1 MONTH OF
FREE
READING

at

www.ForgottenBooks.com

By purchasing this book you are eligible for one month membership to ForgottenBooks.com, giving you unlimited access to our entire collection of over 1,000,000 titles via our web site and mobile apps.

To claim your free month visit:

www.forgottenbooks.com/free1219618

ISBN 978-0-332-69970-7
PIBN 11219618

DEPARTMENT OF THE INTERIOR

UNITED STATES GEOLOGICAL SURVEY

CHARLES D. WALCOTT, Director

REPORT

OF

PROGRESS OF STREAM MEASUREMENTS

FOR

THE CALENDAR YEAR 1903

PREPARED UNDER THE DIRECTION OF F. H. NEWELL

BY

JOHN C. HOYT

PART III.—Western Mississippi River and Western Gulf of Mexico Drainage

WASHINGTON

GOVERNMENT PRINTING OFFICE

1904

659126

CONTENTS.

4 CONTENTS.

ILLUSTRATION.

LETTER OF TRANSMITTAL.

DEPARTMENT OF THE INTERIOR,
UNITED STATES GEOLOGICAL SURVEY,
HYDROGRAPHIC BRANCH,
Washington, D. C., March 23, 1904.

SIR: I have the honor to transmit herewith Water-Supply Paper No. 99, which is Part III of a series of four papers numbered 97 to 100, inclusive. These papers compose the report of progress of stream measurements for the calendar year 1903. Parts I and II of this report contain the results of the data collected from the territory east of the Mississippi River. Parts III and IV are devoted to the data collected from the territory west of the Mississippi River.

The actual work of assembling the original data on which this report is based, and of preparing the same for publication, has been done under the immediate direction of John C. Hoyt, who has been assisted by Frank H. Brundage, L. R. Stockman, R. H. Bolster, II. J. Saunders, and W. A. Brothers. Acknowledgment is due each of these persons and also the various resident hydrographers and engineers under whose direction the data herein given were collected.

Very respectfully,

F. II. NEWELL,
Chief Engineer.

Hon. CHARLES D. WALCOTT,
Director United States Geological Survey.

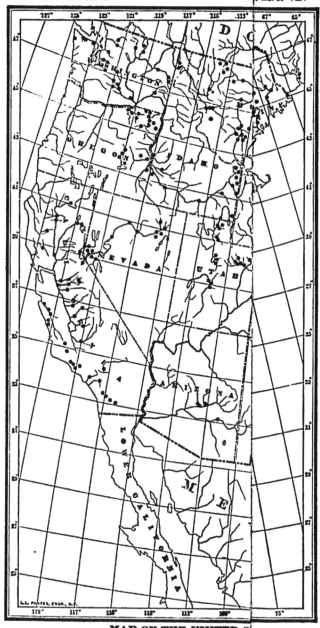

MAP OF THE UNITED S

PROGRESS REPORT OF STREAM MEASUREMENTS FOR THE CALENDAR YEAR 1903.

PART III.

By JOHN C. HOYT.

INTRODUCTION.

During the calendar year 1903 the division of hydrography has continued the work of measuring the flow of streams on the same general lines as in previous years. Special efforts have been made to collect such other information, aside from the flow, as will be of use in general hydrographic studies. Reconnaissances made on many of the important rivers in different portions of the United States have resulted in the collection of many valuable data in regard to floods, water powers, river profiles, etc.

During 1903 the number of regular stations at which stream measurements are being made has been steadily increased, so that at the close of the year systematic measurements were being carried on at approximately 500 stations. These are distributed so as to best cover the needs of the various States and Territories. (See Pl. I for location of principal gaging stations.) This expansion in the work is the result of the constantly increasing demand from the general and engineering public for the stream data collected by the Survey. The requests for information have been so numerous that the supply of publications containing the results has in many cases become exhausted.

The Survey has continued to receive the hearty cooperation of various individuals, corporations, and States, as mentioned hereafter. This cooperation has made possible the publication of many valuable records which could not otherwise have been obtained.

The Report of the Progress of Stream Measurements for the Calendar Year 1903, of which this is Part III, is published in a series of four Water-Supply Papers, Nos. 97–100, inclusive, under the following subtitles:

Part I. Northern Atlantic and Great Lakes Drainage.

Part II. Southern Atlantic, Eastern Gulf of Mexico, and Eastern Mississippi River Drainage.

Part III. Western Mississippi River and Western Gulf of Mexico Drainage.

Part IV. Interior Basin, Pacific, and Hudson Bay Drainage.

The territory covered by each paper is given in the subtitle, and these larger drainages are, for convenience in arrangement, subdivided into smaller ones, under which the data are arranged, as far as practicable, geographically.

These papers contain the data that have been collected at the regular gaging stations, the results of the computations based upon the observations, and such other information as has been collected in the various drainage areas that may be of use in hydrographic studies, including, as far as available, a description of the drainage area and the streams draining it.

For each regular station are given, as far as available, the following data:

1. Description of station.
2. List of discharge measurements.
3. Gage-height table.
4. Rating table.
5. Table of estimated monthly and yearly discharges and run-off.

The descriptions of stations give such general facts about the locality and equipment as would enable the reader to find and use the station. They also contain, as far as possible, a complete history of all the changes that have occurred since the establishment of the station which would be factors in using the data collected.

The discharge-measurement table gives the results of the discharge measurements made during the year. This includes the date, the hydrographer's name, the gage height, and the discharge in second-feet.

The table of daily gage heights gives for each day the fluctuations of the surface of the river as found from the mean of the gage readings taken on that day. At most of the stations the gage is read in the morning and evening.

The rating table gives discharges in second-feet corresponding to each stage of the river as given by the gage heights. It depends on the general law that for streams of practically constant cross section the discharge is a function of the gage height and that like gage heights will have the same discharge. In its preparation the discharge measurements are plotted on cross-section paper to some convenient scale, using gage heights as ordinates and discharges as abscissas. Through these points a smooth curve is drawn, which is the basis for the table. From this curve are tabulated, on forms prepared for the purpose, the discharges corresponding to each tenth of a foot on the gage. The first and second differences between the successive discharges are then taken. These are adjusted on the assumption that there is a gradual increase in the discharge as the gage height increases, and the discharge values in the table are then adjusted according to these revised differences. In preparing the rating table all available data are brought into use, including special conditions which might affect the discharge. For high waters above the stage covered by discharge measurements the general rule is to extend the curve by a tangent line. In case the river overflows its banks a percentage of the discharge is added, depending on the depth and velocity of the overflowed portion. For stages below that portion of the curve which is fixed by discharge measurements the curve has been extended, following the general form of the determined lower portion. Notes under each rating table indicate those portions based on actual observation and those that are estimated.

From the rating table and daily gage heights a table giving the daily discharge of the streams is prepared. From this the table of estimated monthly and yearly discharges and run-off is computed. This latter table gives in condensed form a summary of the results obtained from the observations made during the year at the station. In order to explain this table the following definitions are given:

The term "second-feet" (sec.-ft.) is an abbreviation for "cubic feet per second." It is the number of cubic feet of water flowing by the gaging section every second. The column headed "Maximum" gives the mean flow for the day when the mean gage height was the highest, and it is the flow as given in the rating table for that mean gage height. As the gage height is the mean for the day, there might have been short periods when the water was higher and the corresponding discharge larger than given in this column. Likewise in the column of "Minimum" the quantity given is the mean flow for the day when the mean gage height was lowest. The column headed "Mean" is the average flow for each second during the month. Upon this the computations for the three remaining columns in the table are based.

An "acre-foot" is the quantity of water it would take to cover an acre to the depth of one foot—an amount equivalent to 43,560 cubic feet of water. This quantity is used in making estimates for irrigation projects and is computed only for such streams as may be used for irrigation. The quantities in the column headed "Total in acre-feet" show the number of acres which would be covered one foot by the flow during the month had all the water been impounded.

The expression "second-feet per square mile" means the number of cubic feet of water flowing from every square mile of drainage area for each second.

"Depth in inches" means the depth of water in inches that would have covered the drainage area, uniformly distributed, if all the water could have accumulated on the surface. This quantity is used for comparing run-off with rainfall, which quantity is also given in depth in inches.

It should be noticed that "acre-feet" and "depth in inches" represent the actual quantities of water which are produced during the periods in question, while "second-feet," on the contrary, is merely a rate of flow, into which the element of time does not enter.

The results of stream measurements during previous years by the United States Geological Survey can be found in the following Survey publications:

1888. Tenth Annual Report, Part II.
1889. Eleventh Annual Report, Part II.
1890. Twelfth Annual Report, Part II.
1891. Thirteenth Annual Report, Part III.
1892. Fourteenth Annual Report, Part II.
1893. Bulletin No. 131.
1894. Bulletin No. 131; Sixteenth Annual Report, Part II.
1895. Bulletin No. 140.
1896. Water-Supply Paper No. 11; Eighteenth Annual Report, Part IV.
1897. Water-Supply Papers Nos. 15 and 16; Nineteenth Annual Report, Part IV.
1898. Water-Supply Papers Nos. 27 and 28; Twentieth Annual Report, Part IV.
1899. Water-Supply Papers Nos. 35 to 39, inclusive; Twenty-first Annual Report, Part IV.
1900. Water-Supply Papers Nos. 47 to 52, inclusive; Twenty-second Annual Report, Part IV.
1901. Water-Supply Papers Nos. 65, 66, and 75.
1902. Water-Supply Papers Nos. 82 to 85, inclusive.
1903. Water-Supply Papers Nos. 97 to 100, inclusive.

A limited number of these are for free distribution, and as long as the supply lasts they may be obtained by application to the Director, United States Geological Survey. Aside from these, other copies are filed with the Superintendent of Public Documents, Washington, D. C., from whom they may be had at nominal cost. Copies of Government publications are as a rule furnished to the municipal public libraries in our large cities, where they may be consulted by those interested.

ACKNOWLEDGMENTS.

Most of the measurements presented in this paper have been obtained through local hydrographers. Acknowledgment is due to each of these persons, and thanks are extended to other persons and corporations who have assisted local hydrographers or have cooperated in any way, either by furnishing records of the height of water or by assisting in transportation.

The following list, arranged alphabetically by States, gives the names of the resident hydrographers and others who have assisted in furnishing and preparing the data contained in this report:

Colorado: District hydrographer, M. C. Hinderlider, assisted by Filmore Cogswell. Antoine Jacob, Oro McDermith, R. I. Meeker, W. N. Sammis, and others. Thanks are due A. J. McCune, State engineer of Colorado, for courtesies extended in the form of office privileges from January 1 to April 1, and for other favors; to F. H. Brandenburg, district weather forecast official at Denver, for information of various kinds; to Mr. T. J. Burroughs for records of the flow of the Arkansas at Pueblo, Colo., and to William Huston for similar records at Prowers, Colo. Acknowledgments and thanks are due for assistance in the form of transportation for hydrographers to the Denver and Rio Grande, Rio Grande Southern, Colorado and Southern, and Union Pacific railroads.

Indian Territory: District hydrographers, W. G. Russell and G. H. Matthes, assisted by E. R. Kerby, F. Bonstedt, L. M. Holt, and Charles Gould.

Iowa: District hydrographer, E. Johnson, jr.; assistant engineer, F. W. Hanna, and Carl C. Kastburg, city engineer of Boone, Iowa. Acknowledgment should also be made to Frank Dearborn, Stone City, Iowa, who assisted in establishing a gage at and supplied voluntary readings from Stone City, Iowa.

Kansas: District hydrographer, W. G. Russell. Acknowledgments are due the Union Pacific, Atchison, Topeka and Santa Fe, and Missouri, Kansas and Texas railway companies for transportation furnished the district hydrographer in Kansas, northern Oklahoma, and Indian Territory.

Minnesota: District hydrographer, E. Johnson. jr., assisted by L. R. Stockman, assistant engineer, and W. R. Hoag, of the engineering department of the University of Minnesota.

Missouri: District hydrographer, E. Johnson, jr.; F. W. Hanna, assistant engineer; I. W. McConnell, engineer. Acknowledgment should also be made to the St. Louis and San Francisco Railroad for transportation furnished to F. W. Hanna between St. Louis and Arlington, Mo.

Montana: District hydrographer, C. C. Babb, assisted by C. T. Prall, A. E. Place, and W. B. Freeman.

Nebraska: District hydrographer, J. C. Stevens. Acknowledgments and thanks are due Prof. A. B. Crane, of Brookings, S. Dak., for measurements made on the Big Sioux and neighboring streams; to Adna Dobson, State engineer of Nebraska, and his assistants in office, and to Prof. O. V. P. Stout, of the University of Nebraska, for the use of scientific instruments, office privileges, and other favors. Acknowledgments are also due for transportation furnished the district hydrographer by the officials of the Burlington and Missouri River, Union Pacific, and Chicago and Northwestern railways.

New Mexico: District hydrographers, M. C. Hinderlider and W. M. Reed. All gagings on the Rio Grande were made by F. Cogswell under the direct supervision of the Denver office. Those measurements on the headwaters of Canadian and Pecos rivers were made by M. C. Hinderlider. Acknowledgments are due Hugh Loudon, secretary of the Lacueva Ranch Company, for records of the flow of Mora River at Lacueva, and to the Denver and Rio Grande Railroad for transportation for Filmore Cogswell, hydrographer.

North Dakota: District hydrographer, E. Johnson, jr.. assisted by Prof. E. F. Chandler, of the engineering department of the University of North Dakota.

Oklahoma: District hydrographers, W. G. Russell and G. H. Matthes, assisted by E. R. Kerby, F. Bonstedt, L. M. Holt, and Charles N. Gould.

South Dakota: District hydrographer, R. F. Walter.

Texas: District hydrographer, Thomas U. Taylor. Special acknowledgment is due to W. W. Follett, consulting engineer of the International (Water) Boundary Commission, for the results of the data collected by him at the stations in the Rio Grande drainage.

Wyoming: District hydrographer, A. J. Parshall, assisted by Rex G. Schnitger. Acknowledgments and thanks are due to the Burlington and Missouri River. Union Pacific, Colorado and Southern, and Fremont, Elkhorn and Missouri Valley railroads for transportation for the district hydrographer. Stations maintained in connection with the Shoshone irrigation investigations were carried on under the direction of Jeremiah Ahern, engineer for this project.

WESTERN MISSISSIPPI RIVER DRAINAGE.

For convenience in arrangement, the stations from which the Geological Survey has obtained data during 1903, and which are located on rivers tributary either directly or indirectly to the Mississippi from the West, have been grouped under Crow Wing, Minnesota, Maquoketa, Wapsipinicon, Iowa, Des Moines, Missouri, Platte, Kansas, Meramec, Arkansas, and Red River drainage basins. These are arranged in the report in this order.

CROW WING RIVER DRAINAGE BASIN.

Crow Wing River rises in the eastern part of Becker County, Minn., flows south and then east for a course of about 80 miles, and enters the Mississippi 8 miles below Brainard. This stream in its usual stages is small, and there are no great possibilities for the development of water power.

The drainage area is about 3,000 square miles, and a study of the river is important in connection with the work of regulating the flow of the Mississippi by means of storage basins constructed at the outlets of numerous lakes, from which it takes its source. A station was established at Pillager, about 8 miles from the mouth, but the location was found unsuitable for obtaining accurate results and was finally abandoned.

CROW WING RIVER AT PILLAGER, MINN.

This station was established May 24, 1903, by E. Johnson, jr., assisted by W. R. Hoag. It is located three-fourths of a mile from Pillager and from the railroad station. The gage is a vertical 1½-inch by 6-inch board, secured to a post driven in the river bed and tied to the bank. The observer during 1903 was Olaf Johnson. The gage was read once each day. It is 300 feet below the point at which measurements are made. Discharge measurements are made from a cable and boat. The channel is straight for 300 feet above the station and for 800 feet below. The right bank is low and will overflow at high stages. The left bank is high and is not liable to overflow. The bed of the stream is sandy and shifting. There is but one channel, with a depth at ordinary stages of 4 to 6 feet. The bench mark is on an 18-inch elm tree 40 feet from the water's edge. Its elevation is 15.56 feet above the zero of the gage. The station was discontinued September 25, 1903.

The observations at this station during 1903 have been made under the direction of E. Johnson, jr., district hydrographer.

Discharge measurements of Crow Wing River at Pillager, Minn., in 1903.

Date.	Hydrographer.	Gage height.	Discharge.
		Feet.	*Second-feet.*
May 24	E. Johnson, jr	10.30	3,066
June 30	W. R. Hoag	8.26	888

Mean daily gage height, in feet, of Crow Wing River at Pillager, Minn., for 1903.

Day.	May.	June.	July.	Aug.	Sep.	Day.	May.	June.	July.	Aug.	Sept.
1		9.60	8.40	9.70	8.45	17		8.50	9.00	8.40	
2		9.40	8.30		8.40	18		8.50	8.80	8.40	
3		9.30	8.45		8.30	19		8.60	8.90	8.30	
4		9.10	8.65		8.20	20		8.40	8.70	8.20	
5		9.00	8.60		8.20	21		8.30	8.60	8.20	
6		8.80	8.60			22		8.30	8.60	8.10	
7		8.70	8.55			23		8.40	8.60	8.10	
8		8.50	8.80			24		8.20	8.55	8.15	
9		8.60	8.90	9.60		25	10.30	8.10	8.40	8.10	
10		8.75	8.70	9.00		26	10.30	8.10	8.50	8.20	
11		8.80	8.60	8.90		27	10.35	8.10	8.60	8.20	
12		8.80	8.60	8.80		28	10.30	8.10	8.80	8.30	
13		8.80	8.50	8.70		29	10.10	8.10	9.50	8.60	
14		8.60	8.60	8.60		30	9.90	8.15	9.40	8.60	
15		8.60	8.70	8.60		31	9.70		9.50	8.60	
16		8.50	8.50	8.50							

MINNESOTA RIVER DRAINAGE BASIN.

Minnesota River rises in Bigstone Lake, which forms part of the boundary between South Dakota and Minnesota. The river flows in a southeasterly direction until the city of Mankato is reached, in the northern part of Blue Earth County, when it makes an abrupt turn to the north and continues in a northerly and northeasterly direction, entering Mississippi River at a point midway between Minneapolis and St. Paul.

The course of this river is generally marked by wide bottom lands. It has a sluggish current, affording few opportunities for the development of water power. The United States Geological Survey maintains a gaging station at Mankato, below the junction of Blue Earth River, mainly for the study of the sewage disposal of the city of Mankato and for the purpose of ascertaining the run-off in the southern part of Minnesota.

MINNESOTA RIVER ABOVE MANKATO, MINN.

This station was established May 20, 1903, by E. Johnson, jr., assisted by W. R. Hoag. It is located at Sibley Park, 1 mile below the highway and railroad bridges across Blue Earth River and 1½ miles above the city bridge in Mankato. The drainage area above the station is about 13,400 square miles. Blue Earth River joins the Minnesota about 500 feet above the station. The gage is a vertical timber fastened to a post which is driven into the river bed a few feet from the right bank. It is read once each day by George E. Blake. On February 4, 1904, a new five-sixteenths-inch galvanized cable was stretched across the river between trees on the banks. The old one-fourth-inch wire cable from which measurements were previously made was left undisturbed and will be used as a tag line. The new cable is wound with heavy wire every 25 feet. The first point at which wire is wound is 17.34 feet from the tree to which the cable is attached. Discharge measurements are made from a small rowboat running on this cable. The initial point for soundings is a spike in the base of the willow tree to which the small cable is attached on the right bank. It is about 15 feet from the top of the bank. The channel is straight for 1,000 feet above and 2,000 feet below the station. The current velocity is moderate, though somewhat sluggish near the banks at low stages. The width of the channel is about 300 feet at low water and about 350 feet at high stages. The right bank is low and liable to overflow for a distance of 50 to 75 feet from the gage. The left bank is a steep, rocky bluff. The bed of the stream is composed of sand, gravel, and blue earth, and may shift somewhat at high water. There is but one channel at all stages. The bench mark is on a 20-inch cottonwood tree on the right bank, a short distance above the station and about 30 feet from the water's edge. Its elevation is 14.78 feet above the zero of the gage.

The observations at this station during 1903 have been made under the direction of E. Johnson, jr., district hydrographer.

Discharge measurements of Minnesota River above Mankato, Minn., in 1903.

Date.	Hydrographer.	Gage height.	Discharge.
		Feet.	*Second-feet*
May 20	E. Johnson, jr	a 9.20	7,63.
June 23	W. R. Hoag	b 4.95	2,11.
July 30	....do	5.11	3,01.
September 7	L. R. Stockman	5.15	3,26.
October 15	....do	10.45	12,29.
November 27	....do	3.30	1,27.

a Area is in error.   b Area is probably in error.

Mean daily gage height, in feet, of Minnesota River above Mankato, Minn., for 1903.

Day.	May.	June.	July.	Aug.	Sept.	Oct.	Nov.	Dec.
1		15.90	5.60	4.10	5.00	7.10	5.55	3.35
2		14.45	5.75	4.95	5.65	7.00	5.40	3.25
3		13.00	6.10	4.90	5.40	6.90	5.30	3.20
4		11.85	7.85	5.00	5.90	7.30	5.20	3.25
5		10.90	7.60	5.35	5.95	7.35	5.10	3.25
6		10.00	7.65	5.70	5.95	7.55	5.00	3.70
7		9.20	7.60	5.90	5.00	11.80	4.80	3.60
8		8.10	7.55	5.70	5.40	13.00	4.70	3.50
9		7.70	7.45	5.30	5.50	13.45	4.60	3.40
10		7.70	7.50	5.10	5.70	13.40	4.55	3.30
11		7.20	7.45	4.90	6.00	12.90	4.50	3.30
12		6.90	7.10	4.80	7.00	12.40	4.45	3.25
13		6.70	6.90	4.70	8.50	11.75	4.40	3.25
14		6.40	6.80	4.70	10.95	11.10	4.35	3.20
15		6.20	6.40	5.80	12.60	10.40	4.30	3.15
16		5.80	6.00	6.10	14.25	9.80	4.00	3.15
17		5.60	5.95	6.90	15.65	9.45	3.40	3.15
18		5.20	6.35	6.70	15.00	9.00	3.20	3.10
19		5.10	6.55	6.40	14.20	8.60	3.00	3.10
20	9.20	4.95	6.90	6.00	13.20	8.40	2.90	3.10
21	8.70	4.95	7.30	5.90	12.40	8.10	2.80	3.10
22	8.25	4.90	7.00	5.80	11.45	7.75	3.70	3.10
23	8.20	4.90	6.95	5.40	10.80	7.40	3.75	3.10
24	8.25	4.80	6.60	5.10	10.00	7.20	3.70	3.10
25	9.80	4.75	6.35	4.90	9.45	6.80	3.45	3.10
26	11.50	4.60	6.00	4.70	8.85	6.60	3.40	3.10
27	14.20	4.60	5.90	4.60	8.55	6.40	3.30	3.10
28	17.10	4.55	6.00	4.55	7.70	6.20	3.30	3.10
29	19.60	4.50	5.60	4.95	7.00	6.00	3.25	3.10
30	19.00	4.60	5.40	4.96	7.35	5.80	3.40	3.10
31	17.50	....	5.00	4.96	....	5.65	....	3.10

MAQUOKETA RIVER DRAINAGE BASIN.

Maquoketa River rises in Fayette County, Iowa, flows southeast, and enters the Mississippi near Green Island, Jackson County, Iowa. There are several dams on this stream, affording small water power, mostly used for running grist mills. After a thorough reconnaissance no suitable place could be found at which to establish a gaging station, though the most favorable appears to be the Klondike dam, just below the fork of Maquoketa River at Maquoketa. This dam is 12¼ feet high, with a spillway 300 feet in length, the crest of which appears perfectly level and sharp. The United States Geological Survey maintained a station for a short time at Manchester.

The total drainage area of Maquoketa River is 1,874 square miles, while the area at Manchester is 226 square miles.

MAQUOKETA RIVER AT MANCHESTER, IOWA.

This station was established June 6, 1903, by F. W. Hanna, and was discontinued July 15, 1903. It was located one-fourth of a mile above the city bridge and the Hoag dam. The gage is a 2 by 4 inch pine rod 12 feet long. It is driven into the river bed and anchored to H. S. Weber's boat landing. The gage was read twice each day by H. S. Weber. Discharge measurements can be made by means of a boat and a tagged wire. The soundings were referred to the base of the oak tree to which the tagged wire was attached, 16 feet from the northwest corner of the ice house. The channel is straight for 300 feet above and 1,000 feet below the station. Both banks will overflow at flood stages. The bed of the stream is sandy and the current is sluggish. Bench mark No. 1 is a spike driven into a blaze on an oak tree 12.7 feet from the gage and 16.5 feet from the northwest corner of the ice house. Its elevation is 14.33 feet above the zero of the gage. Bench mark No. 2 is a similar mark on the tree 19 feet from the gage and 14.9 feet from the northwest corner of the ice house. Its elevation is 11.16 feet above the zero of the gage.

The observations at this station during 1903 have been made under the direction of E. Johnson, jr., district hydrographer.

A discharge measurement made by F. W. Hanna on June 6, 1903, gave the following results:

Gage height, 4.20 feet; discharge, 245 second-feet.

Mean daily gage height, in feet, of Maquoketa River at Manchester, Iowa, for 1903.

Day.	June.	July.	Day.	June.	July.	Day.	June.	July.	Day.	June.	July.
1		4.70	9	4.15	4.40	17		4.00	25		4.15
2		4.35	10	4.15	8.70	18		3.85	26		4.00
3		4.10	11	4.15		19		3.95	27		3.95
4		4.35	12	4.05		20		4.05	28		4.90
5		4.60	13	4.05		21		4.00	29		4.65
6		4.40	14	4.00		22		4.25	30		4.70
7	4.10	4.20	15	4.10		23		4.55	31		
8	4.10	4.10	16	4.00		24		4.35			

MISCELLANEOUS MEASUREMENTS IN THE MAQUOKETA RIVER DRAINAGE BASIN.

The following miscellaneous measurements were made during the year 1903 on Maquoketa River:

Miscellaneous measurements in Maquoketa River drainage basin in 1903.

Stream.	Locality.	Date.	Discharge.
			Second-feet.
North Maquoketa River	Cascade, Iowa	July	152
South Maquoketa River	Monticello, Iowa	do	878

WAPSIPINICON RIVER DRAINAGE BASIN.

Wapsipinicon River rises in Mower County, Minn., and flows south-ast, entering the Mississippi near Shaffton, Scott County, Iowa. There are numerous small power sites on this stream at Toronto,)xford Mills, Newport, Anamosa, Central City, Troy Mills, Quasque-ton, Independence, and Littleton. At several of these points, nota-bly at Anamosa, there are good opportunities for building dams.

The river is generally fairly constant in its flow, though its head-waters are sometimes low. As a rule, its banks are fairly high and the stream is well confined. The United States Geological Survey maintains a gaging station at Stone City.

The total drainage area of the Wapsipinicon River is 2,304 square miles; that at Stone City, where the United States Geological Survey has its station, is 1,308 square miles.

WAPSIPINICON RIVER AT STONE CITY, IOWA.

This station was established August 19, 1903, by F. W. Hanna. It is located at the highway bridge just above the Chicago, Milwaukee and St. Paul Railway bridge and near the Dearborn stone quarry. There is a dam at Waubeck, 4 miles above the station, and another at Anamosa, the same distance below. The standard boxed chain gage is attached to the guard rail on the upstream side of the bridge. The length of the chain from the end of the weight to the marker is 36.4 feet. The gage is read once each day by Frank Dearborn. Dis-charge measurements are made from the two-span Pratt truss high-way bridge, which has a length of 225 feet. The initial point for soundings is the end of the lower chord on the upstream side of the bridge at the left bank. The channel is straight for about 400 feet above and below the station, at which points the river makes abrupt turns. Both banks are high and are not subject to overflow. The bed of the stream is of solid rock and sand and is permanent. The channel is broken by one pier, and the current velocity is mod-erate. Bench mark No. 1 is the east end of the south rail west of the first switch east of the railroad station at Stone City. Its elevation is 815.08 feet above sea level and 38.75 feet above gage datum. Bench mark No. 2 is a cross on the northwest corner of the middle pier of the highway bridge at which the station is located. Its eleva-tion is 31.09 feet above gage datum and 807.42 feet above sea level. The center of the gage pulley is at an elevation of 35.58 feet above gage datum and 811.91 feet above sea level. The top of the lower chord, upstream side, at the first cross girder east of the pier is 31.75 feet above gage datum. The elevations of the bench marks above sea level have been determined by Chicago, Milwaukee and St. Paul Railway levels.

The observations at this station during 1903 have been made under the direction of E. Johnson, jr., district hydrographer.

Discharge measurements of Wapsipinicon River at Stone City, Iowa, in 1903

Date.	Hydrographer.	Gage height.	Dischar
		Feet.	*Second-f*
July 13	F. W. Hanna	15.20	10,
August 19	do	3.00	
September 16	do	4.13	
October 15	do	3.53	
November 16	do	2.80	
December 19	do	1.77	1

Mean daily gage height, in feet, of Wapsipinicon River at Stone City, Iowa,
1903.

Day.	Aug.	Sept.	Oct.	Nov.	Dec.	Day.	Aug.	Sept.	Oct.	Nov.	D
1		3.55	3.45	(a)	2.70	17		3.95	3.65	2.80	
2		3.70	3.50	2.92	2.70	18		3.85	3.65	2.75	
3		3.80	3.40	2.92	2.70	19	3.30	3.85	3.60	2.75	
4		3.75	3.45	2.92	2.68	20	3.25	3.95	3.45	2.70	
5		3.73	3.40	2.92	2.65	21	3.30	3.90	3.35	2.70	
6		3.65	3.30	2.92	1.90	22	3.25	3.90	(a)	2.70	
7		3.65	3.60	2.90	1.90	23	3.25	3.85	3.25	2.73	
8		3.60	3.65	2.90	1.90	24	3.20	3.85	3.20	2.70	
9		3.70	3.70	2.87	1.90	25	3.15	3.75	3.15	2.70	
10		4.30	3.95	2.87	1.85	26	3.35	3.75	3.13	2.70	1
11		3.90	4.00	3.12	1.85	27	3.70	3.70	3.10	2.70	
12		3.80	3.10	2.92	1.80	28	3.70	3.60	3.10	2.70	
13		3.95	3.85	2.90		29	(a)	3.55	3.08	2.70	
14		3.90	3.75	2.90		30	3.55	3.50	3.05	2.68	
15		4.60	3.70	2.85	1.67	31	3.53		3.05		
16		4.10	3.70	2.80							

a Missing.

Rating table for Wapsipinicon River at Stone City, Iowa, from August 19
December 31, 1903.

Gage height.	Discharge.	Gage height.	Discharge.	Gage height.	Discharge.	Gage height.	Discharge
Feet.	*Second-feet.*	*Feet.*	*Second-feet.*	*Feet.*	*Second-feet.*	*Feet.*	*Second-feet*
1.5	79	2.4	214	3.3	451	4.2	837
1.6	90	2.5	234	3.4	487	4.3	886
1.7	102	2.6	255	3.5	525	4.4	936
1.8	115	2.7	277	3.6	565	4.5	987
1.9	129	2.8	301	3.7	607	4.6	1,039
2.0	144	2.9	327	3.8	651	4.8	1,146
2.1	160	3.0	355	3.9	696	5.0	1,255
2.2	177	3.1	385	4.0	742	5.2	1,365
2.3	195	3.2	417	4.1	789		

Estimated monthly discharge of Wapsipinicon River at Stone City, Iowa, for 1903.

[Drainage area, 1,303 square miles.]

Month.	Discharge in second-feet.			Run-off.	
	Maximum.	Minimum.	Mean.	Second-feet per square mile.	Depth in inches.
August 19–31 [a]	607	401	483	0.37	0.18
September	1,039	525	664	.51	.57
October [b]	742	370	509	.39	.45
November [c]	385	277	306	.23	.26
December 1–12	277	115	187	.14	.06

[a] August 29 estimated.   [b] October 22 estimated.   [c] November 1 estimated.

MISCELLANEOUS MEASUREMENT IN THE WAPSIPINICON RIVER DRAIN-
AGE BASIN.

At Independence in July the stream was found to have a discharge of 2,423 second-feet.

IOWA RIVER DRAINAGE BASIN.

Iowa River rises in the north-central part of Iowa, flows in a south-easterly direction, and joins Mississippi River about 30 miles above Burlington. Its chief tributary is Cedar River, which rises in Minnesota and flows into the Iowa about 50 miles above the latter's mouth.

The total drainage area of Iowa River is 12,412 square miles, 4,469 square miles of this being tributary to it above its junction with Cedar River at Columbus Junction. Its drainage areas at Marshall-town and Iowa City are, respectively, 1,463 and 3,317 square miles. Cedar River drains 7,597 square miles above its union with Iowa River, and 6,317 square miles at Cedar Rapids.

The chief power plants in the Iowa drainage basin are at Iowa City, Marshalltown, Cedar Rapids, and Waterloo. From the standpoint of power development the Iowa River system is important.

During 1903 stations have been maintained on Cedar River at Cedar Rapids, Iowa, and on Iowa River at Iowa City and Marshalltown, Iowa.

CEDAR RIVER AT CEDAR RAPIDS, IOWA.

This station was established October 26, 1902, by K. C. Kastberg, assisted by J. C. Stevens. It is located near the city gas works and near the plant of the Iowa Windmill and Pump Company. The gage is an inclined 4 by 6 inch timber, graduated to read to vertical tenths and half-tenths feet. It is fastened to the right bank by posts set in the ground. It is read twice each day by R. S. Toogood. Discharge measurements were made during 1903 from a cable and car at a point just above the gage. The points of measurement are marked

tag wire. Discharge measurements have been made since 1903 at the
First Avenue Bridge about one-half mile upstream from the gage. At
low stages the discharge can be measured by wading at the cable.
The initial point for soundings at the cable is the foot of the pole supporting the cable on the right bank of the river. The channel is
straight for 800 feet above and below the cable. The right bank is
high and will overflow only at extreme stages. The left bank is an
earth embankment and will overflow. The bed of the stream is rocky
and permanent. At the cable the cross section is regular and the
channel is about 400 feet wide.

Bench mark No. 1 is a city bench mark at Williams & Huntings office,
on the north corner of Fifth avenue and West First street. It is
marked by a triangle on the stone just north of the first iron post
from the corner of the building. Its elevation above sea level, as
determined by the city of Cedar Rapids from levels of the Northwestern Railroad, is 746.6 feet. Its elevation above the zero of the
gage is 23.57 feet. Bench mark No. 2 is a bolt driven into the
masonry foundation of the east face of the Iowa Windmill and Pump
Company's building on the west bank of the river just north of
Seventh avenue. Its elevation is 740.05 feet above sea level and 17.02
feet above the zero of the gage.

At the First Avenue Bridge the channel has a rock bottom and a
straight course, both above and below, for several hundred feet. It is
divided into five parts by the 5-span Pratt truss bridge, its length
from abutment to abutment being 714 feet. The current is strong
and direct. The initial point for soundings is the inner upstream
face of the left abutment, the measurements being taken from the
upstream side of the bridge.

The observations at this station during 1903 have been made under
the direction of E. Johnson, jr., district hydrographer.

Discharge measurements of Cedar River at Cedar Rapids, Iowa, for 1903.

Date.	Hydrographer.	Gage height.	Discharge
		Feet.	*Second-feet*
March 21	K. C. Kastberg	5.75	9,49
April 25	....do	4.65	5,30
May 10	....do	4.50	5,08
June 5	F. W. Hanna	7.20	15,09
July 8	....do	3.95	3,01
Do	E. C. Murphy	3.95	2,93
August 18	F. W. Hanna	4.10	3,64
September 15	....do	4.80	5,91
October 14	....do	5.00	6,37
November 14	....do	3.60	2,06
December 18	....do	a3.60	1,15

a *Gage affected by ice; river partly frozen over.*

Mean daily gage height, in feet, of Cedar River at Cedar Rapids, Iowa, for 1903.

Day.	Jan.	Feb.	Mar.	Apr.	May.	June.	July.	Aug.	Sept.	Oct.	Nov.	Dec.
1	(a)	(a)	4.80	5.30	4.35	14.03	3.85	4.40	4.80	4.25	3.85	3.60
2	(a)	(a)	4.90	5.25	4.35	11.70	4.10	4.30	4.90	4.23	3.80	3.60
3	(a)	(a)	5.35	5.25	4.45	9.38	4.08	4.33	4.90	4.19	3.80	3.25
4	(a)	(a)	5.70	5.20	4.70	8.05	4.22	4.38	4.80	4.20	3.75	3.50
5	(a)	(a)	5.65	5.20	4.75	7.20	4.20	5.10	4.60	4.15	3.75	3.35
6	(a)	(a)	5.75	5.20	4.85	6.68	4.10	5.63	4.50	4.18	3.75	3.20
7	(a)	(a)	6.00	5.15	4.95	6.20	4.00	5.83	4.30	4.38	3.70	3.25
8	(a)	(a)	5.80	4.95	4.85	5.95	3.90	6.95	4.40	4.48	3.70	3.25
9	(a)	(a)	5.55	4.80	4.70	5.65	4.00	7.30	4.60	4.58	3.70	3.15
10	(a)	(a)	5.55	4.70	4.50	5.40	4.50	6.60	4.70	4.80	3.65	3.25
11	(a)	(a)	5.95	4.80	4.40	5.18	5.80	5.60	4.60	5.00	3.63	3.15
12	(a)	(a)	6.10	5.10	4.85	5.95	6.45	5.20	4.40	5.10	3.60	3.45
13	(a)	(a)	6.15	5.30	4.60	4.83	6.82	4.80	4.35	5.08	3.60	3.80
14	(a)	3.65	6.25	5.55	4.90	4.70	7.15	4.60	4.42	4.98	3.60	3.60
15	(a)	3.75	6.20	5.90	5.05	4.58	7.05	4.40	4.75	4.80	3.60	3.60
16	(a)	3.70	6.10	6.20	5.45	4.48	6.40	4.30	4.88	4.65	3.60	3.50
17	(a)	3.65	6.00	6.35	6.00	4.35	5.72	4.20	5.05	4.53	3.45	3.40
18	(a)	3.50	5.90	6.40	6.20	4.30	5.70	4.10	5.48	4.40	3.35	3.70
19	(a)	3.40	5.70	5.90	6.10	4.23	6.00	4.10	5.85	4.38	3.15	3.40
20	(a)	3.35	5.75	5.95	6.00	4.13	5.90	4.00	6.35	4.38	3.05	3.30
21	(a)	3.40	5.75	5.05	5.25	4.10	5.80	3.90	6.40	4.30	3.20	3.20
22	(a)	3.35	5.70	4.90	5.10	4.28	5.70	3.90	5.70	4.25	3.20	3.10
23	(a)	3.20	5.85	4.75	5.25	4.45	5.50	3.90	5.30	4.18	3.50	3.30
24	(a)	3.15	5.90	4.70	5.25	4.28	5.20	3.80	5.00	4.10	3.45	3.40
25	(a)	3.25	5.70	4.65	5.10	4.13	4.90	3.80	4.78	4.10	3.40	3.40
26	(a)	3.20	5.45	4.55	5.20	4.08	4.60	3.80	4.63	4.00	3.85	3.60
27	(a)	4.30	5.35	4.45	5.55	3.96	4.40	3.90	4.50	3.95	3.85	3.70
28	(a)	4.65	5.35	4.40	5.23	3.90	4.30	3.90	4.43	3.90	3.25	3.90
29	(a)	......	5.30	4.35	9.03	3.88	4.20	4.10	4.33	3.88	3.80	3.70
30	(a)	......	5.30	4.40	14.35	3.85	4.40	4.25	4.28	3.85	3.90	3.60
31	(a)	......	5.35		16.55	......	4.30	4.60	......	3.80	......	3.50

a River frozen.

Rating table for Cedar River at Cedar Rapids, Iowa, from February 22 to December 31, 1903.

Gage height.	Discharge.	Gage height.	Discharge.	Gage height.	Discharge.	Gage height.	Discharge.
Feet.	Second-feet.	Feet.	Second-feet.	Feet.	Second-feet.	Feet.	Second-feet.
3.2	1,110	4.0	3,250	4.8	5,880	5.6	8,730
3.3	1,325	4.1	3,550	4.9	6,190	5.7	9,110
3.4	1,560	4.2	3,855	5.0	6,550	5.8	9,490
3.5	1,815	4.3	4,165	5.1	6,910	5.9	9,870
3.6	2,090	4.4	4,480	5.2	7,270	6.0	10,250
3.7	2,370	4.5	4,800	5.3	7,630	6.1	10,650
3.8	2,660	4.6	5,140	5.4	7,990		
3.9	2,950	4.7	5,480	5.5	8,350		

Tangent at 6 feet gage height; differences above this point, 400 per tenth; approximate above and below 3.6 and 7.2 feet gage heights.

Estimated monthly discharge of Cedar River at Cedar Rapids, Iowa, for 1903.

[Drainage area, 6,817 square miles.]

Month	Discharge in second-feet.			Run-off.	
	Maximum.	Minimum.	Mean.	Second-feet per square mile.	Depth in inches.
February 22–28 a	5,310	1,005	2,224	0.35	0.09
March	11,250	5,830	9,063	1.43	1.65
April	11,850	4,320	7,079	1.12	1.25
May	52,450	4,320	9,909	1.57	1.81
June......................	42,450	2,805	9,153	1.45	1.62
July	14,850	2,805	6,987	1.11	1.28
August	15,450	2,660	5,495	.87	1.00
September	11,850	4,165	6,143	.97	1.08
October	6,910	2,660	4,409	.70	.81
November	3,855	795	2,070	.33	.37
December	2,950	900	1,700	.27	.31

a Frozen to February 21.

IOWA RIVER AT IOWA CITY, IOWA.

This station was established June 11, 1903, by F. W. Hanna. At that time a chain gage was installed on the county bridge directly west of the State University grounds. On September 15, 1903, the standard boxed chain gage was substituted for the gage which was first established and made to read the same, with the same length of chain from the end of the weight to the marker. This length is 24.43 feet. The gage is read twice each day by Arthur G. Smith. Discharge measurements are made from the upstream side of the two-span highway bridge at the foot of Burlington street, about 1,000 feet below the bridge at which the gage is located. The initial point for soundings is the inner face of the right abutment, located at the west end of the bridge, and is marked zero on the bridge floor. The bridge is not quite at right angles with the direction of the current. The channel is straight for about 1,600 feet above and below the station. The channel is 316 feet wide between abutments, and is broken by one pier. The bed of the stream is of soft material, except near the bridge pier, and is slightly shifting. The right bank is high and somewhat rocky and not liable to overflow. The left bank does not overflow, owing to the road embankment. Both banks are covered with willows above ordinary high-water mark, which will interfere with flood measurements. Bench mark No. 1 is a nail driven into the base of an elm tree 16 feet northwest of the bridge at which the gage is located. Its elevation above gage datum is 18.96 feet, and above the Iowa City datum is 61.01 feet. Bench mark No. 2 is on the top stone just north of and against the northwest railing post of the bridge at which the gage is located. Its elevation above gage datum is 22.67 feet, and above the city datum is 64.72 feet. The city bench mark used for reference is on the north side of the base of a maple tree, one block east and one block and 40 feet south of the bridge at which the gage is located. The elevation of this bench mark is 22.53 feet above gage datum, 64.58 feet above the city datum, and 671.92 feet above sea level. The elevation above sea level was obtained from the levels of the Chicago, Rock Island and Pacific Railroad. The bottom of the top hand rail, just above the pulley center, has an elevation of 26.73 feet above gage datum.

The observations at this station during 1903 have been made under the direction of E. Johnson, jr., district hydrographer.

Discharge measurements of Iowa River at Iowa City, Iowa, in 1903.

Date.	Hydrographer.	Gage height.	Discharge.
		Feet.	*Second-feet.*
June 2	F. W. Hanna	13.95	18,299
June 12	....do	6.95	7,226
July 9	E. C. Murphy	1.87	1,816
August 20	F. W. Hanna	.31	707
September 15	....do	2.40	2,481
October 13	....do	.95	1,215
November 14	....do	— .20	501
December 18	....do	—1.06	279

Mean daily gage height, in feet, of Iowa River at Iowa City, Iowa, for 1903.

Day.	June.	July.	Aug.	Sept.	Oct.	Nov.	Dec.
1	12.70	1.20	1.40	1.30	0.90	0.15	−0.45
2	14.60	1.15	3.75	1.20	.85	.25	− .35
3	14.65	1.25	2.55	1.00	1.20	.50	− .55
4	13.80	1.65	1.90	.90	1.15	.40	− .40
5	12.75	1.80	1.90	.80	1.00	.30	− .35
6	11.60	2.20	2.30	.80	.80	.20	ᵃ−1.20
7	10.65	2.50	2.55	.60	1.95	.20	−1.45
8	9.90	2.30	2.15	.60	1.30	.15	−1.45
9	9.25	1.80	1.85	1.00	1.10	.05	−1.30
10	8.55	2.10	1.45	2.40	1.35	.10	−1.15
11	7.75	2.20	1.20	2.30	1.25	.10	−1.10
12	7.55	1.80	1.10	2.20	1.10	.05	−1.00
13	6.20	2.70	1.00	3.10	1.00	.00	−1.00
14	5.10	3.10	.85	2.30	1.00	.00	−1.00
15	4.30	3.00	.65	2.40	1.05	− .15	− .95
16	3.65	3.00	.50	1.90	1.10	−.25	− .95

ᵃ Measurements to top of ice.

Mean daily gage height, in feet, of Iowa River at Iowa City, Iowa, etc.—Cont'd.

Day.	June.	July.	Aug.	Sept.	Oct.	Nov.	Dec.
17	3.30	3.30	.45	1.60	1.20	(a)	-1.00
18	2.95	3.50	.45	1.50	1.05	-.50	-1.00
19	2.65	3.20	.40	1.30	.95	-.45	-1.10
20	2.40	3.40	.35	1.40	.95	-.55	-1.20
21	2.30	3.60	.20	1.50	.85	-.55	-1.30
22	2.20	3.60	.10	1.50	.65	-.45	-1.40
23	2.20	3.50	.10	1.40	.55	-.55	-1.50
24	2.45	3.30	.00	1.40	.40	-.25	-1.55
25	2.45	2.90	.20	1.30	.45	-.25	-1.60
26	2.10	3.10	.40	1.20	.45	-.45	-1.60
27	1.85	2.20	.40	1.05	.40	-.40	-1.70
28	1.65	1.90	.60	.95	.25	-.50	-1.70
29	1.50	1.70	1.10	.90	.25	-.20	-1.80
30	1.35	1.60	1.50	.85	.25	-.15	-1.80
31	....	1.50	1.40	....	.30	....	-1.90

a Missing.

Rating table for Iowa River at Iowa City, Iowa, from June 1 to December 31, 1903

Gage height.	Discharge.	Gage height.	Discharge.	Gage height.	Discharge.	Gage height.	Discharge.
Feet.	*Second-feet.*	*Feet.*	*Second-feet.*	*Feet.*	*Second-feet.*	*Feet.*	*Second-feet.*
-1.9	185	-0.3	433	1.6	1,635	7.5	8,460
-1.8	195	-.2	467	1.8	1,815	8.0	9,160
-1.7	205	-.1	509	2.0	1,995	8.5	9,885
-1.6	217	.0	557	2.2	2,175	9.0	10,610
-1.5	229	.1	609	2.4	2,355	9.5	11,360
-1.4	242	.2	663	2.6	2,540	10.0	12,110
-1.3	256	.3	719	2.8	2,740	10.5	12,860
-1.2	270	.4	779	3.0	2,940	11.0	13,610
-1.1	284	.5	839	3.5	3,460	11.5	14,410
-1.0	298	.6	903	4.0	4,010	12.0	15,210
-.9	314	.7	967	4.5	4,585	12.5	16,010
-.8	330	.8	1,035	5.0	5,160	13.0	16,810
-.7	347	.9	1,105	5.5	5,785	13.5	17,610
-.6	365	1.0	1,175	6.0	6,410	14.0	18,410
-.5	385	1.2	1,315	6.5	7,085	14.5	19,210
-.4	407	1.4	1,475	7.0	7,760	15.0	20,010

Estimated monthly discharge of Iowa River at Iowa City for 1903.

[Drainage area, 3,317 square miles.]

Month.	Discharge in second-feet.			Run off.	
	Maximum.	Minimum.	Mean.	Second-feet per square mile.	Depth in inches.
une	19,450	1,435	7,286	2.20	2.45
July	3,570	1,280	2,441	.74	.85
August	3,735	557	1,333	.40	.46
September	3,040	903	1,525	.46	.1
October	1,950	691	1,099	.33	.33
November a	839	375	527	.16	.18
December 1–5 b	419	375	403	.12	.02

a November 17, estimated.   b Frozen December 6–31.

IOWA RIVER AT MARSHALLTOWN, IOWA.

This station was established October 23, 1902, by K. C. Kastberg, assisted by J. C. Stevens. It is located at the highway bridge 50 feet above a flour-mill dam and a short distance above the city waterworks. The gage is a vertical 2 by 6 inch pine board attached to the piling under the south end of the highway bridge. The station was discontinued August 1, 1903. Up to this time the gage was read once each day by George G. Bassett. Discharge measurements were made from the downstream side of the bridge to which the gage is attached. The initial point for soundings was taken at the piling to which the gage is fastened at the south end of the bridge. The channel makes an abrupt bend immediately above the bridge, and about 50 feet below the bridge the power dam of the flour mill is located, with about 8 feet fall. The current has a moderate velocity at ordinary stages. The right bank is high and will not overflow. The left bank overflows at high stages. There is but one channel. The bench mark is a bolt head in the north rim of the top of the tubular iron pier under the southeast batter post of the bridge. Its elevation is 11.19 feet above the zero of the gage.

The observations at this station during 1903 have been made under the direction of E. Johnson, jr., district hydrographer.

Discharge measurements of Iowa River at Marshalltown, Iowa, in 1903.

Date.	Hydrographer.	Gage height.	Discharge.
		Feet.	Second-feet.
March 7	K. C. Kastberg	3.05	1,325
March 23	......do	3.40	1,964
April 24	......do	2.80	1,150
June 4	F. W. Hanna	5.45	4.790

Mean daily gage height, in feet, of Iowa River at Marshalltown, Iowa, for 1903.

Day.	Jan.	Feb.	Mar.	Apr.	May.	June.	July.	Aug.
1		(a)	4.25	8.10	2.60	6.40	8.10	2.15
2		(a)	8.15	8.15	2.60	5.90	2.70	2.75
3		(a)	2.85	8.25	2.60	5.65	8.20	2.55
4	(a)	(a)	2.90	8.10	2.50	5.45	8.00	1.90
5	(a)	(a)	2.80	8.10	2.50	5.30	2.60	1.90
6	(a)	(a)	2.75	2.95	2.80	4.80	2.90	2.90
7	(a)	(a)	8.00	2.90	2.65	4.75	2.90	2.15
8	(a)	(a)	8.55	2.80	2.50	4.60	2.10	2.15
9	(a)	(a)	8.55	2.60	2.85	8.85	2.05	....
10	(a)	(a)	8.65	2.60	2.45	8.55	2.90	....
11	(a)	(a)	8.80	2.65	2.85	8.30	8.50	....
12	(a)	(a)	4.05	8.20	2.50	8.10	4.15	....
13	(a)	(a)	4.25	8.80	8.05	2.95	4.80	....
14	(a)	(a)	4.25	4.10	8.70	2.90	8.95	....
15	(a)	(a)	4.00	4.15	4.15	2.65	8.45	....
16	(a)	(a)	8.75	8.85	4.25	2.50	8.00	....
17	(a)	(a)	8.60	8.50	4.15	2.40	2.75	....
18	(a)	(a)	8.60	8.20	8.85	2.80	4.55	....
19	(a)	(a)	8.60	8.05	8.60	2.80	4.00	....
20	(a)	(a)	8.80	8.00	8.80	2.25	8.65	....
21	(a)	(a)	8.60	8.05	8.05	2.80	8.65	....
22	(a)	(a)	8.55	2.95	8.15	2.20	8.80	....
23	(a)	1.40	8.40	2.90	8.95	2.10	8.00	....
24	(a)	1.70	8.25	2.85	8.70	2.05	2.70	....
25	(a)	1.80	2.95	2.70	8.50	2.00	2.50	....
26	(a)	1.80	2.95	2.65	4.10	2.00	2.40	....
27	(a)	2.75	8.85	2.50	5.80	1.90	2.90	....
28	(a)	4.25	8.65	2.40	6.65	1.90	2.30	....
29	(a)	....	8.45	2.85	6.75	1.90	2.45	....
30	(a)	....	8.80	2.40	6.75	1.90	2.80	....
31	(a)	....	8.90	....	6.75	....	2.90	....

a River frozen.

DES MOINES RIVER DRAINAGE BASIN.

Des Moines River rises in the southern part of Minnesota and flows south and southeast, entering the Mississippi near Keokuk, Iowa. Its principal tributary is Raccoon River, which enters the Des Moines at the city of Des Moines. Des Moines River is the largest river of Iowa and affords numerous good opportunities for the development of water power. As it drains a large area, it is a natural recipient of the sewage of many towns. The United States Geological Survey maintained stations for a short time at Des Moines on Des Moines and Raccoon rivers. These were after a time discontinued and a permanent station was established at Keosauqua, Iowa.

The total drainage area of Des Moines River is 14,717 square miles. Its drainage above the mouth of Raccoon River is 6,462 square miles, while that of the Raccoon at its confluence with the Des Moines is 3,677 square miles.

During 1903 stations were maintained in Des Moines River drainage basin on Des Moines River at Keosauqua, Iowa, and at Des Moines, Iowa, and on Raccoon River at Des Moines, Iowa.

DES MOINES RIVER AT KEOSAUQUA, IOWA.

This station was established May 30, 1903, by F. W. Hanna. It is located at the county bridge one-fourth mile above the old dam site and Government locks. The standard chain gage has the zero of its scale 1 foot to the left of the sixth strut from the right end of the second span from the right end of the bridge. The length of the chain from the end of the weight to the marker is 36.80 feet. The gage is read twice each day by Oscar McCrary. Discharge measurements are made from the downstream side of the four-span bowstring truss bridge, to which the gage is attached. The initial point for soundings is marked zero on the hand rail on the left bank. This is the edge of the west abutment. The channel is straight for 1,000 feet above and below the station and has a width between abutments of 614 feet, broken by 3 piers. The right bank is high and rocky and is not subject to overflow; the low alluvial bank on the left side is subject to overflow at extremely high stages. The bed of the stream is regular in shape and is composed of sand and gravel on the left and of rock on the right side. A slight shifting of the sand occurs at flood stages. Bench mark No. 1 is a cross on the bridge-seat stone at the northeast corner of the bridge. Its elevation above gage datum is 30.11 feet, and above sea level is 618 feet, as determined from the Iowa State geological survey bench mark in front of the Keosauqua railroad station. Bench mark No. 2 is located on a rock cliff on the right bank of the river south of the road leading across the highway bridge and at a distance of 120 feet from the east end of the bridge. It consists of a notch chiseled into the ledge of rock. Its elevation above gage datum is 36.25 feet. A third bench mark is located on the top of the guard rail, 2 feet east of the zero of the gage. Its elevation is 36.60 feet above gage datum.

The observations at this station have been made under the direction of E. Johnson, jr., district hydrographer.

Discharge measurements of Des Moines River at Keosauqua. Iowa, in 1903.

Date.	Hydrographer.	Gage height.	Discharge.
		Feet.	*Second-feet.*
May 30	F. W. Hanna	15.70	50,803
June 11	do	10.68	35,383
July 10	E. C. Murphy	3.73	8,111
August 24	F. W. Hanna	2.60	5,146
September 19	do	5.16	18,920
October 17	do	4.00	9,564
November 20	do	1.80	2,894
December 12	do	.68	1,455

Mean daily gage height, in feet, of Des Moines River at Keosauqua, Iowa, for 1903.

Day.	May.	June.	July.	Aug.	Sept.	Oct.	Nov.	Dec.
1		27.70	4.05	4.30	9.00	4.35	2.60	1.3
2		27.10	3.75	3.85	7.50	5.55	2.60	1.4
3		24.85	3.80	3.90	5.10	5.25	2.50	1.3
4		21.70	5.15	3.80	3.90	5.05	2.50	1.1
5		19.95	5.50	3.85	3.60	4.40	2.50	.5
6		17.90	5.50	4.35	3.20	4.15	2.50	.5
7		15.60	5.10	4.90	3.00	5.90	2.50	.6
8		14.00	4.50	5.35	2.90	5.05	2.40	.7
9		12.75	4.20	4.90	4.10	4.45	2.30	.6
10		12.00	3.90	4.60	7.40	4.80	2.30	.6
11		10.50	3.90	4.30	7.30	4.85	2.30	.5
12		10.00	4.00	4.60	5.00	4.55	2.25	.5
13		8.85	4.60	4.30	5.00	4.10	2.10	.5
14		7.15	5.60	4.00	4.50	3.85	2.10	.5
15		6.45	6.10	3.80	4.60	3.70	2.10	.5
16		5.85	6.10	3.70	4.70	3.95	2.00	.5
17		5.30	5.70	3.70	5.20	4.00	2.00	.5
18		4.90	5.30	4.00	5.40	3.85	1.95	.50
19		4.80	5.00	4.30	5.20	3.75	1.90	.50
20		5.30	5.10	4.40	5.20	3.65	1.65	.60
21		6.30	5.70	4.30	5.50	3.55	1.40	.6
22		6.75	6.20	4.00	5.80	3.40	1.50	.60
23		7.05	6.30	2.90	6.00	3.25	1.50	.55
24		7.15	6.20	2.60	6.20	3.10	1.50	.50
25		6.90	5.40	2.40	6.20	3.05	1.50	.45
26		5.85	4.70	2.20	6.00	2.95	1.50	.45
27		5.30	4.30	3.00	5.70	2.90	1.40	.40
28		5.10	4.00	6.30	5.30	2.80	1.50	.40
29	12.10	5.00	3.80	8.10	5.90	2.80	1.50	.40
30	15.35	4.45	3.70	9.70	4.60	2.70	1.50	.3
31	22.90		4.20	10.20		2.65		.3

Rating table for Des Moines River at Keosauqua, Iowa, from May 29 to December 31, 1903.

Gage height.	Discharge.	Gage height.	Discharge.	Gage height.	Discharge.	Gage height.	Discharge.
Feet.	*Second-feet.*	*Feet.*	*Second-feet.*	*Feet.*	*Second-feet.*	*Feet.*	*Second-feet.*
0.0	1,030	1.3	2,200	2.6	5,060	3.9	9,140
.1	1,060	1.4	2,360	2.7	5,340	4.0	9,490
.2	1,100	1.5	2,530	2.8	5,630	4.1	9,840
.3	1,150	1.6	2,710	2.9	5,920	4.2	10,200
.4	1,210	1.7	2,900	3.0	6,220	4.3	10,560
.5	1,280	1.8	3,100	3.1	6,520	4.4	10,930
.6	1,360	1.9	3,310	3.2	6,830	4.5	11,300
.7	1,450	2.0	3,530	3.3	7,150	4.6	11,670
.8	1,550	2.1	3,760	3.4	7,470	4.7	12,040
.9	1,660	2.2	4,000	3.5	7,800	4.8	12,410
1.0	1,780	2.3	4,250	3.6	8,130	5.0	13,150
1.1	1,910	2.4	4,510	3.7	8,460		
1.2	2,050	2.5	4,780	3.8	8,800		

Tangent above 4.4 feet gage height.   Differences above this point, 370 per tenth.

ated monthly discharge of Des Moines River at Keosauqua, Iowa, for 1903.

[Drainage area, 14,291 square miles.]

Month.	Discharge in second-feet.			Run-off.	
	Maximum.	Minimum.	Mean.	Second-feet per square mile.	Depth in inches.
?9-31	79,380	39,420	56,748	3.97	0.44
	97,140	11,115	34,425	2.41	2.69
	17,960	8,460	12,735	.89	1.03
ist	32,390	4,000	11,584	.81	.93
mber	27,950	5,920	14,313	1.00	1.12
ber	16,480	5,200	9,445	.66	.76
mber	5,060	2,360	3,608	.25	.28
mber 1–5 a	2,530	1,405	2,081	.15	.03

a River frozen December 6–31.

DES MOINES RIVER AT DES MOINES, IOWA.

his station was established October 2, 1902, by K. C. Kastberg,
sted by J. C. Stevens. It was discontinued August 1, 1903. The
ion was located at the Walnut Street Bridge, one-fourth mile below
Edison Power and Light Company's dam and one-fourth mile
re the forks of Des Moines and Raccoon rivers. The United States
ither Bureau gage is a 2 by 10 inch vertical plank attached to the
tubular pier of the Locust Street Bridge, about 300 feet above the
nut Street Bridge. The observer was E. T. Blagburn. Discharge
surements are made from the upstream side of the footway of the
nut Street Bridge. The initial point for soundings is the face of
masonry abutment at the right-bank end of the bridge. The
inel is straight for 1,200 feet above and below the station. At
. stages the current is very swift. Above the station a sand bar
red with willows affects the velocity of the current at low stages.
i banks are high and are supported by retaining walls. The bed
he stream is composed of sand and gravel and is permanent.
ih mark No. 1 is on the north wing of the east abutment of the
rt Avenue Bridge. Its elevation is 26.69 feet above the zero of the
i and 25.81 feet above the city datum. Bench mark No. 2 is the
h wing of the west abutment of the Walnut Street Bridge. Its
ation is 28.09 feet above gage datum and 27.2 feet above the city
m. Bench mark No. 3 is on the south wing of the west abutment
e Locust Street Bridge. Its elevation is 27.8 feet above the zero
e gage and 26.924 feet above the city datum.
ie observations at this station during 1903 have been made under
lirection of E. Johnson, jr., district hydrographer.

Discharge measurements of Des Moines River at Des Moines, Iowa, in 1902.

Date.	Hydrographer.	Gage height.	Discharge.
		Feet.	Second-feet
October 20	J. C. Stevens..................	7. 10	5, 7
November 12.................	K. C. Kastberg..............	3. 05	1, 7
November 15.................	.....do	5. 00	3, 4

Discharge measurements of Des Moines River at Locust street, Des Moines. Iowa in 1903.

Date.	Hydrographer.	Gage height.	Discharge.
		Feet.	Second-feet
March 16.....................	K. C. Kastberg..............	7. 70	8, 5
May 21.......................	F. W. Hanna	7. 25	6, 7
June 3	.....do	16. 57	22, 1
July 9........................	E. C. Murphy	4. 97	4, 0

Mean daily gage height, in feet, of Des Moines River at Des Moines, Iowa, for 19

Day.	Oct.	Nov.	Dec.	Day.	Oct.	Nov.	Dec.	Day.	Oct.	Nov.	D
1.............	7.80	3.05	3.70	12.............		3.05	2.50	23.............	4.70	6.30	
2.............	7.65	3.65	3.55	13.............	4.60	3.00	2.70	24.............	4.20	6.05	
3.............	7.60	3.85	3.65	14.............	4.35	3.35	2.80	25.............	4.05	5.65	
4.............	7.80	3.75	3.60	15.............	4.10	5.00	2.80	26.............		5.15	
5.............		3.55	3.40	16.............	4.20	5.45	2.80	27.............	3.65	4.80	
6.............	7.35	3.40	3.05	17.............	4.20	6.75	2.70	28.............	3.45	4.50	
7.............	6.75	3.35	2.50	18.............	9.80	7.40	2.70	29.............	3.30	4.20	
8.............	6.00	3.30	2.70	19.............	10.30	7.75	2.65	30.............	3.25	3.90	
9.............	5.55	3.20	2.60	20.............	7.65	7.35	2.55	31.............	3.15		
10.............	4.90	3.10	2.50	21.............	6.20	7.05	3.90				
11.............	4.75	3.10	2.40	22.............	5.25	6.75	4.60				

Mean daily gage height, in feet, of Des Moines River at Locust street, Des Moin Iowa, for 1903.

Day.	Jan.	Feb.	Mar.	Apr.	May.	June.	July.	Au
1....................................	2.75	1.85	4.50	5.20	4.05	19.65	5.05	
2....................................	2.70	1.80	4.30	5.10	4.10	18.25	7.70	
3....................................	2.70	1.80	4.40	5.05	4.70	16.50	7.45	
4....................................	2.55	1.90	4.25	5.10	5.10	15.25	6.65	....
5....................................	2.45	1.95	4.15	5.00	5.25	14.50	5.95	....
6....................................	2.30	1.90	4.40	4.90	5.25	13.60	5.60	....
7....................................	2.70	1.80	4.50	4.75	5.10	13.15	5.20	....
8....................................	2.85	1.80	4.30	4.70	4.80	12.80	5.10	....
9....................................	2.40	1.85	5.15	4.60	4.60	11.10	5.00	....
10....................................	2.30	1.85	6.40	4.45	4.40	10.05	5.35	....
11....................................	2.20	1.90	7.05	4.30	4.10	9.10	5.50	....
12....................................	2.20	1.85	8.25	4.15	5.90	8.15	8.00	....

Mean daily gage height, in feet, of Des Moines River, etc.—Continued.

Day.	Jan.	Feb.	Mar.	Apr.	May.	June.	July.	Aug.
13	2.10	1.80	8.75	4.10	7.60	7.90	8.35	......
14	2.10	1.85	8.85	5.25	8.75	7.70	7.95	......
15	2.00	1.90	8.10	5.05	10.00	7.55	7.80	......
16	2.00	1.80	7.70	4.90	10.30	6.10	7.00	......
17	1.95	1.75	7.45	4.80	9.85	5.80	6.45	......
18	1.90	1.60	7.25	4.60	8.80	5.50	6.75	......
19	1.90	1.45	7.30	4.85	8.85	6.00	7.45	......
20	1.90	1.30	7.30	4.95	7.20	5.90	8.35	......
21	2.00	1.15	7.15	5.10	6.95	6.15	8.30	......
22	2.00	1.25	6.90	5.35	8.05	6.85	7.55	......
23	1.95	1.35	6.55	5.35	8.40	6.80	6.75	......
24	1.90	1.50	6.10	5.05	8.30	6.50	5.90	......
25	1.90	1.70	5.90	4.70	8.50	6.15	5.55	......
26	1.85	1.85	5.65	4.40	9.25	5.80	5.10	......
27	1.85	3.10	5.60	4.20	14.12	5.95	4.95	......
28	1.80	3.70	5.55	4.05	19.71	5.15	4.65	......
29	1.75	......	5.55	3.95	20.44	4.75	4.95	......
30	1.85	......	5.45	3.90	22.88	4.45	4.90	......
31	1.90	......	5.30	......	21.55	......	4.80	......

RACCOON RIVER AT DES MOINES, IOWA.

This station was established November 18, 1902, by K. C. Kastberg, assisted by J. C. Stevens. It was discontinued May 31, 1903, on account of backwater from Des Moines River. The station was located at the South Ninth street highway bridge 1 mile south of the town. This bridge makes an angle of about 19° with the normal to the direction of the current. The gage rod as originally established consisted of a stadia rod. It was attached to an overhanging willow tree 200 feet below the bridge. This gage was replaced by a permanent one fastened to the south abutment and consisting of a heavy plank spiked to the coping. Its zero was placed at the same elevation as that of the original gage.

The gage was read once each day by E. T. Blagburn. Discharge measurements were made from the South Ninth street highway bridge. The initial point for soundings is the face of the south abutment. The channel is straight for 800 feet above and for 500 feet below the station. The right bank is the railroad embankment and does not overflow. The left bank overflows at extreme stages. The bed of the stream is sandy and slightly shifting. The bench mark is a crow's-foot on the top of the south abutment of the bridge. Its elevation is 28.80 feet above the zero of the gage.

The observations at this station during 1903 have been made under the direction of E. Johnson, jr., district hydrographer.

The following discharge measurement was made by K. C. Kastberg in 1903:

March 17: Gage height, 8.05 feet; discharge, 6,082 second-feet.

Mean daily gage height, in feet, of Raccoon River at South Ninth street, Des Moines, Iowa, for 1903.

Day.	Jan.	Feb.	Mar.	Apr.	May.	Day.	Jan.	Feb.	Mar.	Apr.	May.
1	5.15	2.95	5.45	5.80	5.75	17	3.25	3.80	8.05	5.30	10.30
2	5.30	2.90	5.70	5.80	5.90	18	3.40	4.20	7.90	5.15	9.60
3	4.55	2.90	5.20	5.75	5.80	19	3.55	3.95	7.80	5.50	9.15
4	4.40	2.95	5.25	5.70	6.10	20	3.30	3.75	7.85	5.40	8.30
5	4.45	3.05	5.15	5.60	6.20	21	3.30	3.40	7.65	5.70	7.80
6	4.35	3.05	5.30	5.55	6.05	22	3.35	3.45	7.50	5.90	9.85
7	3.75	......	5.60	5.35	5.70	23	3.45	3.65	7.35	5.65	11.30
8	4.00	3.00	5.40	5.20	5.50	24	3.45	3.80	7.10	5.30	......
9	3.90	3.00	6.00	5.00	5.35	25	3.40	3.95	6.60	5.10	......
10	3.90	2.90	6.90	4.95	5.20	26	3.35	4.05	6.40	4.80	......
11	3.90	2.95	7.65	5.05	5.10	27	3.25	......	6.35	5.00	15.00
12	3.95	3.15	8.80	5.00	6.05	28	3.10	5.60	6.20	5.05	19.80
13	3.70	3.30	9.15	4.85	8.00	29	3.00	......	6.10	5.30	20.10
14	3.50	......	9.10	6.55	9.40	30	3.05	......	6.05	5.45	23.00
15	3.45	3.65	8.45	6.00	10.55	31	3.20	......	5.85	......	......
16	3.30	3.50	8.15	5.65	10.85						

MISCELLANEOUS MEASUREMENTS IN THE DES MOINES RIVER DRAINAGE BASIN.

Des Moines River was measured at Ottumwa, Iowa, September 17, 1903, by F. W. Hanna, hydrographer. It had a discharge of 14,913 second-feet; the water surface was 18.3 feet below the floor of Market Street Bridge.

MISSOURI RIVER DRAINAGE BASIN.

Missouri River is formed by the junction of Jefferson, Madison, and Gallatin rivers at Three Forks, Mont. Of its tributaries, Osage River rises in the eastern part of Kansas, flows east, and enters the Missouri below Jefferson City, Mo. Big Sioux River drains the extreme eastern part of South Dakota and flows south into the Missouri at Sioux City, Iowa. Niobrara River drains the extreme northern part of Nebraska and flows east into the Missouri at Niobrara, Nebr. Yellowstone River has its source in a lake of the same name in Yellowstone National Park, from which it takes a general northeasterly course and flows into the Missouri near the Montana-North Dakota boundary. Its principal tributaries are Powder River, of which Clear Creek and Piney Creek in Wyoming are small tributaries near its headwaters, and the Bighorn, which drains northeastern Wyoming and of which the Shoshone is a tributary. Milk River rises on the divide in the northwestern part of Montana, flows northeast into Canada and then, with a general easterly course, flows back into Montana, emptying into the Missouri in the northeastern part of the State. Musselshell River rises in Meagher County, in south-central Montana, flows east, then north into the Missouri, about 100 miles above the mouth of Milk River. Marias River rises on the divide in the northwestern part of Montana and flows east into the Missouri. Two Medicine Creek is a tributary near its head-

ters.   West Gallatin River, the longest fork of the Gallatin, rises in
llowstone Park and flows north to its junction at Three Forks,
nt., with the Madison and Jefferson rivers.   Middle Creek is a small
butary of East Gallatin River.   Madison River rises in Yellow-
ne Park and flows north parallel to the West Gallatin.   Jefferson
iver is formed by Beaverhead and Big Hole rivers.   Its tributaries
rain extreme southwestern Montana.   Gasconade River rises in the
zark Mountains in southern Missouri and flows northeasterly into
lissouri River; it has excellent water-power facilities; its total drain-
ge area is 3,466 square miles, and its drainage area at Arlington is
,725 square miles.   The following is a list of the stations in the Mis-
ouri River drainage basin:

Gasconade River near Arlington, Mo.
Little Piney Creek near Arlington, Mo.
Big Piney Creek near Hooker, Mo.
Osage River at Ottawa, Kans.
Big Sioux River near Watertown, S. Dak.
James River at Lamoure, N. Dak.
Niobrara River near Valentine, Nebr.
Elk Creek, near Piedmont, S. Dak.
Box Elder Creek at Black Hawk, S. Dak.
Rapid Creek at Rapid, S. Dak.
Spring Creek at Blair's ranch, near Rapid, S. Dak.
Battle Creek at Hermosa, S. Dak.
Cheyenne River at Edgemont, S. Dak.
Belle Fourche River at Belle Fourche, S. Dak.
Red Water River at Belle Fourche, S. Dak.
Red Water River at Minnesela, S. Dak.
Spearfish River at Toomey's ranch, near Spearfish, S. Dak.
Cannon Ball River at Stevenson, N. Dak
Heart River near Richardton, N. Dak.
Knife River at Broncho, N. Dak.
Little Missouri River at Medora, N. Dak.
Little Missouri River at Camp Crook, S. Dak.
Clear Creek at Buffalo, Wyo.
Piney Creek at Kearney, Wyo.
Cruez ditch near Story, Wyo.
Prairie Dog ditch near Story, Wyo.
Tongue River at Dayton, Wyo.
Shoshone River near Cody, Wyo.
Shoshone River (South Fork) at Marquette, Wyo.
Grey Bull River at Meeteetse, Wyo.
Bighorn River near Thermopolis, Wyo.
Yellowstone River at Glendive, Mont.
Yellowstone River near Livingston, Mont.
Beaver Creek near Ashfield, Mont.
Beaver Creek overflow near Bowdoin, Mont.
Milk River at Malta, Mont.
Harlem canal (head of) near Zurich, Mont.
Paradise Valley canal near Chinook, Mont.
Fort Belknap canal at Yantic, Mont.
Milk River at Havre, Mont.
Musselshell River near Shawmut, Mont.

Marias River near Shelby, Mont.
Two Medicine Creek at Midvale, Mont.
Sun River near Augusta, Mont.
Missouri River at Cascade, Mont.
Missouri River near Townsend, Mont.
Gallatin River at Logan, Mont.
West Gallatin River near Salesville, Mont.
Middle Creek near Bozeman, Mont.
Madison River (including Cherry Creek) near Norris, Mont.
Jefferson River near Sappington, Mont.

GASCONADE RIVER NEAR ARLINGTON, MO.

This station was established April 11, 1903, by I. W. McCom
It is located about 2 miles below Arlington, at the foot of the
bluff near the river on the right bank below the city. The gage
vertical 2 by 4 inch rod 16 feet long, fastened to a leaning willow
It was read once each day during 1903 by C. B. Louzader. Disch
measurements are made from a cable and boat, with distances ma
on a tag wire. On account of the cable being washed away by fl
an apparatus for raising and lowering the cable was establi
November 28, 1903. The elevation of the gage was verified at
time. The initial point for soundings is a blazed white-oak tre
feet up the bank from the gage in the line of the cross section.
bench mark to which the gage is referred is on the same tree and
an elevation of 27.2 feet above the zero of the gage. The chann
straight for 500 feet above and 2,000 feet below the station. The
rent is swift. The right bank is high and wooded to the edge of
cliff, which rises for a vertical distance of 200 to 250 feet. The
bank is low, but not likely to be overflowed except at extreme
stages. The bed of the stream is of gravel and bed rock, with c
sional bowlders, and is not likely to shift. There is but one cha
at all stages. At ordinary stages the water is about 10 feet deep.

The observations at this station during 1903 have been made u
the direction of E. Johnson, jr., district hydrographer.

Discharge measurements of Gasconade River near Arlington, Mo., in 190

Date.	Hydrographer.	Gage height.	Disch
		Feet.	*Secon*
April 18	W. A. Luther	6.10	
May 9	R. S. Webster	4.38	
June 25	F. W. Hanna	5.50	
July 23	do	2.94	
August 28	do	2.84	
September 23	do	3.42	
October 20	do	3.40	
November 28	do	2.40	
December 28	do	3.20	

ily gage height, in feet, of Gasconade River near Arlington, Mo., for 1903.

Day.	Apr.	May.	June.	July.	Aug.	Sept.	Oct.	Nov.	Dec.
....	....	3.80	14.80	3.90	3.00	2.70	2.80	2.70	2.50
....	....	3.70	13.60	3.70	3.30	2.70	2.80	2.70	2.40
....	....	3.60	14.70	3.60	2.90	2.60	2.80	2.70	2.40
....	....	3.50	13.40	3.50	3.00	2.60	3.10	2.90	2.40
....	....	3.50	12.50	3.40	2.90	2.50	3.00	2.80	2.40
....	....	3.80	11.00	3.30	2.90	2.50	3.00	2.70	2.40
....	....	3.90	9.00	3.20	4.50	2.40	7.50	2.70	2.30
....	....	4.20	8.20	3.20	3.80	2.40	11.30	2.60	2.30
....	....	4.40	7.80	3.10	3.50	2.90	8.00	2.50	2.30
....	....	4.30	6.60	3.00	4.70	3.50	6.60	2.50	2.30
....	....	4.20	6.10	3.00	6.70	3.30	5.50	2.50	2.30
....	5.10	4.00	5.60	3.00	5.00	3.20	4.90	2.50	2.30
....	7.40	4.00	5.20	3.00	7.20	3.10	4.50	2.50	2.40
....	9.40	4.10	5.00	3.00	8.70	6.80	4.20	2.50	2.40
....	9.50	4.60	4.80	3.00	6.50	5.40	4.00	2.50	2.30
....	8.00	6.90	4.60	3.00	5.40	5.20	3.80	2.50	2.30
....	6.90	6.50	4.40	3.00	4.70	4.80	3.70	2.50	2.30
....	6.20	6.40	4.30	2.90	4.20	5.20	3.60	2.50	2.30
....	5.80	6.60	4.20	2.90	3.90	4.50	3.50	2.50	2.30
....	5.90	7.40	4.00	2.90	3.70	4.20	3.40	2.50	2.30
....	5.40	7.40	3.90	2.90	3.70	4.00	3.30	2.50	2.30
....	5.10	8.30	4.60	2.90	3.40	3.70	3.20	2.50	2.30
....	4.90	9.10	9.50	2.90	3.20	3.50	3.10	2.50	2.30
....	4.70	8.00	6.70	3.20	3.10	3.30	3.00	2.50	2.60
....	4.50	6.80	5.60	5.00	3.00	3.20	2.90	2.50	2.70
....	4.30	6.10	5.00	4.50	2.90	3.10	2.80	2.50	2.50
....	4.20	5.60	4.60	3.90	2.90	3.00	2.80	2.50	2.60
....	4.10	5.20	4.40	3.60	2.80	2.90	2.70	2.50	3.30
....	4.00	5.20	4.80	3.30	2.80	2.90	2.70	2.50	3.30
....	3.90	6.00	4.00	3.30	2.80	2.80	2.70	2.50	3.10
....	....	15.00	....	3.10	2.80	....	2.70	....	3.00

table for Gasconade River near Arlington, Mo., from April 12 to December 31, 1903.

Discharge.	Gage height.	Discharge.	Gage height.	Discharge.	Gage height.	Discharge.
Second-feet.	*Feet.*	*Second-feet.*	*Feet.*	*Second-feet.*	*Feet.*	*Second-feet.*
865	3.0	1,112	3.8	1,780	4.6	2,716
874	3.1	1,176	3.9	1,884	4.7	2,842
884	3.2	1,246	4.0	1,992	4.8	2,968
902	3.3	1,322	4.1	2,104	4.9	3,094
928	3.4	1,404	4.2	2,220	5.0	3,220
962	3.5	1,492	4.3	2,340		
1,004	3.6	1,584	4.4	2,464		
1,054	3.7	1,680	4.5	2,590		

mt at 4.5 feet gage height. Differences above this point, 126 per tenth.

Estimated monthly discharge of Gasconade River near Arlington, Mo., for 1903.

Month.	Discharge in second-feet.			Total in acre-feet.
	Maximum.	Minimum.	Mean.	
April 12–30	8,890	1,884	4,174	157,98
May	15,820	1,492	4,105	252,47
June	15,568	1,884	5,824	346,
July	8,220	1,054	1,874	84,
August	7,882	1,004	2,196	135,07
September	5,488	884	1,690	100,58
October	11,158	962	2,246	138,18
November	1,054	902	921	54,68
December	1,822	874	931	57,24

LITTLE PINEY CREEK NEAR ARLINGTON, MO.

This station was established April 10, 1903, by I. W. McConnell, and was discontinued August 21 because it was found to be influenced by back water from the Gasconade and on account of the small size of the stream, which flows through a territory the character of which is very similar to that of Gasconade River itself.

The station was located half a mile above the point where the Little Piney joins the Gasconade. The measurements were made from a boat and tag wire, and the gage consisted of a vertical pine rod graduated to feet and tenths, and was read daily by Charles Courson. The initial point for soundings was at the edge of a vertical rock cliff on the right bank, about 70 feet from the water's edge. The bench mark consisted of a nail in the root of a tree on the right bank. This was 13.92 feet above the zero of the gage.

The observations at this station during 1903 have been made under the direction of E. Johnson, jr., district hydrographer.

Discharge measurements of Little Piney Creek near Arlington, Mo., in 1903.

Date.	Hydrographer.	Gage height.	Discharge.
		Feet.	Second-feet.
April 16	I. W. McConnell	4.05	504
April 25	R. S. Webster	3.50	289
May 9	W. A. Luther	3.20	200
June 24	F. W. Hanna	8.30	265
July 23	....do	2.70	97

Mean daily gage height, in feet, of Little Piney Creek near Arlington, Mo., for 1903.

Day.	Apr.	May.	June.	July.	Day.	Apr.	May.	June.	July.
1	......	3.90	10.20	2.90	17	3.80	3.50	3.40	2.70
2	......	3.20	10.20	2.90	18	3.60	3.40	3.40	2.70
3	......	3.20	10.80	2.90	19	3.60	3.40	3.30	2.70
4	......	3.30	10.00	2.90	20	3.90	3.90	3.80	2.70
5	......	3.60	8.00	2.90	21	3.50	4.50	3.20	2.70
6	......	3.40	5.20	2.90	22	3.40	4.90	4.00	2.70
7	......	3.30	4.30	2.80	23	3.40	5.00	3.80	2.70
8	......	3.30	4.10	2.80	24	3.40	3.40	3.30	2.70
9	......	3.90	3.80	2.70	25	3.40	3.40	3.20	2.70
10	......	3.30	3.70	2.70	26	3.40	3.40	3.20	2.70
11	......	3.40	3.60	2.70	27	3.40	3.40	3.10	2.70
12	2.50	3.40	3.60	2.70	28	3.40	3.40	3.10	2.70
13	5.90	3.60	3.50	2.70	29	3.90	3.50	3.00	2.70
14	5.50	3.50	3.50	2.70	30	3.90	3.80	2.90	2.70
15	4.90	3.20	3.50	2.70	31	......	9.60	......	2.70
16	4.00	3.20	3.50	2.70					

Rating table for Little Piney Creek near Arlington, Mo., from April 12 to August 1, 1903.

Gage height.	Discharge.	Gage height.	Discharge.	Gage height.	Discharge.	Gage height.	Discharge.
Feet.	Second-feet.	Feet.	Second-feet.	Feet.	Second-feet.	Feet.	Second-feet.
2.5	58	3.2	225	3.9	450	4.6	700
2.6	76	3.3	255	4.0	484	4.7	738
2.7	96	3.4	286	4.1	519	4.8	776
2.8	118	3.5	318	4.2	554	4.9	815
2.9	142	3.6	350	4.3	590	5.0	854
3.0	168	3.7	383	4.4	626	5.1	894
3.1	196	3.8	416	4.5	663	5.2	934

Tangent at 5 feet gage height. Differences above this point, 40 per tenth. Limits of accuracy are 2.70 and 4.00. Above and below, table is approximate.

Estimated monthly discharge of Little Piney Creek near Arlington, Mo., for 1903.

Month.	Discharge in second-feet.			Total in acre-feet.
	Maximum.	Minimum.	Mean.	
April 12–30	1,214	58	422	15,903
May	2,694	225	413	25,394
June	3,174	142	746	44,390
July	142	96	106	6,518

BIG PINEY CREEK NEAR HOOKER, MO.

This station was established on April 12, 1903, by I. W. McConnell, and was discontinued August 22 on account of its being influenced by back water from the Gasconade. No station was reestablished on this stream, as its drainage basin is similar to that of the Gasconade.

The station was located half a mile above the wagon ford at Hooker, Mo., and 2 miles above the mouth of the stream. The measurements were made from a boat and tag wire. The gage was an oak timber securely fastened to an overhanging tree and graduated to feet and tenths. It was read daily by P. H. Hooker. The initial point for soundings was a tree at the water's edge on the left bank, to which the tag wire was attached. The bench mark consisted of a spike driven in the tree near the gage. The head of this spike is 18.70 feet above the zero of the gage.

The observations at this station during 1903 have been made under the direction of E. Johnson, jr., district hydrographer.

Discharge measurements of Big Piney Creek near Hooker, Mo., in 1903.

Date.	Hydrographer.	Gage height.	Discharge.
		Feet.	*Second-feet.*
April 19	R. S. Webster	2.90	1,151
April 25	do	2.33	683
May 9	do	2.18	640
June 26	F. W. Hanna	2.70	956
July 22	do	1.70	397

Mean daily gage height, in feet, of Big Piney Creek near Hooker, Mo., for 1903.

Day.	Apr.	May.	June.	July.	Aug.	Day.	Apr.	May.	July.	Aug.
1		2.00	8.10		1.60	17	3.70	3.00	1.90	1.00
2		2.00			1.60	18	3.20	3.90	1.80	2.80
3		2.00	6.90		1.60	19	2.90	4.90	1.80	2.40
4		2.00	5.70		1.60	20	3.00	4.00	1.80	2.00
5		2.10	5.40		1.70	21	2.90	4.50	1.70	2.00
6		2.20	5.20		1.70	22	2.90	5.70	1.70	1.90
7		2.20	4.80		1.60	23	2.90	4.90	1.90	
8		2.30	4.20		1.70	24	2.60	3.80	2.90	
9		2.10	3.90		1.80	25	2.50	3.70	3.10	
10		2.10	3.70		2.00	26	2.40	3.70	2.00	
11		2.10	3.20		2.00	27	2.30	3.50	1.90	
12	2.60	2.20	2.90	2.30	2.10	28	2.30	3.10	1.80	
13	3.50	2.20	2.60	2.10	4.30	29	2.30	3.20	1.70	
14	4.30	2.40	2.60	2.00	5.40	30	2.20	4.30	1.70	
15	4.00	3.10	2.60	2.00	3.90	31		6.90	1.60	
16	3.80	3.80	2.50	2.00	3.20					

ㅈㅈㅌing table for Big Piney Creek near Hooker, Mo., from April 12 to August 22, 1903.

Gage height.	Discharge.	Gage height.	Discharge.	Gage height.	Discharge.	Gage height.	Discharge.
Feet.	*Second-feet.*	*Feet.*	*Second-feet.*	*Feet.*	*Second-feet.*	*Feet.*	*Second-feet.*
1.6	387	2.3	654	3.0	1,242	3.7	1,886
1.7	397	2.4	725	3.1	1,334	3.8	1,978
1.8	417	2.5	802	3.2	1,426	3.9	2,070
1.9	447	2.6	884	3.3	1,518	4.0	2,162
2.0	486	2.7	970	3.4	1,610		
2.1	534	2.8	1,059	3.5	1,702		
2.2	590	2.9	1,150	3.6	1,794		

Tangent at 3 feet gage height. Differences above this point, 92 per tenth.

Estimated monthly discharge of Big Piney Creek near Hooker, Mo., for 1903.

Month.	Discharge in second-feet.			Total in acre-feet.
	Maximum.	Minimum.	Mean.	
April 12–30	2,438	590	1,225	46,165
May	4,830	486	1,482	91,125
June 1–20	5,934	725	2,206	87,512
July 12–31	1,834	387	530	21,025
August 1–22	3,450	387	858	37,440

OSAGE [a] AT OTTAWA, KANS.

This station was established August 26, 1902, by W. G. Russell. It is located at the highway bridge near the center of the town of Ottawa, Kans. The gage is of the old wire type, with its scale spiked to the bridge floor. It is read once each day by W. H. Blacksten. Discharge measurements are made from the highway bridge, to which the gage is attached. The bridge is somewhat oblique to the thread of the stream and has a total span between abutments of 135 feet. At low stages the discharge can be measured by wading at a short distance below the bridge. The channel is slightly curved and the current is swift. Both banks are high and the bed of the stream is rocky.

The observations at this station during 1903 have been made under the direction of W. G. Russell, district hydrographer.

---

[a] Known also by the name Marais des Cygnes.

Discharge measurements of Osage River at Ottawa, Kans., in 1903.

Date.	Hydrographer.	Gage height.	Dischar
		Feet.	Second-f
April 8	W. G. Russell	2.50	
August 21	do	2.20	
September 28	E. C. Murphy	1.50	

Mean daily gage height, in feet, of Osage River at Ottawa, Kans., for 1903.

Day.	Jan.	Feb.	Mar.	Apr.	May.	June.	July.	Aug.	Sept.	Oct.	Nov.	D
1	2.10	2.00	5.30	2.20	2.00	23.50	2.00	1.40	1.60	1.50	11.00	
2	2.10	2.00	2.90	2.20	2.00	22.10	2.10	1.60	1.60	1.60	12.80	
3	2.10	2.10	2.60	3.90	2.00	18.45	2.20	6.60	1.60	2.00	8.00	
4	2.10	2.00	2.50	8.90	2.00	18.85	3.00	2.20	1.60	2.30	4.80	
5	2.20	2.00	2.50	6.30	2.00	11.80	3.20	1.90	1.60	2.00	3.50	
6	2.10	2.00	3.10	3.50	2.10	9.50	2.10	8.55	1.60	1.80	3.10	
7	2.00	2.00	4.50	2.80	2.20	5.50	2.00	7.60	1.60	20.50	2.80	
8	2.40	1.90	6.60	2.50	2.30	3.40	2.00	10.80	1.70	21.50	2.70	
9	2.10	1.90	2.80	2.40	2.20	3.00	1.11	10.30	8.55	11.30	2.60	
10	2.10	1.90	2.60	2.80	2.10	2.90	1.60	3.80	7.00	3.10	2.50	
11	2.10	1.90	3.90	6.20	2.10	2.70	1.00	5.80	5.00	2.50	2.40	
12	2.00	1.90	3.20	4.20	2.70	2.50	2.10	2.70	2.40	2.20	2.70	
13	2.00	1.90	2.80	2.70	9.90	2.40	2.00	5.85	2.30	2.30	2.40	
14	2.00	1.90	2.50	2.70	8.80	2.40	1.10	9.70	2.30	18.75	2.30	
15	2.00	1.90	2.40	3.00	3.70	2.30	1.10	13.30	2.10	17.10	2.30	
16	2.00	1.90	2.40	2.60	3.00	2.30	2.07	15.90	2.00	18.00	2.30	
17	2.00	1.90	2.30	2.40	2.60	2.20	2.00	8.65	1.90	5.40	2.20	
18	2.00	1.90	2.30	2.30	2.50	2.20	2.00	3.30	1.80	3.00	2.20	
19	2.00	1.90	2.40	2.30	2.30	2.20	1.10	2.50	1.80	2.80	2.10	
20	2.00	1.90	6.40	2.30	2.20	2.40	1.80	2.30	1.70	2.50	2.00	
21	1.90	1.90	4.00	2.30	2.20	2.60	1.60	2.20	1.70	2.40	2.00	
22	1.90	1.90	2.70	2.20	2.70	2.30	1.60	2.10	1.70	2.40	2.00	
23	1.90	1.90	2.50	2.30	3.30	2.10	1.50	2.00	1.70	2.30	2.10	
24	2.00	2.00	2.50	2.10	21.90	2.00	1.50	1.90	1.60	2.20	2.10	
25	1.90	2.40	2.60	2.10	24.00	2.00	1.50	1.90	1.60	2.20	2.10	
26	2.00	2.70	2.60	2.00	21.90	3.40	1.50	1.80	1.60	2.10	2.10	
27	2.10	3.50	2.50	2.00	9.90	2.40	1.40	1.80	1.60	2.10	2.00	
28	2.20	3.60	2.40	2.00	10.10	2.30	1.40	1.70	1.50	2.00	2.00	
29	2.30		2.30	2.00	13.30	2.10	1.40	1.70	1.50	2.00	2.00	
30	2.20		2.30	2.00	18.50	2.00	1.40	1.70	1.50	2.00	2.00	
31	2.10		2.20		22.60		1.40	1.70		4.20		

BIG SIOUX RIVER NEAR WATERTOWN, S. DAK.

Big Sioux River rises in Grant County, S. Dak., about 30 mi
north of Watertown. Its principal headwaters drain lands con
tuting part of the Sisseton and Wahpeton Indian Reservation.
general course is southeast, and it empties into Missouri River n
Sioux City, Iowa. The river is of interest on account of its wa
powers, a number of which have been developed, principally at Fl
dreau, Dell Rapids, and Sioux Falls, S. Dak., and at Akron, Iowa

e Poinsett, which lies almost wholly in Hamlin County, S. Dak.,
; outlet in Big Sioux River near Dempster, a short distance
Estelline. Immediately below the outlet of the lake a dam has
onstructed on the Big Sioux to maintain the level of the lake
certain limits.

gaging station was established by O. V. P. Stout, the gage
put in September 15, 1900, by George W. Carpenter, county
or for Codington County. It is located on the farm of L. E.
about 4 miles above Watertown. The gage consists of an
l rod securely fastened on the right bank of the stream. The
r is L. E. Spicer. This station was discontinued January 23,

bservations at this station during 1903 have been made under
ction of J. C. Stevens, district hydrographer.

rge measurements of Big Sioux River near Watertown, S. Dak., in 1903.

Date.	Hydrographer.	Gage height.	Discharge.
		Feet.	*Second-feet.*
....	R. B. Crane................	1.45	28
....	......do	1.60	23
1?	......do	2.18	34

aily gage height, in feet, of Big Sioux River near Watertown, S. Dak., for 1903.

Day.	Mar.	Apr.	May.	June.	July.	Aug.	Sept.	Oct.	Nov.
			1.35	......	......	......	1.85	......	1.75
			......	1.50	......	2.50		2.10	......
		2.30	1.35	......	1.45	......	......	......	1.70
					......	2.70	1.80	2.20	......
		2.10	1.30	1.50	3.30	......	......	......	......
			......	......	4.15	......	1.70	2.30	1.60
		2.00	......	2.65	3.40	2.70	......	......	......
			1.25	......	......	......	2.00	......	1.60
			......	2.60	......	2.55	......	2.50	......
		1.90	1.25	......	2.55	......	......	......	1.60
	2.00		......	......	......	2.35	2.00	2.40	......
	6.90	1.75	1.30	2.00	2.15	......	......	......	......
	6.50			......	......	......	3.75	2.30	1.60
		1.65	......	1.70	2.00	2.20	4.70	......	......
	5.50		1.20	......	......	......	4.25	......	1.60
			......	1.45	......	2.10	......	2.18	......
	4.00	1.55	1.15	......	2.05	......	......	......	1.60
			......	......	......	2.00	3.55	2.10	......
		1.50	1.00	1.45	1.95	......	......	......	......
	3.80		......	......	......	......	3.15	2.00	1.60
		1.45	......	1.35	1.86	1.90	......	......	......
	3.75		1.50	......	......	......	2.90	......	......

Mean daily gage height, in feet, of Big Sioux River, etc.—Continued.

Day.	Mar.	Apr.	May.	June.	July.	Aug.	Sept.	Oct.	Nov.
23				1.25		1.80		1.85	
24	3.55	1.30	1.50		2.20				
25						1.70	2.35	1.80	
26		1.20	1.50	1.10	2.30				
27	3.00		2.20				2.30	1.80	
28		1.20	2.00	1.05	2.40	1.85			
29	2.75						2.25	1.75	
30		1.35	1.60	1.80		1.85			
31	2.50		1.55		2.25				

Rating table for Big Sioux River near Watertown, S. Dak., from January 1 to December 31, 1903.

Gage height.	Discharge.	Gage height.	Discharge.	Gage height.	Discharge.	Gage height.	Discharge.
Feet.	Second-feet.	Feet.	Second-feet.	Feet.	Second-feet.	Feet.	Second-feet.
0.8	2	1.7	27	2.6	91	3.5	200
.9	4	1.8	32	2.7	101	3.6	216
1.0	6	1.9	38	2.8	111	3.7	232
1.1	8	2.0	44	2.9	121	3.8	248
1.2	10	2.1	51	3.0	132	3.9	264
1.3	13	2.2	58	3.1	144	4.0	280
1.4	16	2.3	65	3.2	157		
1.5	19	2.4	73	3.3	171		
1.6	23	2.5	82	3.4	185		

This table was applied indirectly according to the method outlined in Nineteenth Annual Report United States Geological Survey, part 4, page 323.

Estimated monthly discharge of Big Sioux River near Watertown, S. Dak., for 1903.

Month.	Discharge in second-feet.			Total in acre-feet
	Maximum.	Minimum.	Mean.	
March, for 11 days	740	44	302	6,5...
April, for 13 days	74	16	39	1,0...
May, for 16 days	62	8	22	6...
June, for 13 days	96	7	30	7...
July, for 14 days	300	18	75	2,0...
August, for 13 days	74	16	40	1,0...
September, for 14 days	340	16	100	2,7...
October, for 13 days	54	20	33	8...
November, for 9 days	26	22	24	4...

JAMES RIVER AT LAMOURE, N. DAK.

This station was established May 8, 1903, by F. E. Weymouth. The station is located at the highway bridge one-half mile from Lamoure, N. Dak. In July, 1902, the United States Weather Bureau established a vertical timber gage at this bridge. It reads from −3 feet to +17 feet, is well painted, and is securely fastened to one of the bridge piers. The Geological Survey gage readings have been made on this rod. The observer is Carl M. Knudson. Discharge measurements are made from the bridge to which the gage is attached. The initial point for soundings is the end of the hand rail on the upstream side of the bridge on the left bank. The channel is straight for a considerable distance above and below the station and the current is sluggish. Both banks are high. The bed of the stream is sandy and shifting. The bench mark is the top of the plate on the sheet-iron concrete pier on the lower side at the right bank. Its elevation is 14.07 feet above the zero of the gage. During the summer the channel becomes choked with luxuriant plant growth, so that no definite relation between the amount of flow and the height of water surface can be established. Observations were therefore discontinued July 31, 1903.

The observations at this station during 1903 have been made under the direction of E. F. Chandler, district hydrographer.

Discharge measurements of James River at Lamoure, N. Dak., in 1903.

Date.	Hydrographer.	Gage height.	Discharge.
		Feet.	*Second-feet.*
May 8	F. E. Weymouth	−1.50	127
July 29	E. F. Chandler	−0.24	About 5 or 10.

Mean daily gage height, in feet, of James River at Lamoure, N. Dak., for 1903.

Day.	Apr.	May.	June.	July.	Day.	Apr.	May.	June.	July.
1	(a)	−1.60	−0.50	−0.10	17	−0.10	−1.40	−0.20	0.00
2	3.90	−1.60	− .50	− .10	18	− .40	−1.30	− .20	− .10
3	5.40	−1.60	− .50	.00	19	− .60	−1.40	− .20	− .10
4	6.20	−1.60	− .50	.00	20	− .80	−1.50	− .10	− .10
5	6.50	−1.60	− .50	.00	21	− .90	−1.60	− .10	− .10
6	6.50	−1.60	− .50	.00	22	−1.00	−1.10	− .10	− .20
7	6.50	−1.60	− .40	.00	23	−1.10	− .30	− .10	− .20
8	5.40	−1.40	− .40	.00	24	−1.20	−1.40	− .10	− .20
9	4.70	−1.50	− .40	.00	25	−1.30	−1.00	− .10	− .20
10	4.20	−1.60	− .40	.00	26	−1.30	.00	− .10	− .20
11	3.70	−1.60	− .40	.00	27	−1.10	.00	− .10	− .20
12	3.10	−1.60	− .30	.00	28	−1.40	− .20	− .10	− .20
13	2.40	−1.60	− .30	.00	29	−1.50	− .30	− .10	− .20
14	1.50	−1.60	− .30	.00	30	−1.60	− .40	− .10	− .30
15	.80	−1.50	− .30	.00	31		− .40		− .30
16	.20	−1.50	− .20	.00					

a Frozen.

NIOBRARA RIVER NEAR VALENTINE, NEBR.

This station was established July 22, 1897, by O. V. P. Stout, and known as the Fort Niobrara station. On June 26, 1901, it was r tablished at the Borman bridge, which is about 3 miles upstream f the military bridge. The gage is of the wire and weight type, loc about 1,000 feet upstream from the bridge on the left bank. The runs over a pulley supported by a frame, on the arm of which the g heights are read from a scale graduated to half tenths. The dista from the end of the weight to the index or marker on the wire is feet. Measurements are made from a one-span steel bridge, restin; tubular concrete piers, none of which obstruct the channel. The stream hand rail is marked at 10-foot intervals, and the initial p for soundings is the zero mark on the upstream hand rail in line the west face of the east pier. The channel is straight for about feet both above and below the gaging section. The banks are high not liable to overflow, while the bed is rock covered at times wi thin layer of shifting sand. The velocity of the water is always high.

The range of gage heights seldom exceeds 1 foot and the str has a marked constancy of flow.

Bench mark No. 1 is the head of a nail driven in the stump of a l elder tree—one of a clump of four—just east of Mr. Borman's ho Its elevation above gage datum is 17.26 feet. Bench mark No. 2 is by 6 inch pine head block on which the gate rests when closed. elevation is 16.19 feet above gage datum.

The observations at this station during 1903 have been made ur the direction of J. C. Stevens, district hydrographer.

Discharge measurements of Niobrara River near Valentine, Nebr., in 190;

Date.	Hydrographer.	Gage height.	Disch
		Feet.	*Secona*
April 26	J. C. Stevens	1.55	
May 26	....do	1.55	
Do	....do	1.55	
June 25	....do	1.40	
August 25	....do	1.80	
September 17	E. C. Murphy	1.38	
Do	....do	1.38	
December 23	J. C. Stevens	1.75	

aily gage height, in feet, of Niobrara River near Valentine, Nebr., for 1903.

Day.	Feb.	Mar.	Apr.	May.	June.	July.	Aug.	Sept.	Oct.	Nov.	Dec.
....	.....	1.18	1.90	1.50	1.50	1.40	1.56	1.30	1.38	1.52	1.58
....	.....	1.67	1.94	1.52	1.48	1.37	1.95	1.33	1.41	1.50	1.62
....	.....	1.90	1.98	1.50	1.42	1.35	3.20	1.31	1.45	1.51	1.57
....	.....	1.75	1.80	1.55	1.47	1.37	1.60	1.37	1.38	1.55	1.60
....	.....	1.80	1.85	1.55	1.43	1.30	1.51	1.29	1.42	1.46	(a)
....	.....	1.78	1.65	1.57	1.40	1.33	1.25	1.32	1.41	1.50	1.55
....	.....	1.90	1.63	1.62	1.38	1.30	1.26	1.35	1.40	1.50	1.60
....	.....	1.95	1.65	1.60	1.44	1.30	1.27	1.33	1.32	1.50	1.62
....	.....	1.98	1.68	1.61	1.48	1.29	1.20	1.31	1.34	1.52	1.66
....	.....	2.10	1.85	1.72	1.39	1.31	1.22	1.32	1.35	1.50	1.62
....	.....	2.05	1.70	1.68	1.41	1.33	1.24	1.29	1.44	1.52	1.61
....	.....	2.13	1.58	1.55	1.31	1.35	1.28	1.32	1.50	1.55	1.60
....	.....	2.15	1.62	1.58	1.28	1.32	1.30	1.33	1.42	1.58	1.51
....	.....	2.10	1.60	1.48	1.30	1.38	1.42	1.35	1.43	1.55	1.41
....	1.26	2.05	1.52	1.47	1.32	1.40	1.35	1.43	1.45	1.60	1.45
....	1.22	1.98	1.57	1.45	1.30	1.37	1.40	1.40	1.51	(b)	1.65
....	1.18	2.02	1.60	1.50	1.34	2.12	1.39	1.38	1.40	(b)	1.80
....	1.56	2.05	1.68	1.51	1.20	3.37	1.35	1.37	1.41	(b)	1.82
....	1.66	2.08	1.90	1.48	1.42	2.00	1.33	1.35	1.39	(a)	1.88
....	1.75	1.78	1.87	1.45	1.35	1.55	1.34	1.38	1.42	(b)	1.82
....	1.84	1.85	1.75	1.47	1.52	1.41	1.33	1.41	1.45	(b)	1.89
....	1.95	1.72	1.70	1.45	1.35	1.35	1.34	1.45	1.48	1.80	1.78
....	1.70	1.80	1.72	1.40	1.48	1.34	1.31	1.34	1.48	1.75	1.80
....	1.83	1.75	1.58	1.42	1.40	1.32	1.32	1.35	1.47	1.65	1.79
.....·...............	1.76	1.78	1.63	1.38	1.40	1.33	1.29	1.42	1.45	1.62	1.90
....	1.98	1.96	1.55	1.57	1.35	1.35	1.35	1.40	1.40	1.60	2.00
....	1.95	1.95	1.53	1.40	1.36	1.30	1.45	1.34	1.45	1.62	1.80
....	1.42	1.90	1.65	1.40	1.40	1.25	1.37	1.36	1.48	1.63	1.75
....	.....	1.88	1.60	1.50	1.35	1.30	1.33	1.40	1.49	1.57	1.74
....	.....	1.85	1.57	1.41	1.33	1.36	1.27	1.45	1.47	1.52	1.72
....	.....	1.88	......	1.42	......	1.47	1.31	......	1.48	......	1.73

a No record.   *b* Ice.

table for Niobrara River near Valentine, Nebr., from January 1 to December 31, 1903.

Gage height.	Discharge.	Gage height.	Discharge.	Gage height.	Discharge.	Gage height.	Discharge.
Feet.	*Second-feet.*	*Feet.*	*Second-feet.*	*Feet.*	*Second-feet.*	*Feet.*	*Second-feet.*
.0	475	1.4	655	1.8	1,155	2.2	1,985
1	490	1.5	750	1.9	1,330		
2	525	1.6	865	2.0	1,525		
3	580	1.7	1,000	2.1	1,740		

table was applied indirectly according to the methods outlined in Nine-Annual Report of the United States Geological Survey, part 4, page 323.

Estimated monthly discharge of Niobrara River near Valentine, Nebr., for 1903.

[Drainage area, 6,070 square miles.]

Month.	Discharge in second-feet.			Total in acre-feet.	Run-off.	
	Maximum.	Minimum.	Mean.		Second-feet per square mile.	Depth in inches.
February 15–28	1,525	505	998	27,713	0.164	0.065
March	1,860	525	1,357	83,439	.224	.258
April	1,425	750	979	58,255	.161	.180
May	1,075	700	794	48,821	.131	.151
June.............	805	615	692	41,177	.114	.127
July	7,000	550	912	56,077	.150	.173
August............	6,800	525	843	51,834	.139	.160
September	750	580	638	37,964	.105	.117
October	750	580	685	42,119	.113	.130
November 1–14 and 22–30 [a]	....	.....	724	33,028	.119	.102
December, 5th missing..............	1,000	525	723	43,021	.119	.133

[a] Ice.

ELK CREEK NEAR PIEDMONT, S. DAK.

This station was established August 21, 1903, by R. F. Walter, assisted by W. T. Carpenter. It is located on the highway known as the Valley road, and is on the first bridge below Piedmont. The gage is a vertical 1½ by 4 inch pine timber spiked to the downstream pile of the pier at the middle of the span. The foot marks are steel nails projecting about an inch from the timber. The tenths are marked with brass tacks. The gage was read twice daily by Reese Fockler. The discharge measurements are best made by wading, as the bridge crosses the stream at an angle. The initial point for soundings is a steel nail driven into the side of a fence post on the south or downstream side of the bridge. The channel is curved for 100 feet above and below the station. The stream has a swift current. Both banks are low but are not subject to overflow. They are wooded below, but cleared above the station. The bed of the stream is sandy and shifting.

Bench mark No. 1 is the top of a nail surrounded by a circle of brass tacks driven into the top of a pile in the wing. The pile is the most easterly one in the north end of the bridge. Its elevation above the zero of the gage is 6.98 feet. Bench mark No. 2 is the top of a steel spike driven into a root of an elm tree 300 feet north of the bridge. Its elevation is 5.26 feet above the zero of the gage. The drainage area above the station is 100 square miles.

This station was abandoned soon after it was established on accoun of the fact that the observer left and another could not be found near the station. Miscellaneous measurements will be taken here as opportunity affords.

The observations at this station during 1903 have been made under the direction of R. F. Walter, district engineer.

The following discharge measurement was made by W. T. Carpenter in 1903:

August 21: Gage height, 0.65 feet; discharge, 4.5 second-feet.

Mean daily gage height, in feet, of Elk Creek near Piedmont, S. Dak., for 1903.

Day.	Aug.	Sept.	Day.	Aug.	Sept.	Day.	Aug.	Sept.	Day.	Aug.	Sept.
1		0.70	9		0.70	17		1.02	25	0.88	0.90
2		.68	10		.80	18		1.10	26	.80	.90
3		.69	11		.80	19		.95	27	1.05	.90
4		.72	12		.85	20		.92	28	.80	.90
5		.65	13		.88	21	a0.65	.92	29	.78	.90
6		.68	14		.80	22	.75	.90	30	.70	.90
7		.70	15		.82	23	.65	.90	31	.70	
8		.65	16		1.08	24	.62	.90			

a Observer left, so only records for this month were kept.

BOX ELDER CREEK AT BLACKHAWK, S. DAK.

This station was established June 27, 1903, by R. F. Walter. It is located at the bridge on the road leading past the church at Blackhawk. It is about 2 miles downstream from the point at which Box Elder Creek crosses the road from Rapid to Piedmont. The gage is a vertical timber graduated to feet and tenths and securely spiked to a pile on the upstream side of the east end of the bridge. The gage is read once each day by Roy H. Haedt. Discharge measurements are best made by wading. The initial point for soundings is a brass tack driven into the hand rail vertically above the gage. The channel is straight for 150 feet above the station and curved for 75 feet below. Both banks are high and cleared, but may overflow in extremely high water. The bed of the stream is stony, with soft mud along the edges. There is considerable water grass just below the station. Bench mark No. 1 is a steel spike driven in a large fence post, about 6 inches above its base, at the corner of the barnyard, 300 feet east of the station. Its elevation is 16.84 feet above the zero of the gage. Bench mark No. 2 is a steel spike driven into the south gatepost, about 6 inches above its base. This gate is located near a house about 500 feet from the station, along the road toward Rapid. The elevation of the bench mark above the zero of the gage is 16.81 feet. The drainage area above the station is 157 square miles.

The observations at this station during 1903 have been made under the direction of R. F. Walter, district engineer.

Discharge measurements of Box Elder Creek at Blackhawk, S. Dak., in 1903.

Date.	Hydrographer.	Gage height.	Discharge.
		Feet.	*Second-feet.*
June 27	R. F. Walter	1.40	39.00
August 21	W. T. Carpenter	.90	3.40
November 6	R. F. Walter	.60	.43

Mean daily gage height, in feet, of Box Elder Creek at Blackhawk, S. Dak., for 1903.

Day.	June.	July.	Aug.	Sept.	Oct.	Nov.	Dec.
1		1.30	1.00	0.90	0.80	0.60	0.60
2		1.30	1.10	.90	.80	.60	.60
3		1.20	1.10	.90	.80	.60	.60
4		1.30	1.00	.90	.80	.60	.60
5		1.80	1.00	.90	.80	.60	.60
6		1.20	1.10	.90	.80	.60	.60
7		1.10	1.10	.90	.80	.60	.60
8		1.00	1.00	.90	.80	.60	.60
9		1.00	1.00	.90	.80	.60	.60
10		1.00	1.00	.90	.80	.60	.60
11		1.00	1.00	.90	.80	.60	.60
12		1.00	1.00	.90	.80	.60	.50
13		.90	1.00	.90	.80	.60	.50
14		.90	1.00	.90	.80	.60	.50
15		.90	1.00	.90	.80	.60	.50
16		.90	.90	.90	.80	.60	.50
17		1.00	.90	.90	.80	.60	.50
18		1.20	.90	.90	.70	.60	.50
19		1.20	.90	.90	.70	.60	a .50
20		1.10	.90	.90	.70	.60	....
21		1.00	.90	.90	.70	.60	....
22		.90	.90	.90	.70	.60	....
23		.90	.90	.80	.70	.60	....
24		.90	.90	.80	.70	.60	....
25		.90	.90	.80	.70	.60	....
26		.90	.90	.80	.70	.60	....
27		.90	.90	.80	.60	.60	....
28	1.40	.90	.90	.80	.60	.60	....
29	1.30	.90	.90	.80	.60	.60	....
30	1.30	1.00	.90	.80	.60	.60	....
31		1.00	.90		.60		....

a Readings discontinued after December 19 on account of ice.

RAPID CREEK AT RAPID, S. DAK.

This station was established June 10, 1903, by R. F. Walter. I is located at a wagon bridge one-half mile downstream from the Rapid River Milling Company's mill and one-fourth mile north of the Fremont, Elkhorn and Missouri Valley Railway. The gage is a vertical 2 by 4 inch timber, securely spiked to a timber at the south

the bridge. It is graduated to feet and tenths by means
led tacks. The observer, Prof. Mark Ehle, jr., of the
a School of Mines, reads the gage twice each day. Dis-
urements are made from the downstream side of the
₂. The initial point for soundings is a brass-headed tack
ɔy four similar tacks driven into the downstream face of
at the south end of the bridge. The channel is straight
ibove and 100 feet below the station. The banks are 12
igh, and will overflow only at extreme high water. The
ream is muddy, with embedded bowlders. At high water
flows in two channels, one of which is extremely small.
subject to rapid fluctuations in height, owing to the open-
sing of head gates of a ditch above the station. Bench
is the top of a nail driven flush with the surface of the
post at the south end of the bridge. Its elevation above
the gage is 12.73 feet. Bench mark No. 2 is the upper
he lowest timber of the front doorsill of a house owned
upied) by Mr. Feigel. Its elevation above the zero of the
₂ feet. This is the first house on the west side of the road
from the bridge.
·vations at this station during 1903 have been made under
n of R. F. Walter, district engineer.

rge measurements of Rapid Creek at Rapid, S. Dak., in 1903.

ate.	Hydrographer.	Gage height.	Discharge.
		Feet.	*Second-feet.*
....	R. F. Walter..............	2.30	170
....	W. T. Carpenter	1.70	46
....	......do	1.95	91
....	F. C. Magruder	1.75	58

·gage height, in feet, of Rapid Creek at Rapid, S. Dak., for 1903.

Day.	June.	July.	Aug.	Sept.	Oct.	Nov.	Dec.
....	2.10	2.40	2.01	1.70	1.85	1.78	1.75
....	2.10	2.25	1.85	1.82	1.82	1.78	1.80
....	2.10	2.20	2.01	1.88	1.82	1.78	1.80
....	2.15	2.25	2.00·	1.90	1.82	1.80	1.85
....	2.15	2.30	1.98	1.75	1.72	1.80	1.80
....	2.35	2.20	1.98	1.91	1.80	1.80	1.70
....	2.35	2.10	1.98	1.90	1.80	1.78	1.70
....	2.35	2.10	1.95	1.85	1.78	1.78	1.65
....	2.30	2.05	1.92	1.80	1.80	1.72	1.70
....	2.30	2.05	1.85	1.85	1.80	1.72	1.70
....	2.25	2.05	1.95	1.85	1.80	1.75	1.65

Mean daily gage height, in feet, of Rapid Creek, etc.—Continued.

Day.	June.	July.	Aug.	Sept.	Oct.	Nov.	Dec.
12	2.20	2.05	1.88	1.75	1.80	1.73	(*a*)
13	2.20	2.05	1.85	1.90	1.80	1.73	(*a*)
14	2.15	2.02	1.70	1.92	1.82	1.75	(*a*)
15	2.20	2.02	1.82	1.92	1.82	1.73	1.8
16	2.20	2.02	1.82	1.96	1.80	(*a*)	1.8
17	2.20	2.09	1.80	1.95	1.78	(*a*)	1.8
18	2.15	2.12	1.85	1.92	1.80	(*a*)	1.8
19	2.10	2.10	1.78	1.88	1.78	(*a*)	1.8
20	2.15	2.05	1.75	1.85	1.78	(*a*)	(*a*)
21	2.25	2.01	1.72	1.84	1.78	1.70	
22	2.40	1.99	1.86	1.82	1.75	1.65	
23	2.45	1.91	1.78	1.78	1.78	1.85	
24	2.55	1.80	1.82	1.85	1.80	1.80	
25	2.50	1.90	1.78	1.80	1.78	1.80	
26	2.45	1.95	1.81	1.78	1.80	1.85	
27	2.35	1.96	1.78	1.78	1.80	1.85	
28	2.35	1.98	1.75	1.82	1.80	1.90	
29	2.30	2.02	1.80	1.88	1.78	1.80	
30	2.25	2.05	1.78	1.85	1.75	1.80	
31		2.05	1.88		1.78		

a River frozen.

Rating table for Rapid Creek at Rapid, S. Dak., from June 1 to December 31, 1903.

Gage height.	Discharge.	Gage height.	Discharge.	Gage height.	Discharge.	Gage height.	Discharge.
Feet.	*Second-feet.*	*Feet.*	*Second-feet.*	*Feet.*	*Second-feet.*	*Feet.*	*Second-feet.*
1.5	21	1.9	81	2.3	171	2.7	285
1.6	33	2.0	102	2.4	197	2.8	315
1.7	48	2.1	124	2.5	225	2.9	345
1.8	64	2.2	146	2.6	255	3.0	375

Estimated monthly discharge of Rapid Creek at Rapid, S. Dak., for 1903.

[Drainage area, 410 square miles.]

Month.	Discharge in second-feet.			Total in acre-feet.	Run-off.	
	Maximum.	Minimum.	Mean.		Second-feet per square mile.	Depth in inches
June	240	124	163	9,699	0.398	0.4
July	197	64	117	7,194	.285	.3
August	102	48	75	4,612	.183	.2
September	102	48	73	4,344	.178	.1
October	72	48	63	3,874	.154	.1
November *a*	81	40	52	3,094	.127	.1
December 1–19 *b*	72	33	41	1,545	.100	.0

a Nov. 16 to 20, inclusive, estimated.   *b* Dec. 12 to 14, inclusive, estimated.

IG CREEK AT BLAIR'S RANCH, NEAR RAPID, S. DAK.

ion was established June 27, 1903, by R. F. Walter. It is
he highway from Rapid to Hermosa, on the property of Frank
ə gage is an upright 2 by 4 inch timber spiked to a tree at
edge about 100 feet above a small plank footbridge from
surements of discharge are sometimes made. The gage is
ach day by Lyda Blair. The channel is straight for 100 feet
75 feet below the station. The water is sluggish above and
ootbridge, but has a high velocity where measurements are
th banks are low and subject to overflow. At the station
the stream is gravelly; above and below this point it has a
tom. Bench mark No. 1 is a steel nail driven into the tree
he gage is attached. Its elevation is 2.14 feet above the
gage. Bench mark No. 2 is a steel nail driven into the
ree about 50 feet upstream from the gage. Its elevation is
)ove the zero of the gage. Bench mark No. 3 is a steel nail
) the roots of a tree about 75 feet downstream from the
elevation is 3.36 feet above the zero of the gage. The
rea above the station is 205 square miles.
·rvations at this station during 1903 have been made under
)n of R. F. Walter, district engineer.

measurements of Spring Creek at Blair's ranch, S. Dak., in 1903.

)ate.	Hydrographer.	Gage height.	Discharge.
		Feet.	*Second-feet.*
----------------	R. F. Walter ----------------	1.25	28.0
---------- --------	W. T. Carpenter -----------	.95	6.4
}----------------	-----do ----------------------	.97	7.7

gage height, in feet, of Spring Creek at Blair's ranch, S. Dak., for 1903.

Day.	June.	July.	Aug.	Sept.	Oct.	Nov.	Dec.
--------------------------	--------	1.30	1.10	0.90	0.90	0.90	0.80
--------------------------	--------	1.10	1.10	.90	.90	.90	.80
--------------------------	--------	1.10	1.10	.90	.90	.90	.80
--------------------------	--------	1.10	1.10	.90	.90	.90	.70
--------------------------	--------	1.10	1.10	.90	.90	.90	.65
--------------------------	--------	1.05	1.05	.90	.90	.90	.65
--------------------------	--------	1.00	1.00	.90	.90	.90	.70
--------------------------	--------	1.00	1.00	.90	.90	.90	.70
--------------------------	--------	1.00	1.00	.90	.90	.90	.70
--------------------------	--------	1.00	1.00	.90	.90	.90	.85
--------------------------	--------	1.00	1.00	.90	.90	.90	.85
--------------------------	--------	1.00	1.00	.90	.90	.85	.90

Mean daily gage height, in feet, of Spring Creek, etc.—Continued.

Day.	June.	July.	Aug.	Sept.	Oct.	Nov.	Dec.
13		1.00	1.10	0.90	0.90	0.70	1.0
14		1.00	1.00	1.00	.90	.60	.5
15		1.30	1.00	1.00	.90	1.00	.5
16		1.25	1.00	1.00	:90	.70	1.0
17		1.20	.90	1.00	.90	b.40	1.0
18		1.30	.90	.95	.90	b.30	1.0
19		1.20	.90	.95	.90	b.10	a1.0
20		1.10	1.00	.95	.90	(b)	
21		1.05	.90	.90	.90	(b)	
22		1.00	.90	.90	.90	.60	
23		1.00	.90	.90	.90	.85	
24		1.00	.90	.90	.90	.85	
25		1.00	1.10	.90	.90	.70	
26		1.00	.90	.90	.90	.80	
27	1.25	1.10	.90	.90	.90	.80	
28	1.20	1.10	1.00	.90	.90	.70	
29	1.20	1.10	1.00	.90	.90	.70	
30	1.15	1.10	1.00	.90	.90	.90	
31		1.10	.90		.90		

a Readings discontinued after December 19 on account of ice.   *b* Frozen.

BATTLE CREEK AT HERMOSA, S. DAK.

This station was established August 22, 1903, by R. F. Walter, assisted by W. T. Carpenter. It is located at the Chicago and Northwestern Railroad bridge. The gage is a 1½ by 4 inch vertical pine timber nailed securely to a pile in the sixth bent from the north end of the railroad bridge. It is read once each day by A. B. Huff. Discharge measurements are made from the downstream side of the bridge, to which the gage is fastened. At low water measurements can be made by wading. The initial point for soundings is a steel nail surrounded by a circle of brass-headed tacks on the downstream end of the fourth bent from the north end of the bridge. The channel is curved for about 700 feet above the station and is straight for 100 feet below. The current velocity is moderate. The right bank is high and precipitous. The left bank is high, but slopes toward the water's edge. The banks are not wooded. The bed of the stream is composed of sand and gravel, with mud near the banks. There are two channels at low water. Bench mark No. 1 is the initial point for soundings. Its elevation above the zero of the gage is 7.16 feet. Bench mark No. 2 is the top of a fence post in the corner of a wire fence 50 feet east of the north end of the railroad bridge. The point is surrounded by a circle of brass-headed tacks. Its elevation above the zero of the gage is 9.86 feet.

The drainage area above the station is 175 square miles.

The observations at this station during 1903 have been made under the direction of R. F. Walter, district engineer.

Discharge measurements of Battle Creek at Hermosa, S. Dak., in 1903.

Date.	Hydrographer.	Gage height.	Discharge.
		Feet.	*Second-feet.*
August 22	W. T. Carpenter	1.00	7.3
November 7	F. M. Madden	.90	2.0

Mean daily gage height, in feet, of Battle Creek at Hermosa, S. Dak., for 1903.

Day.	Aug.	Sept.	Oct.	Nov.	Dec.	Day.	Aug.	Sept.	Oct.	Nov.	Dec.
		1.00	1.00	0.90	0.90	17		1.20	0.90	0.90	0.90
		1.00	1.00	.90	.90	18		1.20	.90	.90	.90
		1.00	1.00	.90	.90	19		1.00	.90	.90	a.90
		1.00	1.00	.90	.90	20		1.00	.90	.90	.....
		1.00	1.00	.90	.90	21		1.00	.90	.90	.....
		1.00	1.00	.90	.90	22		1.00	.90	.90	.....
		1.00	1.00	.90	.90	23	1.00	1.00	.90	.90	.....
		1.00	1.00	.90	.90	24	1.00	1.00	.90	.90	.....
		1.00	1.00	.90	.90	25	1.00	1.00	.90	.90	.....
		1.00	1.00	.90	.90	26	2.50	1.00	.90	.90	.....
		1.30	1.00	.90	.90	27	2.00	1.00	.90	.90	.....
		1.30	1.00	.90	.90	28	1.00	1.00	.90	.90	.....
		1.30	.90	.90	.90	29	1.10	1.00	.90	.90	.....
		1.30	.90	.90	.90	30	1.00	1.00	.90	.90	.....
		1.20	.90	.90	.90	31	1.00		.90		

a Readings discontinued December 20 on account of ice.

CHEYENNE RIVER AT EDGEMONT, S. DAK.

This station was established June 19, 1903, by R. F. Walter. It is located at the highway bridge just downstream from the Burlington and Missouri River Railroad bridge. It is just above the junction of Cottonwood Creek and Cheyenne River. The gage is a vertical 2 by inch timber graduated to feet and tenths, and fastened by bolts and iron bands to the downstream side of the middle steel pier of the bridge. The observer is Loyd Stewart, who reads the gage once each day. Discharge measurements are made from the highway bridge. The initial point for soundings is a brass-headed tack surrounded by four similar tacks in the first post of the hand rail on the west end of the bridge. The channel is straight above and below the station, and the current velocity is low. The right bank is high enough to prevent overflow, and is cleared. The left bank is low and subject to overflow; it has a few scattered trees. The bed of the stream is sandy and shifting. The water flows in two channels at low water and in one at high stages. Bench mark No. 1 is a nail surrounded by a circle of brass tacks driven into a knot on the south side of a cottonwood tree 250 *feet north of* the north end of the wagon bridge. Its elevation above the zero of the gage is 11.29 feet. Bench mark No. 2 is the corner of the abutment at the north end of the railroad bridge. It is on the fourth step of the abutment and on the corner on the east side away from the bridge. Its elevation above the zero of the gage is 20.23 feet. The drainage area above the station is 7,350 square miles.

The observations at this station during 1903 have been made under the direction of R. F. Walter, district engineer.

Discharge measurements of Cheyenne River at Edgemont, S. Dak., in 1903.

Date.	Hydrographer.	Gage height.	Discharge.
		Feet.	*Second-feet.*
June 19	R. F. Walter	1.60	48
August 24	W. T. Carpenter	1.70	43
September 21	do	2.20	95
November 14	F. M. Madden	1.88	12

Mean daily gage height, in feet, of Cheyenne River at Edgemont, S. Dak., for 1903.

Day.	June.	July.	Aug.	Sept.	Oct.	Nov.	Dec.
1		1.60	4.30	2.70	1.60	1.30	1.80
2		1.50	5.40	2.50	1.50	1.30	1.80
3		1.40	4.40	2.20	1.50	1.30	1.80
4		1.30	4.10	2.10	1.50	1.30	1.80
5		1.25	3.70	2.00	1.50	1.30	1.80
6		1.20	3.40	2.00	1.50	1.30	1.90
7	1.80	1.15	3.10	2.00	1.40	1.30	a1.90
8	1.70	1.10	2.80	1.90	1.40	1.20	
9	2.40	1.10	2.50	2.20	1.40	1.20	
10	2.00	1.10	2.30	2.20	1.40	1.20	
11	1.75	6.40	2.00	2.20	1.40	1.20	
12	1.55	7.00	5.70	2.20	1.40	1.20	
13	1.50	4.30	4.00	7.40	1.40	1.20	
14	1.40	3.10	3.70	4.80	1.40	1.20	
15	1.35	2.70	3.20	3.70	1.40	1.20	
16	1.90	2.30	3.00	3.40	1.40	1.20	
17	2.15	8.00	2.70	3.00	1.40	1.20	
18	1.80	6.30	2.50	2.70	1.40	1.40	
19	1.60	3.90	2.30	2.50	1.40	1.60	
20	1.55	2.50	2.20	2.30	1.40	1.60	
21	1.50	2.30	1.90	2.10	1.40	1.60	
22	2.30	2.00	1.80	2.00	1.30	1.60	
23	1.70	1.80	1.80	2.00	1.30	1.60	
24	1.65	2.30	1.70	1.80	1.30	1.70	
25	1.60	4.10	3.50	1.70	1.30	1.70	
26	1.60	8.00	3.10	1.60	1.30	1.70	
27	1.60	8.50	3.70	1.60	1.30	1.70	
28	1.60	3.20	3.50	1.60	1.30	1.70	
29	1.40	2.80	3.30	1.60	1.30	1.90	
30	1.30	8.00	3.10	1.60	1.30	1.80	
31		3.80	2.90		1.30		

a Observations at station discontinued after December 7 till spring, 1904, on account of ice.

ELLE FOURCHE RIVER AT BELLE FOURCHE; S. DAK.

ation was established May 26, 1903, by R. F. Walter. It is
t the county highway bridge on the western outskirts of
urche, S. Dak. The gage consists of a 2 by 4 inch vertical
piked to a pile on the north side of the bridge on the third
n the east end. It is graduated to feet and tenths. The foot-
e copper figures and the tenths are marked with brass tacks.
e is read once each day by Henry Roberts. Discharge meas-
s are made from the north side of the bridge. The initial
· soundings is the center of the north pile in the first bent on
side. The channel is straight for about 225 feet above the
nd for 300 feet below. The current is swift. The left bank
vith a few scattered trees. The right bank is low and subject
ow at flood stages. It is sparsely wooded. The bed of the
s of gravel and is permanent. The water flows in a single
except at low stages, when it is divided by a gravel bar below
ge. Bench mark No. 1 is a spike in a cottonwood tree on the
k 50 feet downstream from the bridge. Its elevation above
of the gage is 8.016 feet. Bench mark No. 2 is the top of the
kwater on the east pier of the railroad bridge, which is 300
w the highway bridge. Its elevation is 8.21 feet above the
he gage. The drainage area above the station is 3,250 square

servations at this station during 1903 have been made under
tion of R. F. Walter, district engineer.

measurements of Belle Fourche River at Belle Fourche, S. Dak., in 1903.

Date.	Hydrographer.	Gage height.	Discharge.
		Feet.	*Second-feet.*
....	R. F. Walter	3.20	573
....	.....do	1.80	97
....	.....do	2.15	212
....	F. M. Madden...............	3.70	928
....	R. F. Walter	4.00	1,126
....	W. T. Carpenter............	2.20	213
....	.....do ....·..........	8.10	4,072
3	Madden and Magruder.	1.40	79

Mean daily gage height, in feet, of Belle Fourche River at Belle Fourche, S. Dak., for 1903.

Day.	June.	July.	Aug.	Sept.	Oct.	Day.	June.	July.	Aug.	Sept.	Oct.
1	1.75	1.80	4.50	4.00	1.50	17	0.50	1.80	3.00	4.50	1.30
2	1.75	1.20	7.00	2.80	1.40	18	.50	1.50	2.10	3.80	1.30
3	1.65	1.00	4.00	2.80	1.50	19	.50	1.30	1.70	3.60	1.30
4	1.50	.90	3.50	2.50	1.50	20	.50	1.60	1.80	2.40	1.30
5	1.35	.90	2.70	2.40	1.20	21	1.00	2.30	1.60	2.50	1.30
6	1.35	.90	2.30	2.10	1.20	22	1.80	1.70	1.80	2.10	1.30
7	1.35	.80	2.20	2.10	1.20	23	1.40	1.50	1.80	1.90	1.30
8	1.35	.80	2.30	2.10	1.20	24	1.50	1.30	1.75	1.80	1.30
9	1.00	.90	2.80	1.80	1.20	25	2.30	2.50	1.45	1.80	1.30
10	1.00	.70	3.60	1.50	1.20	26	1.80	2.50	2.45	1.60	1.30
11	75	.60	1.70	2.50	1.20	27	1.30	2.70	5.45	1.40	1.30
12	75	.70	3.20	7.90	1.20	28	1.30	2.30	2.80	2.80	1.30
13	75	.70	2.20	4.60	1.20	29	1.40	2.20	6.95	1.50	1.30
14	75	1.00	2.10	3.50	1.20	30	1.70	1.40	8.20	1.30	1.30
15	.50	1.30	1.90	2.50	1.20	31		3.75	5.80		a1.30
16	.50	1.70	2.00	7.90	1.20						

a Observer left November 1 and ice formed at gage before new observer could be secured. Readings discontinued till spring 1904.

Rating table for Belle Fourche River at Belle Fourche, S. Dak., from June 1 to November 30, 1903.

Gage height.	Discharge.	Gage height.	Discharge.	Gage height.	Discharge.	Gage height.	Discharge.
Feet.	*Second-feet.*	*Feet.*	*Second-feet.*	*Feet.*	*Second-feet.*	*Feet.*	*Second-feet.*
0.6	5	1.4	70	2.2	200	3.0	465
.7	10	1.5	82	2.3	225	3.1	515
.8	15	1.6	95	2.4	250	3.2	575
.9	20	1.7	108	2.5	280	3.3	635
1.0	27	1.8	122	2.6	310	3.4	705
1.1	37	1.9	137	2.7	345	3.5	775
1.2	47	2.0	157	2.8	380		
1.3	58	2.1	177	2.9	420		

Tangent above 3.30 feet gage height; differences above this point, 70 per tenth

Estimated monthly discharge of Belle Fourche River at Belle Fourche, S. Dak., for 1903.

[Drainage area, 3,250 square miles.]

Month.	Discharge in second-feet.			Total in acre-feet.	Run-off.	
	Maximum.	Minimum.	Mean.		Second-feet per square mile.	Depth in inches.
1903.						
June [a]	225	1	57	3,392	0.017	0.019
July [b]	950	5	117	7,194	.036	.042
August	4,065	76	751	46,177	.231	.266
September	3,855	58	624	37,131	.192	.214
October	82	47	54	3,320	.017	.020

[a] 21 to 27, inclusive (estimated).   [b] 19 and 20 (estimated).

RED WATER RIVER AT BELLE FOURCHE, S. DAK.

This station was established July 20, 1903, by R. F. Walter. It is located at the county highway bridge in the eastern limits of Belle Fourche, S. Dak. The gage consists of a 2 by 4 inch timber spiked to the north pile of the middle bent of the highway bridge. It is read once each day by Henry Roberts. Discharge measurements are made from the north side of the bridge. The initial point for soundings is the center of the pile on the north side and at the east end of the bridge. The channel is straight for 50 feet above and below the station, and the current is swift. The right bank is high enough to prevent overflow on that side. The left bank, however, is low and subject to overflow. There are trees along both banks. The bed of the stream is rocky and the channel permanent. The water flows in one channel. At very high stages the gage height may be affected by back water from Belle Fourche River. There is considerable "dead water." Bench mark No. 1 is the stone water table at the northeast corner of the public school building. Its elevation above the zero of the gage is 26.05 feet. Bench mark No. 2 is the top of the hydrant 75 feet northwest of the public school building. Its elevation above the zero of the gage is 21.83 feet.

The observations at this station during 1903 have been made under the direction of R. F. Walter, district engineer.

Discharge measurements of Red Water River at Belle Fourche, S. Dak., in 1903.

Date.	Hydrographer.	Gage height.	Discharge.
		Feet.	*Second-feet.*
August 12	W. T. Carpenter	2.85	132
July 25	R. F. Walter	2.45	72
October 28	F. M. Madden	2.85	136
August 3	R. F. Walter	3.00	208

Mean daily gage height, in feet, of Red Water River at Belle Fourche for 1903.

Day.	July.	Aug.	Sept.	Oct.	Day.	July.	Aug.	Sept.	Oct.
1	......	2.90	2.80	2.80	17	......	2.60	3.20	2.8
2	......	3.00	2.80	2.80	18	......	2.50	3.20	2.8
3	......	3.00	2.80	2.80	19	......	2.50	3.20	2.8
4	......	2.90	2.80	2.80	20	......	2.50	3.20	2.8
5	......	2.80	2.80	2.80	21	2.45	2.50	2.80	2.8
6	......	2.80	2.80	2.80	22	2 50	2.50	3.00	2.8
7	......	2.80	2.80	2.80	23	2.45	2.45	2.90	2.8
8	......	3.00	2.80	2.80	24	2.45	2.50	3.20	2.9
9	......	3.45	2.80	2.80	25	2.45	2.45	3.20	2.9
10	......	3.00	2.90	2.80	26	2.60	2.45	3.20	2.9
11	......	2.80	3.00	2.80	27	2.70	3.60	3.20	2.8
12	......	2.80	3.80	2.80	28	2.80	3.20	3.20	2.8
13	......	2.80	3.50	2.80	29	2.70	2.90	2.80	2.8
14	......	2.60	3.80	2.80	30	2.70	2.90	2.80	2.8
15	......	2.80	3.40	2.90	31	2.85	2.90	......	2.8
16	......	2.60	3.10	2.80					

Rating table for Red Water River at Belle Fourche, S. Dak., from July 20 to December 31, 1903.

Gage height.	Discharge.	Gage height.	Discharge.	Gage height.	Discharge.	Gage height.	Discharge.
Feet.	*Second-feet.*	*Feet.*	*Second-feet.*	*Feet.*	*Second-feet.*	*Feet.*	*Second-feet.*
2.4	70	2.8	119	3.2	370	3.6	720
2.5	75	2.9	153	3.3	454	3.7	810
2.6	84	3.0	208	3.4	540	3.8	900
2.7	98	3.1	288	3.5	630		

No measurements made above 3 feet gage height. For gage heights above 3.5 feet discharges should be considered as estimates.

Estimated monthly discharge of Red Water River at Belle Fourche, S. Dak., for 1903.

[Drainage area, 1,015 square miles.]

Month.	Discharge in second-feet.			Total in acre-feet.	Run-off.	
	Maximum.	Minimum.	Mean.		Second-feet per square mile.	Depth in inches
July 21–31. inclusive	136	72	91	1,985	0.090	0.0
August [a]	720	72	162	9,961	.160	.1
September [b]	900	119	291	17,316	.287	.3
October	153	119	123	7,563	.121	.1

[a] 30 and 31 (estimated). [b] 1 to 5, inclusive (estimated).

RED WATER RIVER AT MINNESELA, S. DAK.

This station was established May 12, 1903, by J. R. Hickox at the highway bridge 1 mile west of the town of Minnesela, S. Dak. It was abandoned on account of the difficulty of obtaining an observer and on account of the location of a power canal diverting water just below the station, which wasted the water into the Belle Fourche River above the station on that stream.   When this station was abandoned a new station was established at Belle Fourche, on Red Water River.

The observations at this station during 1903 have been made under the direction of R. F. Walter, district engineer.

Discharge measurements of Red Water River at Minnesela, S. Dak., in 1903.

Date.	Hydrographer.	Gage height.	Discharge.
		Feet.	*Second-feet.*
May 12	J. R. Hickox	3.30	271
May 26	R. F. Walter	2.95	212
August 13	W. T. Carpenter	2.60	141

Mean daily gage height, in feet, of Red Water River at Minnesela, S. Dak., for 1903.

Day.	June.	July.	Aug.	Day.	June.	July.	Aug.	Day.	June.	July.	Aug.
1		2.30	*2.90	12		1.90		23	2.20	2.10	
2		2.30		13		1.90		24	2.50	2.00	
3		2.30		14		1.80		25	2.70	2.10	
4		2.30		15		1.90		26	2.80	2.40	
5		2.30		16		2.00		27	2.60	2.50	
6		2.30		17		2.20		28	2.50	2.50	
7		2.20		18		2.20		29	2.50	2.50	
8		2.20		19		2.20		30	2.40	2.50	
9		2.20		20		2.00		31		2.50	
10		2.10		21		2.10					
11		2.10		22	2.10	2.10					

*Station abandoned August 1 and one established 3 miles below, so as to be below Belle Fourche power canal.

Rating table for Red Water River at Minnesela, S. Dak., from May 12 to August 13, 1903.

Gage height.	Discharge.	Gage height.	Discharge.	Gage height.	Discharge.	Gage height.	Discharge.
Feet.	*Second-feet.*	*Feet.*	*Second-feet.*	*Feet.*	*Second-feet.*	*Feet.*	*Second-feet.*
1.8	55	2.2	85	2.6	142	3.0	214
1.9	60	2.3	97	2.7	160	3.1	233
2.0	67	2.4	110	2.8	178	3.2	252
2.1	75	2.5	125	2.9	196	3.3	271

Estimated monthly discharge of Red Water River at Minnesela, S. Dak., for 1903.

Month.	Discharge in second-feet.			Total in acre-feet.
	Maximum.	Minimum.	Mean.	
June 22-30	173	75	125	2,231
July	125	55	88	5,411

SPEARFISH CREEK AT TOOMEY'S RANCH, NEAR SPEARFISH, S. DAK.

This station is a temporary station, and was established May 14, 1903, by J. R. Hickox. It is about 6 miles by wagon road from Spearfish, S. Dak., at the ranch of D. J. Toomey, and about 1 mile above the junction of the Spring Branch with Spearfish Creek. The gage is a vertical 2 by 4 inch timber graduated to feet and tenths and marked with brass-headed tacks. It is located at the east end of a small plank footbridge about 300 feet southeast of Toomey's ranch house. Allie Toomey, the observer, reads the gage once each day. The discharge measurements are made from the footbridge. The initial point for soundings is a nail driven into the top of a post and surrounded by a circle of tacks and located at the west end of the footbridge. The channel is straight for 100 feet above and below the station. Both banks are low and wooded and are subject to overflow. The bed of the stream is covered with round stones not larger than 6 inches in diameter. The left bank is skirted by a high bluff, about 100 feet back from the stream, which does not overflow. Bench mark No. 1 is a steel spike in a telephone pole 30 feet east of the driveway in front of Toomey's house. Its elevation above the zero of the gage is 16.69 feet. Bench mark No. 2 is a steel nail surrounded by a circle of brass tacks on the corner of the back porch floor of Toomey's house. Its elevation above the zero of the gage is 16.55 feet.

The observations at this station during 1903 have been made under the direction of R. F. Walter, district engineer.

Discharge measurements of Spearfish Creek near Spearfish, S. Dak., in 1903.

Date.	Hydrographer.	Gage height.	Discharge.
		Feet.	*Second-feet.*
May 14...	J. R. Hickox	2.94	106
May 27.............	R. F. Walter................	2.70	61
August 14.....	W. T. Carpenter	2.50	38
September 22..	.....do	2.80	71

daily gage height, in feet, of Spearfish Creek near Spearfish, for 1903.

Day.	May.	June.	July.	Aug.	Sept.	Oct.	Nov.	Dec.
....		2.70	2.65	2.50	2.55	2.80	2.80	2.75
....		2.70	2.55	2.55	2.55	2.80	2.80	2.75
....		2.65	2.60	2.60	2.55	2.75	2.80	2.75
....		2.65	2.60	2.60	2.60	2.75	2.80	2.75
....		2.60	2.60	2.70	2.80	2.75	2.80	2.50
....		2.60	2.65	2.60	2.80	2.75	2.80	2.75
....		2.60	2.60	2.60	2.80	2.75	2.80	2.75
....		2.60	2.60	2.60	2.80	2.80	2.80	2.75
....		2.60	2.60	2.60	2.75	2.80	2.80	2.75
....		2.60	2.60	2.55	2.70	2.75	2.80	2.75
....		2.55	2.60	2.50	2.70	2.80	2.70	2.75
....		2.60	2.50	2.50	2.95	2.80	2.75	2.75
....		2.50	2.50	2.50	2.85	2.80	2.65	2.55
....		2.45	2.50	2.50	2.85	2.80	2.65	2.50
....		2.45	2.50	2.50	2.85	2.80	2.70	2.35
....		2.50	2.50	2.50	2.85	2.80	2.70	2.75
....		2.50	2.60	2.50	2.85	2.80	2.75	2.70
....		2.50	2.60	2.50	2.85	2.80	2.40	2.70
....		2.45	2.60	2.50	2.85	2.75	2.40	2.75
....		2.45	2.60	2.50	2.85	2.80	2.45	2.70
....		2.60	2.60	2.50	2.85	2.80	2.70	(a)
....		2.60	2.60	2.50	2.80	2.70	2.75	....
....		2.65	2.55	2.50	2.80	2.75	2.75	....
....		2.70	2.50	2.50	2.80	2.75	2.75	....
....		2.75	2.40	2.50	2.80	2.75	2.75	....
....		2.80	2.40	2.50	2.80	2.75	2.75	....
....		2.75	2.40	2.55	2.80	2.75	2.75	....
....		2.70	2.40	2.60	2.80	2.75	2.75	....
....		2.70	2.40	2.65	2.80	2.75	2.75	....
....		2.65	2.40	2.60	2.80	2.75	2.75	....
....	2.70	....	2.50	2.60	....	2.80	....	....

a Frozen.

able for Spearfish Creek near Spearfish, S. Dak., from May 29 to December 31, 1903.

Gage height.	Discharge.	Gage height.	Discharge.	Gage height.	Discharge.
Feet.	*Second-feet.*	*Feet.*	*Second-feet.*	*Feet.*	*Second-feet.*
2.4	31	2.6	47	2.8	74
2.5	38	2.7	59	2.9	95

Estimated monthly discharge of Spearfish Creek near Spearfish, S. Dak., for 1[

[Drainage area, 230 square miles.]

Month.	Discharge in second-feet.			Total in acre-feet.	Run-off.	
	Maximum.	Minimum.	Mean.		Second-feet per square mile.	Dept inch
June	74	34	49	2,916	0.21	(
July	52	31	42	2,582	.18	
August	59	38	42	2,582	.18	
September	108	42	73	4,344	.32	
October	74	59	70	4,304	.30	
November.........	74	31	63	3,749	.27	
December 1–20	66	28	59	2,340	.26	

CANNON BALL RIVER AT STEVENSON, N. DAK.

This station was established June 10, 1903, by F. E. Weymou
It is located one-half mile west-northwest of the post-office at Stev
son, in sec. 20, T. 133 N., R. 82 W., and is about 40 miles south
Mandan, N. Dak. The standard chain gage is located on the
bank. The pulley is fastened to a horizontal timber at a point 10
feet from the zero of the scale. The length of the chain from the (
of the weight to the end of the chain and ring is 26.09 feet. It is re
once each day by Donald Stevenson. Discharge measurements
made from a three-fourths-inch wire cable and car about 200 f
above the gage. The left end of the cable is supported by a pos
feet above ground and 10 feet back from the crest of the bank, a
is fastened to a 4-foot log anchor buried in the ground. This end
the cable is about 34 feet above gage datum. The right end of
cable is attached to a 13-inch cottonwood tree at a point 12 feet ab(
ground and 30 feet above the gage datum. A tag wire, consisting
barbed wire with zinc markers every 10 feet, is stretched above
cable. The initial point for soundings is a point 2 feet back from
cable support on the left bank. The channel is straight for 100 f
above the station and for 400 feet below. The current velocit;
sluggish at ordinary stages. The right bank is low and is a grad
slope up from the water's edge. It is covered with timber and bru
The left bank is steep and about 25 feet high. The bench marl
the highest point of a bowlder, whose dimensions are 6 by 8 f(
located in the stream about 600 feet below the gage, toward the ri
bank. Its elevation above gage datum is 4.84 feet. The datun
the gage is about 1,700 feet above sea level, as determined by carry
an aneroid barometer six times between this station and Mand
N. Dak.

The bed of the stream consists of clay, soft mud, and loose stor
and is probably somewhat shifting. The depth varies from 2 to 5
at ordinary stages.

)bservations at this station during 1903 have been made under
:ction of E. F. Chandler, district hydrographer.

rge measurements of Cannon Ball River at Stevenson, N. Dak., in 1903.

Date.	Hydrographer.	Gage height.	Discharge.
		Feet.	Second-feet.
....	F. E. Weymouth...............	3.10	75
....	E. F. Chandler...............	2.62	11
....	F. E. Weymouth...............	2.63	12
....	E. F. Chandler...........	3.10	56
.....do	2.83	19
.....do	2.71	8
26	.....do	3.10	a 35

a Meter in poor adjustment.

*ily gage height, in feet, of Cannon Ball River at Stevenson, N. Dak., for
1903.*

Day.	June.	July.	Aug.	Sept.	Oct.	Nov.	Dec.
....		2.70	3.60	3.90	2.80	2.80	(a)
....		3.20	4.00	3.40	3.10	2.80	(a)
....		3.40	3.60	3.20	3.10	3.10	(a)
....		2.90	3.40	3.00	3.10	3.20	(a)
....		3.00	4.60	3.10	3.00	3.10	(a)
....		2.90	3.50	2.60	2.90	3.00	(a)
....		2.80	3.30	2.70	3.00	3.10	(a)
....		2.90	4.40	2.60	3.00	3.00	(a)
....		3.00	4.90	2.50	3.00	3.10	(a)
....	3.00	3.30	3.70	2.50	3.10	3.10	(a)
....	2.90	2.80	3.60	2.60	2.90	2.90	(a)
....	3.00	2.20	3.30	2.70	2.70	2.90	(a)
....	2.90	2.20	3.50	5.60	2.70	2.70	(a)
....	2.80	2.60	6.10	5.20	2.70	2.80	(a)
....	2.80	2.60	4.60	5.00	2.80	2.90	(a)
....	2.80	2.60	3.80	4.80	2.70	2.90	(a)
....	2.80	2.50	3.40	4.80	2.90	3.10	(a)
....	2.70	2.50	3.20	4.60	2.90	a3.30	(a)
....	2.90	2.40	3.00	4.50	3.00	a3.30	(a)
....	2.70	2.40	2.90	4.10	2.90	a3.50	(a)
....	2.70	2.40	2.90	3.90	3.10	a3.60	(a)
....	2.60	2.50	2.80	3.70	3.20	(a)	(a)
....	2.60	3.20	2.80	3.70	3.20	(a)	(a)
....	2.80	2.90	3.50	3.60	3.10	....	(a)
....	2.90	2.70	5.20	3.60	3.20	....	(a)
....	2.90	2.70	5.00	3.70	3.30	....	(a)
....	2.80	2.70	5.40	3.20	3.10	(a)	(a)
....	2.60	2.50	6.70	3.00	2.90	....	(a)
....	2.60	2.50	5.50	2.90	2.90	....	(a)
....	2.70	3.00	4.70	2.80	2.80	....	(a)
....		3.10	4.00	....	3.00	....	(a)

a Frozen.

HEART RIVER NEAR RICHARDTON, N. DAK.

This station was established June 2, 1903, by F. E. Weymouth. It is located at the iron highway bridge 10 miles south of the Northern Pacific Railroad station at Richardton, N. Dak. The standard chain gage is located on the lower side of the bridge. The scale, reading from 2 to 23 feet, is marked on the foot guard rail with wire staples and white paint marks. The length of the chain is 24.34 feet from the end of the weight to the marker. The gage is read once each day by W. F. Church. Discharge measurements are made from the bridge to which the gage is attached. The initial point for sounding is the end of the guard rail on the lower side of the bridge at the left bank. Distances are marked and numbered with white paint every 10 feet across the entire length of the bridge. The channel is straight for 150 feet above and below the station. The current velocity is moderate. Both banks are high and covered with brush. The bed of the stream is sandy and shifting. The bench mark is the top of the foot guard rail at a distance of 45 feet from the initial point for soundings. Its elevation is 25.58 feet above gage datum. The top of the gage pulley has this same elevation. Gage datum is 2,150 feet above sea level, as determined by carrying an aneroid barometer six times between this point and the railroad station at Richardton.

The observations at this station during 1903 have been made under the direction of E. F. Chandler, district hydrographer.

Discharge measurements of Heart River near Richardton, N. Dak., in 1903.

Date.	Hydrographer.	Gage height.	Discharge
		Feet.	*Second-feet*
May 18	F. E. Weymouth	4.50	
June 2	do	5.00	
June 18	do	4.50	
July 6	D. E. Willard	4.40	
July 21	E. F. Chandler	4.07	
September 5	do	4.71	
October 18	do	4.87	

Mean daily gage height, in feet, of Heart River near Richardton, N. Dak., for 1903.

Day.	May.	June.	July.	Aug.	Sept.	Oct.	Nov.	Dec.
1		5.30	4.40	4.00	6.00	4.50	4.40	(a)
2		5.20	4.40	4.00	5.40	4.40	4.40	(a)
3		4.90	4.50	4.00	5.10	4.40	4.40	(a)
4		4.80	4.50	4.00	4.90	4.40	4.40	(a)
5		4.80	4.40	4.00	4.80	4.40	4.40	(a)
6		4.90	4.40	4.00	4.80	4.40	4.40	(a)
7		4.60	4.40	4.00	4.70	4.40	4.40	(a)
8		4.60	4.30	4.00	4.70	4.40	4.40	(a)
9		4.60	4.30	4.00	4.60	4.40	4.40	(a)
10		4.70	4.30	4.00	4.60	4.40	4.40	(a)
11		4.60	4.40	4.00	4.50	4.40	4.40	(a)
12		4.60	4.30	4.20	4.50	4.40	4.40	(a)
13		4.50	4.20	4.20	4.40	4.40	4.40	(a)
14		4.50	4.20	4.10	4.50	4.40	4.40	(a)
15		4.50	4.10	4.20	4.60	4.40	(a)	(a)
16		4.40	4.10	4.10	4.70	4.40	(a)	(a)
17		4.40	4.10	4.10	4.80	4.40	(a)	(a)
18	4.50	4.40	4.10	4.10	6.20	4.40	(a)	(a)
19	4.50	4.40	4.10	4.10	5.80	4.40	(a)	(a)
20	4.50	4.40	4.10	4.00	5.60	4.40	(a)	(a)
21	4.50	4.40	4.10	4.00	5.30	4.40	(a)	(a)
22	5.00	4.40	4.10	4.00	5.10	4.40	(a)	(a)
23	13.80	4.40	4.10	4.00	4.90	4.40	(a)	(a)
24	16.20	4.40	4.10	4.30	4.70	4.40	(a)	(a)
25	12.40	4.30	4.10	4.30	4.70	4.40	(a)	(a)
26	8.50	4.40	4.00	4.20	4.60	4.40	(a)	(a)
27	7.40	4.40	4.00	4.30	4.60	4.40	(a)	(a)
28	6.50	4.40	4.00	4.20	4.60	4.40	(a)	(a)
29	6.10	4.40	4.00	4.10	4.60	4.40	(a)	(a)
30	5.80	4.40	4.00	4.10	4.50	4.40	(a)	(a)
31	5.50	....	4.00	6.90	....	4.40	(a)	(a)

a Frozen.

Rating table for Heart River near Richardton, N. Dak., from May 18 to December 31, 1903.

Gage height.	Discharge.	Gage height.	Discharge.	Gage height.	Discharge.	Gage height.	Discharge.
Feet.	Second-feet.	Feet.	Second-feet.	Feet.	Second-feet.	Feet.	Second-feet.
4.0	0	4.7	28	5.4	138	6.1	282
4.1	1	4.8	38	5.5	158	6.2	304
4.2	3	4.9	50	5.6	178	6.3	326
4.3	5	5.0	64	5.7	198	6.4	348
4.4	8	5.1	81	5.8	218		
4.5	13	5.2	99	5.9	239		
4.6	19	5.3	118	6.0	260		

Tangent above 6 feet gage height. Differences above this point, 22 per tenth.

Estimated monthly discharge of Heart River near Richardton, N. Dak., for 190

Month.	Discharge in second-feet.			Total in acre-feet
	Maximum.	Minimum.	Mean.	
May 18–31	2,504	13	611	16,9
June	118	5	22	1,3
July	13	0	4	2
August	530	0	18	1,1
September	336	8	67	3,8
October	13	8	8	4
November 1–14	8	8	8	2

KNIFE RIVER AT BRONCHO, N. DAK.

This station was established May 29, 1903, by F. E. Weymouth.
is located about 600 feet east of H. M. Haven's ranch house, where th
post-office is located. The station is about 23 miles north of Hebro
N. Dak., which is on the Northern Pacific Railroad. The standa
chain gage is located on the right bank, with the pulley fastened to
2 by 4 inch timber projecting horizontally. The length of the cha
from the end of the weight to the leather marker is 28.24 feet. Th
total length of the chain from the end of the weight to the end of th
ring is 38.24 feet. The horizontal scale is marked with white pai
and reads from zero to 16 feet. A temporary 2 by 4 inch vertic
board gage is spiked to a tree on the right bank above the ford, thre
fourths of a mile below the ranch. It reads from zero to 19.3 fee
The gage is read once each day by H. M. Haven. Discharge mea
urements are made from the ½-inch cable and car with a span of 1
feet. The tag wire, consisting of barbed wire with zinc markers eve
10 feet, is stretched about 5 feet east of the cable. The initial poi
for soundings is the post to which the right end of the tag wire
fastened. The channel is straight for 100 feet above the station a
for 200 feet below. The current velocity is moderate. Both ban
are high, with a few small trees and some brush. The bed of t
stream is composed of sand and gravel and is probably fairly p
manent. The bench mark for the temporary gage rod is a spike
the tree to which the gage is fastened at an elevation of 19.25 f
above its zero. The gage zero is about 1,870 feet above sea level,
determined by carrying an aneroid barometer four times between th
station and Hebron, N. Dak.

The observations at this station during 1903 have been made und
the direction of E. F. Chandler, district hydrographer.

Discharge measurements of Knife River at Broncho, N. Dak., in 1903.

Date.	Hydrographer.	Gage height.	Discharge.
		Feet.	*Second-feet.*
May 29	F. E. Weymouth	1.45	80
June 20	do	.80	22
July 7	D. E. Willard	.80	9
July 28	E. F. Chandler	2.20	247
October 17	do	.84	18

Mean daily gage height, in feet, of Knife River at Broncho, N. Dak., for 1903.

Day.	May.	June.	July.	Aug.	Sept.	Oct.	Nov.	Dec.
1		1.20	0.80	1.00	2.00	0.50	0.50	(a)
2		1.10	.80	.90	1.80	.50	.50	(a)
3		1.00	.80	.90	1.70	.50	.40	(a)
4		.90	.80	1.00	1.70	.50	.40	(a)
5		1.00	.80	1.00	1.60	.50	.40	(a)
6		1.00	.80	.80	1.00	.50	.30	(a)
7		1.00	.80	.80	1.00	.50	.30	(a)
8		1.30	.80	.70	1.00	.50	.30	(a)
9		1.00	.80	.60	.90	.50	.30	(a)
10		1.00	.80	.80	.90	.50	.40	(a)
11		.90	.80	.80	.80	.50	.40	(a)
12		.90	.70	.80	.80	.50	.40	(a)
13		.80	.70	.90	.80	.50	a.40	(a)
14		.80	.70	.80	1.90	.40	a.40	(a)
15		.80	.70	.80	2.10	.40	(a)	(a)
16		.80	.70	.80	1.50	.40	(a)	(a)
17		.80	.70	.80	1.70	.40	(a)	(a)
18		.80	.70	.80	1.60	.40	(a)	(a)
19		.80	.70	.90	1.00	.40	(a)	(a)
20		.80	.70	.80	1.00	.40	(a)	(a)
21		.80	.70	.80	.80	.40	(a)	(a)
22		.80	.60	.80	.80	.40	(a)	(a)
23		1.00	.60	.80	.70	.40	(a)	(a)
24		.90	.60	.80	.60	.40	(a)	(a)
25		.80	.60	.90	.60	.40	(a)	(a)
26		.80	.70	.90	.60	.40	(a)	(a)
27		.80	3.70	.90	.50	.40	(a)	(a)
28		.80	2.30	1.60	.50	.50	(a)	(a)
29	1.45	.80	1.50	2.40	.50	.50	(a)	(a)
30	1.40	.80	1.00	2.00	.50	.50	(a)	(a)
31	1.30		.70	1.60		.50	(a)	(a)

a Frozen.

Rating table for Knife River at Broncho, N. Dak., from January 1 to December 31, 1903.

Gage height.	Discharge.	Gage height.	Discharge.	Gage height.	Discharge.	Gage height.	Discharge.
Feet.	Second-feet.	Feet.	Second-feet.	Feet.	Second-feet.	Feet.	Second-feet.
0.2	8	1.0	36	1.8	143	2.6	382
.3	9	1.1	44	1.9	165	2.7	419
.4	11	1.2	53	2.0	190	2.8	458
.5	13	1.3	64	2.1	217	2.9	498
.6	16	1.4	76	2.2	247	3.0	540
.7	20	1.5	90	2.3	278		
.8	24	1.6	105	2.4	311		
.9	29	1.7	123	2.5	346		

Table not accurate below 1 foot gage height on account of low velocity of stream.

Estimated monthly discharge of Knife River at Broncho, N. Dak., for 1903.

Month.	Discharge in second-feet.			Total in acre-feet.
	Maximum.	Minimum.	Mean.	
May 29–31	83	64	74	440
June	64	24	30	1,785
July	840	16	60	3,689
August	311	16	46	2,898
September	217	13	62	3,689
October	13	11	12	738
November 1–14	13	9	11	305

LITTLE MISSOURI RIVER AT MEDORA, N. DAK.

This station was established May 12, 1903, by F. E. Weymouth. It is located at the Northern Pacific Railroad bridge, one-third mile west of the railroad station at Medora, N. Dak. The standard chain gage is located on the lower side of the railroad bridge at a point 91 feet from the initial point for soundings. The length of the chain from the end of the weight to the marker is 30.75 feet. There is also a 1 by 6 inch vertical board gage, reading from 1 to 9 feet, nailed to a pile in the river about 200 feet above the bridge. The gage is read once each day by W. A. Brubaker. Discharge measurements are made from the railroad bridge to which the gage is attached. The initial point for soundings is the left end of the guard rail on the lower side of the bridge. This point is 2.9 feet west of the east face f the concrete abutment. The guard rail is marked and numbered very 10 feet with white paint. The channel is straight for 100 feet above the station and for 300 feet below. The right bank is low and overflows at very high stages. The left bank is almost perpendicular and has a height of 30 feet above gage datum. There is but one channel, broken by one to three piers, according to the stage of the river. The bed of the stream is of clay and sand and sometimes scours from 5 to 8 feet during floods. At ordinary stages the bed of the stream changes only slightly. About November 1, 1903, the cross section at the bridge was temporarily changed by the railroad company during repairs to the bridge, continuing several months. The bench mark is the top of the southwest corner of the concrete abutment on the left bank. Its elevation above gage datum is 30.11 feet. The top of the gage pulley is 31.05 feet above gage datum. The gage datum is about 2,330 feet above sea level, as determined by hand level from the railroad station at Medora, N. Dak.

The observations at this station during 1903 have been made under the direction of E. F. Chandler, district hydrographer.

Discharge measurements of Little Missouri River at Medora, N. Dak., in 1903.

Date.	Hydrographer.	Gage height.	Discharge.
		Feet.	*Second-feet.*
April 24	C. C. Babb	2.45	69
May 13	F. E. Weymouth	8.75	48
June 3	do	4.80	281
June 17	J. N. Kerr	4.00	52
July 5	F. E. Weymouth	7.37	2,325
July 20	J. N. Kerr	4.20	105
September 5	E. F. Chandler	6.15	1,192
October 19	do	3.80	95

Mean daily gage height, in feet, of Little Missouri River at Medora, N. Dak., for 1903.

Day.	May.	June.	July.	Aug.	Sept.	Oct.	Nov.	Dec.
1		5.10	5.00	7.70	10.10	4.40	4.00	(a)
2		4.90	7.90	7.90	8.70	4.20	4.10	(a)
3		4.80	7.90	8.00	8.10	4.00	4.00	(a)
4		4.80	8.00	8.00	7.20	4.00	4.00	(a)
5		4.90	8.80	8.70	5.10	4.10	3.90	(a)
6		4.70	7.80	8.90	5.40	4.30	3.90	(a)
7		4.60	6.50	8.70	5.10	4.20	3.90	(a)
8		4.50	6.30	7.80	5.00	4.00	4.00	(a)
9		4.40	5.80	6.70	4.60	3.90	4.10	(a)
10		4.30	5.40	6.50	4.50	4.00	4.00	b 3.90

a Frozen.   b Frozen; height of water surface under ice.

Mean daily gage height, in feet, of Little Missouri River, etc.—Continued.

Day.	May.	June.	July.	Aug.	Sept.	Oct.	Nov.	Dec.
11		4.20	5.20	6.20	4.80	4.10	4.00	(a)
12	3.70	4.20	5.10	5.90	5.60	4.10	4.00	(a)
13	3.60	4.10	5.00	6.80	5.60	4.00	3.90	(a)
14	3.60	4.00	4.90	6.70	5.40	3.90	a4.10	(a)
15	3.50	4.00	4.70	5.20	5.20	3.90	a4.20	(a)
16	4.00	3.90	4.60	5.80	5.80	3.80	a4.20	(a)
17	3.80	3.90	4.50	6.20	5.90	3.80	a4.20	(a)
18	4.10	3.90	4.50	6.40	6.00	3.80	(a)	(a)
19	4.20	3.90	4.40	5.80	6.90	3.70	(a)	(a)
20	3.90	3.80	4.40	6.80	7.60	3.70	(a)	b3.80
21	5.00	3.90	4.30	6.50	7.40	3.90	(a)	(a)
22	7.90	4.20	4.30	6.40	7.30	3.90	(a)	(a)
23	7.70	4.20	5.20	6.80	7.00	3.80	(a)	(a)
24	6.80	4.20	5.50	5.90	6.50	3.70	(a)	(a)
25	7.90	4.10	5.40	7.20	5.30	3.70	(a)	(a)
26	7.30	5.70	5.80	7.50	5.00	3.60	(a)	(a)
27	6.80	5.60	5.90	7.80	5.00	3.60	(a)	(a)
28	6.10	5.50	6.90	7.90	4.80	3.60	(a)	(a)
29	6.00	5.30	5.70	8.80	4.60	3.90	(a)	(a)
30	6.00	5.20	6.90	8.70	4.50	4.00	b3.90	(a)
31	5.40	....	7.20	10.50	....	3.80	....	b3.60

a Frozen.   b Frozen; height of water surface under ice.

Rating table for Little Missouri River at Medora, N. Dak., from May 12 to December 31, 1903.

Gage height.	Discharge.	Gage height.	Discharge.	Gage height.	Discharge.	Gage height.	Discharge.
Feet.	Second-feet.	Feet.	Second-feet.	Feet.	Second-feet.	Feet.	Second-feet.
3.4	36	4.8	290	6.2	1,230	8.2	3,210
3.5	39	4.9	338	6.3	1,320	8.4	3,430
3.6	43	5.0	390	6.4	1,410	8.6	3,660
3.7	48	5.1	444	6.5	1,500	8.8	3,900
3.8	54	5.2	500	6.6	1,590	9.0	4,140
3.9	61	5.3	560	6.7	1,680	9.2	4,380
4.0	71	5.4	620	6.8	1,770	9.4	4,630
4.1	84	5.5	690	6.9	1,860	9.6	4,890
4.2	101	5.6	760	7.0	1,960	9.8	5,150
4.3	122	5.7	830	7.2	2,160	10.0	5,420
4.4	147	5.8	900	7.4	2,360	10.2	5,680
4.5	176	5.9	980	7.6	2,560	10.4	5,960
4.6	209	6.0	1,060	7.8	2,770	10.5	6,100
4.7	247	6.1	1,140	8.0	2,990		

Estimated monthly discharge of Little Missouri River at Medora, N. Dak., for 1903.

Month.	Discharge in second-feet.			Total in acre-feet.
	Maximum.	Minimum.	Mean.	
May 13-31	2,880	39	952	37,765
June	830	54	287	14,102
July	3,900	122	1,083	66,591
August	6,100	500	2,295	141,114
September	5,550	176	1,242	73,904
October	147	43	68	4,181
November 1-17	101	61	76	2,563

LITTLE MISSOURI RIVER AT CAMP CROOK, S. DAK.

This station was established September 2, 1903, by R. F. Walter, assisted by W. T. Carpenter. The station is at the highway bridge on the road from Camp Crook to Belle Forche, about one-half mile from Camp Crook. The gage is a vertical 2 by 4 inch pine timber fastened to the first pier from the west end of the bridge, on the downstream side. It is graduated to feet and tenths with bronze figures marking the foot marks. The gage is read once each day by George A. Lane. Discharge measurements are made from the bridge to which the gage is attached. The initial point for soundings is a brass-headed tack surrounded by four similar tacks driven into the end post of the hand rail at the west end and downstream side of the bridge. The channel is straight for 300 feet above and 200 feet below the station. The current has a moderate velocity. The right bank is high and not subject to overflow. The left bank is subject to overflow at high water. Both banks are sparsely wooded. The bed of the stream is covered with stones, the largest of which are about 1 foot in diameter. The water flows in one channel at medium and high stages. Bench mark No. 1 is a large spike driven into a blaze on a tree 6 inches above the ground. The tree is at the side of the road, 100 feet from the end of the bridge on the right bank. The elevation of the bench mark above the zero of the gage is 10.53 feet. Bench mark No. 2 is a large spike driven into a blaze on a tree about 6 inches above ground. The tree is on the opposite side of the road from the first bench mark and is 175 feet farther downstream. The elevation of the bench mark above the zero of the gage is 12.89 feet. The drainage area above the station is 1,900 square miles.

The observations at this station during 1903 have been made under the direction of R. F. Walter, district engineer.

Discharge measurements of Little Missouri River at Camp Crook, S. Dak., in 190:.

Date.	Hydrographer.	Gage height.	Discharge.
		Feet.	Second-feet.
September 2	W. T. Carpenter	1.90	390
November 11	F. M. Madden	.53	36

Mean daily gage height, in feet, of Little Missouri River at Camp Crook, S. Dak., for 1903.

Day.	Sept.	Oct.	Nov.	Dec.	Day.	Sept.	Oct.	Nov.	Dec.
1			0.40	0.60	17	5.50		(a)	(a)
2	2.00		.40	.55	18	4.20		(a)	(a)
3	1.40		.40	.70	19	3.20		(a)	(a)
4	1.10	0.70	.40	.70	20	2.10	.50	.70	(a)
5	1.00		.40	.70	21	1.50	.45	(a)	(a)
6	.90	.70	.40	.70	22	1.30	.45	(a)	(a)
7	2.90		.40	.70	23	1.20	.45	(a)	(a)
8	3.10		.40	.70	24	1.10		(a)	(a)
9	2.20		.40	.675	25	1.10	.40	(a)	(a)
10	2.10		.40	60	26	1.10		(a)	(a)
11			.60	.50	27	1.10	.40	(a)	(a)
12	2.20		.50	.50	28	.80		.70	(a)
13	6.30		.50	.50	29	.80		.70	(a)
14	9.00		.50	.50	30	.80	.40	.65	(a)
15	9.00	.80	.60	.20	31				(a)
16	7.00		.60	(a)					

a Frozen.

CLEAR CREEK AT BUFFALO, WYO.

This station was established October 24, 1902, by Jeremiah Ahern. It is located at the highway bridge in the town of Buffalo, Johnson County, Wyo. The gage is a plain staff graduated to feet and tenths, spiked to the pier at the northwest end of the bridge. The initial point for soundings is on the left bank. Measurements are made from the bridge. The channel is straight both above and below the station. Both banks are high and rocky. The bed of the stream is also rocky. The gage is read daily by P. A. Gatchell. The bench mark is U. S. G. S. B. M. at the court-house, marked "SHER 4635." Its elevation is 4,635.033 feet above sea level. The elevation of the gage as determined from this bench mark is 4,605.766 feet.

The observations at this station during 1903 have been made under the direction of A. J. Parshall, district hydrographer.

rge measurements of Clear Creek at Buffalo, Wyo., in 1903.

te.	Hydrographer.	Gage height.	Discharge.
		Feet.	*Second-feet.*
....	R. G. Schnitger	1.00	33
....	....do	1.00	56
....	....do	1.10	74
....	....do	1.72	185
....	....do	1.55	155
....	....do	1.35	115
....	....do	1.90	267
....	....do	2.20	399
....	....do	1.71	208
....	....do	2.60	491
....	....do	2.30	419
....	....do	2.25	432
....	....do	1.80	176
....	....do	1.10	55
....	....do	1.53	131
....	....do	1.20	64
....	....do	1.05	50
....	....do	1.10	71

gage height, in feet, of Clear Creek at Buffalo, Wyo., for 1903.

	Jan.	Feb.	Mar.	Apr.	May.	June.	July.	Aug.	Sept.	Oct.	Nov.	Dec.
....	1.50	1.00	0.90	0.80	1.00	1.80	1.90	1.90	1.01	1.35	1.08	1.10
....	1.40	1.00	.90	.85	1.00	2.00	1.85	1.75	1.01	1.30	1.06	1.20
....	1.00	1.50	.80	.90		2.55	1.80	1.45	1.00	1.30	1.04	1.30
....	.90	1.80	.75	.85	1.00	2.55	1.55	1.45	1.00	1.30	1.02	1.30
....	.80	1.50	.75	.80	1.10	2.60	1.50	1.25	1.00	1.25	1.00	1.50
....	.80	1.05	.80	.75	1.10	2.67	1.45	1.10	1.00	1.25	1.00	1.40
....	.70	1.00	.80	.80	1.00	2.67	1.90	1.10	1.00	1.22	1.00	1.30
....	.90	1.00	.80	.85	1.05	2.45	1.50	1.10	1.00	1.20	1.00	1.00
....	.80	1.00	.90	.95	1.10	2.15	1.50	1.10	1.00	1.20	1.01	1.00
....	1.80	1.20	.90	.90	1.10	1.95	1.45	1.07	1.00	1.20	1.02	1.00
....	1.50	1.20	.80	.95	1.20	1.80	1.40	1.40	1.05	1.20	1.02	1.00
....	1.20	1.40	.75	.90	1.30	1.85	1.50	1.25	1.10	1.20	1.03	1.00
....	1.00	1.20	.80	.85	1.40	2.25	1.45	1.10	1.10	1.20	1.04	1.20
....	1.00	1.00	.80	.90	1.50	2.05	1.40	1.10	1.15	1.20	1.05	1.90
....	.80	1.00	.80	.90	1.60	2.00	1.35	1.05	1.15	1.10	1.05	1.80
....	.80	1.00	.80	.90	1.70	2.75	1.35	1.05	1.13	1.10	1.05	1.70
....	.80	.80	.90	.98	1.60	2.95	1.37	1.04	1.17	1.10	1.50	1.60
....	.80	.80	1.00	1.10	1.80	2.95	1.70	1.02	1.35	1.10	1.50	1.50
....	.80	.80	1.10	1.05	1.90	2.65	1.45	1.00	1.40	1.10	1.00	1.50
....	.80	.80	1.20	1.00	1.70	2.45	1.40	1.00	1.40	1.10	1.80	1.50
....	.85	1.00	1.00	1.00	1.60	2.50	1.30	1.00	1.30	1.10	2.00	1.40
....	.80	1.00	1.00	1.00	1.50	2.30	1.35	1.00	1.40	1.10	2.00	1.30
....	.75	1.00	1.00	.90	1.50	2.20	1.40	1.00	1.45	1.10	1.50	1.20
....	.70	.90	.90	.95	1.50	2.05	1.35	1.04	1.45	1.10	1.00	1.20

Mean daily gage height, in feet, of Clear Creek, etc.—Continued.

Day.	Jan.	Feb.	Mar.	Apr.	May	June.	July.	Aug.	Sept.	Oct.	Nov.	1
25	0.80	0.90	0.90	1.05	1.50	1.85	1.30	1.01	1.85	1.10	1.00	
26	.70	.85	.85	1.00	1.40	1.80	1.25	1.01	1.30	1.10	.90	
27	.70	.90	.95	1.00	1.40	1.80	1.10	1.01	1.30	1.10	.90	
28	.75	.90	1.00	.95	1.40	1.85	1.10	1.01	1.25	1.10	1.00	
29	.80	......	.90	.95	1.40	2.00	1.10	1.01	1.25	1.10	1.00	
30	.90	......	.90	1.00	1.50	2.00	1.35	1.01	1.30	1.10	1.00	
31	1.00	......	.85	......	1.50	......	1.95	1.01	......	1.10	......	

Rating table for Clear Creek at Buffalo, Wyo., from January 1 to December 31, 1

Gage height.	Discharge.	Gage height.	Discharge.	Gage height.	Discharge.	Gage height.	Discharg
Feet.	Second-feet.	Feet.	Second-feet.	Feet.	Second-feet.	Feet.	Second-fe
0.7	6	1.3	98	1.9	265	2.5	585
.8	18	1.4	116	2.0	305	2.6	585
.9	30	1.5	136	2.1	345	2.7	635
1.0	46	1.6	160	2.2	390	2.8	685
1.1	62	1.7	190	2.3	435	2.9	735
1.2	80	1.8	225	2.4	485	3.0	785

This table is only an approximate estimate of the discharge of Clear Creek. '
data are so inconsistent that the exact discharges can not be obtained.

Estimated monthly discharge of Clear Creek at Buffalo, Wyo., for 1903.

[Drainage area, 118 square miles.]

Month.	Discharge in second-feet.				Run-off.		
	Maximum.	Minimum.	Mean.	Total in acre-feet.	Second feet per square mile.	Dep in inc
January	225	6	40	2,459	0.339	0.
February	225	18	58	3,221	.492	.
March	80	12	30	1,845	.254	
April	62	12	35	2,083	.297	.
May	265	46	116	7,132	.984	1.
June	760	225	422	25,111	3.576	3.
July	285	62	134	8,239	1.135	1.
August	265	46	73	4,489	.619	.
September	126	46	77	4,582	.652	.
October	107	62	73	4,489	.619	.
November	305	30	83	4,939	.703	.
December	265	46	103	6,333	.874	1.
The year	760	6	104	74,922	.879	11

PINEY CREEK AT KEARNEY, WYO.

itation was established September 6, 1902, by Jeremiah Ahern.
ited at the highway bridge at Kearney, Johnson County, Wyo.,
s on the stage route between Sheridan and Buffalo, 24 miles
heridan. The gage is a vertical staff graduated to feet and
ind spiked to the abutment of the bridge. It is read daily by
ioyce. The initial point for soundings is on the right bank.
rge measurements are made from the bridge. The channel is
t for 600 feet above and 200 feet below the station. The left
high and liable to overflow at extreme high water. The right
high and does not overflow. The bed of the stream is com-
of gravel. The bench mark is the U. S. G. S. bench mark at
ranch, about 600 feet from the gage. It is marked "SHER
Its elevation is 4,661.767 feet above sea level. The elevation
ero of the gage as referred to this bench mark is 4,645.963 feet.
ibservations at this station during 1903 have been made under
ction of A. J. Parshall, district hydrographer.

ischarge measurements of Piney Creek at Kearney, Wyo., in 1903.

Date.	Hydrographer.	Gage height.	Discharge.
		Feet.	*Second-feet.*
....	R. G. Schnitger	1.85	69
....	....do	2.00	86
....	....do	2.15	127
....	....do	2.20	156
....	....do	2.58	336
....	....do	2.62	385
....	....do	2.42	278
....	....do	2.71	398
....	....do	2.81	588
....	....do	3.00	818
....	... .do	2.53	348
....	....do	2.60	412
....	....do	2.80	598
....	....do	2.70	531
....	....do	2.76	545
....	....do	2.58	434
....	....do	2.50	378
....	....do	2.45	329
....	....do	2.82	270
....	....do	2.40	308
....	....do	2.20	182
....	....do	2.00	182
....	....do	1.90	105
....	....do	1.80	80
....	....do	1.95	

Discharge measurements of Piney Creek at Kearney, Wyo., in 1903—Con

Date.	Hydrographer.	Gage height.	
		Feet.	*Second*
August 1	R. G. Schnitger	2.20	
August 3	do	1.92	
August 4	do	1.80	
August 8	do	1.80	
August 10	do	1.85	
August 11	do	1.75	
August 12	do	1.70	
August 17	do	1.40	
August 19	do	1.40	
August 20	do	1.42	
August 24	do	1.60	
August 26	do	1.46	
August 27	do	1.86	

Mean daily gage height, in feet, of Piney Creek at Kearney, Wyo., for 1903.

Day.	Apr.	May.	June.	July.	Aug.	Sept.	Oct.	Nov.	Dec.
1		1.98	2.70	2.30	2.15	1.55	1.85	1.65	1.67
2		1.95	2.71	2.30	1.98	1.55	1.86	1.61	1.60
3		1.97	2.83	2.23	1.92	1.55	1.90	1.59	1.60
4		1.95	2.90	2.17	1.86	1.55	1.92	1.60	1.60
5		2.10	2.90	2.10	1.90	1.50	1.90	1.60	1.60
6		2.09	2.83	2.12	1.80	1.51	1.90	1.60	1.60
7		2.15	2.80	2.42	1.86	1.49	1.87	1.60	1.60
8		2.18	2.87	2.30	1.85	1.49	1.88	1.60	1.60
9		2.20	2.80	2.10	1.82	1.47	1.86	1.67	1.60
10		2.15	2.55	2.03	1.80	1.45	1.85	1.62	1.55
11		2.22	2.49	2.02	1.76	1.53	1.83	1.72	1.55
12		2.20	2.56	2.02	1.70	1.55	1.80	1.67	1.55
13		2.30	2.67	2.01	1.65	1.55	1.80	1.80	1.50
14		2.52	2.62	2.04	1.61	1.53	1.80	1.82	1.55
15		2.50	3.00	2.06	1.57	1.50	1.80	1.85	1.55
16		2.55	2.92	2.02	1.50	1.48	1.80	(a)	1.55
17		2.70	2.85	2.02	1.45	1.46	1.80	(a)	1.55
18		2.62	2.85	2.18	1.41	1.47	1.80	(a)	1.55
19	1.90	2.55	2.85	2.04	1.40	1.47	1.76	(a)	1.55
20	1.88	2.50	2.77	1.98	1.40	1.48	1.70	(a)	1.55
21	1.85	2.50	2.85	1.93	1.45	1.46	1.70	(a)	1.50
22	1.90	2.50	2.67	1.94	1.53	1.46	1.70	(a)	1.50
23	1.91	2.40	2.57	1.97	1.61	1.45	1.70	1.60	1.50
24	1.92	2.40	2.53	1.98	1.62	1.45	1.65	1.65	1.50
25	1.97	2.40	2.45	1.94	1.60	1.45	1.70	1.65	1.50
26	2.60	2.40	2.40	1.75	1.60	1.48	1.70	1.63	1.50
27	2.30	2.42	2.40	1.84	1.82	1.47	1.70	1.60	1.50
28	1.96	2.40	2.47	1.84	1.67	1.85	1.70	1.60	1.50
29	1.95	2.40	2.48	1.88	1.62	1.85	1.70	1.60	1.50
30	1.98	2.50	2.46	1.96	1.59	1.85	1.70	1.62	1.50
31		2.50		2.01	1.56		1.68		1.50

a Ice.

Rating table for Piney Creek at Kearney, Wyo., from April 19 to June 15, 1903.

Gage height.	Discharge.	Gage height.	Discharge.	Gage height.	Discharge.	Gage height.	Discharge.
Feet.	Second-feet.	Feet.	Second-feet.	Feet.	Second-feet.	Feet.	Second-feet.
1.8	61	2.2	151	2.6	358	3.0	818
1.9	71	2.3	193	2.7	430		
2.0	90	2.4	242	2.8	523		
2.1	117	2.5	297	2.9	651		

Rating table for Piney Creek at Kearney, Wyo., from June 16 to December 31, 1903.

Gage height.	Discharge.	Gage height.	Discharge.	Gage height.	Discharge.	Gage height.	Discharge.
Feet.	Second-feet.	Feet.	Second-feet.	Feet.	Second-feet.	Feet.	Second-feet.
1.4	36	1.8	78	2.2	208	2.6	441
1.5	41	1.9	101	2.3	256	2.7	514
1.6	48	2.0	130	2.4	311	2.8	594
1.7	60	2.1	166	2.5	373	2.9	685

Estimated monthly discharge of Piney Creek at Kearney, Wyo., for 1903.

Month.	Discharge in second-feet.			Total in acre-feet.
	Maximum.	Minimum.	Mean.	
April 19–30	358	61	108	2,571
May	430	76	228	14,019
June	818	311	488	29,038
July	311	69	152	9,346
August	186	36	66	4,058
September	89	38	46	2,737
October	101	54	76	4,673
November [a]	89	48	57	3,392
December	54	41	44	2,705

[a] River frozen November 16–27, and discharge estimated.

CRUEZ DITCH NEAR STORY, WYO. .

This station was established May 10, 1903. by R. G. Schnitger. It is located in the flume constructed to divert the water from Piney Creek into Prairie Dog Valley. Is about 1 mile north of the post-office at Story, Wyo. The gage is a vertical 2-foot rod. It is read once each day by Miss Stella Sanders. Discharge measurements are made by wading. The channel is straight for 400 feet below the station, at which point it enters a small tunnel, and is straight also for 100 feet above the station. There is never enough water turned into the ditch to cause it to overflow its banks. The bed of the stream is sandy and to a small extent is covered with vegetation. The bench mark is a notch cut in the flume at the 2-foot mark on the gage rod.

The observations at this station during 1903 have been made under the direction of A. J. Parshall, district hydrographer.

Discharge measurements of Cruez ditch near Story, Wyo., in 1903.

Date.	Hydrographer.	Gage height.	Discharge.
		Feet.	*Second-feet.*
June 25	R. G. Schnitger	1.30	51
July 2	do	.80	36
June 7	do	.35	52
July 9	do	.65	68
July 25	do	.40	45
August 2	do	.40	45
August 13	do	.35	30
August 20	do	.32	19
August 9	do	.40	43
June 14	do	1.30	97

Mean daily gage height, in feet, of Cruez ditch near Story, Wyo., for 1903.

Day.	June.	July.	Aug.	Day.	June.	July.	Aug.	Day.	June	July.	Aug.
1		1.25	0.80	12	1.10	0.40	0.50	23	1.30	0.60	0.40
2		1.30	.60	13	1.20	.60	.35	24	1.30	.50	.40
3		.35	.60	14	1.20	.50	.35	25	1.05	.40	.40
4		.60	.60	15	1.40	.50	.50	26	1.15	.45	.4
5	0.30	.60	.60	16	1.20	.50	.40	27	1.10	.40	.5
6	.50	.80	.60	17	1.20	.55	.35	28	1.30	.40	.5
7	.35	.60	.60	18	1.15	.50	.30	29	1.20	.40	.4
8	.35	.60	.60	19	1.40	.50	.30	30	1.25	.50	.4
9	.70	.40	.40	20	1.40	.50	.40	31		.55	.4
10	.70	.50	.50	21	.60	.50	.70				
11	1.10	.30	.50	22	1.30	.60	.40				

PRAIRIE DOG DITCH NEAR STORY, WYO.

This station was established May 10, 1903, by R. G. Schnitger. It is located at the point where the ditch cuts through the divide between Piney and Prairie Dog creeks. The gage is a 2-foot vertical rod. It

once each day by Miss Stella Sanders.   Discharge measure-
re made by wading.

hannel is straight for 100 feet above and 600 feet below the
Both banks are high and rocky.   The bed of the stream is of
nd sand without vegetation and is shifting.   The bench mark
:h cut in the pier at the 2-foot mark at the top of the gage rod.
bservations at this station during 1903 have been made under
ction of A. J. Parshall, district hydrographer.

large measurements of Prairie Dog ditch near Story, Wyo., in 1903.

Date.	Hydrographer.	Gage height.	Discharge.
		Feet.	*Second-feet.*
....	R. G. Schnitger	1.85	87
....	.....do	2.80	148
....	.....do	2.40	145
....	.....do	2.76	124
....	.....do	2.10	110
....	.....do	1.90	90
....	.....do	1.40	39
....	.....do	1.70	57
....	.....do	1.60	48
....	.....do	1.75	69
3.....................	.....do	1.85	53
0.....................	.....do	1.75	44

ly gage height, in feet, of Prairie Dog ditch near Story, Wyo., for 1903.

	May.	June.	July.	Aug.	Day.	May.	June.	July.	Aug.
..........|..........		0.80	3.11	1.60	17	0.70	2.80	3.60	0.80
..... |..........		.85	3.01	1.40	18	.30	3.60	3.50	.90
..... |..........		1.25	3.00	1.35	19	.20	3.55	3.20	.80
..... |..........		2.05	2.95	1.00	20	.40	3.55	3.00	.90
..... |..........		2.05	2.90	1.30	21	.40	3.55	3.00	1.30
..... |..........		1.25	2.75	1.20	22	.45	1.50	2.30	1.20
..... |..........		2.70	3.05	1.20	23	.50	1.60	2.30	1.20
..... |..........		1.60	2.80	1.00	24	.55	1.50	2.20	.90
..... |..........		1.60	2.80	1.75	25	.55	1.30	2.00	.80
..... |..........		1.60	3.00	1.10	26	.40	1.05	1.70	.80
..... |..........		1.70	3.00	1.10	27	.70	1.05	2.00	1.30
..... |..........		1.70	2.20	1.10	28	.70	3.25	1.75	1.50
..... |..........		1.70	2.30	1.85	29	.65	3.25	1.40	1.20
.....	0.75	1.65	2.30	.95	30	.65	3.20	1.40	1.20
.....	.95	1.70	3.40	.80	31	.90	....	1.45	1.25
.....	.90	3.60	3.40	.80					

TONGUE RIVER AT DAYTON, WYO.

This station was established May 3, 1903, by R. G. Schnitger.
is located at the bridge 20 miles west of Sheridan, Wyo., at Dayt
The gage is read twice each day by T. S. Wilson, the postmaster
Dayton. The gage is a vertical 2 by 6 inch timber 9 feet lo
nailed to the bridge pier. Discharge measurements are made fr
the bridge to which the gage is attached or by wading. The chan
is slightly curved above the station and straight below it. B
banks are low and rocky and are liable to overflow. There is se
brush along the right bank. The bed of the stream is rocky ;
covered with bowlders. There are two channels at low water. '
bench mark is a large spike in the bridge floor opposite the 7-f
mark on the gage rod. Its elevation is 7 feet above the zero of
gage.

The observations at this station during 1903 have been made un
the direction of A. J. Parshall, district hydrographer.

Discharge measurements of Tongue River at Dayton, Wyo., in 1903.

Date.	Hydrographer.	Gage height.	Disch
		Feet.	*Secon*
May 2	R. G. Schnitger	1.60	
May 12	....do	1.95	
May 25	....do	2.31	
May 31	....do	2.60	.
June 9	....do	2.91	
June 16	...do	2.95	
June 27	...do	2.50	
July 5	...do	2.71	
July 11	....do	1.91	
July 30	....do	1.40	
August 7	....do	1.20	
August 15	....do	1.47	
August 22	....do	1.58	
August 30	....do	1.20	

laily gage height, in feet, of Tongue River at Dayton, Wyo., for 1903.

Day.	May.	June.	July.	Aug.	Sept.	Oct.
....	1.60	2.85	2.65	1.75	1.40	1.40
....	1.70	2.85	2.70	1.65	1.40	1.40
....	1.70	2.90	2.65	1.60	1.40	1.40
....	1.70	2.90	2.65	1.50	1.40	1.65
....	1.75	3.00	2.65	1.50	1.40	1.60
....	1.85	3.05	2.55	1.50	1.37	1.50
....	1.85	3.05	2.45	1.50	1.35	1.50
....	1.90	3.00	2.25	1.50	1.35	1.50
....	1.95	2.95	2.10	1.50	1.35	1.50
....	1.90	2.97	2.05	1.50	1.35	1.50
....	1.90	3.00	2.05	1.50	1.35	1.40
....	1.95	2.97	2.00	1.50	1.40	1.40
....	2.05	2.97	2.05	1.45	1.40	1.40
....	2.80	3.00	2.00	1.43	1.40	1.40
....	2.95	2.92	2.00	1.40	1.42	1.40
....	2.90	2.97	1.95	1.40	1.45	1.40
....	2.75	2.62	1.90	1.40	1.50	1.40
....	2.65	2.75	1.95	1.40	1.50	1.40
....	2.70	2.87	1.95	1.40	1.50	1.38
....	2.45	2.75	1.90	1.40	1.45	1.35
....	2.40	2.74	1.80	1.40	1.45	1.35
....	2.55	2.77	1.80	1.40	1.45	1.35
....	2.40	2.75	1.70	1.40	1.40	1.35
....	2.30	2.70	1.70	1.40	1.40	1.35
....	2.35	2.65	1.60	1.40	1.40	1.35
....	2.35	2.65	1.55	1.38	1.40	1.35
....	2.30	2.65	1.60	1.50	1.35	1.35
....	2.35	2.60	1.40	1.50	1.35	1.35
....	2.50	2.70	1.40	1.47	1.35	1.32
....	2.65	2.65	1.40	1.43	1.35	1.30
....	2.75	....	1.65	1.40	....	1.30

SHOSHONE RIVER NEAR CODY, WYO.

station was established April 26, 1902, by A. J. Parshall. It
ed at the wagon bridge 1 mile northwest of Cody, Wyo. The
a plain staff graduated to half tenths and is fastened to the
from which discharge measurements are made. The initial
)r soundings is on the left bank. The channel is straight both
.nd below the station. The current is swift. The right bank
.nd subject to overflow. The left bank is high and does not
v. The bed of the stream is gravel and rock. The bench mark
t in the bridge sleeper 1.15 feet above the 12-foot mark on the
It is indicated by a cross. The observer is W. J. Kissick, who
1e gage twice daily.
)bservations at this station during 1903 have been made under
:ction of A. J. Parshall, district hydrographer.

Discharge measurements of Shoshone River near Cody, Wyo., in 1903.

Date.	Hydrographer.	Gage height.	Dischar
		Feet.	*Second-ft*
March 20	A. J. Parshall	1.95	
April 21	do	2.41	
April 25	do	3.00	1,
May 19	do	3.20	1,
May 20	do	3.00	1.
June 6	Jeremiah Ahern	4.70	2,
June 11	do	5.25	5,
June 16	A. J. Parshall	5.70	7,
June 17	do	6.10	8,
June 18	do	6.20	8,
June 24	do	5.10	4,
July 14	do	4.70	3,
July 15	do	4.50	3,
July 25	do	4.25	3,
October 31	Jeremiah Ahearn	2.20	

Mean daily gage height, in feet, of Shoshone River near Cody, Wyo., for 1903

Day.	Jan.	Feb.	Mar.	Apr.	May.	June.	July.	Aug.	Sept.	Oct.	Nov.	
1	2.15	2.00	2.00	2.50	2.85	4.60	5.40	3.75	2.00	2.00	2.75	
2	2.05	2.00	2.00	2.80	2.80	5.00	5.35	3.85	2.00	2.00	2.00	
3	2.10	2.00	2.00	2.50	2.75	4.75	5.00	3.70	2.10	2.05	2.00	
4	1.95	1.95	1.90	2.50	3.00	4.50	4.25	3.80	1.85	2.00	2.00	
5	2.00	1.95	2.00	2.50	3.00	5.50	4.40	3.80	1.90	2.30	2.00	
6	1.95	1.95	2.00	2.50	3.50	5.50	4.50	3.90	1.95	2.40	1.95	
7	2.00	2.00	2.00	2.50	3.50	5.35	5.00	3.85	1.95	2.30	2.00	
8	2.00	2.00	1.90	2.50	3.50	5.40	4.80	3.80	1.95	2.10	2.00	
9	2.05	1.95	2.00	2.50	3.50	5.40	4.70	3.75	1.95	2.10	2.10	
10	2.00	1.95	2.00	2.50	3.00	5.35	4.45	3.70	1.95	2.00	2.10	
11	2.05	1.95	2.00	2.50	3.50	5.45	4.70	3.80	1.80	2.50	2.20	
12	1.95	2.00	2.00	2.25	3.50	5.35	3.45	1.50	2.30	2.00		
13	2.00	2.00	2.00	2.00	3.80	5.40	4.80	3.50	1.45	2.35	2.30	
14	2.00	2.00	2.00	2.00	4.25	5.60	4.65	3.55	1.35	2.15	2.30	
15	2.00	2.00	2.05	2.00	4.25	5.70	4.55	3.50	1.95	2.00	2.35	
16	1.95	2.00	1.95	2.00	4.30	6.25	4.70	3.50	1.95	2.05	2.45	
17	2.05	2.00	1.95	2.00	3.00	6.25	4.60	3.45	1.95	2.50	2.00	
18	1.95	2.00	2.05	2.00	3.00	6.10	4.55	3.00	1.85	2.45	2.45	
19	2.05	1.95	1.95	2.50	3.00	6.05	4.35	3.00	1.90	2.45	2.50	
20	2.05	2.00	2.00	2.50	3.10	5.90	4.40	2.95	1.80	2.70	2.45	
21	2.00	2.00	2.00	2.50	3.00	5.90	4.55	3.00	1.70	2.60	2.50	
22	1.95	1.90	2.00	2.90	3.00	5.45	4.35	3.75	1.65	2.75	2.50	
23	2.00	1.95	2.00	3.00	3.00	5.30	4.50	2.80	1.90	2.85	2.25	
24	1.95	1.90	2.00	3.20	3.00	5.40	4.40	2.90	1.90	2.95	2.35	
25	2.05	2.00	2.00	3.50	3.00	5.80	4.50	3.00	1.80	2.70	2.50	
26	2.00	2.00	2.00	3.85	3.00	5.75	4.50	2.85	1.60	2.95	2.55	
27	2.00	1.95	2.00	3.95	3.50	5.95	4.50	2.85	1.95	2.40	2.65	
28	2.05	1.95	2.00	3.60	3.50	5.90	4.00	2.85	2.00	2.40	2.55	
29	2.00		2.50	3.00	3.50	6.50	3.90	2.75	1.95	2.40	2.50	
30	2.05		2.50	2.90	3.60	6.25	4.05	2.90	2.00	2.55	2.30	
31	2.00		2.80		4.25		3.80	2.40		2.65		

ating table for Shoshone River near Cody, Wyo., from January 1 to December 31, 1903.

Gage height.	Discharge.	Gage height.	Discharge.	Gage height.	Discharge.	Gage height.	Discharge.
Feet.	*Second-feet.*	*Feet.*	*Second-feet.*	*Feet.*	*Second-feet.*	*Feet.*	*Second-feet.*
1.3	50	2.7	880	4.1	2,610	5.5	6,440
1.4	70	2.8	925	4.2	2,800	5.6	6,810
1.5	95	2.9	1,020	4.3	3,000	5.7	7,180
1.6	125	3.0	1,120	4.4	3,210	5.8	7,550
1.7	160	3.1	1,225	4.5	3,420	5.9	7,920
1.8	200	3.2	1,335	4.6	3,650	6.0	8,290
1.9	245	3.3	1,450	4.7	3,900	6.1	8,660
2.0	295	3.4	1,575	4.8	4,170	6.2	9,030
2.1	355	3.5	1,710	4.9	4,440	6.3	9,400
2.2	425	3.6	1,850	5.0	4,730	6.4	9,770
2.3	500	3.7	1,990	5.1	5,040	6.5	10,140
2.4	575	3.8	2,130	5.2	5,370		
2.5	655	3.9	2,280	5.3	5,710		
2.6	740	4.0	2,440	5.4	6,070		

Estimated monthly discharge of Shoshone River near Cody, Wyo., for 1903.

Month.	Discharge in second-feet.			Total in acre-feet.
	Maximum.	Minimum.	Mean.	
anuary	385	270	303	18,631
'ebruary	295	245	283	15,717
larch	925	245	335	20,598
pril	2,360	295	847	50,400
lay	2,900	875	1,558	95,798
une	10,140	3,420	6,819	405,759
uly	6,060	2,060	3,562	219,019
ngust	2,280	575	1,565	96,228
ptember	355	60	231	13,745
ctober	1,070	295	577	35,478
ovember	875	270	499	29,693
ecember	830	95	437	26,870
The year	10,140	60	1,418	1,027,936

SHOSHONE RIVER (SOUTH FORK) AT MARQUETTE, WYO.

This station was established by A. J. Parshall April 26, 1903. It located at the county bridge at Marquette, Wyo., 15 miles west of ody, Wyo. The gage is a vertical timber set in the stream an^

braced to the bridge. It is read twice each day by E. C. Payne. Discharge measurements are made from the bridge to which the gage is attached. The initial point for soundings is taken at either bank, depending upon the depth of the water. The channel is straight for 150 feet above and 200 feet below the station. The current velocity is swift. Both banks are low and rocky and will overflow until the water reaches the bridge abutments. The bed of the stream is rocky. The bench mark is the lower surface of the iron plate at the bottom of the bridge stringer at the gage rod. Its elevation is 10.6 feet above the zero of the gage.

The observations at this station during 1903 have been made under the direction of A. J. Parshall, district hydrographer.

Discharge measurements of South Fork of Shoshone River at Marquette, Wyo., in 1903.

Date.	Hydrographer.	Gage height.	Discharge.
		Feet.	*Second-feet.*
June 4	P. M. Churchill	2.30	1,397
June 17	A. J. Parshall	3.40	2,740
June 25	do	2.50	1,448
October 27	P. M. Churchill	.30	165
April 26	A. J. Parshall	1.10	534
May 20	do	.95	334
July 15	do	2.30	1,199
July 26	do	1.80	857

Mean daily gage height, in feet, of South Fork of Shoshone River at Marquette, Wyo., for 1903.

Day.	Apr.	May.	June.	July.	Aug.	Sept.	Oct.	Nov.
1	0.65	2.40	3.30	1.60	0.60	0.75	0.50	
2	.65	2.55	2.65	1.55	.50	.70	.50	
3	.60	2.05	1.95	1.40	.50	.70	.50	
4	.70	2.45	2.05	1.40	.50	.70	.50	
5	.75	2.65	2.00	1.40	.50	.70	.50	
6	.80	2.55	2.10	1.35	.75	.70	.50	
7	.95	2.70	2.30	1.25	.70	.70	.50	
8	1.05	2.90	2.00	1.15	.70	.70	.50	
9	.95	2.70	2.05	1.05	.70	.70	.50	
10	.80	2.25	2.15	1.00	.70	.70	.50	
11	.85	2.40	2.00	1.00	.80	.70	.50	
12	.85	2.75	2.40	1.00	.90	.70	.50	
13	1.15	2.75	2.40	1.00	.90	.70	.50	
14	1.85	2.85	2.50	1.00	.90	.70	.50	
15	1.90	3.15	2.05	.95	.90	.70	.50	
16	2.00	3.45	1.95	.90	.90	.70	.50	
17	1.70	3.65	1.95	.90	.80	.70	.50	
18	1.10	3.50	1.90	.80	.80	.60	.60	
19	.95	3.15	1.90	.80	.95	.60	.60	
20	1.00	3.10	1.95	.70	1.00	.60	.60	

ean daily gage height, in feet, of South Fork of Shoshone River, etc.—Cont'd.

Day.	Apr.	May.	June.	July.	Aug.	Sept.	Oct.	Nov.
....	....	0.95	2.80	2.00	0.70	0.95	0.60	0.70
....	....	.85	2.80	1.90	1.75	.90	.60	.65
....	....	.85	2.75	1.65	1.65	.90	.60	.55
....	....	.85	2.50	2.00	1.40	.90	.60	.50
....	....	1.00	2.60	1.90	1.10	.90	.50	.60
....	1.10	.95	2.70	1.90	1.00	.90	.50	.50
....	1.15	.90	3.20	1.95	1.00	.90	.50	.50
....	.95	1.05	3.20	1.90	.90	.90	.50	.50
....	.80	1.10	3.40	1.80	.75	.90	.50	.50
....	.75	1.45	3.15	1.65	.70	.80	.50	.50
....	....	1.95	....	1.65	.70	....	.50	....

ıting table for South Fork of Shoshone River at Marquette, Wyo., from April 26 to December 31, 1903.

Gage height.	Discharge.	Gage height.	Discharge.	Gage height.	Discharge.	Gage height.	Discharge.
Feet.	*Second-feet.*	*Feet.*	*Second-feet.*	*Feet.*	*Second-feet.*	*Feet.*	*Second-feet.*
0.3	165	1.2	435	2.1	1,075	3.0	2,145
.4	181	1.3	485	2.2	1,175	3.1	2,290
.5	202	1.4	545	2.3	1,275	3.2	2,440
.6	227	1.5	605	2.4	1,380	3.3	2,590
.7	255	1.6	670	2.5	1,495	3.4	2,740
.8	285	1.7	740	2.6	1,615	3.5	2,895
.9	317	1.8	815	2.7	1,740	3.6	3,055
1.0	352	1.9	895	2.8	1,870	3.7	3,215
1.1	390	2.0	980	2.9	2,005		

ıtimated monthly discharge of South Fork of Shoshone River at Marquette, Wyo., in 1903.

[Drainage area, 500 square miles.]

Month.	Discharge in second-feet.			Total in acre-feet.	Run-off.	
	Maximum.	Minimum.	Mean.		Second-feet per square mile.	Depth in inches.
pril 26–30	412	270	338	3,352	0.68	0.13
ıy	980	227	417	25,640	.83	.96
ne	3,135	1,027	1,951	116,093	3.90	4.35
ly	2,590	705	1,072	65,915	2.14	2.47
ıgust	777	255	411	25,271	.82	.95
ıtember	352	202	287	17,078	.57	.64
:ober	270	202	237	14,573	.47	.54
vember	255	202	209	12,436	.42	.47

GREY BULL RIVER AT MEETEETSE, WYO.

This station was established April 24, 1903, by A. J. Parshall, assisted by R. G. Schnitger. It is located within the town limits of Meeteetse, Wyo., at the county bridge. The gage is a vertical 2 by 6 inch timber nailed to the bridge pier. It is read twice each day by W. L. Simpson. Discharge measurements are made by wading, at a point about 200 feet below the bridge to which the gage is attached. A cable will be necessary for high-water measurements. The initial point for soundings is on the right bank.

The channel is straight for about 200 feet above the station and for 150 feet below. The current velocity is swift. The right bank is liable to overflow at high water; the left bank is high and level. The bed of the stream is rocky. The bench mark is the top of the nut on the bridge rod. Its elevation is 14.8 feet above the zero of the gage.

The observations at this station during 1903 have been made under the direction of A. J. Parshall, district hydrographer.

Discharge measurements of Grey Bull River near Meeteetse, Wyo., in 1903.

Date.	Hydrographer.	Gage height.	Discharge.
		Feet.	*Second-feet.*
April 24	A. J. Parshall	0.90	160
May 27	do	.85	131
July 16	do	2.25	541

Mean daily gage height, in feet, of Grey Bull River at Meeteetse, Wyo., for 1903.

Day.	Apr.	May.	June.	July.	Aug.	Sept.	Oct.
1		0.55	3.65	3.55	2.60	1.45	1.40
2		.55	3.15	3.65	2.45	1.50	1.40
3		.55	3.05	3.05	2.55	1.50	1.40
4		.65	3.55	2.50	2.55	1.50	1.40
5		.90	3.85	2.50	2.40	1.50	1.40
6		.90	3.75	2.40	2.15	1.50	1.40
7		1.10	3.85	2.80	2.10	1.40	1.35
8		.85	3.85	2.75	2.10	1.40	1.30
9		.95	3.65	2.55	2.05	1.40	1.40
10		.80	3.55	2.75	1.90	1.40	1.40
11		.75	3.45	2.80	1.90	1.40	1.40
12		.75	3.65	2.90	1.75	1.45	1.30
13		1.30	3.90	3.00	1.75	1.50	1.30
14		2.05	3.25	2.75	1.65	1.55	1.30
15		2.15	4.15	2.90	1.60	1.55	1.30
16		2.05	4.90	2.65	1.55	1.60	1.30
17		1.85	5.45	2.55	1.55	1.75	1.30
18		1.60	4.95	2.50	1.60	1.60	1.30
19		1.30	4.35	2.45	1.60	1.55	1.30

Mean daily gage height, in feet, of Grey Bull River, etc.—Continued.

Day.	Apr.	May.	June.	July.	Aug.	Sept.	Oct.
10	....	.80	4.55	2.60	1.45	1.50	(a)
11	....	.45	4.30	2.55	1.45	1.50	(a)
12	....	.95	4.15	2.40	2.40	1.50	(a)
13	....	1.05	4.00	2.50	1.65	1.50	(a)
14	0.80	1.10	3.60	2.50	1.60	1.40	(a)
15	.70	1.15	3.50	2.45	1.55	1.40	(a)
16	1.40	1.05	3.85	2.65	1.55	....	(a)
17	1.35	1.35	4.00	2.45	1.50	....	1.30
18	.95	1.65	3.95	2.40	1.40	1.40	1.30
19	.50	1.85	4.40	2.45	1.45	1.40	1.30
30	.45	2.50	4.30	2.45	1.50	1.40	1.30
31	....	3.40	....	2.75	1.45	....	1.30

a Observer absent.

BIGHORN RIVER NEAR THERMOPOLIS, WYO.

This station was established on May 28, 1900, by A. J. Parshall. It was located about a half mile northeast of Thermopolis, at the ferry crossing the river. The gage used during 1900 was a horizontal rod extending out over the water and fastened to posts firmly set in the ground, with the usual wire attachments.

The bench mark was the head of a nail driven in a stake set in the ground 1 foot south of the post to which the gage rod was fastened, 2.58 feet below the gage frame. Its elevation was 6.50 feet above gage datum. It has been destroyed.

In 1901 a new gage with relative heights, the same as that in use in 1900, was painted on the lower center pier of the recently constructed iron bridge, which is located 400 or 500 feet upstream from the former station. In March, 1903, a new gage, serving as a bench mark, and for the observer's use at times of high water, was painted on the lower pier of the bridge on the right bank of the river. This gage is reached by the water only when it is at a height of 3 feet above datum. It will enable the observer to read heights of high water more accurately than can be done on the gage, which is in the center of the stream and is exposed to the force of the current, causing a rise and fall of several inches on the gage. The channel is straight above and below the station. Both banks are high and not subject to overflow. The bed of the stream is composed of rock and gravel. It shifts very little from year to year. Measurements, which were formerly made from a ferryboat, are now made from the bridge.

The observations at this station during 1903 have been made under the direction of A. J. Parshall, district hydrographer.

Discharge measurements of Bighorn River near Thermopolis, Wyo., in 1903

Date.	Hydrographer.	Gage height.	Discha
		Feet.	Second
March 27	A. J. Parshall	0.80	
May 23	do	1.90	1.
May 25	do	1.75	1.
June 20	do	6.05	9.
June 21	do	5.80	8.
June 22	do	5.30	7.
July 18	do	3.05	3.
July 20	do	3.15	4.

Mean daily gage height, in feet, of Bighorn River near Thermopolis, Wyo.,
1903.

Day.	Mar.	Apr.	May.	June.	July.	Aug.	Sept.	O
1		1.50	1.45	2.05	4.85	2.40	1.70	
2		1.35	1.55	2.30	4.55	2.40	1.60	
3		1.20	1.40	2.75	4.20	2.45	1.55	
4		1.35	1.55	3.25	3.95	2.35	1.50	
5		1.45	1.75	4.15	3.75	2.25	1.50	
6		1.30	1.65	4.55	3.55	2.15	1.50	
7		1.20	1.55	4.65	3.30	2.10	1.35	
8		1.05	1.35	4.55	2.95	2.15	1.25	
9		.90	1.80	5.15	3.20	2.05	1.15	
10		.85	1.80	5.75	3.00	2.00	1.35	
11		.80	1.30	5.90	2.95	1.90	1.55	
12		.80	1.20	5.30	2.90	1.90	1.65	
13		.75	1.20	5.40	3.25	1.80	1.75	
14		.70	1.45	5.35	3.40	1.75	1.65	
15		.70	1.65	5.35	3.35	1.70	1.55	
16		.70	2.25	5.55	3.45	1.80	1.50	
17		.70	2.55	5.65	3.25	1.80	1.40	
18		.70	2.35	5.90	3.05	1.70	1.40	
19		.85	2.25	6.55	2.90	1.70	1.40	
20		.80	2.15	6.05	2.95	1.60	1.40	
21		.70	2.05	5.85	2.95	1.45	1.35	
22		.70	1.95	5.35	2.85	2.10	1.30	
23		.70	1.90	5.10	2.95	2.30	1.30	
24		.70	1.80	4.65	2.95	2.20	1.30	
25		.85	1.70	4.35	2.80	2.30	1.20	
26		.90	1.70	4.15	2.75	2.20	1.20	
27		1.00	1.60	4.00	2.85	2.15	1.20	
28		1.15	1.50	4.30	2.80	2.05	1.20	
29	1.15	1.25	1.50	4.80	2.80	1.95	1.20	
30	1.35	1.35	1.65	5.05	2.60	1.85	1.10	
31	1.40		1.85		2.50	1.75		

Rating table for Bighorn River near Thermopolis, Wyo., from January 1 to December 31, 1903.

Gage height.	Discharge.	Gage height.	Discharge.	Gage height.	Discharge.	Gage height.	Discharge.
Feet.	*Second-feet.*	*Feet.*	*Second-feet.*	*Feet.*	*Second-feet.*	*Feet.*	*Second-feet.*
0.7	550	1.5	1,350	2.2	2,445	2.9	3,630
.8	620	1.6	1,485	2.3	2,610	3.0	3,800
1.0	780	1.7	1,630	2.4	2,780	3.1	3,970
1.1	880	1.8	1,785	2.5	2,950	3.2	4,140
1.2	990	1.9	1,950	2.6	3,120		
1.3	1,105	2.0	2,115	2.7	3,290		
1.4	1,225	2.1	2,280	2.8	3,460		

Tangent above 2.3 feet gage height. Differences above this point, 170 per tenth.

Estimated monthly discharge of Bighorn River near Thermopolis, Wyo., in 1903.

[Drainage area, 8,184 square miles.]

Month.	Discharge in second-feet.			Total in acre-feet.	Run-off.	
	Maximum.	Minimum.	Mean.		Second-feet per square mile.	Depth in inches.
April	1,350	550	785	46,711	0.10	0.11
May	3,035	990	1,654	101,701	.20	.23
June	9,835	2,195	6,846	407,365	.84	.94
July	6,945	2,950	4,159	255,727	.51	.59
August	2,780	1,285	2,133	131,153	.26	.30
September	1,705	880	1,237	73,607	.15	.17
October	840	620	797	49,006	.10	.12

YELLOWSTONE RIVER AT GLENDIVE, MONT.

This station, established in 1893 by the United States War Department, was transferred to the Department of Agriculture, where daily records of the gage are kept.

When a study of the lower Yellowstone Valley was begun in 1903, this station was taken up by the United States Geological Survey, whose records begin August 1, 1903. It is located at the steel highway bridge leading northward out of Glendive, about one-fourth mile from the post-office. The gage is of the wire type and is fastened to the upstream hand rail of the second span, about 200 feet from its right end. The graduations are marked upon the hand rail. The marker is an adjustable pointer near the handle. The distance from the marker to the bottom of the weight is 48.99 feet. The gage is read daily by H. A. Sample, of Glendive, Mont.

The measurements are made from the downstream hand rail of the bridge, upon which the distances are painted. The initial point is over the southeast cement-filled iron pier at the right. The bridge spans the channel in four truss spans of about 300 feet each. The channel above is straight for about 500 feet and below for the same distance. The water, especially at high stages, is very swift. The right bank is high and of clay and gravel. Being riprapped for some distance, it is not subject to inundation. The left is a low bank, beyond which are sandy flats covered with trees and brush and extending to the hills, 1 mile distant. These are liable to become submerged. At low water an island becomes visible between the third and fourth spans from the right, and at flood stages water finds its way through a slough traversing the above-mentioned flats. The bed is of sand and clay and liable to shift with every flood. Old piling and cribwork, remains of a former bridge, obstruct the channel on the upstream side.

Bench mark No. 1 is a point on the surface of the cement at the northwest corner of the girder on the pier under Station "O." Its elevation is 26.26 feet above gage datum and 2,060.6 feet above sea level. Bench mark No. 2 is the top of the west rail of the Northern Pacific track at the right end of the bridge. Its elevation is 30.19 feet above gage datum.

The elevation above sea level of bench mark No. 1 was determined from the Northern Pacific Railway elevation of the rail at Terry, Mont. This is 2,242 feet above sea level.

The gage readings from 1897 to the establishment of the Geological Survey wire gage were referred to a "T" on the top of the southeast anchor bolt in the south caisson at the east end of the bridge. Its elevation when established was 25.08 feet above gage datum. The gage heights up to April 8, 1899, were read from a wire gage. On this date the gage was carried away by ice and the gage heights during the remainder of 1899 and up to the establishment of the Geological Survey gage were determined by measuring down from the bench mark on the anchor bolt. A new bridge was erected in 1901, and it is not known whether this bench mark was disturbed or not.

The observations at this station since August 1, 1903, have been made under the direction of C. C. Babb, district engineer.

Discharge measurements of Yellowstone River at Glendive, Mont., in 1903.

Date.	Hydrographer.	Gage height.	Discharge.
		Feet.	*Second-feet.*
August 1	John N. Kerr	6.50	31,058
August 31	F. E. Weymouth	2.80	9,645
September 28	do	1.95	7,181
October 31	John N. Kerr	1.45	5,966

ily gage height, in feet, of Yellowstone River at Glendive, Mont., for 1897-1903.

Day.	Jan.	Feb.	Mar.	Apr.	May.	June.	July.	Aug.	Sept.	Oct.	Nov.	Dec.
1897.												
							5.1	2.5	1.1	0.2	0.6	0.3
							4.8	2.5	1.1	.2	.5	.3
							4.5	2.4	1.1	.1	.5	.3
							4.5	2.4	1.0	.1	.5	.3
							5.0	2.4	1.0	.1	.4	.4
							5.0	2.3	1.0	.0	.4	.4
							4.5	2.5	.9	.0	.4	.5
							4.4	2.3	.9	.0	.4	.5
							4.1	2.2	.8	.0	.3	.6
							4.0	2.1	.8	.0	.3	.7
							4.0	2.1	.7	.2	.3	.6
							3.9	2.1	.7	.3	.2	.6
							3.8	2.1	.7	.2	.2	.5
							3.7	2.0	.7	.2	.2	.5
							3.6	2.0	.7	.1	.2	.4
							3.5	2.0	.6	.1	.3	.4
							3.6	1.9	.6	.1	.4	.4
							3.7	1.9	.6	.2	.4	.4
							3.6	1.8	.6	.4	.4	.4
							3.6	1.8	.6	.5	.4	.4
							3.6	1.7	.6	.6	.4	.5
							3.4	1.6	.5	.6	.3	.6
							3.1	1.6	.5	.6	.2	.6
							3.0	1.5	.5	.6	.2	.7
							3.1	1.5	.4	.6	.3	.7
							3.0	1.4	.4	.6	.3	.7
							2.9	1.3	.3	.5	.3	.7
							2.7	1.2	.3	.5	.3	.9
							2.6	1.2	.3	.4	.3	1.0
							2.6	1.2	.3	.4	.3	1.1
							1.1					1.3
1898.												
	1.5	1.1	4.3	2.1	2.4	8.4	6.9	2.8	1.4	.4	.5	
	2.1	1.0	4.4	2.3	2.3	7.4	6.9	2.7	1.6	.5	.5	
	2.5	.8	4.3	2.5	2.2	6.5	7.1	2.6	2.0	.6	.5	
	2.7	.8	4.4	4.2	2.4	6.2	6.8	2.5	3.1	.7	.6	
	2.7	.6	4.2	4.1	2.2	6.2	6.2	2.4	2.6	1.2	.4	
	2.5	.8	4.1	3.9	2.0	6.0	5.8	2.3	2.2	1.5	.4	
	2.2	1.3	4.1	3.9	1.8	5.9	5.6	2.2	2.1	1.1	.4	
	2.0	1.7	4.0	4.2	1.7	5.7	5.7	2.2	2.0	.9	.4	
	1.9	1.7	4.0	4.8	1.6	5.5	5.7	2.3	1.9	.9	.4	
	1.8	1.7	4.1	8.0	1.5	5.4	5.7	2.2	1.7	.8	.3	
	1.7	2.2	4.2	4.7	1.4	5.3	5.6	2.1	1.6	.8	.3	
	1.6	2.6	4.0	4.3	1.6	5.5	5.6	2.2	1.4	.9	.3	
	1.8	2.7	3.9	3.5	1.7	5.8	5.6	2.1	1.4	.8	.3	
	1.9	3.3	3.8	2.1	1.9	6.4	5.6	2.0	1.3	.8	.2	
	1.9	5.7	3.7	1.5	2.4	6.8	5.7	2.0	1.2	.7	.1	
	1.5	5.1	3.6	1.5	2.8	7.5	5.9	1.9	1.1	.7		
	1.4	4.7	3.4	1.4	3.2	8.4	5.7	1.8	1.0	.7		
	1.3	3.9	3.3	1.3	3.6	8.8	5.5	1.8	1.0	.6		
	1.4	3.8	3.1	1.2	4.6	9.2	5.1	1.9	.9	.6		
	1.6	3.8	3.0	1.2	5.6	9.4	4.9	2.0	.9	.5		

Mean daily gage height, in feet, of Yellowstone River, etc.—Continued.

Day.	Jan.	Feb.	Mar.	Apr.	May	June	July	Aug.	Sept.	Oct.	Nov.	Dec.
1898.												
21	1.5	3.6	2.7	1.6	5.4	9.6	4.5	2.0	0.8	0.5		
22	1.4	3.4	2.5	1.8	5.6	9.8	4.2	1.9	.8	.6		
23	1.2	3.2	2.2	1.9	5.3	10.2	3.8	1.8	.7	.6		
24	1.0	3.0	2.2	2.0	4.6	10.4	3.7	1.7	.7	.6		
25	1.0	3.0	1.9	1.9	4.4	9.5	3.5	1.7	.6	.6		
26	.9	3.4	1.8	1.5	4.7	9.5	3.4	1.6	.5	.6		
27	1.3	3.7	1.7	1.8	6.4	9.3	3.3	1.6	.5	.7		
28	1.8	3.7	1.6	1.8	7.5	8.7	3.2	1.5	.5	.6		
29	1.9		1.3	1.4	8.5	7.9	3.1	1.4	.5	.6		
30	1.6		1.8	1.9	9.1	7.4	3.0	1.4	.5	.5		
31	1.4		2.0		9.2		2.9	1.4		.5		
1899.												
1					3.1	3.5	11.1	8.7	1.4	.3	0.3	0.0
2					6.4	3.1	3.2	11.4	3.4	1.3	.2	.2
3					6.8	3.4	3.7	11.6	3.1	1.2	.1	.3
4					7.0	3.3	4.7	11.4	2.9	1.3	.1	.1
5					7.5	3.1	5.8	10.8	2.8	1.4	.2	.0
6					8.8	2.7	6.7	10.6	2.9	1.4	.2	.0
7					11.8	2.3	7.1	10.3	2.9	1.3	.2	.0
8					13.4	2.1	6.3	9.8	3.3	1.4	.1	-.1
9					10.0	1.9	5.3	9.3	3.4	1.2	.0	.0
10					10.6	1.7	4.7	8.9	3.2	1.0	.0	.0
11					9.6	1.6	4.3	8.8	2.9	.9	.1	.0
12					8.0	1.9	4.2	8.6	2.8	.8	.1	.0
13					6.0	2.2	4.5	8.4	2.7	.7	.4	.1
14					5.1	2.5	5.7	8.0	2.6	.6	.5	.1
15					5.2	2.8	6.4	7.8	2.5	.5	.7	.0
16					5.0	3.2	6.0	7.4	2.3	.4	.6	.0
17					4.7	3.3	5.7	7.0	2.2	.5	.5	.0
18					3.9	3.2	5.2	6.5	2.0	.7	.6	.1
19					3.0	3.4	5.3	6.1	1.8	.8	.6	.1
20					2.6	3.6	6.7	5.6	1.6	.7	.5	.2
21					2.5	4.1	8.4	5.1	1.5	.6	.5	.1
22					2.8	4.2	9.7	4.8	1.4	.5	.8	.1
23					3.2	4.7	10.5	4.5	1.4	.4	.7	.0
24					3.1	5.0	10.9	4.2	1.3	.4	.6	.0
25					3.0	4.6	9.8	3.8	1.4	.3	.5	.1
26					3.1	3.8	8.7	3.5	1.4	.4	.6	.0
27					2.9	3.5	9.2	3.4	1.3	.4	.6	.0
28					2.6	3.8	10.2	3.2	1.2	.3	.8	.0
29					2.4	3.7	10.8	3.1	1.2	.3	.8	-.1
30					2.8	3.5	11.0	3.2	1.3	.4	.6	-.1
31					3.7		3.6	1.4		.4		.3
1900.												
1	.3	.1	2.0	1.6	6.0	8.8	6.0	1.2	.4	.9	.3	1.7
2	.3	.0	1.7	1.7	5.5	8.2	5.7	1.1	.5	1.1	.3	1.9
3	.2	.0	1.3	1.8	3.8	7.8	5.6	1.0	1.1	1.0	.4	1.8
4	.2	.0	.7	1.7	3.7	7.9	5.1	1.0	.9	.8	.4	1.8
5	.3	.0	.4	1.6	3.3	8.6	4.6	1.0	.7	.7	.3	1.7
6	.4	.0	.4	1.5	3.1	8.9	4.6	1.1	.5	.6	.2	1.6
7	.5	.0	1.0	1.7	3.8	9.4	4.3	1.0	1.0	.5	.2	1.6
8	.5	.6	1.7	1.8	4.8	9.7	4.0	.9	3.1	.4	.2	1.6
9	.6	.5	2.5	1.8	5.1	10.0	3.7	1.2	4.0	.4	.1	1.

daily gage height, in feet, of Yellowstone River, etc.—Continued.

	Jan.	Feb.	Mar.	Apr.	May.	June.	July.	Aug.	Sept.	Oct.	Nov.	Dec.
	0.6	0.5	2.9	1.7	5.2	9.7	3.4	4.8	2.5	0.3	0.1	1.3
	.5	.4	2.1	1.5	5.7	9.3	3.2	2.0	1.0	.4	.2	1.2
	.5	.3	1.5	1.4	6.2	8.6	3.1	1.5	.5	.3	.3	1.1
	.7	.3	5.0	2.3	6.8	8.1	3.0	1.3	.4	.3	.2	1.0
	.5	.2	4.6	2.6	7.4	7.7	2.8	1.2	.5	.2	.2	.9
	.5	.2	3.8	2.3	7.5	4.4	2.7	1.0	.3	.2	.3	.9
	.4	.2	4.8	2.0	7.0	7.2	2.6	.9	.4	.3	.4	.9
	.5	.3	5.3	2.7	6.5	7.1	2.5	.8	.4	.2	.7	.8
	.6	.3	5.0	2.5	6.0	7.0	2.4	.9	.6	.1	1.0	.8
	.6	.2	5.0	2.4	5.8	7.3	2.3	1.0	.9	.1	1.2	.7
	.5	.2	5.0	2.3	5.3	7.1	2.4	.8	.7	.3	1.5	.6
	.5	.3	4.0	2.1	5.1	7.0	2.1	.7	.6	.3	1.6	.6
	.5	.3	8.3	2.0	5.1	7.1	2.0	.6	.5	.2	1.4	.6
	.4	.2	2.8	2.4	4.9	7.5	1.8	.5	.5	.2	1.3	.5
	.8	.4	2.3	2.5	5.4	7.6	1.7	.4	.4	.1	1.0	.5
	.3	.4	1.8	4.6	5.4	7.3	1.6	.4	.4	.1	1.0	.4
	.2	.5	1.3	5.0	5.6	7.2	1.6	.3	.5	.0	1.0	.4
	.2	1.0	1.0	4.0	6.1	7.1	1.5	.3	.4	.0	1.0	.5
	.2	1.5	1.1	3.6	6.0	7.0	1.4	.2	.4	.1	1.4	.6
	.1	......	1.0	4.6	7.2	6.9	1.6	.2	.6	.1	1.5	.7
	.1	......	1.0	5.5	8.3	6.4	1.5	.3	......	.2	1.6	.7
	.1	......	1.5	......	8.7	......	1.4	.2	......	.2	......	......
	.6	.8	1.5	.0	1.8	9.6	4.0	1.7	.5	.6	.4	.2
	.5	.8	1.6	.0	1.7	9.6	4.1	1.8	.6	.5	.3	.2
	.5	.6	1.5	.1	1.6	7.8	4.9	1.8	.6	.4	.2	.1
	.5	.5	1.4	.0	1.5	7.5	6.0	1.9	.8	.3	.1	.1
	.5	.5	1.7	.0	2.2	7.1	5.0	2.5	1.8	.3	.3	.2
	.4	.6	2.0	.1	4.2	7.0	4.4	2.6	1.7	.3	.4	.2
	.4	.7	2.1	.2	4.5	6.3	4.0	2.4	1.6	.5	.4	.4
	.4	.6	2.0	.2	4.6	5.8	3.8	2.0	1.4	.6	.3	.5
	.3	.6	2.2	.3	4.4	5.3	5.0	2.0	1.0	.6	.3	.7
	.2	.4	1.8	.2	4.1	5.2	4.2	1.9	.8	.6	.2	.8
	.4	.6	1.8	.1	4.0	5.3	4.0	2.0	1.1	.5	.4	.8
	.5	.7	2.2	.0	4.4	5.8	3.9	1.7	1.2	.4	.4	.9
	.6	.8	2.4	.1	4.8	5.3	5.0	1.4	1.0	.6	.2	1.0
	.6	.9	2.8	.2	5.1	4.9	4.2	1.2	.6	.8	.1	1.0
	.7	1.0	3.0	.2	5.6	4.5	3.8	1.0	.5	.9	.0	.8
	.6	1.0	3.3	.1	5.9	4.1	3.6	.9	.4	.8	.0	.7
	.5	1.2	8.6	.2	6.3	3.8	3.4	.8	.3	.7	.1	.7
	.6	.9	3.5	.4	6.8	3.6	3.2	.8	.4	.6	.1	.7
	.6	.8	3.4	.5	7.6	3.5	3.7	.7	.4	.5	.1	.6
	.8	.7	3.3	.7	7.8	4.2	3.7	.6	.6	.5	.2	.6
	.9	.7	3.2	.6	8.0	4.9	3.3	.7	.6	.4	.1	.5
	1.0	.7	3.0	.6	8.3	5.0	3.1	.8	.5	.5	.2	.8
	.9	.6	3.0	.5	8.6	5.1	2.9	.9	.4	.5	.3	.9
	.9	.5	2.6	.6	8.8	6.0	2.7	1.5	.5	.4	.4	1.5
	.8	.5	2.3	.6	7.4	6.5	2.4	1.2	1.0	.5	.4	2.0
	.8	.6	2.0	.7	6.8	5.3	2.4	1.0	1.3	.4	.5	2.1
	.9	.8	1.9	.8	6.5	4.8	2.3	.9	1.4	.5	.5	2.3
	.8	1.1	1.4	1.0	7.0	4.6	2.4	1.0	1.3	.4	.5	2.5
	.7	......	1.0	1.4	7.5	4.4	2.3	1.1	1.0	.3	.8	3.0
	.6	......	.5	1.6	8.2	4.3	2.1	1.9	.7	.3	.3	3.2
	.6	......	.1	......	8.7	......	2.0	.8	......	.3	......	3.3

Mean daily gage height, in feet, of Yellowstone River, etc.—Continued.

Day.	Jan.	Feb.	Mar.	Apr.	May.	June.	July.	Aug.	Sept.	Oct.	Nov.	Dec.
1902.												
1	3.4	0.2	2.0	1.0	1.3	5.1	4.0		0.8	0.8		
2	3.3	.2	2.0	1.0	1.7	5.5	3.8		.7	.9		
3	3.0	.1	2.2	1.1	1.8	5.8	3.5		.6	.9		
4	3.0	.3	2.6	.8	1.9	5.6	3.4		.5	.8		
5	3.0	.4	3.2	.7	1.8	5.2	3.4		.5	.8		
6	2.8	.4	3.3	.6	1.6	4.9	3.3		.4	.7		
7	2.8	.3	3.4	.5	1.5	4.7	3.3		.4	.7		
8	2.8	.4	3.7	.5	1.3	4.3	3.1		.4	.6		
9	2.7	.4	4.0	.5	1.2	4.1	3.0		.3	.5		
10	2.6	.5	4.0	.4	1.0	4.0	2.8		.3	.5		
11	2.4	.6	3.7	.3	1.0	4.2	2.6		.3	.6		
12	2.3	.7	3.4	.3	.9	4.4	2.4		.2	.6		
13	2.1	.6	3.0	.4	.8	5.3	2.1		.2	.5		
14	2.0	.5	2.6	.4	.6	5.9	2.0		.1	.5		
15	1.9	.5	2.2	.3	.9	6.3	2.0		.1	.4		
16	1.8	.4	1.9	.2	1.5	6.5	2.3		.0	.4		
17	1.6	.6	1.8	.4	2.0	6.8	2.5		.0	.3		
18	1.2	.7	2.0	.3	2.5	6.9	2.8		.1	.4		
19	1.1	1.0	1.8	.2	3.1	7.0	3.0		.1	.5		
20	1.0	1.6	1.6	.1	3.7	6.7	3.1		.2	.5		
21	1.0	2.0	1.7	.1	4.1	6.2	2.9		.2	.6		
22	.9	4.0	1.8	.0	4.0	5.7	2.7		.3	.5		
23	.8	5.0	2.9	.0	4.0	5.3	2.4		.4	.4		
24	.6	4.8	3.7	.1	3.7	5.0	2.3		.5	.3		
25	.6	4.0	4.1	.2	3.4	5.1	2.1		.5	.2		
26	.4	3.6	3.6	.4	3.3	5.3	2.0		.6	.1		
27	.4	3.0	3.2	.5	3.0	5.4	2.0		.6	.1		
28	.3	2.4	2.4	.6	2.7	5.2	1.9		.7	.0		
29	.3		1.7	.8	2.4	4.9	1.8		.7	.0		
30	.2		1.5	.9	3.0	4.5	1.8		.8	−.1		
31	.2		1.3		4.3		1.7			−.1		
1903.												
1				6.0	0.4	3.0	7.9	5.5	2.7	1.9	1.4	5.7
2				7.0	.5	3.1	8.3	6.1	2.5	1.9	1.4	5.5
3				7.2	.7	3.2	8.9	6.3	2.3	2.0	1.3	6.0
4				7.0	.9	4.3	8.1	5.6	2.0	2.0	1.8	6.0
5				6.5	1.2	5.0	7.9	4.9	1.9	2.0	1.8	6.0
6				4.0	1.5	5.5	6.9	4.9	1.7	2.1	1.3	6.0
7				3.5	1.8	6.3	6.8	4.4	1.6	2.0	1.3	6.
8				3.0	2.0	7.0	6.0	4.2	1.6	2.0	1.3	6.
9				2.7	2.1	7.1	5.8	4.0	1.5	2.0	1.3	6.
10				2.0	2.0	7.2	6.0	3.8	1.5	2.0	1.3	6.
11				1.5	1.8	7.7	5.9	3.6	1.6	1.9	1.3	6
12				1.0	1.6	8.0	5.8	3.7	1.7	2.0	1.3	6
13				1.2	1.8	7.7	5.6	3.5	2.2	2.0	1.3	6
14				1.0	1.9	7.5	5.4	3.4	2.6	2.0	1.3	6
15				.9	2.0	7.5	5.2	3.8	2.65	1.9	1.3	6
16			4.2	.8	2.2	7.6	5.6	3.4	2.9	1.8	1.3	6
17			4.1	.8	2.3	8.3	5.4	3.5	2.9	1.9	1.4	5
18			4.2	.7	2.8	8.4	5.5	3.0	2.75	1.9	1.4	5
19			4.4	.7	2.4	8.5	5.4	2.9	2.55	1.9	1.4	5
20			4.4	.6	2.5	9.0	5.3	2.7	2.4	1.9	1.4	4
21			4.5	.6	4.5	8.8	5.3	2.6	2.3	1.8	1.4	4

Mean daily gage height, in feet, of Yellowstone River, etc.—Continued.

Day.	Jan	Feb.	Mar.	Apr.	May.	June.	July.	Aug.	Sept.	Oct.	Nov.	Dec.
1903.												
22			4.3	0.5	5.0	8.6	5.0	2.4	2.3	1.8	1.4	4.4
23			4.4	.5	5.1	8.6	5.0	2.2	2.2	1.8	1.4	4.4
24			4.4	.4	5.3	8.4	4.8	2.2	2.15	1.7	1.4	4.4
25			4.5	.4	5.3	8.2	4.7	2.5	2.1	1.6	1.4	4.6
26			4.5	.5	5.0	7.8	4.7	2.4	2.05	1.6	a5.6	4.6
27			4.4	.5	4.7	7.5	5.4	2.7	2.0	1.6	5.6	4.6
28			4.5	.4	4.3	7.3	5.2	4.0	1.9	1.6	5.6	4.6
29			4.8	.3	3.9	7.0	4.5	3.9	1.9	1.5	5.7	4.6
30			5.0	.3	3.5	7.0	4.5	2.7	1.9	1.5	5.7	5.1
31			5.0		3.0		4.3	2.6		1.4		5.1

a Ice gorge one-eighth mile below gage cause of sudden rise.

Top of ice, November 16–25; December 1–31. Ice broken, November 26–30.

YELLOWSTONE RIVER NEAR LIVINGSTON, MONT.

This station was established May 2, 1897. It is located at Carter's bridge, 5 miles south of Livingston, Mont., at the mouth of the lower canyon. A vertical rod was first installed, but was replaced by a wire gage, which is located on the lower side of the east span. The scale is fastened to the guard rail. The length of the wire from the end of the weight to the marker is 21.76 feet. The gage is read once each day by Thomas S. Carter. Discharge measurements are made from the downstream side of the bridge. The initial point for soundings is marked on the guard rail, 2 feet east of the center of the northwest pier. The channel is straight for about 300 feet above and for 250 feet below the station. The right bank is low, and at high stages a part of the water escapes through a slough on that side and has to be measured separately. The left bank is high and rocky and will not overflow. The current velocity is swift, and at high water it is difficult to keep the meter submerged. On January 1, 1903, the gage and the datum were lowered 3 feet, increasing all subsequent readings by that amount. Bench mark No. 1 is the head of the 2-inch nut on the center pin at the foot of the end diagonal of the upstream truss at the east pier. Its elevation is 13.44 feet above the old and 16.44 feet above the new datum. Bench mark No. 2 is a white paint mark on the corner of the top of the shoe at the foot of the batter post on the east end of the upstream side of the bridge. The batter post is marked "B. M. 14.20, U. S. G. S." in white paint. It is also marked "17.20" in red paint. Its elevation is 14.20 feet above the old and 17.20 feet above the new gage datum.

The observations at this station during 1903 have been made under the direction of C. C. Babb, district engineer.

Mean daily gage height, in feet, of Yellowstone River near Livingston, Mont., for 1903.

Day.	Jan.	Feb.	Mar.	Apr.	May.	June.	July.	Aug.	Sept.	Oct.	Nov.	Dec.
1	1.60	1.55	1.50	1.85	2.30	6.45	6.90	4.70	3.10	2.75	2.35	1.30
2	1.70	1.10	1.70	1.80	2.25	6.70	6.73	4.57	3.05	2.75	2.30	1.35
3	1.70	1.40	1.60	1.75	2.15	6.85	6.63	4.50	3.05	2.75	2.30	1.30
4	1.80	1.50	1.70	1.80	2.15	7.05	6.45	4.40	3.00	2.80	2.30	1.00
5	1.80	1.50	1.70	1.85	2.25	7.15	6.40	4.32	2.95	2.80	2.30	1.80
6	1.80	1.50	1.75	1.75	2.35	7.15	6.35	4.23	3.00	2.80	2.30	.....
7	1.70	1.40	1.75	1.75	2.60	7.40	6.40	4.17	3.00	2.90	2.30	1.25
8	1.60	1.50	1.75	1.80	2.85	7.63	6.38	4.08	3.00	2.85	2.30	1.20
9	1.60	1.60	1.65	1.85	2.78	7.75	6.23	4.00	3.00	2.90	2.30	2.00
10	1.70	1.70	1.75	2.00	2.62	7.48	6.13	4.00	2.95	2.85	2.10	1.10
11	1.70	1.60	1.75	1.95	2.55	7.40	6.13	3.95	2.95	2.85	2.10	1.00
12	1.60	1.50	1.75	1.80	2.55	7.63	6.10	3.90	2.95	2.80	2.05	1.00
13	1.50	1.45	1.75	1.70	3.00	7.60	6.03	3.83	2.95	2.75	2.05	1.00
14	1.50	(a)	1.65	1.80	3.65	7.83	5.93	3.75	3.00	2.70	2.00	1.00
15	1.60	(a)	1.80	1.85	5.00	8.00	5.83	3.67	2.95	2.70	.....	1.10
16	1.60	(a)	1.70	1.85	5.22	8.10	5.73	3.63	2.90	2.65	.....	1.00
17	1.60	(a)	1.65	1.85	4.60	8.15	5.75	3.57	2.90	2.60	.....	1.10
18	1.60	(a)	1.65	1.85	4.10	8.20	5.73	3.50	2.95	2.55	.....	1.15
19	1.60	1.50	1.60	1.85	3.75	7.80	5.55	3.50	2.95	2.55	.....	1.00
20	1.60	1.60	1.55	1.90	3.50	7.65	5.45	3.48	2.90	2.55	.....	1.80
21	1.70	1.60	1.55	1.95	3.40	7.50	5.40	3.45	2.85	2.55	.....	2.00
22	1.70	1.70	1.60	1.95	3.28	7.30	5.30	3.47	2.85	2.50	.....	1.80
23	1.60	1.70	1.55	2.05	3.20	7.28	5.25	3.50	2.85	2.50	.....	1.90
24	1.75	1.70	1.55	2.10	3.15	7.00	5.15	3.47	2.80	2.50	.....	1.15
25	1.70	1.70	1.60	2.20	3.05	6.85	5.05	3.38	2.75	2.50	.....	1.80
26	1.70	1.70	1.60	2.40	3.10	6.80	5.00	3.33	2.75	2.50	2.45	1.15
27	1.80	1.60	1.70	2.65	3.23	7.10	4.95	3.30	2.75	2.50	2.35	1.80
28	1.60	1.50	1.70	2.55	3.65	7.00	4.83	3.27	2.75	2.45	2.30	1.90
29	1.45	.....	1.75	2.40	3.98	7.00	4.70	3.23	2.75	2.45	.....	1.80
30	1.50	.....	1.80	2.25	4.45	7.05	4.68	3.18	2.75	2.45	2.25	1.55
31	1.45	.....	1.85	.....	5.50	.....	4.70	3.12	.....	2.40	.....	1.70

a Ice.

Discharge measurements of Yellowstone River near Livingston, Mont., in 1903.

Date.	Hydrographer.	Gage height.	Discharge.
		Feet.	Second-feet.
April 4	J. S. Baker	1.86	1,494
May 25	F. L. Travenner	3.01	2,549
June 24	J. H. Sloan	7.10	14,208
July 3	do	6.66	11,763
July 20	do	5.45	7,568
September 1	do	3.12	3,009

le for *Yellowstone River near Livingston, Mont., from January 1 to December 31, 1903.*

)ischarge.	Gage height.	Discharge.	Gage height.	Discharge.	Gage height.	Discharge.
rcond-feet.	*Feet.*	*Second-feet.*	*Feet.*	*Second-feet.*	*Feet.*	*Second-feet.*
880	2.6	2,080	4.1	4,400	5.6	7,780
920	2.7	2,220	4.2	4,580	5.7	8,080
960	2.8	2,360	4.3	4,760	5.8	8,400
1,000	2.9	2,500	4.4	4,940	5.9	8,740
1,060	3.0	2,650	4.5	5,140	6.0	9,080
1,130	3.1	2,800	4.6	5,340	6.1	9,440
1,200	3.2	2,950	4.7	5,540	6.2	9,800
1,280	3.3	3,100	4.8	5,760	6.3	10,180
1,360	3.4	3,260	4.9	5,980	6.4	10,560
1,450	3.5	3,420	5.0	6,200	6.5	10,980
1,540	3.6	3,580	5.1	6,440	6.6	11,460
1,630	3.7	3,740	5.2	6,680	6.7	12,000
1,720	3.8	3,900	5.3	6,940	6.8	12,560
1,830	3.9	4,060	5.4	7,200	6.9	13,120
1,950	4.0	4,220	5.5	7,480	7.0	13,680

it above 6.8 feet gage height, with differences of 560 per tenth.

d *monthly discharge of Yellowstone River near Livingston, Mont., for 1903.*

[Drainage area, 3,580 square miles.]

Month.	Discharge in second-feet.				Run-off.	
	Maximum.	Minimum.	Mean.	Total in acre-feet.	Second-feet per square mile.	Depth in inches.
....	1,280	960	1,152	70,834	0.322	0.371
ry *a* *...........*	1,200	880	1,087	60,369	.304	.317
... ...	1,320	1,060	1,185	72,863	.331	.38?
....	2,150	1,200	1,436	85,448	.401	.447
... ...	7,480	1,585	3,236	198,974	.904	1.040
....	20,400	10,770	15,711	934,870	4.390	4.900
....	13,120	5,400	8,481	521,476	2.370	2.780
....	5,540	2,830	3,852	236,850	1.080	1.250
ber *..........*	2,800	2,290	2,521	150,010	.704	.784
r *...........*	2,500	1,830	2,155	132,506	.602	.694
ber *b* *........*	1,890	1,450	1,673	99,550	.467	.521
ber *c* *........*	1,950	1,200	1,462	89,895	.408	.470
The year *.....*	20,400	880	3,662	2,653,645	1.024	13.906

ary 14 to 18 estimated.   *b* November 15 to 25, also 29, estimated.   *c* December 6 estimated.

BEAVER CREEK NEAR ASHFIELD, MONT.

The original station was established July 5, 1903, by A. E. Place. It was located at railroad bridge No. 455 of the Great Northern Railroad one-fourth mile west of Ashfield, Mont. The gage was read once each day by John Borek, and consisted of a vertical board gage in two sections fastened to the left abutment and to some old piles near the channel. Discharge measurements were made from the railroad bridge. The bench mark was a spike in the pile to the left of the upper part of the gage, and was marked "U. S. G. S. B. M." Its elevation above the zero of the gage was 7 feet. This station was discontinued in December, 1903, on account of the influence of the backwater from a dam completed August 2, 1903, about 1 mile below. A new station was established December 31, 1903, by W. B. Freeman, at a point 2½ miles farther upstream, at the Thomas & Gould ranch. The new station is 18 miles from Malta, Mont. There is a railroad dam at Ashfield, Mont., but the backwater does not affect the new station. The gage is an inclined 2 by 6 inch plank securely fastened to posts set in the ground at each end. Discharge measurements are made from a ½-inch steel cable about 225 yards above the gage. The cable has a span of 82 feet between supports. A car is suspended from the cable, and a tag wire has been installed and marked at intervals of 5 feet. The initial point for soundings is a nail in the cable support on the left bank. The channel is straight for 300 feet above and below the cable. The right bank is sloping and covered with willows and is not liable to overflow. The left bank is steep and is also covered with willows. The stream has a sluggish velocity. The bed of the stream is composed of mud, and there is but one channel. Bench mark No. 1 is the top of a nail driven flush in the top of a 2 by 4 stake 12½ feet south of the gage. Elevation, 11.432 feet above zero of gage. Bench mark No. 2 is the top of a railroad spike driven in the northwest corner of Thomas's house 1 foot above ground. Elevation, 13.346 feet above zero of gage.

The observations at this station during 1903 have been made under the direction of C. C. Babb, district engineer.

Discharge measurements of Beaver Creek near Ashfield, Mont., in 1903.

Date.	Hydrographer.	Gage height.	Discharge.
		Feet.	*Second-feet.*
July 22	A. E. Place	1.21	60
August 16	do		a 21.6
September 12	E. W. Myers		b .5

a Ditch, Blum Cattle Co., 3 miles above station.   b Estimated, 10 miles above station.

ly gage height, in feet, of Beaver Creek near Ashfield, Mont., for 1903.

Day.	July.	Aug.	Sept.	Oct.	Nov.	Dec.
		1.60	2.70	2.40	2.00	1.90
		(2)	2.70	2.80	2.00	1.90
		2.60	2.80	2.70	2.50	1.90
		2.70	2.90	2.30	2.00	1.80
		3.10	3.00	2.30	2.00	1.80
		3.10	3.20	2.20	1.90	1.80
	40.30	3.10	3.20	2.20	1.90	1.80
	1.00	3.20	3.20	2.20	1.90	1.70
	1.20	3.20	3.10	2.20	1.90	1.70
	1.10	3.20	3.00	2.20	1.90	1.70
	1.30	3.30	3.00	2.20	1.00	1.70
	1.30	3.30	2.90	2.20	1.90	1.70
	1.20	3.30	2.90	2.10	1.90	1.70
	1.30	3.20	2.80	2.10	1.90	1.70
	1.20	3.20	2.70	2.10	1.90	1.70
	1.10	3.20	2.70	2.10	1.90	1.70
	1.10	3.10	2.70	2.10	1.90	1.70
	1.00	3.00	2.60	2.10	1.90	1.70
		2.90	2.60	2.10	1.90	1.70
	1.30	2.90	2.60	2.10	1.90	1.70
	1.30	2.80	2.50	2.10	1.90	1.70
	1.20	2.80	2.50	2.10	1.90	1.70
	1.20	2.70	2.50	2.10	1.90	1.70
	1.10	2.60	2.50	2.10	1.90	1.70
	1.00	2.60	2.50	2.10	1.90	1.70
	.30	2.50	2.40	2.10	1.90	1.70
	.20	2.70	2.40	2.10	1.90	1.70
	.30	2.70	2.40	2.10	1.90	1.70
	.30	2.70	2.40	2.10	1.90	1.70
	.40	2.70	2.40	2.00	1.90	1.70
	.50	2.70	....	2.00	....	1.70

*ust 2 a dam was erected about 1 mile below station. This accounts for rise of water

eported not running from July 7 to December 31.
:hes thick.
ga to top of ice November 15 to December 31.

VER CREEK OVERFLOW CHANNEL NEAR BOWDOIN, MONT.

station was established June 29, 1903, by Cyrus C. Babb. It
ed on an overflow channel of Beaver Creek, about 4 miles
st of the Great Northern Railroad station at Bowdoin, Mont. It
ed best by wagon from Malta, Mont., 12 miles distant. The
upply is partly due to natural drainage of irrigated lands and
o water wasted over the spillway at the dam above on Beaver
The gage is a vertical board nailed to a cottonwood tree
.00 feet above the cable. It is graduated to tenths of feet
.rked with black paint. The gage is read once each day by
urmell. Discharge measurements are made by means of a
·ar, and tagged wire, and also by wading. The initial point for

soundings is the east side of a post on the west or left bank near th
wire fence crossing the channel. The channel is straight for 200 fe
above the station and for 100 feet below. The stream is sluggish
During high floods both banks will be submerged, as at that time th
whole valley is flooded. The bed of the stream at the station form
a small bar covered with weeds and swamp grass. Bench mark N
1 is a nail in a hub 1 foot south of the back of the gage. Its elevation is 5.79 feet above the zero of the gage and is 2,213.6 feet abo
sea level. Bench mark No. 2 is a nail in the northwest corner
Turmell's house. Its elevation above the zero of the gage is 17.64 fee
The elevation of these bench marks has been determined from th
bench mark at the Malta schoolhouse. Its elevation is based on th
elevation of the rail in front of the Great Northern Railroad statio
at Malta, Mont.

The observations at this station during 1903 have been made und
the direction of C. C. Babb, district engineer.

Discharge measurements of Beaver Creek overflow near Bowdoin, Mont., in 19

Date.	Hydrographer.	Gage height.	Discharg
		Feet.	*Second-f*
July 21	Babb and Myers	4.15	14
August 10	C. C. Babb	3.30	

Mean daily gage height, in feet, of Beaver Creek overflow near Bowdoin, Mon
for 1903.

Day.	June.	July.	Aug.	Sept.	Oct.	Nov.	De
1			3.28	2.83	2.27	1.67	1
2		2.70	3.25	2.83	2.24	1.65	1
3		2.70	3.40	2.82	2.10	1.63	1
4		2.70	3.60	2.81	2.16	1.62	
5		2.70	3.50	2.70	2.14	1.61	
6		2.70	3.50	2.78	2.10	1.50	
7		2.70	3.45	2.75	2.10	1.50	
8		3.70	3.40	2.73	2.08	1.47	
9		3.60	3.30	2.73	2.05	1.45	
10		4.70	3.20	2.60	2.08	1.44	
11		5.00	3.20	2.58	2.00	1.43	
12		4.50	3.18	2.68	1.98	1.43	
13		3.80	3.16	2.66	1.97		
14		3.60	3.15	2.65	1.95		
15		3.50	3.13	2.63	1.90	(*a*)	
16		3.50	3.10	2.50	1.88	(*a*)	
17		5.80	3.05	2.50	1.87	(*a*)	
18		6.40	3.00	2.48	1.86	(*a*)	
19		5.90	2.98	2.46	1.84	(*a*)	

a Creek frozen from November 15 to December 31; readings taken at top of ice.

Mean gage height, in feet, of Beaver Creek overflow, etc.—Continued.

Day.	June.	July.	Aug.	Sept.	Oct.	Nov.	Dec.
20		5.40	2.96	2.45	1.83	(a)	1.17
21		4.60	2.96	2.45	1.83	(a)	1.15
22		3.70	2.96	2.44	1.81	1.42	1.15
23		3.50	2.95	2.43	1.81	1.42	1.14
24		3.40	2.95	2.43	1.79	1.42	1.13
25		3.43	2.93	2.41	1.78	1.41	1.13
26		3.40	2.91	2.30	1.76	1.41	1.12
27		3.36	2.90	2.36	1.76	1.41	1.12
28		3.38	2.88	2.35	1.74	1.30	1.11
29	2.70	3.31	2.86	2.31	1.73	1.30	1.00
30	2.65	b3.30	2.85	2.30	1.69	c1.30	1.00
31		3.28	2.85	....	1.68	....	c.99

a Creek frozen from November 15 to December 31; readings taken at top of ice.
b There was no flow from July 30 to December 31.
c Thickness of ice November 30, 6 inches; December 31, 10 inches.

MILK RIVER AT MALTA, MONT.

This station was established July 31, 1902, by C. T. Prall. It is located at the highway bridge on the main road one-fourth mile east of the railroad station and post-office at Malta, Mont. The original wire gage has been replaced by a standard boxed chain gage, reading the same as the old one. It is fastened to the lower downstream guard rail of the highway bridge. The length of the chain from the end of the weight to the marker is 28.67 feet. The distance from the zero of the scale to the outside edge of the pulley is 0.5 foot. The marker is a small copper ring 0.8 foot from the extreme end of the ring at the end of the chain. Discharge measurements are made from the downstream side of the bridge to which the gage is attached. The initial point for soundings is a spike on the downstream guard rail at the right bank. It is marked zero with paint. The channel is straight for about 700 feet above and 250 feet below the station. The current velocity is moderate. Both banks are high and sandy and are not liable to overflow. The water passes beneath the bridge at all stages. The bed of the stream is composed of sand and gravel, and there is but one channel. The bench mark is a paint mark on the east end of a plank spiked to the south abutment. It is marked "B. M. U. S. G. S." Its elevation above gage datum is 13.66 feet.

The observations at this station during 1903 have been made under the direction of C. C. Babb, district engineer.

Discharge measurements of Milk River at Malta, Mont., in 1903.

Date.	Hydrographer.	Gage height.	Dischar
		Feet.	*Second-f*
April 25	J. S. Baker	3.46	1.
May 12	C. C. Babb	2.12	
May 14	F. M. Brown	2.23	
Do	do	2.23	
May 24	C. T. Prall	7.15	3
May 28	do	5.60	2
June 2	O. M. Sprigg	7.70	4
June 9	W. B. Freeman	4.25	1
June 16	S. B. Robbins	2.60	
June 22	do	2.20	
June 23	A. E. Place	2.20	
July 8	do	2.24	
July 15	do	2.64	
July 21	E. W. Myers	2.32	
Do	do	2.32	
August 12	P. A. Blair	2.43	
Do	A. E. Place	2.43	
September 8	do	2.16	
October 8	do	1.74	
November 6	W. B. Freeman	1.72	
November 12	do	1.55	
November 26	do	1.33	
December 7	do	1.43	
December 10	do	1.52	

Mean daily gage height, in feet, of Milk River at Malta, Mont., for 1903.

Day.	Mar.	Apr.	May.	June.	July.	Aug.	Sept	Oct.	Nov.
1		6.65	3.60	6.00	2.10	(a)	2.90	70	1.70
2		6.45	3.50	7.70	2.30	(a)	2.70	70	80
3		6.10	3.45	6.00	2.20	(a)	2.60	1.70	1.80
4		5.25	3.30	5.00	2.20	(a)	2.40	1.70	1.80
5		5.25	3.10	4.70	2.20	(a)	2.30	1.70	70
6		6.60	2.90	4.70	2.20	(a)	2.20	1.70	1.70
7		5.30	2.70	4.70	2.10	(a)	2.20	1.70	1.70
8		5.40	2.50	4.65	2.25	2.70	2.20	1.70	1.70
9		5.15	2.40	4.20	2.25	2.70	2.10	1.70	1.60
10		5.15	2.30	3.85	2.30	2.60	2.00	1.75	1.70
11		5.00	2.30	3.65	2.60	2.60	2.00	1.70	1.70
12		5.00	2.20	3.45	2.75	2.40	2.00	1.70	1.60

a Gage stolen.

Mean daily gage height, in feet, of Milk River, etc.—Continued.

Day.	Mar.	Apr.	May.	June.	July.	Aug.	Sept.	Oct.	Nov.	Dec.
....		4.80	2.20	3.25	2.60	2.30	1.90	1.75	1.60	c2.20
....		4.65	2.20	3.15	2.65	2.20	1.90	1.75	1.60	2.20
....		4.30	2.20	2.95	2.60	2.20	1.80	......	1.60	2.20
....		3.95	2.20	2.65	2.65	2.10	1.80	1.90	b1.70	2.20
....		3.65	2.20	2.60	2.60	2.10	1.80	1.90	1.90	2.10
....		3.40	2.20	2.50	2.60	2.00	1.80	1.90	1.90	2.40
....		3.15	2.20	2.40	2.50	2.00	1.90	1.90	1.90	2.20
....		3.00	2.40	2.30	2.35	2.00	2.00	1.90	1.90	2.10
....		3.00	2.70	2.20	2.30	2.00	1.90	1.90	1.90	1.90
....		3.05	3.80	2.10	2.20	2.00	1.90	1.90	1.55	1.70
....		3.20	5.90	2.10	2.10	2.00	1.90	1.85	1.45	1.80
....		3.30	7.20	2.10	2.10	2.00	1.85	1.85	1.35	1.50
....		3.40	7.30	2.10	2.00	1.90	1.85	1.85	1.35	1.50
....		3.40	7.35	2.00	1.90	1.90	1.85	1.85	1.35	1.50
....		3.35	6.95	2.05	1.80	2.20	1.80	1.85	1.33	1.50
....		3.20	5.65	2.00	1.70	2.90	1.80	1.70	1.30	1.50
....		3.55	5.40	2.10	1.70	2.80	1.80	1.70	1.40	1.50
....	5.75	3.70	5.20	2.10	(a)	2.70	1.80	1.70	1.40	1.50
....	5.75	......	5.00	......	(a)	2.90	......	1.70	......	1.50

a Gage stolen.

b River frozen November 16. Readings taken through ice to water surface November 24 to December 12.

c From December 13 to 31 the readings do not represent the flow closely.

Rating table for Milk River at Malta, Mont., from January 1 to December 31, 1903.

Gage height.	Discharge.	Gage height.	Discharge.	Gage height.	Discharge.	Gage height.	Discharge.
Feet.	Second-feet.	Feet.	Second-feet.	Feet.	Second-feet.	Feet.	Second-feet.
1.3	86	2.6	600	3.9	1,430	5.4	2,480
1.4	94	2.7	655	4.0	1,500	5.6	2,620
1.5	104	2.8	710	4.1	1,570	5.8	2,760
1.6	120	2.9	770	4.2	1,640	6.0	2,900
1.7	150	3.0	830	4.3	1,710	6.2	3,040
1.8	185	3.1	890	4.4	1,780	6.4	3,180
1.9	228	3.2	950	4.5	1,850	6.6	3,320
2.0	275	3.3	1,010	4.6	1,920	6.8	3,460
2.1	325	3.4	1,080	4.7	1,990	7.0	3,610
2.2	380	3.5	1,150	4.8	2,060	7.7	4,170
2.3	435	3.6	1,220	4.9	2,130		
2.4	490	3.7	1,290	5.0	2,200		
2.5	545	3.8	1,360	5.2	2,340		

Table not extended above or below measurements. (Gage heights 1.30 to 7.70 feet.)

Estimated monthly discharge of Milk River at Malta, Mont., for 1903.

[Drainage area, 14,044 square miles.]

Month.	Discharge in second-feet.			Total in acre-feet.	Run-off.	
	Maximum.	Minimum.	Mean.		Second-feet per square mile.	Depth in inches
January			a 220	13,527	0.015	0.0
February			a 220	12,218	.015	.0
March			a 380	23,365	.027	.0
April	3,355	830	1,769	105,263	.126	.1
May	3,890	380	1,350	83,008	.096	.1
June	4,170	275	1,136	67,597	.081	.0
July b	710	150	408	25,087	.029	.0
August c	770	185	417	25,640	.030	.0
September	770	185	299	17,792	.021	.0
October	228	150	180	11,068	.013	.0
November	228	86	145	8,628	.010	.0
December	490	86	177	10,883	.013	.0
The year			558	404,076	.040	.5

a Estimated.   b July 30 and 31 estimated.   c August 1 to 7 estimated.

HARLEM CANAL AT HEAD GATES, NEAR ZURICH, MONT.

Harlem canal is owned by a cooperative company of farmers. irrigates large areas of bottom lands near Harlem, Mont.

The station was established in June, 1903, by C. T. Prall. It located about 500 feet below the head gates of the canal and about mile southeast of the Great Northern Railroad section house at Zuric Mont. It is best reached by driving from Harlem or Chinook, Mon The gage is a vertical board driven into the bed of the canal. D charge measurements are made by wading. The initial point f soundings is a stake driven into the left bank opposite the gage. T bench mark is a spike in the downstream face of the 6 by 8 inch p on the left and on the downstream end of the head gate. This spike about 10 inches below the top of the cross piece, and is at an elevati of 11.89 feet above the gage datum.

The observations at this station during 1903 have been made und the direction of C. C. Babb, district engineer, and the following c charge measurements have been made:

Discharge measurements of Harlem canal near Zurich, Mont., in 1903.

- June 27: Gage height, 2.28 feet; discharge, 37.1 second-feet.
  July 27: Gage height, 1.89 feet; discharge, 31.2 second-feet.
  July 27: Gage height, 1.89 feet; discharge, 31.4 second-feet.
  August 19: Gage height, 0.85 feet; discharge, 4 second-feet.

PARADISE VALLEY CANAL AT HEAD GATES, NEAR CHINOOK, MONT.

This station was established in June, 1903, by C. T. Prall. It is located near the South River road, 7 miles east of Chinook, from which place it is reached by driving. The gage is a vertical board driven into the bed of the canal about 500 feet below the head gates, near the house of the observer, Rudolph Friede. Discharge measurements are made by wading. The initial point for soundings is a stake driven into the left bank opposite the gage. The channel is straight, and the section at the station is semicircular. The bed of the stream is composed of mud and sand. This canal carries water only during the irrigating season. Bench mark No. 1 is a spike in the top of a 6 by 6 inch post on the left side of the downstream end of the head gate. Its elevation above gage datum is 7.29 feet. Bench mark No. 2 is a spike in the southwest side of a 12-inch post 40 feet south of the south end of the dam and 28 paces east of bench mark No. 1. Its elevation above gage datum is 7.38 feet.

The observations at this station during 1903 have been made under the direction of C. C. Babb, district engineer, and the following discharge measurements have been made:

Discharge measurements of Paradise Valley canal near Chinook, Mont., in 1903.

June 26: Gage height, 1.58 feet; discharge, 16.8 second-feet.
August 20: Gage height, 0.75 feet; discharge, 0.5 second-feet.

Mean daily gage height, in feet, of Paradise Valley canal near Chinook, Mont., for 1903.

Day.	July.	Aug.	Sept.	Oct.	Day.	July.	Aug.	Sept.	Oct.
1		1.01	0.89	0.90	17		0.75	(a)	1.00
2		1.08	.84	.90	18		.74	(a)	1.15
3		1.18	.81	.98	19		.70	(a)	1.05
4		1.27	.78	1.10	20		.70	(a)	1.00
5		1.30	.75	1.05	21		.65	(a)	1.00
6		1.20	(a)	.85	22		(a)	(a)	1.00
7		1.18	(a)	.90	23		(a)	(a)	1.00
8		1.12	(a)	.82	24		(a)	(a)	.97
9		1.08	(a)	.95	25		(a)	(a)	1.07
10		.97	(a)	1.02	26	0.85	(a)	(a)	1.05
11		.92	(a)	1.12	27	2.23	.90	(a)	1.00
12		.91	(a)	1.12	28	1.50	.93	(a)	1.00
13		.89	(a)	1.10	29	1.33	.91	(a)	1.00
14		.83	(a)	.97	30	1.28	.90	(a)	.95
15		.80	(a)	1.05	31	1.05	.89	(a)	1.05
16		.75	(a)	1.06					

a Dry.

FORT BELKNAP CANAL AT HEAD GATES, AT YANTIC, MONT.

This station was established June 21, 1903, by L. E. Granke. It is located 1 mile below the head gates of the canal at the highway crossing. It is reached by driving from Chinook, Mont. The gage is a vertical board fastened to the downstream chord of the wooden bridge. Discharge measurements are made from the wooden highway bridge, to which the gage is attached. The initial point for soundings is a spike 15 feet from the gage on the right end of the downstream hand rail. Distances from the initial point are marked by spikes every 3 feet. Both banks are sloping, and the bed is composed of sand. The bench mark is a nail at the base of a large cottonwood tree 300 feet southeast of the gage. Its elevation is 3.66 feet above the zero of the gage and 2,427.8 feet above sea level. The elevation above sea level is determined from the elevation of the rail in front of the Great Northern Railroad station at Havre, Mont. The canal carries water only during the irrigating season.

The observations at this station during 1903 have been made under the direction of C. C. Babb, district engineer.

Discharge measurements of Belknap canal at Yantic, Mont., in 1903.

Date.	Hydrographer.	Gage height.	Discharge.
		Feet.	*Second-feet.*
June 28	A. E. Place	3.28	87.0
July 25	E. W. Myers	3.25	76.0
August 21	A. E. Place	.75	.8

Mean daily gage height, in feet, of Belknap canal at Yantic, Mont., for 1903.

Day.	June	July.	Day.	June.	July.	Day.	June.	July.	Day.	June.	July.
1		3.80	9			17			25	3.60	
2		3.30	10			18			26	3.70	
3		3.40	11			19			27	3.70	
4		3.40	12			20			28	3.30	
5			13			21	3.70		29	3.50	
6			14			22	3.60		30	3.50	
7			15			23	3.50		31	3.30	
8			16			24	3.60				

MILK RIVER AT HAVRE, MONT.

This station was originally established by Cyrus C. Babb May 15, 1898. At that time a cable was stretched across the river not far from the highway bridge. The station was reestablished November 4, 1902, by L. V. Branch, when the cable was taken out and the station was transferred to the highway bridge, at which it is now located. The original wire gage was replaced January 12, 1904, by a standard chain gage, which was made to read the same as the old gage. The length of chain from the end of the weight to the marker is 25 feet. The marker is a ring of fine copper wire wrapped around the chain. It is placed 0.85 foot from the end of the cast-iron ring at the

.he chain. The scale is nailed to the floor of the bridge on
nstream side 90 to 100 feet from the initial point for sound-
t is boxed, and reads from 2.5 to 10.5 feet. The gage pulley
d at a point which corresponds to a reading of 2 feet on the
L. H. Ling, who reads the gage, is the Weather Bureau
r. Discharge measurements are made from the downstream
the bridge to which the gage is attached. The initial point
adings is marked on the downstream guard rail at the right
The channel is straight above for a short distance, when a
urve occurs. Below the station it is straight. The current is
1 at ordinary stages. Both banks are high and are not wooded.
d of the stream is shifting and is liable to change after each

The bridge from which the measurements are made makes
e of 20 degrees with the normal to the current. The datum
the establishment of the bench marks is that of the city of
deduced from a destroyed temporary bench mark of the United
Geological Survey. Bench mark No. 2 is a 20-penny nail in a
ood tree 100 feet southwest of the southwest pier of the bridge.
ation is 2,477.44 feet. Bench mark No. 3 is a standard United
Geological Survey bench mark plate in the stone pier of a
supporting the Great Northern Railroad water tank one-half
itheast of the bridge and near the roundhouse. It is marked
The correct elevation is 2,482.46 feet. There is also a tempo-
ich mark on the top of the southeast corner of the lid of the
r. Its elevation is 2,488.60 feet.

bservations at this station during 1903 have been made under
ction of C. C. Babb, district engineer.

ischarge measurements of Milk River at Havre, Mont., in 1903.

Date.	Hydrographer.	Gage height.	Discharge.
		Feet.	*Second-feet.*
....	L. H. Ling	5.90	1,130
....	.....do	5.60	1,034
....	.....do	5.40	895
....	J. S. Baker..........	5.18	912
....	.....do	5.18	841
....	L. H. Ling	4.63	588
....	F. M. Brown	4.70	599
....	.....do	4.70	552
....	L. H Ling	4.16	388
. ...	.....do	5.90	1,786
....	C. T. Prall	6.25	1,987
....	L. H. Ling	7.22	2,784

Discharge measurements of Milk River at Havre, Mont., in 1903—Continued.

Date.	Hydrographer.	Gage height.	Discharg
		Feet.	*Second-f(*
May 29	C. T. Prall	8.30	3,7
May 30	L. H. Ling	8.51	4,
June 24	A. E. Place	4.80	
July 23	E. W. Myers	4.04	
July 24	do	3.93	
July 28	do	4.35	
August 18	A. E. Place	4.21	
August 28	J. H. Sloan	5.35	
September 29	do	3.60	
November 4	W. B. Freeman	3.63	

Mean daily gage height, in feet, of Milk River at Havre, Mont., for 1903.

Day.	Jan.	Feb.	Mar.	Apr.	May.	June.	July.	Aug.	Sept.	Oct.	Nov.	D
1	(a)	(a)	(a)	6.00	5.20	7.20	4.50	4.30	4.35	3.60	3.70	(
2	(a)	(a)	(a)	5.60	5.10	6.90	4.50	4.40	4.20	3.60	3.60	(
3	(a)	(a)	(a)	5.40	4.65	6.90	4.50	4.80	4.10	3.50	3.60	(
4	(a)	(a)	(a)	5.90	4.63	6.60	4.50	4.80	4.10	3.50	3.60	(
5	(a)	(a)	(a)	5.90	4.60	6.30	4.50	4.10	4.10	3.50	3.60	(
6	(a)	(a)	(a)	5.60	4.40	5.80	4.50	4.80	3.80	3.60	3.60	(
7	(a)	(a)	(a)	6.00	4.30	5.70	5.00	4.70	3.90	3.75	3.60	(
8	(a)	(a)	(a)	6.00	4.30	5.50	4.90	4.60	3.80	3.80	3.60	(
9	(a)	(a)	(a)	5.90	4.40	5.40	4.80	4.50	3.70	3.80	3.50	(
10	(a)	(a)	(a)	(b)	4.40	5.20	4.80	4.40	3.70	3.80	3.50	(
11	(a)	(a)	(a)	(b)	4.30	5.10	4.80	4.40	3.70	3.80	3.60	(
12	(a)	(a)	(a)	5.00	4.20	5.00	5.00	4.30	3.70	3.80	3.60	(
13	(a)	(a)	(a)	5.30	4.70	4.70	5.00	4.20	3.70	3.80	(a)	(
14	(a)	(a)	(a)	5.40	4.40	4.90	5.10	4.10	3.70	3.90	(a)	(
15	(a)	(a)	(a)	5.10	4.20	4.80	5.00	4.00	3.90	3.90	(a)	(
16	(a)	(a)	(a)	5.00	4.16	4.70	4.80	4.00	3.90	4.00	(a)	(
17	(a)	(a)	(a)	5.00	4.30	4.70	4.50	4.30	3.90	3.90	(a)	(
18	(a)	(a)	(a)	4.90	4.50	4.60	4.50	4.20	3.90	3.90	(a)	(
19	(a)	(a)	(a)	4.90	5.30	4.60	4.30	4.00	3.90	3.90	(a)	(
20	(a)	(a)	(a)	4.90	5.30	4.80	4.20	4.00	3.90	3.80	(a)	(
21	(a)	(a)	(a)	4.80	5.30	4.70	4.10	3.90	3.90	3.70	(a)	(
22	(a)	(a)	(a)	5.10	5.30	4.70	4.10	3.90	5.80	3.70	(a)	(
23	(a)	(a)	(a)	5.10	5.40	4.65	4.00	3.90	3.75	3.70	(a)	(
24	(a)	(a)	(a)	5.20	5.70	4.70	4.00	3.90	3.70	3.70	(a)	(
25	(a)	(a)	(a)	5.20	5.90	4.60	3.90	3.90	3.70	3.70	(a)	(
26	(a)	(a)	(a)	5.10	6.00	4.60	4.00	3.85	3.60	3.70	(a)	(
27	(a)	(a)	(a)	5.10	6.20	4.70	4.20	3.80	3.65	3.70	(a)	(
28	(a)	(a)	(a)	5.10	7.20	4.70	4.30	5.10	3.60	3.70	(a)	(
29	(a)		(a)	5.00	8.30	4.60	4.30	5.00	3.60	3.70	(a)	(
30	(a)		5.00	5.10	8.50	4.56	4.20	4.90	3.60	3.70	(a)	(
31	(a)		5.20		7.80		4.00	4.60		3.70	(a)	(

a Ice.   b Gage stolen.

Rating table for Milk River at Havre, Mont., from January 1 to December 31, 1903.

Gage height.	Discharge.	Gage height.	Discharge.	Gage height.	Discharge.	Gage height.	Discharge.
Feet.	*Second-feet.*	*Feet.*	*Second-feet.*	*Feet.*	*Second-feet.*	*Feet.*	*Second-feet.*
3.6	118	4.3	339	5.0	784	5.7	1,320
3.7	128	4.4	395	5.1	795	5.8	1,420
3.8	145	4.5	451	5.2	860	5.9	1,520
3.9	168	4.6	507	5.3	935	6.0	1,620
4.0	195	4.7	563	5.4	1,025		
4.1	235	4.8	620	5.5	1,120		
4.2	285	4.9	677	5.6	1,220		

Tangent above gage height 4.5 feet with differences of 100 per tenth. Table is not extended at either end beyond well-determined points.

Estimated monthly discharge of Milk River at Havre, Mont., for 1903.

[Drainage area, 7,300 square miles.]

Month.	Discharge in second-feet.			Total in acre-feet.	Run-off.	
	Maximum.	Minimum.	Mean.		Second-feet per square mile.	Depth in inches.
January			a 200	12,298	0.027	0.031
February			a 200	11,107	.027	.028
March			a 240	14,757	.033	.038
April	1,620	620	996	59,266	.136	.152
May	4,120	285	1,079	66,344	.148	.171
June	2,820	507	975	58,015	.134	.150
July	795	168	445	27,362	.061	.070
August	795	145	378	23,242	.052	.060
September	839	118	164	9,759	.022	.024
October	195	110	138	8,485	.018	.021
November			a 115	6,843	.016	.018
December			a 147	9,038	.020	.023
The year			423	306,516	.058	.786

a Estimated.

MUSSELSHELL RIVER NEAR SHAWMUT, MONT.

This station was established August 12, 1902, by S. B. Robbins. It is located at Crawford's ranch, one-eighth mile west of the post-office at Shawmut, Mont., and 25 miles east of Harlowtown, Mont. It was established to determine the amount of water available for irrigation and storage. The gage is a vertical rod, graduated to feet and tenths with nails, and spiked to a cottonwood stump on the right bank of the river about 200 feet north of Crawford's bunk house. The gage is read once each day by Dwight E. Crawford. A cable will be installed, as there is no bridge from which to make discharge measurements. The channel is straight for 250 feet above and 300 feet below the station. The right bank is high and steep and will overflow at extreme high water. The left bank is low and sloping and will overflow at high water. There is one channel at all stages. At extreme high water the banks may overflow so that the discharge can not be determined accurately, but this location is the best in this vicinity. Bench mark No. 1 is a 30-penny spike in an 18-inch cottonwood tree, below a blaze. This tree stands at the corral fence, 75 feet northeast of the gage rod. The elevation of the bench mark is 5.89 feet above the zero of the gage. Bench mark No. 2 is a 10-penny spike in a 6-inch cottonwood tree, below a blaze. This tree is 60 feet southeast of the gage rod. The elevation of the bench mark is 6.14 feet above the zero of the gage. The barometric elevation of this station is 3,900 feet above sea level.

The observations at this station during 1903 have been made under the direction of C. C. Babb, district engineer.

Mean daily gage height, in feet, of Musselshell River near Shawmut, Mont., for 1903.

Day.	Jan.	Feb.	Mar.	Apr.	May.	June.	July.	Aug.	Sept.	Oct.	Nov.	De
1	1.40	1.30	1.30	1.60	1.60	2.00	1.70	0.80	0.50	1.00	1.10	1.
2	1.30	1.40	1.30	1.60	1.60	2.10	1.70	.80	.50	1.00	1.10	1.
3	1.30	1.40	1.30	1.60	1.60	2.40	1.70	.80	.50	1.00	1.10	1.
4	1.30	1.40	1.40	1.60	1.60	2.50	1.70	.80	.50	1.00	1.10	1.
5	1.30	1.40	1.30	1.50	1.60	2.60	1.70	.80	.50	1.00	1.10	(a
6	1.30	1.40	1.30	1.50	1.60	2.60	1.70	.80	.50	1.00	1.10	(a
7	1.30	1.40	1.30	1.30	1.60	2.60	1.80	.80	.50	1.00	1.10	(a
8	1.40	1.40	1.30	1.30	1.70	2.60	1.90	.80	.50	1.00	1.10	(a
9	1.40	1.40	1.30	1.30	1.70	2.50	1.80	.80	.50	1.00	1.10	1.
10	1.40	1.40	1.40	1.40	1.70	2.50	1.70	.70	.50	1.00	1.10	1.
11	1.40	1.40	1.40	1.40	1.70	2.30	1.70	.70	.50	1.00	1.10	1.
12	1.30	1.40	1.40	1.40	1.60	2.10	1.60	.70	.50	1.10	1.10	1.
13	1.20	1.40	1.40	1.30	1.50	2.00	1.50	.60	.50	1.10	1.10	1.
14	1.20	1.40	1.30	1.30	1.50	2.00	1.40	.60	.50	1.10	1.10	1.
15	1.20	1.40	1.30	1.30	1.50	2.00	1.30	.60	.60	1.10	1.10	1.
16	1.20	1.40	1.30	1.30	1.60	2.00	1.20	.60	.60	1.10	1.10	1.
17	1.20	1.40	1.20	1.30	1.80	1.90	1.00	.60	.60	1.10	1.20	1.
18	1.20	1.40	1.20	1.40	2.00	2.00	1.00	.60	.60	1.10	1.20	1.
19	1.20	1.40	1.30	1.60	2.00	2.00	1.00	.60	.70	1.10	1.20	1.
20	1.20	1.40	1.30	1.60	1.80	2.00	1.00	.60	.80	1.10	1.20	1.
21	1.20	1.40	1.30	1.60	1.80	2.10	1.00	.50	.80	1.10	1.30	1
22	1.20	1.40	1.50	1.70	1.90	2.10	1.00	.50	.80	1.10	1.50	1
23	1.20	1.40	1.30	1.70	1.80	2.10	.80	.50	.80	1.10	1.50	1
24	1.20	1.40	1.40	1.70	1.80	2.00	.80	.50	.80	1.10	1.40	1
25	1.20	1.40	1.40	1.70	1.70	1.90	.80	.50	.90	1.10	1.40	1
26	1.30	1.30	1.50	1.80	1.70	1.80	2.05	.50	1.00	1.10	1.40	1
27	1.30	1.40	1.50	1.80	1.60	1.70	1.40	.50	1.00	1.10	1.40	1
28	1.30	1.50	1.50	1.80	1.70	1.60	.80	.50	1.00	1.10	1.40	1
29	1.30		1.50	1.70	1.80	1.70	.80	.50	1.00	1.10	1.40	:
30	1.30		1.60	1.70	1.80	1.70	.80	.50	1.00	1.10	1.50	
31	1.30		1.60		1.90		.80	.50	1.00			

a River frozen.

MARIAS RIVER NEAR SHELBY, MONT.

This station was established April 4, 1902, by J. S. Baker. It is cated at the highway bridge near J. A. Johnston's ranch, 7 miles uth of Shelby, Mont., from which point it is reached by driving. he gage is of the standard chain type, and the scale board is spiked the upstream hand rail of the bridge. The length of the chain from e weight to the marker is 22.06 feet. It is read once each day by lsie Bailey. Discharge measurements are made from the highway ridge, the lower chord of which is about 15 feet above low water. he initial point for soundings is the east end of the east bridge pier, n the lower side of the bridge, on the left bank. It is marked 0. he channel is straight for 300 feet above and 600 feet below the sta- ion. The current velocity is moderate. The right bank is sloping ind sandy and is liable to overflow. The left bank is high, and is rotected by sheet piling and a plank wall. It may overflow at xtreme flood stages. The bed of the stream is composed of sand and gravel with some cobblestones; it is liable to shift after freshets. There is but one channel, and the current flows toward the left bank. Bench mark No. 1 is a rivet head in the footplate at the foot of the atter post on the top of the southeast pier. The plate is marked 'B. M. 17.54" with black paint. The elevation of the bench mark bove gage datum is 17.54 feet. Bench mark No. 2 is a spike in the outhwest side of a cottonwood stump 25 feet southeast of the south- ast pier. It is marked "B. M. 11.24." Its elevation above gage latum is 11.24 feet.

The observations at this station during 1903 have been made under he direction of C. C. Babb, district engineer.

Discharge measurements of Marias River near Shelby, Mont., for 1903.

Date.	Hydrographer.	Gage height.	Discharge.
		Feet.	*Second-feet.*
May 7	F. M. Brown	3.55	1,721
May 30	C. T. Prall	4.35	3,330
June 25	A. E. Place	4.69	3,969
July 24	E. W. Myers	3.15	1,019
August 18	A. E. Place	2.57	617
October 1	J. H. Sloan	2.95	786

Mean daily gage height, in feet, of Marias River near Shelby, Mont., for 1903.

Day.	Jan.	Feb.	Mar.	Apr.	May.	June.	July.	Sept.	Oct.	Nov.	Dec.
1	2.90	3.10	3.40	6.10	3.40	4.70	4.60		2.90	2.60	3.60
2	2.60	2.95	3.40	5.20	3.20	5.60	4.60		2.90	2.60	3.70
3	2.70	2.90	3.20	4.90	3.20	6.50	4.50		2.90	2.60	3.70
4	2.70	2.90	3.00	3.70	3.30	6.40	4.50		2.90	2.60	3.60
5	2.80	2.60	3.00	3.50	3.50	6.50	4.50		2.90	2.60	3.60
6	3.10	2.50	3.00	3.30	3.60	6.20	4.20		2.80	2.80	3.60
7	3.20	2.50	3.00	3.30	3.70	6.60	4.20		2.80	3.00	3.60
8	3.20	2.40	3.00	2.80	4.00	5.90	4.05		2.80	3.10	3.20
9	3.10	2.50	3.20	2.70	3.90	6.10	4.00		2.80	3.00	3.20
10	3.00	2.50	3.60	2.60	3.60	5.90	3.90		2.80	2.90	3.20
11	3.00	2.90	3.50	2.70	3.60	5.50	a3.90	2.40	2.80	2.80	3.00
12	3.00	3.00	3.40	2.50	3.50	5.20		2.60	2.80	2.70	2.80
13	3.00	3.10	3.00	2.50	3.60	5.50		2.60	2.90	2.60	2.70
14	2.90	3.10	3.00	2.60	4.00	5.70		2.60	3.00	2.40	2.60
15	2.90	3.00	3.00	2.60	4.40	5.80		2.60	3.00	2.40	3.00
16	3.00	2.90	3.00	2.60	4.50	5.50		2.60	3.00	2.30	3.20
17	3.00	3.20	3.10	2.40	4.50	5.40		2.60	3.00	2.30	3.20
18	3.10	2.90	3.00	2.60	4.50	5.50		2.60	2.90	2.80	3.40
19	3.10	3.00	3.00	2.60	4.10	5.40		2.50	2.80	3.10	3.60
20	3.20	3.10	3.00	3.00	4.10	5.30		2.50	2.80	3.20	3.80
21	3.20	3.10	3.00	3.00	4.20	5.40		2.50	2.80	3.30	3.80
22	3.20	6.10	3.50	3.10	4.10	5.20		2.60	2.80	3.50	3.80
23	3.30	3.10	3.50	3.30	4.00	5.00		2.60	2.80	3.50	3.70
24	3.30	3.50	3.90	3.50	4.20	4.90		2.60	2.70	3.60	3.70
25	3.20	3.50	3.70	3.50	4.20	4.70		2.60	2.70	3.60	3.60
26	3.30	3.40	3.20	3.50	4.20	4.60		2.90	2.70	3.90	3.50
27	3.20	3.40	3.00	3.80	4.30	4.50		3.30	2.70	3.60	3.40
28	3.30	3.40	3.10	3.60	4.50	4.50		3.20	2.60	3.60	3.10
29	3.30		3.90	3.50	4.50	4.50		3.10	2.60	3.60	3.00
30	3.10		5.50	3.30	4.60	4.70		3.00	2.60	3.60	3.00
31	3.10		6.40		4.60				2.60		3.10

a Record missing from July 11 to September 11. No observer.

Rating table for Marias River near Shelby, Mont., from January 1 to December 31, 1903.

Gage height.	Discharge.	Gage height.	Discharge.	Gage height.	Discharge.	Gage height.	Discharge.
Feet.	*Second-feet.*	*Feet.*	*Second-feet.*	*Feet.*	*Second-feet.*	*Feet.*	*Second-feet.*
2.3	408	2.9	810	3.5	1,600	4.1	2,800
2.4	460	3.0	895	3.6	1,800	4.2	3,000
2.5	518	3.1	990	3.7	2,000	4.3	3,200
2.6	582	3.2	1,100	3.8	2,200	4.4	3,400
2.7	654	3.3	1,240	3.9	2,400	4.5	3,600
2.8	730	3.4	1,410	4.0	2,600	4.6	3,800

Differences above 4.5 feet, 200 per tenth. Curve extended above gage height 4.69 feet, the highest measurement made in 1903. Measurements made in 1902 at low water determine the lower part of the curve.

Estimated monthly discharge of Marias River near Shelby, Mont.

[Drainage area, 2,610 square miles.]

Month.	Discharge in second-feet.			Total in acre-feet.	Run-off.	
	Maximum.	Minimum.	Mean.		Second-feet per square mile.	Depth in inches.
1903.						
January	1,240	408	968	59,520	0.371	0.431
February	1,600	460	934	51,871	.358	.373
March	7,400	895	1,563	96,105	.598	.689
April	6,800	518	1,516	90,208	.581	.651
May	3,800	1,100	2,582	158,761	.989	1.140
June................	7,800	3,600	5,467	325,309	2.094	2.334
July *a*.............	3,800	800	1,993	122,545	.764	.882
August *b*	800	510	671	41,258	.257	.296
September *c*	1,240	460	618	36,774	.237	.264
October	895	582	738	45,378·	.282	·.325
November..........	2,400	408	1,048	62,360	.401	.451
December	2,200	654	1,506	92,600	.577	.668
The year	7,800	408	1,634	1,182,689	.626	8.504

a July 12 to 23 and 25 to 31 estimated.   *b* August 1 to 17 and 19 to 31 estimated.
c September 1 to 10 estimated.

TWO MEDICINE CREEK AT MIDVALE, MONT.

This station was established September 17, 1902, by C. C. Babb. It is located 200 yards above the great Northern Railroad bridge 1 mile from the railroad station at Midvale, Mont. The gage is inclined and graduated so as to give readings of vertical feet and tenths. It is fastened to a tree and stake. The observer during 1903 was Malcolm Clark. Discharge measurements are made by wading. The initial point for soundings is on the left bank. The channel is straight for 700 feet above and 300 feet below the station. The banks are high and rocky. Bench mark No. 1 is a nail in a rock 13.5 feet below the rod. Its elevation is 2.03 feet above the zero of the gage. Bench mark No. 2 is a nail driven in a spruce tree 40 yards north of the rod. Its elevation is 32 feet above the zero of the gage.

This station was established during the investigation of the feasibility of storing the waters of the river for irrigating purposes. This was found to be impracticable, and the station was accordingly discontinued July 1, 1903.

The observations at this station during 1903 have been made under the direction of C. C. Babb, district engineer.

Mean daily gage height, in feet, of Two Medicine Creek at Midvale, Mont., for 1903.

Day.	Jan.	Apr.	May.	June.	July.	Day.	Jan.	Apr.	May.	June.	July.
1	1.20	1.30	1.00	1.90	1.60	17		1.50	1.50	2.00	
2	1.20	1.30	1.00	2.00	1.60	18		1.50	1.50	2.10	
3	1.20	1.30	1.00	2.20	1.30	19		1.50	1.40	2.10	
4	1.20	1.30	1.10	2.40	1.30	20		1.40	1.40	2.60	
5	1.20	1.30	1.10	2.80	(a)	21		1.30	1.30	2.40	
6	1.00	1.30	1.10	3.00		22		1.30	1.10	2.20	
7	1.00	1.30	1.20	3.00		23		1.30	1.10	2.10	
8	1.00	1.20	1.20	3.00		24		1.30	1.30	2.60	
9	1.00	1.30	1.20	2.80		25		1.30	1.30	1.90	
10	1.00	1.30	1.20	2.80		26		1.10	1.40	1.90	
11	(b	1.30	1.30	2.50		27		1.10	1.40	2.00	
12		1.30	1.30	2.50		28		1.00	1.60	2.00	
13		1.30	1.30	2.60		29		1.00	1.70	1.90	
14		1.30	1.40	2.40		30		1.00	1.50	1.70	
15		1.30	1.40	2.10		31			1.80		
16		1.50	1.00	2.00							

a Station discontinued.   b River frozen January 11 to March 31.

SUN RIVER AT CHRISTIAN'S RANCH, NEAR AUGUSTA, MONT.

This station was established October 31, 1903, by C. T. Prall. It is located below Christian's ranch and below the head of Kilraven ditch, 14 miles northwest of Augusta and 21 miles southwest of Choteau, in Lewis and Clarke County, Mont. The wire gage is on the left bank about 250 feet below the cable. The horizontal scale board is painted white, and is nailed to a plank, which is bolted to three posts set in the ground. The length of the wire from the end of the weight to the marker is 13.15 feet. The distance from the outside edge of the pulley to the zero of the scale is 4 feet. Bench mark No. 1 is the highest point of rock in Sun River, 12 feet upstream from gage. The size of the rock is 4 cubic yards. The elevation of the bench mark is 3.96 feet above the zero of the gage. Bench mark No. 2 is a nail in the root of a cottonwood tree 15 feet east of the gage. Its elevation is 14.45 feet above the zero of the gage. Bench mark No. 3 is a nail in the post which supports the inshore end of the gage. Its elevation is 10.67 feet above the zero of the gage. Discharge measurements are made by means of a cable, car, tag wire, and stay wire. The only observer available is G. B. Christian, who is the owner of several ranches. On account of extensive business interests three gage readings a week are the most that the observer can furnish. The channel is straight for about 300 feet above and 200 feet below the station. The right bank is low and may overflow at flood stages; the left bank is high and rocky. The bed of the stream is of solid rock. The current is sluggish. The initial point for soundings is the left cable strut.

The observations at this station during 1903 have been made under ie direction of C. C. Babb, district engineer.

MISSOURI RIVER AT CASCADE, MONT.

This station was established July 20, 1902, by W. W. Schlecht. It i located at the highway bridge at the east end of the town of Cas- ide, Mont., about one-fourth mile from the railroad. There is an land about 600 feet above the station and one about 300 feet below. here is a fall of about 1 foot at the island below the station. The ire gage first installed has been replaced by a standard chain gage, hich is located on the upstream side of the left span, and is fastened i the floor of the bridge. The distance from the end of the weight i the marker was found to be 31.97 feet in October, 1903. The arker is the end of the wire loop, which serves as a handle. The ige is read twice each day by H. W. Ludwig. Discharge measure- ents are made from the upstream side of the two-span highway ridge. The central pier of this bridge is protected by an ice and rift breaker, which extends about 125 feet upstream. The initial iint for soundings is marked on the guard rail over the right abut- ent. The channel is straight for about 600 feet above and 200 feet low the station. The right bank is low and the left bank is high, ith bushes along both banks. There is a single channel, broken by ie bridge piers. The ice breaker affects the current to some extent. The observations at this station during 1903 have been made under ie direction of C. C. Babb, district engineer.

Discharge measurements of Missouri River at Cascade, Mont., in 1903.

Date.	Hydrographer.	Gage height.	Discharge.
		Feet.	*Second-feet.*
ay 15	F. M. Brown	5.38	7,729
me 18	J. H. Sloan	8.62	19,186
lly 8	G. T. Morris	5.85	8,936
lly 22	J. H. Sloan	4.91	5,054
lly 23	E. W. Myers	4.93	5,189
igust 28	J. H. Sloan	3.62	2,317
tober 2	do	4.05	2,640
rvember 3	W. B. Freeman	4.28	2,786

Mean daily gage height, in feet, of Missouri River at Cascade, Mont., for 1903.

Day.	Mar.	Apr.	May.	June	July.	Aug.	Sept.	Oct.	Nov.	Dec.
1		5.85	5.85	6.08	6.60	4.40	3.62	4.08	4.30	4.9
2		5.82	5.72	6.66	6.60	4.35	3.60	4.07	4.30	4.8
3		5.55	5.62	7.52	6.62	4.20	3.60	4.10	4.30	4.6
4		5.18	5.58	7.72	6.65	4.10	3.60	4.10	4.30	4.5
5		5.05	5.55	8.28	6.45	4.00	3.60	4.15	4.30	4.8
6		5.00	5.55	8.82	6.35	3.98	3.65	4.17	4.30	4.7
7		4.98	5.55	9.12	6.28	3.92	3.70	4.20	4.30	4.6
8		4.90	5.58	9.28	6.25	3.90	3.75	4.25	4.30	4.5
9		4.80	5.72	9.30	6.20	3.90	3.80	4.20	4.30	4.5
10		4.82	5.82	9.28	6.10	3.88	3.85	4.30	4.30	4.3
11		4.90	5.85	9.22	6.05	3.82	3.90	4.30	4.30	4.3
12		5.05	5.78	9.08	5.95	3.80	4.00	4.30	4.30	4.3
13		5.10	5.75	8.95	5.75	3.80	4.00	4.25	4.30	4.3
14		5.10	5.75	8.80	5.65	3.80	3.95	4.25	4.30	4.6
15		5.02	5.85	8.82	5.50	3.80	3.90	4.30	4.40	4.5
16		5.00	6.05	8.82	5.35	3.75	3.90	4.25	4.45	4.6
17		5.00	6.45	8.78	5.15	3.75	3.90	4.25	5.00	4.1
18		5.00	6.75	8.65	4.95	3.75	3.85	4.25	5.00	4.1
19		4.95	6.82	8.45	4.90	3.75	3.85	4.30	5.00	4.1
20		5.02	6.78	8.28	5.05	3.73	3.88	4.30	5.05	4.1
21		5.18	6.68	8.00	5.00	3.80	3.85	4.30	5.05	4.
22		5.28	6.55	7.78	4.95	3.80	3.90	4.30	5.05	4.
23		5.32	6.40	7.58	4.85	3.80	3.90	4.30	5.05	4.
24		5.38	6.40	7.45	4.62	3.82	3.90	4.30	5.10	4.
25		5.45	6.35	7.40	4.32	3.88	3.88	4.30	5.10	4.
26		5.58	6.38	7.28	4.25	3.90	3.90	4.30	5.15	4.
27		5.75	6.50	7.05	4.15	3.85	3.90	4.30	5.15	4.
28		5.85	6.42	6.92	4.15	3.78	3.93	4.30	5.20	4.
29	5.45	5.95	6.42	6.78	4.15	3.72	3.95	4.30	5.08	4.
30	5.58	5.92	6.55	6.65	4.25	3.70	3.97	4.30	5.00	4.
31	5.85		6.55		4.45	3.68		4.30		4.

a To top of ice.   *b* River open.

Rating table for Missouri River at Cascade, Mont., from July 17, 1902, to December 31, 1903.

Gage height.	Discharge.	Gage height.	Discharge.	Gage height.	Discharge.	Gage height.	Discharge.
Feet.	Second-feet.	Feet.	Second-feet.	Feet.	Second-feet.	Feet.	Second-feet.
3.3	1,670	4.1	2,740	4.9	5,110	5.7	8,300
3.4	1,760	4.2	2,950	5.0	5,500	5.8	8,700
3.5	1,860	4.3	3,170	5.1	5,900	5.9	9,100
3.6	1,970	4.4	3,420	5.2	6,300	6.0	9,500
3.7	2,090	4.5	3,700	5.3	6,700	6.1	9,900
3.8	2,230	4.6	4,030	5.4	7,100		
3.9	2,380	4.7	4,370	5.5	7,500		
4.0	2,550	4.8	4,730	5.6	7,900		

Differences above gage height 6 feet, 400 per tenth.

Table is determined by measurements between gage heights 3.6 and 6.6 feet. They have been extended above and below these points.

Estimated monthly discharge of Missouri River at Cascade, Mont., for 1902 and 1903.

[Drainage area, 18,295 square miles.]

Month.	Discharge in second-feet.			Total in acre-feet.	Run-off.	
	Maximum.	Minimum.	Mean.		Second-feet per square mile.	Depth in inches.
1902.						
July 17–31	6,900	3,170	4,847	144,208	0.265	0.148
August	2,950	1,810	2,171	133,490	.119	.137
September	2,305	1,915	2,057	122,400	.112	.125
October	2,950	2,380	2,720	167,247	.149	.172
November	4,030	2,550	3,176	188,985	.174	.194
1903.						
April	9,300	4,730	6,536	388,919	.357	.398
May	12,700	7,700	9,958	612,293	.544	.624
June	22,700	12,100	17,953	1,068,278	.981	1.091
July	12,700	2,845	7,302	448,963	.399	.460
August	3,420	2,090	2,872	145,849	.130	.150
September	2,550	1,970	2,274	135,312	.124	.138
October	3,170	2,645	3,056	187,906	.167	.193
November	6,300	3,170	4,470	265,983	.244	.272
December	9,900	2,740	6,083	374,029	.332	.383

MISSOURI RIVER NEAR TOWNSEND, MONT.

This station was established October 1, 1891, by the Missouri River Commission. The locality is 2,504 miles above the mouth of the river.

A standard wire cable gage was erected on a county-road bridge located about 300 feet below the Northern Pacific Railroad bridge across Missouri River near Townsend, Mont. Its zero is set at an approximate elevation of 3,700 feet above sea level, as determined from a primary line of levels run under the direction of the Commission in 1890 from Three Forks to Fort Benton, Mont., and starting from a bench mark of the Northern Pacific Railroad at Gallatin, Mont. (See Annual Report of Chief of Engineers, 1891.)

The gage reads from 3,785 to 3,799 feet. A record of the daily gage height is maintained by the Department of Agriculture. The observer for both Departments is M. McMahon. A gage reading of 3,785 feet is taken as the zero of the United States Geological Survey observations. Discharge measurements are made from the lower side of the bridge, and distances are marked on the guard rail. The initial point for soundings is over the left pier on the lower side and is marked on the guard rail. The channel is straight for 300 feet above and below the station. The right bank is low, but has been known to be over-flowed but once. The left bank is high and rocky. The bed of the stream is composed of clean gravel and is shifting. The channel is broken by three bridge piers, and the velocity is made uncertain by eddies and backwater. The piers are protected on the upstream side by heaps of riprap and by cofferdams. These cause an appreciable difference in the elevation of the water surface above and below these protections. They cause a contraction of the current, which has eroded large holes under the bridge.

The United States engineers' reference bench mark is B. M. 10 (Townsend), described as "located on the right bank of the Missouri River, about one-half mile north of Townsend railroad station, about one-half mile from river, measured in a perpendicular direction to track, and about three-fourths mile south of the railroad bridge over Missouri River. It is about 60 feet west of a point on the track 30 feet north of railroad bridge No. 392, and about 7 feet west of rail-road fence. Compass reading to milepost 1121 is 318°. Marked by stone and pipe; elevation, 3,795.991 feet." This elevation is erroneous, but is the elevation from which the gage is set.

The tabulated records of the Townsend gage are reductions to the St. Louis directrix datum obtained by subtracting 400.063 feet from the daily means of the gage readings.

In April, 1903, two new bench marks were established by J. S. Baker. The first has been destroyed. The second of these bench marks is a spike driven near the ground in the north side of a 12 by 12 inch post located on the north bank of the river about 200 feet below the bridge. Its elevation is 14.98 feet above the Geological Survey gage datum.

The area drained at this point is approximately 15,000 square miles, comprising, as above stated, the inflow from Gallatin, Madison, and Jefferson rivers. This gage at Townsend is the highest of a series of twenty or more gages maintained permanently by the Missouri River Commission. Descriptions of these gages are given in the annual report of the Missouri River Commission, contained in the Annual Report of the Chief of Engineers, United States Army; that for 1891, on page 3819; for 1892, on page 3271, and for 1893, on page 2316. The distance of this point above Sioux City is 1,703 miles, and above Fort Benton, the next gage below maintained by the Corps of Engineers, 219 miles.

The observations at this station during 1903 have been made under the direction of C. C. Babb, district engineer.

Discharge measurements of Missouri River near Townsend, Mont., in 1903.

Date.	Hydrographer.	Gage height.	Discharge.
		Feet.	*Second-feet.*
April 19	W. B. Freeman	4.40	5,173
April 30	F. M. Brown	4.95	7,020
May 29	W. B. Freeman	5.25	7,786
June 20	J. H. Sloan	6.80	15,679
July 6	G. T. Morris	5.25	8,104
July 23	J. H. Sloan	4.30	4,550
August 24	do	3.40	2,029
September 27	do	3.74	2,649

Mean daily gage height, in feet, of Missouri River near Townsend, Mont., for 1903.

Day.	Mar.	Apr.	May.	June.	July.	Aug.	Sept.	Oct.	Nov.	Dec.
1		4.60	4.90	5.40	4.90	3.80	3.30	3.70	3.90	7.90
2		4.50	4.80	6.10	4.90	3.80	3.30	3.70	3.90	7.90
3		4.30	4.70	6.60	4.50	4.00	3.30	3.80	3.90	7.90
4		4.30	4.70	7.10	4.40	3.90	3.30	3.80	3.90	4.10
5		4.30	4.70	7.60	4.40	3.90	3.30	3.80	3.90	4.10
6		4.30	4.60	7.80	4.40	3.80	3.30	3.80	3.90	3.40
7		4.10	4.70	7.80	5.00	3.80	3.30	3.80	3.90	a 7.90
8		4.10	5.00	7.90	5.20	3.80	3.30	3.90	3.90	7.60
9		4.10	5.00	7.90	5.20	3.70	3.30	3.90	3.90	7.50
10		4.10	5.00	7.90	5.10	3.70	3.30	3.90	3.80	7.40
11		4.50	5.00	7.80	5.00	3.70	3.30	3.90	3.80	7.10
12		4.50	5.00	7.50	4.90	3.60	3.30	3.90	3.70	6.80
13		4.40	5.00	7.50	4.80	3.60	3.50	3.90	3.70	6.70
14		4.30	5.00	7.50	4.70	3.50	3.70	3.90	3.70	6.70
15	6.70	4.30	5.30	7.50	4.50	3.50	3.70	3.90	3.70	6.70
16	6.70	4.30	5.80	7.50	4.40	3.40	3.70	3.90	3.70	6.70
17	6.70	4.30	5.80	7.40	4.30	3.50	3.70	3.90	b 8.60	6.70
18	6.70	4.30	5.90	7.30	4.40	3.40	3.70	3.90	8.00	6.70
19	6.70	4.40	5.80	7.10	4.50	3.40	3.70	3.90	8.60	6.70
20	6.70	4.60	5.80	6.90	4.40	3.40	3.70	3.90	8.00	6.50
21	6.70	4.60	5.40	6.70	4.30	3.40	3.70	3.90	8.60	6.50
22	6.70	4.60	5.30	6.50	4.30	3.40	3.70	3.90	8.90	6.40
23	6.70	4.60	5.30	6.40	4.20	3.40	3.70	3.90	8.90	6.40
24	6.40	4.80	5.10	6.40	4.20	3.40	3.70	3.90	8.90	6.20
25	4.40	4.80	5.00	6.30	4.20	3.40	3.70	3.90	8.60	6.20
26	4.30	4.90	5.60	6.10	4.20	3.30	3.70	3.90	8.30	6.00
27	4.30	5.00	5.00	6.10	4.20	3.30	3.70	3.90	7.90	6.00
28	4.50	5.10	5.00	5.70	4.00	3.30	3.70	3.90	7.90	6.00
29	4.60	5.20	5.20	5.70	4.00	3.30	3.70	3.90	7.90	6.00
30	4.70	5.00	5.30	4.90	4.00	3.30	3.70	3.90	7.90	6.00
31	4.70		5.30		3.90	3.30		3.90		6.00

a Ice gorge below station cause of sudden rise.
b Readings to top of ice, November 17 to December 8.

Rating table for Missouri River near Townsend, Mont., from January 1 to December 31, 1903.

Gage height.	Discharge.	Gage height.	Discharge.	Gage height.	Discharge.	Gage height.	Discharge.
Feet.	*Second-feet.*	*Feet.*	*Second-feet.*	*Feet.*	*Second-feet.*	*Feet.*	*Second-feet.*
3.0	1,200	4.2	4,270	5.4	8,555	7.2	17,040
3.1	1,390	4.3	4,600	5.5	8,965	7.4	18,210
3.2	1,590	4.4	4,930	5.6	9,375	7.6	19,450
3.3	1,800	4.5	5,260	5.7	9,795	7.8	20,760
3.4	2,020	4.6	5,600	5.8	10,220	8.0	22,150
3.5	2,250	4.7	5,940	5.9	10,655	8.2	23,650
3.6	2,490	4.8	6,285	6.0	11,100	8.4	25,410
3.7	2,745	4.9	6,635	6.2	12,015	8.6	27,490
3.8	3,020	5.0	7,000	6.4	12,960	8.8	29,830
3.9	3,315	5.1	7,375	6.6	13,930		
4.0	3,630	5.2	7,760	6.8	14,925		
4.1	3,950	5.3	8,155	7.0	15,950		

1903 rating table same as 1901 and 1902.

Estimated monthly discharge of Missouri River near Townsend, Mont., for 1903.

[Drainage area, 14,500 square miles.]

Month	Discharge in second-feet.			Total in acre-feet.	Run-off.	
	Maximum.	Minimum.	Mean.		Second-feet per square mile.	Depth in inches.
March 15–31	14,425	4,600	10,568	356,342	0.729	0.461
April...............	7,760	3,950	5,295	315,074	.365	.407
May	10,655	5,600	7,590	466,691	.523	.603
June...............	21,445	6,635	15,785	939,273	1.089	1.215
July	7,760	3,315	5,282	324,777	.364	.420
August.............	3,630	1,800	2,398	147,447	.165	.190
September..........	2,745	1,800	2,350	139,835	.162	.181
October	3,315	2,745	3,231	198,666	.223	.257
November *a*	3,315	2,745	2,934	174,585	.202	.225
December *a*.......	....	....	2,745	168,783	.189	.215

a November 17 to December 31 estimated.

GALLATIN RIVER AT LOGAN, MONT.

This station was established August 24, 1893, by F. H. Newell, at the railroad pump house immediately below the Northern Pacific bridge crossing Gallatin River at Logan, Mont. The gage was a vertical rod fastened to the cribwork box sunk in the river for the protection of the inlet pipe of the pump. The bench marks consisted of nails driven into the angle of the pier of the bridge facing the gage, these being placed at the elevation of the 7, 8, 9, and 10 foot marks, and designated by corresponding figures. At the bridge itself measurements of volume can not well be made, as the stream is divided into four channels, being very swift in two of these and obstructed by piles, snags, and sand in the others. Above the bridge, however, is a broad straight course, where measurements can be made by means of a boat and cable.

On March 10, 1894, the gage rod was washed out, together with the crib to which it was attached, no discharge measurements having been made while it was in place.

On November 16, 1894, a new gage was established by Arthur P. Davis, under the northeast corner of the Northern Pacific bridge above mentioned. This gage consisted of timbers partly inclined and partly vertical, the lower inclined portion being graduated from 0.6 foot up to 7.1 feet and the vertical portion from 7 feet up to 12.1 feet. Bench mark No. 1 was on the head of a bridge spike in the top of the pile stump to which the lower end of the inclined gage was fastened. Its elevation was 1.62 feet above the zero of the gage. Bench mark No. 2 was the head of a bridge spike driven horizontally into the first pier east of the river. It was driven into the north end and was marked "B. M." Its elevation was 9.32 feet above the zero of the gage. The measurements were made by means of a cable across the river 100 yards above the bridge.

On September 16, 1896, a wire gage was placed in the east span of the railroad bridge and fastened to the guard rail on the upper side. The distance from the outside edge of the pulley to the end of the rod was 1 foot; from the end of the weight to the index marker was 18.4 feet. Bench mark No. 1 was the top of the northeast corner of the iron plate at the foot of the diagonal end member of the truss at the east end on the upper side, and was 13.7 feet above datum. Bench mark No. 2 was a spike in a pile stump, described and called "Bench mark No. 1" in the description of the 1894 gage.

The present station is located 450 feet northeast of the railroad bridge and 600 feet northwest of the railroad station. It was established in October, 1901, by J. S. Baker. The standard chain gage is placed on a horizontal frame 20 feet east of the south cable support. The gage is read once each day by N. A. Smith. Discharge measurements are made by means of a cable, car, and tagged wire about 450 feet above the railroad bridge. The initial point for soundings is the left cable support. The channel is straight for 200 feet above and below the station. The current has a moderate velocity. The right bank is high and rocky and is not liable to overflow. "The left bank is low, but will overflow only at extremely high stages." Both banks are covered with bushes. The bed of the stream is sandy, with a small amount of vegetation. On July 27, 1901, the datum of the gage was raised 1.23 feet. The bench marks in present use are located as follows: Bench mark No. 1 is a railroad spike in the west side of the post of the gage frame, 3.6 feet from the top of the bridge bar. Its elevation is 8 feet above gage zero. Bench mark No. 2 is a temporary bench mark; it is a point halfway down on the bevel edge at the top of the coping of the northeast abutment of the Northern Pacific Railroad bridge. Its elevation is 11.705 feet above gage zero.

The observations at this station during 1903 have been made under the direction of C. C. Babb, district engineer.

Discharge measurements of Gallatin River at Logan, Mont., in 1903.

Date.	Hydrographer.	Gage height.	Discharge.
		Feet.	*Second-feet.*
April 18	W. B. Freeman	1.26	663
May 2	F. M. Brown	1.49	935
May 27	W. B. Freeman	2.07	1,182
May 28	do	2.30	1,349
May 30	do	2.48	1,552
June 6	J. H. Sloan	4.63	4,013
June 21	do	3.55	3,295
July 9	G. T. Morris	2.20	1,496
July 23	J. H. Sloan	1.28	788
August 24	do	.38	309
September 25	do	1.05	537

Mean daily gage height, in feet, of Gallatin River at Logan, Mont., for 1903.

Day.	Jan.	Feb.	Mar.	Apr.	May.	June.	July.	Aug.	Sept.	Oct.	Nov.	Dec.
1	2.00	1.50	1.20	1.90	1.50	3.60	3.10	0.90	0.50	1.15	2.00	1.3
2	2.00	1.40	1.20	1.60	1.50	4.10	3.05	1.00	.50	1.20	2.00	1.4
3	1.00	1.40	1.00	1.40	1.50	(a)	2.95	1.20	.50	1.20	2.05	1.4
4	.90	1.40	1.10	1.40	1.50	(a)	2.90	1.20	.50	1.20	2.05	1.5
5	1.20	1.40	1.10	1.40	1.50	(a)	2.80	1.15	.50	1.20	2.00	1.
6	1.10	1.50	1.10	1.35	1.50	4.50	2.80	1.15	.50	1.25	2.00	1-
7	1.10	1.50	1.15	1.40	1.60	4.80	2.30	1.10	.50	1.30	2.00	1-
8	1.10	1.50	1.10	1.40	1.60	4.60	2.30	1.10	.60	1.30	2.00	1-

a Gage broken.

Mean daily gage height, in feet, of Gallatin River, etc.—Continued.

Day.	Jan.	Feb.	Mar.	Apr.	May	June.	July.	Aug.	Sept.	Oct.	Nov.	Dec.
....	1.10	1.85	1.10	1.70	1.80	4.80	2.30	1.10	.60	1.30	2.30	1.30
....	1.15	1.50	1.20	1.70	1.85	4.70	2.10	1.00	.65	1.30	2.30	1.30
....	1.10	1.70	1.10	1.75	1.70	4.60	1.90	1.00	0.70	1.30	2.30	1.30
....	1.00	1.60	1.50	1.60	1.65	4.50	1.90	.90	.70	1.35	2.25	1.30
....	1.00	1.60	1.60	1.50	1.80	4.50	1.80	.80	.70	1.40	2.20	1.30
....	1.85	1.90	1.90	1.40	2.30	4.60	1.60	.80	.80	1.40	2.15	1.50
....	1.40	2.10	1.90	1.45	3.00	4.50	1.40	.85	.90	1.40	a2.40	a1.80
....	1.40	2.20	1.30	1.40	3.10	4.10	1.40	.75	.90	1.40	3.00	2.30
....	1.40	2.20	1.90	1.45	3.15	4.10	1.50	.70	.90	1.40	3.70	2.30
....	1.90	2.10	1.40	1.30	2.75	4.15	1.50	.60	.95	1.40	4.50	2.40
....	2.30	2.10	1.20	1.40	2.70	4.10	1.50	.60	.90	1.40	3.40	1.80
....	2.90	2.10	1.10	1.45	2.60	3.50	1.40	.55	.90	1.30	2.80	1.10
....	1.40	2.00	1.10	1.40	2.60	3.55	1.30	.55	.95	1.25	2.00	1.10
....	1.20	1.90	1.10	1.40	2.30	3.35	1.30	.55	.95	1.25	2.10	1.40
....	1.10	1.80	1.10	1.35	2.10	3.25	1.30	.50	.95	1.25	1.50	1.60
....	1.10	1.70	1.20	1.30	2.10	3.15	1.25	.50	1.00	1.25	1.40	1.50
....	1.20	1.60	1.45	1.40	2.10	3.05	1.25	.50	1.05	1.25	1.30	1.30
....	1.10	1.60	1.40	1.50	2.00	3.00	1.25	.50	1.10	1.25	1.30	1.20
....	1.05	1.20	1.60	1.80	2.10	3.00	1.30	.50	1.10	1.35	1.20	1.10
....	1.10	1.00	2.00	1.85	2.50	2.90	1.30	.50	1.10	1.35	1.20	1.20
....	1.15	......	2.10	1.70	2.50	2.90	1.20	.50	1.10	1.35	1.20	1.25
....	1.80	......	2.40	1.60	2.80	3.00	1.00	.50	1.10	1.30	1.30	1.30
....	1.70	......	2.45	......	3.10	......	.95	.50	......	1.25	......	1.40

a Ice gorged November 15 to 20 and December 15 to 19.

...ting table for Gallatin River at Logan, Mont., from January 1 to December 31, 1903.

Gage height.	Discharge.	Gage height.	Discharge.	Gage height.	Discharge.	Gage height.	Discharge.
Feet.	*Second-feet.*	*Feet.*	*Second-feet.*	*Feet.*	*Second-feet.*	*Feet.*	*Second-feet.*
0.1	350	1.3	710	2.5	1,850	3.7	3,380
.2	356	1.4	780	2.6	1,965	3.8	3,520
.3	363	1.5	860	2.7	2,080	3.9	3,670
.4	370	1.6	940	2.8	2,200	4.0	3,830
.5	382	1.7	1,025	2.9	2,320	4.1	3,970
.6	400	1.8	1,120	3.0	2,440	4.2	4,125
.7	423	1.9	1,220	3.1	2,570	4.3	4,275
.8	450	2.0	1,315	3.2	2,700	4.4	4,430
.9	480	2.1	1,415	3.3	2,830	4.5	4,580
1.0	530	2.2	1,520	3.4	2,960	4.6	4,730
1.1	590	2.3	1,630	3.5	3,090	4.7	4,880
1.2	650	2.4	1,735	3.6	3,240	4.8	5,030

Estimated monthly discharge of Gallatin River at Logan, Mont., for 1903.

[Drainage area, 1,805 square miles.]

Month.	Discharge in second-feet.			Total in acre-feet.	Run-off.	
	Maximum.	Minimum.	Mean.		Second-feet per square mile.	Depth in inches.
January	2,320	480	824	50,666	0.456	0.535
February	1,520	580	1,041	57,814	.577	.601
March	1,790	530	788	48,452	.436	.502
April	1,220	710	872	51,888	.483	.539
May	2,635	860	1,515	93,154	.839	.967
June...............	5,030	2,320	3,749	223,030	2.077	2.318
July	2,570	500	1,181	72,616	.654	.754
August	650	383	467	28,714	.259	.299
September	590	382	468	27,848	.259	.289
October	780	620	712	43,778	.394	.454
November.........	4,580	650	1,527	90,862	.846	.944
December	1,735	590	840	51,650	.465	.536
The year	5,030	382	1,165	840,622	.645	8,727

WEST GALLATIN RIVER NEAR SALESVILLE, MONT.

The Salesville station, which has been maintained for a number of
years, was established near Williams's ranch, about 16 miles southwest of Bozeman. A gage rod was erected in July, 1895, and observations were begun on August 1 by Ira T. Williams, a ranchman, living
about 600 feet away. The gage was spiked to a tree. The bench
mark consisted of a 6-inch spike driven in the top of a stump 5 feet
north of the gage post. It was 6.71 feet above the zero of the gage, as
lowered 5 feet from the original position. A second bench mark consisted of a 6-inch spike driven into the east bridge abutment. This
was 9.26 feet above the zero of the gage. The initial point for soundings is marked on the guard rail over the center of the left pier. The
highway bridge, from the lower side of which discharge measurements
are made, is not at right angles to the current. At flood stages the
water flows behind the bridge abutments on the right bank, but at
other times is confined within the channel. The bed of the stream is
composed of bowlders and is not liable to change. The velocity is
high, rendering discharge measurements somewhat difficult. The
channel is nearly straight, with slight curves both above and below.
The West Gallatin canal, carrying about 125 second-feet between
July 1 and September 15, is taken out below the station. The Klein-

lt canal, carrying about 50 second-feet between July 1 and
t 15, and one or two smaller ditches are taken out above the
. In September, 1896, a wire gage was placed on the bridge,
lley being fastened to the end of the rod opposite the 0.15-foot
The distance from the end of the weight to the index marker
0 feet. On October 15, 1903, this length was 15.70 feet. The
,ges were made to read the same. Bench mark No. 1 for the
ıge is the head of the southwest bolt in the rim of the southeast
·ical pier. Its elevation is 13.70 feet above the gage datum.
mark No. 2, established May 22, 1900, by J. S. Baker, is a large
·r set firmly in the ground on the south side of the wagon road,
ıe fence. It is 123 feet east of the center of the southeast
·ical pier and is at an elevation of 13.69 feet above the gage
The observer is Miss Marguerite Williams.
observations at this station during 1903 have been made under
ection of C. C. Babb, district engineer.

ʒe measurements of West Gallatin River near Salesville, Mont., in 1903.

Date.	Hydrographer.	Gage height.	Discharge.
		Feet.	*Second-feet.*
· 3	J. S. Baker	2.72	375
	F. L. Travenner	2.82	405
	J. H. Sloan	3.77	929
	do	6.75	5,006
	do	5.85	3,063
	do	4.25	1,604
)er 5	do	2.95	474
15	do	3.17	448

*ıily gage height, in feet, of West Gallatin River near Salesville, Mont.,
for 1903.*

Day.	Jan.	Feb.	Mar.	Apr.	May.	June.	July.	Aug.	Sept.	Oct.	Nov.	Dec.
	2.75	2.65	2.65	2.80	3.10	5.80	5.20	4.20	......	3.00	3.00	3.00
	2.75	2.65	2.65	2.80	3.00	6.00	5.10	3.90	......	3.00	3.00	3.00
	2.75	2.65	2.65	2.65	3.00	6.80	4.80	3.90	......	3.10	3.00	3.00
	2.70	2.65	2.65	2.65	3.10	6.40	4.80	3.90	......	3.20	3.00	2.70
	2.70	2.65	2.65	2.65	3.20	6.50	4.80	3.80	2.15	3.20	3.00	2.70
	2.70	2.65	2.65	2.65	3.20	6.50	4.70	3.80	......	3.10	3.00	2.90
	2.60	2.65	3.00	2.65	3.50	6.40	4.70	3.80	2.95	3.10	3.00	2.90
	2.60	2.65	2.75	2.70	3.50	6.00	4.60	3.60	2.92	3.10	3.00	2.80
	2.65	2.70	2.65	2.95	3.50	6.40	4.50	3.60	3.20	3.10	3.00	2.90
	2.65	2.70	2.65	2.95	3.40	6.20	4.40	3.60	3.10	3.10	3.05	2.80
	2.65	2.70	2.65	2.85	3.30	6.10	4.40	3.60	3.10	3.10	3.05	2.90
	2.60	2.70	2.65	2.85	3.70	6.10	4.50	3.60	3.10	3.10	2.90	2.90

Mean daily gage height, in feet, of West Gallatin River, etc.—Continued.

Day.	Jan.	Feb.	Mar.	Apr.	May.	June.	July.	Aug.	Sept.	Oct.	Nov.	Dec.
13	2.65	2.70	2.60	2.90	4.00	6.50	4.50	3.50	3.10	3.10	2.90	2.80
14	2.65	2.70	2.65	2.90	4.50	6.85	4.30	3.50	3.10	3.10	2.90	2.80
15	2.65	2.75	2.60	2.95	4.90	6.85	4.20	3.50	3.10	3.10	3.00	2.80
16	2.65	2.80	2.60	2.90	4.80	6.50	4.30	3.40	3.00	3.10	3.00	2.80
17	2.65	2.85	2.60	2.90	4.50	6.50	4.60	3.40	3.00	3.10	3.10	2.80
18	2.70	2.95	2.60	2.80	4.30	6.20	4.30	(a)	3.10	3.10	2.80	2.80
19	2.70	3.00	2.60	2.80	4.00	6.20	4.30	......	3.10	3.10	2.70	2.80
20	2.75	3.10	2.60	2.80	3.90	6.30	4.10	......	3.10	3.10	2.70	2.80
21	2.75	3.10	2.60	2.90	3.70	5.90	4.00	......	3.10	3.10	2.70	2.80
22	2.75	2.80	2.65	3.00	3.70	5.80	4.00	......	3.10	3.10	2.80	2.80
23	2.75	2.75	2.65	3.10	3.70	5.90	4.00	......	3.10	3.10	2.80	2.80
24	2.75	2.65	2.65	3.10	3.50	5.30	4.00	......	3.10	3.00	2.80	2.80
25	2.75	2.65	2.65	3.30	3.70	5.30	4.00	......	3.10	3.00	2.90	2.80
26	2.65	2.65	2.65	3.50	3.70	5.40	4.00	......	3.00	3.00	2.90	2.80
27	2.65	2.65	2.70	3.30	4.00	5.40	4.00	......	3.00	3.00	2.95	2.80
28	2.65	2.65	2.75	3.30	4.00	5.40	3.90	......	3.00	3.00	2.95	2.80
29	2.60	......	2.80	3.10	4.20	5.80	3.90	......	3.10	3.00	3.00	2.80
30	2.60	......	2.80	3.10	4.50	5.20	3.90	......	3.10	3.00	3.00	2.80
31	2.60	......	2.80	......	5.10	......	3.90	......	......	3.00	......	2.80

a Gage broken from August 18 to September 7.

Rating table for West Gallatin River near Salesville, Mont., from January 1 December 31, 1903.

Gage height.	Discharge.	Gage height.	Discharge.	Gage height.	Discharge.	Gage height.	Discharge.
Feet.	Second-feet.	Feet.	Second-feet.	Feet.	Second-feet.	Feet.	Second-feet.
2.6	360	3.5	660	4.4	1,710	5.3	3,000
2.7	370	3.6	730	4.5	1,850	5.4	3,150
2.8	390	3.7	810	4.6	1,990	5.5	3,300
2.9	410	3.8	910	4.7	2,130	5.6	3,450
3.0	440	3.9	1,020	4.8	2,270	5.7	3,600
3.1	470	4.0	1,150	4.9	2,410	5.8	3,750
3.2	510	4.1	1,290	5.0	2,550	5.9	3,900
3.3	550	4.2	1,430	5.1	2,700	6.0	4,050
3.4	600	4.3	1,570	5.2	2,850		

Tangent above 5 feet, with differences of 150 per tenth. Table extended 0.1 f below lowest and 0.15 foot above highest measurement.

*timated monthly discharge of West Gallatin River near Salesville, Mont., for
1903.*

[Drainage area, 860 square miles.]

Month.	Discharge in second-feet.			Total in acre-feet.	Run-off.	
	Maximum.	Minimun.	Mean.		Second-feet per square mile.	Depth in inches.
nuary	380	360	369	22,689	0.429	0.494
bruary	470	365	382	21,215	.444	.462
arch	440	360	370	22,750	.430	.496
pril	660	365	429	25,527	.499	.556
ay	2,700	440	1,066	65,546	1.240	1.430
me	5,325	2,850	4,125	245,454	4.800	5.350
ily	2,850	1,020	1,635	100,532	1.901	2.191
ugust a	1,430	470	684	42,058	.795	.916
ptember	510	415	456	27,134	.530	.590
ctober	510	440	463	28,469	.538	.619
ovember	470	370	422	25,111	.491	.551
ecember	470	370	428	26,317	.498	.574
The year	5,325	360	902	652,802	1.050	14.229

a August 17 to 31 estimated.

MIDDLE CREEK NEAR BOZEMAN, MONT.

The station, established on August 3, 1895, is located 9 miles south
[Bozeman, one-eighth of a mile above the old sawmill dam, and 1½
iles above the mouth of the canyon. Discharge measurements are
ade from a wire cable placed across the stream in 1898. The gage
as about 200 feet below the cable and consisted of a horizontal
ame supporting a wire gage. Bench mark No. 1 consisted of a
ike driven horizontally into a stump 5 feet high about 80 feet east
the gage rod. The middle of this spike was at an elevation of 7.03
et above gage datum. Bench mark No. 2 consisted of an 8-inch
idge spike driven horizontally into a charred stump about 25 feet
rtheast of the gage, with an elevation of 3.58 feet. Bench mark
. 3 consisted of a large rock 93 feet east of the gage, marked "B. M."
black paint, and 4.84 feet above gage datum. The initial point for
undings is on the left bank. The water moves swiftly on one side.
e left bank is low and liable to overflow. The bed of the stream
of gravel and is liable to change. Gage heights were not taken in
)7 and 1901, owing to the impossibility of securing an observer at
derate expense.
A new gage and three new bench marks were established at this
tion on May 8, 1902. The gage is located 50 feet above the cable,

The length of the wire from the end of the weight to the marke
14.76 feet. Bench mark No. 1 is the top of a bowlder 56 feet sou
east of the deadman of the east end of the cable. It is marked in
paint "B. M., U. S. G. S., 8.41." Its elevation is 8.41 feet above
zero of the gage. Bench mark No. 2 is the top of a bowlder 68 1
southeast of the deadman, marked in red paint "B. M., U. S. G.
9.26." Its elevation is 9.26 feet above the zero of the gage. Be
mark No. 3 is a large bowlder on the hillside 150 feet northeast of
deadman, marked "B. M., U. S. G. S., 20.43." Its elevation is 2(
feet above the zero of the gage. The relation of the gage zeros
not been determined.

During 1903 measurements were made by means of the cable
car, although the car was out of repair. Soundings were not refer
to the initial point, as the tag wire had been removed. The strean
too rough to afford accurate measurement and has an extremely I
velocity at flood stages. The gage heights were not read during 1!

The observations at this station during 1903 have been made un
the direction of C. C. Babb, district engineer.

Discharge measurements of Middle Creek near Bozeman, Mont., in 1903.

Date.	Hydrographer.	Gage height.	Discha
		Feet.	*Second*
April 17	F. L. Travenner	0.85	
May 21	G. T. Morris	1.23	
June 3	C. E. Lamme	1.89	
June 13	J. H. Sloan	1.70	
June 28	do	1.68	
July 17	do	1.64	
September 6	do	1.10	
October 14	do	1.10	

MADISON RIVER (INCLUDING CHERRY CREEK) NEAR NORRIS, MO:

This station is located at the ranch of the observer, Mrs. S. A. Bl.
4 miles below the Redbluff county iron bridge over the Madison,
about 1½ miles below the mouth of Cherry Creek. It is also abo1
miles below the location of the old Redbluff station, described
Bulletin No. 131, on page 18. It was established May 2, 1897, at wl
time the one at Threeforks was discontinued. The vertical gag
fastened to a post set firmly in the bed of the river and braced v
crosspieces from the bank. It is about 125 yards west of the observ
house. Discharge measurements are made from the lower side
the iron bridge above. Cherry Creek is measured at the same ti
as it enters between the gage and the bridge. The initial point
soundings is at the left abutment of the bridge. The banks are l

and do not overflow.   The bed of the stream is rocky and the current is quite swift.

The channel is curved both above and below the station, but the bridge is at right angles to the current.

The bench mark established June 6, 1901, by J. S. Baker, is the top of a large granite bowlder 20 feet south of the gage rod and 8 feet east of the river bank.   There is a spot of black paint on the top of the bowlder, and the letters "B. M. U. S. G. S. 6.856."   The elevation of this bench mark is 6.856 feet above the zero of the gage.

The observations at this station during 1903 have been made under the direction of C. C. Babb, district engineer.

Discharge measurements of Madison River near Norris, Mont., in 1903.

Date.	Hydrographer.	Gage height.	Discharge.
		Feet.	*Second-feet.*
April 19	F. L. Travenner	1.46	1,480
May 23	J. H. Sloan	1.68	2,206
June 11	....do	2.90	5,983
June 26	....do	2.15	3,793
July 15	....do	1.68	2,183
September 4	....do	1.30	1,806
October 16	....do	1.35	1,185

Mean daily gage height, in feet, of Madison River near Norris, Mont., for 1903.

Day.	Jan.	Feb.	Mar.	Apr.	May.	June.	July.	Aug.	Sept.	Oct.	Nov.	Dec.
1	(a)	(a)	(a)	1.50	1.50	2.25	2.10	1.40	1.30	1.20	1.30	1.30
2	(a)	(a)	(a)	1.50	1.50	2.68	1.95	1.40	1.30	1.25	1.30	1.30
3	(a)	(a)	(a)	1.50	1.50	2.70	1.90	1.40	1.30	1.25	1.30	1.30
4	(a)	(a)	(a)	1.50	1.50	2.75	1.83	1.40	1.30	1.30	1.30	1.30
5	(a)	(a)	(a)	1.40	1.50	2.88	1.90	1.40	1.30	1.30	1.30	1.30
6	(a)	(a)	(a)	1.40	1.50	2.90	1.88	1.40	1.30	1.30	1.30	(a)
7	(a)	(a)	(a)	1.40	1.55	2.85	1.80	1.40	1.30	1.30	1.30	(a)
8	(a)	(a)	(a)	1.40	1.58	2.88	1.80	1.40	1.30	1.30	1.30	(a)
9	(a)	(a)	(a)	1.45	1.60	2.90	1.75	1.40	1.20	1.30	1.30	(a)
10	(a)	(a)	(a)	1.50	1.60	2.93	1.65	1.40	1.20	1.30	1.30	(a)
11	(a)	(a)	(a)	1.50	1.60	2.90	1.60	1.40	1.20	1.30	1.30	(a)
12	(a)	(a)	(a)	1.50	1.60	2.85	1.60	1.35	1.20	1.30	1.30	(a)
13	(a)	(a)	(a)	1.50	1.62	2.85	1.60	1.30	1.20	1.30	1.30	(a)
14	(a)	(a)	(a)	1.50	1.75	2.90	1.60	1.30	1.20	1.30	1.30	(a)
15	(a)	(a)	(a)	1.50	1.90	2.90	1.60	1.30	1.30	1.30	1.30	(a)
16	(a)	(a)	(a)	1.50	2.00	2.90	1.00	1.30	1.20	1.30	1.30	(a)
17	(a)	(a)	(a)	1.50	2.05	2.80	1.60	1.30	1.20	1.30	1.30	(a)
18	(a)	(a)	(a)	1.50	1.90	2.70	1.60	1.30	1.20	1.30	1.30	(a)
19	(a)	(a)	(a)	1.40	1.87	2.60	1.60	1.30	1.20	1.30	1.30	(a)
20	(a)	(a)	(a)	1.40	1.75	2.55	1.55	1.30	1.20	1.30	1.30	(a)
21	(a)	(a)	(a)	1.45	1.70	2.40	1.50	1.30	1.20	1.30	1.30	(a)
22	(a)	(a)	(a)	1.45	1.70	2.40	1.50	1.30	1.20	1.30	1.30	(a)
23	(a)	(a)	(a)	1.50	1.70	2.40	1.50	1.20	1.20	1.30	1.30	(a)
24	(a)	(a)	(a)	1.50	1.68	2.28	1.50	1.20	1.20	1.30	1.30	(a)

a Ice.

Mean daily gage height, in feet, of Madison River, etc.—Continued.

Day.	Jan.	Feb.	Mar.	Apr.	May.	June.	July.	Aug.	Sept.	Oct.	Nov.	Dec.
25	(a)	(a)	(a)	1.50	1.56	2.20	1.50	1.20	1.20	1.20	1.30	(a)
26	(a)	(a)	(a)	1.45	1.60	2.20	1.50	1.20	1.20	1.20	1.30	(a)
27	(a)	(a)	(a)	1.45	1.65	2.10	1.50	1.20	1.20	1.20	1.30	(a)
28	(a)	(a)	(a)	1.45	1.70	2.10	1.50	1.20	1.20	1.20	1.30	(a)
29	(a)		1.50	1.50	1.83	2.10	1.50	1.20	1.20	1.20	1.30	(a)
30	(a)		1.50	1.50	1.93	2.15	1.50	1.20	1.20	1.20	1.30	(a)
31	(a)		1.50		2.03		1.50	1.20		1.20		(a)

a Ice.

Rating table for Madison River near Norris, Mont., from January 1 to December 31, 1903.

Gage height.	Discharge.	Gage height.	Discharge.	Gage height.	Discharge.	Gage height.	Discharge.
Feet.	Second-feet.	Feet.	Second-feet.	Feet.	Second-feet.	Feet.	Second-feet.
1.2	1,050	1.7	2,300	2.2	3,820	2.7	5,370
1.3	1,220	1.8	2,600	2.3	4,130	2.8	5,680
1.4	1,440	1.9	2,900	2.4	4,440	2.9	5,990
1.5	1,700	2.0	3,200	2.5	4,750	3.0	6,300
1.6	2,000	2.1	3,510	2.6	5,060		

Table extended one-tenth foot above and below extreme measurements made during 1903.

Estimated monthly discharge of Madison River near Norris, Mont., for 1903.

[Drainage area, 2,085 square miles.]

Month.	Discharge in second-feet.			Total in acre-feet.	Run-off.	
	Maximum.	Minimum.	Mean.		Second-feet per square mile.	Depth in inches.
January a			1,300	79,934	0.624	0.71
February a			1,300	72,198	.624	.65
March a			1,300	79,934	.624	.71
April	1,700	1,440	1,622	96,516	.778	.86
May	3,355	1,700	2,281	140,253	1.090	1.26
June	6,145	3,510	5,065	301,389	2.430	2.71
July	3,510	1,700	2,141	131,645	1.030	1.19
August	1,440	1,050	1,258	77,351	.603	.69
September	1,220	1,050	1,095	65,157	.525	.58
October	1,220	1,050	1,209	74,338	.580	.67
November	1,220	1,220	1,220	72,595	.585	.64
December a			1,200	73,785	.576	.66
The year			1,749	1,265,095	.839	11.3

a Estimated.

JEFFERSON RIVER NEAR SAPPINGTON, MONT.

This station was established November 13, 1894, by A. P. Davis. It is located 300 feet above the railroad bridge, 1 mile north of the railroad station at Sappington, and 7 miles above Willow Creek. The chain gage is fastened to the guard rail on the upstream side of the south span of the Northern Pacific Railway bridge. The length of the chain from the end of the weight to the marker is 16.35 feet. It is read twice each day by John Fraser. Discharge measurements are made by means of a cable, car, and tagged wire about 300 feet above the railroad bridge to which the gage is attached. The initial point for soundings has been taken as the cable support on either bank. The channel is straight for 600 feet above and 300 feet below the station. The current velocity is swift. Both banks are composed of clay and are covered with willows and underbrush. The right bank may overflow at extreme high water. The bed of the stream is composed of rocks and gravel and there is but one channel. On November 3, 1897, the gage rod was lowered 0.8 foot. The subsequent years were adjusted to the new datum, but the remainder of 1897 was corrected to agree with the old datum.

Bench mark No. 1 consists of a 6-inch wire nail driven horizontally in the east side of the blocking which forms the south abutment of the railroad bridge and is 12.90 feet above gage datum. Bench mark No. 2 is a 6-inch wire nail in a telegraph pole about 30 feet south and east of the south abutment of the bridge and is at an elevation of 9.93 feet above gage datum. Bench mark No. 3 is the head of the northwest bolt fastening the switch standard to the cross-tie 30 feet east of the bridge; its elevation is 17.70 feet above gage datum.

The observations at this station during 1903 have been made under the direction of C. C. Babb, district engineer.

Discharge measurements of Jefferson River near Sappington, Mont., in 1903.

Date.	Hydrographer.	Gage height.	Discharge.
		Feet.	*Second-feet.*
April 17	W. B. Freeman	2.72	1,998
May 2	F. M. Brown	3.41	3,060
May 27	W. B. Freeman	3.46	3,051
June 7	J. H. Sloan	6.40	9,755
June 22	do	4.68	5,968
July 10	G. T. Morris	3.60	3,436
July 24	J. H. Sloan	2.50	1,874
August 24	do	1.45	535
October 4	do	1.90	779

Mean daily gage height, in feet, of Jefferson River near Sappington, Mont., for 1903.

Day.	Jan.	Feb.	Mar.	Apr.	May.	June.	July.	Aug.	Sept.	Oct.	Nov.	Dec.
1	2.40	3.00	2.70	2.60	3.80	3.90	3.90	2.20	1.50	1.90	2.10	4.5
2	2.40	3.00	2.70	2.60	3.55	4.45	3.75	2.20	1.50	1.90	2.10	3.8
3	2.40	3.45	2.80	2.60	3.50	5.00	3.50	2.30	1.50	1.90	2.10	3.1
4	2.40	3.90	2.80	2.65	3.45	5.65	3.30	2.30	1.50	1.90	2.20	2.5
5	2.45	3.90	2.75	2.70	3.50	6.10	3.40	2.20	1.50	1.95	2.20	2.3
6	2.50	3.80	2.70	2.70	3.50	6.30	3.45	2.10	1.55	2.00	2.20	2.2
7	2.50	3.70	2.70	2.60	3.50	6.40	3.50	2.10	1.60	2.00	2.30	2.
8	2.50	3.70	2.75	2.60	3.65	6.30	3.50	2.05	1.60	2.05	2.30	2.
9	2.50	3.70	2.85	2.60	3.60	6.20	3.65	2.00	1.60	2.10	2.30	2.
10	2.50	3.65	2.90	2.65	3.60	6.10	3.55	2.00	1.60	2.10	2.35	2.
11	2.50	3.60	2.90	2.85	3.60	6.10	3.35	2.00	1.60	2.10	2.40	2.
12	2.50	3.60	2.85	2.90	3.60	6.05	3.30	1.95	1.65	2.20	2.40	2.
13	2.50	3.45	2.70	2.90	3.55	5.95	3.15	1.95	1.75	2.20	2.40	2.
14	2.65	3.25	2.60	2.90	3.50	5.85	2.95	1.85	1.85	2.20	2.40	2.
15	2.75	3.45	2.60	2.90	3.75	5.80	2.75	1.85	1.90	2.20	2.45	2.
16	2.90	3.50	2.60	2.90	4.15	5.75	2.65	1.85	2.00	2.20	2.50	2.
17	3.20	3.90	2.60	2.80	4.60	5.65	2.70	1.90	2.00	2.20	2.50	3.
18	2.60	3.90	2.60	2.80	4.40	5.45	2.70	1.80	2.00	2.10	2.65	2.
19	2.45	2.95	2.60	2.95	4.25	5.35	2.70	1.55	2.00	2.10	3.25	2.
20	2.30	2.80	2.60	3.00	4.15	5.10	2.70	1.50	2.00	2.10	4.45	2.
21	2.30	2.75	2.60	3.15	3.95	4.75	2.60	1.50	2.00	2.20	4.65	3.
22	2.30	2.70	2.60	3.25	3.85	4.70	2.60	1.80	1.95	2.20	4.70	3.
23	2.30	2.70	2.65	3.35	3.70	4.70	2.60	1.80	1.90	2.10	4.70	3.
24	2.30	2.60	2.60	3.45	3.00	4.70	2.50	1.80	1.90	2.10	4.80	3.
25	2.30	2.60	2.55	3.60	3.55	4.65	2.50	1.90	1.90	2.15	4.80	3.
26	2.90	2.70	2.55	3.65	3.50	4.45	2.45	1.80	1.90	2.20	4.90	2.
27	2.80	2.70	2.50	3.85	3.40	4.15	2.30	1.80	1.90	2.20	4.90	4.
28	2.75	2.70	2.50	3.90	3.45	4.05	2.30	1.60	1.90	2.20	4.90	4
29	2.70	....	2.50	3.90	3.50	3.90	2.30	1.60	1.90	2.20	4.95	4
30	2.95	....	2.60	3.85	3.55	3.90	2.30	1.50	1.90	2.20	4.70	3
31	3.00	....	2.60	....	3.65	....	2.30	1.50	....	2.10	....	3

a Gage heights from November 10 to December 31 were increased by ice gorges below.

*Rating table for Jefferson River near Sappington, Mont., from January 1
December 31, 1903.*

Gage height.	Discharge.	Gage height.	Discharge.	Gage height.	Discharge.	Gage height.	Discharge.
Feet.	Second-feet.	Feet.	Second-feet.	Feet.	Second-feet.	Feet.	Second-feet.
1.5	570	2.2	1,330	2.9	2,320	3.6	3,410
1.6	640	2.3	1,470	3.0	2,470	3.7	3,600
1.7	730	2.4	1,610	3.1	2,620	3.8	3,800
1.8	830	2.5	1,750	3.2	2,770	4.0	4,250
1.9	940	2.6	1,890	3.3	2,920		
2.0	1,060	2.7	2,030	3.4	3,070		
2.1	1,190	2.8	2,170	3.5	3,230		

Tangent above 4 feet, with differences of 230 per tenth.

Estimated monthly discharge of Jefferson River near Sappington, Mont., for 1903.

[Drainage area, 8,984 square miles.]

Month.	Discharge in second-feet.			Total in acre-feet.	Run-off.	
	Maximum.	Minimum	Mean.		Second-feet per square mile.	Depth in inches.
January	2,470	1,330	1,797	110,493	0.20	0.22
February	4,020	1,890	2,848	158,169	.32	.33
March	2,320	1,750	1,974	121,376	.22	.25
April	4,020	1,890	2,561	152,390	.28	.81
May	5,630	3,070	3,654	224,675	.41	.47
June	9,770	4,020	7,117	423,491	.79	.88
July	4,020	1,470	2,412	148,308	.27	.31
August	1,470	570	859	52,817	.10	.12
September	1,060	570	828	49,269	.09	.10
October	1,330	940	1,202	73,908	.13	.15
November a	6,435	1,190	3,204	190,651	.36	.40
December a	5,515	1,260	2,712	166,754	.30	.35
The year	9,770	570	2,597	1,872,301	.29	3.89

a Discharge as estimated is too high. Gage heights from November 19 to December 31 were raised by ice gorges below gage.

Miscellaneous measurements in the Missouri River drainage basin in 1903.

Date.	Stream.	Locality.	Discharge.
			Sec.-feet.
July 2	Battle Creek	Hermosa, S. Dak	10.2
August 22	Strater ditch	Battle Creek basin	1.3
June 25	Bear Butte Creek	In canyon above Sturgis, S. Dak.	14.2
July 3	Beaver Creek	Buffalo Gap	11.6
Do	Bolan ditch	Beaver Creek drainage basin.	12.6
June 10	Box Elder	Below railroad bridge	48.0
June 15	Cascade Creek	Cascade Springs	28.4
June 26	Elk Creek	6 miles below Piedmont	17.5
July 1	Fall River	Hot Springs, S. Dak., below power house.	34.9
August 27	Rapid Creek	Mystic, S. Dak	39.4
August 20	Lockhart ditch	Rapid Creek Valley	2.3
Do	Hawthorne ditch		6.0
Do	Rapid Valley ditch		26.0
Do	South ditch		14.9
Do	Cyclone ditch		2.6

Miscellaneous measurements in the Missouri River, etc.—Continued.

Date.	Stream.	Locality.	Dis-charge.
			Sec.-feet.
May 13.........	Red Water	Bridge above Crow Creek...	80.6
May 12.........	Red Water ditch	Red Water Valley at head...	40.5
May 13.........	.....do	...do	43.3
August 13...	.....do	...do	37.6
October 28 ..	.....do	Bridge east of Belle Fourche	16.0
August 13......	Belle Fourche power canal.	.....do	9.8
October 28	.....do	Red Water Valley.........	34.3
May 13...........	.....do	.....do	73.7
May 12.........	False Bottom Creek......	.....do	7.4
August 14......	Walter & McVey ditch ..	Spearfish Valley	7.7
Do ......:..	Cook & Burns ditch	.....do	11.0
August 15......	Smith ditch	.....do	3.3
Do	Walter & Schuler ditch ..	.....do	3.4
August 14......	Cook ditch	.....do	20.3
August 29......	.....do	.....do	27.8
August 15......	Eckles ditch.............	.....do	4.3
Do	Evans ditch	.....do	12.8
August 29......	.....do	.....do	23.3
Do	Concord ditch	.....do	8.9
Do	Tindleys ditch.........	.....do	4.4
Do	Spring Branch ditch.... .... do		6.6
Do	Brown ditch	.do	6.4
August 26......	Spring Creek	Hill City, S. Dak	7.5
June 9	.....do	Waugh ranch	32.6
November 2	.....do	Sink Hole (above) :.........	8.8
Do	.....do	Sink Hole (below).........	5.9

Measurements on Crow Creek and Tongue River.

Date.	Hydrographer.	Stream.	Locality.	Dis-charge.
				Sec.-feet.
May 1...........	F. M. Brown	Crow Creek ..	Radersburg......	21
May 25.........	C. T. Prall.......	Harlem ditch...	12 miles east of Chinook.	36
May 28.........	W. B. Freeman ..	Crow Creek	Radersburg...:...	106
Do	.....do	.....do	.do.............	106
August 30......	J. H. Sloan	Crow Creek	.....do.............	13
August —......	R. G. Schnitger ..	Tongue River ..	P. K. cabins......	142
September 26...	J. H. Sloan	Crow Creek	Radersburg......	26

Measurements of ditches and tributaries from Piney Creek.

Date	Hydrographer.	Dis-charge.	Stream.
	-	*Sec.-feet.*	
ne 17	R. G. Schnitger	6. 6	Meyers ditch.
ine 20	.....do	20. 8	Coffeen ditch.
ine 22	.....do	16. 1	Little Piney Creek.[a]
ine 25	.....do	27. 4	Piney Divide ditch.
uly 9	.....do	87. 2	Do.
Do....	.....do	14. 5	Coffeen ditch.
August 9....	.....do	12. 6	Do.
Do......	.....do	12. 7	Piney Divide ditch.

[a] Little Piney Creek empties into Piney a short distance below gaging station.

PLATTE RIVER DRAINAGE BASIN.

Platte River, one of the largest tributaries of the Missouri, is formed by the junction of North Platte and South Platte rivers in Lincoln County, Nebr., and flows east into Missouri River 18 miles south of Omaha, Nebr. The principal tributaries of the Platte in Nebraska are Elkhorn and Loup rivers. Elkhorn River rises in the northern part of Nebraska and flows southeast into the Platte about 35 miles above its mouth. North Loup, Middle Loup, and South Loup rivers, with their tributaries, drain north-central Nebraska and join Platte River near Columbus, Nebr.

North Platte River rises in northern Colorado, flows north into Wyoming, and then east into Nebraska. Sweetwater and Laramie rivers are its principal tributaries. The Sweetwater joins it from the west in south-central Wyoming. Laramie River rises in the northern part of Colorado and flows north into the North Platte in eastern Wyoming. The Little Laramie is a small tributary of Laramie River in southern Wyoming.

South Platte River rises in central Colorado, flows north, then east to its junction with North Platte River. Its tributaries are mostly small creeks. Big Thompson Creek joins it 8 miles south of Greeley, Colo. St. Vrain Creek joins it about 15 miles south of Greeley. Clear Creek joins it about 3 miles north of Denver, and Bear Creek 8 miles south of Denver.

The following list includes the stations in the Platte River drainage basin:

Platte River at Southbend, Nebr.
Elkhorn River near Arlington, Nebr.
Elkhorn River at Norfolk, Nebr.
Loup River at Columbus, Nebr.
North Loup River near St. Paul, Nebr.
Middle Loup River near St. Paul, Nebr.

Miscellaneous measurements in the Missouri River, etc.—Continued.

Date.	Stream.	Locality.	Discharge.
			Sec.-feet.
May 13	Red Water	Bridge above Crow Creek	80 . 0
May 12	Red Water ditch	Red Water Valley at head	40 . 5
May 13	do	do	43 . 3
August 13	do	do	37 . 6
October 28	do	Bridge east of Belle Fourche	16 . 0
August 13	Belle Fourche power canal.	do	9 . 8
October 28	do	Red Water Valley	34 . 3
May 13	do	do	72 . 7
May 12	False Bottom Creek	do	7 . 4
August 14	Walter & McVey ditch	Spearfish Valley	7 . 7
Do	Cook & Burns ditch	do	11 . 0
August 15	Smith ditch	do	3 . 2
Do	Walter & Schuler ditch	do	2 . 4
August 14	Cook ditch	do	20 . 2
August 29	do	do	27 . 8
August 15	Eckles ditch	do	4 . 2
Do	Evans ditch	do	12 . 8
August 29	do	do	23 . 2
Do	Concord ditch	do	8 . 9
Do	Tindleys ditch	do	4 . 4
Do	Spring Branch ditch	do	6 . 6
Do	Brown ditch	do	6 . 4
August 26	Spring Creek	Hill City, S. Dak	7 . 5
June 9	do	Waugh ranch	32 . 6
November 2	do	Sink Hole (above)	8 . 8
Do	do	Sink Hole (below)	5 . 9

Measurements on Crow Creek and Tongue River.

Date.	Hydrographer.	Stream.	Locality.	Discharge.
				Sec.-feet.
May 1	F. M. Brown	Crow Creek	Radersburg	21
May 25	C. T. Prall	Harlem ditch	12 miles east of Chinook.	36
May 28	W. B. Freeman	Crow Creek	Radersburg	106
Do	do	do	do	106
August 30	J. H. Sloan	Crow Creek	do	13
August —	R. G. Schnitger	Tongue River	P. K. cabins	142
September 26	J. H. Sloan	Crow Creek	Radersburg	26

?4 inch stone set on end in the ground at a point just
> fence on the east side of the road leading to the bridge.
: south of the south end of the bridge, 20 feet north of a
willow trees, and 39 feet west of the second telephone pole
bridge, in the pasture to the east of the road. Its eleva-
feet above the zero of the gage. Bench mark No. 2 is the
orth window of the Burlington and Missouri River Rail-
. Its elevation is 10.47 feet above the zero of the gage.
: No. 3 is the southeast corner of the southeast corner
> masonry abutment at the north end of the old bridge.
i is 13.78 feet above the zero of the gage.
vations at this station during 1903 have been made under
i of J. C. Stevens, district hydrographer.
ving measurements were made by J. C. Stevens in 1903:

Gage height, 3.80 feet; discharge, 18,178 second-feet.
t 31: Gage height, 4.90 feet; discharge, 38.846 second-feet.
iber 14: Gage height, 2.30 feet; discharge, 11.986 second-feet.

rage height, in feet, of Platte River at Southbend, Nebr., for 1903.

Day.	Mar.	Apr.	May.	June.	July.	Aug.	Sept.
......		3.15	3.40	6.50	3.15	2.55	4.50
......		3.05	3.25	5.95	2.90	5.50	4.15
......		3.00	3.25	5.70	2.85	4.55	3.20
......		2.10	3.45	5.20	3.05	4.25	2.90
......		2.10	3.80	4.70	5.00	3.85	2.80
......		2.90	3.55	4.45	4.40	3.45	2.05
......		2.70	3.35	3.80	4.00	3.25	1.85
......		2.55	3.25	3.65	3.65	2.90	1.70
......		2.50	3.15	3.50	3.15	2.45	2.00
......		2.50	3.15	3.20	2.65	2.25	1.95
......		2.50	3.85	3.05	2.75	1.95	1.85
......		2.52	4.70	2.85	2.55	2.00	1.95
......		2.75	4.65	2.70	2.35	1.95	2.00
......		2.60	4.45	2.55	2.25	4.00	2.30
......		2.45	4.15	2.35	2.15	3.60	2.55
......		2.50	4.10	2.25	2.15	3.60	2.75
......		2.55	3.40	2.00	3.05	3.15	2.90
......		2.55	3.15	2.05	3.55	2.75	2.45
......		2.55	3.40	2.35	3.70	2.50	1.90
......		2.55	3.50	2.20	3.45	2.20	1.60
......		2.60	3.60	2.55	2.25	2.20	1.50
......		2.60	3.95	3.00	2.90	1.90	1.40
......		2.65	4.05	3.25	2.45	1.80	1.35
......		2.70	4.00	2.95	2.35	1.70	1.15
......	3.55	2.60	4.50	3.30	1.70	1.45	1.05
......	3.65	2.35	5.40	3.60	1.35	1.75	1.05
......	3.65	2.35	5.15	3.75	1.25	2.30	1.05
......	3.15	2.45	5.40	3.65	1.25	5.55	.95
......	3.25	2.95	5.60	3.60	1.95	5.45	.90
......	3.15	3.50	6.55	3.60	1.75	5.15	.90
......	3.45	......	6.60	......	2.60	4.90	(b)

.gings of Platte River and Salt, Wahoo, and Clear creeks 10 miles above South-
discontinued.

Platte River near Columbus, Nebr.
Platte River near Lexington, Nebr.
North Platte River at North Platte, Nebr.
North Platte River at Bridgeport, Nebr.
North Platte River near Mitchell, Nebr.
North Platte River at Guernsey, Wyo.
Laramie River at Uva, Wyo.
Little Laramie River at Haley's ranch, near Laramie, Wyo.
Little Laramie River near Hatton, Wyo.
Sweetwater River at Devils Gate, near Splitrock, Wyo
North Platte River at Saratoga, Wyo.
South Platte River at Bigspring, Nebr.
South Platte River near Julesburg, Colo.
Middle Crow Creek near Hecla, Wyo.
South Platte River at Kersey, Colo.
Cache la Poudre River near Greeley, Colo.
Big Thompson Creek near Arkins, Colo.
Handy ditch near Arkins, Colo.
St. Vrain Creek near Lyons, Colo.
Supply ditch at Lyons, Colo.
Clear Creek at Forkscreek, Colo.
South Platte River at Denver, Colo.
South Platte River at South Platte, Colo.

PLATTE RIVER AT SOUTHBEND, NEBR.

A temporary gage was established at this point March 25, 1903, by J. C. Stevens. A permanent gage was established April 21, 1903. It is located in the NE. ¼ sec. 13, T. 12 N., R. 10 E. It is one-fourth mile north of the Burlington and Missouri River Railroad station at Southbend, Nebr. The gaging section is 1 mile above the Chicago, Rock Island and Pacific Railroad bridge. The gage is an inclined yellow-pine rod set at an angle of 30° 26' to the horizontal and graduated to read directly to vertical half tenths. The rod is bolted at the water's edge to a large flat stone embedded in the bank. The upper end is fastened to the roots of a large elm tree by lag screws. During 1903 the observer was Mannie Bunker. It was expected to make discharge measurements from a footbridge, which, at the time of establishing the station, needed repairing. As these repairs have not been made, and as there is no other inexpensive means of making measurements, the station was discontinued September 30, 1903. Discharge measurements can be made from the Burlington and Missouri River Railroad bridge 10 miles upstream from the gage by adding the discharges of Salt, Wahoo, and Clear creeks, which empty into the river between the gage and the bridge. The channel is straight for 1,000 feet above and below the station. The water is never sluggish. The right bank is low and sandy and is not liable to overflow. The left bank is a rocky bluff, over 100 feet high and covered with timber. The bed is composed of shifting sand and gravel, free from vegetation. There is one channel at high stages and from three to ten channels at low stages. Bench mark No. 1 is

y 24 inch stone set on end in the ground at a point just
the fence on the east side of the road leading to the bridge.
eet south of the south end of the bridge, 20 feet north of a
six willow trees, and 39 feet west of the second telephone pole
he bridge, in the pasture to the east of the road. Its eleva-
)9 feet above the zero of the gage. Bench mark No. 2 is the
north window of the Burlington and Missouri River Rail-
on. Its elevation is 10.47 feet above the zero of the gage.
ark No. 3 is the southeast corner of the southeast corner
the masonry abutment at the north end of the old bridge.
ion is 13.78 feet above the zero of the gage.
servations at this station during 1903 have been made under
ion of J. C. Stevens, district hydrographer.
lowing measurements were made by J. C. Stevens in 1903:

5: Gage height, 3.80 feet; discharge, 13,178 second-feet.
rust 31: Gage height, 4.90 feet; discharge, 38.846 second-feet.
tember 14: Gage height, 2.30 feet; discharge, 11.986 second-feet.

'y gage height, in feet, of Platte River at Southbend, Nebr., for 1903.

Day.	Mar.	Apr.	May.	June.	July.	Aug.	Sept.
		3.15	3.40	6.50	3.15	2.55	4.50
		3.05	3.25	5.95	2.90	5.50	4.15
		3.00	3.25	5.70	2.85	4.55	3.20
		2.10	3.45	5.20	3.05	4.25	2.90
		2.10	3.80	4.70	5.00	3.85	2.30
		2.90	3.55	4.45	4.40	3.45	2.05
		2.70	3.35	3.80	4.00	3.25	1.85
		2.55	3.25	3.65	3.65	2.90	1.70
		2.50	3.15	3.50	3.15	2.45	2.00
		2.50	3.15	3.20	2.65	2.25	1.95
		2.50	3.85	3.05	2.75	1.95	1.85
		2.52	4.70	2.85	2.55	2.00	1.95
		2.75	4.65	2.70	2.35	1.95	2.00
		2.00	4.45	2.55	2.25	4.00	2.30
		2.45	4.15	2.35	2.15	3.60	2.55
		2.50	4.10	2.25	2.15	3.60	2.75
		2.55	3.40	2.00	3.05	3.15	2.90
		2.55	3.15	2.05	3.55	2.75	2.45
		2.55	3.40	2.35	3.70	2.50	1.90
		2.55	3.50	2.20	3.45	2.20	1.60
		2.60	3.60	2.55	2.25	2.20	1.50
		2.60	3.95	3.00	2.90	1.90	1.40
		2.65	4.05	3.25	2.45	1.80	1.35
		2.70	4.00	2.95	2.35	1.70	1.15
	3.55	2.60	4.50	3.30	1.70	1.45	1.05
	3.65	2.35	5.40	3.60	1.35	1.75	1.05
	3.65	2.35	5.15	3.75	1.25	2.30	1.05
	3.15	2.45	5.40	3.65	1.25	5.55	.95
	3.25	2.95	5.60	3.60	1.95	5.45	.90
	3.15	3.50	6.55	3.60	1.75	5.15	.90
	3.45		6.60		2.60	4.90	(b)

f gagings of Platte River and Salt, Wahoo, and Clear creeks 10 miles above South-
ion discontinued.

ELKHORN RIVER NEAR ARLINGTON, NEBR.

This station was established April 28, 1899, by Glenn E. Smith.
is located at the highway bridge 1 mile south of Arlington.
Northwestern Railroad bridge is just above the station. The o ˙
gage rod was fastened to cross-ties solidly embedded in the
bank. It consisted of a new oak stake 3 inches by 4 inches an
feet long, located 200 feet downstream from the bridge, on the
bank. This rod was washed out May 10, 1899, and on May 29 a
gage, consisting of a 2 by 6 inch oak rod 16 feet long, was set on
same bank 25 feet farther upstream. The gage datum was
changed. High-water gage heights are determined by m
down from a point on the bridge. During 1903 the observer
been Mike Hammang. Discharge measurements are made from
upstream side of the single span highway bridge. This bridge
oblique to the direction of the current. The upstream hand rail
the bridge is marked at intervals of 10 feet, and the initial point
soundings is the zero mark on the hand rail. The channel is st ˙
for 500 feet above and for 300 feet below the station. When the
road bridge was rebuilt in 1902 piles were left in the river bed at
bridge, which caused drift to collect at this point. This caused
channel at the highway bridge to scour, and the present irregu
section causes eddies on both sides of the river for one-third of
width. The channel is 180 feet wide at ordinary stages, width of piers
being deducted. Both banks are high and wooded, but overflow at
high water. The velocity is rapid and poorly distributed. The bed is
composed of sand and silt and is changeable. Bench mark No. 1 is a
large spike driven in a piling 25 feet upstream from the gage. Its elevation is 6.36 feet above the zero of the gage. Bench mark No. 2 is a
large spike driven in a piling 20 feet back and 15 feet downstream from
the gage. Its elevation is 9.12 feet above the zero of the gage. Bench
mark No. 3 is a vertical spike driven into a small leaning tree 30 feet
downstream from the gage. Its elevation is 7.43 feet above the zero
of the gage. This station was discontinued November 21, 1903.

The observations at this station during 1903 have been made under
the direction of J. C. Stevens, district hydrographer.

e measurements of Elkhorn River near Arlington, Nebr., in 1903.

Date.	Hydrographer.	Gage height.	Discharge.
		Feet.	*Second-feet.*
....	J. C. Stevens.........	4.40	2,120
....	.....do	7.43	4,526
....	.....do	7.29	4,134
....	.....do	10.80	8,431
....	.....do	5.40	2,136
....	.....do	4.65	2,093
....	.....do	10.65	8,658
6...............	E. C. Murphy	7.70	4,843

gage height, in feet, of Elkhorn River near Arlington, Nebr., for 1903.

r.	Mar.	Apr.	May.	June.	July.	Aug.	Sept.	Oct.	Nov.
....		6.11	3.44	10.25	4.74	3.35	9.98	3.22	3.15
....		5.87	4.00	10.40	4.43	4.15	9.30	3.43	3.28
....		5.70	4.37	10.33	4.23	5.54	8.18	4.87	3.50
....		5.48	4.80	10.05	5.39	7.87	7.50	4.50	3.86
....		5.26	4.96	9.76	7.29	8.40	6.98	3.75	4.10
....		5.08	4.85	9.47	7.82	8.10	6.52	3.42	3.96
....		4.87	......	9.37	7.91	7.37	5.96	3.27	3.78
....		4.71	4.45	9.25	7.07	6.34	5.42	3.75	3.60
....		4.50	4.25	9.17	6.25	5.74	5.10	4.88	3.55
....		4.40	4.12	9.05	5.85	5.28	4.94	4.86	3.44
....		4.38	4.50	8.64	5.60	4.90	4.85	4.50	3.43
....		4.40	6.41	8.15	5.33	4.68	5.20	4.50	3.34
....		4.50	7.61	7.84	5.12	4.94	6.00	4.62	3.23
....		4.37	7.89	7.48	4.93	4.87	6.42	4.80	3.25
....		4.10	7.36	7.08	4.64	5.05	7.62	4.90	3.22
....		3.91	6.64	6.38	4.30	4.65	7.08	4.88	3.20
....		3.76	6.26	5.84	5.46	4.68	7.12	4.88	3.00
....		3.60	6.05	5.50	7.31	4.72	6.06	4.64	1.70
....		3.45	7.56	5.33	6.98	4.80	5.52	4.35	1.87
....		3.37	7.19	5.20	6.61	4.54	5.38	4.12	2.20
....		3.31	7.67	5.19	6.55	4.47	5.15	3.90	2.54
....		3.30	8.55	5.40	4.86	4.30	5.00	3.72	....
....		3.18	9.45	5.17	4.35	4.00	4.62	3.57	....
....		3.12	8.50	5.43	4.05	3.72	4.20	3.43	....
....	7.53	3.05	8.00	5.97	3.80	3.72	3.85	3.35	....
....	7.16	2.92	8.18	5.68	3.55	....	3.65	3.30	....
....	7.02	2.80	8.85	5.40	3.38	9.65	3.43	3.26	....
....	6.93	2.82	8.40	5.22	3.22	10.42	3.32	3.12	....
....	6.85	2.88	8.25	5.06	3.25	10.65	3.22	3.12	....
....	6.72	3.18	9.20	4.87	3.24	10.63	3.20	3.18	....
....	6.53		10.05	....	3.16	10.23	....	3.20	....

Rating table for Elkhorn River near Arlington, Nebr., from January 1 to December 31, 1903.

Gage height.	Discharge.	Gage height.	Discharge.	Gage height.	Discharge.	Gage height.	Discharge.
Feet.	*Second-feet.*	*Feet.*	*Second-feet.*	*Feet.*	*Second-feet.*	*Feet.*	*Second-feet.*
0.4	238	2.6	1,037	5.4	2,915	8.2	5,480
.5	256	2.8	1,147	5.6	3,085	8.4	5,700
.6	276	3.0	1,260	5.8	3,255	8.6	5,930
.7	298	3.2	1,380	6.0	3,425	8.8	6,170
.8	322	3.4	1,500	6.2	3,595	9.0	6,410
.9	347	3.6	1,620	6.4	3,765	9.2	6,670
1.0	375	3.8	1,745	6.6	3,935	9.4	6,930
1.2	435	4.0	1,875	6.8	4,105	9.6	7,200
1.4	501	4.2	2,015	7.0	4,280	9.8	7,490
1.6	574	4.4	2,155	7.2	4,460	10.0	7,790
1.8	654	4.6	2,295	7.4	4,650	10.5	8,590
2.0	740	4.8	2,440	7.6	4,850	11.0	9,390
2.2	833	5.0	2,595	7.8	5,055		
2.4	932	5.2	2,755	8.0	5,265		

This table was applied indirectly according to the method outlined in Nineteenth Ann. Rept. U. S. Geol. Survey, pt. 4, pp. 323 et seq.

Estimated monthly discharge of Elkhorn River near Arlington, Nebr., for 1903.

[Drainage area, 5,980 square miles.]

Month.	Discharge in second-feet.				Run-off.	
	Maximum.	Minimum.	Mean.	Total in acre-feet.	Second-feet per square mile.	Depth in inches.
March 25–31	4,800	3,890	4,255	59,079	0.712	0.18
April	3,510	1,064	1,936	115,200	.324	.36
May (no record 7th)	7,665	1,410	4,027	239,623	.673	.75
June	8,510	1,945	4,319	256,998	.722	.80
July	4,650	1,178	2,516	154,703	.421	.48
August (no record 26th)	8,670	1,290	3,567	212,251	.596	.66
September	7,565	1,380	3,250	193,388	.543	.60
October	2,515	1,320	1,884	115,843	.315	.36
November 1–21	1,945	613	1,399	58,272	.284	.18

ELKHORN RIVER AT NORFOLK, NEBR.

This station was established July 16, 1896, by O. V. P. Stout. The
ge is located one-fourth mile downstream from the Thirteenth
·eet Bridge, from which discharge measurements are made. About
miles below the station the North Fork of the Elkhorn empties into
khorn River. The gage is an inclined 2 by 4 inch oak rod 12 feet
ig, securely embedded in the left bank at an angle of 30° to the hori-
ital. It is graduated to read directly to vertical half tenths of feet.
ıring 1903 the gage was read once each day by Harold E. Taft. The
irteenth Street Bridge consists of a single span resting on concrete
)ular piers. The downstream edge of the floor of the bridge is
.rked at intervals of 10 feet and the initial point for soundings is the
·o mark on the bridge floor. The channel is straight for 300 feet
ove and 200 feet below the station. It is 120 feet wide at ordi-
ry stages and 225 feet wide at high stages. The right bank is high,
ıdy, and wooded. The left bank is lower and liable to overflow at
;h stages. The bed is composed of sand and mud and is irregular
shape. Snags and brush obstruct the current, which is rapid and
orly distributed. The zero of the gage is 8.21 feet below a small
ike driven horizontally into a tree, near the root, about 20 feet back
d downstream from the gage and 3.96 feet below the head of a lag
rew which is placed vertically in the horizontal trunk of a large
·ing willow tree which overhangs the stream about 15 feet below
.e gage. Bench mark No. 3 is a standard 4-foot iron pipe of the
nited States Geological Survey, located 35 feet west and 7 feet north
f the top of the gage and 15.5 feet west of the ash tree on which
s bench mark No. 1. Its elevation is 10.70 feet above the zero of the
age. The station was discontinued November 21, 1903.

The o'·servations at this station during 1903 have been made under
he direction of J. C. Stevens, district hydrographer.

Discharge measurements of Elkhorn River at Norfolk, Nebr., in 1903.

Date.	Hydrographer.	Gage height.	Discharge.
		Feet.	*Second-feet.*
ıay 25	J. C. Stevens	3.97	1,167
'une 24	do	3.69	1,193
ıugust 26	do	3.13	784
eptember 18	E. C. Murphy	2.37	389

Mean daily gage height, in feet, of Elkhorn River at Norfolk, Nebr., for 1903

Day.	Mar.	Apr.	May.	June.	July.	Aug.	Sept.	Oct.	N
1		4.40	3.54	6.60	3.32	3.52	3.61	1.60	1
2		4.26	3.61	6.46	3.01	3.67	3.53	2.03	1
3		4.21	3.73	6.94	3.38	3.85	3.42	2.17	1
4		4.12	3.65	6.76	7.80	3.87	3.30	2.09	1
5		4.01	3.61	7.22	6.23	3.65	3.23	2.04	1
6		3.97	3.52	7.15	4.21	3.76	3.25	2.17	1
7		3.84	3.31	6.96	3.84	3.75	3.14	3.19	1
8		3.78	3.33	6.49	3.32	3.64	3.01	2.61	1
9	7.00	3.63	3.30	6.96	3.16	3.58	2.97	2.15	1
10	8.30	3.00	5.66	5.89	2.61	3.45	2.91	1.91	1
11	7.50	3.54	5.11	5.58	2.63	3.65	2.85	1.42	
12	5.30	3.51	5.08	5.18	2.21	3.85	2.98	1.62	
13	4.54	3.42	4.38	4.88	2.32	3.90	3.01	1.73	
14	5.00	3.31	3.95	4.53	2.51	4.02	3.00	1.64	
15	5.88	3.20	3.97	4.21	2.37	4.33	2.96	1.51	
16	6.80	3.17	4.07	4.05	2.43	3.97	2.84	1.56	
17	7.95	3.11	4.31	3.84	4.45	3.67	2.62	1.49	
18	7.75	3.04	4.01	3.62	3.71	3.72	2.42	1.47	
19	6.12	2.94	3.85	3.42	3.57	3.75	2.38	1.47	
20	5.87	2.91	3.77	3.21	3.21	3.80	2.27	1.47	
21	5.85	2.81	3.61	4.92	2.95	3.91	2.18	1.43	
22	5.82	2.71	3.61	4.43	2.86	3.69	2.16	1.41	
23	5.80	2.63	3.73	4.21	2.73	3.54	2.15	1.39	
24	5.86	2.59	3.95	3.51	2.65	3.43	2.15	1.35	
25	5.71	2.55	3.97	3.33	2.54	3.43	2.13	1.31	
26	5.54	2.45	3.97	3.20	2.61	3.62	2.10	1.29	
27	5.29	2.42	4.50	3.08	2.72	3.83	1.98	1.36	
28	5.09	2.65	4.43	3.00	2.83	5.00	1.85	1.34	
29	4.97	3.15	6.55	2.94	2.80	4.38	1.78	1.37	
30	4.76	3.35	9.25	2.83	2.97	4.13	1.61	1.33	
31	4.50		7.68		2.99	3.94		1.30	

Rating table for Elkhorn River at Norfolk, Nebr., from January 1 to December 31, 1903.[a]

Gage height.	Discharge.	Gage height.	Discharge.	Gage height.	Discharge.	Gage height.	Discharge
Feet.	Second-feet.	Feet.	Second-feet.	Feet.	Second-feet.	Feet.	Second-fee
0.8	140	2.2	430	3.6	1,085	5.0	1,985
1.0	160	2.4	490	3.8	1,205	5.2	2,135
1.2	190	2.6	570	4.0	1,325	5.4	2,300
1.4	230	2.8	665	4.2	1,445	5.6	2,480
1.6	270	3.0	765	4.4	1,565	5.8	2,660
1.8	310	3.2	865	4.6	1,705	6.0	2,840
2.0	370	3.4	965	4.8	1,845		

[a] This table was applied indirectly according to the method outlined in Nineteenth Ann. R U. S. Geol. Surv., pt. 4, pp. 323 et seq.

Estimated monthly discharge of Elkhorn River at Norfolk, Nebr., for 1903.

[Drainage area, 2,474 square miles.]

Month.	Discharge in second-feet.			Total in acre-feet.	Run-off.	
	Maximum.	Minimum.	Mean.		Second-feet per square mile.	Depth in inches.
March 9–31	4,500	1,570	2,707	123,493	1.094	0.936
April	1,500	430	856	50,936	.346	.386
May	8,000	810	1,610	98,995	.651	.751
June	3,860	720	1,998	118,889	.808	.901
July	5,000	440	1,007	61,918	.407	.469
August	1,920	940	1,177	72,371	.476	.549
September	1,010	210	539	32,073	.218	.243
October	710	180	248	15,249	.100	.115
November 1–16 a	320	180	229	7,267	.093	.055

a Ice for days not included.

LOUP RIVER AT COLUMBUS, NEBR.

·Observations at this station were begun October 13, 1894. The station was established by O. V. P. Stout. It is located about 75 yards above the Union Pacific Railroad bridge and about 6 miles above the mouth of the river. There is an island about 1,000 feet above the gaging section. The gage is a vertical rod 12 feet long bolted to a pile 160 feet above the railroad bridge. The observer is David J. Mowery. Discharge measurements are made by means of a ⅝-inch cable and car. The cable is fastened to a large tree, and is supported by a post on the west bank. On the east bank it is anchored to a timber set in the ground. Distances from the initial point are marked in red paint on the cable. The initial point for soundings is the zero mark on the cable. The channel is straight for 1,000 feet above and 300 feet below the cable. At a point 150 feet below the cable the current is broken by old pilings and masonry piers of a railroad bridge. The section at the gage is broad and shallow, with rapid velocity and a shifting, sandy bed. The right bank is low, sandy, covered with willows and brush, and is liable to overflow at very high stages. The left bank is about 10 feet high and is not liable to overflow. The bench mark is a standard 4-foot iron post of the Geological Survey, located 72 feet east of the gage. Its elevation is 13.27 feet above the zero of the gage. On account of the constantly changing cross section a large number of discharge measurements are necessary to obtain the daily discharge with a fair degree of accuracy.

The observations at this station during 1903 have been made under the direction of J. C. Stevens, district hydrographer.

Discharge measurements of Loup River at Columbus, Nebr., in 1903.

Date.	Hydrographer.	Gage height.	Discharge.
		Feet.	*Second-feet.*
March 20	J. C. Stevens	5.10	6,074
April 8	do	5.00	3,905
May 7	do	5.10	3,400
June 11	do	5.13	3,432
June 27	do	5.05	2,799
July 10	do	4.67	3,022
July 17	do	5.56	5,022
August 1	do	6.78	14,580
August 18	do	5.13	4,177
September 19	E. C. Murphy	4.49	2,644
September 29	J. C. Stevens	4.40	2,396
October 30	do	4.61	2,607
December 20	do	6.50	2,280

Mean daily gage height, in feet, of Loup River at Columbus, Nebr., for 1903.

Day.	Mar.	Apr.	May.	June.	July.	Aug.	Sept.	Oct.	Nov.	Dec.
1		5.00	5.20	5.80	5.00	6.72	4.90	4.40	4.60	4.95
2		4.90	5.20	5.60	4.80	6.60	4.80	4.45	4.60	4.95
3		5.10	5.20	5.50	4.90	5.60	4.70	4.45	4.70	5.00
4		5.00	5.20	5.45	7.00	5.30	4.60	4.50	4.70	5.00
5		5.00	5.10	5.40	6.60	5.00	4.50	4.50	4.70	4.95
6		4.95	5.10	5.30	5.65	4.90	4.50	4.55	4.70	
7		5.00	5.10	5.30	5.00	4.80	4.90	4.55	4.70	
8	a9.03	5.00	5.15	5.30	4.95	4.70	4.80	4.55	4.70	
9		5.10	5.15	5.30	4.70	4.70	4.70	4.50	4.70	
10		5.20	5.15	5.25	4.70	4.70	4.70	4.50	4.65	
11		5.20	5.40	5.12	4.70	4.90	4.75	4.60	4.65	
12		5.10	5.90	5.00	4.50	5.10	4.75	4.60	4.65	
13	5.45	5.00	5.75	4.95	4.50	5.60	4.75	4.75	4.65	
14	5.35	5.20	5.50	4.85	4.55	6.40	4.90	4.95	4.70	
15	5.20	5.25	5.25	4.85	5.55	6.00	4.80	4.90	4.70	
16	5.20	5.10	5.25	4.85	5.95	5.10	4.70	4.80	4.80	
17	5.25	5.10	5.25	4.85	5.56	5.15	4.60	4.80	5.00	
18	5.22	5.20	5.10	4.75	5.25	5.00	4.50	4.70	5.20	
19	5.25	5.20	5.10	4.70	5.10	4.90	4.53	4.75	5.30	
20	5.10	5.30	4.95	4.70	5.05	4.70	4.50	4.75	5.30	
21	4.90	5.30	5.00	4.70	4.90	4.60	4.50	4.70	5.40	
22	4.78	5.30	5.10	4.80	4.80	4.50	4.40	4.70	5.45	
23	4.80	5.10	5.15	4.80	4.70	4.50	4.40	4.70	5.45	
24	4.90	5.20	5.55	4.80	4.60	4.50	4.40	4.68	5.40	
25	4.80	5.10	5.40	4.95	4.60	4.60	4.40	4.65	5.50	
26	4.80	5.05	5.35	5.00	4.60	4.80	4.40	4.65	5.15	
27	4.80	5.00	5.15	5.05	4.60	7.20	4.40	4.63	5.00	
28	4.85	5.00	5.40	5.05	4.60	6.80	4.40	4.62	4.90	
29	5.12	5.00	5.65	5.00	4.70	5.80	4.40	4.65	4.80	
30	5.15	5.00	5.70	4.95	5.60	5.50	4.40	4.61	4.70	
31	5.20		5.80		5.40	5.00		4.60		

a From ice gorge.

Rating table for Loup River at Columbus, Nebr., from January 1 to December 31, 1903.a

Gage height.	Discharge.	Gage height.	Discharge.	Gage height.	Discharge.	Gage height.	Discharge.
Feet.	Second-feet.	Feet.	Second-feet.	Feet.	Second-feet.	Feet.	Second-feet.
3.8	1,140	4.6	2,040	5.4	4,300	6.2	8,750
3.9	1,180	4.7	2,260	5.5	4,740	6.3	9,750
4.0	1,220	4.8	2,500	5.6	5,180	6.4	10,750
4.1	1,300	4.9	2,740	5.7	5,620	6.6	12,750
4.2	1,400	5.0	3,000	5.8	6,060	6.8	14,750
4.3	1,510	5.1	3,260	5.9	6,500	7.0	16,750
4.4	1,660	5.2	3,550	6.0	7,050		
4.5	1,820	5.3	3,890	6.1	7,800		

aThis table was applied indirectly according to the method outlined in Nineteenth Ann. Rept. U. S. Geol. Surv., pt. 4, p. 328 et seq.

Estimated monthly discharge of Loup River at Columbus, Nebr., for 1903.

[Drainage area, 13,542 square miles.]

Month.	Discharge in second-feet.			Total in acre-feet.	Run-off.	
	Maximum.	Minimum.	Mean.		Second-feet per square mile.	Depth in inches.
March 13–31 a	8,250	4,090	5,554	209,308	0.410	0.290
April	4,740	3,400	4,027	239,623	.297	.331
May	6,500	2,870	4,085	251,177	.301	.347
June	6,060	2,150	3,188	189,699	.235	.262
July (4th missing)	12,750	2,040	3,549	211,180	.262	.292
August	20,000	2,260	5,873	330,373	.397	.458
September	3,550	2,380	2,763	164,410	.204	.228
October	3,710	2,380	2,796	171,919	.206	.237
November 1–28	5,840	2,620	3,665	203,544	.271	.282

a Rod washed out March 8-13 by back water from ice gorge.

NORTH LOUP RIVER NEAR ST. PAUL, NEBR.

This station was established May 5, 1895, by O. V. P. Stout. The original station was located at an old bridge about 800 feet above the present station. The present station was established April 15, 1903, by J. C. Stevens. It is located in sec. 22, T. 15 N., R. 10 W. of the sixth principal meridian. The gage consists of a 4 by 4 inch yellow-pine rod 14 feet long, fastened to two 4-foot upright posts at an angle of 32° 57′ to the horizontal, and graduated to read directly to vertical half tenths. The lower end of the rod is driven firmly into the bed of the stream. The gage is located on the right bank about one-half mile below the highway bridge. It is read once each day by C. Schack. Discharge measurements are made from the new highway bridge 3 miles *north of St. Paul.* The bridge consists of six

100-foot spans, resting on tubular concrete piers. The upstream hand rail is marked at intervals of 10 feet, and the initial point for soundings is the zero mark over the center of the first pier on the right bank. The channel is straight for 500 feet above and 1,000 feet below the station. The water is never sluggish. Both banks are covered with willows and trees and are not liable to overflow. The bed of the stream is of shifting sand free from vegetation. The channel is 250 feet wide at ordinary stages. The direction of the current at the bridge is variable, sometimes making an angle of 45° with the bridge. Bench mark No. 1 is the sill of the north window of the brick house belonging to C. Schack. Its elevation is 44.90 feet above the zero of the gage. Bench mark No. 2 is a notch cut in an ash tree 30 feet south of the gage rod. Its elevation is 6.47 feet above the zero of the gage. This station was discontinued November 30, 1903.

The observations at this station during 1903 have been made under the direction of J. C. Stevens, district hydrographer.

Discharge measurements of North Loup River near St. Paul, Nebr., in 1903.

Date.	Hydrographer.	Gage height.	Discharge.
		Feet.	*Second-feet.*
April 15	J. C. Stevens	1.44	1,042
May 14	do	1.70	1,222
June 2	do	1.94	1,843
July 16	do	1.65	1,017
August 11	do	1.44	965
September 19	do	1.49	807
November 28	do	1.60	1,784

Mean daily gage height, in feet, of North Loup River near St. Paul, Nebr., for 1903

Day.	Apr.	May.	June.	July.	Aug.	Sept.	Oct.	Nov.	Dec.
1		1.80	1.85	1.50	2.55	1.45	1.40	1.60	1.30
2		1.70	1.90	1.45	2.00	1.40	1.45	1.65	1.25
3		1.65	1.85	2.50	(a)	1.80	1.60	1.60	1.40
4		1.65	1.80	1.90	(a)	1.35	1.40	1.60	1.30
5		1.80	1.80	1.75	(a)	1.40	1.45	1.55	
6		1.65	1.75	1.55	(a)	1.60	1.45	1.50	
7		1.65	1.75	1.45	(a)	1.50	1.80	1.45	
8		1.65	1.70	1.40	(a)	1.60	1.55	1.60	
9		1.65	1.70	1.35	(a)	1.55	1.50	1.65	
10		2.43	1.70	1.35	(a)	1.50	1.40	1.65	
11		1.95	1.70	1.40	1.45	1.50	1.50	1.60	
12		1.90	1.65	1.45	1.50	1.60	1.50	1.60	
13		1.80	1.65	1.40	1.85	1.50	1.55	1.55	
14		1.70	1.65	1.85	1.65	1.50	1.65	1.55	
15	1.45	1.70	1.60	1.65	1.50	1.55	1.60	1.60	
16	1.45	1.70	1.55	1.65	1.45	1.55	1.55	1.55	
17	1.45	1.70	1.55	1.65	1.45	1.45	1.45	.70	

a Sand on gage rod.

Mean daily gage height, in feet, of North Loup River, etc.—Continued.

Day.	Apr.	May.	June.	July.	Aug.	Sept.	Oct.	Nov.	Dec.
18	1.50	1.65	1.50	1.60	1.50	1.50	1.40	0.80	.....
19	1.55	1.65	1.55	1.75	1.40	1.50	1.45	1.05	.....
20	1.55	1.60	1.50	1.60	1.50	1.55	1.45	1.30	.....
21	1.55	1.55	1.55	1.70	1.40	1.45	1.45	1.70	.....
22	1.60	1.55	1.60	1.60	1.85	1.40	1.50	1.80	.....
23	1.60	1.85	1.60	1.55	1.85	1.40	1.50	1.85	.....
24	1.60	1.60	1.60	1.50	2.50	1.35	1.50	1.60	.....
25	1.65	1.60	1.65	1.50	2.10	1.35	1.55	1.55	.....
26	1.55	1.80	1.65	1.45	2.65	1.35	1.50	1.55	.....
27	1.50	1.70	1.55	1.45	1.90	1.30	1.55	1.50	.....
28	1.60	1.60	1.50	1.40	1.60	1.35	1.60	1.60	.....
29	1.90	1.80	1.50	2.68	1.50	1.40	1.55	1.35	.....
30	1.65	2.00	1.45	.....	1.50	1.50	1.55	1.35	.....
31	.....	1.95	.....	.....	1.48	.....	1.60	.....	.....

Rating table for North Loup River near St. Paul, Nebr., from January 1 to December 31, 1903.

Gage height.	Discharge.	Gage height.	Discharge.	Gage height.	Discharge.	Gage height.	Discharge.
Feet.	*Second-feet.*	*Feet.*	*Second-feet.*	*Feet.*	*Second-feet.*	*Feet.*	*Second-feet.*
1.0	430	1.5	810	2.0	1,980	2.5	3,560
1.1	500	1.6	980	2.1	2,270	2.6	3,900
1.2	540	1.7	1,180	2.2	2,570	2.7	4,240
1.3	610	1.8	1,420	2.3	2,890		
1.4	700	1.9	1,700	2.4	3,220		

This table was applied indirectly, according to the method outlined in Nineteenth Ann. Rept. U. S. Geol. Survey, pt. 4, p. 323.

Estimated monthly discharge of North Loup River near St. Paul, Nebr., for 1903.

[Drainage area, 4,024 square miles.]

Month.	Discharge in second-feet.			Total in acre-feet.	Run-off.	
	Maximum.	Minimum.	Mean.		Second-feet per square mile.	Depth in inches.
April 15–30	1,980	1,070	1,273	40,399	0.316	0.188
May [a]	1,980	890	1,276	75,927	.317	.354
June	1,700	750	1,089	64,800	.271	.302
July 1–29 [b]	4,500	650	1,170	67,299	.291	.314
August 1–2 and 11–31 [b]	4,410	810	1,631	74,406	.405	.346
September	1,180	610	835	49,686	.208	.232
October	1,560	700	997	61,303	.248	.286
November	2,270	400	1,316	78,307	.327	.365

[a] Discharge for May 10 missing.   [b] Sand at gage rod on days not included.

MIDDLE LOUP RIVER NEAR ST. PAUL, NEBR.

This station was originally established May 5, 1895, by Glenn E. Smith. The present gage was installed April 16, 1903, by J. C. Stevens. The station is located in sec. 10, T. 14 N., R. 10 W., at a railroad and highway bridge which is crossed by the Union Pacific Railroad 1 mile south of St. Paul. The gage consists of a 2 by 6 inch vertical pine rod 8 feet long spiked to the last piling downstream in the south row of the bent next to the south abutment. The gage can be read from the bridge. Discharge measurements are made from the upstream side of the railroad highway bridge, the hand rail of which is marked at intervals of 10 feet. The initial point for soundings is the zero mark on the hand rail over the south abutment. The channel bends to the south about 50 feet above the bridge and is straight for 1,000 feet below. It has a width of 740 feet between bridge abutments broken by 5 piers and several rows of short piles of an old bridge. The section is broad and shallow, with a rapid velocity and shifting sandy bed. Both banks are wooded and are not liable to overflow. Bench mark No. 1 is the top of the piling to which the gage rod is fastened. Its elevation is 8.15 feet above the zero of the gage. Bench mark No. 2 is a standard 4-foot United States Geological Survey iron post located 60 feet south of the south end of the bridge and 100 feet east of the end of the downstream hand rail. Its elevation is 7.54 feet above the zero of the gage. This station was discontinued November 21, 1903.

The observations at this station during 1903 have been made under the direction of J. C. Stevens, district hydrographer.

Discharge measurements of Middle Loup River near St. Paul, Nebr., in 1903.

Date.	Hydrographer.	Gage height.	Discharge.
		Feet.	*Second-feet.*
April 7	J. C. Stevens		2,18
April 16	do	2.43	1,36
May 14	do	2.30	1,70
June 2	do	2.75	2,77
July 16	do	2.67	1,73
August 11	do	2.68	1,70
September 21	E. C. Murphy	2.41	1,17

Mean daily gage height, in feet, of Middle Loup River near St. Paul, Nebr., for 1903.

Day.	Apr.	May.	June.	July.	Aug.	Sept.	Oct.	Nov.
1		2.56	2.85	2.55	3.50	2.50	2.40	2.35
2		2.55	2.80	2.55	3.50	2.40	2.53	2.55
3		2.55	2.70	2.90	3.00	2.40	2.45	2.55
4		2.48	2.60	3.88	2.65	2.40	2.50	2.30
5		2.48	2.70	3.05	2.45	2.40	2.55	2.25
6		2.50	2.60	2.70	2.30	2.50	2.00	2.20
7		2.50	2.65	2.55	2.20	2.40	2.60	2.20
8		2.55	2.62	2.45	2.20	2.50	2.60	2.20
9		2.48	2.60	2.40	2.45	2.50	2.50	2.20
10		2.45	2.55	2.45	2.40	2.55	2.40	2.20
11		2.80	2.55	2.45	2.65	2.50	2.25	2.20
12		2.74	2.55	2.35	2.82	2.40	2.25	2.20
13		2.64	2.50	2.35	3.15	2.55	2.65	2.25
14		2.30	2.50	2.72	2.87	2.60	2.50	2.25
15		2.15	2.50	2.77	2.70	2.55	2.45	2.20
16	2.55	2.85	2.55	2.65	2.75	2.50	2.40	2.30
17	2.37	2.85	2.55	2.75	2.95	2.50	2.30	2.30
18	2.37	2.35	2.50	2.65	2.60	2.50	2.25	1.70
19	2.40	2.50	2.50	2.55	2.70	2.45	2.25	1.90
20	2.58	2.40	2.50	2.55	2.50	2.40	2.20	2.00
21	2.70	2.55	2.45	2.45	2.40	2.40	2.20	2.10
22	2.51	2.40	2.58	2.40	2.40	2.40	2.20	....
23	2.50	2.90	2.50	2.40	2.40	2.45	2.20	....
24	2.50	2.87	2.72	2.40	2.40	2.45	2.20	....
25	2.50	2.75	2.98	2.35	3.35	2.40	2.25	....
26	2.50	2.85	2.80	2.80	3.75	2.40	2.25	....
27	2.40	2.97	2.75	2.30	3.85	2.45	2.30	....
28	2.42	2.70	2.65	2.30	3.60	2.40	2.20	....
29	2.69	3.04	2.65	2.45	3.20	2.45	2.20	....
30	2.58	3.98	2.65	2.77	2.90	2.45	2.25	....
31		3.80	....	2.70	2.55	....	2.25	....

Rating table for Middle Loup River near St. Paul, Nebr., from January 1 to December 31, 1903.

Gage height.	Discharge.	Gage height.	Discharge.	Gage height.	Discharge.	Gage height.	Discharge.
Feet.	Second-feet.	Feet.	Second-feet.	Feet.	Second-feet.	Feet.	Second-feet.
1.7	845	2.3	1,100	2.9	3,500	3.5	8,300
1.8	855	2.4	1,250	3.0	4,300	3.6	9,100
1.9	870	2.5	1,430	3.1	5,100	3.7	10,000
2.0	890	2.6	1,670	3.2	5,900	3.8	11,000
2.1	930	2.7	1,970	3.3	6,700	3.9	12,000
2.2	1,000	2.8	2,700	3.4	7,500	4.0	13,000

This table was applied indirectly according to the method outlined in Nineteenth Ann. Rept. U. S. Geol. Survey, pt. 4, p. 323.

Estimated monthly discharge of Middle Loup River near St. Paul, Nebr., for 1903.

[Drainage area, 6,849 square miles.]

Month.	Discharge in second-feet.			Total in acre-feet.	Run-off.	
	Maximum.	Minimum.	Mean.		Second-feet per square mile.	Depth in inches.
April 16–30	3,150	1,160	1,719	51,144	0.251	0.140
May	12,400	1,250	3,209	197,314	.469	.541
June	4,950	1,340	1,938	115,319	.283	.316
July	12,000	1,100	2,028	124,697	.296	.341
August	11,500	1,000	3,365	206,906	.491	.566
September	1,670	1,160	1,330	79,141	.194	.216
October	1,820	1,000	1,234	75,876	.180	.208
November 1–21	1,550	845	1,059	44,110	.155	.121

PLATTE RIVER NEAR COLUMBUS, NEBR.

This station was established June 4, 1895, by O. V. P. Stout, at the Meridian Bridge, 2 miles south of Columbus, Nebr. The gage is an inclined oak timber, fastened to cross-ties embedded in the left bank about 75 feet upstream from Meridian Bridge on the main channel. The rod is graduated to read directly to vertical half-tenths of feet. The river at this point flows in three channels, known as the main, middle, and south channels, having widths of 1,940 feet, 320 feet, and 75 feet, respectively. Each channel is spanned by a pile bridge, from the upstream side of which discharge measurements are made. The main channel is crossed by the Meridian Bridge, which consists of sixty-five 30-foot spans and above which the only gage rod is located. The middle channel bridge consists of sixteen 20-foot spans, and is located 1⅛ miles south of the Meridian Bridge. The south channel is spanned by a pile bridge about 80 feet long, located one-eighth mile south of the middle channel bridge. Discharge measurements are made from the upstream side of all three bridges, whose hand rails are marked at intervals of 10 or 20 feet. The initial point for soundings is the zero mark on the upstream hand rail at the north end of the main channel bridge, and at the south ends of the middle and south channel bridges. Above the bridges the channels are straight for 5,000 feet in the main channel, 200 feet in the middle channel, and 100 feet in the south channel. Below the bridges the channels are straight for 3,000 feet in the main channel, 300 feet in the middle channel, and 500 feet in the south channel. The sections are broad and shallow, with rapid velocity and shifting, sandy bed. At low stages the river flows in many shallow channels, so that the measurement of the discharge is mainly a matter of estimation. The bed is so changeable that the daily discharges are not proportional to the corresponding gage heights. The river is usually dry from August 15 to October 15. The bench mark is a standard iron post of the Geological Survey, located 44.5 feet east of the gage, 60 feet north of the north end of the north bridge truss, and 10 feet west of a cottonwood tree 6 inches in diameter. Its elevation is 7.06 feet above the zero of the gage.

The observations at this station during 1903 have been made under the direction of J. C. Stevens, district hydrographer.

Discharge measurements of Platte River near Columbus, Nebr., in 1903.

Date.	Hydrographer.	Gage height.	Discharge.
		Feet.	*Second-feet.*
April 3	J. C. Stevens	3.28	5,912
May 7	....do	3.54	5,420
June 11	....do	3.55	4,984
June 27	....do	4.64	13,133
July 10	....do	3.96	7,198
July 17	....do	4.10	7,524
August 1	....do	3.87	5,864
August 18	....do	3.62	2,898
September 29	....do	2.00	126
October 30	....do	2.82	878

Mean daily gage height, in feet, of Platte River near Columbus, Nebr., for 1903.

Day.	Mar.	Apr.	May.	June.	July.	Aug.	Sept.	Oct.	Nov.	Dec.
1		3.55	3.70	5.00	3.75	3.94	4.00	2.00	3.10	3.10
2		3.40	3.45	4.75	3.70	3.95	4.10	2.10	3.20	3.10
3		3.30	4.00	4.65	3.70	3.95	4.00	2.10	3.30	3.20
4		3.25	3.90	4.45	4.00	3.95	3.90	2.10	3.30	3.20
5		3.30	3.90	4.25	4.60	3.80	3.60	2.10	3.31	3.20
6		2.85	3.80	4.10	4.65	3.70	3.40	2.10	3.35	
7		2.90	3.55	3.90	4.40	3.50	3.00	2.15	3.40	
8		3.00	3.75	3.70	4.10	3.40	2.95	2.15	3.40	
9	4.11	3.10	4.10	3.60	4.00	3.41	2.90	2.20	3.30	
10		3.20	4.10	3.55	3.96	3.10	2.80	2.80	3.20	
11		3.00	4.35	3.55	3.96	3.00	2.80	2.30	3.25	
12		2.65	4.60	3.45	3.70	3.10	2.80	2.30	3.20	
13		2.55	4.60	3.40	3.65	3.35	2.90	2.35	3.00	
14		2.50	4.40	3.35	3.65	3.40	3.00	2.40	2.90	
15	3.80	2.90	4.35	3.60	3.60	3.80	2.90	2.50	2.90	
16	3.85	3.10	4.15	3.60	3.60	3.45	2.80	2.70	2.90	
17	3.85	3.10	4.10	3.70	4.12	3.65	2.75	2.75	2.80	
18	3.95	3.05	3.85	3.75	4.10	3.60	2.75	2.85	2.70	
19	3.90	3.20	3.55	3.80	4.05	3.60	2.72	2.80	2.60	
20	4.00	3.20	3.25	3.85	4.00	3.55	2.50	2.80	2.55	
21	4.10	3.25	3.10	4.05	3.85	3.40	2.30	2.85	2.55	
22	3.80	3.25	3.20	4.20	3.70	3.20	2.20	2.85	2.60	
23	3.85	3.35	3.25	4.15	3.65	3.10	2.10	2.85	2.60	
24	3.70	3.05	3.30	4.15	3.50	3.10	2.10	2.85	2.70	

Mean daily gage height, in feet, of Platte River, etc.—Continued.

Day.	Mar.	Apr.	May.	June.	July.	Aug.	Sept.	Oct.	Nov.	D
25	3.65	3.05	3.80	4.30	3.30	3.20	2.00	2.82	2.85	
26	3.55	3.05	4.00	4.50	3.20	3.40	2.00	2.80	2.90	
27	3.50	3.10	4.10	4.66	3.20	3.80	2.00	2.80	2.95	
28	3.45	3.00	4.25	4.45	3.20	3.90	2.00	2.80	3.00	
29	3.50	3.00	4.55	4.30	3.10	4.30	2.00	2.80	3.00	
30	3.80	3.40	4.75	4.00	3.20	4.20	2.00	2.83	3.05	
31	3.55	-----	5.00	-----	3.30	4.00	-----	2.80		

Rating table for Platte River near Columbus, Nebr., from January 1 to Decem 31, 1903.[a]

Gage height.	Discharge.	Gage height.	Discharge.	Gage height.	Discharge.	Gage height.	Discharge
Feet.	*Second-feet.*	*Feet.*	*Second-feet.*	*Feet.*	*Second-feet.*	*Feet.*	*Second-fe*
1.0	190	2.0	820	3.0	3,100	4.0	9,150
1.1	220	2.1	930	3.1	3,500	4.2	11,250
1.2	250	2.2	1,040	3.2	4,000	4.4	13,550
1.3	285	2.3	1,155	3.3	4,500	4.6	16,400
1.4	340	2.4	1,285	3.4	5,000	4.8	19,700
1.5	400	2.5	1,415	3.5	5,500	5.0	23,700
1.6	475	2.6	1,680	3.6	6,050	5.2	28,400
1.7	555	2.7	1,980	3.7	6,700	5.4	33,700
1.8	635	2.8	2,320	3.8	7,400	5.5	36,500
1.9	725	2.9	2,700	3.9	8,200		

[a] This table was applied indirectly, according to the method outlined in Nineteenth Ann. U. S. Geol. Survey, pt. 4, p. 323 et seq.

Estimated monthly discharge of Platte River near Columbus, Nebr., for 19

[Drainage area, 56,867 square miles.]

Month.	Discharge in second-feet.			Total in acre-feet.	Run-off.	
	Maximum.	Minimum.	Mean.		Second-feet per square mile.	De inc
March 15–31[a]	13,550	7,050	9,652	325,456	0.170	
April	7,800	1,980	4,339	258,188	.076	
May	21,600	3,300	9,159	563,165	.161	
June	21,600	4,000	8,816	524,588	.155	
July	13,550	2,320	5,847	359,518	.103	
August	6,375	1,350	3,507	215,637	.062	
September	4,750	190	1,175	69,917	.021	
October	1,095	190	613	37,692	.011	
November	3,100	1,095	1,926	114,605	.034	

[a] Frozen for days not included.

PLATTE RIVER NEAR LEXINGTON, NEBR.

This station was established April 2, 1902, by H. O. Smith. It is located in sec. 20, T. 9 N., R. 21 W., at the highway bridge 3 miles south of Lexington, Nebr. The boxed wire gage is fastened to the upstream hand rail of the bridge, about 50 feet from the north end. The marker is the end of the ring on the wire, 15.04 feet from the end of the weight. The gage is read once each day by Charles J. Freeman. Discharge measurements are made from the upstream side of the highway bridge, which consists of one hundred and eighty-seven 20-foot spans supported by pile piers. A second smaller channel, about one-fourth mile south, is measured from a similar pile bridge having a total span of 60 feet. The upstream hand rail at the main bridge is marked at 20-foot intervals. The initial point for soundings is the zero mark on the hand rail. The channel is straight for 1 mile above and below the station. The section is broad and shallow, with a rapid velocity and shifting, sandy bed. At low stages the river flows in as many as 40 channels of varying widths and depths. It is impossible to measure the discharge at such stages with a fair degree of accuracy. The river usually goes dry in August. Both banks are low, but are not subject to overflow. Bench mark No. 1 is a hub on the west side of the bridge approach. Its elevation is 6.25 feet above gage datum and 2,392 feet above sea level. Bench mark No. 2 is the top of the east end of the first cap at the north end of the bridge. Its elevation is 7.66 feet above gage datum.

The observations at this station during 1903 have been made under the direction of J. C. Stevens, district hydrographer.

Discharge measurements of Platte River near Lexington, Nebr., in 1903.

Date.	Hydrographer.	Gage height.	Discharge.
		Feet.	*Second-feet.*
May 8	J. C. Stevens	3.35	7,437
June 12	...do	3.36	4,749
June 28	...do	3.62	9,645
July 9	...do	3.25	5,188
July 23	...do	2.77	2,075
August 15	...do	2.88	1,340
September 22	W. C. Sturdevant	2.60	296
September 30	J. C. Stevens	2.52	250
November 1	...do	2.80	1,218

Mean daily gage height, in feet, of Platte River near Lexington, Nebr., for 1903.

Day.	Jan.	Feb.	Mar.	Apr.	May.	June.	July.	Aug.	Sept.	Oct.	Nov.	Dec.
1	3.05	3.80	4.00	2.90	3.35	3.20	3.45	2.95	2.80	2.55	2.15	14
2	3.00	3.80	4.00	2.75	2.80	3.30	3.55	2.65	2.90	2.70	2.10	12
3	3.15	3.75	4.00	2.70	2.85	3.20	3.55	2.90	2.80	2.75	2.15	12
4	3.20	3.65	4.05	3.00	3.05	3.25	3.40	2.80	2.65	2.75	2.10	15
5	3.20	3.75	4.05	2.90	3.10	3.20	3.50	2.65	2.50	2.80	2.15	14
6	3.25	3.75	4.05	2.75	3.05	3.15	3.40	2.60	2.45	2.85	2.10	15
7	3.15	3.70	4.10	2.75	3.35	3.10	3.35	2.55	2.50	2.75	2.15	16
8	3.25	3.65	4.15	2.75	3.35	3.10	3.25	2.60	2.40	2.75	2.10	14
9	3.35	3.65	4.30	2.80	3.35	3.10	3.10	2.60	2.50	2.75	2.15	14
10	3.35	3.65	4.55	2.90	3.10	3.00	2.90	2.55	2.40	2.90	2.15	14
11	3.40	3.70	4.65	3.00	3.05	3.15	2.90	2.70	2.55	2.80	2.10	15
12	3.40	3.75	4.65	2.70	2.75	3.35	3.00	2.75	2.50	2.70	2.15	15
13	3.50	3.75	4.75	2.75	3.05	3.25	3.00	2.80	2.35	2.60	2.00	
14	3.50	3.80	3.80	2.90	3.15	3.20	3.05	2.65	2.40	2.60	2.05	
15	3.45	3.80	4.90	2.90	3.10	3.20	3.10	2.90	2.40	2.75	2.00	
16	3.50	3.80	4.80	2.90	3.10	3.25	3.00	2.95	2.45	2.75	2.75	
17	3.50	3.80	4.20	3.00	3.35	3.50	3.05	2.90	2.45	2.75	2.55	
18	3.55	3.85	4.50	3.05	3.15	3.40	2.90	2.85	2.50	2.80	2.00	
19	3.60	3.80	4.10	2.80	3.15	3.45	2.85	2.80	2.55	2.80		
20	3.60	3.80	3.50	2.85	3.00	3.55	2.95	2.80	2.55	2.80	2.55	
21	3.65	3.80	3.65	2.95	3.40	3.65	2.85	2.75	2.60	2.70	2.50	
22	3.65	3.80	3.50	2.95	3.35	3.60	2.85	2.70	2.60	2.75	2.50	
23	3.70	3.85	3.10	3.05	3.15	3.65	2.60	2.60	2.55	2.80	2.70	
24	3.65	3.85	3.15	2.95	3.10	3.85	2.80	2.55	2.55	2.80	2.75	
25	3.65	3.90	3.00	2.90	3.05	3.80	2.75	2.50	2.50	2.75	2.60	
26	3.60	3.95	3.00	3.40	3.25	3.80	2.75	2.70	2.35	2.80	2.65	
27	3.65	3.90	3.40	3.40	3.30	3.85	2.95	2.70	2.30	2.75	2.70	
28	3.70	3.95	3.20	2.70	3.40	3.65	2.70	2.60	2.40	2.80	3.05	
29	3.75		3.35	3.10	2.80	3.55	2.80	2.65	2.55	2.95	2.90	
30	3.75		3.10	3.50	3.15	3.65	2.85	2.60	2.60	2.85	2.80	
31	3.70		3.00		3.15		2.90	2.75		2.80		

Rating table for Platte River near Lexington, Nebr., from January 1 to December 31, 1903.[a]

Gage height.	Discharge.	Gage height.	Discharge.	Gage height.	Discharge.	Gage height.	Discharge.
Feet.	*Second-feet.*	*Feet.*	*Second-feet.*	*Feet.*	*Second-feet.*	*Feet.*	*Second-feet.*
2.1	160	2.6	1,680	3.1	5,140	3.6	10,240
2.2	390	2.7	2,220	3.2	6,060	3.7	11,520
2.3	640	2.8	2,820	3.3	7,000	3.8	12,860
2.4	920	2.9	3,500	3.4	8,020	3.9	14,240
2.5	1,260	3.0	4,290	3.5	9,060	4.0	15,680

[a] This table was applied indirectly, according to the method outlined in Nineteenth Ann. Rept U. S. Geol. Survey, pt. 4, p. 323 et seq.

Estimated monthly discharge of Platte River near Lexington, Nebr., for 1903.

[Drainage area, 53,800 square miles.]

Month.	Discharge in second-feet.			Total in acre-feet.	Run-off.	
	Maximum.	Minimum.	Mean.		Second-feet per square mile.	Depth in inches.
March 20-31 a	10,880	4,290	6,612	157,377	0.124	0.055
April	9,060	2,220	3,893	231,650	.073	.081
May	7,510	1,950	3,749	230,517	.070	.081
June..............	12,860	2,520	6,471	385,051	.121	.135
July	9,060	1,090	3,709	228,057	.070	.081
August	2,220	275	1,009	62,041	.019	.022
September	1,260	0	237	14,102	.004	.004
October	1,680	390	1,140	70,096	.021	.024
November 1-16 a....	2,220	1,090	1,499	47,572	.028	.017

a Days river was frozen not included.

NORTH PLATTE RIVER AT NORTH PLATTE, NEBR.

The lowest gaging station on this river is located at the wagon bridge just north of North Platte, Nebr., and was established October 5, 1894. It is 3.5 miles above the junction with South Platte River, in sec. 28, T. 14 N., R. 30 W. The bridge is a long, low, pile bridge, having 93 spans of approximately 20 feet each, crossing the main channel of the river. North of this, at a distance of about 440 feet, is another bridge crossing a smaller branch or slough and having six spans of about 20 feet each. The water, except in times of flood, does not pass under all of the spans of the long bridge. Usually the greater part flows under two or three of the spans, spreading out in shallow pools or streamlets under others. Beneath the greater number of spans is a dry, sandy bed at ordinary stages. The initial point for soundings is on the right bank and consists of a mark on the railing on the upstream side of the bridge. The channel is nearly straight for about 500 feet both above and below the station. The banks are low, but are rarely, if ever, overflowed. The current is moderately rapid, and the bed is sandy and shifting.

The observations of river height are made at the Union Pacific Railroad bridge, about 2 miles below the wagon-road bridge. The railroad bridge is 2 miles above the junction of North and South Platte rivers. The gage is vertical, marked to tenths of a foot, and is fastened by screws to the piling under the bridge. The gage is read twice each day by H. E. Dress, the railroad bridge watchman.

The bench mark is the top of the east rail directly over the gage rod. Its elevation is 12 feet above the zero of the gage. The river usually goes dry some time in August. In measuring depths care must be taken not to sound too close to the piling, as there is usually a hole washed *out around the base of* each, in which there is no curren

The observations at this station during 1903 have been made under the direction of J. C. Stevens, district hydrographer.

Discharge measurements of North Platte River at North Platte, Nebr., in 1903.

Date.	Hydrographer.	Gage height.	Discharge.
		Feet.	*Second-feet.*
April 2	J. C. Stevens	2.30	2,143
May 8	do	2.85	5,004
June 13	do	2.90	6,490
June 29	do	3.25	9,114
July 8	do	2.80	4,768
July 31	do	1.85	787
August 14	do	1.80	791
September 17	W. C. Sturdevant	1.60	492
October 31	J. C. Stevens	2.00	1,145

Mean daily gage height, in feet, of North Platte River at North Platte, Nebr., for 1903.

Day.	Mar.	Apr.	May.	June.	July.	Aug.	Sept.	Oct.	Nov.	Dec.
1		2.40	2.70	2.70	3.15	2.00	1.75	1.85	2.00	2.40
2		2.30	2.75	2.60	3.05	1.90	1.60	1.90	2.10	2.40
3		2.35	2.85	2.55	3.10	1.85	1.55	1.90	2.10	2.25
4		2.45	2.90	2.50	3.20	1.75	1.50	2.00	2.10	2.20
5		2.50	2.90	2.45	3.10	1.65	1.50	2.00	2.10	2.05
6		2.40	3.00	2.50	3.00	1.60	1.60	2.00	2.10	2.05
7		2.40	2.90	2.50	2.85	1.50	1.60	2.00	2.05	1.85
8		2.45	2.85	2.55	2.80	1.45	1.55	1.90	2.05	2.00
9		2.70	2.75	2.60	2.75	1.45	1.60	1.95	2.10	2.10
10		2.70	2.80	2.75	2.65	1.40	1.65	1.90	2.10	2.10
11		2.65	2.75	2.85	2.60	1.85	1.65	1.90	2.10	2.30
12		2.60	2.80	2.90	2.70	1.80	1.55	1.85	2.10	2.05
13		2.55	2.65	2.90	2.55	1.80	1.40	1.95	2.10	2.20
14		2.50	2.65	3.00	2.50	1.80	1.40	2.00	2.10	2.25
15	3.05	2.50	2.60	3.00	2.45	1.70	1.55	2.00	2.10	2.30
16	3.70	2.55	2.70	3.10	2.40	1.60	1.60	2.00	2.05	2.40
17	3.55	2.55	2.70	3.20	2.35	1.65	1.60	2.05	1.55	2.50
18	3.55	2.60	2.70	3.20	2.30	1.75	1.70	2.05	1.65	2.50
19	3.45	2.50	2.75	3.30	2.40	1.90	1.70	1.95	2.05	2.50
20	3.40	2.60	2.85	3.30	2.30	1.75	1.70	1.95	2.20	2.50
21	3.10	2.60	2.85	3.40	2.25	1.60	1.70	1.95	2.25	2.50
22	2.80	2.60	2.90	3.40	2.40	1.60	1.70	1.90	2.35	2.60
23	2.65	2.60	2.90	3.40	2.15	1.60	1.65	2.00	2.50	2.60
24	2.55	2.55	2.90	3.40	2.05	1.70	1.70	2.00	2.65	2.60
25	2.60	2.55	3.00	3.55	2.15	1.70	1.70	2.00	2.75	2.60
26	2.45	2.65	3.05	3.50	2.00	1.70	1.65	2.00	2.80	2.60
27	2.60	2.60	3.00	3.40	1.85	2.05	1.65	2.00	2.75	2.70
28	2.55	2.50	2.95	3.35	1.70	1.75	1.75	2.00	2.40	2.70
29	2.50	2.65	2.80	3.25	1.70	1.95	1.80	2.00	2.40	2.65
30	2.40	2.85	2.80	3.15	1.85	2.00	1.85	2.00	2.40	2.65
31	2.45		2.80		1.90	1.80		2.00		2.70

Rating table for North Platte River at North Platte, Nebr., from January 1 to December 31, 1903.[a]

Gage height.	Discharge.	Gage height.	Discharge.	Gage height.	Discharge.	Gage height.	Discharge.
Feet.	*Second-feet.*	*Feet.*	*Second-feet.*	*Feet.*	*Second-feet.*	*Feet.*	*Second-feet.*
0.5	15	1.5	565	2.5	2,950	3.5	11,100
.6	30	1.6	670	2.6	3,440	3.6	12,700
.7	50	1.7	800	2.7	4,050	3.7	14,300
.8	70	1.8	950	2.8	4,700	3.8	16,000
.9	100	1.9	1,120	2.9	5,350	3.9	17,700
1.0	150	2.0	1,320	3.0	6,100	4.0	19,400
1.1	210	2.1	1,560	3.1	6,900		
1.2	280	2.2	1,800	3.2	7,800		
1.3	370	2.3	2,150	3.3	8,800		
1.4	465	2.4	2,520	3.4	9,900		

[a] This table was applied indirectly, according to the method outlined in Nineteenth Ann. Rept. U. S. Geol. Survey, pt. 4, p. 323 et seq.

Estimated monthly discharge of North Platte River at North Platte, Nebr., for 1903.

[Drainage area, 23,517 square miles.]

Month.	Discharge in second-feet.			Total in acre-feet.	Run-off.	
	Maximum.	Minimum.	Mean.		Second-feet per square mile.	Depth in inches.
March 15–31 [a]	14,300	2,520	6,562	221,263	0.230	0.145
April	5,025	2,150	3,223	191,782	.113	.126
May	6,500	3,440	4,866	299,199	.171	.197
June	12,700	2,735	6,825	406,116	.239	.267
July	8,300	670	3,348	205,861	.117	.135
August	1,220	370	711	43,718	.025	.029
September	800	325	545	32,430	.019	.021
October	1,220	875	1,087	66,837	.038	.044
November 1–18 [a]	1,440	565	1,316	46,985	.046	.031

[a] River frozen on days not included.

NORTH PLATTE RIVER AT BRIDGEPORT, NEBR.

This station was established May 4, 1902, by R. H. Willis. It is located at the highway bridge on the public road due north of Bridgeport. The Burlington and Missouri River Railroad bridge crosses the river 1 mile above the gaging station. The wire gage, with inclosed scale, is fastened to the upstream hand rail of the bridge, at the south end. The length of the wire from the end of the weight to the marker is 12.80 feet. The gage is read once each day by Porter Hannawald. Discharge measurements are made from the upstream side of the highway bridge, which is supported by pile bents 20 feet apart. The hand rail is marked at 10-foot intervals with black paint, and the initial point for soundings is the zero mark on the hand rail at the south end of the bridge. The channel widens just above and below the station and is then straight for 1 mile above and one-half mile below. The water is never sluggish. Both banks are low and sandy, but are not liable to overflow. The bed is of shifting sand, free from vegetation. There is one channel at ordinary or high stages, but at low water the river is divided into many winding channels. The river generally goes dry in August. In sounding, during discharge measurements, care must be taken not to get too close to the piles of the bridge supports, as there is usually a hole near the base of each, filled with dead water. Bench mark No. 1 is a 6 by 6 inch stone, marked "U. S. C. & G. S.," located in the NE. ¼ sec. 32, T. 20 N., R. 50 W., of the sixth principal meridian. It is 130 feet east of the east gate of the stock yards and 130 feet northwest of the northwest corner of the public school building. Its elevation is 9.94 feet above gage datum. Bench mark No. 2 is a standard aluminum Geological Survey bench-mark cap set in a 20 by 12 by 6 inch stone, the top of which is built up with concrete to form a truncated pyramid. It is located about 50 feet south and a little east of the northeast corner of lot No. 4, block No. 2, of the Riverside Addition to Bridgeport, Nebr. Its elevation is 11.32 feet above gage datum. Bench mark No. 3 is the head of a nail driven in the top of the west end of the cap of the south bent of the bridge. Its elevation is 10.14 feet above gage datum. Gage datum is the water surface when the wire gage reads zero.

The observations at this station during 1903 have been made under the direction of J. C. Stevens, district hydrographer.

Discharge measurements of North Platte River at Bridgeport, Nebr., in 1903.

Date.	Hydrographer.	Gage height.	Discharge.
		Feet.	*Second-feet*
March 31	J. C. Stevens	5.32	1,959
June 5	R. H. Willis	5.67	3,789
June 9	do	6.06	6,174
June 16	do	6.41	8,444
June 19	J. C. Stevens	6.40	10,464
July 4	R. H. Willis	5.77	5,544
July 7	J. C. Stevens	5.65	a 4,234
July 13	R. H. Willis	5.87	2,494
July 27	do	5.10	914

a Mean of two gagings.

Discharge measurements of North Platte River, etc.—Continued.

Date.	Hydrographer.	Gage height.	Discharge.
		Feet.	Second-feet.
July 30	J. C. Stevens	4.98	824
August 4	R. H. Willis	4.95	699
August 13	J. C. Stevens	4.90	331
August 18	R. H. Willis	5.12	364
August 29	do	5.05	183
September 7	O. V. P. Stout	5.10	284
October 19	R. H. Willis	5.10	592

Mean daily gage height, in feet, of North Platte River at Bridgeport, Nebr., for 1903.

Day.	Apr.	May.	June.	July.	Aug.	Sept.	Oct	Nov.	Dec.
1	5.88	6.00	5.70	5.98	5.13	5.10	5.00	5.07	5.32
2	5.36	5.98	5.65	5.86	5.00	5.10	5.05	5.08	5.30
3	5.22	5.95	5.73	5.83	5.06	5.08	5.05	5.10	5.32
4	5.32	5.90	5.85	5.81	5.14	5.10	5.07	5.05	5.30
5	5.58	5.95	5.67	5.85	5.13	5.05	4.95	5.07	5.30
6	5.73	5.98	5.73	5.72	5.09	5.08	5.10	5.10	5.32
7	5.60	5.75	5.92	5.68	5.10	5.10	5.20	5.08	5.50
8	5.60	5.65	6.00	5.63	4.90	5.00	5.15	5.05	5.58
9	5.60	5.77	6.06	5.41	4.95	5.00	5.10	5.05	5.48
10	5.55	5.78	6.20	5.45	5.05	4.98	5.10	5.00	5.40
11	5.52	5.67	6.21	5.54	5.00	4.85	5.22	5.00	5.42
12	5.42	5.60	6.23	5.46	4.95	4.95	5.10	5.02	5.40
13	5.43	5.63	6.30	5.35	5.08	4.90	5.10	5.30	5.38
14	5.41	5.57	6.38	5.29	4.95	5.00	5.15	5.35	......
15	5.46	5.70	6.38	5.31	4.90	5.15	5.10	5.30	......
16	5.47	5.65	6.41	5.37	5.08	5.12	5.08	5.40	......
17	5.50	5.80	6.35	5.45	5.00	4.98	5.05	5.40	......
18	5.50	5.83	6.34	5.22	5.10	4.96	5.08	5.40	......
19	5.83	5.75	6.39	5.17	5.15	5.00	5.08	5.40	......
20	5.53	5.92	6.40	5.30	5.10	5.00	5.05	5.40	......
21	5.55	6.05	6.36	5.17	5.08	5.00	5.05	5.35	5.32
22	5.60	6.05	6.33	5.13	5.00	5.10	5.10	5.30	5.35
23	5.58	6.10	6.45	5.12	4.98	5.20	5.12	5.35	5.32
24	5.57	6.15	6.37	4.99	5.30	5.18	5.10	5.37	5.54
25	5.60	5.95	6.37	5.08	5.10	5.18	5.08	5.30	5.50
26	5.80	5.87	6.31	5.27	5.10	5.10	5.05	5.37	5.60
27	5.97	5.85	6.27	5.22	5.20	5.15	5.05	5.35	5.65
28	5.85	5.80	6.16	5.22	5.00	5.07	5.07	5.37	5.75
29	5.85	5.93	6.10	5.11	5.05	5.07	5.06	5.37	5.80
30	5.80	5.90	5.97	5.19	5.08	5.05	5.07	5.37	5.67
31	......	5.72	......	5.12	5.10	......	5.08	......	5.65

Rating table for North Platte River at Bridgeport, Nebr., from January 1 to December 31, 1903.[a]

Gage height.	Discharge.	Gage height.	Discharge.	Gage height.	Discharge.	Gage height.	Discharge.
Feet.	*Second-feet.*	*Feet.*	*Second-feet.*	*Feet.*	*Second-feet.*	*Feet.*	*Second-feet.*
4.5	0	5.1	1,170	5.7	4,450	6.3	9,190
4.6	45	5.2	1,575	5.8	5,160	6.4	10,060
4.7	150	5.3	2,040	5.9	5,910	6.5	10,950
4.8	315	5.4	2,565	6.0	6,700	6.6	11,860
4.9	540	5.5	3,150	6.1	7,510	6.7	12,780
5.0	825	5.6	3,780	6.2	8,340	6.8	13,710

[a] This table was applied indirectly, according to the method outlined in Nineteenth Am. Rept. U. S. Geol. Survey, pt. 4, p. 328 et seq.

Estimated monthly discharge of North Platte River at Bridgeport, Nebr., for 1903.

[Drainage area, 23,190 square miles.]

Month.	Discharge in second-feet.			Total in acre-feet.	Run-off.	
	Maximum.	Minimum.	Mean.		Second-feet per square mile.	Depth in inches.
April [a]	6,300	1,575	3,664	218,023	0.158	0.176
May	7,920	3,460	5,526	339,780	.238	.274
June	10,500	3,780	7,665	456,099	.331	.369
July	6,700	825	2,837	174,440	.122	.141
August	1,365	90	589	36,216	.025	.029
September	420	200	240	14,281	.013	.014
October	675	90	487	29,944	.021	.024
November 1–12 [a]	1,170	675	883	21,017	.038	.017

[a] River frozen for days not included.

NORTH PLATTE RIVER NEAR MITCHELL, NEBR.

This station was established June 3, 1902, by O. V. P. Stout, assisted by R. H. Willis. It is located at the highway bridge 1 mile south of Mitchell, Nebr. It replaces the station at Gering, Nebr., which was discontinued, as the narrower channel at Mitchell seemed favorable to increased accuracy of gagings, and, being nearer the Wyoming line, it serves better as a State-line gaging station. The gage consists of a sash weight hung by a brass chain running over a pulley. The scale is inclosed in a box fastened to the upstream hand rail near the north end of the bridge. The length of the chain from the end of the weight to the marker is 10.60 feet. The gage is read once each day by Purly Eytchison.

On April 4, 1902, a temporary gage rod was set whose zero mark was 3.79 feet lower than the old rod of 1901. On May 3, 1902, a new

ent gage was put in whose zero mark was 1 foot lower than the
of 1901, in order to avoid negative gage heights. Hence,
1 April 4 and May 3, 1902, 2.79 feet was subtracted from the
iights reported by the observer. To make the gage heights
1 during 1901 comparable with those of 1902, 1 foot must be

large measurements are made from the upstream side of the
y bridge, which is supported by pile bents 20 feet apart. The
m hand rail is marked at intervals of 12½ feet with blue paint
intervals of 10 feet with white paint. The initial point for ·
igs is the zero mark at the north end of the bridge. The chan-
traight for 2,000 feet above and below the station. The mean
r varies from 1 foot at low stages to 4 feet per second at high
Both banks are low and sandy, but are not liable to over-
There is a rather large island in the gaging section. At high
here are three channels, and at low stages there are from three
channels. The bridge has a total span of 1,565 feet. At low
he channels are too shallow for current-meter measurements,
such times velocities must be estimated or determined with
Bench mark No. 1 is a cross cut in the floor of the bridge at
e. Its elevation is 9.74 feet above gage datum. Bench mark
the head of a nail driven in the top at the west end of the cap
irst bent at the north end of the bridge. Its elevation is 8.56
ove gage datum. Bench mark No. 3 is a standard aluminum
mark cap of the Geological Survey leaded into the top of a
gas pipe 4 feet long, with a plate at the bottom. It is located
; north of the north end of the bridge and 30 feet east of the
the downstream hand rail extended. It is 2 feet outside of a
nce. Its elevation is 8.64 feet above the gage datum. The
tum is the water surface when the chain gage reads zero.
bservations at this station during 1903 have been made under
ction of J. C. Stevens, district hydrographer.

irge measurements of North Platte River near Mitchell, Nebr., in 1903.

Date.	Hydrographer.	Gage height.	Discharge.
		Feet.	*Second-feet.*
....	J. C. Stevens	1.76	1,735
....do		2.60 .	3,580
....do		3.68	9,637
.... .do		2.72	4,230
.... .do		1.50	1,067
2 ....do		1.20	456
1 ....do		.87	184
2 ....do		.88	215
er 3 .... O. V. P. Stout....		.82	158

Mean daily gage height, in feet, of North Platte River near Mitchell, Nebr., for 1903.

Day.	Mar.	Apr.	May.	June.	July.	Aug.	Sept.	Oct.	Nov.	Dec.
1		1.76	2.95	2.50	2.95	1.47	0.85	1.80	1.88	2.14
2		1.70	2.80	2.52	2.95	1.47	.82	1.72	1.86	2.10
3		1.80	2.83	2.51	2.95	1.40	.80	1.69	1.86	1.81
4		2.15	2.81	2.56	2.95	1.40	.80	1.71	1.85	1.97
5		2.44	2.71	2.65	2.80	1.40	.82	1.67	1.89	2.12
6		2.30	2.71	2.90	2.70	1.40	.72	1.61	1.87	2.20
7		2.21	2.68	3.05	2.72	1.80	.77	1.65	1.90	2.01
8		2.23	2.63	3.16	2.65	1.30	.77	1.70	1.86	1.81
9		2.15	2.59	3.86	2.60	1.25	.95	1.75	1.91	1.80
10		2.14	2.60	3.39	2.35	1.25	.95	1.80	1.89	1.67
11		2.14	2.58	3.40	2.32	1.25	.72	1.76	1.81	1.60
12		2.15	2.60	3.44	2.32	1.25	.82	1.77	2.02	1.65
13		2.25	2.65	3.60	2.25	1.25	.81	1.74	2.01	1.68
14		2.25	2.69	3.50	2.15	1.30	1.35	1.76	2.10	1.80
15		2.26	2.56	3.50	2.10	1.15	1.50	1.76	2.07	1.66
16		2.21	2.85	3.46	2.05	1.15	1.60	1.79	2.08	1.67
17		2.30	2.87	3.59	2.05	1.28	1.60	1.80	2.78	1.44
18		2.34	2.90	3.65	2.00	1.18	1.65	1.84	2.67	1.65
19		2.33	3.02	3.71	1.87	1.05	1.65	1.87	2.18	1.70
20		2.35	3.11	3.65	1.80	.95	1.66	1.82	1.65	1.68
21		2.34	3.20	3.78	1.70	.98	1.65	1.81	2.00	1.76
22		2.35	3.24	3.91	1.68	.90	1.65	1.85	2.30	1.85
23		2.36	3.17	3.80	1.67	.72	1.68	1.85	2.00	1.81
24		2.46	2.97	3.75	1.82	1.10	1.70	1.80	2.01	1.71
25		2.40	2.95	3.70	1.87	1.30	1.70	1.80	1.92	1.82
26		2.41	2.90	3.61	1.75	1.10	1.75	1.80	2.04	1.88
27		2.04	2.80	3.42	1.60	.87	1.74	1.81	2.08	2.00
28		2.70	2.84	3.30	1.55	.85	1.80	1.82	2.08	1.82
29		2.83	2.76	3.31	1.50	.85	1.77	1.82	2.10	1.79
30		2.93	2.60	3.25	1.50	.82	1.72	1.88	2.16	1.80
31	1.78				1.50	.87		1.89		2.00

Rating table for North Platte River near Mitchell, Nebr., from January 1 to December 31, 1903.[a]

Gage height.	Discharge.	Gage height.	Discharge.	Gage height.	Discharge.	Gage height.	Discharge.
Feet.	Second-feet.	Feet.	Second-feet.	Feet.	Second-feet.	Feet.	Second-feet.
0.7	165	1.6	1,100	2.5	3,250	3.4	7,450
.8	180	1.7	1,300	2.6	3,580	3.5	8,080
.9	200	1.8	1,500	2.7	3,950	3.6	8,740
1.0	250	1.9	1,710	2.8	4,370	3.7	9,430
1.1	350	2.0	1,930	2.9	4,830	3.8	10,150
1.2	460	2.1	2,170	3.0	5,310	3.9	10,870
1.3	580	2.2	2,410	3.1	5,800	4.0	11,610
1.4	720	2.3	2,670	3.2	6,310	4.2	13,090
1.5	900	2.4	2,950	3.3	6,860		

[a] This table was applied indirectly, according to the method outlined in Nineteenth Ann. Rept. U. S. Geol. Survey, pt. 4, p. 323.

d monthly discharge of North Platte River near Mitchell, Nebr., for 1903.

[Drainage area, 24,400 square miles.]

	Discharge in second-feet.			Total in acre-feet.	Run-off.	
fonth.	Maximum.	Minimum.	Mean.		Second-feet per square mile.	Depth in inches.
....	5,070	1,600	2,862	170,301	0.117	0.131
....	6,580	3,400	4,471	274,911	.183	.211
....	10,870	3,250	7,340	436,760	.301	.336
....	5,310	1,000	2,632	161,835	.108	.125
....	900	165	472	29,022	.019	.022
er	1,500	165	765	45,521	.031	.035
....	1,710	1,100	1,449	89,095	.059	.068
er 1 to 16a..	2,170	1,500	1,762	55,918	.072	.043

a River frozen on days not included.

NORTH PLATTE RIVER AT GUERNSEY, WYO.

station was established June 14, 1900, by A. J. Parshall. It
ed at the county bridge about a half mile northwest of Guern-
'he bridge has eight piers, the sides are planked, and there is
1 flow under each span. The rod originally consisted of a 4
ch by 12-foot scantling firmly attached to one of the piers of
lge. As the station was to be a temporary one, a metallic tape,
into feet and tenths, was securely fastened to the rod.
s found that considerable inconvenience was caused by sand
dating about the rod as the high water subsided. With the
; of the season of 1902 a new rod was placed about 200 feet
he first location. It was fastened to one of the piers of the
l bridge, and was placed 1 foot lower in the water.
.902 bench mark is a spike driven into a sleeper of the bridge
rom the rod and at an elevation of 10.04 feet above the zero.
annel is straight for a distance above and below the station.
inks are high and do not overflow at high stages. The bed of
am is sandy, but probably does not shift much.
ig the latter part of the season of 1903 the 1902 gage was
d by a new rod, with its zero at the same elevation as that of
gage. The new rod is attached to a pile of the railroad bridge,
ias been recently reconstructed. In February, 1904, the eleva-
the new rod was checked with the temporary bench mark, and
parison with the readings of the old rods which are still in
This comparison shows that the rod has settled 0.076 foot.
03 bench mark is a monument set 125 feet southeast of the rod
evation of 13.65 feet above the zero of the gage and 4,331 feet
ea level.
urements are still *made from* the county bridge, which, whil'

it does not furnish a perfect location for measurements, is the best yet found on the river in Wyoming below the Saratoga station.

The observations at this station during 1903 have been made under the direction of A. J. Parshall, district hydrographer.

Discharge measurements of North Platte River at Guernsey, Wyo., in 1903.

Date.	Hydrographer.	Gage height.	Discharge.
		Feet.	Second-feet.
April 3	A. J. Parshall	2.60	4,291
April 22	John E. Field	2.10	2,671
April 24	L. V. Branch	2.37	3,306
May 17	John E. Field	3.20	4,966
June 17	H. G. Stokes	5.00	10,818
August 1	Chas. Carpenter	1.00	1,137
August 13	do	.45	544
August 25	do	.30	454
September 3	do	.20	369
June 12	John E. Field	4.85	9,298

Mean daily gage height, in feet, of North Platte River at Guernsey, Wyo., for 1903.

Day.	Jan.	Feb.	Mar.	Apr.	May.	June.	July.	Aug.	Sept.	Oct.	Nov.	Dec.
1	0.80	0.70	0.90	1.30	2.90	2.70	3.70	1.00	0.20	0.80	0.70	0.80
2	.80	.70	1.65	2.80	2.75	3.55	.95	.20	1.00	.70	.80	
3	.80	.70	.90	2.50	2.70	2.80	3.40	.90	.20	.90	.70	.80
4	.80	.70	1.00	2.25	2.80	2.90	3.15	.80	.20	.90	.70	.80
5	1.00	.70	1.00	2.05	2.55	3.10	3.00	.70	.20	.90	.70	.80
6	1.00	.70	1.00	1.95	2.45	3.50	2.90	.70	.20	.95	.70	.80
7	1.00	.70	1.00	1.85	2.40	3.75	2.75	.70	.20	.90	.70	.80
8	1.00	.70	1.00	1.70	2.40	4.10	2.60	.70	.20	.80	.70	.80
9	1.00	.70	1.00	1.60	2.45	4.25	2.50	.70	.15	.85	.70	.60
10	1.00	.70	1.00	1.50	2.55	4.40	2.40	.70	.15	.90	.70	.70
11	.80	.80	1.00	1.50	2.65	4.55	2.20	.60	.15	.80	.70	.70
12	.90	.80	1.00	1.65	2.70	4.75	2.05	.60	.15	.80	.70	.70
13	.90	.80	1.10	1.70	2.90	4.80	2.00	.50	.20	.80	.70	.70
14	.95	.80	1.20	1.75	2.95	4.80	1.90	.50	.20	.80	.70	.50
15	.95	.80	1.70	1.85	3.00	4.70	1.80	.40	.55	.90	.65	.50
16	.95	.80	1.60	1.90	3.15	4.90	1.65	.40	.60	.90	.60	.60
17	.95	.80	1.60	.90	3.40	5.00	1.50	.40	.70	1.00	.60	.60
18	.95	.80	1.80	2.00	3.50	5.10	1.40	.30	.80	1.00	.60	.60
19	.95	.80	1.70	2.10	3.65	5.00	1.30	.30	.80	1.00	.60	1.00
20	.90	.90	.90	2.00	3.75	5.00	1.30	.30	.65	1.00	.60	1.10
21	.90	.90	.80	2.00	3.80	5.25	1.20	.30	.90	1.00	.60	1.00
22	.90	.90	.60	2.15	3.85	5.10	1.20	.40	.90	.95	.50	1.10
23	.90	.90	.40	2.25	3.75	5.00	1.30	.40	.90	.90	.50	1.00
24	.90	.90	1.35	2.40	3.55	4.90	1.35	.85	.90	.90	1.70	1.00
25	.80	.90	2.50	2.50	3.35	4.75	1.15	.40	.90	.85	1.25	1.00
26	.80	.90	.30	2.50	3.15	4.55	1.10	.80	.90	.80	.95	1.00
27	.80	.90	.05	2.60	3.10	4.35	1.10	.20	.90	.80	.95	1.00
28	.70	.90	1.00	2.80	2.85	4.15	1.00	.45	.90	.80	1.00	1.00
29	.70		1.10	2.90	2.90	3.95	1.00	.35	.90	.70	.95	1.00
30	.70		1.25	2.90	2.75	3.75	1.00	.30	.80	.70	.90	.90
31	.70		1.30		2.60		1.00	.30		.70		.90

ing table for North Platte River at Guernsey, Wyo., from January 1 to December 31, 1903.

Gage height.	Discharge.	Gage height.	Discharge.	Gage height.	Discharge.	Gage heigth.	Discharge.
Feet.	*Second-feet.*	*Feet.*	*Second-feet.*	*Feet.*	*Second-feet.*	*Feet.*	*Second-feet.*
0.2	365	1.2	1,365	2.2	2,915	3.4	5,595
.3	435	1.3	1,500	2.3	3,105	3.6	6,145
.4	513	1.4	1,635	2.4	3,300	3.8	6,750
.5	600	1.5	1,775	2.5	3,500	4.0	7,400
.6	695	1.6	1,920	2.6	3,705	4.2	8,070
.7	795	1.7	2,070	2.7	3,915	4.4	8,740
.8	900	1.8	2,225	2.8	4,135	4.6	9,410
.9	1,007	1.9	2,385	2.9	4,360	4.8	10,080
1.0	1,116	2.0	2,555	3.0	4,595	5.0	10,750
1.1	1,235	2.1	2,730	3.2	5,080	5.2	11,420

'urve well defined.

'imated monthly discharge of North Platte River at Guernsey, Wyo., for 1903.

[Drainage area, 16,243 square miles.]

Month.	Discharge in second-feet.			Total in acre-feet.	Run-off.	
	Maximum.	Minimum.	Mean.		Second-feet per square mile.	Depth in inches.
ıuary	1,116	795	987	60,688	0.061	0.070
ɔrnary	1,007	795	897	49,817	.055	.057
rch	2,385	1,007	1,454	89,403	.090	.104
ril	4,360	1,500	2,711	161,316	.167	.186
y	6,910	3,300	4,671	287,209	.288	.332
ıe	11,588	3,915	8,485	504,898	.522	.582
y	6,440	1,116	2,634	161,958	.162	.187
gust	1,116	435	635	39,045	.039	.045
ıtember	1,007	333	665	39,570	.041	.046
ɔober	1,116	795	980	60,258	.060	.069
vember	2,070	600	869	51,709	.053	.059
ɔember	1,235	435	882	54,232	.054	.062
The year	11,588	333	2,156	1,560,098	.133	1.799

LARAMIE RIVER AT UVA, WYO.

This station was established April 14, 1903, by R. G. Schnitger. It located at the Colorado and Southern Railroad bridge one-half e south of Uva, Wyo. The vertical gage is attached to the piles

which support the bridge. It is read twice each day by J. A. Carl
Discharge measurements are made by wading at a point one-four
mile below. The initial point for soundings is on the north ban
The channel is straight for 200 feet above and below the station, a
the current is sluggish. The right bank is high and not liable
overflow. The left bank is liable to overflow. The bed of the stre
is sandy and shifting. The bench mark is a spike driven in t
bridge timber marked with a cross. Its elevation is 14.1 feet ab
the zero of the gage, and 7.1 feet above the top of the rod.

The observations at this station during 1903 have been made un
the direction of A. J. Parshall, district hydrographer.

Discharge measurements of Laramie River at Uva, Wyo., in 1903.

Date.	Hydrographer.	Gage height.	Dischar
		Feet.	Second-
April 14	R. G. Schnitger	0.70	19
June 5	A. J. Parshall	.85	21
July 4	do	.25	6
August 3	do	.10	3
September 1	do	−.85	

Mean daily gage height, in feet, of Laramie River at Uva, Wyo., for 1903.

Day.	Apr.	May.	June.	July.	Aug.	Sept.	O
1		0.90	0.60	0.40	0.20	−0.35	
2		.80	.60	.30	.10	− .85	
3		.90	.60	.30	.10	− 40	−
4		.72	1.00	.20	.10	− .40	−
5		.70	1.00	.20	.10	− .40	
6		.80	.90	.20	.10	− .40	
7		1.00	.80	.20	.10	− .40	
8		1.00	.70	.20	.10	. 40	
9		.90	.60	.20	.05	− .40	
10		1.00	.60	.20	.05	− .40	
11		1.10	.60	.20	.00	− .40	
12		1.10	.60	.20	.00	− .40	
13		1.10	.60	.20	.00	.00	
14	0.75	1.00	.60	.20	− .10	.00	
15	.75	1.00	.50	.20	− .10	.00	
16	.70	.90	.50	.20	− .10	.00	
17	.72	.90	.50	.40	− .10	.00	
18	.75	.90	.50	.35	− .10	.00	
19	.70	.90	.50	.25	− .20	.00	
20	.70	.80	.80	.20	− .20	.00	
21	.65	.80	.70	.20	− .20	− .10	
22	.72	.80	.60	.20	− .20	− .05	
23	.70	.80	.60	.20	− .20	− .05	
24	1.00	.80	.60	.20	− .20	− .20	
25	1.00	.80	.60	.20	− .20	− .20	
26	.90	.70	.60	1.30	− .30	− .20	
27	1.00	.70	.50	1.00	− .30	− .20	−
28	1.00	.60	.50	.90	− .30	− .20	
29	1.10	.60	.50	.70	− .30	− .20	
30	.90	.60	.50	.40	− .33	− .20	
31		.60		.30	− .30		−

Rating table for Laramie River at Uva, Wyo., from April 14 to October 31, 1903.

Gage height.	Discharge.	Gage height.	Discharge.	Gage height.	Discharge.	Gage height.	Discharge.
Feet.	*Second-feet.*	*Feet.*	*Second-feet.*	*Feet.*	*Second-feet.*	*Feet.*	*Second-feet.*
- 0.4	3	0.2	50	0.8	202	1.4	480
− .3	6	.3	70	.9	240	1.5	530
− .2	11	.4	92	1.0	284	1.6	580
− .1	18	.5	116	1.1	330		
+ .0	26	.6	142	1.2	380		
.1	36	.7	170	1.3	430		

Tangent above 1.1 feet gage height. Differences above this point. 50 per tenth.

Estimated monthly discharge of Laramie River at Uva, Wyo., for 1903.

[Drainage area, 3,179 square miles.]

Month.	Discharge in second-feet.			Total in acre-feet.	Run-off.	
	Maximum.	Minimum.	Mean.		Second-feet per square mile.	Depth in inches.
April 14–30	330	156	217	7,317	0.068	0.043
May	330	142	225	13,835	.071	.082
June	580	116	163	9,699	.051	.057
July	430	50	130	7,993	.041	.047
August	50	5	21	1,291	.007	.008
September	26	3	13	774	.004	.004
October	11	11	11	676	.003	.003

LITTLE LARAMIE RIVER AT HALEY'S RANCH, NEAR LARAMIE, WYO.

This station was established in March, 1903, by B. P. Fleming. It was located at Haley's ranch, near Laramie, Wyo. The bench mark is a cross on the top of the railing post on the upstream side of the bridge at the left end. Its elevation is 10.3 feet above the zero of the gage. This station was discontinued September 30, 1903.

The observations at this station during 1903 have been made under the direction of A. J. Parshall, district hydrographer.

Discharge measurements of Little Laramie River at Haley's ranch, near Laramie, Wyo., in 1903.

Date.	Hydrographer	Gage height.	Discharge.
		Feet.	*Second-feet.*
May 5	B. P. Fleming	1.70	40.54
July 8	...do	1.97	46.47
August 12	...do	1.38	8.34

Mean daily gage height, in feet, of Little Laramie River near Laramie, Wyo., for 1903.

Day.	Mar.	Apr.	May.	June.	July.	Aug.	Sept.
1		2.20	1.70	1.70	2.70	1.70	0.90
2		2.20	1.70	1.70	2.60	1.60	.90
3		2.20	1.80	1.80	2.50	1.60	.90
4		2.60	1.70	2.00	2.50	1.50	.90
5		2.60	1.70	2.70	2.40	1.50	.90
6		2.80	1.70	3.20	2.30	1.50	.90
7		2.60	1.60	3.60	2.30	1.40	.90
8		2.40	1.50	3.60	2.20	1.40	.90
9		2.40	1.50	4.30	2.20	1.30	.00
10		2.20	1.50	5.00	2.20	1.30	.00
11		2.20	1.60	4.80	2.20	1.30	.00
12		2.00	1.60	4.50	2.10	1.30	.00
13		2.00	1.50	4.00	2.10	1.30	1.10
14		1.90	1.50	4.30	2.10	1.30	1.20
15		1.80	1.60	4.40	2.10	1.30	1.30
16		1.80	1.70	4.60	2.10	1.20	1.80
17		1.80	2.00	4.50	2.00	1.10	2.00
18		1.80	2.40	4.80	2.00	1.10	2.10
19		1.90	2.20	4.20	1.90	1.10	2.20
20		1.80	2.00	4.10	1.90	1.10	2.30
21		1.80	1.80	4.00	1.90	1.10	2.30
22		1.70	1.90	3.90	1.80	1.10	2.30
23		1.70	1.80	3.80	1.80	1.00	2.30
24		1.70	1.70	3.60	1.80	1.00	2.30
25	1.80	1.70	1.70	3.50	1.70	1.00	2.10
26	1.80	1.70	1.70	3.30	1.80	1.00	2.10
27	1.80	1.70	1.70	3.10	1.80	1.00	2.10
28	1.80	1.70	1.60	2.90	1.80	1.00	2.10
29	1.80	1.70	1.70	2.70	1.80	1.00	2.10
30	2.00	1.70	1.70	2.80	1.80	1.00	2.00
31	2.00		1.70		1.80	1.00	

LITTLE LARAMIE RIVER NEAR HATTON, WYO.

This station was established in 1902 by B. P. Fleming. It is located at an old bridge 25 miles northwest of Laramie, Wyo. The gage is a plain staff graduated to feet and tenths and secured to the bridge. It is read twice daily by J. M. May, a ranchman, who lives about 75 yards from the gage. The initial point for soundings is a nail driven in the bridge stringer at the left bank. The channel is straight above the station and curved below. The water is quite swift. The right bank is liable to overflow. The left bank is high and rocky. There is an old channel at the right of the station in which there is some flow at high water. The bed of the stream is rocky and very rough.

The bench mark is a cross on the top of the first post in the fence which forms the approach to the bridge from the left bank. Its elevation is 10.27 feet above the zero of the gage.

The observations at this station during 1903 have been made under the direction of A. J. Parshall, district hydrographer.

ischarge measurements of Little Laramie River near Hatton, Wyo., in 1903.

Date.	Hydrographer.	Gage height.	Discharge.
		Feet.	Second-feet.
r 9	B. P. Fleming	1.15	129
y 4	do	1.90	386
ober 27	do	.95	86

in daily gage height, in feet, of Little Laramie River near Hatton Wyo., for 1903.

Day.	Apr.	May.	June.	July.	Aug.	Day.	Apr.	May.	June.	July.	Aug.
		1.25	1.80	2.50	1.00	17	1.40	1.55	3.75	1.55	
		1.00	1.70	2.10	.95	18	1.35	1.70	3.95	1.45	
		1.00	2.10	1.70	.90	19	1.15	2.10	4.10	1.35	
		1.25	2.55	1.55	.90	20	1.30	2.30	4.00	1.25	
		1.15	2.25	1.50	.90	21	1.05	2.05	3.85	1.25	
		.90	2.90	1.50	.80	22	1.00	1.60	3.50	1.20	
		.80	3.20	1.50	.80	23	1.10	1.85	3.50	1.20	
		.80	3.35	1.50	.80	24	1.10	1.70	3.55	1.25	
		.80	3.80	1.40	.80	25	1.10	1.80	4.00	1.20	
		1.35	4.00	1.40	.90	26	1.10	1.65	3.80	1.20	
		1.30	4.20	1.35	1.00	27	1.30	1.55	3.00	1.25	
	1.15	1.30	3.70	1.35	.80	28	1.10	1.60	3.00	1.15	
	1.25	1.25	3.80	1.35	.80	29	1.05	1.50	2.85	1.10	
	1.40	1.30	3.80	1.40	.80	30	1.35	1.50	2.70	1.00	
	1.15	1.25	3.95	1.35	.80	31		1.65		1.00	
	1.10	1.10	3.90	1.40							

Rating table for Little Laramie River near Hatton, Wyo., from April 12 to August 15, 1903.

Gage height.	Discharge.	Gage height.	Discharge.	Gage height.	Discharge.	Gage height.	Discharge.
Feet.	Second-feet.	Feet.	Second-feet.	Feet.	Second-feet.	Feet.	Second-feet.
0.8	64	1.7	302	2.6	760	3.5	1,800
.9	80	1.8	343	2.7	820	3.6	1,860
1.0	96	1.9	386	2.8	880	3.7	1,420
1.1	118	2.0	433	2.9	940	3.8	1,480
1.2	141	2.1	482	3.0	1,000	3.9	1,540
1.3	167	2.2	533	3.1	1,060	4.0	1,600
1.4	196	2.3	586	3.2	1,120	4.1	1,660
1.5	229	2.4	642	3.3	1,180	4.2	1,720
1.6	264	2.5	700	3.4	1,240		

Estimated monthly discharge of Little Laramie River near Hatton, Wyo., for 1903.

Month.	Discharge in second-feet.			Total in acre-feet.
	Maximum.	Minimum.	Mean.	
April 12–30	196	98	140	5,276
May	586	64	222	13,650
June	1,720	302	1,217	72,417
July	700	98	207	12,728
August 1–15	98	64	74	2,202

SWEETWATER RIVER AT DEVILS GATE, NEAR SPLITROCK, WYO.

This station was established in October, 1902, by W. W. Schlecht. It is located about one-fourth mile above Devils Gate and is reached by team from Rawlins or from Casper, Wyo. The gage is of the wire type, with the scale board graduated to feet and tenths. It is read twice daily by Tom Sun, jr., a rancher, who lives about 300 yards from the gage. The gagings are made temporarily from a foot bridge at Tom Sun's house. It is intended later to establish a cable at this point.

The observations at this station during 1903 have been made under the direction of A. J. Parshall, district hydrographer.

Discharge measurements of Sweetwater River at Devils Gate, near Splitrock, Wyo. in 1903.

Date.	Hydrographer.	Gage height.	Discharge
		Feet.	Second-feet.
June 3	L. V. Branch	5.20	20
June 7	do	5.65	3
June 18	do	5.85	3
June 27	do	5.20	1
June 29	do	5.10	1
July 4	A. Weiss	4.85	1
July 12	L. V. Branch	4.46	
July 16	do	4.30	
July 24	do	4.14	
May 7	A. J. Parshall	4.85	1

an daily gage height, in feet, of Sweetwater River at Devils Gate, near Splitrock, Wyo., for 1903.

Day.	Jan.	Feb.	Mar.	Apr.	May.	June.	July.	Aug.	Sept.	Oct.	Nov.
....	4.50	4.70	4.90	5.18	5.30	5.15	4.95	4.10	4.00	4.15	4.45	4.45
....	4.50	4.62	4.90	5.00	5.20	5.15	4.90	4.10	4.00	4.15	4.45	4.40
... .	4.50	4.70	4.90	4.98	5.15	5.15	4.90	4.10	3.95	4.20	4.45	4.40
... ..	4.50	4.60	4.90	5.00	4.90	5.25	4.80	4.05	3.95	4.20	4.45	4.40
....:.	4.55	4.60	4.90	4.88	4.90	5.43	4.80	4.00	3.95	4.20	4.45	4.40
.....•....	4.60	4.65	4.92	4.80	4.80	5.55	4.80	4.00	3.95	4.20	4.45	4.40
....:	4.60	4.70	4.95	4.80	4.82	5.67	4.80	4.00	3.95	4.25	4.45	4.40
....	4.60	4.70	4.90	4.70	4.90	5.72	4.67	4.00	3.95	4.25	4.45	4.42
....	4.60	4.70	4.90	4.70	4.98	5.70	4.50	4.05	3.95	4.30	4.40	4.40
....	4.60	4.70	4.80	4.70	4.90	5.70	4.50	4.50	3.95	4.30	4.42	4.40
....	4.60	4.70	4.78	4.62	4.90	5.80	4.45	4.30	3.95	4.30	4.42	4.40
....	4.58	4.70	4.55	4.65	4.85	5.85	4.40	4.10	3.95	4.30	4.50	4.40
....	4.55	4.70	4.45	4.72	4.85	5.85	4.40	4.00	4.15	4.30	4.50	4.40
.......•.........................	4.60	4.62	4.55	4.78	4.80	5.80	4.40	4.00	4.10	4.35	4.50	4.42
....'	4.60	4.60	4.58	4.78	4.80	5.73	4.30	4.00	4.05	4.35	4.50	4.48
....	4.58	4.70	4.40	4.75	4.80	5.80	4.30	4.00	4.02	4.35	4.50	4.55
....	4.50	4.70	4.40	4.85	4.85	5.90	4.30	4.00	4.00	4.35	4.50	4.62
....	4.50	4.70	4.40	4.80	4.92	5.80	4.27	4.00	4.00	4.40	4.50	4.65
....	4.58	4.70	4.40	4.82	5.10	5.80	4.23	4.00	4.00	4.40	4.50	4.70
....	4.60	4.70	4.42	4.72	5.10	5.78	4.20	4.00	4.00	4.40	4.50	4.70
....	4.62	4.70	4.40	4.75	5.10	5.73	4.15	4.00	4.00	4.40	4.50	4.70
....	4.65	4.70	4.40	4.70	5.10	5.73	4.15	4.00	4.00	4.40	4.50	4.70
.....:	4.68	4.72	4.40	4.70	5.10	5.63	4.13	4.00	4.00	4.40	4.50	4.70
....	4.70	4.75	4.40	4.70	5.05	5.45	4.10	4.00	4.00	4.40	4.50	4.70
....	4.70	4.80	4.38	4.78	5.10	5.35	4.10	4.00	4.05	4.40	4.50	4.70
....	4.68	4.88	4.33	4.92	5.10	5.28	4.10	4.00	4.07	4.40	4.50	4.70
....	4.70	4.90	4.52	5.00	5.10	5.20	4.10	4.00	4.10	4.40	4.50 ·	4.70
....	4.70	4.90	4.50	5.10	5.10	5.10	4.10	4.00	4.10	4.45	4.50 ·	4.70
....	4.70	......	4.58	5.20	5.10	5.10	4.10	4.00	4.10	4.45	4.50 ·	4.70
....	4.70	......	4.58	5.20	5.13	4.97	4.10	4.00	4.10	4.47	4.50	4.70
....	4.70	......	4.60	......	5.10	......	4.10	4.00	......	4.45	......	4.70

ating table for Sweetwater River at Devils Gate, near Splitrock, Wyo., from January 1 to December 31, 1903.

Gage height.	Discharge.	Gage height.	Discharge.	Gage height.	Discharge.	Gage height.	Discharge.
Feet.	*Second-feet.*	*Feet.*	*Second-feet.*	*Feet.*	*Second-feet.*	*Feet.*	*Second-feet.*
4.0	14	4.5	54	5.0	189	5.5	270
4.1	19	4.6	67	5.1	162	5.6	300
4.2	25	4.7	82	5.2	187	5.7	332
4.3	33	4.8	99	5.3	213	5.8	365
4.4	43	4.9	118	5.4	241	5.9	400

*Estimated monthly discharge of Sweetwater River at Devils Gate, near Splitrock,
Wyo., for 1903.*

Month.	Discharge in second-feet.			Total in acre-feet.
	Maximum.	Minimum.	Mean.	
January	82	54	69	4,28
February	118	67	84	4,66
March	128	38	73	4,49
April	187	67	110	6,58
May	218	99	140	8,60
June	400	188	287	17,078
July	128	19	49	3,018
August	54	14	17	1,045
September	22	12	15	896
October	48	22	37	2,275
November	54	43	51	3,035
December	82	43	62	3,812
The year	400	12	83	59,701

NORTH PLATTE RIVER AT SARATOGA, WYO.

This station was established June 9, 1903, by A. J. Parshall. The gage consists of a flexible Gurley rod fastened securely to a 3 by 6 inch timber, spiked and braced to the crib work at the northeast corner of Harry Kuykendal's residence, 100 yards below the bridge, on the left bank of the river. The gage is read twice each day by J. M. Sterrett. Discharge measurements are made from the bridge. The initial point for soundings is on the east end of the bridge, at the pier. The channel is straight for 500 feet above and 400 feet below the station. The current has a measurable velocity at all stages. The right bank is high, but overflows at a gage height of 7 feet. The left bank overflows only at extreme high water. The river has a permanent bed of gravel. There is but one channel, broken by the center pier of the bridge. Bench mark No. 1 is a cross on the south side of the southwest pier at the bottom of the top section. Its elevation is 10.7 feet above the zero of the gage. Bench mark No. 2 is a spike driven in a cottonwood tree standing about 30 feet north west of the gage. Its elevation is 9.84 feet above the zero of the gage.

The observations at this station during 1903 have been made under the direction of A. J. Parshall, district hydrographer.

arge measurements of North Platte River at Saratoga, Wyo., in 1903.

Date.	Hydrographer.	Gage height.	Discharge.
		Feet.	*Second-feet.*
....	A. J. Parshall...............	5.20	7,954
....	C. B. Sterrett...............	5.00	7,089
....	....do	4.70	5,708
....	....do	4.20	4,511
....	....do	3.45	1,973
....	....do	2.35	784
r 26...................	....do	2.10	728

ly gage height, in feet, of North Platte River at Saratoga, Wyo., for 1903.

Day.	June.	July.	Aug.	Sept.	Oct.	Nov.	Dec.
....		3.73	1.98	1.51	2.13	1.96	1.95
....		3.60	1.95	1.46	2.17	1.96	1.97
....		3.45	1.90	1.42	2.21	1.97	1.84
....		3.45	1.81	1.41	2.23	1.97	1.75
....		3.37	1.74	1.40	2.23	2.01	1.65
....		3.17	1.69	1.47	2.25	2.03	1.72
....		3.03	1.66	1.85	2.26	2.03	1.80
....		2.97	1.62	1.90	2.21	2.04	1.80
....		2.93	1.59	1.98	2.20	2.01	1.88
....	5.02	2.85	1.59	1.87	2.23	1.93	1.88
....	4.92	2.73	1.57	1.87	2.24	1.67	1.88
....	4.97	2.68	1.53	1.97	2.24	1.60	1.88
....	4.97	2.63	1.53	2.02	2.23	1.80	1.86
....	5.13	2.63	1.54	2.19	2.24	1.94	1.86
....	5.08	2.57	1.58	2.29	2.23	2.07	1.88
....	5.07	2.55	1.56	2.32	2.23	2.06	1.86
....	5.17	2.80	1.54	2.30	2.22	1.85	1.90
....	5.25	2.73	1.50	2.29	2.18	1.75	1.88
....	5.12	2.64	1.51	2.29	2.17	2.00	1.87
....	5.08	2.50	1.48	2.31	2.15	2.05	1.86
....	4.92	2.43	1.45	2.30	2.13	2.00	1.84
....	4.80	2.40	1.46	2.25	2.10	2.01	1.84
....	4.70	2.40	1.51	2.20	2.07	2.06	1.86
....	4.62	2.33	1.60	2.16	2.05	2.04	1.88
....	4.37	2.33	1.61	2.11	2.04	2.09	1.88
....	4.20	2.33	1.66	2.09	2.03	2.12	1.90
....	4.07	2.38	1.71	2.08	2.02	2.08	1.93
....	4.10	2.45	1.80	2.05	1.99	1.93	1.97
....	4.05	1.98	1.70	2.04	2.00	1.83	1.84
....	3.93	1.95	1.62	2.08	2.02	1.87	1.74
....	....	2.00	1.56	....	1.99	....	1.72

Rating table for North Platte River at Saratoga, Wyo., from June 9 to December 31, 1903.

Gage height.	Discharge.	Gage height.	Discharge.	Gage height.	Discharge.	Gage height.	Discharge.
Feet.	Second-feet.	Feet.	Second-feet.	Feet.	Second-feet.	Feet.	Second-feet.
1.4	865	2.4	815	3.4	2,245	4.4	4,830
1.5	400	2.5	905	3.5	2,455	4.5	5,165
1.6	435	2.6	1,010	3.6	2,680	4.6	5,515
1.7	470	2.7	1,125	3.7	2,915	4.7	5,885
1.8	505	2.8	1,250	3.8	3,160	4.8	6,270
1.9	540	2.9	1,390	3.9	3,410	4.9	6,675
2.0	580	3.0	1,540	4.0	3,665	5.0	7,090
2.1	630	3.1	1,700	4.1	3,930	5.1	7,515
2.2	685	3.2	1,870	4.2	4,210	5.2	7,955
2.3	745	3.3	2,050	4.3	4,510		

Table fairly well defined.

Estimated monthly discharge of North Platte River at Saratoga, Wyo., for 1903.

Month.	Discharge in second-feet.			Total in acre-feet.
	Maximum.	Minimum.	Mean.	
June 10–30	8,180	3,538	6,158	256,531
July	3,038	560	1,267	77,905
August	580	382	446	27,428
September	745	365	592	35,296
October	715	580	662	40,705
November	630	435	567	33,739
December	560	452	524	32,220

SOUTH PLATTE RIVER AT BIGSPRING, NEBR.

This station was established September 5, 1902, by J. C. Stevens. It is located at the highway bridge one-fourth mile south of the Union Pacific Railroad station at Bigspring, Nebr. The station was established as a State-line station, and was discontinued November 21, 1903, for the reason that the desired information can be obtained at the Julesburg, Colo., station. The wire gage established in 1902 was replaced by a chain gage fastened to the upstream girder of the bridge near the north end. The length of the chain from the end of the weight to the marker is 9.45 feet. During 1903 the gage was read once each day by Victor Root. Discharge measurements are made from the upstream side of the highway bridge, which consists of 88 20-foot spans, supported by pile bents. The hand rail is marked at atervals of 20 feet and the initial point for soundings is the zero mark at the north end of the bridge on the upstream side. The channel is straight for 500 feet above and for 1 mile below the station. It is broad and shallow, with a shifting, sandy bed full of vegetation. The river is nearly dry from April to November. Both banks are low, but are not liable to overflow. Bench mark No. 1 is a Geological Survey 4-foot iron post located 200 feet north of the Union Pacific Railroad station on the west side of the street. Its elevation is 5.15 feet above gage datum and 3,370 feet above sea level. Bench mark No. 2 is the floor of the bridge over the zero of the gage scale. Its elevation is 9.10 feet above gage datum. Gage datum is the water surface when the gage reads zero.

The observations at this station during 1903 have been made under the direction of J. C. Stevens, district hydrographer.

Discharge measurements of South Platte River at Bigspring, Nebr., in 1903.

Date.	Hydrographer.	Gage height.	Discharge.
		Feet.	*Second-feet.*
April 1	J. C. Stevens	2.85	1,299
May 9	....do	1.98	20
July 8	....do	1.80	1
July 28	....do	1.80	2
August 20	....do	1.96	10

Mean daily gage height, in feet, of South Platte River at Bigspring, Nebr., for 1903.

Day.	Mar.	Apr.	May.	June.	July.	Aug.	Sept.	Oct.	Nov.
1		2.90	2.00	2.00	1.80	1.85	1.90	2.00	2.12
2		2.95	2.00	2.00	1.75	1.85	1.85	2.00	2.15
3	3.75	2.65	2.00	2.00	1.85	1.80	1.85	2.00	......
4	(a)	2.65	2.00	2.00	1.85	1.75	1.85	2.05	2.15
5	(a)	2.75	2.00	2.05	1.80	1.75	1.90	2.05	2.12
6	(a)	2.60	2.00	2.00	1.75	1.70	1.90	2.10	2.12
7	(a)	2.65	2.00	......	1.85	1.70	1.90	2.05	2.12
8	(a)	2.65	1.95	......	1.80	1.75	1.90	2.00	2.11
9	(a)	2.60	2.00	1.90	1.75	1.75	1.90	2.05	2.10
10	4.00	2.50	2.05	1.85	1.85	1.75	1.90	2.05	2.10
11	3.90	2.40	2.00	1.80	1.80	1.80	......	2.05	2.15
12	3.90	2.40	2.00	1.80	1.80	1.80	1.90	2.10	2.10
13	3.75	2.40	1.75	1.80	1.80	1.75	1.90	2.10	2.12
14	3.55	2.00	1.80	1.80	1.90	1.75	1.90	2.13	2.13
15	3.25	2.30	1.75	1.80	1.80	1.70	1.90	2.11	2.10
16	3.25	2.00	1.75	1.80	1.80	1.85	1.90	2.10	2.10
17	2.80	2.00	1.80	1.80	1.80	2.45	1.90	2.10	....

a Frozen at gage.

Mean daily gage height, in feet, of South Platte River, etc.—Continued.

Day.	Mar.	Apr.	May	June	July	Aug.	Sept.	Oct.	Nov.
18	2.60	2.00	1.80	1.80	1.80	2.35	1.95	2.11	1.18
19	2.25	2.00	1.90	1.90	1.80	2.00	1.90	2.10	1.18
20	2.25	2.00	1.95	1.90	1.75	1.98	1.90	2.10	1.18
21	2.30	2.00	1.90	1.90	.....	1.85	.....	2.10	1.18
22	2.30	2.00	1.95	1.95	1.80	1.85	1.95	2.08	.....
23	2.30	2.00	1.95	1.95	1.80	2.25	1.95	2.07	.....
24	2.15	2.00	2.00	1.95	1.75	2.05	1.95	2.07	.....
25	2.15	2.75	2.00	1.90	1.75	2.00	1.95	2.11	.....
26	2.25	2.00	1.95	1.85	1.75	2.00	1.95	2.11	.....
27	2.30	2.00	1.95	1.85	1.80	2.13	1.95	2.15	.....
28	2.15	2.00	1.90	1.80	1.75	2.00	2.00	2.15	.....
29	2.20	3.20	2.00	1.80	1.75	2.00	2.00	2.16	.....
30	2.15	3.10	2.00	1.80	1.80	1.95	2.00	2.19	.....
31	2.15	.....	.....	.....	1.90	1.95	.....	2.19	.....

Rating table for South Platte River at Bigspring, Nebr., from January 1 to December 31, 1903.

Gage height.	Discharge.	Gage height.	Discharge.	Gage height.	Discharge.	Gage height.	Discharge.
Feet.	Second-feet.	Feet.	Second-feet.	Feet.	Second-feet.	Feet.	Second-feet.
1.9	5	2.5	660	3.1	1,740	3.7	2,820
2.0	20	2.6	840	3.2	1,920	3.8	3,000
2.1	70	2.7	1,020	3.3	2,100	3.9	3,180
2.2	160	2.8	1,200	3.4	2,280	4.0	3,360
2.3	310	2.9	1,380	3.5	2,460		
2.4	480	3.0	1,560	3.6	2,640		

Estimated monthly discharge of South Platte River at Bigspring, Nebr., for 1903.

Month.	Discharge in second-feet.			Total in acre-feet.
	Maximum.	Minimum.	Mean.	
March 17-31	1,200	110	272	8,093
April	1,920	20	560	33,322
May 1-30	40	0	13	774
June 1-6 and 9-30	40	0	7	389
July (missed 21st)	5	0	Trace.	Trace.
August	570	0	42	2,582
September (missed 11th and 21st)	20	0	7	389
October	110	20	61	3,751
November 1-21 (missed 3d and 17th)	230	70	87	3,279

SOUTH PLATTE RIVER NEAR JULESBURG, COLO.

This station was established April 2, 1902, by John E. Field, at the wagon bridge crossing South Platte River, about 1 mile southeast of Julesburg, Colo., a station at the junction of the main line and Denver branch of the Union Pacific Railroad. As this is the last station on South Platte River in Colorado, and is also below all irrigation ditches in Colorado taking water from the South Platte, with the exception of one, it is of considerable importance for securing data relative to the flow of return waters as well as the natural flow in the main channel. However, the conditions at this station are not as desirable as they should be, owing to the great width of the channel and the general instability of the bed of the stream, features characteristic of the whole course of South Platte River through this part of the State. The channel at this point is about one-half mile wide, and at low stages of the river it is badly broken up by islands, causing the stream to flow in several different channels. The banks are low, but do not overflow. The bed of the river is dry during the greater part of the year. The gage heights from March 15 to March 31, and from November 25 to December 31, 1903, were affected by ice gorges and anchor ice, so that no rating table could be constructed for those periods. The gage is a 2 by 4 inch vertical timber spiked to a pile of the bridge about 600 feet from the north end. It is read once each day by T. W. Jenkins. Discharge measurements are made at high water from the bridge, which is one-half mile long and is supported on piles at intervals of 20 feet. The initial point for soundings is the north abutment on the left bank. The channel is straight for 600 feet above and 400 feet below the station. The right bank is low, but is not liable to overflow. The left bank overflows only at very high stages. The section is broad and shallow, with a sandy, changeable bed. The channel is broken by islands at ordinary stages, and is covered with undergrowth and vegetation. At low water it is impossible to measure the velocity in the many small channels and the discharge has to be estimated. The bench mark is a 20-penny nail driven into the south face of the cap on the piling to which the gage is spiked. It is directly over the gage rod and is inclosed in a circle of white paint and marked " B. M." Its elevation is 8 feet above the zero of the gage.

The observations at this station during 1903 have been made under the direction of M. C. Hinderlider, district hydrographer.

Discharge measurements of South Platte River near Julesburg, Colo., in 1903.

Date.	Hydrographer.	Gage height.	Discharge.
		Feet.	*Second-feet.*
April 1	J. C. Stevens	2.35	1,145
May 9	do	1.00	35
September 22 *a*	E. C. Murphy	1.00	4
April 17	M. C. Hinderlider	1.22	40
May 11 *b*	R. I. Meeker	1.09	36
Do	do		*c* 42
July 23 *a*	M. C. Hinderlider	.95	*c* 1
June 13 *a*	do	1.00	*c* 11
December 23	do	2.68	*d* 536

a Estimated.

b Water in 8 channels.

c Made 75 feet above head gate Western Irrigation District ditch. At this point the water is confined to one channel by sand and brush wing dam upstream for one-fourth mile. Head gate of canal about 2½ miles east of Julesburg.

d Very rough measurement owing to great quantity of ice in the channel, ice gorge at bridge greatly affecting the gage reading.

Mean daily gage height, in feet, of South Platte River at Julesburg, Colo., for 1903.

Day.	Mar.	Apr.	May.	June.	July.	Aug.	Sept.	Oct.	Nov.	Dec.
1		2.30	1.00	1.00	1.00	0.90	1.00	1.05		1.70
2		2.30	1.00	1.05	.95	.90	1.00	1.05		1.60
3		2.30	1.00	1.05	.95	.90	1.00	1.05		1.55
4		2.30	1.00	1.05	.95	.90	1.00	1.05		1.50
5		2.25	1.00	1.05	.95	.90	1.00	1.05		1.45
6		2.15	1.00	1.05	.95	.90	1.00	1.05		1.75
7		2.50	1.10	1.05	.95	.90	1.00	1.05		1.80
8		2.00	(*a*)	1.05	.95	.95	1.00	1.05		1.90
9		2.00	(*a*)	1.05	.95	.95	1.00	1.05		1.10
10		1.95	1.00	1.10	.95	.95	1.00	1.05	1.05	2.00
11		1.85	1.50	1.05	.95	.95	1.00	1.05	1.05	2.10
12		1.80	1.00	1.05	.95	.95	1.00	1.05	1.05	2.15
13		1.50	1.00	1.00	.95	.95	1.00	1.05	1.05	2.15
14		1.50	1.00	1.00	.95	.95	1.00	1.05	1.05	2.15
15		*b* 3.30	1.30	1.00	1.00	.95	.95	1.00	1.05	1.05	2.20
16		1.20	1.00	1.00	.95	*c* 2.28	1.00	1.05	1.10	2.40
17		1.15	1.15	1.00	.95	1.73	1.05	1.05	1.10	2.40
18	2.65	1.15	1.10	1.00	.95	1.20	1.05	1.00	1.10	2.40
19	2.60	1.25	1.00	1.00	.95	1.10	1.05	1.00	1.10	2.45
20	2.55	1.30	1.00	1.00	.95	1.10	1.05	1.00	1.10	2.70
21	2.50	1.20	1.00	1.00	.95	1.05	1.05	1.00	1.10	2.70
22	2.50	1.10	1.00	1.00	.95	1.05	1.05	1.00	1.10	2.9
23	2.50	1.10	1.00	1.00	.95	*c* 2.30	1.05	1.00	1.10	2.7
24	2.65	1.00	1.00	1.00	.95	1.95	1.05	1.00	1.10	2.6
25	2.40	.00	1.00	1.00	.90	1.35	1.05	1.00	*b* 1.80	2.6
26	2.40	.95	1.00	1.00	.90	1.10	1.05	1.00	1.80	2.5
27	2.30	.95	1.00	1.00	.90	1.05	1.05	1.00	1.85	2.4
28	2.30	.95	1.00	1.00	.90	1.05	1.05	1.00	1.85	2.4
29	2.30	1.00	1.10	1.00	.90	1.05	1.05	1.00	1.85	2.4
30	2.50	1.50	1.50	1.00	.90	1.00	1.05	1.00	1.75	2.4
31	2.30		1.00		.90	1.00		1.00		2.4

a Observer absent.

c Heavy rain.

b March 15 to 31 and November 25 to December 31, 1903, affected by ice gorges.

Rating table for South Platte River near Julesburg, Colo., from April 1 to November 24, 1903.

Gage height.	Discharge.	Gage height.	Discharge.	Gage height.	Discharge.	Gage height.	Discharge.
Feet.	Second-feet.	Feet.	Second-feet.	Feet.	Second-feet.	Feet.	Second-feet.
0.9	1	1.6	452	2.3	1,089	3.0	1,726
1.0	9	1.7	543	2.4	1,180	3.1	1,817
1.1	42	1.8	634	2.5	1,271	3.2	1,908
1.2	96	1.9	725	2.6	1,362	3.3	1,999
1.3	180	2.0	816	2.7	1,453	3.4	2,090
1.4	270	2.1	907	2.8	1,544	3.5	2,181
1.5	361	2.2	998	2.9	1,635	....	....

Discharge measurements have been made between 0.95 and 2.35 feet gage heights. On account of the shifting character of the river bed, the entire table is approximate.

Estimated monthly discharge of South Platte River near Julesburg, Colo., for 1903.

[Drainage area, 20,598 square miles.]

Month.	Discharge in second-feet.			Total in acre-feet.	Run-off.	
	Maximum.	Minimum.	Mean.		Second-feet per square mile.	Depth in inches.
April a	1,271	3	445	26,479	0.0216	0.0241
May	361	9	38	2,337	.0018	.0021
June	42	9	14	833	.0007	.0008
July	9	1	3	184	.00015	.00017
August	1,071	1	130	7,993	.0063	.0073
September	22	9	15	893	.0007	.0008
October	22	9	16	984	.0008	.0009
November 1–24	42	9	27	1,285	.0013	.0012
The period				40,988		

a Ice March 15 to 31 and November 25 to December 31.

MIDDLE CROW CREEK NEAR HECLA, WYO.

This station was established April 8, 1903, by G. W. Zorn, the constructing engineer of the Cheyenne dam. The gage is located one-fourth mile above the dam and 4 miles northwest of Hecla, Wyo. It is a vertical rod driven into the bed of the stream at its deepest part and securely fastened to the timber which spans the stream, and which, in turn, is fastened to stumps of trees on one bank and a post planted in the ground on the other. The gage is read twice each day by G. W. Zorn. Discharge measurements are made from the timber which spans the creek at the gage rod. The right bank has

been taken as the initial point for soundings at each measureme[...]
The channel is very winding. It is straight for only 10 feet abo[...]
and below the station. The banks are low, but are not liable to ov[...]
flow. There is but one channel, and the current velocity is swi[...]
The bed of the stream is composed of clean gravel, and is permane[...]
No bench marks have been established. This station has no conn[...]
tion with the station maintained during 1902.

The observations at this station during 1903 have been made und[...]
the direction of A. J. Parshall, district hydrographer.

Discharge measurements of Middle Crow Creek near Hecla, Wyo., in 1903.

Date.	Hydrographer.	Gage height.	Dischar[
		Feet.	*Second-f[*
May 13	A. J. Parshall	1.90	4[
June 7	do	1.20	2[
July 5	do	.75	[
July 30	do	.60	

Mean daily gage height, in feet, of Middle Crow Creek near Hecla, Wyo., for 19[

Day.	Apr.	May.	June.	July.	Aug.	Sept.	Oct.	N[
1		1.97	1.30	0.75	0.60	0.50	0.70	
2		1.85	1.25	.75	.60	.50	.60	
3		1.77	1.20	.75	.60	.50	.60	
4		1.75	1.30	.85	.55	.45	.60	
5		1.77	1.27	.75	.55	.45	.60	
6		1.82	1.32	.75	.55	.45	.60	
7		1.85	1.20	.70	.55	.45	.55	
8	0.85	2.05	1.18	.70	.55	.45	.55	
9	1.12	2.02	1.15	.70	.62	.50	.55	
10	1.40	2.00	1.20	.70	.60	.50	.55	
11	1.70	2.02	1.10	.65	.60	.50	.55	
12	1.47	1.95	1.05	.65	.60	.55	.55	
13	1.57	1.87	1.05	.65	.55	.60	.60	
14	1.40	1.90	1.42	.70	.55	.65	.60	
15	1.30	1.87	1.22	.72	.55	.65	.60	
16	1.22	1.87	1.10	.67	.55	.60	.60	
17	1.25	1.90	1.07	.78	.55	.60	.60	
18	1.20	1.80	1.00	.77	.55	.65	.55	
19	1.00	1.65	.95	.77	.55	.65	.55	
20	1.15	1.52	.95	.72	.55	.60	.55	
21	1.45	1.50	1.00	.67	.55	.60	.55	
22	1.77	1.50	.97	.65	.55	.60	.55	...
23	2.00	1.45	.95	.65	.57	.55	.55	...
24	2.05	1.40	.92	.77	1.00	.55	.55	...
25	2.20	1.40	.85	.77	.72	.50	.55	...
26	2.42	1.35	.85	.77	.60	.50	.55	...
27	2.60	1.30	.82	.70	.60	.50	.55	...
28	2.45	1.32	.80	.65	.55	.50	.55	...
29	1.92	1.45	.77	.62	.55	.55	.55	...
30	1.90	1.32	.77	.60	.55	.65	.55	...
31		1.27		.60	.50		.60	...

Rating table for Middle Crow Creek near Hecla, Wyo., from April 8 to December 31, 1903.

Gage height.	Discharge.	Gage height.	Discharge.	Gage height.	Discharge.	Gage height.	Discharge.
Feet.	*Second-feet.*	*Feet.*	*Second-feet.*	*Feet.*	*Second-feet.*	*Feet.*	*Second-feet.*
0.5	1.5	1.1	18.6	1.7	40.5	2.3	64.0
.6	2.7	1.2	22.2	1.8	44.3	2.4	68.0
.7	4.8	1.3	25.8	1.9	48.1	2.5	72.0
.8	7.9	1.4	29.4	2.0	52.0	2.6	76.0
.9	11.4	1.5	33.1	2.1	56.0		
1.0	15.0	1.6	36.8	2.2	60.0		

Table well defined.

Estimated monthly discharge of Middle Crow Creek near Hecla, Wyo., for 1903.

Month.	Discharge in second-feet.			Total in acre-feet.
	Maximum.	Minimum.	Mean.	
April 8–30	76.0	9.6	38.0	1,734
May	54.0	24.7	40.4	2,484
June	30.1	.7.0	17.4	1,035
July	9.6	2.7	5.2	320
August	15.0	1.5	2.8	172
September	3.7	1.1	2.1	125
October	4.8	2.1	2.4	148
November 1–21	2.7	2.1	2.6	108

SOUTH PLATTE RIVER AT KERSEY, COLO.

This station was established April 27, 1901, by A. L. Fellows, assisted by John E. Field, at Kersey, a station on the Union Pacific Railroad about 6 miles east of Greeley, at a bridge 1½ miles north of the railroad station itself. This station was intended to take the place of the one previously maintained at Orchard, Colo. This point was selected on account of the regularity of the channel and the fact that there was a wagon bridge crossing the river from which gagings could be made at high water. It is of particular value owing to the fact that the point is just below all the important tributaries of the South Platte, which derive their supply from the mountain region. It is also at about the point where water could be used to the best advantage for storage in reservoirs along South Platte River in northeastern Colorado. The observer is Edward K. Plumb, who lives near the bridge and reads the rod daily. The gage rod consists of a vertical 2 by 6 inch timber, 8 feet long, marked in feet and tenths and spiked to the downstream side of the third pile of the bridge from its south end.

Discharge measurements are made from the bridge at high water and by wading at low water. The initial point for soundings is the right end of the bridge on the downstream side. The channel is straight for 300 feet above and 400 feet below the station. The current is swift. The right bank is high and not liable to overflow. The left bank is low and not well defined. It extends into high timber and overflows at extreme high stages. This has caused very little trouble, as the water seldom gets high enough to overflow. The bed of the stream is composed of shifting sand and gravel. There is one channel at low water and two channels at high water. In making discharge measurements from the bridge great care must be taken, as the piles under the bridge break up the current and cause eddies. No bench mark has been established, but a spike has been driven at each foot mark into the pile to which the gage is attached.

Discharge measurements of South Platte River at Kersey, Colo., in 1903.

Date.	Hydrographer.	Gage height.	Discharge
		Feet.	*Second-fee*
January 8	M. C. Hinderlider	2.58	57
March 24	do	2.95	88
April 16	do	2.38	50
May 10	R. I. Meeker	2.00	31
June 14	M. C. Hinderlider	3.72	1,51
June 18	do	4.80	3,34
July 3	do	1.85	2
July 22	do	1.38	1:
August 14	R. I. Meeker	1.15	8
October 8	E. C. Murphy	1.55	11

Mean daily gage height, in feet, of South Platte River at Kersey. Colo., for 190

Day.	Jan.	Feb.	Mar.	Apr.	May.	June.	July.	Aug.	Sept.	Oct.	Nov.	De
1	2.40	2.40	2.70	3.40	2.50	1.00	2.30	1.30	1.30	1.50	2.00	2.
2	2.40	2.50	2.75	3.25	2.40	1.00	2.00	1.30	1.30	1.50	2.00	2.
3	2.40	2.55	2.75	2.90	2.40	1.00	1.85	1.30	1.30	1.50	2.10	2.
4	2.40	2.45	2.80	2.90	2.40	1.00	1.80	1.30	1.30	1.50	2.20	2.
5	2.50	2.40	2.90	2.80	2.50	1.50	2.30	1.30	1.30	1.50	2.20	2.
6	2.50	2.40	3.00	2.60	2.40	1.50	2.00	1.30	1.30	1.50	2.00	2.
7	2.50	2.40	3.00	2.50	2.20	1.90	2.00	2.30	1.20	1.50	2.00	2.
8	2.60	2.40	3.10	2.50	2.00	2.00	1.90	2.30	1.20	1.50	2.00	2.
9	2.60	2.40	3.20	2.45	1.90	2.90	1.85	1.50	1.20	1.50	2.10	2.

Mean daily gage height, in feet, of South Platte River, etc.—Continued.

Day.	Jan.	Feb.	Mar.	Apr.	May.	June.	July.	Aug.	Sept.	Oct.	Nov.	Dec.
10	2.60	2.40	3.20	2.40	1.85	4.00	1.80	1.30	1.20	1.55	2.10	2.20
11	2.60	2.40	3.20	2.40	1.80	4.20	1.50	1.20	1.20	1.60	2.05	2.20
12	5.60	2.40	3.40	2.40	1.50	4.00	1.50	1.20	1.20	1.75	2.05	2.20
13	2.50	2.40	3.20	2.40	1.50	3.80	1.40	1.20	1.20	1.70	2.00	2.30
14	2.45	2.40	3.10	2.40	1.45	3.60	1.40	1.20	1.25	1.65	2.00	2.30
15	2.45	2.50	3.00	2.35	1.40	3.80	1.40	1.20	1.25	1.70	2.10	2.35
16	2.40	2.50	3.00	2.30	1.40	4.00	1.40	1.20	1.30	1.65	2.10	2.35
17	2.40	2.50	3.00	2.20	1.30	4.30	1.40	1.20	1.40	1.80	2.20	2.35
18	2.40	2.50	3.00	2.20	1.30	4.70	1.35	1.20	1.40	1.80	2.25	2.40
19	2.40	2.60	3.00	2.20	1.25	5.00	1.35	1.20	1.40	1.75	2.30	2.40
20	2.40	2.60	3.00	2.20	1.20	4.60	1.30	1.20	1.40	1.70	2.40	2.40
21	2.40	2.70	3.00	2.20	1.25	4.00	1.30	1.20	1.40	1.70	2.40	2.40
22	2.40	2.80	3.00	2.15	1.25	4.00	1.30	1.20	1.45	1.70	2.40	2.35
23	2.40	2.80	2.95	2.15	1.25	3.85	1.30	1.20	1.45	1.75	2.35	2.35
24	2.40	2.90	2.95	2.10	1.25	3.75	1.30	1.20	1.50	1.70	2.30	2.30
25	2.40	2.80	2.95	2.00	1.25	3.43	1.30	1.40	1.50	1.70	2.30	2.30
26	2.40	2.75	2.95	2.00	1.20	3.23	1.30	1.50	1.50	1.65	2.30	2.30
27	2.40	2.70	2.95	2.00	1.20	2.75	1.30	1.40	1.50	1.70	2.30	2.30
28	2.40	2.70	3.00	2.00	1.10	2.70	1.30	1.30	1.50	1.70	2.30	2.30
29	2.40	......	3.20	2.00	1.05	2.50	1.30	1.30	1.50	1.75	2.30	2.30
30	2.40	......	3.50	2.50	1.00	2.50	1.30	1.30	1.50	1.80	2.30	2.30
31	2.40	......	3.70	......	1.00	......	1.30	1.30	......	2.00	......	2.30

Rating table for South Platte River at Kersey, Colo., from January 1 to December 31, 1903.

Gage height.	Discharge.	Gage height.	Discharge.	Gage height.	Discharge.	Gage height.	Discharge.
Feet.	Second-feet.	Feet.	Second-feet.	Feet.	Second-feet.	Feet.	Second-feet.
1.0	67	2.0	320	3.0	875	4.0	1,865
1.1	83	2.1	359	3.1	951	4.1	2,005
1.2	100	2.2	402	3.2	1,030	4.2	2,155
1.3	121	2.3	449	3.3	1,115	4.3	2,320
1.4	144	2.4	496	3.4	1,205	4.4	2,500
1.5	170	2.5	550	3.5	1,300	4.5	2,695
1.6	196	2.6	605	3.6	1,400	4.6	2,900
1.7	224	2.7	665	3.7	1,505	4.7	3,120
1.8	254	2.8	730	3.8	1,615	4.8	3,350
1.9	285	2.9	800	3.9	1,735	5.0	3,850

Curve well defined.

Estimated monthly discharge of South Platte River at Kersey, Colo., for 19

[Drainage area, 9,470 square miles.]

Month.	Discharge in second-feet.			Total in acre-feet.	Run-off.	
	Maximum.	Minimum.	Mean.		Second-feet per square mile.	Dep in
January	605	498	524	32,220	0.055	(
February	800	498	579	32,156	.061	
March	1,505	665	918	56,446	.097	
April	1,205	320	518	30,828	.055	
May	550	67	228	13,712	.024	
June	3,850	67	1,278	76,046	.135	
July	449	121	192	11,806	.020	
August	449	100	137	8,424	.014	
September	170	100	133	7,914	.014	
October	320	170	213	13,097	.022	
November	498	320	397	23,628	.042	
December	498	402	448	27,289	.047	
The year	3,850	67	463	333,506	.049	

CACHE LA POUDRE RIVER NEAR GREELEY, COLO.

This station was established February 5, 1903, by M. C. Hin
lider, assisted by J. J. Armstrong. The gage is located at a p
1 mile above the junction of the Cache la Poudre with South Pl
River. It is located on the right bank and consists of two secti
The lower section, reading from zero to 4.5 feet, consists of a 4
inch inclined fir timber, fastened to the bank at its upper and le
ends by means of stakes. One foot vertical equals 1.86 feet on
rod. A vertical 2 by 4 inch timber for higher readings is set in
bank at the upper end of the inclined section. Discharge meas
ments are made by wading at the gage rod. Approximate f
measurements may be made by gaging at a wagon bridge just al
the mouth of the river and at the Kersey station just below.
initial point for soundings has been taken at the water's edge.
channel is straight for 150 feet above and 100 feet below the sta1
The current has a moderate velocity at ordinary stages. The 1
bank is high and partly wooded and seldom overflows. The left I
consists of a sand and gravel bar rising gently from the main cha
of the stream. The bed of the stream is composed of sand and gr
and appears to be permanent. The cross section is regular, and t
is but one channel. The bench mark is a spike driven into the
root of a 30-inch cottonwood tree 63 feet southwest of the gage.
tree is blazed on the face next to the gage. The elevation of

bench mark is 7.64 feet above the zero of the gage. The gage is read twice each day by John Nastrom.

Discharge measurements of Cache la Poudre River near Greeley, Colo., in 1903.

Date.	Hydrographer.	Gage height.	Discharge.
		Feet.	*Second-feet.*
February 5	M. C. Hinderlider	1.30	59
March 24	do	1.50	108
April 16	do	1.68	141
May 10	R. I. Meeker	1.39	76
June 18	M. C. Hinderlider	4.50	1,676
July 22	do	1.12	45
August 14	R. I. Meeker	1.05	31
October 8	E. C. Murphy	1.40	91

Mean daily gage height, in feet, of Cache la Poudre River near Greeley, Colo., for 1903.

Day.	Mar.	Apr.	May.	June.	July.	Aug.	Sept.	Oct.
1		1.68	1.98	1.13	1.48	1.08	1.08	1.28
2		1.80	1.98	1.08	1.38	1.08	1.08	1.25
3		1.85	1.95	1.05	1.43	1.08	1.08	1.23
4		1.75	1.65	1.08	1.40	1.08	1.08	1.23
5		1.68	1.65	a 1.38	1.43	1.13	1.08	1.28
6		1.60	1.53	1.85	1.50	1.13	1.03	1.30
7		1.60	1.25	2.50	1.53	1.13	1.08	1.33
8		1.58	1.28	a 3.18	1.43	1.08	1.08	1.38
9		1.60	1.28	a 3.45	1.38	1.08	1.08	1.45
10		1.68	1.18	4.80	1.15	1.08	1.08	1.53
11		1.63	1.15	4.96	1.10	1.08	1.08	1.60
12		1.78	1.18	4.18	1.13	1.08	1.08	1.53
13		1.83	1.20	4.08	1.13	1.08	1.05	1.50
14		1.68	1.18	3.98	1.15	1.08	1.13	1.48
15		1.65	1.15	a 3.95	1.85	1.08	1.28	1.48
16		1.63	1.18	4.45	1.18	1.08	1.28	1.43
17		1.60	1.20	4.55	1.13	1.08	1.28	1.28
18		1.58	1.43	4.63	1.10	1.08	1.23	1.23
19		1.60	1.53	4.58	1.10	1.08	1.23	1.25
20		1.50	1.35	4.58	1.13	1.08	1.20	1.30
21		1.45	1.15	1.56	1.15	1.08	1.28	1.33
22		1.45	1.15	1.68	1.15	1.08	1.38	1.33
23		1.40	1.18	3.75	1.15	1.08	1.43	1.38
24	1.50	1.40	1.13	2.55	1.18	1.10	1.28	1.48
25	1.55	1.43	1.13	2.43	1.18	1.15	1.28	1.48
26	1.58	1.48	1.10	1.88	1.15	1.20	1.30	1.48
27	1.58	1.63	1.05	1.70	1.13	1.10	1.28	1.48
28	2.08	1.78	1.05	2.08	1.10	1.08	1.28	1.48
29	2.20	2.20	1.05	1.85	1.08	1.08	1.30	1.48
30	1.85	2.33	1.08	1.73	1.08	1.08	1.28	1.48
31	1.75		1.08		1.08	1.08		1.48

a Heavy rain.

Rating table for Cache la Poudre River near Greeley, Colo., from March 24 to December 31, 1903.

Gage height.	Discharge	Gage height.	Discharge.	Gage height.	Discharge.	Gage height.	Discharge.
Feet.	*Second-feet.*	*Feet.*	*Second-feet.*	*Feet.*	*Second-feet.*	*Feet.*	*Second-feet.*
1.0	25	2.0	218	3.0	576	4.0	1,204
1.1	39	2.1	245	3.1	625	4.1	1,285
1.2	55	2.2	278	3.2	676	4.2	1,370
1.3	72	2.3	308	3.3	731	4.3	1,465
1.4	90	2.4	335	3.4	789	4.4	1,570
1.5	108	2.5	369	3.5	849	4.5	1,690
1.6	127	2.6	406	3.6	912	4.6	1,825
1.7	147	2.7	445	3.7	979	4.7	1,975
1.8	169	2.8	486	3.8	1,050	4.8	2,140
1.9	193	2.9	530	3.9	1,125	5.0	2,520

Estimated monthly discharge of Cache la Poudre River near Greeley, Colo., fo 1903.

Month.	Discharge in second-feet			Total in acre-feet
	Maximum.	Minimum.	Mean.	
March 24–31	273	108	164	2,6
April	313	90	142	8,4
May	213	29	75	4,6
June....	2,480	32	785	46,7
July	181	36	64	3,9
August	55	29	36	2,2
September	95	29	51	3,0
October	127	60	89	5,4
The period	2,480	29	....	77,0

BIG THOMPSON CREEK NEAR ARKINS, COLO.

This stream drains the country immediately north of that drain
by the headwaters of St. Vrain Creek, and is one of the largest trib
taries of South Platte River, into which it empties about 4 miles abo
the town of Evans. Little Thompson Creek is an important tributa
of Big Thompson Creek, and the country drained by these two strear
makes up Irrigation District No. 4. The junction of these creeks
near the lower end of the district, a short distance above the poi
where their combined waters enter the South Platte.

Records of the flow of this stream were begun in April, 1888. The station was established at its present location on April 1, 1899. The only diversion above the gaging station is Handy ditch, a record of the gage heights of which is kept by the water commissioner of that district, J. M. Wolaver, who also kept the records of Big Thompson Creek at this point during the year 1900. It is necessary to include the discharge of Handy ditch in order to obtain the total run-off of the basin.

The station is located 4 miles southwest of Arkins, Colo., and 12 miles west of Loveland, Colo., on the ranch of John Chasteen. The gage is a vertical 2 by 4 inch timber fastened to the downstream side of the wagon bridge at the right bank. This bridge is located about one-fourth mile below the mouth of the canyon. The gage is read twice each day by L. R. Stone. Discharge measurements are made from the wagon bridge, and at low water by wading below the bridge. The initial point for soundings is the south abutment, on the right bank of the river. The channel is straight for 600 feet above and 300 feet below the station. The current is rapid. Both banks are high and wooded, and are not liable to overflow. There is but one channel, broken by the center pier of the bridge. The bed of the stream is lined with bowlders and is very rough, but is not subject to change. The bench mark is a nail in the root of a cottonwood stump 25 feet south of the gage. Its elevation is 9.35 feet above the zero of the gage.

Discharge measurements of Big Thompson Creek near Arkins, Colo., in 1903.

Date.	Hydrographer.	Gage height.	Discharge.
		Feet.	*Second-feet.*
October 7	E. C. Murphy	0.80	76
April 13	M. C. Hinderlider	.78	83
May 9	R. I. Meeker	1.35	285
June 17	M. C. Hinderlider	2.88	1,063
July 2	do	2.23	698
July 21	do	1.78	429
August 13	R. I. Meeker	.91	114
September 10	do	.75	86

Mean daily gage height, in feet, of Big Thompson Creek near Arkins, Colo., for 1903.

Day.	Apr.	May.	June.	July.	Aug.	Sept.
1		0.88	1.73	2.20	1.28	0.73
2		.85	1.65	2.20	1.13	.70
3		.78	1.90	2.18	1.13	.70
4		.75	2.08	1.98	1.08	.65
5		.78	a2.95	1.98	1.08	.65
6		.75	2.20	1.85	1.08	.65
7		.98	a2.25	2.05	1.10	.65
8		1.10	2.45	2.15	1.13	.65
9		1.38	2.45	2.10	1.10	.65
10		1.43	2.35	2.13	1.10	.73
11		1.40	2.25	2.10	1.00	.70
12		1.38	2.33	2.05	.95	.65
13		1.41	2.45	2.08	.95	.80
14		1.58	2.38	2.05	.95	.65
15	0.83	1.73	a2.55	2.08	.93	.65
16	.85	1.83	2.48	2.03	1.00	.90
17	.88	1.90	2.68	2.25	1.03	.85
18	.78	1.70	2.80	2.03	.98	.73
19	b.50	1.58	2.68	1.90	1.00	.75
20	.53	1.40	2.65	1.73	.98	.70
21	.49	1.45	2.45	1.70	.95	.70
22	.54	1.25	2.30	1.83	1.03	(c)
23	.63	1.20	2.33	1.90	1.03	
24	.73	1.20	2.33	1.88	1.15	
25	.83	1.15	2.20	1.83	1.13	
26	.85	1.18	2.18	1.78	1.05	
27	.90	1.27	2.25	1.70	.98	
28	.95	1.33	2.40	1.65	.93	
29	.88	1.35	2.50	1.50	.88	
30	.75	1.43	2.40	1.40	.68	
31		a1.60		1.35	.70	

a Heavy rain.   b Fall due to water being taken out by Handy ditch.   c Observer moved away.

Rating table for Big Thompson Creek near Arkins, Colo., from January 1 to December 31, 1903.

Gage height.	Discharge.	Gage height.	Discharge.	Gage height.	Discharge.	Gage height.	Discharge.
Feet.	Second-feet.	Feet.	Second-feet.	Feet.	Second-feet.	Feet.	Second-feet.
0.4	27	1.0	138	1.6	349	2.2	669
.5	37	1.1	167	1.7	392	2.3	727
.6	52	1.2	198	1.8	440	2.4	785
.7	69	1.3	231	1.9	495	2.5	843
.8	88	1.4	268	2.0	553	2.6	901
.9	111	1.5	308	2.1	611	2.7	959

Estimated monthly discharge of Big Thompson Creek near Arkins, Colo., for 1903.[a]

[Drainage area, 305 square miles.]

Month.	Discharge in second-feet.			Total in acre-feet.	Run-off.	
	Maximum.	Minimum.	Mean.		Second-feet per square mile.	Depth in inches.
April 15–30.........	195	60	114	3,618	0.37	0.22
May	526	150	274	16,848	.90	1.04
June..	1,182	402	885	52,661	2.90	3.24
July	819	281	568	34,925	1.86	2.14
August	256	86	170	10,453	.56	.65
September 1–2 [b]	139	81	102	4,249	.33	.26
The period ...				122,754		

[a] Monthly discharge of Big Thompson Creek includes water in Handy ditch.
[b] Observer moved.

HANDY DITCH NEAR ARKINS, COLO.

Handy ditch diverts water for irrigating purposes, taking its supply from Big Thompson Creek where the latter emerges from the
mountains, at a point about one-fourth mile above the gaging station on the Big Thompson. The capacity of this ditch is about 160
second-feet. The rating flume on the ditch is located about one-
fourth mile below the head gates and within about 200 yards of the
gaging station on the river. The flume is of trapezoidal form and
equipped with a gage. A complete record was kept of the amount of
water passing through the flume during the year 1903. The amount
of water passing through the flume added to the amount passing
the gaging station on the creek gives the total discharge of Big
Thompson Creek where it leaves the mountains. Above this point
little or no water is diverted.

Discharge measurements of Handy ditch near Arkins, Colo., in 1903.

Date.	Hydrographer.	Gage height.	Discharge.
		Feet.	*Second-feet.*
July 2...................... .	M. C. Hinderlider	2.14	157
July 21........................ .....do		.70	37
August 13 . ... R. I. Meeker..................		.62	33
September 10 ....do		.32	17
October 7 . ... E. C. Murphy		.30	15
Do ....do		.30	16

Mean daily gage height, in feet, of Big Thompson Creek near Arkins, Colo., for 1903.

Day.	Apr.	May.	June.	July.	Aug.	Sept.
1		0.88	1.73	2.20	1.28	0.7
2		.85	1.65	2.20	1.13	.7
3		.78	1.90	2.18	1.13	.7
4		.75	2.08	1.96	1.08	.6
5		.78	a2.25	1.98	1.08	.6
6		.75	2.20	1.85	1.08	.6
7		.96	a2.25	2.05	1.10	.6
8		1.10	2.45	2.15	1.13	.9
9		1.38	2.45	2.10	1.10	.8
10		1.43	2.35	2.13	1.10	.7
11		1.40	2.35	2.10	1.00	.7
12		1.38	2.33	2.05	.95	.8
13		1.41	2.45	2.08	.95	.8
14		1.58	2.38	2.05	.95	.9
15	0.83	1.73	a2.55	2.08	.93	.8
16	.85	1.83	2.48	2.08	1.00	.9
17	.88	1.90	2.68	2.25	1.08	.8
18	.73	1.70	2.80	2.03	.98	.7
19	b.50	1.53	2.68	1.90	1.00	.7
20	.53	1.40	2.65	1.73	.98	.7
21	.49	1.45	2.45	1.70	.95	.7
22	.54	1.25	2.30	1.83	1.08	(c)
23	.63	1.20	2.38	1.90	1.08	......
24	.73	1.20	2.33	1.88	1.15	......
25	.88	1.15	2.20	1.83	1.13	......
26	.85	1.18	2.18	1.78	1.05	......
27	.90	1.27	2.25	1.70	.98	......
28	.95	1.33	2.40	1.65	.93	......
29	.88	1.35	2.50	1.50	.88	......
30	.75	1.43	2.40	1.40	.68	......
31		a1.60		1.35	.70	......

a Heavy rain.   *b* Fall due to water being taken out by Handy ditch.   *c* Observer moved away

Rating table for Big Thompson Creek near Arkins, Colo., from January 1 to December 31, 1903.

Gage height.	Discharge.	Gage height.	Discharge.	Gage height.	Discharge.	Gage height.	Discharge.
Feet.	Second-feet	Feet.	Second-feet.	Feet.	Second-feet.	Feet.	Second-feet.
0.4	27	1.0	138	1.6	349	2.2	669
.5	37	1.1	167	1.7	392	2.3	727
.6	52	1.2	198	1.8	440	2.4	785
.7	69	1.3	231	1.9	495	2.5	843
.8	88	1.4	268	2.0	553	2.6	901
.9	111	1.5	308	2.1	611	2.7	959

Estimated monthly discharge of Big Thompson Creek near Arkins, Colo., for 1903.[a]

[Drainage area, 305 square miles.]

Month.	Discharge in second-feet.			Total in acre-feet.	Run-off.	
	Maximum.	Minimum.	Mean.		Second-feet per square mile.	Depth in inches.
April 15–30	195	60	114	3,618	0.37	0.22
May	526	150	274	16,848	.90	1.04
June	1,182	402	885	52,661	2.90	3.24
July	819	281	568	34,925	1.86	2.14
August	256	86	170	10,453	.56	.65
September 1–2 [b]	139	81	102	4,249	.33	.26
The period				122,754		

[a] Monthly discharge of Big Thompson Creek includes water in Handy ditch.
[b] Observer moved.

HANDY DITCH NEAR ARKINS, COLO.

Handy ditch diverts water for irrigating purposes, taking its supply from Big Thompson Creek where the latter emerges from the mountains, at a point about one-fourth mile above the gaging station on the Big Thompson. The capacity of this ditch is about 160 second-feet. The rating flume on the ditch is located about one-fourth mile below the head gates and within about 200 yards of the gaging station on the river. The flume is of trapezoidal form and equipped with a gage. A complete record was kept of the amount of water passing through the flume during the year 1903. The amount of water passing through the flume added to the amount passing the gaging station on the creek gives the total discharge of Big Thompson Creek where it leaves the mountains. Above this point little or no water is diverted.

Discharge measurements of Handy ditch near Arkins, Colo., in 1903.

Date.	Hydrographer.	Gage height.	Discharge.
		Feet.	*Second-feet.*
July 2	M. C. Hinderlider	2.14	157
July 21	do	.70	37
August 13	R. I. Meeker	.62	33
September 10	do	.32	17
October 7	E. C. Murphy	.30	15
Do	do	.30	16

Mean daily gage height, in feet, of Handy ditch near Arkins, Colo., for 1903.

Day.	Apr.	May.	June.	July.	Aug.	Sept.
1		1.20	0.60	2.00	0.60	0.40
2		1.20	.60	2.06	.60	.40
3		1.20	.60	1.85	.60	.40
4		1.20	.83	1.70	.60	.30
5		1.20	1.28	1.20	.60	.30
6		1.20	1.75	.70	.60	.30
7		.95	2.05	.70	.60	.30
8		.85	2.05	.90	.60	.30
9		(a)	2.00	1.25	.60	.30
10		(a)	2.00	1.20	.60	.30
11		(a)	2.00	.83	.60	.30
12		(a)	2.00	.60	.60	.30
13		(a)	2.30	.60	.60	.30
14		(a)	2.30	.60	.50	.30
15		.28	2.30	.75	.40	.30
16		.28	2.30	.60	.40	.30
17		.58	2.25	.60	.40	.30
18	0.40	.70	2.28	1.00	.40	.30
19	.45	.70	2.28	.80	.40	.30
20	.45	.70	2.30	.60	.40	.30
21	.53	.70	2.30	.60	.40	.30
22	.60	.65	2.30	.60	.30	(b)
23	.60	.70	2.30	.60	.30	
24	.60	.65	2.30	.60	.30	
25	.60	.60	2.30	.60	.50	
26	1.10	.60	2.30	.60	.50	
27	1.10	.60	2.20	.60	.40	
28	1.18	.60	2.15	.60	.40	
29	1.20	.60	2.00	.60	.40	
30	1.15	.60	2.10	.60	.40	
31		.60		.60	.40	

a Water shut out for repairs.   b Observer moved away.

Rating table for Handy ditch near Arkins, Colo., from January 1 to December 31, 1903.

Gage height.	Discharge.	Gage height.	Discharge.	Gage height.	Discharge.	Gage height.	Discharge.
Feet.	Second-feet.	Feet.	Second-feet.	Feet.	Second-feet.	Feet.	Second-feet.
0.2	11	0.8	44	1.4	87	2.0	142
.3	15	.9	51	1.5	95	2.1	152
.4	20	1.0	58	1.6	104	2.2	162
.5	26	1.1	65	1.7	113	2.3	172
.6	32	1.2	72	1.8	122		
.7	38	1.3	79	1.9	132		

Curve fairly well defined.

Estimated monthly discharge of Handy ditch near Arkins, Colo., for 1903.

Month.	Discharge in second-feet.			Total in acre-feet.
	Maximum.	Minimum.	Mean.	
pril 18–30	72	20	43	1,109
[ay 1–8 ª and 15–31	72	12	42	2,083
une	172	32	140	8,331
uly	150	32	52	3,197
August	32	15	25	1,537
September 1–21	20	15	16	666
The period				16,923

ª Water shut out for repairs to ditch.

ST. VRAIN CREEK NEAR LYONS, COLO.

The town of Lyons is situated between the north and south forks of St. Vrain Creek, the forks uniting at a point about 1 mile east of the center of the town. Records of the flow of the creek at or near Lyons have been kept since April, 1888, except during the years 1893 and 1894, but the station was not put in its present condition until May 11, 1895, since which date records have been kept throughout each irrigation season. To obtain the total run-off of the drainage basin of the creek at this station the amount of water in the ditch must be added to the discharge of the creek. The present station is located one-half mile southeast of Lyons, Colo., below the junction of the north and south forks of St. Vrain Creek. It is opposite the switch of the Burlington and Missouri River Railroad and is about 500 feet below the head of the supply ditch which takes water out on the left side of the stream. The gage is an inclined 2 by 4 inch timber graduated to vertical tenths of a foot with brass-headed nails. It is embedded in the slag on the slope of the railroad embankment. The gage is read twice each day by Miss Bessie Sites. When the gage height is 3 feet or less discharge measurements are made by wading near the gage. At high water measurements are made from a wagon bridge about 600 feet below the gage. The initial point for soundings is the gage rod at low water and is the left end of the bridge at the north abutment when measurements are made from the bridge. The channel is straight for 300 feet above and 200 feet below the station. The current is rapid. The bed of the stream is composed of bowlders and cobblestones and is rough. The right bank is low, rocky, partly wooded, and liable to overflow. The left bank is the railroad embankment, which is partly protected by a covering of slag and some bowlders. There is but one channel at all stages. At high water the current is somewhat affected by the bridge piers and the abutments. The bench mark is a spike in the root of a large cottonwood tree 150 feet north of the gage. Its elevation, when established, was 6.51 feet above the zero of the gage. On November 8, 1902, its elevation was found to be 6.47 feet above the zero of the gage.

Discharge measurements of St. Vrain Creek near Lyons, Colo., in 1903.

Date.	Hydrographer.	Gage height.	Discharge
		Feet.	Second-feet
April 14	M. C. Hinderlider	2.38	9
Do	do	2.35	7
May 8	R. I. Meeker	2.85	2
Do	do	2.85	3
June 16	M. C. Hinderlider	4.01	8
July 1	do	3.55	4
July 30	do	3.20	1
August 12	R. I. Meeker	2.50	
September 9	do	2.27	
October 6	E. C. Murphy	2.08	

Mean daily gage height, in feet, of St. Vrain Creek near Lyons, Colo., for 1903.

Day.	Apr.	May.	June.	July.	Aug.	Sept.	Oc
1		2.85	3.10	3.65	2.73	2.38	
2	2.30	2.95	3.00	3.65	2.65	2.33	
3	2.48	2.90	3.15	3.60	2.63	2.33	
4	2.48	3.00	3.30	3.35	2.63	2.30	
5	2.48	2.88	3.40	3.10	2.60	2.28	
6	2.48	2.83	3.48	3.13	2.63	2.33	
7	2.48	2.83	3.53	3.25	2.63	2.33	
8	2.43	2.88	3.73	3.40	2.63	2.35	
9	2.43	2.83	3.80	3.35	2.63	2.30	
10	2.38	2.95	3.75	3.40	2.56	2.25	
11	2.35	2.85	3.70	3.38	2.50	2.30	
12	2.30	2.88	3.83	3.30	2.50	2.30	
13	2.60	2.88	3.85	3.25	2.50	2.30	
14	2.65	2.85	3.73	3.25	2.43	2.40	
15	2.73	2.85	3.75	3.30	2.45	2.38	
16	2.55	3.13	4.05	3.45	2.40	2.33	
17	2.65	3.30	4.15	3.65	2.45	2.30	
18	2.65	3.35	4.15	3.65	2.43	2.33	
19	2.73	2.85	4.10	3.20	2.43	2.30	
20	2.73	2.83	4.15	3.15	2.40	2.25	
21	2.48	2.83	4.05	3.33	2.45	2.23	
22	2.43	2.73	3.90	3.38	2.50	2.30	
23	2.53	2.73	4.25	3.30	2.65	2.20	
24	2.48	2.88	3.90	3.25	2.55	2.20	
25	2.48	2.88	3.77	3.15	2.60	2.18	
26	2.45	2.73	3.65	2.85	2.55	2.20	
27	2.45	2.73	3.70	3.05	2.30	2.20	
28	2.55	2.40	3.73	2.92	2.30	2.10	
29	2.45	2.85	3.85	2.92	2.30	2.08	
30	2.53	2.95	3.90	2.85	2.43	2.10	
31		2.85		2.90	2.40		

Mean storage in Clear Creek Park reservoir.

Rating table for St. Vrain Creek near Lyons, Colo., from January 1 to December 31, 1903.

Gage height.	Discharge.	Gage height.	Discharge.	Gage height.	Discharge.	Gage height.	Discharge.
Feet.	*Second-feet.*	*Feet.*	*Second-feet.*	*Feet.*	*Second-feet.*	*Feet.*	*Second-feet.*
2.0	23	2.6	146	3.2	334	3.8	660
2.1	34	2.7	174	3.3	372	3.9	752
2.2	50	2.8	203	3.4	414	4.0	860
2.3	70	2.9	234	3.5	461	4.1	980
2.4	93	3.0	266	3.6	516	4.2	1,110
2.5	119	3.1	299	3.7	582	4.3	1,250

Curve well defined up to 4 feet gage height.

Estimated monthly discharge of St. Vrain Creek near Lyons, Colo., for 1903.[a]

[Drainage area, 209 square miles.]

Month.	Discharge in second-feet.			Total in acre-feet.	Run-off.	
	Maximum.	Minimum.	Mean.		Second-feet per square mile.	Depth in inches.
April 2–30	308	82	175	10,066	0.84	0.91
May	489	173	269	16,540	1.29	1.49
June	1,280	271	769	45,759	3.68	4.11
July	638	210	412	25,333	1.97	2.37
August	190	100	139	8,547	.67	.77
September	100	44	72	4,284	.34	.38
October	78	35	50	3,074	.24	.28
The period				113,603		

[a] Supply ditch included in computations.

SUPPLY DITCH AT LYONS, COLO.

The supply ditch takes its supply of water for irrigating purposes from St. Vrain Creek at a point about 500 feet above the gage rod in the creek. The capacity of the ditch is approximately 90 second feet. The rating flume on the ditch is located about 200 feet from the head gates, the same being of rectangular cross section, in good repair and equipped with a gage. Observations were made twice daily of the amount of water passing through this flume during 1903, and discharge measurements made each time gagings were made of the creek. The amount of water diverted by this canal, added to the amount of water in St. Vrain Creek at the gaging station gives the total amount of water in the creek at this point. This is almost the entire discharge of the creek, small amounts being diverted above this point

Discharge measurements of supply ditch at Lyons, Colo., in 1903.

Date.	Hydrographer.	Gage height.	Discharge.
		Feet.	*Second-feet.*
May 8	R. I. Meeker	0.80	44
June 16	M. C. Hinderlider	.95	60
July 1	....do	1.18	87
July 20	....do	.70	31
August 12	R. I. Meeker	.30	7
September 9	....do	.30	7
October 6	E. C. Murphy	.50	18

Mean daily gage height, in feet, of supply ditch at Lyons, Colo., for 1903.

Day.	Apr.	May.	June.	July.	Aug.	Sept.	Oct.
1		0.35	0.25	1.20	0.30	0.30	0.50
2		.60	.25	1.20	.30	.30	.50
3		.60	.25	1.10	.30	.30	.50
4		.60	1.20	1.10	.30	.30	.50
5		.60	1.15	.80	.30	.30	.45
6		.70	1.15	.70	.30	.30	.30
7		.80	1.15	.80	.30	.30	.30
8		.90	1.50	.90	.30	.30	.30
9		1.05	1.50	.90	.30	.30	.30
10		1.10	1.50	.80	.30	.30	.30
11		1.10	1.40	.70	.30	.30	.30
12		1.00	1.40	.70	.30	.30	.30
13		1.00	1.50	.70	.30	.30	.30
14		1.00	1.55	.20	.30	.30	.30
15		1.00	1.70	.90	.30	.30	.30
16		.90	1.00	.75	.30	.30	.30
17		1.00	1.00	.75	.30	.30	.30
18		1.15	1.20	.75	.30	.30	.38
19	0.20	.95	1.30	1.05	.30	.50	.30
20	.30	.80	1.30	.75	.30	.30	.30
21	.40	.68	1.30	.75	.60	.30	.30
22	.50	.58	1.30	.75	.60	.30	.30
23	.60	.53	1.30	.75	.65	.30	.35
24	.60	.50	1.40	.75	.65	.30	.40
25	.35	.25	1.30	.75	.60	.30	.40
26	.10	.25	1.20	.75	.60	.30	.40
27	.60	.25	1.20	.75	.60	.30	.40
28	.60	.25	1.20	.75	.35	.40	.40
29	.10	.25	1.20	.55	.35	.40	.40
30	.10	.25	1.20	.30	.35	.40	.40
31		.25		.30	.35		.40

Rating table for supply ditch at Lyons, Colo., from January 1 to December 31, 1903.

Gage height.	Discharge.	Gage height.	Discharge.	Gage height.	Discharge.	Gage height.	Discharge.
Feet.	*Second-feet.*	*Feet.*	*Second-feet.*	*Feet.*	*Second-feet.*	*Feet.*	*Second-feet.*
0.1	1	0.6	25	1.1	78	1.6	133
.2	3	.7	34	1.2	89	1.7	144
.3	7	.8	45	1.3	100		
.4	12	.9	56	1.4	111		
.5	18	1.0	67	1.5	122		

Curve well defined.

Estimated monthly discharge of supply ditch at Lyons, Colo., for 1903.

Month.	Discharge in second-feet.			Total in acre-feet.
	Maximum.	Minimum.	Mean.	
April 19–30	25	1	13	309
May	83	5	38	2,337
June	144	5	90	5,355
July	89	3	44	2,705
August	30	7	12	738
September	12	7	7.50	446
October	18	7	10	615
The period				12,505

CLEAR CREEK AT FORKSCREEK, COLO.

This station was established May 29, 1899, by John E. Field. The station is located at the footbridge just below the Colorado and Southern Railroad bridge near the railroad station. The chain gage is fastened to the wooden cribwork on the left bank 30 feet below the footbridge. The length of the chain from the end of the weight to the marker is 20.25 feet. Discharge measurements are made from the single-span footbridge. The initial point for soundings is a brass-headed nail on the rail of the bridge at the north end. It is marked zero with white paint. The channel is straight for 30 feet above and 150 feet below the station. Both banks are high and are not liable to overflow. The right bank is the side of the mountain and the left bank is the retaining wall of the railroad embankment. The bed of the stream is composed of small cobblestones and bowlders, with some silt at low stages. The current is rapid and it is necessary to use a stay wire on the meter at high stages. The bench mark, established October 8, 1903, is a cross cut in the top of the downstream stringer over the first strut from the south end of the footbridge. When established its elevation above the water surface was 12.73 at a gage reading of 2 feet. The gage is read twice each day by C. W Hoisington.

Discharge measurements of Clear Creek at Forkscreek, Colo., in 1903.

Date.	Hydrographer.	Gage height.	Dischar;
		Feet.	*Second-f*
May 1	M. C. Hinderlider	2.10	1
May 14	....do	2.65	;
May 29	....do	2.58	;
June 20	R. I. Meeker	3.60	1,
June 25	M. C. Hinderlider	3.58	;
July 17	....do	3.24	(
September 7	R. I. Meeker	2.30	;
October 8	M. C. Hinderlider	2.03	;
July 16	....do	2.15	;

Mean daily gage height, in feet, of Clear Creek at Forkscreek, Colo., for 1903

Day.	Apr.	May.	June.	July.	Aug.	Sept.	Oc
1	1.60	2.05	2.75	3.55	2.70	2.18	;
2	1.55	2.15	2.75	3.55	2.65	2.15	;
3	1.40	2.05	3.00	3.48	2.58	2.18	;
4	(a)	2.05	2.95	3.25	2.55	2.10	;
5	(a)	2.15	2.98	3.15	2.58	2.13	
6	(a)	2.23	3.00	3.18	2.55	2.15	
7	1.30	2.25	3.08	3.15	2.48	2.28	
8	1.40	2.35	3.10	3.23	2.43	2.18	
9	1.50	2.40	3.13	3.18	2.40	2.15	
10	1.55	2.65	3.10	3.20	2.38	2.10	
11	1.55	2.55	3.00	·3.18	2.35	2.10	
12	1.50	2.55	3.15	3.10	2.35	2.13	
13	1.40	2.45	3.20	3.10	2.38	2.10	
14	1.40	2.55	3.18	3.15	2.35	2.18	
15	1.45	2.73	3.23	3.15	2.33	2.10	
16	1.40	2.75	3.53	3.30	2.45	2.03	
17	1.40	2.85	3.55	3.25	2.38	2.10	
18	1.45	2.65	3.60	3.18	2.30	2.13	
19	1.55	2.50	3.70	3.03	2.28	2.10	
20	1.50	2.50	3.80	2.95	2.25	2.10	
21	1.55	2.55	3.68	2.93	2.25	2.03	
22	1.75	2.48	3.73	2.95	2.23	2.00	
23	1.75	2.43	3.68	2.95	2.53	2.00	
24	1.95	2.43	3.65	3.10	2.55	2.00	
25	2.00	2.45	3.60	3.05	2.38	1.98	
26	2.05	2.63	3.55	2.93	2.40	1.95	
27	2.15	2.55	3.53	2.88	2.33	1.95	
28	2.15	2.60	3.58	2.80	2.28	1.90	
29	2.05	2.60	3.60	2.78	2.25	2.00	
30	2.00	2.63	3.55	2.68	2.23	2.00	
31		2.65		2.70	2.30		

a Water below weight.

Rating table for Clear Creek at Forkscreek, Colo., from April 1 to December 31, 1903.

Gage height.	Discharge.	Gage height.	Discharge.	Gage height.	Discharge.	Gage height.	Discharge.
Feet.	Second-feet.	Feet.	Second-feet.	Feet.	Second-feet.	Feet.	Second-feet.
1.3	30	2.0	104	2.7	338	3.4	784
1.4	34	2.1	127	2.8	385	3.5	898
1.5	39	2.2	154	2.9	436	3.6	1,042
1.6	48	2.3	183	3.0	491	3.7	1,220
1.7	58	2.4	217	3.1	550	3.8	1,438
1.8	70	2.5	254	3.2	616		
1.9	85	2.6	295	3.3	692		

Estimated monthly discharge of Clear Creek at Forkscreek, Colo., for 1903.

[Drainage area, 345 square miles.]

Month.	Discharge in second-feet.			Total in acre-feet.	Run-off.	
	Maximum.	Minimum.	Mean.		Second-feet per square mile.	Depth in inches.
April (4, 5, and 6 missing)a	140	30	60	3,213	0.17	0.17
May	410	115	248	15,249	.72	.83
June.................	1,438	361	804	47,841	2.33	2.60
July	966	329	567	34,863	1.64	1.89
August	338	154	219	13,466	.63	.73
September	177	85	124	7,379	.36	.40
October	140	77	108	6,641	.31	.36
The period.. ..	1,438	30	....	128,652	....	....

a Water below weight.

SOUTH PLATTE RIVER AT DENVER, COLO.

In the spring of 1895 a river station was established at the Twenty-third Street Viaduct in the city of Denver, but observations were discontinued on June 18, as the location was found to be unfavorable for accurate measurements, and the water had fallen below the gage. In July a station was established at the Fifteenth Street Bridge and observations were begun, these being made in the morning and in the afternoon. Stream measurements were then made from the lower side of the bridge. The original gage consisted of two 6 by 2 inch

timbers spiked together, inclined and graduated to vertical tenths (
a foot. The space between the marks was 0.156 foot. The timber
were fastened to posts driven into the bank. The bench mark is 10
feet southwest from the gage and is a cross mark on top of the ea
abutment of the Fifteenth Street Bridge on the north corner. It
marked "B. M." and is 6.15 feet above the 9-foot mark of the origin:
Fifteenth street gage rod, which has since been destroyed.

Another inclined gage rod, reading the same as the one on the rigl
bank, was placed on the left side in August, 1898. It consisted of
4 by 4 inch by 12-foot timber fastened to posts driven into the le
bank and graduated to vertical feet and tenths.

This rod was washed out by the high water of June, 1900, whic
also removed the sand bar in front of the rod on the right-hand sid
making it available at low-water stages, and since that time the rea
ings have been taken from the latter rod, or from rods which replace
it and which were located at the same point.

May 15, 1901, a T-rail was placed on the site of the latter rod, whic
was stolen. The rail was embedded in an inclined position in th
slag bank. All readings were taken from this rod from May 15, 190
to June 9, 1903, when the rod was again stolen. On June 10, 1903,
vertical 4 by 4 inch timber was placed at the same point and fastene
to a cottonwood tree. The zero of the rod was placed at the sam
elevation as the zero of the old gage. It is read twice each day b
Clarence Crisman.

The present gage and the bridge from which measurements are mac
are located at a point immediately below the mouth of Cherry Creel
which enters between the Fourteenth and Fifteenth Street bridge
The stream at this point is confined between artificial embankmen
of furnace slag.

Discharge measurements are made by wading at a point below
above the gage at low water and from the Fifteenth Street Bridge
high water. The initial point for soundings is the edge of the rig
abutment of the Fifteenth Street Bridge. The channel is straight f
100 feet above and for 500 feet below the station. The current
swift at high stages, but becomes sluggish at low stages. The ban
are slag embankments and can only overflow at extreme flood stag
The bed is composed of sand, free from vegetation, but very shiftir
The bench mark is the top of the capstone on the south end of t
southeast abutment of the Sixteenth Street Viaduct on the right ba
below the gage. Its elevation is 12 feet above the zero of the gaj
The channel scoured during the flood about June 15, 1903, neces
tating two rating tables for this year.

ischarge measurements of South Platte River at Denver, Colo., in 1903.

Date.	Hydrographer.	Gage height.	Discharge.
		Feet.	*Second-feet.*
y 5	M. C. Hinderlider	1.29	78
y 27	do	1.21	75
ry 17	do	1.00	49
11	do	1.61	150
28	F. Cogswell	1.10	74
3	M. C. Hinderlider	1.06	85
6	R. I. Meeker	1.16	83
1	do	.96	46
	do	2.15	312
)	G. B. Monk	2.80	788
3	do	3.00	1,240
	do	2.20	507
	do	1.40	263
	T. E. Brick	1.40	276
	G. B. Monk	1.20	186
	R. I. Meeker	.70	77
	G. B. Monk	.70	68
t 7	do	.50	61
t 27	do	1.30	174
iber 17	T. E. Brick	1.20	189
	do	1.20	171
r 6	R. I. Meeker	.87	85

laily gage height, in feet, of South Platte River at Denver, Colo., for 1903.

Day.	Jan.	Feb.	Mar.	Apr.	May	June.	July.	Aug.	Sept.	Oct.	Nov.	Dec.
	1.05	1.20	1.20	0.83	1.43	1.23	2.20	0.78	0.95	0.83	1.16	0.98
	1.05	1.25	1.38	1.23	1.33	1.45	1.95	.60	.90	.88	1.08	.83
	1.30	1.23	1.43	1.30	1.33	1.63	a2.25	.65	1.05	.85	1.00	.78
	1.78	1.05	1.38	1.10	1.70	1.75	2.10	.65	1.08	.83	.90	.88
	1.25	.98	1.40	.90	1.33	1.90	2.25	.65	1.20	.90	.85	.70
	1.48	.98	1.40	1.23	1.43	2.05	1.85	.65	.95	.90	.85	.73
	1.58	1.00	2.00	1.20	1.53	a2.10	1.75	.65	.75	.90	.85	.75
	1.20	.95	1.60	1.15	1.18	2.30	1.55	.65	1.05	.85	1.35	.80
	1.50	1.10	1.48	1.15	1.20	a2.60	1.35	.70	.75	.85	1.40	.70
	1.25	1.40	1.38	1.05	a1.18	2.90	1.35	.70	.80	.85	1.38	.73
	1.10	1.45	1.55	1.05	1.30	2.70	1.15	.70	.80	.85	1.33	.63
	1.13	1.40	1.43	.95	1.10	a2.65	1.35	.68	.90	.80	1.05	.70
	1.08	1.18	1.50	1.10	.95	2.70	1.40	.65	.95	.90	.95	.68
	1.08	1.00	1.43	1.05	1.05	2.73	1.40	.60	1.00	.70	.93	.85
	1.28	.95	1.55	1.20	.95	3.00	1.45	.80	1.15	.90	.75	.85
	1.10	.98	1.18	1.13	.98	2.95	1.30	.70	1.20	.85	.73	.95
	1.15	1.00	1.03	1.20	.90	3.00	1.65	.63	1.40	1.00	.75	.68

a Rain.

Mean daily gage height, in feet, of South Platte River, etc.—Continued.

Day.	Jan.	Feb.	Mar.	Apr.	May.	June.	July.	Aug.	Sept.	Oct.	Nov.	Dec.
18	1.08	1.00	1.25	1.23	.98	2.90	1.85	.80	1.20	.95	.70	.73
19	1.10	1.20	1.00	.95	.98	2.50	1.75	1.25	1.05	.95	.68	.73
20	1.10	1.35	.98	1.13	.90	2.40	1.75	.88	.90	1.03	.73	.73
21	1.10	1.35	.95	1.00	.85	2.30	1.45	.95	1.20	.75	.75	.6
22	1.10	1.30	1.25	.75	.88	2.20	1.25	1.05	.83	1.00	.75	.73
23	1.08	1.45	1.45	.88	.88	1.95	1.23	1.45	.80	.95	.68	.73
24	1.60	1.50	1.10	.88	.90	1.75	1.20	1.35	.80	1.08	.80	.70
25	1.25	1.50	1.10	.90	1.00	1.65	1.35	.85	.83	1.00	.78	.73
26	1.25	1.35	1.10	.75	1.05	2.10	1.23	1.15	.80	.95	.80	.73
27	1.25	1.30	1.20	.90	1.08	2.35	1.10	1.20	.75	.90	.80	.6
28	1.20	1.28	1.05	a.23	1.15	2.20	.95	1.18	.70	.90	.75	.73
29	.95		.95	1.25	1.20	2.10	.98	1.03	.83	1.18	.68	.73
30	1.08		1.00	1.25	1.20	2.35	.65	1.05	.83	1.20	.88	.73
31	1.15		.98		1.10		.80	1.18		1.20		.73

a Rain.

Rating table for South Platte River at Denver. Colo., from January 1 to June 15, 1903.

Gage height.	Discharge.	Gage height.	Discharge.	Gage height.	Discharge.	Gage height.	Discharge.
Feet.	*Second-feet.*	*Feet.*	*Second-feet.*	*Feet.*	*Second-feet.*	*Feet.*	*Second-feet.*
0.7	26	1.3	88	1.9	225	2.5	508
.8	32	1.4	106	2.0	256	2.6	597
.9	40	1.5	125	2.1	292	2.7	705
1.0	49	1.6	148	2.2	334	2.8	845
1.1	59	1.7	172	2.3	382	2.9	1,025
1.2	72	1.8	197	2.4	440	3.0	1,240

Rating table for South Platte River at Denver, Colo., from June 16 to December 31, 1903.

Gage height.	Discharge.	Gage height.	Discharge.	Gage height.	Discharge.	Gage height.	Discharge.
Feet.	*Second-feet.*	*Feet.*	*Second-feet.*	*Feet.*	*Second-feet.*	*Feet.*	*Second-feet.*
0.5	60	1.2	183	1.9	523	2.6	954
.6	65	1.3	224	2.0	580	2.7	1,022
.7	72	1.4	268	2.1	640	2.8	1,093
.8	80	1.5	315	2.2	702	2.9	1,166
.9	94	1.6	365	2.3	764	3.0	1,240
1.0	116	1.7	415	2.4	826		
1.1	146	1.8	468	2.5	889		

Both tables are somewhat approximate in the upper part, depending upon on measurement when channel was scouring.

Estimated monthly discharge of South Platte River at Denver, Colo., for 1903.

[Drainage area, 3,840 square miles.]

Month.	Discharge in second-feet.			Total in acre-feet.	Run-off.	
	Maximum.	Minimum.	Mean.		Second-feet per square mile.	Depth in inches.
January..............	192	44	78	4,796	0.020	0.023
February	125	44	78	4,332	.020	.021
March	256	42	94	5,780	.024	.028
April	88	29	58	3,451	.015	.017
May	172	36	68	4,181	.018	.021
June................	1,240	77	637	37,904	.166	.185
July	733	68	328	20,168	.085	.098
August	291	65	108	6,641	.028	.032
September	268	72	112	6,664	.029	.032
October	183	72	108	6,641	.028	.032
November	268	70	111	6,605	.029	.032
December	111	67	78	4,796	.020	.023
The year	1,240	29	155	111,959	.040	.544

SOUTH PLATTE RIVER AT SOUTH PLATTE, COLO.

This station is located at the junction of the North and South forks of South Platte River, at the town of South Platte, located on the Colorado and Southern Railroad, about 9 miles above the mouth of the canyon. The station was established March 28, 1902, by John E. Field, at the wagon bridge crossing the main stream about 150 feet below the junction of the two forks. This station is of special importance, its location being above the diverting gates of all irrigating ditches, and also above the intake of the Denver Union Water Company, which derives the greater part of its supply of water from South Platte River a few miles below this station. The location of the Cheesman storage reservoir on the South Fork, 20 miles above this station, and the contemplated installation of large power plants of the two forks above, also add to the importance of this station as a point from which to secure data. Bridges across either fork above the main station allow of measurements being made on these streams, thereby checking all gagings on the main stream below. This station is located only about 200 feet from the Colorado and Southern Railroad station and is easily accessible from Denver. The gage is a vertical 2 by 6 inch pine timber reading from zero to 7 feet and is fastened to the upper side of the center pier of the highway bridge. When the gage reads 2.3 feet or less, measurements are made by wading about 250 feet below the bridge. At high water discharge measurements are made from the lower side of the two-span highway bridge to which the gage is attached.

The gage is read twice each day by Mrs. Mata Wallbrecht. The initial point for soundings is the river face of the right abutment. The channel is straight for 250 feet above and 400 feet below the station. The current is rough and rapid beneath the bridge, but becomes more even below. The right bank is the side of the mountain. The left bank, at ordinary stages, is part of the old river bed about 100 feet wide. At extremely high water the Colorado and Southern Railroad roadbed is the left bank. At the bridge the river is divided into two channels by the center pier. The right channel is the deeper and carries most of the water. An old abutment foundation directly under the left end of the bridge interferes with the current during high stages. The bench mark is a nail driven horizontally in the right side of the upper end of a 12 by 12 inch cross timber from the floor of the bridge. It is inclosed in a circle of white paint and has four other nails driven about it. It is directly over the gage rod and has an elevation of 10.15 feet above the zero of the gage. Bench mark No. 2 is a regulation metal tablet in the cliff about 75 feet from the south end of the bridge. The elevation is 18.43 feet above zero of gage.

Discharge measurements of South Platte River at South Platte, Colo., in 1903.

Date.	Hydrographer.	Gage height.	Discharge
		Feet.	Second-feet
March 28	M. C. Hinderlider	1.40	7
Do	do	1.40	7
May 19	R. I. Meeker	1.92	2
June 23	M. C. Hinderlider	4.60	1,1
June 27	do	3.60	6
August 8	do	1.20	1
August 26	G. B. Monk	2.00	8
October 9	E. C. Murphy	1.23	

Mean daily gage height, in feet, of South Platte River at South Platte, Colo., 1903.

Day.	Mar.	Apr.	May.	June.	July.	Aug.	Sept.	Oct.	Nov.	D
1	46	1.75	2.20	3.40	1.50	2.13	1.50	1.45		1
2	45	70	2.15	3.15	1.40	2.05	1.48	1.50		1
3	35	1.73	2.33	3.05	1.33	2.00	1.43	1.50		1
4	50	73	2.48	3.05	1.25	2.00	1.45	1.65		1
5	56	.68	2.48	2.68	1.35	1.95	1.50	1.68		2
6	40	70	2.58	2.60	1.30	1.90	1.45	2.05		1

Mean daily gage height, in feet, of South Platte River, etc.—Continued.

Day.	Mar.	Apr.	May	June	July	Aug.	Sept.	Oct.	Nov.	Dec.
7		1.43	1.80	2.73	2.45	1.25	1.90	1.35	2.58	1.30
8		1.40	1.90	2.88	2.30	1.20	1.88	1.33	2.58	1.35
9		1.48	1.98	3.15	2.45	1.15	1.83	1.40	2.55	1.40
10		1.48	2.00	3.50	2.33	1.35	1.85	1.30	2.28	1.35
11		1.50	2.13	3.70	2.33	1.48	1.85	1.25	2.00	1.25
12		1.88	2.03	4.50	2.30	1.33	1.90	1.25	1.78	1.20
13		1.55	2.15	4.00	2.23	1.30	1.95	1.23	1.65	1.30
14		1.50	2.23	5.15	2.25	1.28	2.05	1.23	1.50	1.33
15		1.53	2.28	5.70	2.38	1.28	2.10	1.30	1.38	1.20
16		1.50	2.30	5.65	2.23	1.50	2.05	1.43	1.33	1.30
17		1.53	2.28	5.78	2.30	1.68	1.93	1.50	1.40	1.35
18		1.78	2.13	5.70	2.43	2.00	1.65	1.30	1.50	1.40
19		1.70	2.00	5.50	2.45	1.93	1.60	1.53	1.55	1.30
20		1.78	1.85	4.80	2.38	1.68	1.65	1.70	1.83	1.20
21		1.78	1.73	4.75	2.30	1.78	1.60	1.60	1.60	1.18
22		1.78	1.80	4.65	2.20	c2.90	1.65	1.60	1.50	1.18
23		1.75	1.73	4.48	2.05	(b)	1.58	1.60	1.48	1.23
24		1.63	1.68	4.15	2.03	(b)	1.53	1.50	1.40	1.30
25		1.73	2.03	4.20	1.90	2.00	1.50	1.50	1.43	1.20
26		1.70	1.88	3.80	1.80	2.00	1.40	1.55	1.45	1.15
27		1.75	2.25	3.50	1.95	2.03	1.50	1.50	1.48	1.10
28	1.35	1.78	1.80	3.00	1.78	2.03	1.55	1.40	1.33
29	1.50	1.80	1.88	3.30	1.75	2.05	1.60	1.48	1.73
30	c1.42	1.63	1.75	3.50	1.63	2.20	1.53	1.48	1.43	1.95
31	1.40	------	1.88	------	1.58	2.15	------	1.50	------	1.80

a Heavy rain.
b No measurements.
c Drift about gage rod removed Mar. 30, causing lowering of reading on gage rod of 0.08 foot.
Gage blocked by drift.

Rating table for South Platte River at South Platte, Colo., from March 28 to June 17, 1903, and from September 17 to December 31, 1903.

Gage height.	Discharge.	Gage height.	Discharge.	Gage height.	Discharge.	Gage height.	Discharge.
Feet.	Second-feet.	Feet.	Second-feet.	Feet.	Second-feet.	Feet.	Second-feet.
1.0	40	2.0	228	3.0	480	4.0	810
1.1	54	2.1	252	3.1	508	4.2	896
1.2	70	2.2	276	3.2	536	4.4	988
1.3	87	2.3	300	3.3	565	4.6	1,084
1.4	105	2.4	324	3.4	595	4.8	1,184
1.5	123	2.5	350	3.5	627	5.0	1,284
1.6	142	2.6	376	3.6	660	5.2	1,384
1.7	162	2.7	402	3.7	695	5.4	1,486
1.8	184	2.8	428	3.8	732	5.6	1,590
1.9	206	2.9	454	3.9	770	5.8	1,694

Rating table for South Platte River at South Platte, Colo., from June 18 to September 16, 1903.

Gage height.	Discharge.	Gage height.	Discharge.	Gage height.	Discharge.	Gage height.	Discharge.
Feet.	*Second-feet.*	*Feet.*	*Second-feet.*	*Feet.*	*Second-feet.*	*Feet.*	*Second-feet.*
1.0	105	2.0	280	3.0	513	4.0	845
1.1	121	2.1	301	3.1	539	4.2	937
1.2	137	2.2	323	3.2	567	4.4	1,029
1.3	153	2.3	345	3.3	595	4.6	1,121
1.4	169	2.4	367	3.4	623	4.8	1,213
1.5	186	2.5	389	3.5	654	5.0	1,305
1.6	204	2.6	413	3.6	689	5.2	1,399
1.7	222	2.7	437	3.7	725	5.4	1,495
1.8	240	2.	461	3.8	762	5.6	1,594
1.9	260	2.9	487	3.9	801	5.8	1,694

Estimated monthly discharge of South Platte River at South Platte, Colo., for 1903

[Drainage area, 2,612 square miles.]

Month.	Discharge in second-feet.			Total in acre-feet.	Run-off.	
	Maximum.	Minimum.	Mean.		Second-feet per square mile.	Depth in inches.
March 28–31........	123	96	108	857	0.041	0.00(
April	202	96	144	8,569	.055	.06
May	295	158	212	13,035	.082	.09
June...............	1,684	264	861	51,233	.330	.36
July	623	200	353	21,705	.135	.1?
August (23 and 24 missing)	487	129	217	12,482	.083	.0(
September	308	105	211	12,555	.081	.0?
October	162	75	113	6,948	.043	.0?
November..........	371	92	163	9,699	.062	.0(
December	217	54	97	5,964	.037	.0·

MISCELLANEOUS MEASUREMENTS IN THE SOUTH PLATTE DRAINAGE BASIN.

The following miscellaneous measurements were made in South Platte basin in 1903.

Miscellaneous measurements in South Platte basin, 1903.

Date.	Stream.	Locality.	Gage height.	Discharge.
			Feet.	*Sec. feet.*
June 14	Latham ditch	Kersey, Colo	....	120
May 1	North Fork Clear Creek.	Forkscreek, Colo	....	22
May 14	.....do	.....do	*a* 2.65	57
June 25	.....do	.....do		34
June 12	North Fork South Platte.	South Platte, Colo	*a* 4.50	543
June 23	.....do	.....do	*a* 4.60	412
June 27	.....do	.....do	*a* 2.60	439
July 14	.....do	Rhodes Station, Colo..	1.90	259
July 16	.....do	.....do	2.05	217
Do	.....do	¼ mile below Rhodes Station, Colo.	....	247
August 8	.....do	South Platte, Colo	*a* 1.20	103
August 26	.....do	.....do	*a* 2.00	156
May 14	South Fork Clear Creek.	Forkscreek, Colo	*a* 2.65	253
July 16	South Fork South Platte.	South Platte, Colo	*a* 2.15	101
April 17	South Platte	Sterling, Colo	(*b*)	18
June 18	.....do	Bridge on Kersey road, Colo.	....	1,690
June 19	.....do	Sterling, Colo	(*b*)	81
July 21	.....do	Platte Canyon, Colo...	.80	271
July 23	.....do	.....do	.70	211
July 28	.....do	.....do	.60	166
July 29	.....do	.....do	.40	129
August 6	.....do	.....do	.80	98
August 20	.....do	.....do	.50	147

a Gage read on main stream.   *b* Estimated.

Mean daily gage height, in feet, of South Platte River at Platte Canyon for 1903.

Day.	July.	Aug.	Sept.	Day.	July.	Aug.	Sept.	Day.	July.	Aug.	Sept.
1		0.40	0.80	12		0.40		23	0.70		......
2		.40	.80	13		.40		24			......
3		.40	.80	14		.40		25			......
4		.30	.70	15		.40		26	.70		......
5		.30	.60	16		.60		27	.60		......
6		.30		17		.70		28	.60		......
7		.60		18		.80		29	.40		......
8		.40		19		.70		30	.40	0.90	......
9		.30		20		.50		31	.40	.80	......
10		.30		21	0.80	.60					
11		.40		22		.60					

KANSAS RIVER DRAINAGE BASIN.

The drainage basin of Kansas River lies between those of Platte and Arkansas rivers, being entirely within the region of the Great Plains, and principally within the arid or semiarid area.   It has no mountain tributaries, but depends entirely for its water supply upon the water which, falling within or near the basin, percolates slowly to the drainage channels.   The catchment area extends from eastern Colorado to Missouri River, a distance from east to west of 485 miles. Its extreme width is about 200 miles.   The main stream of Kansas River is formed at Junction, Kans., by Republican and Smoky Hill rivers, and flows east into the Missouri at Kansas City.   The Republican, its principal tributary, drains southern Nebraska and northern Kansas by means of many small tributary creeks.   Solomon, Saline, and Smoky Hill rivers drain the plains of northwestern Kansas. Blue River is a tributary in northeastern Kansas, flowing south into Kansas River near Manhattan, Kans.   The following is a list of the stations in the Kansas River drainage basin maintained during 1903:

Kansas, or Kaw, River at Lecompton, Kans.
Blue River near Manhattan, Kans.
Republican River at Junction, Kans.
Republican River and the mill race near Superior, Nebr.
Republican River at Benkelman, Nebr.
Republican River (South Fork) at Benkelman, Nebr.
Solomon River at Niles, Kans.
Saline River near Salina, Kans.
Smoky Hill River at Ellsworth, Kans.

KANSAS RIVER AT LECOMPTON, KANS.

The gaging station at Lecompton was established April 16, 1899, at the new wagon bridge.   On June 24, 1900, a new gage was established, the old gage having been broken.   The present gage, a pine board 1 inch by 6 inches by 10 feet long, graduated to feet and tenths, is spiked on top of the old gage, and is at the same elevation.   On October 26, 1900, a bench mark was established on top of the bottom flange of the iron strut connecting the two iron cylinders at the south end of the highway bridge over the river.   The bench mark is at the west end of the strut, next to the cylinder.   Its elevation is 12.19 feet above the zero of the gage.   The observer is A. D. McAdow.   The channel is somewhat curved at the bridge, and the bridge is slightly oblique to the direction of the current.   The channel has a width of 800 feet, broken by four metal piers.   The left bank is low and subject to overflow during high water.   The bed of the stream is sandy, with some rock, and changes slightly.

The observations at this station during 1903 have been made under the direction of W. G. Russell, district hydrographer.

Discharge measurements of Kansas River at Lecompton, Kans., in 1903.

Date.	Hydrographer.	Gage height.	Discharge.
		Feet.	*Second-feet.*
May 19	W. G. Russell	9.30	24,298
May 20	do	9.85	28,509
May 29	do	21.10	95,017
May 31	do	29.50	a 280,000
June 27	do	9.20	28,051
August 19	do	9.75	32,781
August 20	do	8.90	28,782
August 22	do	7.85	22,252
August 29	do	11.15	47,649

a Estimated.

Mean daily gage height, in feet, of Kansas River at Lecompton, Kans., for 1903.

Day.	Jan.	Feb.	Mar.	Apr.	May.	June.	July.	Aug.	Sept.	Oct.	Nov.	Dec.
1	6.90	5.90	6.80	5.75	4.00	27.20	8.10	6.30	5.95	3.00	4.25	5.10
2	6.90	5.85	6.90	5.70	4.00	26.25	7.70	6.30	5.85	3.10	5.30	5.10
3	6.85	5.80	7.40	5.60	6.90	25.35	7.40	6.30	5.50	3.50	6.40	5.00
4	6.80	5.70	7.85	5.60	6.70	24.35	7.05	6.40	5.25	3.50	6.40	5.00
5	6.80	5.70	7.65	5.50	6.40	23.05	7.00	6.60	5.05	3.40	6.40	4.95
6	6.80	5.60	6.80	5.45	6.40	20.50	6.90	6.70	4.80	3.40	6.30	4.96
7	6.75	5.60	6.25	5.40	6.30	18.20	6.80	6.90	4.80	3.35	6.30	4.90
8	6.70	5.55	5.95	5.10	6.15	16.50	6.75	7.55	4.70	3.30	6.25	4.85
9	6.60	5.50	5.55	4.80	5.90	14.80	6.70	7.95	4.65	3.25	6.20	4.80
10	6.60	5.45	7.05	4.80	5.95	13.85	6.70	10.30	4.60	3.20	6.15	4.70
11	6.60	5.40	8.15	4.70	6.30	12.70	6.60	9.25	4.60	3.15	6.10	4.70
12	6.60	5.65	8.90	4.70	7.50	11.55	6.55	9.25	4.50	3.10	6.00	4.70
13	6.50	6.10	8.90	4.70	9.25	10.45	6.45	10.40	4.50	3.00	5.90	4.60
14	6.50	6.40	9.45	4.65	10.80	9.85	6.40	11.45	4.85	3.15	5.85	4.60
15	6.50	6.55	9.65	4.60	13.10	9.30	6.30	11.70	7.20	3.25	5.80	4.50
16	6.50	6.60	10.00	4.60	12.45	8.90	6.20	12.15	7.50	3.40	5.80	4.50
17	6.40	6.50	9.85	4.55	10.25	8.50	6.40	12.50	7.65	3.55	5.75	4.40
18	6.40	6.35	8.95	4.50	9.65	8.05	6.90	11.80	7.20	3.85	5.70	4.40
19	6.40	6.25	8.70	4.50	9.40	7.80	7.15	9.60	6.60	4.00	5.60	4.30
20	6.35	6.20	8.20	4.50	9.80	7.45	7.05	8.70	6.25	4.00	5.60	4.30
21	6.30	6.00	7.50	4.45	9.60	7.00	6.80	8.25	5.70	3.95	5.50	4.30
22	6.25	6.00	7.40	4.40	9.95	6.65	6.75	7.80	5.60	3.90	5.50	4.30
23	6.20	5.80	7.20	4.40	12.90	6.40	6.70	7.40	5.30	3.85	5.40	4.20
24	6.20	5.75	6.95	4.30	14.20	6.00	6.70	7.05	4.70	3.80	5.40	4.20
25	6.10	5.70	6.65	4.30	14.15	6.20	6.60	6.90	4.05	3.80	5.40	4.20
26	6.10	6.10	6.60	4.30	12.15	7.80	6.60	7.25	3.60	3.80	5.30	4.10
27	6.10	6.50	6.35	4.20	16.60	9.30	6.60	10.80	3.30	3.70	5.40	4.10
28	6.05	6.80	6.05	4.20	18.00	9.30	6.50	12.75	3.15	3.70	5.20	4.10
29	6.00		5.95	4.15	21.10	8.90	6.50	10.70	3.10	3.60	5.20	4.10
30	6.00		5.90	4.10	24.00	8.55	6.40	7.15	3.10	3.60	5.10	4.05
31	6.00		5.85		28.75		6.40	6.05		3.55		4.00

Rating table for Kansas River at Lecompton, Kans., from January 1 to May 25, 1903.

Gage height.	Discharge.	Gage height.	Discharge.	Gage height.	Discharge.	Gage height.	Discharge.
Feet.	*Second-feet.*	*Feet.*	*Second-feet.*	*Feet.*	*Second-feet.*	*Feet.*	*Second-feet.*
4.0	5,000	5.6	9,770	7.2	15,200	8.8	20,800
4.2	5,550	5.8	10,430	7.4	15,900	9.0	21,500
4.4	6,120	6.0	11,110	7.6	16,600	9.2	22,200
4.6	6,710	6.2	11,790	7.8	17,300	9.4	22,900
4.8	7,300	6.4	12,470	8.0	18,000	9.6	23,600
5.0	7,900	6.6	13,150	8.2	18,700	10.0	25,000
5.2	8,510	6.8	13,830	8.4	19,400		
5.4	9,180	7.0	14,510	8.6	20,100		

Rating table for Kansas River at Lecompton, Kans., from May 25 to December 31, 1903.

Gage height.	Discharge.	Gage height.	Discharge.	Gage height.	Discharge.	Gage height.	Discharge.
Feet.	*Second-feet.*	*Feet.*	*Second-feet.*	*Feet.*	*Second-feet.*	*Feet.*	*Second-feet.*
3.0	2,375	4.3	5,835	6.2	11,750	8.8	25,400
3.1	2,635	4.4	6,120	6.4	12,500	9.0	26,800
3.2	2,895	4.5	6,415	6.6	13,300	9.2	28,250
3.3	3,155	4.6	6,710	6.8	14,150	9.4	29,750
3.4	3,415	4.7	7,005	7.0	15,050	9.6	31,300
3.5	3,675	4.8	7,300	7.2	16,000	9.8	32,900
3.6	3,940	4.9	7,600	7.4	17,000	10.0	34,550
3.7	4,205	5.0	7,900	7.6	18,050	10.4	38,000
3.8	4,470	5.2	8,510	7.8	19,150	11.0	43,550
3.9	4,735	5.4	9,130	8.0	20,300	a 11.2	45,500
4.0	5,000	5.6	9,770	8.2	21,500	11.4	47,500
4.1	5,275	5.8	10,430	8.4	22,750		
4.2	5,550	6.0	11,000	8.6	24,050		

Tangent above 11.2 feet gage height, with differences of 1,000 per tenth.

Estimated monthly discharge of Kansas River at Lecompton, Kans., for 1903.

[Drainage area, 58,550 square miles.]

Month.	Discharge in second-feet.			Total in acre-feet.	Run-off.	
	Maximum.	Minimum.	Mean.		Second-feet per square mile.	Depth in inches.
January	14,170	11,110	12,613	775,542	0.22	0.25
February	13,830	9,130	11,085	612,853	.19	.20
March	25,000	9,610	16,156	993,394	.28	.32
April	10,265	5,275	7,182	427,359	.13	.13
May	221,000	5,000	45,018	2,768,049	.77	.89
June	205,500	11,000	65,268	3,888,715	1.11	1.24
July	20,900	11,750	14,085	866,052	.24	.28
August	58,500	11,187	27,568	1,695,090	.47	.54
September	18,325	2,635	8,631	513,580	.15	.17
October	5,000	2,375	3,698	227,381	.063	.073
November	12,500	5,692	10,182	605,871	.17	.19
December	8,205	5,000	6,472	397,948	.11	.13
The year	221,000	2,375	18,992	13,766,834	.324	4.413

BLUE RIVER NEAR MANHATTAN, KANS.

The gaging station, established April 12, 1895, is at the county bridge 4 miles north of Manhattan. The gage rod consists of three sections. The low-water gage is a chain gage on the hand railing of the bridge on the east side and south end of the bridge. It reads from zero to 14 feet from a scale spiked to the hand railing. The other two sections of the rod are bolted to the south bridge pier. The bench mark is a cross cut in the capstone of the south bridge pier immediately above the upper gage, and is 32.14 feet above gage datum. The observer is J. M. Deckert.

The channel is straight for several hundred feet above and below the station, and has a width of 225 feet at ordinary stages. Both banks are subject to overflow during floods, the left bank being the lower of the two. The bed is composed of clay and silt and changes slightly. The velocity is rapid.

The observations at this station during 1903 have been made under the direction of W. G. Russell, district hydrographer.

Discharge measurements of Blue River near Manhattan, Kans., in 1903.

Date.	Hydrographer.	Gage height.	Discharge.
		Feet.	*Second-feet.*
March 10	W. G. Russell	13.00	9,10(
May 15	....do	18.15	16,33
May 16	....do	12.15	7,10?
May 31	....do	36.50	a 98,00
June 25	....do	9.40	4,09
August 27	....do	30.80	32,18
August 28	....do	26.60	34,84
September 25	E. C. Murphy	6.25	1,3?

a Estimated.

Mean daily gage height, in feet, of Blue River near Manhattan, Kans., for 190

Day.	Jan.	Feb.	Mar.	Apr.	May.	June.	July.	Aug.	Sept.	Oct.	Nov.	De
1	7.50	5.70	13.90	6.50	11.00	......	8.70	6.90	9.10	6.10	6.20	5.
2	7.70	5.60	12.10	6.40	11.50	......	8.50	8.80	8.80	6.00	7.70	5.
3	7.00	5.70	10.00	6.40	9.50	32.50	8.50	9.80	8.70	5.90	8.10	5.
4	6.80	5.90	9.20	5.80	7.20	30.00	8.20	11.20	8.50	5.90	10.70	5.
5	6.60	6.50	9.10	6.20	6.80	27.00	7.80	10.80	8.30	5.80	11.00	5.
6	6.40	6.80	9.00	6.10	6.60	25.50	7.80	10.40	7.90	5.80	9.40	5.
7	5.60	5.20	10.10	6.00	6.60	21.50	7.60	10.10	7.80	6.30	8.10	5.
8	8.00	5.50	11.10	5.90	6.40	17.50	7.40	10.30	7.40	6.20	7.50	8.
9	7.50	5.40	13.00	5.80	6.20	14.60	7.80	10.70	7.90	6.10	7.10	5.
10	6.70	5.70	13.30	5.80	6.00	13.30	7.20	10.70	9.60	5.90	6.80	5.
11	6.70	7.70	13.30	5.80	5.90	12.40	7.10	9.20	9.00	5.80	6.70	5
12	6.60	10.10	13.30	5.70	6.30	11.90	7.00	8.40	7.80	5.90	6.60	5
13	6.90	7.60	12.30	5.50	11.90	11.30	7.40	14.50	8.40	6.00	6.50	5
14	7.00	6.70	12.90	5.90	18.30	10.70	7.40	14.40	11.30	6.40	6.30	5
15	7.20	6.10	12.80	5.50	18.60	10.40	8.60	18.20	9.30	7.20	6.20	7
16	6.70	6.00	12.70	5.60	12.30	10.10	7.50	21.80	10.40	8.00	6.10	7
17	6.70	6.90	11.50	5.90	10.40	9.80	9.50	22.40	7.80	7.80	6.00	6
18	6.10	7.60	9.50	5.90	9.70	9.50	7.80	15.20	7.20	6.90	6.00	6
19	6.00	7.50	9.00	5.70	9.00	9.30	9.90	12.90	7.00	6.50	5.90	6
20	5.90	6.50	8.00	5.50	8.60	11.10	9.50	11.00	6.80	6.30	5.80	6
21	5.50	6.50	8.40	5.40	11.70	10.60	8.70	9.70	6.70	6.20	5.80	6
22	6.20	7.00	7.90	5.40	22.00	10.00	8.50	9.00	6.50	6.10	5.80	6
23	6.30	7.50	7.70	5.50	23.00	9.50	8.20	8.60	6.40	5.90	5.80	6
24	6.00	8.50	7.80	5.80	25.00	9.10	8.10	8.20	6.30	5.90	5.90	6
25	5.40	8.80	7.80	5.30	17.50	9.30	8.00	7.90	6.20	5.80	5.90	6
26	5.40	10.00	7.40	5.80	25.00	12.80	7.40	23.30	6.20	5.70	5.90	6
27	5.50	10.20	7.30	5.20	30.50	12.20	7.10	30.50	6.00	5.70	5.80	6
28	5.80	15.00	7.30	5.20	32.00	9.80	6.90	27.20	6.00	5.60	5.80	6
29	5.80	......	6.90	5.30		9.30	6.90	17.20	5.90	5.60	5.80	6
30	5.70	......	6.00	5.30	......	8.80	6.80	11.70	6.10	5.50	5.70	6
31	5.70	......	6.60	......	a36.50	......	6.70	9.80	......	5.90	......	6

a Extreme height.

Rating table for Blue River at Manhattan, Kans., from January 1 to December 31, 1903.

Gage height.	Discharge.	Gage height.	Discharge.	Gage height.	Discharge.	Gage height.	Discharge.
Feet.	*Second-feet.*	*Feet.*	*Second-feet.*	*Feet.*	*Second-feet.*	*Feet.*	*Second-feet.*
5.0	705	6.3	1,420	8.2	2,880	10.8	5,620
5.1	745	6.4	1,490	8.4	3,060	11.0	5,860
5.2	785	6.5	1,560	8.6	3,240	11.2	6,110
5.3	825	6.6	1,630	8.8	3,430	11.4	6,360
5.4	875	6.7	1,700	9.0	3,630	11.6	6,620
5.5	925	6.8	1,770	9.2	3,830	11.8	6,880
5.6	975	6.9	1,840	9.4	4,040	12.0	7,150
5.7	1,035	7.0	1,910	9.6	4,250	12.5	7,840
5.8	1,095	7.2	2,060	9.8	4,470	13.0	8,540
5.9	1,155	7.4	2,210	10.0	4,690	13.5	9,240
6.0	1,220	7.6	2,370	10.2	4,920	14.0	9,960
6.1	1,285	7.8	2,540	10.4	5,150		
6.2	1,350	8.0	2,710	10.6	5,380		

Tangent above 14 feet gage height. Differences above this point, 150 per tenth. Above 25 feet gage height all discharges are increased 30 per cent to get actual discharge.

Estimated monthly discharge of Blue River near Manhattan, Kans., for 1903.

[Drainage area, 9,490 square miles.]

Month.	Discharge in second-feet.			Total in acre-feet.	Run-off.	
	Maximum.	Minimum.	Mean.		Second-feet per square mile.	Depth in inches.
January	2,710	875	1,542	94,814	0.16	0.18
February	11,460	785	2,373	131,790	.25	.26
March	9,815	1,220	4,929	303,072	.52	.60
April	1,560	785	1,046	62,241	.11	.12
May	68,770	1,155	19,092	1,173,920	2.01	2.32
June	66,170	3,430	16,427	977,474	1.73	1.93
July	4,580	1,700	2,644	162,573	.28	.32
August	34,710	1,840	9,626	591,880	1.01	1.16
September	6,235	1,155	2,576	153,282	.27	.30
October	2,710	925	1,337	82,209	.14	.16
November	5,860	1,035	1,849	110,023	.19	.21
December	2,290	785	1,044	64,193	.11	.13
The year	68,770	785	5,374	3,907,471	.565	7.69

REPUBLICAN RIVER AT JUNCTION, KANS.

The gaging station at this point, established by Arthur P. Davis, April 26, 1895, is located at the wagon bridge at the north end of Washington street, just above the mouth of the river. The gage consists of two oak timbers bolted to a post and to a cottonwood tree. The observer is J. H. Rathert. The channel is straight for 300 feet above and below the station, broken by three piers. The right bank is high, but the left is low and may overflow at high water. The bed of the stream is sandy and liable to change. The flow is moderately rapid. One bench mark consists of a 60-penny spike driven into the base of the abutment of the bridge at an elevation of 10.67 feet on the rod. The second bench mark is the top of a stone in the base of the bridge abutment, 18 feet south of the gage and at an elevation of 14.51 feet above gage datum.

On October 23, 1900, a new bench mark was established, at an elevation of 12.35 feet above the zero of the old gage. It is a spike driven in the west side of a cottonwood tree 18 inches in diameter and 10 feet west of the bridge. The spike is about 2 feet above the ground.

The observations at this station during 1903 have been made under the direction of W. G. Russell, district hydrographer.

Discharge measurements of Republican River at Junction, Kans., in 1903.

Date.	Hydrographer.	Gage height.	Discharge
		Feet.	*Second-feet*
March 9	W. G. Russell	7.75	6,19
May 12	....do	9.00	9,66
May 13	....do	10.45	18,01
May 14	....do	12.35	28,18
May 30	....do	18.20	71,00
June 24	....do	4.60	1,9
July 15	....do	7.60	6,2
July 16	....do	8.40	8,7
August 8	....do	6.80	4,5
September 29	E. C. Murphy	8.40	5

Mean daily gage height, in feet, of Republican River at Junction, Kans., for 1903.

Day.	Jan.	Feb.	Mar.	Apr.	May.	June.	July.	Aug.	Sept.	Oct.	Nov.	Dec.
1	3.50	3.55	5.47	4.25	7.17	17.32	5.45	4.32	5.72	3.40	3.72	3.90
2	3.50	3.55	4.52	4.20	5.90	16.22	5.40	6.80	5.25	3.40	3.80	3.97
3	3.50	3.55	4.80	4.20	4.87	13.85	4.95	8.10	4.80	3.40	4.12	4.00
4	3.50	3.70	4.95	4.17	4.82	11.38	4.67	8.45	4.50	3.17	4.37	4.00
5	3.48	4.15	5.05	4.00	4.12	10.60	4.47	7.17	4.17	3.40	4.10	4.00
6	3.50	4.57	5.70	4.00	3.95	9.35	4.40	6.42	4.05	3.32	4.15	4.00
7	3.38	4.47	5.80	4.00	3.92	8.22	4.22	6.15	3.85	4.45	4.25	4.00
8	3.25	4.45	6.05	3.92	3.90	7.35	4.30	6.00	3.80	3.52	4.12	3.97
9	3.45	4.40	5.07	3.80	3.90	6.80	4.22	4.95	3.92	3.30	4.02	3.85
10	3.43	4.40	8.70	3.80	3.85	6.47	4.60	4.70	5.17	3.90	3.92	3.80
11	4.25	4.40	8.55	3.80	4.40	6.20	5.10	4.45	4.12	3.05	3.90	3.80
12	4.10	4.35	8.45	3.80	8.42	5.97	4.75	4.30	4.10	3.10	3.90	3.72
13	4.10	4.15	9.15	3.90	10.85	5.72	4.67	5.90	4.17	3.40	3.90	4.10
14	4.10	4.00	9.35	3.80	12.27	5.45	4.87	6.87	4.20	3.55	3.90	4.10
15	4.05	4.00	9.50	3.72	10.05	5.40	3.72	6.45	3.80	3.65	3.90	4.10
16	4.05	4.05	8.80	3.65	8.55	5.25	8.27	6.50	3.70	3.60	3.90	4.10
17	4.00	4.30	7.32	3.62	7.62	5.15	7.62	8.22	3.62	3.60	3.87	4.10
18	4.05	4.30	6.65	3.55	7.20	5.08	7.40	6.40	3.60	3.60	3.80	4.25
19	4.10	4.30	6.30	3.60	7.87	4.98	6.45	6.10	3.55	3.60	3.77	4.10
20	4.15	4.30	6.05	3.60	6.52	4.75	5.80	5.00	3.50	3.60	3.50	3.80
21	4.20	4.30	6.25	3.52	6.37	4.77	5.25	4.57	3.50	3.50	3.60	3.80
22	3.92	4.30	6.15	3.45	6.72	4.67	5.82	4.50	3.50	3.50	3.72	3.80
23	4.05	4.30	6.05	3.40	8.90	4.60	5.72	4.65	3.50	3.50	3.77	3.80
24	4.12	4.70	5.85	3.45	8.17	4.52	5.12	4.37	3.40	3.50	3.82	3.80
25	4.10	4.55	5.50	3.40	7.35	4.60	4.70	4.32	3.40	3.50	3.85	3.80
26	3.60	4.50	5.17	3.40	9.20	5.48	4.62	4.30	3.40	3.50	3.90	3.80
27	3.55	5.30	4.95	3.35	11.50	6.07	4.45	4.17	3.40	3.50	3.90	3.80
28	3.55	6.02	4.70	3.35	13.10	5.47	4.32	4.00	3.40	3.50	3.90	3.80
29	3.55	......	4.52	3.45	18.02	5.20	4.25	3.95	3.45	3.50	3.90	3.80
30	3.55	......	4.35	3.70	18.20	5.50	4.17	3.90	3.40	3.70	3.90	3.82
31	3.55	......	4.30	......	17.75	......	4.25	4.40	......	3.80	......	3.90

Rating table for Republican River at Junction, Kans., from January 1 to December 31, 1903.

Gage height.	Discharge.	Gage height.	Discharge.	Gage height.	Discharge.	Gage height.	Discharge.
Feet.	*Second-feet.*	*Feet.*	*Second-feet.*	*Feet.*	*Second-feet.*	*Feet.*	*Second-feet.*
2.2	20	4.6	1,480	7.0	6,000	9.4	13,200
2.4	30	4.8	1,720	7.2	6,600	9.6	13,800
2.6	70	5.0	1,960	7.4	7,200	9.8	14,400
2.8	140	5.2	2,210	7.6	7,800	10.0	15,000
3.0	220	5.4	2,520	7.8	8,400	10.2	15,600
3.2	300	5.6	2,840	8.0	9,000	10.4	16,200
3.4	400	5.8	3,200	8.2	9,600	10.6	16,800
3.6	510	6.0	3,600	8.4	10,200	10.8	17,400
3.8	630	6.2	4,000	8.6	10,800	11.0	18,000
4.0	840	6.4	4,400	8.8	11,400	11.5	19,500
4.2	1,050	6.6	4,900	9.0	12,000	12.0	21,000
4.4	1,260	6.8	5,400	9.2	12,600	18.0	24,000

Twenty per cent added to discharge of 1902 for gage heights above 10 feet.

Estimated monthly discharge of Republican River at Junction, Kans., for 1903.

[Drainage area, 25,837 square miles.]

Month.	Discharge in second-feet.			Total in acre-feet.	Run-off.	
	Maximum.	Minimum.	Mean.		Second-feet per square mile.	Depth in inches.
January	1,102	325	684	42,058	0.026	0.030
February	3,600	482	1,217	67,589	.047	.041
March	13,500	1,155	5,099	313,525	.200	.23
April	1,102	375	624	37,130	.024	.02
May	47,520	682	12,112	744,738	.470	.54
June	44,280	1,370	8,162	485,673	.310	.34
July	9,750	997	2,578	158,515	.100	.12
August	10,350	735	3,254	200,081	.120	.14
September	3,020	400	859	51,114	.033	.03
October	1,315	240	466	23,653	.018	.02
November	1,207	455	746	44,390	.029	.03
December	1,102	570	757	46,546	.029	.03
The year	47,520	240	3,046	2,220,012	.117	1.5

REPUBLICAN RIVER AND THE MILL RACE NEAR SUPERIOR, NEBR.

This station was established June 20, 1896, about 1 mile west of
Superior, Nebr. The old gage rod was first placed just above the
highway bridge, which is itself 75 yards above the dam that diverts
water into the mill race.

This gage consists of an oak piece 2 by 4 inches, 10 feet long. The
face is inclined 30 degrees to the horizontal, and the footmarks are
placed 2 feet apart to correspond to this inclination. The rod is
fastened to cross-ties, which are bedded in the bank of the river. The
location is on the outside of an easy bend in the river. The bed of
the river is of mud and sand. Bench mark No. 1 is the standard 4-foot
iron pipe of the United States Geological Survey. It is 83 feet north
of the upstream cylinder of the north pier of the bridge and is 10 feet
west of the line of the upstream truss of the bridge. It is 1 foot inside
a wire fence. The top of the pipe is 4 inches above the ground, and
the elevation is 4.88 feet above zero of the gage.

In the spring of 1898 two new gages were established. The first
gage was placed in the river a few feet upstream from the crest of the
dam, the zero being at the same elevation as the crest. The second
in the mill race, which is crossed by a wagon bridge about 50 yards
below its head. Discharge measurements of the river are made from
the highway bridge, thus determining at once the discharge through
the mill race and over the dam. The discharge from the mill race

d, and is deducted from the total discharge of the river, to give
unt passing over the dam. The bench mark for this river gage
mark No. 1 of the old gage described above. Its elevation is
above the datum of the gage at the dam. The gage of the
e reads 2 feet higher than that of the river gage, so that its
i.92 feet below the same bench mark. Bench mark No. 2 is
of a 6 by 8 inch oak piling in direct line with crest of dam
eet east of the west face of retaining wall at east end of dam.
tion is 9.07 feet above zero of gage. Bench mark No. 3 is the
a 60-penny spike driven vertically into a root at the north
i large cottonwood tree, 6 feet in circumference, located 36
: of retaining wall spoken of above. Its elevation is 5.67 feet
ro of gage. Bench marks Nos. 2 and 3 were established Sep-
11, 1903, on which date bench mark No. 1 could not be found,
indoubtedly been removed by parties repairing the bridge.

found, however, that the discharge of the mill race was reg-
nore or less by the mill below, so that there was no relation
r between the gage height and the discharge. To remedy
iculty, the observer records the depth of water in midstream
agon bridge and immediately thereafter notes the number of
required for a float to pass over a 50-foot range in midstream.
o of the time to the depth was found to bear a constant rela-
he discharge.

g 1899 Glenn E. Smith made an examination of a portion of
epublican River. On September 6, 1899, at Oxford, Nebr., the
annel was dry and was reported to have been in this condition
days. At Orleans, about 12 miles below, there was an esti-
ischarge of 0.3 second-foot—this small amount coming from
iver, which enters at this point.

tation was discontinued November 30, 1903.

bservations at this station during 1903 have been made under
tion of J. C. Stevens, district hydrographer.

rge measurements of Republican River near Superior, Nebr., in 1903.

Date.	Hydrographer.	Gage height.	Discharge.
		Feet.	*Second-feet.*
....	J. C. Stevens	0.80	950
....	....do	1.35	2,198
....	....do	1.76	2,896
....	....do	1.05	1,008
....	....do	3.20	7,662
r 12....................	....do	0.42	364

Mean daily gage height, in feet, of Republican River near Superior, Nebr., for 1903.

Day.	Mar.	Apr.	May.	June.	July.	Aug.	Sept.	Oct.	Nov.
1		1.18	0.80	2.60	1.05	2.10	0.90	0.39	0.79
2		1.12	.86	2.10	.96	2.08	.75	.38	.73
3		1.11	.96	1.89	.92	1.70	.61	.38	.6
4		1.11	.80	1.76	.88	1.88	.65	.19	.8
5		1.00	.95	1.58	.80	1.28	.62	.40	.6
6		1.03	.99	1.45	.79	1.10	.56	.42	
7		1.01	1.00	1.33	1.55	1.03	.52	.46	
8		1.00	2.02	1.44	1.30	.95	.50	.34	
9		(a)	.97	2.08	1.41	.85	.48	.40	
10		(a)	.95	2.59	1.20	.99	.83	.49	.35
11		(a)	.80	3.20	1.19	1.80	.78	.46	.45
12		(a)	.86	2.75	1.21	2.41	.89	.42	.45
13		(a)	.81	2.60	1.13	2.76	1.12	.40	.50
14		3.10	.80	2.32	1.03	3.18	1.84	.39	.51
15		2.81	.90	2.08	1.07	2.90	1.30	.38	.50
16		2.21	.88	1.99	1.04	2.78	1.10	.40	.47
17		2.08	.92	1.68	1.00	1.76	.92	.40	.49
18		1.98	.90	1.42	.99	1.42	.80	.40	.55
19		1.95	.75	1.41	.96	2.10	1.02	.38	.49
20		1.88	.83	1.33	.91	1.75	1.02	.43	.50
21		1.60	.85	1.20	.96	1.47	1.09	.41	.50
22		1.59	.87	1.54	.94	1.30	1.05	.38	.51
23		1.42	.80	1.41	1.09	1.11	.81	.40	.49
24		1.44	.78	1.30	1.08	1.00	.82	.42	.50
25		1.35	.83	3.48	1.16	.91	.75	.36	.53
26		1.30	.81	2.80	1.29	.88	.73	.31	.50
27		1.29	.72	2.32	1.76	.87	.78	.47	.50
28		1.28	.76	1.96	1.58	.80	1.42	.37	.51
29		1.18	.88	3.88	1.41	.80	1.90	.38	.50
30		1.20	.92	3.76	1.19	.78	1.46	.40	.53
31		1.21		2.88		.76	1.20		.52	

a Gage covered with ice, trash, etc.

Rating table for Republican River near Superior, Nebr., from January 1 to December 31, 1903.

Gage height.	Discharge.	Gage height.	Discharge.	Gage height.	Discharge.	Gage height.	Discharge.
Feet.	Second-feet.	Feet.	Second-feet.	Feet.	Second-feet.	Feet.	Second-feet.
0.1	45	1.2	1,190	2.3	4,180	3.4	9,445
.2	95	1.3	1,390	2.4	4,555	3.5	10,050
.3	150	1.4	1,610	2.5	4,950	3.6	10,675
.4	215	1.5	1,840	2.6	5,365	3.7	11,320
.5	285	1.6	2,085	2.7	5,800	3.8	11,985
.6	360	1.7	2,340	2.8	6,255	3.9	12,670
.7	460	1.8	2,605	2.9	6,730	4.0	13,375
.8	570	1.9	2,880	3.0	7,225	4.1	14,100
.9	690	2.0	3,175	3.1	7,750		
1.0	845	2.1	3,490	3.2	8,295		
1.1	1,005	2.2	3,825	3.3	8,860		

This table was applied indirectly, according to the method outlined in Nineteenth Ann. Rept. U. S. Geol. Survey, p. 323.

Estimated monthly discharge of Republican River near Superior, Nebr., in 1903.

[Drainage area, 22,347 square miles.]

Month.	Discharge in second-feet.			Total in acre-feet.	Run-off.	
	Maximum.	Minimum.	Mean.		Second-feet per square mile.	Depth in inches.
March 14-31	7,750	1,260	2,692	96,111	0.120	0.080
April	1,260	705	918	54,625	.041	.046
May	14,100	845	4,491	276,141	.201	.232
June	6,455	835	1,921	114,307	.086	.096
July	7,825	465	1,948	119,778	.087	.100
August	3,285	520	1,230	75,630	.055	.063
September	700	270	389	23,147	.017	.019
October	390	245	326	20,045	.015	.017
November	690	130	417	24,813	.019	.021

Discharge measurements of mill race near Superior, Nebr., in 1903.

Date.	Hydrographer.	Discharge.
		Second-feet.
April 11	J. C. Stevens	87
May 18	do	244
June 4	do	85
June 23	do	78
September 12	do	158

Depth of water, in feet, in midstream of mill race near Superior, Nebr., for 1903.

Day.	Mar.	Apr.	May.	June.	July.	Aug.	Sept.	Oct.	Nov.
1		4.01	4.18	5.34	3.13	4.75	2.25	4.01	4.50
2		3.85	4.02	4.39	3.00	4.15	2.55	4.00	4.42
3		4.20	4.10	3.83	3.00	3.82	2.22	4.00	4.22
4		3.65	4.01	3.60	2.75	3.80	2.18	3.90	4.43
5		3.62	4.24	3.20	3.10	3.15	2.05	4.15	4.29
6		3.70	4.18	3.08	2.95	2.86	1.80	4.27	4.38
7		3.75	4.19	2.05	4.00	3.00	1.52	4.12	4.30
8	3.20	3.72	5.95	3.70	3.60	2.90	3.40	4.02	4.40
9	7.45	3.70	6.75	3.75	3.32	2.52	3.35	4.12	4.42
10	7.30	3.68	5.80	3.55	3.14	2.48	2.88	3.95	4.27
11	7.65	3.70	6.84	3.40	3.12	2.43	2.96	4.22	4.36
12	7.60	3.85	5.40	3.20	4.96	2.51	2.80	4.03	4.29
13	7.50	3.50	5.00	3.25	5.10	2.80	3.00	4.10	3.72
14	3.60	3.52	4.21	2.60	5.40	3.00	3.32	4.25	3.82
15	3.00	3.58	3.79	3.80	4.95	2.71	3.40	4.12	3.84
16	2.20	3.50	3.72	3.70	4.67	2.55	3.32	4.02	4.06

Depth of water, in feet, in midstream of mill race near Superior, Nebr., etc.—
Continued.

Day.	Mar.	Apr.	May.	June.	July.	Aug.	Sept.	Oct	Nov.
17	2.05	3.62	2.89	3.20	3.40	2.35	2.52	4.15	3.84
18	1.78	3.50	3.15	3.38	2.90	2.35	3.63	4.35	
19	1.63	3.10	3.55	3.31	3.85	2.50	3.70	4.23	4.0
20	1.40	4.00	3.52	3.25	3.05	2.58	3.93	4.26	1.8
21	1.48	4.01	3.56	3.40	2.65	2.61	3.91	4.32	4.0
22	1.38	3.98	3.83	3.48	2.58	2.60	3.90	4.24	4.5
23	2.30	3.95	3.79	3.88	2.42	2.35	3.95	4.15	4.5
24	3.30	3.90	3.65	3.64	1.75	2.38	3.91	4.20	4.0
25	3.20	4.10	7.70	3.70	2.45	2.30	4.00	4.30	4.3
26	3.10	4.06	6.45	3.02	2.85	2.22	3.86	4.15	4.3
27	3.09	3.94	5.36	4.45	3.40	2.25	4.16	4.15	4.3
28	3.05	3.80	4.80	4.02	3.12	2.90	4.00	4.21	4.2
29	3.02	4.10	6.80	3.50	2.90	3.65	4.03	4.20	4.6
30	4.20	4.15	7.20	3.30	2.82	3.04	4.02	4.25	
31	4.23		6.10		2.85	2.60		4.28	

REPUBLICAN RIVER AT BENKELMAN, NEBR.

This station was established May 20, 1903, by J. C. Stevens. It is
located at a highway bridge between secs. 17 and 20, T. 1 N., R. 37 W.,
and about one-half mile east of Benkelman, Nebr., which is on the
main line of the Burlington and Missouri River Railroad. The
gage is a vertical 2 by 4 inch rod 6 feet long, spiked to the downstream side of the second bent from the west end of the bridge. It
is read once each day by Leon L. Hines. Discharge measurements
are made from the upstream side of the highway bridge, the upstream
hand rail of which is marked at 5-foot intervals by notches cut into
the rail. This bridge is one-fourth mile above the mouth of the
South Fork of Republican River. The initial point for soundings
is the zero mark on the upstream hand rail at the east end of the
bridge. The channel is straight for 200 feet above and 500 feet below
the station. Both banks are low and sandy, free from timber and
liable to overflow at very high stages. The bed is composed of shifting sand, and there is but one channel, except at extreme low water.
The river sometimes goes dry during the summer. Bench mark No. 1
is the top of the south end of the concrete foundation for the first
or west upright bent of the elevated track in the Burlington and Missouri River Railroad yards just east of the railroad station. To this
bench mark the gage rods of both the Republican and South Fork of
the Republican stations were referred. Its elevation above the zero
of the Republican River gage is 16.14 feet. Bench marks Nos. 2, 3, 4
and 5 are, respectively, the tops of the south ends of the caps of the
first, second, third, and fourth bents from the west end of the Republican River bridge. Their elevations above the zero of the Republican River gage are 6.88, 6.74, 6.34, and 6.19 feet, respectively.

;ion was maintained on this stream from November 1, 1894, to
)er 7, 1895, the gagings being made several miles farther
n. A description of this station and the results of the obser-
are given on page 125 of bulletin No. 140.

bservations at this station during 1903 have been made under
ction of J. C. Stevens, district hydographer.

rge measurements of Republican River at Benkelman, Nebr., in 1903.

Date.	Hydrographer.	Gage height.	Discharge.
		Feet.	*Second-feet.*
-------------------	J. C. Stevens	1.15	57
-------------------	----.do -------------------	1.20	82
r 7.--------------	----.do -------------------	.80	33

ily gage height, in feet, of Republican River at Benkelman, Nebr., for 1903.

Day.	May.	June.	July.	Aug.	Sept.	Oct.	Nov.	Dec.
-------------------	--------	1.15	0.90	1.22	0.60	1.00	1.10	1.10
-------------------	--------	1.20	.85	1.00	.60	1.05	1.10	1.15
-------------------	--------	1.20	.75	.90	.60	1.00	1.15	1.10
-------------------	--------	1.20	.95	.90	.60	1.00	1.10	1.00
-------------------	--------	1.20	1.00	.80	.65	1.05	1.10	1.10
-------------------	--------	1.20	.80	.70	.85	1.00	1.10	1.85
-------------------	--------	1.50	.75	.70	.85	1.00	1.15	1.05
-------------------	--------	1.45	.70	⁻70	.80	1.00	1.10	1.25
-------------------	--------	1.35	.70	.70	.80	1.05	1.10	1.25
-------------------	--------	1.30	.75	.80	.90	1.05	1.10	1.25
-------------------	--------	1.35	1.98	.70	.90	1.05	1.05	1.05
-------------------	--------	1.30	1.20	.75	.90	1.05	1.10	1.35
-------------------	--------	1.30	.90	.75	.90	1.00	1.10	1.25
-------------------	--------	1.30	.80	.70	.90	1.05	1.15	1.50
-------------------	--------	1.25	1.10	.70	.90	1.00	1.10	1.50
-------------------	--------	1.25	1.25	.70	.95	.95	1.15	1.50
-------------------	--------	1.20	1.00	.90	.90	1.00	1.20	1.50
-------------------	--------	1.20	1.00	.90	.95	1.00	1.15	1.60
-------------------	--------	1.10	1.00	.85	.95	1.00	1.15	1.70
-------------------	1.15	1.10	1.10	.85	.95	1.05	1.25	1.60
-------------------	1.15	1.10	1.00	.90	1.00	1.05	1.60	1.70
-------------------	1.00	1.00	.90	.75	1.00	1.05	1.60	1.50
-------------------	1.00	1.00	.90	.75	1.00	1.05	1.20	1.50
-------------------	1.05	1.00	.85	.75	1.00	1.05	1.10	1.40
-------------------	1.00	1.00	.70	.70	1.00	1.00	1.15	1.40
-------------------	1.25	1.05	.70	.75	.95	1.00	1.20	1.30
-------------------	1.25	1.00	.65	.85	1.00	1.05	1.20	1.60
-------------------	1.30	1.00	.80	.75	1.00	1.05	1.20	1.50
-------------------	1.25	1.00	.80	.75	1.00	1.05	f.10	1.50
-------------------	1.30	.95	.60	.70	1.00	1.10	1.05	1.40
-------------------	1.25	....	.65	.70	....	1.10	....	1.40

*Rating table for Republican River at Benkelman, Nebr., from January 1, 1
December 31, 1903.*

Gage height.	Discharge.	Gage height.	Discharge.	Gage height.	Discharge.	Gage height.	Dischar;
Feet.	*Second-feet.*	*Feet.*	*Second-feet.*	*Feet.*	*Second-feet.*	*Feet.*	*Second-f*
0.4	6	0.8	33	1.2	61	1.6	8
.5	12	.9	40	1.3	68	1.7	9
.6	19	1.0	47	1.4	75	1.8	10
.7	26	1.1	54	1.5	81	1.9	10

Estimated monthly discharge of Republican River at Benkelman, Nebr., for .

Month.	Discharge in second-feet.			Tota acre-f
	Maximum.	Minimum.	Mean.	
May 20–31	71	47	58	1
June	81	43	59	5
July	112	19	39	2
August	61	26	32	1
September	47	19	39	2
October	54	43	49	3
November 1–20 [a]	64	50	56	5

a River frozen on days not included.

REPUBLICAN RIVER (SOUTH FORK) AT BENKELMAN, NEBR.

This station was established May 20, 1903, by J. C. Stevens.
located at a highway bridge between secs. 17 and 20, T. 1 N., R. 37
and about three-fourths of a mile east of Benkelman, Nebr.
South Fork empties into Republican River about one-fourth (
mile below this station. A station is also maintained on Republ
River one-fourth of a mile above the junction of the two strea
The gage is a vertical 2 by 4 inch rod 5½ feet long, spiked to
upstream side of the first bent in the channel from the east b
of the stream. It is read once each day by Leon L. Hines.
charge measurements are made from the upstream side of the h
way bridge to which the gage is attached. The upstream b
rail of this bridge is marked at 5-foot intervals by notches cut
the rail, and the initial point for soundings is the zero mark on
rail at the east end of the bridge. The channel is straight for 1
feet above and 500 feet below the station. Both banks are low
sandy, free from timber, and liable to overflow at very high sta
The bed of the stream is sandy and shifting, and there is but
channel except at extreme low stages. The river usually goes

the summer. Bench mark No. 1 is the top of the south end
concrete foundation for the first or west upright bent of the
d track in the Burlington and Missouri River Railroad yards
st of the station. Its elevation above the zero of the South
age is 18.29 feet. This is bench mark No. 1 of the Republican
gage. Bench marks Nos. 2, 3, 4, and 5 are, respectively, the
the south ends of the first, second, third, and fourth bents
1e west end of the South Fork River bridge. Their elevations
:he zero of the South Fork River gage are 6.62, 6.48, 6.38, and
et, respectively.

ition was maintained on this river from November 1, 1894, to
iber 7, 1895, the gagings being made near the present station.
ription of the old station and the results of the observations are
n page 130 of Bulletin No. 140.

observations at this station during 1903 have been made under
ection of J. C. Stevens, district hydrographer.

*ge measurements of South Fork of Republican River at Benkelman, Nebr.,
in 1903.*

Date.	Hydrographer.	Gage height.	Discharge.
		Feet.	*Second-feet.*
....	J. C. Stevens	1.15	58
....	.....do	1.05	37
)er 7...	.....do	.76	2

*ily gage height, in feet, of South Fork of Republican River at Benkelman,
,Nebr., for 1903.*

Day.	May.	June.	July.	Aug.	Sept.	Oct.	Nov.	Dec.
....	....	1.05	0.80	0.98	0.80	0.90	1.20	1.20
....	....	1.05	.80	1.30	.80	.90	1.15	1.20
....	....	1.05	.80	1.10	.80	.90	1.15	1.15
....	....	1.05	.85	1.00	.80	1.00	1.15	1.05
....	....	1.05	.90	.95	.80	.90	1.10	1.10
....	....	1.05	.80	.90	.80	.90	1.15	1.10
....	....	1.20	.80	.90	.80	.90	1.10	1.20
....	....	1.20	.80	.85	(a)	1.00	1.15	1.30
....	....	1.20	.80	1.00	(a)	1.00	1.15	1.40
....	....	1.20	.80	.85	(a)	1.00	1.15	1.30
....	1.15	.85	.80	(a)	1.00	1.20	1.10
....	1.10	1.00	.80	(a)	1.00	1.15	1.15
....	1.05	.80	1.00	.85	1.05	1.15	1.25
....	1.05	1.00	1.00	.85	1.05	1.15	1.30
....	1.00	1.00	1.10	.85	1.00	1.10	1.30
....	1.00	.95	1.00	.85	1.05	1.20	1.40
....	1.00	.90	1.00	.85	1.05	1.20	1.40

Mean daily gage height, in feet, of South Fork of Republican River, etc.—Cont'

Day.	May.	June.	July.	Aug.	Sept.	Oct.	Nov.	De
19.	......	1.00	0.90	1.00	0.85	1.05	1.10	1
20.	1.15	.95	.95	1.00	.85	1.00	1.15	1
21.	1.15	.95	.90	.90	.85	1.05	1.10	1
22.	1.10	.90	.90	.90	.85	1.05	1.30	
23.	1.10	.90	.85	.85	.85	1.05	1.40	
24.	1.10	.90	.80	.80	.90	1.10	1.30	
25.	1.05	.90	.80	.85	.90	1.10	1.20	
26.	1.15	.90	.80	.80	.80	1.10	1.20	
27.	1.05	.85	.80	.80	.90	1.10	1.25	
28.	1.05	.80	.80	.80	.90	1.10	1.25	
29.	1.00	.80	.80	.80	.90	1.10	1.20	
30.	1.05	.80	.80	.80	.90	1.10	1.10	
31.	1.10	......	.80	.80	......	1.10	......	

Rating table for South Fork of Republican River at Benkelman, Nebr., fro January 1 to December 31, 1903.

Gage height.	Discharge.	Gage height.	Discharge.	Gage height.	Discharge.	Gage height.	Dischar
Feet.	Second-feet.	Feet.	Second-feet.	Feet.	Second-feet.	Feet.	Second-fe
0.8	7	1.1	50	1.4	93	1.7	13
.9	22	1.2	65	1.5	107	1.8	15
1.0	36	1.3	79	1.6	121	1.9	16

Estimated monthly discharge of South Fork of Republican River at Benkel Nebr., for 1903.

Month.	Discharge in second-feet.			Tota acre-f
	Maximum.	Minimum.	Mean.	
May 20–31	57	36	48	1
June	65	7	37	2
July	36	7	15	
August	79	7	25	1
September 1–5 and 14–30 [a]	22	7	15	
October	50	22	39	5
November 1–20 [b]	65	50	57	5

[a] Sand at gage rod.   [b] River frozen on days not included.

SOLOMON RIVER NEAR NILES, KANS.

The station at Niles was established May 5, 1897, and was loc at a bridge one-half mile west of the town and 7 miles above mouth of the river. The rod of the wire gage is spiked to the t of the bridge. The bench mark is the uppermost of three nails driven into a cottonwood tree 18 inches in diameter, on the north side of the river and 25 feet east of the bridge, at an elevation of 24.96 feet above gage datum. The channel is straight for about 100 feet above and below the section. The current is sluggish; the right bank is high, and the left bank overflows only at very high stages. The bed of the stream is muddy.

The observer during 1903 was J. J. Little. The station was discontinued November 30, 1903.

The observations at this station during 1903 have been made under the direction of W. G. Russell, district hydrographer.

Discharge measurements of Solomon River near Niles, Kans., in 1903.

Date.	Hydrographer.	Gage height.	Discharge.
		Feet.	*Second-feet.*
March 7	W. G. Russell	10.90	973
May 8	do	10.70	1,118
May 9	do	14.20	2,168
May 12	do	22.70	4,825
June 3	do	33.80	*a* 41,000
June 23	do	13.60	1,358

a Estimated.

Mean daily gage height, in feet, of Solomon River near Niles, Kans., for 1903.

Day.	Jan.	Feb.	Mar.	Apr.	May.	June.	July.	Aug.	Sept.	Oct.	Nov.
1	7.00	6.70	9.10	8.80	15.60	32.90	15.20	10.70	15.50	8.40	8.20
2	7.00	6.50	8.90	8.60	15.60	33.30	14.60	18.80	13.00	8.40	8.20
3	6.80	6.70	9.10	8.60	12.80	33.80	13.90	23.10	12.80	8.40	8.40
4	6.80	7.00	8.70	8.50	10.50	33.60	14.40	24.80	11.90	8.20	8.80
5	6.70	6.40	9.80	8.40	10.20	32.80	13.40	25.90	11.40	8.30	9.50
6	6.70	6.20	11.60	8.30	9.10	32.00	12.40	25.30	10.70	8.40	9.60
7	6.60	6.10	10.90	8.20	10.40	30.00	12.00	16.80	10.60	8.20	9.50
8	6.60	6.20	11.60	8.20	10.70	25.60	11.80	25.70	10.20	8.20	9.00
9	6.80	6.20	13.80	8.00	16.80	20.80	11.50	18.80	11.10	8.10	8.80
10	6.90	6.40	17.10	8.00	20.00	18.80	11.50	12.70	11.30	8.00	8.60
11	6.60	7.00	19.10	8.00	21.80	18.10	11.20	11.60	13.60	8.20	8.50
12	6.50	6.60	20.30	7.80	23.10	18.50	10.90	11.20	12.20	8.10	8.40
13	6.80	7.10	21.20	8.00	23.20	18.00	10.70	15.40	11.20	8.20	8.20
14	7.00	7.20	22.20	8.30	25.10	16.60	11.30	18.40	10.50	8.30	8.10
15	6.90	6.70	22.70	7.90	28.20	15.90	16.70	24.70	9.80	8.40	8.20
16	6.80	7.00	21.70	7.70	30.00	15.50	20.20	24.70	9.50	8.50	8.10
17	6.80	6.70	16.80	7.60	32.10	15.10	19.60	26.60	9.40	8.30	8.10
18	6.50	7.00	13.30	7.50	31.00	14.70	17.70	28.00	9.20	8.30	8.10
19	6.70	6.90	12.20	7.60	25.00	14.40	15.60	28.20	9.00	8.20	7.90
20	6.70	6.70	12.70	7.50	18.20	14.20	13.00	25.40	9.20	8.20	7.90
21	6.70	6.80	11.70	7.40	17.50	14.00	12.20	16.70	9.00	8.20	7.90
22	6.70	6.70	13.50	7.30	21.50	13.70	11.60	15.70	9.20	8.10	7.80

Mean daily gage height, in feet, of Solomon River, etc.—Continued.

Day.	Jan.	Feb.	Mar.	Apr.	May.	June.	July.	Aug.	Sept.	Oct.	Nov.
23	6.80	7.20	11.80	7.00	20.20	13.40	11.20	16.70	9.00	8.10	7.80
24	6.50	7.10	10.60	7.60	17.70	13.20	10.80	15.00	8.90	8.10	7.90
25	6.40	7.00	10.00	7.40	18.00	13.50	10.60	13.20	8.80	8.10	7.90
26	6.60	6.80	9.70	7.10	20.30	13.80	10.40	12.40	8.50	7.90	7.90
27	6.80	6.70	9.50	6.90	25.00	13.80	10.30	12.00	8.50	8.00	7.90
28	7.10	9.00	9.30	6.60	23.20	15.70	10.10	11.20	8.60	8.00	8.00
29	6.60	......	9.20	7.10	32.80	15.60	10.00	14.40	8.40	8.00	8.30
30	6.60	......	9.00	7.90	31.90	17.20	10.00	15.60	8.50	8.10	8.10
31	6.90	......	8.90	......	32.20	......	9.90	16.90	......	8.20	......

Rating table for Solomon River near .Niles, Kans., from January 1 to March 5, 1903.

Gage height.	Discharge.	Gage height.	Discharge.	Gage height.	Discharge.	Gage height.	Discharge.
Feet.	Second-feet.	Feet.	Second-feet.	Feet.	Second-feet.	Feet.	Second-feet.
6.0	222	7.0	390	8.1	623	9.1	855
6.1	237	7.1	410	8.2	646	9.2	880
6.2	252	7.2	430	8.3	669	9.3	905
6.3	267	7.3	450	8.4	692	9.4	930
6.4	282	7.4	470	8.5	715	9.5	955
6.5	299	7.5	490	8.6	738	9.6	980
6.6	316	7.6	512	8.7	761	9.7	1,005
6.7	333	7.8	556	8.8	784	9.8	1,030
6.8	352	7.9	578	8.9	807	9.9	1,055
6.9	371	8.0	600	9.0	830	10.0	1,080

Rating table for Solomon River near Niles, Kans., from March 6 to November 30, 1903.

Gage height.	Discharge.	Gage height.	Discharge.	Gage height.	Discharge.	Gage height.	Discharge.
Feet.	Second-feet.	Feet.	Second-feet.	Feet.	Second-feet.	Feet.	Second-feet.
10.0	860	12.0	1,410	14.0	2,005	19.0	3,640
10.2	915	12.2	1,465	14.5	2,155	19.5	3,820
10.4	970	12.4	1,525	15.0	2,305	20.0	4,000
10.6	1,025	12.6	1,585	15.5	2,460	21.0	4,370
10.8	1,080	12.8	1,645	16.0	2,615	22.0	4,740
11.0	1,135	13.0	1,705	16.5	2,775	23.0	5,115
11.2	1,190	13.2	1,765	17.0	2,945	24.0	5,495
11.4	1,245	13.4	1,825	17.5	3,115	25.0	5,875
11.6	1,300	13.6	1,885	18.0	3,285		
11.8	1,355	13.8	1,945	18.5	3,460		

ited monthly discharge of Solomon River near Niles, Kans., for 1903.

[Drainage area, 6,815 square miles.]

nth.	Discharge in second-feet.			Total in acre-feet.	Run-off.	
	Maximum.	Minimum.	Mean.		Second-feet per square mile.	Depth in inches.
....	410	282	341	20,967	0.050	. 0.058
....	830	237	358	19,882	.052	.054
.. ..	5,002	587	1,866	114,736	.270	.310
....	564	120	345	20,529	.051	.057
....	9,946	635	4,835	297,292	.710	.820
....	10,602	1,765	4,447	264,615	.650	.720
....	4,074	835	1,658	101,946	.240	.280
....	7,091	1,052	3,625	222,893	.530	.610
. ...	2,460	472	970	57,719	.140	.160
....	495	363	427	26,255	.063	.073
. ...	760	343	456	27,134	.067	.075

SALINE RIVER NEAR SALINA, KANS.

.ation, established May 4, 1897, is located at a bridge 4.5 miles
it of Salina, near the mouth of the river. The rod of the wire
ipiked to the floor of the bridge. Bench mark No. 1 is a nail
n tree 2 feet in diameter on the north side of the river and 6
t of the bridge. Its elevation is 22.90 feet above gage datum.
iark No. 2 is six nails driven into a 16-inch boxelder tree on
h side of the river and 35 feet east of the bridge. Its eleva-
!2.90 feet above gage datum. The channel is straight for a
stance above and below the station and has a width of about
it low stages. Both banks are high and not liable to over-
ipt during floods. The bed of the stream is composed of sand
d and the flow is sluggish. The station was discontinued
er 30, 1902.

bservations at this station during 1903 have been made under
:tion of W. G. Russell, district hydrographer.

Discharge measurements of Saline River near Salina, Kans.

Date.	Hydrographer.	Gage height.	Dischar
		Feet.	*Second-*
March 6	W. G. Russell	8.40	
May 2	do	33.90	a 27.
May 6	do	13.00	1.
July 7	do	11.10	
August 4	do	24.60	2
August 5	do	19.10	1
September 30	E. C. Murphy	7.96	

a Estimated.

Mean daily gage height, in feet, of Saline River near Salina, Kans., for 19.

Day.	Jan.	Feb.	Mar.	Apr.	May.	June.	July.	Aug.	Sept.	Oct.
1	5.60	5.80	6.00	7.00	11.50	a33.50	13.40	16.90	9.60	8.00
2	5.70	5.60	6.00	6.90	15.90	a33.30	12.50	23.20	9.50	7.90
3	5.60	5.40	5.70	6.90	13.90	a33.00	12.10	27.80	9.40	7.90
4	5.80	6.20	6.80	6.70	13.30	32.70	11.80	21.70	9.30	7.80
5	5.60	6.10	8.80	6.80	11.20	32.00	11.70	16.70	9.20	7.90
6	5.80	5.90	8.60	6.80	13.00	31.20	11.80	18.00	9.10	7.90
7	5.70	5.60	7.60	6.70	11.80	30.60	11.00	13.10	8.90	7.80
8	5.60	5.00	7.40	6.60	13.30	28.50	10.80	12.00	8.80	7.80
9	6.00	5.10	9.80	6.50	15.80	24.70	10.60	12.90	8.90	7.80
10	6.00	5.20	12.10	6.50	16.70	20.20	10.40	11.70	9.80	7.80
11	5.60	5.40	11.80	6.40	16.80	17.80	10.30	11.40	9.50	7.80
12	5.40	5.80	11.20	6.30	15.40	17.20	10.40	10.00	9.50	7.80
13	5.50	5.60	11.20	6.40	13.90	16.50	10.10	15.80	9.80	7.90
14	5.40	6.00	10.90	6.20	16.10	15.80	10.30	20.40	8.90	8.10
15	5.40	6.20	12.90	6.20	17.30	15.30	10.30	24.90	8.80	8.10
16	5.40	6.40	13.80	6.20	17.50	14.90	10.20	27.50	8.50	8.20
17	5.00	6.00	12.80	6.10	18.00	14.50	10.00	25.50	8.60	8.30
18	5.80	5.60	11.40	6.10	17.20	14.10	9.90	24.50	8.40	8.30
19	5.60	5.20	11.20	6.00	18.60	13.80	9.80	21.30	8.40	8.30
20	5.00	5.20	9.90	6.00	12.30	13.50	9.60	14.40	8.30	8.20
21	5.60	5.30	9.00	6.00	13.30	14.00	9.50	15.10	8.30	8.30
22	5.60	5.40	8.60	5.90	20.10	14.10	9.50	16.60	8.20	7.90
23	5.00	5.00	8.40	6.00	23.10	13.50	9.40	15.60	8.10	7.80
24	5.70	5.80	8.00	5.80	20.20	13.60	9.30	12.80	8.30	7.90
25	5.60	5.70	7.80	5.80	18.00	13.50	9.20	11.60	8.00	7.80
26	5.70	5.80	7.60	5.80	28.70	13.20	9.00	11.00	8.00	7.80
27	5.80	5.80	.50	6.00	32.60	12.70	9.00	10.60	8.00	7.60
28	5.70	6.00	†	6.00	30.70	12.40	9.00	10.20	7.90	7.70
29	5.80		30	6.00	a31.40	12.70	8.80	10.00	7.90	7.60
30	5.80		7.20	6.80	a32.10	13.80	9.00	9.90	7.90	7.90
31	6.00		7.10		a32.80		9.00	9.70		8.00

a Estimated by interpolation.

Rating table for Saline River near Salina, Kans., from January 1 to May 25, 1903.

Gage height.	Discharge.	Gage height.	Discharge.	Gage height.	Discharge.	Gage height.	Discharge.
Feet.	Second-feet.	Feet.	Second-feet.	Feet.	Second-feet.	Feet.	Second-feet.
5.0	140	6.4	280	7.8	436	13.0	1,200
5.1	150	6.5	290	7.9	448	14.0	1,370
5.2	160	6.6	300	8.0	460	15.0	1,550
5.3	170	6.7	310	8.2	488	16.0	1,730
5.4	180	6.8	320	8.4	516	17.0	1,910
5.5	190	6.9	330	8.6	544	18.0	2,090
5.6	200	7.0	340	8.8	572	19.0	2,270
5.7	210	7.1	352	9.0	600	20.0	2,450
5.8	220	7.2	364	9.5	670	21.0	2,640
5.9	230	7.3	376	10.0	740	22.0	2,860
6.0	240	7.4	388	10.5	810	23.0	3,170
6.1	250	7.5	400	11.0	880	24.0	3,620
6.2	260	7.6	412	11.5	955	25.0	4,220
6.3	270	7.7	424	12.0	1,030		

Rating table for Saline River near Salina, Kans., from June 10 to November 30, 1903.

Gage height.	Discharge.	Gage height.	Discharge.	Gage height.	Discharge.	Gage height.	Discharge.
Feet.	Second-feet.	Feet.	Second-feet.	Feet.	Second-feet.	Feet.	Second-feet.
7.0	90	7.9	207	11.0	640	20.0	2,000
7.1	103	8.0	220	12.0	780	21.0	2,180
7.2	116	8.2	248	13.0	925	22.0	2,360
7.3	129	8.4	276	14.0	1,075	23.0	2,540
7.4	142	8.6	304	15.0	1,225	24.0	2,720
7.5	155	8.8	332	16.0	1,375	25.0	2,900
7.6	168	9.0	360	17.0	1,525	26.0	3,090
7.7	181	9.5	430	18.0	1,675	27.0	3,290
7.8	194	10.0	500	19.0	1,830		

Estimated monthly discharge of Saline River near Salina, Kans., for 1903.

[Drainage area, 3,311 square miles.]

Month.	Discharge in second-feet.			Total in acre-feet.	Run-off.	
	Maximum.	Minimum.	Mean.		Second-feet per square mile	Depth in inches.
January	240	180	205	12,605	0.062	0.07
February	280	140	207	11,496	.032	.06
March	1,336	210	628	38,614	.190	.22
April	340	220	271	16,126	.082	.09
May *a*	7,580	910	2,685	165,094	.810	.93
June *b*	7,895	835	2,878	170,955	.870	.97
July	985	332	533	32,773	.160	.1
August	3,410	458	1,434	88,178	.430	.5
September	472	207	322	19,160	.097	.1
October	262	168	212	13,035	.064	.0
November	304	168	223	13,270	.067	.0

a May 29, 30, and 31 estimated.   *b* June 1, 2, and 3 estimated.

SMOKY HILL RIVER AT ELLSWORTH, KANS.

The gaging station, established April 17, 1895, is located at the highway bridge on Douglass avenue, Ellsworth, Kans. The gage an inclined ash timber spiked to a post driven in the bed of the river and bolted to an iron post on the bridge pier. The bench mark is nail driven into the base of a large box-elder tree near the southeast corner of the bridge, 90 feet from the gage, and its elevation is 13.0 feet above the zero of the gage. A slope gage is spiked to the S Louis and San Francisco Railroad bridge, 2,536 feet upstream, and referred to the same datum. The channel is nearly straight above and below the gage, and the bed is sandy and shifting. The observer Thomas Coyne.

The observations at this station during 1903 have been made under the direction of W. G. Russell, district hydrographer.

Discharge measurements of Smoky Hill River at Ellsworth, Kans., in 1903.

Date.	Hydrographer.	Gage height.	Discharge.
		Feet.	Second-feet.
February 23	W. G. Russell	1.60	50
April 6	do	1.85	194
May 7	do	6.30	2,812
June 4	do	4.90	1,697
June 8	do	3.60	821

Mean daily gage height, in feet, of Smoky Hill River at Ellsworth, Kans., for 1903.

Day.	Jan.	Feb.	Mar.	Apr.	May.	June.	July.	Aug.	Sept.	Oct.	Nov.	Dec.
1	(a)	1.50	1.50	2.00	3.50	7.40	2.40	5.00	1.85	1.30	1.50	1.25
2	(a)	1.35	1.50	2.00	3.00	7.00	2.30	2.60	1.85	1.30	1.60	1.25
3	(a)	1.30	1.40	2.20	3.00	5.40	2.20	2.40	1.80	1.30	1.70	1.30
4	1.60	1.30	1.40	2.10	2.90	4.80	2.10	4.45	1.80	1.30	1.75	1.30
5	1.80	(a)	1.50	1.90	5.00	4.40	2.10	4.00	1.70	1.30	1.75	1.25
6	1.35	(a)	1.60	1.75	5.30	4.30	2.10	3.40	1.65	1.30	1.75	1.25
7	1.80	(a)	2.40	1.80	6.30	3.90	2.00	3.00	1.60	1.30	1.70	1.25
8	(a)	1.50	2.00	1.80	5.90	3.60	2.00	2.75	1.60	1.30	1.65	1.25
9	(a)	1.40	2.00	1.75	5.50	3.50	2.00	2.50	1.60	1.30	1.60	1.25
10	(a)	1.35	1.90	1.70	5.00	3.40	1.95	2.30	1.55	1.30	1.55	1.20
11	(a)	1.30	2.70	1.70	4.90	3.30	1.90	2.25	1.55	1.25	1.45	1.20
12	(a)	1.30	2.80	1.70	4.40	3.20	1.90	2.30	1.50	1.25	1.45	1.20
13	(a)	(a)	4.00	1.65	5.50	3.20	2.00	2.10	1.50	1.40	1.40	(a)
14	(a)	(a)	4.20	1.60	5.60	3.10	2.10	2.00	1.50	1.45	1.40	(a)
15	(a)	1.30	4.40	1.60	4.50	3.00	2.10	6.00	1.50	1.60	1.40	(a)
16	(a)	1.30	4.30	1.60	4.00	2.90	2.00	4.30	1.50	1.60	1.35	(a)
17	1.50	1.30	4.20	1.55	3.80	2.80	2.00	6.15	1.45	1.55	1.35	1.20
18	1.50	1.30	3.40	1.55	3.60	2.70	1.90	5.20	1.45	1.50	1.35	1.20
19	1.45	1.30	3.00	1.60	3.60	2.60	1.90	4.80	1.45	1.45	1.35	1.20
20	1.40	1.30	3.00	1.60	3.20	2.80	2.00	3.90	1.45	1.40	1.30	(a)
21	1.40	1.30	2.70	1.50	3.00	2.80	2.20	3.40	1.40	1.40	1.30	1.20
22	1.35	1.30	2.50	1.50	3.10	2.60	2.00	2.70	1.40	1.40	1.30	1.15
23	1.30	1.60	2.40	1.50	3.50	2.80	2.00	2.50	1.40	1.35	1.30	1.20
24	1.90	1.65	2.40	1.50	3.20	2.80	1.90	2.25	1.40	1.35	1.30	1.20
25	1.35	1.70	2.30	1.45	3.00	2.70	1.90	2.20	1.40	1.35	1.30	1.20
26	1.40	1.70	2.30	1.40	5.00	3.40	1.80	2.10	1.35	1.30	1.30	1.60
27	1.45	1.60	2.30	1.40	3.30	3.90	1.80	2.00	1.35	1.30	1.30	1.30
28	1.45	1.60	2.00	1.40	15.40	3.00	1.75	1.90	1.35	1.30	1.30	1.25
29	1.50		2.00	3.00	16.00	2.80	1.70	1.90	1.35	1.30	1.25	1.20
30	1.50		2.00	5.00	13.80	2.50	1.80	1.90	1.30	1.35	1.25	1.20
31	1.40		1.90		10.20		1.80	1.90		1.40		1.20

a River frozen.

Rating table for Smoky Hill River at Ellsworth, Kans., from January 1 to December 31, 1903.

Gage height.	Discharge.	Gage height.	Discharge.	Gage height.	Discharge.	Gage height.	Discharge.
Feet.	*Second-feet.*	*Feet.*	*Second-feet.*	*Feet.*	*Second-feet.*	*Feet.*	*Second-feet.*
0.5	5	1.8	131	3.2	657	5.8	2,434
.6	9	1.9	155	3.4	765	6.0	2,590
.7	13	2.0	181	3.6	883	6.2	2,748
.8	17	2.1	209	3.8	1,010	6.4	2,910
.9	21	2.2	239	4.0	1,141	6.6	3,075
1.0	26	2.3	271	4.2	1,275	6.8	3,241
1.1	32	2.4	305	4.4	1,410	7.0	3,410
1.2	40	2.5	342	4.6	1,549	7.2	3,580
1.3	50	2.6	381	4.8	1,690	7.4	3,750
1.4	62	2.7	422	5.0	1,834	7.6	3,920
1.5	76	2.8	465	5.2	1,980	7.8	4,090
1.6	92	2.9	510	5.4	2,129	8.0	4,260
1.7	110	3.0	557	5.6	2,280		

Tangent above 7 feet gage height. Differences above this point, 85 feet per tenth. 3 per cent added to discharges greater than 8,510 (for overflow).

Estimated monthly discharge of Smoky Hill River at Ellsworth, Kans., for 1903.

[Drainage area, 7,980 square miles.]

Month.	Discharge in second-feet.			Total in acre-feet.	Run-off.	
	Maximum.	Minimum.	Mean.		Second-feet per square mile.	Depth in inches.
January	131	50	69	4,243	0.0086	0.0099
February	110	50	65	3,610	.0081	.0084
March	1,410	62	431	26,501	.0540	.0620
April	1,834	62	182	10,830	.0230	.0260
May	11,392	510	2,356	144,865	.3000	.3400
June	3,750	342	937	55,755	.1200	.1300
July	305	110	180	11,068	.0230	.0260
August	2,708	155	735	45,193	.0920	.1100
September	143	50	81	4,820	.0100	.0110
October	92	45	58	3,566	.0073	.0084
November	120	45	70	4,165	.0088	.0098
December	92	36	44	2,705	.0055	.0063
The year	11,392	36	434	317,321	.0550	.7478

MISCELLANEOUS MEASUREMENTS IN THE KANSAS RIVER DRAINAGE BASIN.

Kansas River was measured by W. G. Russell at Topeka, Kans., on August 19, 1903. The measurement was made from the Melan Arch Bridge, on Kansas avenue. The stream had a discharge of 30,609 second-feet.

MERAMEC RIVER DRAINAGE BASIN.

Meramec River rises in Dent County, Mo., flows northeast, and enters the Mississippi near St. Louis. This river drains a rugged, hilly, and comparatively thinly populated country. There are, however, numerous good sites for dams, and the United States Geological Survey is studying the river in connection with the possible water-power developments and for its possible use as a future water supply for the city of St. Louis.

The total drainage area of Meramec River is 3,619 square miles; at Eureka it is 3,497 square miles. The drainage area above Dry Fork is 340 square miles, while that of Dry Fork is 360 square miles.

The following stations have been maintained in Meramec River drainage basin during 1903:

Meramec River near Meramec, Mo.
Meramec Spring near Meramec, Mo.
Meramec River (Station No. 1) at Fenton, Mo.
Meramec River (Station No. 2) below Fenton, Mo.
Meramec River near Eureka, Mo.
Meramec River (Dry Fork) near St. James, Mo.

MERAMEC RIVER NEAR MERAMEC, MO.

This station was established February 28, 1903, by I. W. McConnell. It is located about 600 feet below the mouth of the Spring branch and about 1 mile from the post-road between Meramec, Mo., and St. James, Mo. The nearest railroad station is St. James, Mo., 7 miles northwest of the station. The gage is a 2 by 3 inch pine stick, 10 feet long, graduated to feet and tenths by means of brass-headed nails. The gage is driven into the bed of the river and is nailed at the top to a leaning tree. The gage is read once each day by C. C. Smallwood. The equipment by means of which discharge measurements are made consists of a cable, boat, and tagged wire. The initial point for soundings has been taken as the tree to which the cable is attached on the left bank. The channel is straight for about 300 feet above the station and for 2,000 feet below. The current velocity is sufficient for accurate measurement except at low stages, when it becomes sluggish. Both banks are low, and at high water the river spreads over wide flats. The bed of the stream is of cemented gravel and is not liable to shift. The bench mark is a point on a bowlder on the left bank at the foot of the cliff near the mouth of the Spring branch. Its elevation is 10.44 feet above the zero of the gage.

The observations at this station during 1903 have been made under the direction of E. Johnson, jr., district hydrographer.

Discharge measurements of Meramec River near Meramec, Mo., in 1903.

Date.	Hydrographer.	Gage height.	Discharge.
		Feet.	*Second-feet.*
March 28	I. W. McConnell	3.60	564
March 10	do	5.30	1,390
April 24	do	3.50	50
May 26	F. W. Hanna	3.20	40
June 27	do	3.00	3?
July 20	do	2.50	1?
August 27	do	2.45	1?
September 22	do	2.50	1?
October 20	do	2.60	1?
November 29	do	2.40	1?
December 30	do	2.52	1?

Mean daily gage height, in feet, of Meramec River near Meramec, Mo., for 190?

Day	Mar.	Apr.	May.	June.	July.	Aug.	Sept.	Oct.	Nov.	D?
1	4.60	3.40	3.10	8.40	2.90	2.50	2.50	2.40	2.50	?
2	4.30	3.40	3.00	7.20	2.80	7.30	2.50	2.50	2.50	?
3	4.20	3.40	3.00	6.10	2.80	3.70	2.50	2.50	2.50	?
4	4.60	3.30	2.90	5.60	2.80	3.50	2.50	2.50	2.50	?
5	9.30	3.20	2.90	5.40	2.80	3.40	2.50	2.60	2.50	
6	8.20	3.20	3.00	5.00	2.70	3.00	2.50	2.60	2.50	
7	7.60	3.20	3.10	4.90	2.70	2.90	2.50	5.00	2.50	
8	7.00	3.40	3.10	4.70	2.70	3.00	2.50	4.70	2.50	
9	6.40	3.30	3.00	4.60	2.60	3.80	2.60	3.80	2.40	
10	5.30	3.30	3.00	4.50	2.60	3.60	2.70	3.40	2.40	
11	4.90	3.30	2.90	4.50	2.60	3.20	2.60	3.10	2.40	
12	4.80	3.30	2.80	4.40	2.60	2.90	2.50	3.00	2.40	
13	4.50	6.10	2.80	4.20	2.60	2.80	2.60	2.90	2.40	
14	4.30	5.70	3.40	3.50	2.60	2.80	3.40	2.90	2.40	
15	4.10	4.80	3.60	3.30	2.60	2.80	2.80	2.80	2.40	
16	3.90	4.50	3.50	3.20	2.60	2.80	2.80	2.80	2.40	
17	3.80	4.00	4.70	3.10	2.60	2.80	2.90	2.80	2.40	
18	3.70	3.90	4.60	3.10	2.60	2.70	2.80	2.70	2.40	
19	3.70	3.70	4.30	2.90	2.50	2.70	2.60	2.60	2.40	
20	5.30	5.70	4.00	2.90	2.50	2.70	2.50	2.60	2.40	
21	5.00	4.50	4.00	2.80	2.50	2.60	2.50	2.50	2.40	
22	4.20	3.80	3.80	2.90	2.50	2.60	2.50	2.50	2.40	
23	4.30	3.70	3.50	2.90	2.50	2.60	2.50	2.50	2.40	
24	4.30	3.60	3.30	2.80	2.50	2.50	2.50	2.50	2.40	
25	4.20	3.40	3.20	3.00	2.50	2.50	2.40	2.50	2.40	
26	4.00	3.30	3.20	3.10	2.50	2.50	2.40	2.60	2.40	
27	3.80	3.30	3.10	3.00	2.50	2.50	2.40	2.50	2.40	
28	3.60	3.20	3.00	3.00	2.50	2.50	2.40	2.50	2.40	
29	3.50	3.10	3.00	2.90	2.50	2.50	2.40	2.50	2.40	
30	3.50	3.10	3.10	2.90	2.50	2.50	2.40	2.50	2.40	
31	3.40		9.40		2.50	2.50		2.50		

Rating table for Meramec River near Meramec, Mo., from March 1 to December
31, 1903.

Gage height.	Discharge.	Gage height.	Discharge.	Gage height.	Discharge.	Gage height.	Discharge.
Feet.	Second-feet.	Feet.	Second-feet.	Feet.	Second-feet.	Feet.	Second-feet.
2.3	103	3.2	401	4.1	777	5.0	1,227
2.4	134	3.3	439	4.2	823	5.1	1,282
2.5	165	3.4	478	4.3	870	5.2	1,338
2.6	197	3.5	518	4.4	918	5.3	1,395
2.7	229	3.6	559	4.5	967	5.4	1,452
2.8	262	3.7	601	4.6	1,017	5.5	1,510
2.9	295	3.8	644	4.7	1,068		
3.0	329	3.9	688	4.8	1,120		
3.1	364	4.0	732	4.9	1,173		

Tangent above 5.50 feet gage height.   Differences above this point, 60 per tenth.
Table is approximate above 5.30.

Estimated monthly discharge of Meramec River near Meramec, Mo., for 1903.

Month.	Discharge in second-feet.			Total in acre-feet.
	Maximum.	Minimum.	Mean.	
March	3,790	478	1,197	73,601
April	1,870	364	660	39,273
May	3,850	262	573	35,232
June	3,250	262	819	48,734
July	295	165	198	12,175
August	2,590	165	358	22,013
September	478	134	190	11,306
October	1,570	134	300	18,446
November	165	134	142	8,450
December	262	134	153	9,408

MERAMEC SPRING NEAR MERAMEC, MO.

This station was established February 28, 1903, by I. W. McConnell. It is located on the Spring branch 500 feet from the spring, at a footbridge. This point is about 1 mile from the mouth of the Spring branch and 2 miles above the mouth of Dry Fork. The gage is a 2 by 3 inch pine rod 10 feet long. It is read once each day by C. C. Smallwood. Discharge measurements are made from the footbridge. The initial point for soundings is at the tree at the end of the footbridge on the right bank. The channel is straight for 50 feet above and for 500 feet below the station. Both banks are of clay and gravel and are about 6 feet high. They will overflow only at unusual flood stages. The bed of the stream is composed of gravel with some bowlders and is clean. The water is very swift, making accurate gage readings very difficult. The bench mark is located on the left corner stone of the breast wheel of the tail race just below the old breast wheel and is designated by a cross on the top surface of the stone. Its elevation above the zero of the gage is 4.68 feet.

The observations at this station during 1903 have been made under the direction of E. Johnson, jr., district hydrographer.

Discharge measurements of Meramec Spring near Meramec, Mo., in 1903.

Date.	Hydrographer.	Gage height.	Discharge
		Feet.	Second-feet
February 28	I. W. McConnell	1.30	38
March 10	....do	1.85	46
March 27	....do	1.00	27
April 24	....do	.90	24
May 26	F. W. Hanna	.90	23
June 27	....do	.80	18
July 21	....do	.50	13
August 27	....do	.40	1
September 22	....do	.40	1
October 20	....do	.39	1
November 29	....do	.275	
December 30	....do	.35	

Mean daily gage height, in feet, of Meramec Spring near Meramec, Mo., for 1903.

Day.	Mar.	Apr.	May.	June.	July.	Aug.	Sept.	Oct.	Nov.	Dec.
1	1.40	0.90	0.70	3.10	0.70	0.50	0.40	0.40	0.40	0.
2	1.30	.90	.70	2.80	.70	2.00	.40	.40	.40	
3	1.10	.90	.70	2.60	.70	1.40	.40	.40	.40	
4	1.00	.90	.70	2.50	.70	1.00	.40	.40	.40	
5	2.10	.80	.70	2.20	.70	.90	.40	.50	.40	
6	2.10	.80	.80	2.00	.70	.80	.40	.50	.40	
7	2.40	.80	.80	1.90	.70	.70	.40	1.00	.40	
8	2.60	.80	.80	1.60	.60	.70	.40	1.20	.40	
9	2.40	.90	.80	1.40	.60	1.00	.40	1.00	.30	
10	2.10	.90	.70	1.30	.60	.90	.50	.80	.30	
11	2.00	.90	.70	1.20	.50	.90	.50	.70	.30	
12	1.80	.90	.70	1.20	.50	.70	.50	.60	.30	
13	1.60	1.50	.70	1.10	.50	.70	.40	.50	.30	
14	1.40	1.50	.90	1.10	.50	.70	.60	.60	.30	
15	1.30	1.40	.90	1.10	.50	.60	.50	.60	.30	
16	1.20	1.20	.90	.90	.50	.60	.50	.60	.30	
17	1.10	1.10	1.10	.90	.50	.60	.60	.60	.30	

Mean daily gage height, in feet, of Meramec Spring, etc.—Continued.

Day.	Mar.	Apr.	May.	June.	July.	Aug.	Sept.	Oct.	Nov.	Dec.
....	1.10	1.10	1.20	0.90	0.50	0.60	0.50	0.40	0.30	0.30
....	1.10	1.00	1.20	.90	.50	.60	.50	.40	.30	.30
....	1.70	1.10	1.20	.90	.50	.60	.50	.40	.30	.30
....	1.80	1.20	1.10	.90	.50	.50	.50	.40	.30	.30
....	1.70	1.10	1.20	.90	.50	.50	.50	.40	.30	.30
....	1.50	1.00	1.10	.90	.50	.50	.50	.40	.30	.30
....	1.40	.90	1.00	.90	.50	.50	.50	.40	.30	.30
....	1.20	.90	.90	.90	.50	.50	.40	.40	.30	.40
....	1.20	.90	.90	.80	.50	.50	.40	.40	.30	.40
....	1.10	.80	.80	.80	.50	.40	.40	.40	.30	.40
....	.90	.80	.80	.80	.50	.40	.40	.40	.30	.40
....	.90	.80	.80	.70	.50	.40	.40	.40	.30	.35
....	.90	.80	.90	.70	.50	.40	.40	.40	.30	.35
....	.90	......	3.20	......	.50	.40	......	.40	......	.30

Rating table for Meramec Spring near Meramec, Mo., from February 28 to December 31, 1903.

Gage height.	Discharge.	Gage height.	Discharge.	Gage height.	Discharge.	Gage height.	Discharge.
Feet.	*Second-feet.*	*Feet.*	*Second-feet.*	*Feet.*	*Second-feet.*	*Feet.*	*Second-feet.*
0.3	94	1.1	305	1.9	608	2.7	912
.4	112	1.2	342	2.0	646	2.8	950
.5	133	1.3	380	2.1	684	2.9	988
.6	156	1.4	418	2.2	722	3.0	1,026
.7	180	1.5	456	2.3	760	3.1	1,064
.8	206	1.6	494	2.4	798	3.2	1,102
.9	237	1.7	532	2.5	836		
1.0	270	1.8	570	2.6	874		

Estimated monthly discharge of Meramec River near Meramec, Mo., for 1903.

Month.	Discharge in second-feet.			Total in acre-feet.
	Maximum.	Minimum.	Mean.	
March	874	237	462	28,407
April	456	206	268	15,947
May	1,102	180	263	16,171
June	1,064	180	398	23,683
July	180	133	146	8,977
August	646	112	189	11,621
September	156	112	123	7,319
October	342	112	144	8,854
November	112	94	99	5,891
December	112	94	97	5,964

MERAMEC RIVER (STATION NO. 1) AT FENTON, MO.

This station was established April 18, 1903, by M. O. Leighton, assisted by I. W. McConnell. The station is located on the wagon bridge crossing the river on the road leading from Fenton to St. Louis. A chain gage was located on the wagon bridge, and discharge measurements were made from this bridge. The gage was read twice each day by J. F. Stahlhuth. The station was discontinued August 25, 1903, on account of backwater from Mississippi River, and a station was established at Eureka, Mo., 12 miles up the river, to take the place of the Fenton stations. At the Fenton wagon bridge the channel is straight for 100 feet above and 2,000 feet below the station. The right bank will not overflow; the left bank overflows at extreme high water. The bed of the stream is of rock and gravel and is probably permanent. The bench mark is the top surface of the pier on the downstream side. Its elevation is 33.30 feet above the zero of the gage.

The observations at this station during 1903 have been made under the direction of E. Johnson, jr., district hydrographer.

Discharge measurements of Meramec River at station No. 1, near Fenton, Mo., in 1903.

Date.	Hydrographer.	Gage height.	Discharge.
		Feet.	*Second-feet.*
April 18	I. W. McConnell	15.10	5,003
May 9	....do	10.90	1,609
June 23	F. W. Hanna	27.60	1,501
July 27	....do	23.40	717

Mean daily gage height, in feet, of Meramec River at station No. 1, at Fenton, Mo., for 1903.

Day.	Apr.	May.	June.	July.	Aug.	Day.	Apr.	May.	June.	July.	Aug.
1		29.70	31.40	24.10	22.40	17		30.70	33.75	24.10	23.90
2		29.60	34.15	23.90	22.40	18		30.40	31.70	24.00	23.60
3		29.50	36.40	23.50	22.40	19	33.05	30.70	30.88	24.60	23.30
4		29.40	37.65	23.20	22.65	20	33.00	30.75	29.75	24.70	23.10
5		29.30	36.65	23.10	24.50	21	32.60	30.60	28.40	23.00	23.10
6		29.20	36.70	23.10	23.70	22	32.50	30.35	28.50	22.90	22.90
7		29.20	37.65	23.00	23.30	23	32.80	30.00	27.75	22.80	
8		29.20	38.05	22.90	23.25	24	32.30	30.00	27.40	22.80	
9		29.30	38.35	22.80	23.10	25	30.70	30.65	27.05	22.90	
10		29.20	39.35	22.80	23.10	26	31.00	27.75	26.70	23.20	
11		29.20	38.70	22.80	23.00	27	30.40	27.85	26.10	23.40	
12		29.20	38.00	22.90	23.20	28	30.25	27.70	24.90	22.90	
13		29.20	38.10	23.10	23.10	29	30.15	27.40	24.30	22.60	
14		29.40	36.70	23.00	23.00	30	30.00	27.60	24.10	22.50	
15		29.90	36.20	23.30	23.90	31		28.60		22.40	
16		30.70	34.95	26.20	24.00						

a 18.4 added to gage heights from April 19 to May 23, inclusive, to agree with a new gage set May 22.

b All readings from May 26 to June 27, inclusive, are affected by backwater from Mississippi River.

MERAMEC RIVER AT STATION NO. 2, BELOW FENTON, MO.

This station was established May 10, 1903, by I. W. McConnell. The station is located about 1,200 feet downstream from the wagon bridge. The gage is a vertical 2 by 4 inch rod 16 feet long, fastened to a leaning birch tree on the left bank. The gage was read twice each day by J. F. Stahlhuth. Discharge measurements were made by means of a boat and tag wire. The initial point for soundings was taken at a large soft-maple tree 50 feet west of the gage, on the left bank. The bench mark is located on this tree, and is a notch near the foot of the tree at an elevation of 14.30 feet above the zero of the gage. The channel is straight for 1,300 feet above and 2,000 feet below the station. The current is swift. The right bank is high and rocky and never overflows. The left bank is low and of alluvial soil, and overflows only at flood stage. The bed of the stream is of rock and gravel and is probably permanent. The station was discontinued in August, 1903, on account of the gage heights being affected by back water from the Mississippi. The station was replaced by the station at Eureka, 12 miles upstream.

The observations at this station during 1903 have been made under the direction of E. Johnson, jr., district hydrographer.

The following discharge measurement was made by I. W. McConnell at this station in 1903:

May 10: Gage height, 6.50 feet; discharge, 1,537 second-feet.

Mean daily gage height, in feet, of Meramec River at Station No. 2, below Fenton, Mo., for 1903.

Day.	May.	June.	July.	Aug.	Day.	May.	June.	July.	Aug.
1		14.15	7.15	5.50	17	8.30	16.35	7.80	6.90
2		16.75	6.95	5.50	18	8.00	13.80	7.20	6.70
3		19.00	6.60	5.50	19	8.30	12.90	7.10	6.30
4		20.15	6.85	6.10	20	8.35	12.25	6.40	6.10
5		19.25	6.20	7.50	21	8.20	11.00	6.20	6.10
6		19.30	6.20	6.75	22	7.95	11.45	6.00	5.90
7		20.25	6.10	6.40	23	7.60	11.15	5.90	
8		20.65	6.00	6.35	24	7.40	11.80	5.90	
9		20.95	5.90	6.10	25	7.45	10.45	6.00	
10	6.50	21.20	5.90	6.10	26	9.00	10.15	6.40	
11	6.50	21.30	5.80	6.10	27	11.05	9.60	6.50	
12	6.50	21.20	6.00	6.30	28	10.45	8.50	6.00	
13	6.50	20.70	5.80	6.10	29	10.15	7.95	5.80	
14	6.65	20.00	6.20	6.00	30	10.40	7.80	5.70	
15	7.20	18.80	6.50	6.90	31	11.10		5.50	
16	8.05	17.55	9.30	7.00					

MERAMEC RIVER NEAR EUREKA, MO.

This station was established August 26, 1903, by F. W. Hanna. It is located at the highway bridge on the road between Crescent, Mo., and Eureka, Mo., about 1½ miles from Eureka, 2 miles below the mouth of Big River, and 2 miles above the 'Frisco Railroad bridge. The standard boxed chain gage is attached to the floor of the bridge. The length of the chain from the end of the weight to the marker is 42.46 feet. The gage is read once each day by Rhoda Hilderbran. Discharge measurements are made from the bridge to which the gag is attached. This is a two-span Pratt truss highway bridge, with length between abutments of 450 feet. The initial point for sound ings is the inner face of the right abutment, on the upstream side. The channel is straight for about 250 feet above and 1,000 feet belo the station. The current is never sluggish. The right bank is high and rocky and not subject to overflow. It is wooded above the high water line. The left bank is somewhat lower and is composed of allu vial soil, and it also is wooded above the high-water line. The bed of the stream is composed of coarse gravel and stones. There is but one channel, broken by the center pier of the bridge. Bench mar No. 1 is a notch cut into the high rock cliff just below the bridg It is marked by a painted cross and the letters "U. S. G. S." Its elevation is 38.26 feet above gage datum. Bench mark No. 2 is a mark on the top of the downstream guard rail, just above the ga box. Its elevation is 42.56 feet above gage datum. Bench ma No. 3 is a painted cross, surrounded by the letters "U. S. G. S.," the lower wing of the left abutment. Its elevation is 36.29 feet abo gage datum.

The observations at this station during 1903 have been made und the direction of E. Johnson, jr., district hydrographer.

Discharge measurements of Meramec River near Eureka, Mo., in 1903.

Date.	Hydrographer.	Gage height.	Dischar
		Feet.	*Second-f*
August 26	F. W. Hanna	3.00	
September 21	do	3.90	1.
October 19	do	3.40	1.
November 30	do	2.80	
December 31	do	3.35	

Mean daily gage height, in feet, of Meramec River near Eureka, Mo., for 1903.

Day.	Aug.	Sept.	Oct.	Nov.	Dec.	Day.	Aug.	Sept.	Oct.	Nov.	Dec.
1		3.10	2.80	2.90	2.80	17		5.40	3.00	2.80	2.50
2		3.00	2.80	2.90	2.70	18		4.80	3.50	2.80	2.80
3		2.90	2.80	2.90	2.70	19		5.00	3.40	2.90	2.60
4		2.80	2.60	3.00	2.70	20		4.40	3.30	2.90	2.75
5		2.80	2.80	3.20	2.70	21		4.00	3.20	2.80	2.90
6		2.70	5.80	3.10	2.60	22		3.70	3.20	2.80	2.80
7		2.70	3.90	3.00	2.60	23		3.40	3.10	2.80	2.70
8		2.70	4.50	3.00	2.60	24		3.30	3.10	2.80	2.70
9		3.50	8.30	3.00	2.50	25		3.20	3.00	2.80	2.70
10		3.00	7.50	3.00	2.50	26	3.00	3.10	3.00	2.70	2.60
11		3.10	5.50	2.90	2.50	27	2.90	3.00	3.00	2.70	3.00
12		3.80	4.80	2.90	2.60	28	3.70	2.90	2.90	2.70	3.60
13		4.30	4.30	2.80	2.60	29	3.70	2.90	2.90	2.80	3.40
14		3.80	4.10	2.90	2.60	30	3.60	2.90	2.90	2.80	3.00
15		2.90	3.80	2.90	2.60	31	3.30		2.90		3.20
16		5.30	8.70	2.90	2.70						

Rating table for Meramec River near Eureka, Mo., from August 26 to December 31, 1903.

Gage height.	Discharge.	Gage height.	Discharge.	Gage height.	Discharge.	Gage height.	Discharge.
Feet.	*Second-feet.*	*Feet.*	*Second-feet.*	*Feet.*	*Second-feet.*	*Feet.*	*Second-feet.*
2.5	377	3.8	1,228	5.1	2,103	6.8	3,283
2.6	442	3.9	1,294	5.2	2,171	7.0	3,425
2.7	507	4.0	1,361	5.3	2,239	7.2	3,567
2.8	572	4.1	1,428	5.4	2,308	7.4	3,709
2.9	637	4.2	1,495	5.5	2,377	7.6	3,853
3.0	702	4.3	1,562	5.6	2,446	7.8	3,997
3.1	767	4.4	1,629	5.7	2,515	8.0	4,141
3.2	832	4.5	1,696	5.8	2,584	8.2	4,286
3.3	898	4.6	1,763	5.9	2,653	8.4	4,432
3.4	964	4.7	1,831	6.0	2,722	8.6	4,578
3.5	1,030	4.8	1,899	6.2	2,862	8.8	4,724
3.6	1,096	4.9	1,967	6.4	3,002	9.0	4,872
3.7	1,162	5.0	2,035	6.6	3,142		

Estimated monthly discharge of Meramec River near Eureka, Mo., for 1903.

Month.	Discharge in second-feet.			Total in acre-feet.
	Maximum.	Minimum.	Mean.	
August 26–31	1,162	637	943	11,223
September	2,446	507	1,057	62,896
October	4,359	572	1,228	75,307
November	832	507	624	37,131
December	1,096	377	520	33,203

MERAMEC RIVER (DRY FORK) NEAR ST. JAMES, MO.

This station was established March 28, 1903, by I. W. McConnell, and was discontinued May 31, inasmuch as the gage was washed out and the flow of the stream at ordinary stages was less than 50 second-feet. As it has no power possibilities and as the discharge is measured at Eureka station, it was considered best to discontinue the observations at this point.

The observations at this station during 1903 have been made under the direction of E. Johnson, jr., district hydrographer.

Discharge measurements of Meramec River (Dry Fork) near St. James, Mo., in 1903.

Date.	Hydrographer.	Gage height.	Discharge.
		Feet.	Second-feet.
April 25	I. W. McConnell	1.40	4
May 25	F. W. Hanna	1.30	

Mean daily gage height, in feet, of Meramec River (Dry Fork) near St. James, Mo., for 1903.

Day.	Apr.	May.	June.	Day.	Apr.	May.	June.	Day.	Apr.	May.	June.
1	1.40	1.20		12	1.60	1.30		23		1.50	1.60	
2	1.30	1.20		13	7.00	1.30	1.60	24		1.40	1.60	
3	1.40	1.20		14	5.00	1.60	1.50	25		1.40	1.60	
4	1.50	1.30		15	3.00	1.50	1.30	26		1.30	1.80	
5	1.40	1.30		16	1.90	1.30		27		1.30	1.40	
6	1.30	1.80		17	1.60	5.00	5.00	28		1.30	1.30	
7	1.30	1.60		18	1.40	3.40	3.40	29		1.30	2.10	
8	1.80	1.50		19	1.70	2.00		30		1.30	1.80	
9	1.50	1.30		20	2.40	1.70		31		10.00		
10	1.70	1.20		21	1.90	1.60						
11	1.70	1.20		22	1.60	1.70						

ARKANSAS RIVER DRAINAGE BASIN.

Arkansas River rises in the central part of Colorado, flows south
about 70 miles, then east for 50 miles, increased by many small
untain streams.   At Canyon, Colo., it emerges from the Rocky
untain front and takes a general easterly direction across the great
ins of Colorado, where most of the water is diverted for irrigation.
flows east across Kansas to the center of the State, thence south-
st through Indian Territory and Arkansas into Mississippi River
out 25 miles north of Greenville, Miss.   Throughout the mountain-
s area above Canyon the discharge increases, but as soon as the river
erges onto the Great Plains the water is gradually diverted by lines
canals, so that by the time the Kansas line is reached the river is
ually dry during the summer.   Among its tributaries, Neosho
alled the Grand in Indian Territory) and Verdigris rivers drain
utheastern Kansas and flow south in nearly parallel courses, joining
e Arkansas about 3 miles apart, west of Talequah, Ind. T.   Walnut
iver is a small tributary from the north, flowing into the Arkansas
Arkansas City, Kans.   The North Fork of Canadian River rises in
e northeastern part of New Mexico and flows east through Oklahoma
d Indian Territory into Canadian River about 40 miles above its
outh.   The following is a list of the stations in the Arkansas River
ainage basin:

Canadian River (North Fork) near Elreno, Okla.
Canadian River (North Fork) near Woodward, Okla.
Cimarron River near Waynoka, Okla.
Cimarron River near Arkalon, Kans.
Mora River at Weber, N. Mex.
Mora River at La Cueva, N. Mex.
Sapello River at Los Alamos, N. Mex.
Sapello River at Sapello, N. Mex.
Mill tailrace at Sapello, N. Mex.
Mannelitos River at Sapello, N. Mex.
Gallinas River at Hot Springs, near Las Vegas, N. Mex.
Neosho River near Iola, Kans.
Grand River near Fort Gibson, Ind. T.
Verdigris River near Liberty, Kans.
Verdigris River near Catoosa, Ind. T.
Walnut River near Arkansas City, Kans.
Arkansas River near Arkansas City, Kans.
Arkansas River at Hutchinson, Kans.
Arkansas River at Dodge, Kans.
Arkansas River near Syracuse, Kans.
Arkansas River near Coolidge, Kans.
Arkansas River near Granada, Colo.
Arkansas River near Prowers, Colo.
Arkansas River at La Junta, Colo.
Arkansas River near Rocky Ford, Colo.
Arkansas River near Nepesta, Colo.

Oxford Farmers' canal near Nepesta, Colo.
Arkansas River at Pueblo, Colo.
Arkansas River near Canyon, Colo.
Arkansas River at Salida, Colo.

CANADIAN RIVER (NORTH FORK) NEAR ELRENO, OKLA.

This station was established October 27, 1902, by W. G. Rus
and is located at the highway bridge, 2 miles north of Elreno, Ok
The gage is of the usual wire type, with a scaleboard graduated
feet and tenths and nailed to the railing of the bridge. The dista
from the end of the weight to the marker is 17.06 feet. The be
mark is the top of a steel cylinder on the north side of the brid
Its elevation is 11.3 feet above the zero of the gage. The initial po
for soundings is on the right bank. The channel both above a
below the station is straight for about 200 feet and has a width o
feet. The right bank is high, and the left bank is low. Both ba
are liable to overflow. The bed of the stream is sandy and somew
shifting. The observer is Kenneth A. Killion, who reads the g
once daily. The gage was destroyed July 6, 1903, but was reest
lished July 10.

The observations at this station during 1903 have been made un
the direction of G. H. Matthes, district engineer.

Discharge measurements of North Fork of Canadian River at Elreno, Okla.
1903.

Date.	Hydrographer.	Gage height.	Discha
		Feet.	*Second*
August 11	L. M. Holt	1.60	1
August 26 .	A. C. Redman .	1.90	!
September 20 . .	Ferd. Bonstedt	1.30	
October 20 .	E. R. Kerby	1.30	
November 24 .	. . . do	1.40	
December 20 . . .	. . . . do	1.60	!

Mean daily gage height, in feet, of North Fork of Canadian River at Elre
Okla., for 1903.

Day.	Jan.	Feb.	Mar.	Apr.	May.	June.	July.	Aug.	Sept.	Oct.	Nov.	I
1.	2.60	2.90	3.50	3.60	3.60	5.40	3.50	1.80	1.60	1.40	1.50	
2.	2.60	2.90	3.40	3.60	3.60	5.00	3.30	(a)	1.50	1.40	1.50	
3	2.70	2.90	3.30	3.50	3.30	4.70	3.10		1.50	1.40	1.50	
4.	2.80	2.90	3.20	3.50	3.30	4.50	3.10		1.40	1.40	1.50	
5.	2.90	2.90	3.10	3.50	3.20	4.40	3.10		1.40	1.30	1.50	
6.	2.90	2.80	3.30	3.40	3.50	4.20	(a)		1.40	1.30	1.50	
7.	2.90	2.80	3.30	3.40	4.40	4.10			1.40	1.30	1.40	

a Gage destroyed.

Mean daily gage height, in feet, of North Fork of Canadian River, etc.—Continued.

Day.	Jan.	Feb.	Mar.	Apr.	May.	June.	July.	Aug.	Sept.	Oct.	Nov.	Dec.
8	2.90	2.80	5.20	3.30	4.90	4.20	......	......	1.40	1.30	1.30	1.80
9	2.90	2.80	4.30	3.30	7.30	4.10	......	......	1.30	.70	1.30	1.30
10	3.00	2.80	4.10	3.40	5.70	4.00	......	......	1.40	.50	1.30	1.30
11	2.90	2.80	4.20	3.30	5.40	4.00	2.50	1.60	1.40	.50	1.30	1.50
12	2.50	2.80	4.40	3.20	5.40	4.00	2.50	1.50	1.40	.50	1.30	1.50
13	2.70	2.90	4.40	3.20	5.50	4.20	2.40	1.50	1.40	1.40	1.30	1.50
14	2.80	3.00	4.50	3.20	5.10	4.00	2.40	2.20	1.50	1.40	1.30	1.50
15	2.70	3.00	4.40	3.10	4.80	3.90	2.30	1.90	1.50	1.40	1.30	1.50
16	2.80	3.00	4.20	3.10	4.90	3.80	2.20	1.60	1.50	1.40	1.30	1.60
17	2.80	2.90	4.10	3.10	4.80	3.80	2.10	1.50	1.40	1.40	1.30	1.60
18	2.80	3.00	4.00	3.00	4.60	3.70	2.10	2.90	1.40	1.30	1.30	1.60
19	2.80	2.90	4.30	3.00	5.00	3.70	2.10	2.60	1.30	1.30	1.30	1.60
20	2.80	2.90	4.10	3.00	4.60	3.70	2.10	2.50	1.30	1.30	1.30	1.60
21	2.90	2.90	4.20	3.00	4.60	3.60	2.00	2.40	1.20	1.30	1.30	1.60
22	3.00	2.80	4.20	3.00	5.60	3.40	2.00	2.30	1.30	1.30	1.30	1.60
23	2.90	2.80	3.80	3.00	5.10	3.30	1.90	2.30	1.30	1.30	1.40	1.70
24	2.90	2.90	3.70	2.90	8.90	3.30	1.90	2.00	1.40	1.30	1.40	1.70
25	2.90	3.00	3.70	2.90	6.70	3.20	1.90	1.90	1.30	1.30	1.40	1.70
26	2.90	3.60	3.60	2.90	6.90	3.30	1.80	1.80	1.10	1.30	1.30	1.90
27	3.00	3.60	3.60	2.80	6.20	3.20	1.80	1.80	1.40	1.30	1.30	1.90
28	2.90	3.50	3.50	2.80	9.30	3.70	1.80	1.60	1.40	1.30	1.30	1.90
29	2.90	......	3.50	2.90	8.10	3.70	1.80	1.70	1.40	1.30	1.30	1.90
30	2.90	......	3.50	2.90	6.50	3.80	1.80	1.70	1.40	1.30	1.30	1.90
31	2.90	......	3.50	......	6.00	......	1.80	1.60	......	1.60	......	1.90

CANADIAN RIVER (NORTH FORK) NEAR WOODWARD, OKLA.

This station was established September 13, 1903, by W. G. Russell. It is located 7 miles east of Woodward, at the railroad bridge. The gage is painted on the west face of the second pier from the west end of the bridge. It reads from zero to 7 feet. The observer is John E. Watkins, the section foreman. Discharge measurements are made from the six-span railroad bridge, which has a total length between abutments of 360 feet. The initial point for soundings is the west end of the bridge. The channel is straight for 200 feet above and below the station. The right bank is high and not liable to overflow. The left bank is low and liable to overflow, with scattering trees along the bank. The bed of the stream is sandy and shifting. There is one channel, broken by two piers at low water and five piers at high water. The bench mark is the bottom of the coping stone of the pier at the top of the gage. Its elevation is 7 feet above the zero of the gage.

The observations at this station during 1903 have been made under the direction of W. G. Russell, district hydrographer.

Mean daily gage height, in feet, of North Fork of Canadian River near Woodward,
Okla., for 1903.

Day.	Oct.	Nov.	Dec.	Day.	Oct.	Nov.	Dec.	Day.	Oct.	Nov.	Dec.
1		2.30	2.40	12		2.30	2.40	23	1.80	2.40	2.50
2		2.40	2.40	13		2.20	2.50	24	1.80	2.40	2.50
3		2.40	2.40	14		2.20	2.50	25	1.80	2.40	2.50
4		2.60	2.40	15		2.10	2.40	26	1.80	2.40	2.50
5		2.70	2.30	16		2.20	2.50	27	1.80	2.40	2.60
6		2.60	2.60	17		2.10	2.50	28	1.90	2.40	2.50
7		2.60	2.60	18	1.20	2.10	2.50	29	1.90	2.30	2.50
8		2.40	2.40	19	1.20	2.20	2.40	30	1.90	2.30	2.30
9		2.40	2.50	20	1.10	2.20	2.30	31	2.00		2.50
10		2.30	2.60	21	1.80	2.40	2.30				
11		2.30	2.50	22	1.90	2.40	2.50				

CIMARRON RIVER NEAR WAYNOKA, OKLA.

This station was established September 11, 1903, by W. G. Russell.
It is located at the railway bridge 2¼ miles southwest of Waynoka.
The gage is a 1 by 4 inch pine board 12 feet long, spiked to the east
side of the eighth pile bent from the east end of the bridge. The
observer is M. J. Sunden, the section foreman. Discharge measurements are made from the railway bridge, which is supported on piles
and has a total span of 2,158 feet. The initial point for soundings is
the east end of the bridge. The channel is straight for 1,000 feet
above and below the station. Both banks are liable to overflow at
very high water. At low water there are several channels. At high
water there is one channel, broken by the bridge supports. The
bench mark is the top of the cap of the bench to which the gage is
spiked. Its elevation is 10.93 feet above the zero of the gage.

The observations at this station during 1903 have been made under
the direction of W. G. Russell, district hydrographer.

Mean daily gage height, in feet, of Cimarron River near Waynoka, Okla., for 1903.

Day.	Oct.	Nov.	Dec.	Day.	Oct.	Nov.	Dec.	Day.	Oct.	Nov.	Dec.
1		0.50	0.30	12		0.60	0.30	23		0.20	1
2		.40	.30	13		.40	.30	24		.30	1
3		.30	.20	14		.20	.30	25		.20	1
4		1.50	.20	15		.20	.30	26		.30	1
5		1.40	.20	16		.20	.30	27		.30	1
6		1.30	.30	17		.20	.30	28		.30	1
7		1.00	.30	18		.20	.30	29		.30	
8		1.00	.30	19		.20	.30	30		.30	
9		1.00	.30	20		.20	.30	31	1.00		.5
10		.90	.30	21		.20	.50				
11		.80	.30	22		.20	1.00				

CIMARRON RIVER NEAR ARKALON, KANS.

This station was established August 15, 1903, by W. G. Russell. It is located about a half mile north of Arkalon, Kans., at the bridge of the Chicago, Rock Island and Pacific Railway. The gage is a 1 by 4 inch pine timber, 12 feet long, graduated to feet and tenths and spiked to a pile bent on the downstream side of the bridge. This is the fifth bent from the south end of the bridge. The gage is read once each day by Elwin Singer. Discharge measurements are made from the bridge, but can be made at low water by wading. The initial point for soundings is at the south end of the bridge, or the right bank of the river. The channel is straight for about 100 feet above and below the station. The stream is sluggish at low water, but will probably have a good current when high. Both banks are low and liable to overflow. The railroad embankment is high at each end of the bridge, forcing the water to pass beneath the bridge at all stages. There is but one channel, which is shallow except at high water. The bridge rests upon piles, and drift may at times affect the measurements. The bridge is 245 feet long and 15 feet above the bed of the stream. No bench mark has been established.

The observations at this station during 1903 have been made under the direction of W. G. Russell, district hydrographer.

The following discharge measurement was made by W. G. Russell in 1903:

August 15: Gage height, 1 foot; discharge, 51 second-feet.

Mean daily gage height, in feet, of Cimarron River near Arkalon, Kans., for 1903.

Day.	Aug.	Sept.	Oct.	Nov.	Dec.	Day.	Aug.	Sept.	Oct.	Nov.	Dec.
1		0.80	0.80	0.80	0.80	17	1.30	.80	.80	.80	.90
2		.80	.80	.80	.80	18	1.20	.80	.80	.80	.80
3		.80	.80	.80	.80	19	1.10	.80	.80	.80	.80
4		.80	.80	.80	.80	20	1.20	.80	.80	.80	.90
5		.80	.80	.80	.80	21	1.20	.80	.80	.80	.90
6		.80	.80	.80	.80	22	1.20	.80	.80	.80	.80
7		.80	.80	.80	.80	23	.80	.80	.80	.80	.80
8		.80	.80	.80	.80	24	.80	.80	.80	.80	.80
9		.80	.80	.80	.80	25	.80	.80	.80	.80	.80
10		.80	.80	.80	.80	26	.80	.80	.80	.80	.80
11		.80	.80	.80	.80	27	.80	.80	.80	.80	.80
12		.80	.80	.80	.80	28	.80	.80	.80	.80	.80
13		.80	.80	.80	.80	29	.80	.80	.80	.80	.80
14		.80	.80	.80	1.00	30	.80	.80	.80	.80	.80
15	1.00	.80	.80	.80	.90	31	.80		.80		.80
16	1.00	.80	.80	.80	.90						

MORA RIVER AT WEBER, N. MEX.

This station was established August 21, 1903, by M. C. Hinderlider. It is located at the highway bridge 150 feet north of the post-office at

Weber, N. Mex., and is about 15 miles west of Watrous, N. Mex. The gage is a vertical 2 by 6 inch timber 14 feet long, fastened to the upstream end of the south abutment of the wagon bridge. The zero of the gage rests on the bed of the river. It is read twice each day by Mrs. Emily Biernbaum, the postmistress at Weber. Discharge measurements are made during the greater part of the year by wading below the bridge. At high water they are made from the highway bridge, which consists of a single span of 32 feet. The initial point for soundings is the edge of the left abutment. The channel is straight for 20 feet above and for 75 feet below the bridge. Both banks are of gravel and overflow only at high stages. The bed of the stream is composed of bowlders and mud, overlying shale or stone, with a sand bar near the right bank just below the bridge during low water. A few feet below the bridge the channel is filled across with riprap, is shifting, and free from vegetation. Bench mark No. 1 is a 20-penny nail driven horizontally into the northwest face of a large cottonwood tree 75 feet southeast of the gage. A spike is driven on each side of the bench mark. The elevation of the bench mark is 12.915 feet above the zero of the gage. Bench mark No. 2 is a 20-penny nail driven vertically into the top side of the east end of the cap timber of the north abutment of the bridge. Its elevation is 10.155 feet above the zero of the gage.

Discharge measurements of Mora River at Weber, N. Mex., in 1903.

Date.	Hydrographer.	Gage height.	Discharge.
		Feet.	*Second-feet.*
August 21	M. C. Hinderlider...........	2.40	12
Do..	....do	2.40	13

Mean daily gage height, in feet, of Mora River at Weber, N. Mex., for 1903.

Day.	Aug.	Sept.	Oct.	Nov.	Dec.	Day.	Aug.	Sept.	Oct.	Nov.	Dec.
1		2.10	2.35		2.25	17		2.15	2.50	2.45	2.30
2		2.00	2.35		2.20	18		2.15	2.45	2.45	2.25
3		2.10	2.35		2.30	19		2.20	2.45	2.30	2.40
4		1.75	2.40		2.30	20		2.20	2.45	2.40	2.40
5		1.50	2.40		2.35	21	2.40	2.25	2.50	2.45	2.40
6		2.15	2.40	2.45	2.25	22	2.35	2.15	2.45	2.30	2.40
7		2.15	2.30	2.50	2.20	23	2.40	2.10	2.50	2.20	2.40
8		2.00	2.50	2.45	2.20	24	2.45	2.15	2.50	2.25	2.40
9		2.10	2.50	2.35	2.30	25	2.35	2.35	2.50	2.45	2.40
10		2.05	2.50	2.35	2.30	26	2.25	2.40	2.50	2.20	2.40
11		2.00	2.40	2.30	2.30	27	2.20	2.30	2.50	2.30	2.40
12		1.95	2.50	2.35	2.30	28	2.15	2.45	2.50	2.30	2.50
13		2.00	2.45	2.40	2.30	29	2.00	2.25	2.50	2.30	2.50
14		2.00	2.50	2.40	2.30	30	2.00	2.30	2.50	2.20	2.50
15		2.00	2.50	2.50	2.30	31	2.20		2.50		2.50
16		2.15	2.45	2.40	2.30						

MORA RIVER AT LA CUEVA, N. MEX.

Discharge measurements of Mora River at La Cueva, N. Mex., in 1903.

Date.	Hydrographer.	Gage height.	Discharge.
		Feet.	*Second-feet.*
August 25	M. C. Hinderlider	0.80	2
August 26	do	1.20	12

Mean daily gage height, in feet, of Mora River at La Cueva, N. Mex., for 1903.

Day.	Aug.	Sept.	Oct.	Nov.	Dec.	Day.	Aug.	Sept.	Oct.	Nov.	Dec.
1		0.80	1.03	1.15	1.20	17		0.90		1.15	1.30
2		.80	1.20	.75	1.15	18		.90		1.10	1.30
3		.80	1.15	.75	.85	19		.90		1.10	1.15
4		.83	1.15	1.15	1.20	20		1.22			1.20
5		.83	1.13	1.15	1.00	21		.88			.80
6		1.20	1.00	.70	1.20	22		.88			.85
7		.98	1.13	1.15	1.20	23		.90			
8		.90	1.13	1.15	1.25	24		1.00	1.10		
9		.90	1.15	1.10	1.20	25		1.15	1.10		
10				.80	1.20	26		1.13	.80		
11					1.20	27	0.80	1.20	1.15	1.20	
12				.80	1.35	28	.80	.80	1.10	1.80	
13		1.20		1.10	1.20	29	.80	1.10	.75	.80	
14		.80		.80	1.25	30	1.05	1.15	1.20	.90	
15		.90		1.15		31	.80		1.20		
16		.90		1.10	1.20						

SAPELLO RIVER AT LOS ALAMOS, N. MEX.

This station was established August 22, 1903, by M. C. Hinderlider. It is located at a ford crossing Sapello River at a point due north from Los Alamos and about one-fourth mile distant. The gage consists of a 2 by 6 inch timber, 16 feet long, placed in an inclined position on the right bank of the river at the ford. It is graduated to read directly to vertical tenths of feet from 0.5 foot to 7.5 feet. There is no bridge and no other means of making high-water measurements. All measurements are made by wading. The initial point for soundings is the right bank. The channel is straight for 200 feet above and below the station. The current is moderate at ordinary stages and rapid at flood stages. The right bank is high and steep and does not overflow at the gage. The left bank is a low sand and gravel bar extending for about 100 feet and then rising to above high-water mark. The bed of the stream is composed of small cobblestones, gravel, and sand, and is apparently permanent. The cross section is regular, and there is one channel at all stages. The bench mark is a

20-penny spike driven horizontally into the northwest face of a 10-inch willow tree about 3½ feet above the ground. The tree is about 100 feet southeast of the gage, and there is a spike driven on each side of the bench mark. The elevation of the bench mark is 17.742 feet above the zero of the gage. The gage is read twice each day by W. B. Hogin.

A cable was placed across this stream about 500 feet above the gage rod, equipped with car and tag wire, in March, 1904, from which gagings at high stages are made.

This station was established for the purpose of determining the available amount of water for prospective diversion into the Sanguyjuella basin for storage. This basin lies about 6 miles northwest of Las Vegas.

The following discharge measurement was made by M. C. Hinderlider in 1903:

August 22: Gage height, 1.30 feet; discharge, 7 second-feet.

Mean daily gage height, in feet, of Sapello River at Los Alamos, N. Mex., for 1903.

Day.	Aug.	Sept.	Oct.	Nov.	Dec.	Day.	Aug.	Sept.	Oct.	Nov.	Dec.
1		1.00	1.00	1.00	1.00	17		1.00	1.00	(a)	(a)
2		1.00	1.00	1.00	1.00	18		1.00	1.00	(a)	(a)
3		1.00	1.00	1.00	1.00	19		1.00	1.00	(a)	(a)
4		1.00	1.00	1.00	(a)	20		1.00	1.00	(a)	(a)
5		1.00	1.00	1.00	(a)	21		1.00	1.00	(a)	(a)
6		1.00	1.00	1.00	(a)	22		1.00	1.00	1.00	(a)
7		1.00	1.00	1.00	(a)	23	1.30	1.00	1.00	1.00	(a)
8		1.00	1.00	1.00	(a)	24	1.30	1.00	1.00	1.00	(a)
9		1.00	1.00	1.00	(a)	25	1.30	1.00	1.00	1.00	(a)
10		1.00	1.00	1.00	(a)	26	1.30	1.00	1.00	1.00	(a)
11		1.00	1.00	1.00	(a)	27	1.00	1.00	1.00	1.00	(a)
12		1.00	1.00	1.00	(a)	28	1.00	1.00	1.00	1.00	(a)
13		1.00	1.00	1.00	(a)	29	1.00	1.00	1.00	1.00	(a)
14		1.00	1.00	1.00	(a)	30	1.00	1.00	1.00	1.00	(a)
15		1.00	1.00	1.00	(a)	31	1.00		1.00		(a)
16		1.00	1.00	1.00	(a)						

a Ice.

SAPELLO RIVER AT SAPELLO, N. MEX.

This station was established August 12, 1903, by E. G. Marsh. It is located about one-half mile above the junction of Sapello and Manuelitos rivers, and is about 12 miles from Las Vegas, N. Mex. The gage is an inclined 1 by 8 inch pine board 10 feet long, bolted to the rock and braced with timbers. It is read once each day by Horace R. Titlow. Discharge measurements may be made by wading but at high stages a cable, car, tagged wire, and stay wire will be necessary. The initial point for soundings is a cross in the ledge of rock on the east bank. The channel is winding above and below the

,nd will overflow both banks except at the station. The bed
eam is composed of gravel above and below the station. At
n the bed is an outcrop of rock covered with about 6 inches
ig sand. The bench mark consists of two arrows and the
M. cut in the ledge of rock on the east bank 10 feet down-
om the gaging section. Its elevation is 12.57 feet above the
ie gage.

lowing discharge measurement was made by E. G. Marsh in

ist 12: Gage height, 0.66 foot; discharge, 2.2 second-feet.

ı gage height, in feet, of Sapello River at Sapello, N. Mex., for 1903.

	Aug.	Sept.	Oct.	Nov.	Dec.	Day.	Aug	Sept.	Oct.	Nov.	Dec
.....		0.60	0.50	0.50	0.50	17.................	0.80	0.50	0.50	0.50	0.60
.....		.60	.50	.50	.50	18.................	.70	.50	.50	.50	.60
.....		.60	.50	.50	.50	19.................	.70	.50	.50	.50	.60
.....		.60	.50	.50	.50	20.................	.70	.50	.50	.50	.60
.....		.60	.50	.50	.50	21.................	.70	.50	.50	.50	.60
.....		.60	.50	.50	.50	22.................	.70	.50	50	.50	.60
.....		.50	.50	.50	.50	23.................	.70	.50	.50	.50	.60
.....		.50	.50	.50	.50	24.................	60	.60	.50	.50	.60
.....		.50	.50	.50	.50	25.................	.60	.60	.50	.50	.60
.....		.50	.50	.50	.50	26.................	.60	.60	.50	.50	.60
.....		.50	.50	.50	.50	27.................	.60	.60	50	.50	.60
.....	0.70	.50	.50	.50	.50	28.................	.60	.60	.50	.50	.60
.....	.70	.50	.50	.50	.50	29.................	.60	.60	.50	.50	.60
.....	.70	.50	.50	.50	.60	30.................	.60	.60	.50	.50	.60
.....	.70	.50	.50	.50	.60	31.................	.60		.50	.50	.60
.....	.80	.50	.50	.50	.60						

MILL TAILRACE AT SAPELLO, N. MEX.

rod was placed in the mill tailrace of the flouring mill at
ır the purpose of determining the amount of water passing
he mill wheel. This water is diverted by two ditches from
ello and Manuelitos rivers at points above the gage rods
streams, and reenters Sapello River below the junction of
nd Manuelitos rivers. At low stages these ditches take the
pply in the two rivers before the water reaches the gage
ie same, and as this water is returned to the river below the
s through the mill tailrace it was necessary to place a gage
ill tailrace. This gage consists of a 2 by 4 inch by 5-
er graduated to vertical feet and tenths and placed in the
ace about 150 feet below the mill. The amount of water
his gage added to the amount passing the gages in Sapello
ielitos rivers gives the total discharge of the Sapello proper.

Mean daily gage height, in feet, of mill tailrace at Sapello, N. Mex., for 1903.

Day.	Aug.	Sept.	Oct.	Nov.	Dec.	Day.	Aug.	Sept.	Oct.	Nov.	De
1		0.10	0.10	0.20	0.25	17		0.10	0.20	0.25	0
2		.00	.10	.20	.25	18		.10	.20	.25	
3		.15	.10	.20	.25	19		.10	.20	.25	
4		.15	.10	.20	.25	20		.10	.20	.25	
5		.10	.10	.20	.25	21		.10	.20	.25	
6		.16	.10	.20	.25	22		.10	.20	.25	
7		.10	.10	.20	.25	23		.10	.20	.25	
8		.10	.10	.25	.25	24	0.10	.10	.20	.25	
9		.10	.10	.25	.25	25	.15	.10	.20	.25	
10		.10	.10	.25	.25	26	.15	.10	.20	.25	
11		.10	.15	.25	.25	27	.15	.10	.20	.25	
12		.10	.15	.25	.25	28	.15	.10	.20	.25	
13		.10	.20	.25	.25	29	.15	.10	.20	.25	
14		.10	.20	.25	.30	30	.15	.10	.20	.25	
15		.10	.20	.25	.30	31	.10		.20		
16		.10	.20	.25	.30						

MANUELITOS RIVER AT SAPELLO, N. MEX.

This station was established August 11, 1903, by E. G. Mars
It is located two-fifths of a mile above the junction of Sapello a
Manuelitos rivers and about 12 miles from Las Vegas, N. Me
There is a riprap wing dam 500 feet below the station. The gage
a 1 by 8 inch pine board 10 feet long, bolted in an inclined positi
to the rocky bluff on the left bank, and is graduated to read to ver
cal tenths of feet. It is read once each day by Horace R. Titlo
A cable, car, and tagged wire will be necessary in order to obta
measurements at high water. The initial point for soundings is
cross cut in the face of the rock 5 feet downstream from the gage,
the east bank. The channel is straight for 100 feet above and 2
feet below the station. At flood stages the current is very swift a
there is but one channel. At very low water there are several sm
channels. The right bank is low, brush lined, and overflows at hi
water until it reaches a second higher bank. The left bank is
steep, rocky bluff, and does not overflow. The bed of the stream
composed of quicksand, which is prevented from shifting by the r
rap dam below. The bench mark consists of two arrows cut in t
rocky bluff 39 feet downstream from the gage. Its elevation is 7.
feet above the zero of the gage.

Mean daily gage height, in feet, of Manuelitos River at Sapello, N. Mex., for 1903.

Day.	Aug.	Sept.	Oct.	Nov.	Dec.	Day.	Aug.	Sept.	Oct.	Nov.	Dec.
1		0.40	0.30	0.30	0.30	17	.60	.30	.30	.30	.30
2		.40	.30	.30	.20	18	.50	.30	.30	.30	.30
3		.40	.30	.30	.30	19	.50	.30	.30	.30	.30
4		.30	.30	.30	.30	20	.60	.30	.30	.30	.30
5		.30	.30	.30	.30	21	.50	.30	.30	.30	.30
6		.30	.30	.30	.30	22	.50	.30	.30	.30	.30
7		.30	.30	.30	.30	23	.50	.30	.30	.30	.30
8		.40	.30	.30	.30	24	.40	.40	.30	.30	.30
9		.30	.30	.30	.30	25	.40	.40	.30	.30	.30
10		.30	.30	.30	.30	26	.40	.30	.30	.30	.30
11		.40	.30	.30	.30	27	.40	.30	.30	.30	.30
12	0.50	.40	.30	.30	.30	28	.40	.30	.30	.30	.30
13	.50	.30	.30	.30	.30	29	.40	.30	.30	.30	.30
14	.50	.30	.30	.30	.30	30	.40	.30	.30	.30	.30
15	.50	.30	.30	.30	.30	31	.40		.30		.30
16	.60	.30	.30	.30	.30						

GALLINAS RIVER AT HOT SPRINGS, NEAR LAS VEGAS, N. MEX.

This station was established August 13, 1903, by E. G. Marsh. The station is located at the Hot Springs, 6 miles from Las Vegas, N. Mex. There are four footbridges one-fourth mile below the station, but a dam one-fourth mile farther downstream makes this point unsuitable for a station. There is an adjustable ice dam with a 12-foot fall 200 feet above the station. The gage is a vertical 1 by 8 inch pine board 10 feet long, bolted to the masonry wall on the right bank, which protects Hot Springs Nos. 16 and 17. At low water discharge measurements can be made by wading about 50 feet above the bridge. For high-water measurements a cable, car, tagged wire, and stay wire will be necessary, or they may be made from the various bridges. The initial point for soundings is a cross cut in the rock on the south bank about 1,000 feet above the upper bridge. The channel is straight for 200 feet above and 400 feet below the station. The current is rapid and the banks are high. The bed of the stream and the banks at the station are of solid rock, and there is one channel at all stages. The bench mark is a bolt leaded into a rock on the right bank 400 feet downstream from the gaging section and 200 feet upstream from the gage. Its elevation is 18.46 feet above the zero of the gage. The observer is William Preger, who reads the gage daily, or oftener during sudden fluctuations in the stage of the river.

This station was established for the purpose of determining the amount of water available for storage in the Sanguyjuella basin, located about 6 miles northwest of Las Vegas.

The following discharge measurement was made by E. G. Marsh in 1903:

August 13: Gage height, 0.50 foot; discharge, 2.94 second-feet.

Mean daily gage height, in feet, of Gallinas River at Hot Springs, near Las Vegas N. Mex., for 1903.

Day.	Aug.	Sept.	Oct.	Nov.	Dec.	Day.	Aug.	Sept.	Oct.	Nov.	D.
1		0.50	0.20	0.20	0.20	17	0.50	0.20	0.10	0.20	1
2		.50	.20	.20	.20	18	.70	.20	.10	.20	
3		.50	.20	.20	.20	19	.60	.20	.10	.20	
4		.40	.20	.20	.20	20	.60	.20	.10	.20	
5		.40	.20	.20	.20	21	.50	.20	.10	.20	
6		.40	.10	.20	.20	22	.50	.20	.10	.20	
7		.30	.10	.20	.20	23	.50	.20	.10	.20	
8		.25	.10	.20	.20	24	.50	.20	.10	.20	
9		.20	.10	.20	.20	25	.50	.10	.10	.20	
10		.20	.10	.20	.20	26	.50	.10	.20	.20	
11		.20	.10	.20	.20	27	.50	.20	.20	.20	
12		.20	.10	.20	.20	28	.50	.20	.20	.20	
13		.20	.10	.20	.20	29	.40	.20	.20	.20	
14	0.60	.20	.10	.20	.20	30	.40	.20	.20	.20	
15	.40	.20	.10	.20	.20	31	.50		.20		
16	.60	.20	.10	.20	.20						

NEOSHO RIVER NEAR IOLA, KANS.

The station was established in July, 1895, and was located at the highway bridge 1 mile west of the city of Iola, Kans. The station was discontinued November 30, 1903. The gage is fastened to the head gates of the flume, about 90 feet above the bridge. The bench mark is the heads of three large nails driven into the crosspiece the flume, and is 13.30 feet above the datum of the gage.

The observer during 1903 was Thomas J. Staley.

The observations at this station during 1903 have been made under the direction of W. G. Russell, district hydrographer.

ischarge measurements of Neosho River near Iola, Kans., in 1903.

Date.	Hydrographer.	Gage height.	Discharge.
		Feet.	*Second-feet.*
....	W. G. Russell	5.40	2,324
....	....do	12.40	13,566
....	....do	9.85	10,884
....	....do	12.90	15,614
0..................	....do	3.80	765

laily gage height, in feet, of Neosho River near Iola, Kans., for 1903.

Day.	Jan.	Feb.	Mar.	Apr.	May.	June.	July.	Aug.	Sept.	Oct.	Nov.
....	3.80	3.60	6.15	3.80	3.50	18.95	4.00	4.00	3.50	2.00	12.70
....	3.80	3.60	5.50	3.80	3.50	21.75	4.60	4.10	4.00	2.00	14.30
....	3.80	3.60	4.85	8.05	3.50	22.00	4.50	4.60	3.50	2.00	13.20
....	3.80	8.85	4.60	10.70	4.00	21.95	4.40	4.70	3.40	5.50	12.00
....	3.80	3.65	4.40	6.70	4.10	21.40	6.50	4.50	3.40	4.30	12.00
....	3.80	3.95	4.50	6.25	4.20	20.45	4.50	11.50	3.40	3.60	10.00
....	3.75	2.95	8.75	5.85	3.90	18.45	4.30	10.50	8.30	3.70	8.00
....	3.70	2.90	9.50	5.25	4.10	15.90	4.20	9.80	3.60	12.00	5.80
....	3.70	8.00	5.15	4.75	4.25	7.75	4.20	11.50	4.60	6.00	5.20
....	3.70	3.10	5.00	8.00	4.30	6.30	4.10	9.50	9.60	5.50	5.00
....	3.70	3.10	6.65	7.05	4.15	6.05	4.00	8.40	11.90	4.30	4.90
....	3.70	3.10	5.00	5.35	5.80	5.95	3.90	6.30	5.00	3.90	5.00
....	3.55	3.10	4.50	5.25	9.05	5.75	3.80	6.00	4.30	4.00	4.90
....	3.50	3.10	4.30	5.05	10.80	5.10	3.80	7.10	4.30	6.40	4.90
....	3.50	3.10	4.05	4.70	11.90	5.00	3.80	9.20	4.00	7.60	4.60
....	3.60	3.10	4.00	4.60	7.70	4.90	5.80	14.30	3.90	8.40	4.60
....	3.60	3.10	3.90	4.45	5.90	4.80	4.40	12.10	3.70	6.60	4.60
....	3.60	3.05	3.85	4.25	5.25	4.70	4.10	10.20	3.70	5.50	4.40
....	3.50	3.00	3.80	4.05	4.95	4.70	4.10	10.20	3.50	4.50	4.30
....	3.55	3.00	7.85	4.00	4.85	4.60	4.00	9.80	3.50	4.20	4.30
....	3.40	3.00	6.30	3.95	10.45	4.60	4.00	8.20	3.40	4.00	4.30
....	3.40	3.00	5.55	3.90	12.25	4.50	4.10	7.40	3.40	3.90	4.30
....	3.40	3.15	5.15	3.80	10.35	4.35	4.00	6.30	3.30	3.70	4.30
....	3.40	3.55	4.70	8.80	13.50	5.15	3.90	6.20	3.30	3.60	4.10
....	3.40	4.05	4.40	3.70	14.90	5.45	3.80	6.00	3.20	4.40	3.90
....	3.45	4.30	4.20	3.60	14.80	5.45	3.80	5.70	3.20	3.50	3.90
....	3.50	9.00	4.05	3.50	15.40	5.10	3.80	5.40	3.10	3.50	3.80
....	3.60	8.35	4.00	3.50	15.85	5.00	3.70	5.30	2.00	3.40	3.80
....	3.60	......	4.00	3.50	17.15	4.80	3.70	5.00	2.00	3.40	3.80
....	3.60	......	3.90	3.50	17.35	4.70	3.70	3.80	2.00	3.70	3.70
....	3.60	......	3.90	......	18.50	......	4.00	3.60	......	3.80	......

Rating table for Neosho River near Iola, Kans., from January 1 to December 1903.

Gage height.	Discharge.	Gage height.	Discharge.	Gage height.	Discharge.	Gage height.	Dischar
Feet.	*Second-feet.*	*Feet.*	*Second-feet.*	*Feet.*	*Second-feet.*	*Feet.*	*Second-*
2.0	120	3.9	845	5.8	2,800	9.4	9,21
2.1	145	4.0	920	5.9	2,960	9.6	9,57
2.2	170	4.1	995	6.0	3,120	9.8	9,95
2.3	195	4.2	1,070	6.2	3,460	10.0	10,30
2.4	220	4.3	1,150	6.4	3,810	10.5	11,2
2.5	245	4.4	1,230	6.6	4,170	11.0	12,2
2.6	270	4.5	1,315	6.8	4,530	11.5	13,2
2.7	295	4.6	1,400	7.0	4,890	12.0	14,2
2.8	320	4.7	1,490	7.2	5,250	13.0	16,3
2.9	350	4.8	1,580	7.4	5,610	14.0	18,5
3.0	380	4.9	1,680	7.6	5,970	15.0	20,8
3.1	415	5.0	1,780	7.8	6,330	16.0	23,2
3.2	450	5.1	1,890	8.0	6,690	17.0	25,7
3.3	490	5.2	2,000	8.2	7,050	18.0	28,3
3.4	530	5.3	2,120	8.4	7,410	19.0	31,0
3.5	580	5.4	2,240	8.6	7,770	20.0	33,7
3.6	630	5.5	2,370	8.8	8,130	21.0	36,4
3.7	700	5.6	2,500	9.0	8,490	22.0	39,1
3.8	770	5.7	2,650	9.2	8,850		

Estimated monthly discharge of Neosho River near Iola, Kans., for 190.

[Drainage area, 3,670 square miles.]

Month.	Discharge in second-feet.				Run-off.	
	Maximum.	Minimum.	Mean.	Total in acre-feet.	Second-feet per square mile.	De in
January	770	530	646	39,721	0.18	
February	8,490	350	1,041	57,814	.28	
March	9,390	770	2,109	135,211	.60	
April	11,634	580	2,208	131,385	.60	
May	29,670	580	9,157	563,042	2.50	
June	39,120	1,190	10,725	638,182	2.92	
July	3,990	700	1,123	69,051	.31	
August	19,210	630	6,112	375,812	1.66	
September	14,010	120	1,396	83,068	.38	
October	14,220	120	1,974	121,376	.54	
November	19,210	700	4,246	252,654	1.16	

GRAND RIVER NEAR FORT GIBSON, IND. T.

This station was established September 22, 1903, by W. G. Russell. It is located at the Missouri Pacific Railway bridge three-fourths mile northwest of Fort Gibson, Ind. T. The gage is a vertical 2 by 6 inch plank 18 feet long bolted to the east face of the first stone pier from the left bank. It reads from 9 to 27 feet. The 9-foot mark is the top of the steel caisson, and below this point gage heights are determined by measuring down to the water surface. The gage is read once each day by W. L. Blackwell. Discharge measurements are made from the railroad bridge, to which the gage is attached. The initial point for soundings is the east end of the bridge. The channel is straight for 300 feet above and 1,000 feet below the station. The current is sluggish. The right bank is low and liable to overflow. The left bank is high and rocky. The bed of the stream is composed of rock and is free from vegetation. The channel is broken by two piers, which cause eddies at high water. The bench mark is a cross on a hard limestone rock in a wall 4 feet above the upstream side of the bridge on the left bank of the river. Its elevation is 24.90 feet above the zero of the gage.

The observations at this station during 1903 have been made under the direction of W. G. Russell, district hydrographer.

Mean daily gage height, in feet, of Grand River near Fort Gibson, Ind. T., for 1903.

Day.	Sept.	Oct.	Nov.	Dec.	Day.	Sept.	Oct.	Nov.	Dec.
1		9.20	11.40	9.20	17		11.80	9.80	9.00
2		9.20	11.60	9.20	18		11.70	9.80	9.00
3		9.20	11.60	9.20	19		11.70	9.80	9.00
4		9.30	11.80	9.20	20		11.70	9.60	9.10
5		9.40	13.50	9.20	21		11.60	9.60	9.10
6		9.50	13.80	9.20	22	9.30	11.60	9.50	9.10
7		9.60	14.00	9.20	23	9.30	11.50	9.50	9.10
8		9.60	13.80	9.20	24	9.20	11.40	9.40	9.20
9		9.60	13.40	9.10	25	9.20	11.40	9.40	9.20
10		9.60	13.20	9.10	26	9.20	11.40	9.40	9.20
11		9.80	13.00	9.10	27	9.20	11.40	9.40	9.30
12		10.00	12.50	9.10	28	9.20	11.40	9.40	9.50
13		10.40	11.20	9.00	29	9.20	11.20	9.20	9.80
14		10.80	10.00	9.00	30	9.20	11.20	9.20	9.80
15		10.90	10.00	9.00	31		11.20		9.70
16		11.20	10.00	9.00					

VERDIGRIS RIVER NEAR LIBERTY, KANS.

The station was established in August, 1895, and was located at a wagon bridge about 250 feet below McTaggart's mill dam, about 3 miles southwest of the town of Liberty, Kans. The gage is a vertical timber fastened to the floor of the mill. Bench mark No. 1 is the heads of three large nails in the flume and is at an elevation of 12.46 feet above the zero of the gage. Bench mark No. 2 is the head of a spike in the root of a cottonwood tree 40 feet south of the gage and is at an elevation of 10.98 feet above gage datum. The bed is rocky, composed of gravel and subject to very little change. In 1900 a new bench mark was established, consisting of three nails driven horizontally into a root on the river side of a cottonwood tree 40 feet south of the gage, the nails being 8 inches below a sandstone rock which protrudes from a hollow in the tree. Its elevation is 11.88 feet above the zero of the old gage. The observer during 1903 was C. Fienen.

The station was discontinued November 30, 1903.

The observations at this station during 1903 have been made under the direction of W. G. Russell, district hydrographer.

Discharge measurements of Verdigris River near Liberty, Kans., in 1903.

Date.	Hydrographer.	Gage height.	Discharge.
		Feet.	*Second-feet.*
March 26	W. G. Russell	5.18	1,677
May 26	do	18.00	a 11,089
May 25	do	27.10	a 19,117

a These measurements are probably too small, as shown by previous measurements. The bed is constant.

Mean daily gage height, in feet, of Verdigris River near Liberty, Kans., for 1903.

Day.	Jan.	Feb.	Mar.	Apr.	May.	June.	July.	Aug.	Sept.	Oct.	Nov.
1	4.80	3.80	16.75	4.60	4.10	28.70	4.40	4.90	3.30	3.30	11.80
2	4.35	3.50	6.90	4.60	4.10	27.20	4.40	7.50	3.30	3.20	17.00
3	4.40	6.10	5.70	15.00	4.10	20.00	5.40	7.20	3.20	3.30	13.60
4	4.30	6.25	5.40	24.00	4.10	11.50	13.00	6.80	3.20	3.30	10.60
5	4.25	4.70	5.15	17.00	4.10	11.40	12.60	6.60	3.10	3.40	8.40
6	4.20	4.15	5.10	7.60	4.10	10.40	5.80	8.70	3.00	5.20	7.50
7	4.10	4.00	20.00	6.40	4.10	7.35	5.40	6.50	2.90	7.60	5.20
8	4.00	3.85	22.50	5.70	5.10	7.10	4.50	5.20	2.80	6.90	5.40
9	3.95	3.70	18.45	5.70	5.70	6.50	4.30	4.80	14.60	7.90	5.20
10	3.90	3.70	10.75	5.70	5.20	6.20	4.30	4.80	12.40	5.60	4.90
11	3.80	3.60	9.95	8.20	14.60	5.60	5.40	4.60	8.60	4.50	6.40
12	3.75	3.60	7.95	7.80	22.60	5.20	7.40	4.30	5.90	4.30	5.90
13	3.65	3.60	6.95	8.40	22.50	5.60	6.80	4.40	5.40	4.00	5.20
14	3.55	3.55	6.35	10.30	19.00	4.80	5.40	4.00	5.20	5.60	3.60
15	3.50	3.45	5.70	8.60	8.90	3.70	5.20	6.00	4.80	4.80	3.60
16	3.50	3.30	5.30	5.70	7.40	3.60	4.90	6.20	4.10	4.70	3.60
17	3.50	3.25	4.95	5.50	6.50	3.60	4.40	8.40	4.00	4.60	3.60
18	3.50	3.20	4.65	4.80	5.90	3.40	4.20	6.10	3.90	4.60	3.60
19	3.60	3.10	5.85	4.90	5.50	4.70	4.10	4.70	3.80	4.40	3.60
20	3.60	3.05	11.35	4.70	12.00	4.65	4.30	4.40	3.80	4.20	8.90
21	3.60	3.35	13.35	4.60	27.50	4.50	3.90	4.00	3.60	4.00	5.80

Mean daily gage height, in feet, of Verdigris River, etc.—Continued.

Day.	Jan.	Feb.	Mar.	Apr.	May.	June.	July.	Aug.	Sept.	Oct.	Nov.
22	3.55	4.75	10.90	4.40	36.70	4.30	3.70	4.10	3.50	3.80	5.70
23	3.50	5.75	8.80	4.40	39.00	4.20	3.50	3.70	3.40	3.60	5.40
24	3.50	5.45	5.40	4.30	36.00	5.90	3.30	3.60	3.40	3.60	5.20
25	3.50	5.70	5.30	4.30	27.25	6.40	3.30	3.60	3.40	3.60	4.90
26	3.50	8.75	5.00	4.20	18.20	5.90	3.30	3.50	3.40	3.60	4.20
27	3.50	22.00	5.50	4.20	7.50	5.10	3.30	3.50	3.40	3.60	4.20
28	3.55	24.80	5.30	4.20	10.50	4.70	3.30	3.60	3.30	3.60	4.20
29	3.70	.....	4.70	4.10	19.60	4.60	3.20	3.50	3.30	3.60	4.20
30	3.70	.....	4.60	4.10	27.70	4.60	3.90	3.50	3.30	3.60	4.20
31	3.70	.....	4.60	.....	28.50	.....	4.90	3.40	.....	3.60	.....

Rating table for **Verdigris River** *near Liberty, Kans., from January 1 to December 31, 1903.*

Gage height.	Discharge.	Gage height.	Discharge.	Gage height.	Discharge.	Gage height.	Discharge.
Feet.	Second-feet.	Feet.	Second-feet.	Feet.	Second-feet.	Feet.	Second-feet.
1.2	2	5.2	1,520	9.2	4,375	18.0	12,300
1.4	5	5.4	1,650	9.4	4,525	19.0	13,500
1.6	10	5.6	1,780	9.6	4,680	20.0	14,700
1.8	27	5.8	1,920	9.8	4,840	21.0	15,900
2.0	55	6.0	2,060	10.0	5,000	22.0	17,150
2.2	90	6.2	2,200	10.2	5,160	23.0	18,500
2.4	140	6.4	2,340	10.4	5,320	24.0	19,900
2.6	205	6.6	2,480	10.6	5,480	25.0	21,300
2.8	280	6.8	2,620	10.8	5,640	26.0	22,700
3.0	368	7.0	2,760	11.0	5,801	27.0	24,100
3.2	464	7.2	2,901	11.2	5,963	28.0	25,500
3.4	560	7.4	3,043	11.4	6,126	29.0	26,900
3.6	660	7.6	3,188	11.6	6,290	30.0	28,300
3.8	760	7.8	3,334	11.8	6,454	31.0	29,700
4.0	860	8.0	3,480	12.0	6,619	32.0	31,100
4.2	960	8.2	3,626	13.0	7,458	33.0	32,500
4.4	1,060	8.4	3,775	14.0	8,300	34.0	33,950
4.6	1,164	8.6	3,925	15.0	9,150	35.0	35,450
4.8	1,280	8.8	4,075	16.0	10,100	36.0	36,950
5.0	1,400	9.0	4,225	17.0	11,150		

Estimated monthly discharge of Verdigris River near Liberty, Kans., in 1903.

[Drainage area, 3,067 square miles.]

| Month. | Discharge in second-feet. | | | Total in acre-feet. | Run-off. | |
	Maximum.	Minimum.	Mean.		Second-feet per square mile.	Depth in inches.
January	1,060	610	747	45,931	0.24	0.28
February	21,020	392	2,348	130,401	.76	.79
March	17,800	1,164	4,112	252,837	1.34	1.54
April	19,900	910	3,031	180,357	.99	1.10
May	41,450	910	10,865	668,063	3.54	4.08
June	26,480	560	3,981	236,886	1.30	1.45
July	7,458	464	1,551	95,367	.51	.59
August	4,000	560	1,514	93,092	.49	.56
September	8,810	280	1,306	77,712	.43	.48
October	3,407	464	1,100	67,636	.36	.42
November	11,150	660	2,367	140,846	.77	.87

VERDIGRIS RIVER NEAR CATOOSA, IND. T.

This station was established September 25, 1903, by W. G. Russell. It is located at the Frisco Railway bridge 2 miles northeast of Catoosa, Ind. T. The wire gage is fastened to the guard rail of the railway bridge. The observer is John L. Calloway, the section foreman. Discharge measurements are made from the single span steel railway bridge and its approaches. The initial point for soundings is at the west end of the bridge, 338 feet from the zero of the gage scale, which is marked on the guard rail. The channel is straight for 200 feet above and below the bridge. The current is sluggish. Both banks are low, wooded, and subject to overflow, but all water passes beneath the bridge and its approaches. At low water there are two channels, and at high water the channel is broken by the two stone piers in the pile supports of the bridge. The gaging section is obstructed by broken piles, upon which drift collects. No bench mark has been established.

The observations at this station during 1903 have been made under the direction of W. G. Russell, district hydrographer.

Mean daily gage height,.in feet, of Verdigris River near Catoosa, Ind. T., for 1903.

Day.	Sept.	Oct.	Nov.	Dec.	Day.	Sept.	Oct.	Nov.	Dec.
1		3.20	4.00	3.80	17		5.60	5.10	3.30
2		3.00	8.70	3.90	18		5.60	4.90	3.30
3		2.80	10.00	3.80	19		5.70	4.70	3.30
4		3.60	10.30	3.70	20		5.40	4.40	3.30
5		3.30	18.70	3.50	21		4.50	4.30	3.30
6		11.20	16.60	3.50	22		4.10	4.20	3.30
7		7.40	9.10	3.50	23		3.90	4.00	3.30
8		6.00	7.30	3.50	24		8.60	4.00	3.40
9		4.90	6.50	3.50	25	3.00	3.40	3.90	3.50
10		7.30	6.20	3.40	26	3.40	3.30	3.70	3.50
11		8.50	5.80	3.40	27	3.10	3.30	3.60	3.40
12		5.50	5.40	3.30	28	2.90	3.20	3.50	3.40
13		5.10	5.20	3.30	29	3.00	3.10	3.50	3.40
14		4.30	6.70	3.30	30	2.90	3.00	3.80	3.40
15		4.30	6.50	3.30	31		3.40		3.50
16		4.20	6.00	3.30					

WALNUT RIVER NEAR ARKANSAS CITY, KANS.

This station was established September 21, 1902, by W. G. Russell, and was located on the Madison avenue highway bridge one-half mile east of Arkansas City, Kans. The gage is of the usual wire type, with the scaleboard graduated to feet and tenths; it was read daily by D. T. Burton. The bench mark is the lower edge of the fifth rivet from the top in the upper row on the fourth section of the west side of the south pier. The initial point for soundings is on the right bank. The channel both above and below the station is straight, has a width of 140 feet at ordinary stage, broken by one pier, and the current is sluggish. Both the right and left banks are high. The bed of the stream is sandy and shifting. The station was discontinued November 30, 1903.

The observations at this station during 1903 have been made under the direction of W. G. Russell, district hydrographer.

Discharge measurements of Walnut River near Arkansas City, Kans., in 1903.

Date.	Hydrographer.	Gage. height.	Discharge.
		Feet.	Second-feet.
March 25	W. G. Russell	6.50	599
May 27	do	9.60	2,984
August 17	do	5.30	275

Mean daily gage height, in feet, of Walnut River near Arkansas City, Kans., fo 1903.

Day.	Jan.	Feb.	Mar.	Apr.	May.	June.	July.	Aug.	Sept.	Oct.	Nov
1	5.70	5.10	11.20	5.80	5.50	15.80	6.10	4.70	4.20	4.40	7.
2	5.60	5.00	10.10	5.70	5.70	14.20	6.00	4.70	4.20	4.50	13.
3	5.55	5.20	7.20	5.70	5.60	12.80	6.00	4.60	4.30	4.50	16.
4	5.50	4.70	6.70	5.60	5.60	10.10	6.10	4.50	4.20	4.60	12.
5	5.40	4.70	6.60	8.90	5.50	9.80	6.00	4.70	4.10	4.70	10.
6	5.45	4.60	6.70	8.00	5.50	8.20	5.80	4.90	4.30	4.60	8.
7	5.40	4.50	6.60	7.60	5.60	8.00	5.70	5.40	4.40	4.50	6
8	5.50	4.60	6.90	7.40	5.40	8.10	5.80	5.30	4.60	4.30	6
9	5.20	4.50	7.20	7.20	5.10	8.00	5.50	5.20	5.30	4.20	6
10	5.10	4.60	6.90	7.00	6.20	7.40	5.40	5.30	12.70	4.10	6
11	5.00	4.80	6.80	6.20	7.40	7.00	5.30	5.40	8.10	4.20	6
12	4.80	5.40	6.70	6.30	9.70	6.70	5.40	5.60	6.60	4.20	6
13	4.60	5.30	6.80	6.20	13.90	6.50	5.40	5.70	6.70	4.30	5
14	4.50	5.20	6.90	6.10	13.20	6.60	5.30	5.60	6.60	4.60	5
15	4.50	5.10	6.70	6.00	11.00	6.70	5.10	5.40	6.50	5.10	5
16	4.40	4.80	6.50	5.90	8.20	7.30	4.90	4.90	6.30	5.50	5
17	4.40	4.60	6.40	5.80	7.10	7.40	4.80	5.20	6.40	5.60	5
18	4.50	4.50	6.40	5.70	7.70	7.20	4.70	5.20	6.20	5.50	5
19	4.70	4.60	7.50	5.40	8.40	7.00	4.60	5.40	6.10	5.30	5
20	4.80	4.70	10.30	5.40	9.50	6.80	4.70	5.30	4.90	5.00	5
21	4.90	4.80	12.30	5.30	16.70	6.90	4.90	5.20	4.70	5.00	5
22	5.30	4.90	8.70	5.30	18.70	7.00	4.90	5.00	4.60	4.80	5
23	5.20	4.90	7.50	5.30	18.00	6.80	4.70	5.00	4.60	4.80	5
24	5.10	5.00	6.40	5.40	19.20	6.70	4.60	4.90	4.50	4.70	5
25	5.00	5.10	6.40	5.30	18.30	6.60	4.40	4.90	4.50	4.70	5
26	5.10	6.00	6.30	4.80	16.00	6.40	4.50	5.00	4.40	4.60	5
27	5.10	8.40	6.20	4.80	9.60	6.20	4.60	4.90	4.30	4.70	5
28	5.10	8.90	6.00	5.10	9.70	6.30	4.70	4.80	4.30	4.60	5
29	5.00	......	6.10	5.20	9.80	6.20	4.70	4.60	4.40	4.50	5
30	5.00	......	6.00	5.40	10.30	6.10	4.60	4.40	4.50	4.50	5
31	4.90	......	6.00	......	17.20	......	4.70	4.30	......	4.60	...

Rating table for Walnut River near Arkansas City, Kans., from September 24, 19 to December 31, 1903.

Gage height.	Discharge.	Gage height.	Discharge.	Gage height.	Discharge.	Gage height.	Discharge
Feet.	Second-feet.	Feet.	Second-feet.	Feet.	Second-feet.	Feet.	Second-feet
4.0	45	5.2	260	6.4	570	8.2	1,510
4.1	60	5.3	285	6.5	600	8.4	1,670
4.2	75	5.4	310	6.6	630	8.6	1,850
4.3	90	5.5	335	6.7	665	8.8	2,040
4.4	105	5.6	360	6.8	700	9.0	2,240
4.5	122	5.7	385	6.9	740	9.2	2,460
4.6	140	5.8	410	7.0	780	9.4	2,700
4.7	160	5.9	435	7.2	870	9.6	2,960
4.8	180	6.0	460	7.4	970	9.8	3,240
4.9	200	6.1	485	7.6	1,090	10.0	3,540
5.0	220	6.2	510	7.8	1,220	10.2	3,840
5.1	240	6.3	540	8.0	1,360	10.4	4,140

Tangent above 10 feet gage height. Differences above this point, 150 per ten

Estimated monthly discharge of Walnut River near Arkansas City, Kans., for 1902, and 1903.

Month.	Discharge in second-feet.			Total in acre-feet.
	Maximum.	Minimum.	Mean.	
1902.				
September 24–30	4,140	105	1,524	21,160
October	2,760	45	307	18,877
November	260	75	147	8,747
December	510	90	265	16,294
1903.				
January	385	105	233	14,327
February	2,140	122	323	17,938
March	6,990	460	1,261	77,536
April	2,140	180	523	31,121
May	17,340	240	5,083	312,541
June	11,490	485	1,831	108,952
July	485	105	261	16,048
August	385	90	232	14,265
September	7,590	60	534	31,775
October	360	60	162	9,961
November	12,840	220	1,613	95,980

ARKANSAS RIVER NEAR ARKANSAS CITY, KANS.

This station was established September 23, 1902, by W. G. Russell. It is located on the Chestnut avenue highway bridge, one-half mile west of Arkansas City, Kans. The gage is a painted staff graduated to feet and tenths, spiked to the west side of the south pile of the second bent on Chestnut avenue. The bench mark is the top of the cap on the pile which carries the gage. Its elevation is 17.2 feet above the zero of the gage. The initial point for soundings is on the left bank. The channel is straight for about 200 feet both above and below the station and has a width of 550 feet, broken by 36 pile piers. Both banks are low and liable to overflow. The bed of the river is sandy and shifting. The current is moderately rapid. The observer is D. T. Burton, who reads the gage daily.

The observations at this station during 1903 have been made under the direction of W. G. Russell, district hydrographer.

Discharge measurements of Arkansas River near Arkansas City, Kans., in 1903.

Date.	Hydrographer.	Gage height.	Discharge.
		Feet.	*Second-feet.*
March 25	W. G. Russell	4.70	842
May 27	do	6.00	3,601
June 17	do	8.35	7,396
June 18	do	7.80	6,410
June 19	do	7.20	4,738
June 20	do	7.00	4,333
August 17	do	4.60	928
October 2	E. C. Murphy	3.45	150

Mean daily gage height, in feet, of Arkansas River near Arkansas City, Kans., for 1903.

Day.	Jan.	Feb.	Mar.	Apr.	May.	June.	July.	Aug.	Sept.	Oct.	Nov.	Dec.
1	2.85	3.90	4.70	4.20	4.20	9.40	6.20	3.80	3.80	3.30	4.30	4.10
2	3.20	3.80	4.60	4.20	4.30	10.60	6.10	3.80	3.40	6.80	4.00
3	3.60	3.80	4.40	4.30	4.40	11.00	6.10	3.90	3.70	3.50	7.10	4.00
4	3.50	3.70	4.20	4.40	4.50	10.10	6.00	4.20	3.60	3.60	6.30	3.90
5	3.60	3.00	4.30	4.70	4.50	9.00	6.00	4.80	3.60	3.50	6.80	3.80
6	3.80	3.10	4.60	4.60	4.60	7.20	5.80	4.80	3.70	3.60	6.10	3.90
7	3.80	3.00	4.50	4.60	4.50	7.00	5.70	5.00	3.70	3.70	5.20	3.90
8	3.90	3.80	4.60	4.50	4.40	6.20	5.60	5.10	3.70	3.60	5.20	4.10
9	4.10	3.70	4.70	4.40	4.30	6.00	5.40	5.00	4.00	3.50	5.00	4.10
10	4.00	3.50	4.90	4.40	4.40	5.80	5.00	5.10	4.60	8.40	4.90	4.20
11	3.70	3.70	4.90	4.30	5.10	5.60	5.00	5.20	5.10	3.50	4.80	4.10
12	3.60	3.90	4.90	4.30	5.60	5.30	5.00	5.30	4.70	3.50	4.60	4.00
13	3.40	3.80	4.80	4.20	5.50	5.20	5.10	5.20	3.60	3.60	4.50	4.00
14	3.20	3.80	4.80	4.10	5.45	5.30	5.00	5.20	3.70	3.70	4.40	4.10
15	3.10	3.80	4.90	4.10	5.30	6.50	5.00	5.30	3.70	3.90	4.50	4.10
16	3.00	4.00	5.00	4.00	5.00	9.00	4.80	5.00	3.60	4.40	4.60	4.00
17	3.00	3.80	5.40	4.00	5.20	8.40	5.40	5.00	3.00	4.60	4.50	4.00
18	3.20	3.80	5.20	4.10	5.60	7.80	4.40	4.70	3.50	4.50	4.40	4.00
19	3.30	3.70	5.10	4.20	5.80	7.20	4.30	5.10	3.50	4.40	4.40	3.90
20	3.40	3.80	5.20	4.20	6.10	7.00	4.30	5.00	3.60	3.80	4.30	4.00
21	3.60	3.60	5.60	4.30	7.40	6.80	4.20	4.60	3.60	3.70	4.10	4.00
22	3.70	3.70	5.40	4.20	8.80	7.00	4.20	4.40	3.50	3.70	4.20	4.10
23	2.80	3.80	5.10	4.20	8.40	8.00	4.10	4.30	3.40	3.60	4.10	4.00
24	2.70	3.90	5.00	4.10	9.10	7.80	4.00	4.30	3.40	3.50	4.10	4.00
25	3.80	4.00	4.70	4.00	9.10	7.40	4.00	4.30	3.40	3.50	4.20	3.90
26	3.80	4.60	4.60	4.00	9.00	7.20	4.00	4.50	3.30	3.60	4.20	3.80
27	3.40	4.80	4.50	3.80	6.60	7.00	3.90	4.50	3.30	3.60	4.30	3.80
28	3.50	4.90	4.50	3.80	6.50	6.80	3.90	4.40	3.40	3.40	4.30	3.80
29	3.60		4.80	4.00	6.60	6.60	3.80	4.30	3.40	3.50	4.20	3.90
30	3.70		4.40	4.10	7.10	6.50	3.80	3.90	3.30	3.60	4.20	3.80
31	3.80		4.30		9.60		3.90	3.80		3.70		3.80

nting table for Arkansas River near Arkansas City, Kans., from September 23, 1902, to December 31, 1903.

Gage height.	Discharge.	Gage height.	Discharge.	Gage height.	Discharge.	Gage height.	Discharge.
Feet.	*Second-feet.*	*Feet.*	*Second-feet.*	*Feet.*	*Second-feet.*	*Feet.*	*Second-feet.*
2.6	30	4.0	375	5.4	1,685	7.6	5,675
2.7	37	4.1	445	5.5	1,810	7.8	6,155
2.8	45	4.2	515	5.6	1,935	8.0	6,655
2.9	59	4.3	595	5.7	2,070	8.2	7,175
3.0	70	4.4	675	5.8	2,205	8.4	7,715
3.1	87	4.5	765	5.9	2,355	8.6	8,275
3.2	105	4.6	855	6.0	2,505	8.8	8,855
3.3	125	4.7	950	6.2	2,835	9.0	9,455
3.4	145	4.8	1,045	6.4	3,185	9.2	10,055
3.5	170	4.9	1,145	6.6	3,555	9.4	10,655
3.6	195	5.0	1,245	6.8	3,945	9.6	11,255
3.7	230	5.1	1,350	7.0	4,355	10.0	12,455
3.8	265	5.2	1,455	7.2	4,775	11.0	15,455
3.9	320	5.3	1,570	7.4	5,215		

Tangent above 9 feet gage height. Differences above this point, 300 per tenth.

Estimated monthly discharge of Arkansas River near Arkansas City, Kans., for 1902 and 1903.

Month.	Discharge in second-feet.			Total in acre-feet.
	Maximum.	Minimum.	Mean.	
1902.				
September 23 to 30	950	375	523	8,299
October	810	70	477	29,330
November	675	125	209	12,436
December	265	33	142	8,731
1903.				
January......................	445	37	195	11,990
February	1,145	70	327	18,161
March	1,935	515	1,056	64,931
April	950	265	536	31,894
May	11,255	515	3,286	202,048
June.........................	15,455	1,455	5,598	333,104
July	2,835	265	1,178	72,432
August	1,570	265	922	56,692
September	1,350	125	280	16,661
October	855	125	264	16,233
November....................	4,565	445	1,227	73,012
December	515	265	365	22,443

ARKANSAS RIVER AT HUTCHINSON, KANS.

This station was established May 13, 1895, and is located at the wagon bridge at the south end of Main street. The gage consists of an oak timber spiked to a pile a few feet above the bridge. Bench mark No. 1 is the upper crosspiece of the pier guard. Its elevation is 8.35 feet above the zero of the gage. Bench mark No. 2 is the top of the iron doorsill of the first brick building next to the river. Its elevation is 8.12 feet above gage datum. The channel is straight for some distance above and below the bridge and has a width of 1,020 feet, broken by 11 steel piers. The bed is sandy and very shifting, necessitating frequent discharge measurements and soundings. At low water the stream subdivides into a number of small channels. Measurements of discharge are made from the bridge at high water, and at low water they can be made by wading.

The observer is Daniel Lauer.

The observations at this station during 1903 have been made under the direction of W. G. Russell, district hydrographer.

Discharge measurements of Arkansas River at Hutchinson, Kans., in 1903.

Date.	Hydrographer.	Gage height.	Discharge.
		Feet.	*Second-feet.*
March 24	W. G. Russell	1.85	452
June 16	....do	4.50	6,688
September 8	....do	1.05	138
October 1	E. C. Murphy	.84	78

Mean daily gage height, in feet, of Arkansas River at Hutchinson, Kans., for 1903.

Day.	Jan.	Feb.	Mar.	Apr.	May.	June.	July.	Aug.	Sept.	Oct.	Nov.	Dec.
1	1.40	1.60	1.50	1.55	1.55	2.70	3.00	1.75	1.25	0.90	1.00	1.20
2	1.40	1.65	1.50	1.50	1.50	2.55	2.80	1.85	1.20	.90	1.10	1.25
3	1.40	1.60	1.55	1.60	1.50	2.50	2.70	1.95	1.15	.90	1.20	1.25
4	1.40	1.55	1.70	1.55	1.60	2.40	2.60	2.10	1.10	.90	1.20	1.20
5	1.40	1.50	1.70	1.50	1.70	2.30	2.50	2.25	1.10	.85	1.20	1.20
6	1.30	1.40	1.80	1.45	1.80	2.25	2.45	2.40	1.10	.85	1.30	1.20
7	1.30	1.40	1.90	1.45	1.90	2.20	2.40	2.60	1.10	.85	1.35	1.20
8	1.30	1.40	2.00	1.40	2.00	2.05	2.40	2.70	1.05	.85	1.20	1.20
9	1.30	1.45	2.00	1.40	2.10	1.95	2.30	2.75	1.05	.85	1.20	1.20
10	1.30	1.45	2.05	1.40	2.05	1.95	2.20	2.65	1.05	.85	1.25	1.20
11	1.30	1.45	2.05	1.45	2.30	1.90	2.10	2.55	1.05	.80	1.25	1.20
12	1.30	1.45	2.10	1.45	2.35	1.80	2.00	2.15	1.10	1.00	1.20	1.20
13	1.30	1.45	2.10	1.40	2.30	1.80	1.90	2.00	1.10	1.10	1.20	1.20
14	1.30	1.45	2.50	1.40	2.10	4.95	1.80	1.80	1.00	1.15	1.20	1.20
15	1.30	1.45	2.30	1.40	2.20	5.00	1.75	1.65	.95	1.10	1.10	1.25
16	1.40	1.45	2.25	1.40	2.30	4.50	1.75	1.70	.95	1.10	1.20	1.25
17	1.40	1.45	2.20	1.40	2.30	4.10	1.70	1.75	1.00	1.05	1.25	1.30
18	1.40	1.45	2.25	1.40	2.30	3.90	1.65	1.80	1.00	1.05	1.20	1.30
19	1.40	1.45	2.20	1.35	2.30	3.75	1.65	1.75	1.00	1.00	1.20	1.25
20	1.40	1.45	2.10	1.35	2.10	3.60	1.55	1.70	1.00	1.00	1.20	1.20
21	1.40	1.45	2.00	1.30	1.95	4.50	1.50	1.70	1.00	1.00	1.20	1.15
22	1.35	1.45	1.95	1.30	1.95	4.30	1.50	1.70	1.00	1.00	1.20	1.15

Mean daily gage height, in feet, of Arkansas River, etc.—Continued.

Day.	Jan.	Feb.	Mar.	Apr.	May.	June.	July.	Aug.	Sept.	Oct.	Nov.	Dec.
1........................	1.35	1.45	1.90	1.30	1.90	4.05	1.45	1.70	1.00	1.00	1.90	1.10
2........................	1.30	1.45	1.90	1.45	1.80	4.00	1.45	1.60	1.00	1.00	1.90	1.10
3........................	1.30	1.45	1.85	1.40	1.70	4.00	1.40	1.50	.95	1.00	1.90	1.00
4........................	1.30	1.45	1.85	1.35	1.85	3.90	1.40	1.50	.95	1.00	1.90	1.00
5........................	1.30	1.45	1.80	1.30	2.20	3.80	1.35	1.45	.95	1.00	1.90	1.00
6........................	1.35	1.45	1.80	1.30	2.50	3.75	1.30	1.40	.95	1.00	1.90	1.00
7........................	1.40	....	1.75	1.50	2.90	3.40	1.30	1.30	.95	1.00	1.90	1.00
8........................	1.50	....	1.70	1.60	3.00	3.15	1.25	1.30	.90	1.00	1.90	.95
9........................	1.50	....	1.70	....	2.90	....	1.20	1.30	....	1.00	....	.95

Rating table for Arkansas River at Hutchinson, Kans., from January 1 to December 31, 1903.

Gage height.	Discharge.	Gage height.	Discharge.	Gage height.	Discharge.	Gage height.	Discharge.
Feet.	Second-feet.	Feet.	Second-feet.	Feet.	Second-feet.	Feet.	Second-feet.
0.8	57	1.6	300	2.4	980	3.2	2,550
.9	75	1.7	350	2.5	1,110	3.3	2,810
1.0	100	1.8	410	2.6	1,260	3.4	3,070
1.1	130	1.9	480	2.7	1,430	3.5	3,330
1.2	160	2.0	560	2.8	1,620	3.6	3,590
1.3	190	2.1	650	2.9	1,830		
1.4	220	2.2	750	3.0	2,060		
1.5	260	2.3	860	3.1	2,300		

Tangent above 3.30 feet gage height. Differences above this point, 260 per tenth.

Estimated monthly discharge of Arkansas River at Hutchinson, Kans., in 1903.

[Drainage area, 34,000 square miles.]

Month.	Discharge in second-feet.			Total in acre-feet.	Run-off.	
	Maximum.	Minimum.	Mean.		Second-feet per square mile.	Depth in inches.
January............	260	190	208	12,789	0.0061	0.0070
February	325	220	247	13,718	.0073	.0076
March	1,110	260	535	32,896	.0160	.0180
April	300	190	230	13,686	.0068	.0076
May	2,060	260	730	44,886	.0220	.0250
June..............	7,230	410	3,016	179,464	.0890	.0990
July	2,060	160	598	36,770	.0180	.0210
August	1,525	190	556	34,187	.0160	.0180
September	175	75	110	6,545	.0032	.0036
October...........	145	65	94	5,780	.0028	.0032
November	205	100	162	9,640	.0048	.0054
December	190	87	147	9,039	.0043	.0050
The year	7,230	65	553	399,400	.0164	.2204

ARKANSAS RIVER AT DODGE, KANS.

This station was established November 28, 1902, by W. G. Russell, and is located one-fourth mile south of Dodge, on the highway bridge. The gage is a plain staff graduated to feet and tenths and nailed to the upstream pile of the twelfth bent at the north end of the bridge. The initial point for soundings is on the left bank. The channel both above and below the station is straight for about 100 feet; both banks are low and liable to overflow; the bed of the stream is sandy and shifting. The observer is Alexander Alter, who reads the gage once each day. No bench marks have been established.

The observations at this station during 1903 have been made under the direction of W. G. Russell, district hydrographer.

Discharge measurements of Arkansas River at Dodge, Kans., in 1903.

Date.	Hydrographer.	Gage height.	Discharge.
		Feet.	*Second-feet.*
March 16	W. G. Russell	2.90	630
June 15	do	4.20	5,409
August 14	do	1.15	9

Mean daily gage height, in feet, of Arkansas River at Dodge, Kans., for 1903.

Day.	Jan.	Feb.	Mar.	Apr.	May.	June.	July.	Aug.	Sept.	Oct.	Nov.	Dec.
1	1.45	2.00	1.55	1.25	1.12	0.45	1.55	1.25	(a)	(a)	0.20	1.00
2	1.22	1.65	1.55	1.25	1.35	.55	.35	1.25	(a)	(a)	.25	1.00
3	1.55	1.35	1.55	1.15	1.45	.55	1.25	1.20	(a)	(a)	.40	1.00
4	1.55	1.00	1.55	1.15	1.35	.45	1.45	1.15	(a)	(a)	.40	1.00
5	1.82	1.00	1.55	1.15	1.25	.45	.55	1.15	(a)	(a)	.40	1.00
6	1.82	1.12	1.65	1.15	1.15	.35	1.35	1.30	(a)	(a)	.40	1.00
7	1.65	1.25	1.75	1.00	1.00	.35	1.55	1.45	(a)	(a)	.40	1.00
8	2.00	1.25	1.75	.55	1.25	.35	1.35	1.35	(a)	(a)	.40	1.05
9	2.00	1.35	1.75	.65	1.25	.45	1.35	1.25	(a)	(a)	.40	1.10
10	2.00	1.55	2.00	.55	1.45	1.90	1.45	1.15	(a)	(a)	.40	1.10
11	2.00	1.45	2.00	.45	1.45	2.50	1.55	1.15	(a)	(a)	.40	1.10
12	2.00	1.45	2.00	.45	1.55	5.05	1.45	1.15	(a)	(a)	.40	1.00
13	2.00	1.00	1.75	.45	1.55	5.00	1.35	1.25	(a)	(a)	.40	1.00
14	2.00	1.00	1.82	.45	1.55	3.55	1.25	1.35	(a)	(a)	.40	1.00
15	2.00	1.00	2.00	.55	1.45	3.85	1.15	1.27	(a)	(a)	.70	1.00
16	1.65	1.00	2.00	.55	1.45	3.00	1.15	1.27	(a)	(a)	1.00	1.00
17	1.55	1.00	2.00	.45	1.00	2.82	1.00	1.35	(a)	(a)	1.00	1.00
18	1.75	1.00	2.12	.45	1.00	3.12	1.00	1.50	(a)	(a)	1.00	1.10
19	1.65	1.00	2.25	.45	.85	3.22	.85	1.55	(a)	(a)	1.00	1.10
20	1.65	1.12	2.12	.35	.55	3.25	1.00	1.50	(a)	(a)	1.00	1.10
21	1.65	1.12	2.00	.45	.55	3.55	1.00	1.45	(a)	(a)	1.00	1.10
22	1.75	1.12	1.65	.35	.45	3.55	.85	1.35	(a)	(a)	1.00	1.10
23	1.55	1.12	1.55	.35	.45	3.00	.65	1.30	(a)	(a)	1.00	1.20
24	1.75	1.12	1.55	.35	.45	2.55	.45	1.15	(a)	(a)	1.00	1.20

a Water not running.

*Mean daily gage height, in feet, of Arkansas River, etc.—*Continued.

Day.	Jan.	Feb.	Mar.	Apr.	May.	June.	July.	Aug.	Sept.	Oct.	Nov.	Dec.
....	1.75	1.35	1.55	0.35	0 45	3.00	0.55	1.15	(a)	(a)	1.00	1.20
....	1.55	1.45	1.55	.35	.45	2.75	1.45	1.15	(a)	(a)	1.00	1.20
....	2.00	1.45	1.25	.35	.35	2.55	1.30	1.15	(a)	(a)	1.10	1.20
....	2.00	1.55	1.35	.35	.35	2.00	1.25	1.15	(a)	(a)	1.10	1.20
....	2.00	......	1.35	.25	.45	2.00	1.25	1.15	(a)	(a)	1.00	1.20
....	2.00	......	1.35	.35	.45	1.55	1.25	1.15	(a)	(a)	1.00	1.10
....	2.00	......	1.35	......	.55	......	1.35	1.15	......	(a)	......	1.20

a Water not running.

ARKANSAS RIVER NEAR SYRACUSE, KANS.

This station, established August 21, 1902, by W. G. Russell, is located on the highway bridge 1 mile south of Syracuse, Kans. The gage is a plain staff graduated to feet and tenths, fastened to the east pile of the bent 283 feet south of the north end of the bridge. The initial point for soundings is on the left bank. The channel above and below the station is straight and the water is sluggish. The right bank is low and liable to overflow; the left bank is high, and the bed of the stream is sandy and shifting. The bench mark is on the top of the east end of the first sill at the north end of the bridge. Its elevation is 13.45 feet above the zero of the gage. The gage is read daily by Clyde S. Welborn.

The observations at this station during 1903 have been made under the direction of W. G. Russell, district hydrographer.

Discharge measurements of Arkansas River near Syracuse, Kans., in 1903.

Date.	Hydrographer.	Gage height.	Discharge.
		Feet.	*Second-feet.*
March 16....................	W. G. Russell..............	2.20	181
June 10	.....do	4.45	3,823
June 11	.....do	6.40	15,368
June 12	.....do	7.10	22,695
June 13	.....do	5.50	8,773
August 10	.....do	2.10	28
August 11	.....do	3.65	1,034

Mean daily gage height, in feet, of Arkansas River near Syracuse, Kans., for 1903.

Day.	Jan.	Feb.	Mar.	Apr.	May.	June.	July.	Aug.	Sept.	Oct.	Nov.	Dec.
1	2.00	2.30	3.40	1.80	1.70	1.90	3.60	2.10	2.00	1.90	1.90	2.00
2	2.00	2.30	3.40	1.75	1.70	1.70	3.50	2.10	2.00	1.90	1.90	2.00
3	2.00	2.40	3.50	1.75	1.70	1.60	3.50	2.10	2.00	1.90	1.90	2.00
4	2.00	2.30	3.50	1.70	1.60	1.60	3.50	2.10	2.00	1.90	2.00	2.00
5	2.00	2.10	3.50	1.70	1.60	1.60	3.00	2.10	2.00	1.90	2.00	2.00
6	2.00	2.10	3.60	1.70	1.60	1.60	2.90	2.00	2.00	1.90	1.90	2.00
7	2.00	2.10	3.60	1.70	1.80	3.20	2.80	2.00	2.00	1.90	1.90	2.00
8	2.10	2.20	3.50	1.70	1.80	4.20	2.70	2.00	2.00	1.90	1.90	2.00
9	2.10	2.10	3.40	1.60	1.70	4.60	2.70	2.10	2.00	1.90	1.90	2.00
10	2.10	2.15	3.00	1.60	2.15	4.50	2.70	2.80	2.00	1.90	1.90	2.00
11	2.00	2.10	2.70	1.60	2.20	7.50	3.10	3.00	2.00	1.90	1.90	2.00
12	2.00	2.10	2.70	1.60	2.00	6.80	3.10	3.00	2.00	1.90	1.90	2.00
13	2.20	2.45	2.60	1.55	2.00	5.40	3.10	3.10	2.00	1.90	1.90	2.00
14	2.10	2.30	2.60	1.60	1.90	5.00	3.10	2.80	2.00	1.90	1.90	2.00
15	2.10	2.20	2.40	1.60	1.80	4.50	2.70	3.20	2.00	1.90	1.90	1.90
16	2.00	2.20	2.20	1.60	1.70	5.40	2.70	3.00	2.00	1.90	1.90	1.90
17	2.10	2.20	2.20	1.60	1.70	5.50	2.50	2.60	2.00	1.90	2.00	1.90
18	2.20	2.25	2.15	1.60	1.70	5.50	2.50	2.00	2.00	1.90	2.00	1.90
19	2.20	2.30	2.00	1.50	1.70	5.00	2.50	2.40	2.00	1.90	2.00	1.90
20	2.10	2.40	2.00	1.60	1.70	5.00	3.00	2.30	2.00	1.90	2.00	1.90
21	2.10	2.40	2.00	1.60	1.70	4.80	3.00	2.30	2.00	1.90	2.00	1.90
22	2.30	2.40	2.00	1.60	1.60	4.80	2.80	2.30	2.00	1.90	2.00	1.90
23	2.25	2.50	2.00	1.60	1.60	4.50	2.70	2.10	2.00	1.90	2.00	1.90
24	2.30	2.50	1.90	1.60	1.60	4.50	2.00	2.10	2.00	1.90	2.00	2.00
25	2.40	2.80	1.90	1.60	1.60	4.20	2.60	2.10	2.00	1.90	2.00	2.00
26	2.60	3.00	2.00	1.60	2.10	4.20	2.50	2.00	2.00	1.90	2.00	2.00
27	2.60	3.70	2.00	1.60	2.80	4.00	2.50	2.00	2.00	1.90	2.00	2.00
28	2.40	3.40	1.90	1.70	2.40	3.80	2.40	2.00	1.95	1.90	2.00	2.00
29	2.30	......	1.90	1.80	2.60	3.70	2.30	2.00	1.90	1.90	2.20	2.00
30	2.00	......	1.80	1.80	2.30	3.70	2.20	2.00	1.90	1.90	2.10	2.10
31	2.30	......	1.80	......	2.00	......	2.10	2.00	......	1.90	......	2.10

Rating table for Arkansas River near Syracuse, Kans., from January 1 to June ?, 1903.

Gage height.	Discharge.	Gage height.	Discharge.	Gage height.	Discharge.	Gage height.	Discharge.
Feet.	Second-feet.	Feet.	Second-feet.	Feet.	Second-feet.	Feet.	Second-feet.
1.6	20	2.2	190	2.8	680	3.4	1,315
1.7	30	2.3	250	2.9	780	3.5	1,430
1.8	40	2.4	330	3.0	880	3.6	1,545
1.9	60	2.5	410	3.1	980	3.7	1,660
2.0	80	2.6	500	3.2	1,090	3.8	1,775
2.1	130	2.7	590	3.3	1,200	3.9	1,890

Rating table for Arkansas River near Syracuse, Kans., from June 8 to December 31, 1903.

Gage height.	Discharge.	Gage height.	Discharge.	Gage height.	Discharge.	Gage height.	Discharge.
Feet.	*Second-feet.*	*Feet.*	*Second-feet.*	*Feet.*	*Second-feet.*	*Feet.*	*Second-feet.*
1.9	8	3.0	545	4.1	2,460	5.4	8,350
2.0	20	3.1	645	4.2	2.770	5.6	9,590
2.1	28	3.2	755	4.3	3,100	5.8	10,910
2.2	50	3.3	875	4.4	3,450	6.0	12,310
2.3	80	3.4	1,000	4.5	3,830	6.2	13,790
2.4	120	3.5	1,140	4.6	4,245	6.4	15,360
2.5	170	3.6	1,300	4.7	4,680	6.6	17,200
2.6	230	3.7	1,480	4.8	5,140	6.8	19,300
2.7	300	3.8	1,680	4.9	5,620	7.0	21,600
2.8	375	3.9	1,910	5.0	6,120	7.2	24,100
2.9	455	4.0	2,170	5.2	7,190	7.4	26,800

Estimated monthly discharge of Arkansas River near Syracuse, Kans., in 1903.

[Drainage area, 24,960 square miles.]

Month.	Discharge in second-feet.			Total in acre-feet.	Run-off.	
	Maximum.	Minimum.	Mean.		Second-feet per square mile.	Depth in inches.
January	500	80	169	10,391	0.0068	0.0078
February	1,660	130	365	20,271	.0150	.0160
March	1,545	40	600	36,892	.0240	.0280
April	40	10	245	14,578	.0098	.0109
May	680	20	100	6,149	.0040	.0046
June	28,300	20	4,893	291,154	.2000	.2200
July	1,300	28	418	25,702	.0170	.0200
August	1,000	20	178	10,945	.0072	.0083
September	20	8	19	1,130	.00076	.00085
October	8	8	8	492	.00032	.00037
November	50	8	16	952	.00064	.00071
December	28	8	17	1,045	.00068	.00078
The year	28,300	8	586	419,701	.02385	.31831

ARKANSAS RIVER NEAR COOLIDGE, KANS.

This station was established May 5, 1903, by A. Jacob. It is located at the highway bridge on the road running south from Coolidge, Kans., and about 1¼ miles east of the Colorado-Kansas boundary. The bridge is about 1 mile 'from' the Atchison, Topeka and Santa Fe Railway station at Coolidge, Kans. The gage is a 1 by 5 inch pine board nailed to the upper pile of the thirteenth bent from the south end of the bridge. It is read twice each day by B. N. Johnson. Discharge measurements are made from the bridge at high water and by wading at low water. The initial point for soundings is the second set of piles from the south end of the bridge. The channel is straight for 3,000 feet above and 2,000 feet below the station. The velocity is well distributed and the current is smooth. The right bank is high and covered with grass. The left bank is low, covered with grass and overflows frequently. The channel is broad and shallow, and the bed is sandy. After each flood the bed is inclined to become quicksand, with considerable change in the cross section. Bench mark No. 1 is the top of an iron pipe, 2 inches in diameter and 4 feet long, resting on a stone 3½ feet below the surface of the ground. The pipe projects about one-half foot above the ground and is painted white. It is located 18 feet west and 8 feet south of the second set of piles from the south end of the bridge. Its elevation above the zero of the gage is 5.55 feet. Bench mark No. 2 is a 60-penny nail driven flush in a cross cut on the top of the west end of the fourteenth cap from the south end of the bridge. Its elevation is 8.31 feet above the zero of the gage. This station was discontinued October 31, 1903, the station at Syracuse, Kans., to be continued in its stead.

Mean daily gage height, in feet, of Arkansas River near Coolidge, Kans., for 1903

Day.	May.	June.	July.	Aug.	Sept.	Oct.
1	....	0.60	2.25	1.00	0.70	0.8
2	....	.60	2.25	1.05	.55	.5
3	....	.50	2.05	.95	.50	.3
4	....	.50	1.85	.90	.60	.3
5	....	.45	1.70	.80	.55	.4
6	....	a.85	1.60	.65	.45	.4
7	0.65	1.35	1.55	a.85	.35	.4
8	.55	2.70	1.45	.60	.30	.5
9	.45	2.85	1.40	1.05	.30	.5
10	.85	2.65	1.30	2.30	.20	.6
11	.90	4.50	1.40	2.10	.20	.6
12	.70	4.30	1.90	1.75	.20	.6
13	.60	3.35	1.85	1.75	.20	.7
14	.55	3.00	1.55	1.65	.15	.7
15	.50	2.85	1.45	1.80	.10	.7
16	.45	3.30	1.40	1.55	.10	.7
17	.35	3.40	1.35	1.45	.10	.7
18	.30	3.40	1.30	1.35	.15	.7

a Rain.

Mean daily gage height, in feet, of Arkansas River, etc.—Continued.

Day.	May.	June.	July.	Aug.	Sept.	Oct.
....	0.35	3.05	1.25	1.30	.85	1.00
....	.40	3.10	1.20	1.25	.50	1.00
....	.35	2.95	1.35	1.20	.50	1.00
....	.30	2.90	1.55	1.10	.50	1.00
....	.25	2.90	1.50	1.05	.35	1.00
....	.25	2.90	1.50	1.00	.45	1.00
....	.30	2.85	1.40	.90	.50	1.10
....	a.50	2.60	1.15	.90	.50	1.10
....	1.35	2.40	1.05	.75	.35	1.10
....	a1.00	2.20	1.00	.70	.30	1.10
....	1.15	2.05	1.10	.70	.30	1.10
....	.85	1.95	1.05	.80	.35	a1.25
....	.75	....	1.00	.75	....	1.20

a Rain.

Discharge measurements of Arkansas River near Coolidge, Kans., in 1903.

Date.	Hydrographer.	Gage height.	Discharge.
		Feet.	*Second-feet.*
May 4....................	A. Jacob	....	16
May 5....................	......do....................	0.20	10
May 6....................	......do....................	.71	70
May 18...................	......do....................	.40	11
June 2...................	......do....................	.60	32
June 3...................	......do....................	.50	29
June 9...................	......do....................	2.70	4,486
June 11	......do....................	4.40	27,361
June 26	......do....................	2.60	2,670
July 13..................	......do....................	1.80	375
July 25..................	......do....................	1.35	45
July 27..................	......do....................	1.15	7
August 11...............	......do....................	2.13	971
August 13...............	......do....................	1.78	262

Rating table for Arkansas River near Coolidge. Kans., from May 7 to June 7, 1903.

Gage height.	Discharge.	Gage height.	Discharge.	Gage height.	Discharge.	Gage height.	Discharge.
Feet.	*Second-feet.*	*Feet.*	*Second-feet.*	*Feet.*	*Second-feet.*	*Feet.*	*Second-feet.*
0.2	7	0.6	36	1.0	155	1.4	344
.3	9	.7	57	1.1	199		
.4	13	.8	83	1.2	245		
.5	21	.9	115	1.3	294		

Rating table for Arkansas River near Coolidge, Kans., from June 8 to October 31, 1903.

Gage height.	Discharge.	Gage height.	Discharge.	Gage height.	Discharge.	Gage height.	Discharge.
Feet.	*Second-feet.*	*Feet.*	*Second-feet.*	*Feet.*	*Second-feet.*	*Feet.*	*Second-feet.*
0.6	1	1.6	114	2.6	3,130	3.6	16,480
.7	2	1.7	220	2.7	4,310	3.7	17,840
.8	3	1.8	335	2.8	5,630	3.8	19,200
.9	4	1.9	490	2.9	6,970	3.9	20,560
1.0	5	2.0	680	3.0	8,320	4.0	21,920
1.1	7	2.1	900	3.1	9,680	4.1	23,280
1.2	13	2.2	1,170	3.2	11,040	4.2	24,640
1.3	29	2.3	1,470	3.3	12,400	4.3	26,000
1.4	55	2.4	1,830	3.4	13,760	4.4	27,360
1.5	92	2.5	2,330	3.5	15,120	4.5	28,720

Estimated monthly discharge of Arkansas River near Coolidge, Kans., for 1903.

[Drainage area, 24,800 square miles.]

Month.	Discharge in second-feet.			Total in acre-feet.	Run-off.	
	Maximum.	Minimum.	Mean.		Second-feet per square mile.	Depth in inches.
May [a]	319	8	57	2,826	0.0023	0.0022
June	28,720	17	6,608	393,203	.2700	.3000
July	1,320	5	211	12,974	.0086	.0099
August	1,470	1	122	7,501	.0050	.0058
September	2	0	Trace.			
October	21	0	4	246	.00016	.00018
The period				416,750		

[a] May 7 to 31, inclusive.

ARKANSAS RIVER NEAR GRANADA, COLO.

This station was established July 24, 1903, by A. Jacob. It is located at the highway bridge 2½ miles north of Granada, Colo., three-fourths of a mile below the head gate of Buffalo canal and 1 mile above the mouth of Buffalo Creek. The gage is painted in black on the lower cylinder of the third pier of the bridge from the left bank. The observer is Ben Riley, the gateman at the canal. Discharge measurements are made at high water from the wagon bridge and at low water by wading. The initial point for soundings is the left bridge abutment. The channel is straight for 2,000 feet above

d below the station. The velocity is fairly uniform. The right
nk is low, covered with grass, and liable to overflow. The left
nk is low and marshy and overflows at nearly all floods. The bed
the stream is sandy, free from vegetation, and shifting. At low
.ter there is one channel, at medium stages two channels, and at
;h water one channel. As the gage is painted on the bridge pier,
bench mark has been established.

Discharge measurements of Arkansas River near Granada, Colo., in 1903.

Date.	Hydrographer.	Gage height.	Discharge.
		Feet.	Second-feet.
1gust 10 a	A. Jacob	2.18	899
1gust 14	do	1.30	81
1gust 20	do	.90	b 5

For measurements prior to August 10 see list of miscellaneous measurements on page 306.
Estimated.

an daily gage height, in feet, of Arkansas River near Granada, Colo., for 1903.

Day.	Aug.	Sept.	Oct.	Day.	Aug.	Sept.	Oct.	Day.	Aug.	Sept.	Oct.
	0.70	0.50	0.50	12	2.00	0.50	0.50	23	0.80	0.50	0.50
	.70	.60	.50	13	1.50	.50	.50	24	.70	.50	.50
	.70	.60	.50	14	1.45	.50	.50	25	.70	.50	.50
	.70	.60	.50	15	1.35	.50	.50	26	.70	.50	.50
	.70	.60	.50	16	1.20	.50	.50	27	.70	.50	.50
	.70	.60	.50	17	1.20	.50	.50	28	.80	.50	.50
	.80	.60	.50	18	1.20	.50	.50	29	.70	.50	.50
	.80	.60	.50	19	1.00	.50	.50	30	.70	.50	.50
	1.80	.60	.50	20	.90	.50	.50	31	.70		.60
	2.20	.55	.50	21	.90	.50	.50				
	1.80	.50	.50	22	.80	.50	.50				

ting table for Arkansas River near Granada, Colo., from August 1 to October
31, 1903.

Gage height.	Discharge.	Gage height.	Discharge.	Gage height.	Discharge.	Gage height.	Discharge.
Feet.	Second-feet.	Feet.	Second-feet.	Feet.	Second-feet.	Feet.	Second-feet.
0.5	1	1.0	14	1.5	195	2.0	662
.6	2	1.1	28	1.6	267	2.1	792
.7	3	1.2	50	1.7	352	2.2	950
.8	5	1.3	83	1.8	448		
.9	7	1.4	133	1.9	548		

Estimated monthly discharge of Arkansas River near Granada, Colo., for 1903.

[Drainage area, 28,478 square miles.]

Month.	Discharge in second-feet.			Total in acre-feet.	Run-off.	
	Maximum.	Minimum.	Mean.		Second-feet per square mile.	Depth in inches.
August [a]	950	8	107.0	6,579	0.0046	0.0053
September	2	1	1.8	77	.00006	.00007
October	2	1	1.0	61	.00004	.00005
The period ...	....	....	....	6,717	....	....

[a] Records from April 26 to August 1 thrown out as unreliable.

ARKANSAS RIVER NEAR PROWERS, COLO.

This station was established May 22, 1903, by A. Jacob. It is located 1 mile east of Martin, Colo., and about 5 miles upstream from Prowers, Colo. The station is 100 yards north of the Atchison Topeka and Santa Fe Railway, and just below the mouth of Mud Creek. The gage is a hard pine board nailed to the first abutment pile above the dam at the right bank. It is read twice each day by William Hustan, the gate keeper. At low water discharge measurements are made by wading. At high water measurements are made either from the Prowers bridge, 5½ miles downstream, or from the Caddoa bridge, 6 miles upstream. At the Prowers bridge the initial point for soundings is the left pier. At the Caddoa bridge the initial point for soundings is the south pier. The channel is straight for 50 feet above and 300 feet below the station. At the dam the right bank is high, wooded, and not liable to overflow. The left bank is high, sodded, and liable to overflow at high water. The cross section of the stream at the gage is the crest of the dam.

The bench mark is a nail in the oak pile to which the gage attached. It is at the same elevation as the 6-foot mark on the gage.

The Keesee ditch and the Colorado and Kansas canal are taken out just above the dam. Their discharge must be added to that at the dam to obtain the total for the river at this point.

The cap which forms the crest of the dams was washed away by floods in June and August, 1903. Two rating tables are necessary on this account—one for the periods while the cap was in place and one for the periods when it was washed away.

Discharge measurements of Arkansas River near Prowers, Colo., in 1903.

Date.	Hydrographer.	Gage height.	Discharge.
		Feet.	*Second-feet.*
June 30	A. Jacob	0.80	1,819
July 10	....do	.20	140
July 15	....do	.55	494
July 23	....do	.40	248
July 30	....do	.24	106
August 7	....do	.23	70
August 9	....do	1.40	3,988
August 15	....do	.10	111
August 19	....do	.35	333

Mean daily gage height, in feet, of Arkansas River near Prowers, Colo., for 1903.

Day.	June.	July.	Aug.	Nov.	Dec.	Day.	June.	July.	Aug.	Nov.	Dec.
1	(a)	0.90	0.30	......	0.10	17	2.65	0.40	0.10	(a)	......
2	(a)	.70	.25	......	.10	18	1.90	.30	.43	(a)	......
3	(a)	.50	.10	......	.10	19	1.70	.30	.35	(a)	......
4	(a)	.45	.00	......	.10	20	2.55	.90	.41	0.10	......
5	0.15	.35	.00	......	.10	21	1.85	.55	.40	.10	......
6	2.10	.30	.00	......	.20	22	1.30	.40	.29	.10	......
7	3.10	.25	.20	......	.30	23	1.10	.35	.22	.10	......
8	1.30	.20	.40	......	.20	24	.90	.30	.16	.10	......
9	1.35	.20	1.10	......	.10	25	.85	.30	.10	.20	......
10	1.65	.10	.00	......	.10	26	.75	.40	.13	.20	......
11	2.60	.10	.65	......	.10	27	.00	.40	(a)	.20	......
12	4.80	.05	.48	......	.10	28	.80	.30	(a)	.20	......
13	2.60	.20	.30	......	.10	29	.80	.20	(a)	.20	......
14	1.25	.40	.20	......	.20	30	.80	.20	(a)	.20	......
15	1.50	.50	.10	......	.30	31	......	.20	......	......	......
16	1.45	.50	.10	......	......						

a No water going over dam.

NOTE.—During September and October and the fore part of November there may have been a small amount of water passing over the dam, but there is no record of such flow. If there was any it was very slight.

Rating table for Arkansas River near Prowers, Colo., from June 5 to July from August 9 to August 26, 1903.

Gage height.	Discharge.	Gage height.	Discharge.	Gage height.	Discharge.	Gage height.	Disch
Feet.	*Second-feet.*	*Feet.*	*Second-feet.*	*Feet.*	*Second-feet.*	*Feet.*	*Secon*
0.0	65	1.0	2,550	2.0	6,450	3.0	10,
.1	95	1.1	2,940	2.1	6,840	3.2	11.
.2	145	1.2	3,330	2.2	7,230	3.4	11,
.3	260	1.3	3,720	2.3	7,620	3.6	12,
.4	425	1.4	4,110	2.4	8,010	3.8	13,
.5	655	1.5	4,500	2.5	8,400	4.0	14,
.6	990	1.6	4,890	2.6	8,790	4.2	15,
.7	1,380	1.7	5,280	2.7	9,180	4.4	15,
.8	1,770	1.8	5,670	2.8	9,570	4.6	16,
.9	2,160	1.9	6,060	2.9	9,960	4.8	17,

Rating table for Arkansas River near Prowers, Colo., from July 13 to A .and from November 20 to December 31, 1903.

Gage height.	Discharge.	Gage height.	Discharge.	Gage height.	Discharge.	Gage height.	Disch
Feet.	*Second-feet.*	*Feet.*	*Second-feet.*	*Feet.*	*Second-feet.*	*Feet.*	*Secon*
0.0	5	0.3	140	0.6	690	0.9	1,
.1	25	.4	250	.7	1,005		
.2	65	.5	435	.8	1,380		

Estimated monthly discharge of Arkansas River at Colorado and Kansa near Prowers, Colo., for 1903.

[Drainage area, 19,125 square miles.]

Month.	Discharge in second-feet.			Total in acre-feet.[a]	Run-off	
	Maximum.[a]	Minimum.[a]	Mean.[a]		Second-feet per square mile.	D
June (26 days)	17,446	131.5	4,296	255,630	0.2246	
July	2,202	90.5	443	27,239	.0232	
August	3,003	32	346	21,275	.0181	
September	29.5	26	26	1,547	.0014	
October	23	23	23	1,414	.0012	
November	86.2	46.2	40	2,380	.0021	
December 1-15	148	30.5	60	1,785	.0031	
The period				311,270		

[a] Maximum and minimum include water in Keesee ditch and Colorado and Kansas c

Note. —From August 27 to November 19 there are no definite records for the river, but ing to information from the State engineer's office there was little or no flow over the d ing September, and during October and November some water passed over the dam, much is not known.

Discharge measurements of Colorado and Kansas canal near Prowers, Colo., in 1903.

Day.	Hydrographer.	Gage height.	Discharge.
		Feet.	*Second-feet.*
April 23	A. Jacob	1.07	28
May 1	do	1.32	37
May 9	do	1.04	21
May 20	do	1.42	31
Do	do	1.11	21
Do	do	1.64	41
Do	do	.53	5
June 5	do	.94	21
June 23	do	1.70	49
June 30	do	1.85	54
July 10	do	1.05	22
July 23	do	1.33	38
July 30	do	.85	18
August 7	do	1.00	23
August 15	do	1.15	29

Mean daily gage height, in feet, of Colorado and Kansas canal at Martin, near Prowers, Colo., for 1903.

Day.	June.	July.	Aug.	Sept.	Oct.	Nov.	Dec.
1	0.80	1.20	1.10	1.10	1.00	0.95	0.90
2	.80	2.00	1.25	1.10	1.00	.95	.90
3	.80	2.10	1.10	1.10	1.00	.95	.90
4	.80	2.00	1.00	1.10	1.00	.95	.90
5	.80	2.00	1.00	1.10	1.00	1.00	.80
6	.90	1.80	1.00	1.10	1.00	1.00	.80
7	1.10	1.35	1.10	1.10	1.00	1.00	.50
8	2.10	.95	1.10	1.10	1.00	1.00	.50
9	2.85	.90	1.80	1.10	1.00	1.00	.50
10	2.90	.90	2.15	1.10	1.00	1.00	.50
11	3.10	.90	1.00	1.10	1.00	1.10	.50
12	2.50	1.55	1.50	1.10	1.00	1.10	.50
13	2.50	2.20	1.30	1.10	1.00	1.10	.40
14	3.10	1.90	1.10	1.10	1.00	1.10	.40
15	2.20	2.20	1.10	1.10	1.00	1.10	.40
16	2.30	2.20	1.10	1.10	1.00	1.10	(a)
17	1.95	2.20	1.10	1.00	1.00	1.00	
18	2.10	1.60	1.10	1.00	1.00	1.00	
19	2.20	1.10	1.40	1.00	1.00	1.00	
20	2.10	2.50	1.70	1.00	1.00	1.00	
21	2.10	2.45	1.75	1.00	1.00	1.00	
22	1.90	2.00	1.10	1.00	1.00	1.00	

a Canal dry from December 16 to 31, inclusive.

Mean daily gage height, in feet, of Colorado and Kansas canal, etc.—Continued.

Day.	June.	July.	Aug.	Sept.	Oct.	Nov.	Dec.
23	1.90	1.55	1.10	1.00	1.00	0.95	....
24	1.70	.90	1.10	1.00	1.00	.95	....
25	1.70	1.15	1.10	1.00	1.00	.95	....
26	1.70	1.40	1.10	1.00	1.00	.95	....
27	1.70	1.40	1.10	1.00	1.00	.95	....
28	2.10	.90	1.10	1.00	1.00	.90	....
29	2.10	.80	1.10	1.00	1.00	.90	....
30	2.10	.80	1.10	1.00	1.00	.90	....
31		.80	1.10	....	1.00	....	....

Rating table for Colorado and Kansas canal near Prowers, Colo., from June 1 to December 31, 1903.

Gage height.	Discharge.	Gage height.	Discharge.	Gage height.	Discharge.	Gage height.	Discharge.
Feet.	Second-feet.	Feet.	Second-feet.	Feet.	Second-feet.	Feet.	Second-feet.
0.4	5.5	1.1	26.5	1.8	51.0	2.5	76.0
.5	8.0	1.2	30.0	1.9	54.5	2.6	80.0
.6	10.5	1.3	33.5	2.0	58.0	2.7	84.0
.7	13.5	1.4	37.0	2.1	61.5	2.8	88.0
.8	16.5	1.5	40.5	2.2	65.0	2.9	92.0
.9	19.5	1.6	44.0	2.3	68.5	3.0	96.0
1.0	23.0	1.7	47.5	2.4	72.0	3.1	100.0

Estimated monthly discharge of Colorado and Kansas canal near Prowers, Colo. in 1903.

Month.	Discharge in second-feet.			Total in acre-feet
	Maximum.	Minimum.	Mean.	
June	100.0	16.5	54.9	3,2?
July	76.0	16.5	41.9	2,5?
August	63.2	23.0	30.6	1,8?
September	26.5	23.0	24.9	1,4?
October	23.0	23.0	23.0	1,4?
November	26.5	19.5	22.8	1,3?
December 1-15	19.5	5.5	11.7	3?
The period				12,3?

Discharge measurements of Keesee ditch near Prowers, Colo., in 1903.

Date.	Hydrographer.	Gage height.	Discharge.
		Feet.	*Second-feet.*
ᵣil 23	A. Jacob		8.0
ᵧ 1	do		8.9
ᵧ 9	do	1.32	8.2
ᵧ 20	do	1.35	7.4
Do	do	1.74	13.7
ne 5	do	.75	9.0
ne 23	do	(ᵃ)	(ᵃ)
ne 30	do	(ᵃ)	(ᵃ)
ly 10	do	1.61	9.0
ily 23	do	1.75	13.0
ily 30	do	1.75	12.0
ngust 7	do	.82	13.0
ngust 15	do	1.80	12.0

ᵃ Not drawing.

Mean daily gage height, in feet, of Keesee ditch near Prowers, Colo., for 1903.

Day.	July.	Aug.	Sept.	Day.	July.	Aug.	Sept.	Day.	July	Aug.	Sept.
....	1.75	1.75	1.00	12	1.75	1.75	1.00	23	1.75	1.75	.....
....	1.75	1.75	1.00	13	1.75	1.75	1.00	24	1.80	1.75	.....
....	1.75	1.75	1.00	14	1.75	1.75	1.00	25	1.85	1.75	.....
....	1.75	1.50	1.00	15	1.75	1.75	1.00	26	1.55	1.75	.....
....	1.75	1.50	1.00	16	1.75	1.75	.....	27	1.75	1.75	.....
....	1.75	1.50	1.00	17	1.75	1.75	.....	28	1.75	1.75	.....
....	1.75	1.50	1.00	18	1.75	1.75	.....	29	1.63	1.75	.....
....	1.75	1.75	1.00	19	1.75	1.75	.....	30	1.50	1.75	.....
....	1.75	1.75	1.00	20	1.75	1.75	.....	31	1.75	1.75	.....
....	1.75	1.75	1.00	21	1.75	1.75	.....				
....	1.75	1.75	1.00	22	1.75	1.75	.....				

ating table for Keesee ditch near Prowers, Colo., from July 1 to September 15, 1903.

Gage height.	Discharge.	Gage height.	Discharge.	Gage height.	Discharge.	Gage height.	Discharge.
Feet.	*Second-feet.*	*Feet.*	*Second-feet.*	*Feet.*	*Second-feet.*	*Feet.*	*Second-feet.*
1.0	3.0	1.2	5.4	1.4	7.8	1.6	10.2
1.1	4.2	1.3	6.6	1.5	9.0	1.7	11.4

Estimated monthly discharge of Keesee ditch near Prowers, Colo., for 1903.

Month.	Discharge in second-feet.			Total in acre-feet.
	Maximum.	Minimum.	Mean.	
July	12.6	7.2	11.6	71?
August	12.0	9.0	11.6	71?
September 1–15....................	3.0	3.0	3.0	8?
The period	....	....	....	1,51?

ARKANSAS RIVER NEAR LA JUNTA, COLO.

This station was established April 7, 1903, by M. C. Hinderlider
It is located 1 mile east of La Junta, Colo., 200 feet east of the statio
signal on the Atchison, Topeka and Santa Fe Railway and one-fourt
mile west of the stock yards. The gage is read twice each day t
J. M. Hoskins. It is an inclined 4 by 6 inch pine timber 18 feet lon,
bolted at the lower end to the shale bed, and with its upper end burie
in the shale and clay bank. It reads from 0.2 to 6.5 feet. Dischar;
measurements are made at high water from the downstream side
the highway bridge one-fourth mile upstream. At low water mea
urements are made by wading. The initial point for soundings is t
south end of the wagon bridge. The channel is straight for 50 fe
above the gage and for 600 feet below. The current is moderate a
there is but one channel at all stages. The right bank is hi;
and steep and is not liable to overflow. The left bank is a grav
and sand bar sloping gently, over which the water rises for about 3
feet. The bed of the stream is composed of shale and shifting san
At low water the measurements are made by wading at the gage ro
and a good cross section of smooth shale can be obtained, givin
accurate results. At high water, when measurements are made fro
the bridge, the current is broken by old piles under the bridge, whic
decreases the accuracy of the measurements at these stages. Th
bench mark is a 2-inch iron pipe 4 feet long driven 3 feet into th
shale and set in cement. The top is painted white. The elevatio
of the top is 2.37 feet above the zero of the gage.

Discharge measurements of Arkansas River near La Junta, Colo., in 1903.

Date.	Hydrographer.	Gage height.	Discharge
		Feet.	Second-fee
April 7	M. C. Hinderlider	0.40	
April 25........................	A. Jacob	.80	?
April 30....	....do	.90	?
May 12........................	....do	.75	?
May 13....	....do	.50	
May 23........................	....do	.70	

Discharge measurements of Arkansas River near La Junta, Colo., in 1903—Cont'd.

Date.	Hydrographer.	Gage height.	Discharge.
		Feet.	*Second-feet.*
May 28	A. Jacob	1.05	86
June 6	...do	4.60	5,091
June 18	...do	4.85	4,575
July 2	...do	.40	51
July 9	...do	.80	28
July 17	...do	1.15	259
July 22	...do	.45	40
August 1	...do	.78	87
August 6	...do	.43	22
August 8	...do	1.71	610
October 17	E. C. Murphy	.83	9

Mean daily gage height, in feet, of Arkansas River near La Junta, Colo., for 1903.

Day.	Apr.	May.	June.	July.	Aug.	Sept.	Oct.
1		0.85	1.15	1.20	0.80	0.75	0.40
2		.90	1.05	.45	.55	.70	.45
3		.95	1.10	.30	.55	.70	.45
4		.90	1.45	.30	.50	.75	.50
5	0.40	.85	1.65	.40	.40	.70	.50
6	.40	.80	a 2.05	.55	.45	.70	.40
7	.40	(b)	4.50	.60	1.45	.70	.40
8	.40	(b)	4.55	.45	2.00	.70	.60
9	.30	(b)	2.50	.40	1.40	.70	.45
10	.40	(b)	c 5.75	.35	.85	.70	.40
11	.35	(b)	c 7.00	.95	.55	.50	.40
12	.40	(b)	c 6.50	1.30	.55	.40	.40
13	.40	.50	c 6.00	1.80	.50	.40	.40
14	.40	.95	c 5.40	1.85	.50	.55	.40
15	.40	.80	c 5.00	1.85	.40	.75	.40
16	.40	1.20	c 4.70	1.35	.40	.70	.40
17	.40	1.05	c 4.50	1.20	.60	.70	.40
18	.30	.90	c 4.35	1.40	1.10	.80	.40
19	.70	1.00	c 4.25	2.20	1.50	.75	.45
20	.55	1.25	3.90	1.05	1.20	.60	.35
21	.45	.75	3.60	.60	.90	.50	.30
22	.50	.80	3.50	.50	.50	.40	.30
23	.50	.70	2.85	.45	.45	.40	.30
24	.50	.75	2.65	1.45	.40	.40	.30
25	.70	1.10	2.25	1.20	.45	.40	.30
26	.75	1.25	1.80	1.35	.75	.40	.30
27	.90	1.05	1.65	1.25	.85	.40	.40
28	.90	1.10	2.00	1.10	.80	.40	.40
29	.80	1.15	2.55	1.35	.50	.40	.40
30	.85	1.10	3.15	1.50	.45	.40	.50
31		1.05	....	1.20	.90	....	.75

a Rain.

b Observer changed. No records.

c Records for June 10 to 19, both inclusive, are hypothetical and were sent in by hydrographer, being taken from drift.

Rating table for Arkansas River near La Junta, Colo., from April 5 to June 5, 1903.

Gage height.	Discharge.	Gage height.	Discharge.	Gage height.	Discharge.	Gage height.	Discharge.
Feet.	*Second-feet.*	*Feet.*	*Second-feet.*	*Feet.*	*Second-feet.*	*Feet.*	*Second-feet.*
0.3	3	0.7	26	1.1	96	1.5	176
.4	7	.8	38	1.2	116		
.5	12	.9	56	1.3	136		
.6	18	1.0	76	1.4	156		

Three rating curves owing to change in channel at high and low water.

Rating table for Arkansas River near La Junta, Colo., from June 6 to July 17, 1903.

Gage height.	Discharge.	Gage height.	Discharge.	Gage height.	Discharge.	Gage height.	Discharge.
Feet.	*Second-feet.*	*Feet.*	*Second-feet.*	*Feet.*	*Second-feet.*	*Feet.*	*Second-feet.*
0.3	32	1.3	340	2.6	1,495	4.6	5,125
.4	43	1.4	400	2.8	1,745	4.8	5,555
.5	57	1.5	460	3.0	2,020	5.0	5,985
.6	73	1.6	530	3.2	2,310	5.2	6,415
.7	94	1.7	600	3.4	2,635	5.4	6,845
.8	120	1.8	680	3.6	3,010	5.6	7,275
.9	152	1.9	765	3.8	3,420	5.8	7,705
1.0	191	2.0	855	4.0	3,840	6.0	8,135
1.1	234	2.2	1,045	4.2	4,265	6.5	9,210
1.2	285	2.4	1,260	4.4	4,695	7.0	10,285

Rating table for Arkansas River near La Junta, Colo., from July 18 to October 31, 1903.

Gage height.	Discharge.	Gage height.	Discharge.	Gage height.	Discharge.	Gage height.	Discharge.
Feet.	*Second-feet.*	*Feet.*	*Second-feet.*	*Feet.*	*Second-feet.*	*Feet.*	*Second-feet.*
0.3	6	0.8	90	1.3	352	1.8	667
.4	18	.9	124	1.4	415	1.9	730
.5	32	1.0	173	1.5	478	2.0	793
.6	48	1.1	229	1.6	541	2.1	856
.7	66	1.2	289	1.7	604	2.2	919

Estimated monthly discharge of Arkansas River near La Junta, Colo., for 1903.

[Drainage area, 12,200 square miles.]

Month.	Discharge in second-feet.			Total in acre-feet.	Run-off.	
	Maximum.	Minimum.	Mean.		Second-feet per square mile.	Depth in inches.
April 5–30	56	3	17	877	0.0014	0.0014
May 1–6 and 13–31 *a*.	126	12	68	3,372	.0056	.0052
June	10,285	86	3,328	198,030	.2728	.3043
July	919	32	250	15,372	.0205	.0236
August	793	18	126	7,747	.0103	.0119
September	90	18	47	2,797	.0039	.0044
October	78	6	21	1,291	.0017	.0020
The period ...	....	....	....	229,486	....	....

a May 7–12, no record.

ARKANSAS RIVER NEAR ROCKY FORD, COLO.

The old station was located 2 miles northeast of Rocky Ford, Colo., and was established May 3, 1897. The gage consisted of a vertical 1 by 3 inch timber notched in tenths and securely nailed to the pile protection to the abutment of the wagon bridge at the left side of the stream, upper side of the bridge. The initial point for soundings was on the left bank at the water's edge. Both banks are high and liable to overflow only at very high water. The channel is straight for about 300 feet above and below the station. The bed is sandy and shifting. This station was abandoned April 7, 1900.

The new station was established by R. W. Hawley April 19, 1901. It is located 2 miles west of the wagon bridge at Rocky Ford, 4½ miles northwest of the Rocky Ford station on the Atchison, Topeka and Santa Fe Railway, and one-fourth mile below the old ford. The gage is an inclined 4 by 6 inch pine timber on the left bank at the end of the cable. It is fastened to bed rock by bolts 2 feet long, which are cemented into the rock. During the high water in June, 1903, the upper section of the gage rod was carried away. The rest of the readings during that flood were read on a temporary rod. They were corrected to reduce them to the datum of the previous gage. After the flood had subsided a new gage was established on the old gage datum. The gage is read twice each day by L. Enyart. Discharge measurements are made by means of a cable and car laocted at the gage. At low water measurements may be made by wading. The initial point for soundings is the crest of the shale on the north bank of the river at the top of the gage. The channel is straight for 1,000 feet above and 800 feet below the station. It has a width of 260 feet at ordinary high water and 25 to 35 feet at low water. The velocity is well distributed and never sluggish. The right bank overflows at a gage height of 7 feet. The left bank is composed of shale to a height of 9 feet, above which there is a clay bank extending to a height of 30 feet above the river bed. It will never overflow. The bed of the stream is composed at times of smooth, level, slaty shale. More frequently, however, there is a sand bar in the middle of the stream, forming two channels. The bench mark is a nail driven at the edge of a notch cut on the north side of a cottonwood tree on the right bank of the river. It is about 1½ feet above the ground and is 7.52 feet above the zero of the gage. This station is of especial importance owing to the fact that it is one of the few places on Arkansas River where the channel is at all permanent.

Discharge measurements of Arkansas River near Rocky Ford, Colo., in 1903.

Date.	Hydrographer.	Gage height.	Discharge.
		Feet.	*Second-feet.*
April 6	M. C. Hinderlider	0.28	108
April 27	A. Jacob	− .10	14
May 12	do	− .10	14
May 25	do	.70	328
May 27	do	.55	249
June 8	do	2.55	2,412
June 16	do	8.80	6,712
July 6	do	1.40	555
July 8	do	1.10	341
July 18	do	2.00	954
July 19	do	2.88	2,998
July 21	do	1.38	496
August 2	do	.70	129
August 5	do	.43	61
August 8	do	1.50	578
August 16	do	1.31	439
October 16	E. C. Murphy	.49	110

Mean daily gage height, in feet, of Arkansas River near Rocky Ford, Colo., for 1903.

Day.	Apr.	May.	June.	July.	Aug.	Sept.	Oct.
1		-0.10	0.45	2.10	0.60	0.40	0.45
2		- .10	.75	1.90	.70	.45	.55
3		.10	.83	1.40	.58	.60	.45
4		.00	1.00	1.25	.50	.58	.30
5		- .10	1.50	1.65	.45	.45	.40
6	0.20	- .10	a2.50	1.40	.40	.38	.33
7	.25	- .10	3.35	1.20	2.60	.40	.30
8	.23	- .10	2.60	1.10	1.65	.35	.33
9	.17	- .10	2.85	.75	2.65	.33	.38
10	.18	- .10	6.50	1.10	1.20	.45	.38
11	.20	- .10	3.60	1.50	.55	.80	.38
12	.20	- .10	2.80	2.10	.40	.75	.38
13	.18	- .10	2.60	2.05	.30	.53	.33
14	.18	- .10	2.75	2.00	.80	.60	.40
15	.18	.30	2.85	1.55	.80	.65	.45
16	.13	.80	3.70	1.60	1.85	.60	.50
17	.15	.38	3.90	1.65	1.65	.80	.60
18	.10	.60	3.50	2.00	1.65	.78	.55
19	.10	.60	3.40	2.75	1.05	.60	.60
20	.13	.35	3.30	1.45	1.05	.53	.60
21	.13	.30	3.20	1.30	.45	.53	.50
22	.10	.80	3.10	1.65	.30	.43	.50
23	.10	.30	2.90	1.65	.60	.38	.55
24	.10	.53	2.80	1.80	1.15	.40	.60
25	.10	.70	2.60	1.60	1.08	.55	.60
26	.10	.55	2.50	1.30	.65	.45	. .60
27	.00	.55	2.40	1.05	1.10	.48	.60
28	- .10	.50	2.25	.85	.65	.63	.60
29	- .10	.40	2.35	1.73	.40	.55	.60
30	- .10	.55	2.35	1.15	.85	.48	.65
31		.20		.95	.65		.65

a Rain.

Rating table for Arkansas River near Rocky Ford, Colo., from April 6 to June 10, 1903.

Gage height.	Discharge.	Gage height.	Discharge.	Gage height.	Discharge.	Gage height.	Discharge.
Feet.	Second-feet.	Feet.	Second-feet.	Feet.	Second-feet.	Feet.	Second-feet.
-0.1	14	0.9	458	1.9	1,370	2.9	3,220
.0	38	1.0	526	2.0	1,490	3.0	3,510
.1	69	1.1	600	2.1	1,620	3.1	3,820
.2	102	1.2	680	2.2	1,770	3.2	4,150
.3	137	1.3	760	2.3	1,940	3.3	4,510
.4	176	1.4	840	2.4	2,120	3.4	4,900
.5	222	1.5	930	2.5	2,310	3.8	6,710
.6	273	1.6	1,030	2.6	2,510	6.5	a 19,400
.7	328	1.7	1,140	2.7	2,720		
.8	392	1.8	1,250	2.8	2,950		

a Estimated.

Two rating tables owing to change of river bed.

Rating table for Arkansas River near Rocky Ford, Colo., from June 11 to October 31, 1903.

Gage height.	Discharge.	Gage height.	Discharge.	Gage height.	Discharge.	Gage height.	Discharge.
Feet.	*Second-feet.*	*Feet.*	*Second-feet.*	*Feet.*	*Second-feet.*	*Feet.*	*Second-feet.*
0.0	10	1.0	256	2.0	1,060	3.0	3,390
.1	19	1.1	310	2.1	1,190	3.1	3,730
.2	30	1.2	370	2.2	1,350	3.2	4,080
.3	44	1.3	435	2.3	1,540	3.3	4,440
.4	62	1.4	504	2.4	1,750	3.4	4,870
.5	84	1.5	579	2.5	1,980	3.5	5,310
.6	110	1.6	662	2.6	2,220	3.6	5,770
.7	140	1.7	752	2.7	2,480	3.7	6,240
.8	172	1.8	848	2.8	2,760	3.8	6,710
.9	210	1.9	950	2.9	3,060	3.9	7,180

Curve extended below 0.43 gage height on this table.

Estimated monthly discharge of Arkansas River near Rocky Ford, Colo., for 1903.

[Drainage area, 11,440 square miles.]

Month.	Discharge in second-feet.			Total in acre-feet.	Run-off.	
	Maximum.	Minimum.	Mean.		Second-feet per square mile.	Depth in inches.
April 6–30	119	14	76	3,769	0.007	0.007
May	328	14	117	7,194	.010	.012
June	19,400	198	3,436	204,456	.300	.335
July	2,620	156	684	42,057	.060	.069
August	2,350	44	356	21,890	.031	.036
September	172	49	95	5,653	.008	.009
October	125	44	84	5,165	.007	.008
The period	19,400	14				

ARKANSAS RIVER NEAR NEPESTA, COLO.

The original station, established September 8, 1897, is located 1,000 feet north of Nepesta, Colo., at a wagon bridge, 200 feet below the Atchison, Topeka and Santa Fe Railroad. The gage consists of a vertical timber graduated to feet and tenths, securely fastened to the upstream cylinder of the bridge, on the left side of the river. The channel above and below the station is straight for several hundred

eet, while the bed is sandy and shifting, and the results, therefore, are
lot altogether satisfactory for the purpose of making a rating table.
This station was maintained by the Great Plains Water Company until
December, 1900. May 1, 1901, another station was established by
A. L. Fellows, assisted by C. W. Beach, at the dam and head gate of
he Oxford Farmers' canal, 1½ miles west of Nepesta. The station
onsists of the dam crossing the river at the head gate, forming a weir,
nd a gage rod, consisting of a 2 by 6 inch timber, 8 feet long, fastened
o an oak pile at the south end of the dam; the location being marked
y spikes driven into the pile at each foot mark. The rod is gradu-
ted to feet and tenths vertically. The observer is Z. Swallow, head-
ate keeper of the Oxford Farmers' canal.

At low stages the stream may be measured by wading, but at high
tages it is necessary to use the wagon bridge at Nepesta, making the
ecessary allowances for inflow between the two points; this can
eadily be done, as the distance is only about a mile. Measurements
nade at the highway bridge are not as accurate as those made by
rading, on account of the drift, which collects in front of the pile
pproaches of the bridge, and on account of the interference of the
ridge supports with the current. The initial point for soundings is
he edge of the stone abutment at the east end of the bridge. The
hannel is straight for 300 feet above and 1,000 feet below the station.
The water has a moderate velocity at low stages and becomes rapid at
lood stages. At the gage both banks are the wings of the dam. The
iver overflows just below the dam. The bed of the stream is sandy
nd shifting, except at the dam. The cross section is fairly stable at
he bridge below. No bench mark has been established, as the gage
s spiked firmly to the oak pile of the wing of the dam and is well pro-
ected. The station is of particular value, as it is near the head of
rrigation District No. 17, one of the most important on Arkansas
River.

In computing the discharge of Arkansas River at this point, the
lischarge of the Oxford Farmers' canal is added, as its head gate is
ust above the station.

This station was abandoned October 31, 1903, at least temporarily,
ufficient data on the flow of the river at this point having been
btained.

Discharge measurements of Arkansas River near Nepesta, Colo., in 1903.

Date.	Hydrographer.	Gage height.	Discharge.
		Feet.	*Second-feet.*
April 4	M. C. Hinderlider	0.85	148
Do	do	.85	166
May 26	A. Jacob	.40	358
June 15	do	2.50	6,002
July 7	C. W. Beach	.85	615
Do	A. Jacob	.85	536
July 20	do	.78	540
August 3	do	.62	152
October 15	E. C. Murphy	.55	150
Do	M. C. Hinderlider	.55	144

Mean daily gage height, in feet, of Arkansas River near Nepesta, Colo., for 1903.

Day.	Apr.	May.	June.	July.	Aug.	Sept.	Oct.
1	0.70	0.76	0.68	1.42	0.75	0.58	0.45
2	.75	.82	.93	1.30	.65	.53	.43
3	.73	.76	1.10	1.15	.60	.43	.43
4	.85	.74	1.30	1.30	.55	.43	.43
5	.81	.73	1.33	1.20	.48	.38	.45
6	.80	.70	1.95	1.00	2.00	.40	.45
7	.79	.71	2.00	.75	1.55	.50	.45
8	.75	.70	1.38	.64	1.00	.48	.50
9	.81	.76	a 3.25	1.10	1.30	.63	.50
10	.86	.70	5.25	1.04	.50	.68	.50
11	.83	.71	2.75	1.20	.55	.65	.50
12	.79	.70	2.35	1.28	.60	.53	.55
13	.70	.75	2.08	1.35	.60	.63	.53
14	.82	.89	a 2.65	1.25	.58	.58	.55
15	.77	.84	2.55	1.08	.95	.58	.55
16	.78	.91	2.90	.98	1.20	.68	.55
17	.74	.97	2.80	.96	1.15	.63	.55
18	.70	.96	2.50	1.20	.88	.63	.60
19	.76	.94	2.35	1.35	.83	.58	.54
20	.79	.86	2.28	.85	.40	.58	.58
21	.71	.88	2.20	.95	.45	.58	.58
22	.70	.68	2.05	.95	.48	.58	.55
23	.69	.65	1.95	1.20	.83	.53	.55
24	.67	b .60	1.75	.98	.80	.48	.55
25	.65	.47	1.63	.98	.75	.50	.60
26	.65	.54	1.54	.93	.98	.50	.60
27	.68	.44	1.40	.88	.68	.48	.60
28	.70	.51	1.41	1.10	.53	.48	.58
29	.69	.65	1.58	.90	.48	.45	.55
30	.70	.58	1.55	.88	.35	.43	.55
31		.58		.78	.48		.60

a Heavy rain.   b See appended note.

*Rating table for **Arkansas River** near Nepesta, Colo., from January 1 to May 23, 1903.*

Gage height.	Discharge.	Gage height.	Discharge.	Gage height.	Discharge.
Feet.	*Second-feet.*	*Feet.*	*Second-feet.*	*Feet.*	*Second-feet.*
0.60	39	0.75	84	0.90	208
.65	52	.80	108	.95	292
.70	66	.85	148	1.00	410

Curve poorly defined.

*Rating table for **Arkansas River** near Nepesta. Colo., from May 24 to December 31, 1903.*

Gage height.	Discharge.	Gage height.	Discharge.	Gage height.	Discharge.	Gage height.	Discharge.
Feet.	*Second-feet.*	*Feet.*	*Second-feet.*	*Feet.*	*Second-feet.*	*Feet.*	*Second-feet.*
0.3	27	1.3	2,040	2.3	5,340	3.6	9,630
.4	51	1.4	2,370	2.4	5,670	3.8	10,290
.5	92	1.5	2,700	2.5	6,000	4.0	10,950
.6	158	1.6	3,030	2.6	6,330	4.2	11,610
.7	290	1.7	3,360	2.7	6,660	4.4	12,270
.8	490	1.8	3,690	2.8	6,990	4.6	12,930
.9	740	1.9	4,020	2.9	7,320	4.8	13,590
1.0	1,050	2.0	4,350	3.0	7,650	5.0	14,250
1.1	1,380	2.1	4,680	3.2	8,310	5.2	14,910
1.2	1,710	2.2	5,010	3.4	8,970		

Curve very uncertain above 0.90 foot gage height, depending upon one high-water measurement.

Estimated monthly discharge of Arkansas River near Nepesta, Colo., for 1903.

[Drainage area, 9,130 square miles.]

Month.	Discharge in second-feet.			Run-off.		
	Maximum.	Minimum.	Mean.	Total in acre-feet.	Second-feet per square mile.	Depth in inches.
April a	160	52	120	7,140	0.013	0.015
May a	339	52	141	8,670	.015	.017
June a	15,075	259	4,603	273,898	.504	.562
July a	2,436	201	1,390	85,468	.152	.175
August a	4,350	38	726	44,640	.080	.092
September a	259	46	135	8,033	.015	.017
October a	158	61	120	7,379	.013	.015
The period ...	15,075	38				

a Water in Oxford Farmers' canal included in all means and computations of run-offs.

Discharge measurements of Oxford Farmers' canal at Nepesta, Colo., in 1903.

Date.	Hydrographer.	Gage height.	Discharge.
		Feet.	*Second-feet*
April 4	M. C. Hinderlider	1.10	?
Do	....do	2.00	7
Do	....do	2.50	10
June 15	A. Jacob	1.90	?
July 7	....do	3.20	1?
July 20	....do	3.05	1
August 3	....do	.75	

Mean daily gage height, in feet, of Oxford Farmers' canal at Nepesta, Col
for 1903.

Day.	Apr.	May.	June.	July.	Aug.	Sept.	Oc
1	1.70	0.75	0.75	3.15	0.85	0.70	?
2	2.00	.75	.73	3.15	.70	.70	
3	1.50	.75	.75	3.15	.70	.00	
4	2.30	.75	.75	2.90	.70	.70	
5	2.00	.75	2.90	3.15	.70	.70	
6	2.50	.75	2.90	3.15	1.85	.70	
7	2.70	.75	2.90	3.15	3.00	.70	
8	2.70	.75	2.90	3.15	3.08	.70	
9	.75	.75	2.90	3.15	3.15	.70	
10	.75	.75	2.90	3.15	3.15	.70	
11	.75	.75	2.90	3.15	3.00	.70	
12	.75	.73	2.90	3.15	3.00	.70	
13	.75	.75	2.90	2.55	3.00	.70	
14	.75	.00	2.90	3.00	1.95	.70	
15	.75	.00	2.90	3.00	1.98	.70	
16	.75	.00	2.90	3.08	2.05	.70	
17	.75	.73	2.90	3.00	3.15	.68	
18	.75	.75	2.90	3.00	3.15	.70	
19	.75	.75	2.90	3.05	2.15	.70	
20	.75	.78	2.90	3.00	2.85	.68	
21	.75	.75	2.90	3.13	2.90	.70	
22	.75	.75	2.90	3.15	b 1.45	.70	
23	.75	.73	2.90	3.15	.35	.70	
24	.75	.75	2.90	3.05	.70	.70	
25	.75	.78	2.90	3.00	.70	.70	
26	.75	.78	2.90	3.00	.70	.70	
27	.75	.75	2.90	3.00	.70	.70	
28	.75	.75	2.90	b .43	.70	.70	
29	.75	.75	2.90	.85	.70	.70	
30	.75	.75	2.90	.85	.70	.70	
31		.75		.85	.70		

a Water out of canal for ten days.   *b* Water turned out this day.

Rating table for Oxford Farmers' canal at Nepesta, Colo., from January 1 to December 31, 1903.

Gage height.	Discharge.	Gage height.	Discharge.	Gage height.	Discharge.	Gage height.	Discharge.
Feet.	*Second-feet.*	*Feet.*	*Second-feet.*	*Feet.*	*Second-feet.*	*Feet.*	*Second-feet.*
0.0	0	0.9	19	1.8	63	2.7	109
.1	1	1.0	23	1.9	68	2.8	115
.2	2	1.1	28	2.0	73	2.9	121
.3	3	1.2	33	2.1	78	3.0	127
.4	4	1.3	38	2.2	83	3.1	133
.5	6	1.4	43	2.3	88	3.2	139
.6	8	1.5	48	2.4	93		
.7	11	1.6	53	2.5	98		
.8	15	1.7	58	2.6	103		

Estimated monthly discharge of Oxford Farmers' canal at Nepesta, Colo., for 1903.

Month.	Discharge in second-feet.			Total in acre-feet.
	Maximum.	Minimum.	Mean.	
April	109	13	31	1,845
May	14	0	12	788
June	121	12	107	6,367
July	136	5	116	7,133
August	136	3	64	3,935
September	11	0	11	655
October	11	0	7	430

ARKANSAS RIVER AT PUEBLO, COLO.

This station is an important one, being located near the head of the principal irrigated portion of the valley. Only one ditch of importance is taken out above it in the Pueblo district, although considerable water is used in the ditches in the neighborhood of Canyon, which is in another water district. It is upon the gagings made at this point that the water superintendents and commissioners depend for distribution of water to ditches below.

This station was established in September, 1894, by A. P. Davis. Originally there were two gage rods. The main gage was located at the Santa Fe Avenue Bridge and consisted of a vertical 6 by 6 inch timber and a 1 by 6 inch scale bolted to the abutment of the Denver and Rio Grande Railroad bridge on the left-hand side of the river, graduated to tenths of a foot. There was also a short vertical rod for

extreme low water spiked to a pile about 20 feet out in the strea
reading the same as the main gage. The 12-foot mark of this ga
was opposite the top of the large capstone. The rod at Victor
Avenue Bridge consisted of inclined 4 by 4 inch timbers fastened
posts set in the right bank of the stream, graduated to vertical tent
of a foot, the space between the marks being 0.242 of a foot. Tl
rod was placed in June, 1895, for the purpose of noting the change
the slope of the water surface. The rods were read until July
1898, when on account of the shifting of the bed of the river th
were abandoned and a new gage was installed on the east side of t
Main Street Bridge. From this bridge all the discharge measureme
were made until the season of 1902.

Readings were made at Main Street Bridge until March 3, 19
when, owing to the scouring of the channel, it became necessary
replace the gage by one about 60 feet below the south end of t
Union Avenue Bridge. This gage was a 2 by 6 inch vertical timb
bolted like the former rod to the masonry wall and graduated to f
and tenths. On June 13, 1900, this rod was connected with a ben
mark on the coping at the northwest corner of the Union Aven
Bridge, which was found to be 19.79 feet above the zero of the r
In March, 1902, another rod exactly similar to the one last mention
was bolted in a vertical position to the masonry wall on the rig
hand side of the river about 30 feet above the south end of the Uni
Avenue Bridge. All gage heights up to July 14, 1902, were tak
from the rod just below the Union Avenue Bridge. After July
1902, the readings were taken from the rod above the bridge, t
graduations on the former rod being too dim to be read at low wat
A difference of 0.2 of a foot existed between the two rods when t
new one was set, the new rod reading 0.2 of a foot higher than t
old one. In all the discharge measurements made during 1902 t
gage height was taken from the new rod above the Union Aven
Bridge. The gage is read twice each day by S. N. Rutherford. D
charge measurements are made from the Union Avenue Bridge, whi
is marked every 10 feet on the downstream side. The initial po
for soundings is the edge of the masonry retaining wall on the rig
side of the channel at the south end of the bridge. The channel
straight for 500 feet above and 800 feet below the bridge, has a wid
of 150 feet, and is confined by high masonry walls. The bed is co
posed of bowlders and gravel, and there is little change in the bed
the channel, except that it fills during low water and scours out d
ing high water. The flow of the stream is rapid, but not too swift
accurate measurements. No bench mark has been established
the new gage, but the same is securely bolted to the masonry reta
ing wall.

Discharge measurements of Arkansas River at Pueblo, Colo., in 1903.

Date.	Hydrographer.	Gage height.	Discharge.
		Feet.	*Second-feet.*
March 4	M. C. Hinderlider	2.20	307
April 3	....do	2.50	470
May 6	F. Cogswell	1.90	157
May 26	....do	2.30	448
June 15	....do	5.00	3,386
July 7	C. W. Beach	2.90	798
July 14	F. Cogswell	3.35	1,588
August 3	A. Jacob	2.35	393
Do	....do	2.35	421
August 5	F. Cogswell	2.20	332
August 27	....do	2.25	412
September 23	....do	1.90	229
October 14	E. C. Murphy	1.82	137
November 17	F. Cogswell	2.10	294
December 27	R. I. Meeker	1.85	234

Mean daily gage height, in feet, of Arkansas Riv r at Pueblo, Colo., for 1903.

Day.	Jan.	Feb.	Mar	Apr	May.	June.	July.	Aug.	Sept.	Oct.	Nov.
1	2.10	2.20	2.10	2.50	2.10	3.10	3.90	2.70	2.00	1.90	2.60
2	2.10	2.20	2.15	2.50	2.00	3.20	3.50	2.80	2.00	2.00	2.10
3	2.10	2.10	2.20	2.50	2.00	3.50	3.80	2.50	2.00	2.00	2.00
4	2.20	2.00	2.20	2.50	2.00	3.70	4.00	2.40	2.00	2.00	2.10
5	2.20	2.00	2.30	2.45	2.00	3.45	3.60	2.30	2.30	2.00	2.10
6	2.20	2.10	2.20	2.50	1.95	3.80	3.00	4.50	2.30	2.00	2.20
7	2.20	2.00	2.20	2.55	1.95	4.30	2.55	3.00	2.20	(a)	2.20
8	2.35	2.05	2.20	2.45	2.15	4.00	3.00	3.30	2.50	(a)	......
9	2.30	2.10	2.20	2.55	2.20	6.80	3.00	2.00	2.65	(a)	......
10	2.20	2.20	2.20	2.80	2.25	4.90	3.00	2.60	2.65	(a)	......
11	2.20	2.30	2.20	2.70	2.30	5.00	3.70	2.60	2.30	(a)	......
12	2.20	2.30	2.20	2.60	2.35	4.30	3.70	2.50	2.10	(a)	......
13	2.10	2.00	2.50	2.65	2.40	4.50	2.80	2.40	2.15	(a)	......
14	2.00	2.00	2.50	2.60	2.35	4.90	3.60	2.50	2.10	1.80	......
15	2.20	2.00	2.60	2.55	2.50	5.00	3.50	3.90	2.25	2.10	......
16	2.20	2.00	2.60	2.00	2.85	5.00	3.50	3.20	2.25	1.95	......
17	2.10	2.05	2.50	2.15	3.00	5.00	3.60	2.80	2.15	2.00	......
18	2.20	2.10	2.50	2.00	2.95	5.00	3.80	2.70	2.10	2.50	......
19	2.20	2.20	2.30	2.00	2.90	5.20	3.50	2.60	2.15	2.25	......
20	2.20	2.20	2.20	2.00	2.65	90	3.10	2.70	2.30	2.25	......
21	2.20	2.20	2.20	2.00	2.55	4.90	3.20	2.70	2.15	2.20	......
22	2.20	2.20	2.20	2.00	2.40	60	4.10	2.60	2.00	2.45	......
23	2.20	2.30	2.10	1.90	2.70	80	3.00	2.60	2.15	2.45	......
24	2.20	2.10	2.10	1.80	2.50	4.00	3.00	2.65	1.90	2.55	......
25	2.20	2.20	2.10	1.80	2.50	00	3.05	2.60	1.80	2.10	......

a Mud accumulated about gage and observer failed to notify office or clear it.

Mean daily gage height, in feet, of Arkansas River, etc.—Continued.

Day.	Jan.	Feb.	Mar.	Apr.	May.	June.	July.	Aug.	Sept.	Oct.	Nov.
26	2.20	2.10	2.10	1.90	2.50	4.00	3.00	2.55	1.75	2.00	
27	2.20	2.20	2.10	2.00	2.60	3.90	3.20	2.50	1.70	2.10	
28	2.20	2.20	2.10	2.00	2.70	3.55	3.40	2.40	1.50	2.00	
29	2.20		2.10	2.00	2.70	3.55	3.00	2.45	1.80	2.00	
30	2.20		2.10	2.10	2.50	3.55	2.80	2.30	1.85	2.20	
31	2.20		2.20		2.60		2.60	2.30		2.00	

Rating table for Arkansas River at Pueblo, Colo., from January 1 to December 31, 1903.

Gage height.	Discharge.	Gage height.	Discharge.	Gage height.	Discharge.	Gage height.	Discharge.
Feet.	Second-feet.	Feet.	Second-feet.	Feet.	Second-feet.	Feet.	Second-feet.
1.7	80	2.7	664	3.7	1,690	5.4	3,960
1.8	126	2.8	741	3.8	1,810	5.6	4,260
1.9	175	2.9	822	3.9	1,940	5.8	4,560
2.0	227	3.0	908	4.0	2,070	6.0	4,860
2.1	281	3.1	1,000	4.2	2,330	6.2	5,160
2.2	337	3.2	1,100	4.4	2,590	6.4	5,460
2.3	396	3.3	1,210	4.6	2,850	6.6	5,780
2.4	457	3.4	1,330	4.8	3,110	6.8	6,100
2.5	522	3.5	1,450	5.0	3,390		
2.6	591	3.6	1,570	5.2	3,670		

Estimated monthly discharge of Arkansas River at Pueblo, Colo., for 1903.

[Drainage area, 4,600 square miles.]

Month.	Discharge in second-feet.			Total in acre-feet.	Run-off.	
	Maximum.	Minimum.	Mean.		Second-feet per square mile.	Depth in inches.
January	426	227	327	20,106	0.071	0.082
February	396	227	298	16,550	.065	.068
March	591	281	364	22,381	.079	.091
April	741	126	402	23,921	.087	.09
May	908	201	487	29,944	.106	.12
June	6,100	1,000	2,576	153,283	.560	.6
July	2,200	556	1,295	79,626	.282	.3
August	2,720	396	720	44,271	.157	.1
September	627	80	295	17,554	.064	.0
October	591	126	301	18,508	.065	.0
November 1-7	591	227	334	4,638	.073	.01
The period	6,100	80				

ARKANSAS RIVER NEAR CANYON, COLO.

This station is located at the suspension bridge at the Hot Springs Hotel. It is 1¼ miles above the Strathmore Hotel and 1 mile above the State penitentiary, and a short distance below the mouth of Grape Creek and at a point immediately below where the river leaves the mountains. Observations at this point were begun on April 17, 1889, the station being established here by Robert Robertson. The record has been maintained since that time, with occasional breaks due to absence or change of observer. The station is of special importance, being located at the mouth of the canyon and at a point practically above all of the irrigating ditches except the Canyon ditch (sometimes called the "North Side ditch") and the South Canyon ditch (sometimes called the "South Side ditch"), both of which head above the station. During the irrigation season each of these ditches carries from 25 to 60 cubic feet of water per second, according to the needs of the irrigators, and their discharge should be added to the discharge at the station in order to obain the total run-off at the mouth of the canyon. No accurate records have been kept of the amount of water passing through these canals, although miscellaneous measurements have been made when measurements were made at the regular station. The Canyon estimated monthly discharges do not include water taken out by these canals. This site was used in 1888 for a gaging station by the State engineer of Colorado, and is favorable for obtaining fairly accurate measurements.

The gage rod established by Mr. Robertson was of 2 by 6 inch timber, inclined, and attached to the crib of an old bridge on the south or right-hand side of the river, almost directly in front of the hotel. There were two bench marks: No. 1 on the top of a log of the crib—elevation, 10.01 feet above the datum; No. 2, in the cleft of a red bowlder at the foot of a charred stump 50 feet downstream and on the same side of the river—elevation, 9.60 feet. On April 13, 1891, the station was inspected by Frank Tweedy, and a third bench mark was established, this being a bedded rock 40 feet from the north end of the cable, toward Hot Springs Hotel, and 10 feet from the river bank. It is marked "B. M. No. 3, U. S. G. S.," and is 15.98 feet above the zero of the gage.

On October 4, 1895, it was found that the top of the gage had been broken off, necessitating its renewal for readings during high water. The channel was found to be filled with sand and gravel in front of the gage, requiring considerable work in order to make the water flow to the rod. It was decided, therefore, to put in a new gage where the stream could not deposit material. A point was chosen about 100 feet below the bridge, on the left bank, and a crib was built, anchored in place by rocks and bolts, the lower end of the gage being fastened to it. The upper end of the inclined portion was attached to a juniper

tree. On December 27, 1895, the station was inspected, and it v
found that readings had been made from the old rod, which, at
stage of water prevailing, recorded about 0.40 of a foot above the r
rod on the opposite side of the river.   When the water is high and
tends with unbroken surface from bank to bank the readings are
same, but at low water the observations on the old rod are r
leading, owing to the accumulation of sand and gravel in front o

On August 26, 1902, owing to the shifting of the channel, a new r
consisting of 4 by 4 inch timber, was placed in an inclined posi
on the north or left bank of the river, at the site of the previous :
just below the north end of the suspension bridge.   The rod on
date read practically the same as the old rod.

A new gage was established September 2, 1903, by the State e
neer.   It is located on the right bank just below the bridge, and
sists of an inclined and a vertical section.   The inclined section r
from zero to 7.3 feet.   The vertical section reads from 7.3 to 12 f
The present observer is Dr. J. L. Prentiss.

The measurements were at first made from a car suspended fro
cable stretched across the river, the bridge from which measurem
were originally made by the State engineer having been destro,
Later a new suspension bridge was constructed in front of the h
necessitating the removal and replacement of the gage, and subseq
measurements were made from this bridge, which, having a clear s
offers no obstruction to the current.

The initial point for soundings is the first post of the hand rai
the right bank.   The channel is straight for 500 feet above and
feet below the station and has a width of about 135 feet.   The cur
is very swift at high water and not too sluggish at low stages
accurate measurement.   Both banks are high and rocky, and are
liable to overflow.   The bed of the stream is composed of coarse gr
and small bowlders.   It is not subject to change except on the r
bank near the old gage, where a sand bar is formed at times of
water.   The bed is rough, but is regular in shape.   The bench n
in present use is a bedded bowlder 80 feet southeast of the south
of the suspension bridge, near the river bank.   It is marked wi
cross cut in the rock.   Its elevation is 16.07 feet above the zero of
gage.   This bench mark was checked December 16, 1903.

The head gates of North and South Canyon ditches are above
rod, and no accurate records have been kept by this office concer
the amount of water passing through these canals, although mi
laneous measurements have been made when measurements v
made at the Canyon station.   Therefore these records do not inc
any water taken out by those canals.

Discharge measurements of Arkansas River near Canyon, Colo., in 1903.

Date.	Hydrographer.	Gage height.	Discharge.
		Feet.	*Second-feet.*
April 6	F. Cogswell	3.00	447
May 7	....do	2.30	267
May 27	....do	2.80	494
June 16	....do	5.85	3,862
July 15	....do	5.05	1,606
August 6	....do	3.50	451
August 28	....do	3.65	459
September 24	....do	3.70	284
October 19	E. C. Murphy	3.60	·218
November 18	F. Cogswell	3.60	250
December 16	R. I. Meeker	3.70	298

Mean daily gage height, in feet, of Arkansas River near Canyon, Colo., for 1903.

Day.	Jan.	Feb.	Mar.	Apr.	May.	June.	July.	Aug.	Sept.	Oct.	Nov.	Dec.
1	2.30	2.50	2.25	3.05	2.40	3.40	5.50	3.70	3.40	3.60	3.60	3.60
2	2.35	2.45	2.30	3.00	2.25	3.60	5.25	3.65	3.60	3.60	3.60	3.60
3	2.35	2.35	2.35	3.00	2.30	3.80	5.35	3.55	3.70	3.60	3.60	3.60
4	2.45	2.30	2.45	2.90	2.30	4.10	5.25	3.40	3.70	3.60	3.60	3.60
5	2.50	2.15	2.55	2.95	2.30	4.30	5.00	3.30	3.93	3.60	3.60	3.60
6	2.50	2.30	2.60	3.00	2.25	4.20	4.75	3.45	3.95	3.60	3.60	3.60
7	2.55	2.35	2.70	2.95	2.30	4.15	4.65	3.65	4.00	3.60	3.60	3.60
8	2.40	2.40	2.65	3.10	2.70	4.50	4.95	3.50	4.33	3.60	3.60	3.60
9	2.45	2.40	2.70	3.10	2.70	4.75	4.90	3.65	4.25	3.60	3.60	3.60
10	2.45	2.55	2.70	3.00	2.60	5.05	5.25	3.70	4.08	3.50	3.55	3.60
11	2.40	2.50	2.80	3.00	2.65	5.10	5.30	3.75	3.83	3.55	3.50	3.60
12	2.35	2.45	2.85	3.00	2.75	4.95	5.35	3.75	3.80	3.60	3.60	3.60
13	2.45	1.95	2.85	2.90	2.60	5.05	5.30	3.73	3.80	3.60	3.60	3.60
14	2.40	2.30	2.85	2.80	2.70	5.30	5.20	3.70	3.75	3.60	3.60	3.80
15	2.50	2.20	3.00	2.65	3.00	5.60	5.15	3.95	3.80	3.60	3.60	3.70
16	2.35	2.25	2.95	2.60	3.30	5.90	5.10	4.65	3.80	3.50	3.60	3.75
17	2.40	2.30	2.95	2.45	3.35	6.20	5.35	4.25	3.80	3.55	3.60	3.70
18	2.40	2.25	2.90	2.30	3.35	6.65	5.15	4.10	3.80	3.60	3.60	3.70
19	2.50	2.35	2.85	2.30	3.30	6.70	5.10	3.90	3.75	3.60	3.60	3.75
20	2.35	2.50	2.65	2.20	3.10	6.60	4.95	3.90	3.70	3.60	3.60	3.75
21	2.35	2.50	2.65	2.20	3.00	6.55	4.80	3.80	3.70	3.60	3.60	3.75
22	2.45	2.45	2.60	2.30	2.80	6.45	4.85	4.00	3.70	3.60	3.60	3.70
23	2.50	2.50	2.65	2.30	2.80	6.30	4.80	4.00	3.70	3.60	3.60	3.70
24	2.50	2.50	2.60	2.35	2.75	6.15	4.65	3.65	3.65	3.50	3.60	3.80
25	2.50	2.45	2.60	2.50	2.70	6.00	4.55	3.70	3.60	3.50	3.60	3.70
26	2.50	2.50	2.70	2.55	2.70	5.85	4.50	3.60	3.60	3.50	3.60	3.70
27	2.50	2.40	2.70	2.50	2.70	5.80	4.50	3.60	3.60	3.50	3.60	3.65
28	2.50	2.45	2.60	2.50	2.75	5.65	4.50	3.65	3.60	3.55	3.60	3.80
29	2.50		2.65	2.55	2.80	5.80	4.20	3.55	3.60	3.70	3.60	3.75
30	2.25		2.60	2.50	2.80	5.65	3.85	3.40	3.60	3.80	3.60	3.70
31	2.35		3.00		2.95		3.70	3.30		3.85		3.70

Rating table for Arkansas River near Canyon, Colo., from January 1 to June 26, 1903.

Gage height.	Discharge.	Gage height.	Discharge.	Gage height.	Discharge.	Gage height.	Discharge.
Feet.	*Second-feet.*	*Feet.*	*Second-feet.*	*Feet.*	*Second-feet.*	*Feet.*	*Second-feet.*
1.9	180	2.9	534	3.9	1,438	4.9	2,776
2.0	200	3.0	596	4.0	1,560	5.0	2,880
2.1	222	3.1	664	4.1	1,684	5.2	3,050
2.2	248	3.2	738	4.2	1,808	5.4	3,300
2.3	276	3.3	818	4.3	1,932	5.6	3,550
2.4	308	3.4	906	4.4	2,056	5.8	3,800
2.5	344	3.5	1,000	4.5	2,180	6.0	4,050
2.6	384	3.6	1,100	4.6	2,304	6.2	4,300
2.7	428	3.7	1,206	4.7	2,428	6.4	4,550
2.8	478	3.8	1,320	4.8	2,552	6.6	4,800

Rating table for Arkansas River near Canyon, Colo., from June 27 to September 2, 1903.

Gage height.	Discharge.	Gage height.	Discharge.	Gage height.	Discharge.	Gage height.	Discharge.
Feet.	*Second-feet.*	*Feet.*	*Second-feet.*	*Feet.*	*Second-feet.*	*Feet.*	*Second-feet.*
3.3	320	4.0	722	4.7	1,280	5.4	2,015
3.4	364	4.1	794	4.8	1,370	5.5	2,150
3.5	416	4.2	870	4.9	1,460	5.6	2,390
3.6	472	4.3	950	5.0	1,555	5.7	2,440
3.7	530	4.4	1,030	5.1	1,655	5.8	2,590
3.8	590	4.5	1,110	5.2	1,765		
3.9	654	4.6	1,195	5.3	1,885		

Rating table for Arkansas River near Canyon, Colo., from September 3 to December 31, 1903.

Gage height.	Discharge.	Gage height.	Discharge.	Gage height.	Discharge.	Gage height.	Discharge.
Feet.	*Second-feet.*	*Feet.*	*Second-feet.*	*Feet.*	*Second-feet.*	*Feet.*	*Second-feet.*
3.4	131	3.7	284	4.0	437	4.3	590
3.5	182	3.8	335	4.1	488	4.4	641
3.6	233	3.9	386	4.2	539		

ted monthly discharge of Arkansas River near Canyon, Colo., for 1903.

[Drainage area, 3,060 square miles.]

	Discharge in second-feet.			Total in acre-feet.	Run-off.	
lonth.	Maximum.	Minimum.	Mean.		Second-feet per square mile.	Depth in inches.
----------	364	248	318	19,553	0.104	0.120
y . --------	364	190	300	16,661	.098	.102
----------	596	248	430	26,440	.141	.163
----------	664	248	444	26,420	.145	.162
----------	862	248	472	29,022	.154	.178
----------	4,925	906	3,064	182,321	1.001	1.117
----------	2,150	530	1,491	91,678	.487	.562
----------	1,237	320	560	34,433	.183	.211
er	605	233	337	20,053	.110	.123
----------	360	182	230	14,142	.075	.086
er	233	182	230	13,686	.075	.084
?r	335	233	271	16,663	.089	.108
le year	4,925	182	679	491,072	.222	3.011

ARKANSAS RIVER AT SALIDA.

station is located at the footbridge near the railroad shops at
It was established April 11, 1895, and has been maintained
)rtion of each year since that time. The gage rod has been
d to the north side of the footbridge, but considerable diffi-
as been experienced owing to the fact that ice and logs con-
interfere with the rod, three new rods having been required
. The banks are high and are not subject to overflow, but
re large bowlders in the stream, which interfere with the accu-
the results of measurements.
w gage was placed on the site of the former gage on August 27,
hich has remained continuously ever since. This new gage
i of an inclined 6 by 7 inch pine timber, having its lower end
y bolted to a large embedded granite bowlder, and its upper
ted to the north pier of the footbridge. The gage is painted and
ted to vertical feet and tenths, the distance between footmarks
.24 feet. The upper portion of the gage is painted on the south
the vertical timber pier, to which the gage rod is bolted.
iew rod read 1.30 when the old rod read 0.2.
)ench mark is a circle on the top of a 2 by 2 foot granite
r almost buried in the earth, and located 3 feet northwest
ie upper main post of the north pier of the footbridge on
'th bank of the river. Bench mark is marked with chisel,
U. S. G. S."
bench mark is 2.224 feet above the 3-foot mark on the gage.

The importance of the maintenance of the station lies in the info
mation furnished as to the time required for water to flow from Gra
ite to Salida and again from Salida to Canyon, this question having
bearing upon the distribution of the use of the water turned out fr
Twin Lakes; and it is, moreover, valuable from the point of view th
it is extremely probable that the entire discharge of the Arkansas
this point may eventually be used for power purposes in the Gra
Canyon of the Arkansas.[a]

Discharge measurements of Arkansas River at Salida, Colo., in 1903.

Date.	Hydrographer.	Gage height.	Dischar
		Feet.	*Second-f*
April 6	F. Cogswell	1.35	
May 7	do	1.75	
May 27	do	1.80	
June 16	do	4.20	2,
July 15	do	3.30	1,
August 6	do	1.65	
August 28	do	1.60	
September 24	do	1.50	
October 19	E. C. Murphy	1.29	
November 18	F. Cogswell	1.15	

Mean daily gage height, in feet, of Arkansas River at Salida, Colo., for 190

Day.	Apr.	May.	June.	July.	Aug.	Sept.	Oc
1	1.50	1.15	2.25	3.40	1.95	1.55	
2	1.50	1.25	2.65	3.20	1.80	1.50	
3	1.45	1.15	2.85	3.80	2.00	1.50	
4	1.40	1.25	3.35	3.60	2.40	1.65	
5	1.40	1.30	3.40	b3.95	1.90	1.80	
6	1.40	1.30	3.30	3.85	1.80	1.80	
7	1.45	1.75	3.40	3.45	1.90	1.95	
8	1.65	1.60	3.40	3.05	c2.95	2.00	
9	1.70	1.60	3.30	3.00	b3.00	2.00	
10	1.65	1.60	3.40	b3.65	b3.25	2.00	
11	1.65	1.75	3.25	3.65	b2.85	2.00	
12	1.75	1.80	3.00	3.60	c2.80	2.00	
13	1.70	1.70	3.25	3.45	3.80	1.95	
14	1.65	1.90	b3.45	3.40	3.20	1.80	
15	1.35	2.15	3.80	3.30	3.40	1.75	
16	1.15	2.30	4.10	3.25	3.20	1.70	
17	1.05	2.45	4.45	3.35	3.00	1.70	

[a]For more detailed information concerning this station, see Biennial Reports of the S
Engineers of Colorado: Eighth, p. 480; Ninth, p. 361; Tenth, p. 298. Also publications U. S.
logical Survey: Eighteenth Annual Report, Part IV, p. 224; Nineteenth, Part IV, p. 355; T
tieth, Part IV, p. 381; Twenty-first, Part IV, p. 230; Bulletin No. 140, p. 155; Water-Supply
Irrigation Papers, No. 10, p. 118; No. 28, pp. 110, 116, and 117; No. 37, p. 258; No. 39, p. 480;
No. 50, p. 322. Also report on Agriculture by Irrigation, Eleventh Census, by F. H. Newell, p

b Rain.   c Cloudburst.

Mean daily gage height, in feet, of Arkansas River, etc.—Continued.

Day.	Apr.	May.	June.	July.	Aug.	Sept.	Oct.
....	1.00	2.10	4.65	3.40	3.00	1.60	1.35
....	1.00	2.00	4.55	2.90	2.75	1.60	1.30
....	1.00	2.00	4.45	3.35	2.60	1.60	1.20
....	.95	1.85	3.90	3.00	2.00	1.55	1.20
....	1.10	1.80	3.80	2.90	2.00	1.50	1.20
....	1.35	1.70	3.80	3.00	1.85	1.50	1.20
....	1.40	1.70	3.80	2.80	1.80	1.50	1.20
....	1.35	1.70	3.55	3.05	1.70	1.50	1.20
....	1.35	1.75	3.45	2.95	1.70	1.50	1.20
....	1.30	1.80	3.40	2.90	1.70	1.50	1.20
....	1.80	1.80	3.40	3.00	1.70	1.50	1.30
....	1.20	1.80	3.55	2.90	1.80	1.40	1.70
....	1.15	1.85	3.50	2.90	1.75	1.40	1.60
....	....	1.95	....	2.90	1.60	....	1.35

ating table for Arkansas River at Salida, Colo., from January 1 to June 20, 1903.

Gage height.	Discharge.	Gage height.	Discharge.	Gage height.	Discharge.	Gage height.	Discharge.
Feet.	*Second-feet.*	*Feet.*	*Second-feet.*	*Feet.*	*Second-feet.*	*Feet.*	*Second-feet.*
0.9	200	1.9	676	2.9	1,325	3.9	1,975
1.0	234	2.0	740	3.0	1,390	4.0	2,040
1.1	272	2.1	805	3.1	1,455	4.1	2,105
1.2	312	2.2	870	3.2	1,520	4.2	2,170
1.3	354	2.3	935	3.3	1,585	4.3	2,235
1.4	398	2.4	1,000	3.4	1,650	4.4	2,300
1.5	446	2.5	1,065	3.5	1,715	4.5	2,365
1.6	498	2.6	1,130	3.6	1,780	4.6	2,430
1.7	554	2.7	1,195	3.7	1,845	4.7	2,495
1.8	614	2.8	1,260	3.8	1,910		

Curve poorly defined; extended below 1.35 feet gage height.

lating table for Arkansas River at Salida, Colo., from June 21 to December 31, 1903.

Gage height.	Discharge.	Gage height.	Discharge.	Gage height.	Discharge.	Gage height.	Discharge.
Feet.	*Second-feet.*	*Feet.*	*Second-feet.*	*Feet.*	*Second-feet.*	*Feet.*	*Second-feet.*
1.2	218	1.9	646	2.6	1,108	3.3	1,570
1.3	266	2.0	712	2.7	1,174	3.4	1,636
1.4	322	2.1	778	2.8	1,240	3.5	1,702
1.5	384	2.2	844	2.9	1,306	3.6	1,768
1.6	448	2.3	910	3.0	1,372	3.7	1,834
1.7	514	2.4	976	3.1	1,438	3.8	1,900
1.8	580	2.5	1,042	3.2	1,504	3.9	1,966

Curve poorly defined.

Estimated monthly discharge of Arkansas River at Salida, Colo., for 1903.

[Drainage area, 1,160 square miles.]

Month.	Discharge in second-feet.			Total in acre-feet.	Run-off.	
	Maximum.	Minimum.	Mean.		Second-feet per square mile.	Depth in inches.
April	584	217	389	28,147	0.34	0.38
May	1,032	292	589	36,216	.51	.59
June................	2,462	902	1,787	108,359	1.50	1.67
July	1,999	1,240	1,540	94,691	1.33	1.53
August	1,900	448	973	59,827	.84	.97
September	712	322	500	29,752	.43	.48
October	514	218	286	17,585	.25	.29

MISCELLANEOUS MEASUREMENTS IN THE ARKANSAS DRAINAGE BASIN.

The following miscellaneous measurements were made in the Arkansas drainage basin in 1903:

Miscellaneous measurements in Arkansas drainage basin in 1903.

Date.	Stream.	Locality.	Gage height.	Discharge.
			Feet.	*Sec.-feet.*
Apr. 25	Adobe Creek	At mouth		0.55
30	.....do	Wagon bridge, one-half mile above mouth.	....	.45
May 11	.....do	At mouth	....	.59
14	.....do	.....do	....	.59
23	.....do	.....do	....	.66
28	.....do	.....do	....	.60
June 6	.....do	.....do	....	(a)
18	.....do	... .do	....	2.00
July 1	.....do	.....do	....	b1.00
9	.....do	.....do	....	b1.00
17	.....do	.....do		b1.00
22	.....do	.....do		b1.00
Aug. 1	.....do	.....do		b1.00
6	.....do	.....do		b1.00
Apr. 22	Amity canal	At rating flume	0.30	5.49
May 1	.....do	.....do	.20	2.99
8	... ..do	At head gate	....	2.56
9	.....do	At rating flume	.62	17.89
25	Apishapa River	At mouth, railroad bridge	.60	4.17

a Less than 1 second-foot.   b Discharge was estimated.

Miscellaneous measurements in Arkansas drainage basin in 1903—Continued.

Date.	Stream.	Locality.	Gage height.	Discharge.
			Feet.	*Sec.-feet.*
ly 27	Apishapa River	At mouth, railroad bridge	0.60	4.00
ne 15	....do	....do		a100.00
ly 1	....do	....do	1.23	16.60
g. 3	....do	....do		a15.00
r. 24	Arkansas River	Las Animas		9.11
30	....do	....do		8.46
iy 9	....do	....do		11.76
11	....do	....do	1.00	11.60
14	....do	....do		9.16
23	....do	....do		13.89
28	....do	....do		16.00
ne 6	....do	....do	1.40	1,616.00
20	....do	....do		5,973.00
ly 9	....do	....do		89.00
17	....do	....do		291.00
22	....do	....do		115.00
31	....do	....do		209.00
ig. 6	....do	....do	1.10	22.00
18	....do	....do	1.05	30.00
r. 23	....do	Caddoa bridge		29.00
iy 1	....do	....do		37.95
9	....do	....do		34.90
15	....do	....do		24.00
21	....do	....do		b30.00
29	....do	....do		33.00
ne 5	....do	....do		171.00
22	....do	....do		4,042.00
ly 15	....do	....do		23.72
r. 22	....do	Lamar bridge		15.00
iy 2	....do	....do		14.16
16	....do	....do		b1.00
19	....do	....do		b1.00
21	....do	....do		b1.50
30	....do	....do		b1.50
ine 5	....do	....do		b2.00
ily 15	....do	....do		52.00
24	....do	....do		44.00
29	....do	....do		4.00
r. 21	....do	Byron		12.12

aIndicates that discharge was estimated from train.   b Estimated.

Miscellaneous measurements in Arkansas drainage basin in 1903—Continued.

Date.	Stream.	Locality.	Gage height.	Discharge.
			Feet.	*Sec.-feet.*
May 2	Arkansas River	Byron		17.00
2	Big Sandy Creek	At mouth		*a* .20
7	....do	....do		*a* .30
16	....do	....do		*a* .20
24	....do	....do		.00
June 29	....do	....do		*a* .50
July 11	....do	....do		.00
14	....do	....do		.00
24	....do	....do		*a* 1.00
29	....do	....do		.00
Aug. 10	....do	....do		*a* .50
14	....do	....do		.00
20	....do	....do		.00
Apr. 21	Buffalo canal	Below head gate	.95	6.69
May 7	....do	At rating flume	.10	1.35
7	....do	....do	.28	5.13
7	....do	....do	.40	14.00
7	....do	....do	.48	17.56
June 25	....do	Granada	1.50	39.00
29	....do	....do	1.30	15.00
July 11	....do	....do		.00
14	....do	....do		.00
24	....do	....do	1.30	12.00
28	....do	....do	1.65	28.00
29	....do	....do	1.80	37.00
Aug. 10	....do	....do	1.97	67.00
14	....do	....do	1.15	13.00
May 7	Buffalo Creek	At mouth		*a* .30
16	....do	....do		.00
18	....do	....do		*a* .10
30	....do	....do		2.00
June 4	....do	....do		*a* .50
25	....do	Granada		6.00
29	....do	....do		6.00
July 11	....do	....do		*a* .50
14	....do	....do		*a* .50
24	....do	....do		.00
28	....do	....do		*a* .50
Aug. 10	....do	....do		*a* .50
14	....do	....do		.00

a Estimated.

Miscellaneous measurements in Arkansas drainage basin in 1903—Continued.

Date.	Stream.	Locality.	Gage height.	Discharge.
			Feet.	*Sec.-feet.*
Aug. 20	Buffalo Creek	Granada	....	*a* 1.00
May 1	Caddoa Creek	At mouth	....	.00
9	....do	....do	....	.00
29	....do	....do	....	.00
June 5	....do	....do	....	.00
30	....do	Caddoa	....	.00
July 10	....do	....do	....	.00
15	....do	....do	....	.00
23	....do	....do	....	.00
30	....do	....do	....	.00
Aug. 7	....do	....do	....	*a* .50
15	....do	....do	....	.00
19	....do	....do	....	.00
June 16	Canyon ditch......	Canyon	....	85.00
July 15	....do	....do	....	78.00
Aug. 6	....do	....do	....	16.00
28	....do	....do	....	87.00
Sept. 24	....do	....do	....	70.00
Oct. 19	....do	....do	....	42.00
Nov. 18	....do	....do	....	56.00
May 16	Clay Creek	At mouth	....	1.24
19	....do	....do	....	72.00
June 24	....do	....do	....	*a* 2.00
29	....do	....do	....	*a* 2.00
July 14	....do	At mouth (Lamar)..............	....	4.00
24	....do	At mouth	....	.00
Aug. 10	....do	....do	....	*a*.50
20	....do	....do	....	*a* 1.00
May 13	Crooked Arroyo ...	Near La Junta	....	*a*.30
27	....do	....do	....	.50
June 17	....do	Near La Junta railroad bridge.	....	3.00
July 2	....do	....do	....	*a* 1.00
8	....do	Near La Junta	....	.00
Aug. 5	....do	....do	....	*a*.50
Apr. 5	Flume at sugar factory.	Rocky Ford	....	93.00
6	....do	....do	....	50.00
27	Fort Lyon canal	At rating flume	1.36	62.00
30	....do	....do	1.38	56.00
July 31	....do	....do	....	146.00

a Estimated.

Miscellaneous measurements in Arkansas drainage basin in 1903—Continued.

Date.	Stream.	Locality.	Gage height.	Discharge.
			Feet.	*Sec.-feet.*
May 2	Arkansas River ...	Byron		17.00
2	Big Sandy Creek ..	At mouth		a.20
7	.....do	.....do		a.30
16	.....do	.....do		a.20
24	.....do	.....do		.00
June 29	.....do	.....do		a.50
July 11	.....do	.....do		.00
14	.....do	.....do		.00
24	.....do	.....do		a1.00
29	.....do	.....do		.00
Aug. 10	.....do	.....do		a.50
14	.....do	.....do		.00
20	.....do	.....do		.00
Apr. 21	Buffalo canal	Below head gate	.95	6.69
May 7	.....do	At rating flume	.10	1.35
7	.....do	.....do	.23	5.12
7	.....do	.....do	.40	14.00
7	.....do	.....do	.48	17.56
June 25	.....do	Granada	1.50	39.00
29	.....do	.....do	1.30	15.00
July 11	.....do	.....do		.00
14	.....do	.....do		.00
24	.....do	.....do	1.30	12.00
28	.....do	.....do	1.65	28.00
29	.....do	.....do	1.80	37.00
Aug. 10	.....do	.....do	1.97	67.00
14	.....do	.....do	1.15	13.00
May 7	Buffalo Creek	At mouth		a.30
16	.....do	.....do		.00
18	.....do	.....do		a.10
30	.....do	.....do		2.00
June 4	.....do	.....do		a.50
25	.....do	Granada		6.00
29	.....do	.....do		6.00
July 11	.....do	.....do		a.50
14	.....do	.....do		a.50
24	.....do	.....do		.00
28	.....do	.....do		a.50
Aug. 10	.....do	.....do		a.50
14	.....do	.....do		.00

a Estimated.

Miscellaneous measurements in Arkansas drainage basin in 1904—Continued.

		Stream.		Gage height.	Discharge.
y	23	Mud Creek	At mouth		
	30	...do	...do		.00
g.	7	...do	Prowers, one-fourth mile from mouth.		3.00
	15	...do	...do		.00
	19	...do	...do		.00
r.	23	Purgatory River	At mouth, Las Animas		2.27
y	1	...do	...do		2.87
	9	...do	...do		2.30
	14	...do	...do		3.00
	22	...do	...do		4.31
	28	...do	...do		4.00
1e	6	...do	...do		15.00
	20	...do	...do		956.00
	22	...do	...do		956.00
y	1	...do	...do		274.00
	6	...do	...do		109.00
	10	...do	...do		43.00
	16	...do	...do	1.70	156.00
	22	...do	...do		242.00
	31	...do	...do		19.00
g.	6	...do	...do		4.00
	17	...do	...do		492.00
	18	...do	...do		489.00
y	1	Rule Creek	Three-fourths mile from mouth.		a.80
	9	...do	At mouth		a.20
	15	...do	...do		.00
	29	...do	...do		.00
1e	5	...do	...do		.00
	6	...do	...do		.00
	22	...do	...do		a1.00
	30	...do	...do		.00
ly	10	...do	...do		.00
	15	...do	...do		.00
	23	...do	...do		.00
	30	...do	...do		.00
g.	7	...do	...do		a1.00
	15	...do	...do		.00
	19	...do	...do		.00
1e	16	South Canyon ditch	Canyon		13.00
y	15	South Canyon ditch	Canyon		49.00

a Estimated.

Miscellaneous measurements in Arkansas drainage basin in 1903—Continued

Date.	Stream.	Locality.	Gage height.	Dischar
			Feet.	*Sec.-fe*
June 22	Gagely Arroyo	Wagon bridge, near mouth	....	153
30	.....do	At mouth		a
July 15	.....do	.....do		
28	.....do	.....do		
80	.....do	.....do		
Aug. 15	.....do	.....do		
19	.....do	.....do		
Apr. 21	Graham ditch.....	At head gate		
May 2	.....do	.....do		
7	.....do	.....do		
Apr. 25	Horse Creek........	At mouth		
30	.....do	.....do		
May 11	.....do	.....do	.	
14	.....do	.....do		
23	.....do	.....do		
28	.....do	.....do		
June 6	.....do	.....do		
18	.....do	.....do		2
July 1	.....do	.....do		a
9	.....do	.....do		—
17	.....do	.....do		
22	.....do	.....do		
Aug. 1	.....do	.....do		
6	.....do	.....do		
Apr. 22	Hyde ditch........	At rating flume		1
May 2	.....do	.....do		1
8	.....do	.....do		
Apr. 25	James canal.......	.....do	1.42	2
30	.....do	.....do	1.80	3
25	Las Animas canal .	One-half mile below head gate .		2
30	.....do	In rating flume..........	2.08	2
May 1	Mud Creek........	At mouth		
9	.....do	.....do		
15	.....do	.....do		
29	.....do	.....do		
June 5	.....do	.....do		
23	.....do	.....do		
30	.....do	.....do		—
July 10	.....do	.....do		
15	.....do	.....do		

a Estimated.

Miscellaneous measurements in Arkansas drainage basin in 1903—Continued.

Date.	Stream.	Locality.	Gage height.	Discharge.
			Feet.	*Sec.-feet.*
y 23	Mud Creek	At mouth		0.00
30	....do	do		.00
g. 7	....do	Prowers, one-fourth mile from mouth.		3.00
15	....do	do		.00
19	....do	do		.00
r. 23	Purgatory River	At mouth, Las Animas		2.27
y 1	....do	do		2.87
9	....do	do		2.80
14	....do	do		3.00
22	....do	do		4.31
28	....do	do		4.00
ie 6	....do	do		15.00
20	....do	do	2.50	956.00
22	....do	do	2.50	956.00
y 1	....do	do	2.00	274.00
6	....do	do	1.50	109.00
10	....do	do	1.35	43.00
16	....do	do	1.70	156.00
22	....do	do	1.60	242.00
31	....do	do	.80	19.00
g. 6	....do	do	.55	4.00
17	....do	do	1.65	492.00
18	....do	do	1.70	489.00
y 1	Rule Creek	Three-fourths mile from mouth.		a.30
9	....do	At mouth		a.20
15	....do	do		.00
29	....do	do		.00
ie 5	....do	do		.00
6	....do	do		.00
22	....do	do		a1.00
30	....do	do		.00
ly 10	....do	do		.00
15	....do	do		.00
23	....do	do		.00
30	....do	do		.00
g. 7	....do	do		a1.00
15	....do	do		.00
19	....do	do		.00
ie 16	South Canyon ditch	Canyon		13.00
y 15	South Canyon ditch	Canyon		48.00

a Estimated.

Miscellaneous measurements in Arkansas drainage basin in 1903—Continued.

Date.	Stream.	Locality.	Gage height.	Discharge.
			Feet.	*Sec.-fet.*
Aug. 6	South Canyon ditch	Canyon		.00
28	....do	...do		39.00
Sept. 24	....do	...do		41.00
Nov. 18	....do	...do		.00
Apr. 27	Timpas Creek	At railroad bridge		29.75
28	...do	...do		11.21
May 12	....do	At Swink		13.80
18	...do	...do		30.30
24	...do	...do		9.17
27	...do	...do		12.10
June 7	....do	...do		661.00
17	...do	...do	3.90	66.00
July 1	....do	...do	1.65	a 36.50
2	...do	...do	1.60	b 35.00
8	....do	...do	1.40	30.00
18	....do	...do	1.60	41.00
21	....do	...do	1.50	33.00
Aug. 1	....do	...do	1.30	30.00
5	...do	...do	1.32	21.00
July 25	Two Butte Creek	Holly		.00
Aug. 12	....do	...do		.00
21	....do	...do		.00
May 5	Wild Horse Creek	At mouth		.00
6	.....do	...do		24.30
18	.....do	...do		.00
July 11	....do	...do		.00
12	....do	...do		500.00
24	....do	...do		.00
27	....do	...do		.00
Aug. 13	....do	...do		.00
21	....do	...do		.00
Apr. 21	X Y canal	At road bridge		26.00
May 2	.....do	One-fourth mile from head gate.		24.10
7	....do	At head gate		10.60
Apr. 8	Arkansas River	Holly		7.33
20	...do	1 mile above State line		6.82
20	.....do	At Amity		5.64
21	.....do	At Granada bridge		.70
May 16	.....do	Grote		.00
May 29	.....do	At Colorado and Kansas dam		0.00
June 5	.....do	...do		.00

a By C. W. Beach at railroad bridge.   b By Antoine Jacob at Swink.

Miscellaneous measurements in Arkansas drainage basin in 1903—Continued.

Date.	Stream.	Locality.	Gage height.	Discharge.
			Feet.	*Sec.-feet.*
May 2	Markham Arroyo..	At mouth	....	0.82
7	Manville canal	At head gate	....	13.51
8	Lamar canal	....do	....	12.51
June 21	Kicking Bird canal	At rating flume	6.35	807.00
July 29	Boggs Creek.......	At mouth	....	1.80
June 22	East Fork of Arkansas.[a]	Leadville Junction	2.52	173.00
July 2	....do	....do	2.38	125.00
7	....do	....do	2.13	91.00
13	....do	....do	2.01	76.00
23	....do	....do	1.88	52.00
Aug. 4	....do	....do	1.52	16.00
12	....do	....do	1.48	13.00
June 24	Lake Fork of Arkansas.[a]	Arkansas Junction, Colo......	1.71	170.00
July 1	....do	....do	1.40	133.00
5	....do	....do	.97	84.00
17	....do	....do	2.64	55.00
30	....do	....do	2.27	33.00
Aug. 4	....do	....do	2.07	8.00
12	....do	....do	1.92	6.00
Sept. 5	....do	....do	1.77	10.00
Nov. 25	Deer Creek........	Near Hydro, Okla	....	.23
Dec. 21	....do	....do	....	.19

[a] See following list of gage heights.

Mean daily gage height, in feet, of East Fork of Arkansas River at Leadville Junction for 1903.

Day.	June.	July.	Aug.	Sept.	Oct.	Day.	June.	July.	Aug.	Sept.	Oct.
1		.....	.....	.....	1.42	17		2.04	.....	1.48	.....
2		2.38	1.50	1.51	.....	18	2.60	.....	1.44	...	.....
3		.....	.....	.....	1.42	19		2.00	.....	1.47	.....
4		2.20	1.52	1.50	.....	20	2.50	.....	1.48	1.46	.....
5		.....	.....	.....	1.39	21		1.73	.....	.....	.....
6		2.17	1.47	1.50	.....	22	2.52	.....	1.46	1.42	.....
7		.....	.....	1.67	1.39	23		1.88	.....	.....	.....
8		2.22	1.48	.....	.....	24	2.40	.....	1.94	1.39	.....
9		.....	.....	1.57	1.39	25		.....	1.75	.....	.....
10		2.13	1.48	.....	.....	26		1.76	1.70	1.41	.....
11		.....	.....	1.56	.....	27	2.40	1.72	.....	.....	.....
12		2.22	1.48	.....	1.43	28		.....	1.60	.....	.....
13		2.01	.....	1.55	.....	29		.....	.....	1.47	.....
14		.....	1.44	.....	1.54	30	2.40	.....	1.57	.....	.....
15		2.08	1.44	1.49	.....	31		1.60	.....	.....	.....
16		.....	.....	.....	1.50						

Mean daily gage height, in feet, of Lake Fork of Arkansas River at Arkansas Junction for 1903.

Day.	June.	July.	Aug.	Sept.	Day.	June.	July.	Aug.	Sept.
1	2.00	1.40	2.20	1.82	17	......	2.64	1.98	1.75
2	1.86	1.26	2.16	1.82	18	2.42	2.68	1.94	1.74
3	2.10	1.20	2.12	1.82	19	2.30	2.68	1.90	1.72
4	1.73	1.08	2.07	1.80	20	2.20	2.60	1.87	1.70
5	1.85	.97	2.08	1.77	21	2.00	2.64	1.89	1.88
6	1.68	.87	2.00	1.75	22	2.04	2.60	1.89	1.88
7	2.13	.98	1.98	1.95	23	1.94	2.59	1.90	1.84
8	2.34	1.00	1.97	1.95	24	1.71	2.56	2.00	1.82
9	1.96	1.20	1.97	1.95	25	1.64	2.56	2.00	1.80
10	2.10	1.22	1.96	1.95	26	1.54	2.53	2.00	1.80
11	1.84	1.16	1.92	1.95	27	1.54	2.50	1.97	(a)
12	1.75	1.08	1.92	1.96	28	1.50	2.46	1.94	......
13	1.88	1.06	1.92	1.96	29	1.50	2.34	1.80	......
14	2.28	1.00	1.92	1.96	30	1.46	2.27	1.86	......
15	2.09	.90	1.90	1.96	31	......	2.28	1.84	......
16	2.95	......	1.92	1.96					

a Water shut off during construction at outlet of reservoir.

Mean daily gage height, in feet, of Tennessee Fork of Arkansas River at Leadville Junction for 1903.

Day.	June.	July.	Aug.	Sept.	Oct.	Day.	June.	July.	Aug.	Sept.	Oct.
1					1.26	17				1.26	
2		1.95	1.38	1.25		18	2.50			1.30	
3					1.24	19		1.68		1.24	
4		1.89	1.37	1.25		20	2.40		1.29	1.23	
5					1.23	21		1.58			
6		1.72	1.37	1.25		22	2.29		1.29	1.23	
7			1.38	1.25		23		1.55			
8		1.81	1.38			24	2.15		1.38	1.22	
9				1.32	1.22	25		1.81			
10		1.75	1.27			26		1.55	1.81	1.21	
11				1.31		27	2.10				
12		1.81	1.27		1.21	28		1.48	1.29		
13		1.70		1.30		29				1.26	
14			1.27		1.21	30	2.10		1.27		
15		1.75	1.27	1.30		31	1.38				
16					1.20						

RED RIVER DRAINAGE BASIN.

The headwaters of Red River include several forks, all of which have their sources in northern Texas. Red River takes a general easterly direction along the northern boundary of Texas, and then turns toward the southeast and flows through a low, swampy region in Louisiana into the Mississippi, not far from the southern boundary of Mississippi. Washita River rises in northern Texas, crosses southern Oklahoma, and flows into Red River in the southern part of Indian Territory, about 10 miles from Denison, Tex.

Sulphur Fork of Red River has its headwaters in Hunt and Fannin counties, Tex., flows eastward, forming the boundary between Delta, Red River, and Bowie counties on the north, and Hopkins, Franklin, Titus, Morris, and Cass counties on the south, and empties into Red River in Arkansas about 7 miles north of the boundary line between that State and Louisiana. The flow of this river is very unreliable, changing with the rainfall. If the summer is at all dry it ceases to flow altogether, but there always remains enough water standing in pools to water stock. During or immediately after protracted or unusually heavy rains the river becomes very wide and deep; floods its bottoms, and often occasions considerable loss of stock and damage to planters and the railroads.

Big Cypress River has its headwaters in Franklin and Titus counties, Tex., flows in a general easterly direction, and empties into Red River. The flow of the river is unreliable, varying with the rainfall. In the summer it ceases to flow, becoming dry, except in those places where the water stands in holes. After long or heavy rains the stream is liable to overflow its banks.

The following stations were maintained during 1903 in the Red River drainage basin: Ouachita River near Malvern, Ark.; Washita River at Anardarko, Okla.; Otter Creek near Mountain Park, Okla.; and Red River (North Fork) near Granite, Okla.

OUACHITA RIVER NEAR MALVERN, ARK.

This station is located at the fall line on the river, at the Rockport Bridge, 1¾ miles northwest of Malvern, Ark. The vertical gage, which is fastened to the web between the cylinders of the first pier from the left bank, was installed March 3, 1903, by the General Bauxite Company. It is read once daily by A. M. Baker. Discharge measurements are made from a three-span iron highway bridge 500 feet long and from 30 feet of trestle approach on the right bank, and the first was made by J. M. Giles, May 15, 1903, at which date bench marks were established. The initial point for soundings is the end of the iron bridge on the downstream side at the left bank. The channel is straight for 1,000 feet above and 800 feet below the station. The right bank is high and wooded and overflows only during extreme floods. The left bank does not overflow. The bed of the stream is composed mainly of rock and is permanent. An old dam just below the bridge is the cause of the sluggish current at low water. At this stage, however, measurements can be made by wading at gaps in the dam, where a good velocity may be obtained. Bench mark No. 1 is the top of the first cylindrical pier from the left bank on the downstream side. Its elevation is 27.08 feet above the zero of the gage. Bench mark No. 2 is a copper plug set in the northwest corner of the left concrete abutment. Its elevation is 26.23 feet above the zero of the gage. Bench mark No. 3 is the top of an eyebolt set in the solid rock about 25 feet south of the bridge, at a point 90 feet from the initial point for soundings. Its elevation is 11.86 feet above the zero of the gage.

The observations at this station during 1903 have been made under the direction of M. R. Hall, district hydrographer.

Discharge measurements of Ouachita River near Malvern, Ark., in 1903.

Date.	Hydrographer.	Gage height.	Discharge.
		Feet.	*Second-feet.*
May 15	J. M. Giles	5.60	4,395
May 16	do	5.60	3,455
July 9	do	.65	171
Do	do	.65	196
July 10	do	.80	243

Mean daily gage height, in feet, of Ouachita River near Malvern, Ark., for 1903.

Day.	Mar.	Apr.	May.	June.	July.	Aug.	Sept.	Oct.	Nov.	Dec.
1		3.30	(a)	6.40	1.10	2.60	4.10	1.00	1.00	0.70
2		3.20	(a)	6.00	.90	1.70	3.20	1.00	.90	.70
3	5.70	3.00	1.80	5.00	.90	1.30	2.50	4.40	.90	.70
4	5.50	2.80	1.80	4.30	.90	1.10	2.30	4.00	.80	.70
5	5.50	2.80	1.80	4.00	.90	1.00	2.10	15.00	.80	.70
6	6.40	2.70	2.00	3.50	.90	1.00	1.60	8.70	.80	.70
7	11.30	2.60	2.80	3.00	.90	1.00	1.40	7.00	.80	.70
8	10.90	2.50	9.20	3.00	.80	1.00	1.40	5.60	.80	.70
9	9.90	3.00	20.20	2.70	.70	1.00	1.40	4.30	.80	.70
10	19.40	2.80	10.60	2.90	.80	1.20	1.30	3.70	.80	1.0
11	20.00	2.60	7.20	2.80	.80	3.30	1.00	3.30	.70	1.0
12	18.30	2.80	7.80	2.20	.90	5.00	1.00	3.00	.70	1.0
13	8.80	2.60	7.60	2.00	.80	6.40	1.00	2.70	.70	1.0
14	7.30	2.60	7.20	1.80	.70	5.00	1.00	2.70	.70	1.0
15	6.40	2.40	5.80	1.50	.70	4.80	1.00	2.40	.70	1.0
16	5.70	2.50	5.10	1.40	.70	5.30	1.00	2.00	.70	1.0
17	5.40	2.40	4.60	1.40	.60	4.10	1.90	1.90	.70	1.0
18	5.00	2.30	4.30	1.20	.60	4.00	6.40	1.90	.70	1.0
19	10.60	2.00	4.40	1.00	.60	7.80	4.90	1.80	.70	1.0

a Missing.

Mean daily gage height, in feet, of Ouachita River, etc.—Continued.

Day.	Mar.	Apr.	May.	June.	July.	Aug.	Sept.	Oct.	Nov.	Dec.
20	6.60	1.90	9.50	1.00	0.50	6.10	4.60	1.80	0.70	1.00
21	11.00	1.80	8.50	1.00	.80	3.90	4.60	1.80	.70	1.00
22	8.50	1.80	6.50	.90	.80	3.30	4.60	1.60	.70	1.00
23	6.80	1.80	5.00	.90	.70	2.80	3.60	1.60	.70	1.60
24	5.90	1.80	4.40	.80	.70	2.50	2.80	1.60	.70	1.80
25	5.20	1.80	3.40	.80	.70	2.20	2.00	1.40	.80	3.20
26	4.80	1.80	3.20	1.00	.70	1.90	1.80	1.40	.80	3.00
27	4.60	1.80	2.90	1.50	.70	1.50	1.60	1.40	.80	......
28	4.20	1.80	2.80	1.20	.70	.90	1.30	1.40	.80	......
29	4.00	1.80	2.80	1.20	.60	2.50	1.00	1.30	.70	......
30	3.70	1.80	4.80	1.10	.50	8.00	1.00	1.30	.70	......
31	3.50	......	7.60	......	2.20	6.90	......	1.30	......	......

WASHITA RIVER AT ANADARKO, OKLA.

This station, established October 25, 1902, by W. G. Russell, is located at the highway bridge one-half mile north of the Anadarko railroad depot. The gage is of the wire type, with the scaleboard graduated to feet and tenths, and spiked to the hand rail of the bridge. The initial point for soundings is on the right bank. The channel both above and below the station is straight for 200 feet; the right bank is high, and the left bank is low; both banks are liable to overflow; the bed of the stream is sandy and constant. The observer is James H. Dunlap, who reads the gage once each day.

Bench mark No. 1 is a large spike in the stringer above the second floor beam from the south end of the bridge on the downstream side. Its elevation is 22.92 feet above gage datum. Bench mark No. 2 is a spike driven 4 feet above the ground in the trunk of a cottonwood tree near the south end of the bridge. Its elevation is 26.74 feet above gage datum.

The observations at this station during 1903 have been made under the direction of G. H. Matthes, district engineer.

Discharge measurements of Washita River at Anadarko, Okla., in 1903.

Date.	Hydrographer.	Gage height.	Discharge.
		Feet.	*Second-feet.*
August 10	L. M. Holt	3.50	167
September 19	F. Bonstedt	3.30	94
October 19	E. R. Kerby	2.70	82
November 23	...do	3.20	94
December 19	...do	3.30	108

Mean daily gage height, in feet, of Washita River at Anadarko, Okla., for 1903.

Day.	Jan.	Feb.	Mar.	Apr.	May.	June.	July.	Aug.	Sept.	Oct.	Nov.	Dec.
1	3.10	2.80	13.30	4.30	6.60	13.40	6.10	3.80	3.20	3.00	3.00	3.10
2	3.10	2.80	11.70	4.30	6.60	14.40	5.70	3.70	3.20	3.00	3.00	3.10
3	3.10	2.80	5.70	4.30	6.40	14.40	5.50	3.70	3.20	3.10	3.00	3.10
4	3.10	2.80	5.00	4.10	5.20	10.30	5.40	3.70	3.20	3.00	3.00	3.10
5	3.10	2.80	4.70	4.10	4.70	9.00	5.30	3.70	3.20	3.00	3.00	3.10
6	3.00	2.80	4.60	4.20	4.20	8.40	5.20	3.60	3.20	3.10	3.00	3.10
7	3.00	2.80	4.40	4.20	4.00	7.90	5.10	3.60	3.20	3.00	3.00	3.10
8	2.90	2.80	4.30	4.30	4.30	7.70	5.00	3.60	3.20	3.00	3.10	3.10
9	2.90	2.80	4.20	3.90	4.30	7.80	4.90	3.50	3.10	3.00	3.10	3.10
10	2.90	2.80	4.20	4.70	4.60	7.90	4.80	3.50	3.10	3.00	3.20	3.10
11	2.90	2.80	4.70	5.30	6.50	7.90	4.70	3.50	3.20	3.05	3.20	3.10
12	2.90	2.90	5.10	5.10	8.00	7.90	4.60	3.50	3.10	3.00	3.20	3.10
13	2.90	3.00	4.40	4.90	7.40	7.50	4.60	3.60	3.10	3.00	3.20	3.10
14	2.90	3.00	5.10	4.20	5.90	7.20	4.50	3.70	3.10	3.00	3.20	3.10
15	2.80	2.90	4.80	4.00	6.90	4.40	3.60	3.00	3.00	3.00	3.20	3.10
16	2.80	2.90	4.40	3.80	4.80	6.60	4.40	3.50	3.00	3.00	3.20	3.10
17	2.80	2.80	4.00	3.70	6.40	6.00	4.30	3.50	3.00	3.00	3.20	3.10
18	2.80	2.80	4.00	3.70	4.50	6.40	4.20	3.40	3.00	3.00	3.15	3.10
19	2.80	2.80	6.00	3.60	4.60	6.20	4.20	3.50	3.00	3.00	3.10	3.10
20	2.80	2.80	5.20	3.60	6.10	6.10	4.20	3.40	3.10	3.00	3.10	3.10
21	2.80	3.00	9.90	3.60	5.50	6.10	4.10	3.40	3.10	3.00	3.10	3.10
22	2.80	3.20	9.00	3.50	6.70	6.00	4.10	3.80	3.10	3.00	3.10	3.10
23	2.90	3.70	7.30	3.50	9.30	6.00	4.10	3.30	3.10	3.00	3.20	3.10
24	2.90	4.80	5.60	3.50	14.80	6.10	4.00	3.30	3.00	3.00	3.20	3.10
25	2.90	5.10	5.00	3.50	26.80	6.20	4.00	3.30	3.00	3.00	3.20	3.10
26	2.90	6.90	4.70	3.50	22.80	7.10	4.00	3.60	3.00	3.00	3.30	3.10
27	2.90	8.80	4.50	3.50	21.70	7.60	3.90	3.30	3.00	3.00	3.30	3.10
28	2.90	10.60	4.40	3.50	21.10	7.80	3.80	3.30	3.00	3.00	3.30	3.10
29	2.90	......	4.40	3.60	20.80	8.10	3.80	4.30	3.00	3.00	3.30	3.10
30	2.90	......	4.80	4.20	18.70	7.60	3.80	3.60	3.00	3.00	3.20	3.10
31	2.90	......	4.80	......	11.80	......	3.80	3.30	......	3.00	......	3.10

OTTER CREEK NEAR MOUNTAIN PARK, OKLA.

This station was established April 2, 1903, by G. H. Matthes,
assisted by Ferd. Bonstedt. It is located on G. M. Dale's homestead,
SE. ¼ sec. 21, T. 3 N., R. 17 W., Indian meridian. It is 2 miles west
and 1 mile north of Mountain Park, Okla. The gage consists of a
2 by 6 inch board 26 feet long, bolted to a cottonwood tree which
stands at the water's edge. It is graduated to feet and tenths and
marked with brass figures. The gage is read daily by G. M. Dale.
The channel is slightly curved both above and below the station.
Both banks are about 20 feet high, of sandy loam, covered with vegetation, and subject to overflow. The bed of the stream is of sand,
and is liable to shift. The water flows in one chanel at normal stages
but when about to overflow its banks part of the water is diverted
through a slough into Horse Creek. Measurements of discharge are
made by wading, as there are no bridges in this locality. Bench mark

No. 1 is a 20-penny nail driven into the cottonwood tree to which the gage is attached. It is at the elevation of the 10-foot mark on the gage. Bench mark No. 2 is a nail driven into a mesquite tree 150 feet southwest of G. M. Dale's house. It is 28.62 feet above the zero of the gage. Bench mark No. 3 is a nail driven into a hackberry tree 50 feet east of the creek and 20 feet north of the gage. It is 8.94 feet above the zero of the gage.

The observations at this station during 1903 have been made under the direction of G. H. Matthes, district engineer.

Discharge measurements of Otter Creek near Mountain Park, Okla., in 1903.

Date.	Hydrographer.	Gage height.	Discharge.
		Feet.	*Second-feet.*
April 20	Ferd. Bonstedt	1.35	7.0
May 1	....do	2.20	42.0
May 19	L. M. Holt	5.70	363.0
May 24	....do	9.90	879.0
August 13	Holt and Kerby	.95	.7
September 18	F. Bonstedt	.80	.2
October 22	E. R. Kerby	.90	.3
November 26	....do	1.40	3.0
December 23	....do	1.35	2.0

Mean daily gage height, in feet, of Otter Creek near Mountain Park, Okla., for 1903.

Day.	Apr.	May.	June.	July.	Aug.	Sept.	Oct.	Nov.	Dec.
1		1.80	1.40	20	1.00	0.90	1.00	1.20	1.40
2		1.60	1.40	1.20	1.00	.90	1.00	1.10	1.40
3		1.50	1.40	.20	1.00	.90	.90	1.10	1.40
4		1.40	1.40	20	1.00	.90	.90	1.10	1.40
5		1.40	1.40	1.20	1.00	.90	.90	1.20	1.40
6		1.40	1.40	20	1.00	.90	.90	1.10	1.40
7		1.40	1.40	1.10	.90	.90	.90	1.20	1.40
8		1.40	1.40	10	.90	.90	.90	1.20	1.40
9		1.40	1.60	1.10	.90	.90	.90	1.20	1.40
10		1.90	1.40	10	1.00	.90	.90	1.30	1.40
11		1.90	1.40	1.10	1.00	.90	.90	1.30	1.40
12		1.90	1.30	.00	1.00	.90	.80	1.30	1.50
13		1.60	1.30	1.00	1.00	.90	.80	1.30	1.40
14		1.50	1.40	1.00	.90	.90	.90	1.30	1.40
15		1.50	1.30	00	.90	.90	.90	1.40	1.40
16		1.40	1.30	1.00	.90	.90	.90	1.40	1.40
17		1.40	1.30	.00	.90	.90	.90	1.40	1.40
18		1.40	1.30	1.00	.90	.80	.90	40	1.40
19		6.50	1.20	.00	.90	.90	.90	1.40	1.40
20	1.40	2.70	1.20	00	.90	.90	.90	1.40	1.40
21	1.30	2.00	1.40	00	.90	.90	.90	1.50	1.40

Mean daily gage height, in feet, of Otter Creek, etc.—Continued.

Day.	Apr.	May.	June.	July.	Aug.	Sept.	Oct.	Nov.	Dec.
22	1.40	1.90	1.30	1.00	0.90	0.90	0.90	1.50	1.40
23	1.40	3.40	1.30	1.00	.90	.90	.90	1.50	1.40
24	1.40	2.10	1.20	1.00	.90	.90	.90	1.50	1.40
25	1.40	4.00	1.20	1.00	.90	.90	.90	1.40	1.40
26	1.30	2.60	1.40	1.00	.90	.80	.90	1.40	1.40
27	1.40	1.80	1.40	1.00	.90	.80	.90	1.40	1.40
28	1.40	1.60	1.30	1.00	.90	.80	1.00	1.40	1.40
29	3.90	1.80	1.30	1.00	.90	.90	1.00	1.40	1.30
30	2.70	1.50	1.30	1.00	.90	1.00	1.00	1.40	1.30
31		1.40		1.00	.90		1.10		1.40

Rating table for Otter Creek near Mountain Park, Okla., from April 20 to May 31, 1903.

Gage height.	Discharge.	Gage height.	Discharge.	Gage height.	Discharge.	Gage height.	Discharge.
Feet.	*Second-feet.*	*Feet.*	*Second-feet.*	*Feet.*	*Second-feet.*	*Feet.*	*Second-feet.*
1.3	6	2.2	48	3.1	109	4.0	188
1.4	8	2.3	49	3.2	117	4.2	207
1.5	11	2.4	56	3.3	125	4.4	227
1.6	14	2.5	63	3.4	134	4.6	247
1.7	18	2.6	70	3.5	143	4.8	267
1.8	22	2.7	77	3.6	152	5.0	287
1.9	27	2.8	85	3.7	161	6.0	397
2.0	32	2.9	93	3.8	170		
2.1	37	3.0	101	3.9	179		

Rating table for Otter Creek near Mountain Park, Okla., from June 1 to December 31, 1903.

Gage height.	Discharge.	Gage height.	Discharge.	Gage height.	Discharge.	*Gage height.	Discharge.
Feet.	*Second-feet.*	*Feet.*	*Second-feet.*	*Feet.*	*Second-feet.*	*Feet.*	*Second-feet.*
0.8	0.2	1.1	1.8	1.3	2.3	1.5	3.4
.9	.4	1.2	1.8	1.4	2.8	1.6	4.0
1.0	.8						

Tangent at 6 feet gage height, with differences of 12 per tenth.

Estimated monthly discharge of Otter Creek, near Mountain Park, Okla., for 1903.

[Drainage area, 126 square miles.]

Month.	Discharge in second-feet.			Total in acre-feet.	Run-off.	
	Maximum.	Minimum.	Mean.		Second-feet per square mile.	Depth in inches.
April 20–30	179.0	6.0	29.5	644	0.23	0.09
May	457.0	6.0	41.4	2,546	.33	.38
June	4.0	1.8	2.5	149	.02	.02
July	1.8	.8	1.1	68	.009	.01
August	.8	.4	.5	31	.004	.005
September	.8	.2	.4	24	.003	.003
October	1.8	.2	.5	31	.004	.005
November	3.4	1.3	2.4	143	.019	.021
December	2.8	2.3	2.8	172	.022	.025

RED RIVER (NORTH FORK) NEAR GRANITE, OKLA.

This station was established June 23, 1903, by Ferd. Bonstedt. It is located at the highway bridge 2 miles east and one-half mile north of Granite, Okla. The Chicago, Rock Island and Pacific Railroad crosses the river near this point. The gage is of the wire type and is located on the downstream side of the bridge near the west end. The length from the bottom of the weight to the marker is 20.32 feet. The marker reads 17 feet when the weight is pulled up against the bottom of the bridge. The gage is read daily by Elmer O. Tompkins. Discharge measurements are made from the bridge. The channel is straight for about 500 feet above the station and for about 300 feet below. Both banks are subject to overflow at flood stages. The bed of the stream is sandy and shifting. The water flows in three channels at low water. The gage is referred to a United States Geological Survey standard bench mark (cast-iron post) which is 292 feet east of the bridge near the south line of the highway. Its elevation above gage datum is 10.64 feet and above sea level 1,539.8 feet.

The observations at this station during 1903 have been made under the direction of G. H. Matthes, district engineer.

Discharge measurements of North Fork of Red River near Granite, Okla., in 1903.

Date.	Hydrographer.	Gage height.	Discharge.
		Feet.	Second-feet.
August 12	Holt and Kerby	4.60	61
September 19	F. Bonstedt	3.70	(a)
June 27	do	4.80	72

a No flow; water in pools.

Mean daily gage height, in feet, of North Fork of Red River near Granite, (
for 1903.

Day.	June.	July.	Aug.	Sept.	Oct.	Nov.
1		4.60	4.60	4.60	4.10	(b)
2		4.50	4.10	4.40	4.00	(b)
3		4.50	3.80	4.30	3.90	(b)
4		4.40	(a)	4.00	3.90	(b)
5		4.40	(a)	3.80	3.90	(b)
6		4.40	(a)	3.70	3.80	(b)
7		4.30	(a)	(b)	(b)	(b)
8		4.20	(a)	(b)	(b)	(b)
9		4.20	(a)	(b)	(b)	(b)
10		4.10	(a)	(b)	(b)	(b)
11		4.00	(a)	(c)	(b)	(b)
12		4.10	(a)	(c)	(b)	(b)
13		3.90	(a)	(b)	(b)	(b)
14		3.90	(a)	3.80	(b)	(b)
15		3.80	(a)	(b)	(b)	(b)
16		3.70	4.30	3.80	(b)	(b)
17		3.60	4.10	3.60	(b)	(b)
18		4.00	3.90	3.70	(b)	(b)
19		3.60	3.80	(b)	(b)	(b)
20		4.00	3.70	(b)	(b)	(b)
21		3.80	3.70	(b)	(c)	(b)
22		3.80	3.70	(b)	(b)	(b)
23	5.00	3.70	3.70	(b)	(b)	(b)
24	4.90	3.60	3.70	(b)	(b)	(b)
25	4.90	3.60	3.70	(b)	(b)	(b)
26	4.80	3.60	3.70	(b)	(b)	(b)
27	4.80	3.60	....	(b)	(b)	(b)
28	5.00	3.60	....	(b)	(b)	(b)
29	4.90	3.60	....	3.90	(b)	(b)
30	4.70	3.60	4.90	3.80	(b)	(b)
31		3.50	4.80	....	(b)	....

a Gage chain broke.   *b* Water standing in pools.   *c* Observer absent.

Miscellaneous measurements in Red River drainage basin.

Date.	Hydrographer.	Stream.	Locality.	cl
1903.				Se
June 30	Ferd. Bonstedt ..	Turkey Creek .	Near Victory, Okla	
Sept. 18	.....do	Elk Creek	Near Hobart, Okla.......	
Oct. 22	E. R. Kerby	.....do	.....do	
Nov. 26	.....do	.....do	.....do	
Dec. 22	.....do	.....do	.....do	
Oct. 27	G. H. Matthes ...	Little Elk Creek	.....do	
Dec. 31	E. R. Kerby	Medicine Bluff Creek.	Near Meers, Okla........	
31	.....do	Blue Creek....	T. 5 N., R. 13 W, I. M., Okla.	
31	.....do	Chandler Creek	Near Richards, Okla.....	

Measurements of springs in Wichita Mountains.

Date.	Hydrographer.	Spring.	Locality.	Discharge.
903.				*Second-ft.*
g. 9	G. H. Matthes	Jimmie	SE. ¼ sec. 17, T. 4 N., R. 13 W., I. M.	1.00
)t. 13	....do	do	....do	1.00
t. 22	....do	do	....do	.60
30	....do	do	....do	.50
w. 28	E. R. Kerby	do	....do	.70
c. 8	....do	do	....do	.52
8	....do	do	....do	.54
29	....do	do	....do	.52
29	....do	Frizzle Head	NW. ¼ sec. 13, T. 4 N., R. 14 W., I. M.	.40
29	....do	Springfield	NE. ¼ sec. 14, T. 4 N., R. 14 W., I. M.	.20
29	....do	Thompson No.1	SE. ¼ sec. 10, T. 4 N., R. 14 W., I. M.	.35
29	....do	Thompson No.2	NE. ¼ sec. 10, T. 4 N., R. 11 W., I. M.	.15
30	....do	Podlequah	SE. ¼ sec. 4, T. 4 N., R. 14 W., I. M.	.25
30	....do	Saddle Mountain No. 1.	Saddle Mountain Mission, Okla.	.53
30	...do	Saddle Mountain No. 2.	......do	.20
30	....do	No name	NE. ¼ sec. 15, T. 5 N., R. 14 W., I. M.	.25
31	....do	Eagle Heart	NE. ¼ sec. 5, T. 5 N., R. 13 W., I. M.	.20
31	....do	Custer	SW. ¼ sec. 13, T. 4 N., R. 12 W., I. M.	.50
31	....do	No name	NW. ¼ sec. 36, T. 4 N., R. 12 W., I. M.	.20
5	A. C. Redman	Rock Creek	Near Sulphur, Ind. T	13.00
5	....do	Sulphur Creek	....do	6.00
5	....do	Seven Springs	do	.13
5	....do	Beach Springs	......do	.11
5	....do	Hillside	......do	.08
5	....do	Wilson	do	.01

WESTERN GULF DRAINAGE.

Those stations from which the United States Geological Survey has obtained data during 1903 and which are located on rivers which drain into that portion of the Gulf of Mexico west of the Mississippi River, have been grouped into the following drainage basins: Sabine, Neches, Trinity, Brazos, Colorado (of Texas), Guadalupe, and Rio Grande.

SABINE RIVER DRAINAGE BASIN.

Sabine River has its headwaters in Collin and Hunt counties, flows in a southeasterly direction to the State line, then south, forming the boundary between Texas and Louisiana, and empties into Sabine Lake, an arm of the Gulf, near Orange, Tex. The small tributaries in east Texas support many small water mills, and the Sabine itself is navigable for several hundred miles. The drainage area of the Sabine in Texas above Orange is 7,500 square miles, and its total drainage area above Orange in Louisiana and Texas is 10,400 square miles.

SABINE RIVER NEAR LONGVIEW, TEX.

This station was established December 31, 1903, by Thomas U. Taylor. It is located at the bridge of the International and Great Northern Railway, about 3 miles southwest of Longview Junction, Tex. The zero of the gage is 45.00 feet below the top of the tie at the railroad bridge. The observer is John Wadsock, the pumper for the railroad company. A measurement made at this point by Edward C. H. Bantel gave a discharge of 249 second-feet.

TRINITY RIVER DRAINAGE BASIN.

Trinity River rises in a network of small streams in the counties of Montague, Wise, and Parker, but their combined capacity at Dallas is not sufficient to keep the bottom or bed of the stream moist. The United States Geological Survey maintained a station at Dallas for a time, but it was abandoned on account of the small discharge. Below Dallas the Trinity flows through a wooded country, and consequently it is not subject to the sudden floods with their quick run-offs.

TRINITY RIVER AT RIVERSIDE, TEX.

The drainage area of Trinity River above Riverside, Tex., is 16,000 square miles. A gaging station was established on the Trinity at Riverside, Tex., in December, 1902, by Thomas U. Taylor. The zero of the gage is 66 feet below the top of the ties (or base of rail) in the north arm of the draw span of the International and Great Northern Railroad bridge. The elevation of the top of the pivot pier above

;age datum is 56.5 feet, and that of the top of the channel of the lower
:hord of the arms of the draw span of the bridge is 62.9 feet. Accord-
ng to the survey of the United States Army engineers the elevation
)f the top of the tie with reference to mean low tide of Gulf is 148.70
'eet. The gage consists of a tagged plumber's chain, to which is
attached a lead weight in the form of a frustum of a cone. The bot-
;om of the lead weight is marked 66, and every foot above this is
marked with a brass tag giving its distance in feet above the bottom
)f the weight. The observer at Riverside is G. W. Higdon, who is
in charge of the pumping plant of the International and Great
Northern Railroad. In reading the gage it is only necessary to let
the lead weight touch the water and then read off the distance the
mark or point is from the upper end or zero mark of the chain.

The observations at this station during 1903 have been made under
the direction of Thomas U. Taylor, district hydrographer.

Discharge measurements of Trinity River at Riverside, Tex., in 1902 and 1903.

Date.	Hydrographer.	Gage height.	Discharge.
1902.		*Feet.*	*Second-feet.*
September 10...............	T. U. Taylor................	6.00	160
1903.			
March 18..........	.....do	43.30	24,650
July 1.............	.....do	13.30	3,730
July 2	.....do	14.60	4,500
Do	.....do	15.20	5,160
July 3..............	.....do	18.30	7,200
July 12..........	.....do	22.20	8,800
July 13............	.....do	22.80	9,200
July 15............	.....do	23.60	10,800
June 17	.....do	10.20	980
December 7	G. W. Higdon	8.30	420
December 12	.....do	9.30	500

Mean daily gage height, in feet, of Trinity River at Riverside, Tex., for 1903.

Day.	Jan.	Feb.	Mar.	Apr.	May.	June.	July.	Aug.	Sept.	Oct.	Nov.	Dec.
1............	28.30	19.55	43.90	41.10	10.90	15.75	12.75	15.60	9.60	8.05	9.60	8.30
2............	27.90	17.15	46.20	41.10	10.95	13.40	15.10	19.65	10.55	8.25	9.65	8.30
3..........	25.80	15.15	43.60	41.10	11.00	10.75	18.45	22.40	10.20	8.55	10.00	8.30
4............	24.10	14.20	44.60	41.00	10.90	9.90	18.90	25.05	9.65	9.00	9.80	8.30
5............	21.05	13.75	44.70	40.70	10.80	9.65	22.05	26.65	9.45	10.85	9.80	8.30
6............	18 50	13.70	44.70	40.25	10.55	9.45	20.05	27.65	9.20	14.05	10.20	8.30
7............	13.35	17.40	44.80	39.05	10.00	9.30	19.30	28.70	9.15	17.25	11.35	8.30
8............	15.65	19.05	45.05	38.65	10.00	9.65	19.80	29.75	9.00	19.15	11.85	8.30

Mean daily gage height, in feet, of Trinity River, etc.—Continued.

Day.	Jan.	Feb.	Mar.	Apr.	May.	June.	July.	Aug.	Sept.	Oct.	Nov.	Dec.
9	15.05	20.70	45.25	37.15	10.00	9.20	20.45	30.60	8.65	20.05	11.55	8.30
10	14.30	21.70	47.00	34.70	10.00	9.25	21.25	31.35	8.45	20.65	11.00	8.30
11	13.75	23.50	46.90	29.10	10.00	10.15	21.80	31.70	8.30	21.30	10.50	8.30
12	13.35	23.85	46.45	20.85	10.00	9.50	22.25	31.50	8.20	21.65	10.00	10.50
13	13.05	24.60	46.10	15.45	10.35	9.20	22.50	29.60	8.10	21.80	9.65	13.60
14	12.75	24.30	45.70	14.25	10.75	10.20	23.00	28.15	8.05	21.60	9.45	14.65
15	12.35	23.80	45.25	13.65	10.90	10.20	23.35	14.45	8.00	21.20	9.15	14.15
16	12.45	28.80	44.15	13.05	10.95	10.20	23.65	10.70	7.90	20.75	9.05	13.35
17	13.85	30.90	44.05	12.85	11.45	10.20	23.85	9.35	8.10	19.70	8.90	11.60
18	13.60	32.40	43.45	12.80	11.55	10.20	24.05	9.50	8.70	18.75	8.80	9.85
19	13.40	33.20	43.05	12.80	11.10	10.30	24.25	9.55	8.10	18.15	8.75	9.50
20	13.15	34.60	44.10	13.50	10.60	10.25	24.45	10.10	7.90	18.05	8.70	9.65
21	20.75	34.50	44.40	13.95	10.10	10.15	26.15	10.00	7.80	18.25	8.60	10.90
22	21.05	34.50	44.70	13.40	9.80	10.25	26.35	9.95	7.75	17.70	8.50	10.50
23	21.05	34.70	40.65	12.35	9.70	11.20	25.00	9.65	7.70	16.30	8.50	10.35
24	21.05	34.85	44.30	11.90	9.55	10.85	24.70	9.50	7.60	14.70	8.40	9.80
25	19.35	35.10	43.70	11.55	9.45	10.50	22.40	9.50	7.55	12.80	8.40	10.65
26	18.30	40.10	43.10	11.40	9.30	13.40	17.40	9.45	7.50	11.50	8.40	12.40
27	19.60	42.85	42.55	11.25	9.25	12.95	12.45	9.40	7.50	10.60	8.40	12.70
28	22.20	43.60	42.10	11.05	9.15	11.90	10.45	9.15	7.50	10.05	8.40	12.25
29	24.10	......	41.60	11.00	9.10	11.15	10.30	8.85	7.50	9.65	8.30	11.90
30	24.25	......	41.40	11.00	13.50	11.30	10.40	8.65	7.70	9.45	8.30	10.00
31	22.30	......	41.20	......	16.30	......	10.95	8.60	......	9.45	......	9.55

Rating table for Trinity River at Riverside, Tex., from January 1 to December 31, 1903.

Gage height.	Discharge.	Gage height.	Discharge.	Gage height.	Discharge.	Gage height.	Discharge.
Feet.	Second-feet.	Feet.	Second-feet.	Feet.	Second-feet.	Feet.	Second-feet.
7.5	225	11.0	2,070	21.0	9,070	36.0	19,570
7.6	245	11.5	2,420	22.0	9,770	37.0	20,270
7.8	285	12.0	2,770	23.0	10,470	38.0	20,970
8.0	335	12.5	3,120	24.0	11,170	39.0	21,670
8.2	390	13.0	3,470	25.0	11,870	40.0	22,370
8.4	455	13.5	3,820	26.0	12,570	41.0	23,070
8.6	530	14.0	4,170	27.0	13,270	42.0	23,770
8.8	620	14.5	4,520	28.0	13,970	43.0	24,470
9.0	725	15.0	4,870	29.0	14,670	44.0	25,170
9.2	840	15.5	5,220	30.0	15,370	45.0	25,870
9.4	965	16.0	5,570	31.0	16,070	46.0	26,570
9.6	1,095	17.0	6,270	32.0	16,770	47.0	27,270
9.8	1,230	18.0	6,970	33.0	17,470		
10.0	1,370	19.0	7,670	34.0	18,170		
10.5	1,720	20.0	8,370	35.0	18,870		

Table uncertain for low gage heights; tangent above 10 feet gage height.

Estimated monthly discharge of Trinity River at Riverside, Tex., for 1903.

[Drainage area, 16,000 square miles.]

Month.	Discharge in second-feet.			Total in acre-feet.	Run-off.	
	Maximum.	Minimum.	Mean.		Second-feet per square mile.	Depth in inches.
January.............	14,180	3,015	7,302	448,982	0.46	0.53
February	24,890	3,960	13,189	732,480	.82	.85
March	27,270	22,825	25,276	1,554,161	1.58	1.82
April	23,140	2,070	10,049	597,957	.63	.70
May	5,780	780	1,780	109,448	.11	.13
June................	5,395	840	1,849	110,023	.12	.13
July	12,815	1,580	8,320	511,577	.52	.60
August	16,560	530	6,572	404,097	.41	.47
September	1,755	225	537	31,954	.03	.03
October	9,630	347	5,219	320,904	.33	.38
November..........	2,665	420	1,031	61,349	.06	.07
December	4,625	420	1,642	100,963	.10	.12
The year	27,270	225	6,897	4,983,895	.43	5.83

BRAZOS RIVER DRAINAGE BASIN.

This river has its source in the Staked Plains region of western Texas and has a general southeasterly course, emptying into the Gulf of Mexico south of the mouth of Trinity River. Its drainage basin is entirely within the State of Texas.

Under the direction of Thomas U. Taylor the United States Geological Survey is maintaining stations in this basin at Waco and Richmond, Tex.

BRAZOS RIVER AT WACO, TEX.

On September 14, 1898, a gage was established on the southwest bank of Brazos River at Waco. It consists of an inclined iron bar, 3 inches by 1 inch, reading from 0 to 4.3 feet, bolted to a hard-pine stick, 16 feet long, embedded in cement in the sloping limestone of the bank, flush with the surface, on which are painted the graduations above 4.3 feet.

This part of the gage is inclined to the horizontal at a slope of 27 horizontal to 5 vertical. In the summer of 1903 another section was added, with its lower end connected to the upper end of the first gage. It is similar to the first gage in construction, but is inclined at a slope of 9 horizontal to 4 vertical. It reads from 4.4 to 12 feet.

Three bench marks have been established. The first is on the low-

est water table on the southwest pier of the suspension bridge an
marked "U. S. G. S. 44.33 B. M." It is about on the level of
floor of the suspension bridge.

The hydrant at the corner of First and Austin streets is at an
vation (by gage) of 43.32 feet, while the top of the rail of the
Antonio and Aransas Pass Railroad a few feet from the hydrant i
an elevation of 41.12 feet. The bed of the river is shifting sand,
nearly every freshet modifies the cross section, so that at the s
gage heights the river sometimes flows in one and sometimes in
channels under the suspension bridge from which the measureme
are made.

At high water the gage reading is obtained by measuring to
water surface from the top rail of the stiffening truss of the sus
sion bridge at a certain point when there is no load upon the bri
and by taking this distance from 47.8 feet.

In the early part of 1902 a new camel-back truss bridge of one s
was erected across the Brazos at Waco a few hundred feet above
suspension bridge. This new bridge crosses the river at an angl
76°. It has a footway on the east or downstream side that aff
excellent facilities for measuring the flow of the stream, and there
no midstream piers to render measurements troublesome or doubt

At the north end these bridges are 280 feet apart, and at the so
end they are 380 feet apart. When the river is rising and d
prevents the use of the meter, good float measurements can be m
by timing the drift as it passes from the upper to the lower bridg
the different panel points.

On the north pier of the new bridge a gage has been marked of
the city engineer to agree with the United States Geological Su
gage at the suspension bridge. The top of the cement floor of
new bridge at the southeast batter brace is at an elevation of
feet with respect to the United States Geological Survey gage. H
water gage heights can be read directly from the gage on the n
pier, or the distance of the water surface can be measured from
cement floor, and this subtracted from 45.4 feet will give the heigl
the river referred to the gage.

The channel is straight for several hundred feet above and b
the station and has a width at low stages of about 175 feet witl
piers. The bed is composed of firm sand, subject to some cha
The current is rapid.

The observations at this station during 1903 have been made u
the direction of Thomas U. Taylor, district hydrographer.

Discharge measurements of Brazos River at Waco, Tex., in 1903.

Date.	Hydrographer.	Gage height.	Discharge.
		Feet.	*Second-feet.*
June 4	Thomas U. Taylor	3.00	426
June 15	....do	5.80	2,970
June 17	....do	6.00	3,240
Do	....do	6.00	3,150
June 25	....do	8.10	6,700
Do	....do	4.20	1,200
June 26	....do	6.70	4,000
July 14	....do	3.20	490
July 22	....do	2.80	245
July 23	....do	2.75	235
July 24	....do	2.70	210
July 30	....do	3.70	810
August 1	....do	5.70	2,670
December 18	E. C. Bantel	2.40	144

Mean daily gage height, in feet, of Brazos River at Waco, Tex., for 1903.

Day.	Jan.	Feb.	Mar.	Apr.	May.	June.	July.	Aug.	Sept.	Oct.	Nov.	Dec.
1	3.40	3.70	13.55	5.40	3.20	3.30	5.10	5.65	2.55	14.70	3.15	2.60
2	3.80	3.70	10.95	5.35	3.20	3.20	5.35	5.35	2.50	8.75	3.65	2.60
3	3.80	3.70	9.10	5.30	3.20	3.15	7.00	4.45	2.75	6.85	3.45	2.60
4	3.60	3.65	9.50	5.00	3.20	3.00	6.05	3.75	3.65	6.30	3.30	2.60
5	3.60	3.60	9.35	4.55	3.10	3.00	9.00	3.65	2.90	7.75	3.25	2.60
6	3.60	3.60	8.20	4.35	3.15	3.50	5.45	3.45	2.55	9.90	3.30	2.55
7	3.50	3.60	8.20	4.35	3.20	4.55	3.20	2.50	9.60	3.60	2.50
8	3.40	3.60	7.60	4.45	3.20	4.65	4.40	3.15	2.50	10.35	3.70	2.50
9	3.40	3.65	6.55	4.25	3.20	4.70	4.15	3.00	2.40	8.35	3.60	2.50
10	3.40	3.90	6.10	4.15	3.20	3.90	3.80	3.00	2.40	7.55	3.50	2.45
11	3.40	4.40	11.10	4.10	3.50	3.70	3.65	2.95	2.40	8.90	3.35	2.40
12	3.40	4.10	7.50	4.10	3.65	3.75	3.45	4.65	2.40	7.10	3.20	2.40
13	3.40	4.00	8.95	4.10	3.35	4.05	3.35	5.45	2.35	6.00	3.15	2.40
14	3.40	4.00	9.50	4.05	3.25	4.15	3.25	4.20	2.30	5.45	3.00	2.40
15	3.30	8.90	8.35	4.00	3.20	4.50	3.15	4.05	2.30	4.55	3.10	2.40
16	3.50	11.85	7.00	3.90	3.15	6.70	3.10	4.00	2.30	4.50	3.45	2.40
17	3.60	6.15	6.90	3.90	3.10	5.80	3.00	4.05	2.30	4.40	3.55	2.40
18	3.60	6.25	6.70	3.80	3.10	4.45	2.95	3.50	2.25	4.15	3.35	2.40
19	3.60	6.30	5.95	3.75	3.10	5.20	2.90	3.35	2.20	4.00	3.15	2.30
20	3.60	6.30	11.20	3.65	3.15	4.65	2.90	3.30	3.15	3.80	3.00	2.30
21	4.80	7.10	13.10	3.55	3.60	4.00	2.90	3.25	3.15	3.65	2.95	2.30
22	6.30	6.90	7.20	3.50	3.50	4.65	2.85	3.15	3.00	3.55	2.90	2.30
23	5.95	6.75	6.05	3.50	3.50	4.50	2.75	2.95	3.55	3.50	2.90	2.30
24	4.50	6.65	5.95	3.45	3.50	4.15	2.70	2.75	4.50	3.40	2.90	2.30
25	4.15	6.60	5.90	3.40	3.40	6.20	2.70	2.70	4.35	3.30	2.85	2.35
26	4.00	6.60	5.80	3.40	3.50	6.30	2.70	2.60	4.00	3.10	2.75	2.40
27	4.20	20.45	6.10	3.35	3.15	5.15	2.70	2.50	3.95	3.05	2.70	2.35
28	4.10	13.25	6.30	3.30	3.10	5.95	2.70	2.50	3.65	3.00	2.70	2.30
29	4.00		6.15	3.30	5.70	5.00	4.10	2.50	3.35	3.00	2.70	2.30
30	3.90		5.90	3.25	4.75	4.80	3.65	2.55	7.80	3.00	2.60	2.30
31	3.85		5.45		3.45		8.90	2.60		3.10		2.30

Rating table for Brazos River at Waco, Tex., from January 1 to December 31, 1903.

Gage height.	Discharge.	Gage height.	Discharge.	Gage height.	Discharge.	Gage height.	Discharge.
Feet.	*Second-feet.*	*Feet.*	*Second-feet.*	*Feet.*	*Second-feet.*	*Feet.*	*Second-feet.*
2.2	90	4.0	1,055	6.6	3,850	10.5	11,160
2.3	125	4.1	1,140	6.8	4,150	11.0	12,285
2.4	160	4.2	1,225	7.0	4,470	11.5	13,450
2.5	195	4.3	1,310	7.2	4,800	12.0	14,660
2.6	235	4.4	1,400	7.4	5,135	12.5	15,980
2.7	275	4.5	1,490	7.6	5,475	13.0	17,400
2.8	320	4.6	1,580	7.8	5,825	13.5	18,850
2.9	365	4.7	1,670	8.0	6,180	14.0	20,300
3.0	410	4.8	1,760	8.2	6,540	14.5	21,800
3.1	460	4.9	1,860	8.4	6,910	15.0	23,300
3.2	510	5.0	1,960	8.6	7,285	15.5	24,800
3.3	565	5.2	2,160	8.8	7,665	16.0	26,300
3.4	620	5.4	2,380	9.0	8,055	17.0	29,300
3.5	680	5.6	2,600	9.2	8,450	18.0	32,300
3.6	745	5.8	2,820	9.4	8,850	19.0	35,300
3.7	815	6.0	3,060	9.6	9,255	20.0	38,300
3.8	890	6.2	3,310	9.8	9,665		
3.9	970	6.4	3,570	10.0	10,085		

Table well determined up to 7 feet gage height; above this point table for 1903 is the same as table for 1902.

Estimated monthly discharge of Brazos River at Waco, Tex., for 1903.

[Drainage area, 30,750 square miles.]

Month.	Discharge in second-feet.			Total in acre-feet.	Run-off.	
	Maximum.	Minimum.	Mean		Second-feet per square mile.	Depth in inches.
January	3,440	565	1,004	61,734	0.033	0.038
February	39,650	745	5,860	325,448	.191	.199
March	18,995	2,435	6,532	401,637	.212	.244
April	2,380	535	1,111	66,109	.036	.040
May	2,710	460	662	40,705	.022	.025
June	4,000	410	1,580	94,017	.051	.057
July	8,055	275	1,440	88,542	.047	.054
August	2,655	195	791	48,637	.026	.030
September	5,825	90	604	35,940	.020	.022
October	22,400	410	3,775	232,116	.123	.142
November	815	235	501	29,812	.016	.018
December	235	125	165	10,145	.005	.006
The year	39,650	90	2,002	1,434,842	.065	.875

BRAZOS RIVER AT RICHMOND, TEX.

'his station was established in the latter part of December, 1902,
Thomas U. Taylor.   The gage consists of a plumber's chain, with
ss tags for foot marks.   It is not attached to the bridge, but is
d to measure down to the water's surface from the top of the guard
l in the middle of the sixth panel from the west end of the middle
.n.   Gage datum is 50 feet below the top of the guard rail, where
gage heights are determined.   The channel is straight for 300 feet
)ve and below the station, has a width of about 175 feet at low
ter without piers, and about 500 feet at ordinary high water broken
three piers.   During very high floods the left bank overflows and
ı width of the stream is 900 feet.   The bed is soft except around
ı piers, where it is stony.   It probably does not change much.
e velocity is moderately rapid.   There are some short piles in the
ging section.   The elevation of the top of the guard rail, from
ıich point gage heights are determined, is 93.2 feet above mean
ʋ sea level, so that the gage datum has an elevation of 43.2 feet
ferred to the same plane of reference.   The lowest gage height yet
corded is 1.80 feet above gage datum, or 45 feet above mean low
a level.

Above and at Waco the river rises rapidly, and when it gets above
e gage height of 30 feet overflows the bottom lands below the town.
'hen the floods spread out over the bottom lands, as they do from
'aco to Richmond, the river stays up longer in its lower stretches
an it does in the upper sections, as the bottoms and the lowlands
rve as storage reservoirs for the backwater and are drained slowly
ı the river recedes.   Above Waco the surface water rushes off into
ıe stream more rapidly, and the river rises more suddenly and falls
most as suddenly.   For this reason it is possible for the maximum
scharge at Waco to be greater than it is at Richmond.

At Hearne and below the river in 1899 was several miles in width.
s maximum height at Richmond occurred on July 7, 1899, when it
ood 4 feet below the top of the guard rail, or at a gage height of 46
et.   The water was out over the bottoms above, and in Richmond
covered the tracks of the Southern Pacific Railroad.

The observations at this station during 1903 have been made under
he direction of Thomas U. Taylor, district hydrographer.

Discharge measurements of Brazos River at Richmond, Tex., for 1902 and 1903.

Date.	Hydrographer.	Gage height.	Discharge.
1902.		*Feet.*	*Second-feet.*
August 2	T. U. Taylor	28.10	48,490
August 6	....do	31.90	60,470
August 19	....do	7.10	5,800
1903.			
March 13	....do	11.50	13,00
July 1	....do	5.80	3,80
July 10	....do	8.70	8,90
July 11	....do	7.80	7,73
July 18	....do	4.40	3,30
July 19	....do	4.40	3,30
July 24	....do	3.40	1,91
July 27	....do	2.90	1,63
August 3	....do	19.80	34,00
August 4	....do	21.50	38,50
August 6	H. H. Fox	16.20	25,90
August 10	....do	12.20	15,80
August 11	....do	11.10	13,86
August 13	....do	10.00	11,52
August 14	....do	9.20	9,89
August 15	....do	8.70	8,82
December 10	....do	1.85	1,01

Mean daily gage height, in feet, of Brazos River at Richmond, Tex., for 1903.

Day.	Jan.	Feb.	Mar.	Apr.	May.	June.	July.	Aug.	Sept.	Oct.	Nov.	Dec
1	6.00	6.80	26.20	11.20	8.00	4.10	6.10	11.00	3.10	3.00	3.30	2.0
2	12.00	6.30	27.10	11.00	6.60	4.10	6.00	18.45	3.00	3.00	3.10	2.0
3	14.50	5.90	28.40	10.40	5.60	4.20	6.50	20.98	2.90	3.00	3.00	2.0
4	13.20	5.60	30.10	10.10	5.20	5.50	6.30	20.50	2.80	12.60	2.90	2.0
5	10.30	5.30	31.90	9.70	5.20	5.10	6.30	17.20	3.00	10.60	2.80	2.0
6	9.70	5.20	33.10	9.30	5.40	4.80	6.50	14.60	3.00	8.00	2.70	2.0
7	8.50	5.70	33.40	8.70	7.00	4.50	9.10	13.90	2.90	7.70	2.80	2.0
8	7.90	6.00	32.95	8.20	6.90	4.50	11.50	12.60	2.80	8.20	2.80	2.0
9	7.40	6.50	30.85	8.00	5.90	4.40	10.60	13.00	2.80	10.80	3.00	1.5
10	6.00	7.30	26.05	7.60	5.40	4.20	6.10	12.60	2.80	10.30	3.00	1.5
11	6.00	8.30	22.20	7.40	5.60	4.10	7.80	11.90	2.70	11.90	3.00	1.5
12	5.70	10.40	20.60	7.20	5.40	4.10	6.80	11.10	2.70	10.50	2.90	1.
13	5.30	11.10	19.20	7.00	5.40	4.60	6.20	10.30	2.60	9.50	2.90	1.
14	5.00	11.60	17.90	6.90	5.20	5.00	5.90	9.50	2.60	9.50	3.00	3.
15	4.80	10.50	16.00	6.80	5.00	6.00	5.40	9.10	2.50	8.90	3.00	4.
16	4.80	9.80	15.40	6.60	7.30	6.00	5.20	8.10	2.40	7.60	2.90	3
17	5.40	13.20	14.20	6.60	7.00	5.50	5.00	7.60	2.40	6.30	2.70	3

Mean daily gage height, in feet, of Brazos River, etc.—Continued.

Day.	Jan.	Feb.	Mar.	Apr.	May.	June.	July.	Aug.	Sept.	Oct.	Nov.	Dec.
....	6.00	19.20	12.00	6.60	6.30	5.30	5.00	7.80	2.30	6.00	2.60	3.30
....	6.70	21.00	12.00	6.50	5.60	5.00	4.90	7.00	2.30	6.00	2.50	3.30
....	5.90	21.20	11.80	6.30	5.20	6.20	4.60	6.30	2.20	5.90	2.40	3.40
....	6.00	20.60	15.90	6.00	5.00	6.40	4.60	6.10	2.20	5.40	2.30	3.20
....	6.50	19.80	21.90	5.80	4.60	6.00	4.30	5.60	2.10	5.50	2.30	3.20
....	6.60	18.40	23.60	5.70	4.30	5.70	4.00	4.90	2.10	5.00	2.30	2.90
....	6.30	16.90	22.10	5.50	4.30	5.30	3.90	4.50	2.10	4.60	2.20	2.80
....	7.00	15.20	18.50	5.60	4.20	5.00	3.60	4.10	2.10	4.80	2.30	2.40
....	7.90	15.20	15.00	5.90	4.20	5.00	3.40	3.90	2.10	4.00	2.30	2.20
....	7.50	20.60	13.00	5.30	4.30	4.80	3.30	3.70	2.10	3.80	2.20	2.60
....	7.30	24.90	12.20	5.20	4.30	4.50	2.90	3.60	2.40	3.80	2.20	2.80
....	7.60	....	11.70	5.60	4.40	4.60	3.00	3.80	2.40	3.60	2.10	2.70
....	7.10	....	11.40	5.80	4.30	5.80	3.20	3.40	2.50	3.50	2.10	2.60
....	7.10	....	11.50	....	4.20	....	5.40	3.20	....	3.60	....	2.60

'ating table for Brazos River at Richmond, Tex., from January 1 to December 31, 1903.

Gage height.	Discharge.	Gage height.	Discharge.	Gage height.	Discharge.	Gage height.	Discharge.
Feet.	Second-feet.	Feet.	Second-feet.	Feet.	Second-feet.	Feet.	Second-feet.
1.8	945	3.4	2,000	6.0	5,020	9.2	9,790
1.9	985	3.5	2,100	6.2	5,280	9.4	10,110
2.0	1,025	3.6	2,200	6.4	5,540	9.6	10,440
2.1	1,065	3.7	2,300	6.6	5,810	9.8	10,780
2.2	1,105	3.8	2,410	6.8	6,090	10.0	11,120
2.3	1,155	3.9	2,520	7.0	6,370	10.5	12,020
2.4	1,205	4.0	2,630	7.2	6,670	11.0	13,020
2.5	1,265	4.2	2,860	7.4	6,970	11.5	14,120
2.6	1,325	4.4	3,100	7.6	7,270	12.0	15,270
2.7	1,395	4.6	3,340	7.8	7,570	12.5	16,420
2.8	1,465	4.8	3,580	8.0	7,870	13.0	17,590
2.9	1,545	5.0	3,820	8.2	8,190	13.5	18,790
3.0	1,625	5.2	4,060	8.4	8,510	14.0	19,990
3.1	1,715	5.4	4,300	8.6	8,830		
3.2	1,805	5.6	4,540	8.8	9,150		
3.3	1,900	5.8	4,780	9.0	9,470		

Tangent above 13 feet gage height. Table well defined.

Estimated monthly discharge of Brazos River at Richmond, Tex., in 1903.

[Drainage area, 44,000 square miles.]

Month.	Discharge in second-feet.			Total in acre-feet.	Run-off.	
	Maximum.	Minimum.	Mean.		Second-feet per square mile.	Depth in inches.
January.............	21,190	3,580	7,343	451,503	0.17	0.23
February	46,150	4,060	17,910	994,670	.41	.43
March	66,550	13,900	35,690	2,194,492	.81	.93
April	13,460	4,060	6,974	414,982	.16	.18
May	7,870	2,860	4,365	268,393	.10	.11
June...............	5,540	2,740	3,834	228,139	.09	.10
July	14,120	1,545	4,816	296,124	.11	.13
August	36,670	1,805	11,680	718,176	.27	.31
September	1,715	1,065	1,308	77,831	.03	.03
October	16,650	1,625	6,465	397,517	.15	.17
November.....	1,900	1,065	1,390	82,711	.03	.04
December	3,700	945	1,440	88,542	.03	.04
The year	66,550	945	8,601	6,213,080	.20	2.67

COLORADO RIVER (OF TEXAS) DRAINAGE BASIN.

Colorado River rises in the extreme western portion of the State, within a few miles of the eastern boundary of New Mexico, and flows in a general southeasterly direction, emptying into the Gulf of Mexico in Matagorda County. The drainage area above Austin is 37,000 square miles and above Columbus 40,000 square miles, and it extends into the corner of New Mexico. Its main tributaries are the Concho, the San Saba, and the Llano. The Concho has a reliable flow above its junction with the Colorado, and if the stream below the junction were to receive its name from the one that contributed the most water the river below the junction would be known as the Concho instead of the Colorado. The Concho furnishes water for irrigation and water power, and supports in Irion and Tom Green counties some excellent irrigation systems, described in Water-Supply Paper No. 71. San Saba and Llano rivers are described in the same paper.

The Colorado at Austin emerges from a canyon. From Austin to the Gulf it traverses a rather flat country, and its waters are utilized for many power plants; 60,000 acres of rice were sowed during the season of 1902 in the counties of Colorado, Wharton, and Matagorda, under canals that obtained their water from the Colorado.

Under the direction of Thomas U. Taylor the United States Geological Survey is maintaining gaging stations in this basin at Columbus and Austin, Tex.

COLORADO RIVER AT COLUMBUS, TEX.

This station was established in December, 1902, by Thomas U. Taylor. There is a gage marked on the downstream side of the pier ᴏɴ the west side of the river. Gage datum is taken as 50 feet below the top of this pier, and the observer measures down from this point with a tagged chain and lead weight. The channel is straight for 200 feet above and 600 feet below the stream and has a width of 140 feet at low water unobstructed by piers, and a width of 450 feet at ordinary high water broken by two piers. At very high stages the left bank overflows for several hundred feet, but the water passes under the iron trestle approach to the bridge. The bed is composed of gravel and sand and is fairly permanent.

The observations at this station during 1903 have been made under the direction of Thomas U. Taylor, district hydrographer.

Discharge measurements of Colorado River at Columbus, Tex., in 1902 and 1903.

Date.	Hydrographer.	Gage height.	Discharge.
1902.		*Feet.*	*Second-feet.*
August 2	T. U. Taylor	32.0	36,000
August 4	do	13.4	6,600
August 6	do	13.4	6,800
August 7	do	12.6	4,300
December 27	do	3.5	1,200
1903.			
January 2)	A. A. Cother	8.3	2,056
January 30	do	8.1	1,996
March 20	T. U. Taylor	17.6	15,600
July 28	do	8.5	1,400
June 19	do	12.9	8,500
July 25	do	6.7	800
December 10	H. H. Fox	5.2	516

Mean daily gage height, in feet, of Colorado River at Columbus, Tex., for 1903.

Day.	Jan.	Feb.	Mar.	Apr.	May.	June.	July.	Aug.	Sept.	Oct.	Nov.	Dec.
1	3.50	7.90	35.15	10.50	9.70	11.20	8.50	34.15	7.30	7.30	7.00	5.60
2	3.50	7.90	31.50	10.40	9.05	14.75	11.75	28.00	7.25	7.25	7.00	5.50
3	3.50	7.80	23.25	10.15	8.55	12.95	23.80	16.65	7.20	7.05	6.70	5.50
4	3.40	7.65	20.60	9.95	8.35	11.40	14.80	15.70	7.10	13.20	7.15	5.40
5	3.40	7.50	18.00	9.85	8.80	10.20	11.35	14.95	7.10	23.65	7.60	5.40
6	3.40	7.60	16.40	9.65	12.20	9.55	12.00	13.05	7.00	25.65	6.75	5.40
7	3.40	9.60	15.30	9.55	11.30	9.15	15.20	11.80	7.00	19.50	6.90	5.60
8	3.50	10.40	14.60	9.50	9.25	8.90	13.25	10.80	8.05	15.05	6.00	5.80
9	3.50	8.65	17.35	9.45	8.70	8.85	12.18	10.25	8.05	12.55	6.00	5.80
10	3.40	11.50	14.25	9.25	8.40	8.65	10.80	9.85	8.00	11.20	6.00	5.80
11	3.40	14.60	13.40	9.15	8.30	8.40	9.85	9.50	8.45	12.65	6.00	6.50
12	3.40	12.90	12.90	9.05	8.55	8.30	9.15	9.30	8.15	12.30	6.00	8.00
13	3.40	9.65	12.35	9.00	8.55	8.40	8.80	9.05	7.95	11.50	5.90	8.65
14	3.40	8.95	12.10	8.95	8.25	11.15	8.60	9.35	7.65	10.45	5.80	6.65
15	8.50	9.60	11.65	10.45	10.25	10.15	8.45	8.65	7.40	8.95	5.70	6.50
16	3.60	16.33	11.35	14.35	12.05	9.65	8.35	8.55	7.35	8.75	6.00	6.50

Mean daily gage height, in feet, of Colorado River, etc.—Continued.

Day.	Jan.	Feb.	Mar.	Apr.	May.	June.	July.	Aug.	Sept.	Oct.	Nov.	Dec.
17	3.60	16.10	11.05	13.05	12.95	12.80	8.30	8.35	7.40	8.55	6.00	4.3
18	3.70	15.60	10.80	11.80	11.45	15.05	7.95	8.20	7.00	8.35	6.00	4.3
19	3.80	15.70	10.95	10.60	10.70	15.30	7.80	8.15	7.00	8.30	5.80	4.3
20	4.00	23.60	17.10	10.10	10.20	13.60	7.70	8.20	6.95	8.00	5.90	4.3
21	4.50	18.25	17.15	9.55	9.55	10.70	7.65	8.80	6.90	8.05	6.00	4.0
22	4.60	13.40	14.30	9.25	9.75	10.60	7.60	8.45	9.95	8.25	5.90	4.0
23	5.00	12.70	13.50	9.05	13.20	10.15	7.50	8.30	9.90	7.95	5.80	4.5
24	5.80	11.95	12.50	8.90	11.80	9.65	7.50	8.05	9.10	7.65	5.80	4.5
25	6.40	12.85	11.75	9.55	10.70	9.25	7.45	8.00	8.20	7.45	5.70	4.3
26	7.30	26.90	11.25	8.65	9.75	9.10	7.25	7.85	8.00	7.15	5.60	4.3
27	8.60	32.60	11.70	8.45	9.45	8.45	7.60	7.75	7.90	6.70	5.60	4.3
28	8.40	34.35	11.60	8.35	9.10	8.60	9.55	7.55	7.75	6.50	5.60	4.0
29	8.30	....	11.20	8.40	8.90	8.60	13.55	7.50	7.55	6.45	5.60	4.0
30	8.00	....	10.95	11.15	8.65	8.00	27.25	7.35	7.45	6.30	5.60	4.0
31	8.00	....	10.65	....	9.05	....	33.80	7.35	....	6.10	....	4.0

COLORADO RIVER AT AUSTIN, TEX.

The flood of April 7, 1900, carried away the great masonry dam at Austin. This flood was general over southwest Texas, but its only disaster was limited to the demolition of that structure. A full discussion of this failure will be found in Water-Supply Paper No. 40.

Prior to the flood of April 7, 1900, the discharge of the river at the station below the dam was at low stages absolutely under the control of the turbines at the power house at the dam; and measurements made opposite the city, at the station between the two bridges, did not give the unobstructed flow of the river.

Gage heights were first taken on the crest of the Austin dam on August 13, 1895, and were continued from that date until the failure of the dam occurred in April, 1900. The first discharge measurement was made on December 21, 1897. In February, 1899, the gage was placed on the Congress Avenue Bridge, south of the city. This gage consists of a plain staff attached to a bath house. The observer is W. Peterson. The bench mark is on the first flange above the crib-work of the north pier of the highway bridge. Its elevation is 4.78 feet above the zero of the gage and 424.9 feet above sea level.

Discharge measurements are made at high water from a cable and car about 2 miles above the Congress Avenue Bridge, where the gage is located, and about one-eighth mile above the ruins of the Austin dam and power house. The cable has a span of about 730 feet, but the width of the river at low water is less than half this distance. Good low-water measurements can be made by wading about one-fourth mile above the Congress Avenue Bridge. The bed is composed of gravel and sand and is slightly shifting. The velocity is moderately rapid.

The observations at this station during 1903 have been made under the direction of Thomas U. Taylor, district hydrographer.

Discharge measurements of Colorado River at Austin, Tex., in 1903.

Date.	Hydrographer.	Gage height.	Discharge.
		Feet.	*Second-feet.*
pril 1	A. A. Cother	3.00	1,998
pril 9	....do	2.65	1,363
pril 13	....do	5.22	8,561
pril 18	....do	2.90	1,952
pril 30	....do	2.30	955
ay 30	....do	5.10	8,218
me 1	....do	3.70	3,778
me 9	....do	2.35	1,106
ovember 14	E. C. H. Bantel	1.30	347
ovember 20	....do	1.22	313
ovember 28	....do	1.22	377
ecember 4	....do	1.22	354
ecember 11	....do	1.20	332
ecember 31	....do	1.25	376

Mean daily gage height in feet of Colorado River at Austin, Tex., for 1903.

Day.	Jan.	Feb.	Mar.	Apr.	May.	June.	July.	Aug.	Sept.	Oct.	Nov.	Dec.
	2.20	2.35	7.55	3.00	2.30	3.90	2.25	3.35	1.20	1.40	1.40	1.20
	2.30	2.30	6.60	3.00	2.38	3.20	2.35	3.10	1.20	4.90	1.40	1.20
	2.30	2.30	5.75	2.90	2.30	2.80	2.25	2.90	1.25	9.30	1.40	1.20
	2.40	2.25	5.10	2.95	2.25	2.65	2.10	2.55	2.45	10.40	1.40	1.20
	2.40	2.23	7.25	2.80	2.30	2.55	3.65	2.45	2.30	9.55	1.50	1.20
	2.30	2.35	4.40	2.75	2.30	2.40	3.40	2.35	2.10	4.25	1.60	1.20
	2.25	2.20	4.30	2.73	2.27	2.30	2.95	2.25	2.00	4.15	1.70	1.20
	2.15	2.20	4.00	2.70	2.20	2.30	2.65	2.10	2.00	4.35	1.70	1.20
	2.10	2.20	3.40	2.65	2.60	2.30	2.45	2.20	2.00	4.20	1.70	1.20
	2.05	2.40	3.85	2.62	2.30	2.30	2.30	1.85	1.90	3.60	1.70	1.20
	2.00	2.30	3.70	2.60	2.20	2.20	2.15	1.70	1.75	3.35	1.60	1.20
	2.00	2.50	3.55	2.60	3.60	2.45	2.05	1.60	1.55	3.05	1.50	1.20
	2.00	2.60	3.40	4.80	3.70	2.60	2.20	1.55	1.50	2.70	1.50	1.20
	2.00	2.65	3.30	5.50	4.35	3.05	1.95	1.50	1.40	2.00	1.50	1.20
	2.10	2.90	3.20	4.00	3.95	4.70	1.90	1.50	1.40	2.45	1.40	1.20
	2.20	6.35	3.15	3.45	3.35	6.15	1.80	1.50	1.40	2.30	1.40	1.20
	2.75	4.00	3.10	3.05	3.15	4.45	1.80	2.35	1.30	2.25	1.40	1.20
	3.40	3.25	3.10	2.92	3.05	3.70	1.80	2.15	1.30	2.15	1.40	1.20
	3.15	3.20	3.40	2.85	2.75	3.35	1.73	1.95	3.50	2.00	1.40	1.20
	2.95	3.10	3.95	2.75	4.05	3.15	1.60	1.80	3.15	1.90	1.40	1.20
	2.95	3.50	4.30	2.60	4.25	2.75	1.50	1.70	2.45	1.80	1.40	1.20
	2.90	8.45	3.50	2.50	3.40	2.60	1.47	1.65	2.20	1.70	1.40	1.20
	2.80	3.35	3.30	2.40	3.05	2.60	1.42	1.60	2.10	1.70	1.40	1.20
	2.70	3.10	3.45	2.40	2.85	2.40	1.30	1.50	2.05	1.70	1.40	1.20
	2.70	3.15	3.25	2.33	2.55	2.40	1.30	1.40	1.75	1.70	1.40	1.20
	2.65	11.10	3.20	2.30	2.55	2.50	1.40	1.35	1.45	1.55	1.40	1.20
	2.60	11.30	3.15	2.30	2.70	2.45	1.50	1.30	1.55	1.40	1.40	1.20
	2.50	8.35	3.10	2.25	2.80	2.40	2.15	1.50	1.70	1.40	1.40	1.20
	2.45		3.10	2.20	2.80	2.55	2.65	1.40	1.70	1.40	1.40	1.20
	2.40		3.10	2.25	4.30	2.45	4.50	1.30	1.60	1.40	1.30	1.20
	2.40		3.00		4.95		5.40	1.20		1.35		1.20

Rating table for Colorado River at Austin, Tex., from January 1 to December 31, 1903.

Gage height.	Discharge.	Gage height.	Discharge.	Gage height.	Discharge.	Gage height.	Discharge.
Feet.	Second-feet.	Feet.	Second-feet.	Feet.	Second-feet.	Feet.	Second-feet.
1.2	320	2.4	1,140	3.6	3,365	4.8	7,085
1.3	360	2.5	1,260	3.7	3,615	4.9	7,475
1.4	405	2.6	1,390	3.8	3,870	5.0	7,870
1.5	455	2.7	1,530	3.9	4,130	5.5	9,870
1.6	510	2.8	1,680	4.0	4,395	6.0	11,870
1.7	565	2.9	1,840	4.1	4,665	7.0	15,870
1.8	625	3.0	2,020	4.2	4,955	8.0	19,870
1.9	690	3.1	2,220	4.3	5,265	9.0	23,870
2.0	760	3.2	2,430	4.4	5,595	10.0	27,870
2.1	840	3.3	2,650	4.5	5,945	11.0	31,870
2.2	930	3.4	2,880	4.6	6,315	12.0	35,870
2.3	1,030	3.5	3,120	4.7	6,695		

Table constructed from measurements between gage heights 1.22 feet and 5.22 feet. Above 5.22 feet table is not well determined. Extended parallel to 1902 curve.

Estimated monthly discharge of Colorado River at Austin, Tex., for 1903.

[Drainage area, 37,000 square miles.]

Month.	Discharge in second-feet			Total in acre-feet.	Run-off.	
	Maximum.	Minimum.	Mean.		Second-feet per square mile.	Depth in inches
January	2,880	760	1,276	78,458	0.034	0.039
February	33,070	930	5,066	281,351	.137	.143
March	18,070	2,020	4,854	298,461	.131	.151
April	9,870	930	2,079	123,709	.056	.062
May	7,670	930	2,382	146,463	.064	.074
June	12,470	930	2,279	135,610	.062	.069
July	9,470	360	1,404	86,329	.038	.044
August	2,765	320	801	49,252	.022	.025
September	3,120	320	763	45,402	.021	.023
October	29,470	380	4,222	259,601	.114	.131
November	565	360	439	26,122	.012	.013
December	320	320	320	19,676	.009	.010
The year	33,070	320	2,157	1,550,434	.058	.784

GUADALUPE RIVER DRAINAGE BASIN.

Guadalupe River rises in the southern central part of Texas, flows ȿutheast, and empties into San Antonio Bay. During the summer ′ 1902 its discharge was the least in its observed history, causing ȿuch loss above New Braunfels, where half a dozen power plants ȿre forced to shut down or to run on short time. The flow at this ȿme was so low that special efforts were made to obtain measurements ȿ several points along its course. The results of these measurements ȿe shown in the accompanying table.

Under the direction of Thomas U. Taylor the United States Geoȿgical Survey is maintaining a station in this basin near Cuero, Tex.

GUADALUPE RIVER NEAR CUERO, TEX.

The Guadalupe, while the best water-power stream in Texas, has a ȿrainage area above Cuero of only 5,100 square miles. Its efficiency ȿ due almost entirely to the canal at New Braunfels. Below New ȿraunfels the largest tributary is San Marcos River.

This station was established by Thomas U. Taylor December 26, ȿ02. The original location of the gage was at the dam at Carl ȿuchel's power house, 3 miles north of Cuero, Tex. This gage is a ȿertical staff mounted on the wall of the power house near the dam, ȿnd was read twice each day by Carl Buchel.

For the old station, the initial point for soundings was on the left ȿank. The channel is straight for about one-fourth of a mile above ȿnd 400 feet below the station. The right bank is low and liable to ȿverflow; the left bank is high and rocky. The bed of the stream is ȿf clay. The bench mark is on the crest of the dam. Its elevation ȿ the same as the zero of the gage.

As it proved impossible to measure flood discharges at this point, a ȿew station was established in July, 1903, at the bridge of the San ȿntonio and Aransas Pass Railroad 3 miles west of Cuero. The ȿage datum is 50 feet below the top of the tie in the third panel from ȿhe east end of the bridge. A tagged plumber's chain is used in ȿetermining gage heights by measuring down from the bridge. The ȿbserver is John Hughes, who has charge of the pumping plant ȿelonging to the railroad. Discharge measurements are made from ȿhe highway bridge, 200 feet below the railway bridge. The channel ȿ straight and has a width of 125 feet at low stages. The right bank ȿ low and overflows for several hundred feet at high stages. The ȿection is deep and the flow is sluggish. The bed is composed of soft ȿaterial and may change somewhat.

The observations at this station during 1903 have been made under ȿhe direction of Thomas U. Taylor, district hydrographer.

Discharge measurements of Guadalupe River near Cuero, Tex., in 1903.

Date.	Hydrographer.	Gage height.	Discharge.
		Feet.	*Second-feet.*
June 29	T. U. Taylor	8.70	1,700
June 30	do	8.20	1,400
July 4	do	35.80	33,700
Do	do	33.80	25,600
July 5	do	31.80	20,300
Do	do	30.80	17,800
July 6	do	28.70	15,900
Do	do	28.30	15,300
July 9	do	26.70	12,700
Do	H. H. Fox	25.70	11,300
July 10	do	20.50	6,600
Do	do	17.30	4,900
Do	do	15.70	3,850
Do	do	13.50	3,360
Do	do	12.20	2,700
July 11	do	10.70	2,300
July 12	do	9.70	1,840
July 14	do	8.90	1,570
July 18	T. U. Taylor	8.40	1,370
July 29	do	12.10	2,900
December 11	H. H. Fox	6.80	778

Mean daily gage height, in feet, of Guadalupe River near Cuero, Tex., for 1903.

Day.	Jan.	Feb.	Mar	Apr.	May.	June.	July.	Aug.	Sept.	Oct.	Nov.	Dec.
1	10.50	7.20	43.00	10.40	10.80	7.00	6.90	27.50	7.75	7.10	7.85	6.80
2	10.20	7.20	31.80	10.30	10.30	6.70	13.40	20.40	7.65	7.10	10.40	6.80
3	10.00	7.10	21.30	10.20	10.50	6.70	37.00	28.80	7.55	7.25	8.95	6.75
4	10.00	7.20	16.20	10.20	10.00	6.70	34.80	27.25	7.50	7.15	7.85	6.90
5	10.00	7.10	14.10	10.00	10.00	6.60	31.30	17.80	7.35	7.10	7.60	6.80
6	9.70	7.00	13.60	10.00	10.20	6.80	28.50	12.75	7.35	8.25	8.50	6.80
7	9.60	7.20	13.40	9.80	10.20	6.70	28.40	11.20	7.40	7.35	7.60	6.80
8	9.30	10.00	13.50	9.80	10.20	6.70	26.00	10.70	7.45	7.25	7.10	6.80
9	9.40	9.30	13.20	9.70	10.10	6.60	24.80	9.65	7.30	7.20	7.15	6.75
10	9.40	10.50	13.50	9.70	10.20	6.50	17.50	9.10	7.25	7.25	7.00	6.75
11	9.30	13.20	13.20	9.70	10.10	6.50	10.10	8.95	7.30	6.95	6.95	6.75
12	9.10	14.20	13.00	9.70	9.60	6.70	9.40	8.85	7.10	7.00	6.90	6.80
13	9.10	13.20	12.70	9.40	9.40	6.80	9.20	8.70	7.35	6.95	6.85	6.85
14	9.00	9.60	12.10	9.50	9.20	9.10	8.80	8.50	7.20	7.00	6.80	7.2
15	9.00	10.40	12.10	9.50	9.30	9.60	8.00	8.35	7.20	7.25	6.85	7.1
16	9.00	13.00	11.00	9.40	9.10	10.10	8.00	8.40	7.20	7.00	6.95	7.0
17	9.00	14.60	10.90	9.40	9.40	12.00	6.90	8.25	7.10	10.25	6.80	6.9
18	9.60	14.20	10.80	9.40	9.10	11.00	7.00	8.20	7.00	8.60	6.85	6.8

Mean daily gage height, in feet, of Guadalupe River, etc.—Continued.

Day.	Jan.	Feb.	Mar.	Apr.	May.	June.	July.	Aug.	Sept.	Oct.	Nov.	Dec.
19	9.00	13.30	10.80	9.50	9.00	13.40	8.70	8.60	7.05	7.30	6.90	6.80
20	9.00	16.30	10.80	9.40	9.00	9.20	6.90	8.40	7.05	7.15	6.70	6.80
21	9.00	19.00	13.20	9.30	9.00	9.30	6.90	8.30	7.20	6.95	6.70	6.85
22	9.00	23.20	13.20	9.20	8.80	7.20	6.85	8.20	7.15	6.80	6.80	6.80
23	9.00	18.90	12.10	9.10	9.00	7.00	6.75	8.10	7.20	6.90	6.85	6.75
24	9.00	12.20	11.20	9.00	9.20	7.20	6.70	8.05	7.10	6.85	6.85	6.85
25	9.00	11.10	10.80	9.20	7.00	7.10	6.75	8.00	7.20	6.85	6.80	6.75
26	9.00	16.80	10.80	9.10	7.20	7.00	7.10	8.10	7.15	6.90	6.85	6.70
27	8.80	26.00	10.80	9.10	7.30	6.70	6.80	7.95	7.20	6.85	6.80	6.70
28	8.60	37.00	10.60	9.10	7.20	7.10	6.95	7.95	7.15	6.90	6.70	6.70
29	7.00	....	10.50	9.20	7.20	7.20	7.20	13.40	7.75	7.20	6.90	6.70
30	7.00	....	10.50	9.30	7.20	6.90	18.65	7.65	7.20	6.90	6.95	6.70
31	7.00	....	10.50	....	7.00	....	24.65	7.80	....	7.45	....	6.70

Rating table for Guadalupe River near Cuero, Tex., from January 1 to December 31, 1903.

Gage height.	Discharge.	Gage height.	Discharge.	Gage height.	Discharge.	Gage height.	Discharge.
Feet.	Second-feet.	Feet.	Second-feet.	Feet.	Second-feet.	Feet.	Second-feet.
6.5	755	11.0	2,330	15.6	4,040	23.5	9,130
6.6	790	11.2	2,400	15.8	4,120	24.0	9,630
6.8	860	11.4	2,470	16.0	4,200	24.5	10,140
7.0	930	11.6	2,540	16.2	4,280	25.0	10,690
7.2	1,000	11.8	2,610	16.4	4,360	25.5	11,240
7.4	1,070	12.0	2,680	16.6	4,450	26.0	11,840
7.6	1,140	12.2	2,750	16.8	4,550	26.5	12,440
7.8	1,210	12.4	2,820	17.0	4,650	27.0	13,070
8.0	1,280	12.6	2,890	17.2	4,750	27.5	13,720
8.2	1,350	12.8	2,960	17.4	4,850	28.0	14,370
8.4	1,420	13.0	3,030	17.6	4,950	29.0	15,730
8.6	1,490	13.2	3,100	17.8	5,050	30.0	17,200
8.8	1,560	13.4	3,170	18.0	5,150	31.0	18,950
9.0	1,630	13.6	3,240	18.5	5,440	32.0	21,000
9.2	1,700	13.8	3,320	19.0	5,740	33.0	23,320
9.4	1,770	14.0	3,400	19.5	6,040	34.0	26,000
9.6	1,840	14.2	3,480	20.0	6,340	35.0	29,690
9.8	1,910	14.4	3,560	20.5	6,690	36.0	34,200
10.0	1,980	14.6	3,640	21.0	7,040	37.0	39,500
10.2	2,050	14.8	3,720	21.5	7,440	38.0	44,800
10.4	2,120	15.0	3,800	22.0	7,840	39.0	50,100
10.6	2,190	15.2	3,880	22.5	8,240	40.0	55,400
10.8	2,260	15.4	3,960	23.0	8,680		

Table well defined. Curve extended below 6.8 and above 35.8.

Estimated monthly discharge of Guadalupe River near Cuero, Tex., for 1903.

[Drainage area, 5,100 square miles.]

Month.	Discharge in second-feet.			Total in acre-feet.	Run-off.	
	Maximum.	Minimum.	Mean.		Second-feet per square mile.	Depth in inches.
January	2,155	930	1,659	102,008	0.33	0.38
February	39,500	930	4,474	248,473	.88	.92
March	71,300	2,155	5,659	347,958	.11	1.28
April	2,120	1,630	1,825	108,595	.36	.40
May	2,260	930	1,660	102,069	.33	.38
June	8,170	755	1,212	72,119	.24	.27
July	39,500	825	6,096	374,828	1.20	1.38
August	16,300	1,155	3,343	205,553	.66	.76
September	1,190	930	1,021	60,754	.20	.22
October	2,065	875	1,018	62,594	.20	.23
November	2,120	825	1,008	59,980	.20	.22
December	1,015	825	864	53,125	.17	.20
The year	71,300	755	2,487	1,798,056	.41	6.64

RIO GRANDE DRAINAGE BASIN.

The Rio Grande rises in southern Colorado, in the Rocky Mountains, flows south through New Mexico and thence southeast, forming the boundary between Texas and Mexico. Pecos River, which rises in northern New Mexico and flows south across eastern New Mexico and western Texas, is its longest tributary from the north, although Devils River delivers to the Rio Grande about the same amount of water as does the Pecos. Conchos River is its principal tributary from the Mexican side. The determination of the amount of water in the Rio Grande is of importance, both on account of its use in irrigation and from its bearing upon interstate and international distribution of water. Most of the New Mexico and Texas stations are maintained by the International (Water) Boundary Commission, and the data are furnished by W. W. Follett, consulting engineer for the Commission. On account of the shifting character of the river beds at the international (water) boundary stations, no rating tables have been prepared. The estimated monthly discharges are from daily discharges computed by Mr. Follett directly from the discharge measurements. The following list includes the stations in the Rio Grande drainage basin. Those maintained by the Boundary Commission are marked "B. C."

> Rio Grande at Eagle Pass, Tex. B. C.
> Rio Grande below mouth of Devils River, Tex. B. C.
> Devils River at Devils River, Tex. B. C.
> Pecos River near Moorhead, Tex. B. C.

Pecos River and Margueretta flume near Pecos, Tex.
Pecos River at Carlsbad, N. Mex.
Pecos River near Roswell, N. Mex.
Hondo River at Roswell, N. Mex.
Hondo River at reservoir site, near Roswell, N. Mex.
Pecos River at Santa Rosa, N. Mex.
Rio Grande near Langtry, Tex. B. C.
Rio Grande below Presidio, Tex. B. C.
Rio Grande above Presidio, Tex. B. C.
Rio Grande near Fort Hancock, Tex. B. C.
Rio Grande near El Paso, Tex. B. C.
Rio Grande near San Marcial, N. Mex. B. C.
Rio Grande at water tank near Rio Grande, or Buckman, N. Mex.
Rio Grande at Embudo, N. Mex.
Rio Grande near Cenicero, Colo.
Conejos River near Mogote, Colo.
Rio Grande near Del Norte, Colo.

RIO GRANDE AT EAGLE PASS, TEX.

This station was established in April, 1900, by the International (Water) Boundary Commission. It is a half mile above the highway bridge between Eagle Pass and Ciudad Porfirio Diaz, Mexico, and about 540 miles below El Paso, Tex.

Discharge measurements of Rio Grande at Eagle Pass, Tex., in 1903.

Date.	Hydrographer.	Gage height.	Discharge.
		Feet.	*Second-feet.*
January 2	Robert F. Dowe	2.1	2,385
January 5	..do	2.1	2,332
January 7	..do	2.1	2,381
January 9	..do	2.0	2,047
January 12	..do	2.0	2,129
January 16	..do	2.55	2,985
January 18	..do	2.4	2,688
January 20	..do	2.3	2,454
January 22	..do	2.3	2,451
January 24	..do	2.3	2,428
January 27	..do	2.3	2,458
January 29	..do	2.25	2,391
January 31	..do	2.2	2,262
February 3	..do	2.2	2,273
February 5	..do	2.3	2,267
February 7	..do	2.3	2,280
February 10	..do	2.4	2,379
February 12	J. D. Dillard	2.2	2,353
February 13	Robert F. Dowe	2.2	2,289
February 17	..do	2.8	3,598
February 20	..do	2.6	2,988

Discharge measurements of Rio Grande at Eagle Pass, Tex., in 1903—Continued.

Date.	Hydrographer.	Gage height.	Discharge.
		Feet.	*Second-feet.*
February 23	Robert F. Dowe	2.5	2,689
February 25	do	2.6	2,935
February 27	do	2.6	2,865
March 3	do	2.8	3,208
March 5	do	2.7	2,931
March 7	do	2.6	2,889
March 10	do	2.5	2,769
March 12	do	2.4	2,622
March 19	do	2.2	2,322
March 25	J. D. Dillard	2.0	2,087
March 30	W. W. Sentman	1.9	1,882
April 1	do	1.9	2,008
April 3	do	1.9	1,794
April 6	do	1.8	1,714
April 8	do	1.8	1,692
April 10	do	1.7	1,761
April 14	do	1.7	1,692
April 17	do	1.6	1,65
April 21	do	1.8	1,85
April 23	do	1.8	1,94
April 27	do	1.8	1,706
April 30	do	2.7	3,614
May 7	do	2.0	2,143
May 11	do	2.1	2,273
May 14	do	3.0	4,400
May 16	do	2.3	2,274
May 19	do	3.3	4,694
May 26	do	2.6	3,250
May 28	do	2.6	3,516
May 31	do	2.6	3,170
June 2	do	3.0	4,191
June 6	do	2.6	3,037
June 8	do	3.45	5,339
June 15	do	7.35	20,209
June 18	do	4.55	8,053
June 28	do	4.05	7,021
June 30	do	4.15	7,46
July 2	J. K. Wilson	3.9	7,10
July 6	do	3.9	6,63
July 9	do	4.4	8,76
July 13	do	4.5	8,14

ye measurements of Rio Grande at Eagle Pass, Tex., in 1903—Continued.

Date.	Hydrographer.	Gage height.	Discharge.
		Feet.	*Second-feet.*
....	J. K. Wilson	4.1	7,175
....	.....do	8.65	5,964
....	.....do	8.2	4,571
....	.....do	2.9	8,976
....	.....do	2.7	8,457
....	.....do	2.7	8,501
4	.....do	2.6	3,827
7	.....do	3.0	4,284
11	.....do	2.6	3,274
14	.....do	2.8	2,585
18	.....do	2.7	8,668
22	.....do	2.5	8,029
26	.....do	3.0	4,231
28	.....do	2.7	8,672
31	.....do	3.4	4,885
ber 4	.....do	4.2	7,888
ber 7	.....do	3.9	6,671
ber 9	.....do	3.7	6,407
ber 11	.....do	5.0	10,291
ber 14	.....do	2.8	8,511
ber 17	.....do	3.55	5,011
ber 19	.....do	3.3	4,315
ber 22	.....do	3.1	4,581
ber 25	.....do	3.25	5,081
ber 29	.....do	4.05	7,427
2	.....do	4.85	8,286
5	.....do	4.9	11,784
8	.....do	8.95	7,421
10	.....do	8.5	5,494
13	.....do	8.1	4,456
16	.....do	2.8	8,808
19	.....do	2.7	8,487
22	.....do	2.6	8,227
24	.....do	2.6	8,230
27	.....do	2.4	8,052
31	.....do	2.4	8,017
ber 4	.....do	2.3	2,632
ber 7	.....do	2.2	2,524
ber 10	.....do	2.2	2,518
ber 14	.....do	2.2	2,547
ber 17	.....do	2.2	2,585

Discharge measurements of Rio Grande at Eagle Pass, Tex., in 1903—Continued.

Date.	Hydrographer.	Gage height.	Discharge.
		Feet.	Second-feet.
November 20	J. K. Wilson	2.0	2,353
November 23	......do	2.0	2,390
November 26	......do	2.0	2,335
November 30	......do	2.0	2,414
December 3	......do	2.0	2,328
December 6	......do	2.0	2,387
December 10	......do	2.0	2,343
December 14	......do	2.0	2,350
December 17	......do	2.0	2,292
December 21	......do	2.0	2,416
December 24	......do	2.0	2,363
December 28	......do	2.0	2,063
December 31	......do	2.0	2,079

Mean daily gage height, in feet, of Rio Grande at Eagle Pass, Tex., for 1903.

Day.	Jan.	Feb.	Mar.	Apr.	May.	June.	July.	Aug.	Sept.	Oct.	Nov.	Dec.
1	2.10	2.05	2.90	1.90	2.80	2.40	4.05	3.05	2.95	3.95	2.35	2.00
2	2.10	2.00	2.85	1.90	2.50	2.70	3.90	2.90	2.90	4.25	2.30	2.00
3	2.10	2.15	2.80	1.90	2.30	2.75	3.80	2.70	2.90	4.70	2.30	2.00
4	2.10	2.30	2.75	1.90	2.20	3.05	4.35	2.60	4.20	4.50	2.30	2.00
5	2.10	2.30	2.70	1.75	2.05	2.60	3.85	2.50	3.95	4.90	2.30	2.00
6	2.10	2.30	2.60	1.80	2.00	2.60	3.90	3.00	4.00	4.85	2.30	2.00
7	2.10	2.30	2.60	1.80	1.95	2.50	3.95	2.95	3.85	4.20	2.30	2.00
8	2.00	2.30	2.55	1.80	2.00	3.20	4.25	2.60	3.60	3.95	2.30	2.00
9	2.00	2.30	2.45	1.80	1.85	2.70	4.40	2.40	3.60	3.75	2.30	2.00
10	2.00	2.35	2.45	1.70	1.80	3.45	4.55	2.40	3.50	3.55	2.30	2.00
11	2.00	2.25	2.45	1.70	2.05	3.25	4.00	2.50	5.15	3.45	2.30	2.00
12	2.00	2.20	2.40	1.70	3.65	3.50	4.65	2.40	3.20	3.30	2.30	2.00
13	2.00	2.10	2.40	1.75	4.30	10.55	4.45	2.35	3.00	3.15	2.30	2.00
14	2.05	2.75	2.30	1.65	2.85	12.00	4.35	2.30	2.85	3.05	2.30	2.00
15	2.15	2.95	2.30	1.65	2.40	7.55	4.10	2.45	2.80	2.90	2.30	2.00
16	2.55	2.90	2.30	1.60	2.35	5.95	4.05	2.70	3.20	2.85	2.30	2.00
17	2.55	2.80	2.30	1.60	6.10	5.10	3.85	2.70	3.60	2.70	2.30	2.00
18	2.40	2.75	2.30	1.60	4.45	4.65	3.65	2.65	3.20	2.70	2.10	2.00
19	2.40	2.70	2.20	1.60	3.20	4.40	3.55	2.40	3.30	2.70	2.05	2.00
20	2.30	2.65	2.30	1.60	2.50	4.20	3.40	2.40	3.15	2.65	2.00	2.00
21	2.30	2.60	2.40	1.80	2.85	4.25	3.30	2.50	3.05	2.60	2.00	2.00
22	2.30	2.60	2.20	1.85	3.30	4.30	3.15	2.50	3.10	2.60	2.00	2.00
23	2.30	2.50	2.05	1.80	2.90	4.35	3.15	3.35	3.20	2.60	2.00	2.00
24	2.30	2.50	2.00	1.75	2.90	4.35	3.05	3.10	2.95	2.60	2.00	2.00
25	2.30	2.55	2.00	1.70	2.75	4.30	2.90	3.00	3.00	2.55	2.00	2.00
26	2.30	2.70	2.00	1.65	2.60	4.20	2.90	3.00	3.20	2.45	2.00	2.00
27	2.30	2.60	1.95	1.80	2.60	4.15	2.75	2.85	3.95	2.40	2.00	2.00
28	2.25	2.90	1.90	1.85	2.60	4.15	2.60	2.70	4.30	2.40	2.00	2.00
29	2.20		1.90	5.40	2.60	4.15	2.60	3.05	4.15	2.40	2.00	2.00
30	2.20		1.90	3.55	3.05	4.10	3.00	3.80	3.95	2.40	2.00	2.00
31	2.20		1.90		2.55		2.80	3.20		2.40		2.00

tted monthly discharge of Rio Grande at Eagle Pass, Tex., for 1903.

Month.	Discharge in second-feet.			Total in acre-feet.
	Maximum.	Minimum.	Mean.	
----------------------------	2,990	2,050	2,385	146,638
----------------------------	3,900	2,000	2,761	153,362
----------------------------	3,460	1,880	2,508	154,215
----------------------------	11,200	1,640	2,174	129,362
----------------------------	15,500	1,700	3,880	238,592
----------------------------	47,400	2,670	9,225	548,906
----------------------------	9,010	3,350	6,156	378,506
----------------------------	6,190	2,590	3,630	223,180
----------------------------	10,750	2,480	5,408	321,818
----------------------------	11,780	3,020	5,248	322,691
----------------------------	2,880	2,330	2,490	148,165
----------------------------	2,410	2,060	2,256	138,744
year ----------------------	47,400	1,640	4,010	2,904,179

O GRANDE BELOW MOUTH OF DEVILS RIVER, TEXAS.

ation was established in April, 1900, by the International
Boundary Commission. It is alongside the Southern Pacific
;rack, about a half mile below the mouth of Devils River and
' miles below El Paso.

*measurements of Rio Grande below mouth of Devils River, Texas, in
1903.*

Date.	Hydrographer.	Gage height.	Discharge.
		Feet.	*Second-feet.*
------------------	J. D. Dillard ----------------	3.85	1,957
------------------	----do ----------------	3.9	2,007
------------------	----do ---	4.1	2,306
------------------	----do ---	3.95	2,096
------------------	----do ---	3.95	2,069
------------------	----do ---	3.9	1,980
------------------	----do ---	4.0	2,141
------------------	----do ---	3.95	2,102
------------------	----do ---	4.4	2,859
------------------	---- do ---	4.2	2,483
------------------	----do ---	4.5	3,196
------------------	---- do ---	4.3	2,730
------------------	-- do ----------------	4.1	2,319

Discharge measurements of Rio Grande below mouth of Devils River, etc.—Cont

Date.	Hydrographer.	Gage height.	Dischar
		Feet.	*Second-*
March 15	J. D. Dillard	4.0	2,
March 21	do	4.0	2,
March 28	do	3.75	1,
April 2	do	3.7	1,
April 7	do	3.6	1,
April 13	do	3.5	1,
April 18	do	3.45	1,
April 23	do	3.7	1,
April 27	do	3.7	1,
May 2	do	4.0	2,
May 7	do	3.95	2,
May 11	do	4.2	2,
May 16	do	4.25	2,
May 21	do	4.65	3,
May 27	do	4.8	2
June 1	do	4.2	2
June 5	do	4.3	2
June 10	do	5.05	4
June 15	do	7.5	12
June 20	do	5.2	4
June 25	do	5.4	5
July 1	do	5.4	5
July 7	do	5.6	5
July 11	do	6.3	8
July 15	do	5.7	6
July 18	do	5.1	4
July 23	do	4.6	3
July 29	do	4.5	3
August 4	do	4.1	2
August 8	do	4.05	2
August 12	do	3.9	2
August 16	do	4.4	2
August 21	do	4.2	2
August 25	do	4.75	3
August 29	do	5.8	6
September 4	E. E. Winter	5.9	6
September 8	do	5.0	5
September 11	do	4.7	
September 16	do	4.7	
September 19	do	4.7	5

e measurements of Rio Grande below mouth of Devils River, etc.—Cont'd.

Date.	Hydrographer.	Gage height.	Discharge.
		Feet.	*Second-feet.*
r 23	E. E. Winter	4.5	8,143
r 28	do	5.3	5,404
3	J. D. Dillard	6.3	8,116
3	do	5.4	5,499
14	do	4.5	3,399
18	do	4.35	2,997
22	do	4.25	2,796
26	do	4.1	2,417
30	do	4.0	2,271
r 4	do	3.9	2,141
r 8	do	3.8	2,024
r 12	do	3.8	2,007
r 17	do	3.8	1,989
r 21	do	3.75	1,896
r 25	do	3.75	1,856
r 29	do	3.7	1,820
r 4	do	3.7	1,766
r 8	do	3.7	1,761
r 12	do	3.7	1,759
r 16	do	3.7	1,748
r 23	do	3.7	1,731
r 29	do	3.65	1,688

ily gage height, in feet, of Rio Grande below mouth of Devils River,
Texas, for 1903.

Day.	Jan.	Feb.	Mar.	Apr.	May.	June.	July.	Aug.	Sept.	Oct.	Nov.	Dec.
	3.90	3.90	4.50	3.75	4.15	4.20	5.35	4.55	4.40	5.80	4.10	3.70
	3.90	4.05	4.40	3.70	4.00	4.45	5.25	4.25	4.80	5.85	4.00	3.70
	3.90	4.05	4.35	3.70	3.85	4.70	5.15	4.10	5.95	6.30	3.90	3.70
	3.90	4.00	4.30	3.65	3.80	4.35	5.10	4.20	6.00	6.70	3.90	3.70
	3.90	3.95	4.30	3.60	3.80	4.30	5.25	4.80	5.35	6.55	3.90	3.70
	3.90	3.95	4.25	3.60	3.75	4.30	5.25	4.55	5.10	6.05	3.85	3.70
	3.90	3.95	4.20	3.60	3.90	4.85	5.65	4.15	5.30	5.65	3.80	3.70
	3.90	3.95	4.15	3.60	3.70	4.45	5.90	4.00	5.15	5.35	3.80	3.70
	3.90	3.95	4.15	3.60	3.90	4.45	6.05	4.00	4.90	5.10	3.80	3.70
	3.90	3.95	4.10	3.60	3.90	5.00	6.25	4.10	4.75	4.85	3.80	3.70
	3.90	3.95	4.10	3.60	4.20	4.90	6.30	4.05	4.65	4.80	3.80	3.70
	3.90	3.90	4.10	3.60	6.00	7.80	6.25	3.90	4.45	4.70	3.80	3.70
	3.90	4.65	4.05	3.50	4.50	11.50	6.05	3.90	4.05	4.55	3.80	3.70
	3.95	4.65	4.05	3.50	4.15	7.05	5.85	4.15	4.75	4.50	3.80	3.70
	4.15	4.65	4.00	3.45	4.30	7.35	5.65	4.40	4.55	4.45	3.80	3.70

Mean daily gage height, in feet, of Rio Grande below mouth of Devils River, Texas, for 1903—Continued.

Day.	Jan.	Feb.	Mar.	Apr.	May.	June.	July.	Aug.	Sept.	Oct.	Nov.	Dec.
16.	4.10	4.50	4.00	3.45	4.25	6.40	5.50	4.40	4.60	4.40	3.80	3.70
17.	4.00	4.45	3.90	3.45	5.20	5.50	5.15	4.25	4.65	4.35	3.80	3.70
18.	4.00	4.35	3.90	3.45	4.60	5.25	5.10	4.10	4.75	4.35	3.80	3.70
19.	4.00	4.30	4.10	3.55	4.40	5.20	4.90	4.15	4.70	4.30	3.80	3.70
20.	4.00	4.30	3.95	3.80	4.55	5.20	4.75	4.20	4.60	4.25	3.80	3.70
21.	3.95	4.25	4.00	3.70	4.65	5.10	4.75	4.20	4.70	4.25	3.75	3.70
22.	3.95	4.20	3.90	3.70	4.30	5.45	4.75	5.00	4.75	4.30	3.80	3.70
23.	3.95	4.20	3.85	3.70	4.40	5.45	4.55	5.00	4.45	4.25	3.80	3.70
24.	3.95	4.10	3.80	3.65	4.30	5.40	4.50	4.85	4.75	4.15	3.80	3.70
25.	3.95	4.10	3.80	3.60	4.40	5.40	4.40	4.75	4.85	4.10	3.80	3.70
26.	3.90	4.25	3.80	3.70	4.25	5.45	4.30	4.55	5.65	4.05	3.75	3.70
27.	3.95	4.55	3.80	3.70	4.30	5.50	4.25	4.40	5.50	4.05	3.70	3.65
28.	3.90	4.65	3.75	3.70	4.30	5.40	4.20	4.80	5.25	4.05	3.70	3.65
29.	3.90		3.75	3.70	4.25	5.45	4.45	6.15	5.20	4.05	3.70	3.65
30.	3.90		3.70	4.35	4.20	5.55	4.40	6.00	5.15	4.00	3.70	3.65
31.	3.90		3.70		4.25		4.60	5.45		4.00		3.65

Estimated monthly discharge of Rio Grande below mouth of Devils River, Texas, for 1903.

Month.	Discharge in second-feet.			Total in acre-feet.
	Maximum	Minimum.	Mean.	
January	2,390	1,980	2,068	127,180
February	3,460	1,986	2,544	141,283
March	3,200	1,900	2,292	140,906
April	2,800	1,330	1,640	97,567
May	7,000	1,610	2,745	168,793
June	28,180	2,480	6,214	369,739
July	8,160	2,440	4,829	296,926
August	7,880	2,090	3,241	199,260
September	7,220	2,850	4,392	261,342
October	9,320	2,270	4,136	254,301
November	2,410	1,820	2,009	119,544
December	1,800	1,690	1,744	107,246
The year	28,180	1,330	3,155	2,384,087

DEVILS RIVER AT DEVILS RIVER, TEX.

This station was established in April, 1900, by the International (Water) Boundary Commission. It is opposite the Southern Pacific Railway station at Devils River. The river is about 50 miles in length, has a perennial flow, and during flood periods is subject to great fluctuations.

e measurements of Devils River near Devils River station, Tex., in 1903.

Date.	Hydrographer.	Gage height.	Discharge.
		Feet.	*Second-feet.*
5......................	J. D. Dillard..................	2.05	412
10.....................	.....do	2.05	406
16.....................	.....do	2.4	610
21.....................	.....do	2.2	495
26.....................	.....do	2.15	475
31.....................	.....do	2.1	428
y 4....................	.....do	2.1	439
y 10...................	.....do	2.1	424
y 18...................	.....do	2.1	442
y 23...................	.....do	2.1	428
....	.....do	2.25	533
....	.....do	2.2	495
)...................	.....do	2.2	484
i...................	.....do	2.15	450
....	.....do	2.2	483
,...................	.....do	2.15	438
....	.....do	2.1	414
....	.....do	2.1	428
....	.....do	2.1	417
....	.....do	2.1	420
....	.....do	2.1	414
....	.....do	2.2	471
....	.....do	2.1	405
....	.....do	2.15	439
....	.....do	2.15	447
....	.....do	2.25	540
....	.....do	2.2	491
....	.....do	2.2	482
....	.....do	2.2	490
....	.....do	2.55	732
....	.....do	3.95	2,266
....	.....do	2.7	908
....	.....do	2.55	747
....	.....do	2.5	687
....	.....do	2.45	629
....	.....do	2.4	602
....	.....do	2.35	577
....	.....do	2.35	570
....	.....do	2.35	562
....	.....do	2.3	524

Discharge measurements of Devils River near Devils River station, etc.—Cont'd.

Date.	Hydrographer.	Gage height.	Discharge.
		Second-feet.	*Feet.*
August 15	J. D. Dillard	2.3	529
August 21	....do	2.3	521
August 28	....do	2.3	518
September 7	E. E. Winter	2.3	531
September 10	....do	2.3	521
September 15	....do	2.3	519
September 18	....do	2.3	518
September 22	....do	2.35	540
September 27	....do	2.3	535
October 2	J. D. Dillard	2.5	702
October 7	....do	2.4	619
October 13	....do	2.4	591
October 21	....do	2.35	561
November 3	....do	2.35	587
November 11	....do	2.35	574
November 20	....do	2.35	576
November 28	....do	2.35	566
December 7	....do	2.3	560
December 15	....do	2.3	550
December 22	....do	2.3	547
December 28	....do	2.3	530

Mean daily gage height, in feet, of Devils River near Devils River station, Tex., for 1903.

Day.	Jan.	Feb	Mar.	Apr.	May.	June.	July.	Aug.	Sept.	Oct.	Nov.	Dec.
1		2.10	2.10	2.25	2.10	2.10	2.20	2.50	2.30	2.30	2.50	2.35	2.35
2		2.10	2.10	2.20	2.10	2.10	2.65	2.50	2.30	2.30	2.50	2.35	2.35
3	2.10	2.15	2.20	2.10	2.05	2.35	2.50	2.30	2.30	2.45	2.35	2.35
4	2.05	2.10	2.20	2.10	2.05	2.25	2.45	2.35	2.30	2.45	2.35	2.35
5	2.05	2.10	2.20	2.10	2.10	2.20	2.50	2.35	2.30	2.45	2.35	2.35
6	2.05	2.10	2.20	2.10	2.15	2.20	2.50	2.35	2.30	2.45	2.35	2.30
7	2.05	2.10	2.20	2.10	2.15	2.20	2.50	2.35	2.30	2.40	2.35	2.30
8	2.05	2.10	2.20	2.10	2.15	2.15	2.45	2.30	2.30	2.40	2.35	2.30
9	2.05	2.10	2.20	2.10	2.10	2.60	2.45	2.30	2.30	2.40	2.35	2.30
10	2.05	2.10	2.20	2.15	2.20	2.65	2.40	2.30	2.30	2.40	2.35	2.30
11	2.05	2.10	2.20	2.10	2.20	2.35	2.40	2.50	2.30	2.40	2.35	2.30
12	2.05	2.10	2.15	2.10	2.90	3.40	2.40	2.30	2.30	2.40	2.35	2.30
13	2.10	2.10	2.15	2.10	2.15	7.30	2.40	2.30	2.35	2.40	2.35	2.30
14	2.10	2.10	2.15	2.10	2.10	4.40	2.40	2.30	2.30	2.40	2.35	2.30
15	2.35	2.10	2.15	2.10	2.10	4.05	2.40	2.30	2.30	2.40	2.35	2.30
16	2.30	2.10	2.15	2.10	2.10	3.75	2.40	2.30	2.35	2.40	2.35	2.30
17	2.15	2.10	2.20	2.10	2.10	3.15	2.40	2.30	2.45	2.40	2.35	2.30

Mean daily gage height, in feet, of Devils River near Devils River station, Tex., for 1903—Continued.

Day.	Jan.	Feb.	Mar.	Apr.	May.	June.	July.	Aug.	Sept.	Oct.	Nov.	Dec.
18	2.20	2.10	2.15	2.10	2.45	2.85	2.40	2.30	2.35	2.40	2.35	2.30
19	2.20	2.10	2.55	2.10	2.35	2.75	2.40	2.30	2.35	2.40	2.25	2.30
20	2.20	2.10	2.25	2.10	2.25	2.70	2.40	2.30	2.35	2.40	2.35	2.30
21	2.20	2.10	2.20	2.10	2.20	2.70	2.40	2.30	2.35	2.40	2.35	2.30
22	2.15	2.10	2.15	2.10	2.25	2.70	2.40	2.30	2.35	2.40	2.35	2.30
23	2.15	2.10	2.15	2.10	2.50	2.70	2.35	2.30	2.35	2.40	2.35	2.30
24	2.15	2.10	2.15	2.10	2.30	2.60	2.30	2.30	2.35	2.35	2.35	2.30
25	2.15	2.10	2.15	2.10	2.20	2.55	2.30	2.30	2.30	2.35	2.35	2.30
26	2.15	2.20	2.15	2.10	2.20	2.55	2.30	2.30	2.30	2.35	2.35	2.30
27	2.15	2.20	2.15	2.20	2.20	2.55	2.35	2.30	2.30	2.35	2.35	2.30
28	2.15	2.25	2.15	2.10	2.20	2.55	2.35	2.30	3.15	2.35	2.35	2.30
29	2.10	......	2.10	2.10	2.20	2.50	2.45	2.30	2.80	2.35	2.35	2.30
30	2.10	......	2.10	2.10	2.20	2.50	2.35	2.30	2.60	2.35	2.35	2.30
31	2.10	......	2.10	......	2.20	......	2.30	2.30	......	2.35	......	2.30

Estimated monthly discharge of Devils River near Devils River station, Tex., for 1903.

Month.	Discharge in second-feet.			Total in acre-feet.
	Maximum.	Minimum.	Mean.	
January	580	405	453	27,828
February	535	425	443	24,625
March	760	420	476	29,296
April	470	415	421	25,071
May	1,240	380	523	32,152
June	10,400	440	1,282	76,264
July	690	540	608	37,408
August	560	520	530	32,579
September	1,370	520	585	34,810
October	700	560	603	37,071
November	585	565	575	34,195
December	570	530	550	23,838
The year	10,400	380	587	415,137

PECOS RIVER NEAR MOORHEAD, TEX.

This station was established by the International (Water) Boundary Commission in April, 1900. It is near Moorhead, immediately above the high bridge of the Southern Pacific Railway.

Discharge measurements of Pecos River near Moorhead, Tex., in 1903.

Date.	Hydrographer.	Gage height.	Disch
		Feet.	*Secon*
January 2	J. D. Dillard	1.45	
January 8	....do	1.5	
January 14	....do	1.7	
January 19	....do	1.7	
January 24	....do	1.7	
January 29	....do	1.7	
February 3	....do	1.7	
February 7	....do	1.6	
February 17	....do	1.6	
February 21	....do	1.55	
February 27	....do	1.8	
March 4	....do	1.6	
March 9	....do	1.6	
March 14	....do	1.55	
March 19	....do	1.5	
March 27	....do	1.25	
March 31	....do	1.15	
April 4	....do	1.1	
April 10	....do	1.0	
April 15	....do	.9	
April 20	....do	.8	
April 25	....do	.8	
April 30	....do	.8	
May 5	....do	1.25	
May 9	....do	.9	
May 13	....do	1.4	
May 19	....do	1.8	
May 23	....do	1.15	
May 29	....do	1.1	
June 3	....do	.9	
June 8	....do	.9	
June 13	....do	2.9	
June 19	....do	1.5	
June 23	....do	3.4	
June 29	....do	3.6	
July 3	....do	2.15	
July 8	....do	1.85	
July 13	....do	1.5	
July 17	....do	1.3	

rge measurements of Pecos River near Moorhead, Tex., in 1903—Cont'd.

Date.	Hydrographer.	Gage height.	Discharge.
		Feet.	*Second feet.*
5	J. D. Dillard	1.0	274
)	do	.95	258
t 5	do	.9	288
t 10	do	.85	207
t 14	do	.8	200
t 19	do	.85	214
t 23	do	.8	202
t 27	do	.8	195
t 31	do	.75	182
aber 5	E. E. Winter	.7	354
aber 9	do	.7	258
aber 12	do	.7	286
aber 17	do	1.3	434
aber 20	do	.9	304
aber 24	do	.7	279
aber 30	J. D. Dillard	.9	238
r 5	do	.8	210
r 9	do	.8	199
r 15	do	.8	186
r 20	do	.75	177
r 24	do	.75	174
r 28	do	.8	185
iber 2	do	.8	204
iber 6	do	.85	211
iber 10	do	.85	218
iber 14	do	.9	243
iber 19	do	.9	235
iber 23	do	.8	213
iber 27	do	.9	232
ber 2	do	.9	226
ber 6	do	.9	224
ber 10	do	.9	235
ber 14	do	.9	231
ber 19	do	1.0	267
ber 24	do	.9	236
ber 30	do	.9	229

Mean daily gage height, in feet, of Pecos River near Moorhead, Tex., for 190

Day.	Jan.	Feb.	Mar.	Apr.	May.	June.	July.	Aug.	Sept.	Oct.	Nov.	De
1	1.50	1.70	1.60	1.10	0.80	1.05	3.25	1.00	0.80	0.90	0.80	0.
2	1.40	1.70	1.60	1.10	.80	1.20	2.60	1.00	.75	.80	.80	.
3	1.40	1.70	1.60	1.10	.80	.95	2.20	1.00	.70	.80	.80	.
4	1.40	1.60	1.60	1.10	.80	.90	2.05	.95	.70	.80	.80	.
5	1.40	1.60	1.60	1.00	1.00	.90	2.00	.90	.70	.80	.80	.
6	1.45	1.60	1.60	1.00	1.55	.90	2.00	.90	.70	.85	.80	.
7	1.50	1.60	1.60	1.00	1.20	.90	1.90	.90	.70	.90	.80	.
8	1.50	1.65	1.60	1.00	1.05	.90	1.90	.90	.70	.90	.85	.
9	1.50	1.60	1.60	1.00	.90	.95	1.85	.90	.70	.80	.85	.
10	1.50	1.60	1.60	1.00	.90	1.00	1.70	.90	.70	.80	.85	.
11	1.50	1.60	1.60	1.00	.90	2.20	1.65	1.00	.70	.80	.85	.
12	1.50	1.60	1.60	1.00	2.70	3.40	1.55	.90	.70	.80	.90	.
13	1.50	1.60	1.60	1.00	1.45	3.40	1.45	.85	.70	.80	.90	.
14	1.65	1.60	1.55	.90	1.30	1.50	1.40	.80	1.75	.80	.90	.
15	1.75	1.60	1.50	.90	1.10	1.40	1.40	.80	1.70	.80	.90	.
16	1.60	1.60	1.50	.90	1.45	1.40	1.40	.80	1.45	.80	.85	.
17	1.60	1.60	1.50	.85	1.35	1.30	1.35	.80	1.25	.75	.90	.
18	1.70	1.60	1.50	.80	1.30	1.30	1.25	.80	1.15	.80	.90	.
19	1.70	1.60	1.55	.80	2.05	1.40	1.20	.80	1.00	.80	.85	1.
20	1.70	1.60	1.50	.80	1.70	1.90	1.20	.80	.90	.75	.80	1.
21	1.65	1.55	1.50	.80	1.50	3.30	1.20	.80	.90	.70	.80	1.
22	1.60	1.50	1.50	.80	1.35	3.30	1.10	.80	.90	.70	.80	.
23	1.65	1.50	1.50	.80	1.15	3.40	1.05	.80	.75	.70	.80	.
24	1.70	1.50	1.45	.75	1.00	3.40	1.00	.80	.70	.75	.85	.
25	1.70	1.50	1.40	.75	.90	3.40	1.15	.80	.70	.80	.85	.
26	1.70	1.50	1.35	.80	.90	3.50	1.10	.80	.70	.80	.90	.
27	1.70	1.80	1.25	.80	.90	3.50	1.00	.80	.75	.80	.90	.
28	1.60	1.80	1.20	.80	.90	3.50	1.00	.80	.80	.80	.90	.
29	1.65	......	1.20	.80	1.15	3.60	1.00	.80	.85	.80	.90	.
30	1.70	......	1.20	.80	1.05	3.50	1.00	.80	.85	.80	.90	.
31	1.70	......	1.20	......	1.00	......	1.00	.75	......	.80	......	.

Estimated monthly discharge of Pecos River near Moorhead, Tex., for 1903.

Month.	Discharge in second-feet.			Total in acre-feet
	Maximum.	Minimum.	Mean.	
January	610	400	524	32,1￨
February	665	490	545	30,2
March	530	340	483	29,6￨
April	300	200	244	14,5
May	1,260	210	368	22,6
June	2,140	235	1,011	60,1
July	1,850	275	552	33,9
August	275	180	218	13,3
September	660	205	304	18,0
October	230	170	193	11,8
November	245	195	219	13,0
December	315	225	235	14,4
The year	2,140	170	408	294,3

PECOS RIVER AND MARGUERETTA FLUME NEAR PECOS, TEX.

The summer flow of Pecos River is largely dependent upon numerous springs, which occur in the limestone country in the vicinity of Roswell and Carlsbad. Owing to the numerous diversions for irrigating purposes, however, the river would be dry in the summer where it crosses into Texas were it not for the waters which are gradually returning to the river through seepage and for the various springs that occur below Carlsbad, N. Mex. This water, unfortunately, is impregnated to a considerable extent with alkali, which renders it undesirable for irrigating purposes.

The station on Pecos River was established January 1, 1898, by Thomas U. Taylor, and is located about 6 miles above Pecos, Tex., at the flume of the Barstow Irrigation Company (old Margueretta Canal Company). This canal diverts the water from Pecos River 3 miles above the flume from the west side of the river. The main canal flows for 3 miles on the west side of the river and then is taken by the flume across Pecos River to the east side. However, before it reaches the flume the West Valley canal is taken out of the main canal and is made to carry water to the alfalfa farms on the west side of the river. The gage consists of a graduated strip of wood attached to one of the vertical bents of the flume on the upper side of the same. The bench mark is on the top of the west abutment or pier on the north side, and its elevation is 20.70 feet above the datum of the gage. The channel is straight for 75 feet above and several hundred feet below the station. Its width at low water is about 20 feet and at high water is 200 feet. The bed is sandy and shifting. The velocity is poorly distributed and affected by the aqueduct. In the flume, conditions are favorable for accurate measurement. The gage heights on the flume are obtained by measuring the depths of the water in the flume at the west end. The zero of this gage is at the bottom of the flume. The gages of the river have no connection with each other except that they are geographically at the same place on the river and have the same observer, Willard H. Denis, who reads both gages. For the years 1901, 1902, and 1903 Mr. Denis has also taken the measurements of the flow of the Pecos and the flume at this station. The measurements are made above the flume by wading, a wire being stretched across the river and tagged every 4 feet. Flood discharge measurements are made at a highway bridge east of Pecos.

The observations at this station during 1903 have been made under the direction of Thomas U. Taylor, district hydrographer.

Mean daily gage height, in feet, of Pecos River near Moorhead, Tex., for 190

Day.	Jan.	Feb.	Mar.	Apr.	May.	June.	July.	Aug.	Sept.	Oct.	Nov.	De
1	1.50	1.70	1.60	1.10	0.80	1.05	3.25	1.00	0.80	0.90	0.80	0
2	1.40	1.70	1.60	1.10	.80	1.20	2.60	1.00	.75	.80	.80	
3	1.40	1.70	1.60	1.10	.80	.95	2.20	1.00	.70	.80	.80	
4	1.40	1.60	1.60	1.10	.80	.90	2.05	.95	.70	.80	.80	
5	1.40	1.60	1.60	1.00	1.00	.90	2.00	.90	.70	.80	.80	
6	1.45	1.60	1.60	1.00	1.55	.90	2.00	.90	.70	.85	.80	
7	1.50	1.60	1.60	1.00	1.20	.90	1.90	.90	.70	.90	.80	
8	1.50	1.65	1.60	1.00	1.05	.90	1.90	.90	.70	.90	.85	
9	1.50	1.60	1.60	1.00	.90	.95	1.85	.90	.70	.80	.85	
10	1.50	1.60	1.60	1.00	.90	1.00	1.70	.90	.70	.80	.85	
11	1.50	1.60	1.60	1.00	.90	2.20	1.65	1.00	.70	.80	.85	
12	1.50	1.60	1.60	1.00	2.70	3.40	1.55	.90	.70	.80	.90	
13	1.50	1.60	1.60	1.00	1.45	3.40	1.45	.85	.70	.80	.90	
14	1.65	1.60	1.55	.90	1.30	1.50	1.40	.80	1.75	.80	.90	
15	1.75	1.60	1.50	.90	1.10	1.40	1.40	.80	1.70	.80	.90	
16	1.80	1.60	1.50	.90	1.45	1.40	1.40	.80	1.45	.80	.85	
17	1.60	1.60	1.50	.85	1.35	1.30	1.35	.80	1.25	.75	.90	
18	1.70	1.60	1.50	.80	1.30	1.30	1.25	.80	1.15	.80	.90	
19	1.70	1.60	1.55	.80	2.05	1.40	1.20	.80	1.00	.80	.85	1
20	1.70	1.60	1.50	.80	1.70	1.90	1.20	.80	.90	.75	.80	1
21	1.65	1.55	1.50	.80	1.50	3.30	1.20	.80	.90	.70	.80	1
22	1.60	1.50	1.50	.80	1.35	3.30	1.10	.80	.90	.70	.80	
23	1.65	1.50	1.50	.80	1.15	3.40	1.05	.80	.75	.70	.80	
24	1.70	1.50	1.45	.75	1.00	3.40	1.00	.80	.70	.75	.85	
25	1.70	1.50	1.40	.75	.90	3.40	1.15	.80	.70	.80	.85	
26	1.70	1.50	1.35	.80	.90	3.50	1.10	.80	.70	.80	.90	
27	1.70	1.80	1.25	.80	.90	3.50	1.00	.80	.75	.80	.90	
28	1.60	1.80	1.20	.80	.90	3.50	1.00	.80	.80	.80	.90	
29	1.65		1.20	.80	1.15	3.60	1.00	.80	.85	.80	.90	
30	1.70		1.20	.80	1.05	3.50	1.00	.80	.85	.80	.90	
31	1.70		1.20		1.00		1.00	.75		.80		

Estimated monthly discharge of Pecos River near Moorhead, Tex., for 1903.

Month.	Discharge in second-feet.			Total in acre-feet
	Maximum.	Minimum.	Mean.	
January	610	400	524	32,1
February	665	490	545	30,:
March	530	340	483	29,0
April	300	200	244	14,.
May	1,260	210	368	22,0
June	2,140	235	1,011	60.
July	1,850	275	552	33.
August	275	180	218	13.
September	660	205	304	18.
October	230	170	193	11.
November	245	195	219	13.
December	315	225	235	14.
The year	2,140	170	408	294.

PECOS RIVER AND MARGUERETTA FLUME NEAR PECOS, TEX.

The summer flow of Pecos River is largely dependent upon numerous springs, which occur in the limestone country in the vicinity of Roswell and Carlsbad. Owing to the numerous diversions for irrigating purposes, however, the river would be dry in the summer where it crosses into Texas were it not for the waters which are gradually returning to the river through seepage and for the various springs that occur below Carlsbad, N. Mex. This water, unfortunately, is impregnated to a considerable extent with alkali, which renders it undesirable for irrigating purposes.

The station on Pecos River was established January 1, 1898, by Thomas U. Taylor, and is located about 6 miles above Pecos, Tex., at the flume of the Barstow Irrigation Company (old Margueretta Canal Company). This canal diverts the water from Pecos River 3 miles above the flume from the west side of the river. The main canal flows for 3 miles on the west side of the river and then is taken by the flume across Pecos River to the east side. However, before it reaches the flume the West Valley canal is taken out of the main canal and is made to carry water to the alfalfa farms on the west side of the river. The gage consists of a graduated strip of wood attached to one of the vertical bents of the flume on the upper side of the same. The bench mark is on the top of the west abutment or pier on the north side, and its elevation is 20.70 feet above the datum of the gage. The channel is straight for 75 feet above and several hundred feet below the station. Its width at low water is about 20 feet and at high water is 200 feet. The bed is sandy and shifting. The velocity is poorly distributed and affected by the aqueduct. In the flume, conditions are favorable for accurate measurement. The gage heights on the flume are obtained by measuring the depths of the water in the flume at the west end. The zero of this gage is at the bottom of the flume. The gages of the river have no connection with each other except that they are geographically at the same place on the river and have the same observer, Willard H. Denis, who reads both gages. For the years 1901, 1902, and 1903 Mr. Denis has also taken the measurements of the flow of the Pecos and the flume at this station. The measurements are made above the flume by wading, a wire being stretched across the river and tagged every 4 feet. Flood discharge measurements are made at a highway bridge east of Pecos.

The observations at this station during 1903 have been made under the direction of Thomas U. Taylor, district hydrographer.

Discharge measurements of Pecos River near Pecos, Tex., in 1903.

Date.	Hydrographer.	Gage height.	Discharge.
		Feet.	*Second-feet.*
January 17	W. H. Denis	3.70	438
April 16	do	.80	22
June 15	do	8.70	2,460
June 17	do	8.60	2,385
Do	do	8.40	2,273
June 18	do	8.00	2,165
June 22	do	9.20	2,500
June 25	do	7.50	1,811
Do	do	7.00	1,708
June 26	do	6.20	1,362
Do	do	5.10	957
June 27	do	4.00	537

Mean daily gage height, in feet, of Pecos River near Pecos, Tex., for 1903.

Day.	Jan.	Feb.	Mar.	Apr.	May.	June.	July.	Aug.	Sept.	Oct.	Nov.	Dec.
1	3.05	3.30	3.15	1.00	1.10	0.95	4.50	0.90	1.20	1.40	1.40	1.60
2	3.10	3.30	3.25	.85	1.10	1.00	3.70	.90	1.20	1.30	1.40	1.60
3	3.10	3.20	3.30	.80	1.10	1.00	3.25	.90	1.20	1.30	1.40	1.60
4	3.10	3.20	3.40	.80	1.00	1.00	3.25	.90	1.20	1.30	1.40	1.60
5	3.15	3.10	3.20	.80	1.00	1.00	3.05	1.30	1.20	1.30	1.40	1.60
6	3.20	3.10	3.05	.80	1.00	1.00	2.85	1.45	1.20	1.30	1.40	1.60
7	3.25	3.10	3.00	.80	1.00	2.45	2.30	1.25	1.20	1.30	1.40	1.60
8	3.40	3.10	3.00	.80	.95	2.05	2.15	1.20	1.20	1.30	1.40	1.60
9	3.40	3.10	2.90	.80	.90	1.70	1.85	1.20	1.20	1.30	1.40	1.60
10	3.45	3.10	2.90	.80	1.05	1.55	1.75	1.20	1.20	1.30	1.40	1.80
11	3.50	3.10	2.90	.80	1.00	1.35	1.55	1.20	1.20	1.30	1.40	1.80
12	3.50	3.10	2.90	.80	1.00	1.75	1.60	2.80	1.20	1.30	1.40	1.80
13	3.45	3.10	2.95	.80	1.00	2.35	1.55	1.55	1.20	1.40	1.40	1.80
14	3.45	3.00	2.95	.80	1.00	5.20	1.60	1.50	1.20	1.30	1.40	1.80
15	3.60	3.05	2.85	.80	1.00	8.75	1.60	1.50	1.20	1.30	1.40	1.80
16	3.65	3.10	2.45	.80	1.00	8.85	1.30	1.35	1.20	1.30	1.40	1.80
17	3.70	3.10	2.05	.80	1.00	8.50	1.10	1.30	1.20	1.30	1.40	1.70
18	3.70	3.15	2.00	.80	1.00	7.95	1.00	1.25	1.20	1.30	1.40	1.70
19	3.70	3.30	1.90	.80	1.00	7.80	.90	1.20	1.20	1.30	1.40	1.70
20	3.70	3.35	1.80	.80	1.00	8.65	.90	1.15	1.20	1.30	1.40	1.70
21	3.65	3.50	1.65	.80	1.00	8.95	1.05	1.10	1.20	1.40	1.40	1.70
22	3.50	3.40	1.50	.80	1.00	9.20	1.00	1.10	1.20	1.40	1.40	1.70
23	3.35	3.30	1.00	.80	1.00	9.15	.90	1.10	1.30	1.40	1.40	1.70
24	3.30	3.30	1.60	.80	1.00	8.55	.90	1.10	1.20	1.40	1.70	1.70
25	3.30	3.30	1.60	.80	1.00	7.40	.90	1.10	1.25	1.40	1.70	1.70
26	3.30	3.25	1.30	.75	1.00	5.75	.90	1.10	1.30	1.40	1.40	1.70
27	3.30	3.10	1.15	3.75	1.00	4.20	.90	1.10	1.30	1.40	1.70	1.70
28	3.30	3.10	1.30	3.20	1.00	3.50	.90	1.10	1.30	1.40	1.60	1.70
29	3.20		1.50	1.80	1.00	5.60	.90	1.10	1.30	1.40	1.60	1.70
30	3.35		1.40	1.40	1.00	5.00	.90	1.20	1.65	1.40	1.60	1.70
31	3.80		1.25		1.00		.90	1.20		1.40		1.70

Mean daily gage height, in feet, of flume across Pecos River near Pecos, Tex., for 1903.

Day.	Jan.	Feb.	Mar.	Apr.	May.	June.	July.	Aug.	Sept.	Oct.	Nov.	Dec.
1	2.80	2.10	1.40	1.55	2.05	1.85	2.45	2.35	1.50	1.95	2.00	2.10
2	2.10	2.10	1.00	1.50	1.90	2.10	2.55	2.15	1.30	1.80	2.10	2.00
3	2.10	2.10	1.50	1.50	1.80	2.10	2.70	2.30	1.60	1.80	2.10	2.00
4	2.10	2.15	1.50	1.50	1.80	1.20	2.70	2.30	1.85	1.70	2.10	2.00
5	2.25	2.20	1.85	1.50	1.95	1.30	2.70	2.85	1.70	1.80	2.10	2.00
6	2.30	2.10	1.95	1.65	2.15	1.70	2.65	2.60	1.80	1.70	2.10	2.00
7	2.30	2.20	2.10	2.10	2.20	2.45	2.85	2.25	1.40	1.60	2.10	2.00
8	2.30	2.20	2.15	2.20	2.20	.85	2.90	2.00	1.30	1.60	2.10	2.10
9	2.30	2.20	2.25	2.20	2.20	1.10	2.90	1.95	1.30	1.60	2.10	2.10
10	2.10	2.20	2.20	1.90	2.40	1.50	2.75	1.85	1.30	1.60	2.10	2.00
11	1.40	2.20	2.00	1.80	2.35	2.05	2.85	1.90	1.30	1.50	2.10	2.00
12	1.80	2.20	2.15	1.80	2.30	2.15	3.00	2.65	1.30	1.60	2.10	2.00
13	2.20	2.20	1.85	1.80	2.20	2.15	2.90	2.35	1.30	1.65	2.10	2.00
14	1.70	2.20	1.80	1.75	1.70	1.95	3.00	2.05	1.45	1.75	2.10	2.00
15	.95	2.20	1.85	1.95	1.60	1.90	2.95	1.95	1.65	1.80	2.10	2.00
16	.90	2.20	1.80	2.00	1.60	1.80	2.80	1.75	1.60	1.80	2.10	2.00
17	.90	2.20	1.80	2.00	1.50	1.80	2.65	1.70	1.60	1.90	2.10	2.00
18	.90	2.25	1.80	2.00	1.45	1.90	2.55	1.70	1.65	1.00	2.10	2.00
19	.90	1.80	1.55	2.00	1.95	2.30	2.65	1.75	1.25	2.00	2.10	2.00
20	1.05	1.40	1.60	1.65	2.00	2.35	2.60	1.65	1.20	1.95	2.10	2.10
21	1.20	1.00	1.55	1.70	2.00	2.30	2.95	1.45	1.20	1.80	2.10	2.10
22	1.65	1.85	1.60	1.70	1.60	2.00	2.55	1.40	1.20	1.90	2.10	2.10
23	2.20	1.70	1.40	1.70	1.25	2.40	2.45	1.35	1.65	1.90	2.10	2.10
24	2.10	1.70	1.30	1.80	1.90	2.05	2.35	1.45	1.70	1.90	2.00	2.10
25	2.10	1.70	1.30	1.95	1.90	2.05	2.40	1.80	2.00	1.30	2.05	2.10
26	2.10	1.70	1.30	2.00	1.50	2.25	2.40	1.85	2.20	1.90	2.10	2.10
27	2.10	1.60	1.50	2.40	1.10	2.25	2.35	1.80	2.15	2.00	2.10	2.00
28	2.10	2.00	1.35	2.50	1.00	2.25	2.25	1.80	2.20	2.00	2.10	2.00
29	2.10	.....	1.30	2.40	1.00	2.30	2.40	1.20	2.20	2.00	2.10	2.00
30	2.10	.....	1.30	2.25	1.30	2.30	2.30	2.20	1.85	2.00	2.10	2.10
31	2.10	.....	1.35	.....	1.20	.....	2.90	1.85	.....	2.00	.....	2.10

Rating table for Pecos River near Pecos, Tex., from January 1 to December 31, 1903.

Gage height.	Discharge.	Gage height.	Discharge.	Gage height.	Discharge.	Gage height.	Discharge.
Feet.	Second-feet.	Feet.	Second-feet.	Feet.	Second-feet.	Feet.	Second-feet.
0.8	22	2.0	106	3.2	304	4.4	655
.9	26	2.1	117	3.3	328	4.5	690
1.0	30	2.2	129	3.4	352	4.6	725
1.1	36	2.3	142	3.5	376	4.7	760
1.2	42	2.4	156	3.6	404	4.8	800
1.3	48	2.5	170	3.7	432	4.9	840
1.4	54	2.6	186	3.8	460	5.0	880
1.5	62	2.7	204	3.9	490	6.0	1,280
1.6	70	2.8	224	4.0	520	7.0	1,680
1.7	78	2.9	244	4.1	550	8.0	2,130
1.8	86	3.0	264	4.2	585	9.0	2,580
1.9	96	3.1	284	4.3	620		

Between gage heights 0.80 and 3.70 feet the table is constructed from measurements of 1902. This table was based on measurements of the river only: it does not include ditch or flume

Rating table for flume across Pecos River near Pecos, Tex., from January 1 to December 31, 1903.

Gage height.	Discharge.	Gage height.	Discharge.	Gage height.	Discharge.	Gage height.	Discharge.
Feet.	*Second-feet.*	*Feet.*	*Second-feet.*	*Feet.*	*Second-feet.*	*Feet.*	*Second-feet.*
0.9	23	1.5	54	2.1	111	2.7	188
1.0	27	1.6	62	2.2	123	2.8	202
1.1	31	1.7	70	2.3	135	2.9	216
1.2	35	1.8	80	2.4	147	3.0	230
1.3	41	1.9	90	2.5	160		
1.4	47	2.0	100	2.6	174		

Table not well defined.   Extended above 2.15 feet gage height.

Estimated monthly discharge of Pecos River near Pecos, Tex., for 1903.[a]

Month.	Discharge in second-feet.			Total in acre-feet.
	Maximum.	Minimum.	Mean.	
January	487	395	442	27,177
February	439	346	405	22,493
March	406	89	242	14,880
April	593	76	140	8,331
May	180	57	113	6,948
June	2,792	65	1,167	69,441
July	843	155	293	18,016
August	405	71	146	8,977
September	171	77	109	6,486
October	154	102	132	8,116
November	189	154	169	10,056
December	189	170	182	11,191
The year	2,792	57	295	212,112

[a] This includes discharge of flume.   It does not include discharge of West Valley ditch.

PECOS RIVER AT CARLSBAD, N. MEX.

This station was established May 20, 1903, by V. L. Sullivan, acting under the direction of W. M. Reed.   It is located at the Green street highway bridge, Carlsbad, N. Mex., and is about 500 feet below the station of the Pecos Valley and Northeastern Railroad and 2,000 feet below the Hagerman power dam.   The gage consists of a 2 by 8 inch plank securely spiked at an inclination of 10° with the vertical to the timbers of the third bent from the west end of the bridge.   It is painted white and graduated in black to vertical feet and tenths.

gs are taken twice daily by V. L. Sullivan, a civil engineer in
ploy of the Pecos Valley Irrigation Company. Discharge
ements are made by wading when the stage of the river will
and from the lower side of the bridge during floods. The
)oint for soundings is on the south side of the bridge at the
utment, and 10-foot intervals are marked on the rail. Both
ire high and not subject to overflow. The bed of the river is
ick, much corrugated, which makes low-water measurements
to considerable inaccuracy. The channel is straight for some
e above and below the station. The current is swift at the
but sluggish both above and below. The only bench mark
iitial point for soundings, which is 22.3 feet above the zero of
e.

)bservations at this station during 1903 have been made under
·ction of W. M. Reed, district hydrographer.

charge measurements of Pecos River at Carlsbad, N. Mex., in 1903.

Date.	Hydrographer.	Gage height.	Discharge.
		Feet.	Second-feet.
....	W. M. Reed	0.72	127
16....................	E. G. Marsh	.62	112

aily gage height, in feet, of Pecos River at Carlsbad, N. Mex., for 1903.

Day.	May.	June.	July.	Aug.	Sept.	Oct.	Nov.	Dec.
....		0.81	1.35	0.75	0.65	0.65	0.64	0.65
....		.78	1.20	.75	.65	.65	.64	.65
..................................|.......		.80	1.13	.75	.65	.65	.65	.65
....		.79	.98	.75	.65	.65	.65	.65
..................................|........		.80	.98	.75	.65	.66	.65	.65
....		.79	.98	.74	.65	.65	.65	.64
..................................,........		.80	.98	.74	.66	.65	.65	.64
....		.80	.99	.74	.65	.65	.65	.64
....		.81	.98	.74	.66	.65	.65	.64
....		.81	.97	.73	.65	.65	.65	.64
....		.81	1.08	.73	.65	.65	.65	.64
....		2.15	.84	.73	.66	.65	.65	.64
....		4.65	.83	.73	.65	.65	.66	.64
....		4.78	.82	.72	.65	.65	.65	.64
....		4.28	.82	.72	.66	.65	.65	.64
....		3.98	.83	.71	.66	.65	.65	.64
....		3.76	.81	.70	.66	.65	.64	.64
....		8.05	.79	.69	.66	.65	.64	.64
....		4.90	.81	.68	.66	.65	.64	.64
....		4.80	.81	.67	.65	.65	.65	.64
....		4.55	.81	.67	.65	.65	.65	.64
....		4.25	.82	.66	.65	.65	.65	.64
....		3.65	.80	.66	.65	.65	.65	.64

Mean daily gage height, in feet, of Pecos River, etc.—Continued.

Day.	May.	June.	July.	Aug.	Sept.	Oct.	Nov.	Dec.
24		2.60	0.79	0.65	0.65	0.65	0.65	0.65
25		.88	.78	.65	.65	.64	.65	.65
26		1.49	.78	.65	.65	.64	.65	.65
27		2.45	.75	.65	.65	.64	.65	.65
28		2.34	.75	.65	.65	.64	.65	.65
29	0.79	1.95	.75	.65	.65	.64	.65	.65
30	.74	1.73	.75	.65	.65	.64	.65	.65
31	.78		.75	.65		.64		.65

PECOS RIVER BELOW MOUTH OF HONDO RIVER, NEAR ROSWELL, N. MEX.

This station was established April 24, 1903, by W. M. Reed. It is located at the highway bridge 8 miles southeast of Roswell, N. Mex., and about 200 feet below the mouth of Hondo River. The gage is painted on the right side of the right pier of the bridge. It is read twice each day by Miss Dovie Goldsmith. Discharge measurements are made from the highway bridge at the gage. The initial point for soundings is a zero marked on the guard rail at the west end and north side of the bridge. The channel is straight for one-half mile above and below the station, and has a width at low water of about 50 feet and at ordinary high water of 430 feet. The channel is broken by three iron piers. The current is rapid except near the mouth of Hondo River, where it becomes sluggish. At high water the Pecos and the Hondo join above the bridge. The gage heights on the Pecos may be effected by back water at periods when the Pecos is low and the Hondo is high. Both banks are high and free from timber, but overflow at extreme flood stages. The bed is sandy and shifting and the cross section changes during each flood. The bench mark is the top of the pier upon which the gage is painted. Its elevation is 20 feet above the zero of the gage.

The observations at this station during 1903 have been made under the direction of W. M. Reed, district hydrographer.

Discharge measurements of Pecos River below mouth of Hondo River, near Roswell, N. Mex., in 1903.

Date.	Hydrographer.	Gage height.	Discharge.
		Feet.	*Second-feet.*
April 23	W. M. Reed	2.60	67
May 4	do	3.45	200
June 11	W. A. Wilson	10.00	15,992

ily gage height, in feet, of Pecos River below mouth of Hondo River, near Roswell, N. Mex., for 1903.

Day.	May.	June.	July.	Aug.	Sept.	Oct.	Nov.	Dec.
---------------------------- --------		3.00	4.00	3.20	3.30	3.80	3.70	4.00
---------------------------- --------		3.00	4.00	3.20	3.30	3.60	3.70	4.00
---------------------------- --------		3.00	4.00	3.10	3.25	3.60	3.70	4.00
----------------------------	3.45	3.00	4.00	3.10	3.25	3.60	3.70	4.20
----------------------------	3.45	3.00	3.70	3.20	3.20	3.55	3.70	4.00
----------------------------	3.45	3.80	3.70	3.45	3.20	3.50	3.70	4.00
----------------------------	3.30	3.35	3.70	3.70	3.20	3.45	3.70	4.00
----------------------------	3.30	4.15	3.65	3.70	3.20	3.45	3.65	4.00
----------------------------	3.30	4.00	3.45	3.80	3.25	3.45	3.63	4.00
----------------------------	3.40	9.25	3.45	3.70	3.20	3.45	3.80	4.00
----------------------------	3.40	9.75	3.45	3.60	3.25	3.40	4.50	4.00
----------------------------	4.00	6.00	3.45	3.45	3.20	3.35	4.50	4.00
----------------------------	3.45	5.00	3.40	5.00	3.20	3.35	4.50	4.00
----------------------------	3.40	5.00	3.30	4.00	3.20	3.30	4.50	4.00
----------------------------	3.30	5.45	3.00	4.25	3.20	3.30	4.00	4.00
----------------------------	4.00	7.25	3.00	4.25	3.20	3.30	4.00	4.00
----------------------------	3.00	8.45	3.00	4.20	3.25	3.30	4.00	4.00
----------------------------	3.45	6.00	3.00	4.20	3.25	3.30	4.00	4.00
----------------------------	3.45	5.45	2.70	4.10	3.25	3.30	4.00	4.00
----------------------------	3.35	5.45	2.70	4.00	3.25	3.35	4.00	4.00
----------------------------	3.30	5.00	2.70	3.70	3.25	3.35	4.00	4.00
----------------------------	3.45	5.00	2.70	3.45	3.50	3.35	4.00	4.00
----------------------------	3.45	5.00	3.00	3.45	3.60	3.35	4.00	4.00
----------------------------	3.30	4.45	3.20	3.45	4.20	3.35	4.00	4.00
----------------------------	3.30	4.45	3.70	3.45	4.00	3.45	4.00	4.00
----------------------------	3.25	4.45	3.70	3.45	3.80	3.45	4.00	4.00
----------------------------	3.25	4.45	3.60	3.45	3.80	3.50	4.00	4.00
----------------------------	3.20	4.45	3.45	3.45	3.75	3.50	4.00	4.00
----------------------------	3.00	4.00	3.45	3.45	3.80	3.70	4.00	4.00
----------------------------	3.00	4.00	3.35	3.30	3.80	3.70	4.00	4.00
----------------------------	3.00	....	3.30	3.30	....	3.70	....	4.00

HONDO RIVER AT ROSWELL, N. MEX.

station was established April 25, 1903, by W. M. Reed. It is at the bridge at the intersection of Main and Vegas streets, l, N. Mex. The gage is a 4 by 4 inch inclined timber set on it bank 150 feet below the bridge. It is read by members of logical Survey office force at Roswell. Discharge measure- are made from the highway bridge. The initial point for igs is a zero marked on the east stringer at the north end of lge. The channel is nearly straight for 50 feet above and 450 low the bridge and has a width at ordinary high stages of 40 The current has a moderate velocity. Both banks are low and wn with weeds, but are not liable to overflow. The bed of the is sandy loam, fairly permanent, and free from vegetation. s but one channel at all stages. The bench mark is the initial r soundings. Its elevation is 8.50 feet above the zero of the

The observations at this station during 1903 have been made under the direction of W. M. Reed, district hydrographer.

Discharge measurements of Hondo River at Roswell, N. Mex., in 1903.

Date.	Hydrographer.	Gage height.	Discharge.
		Feet.	*Second-feet.*
June 11	W. A. Wilson	2.50	71.5
June 12	do	2.40	73.5
June 14	do	3.80	245.5
June 15	do	4.10	310.0
June 20	do	2.20	96.0

Mean daily gage height, in feet, of Hondo River at Roswell, N. Mex., for 1903.

Day.	June.	Day.	June.	Day.	June.	Day.	June.	Day.	June.
8	3.80	12	2.40	16	3.80	20	2.20	24	0.0
9	1.20	13	2.70	17	2.90	21	1.90	25	(a)
10	1.40	14	4.20	18	2.90	22	1.70		
11	2.50	15	4.10	19	2.40	23	1.50		

a No water after June 25.

HONDO RIVER AT HONDO RESERVOIR SITE, NEW MEXICO.

This station was established March 9, 1903, by W. A. Wilson. It is located at the First New Mexico Reservoir dam site, 12 miles southwest of Roswell, N. Mex. A footbridge has been constructed 75 feet below the dam for the purpose of making discharge measurements. The gage is a 4 by 4 inch inclined timber which is located 10 feet north of this bridge. The gage is read twice each day by Lee Hall. The initial point for soundings is 1 foot south of the north end of the west stringer of the bridge. The channel is straight for 200 feet above and below the station. The current is swift at high water and sluggish at low water. Both banks are high, without trees, and not liable to overflow. There is but one channel at all stages. The bed is composed of shifting sand, and the cross section changes during each flood. Bench mark No. 1 is the upper surface of the crosspiece which supports the stringer at the north end of the bridge. Its elevation is 8.50 feet above the zero of the gage. Bench mark No. 2 is on a ledge of rock which bears S. 45° W. and is 650 feet distant from the gage. Its elevation is 19.10 feet above the zero of the gage.

The observations at this station during 1903 have been made under the direction of W. M. Reed, district hydrographer.

rge measurements of Hondo River at Hondo reservoir site, New Mexico, in 1903.

Date.	Hydrographer.	Gage height.	Discharge.
		Feet.	*Second-feet.*
)	W. A. Wilson................	5.30	561
....do	5.30	561
} ....do	4.40	489
; 21	E. G. Marsh	.95	11
ber 10	F. S. Dobson...............	1.10	12

laily gage height, in feet, of Hondo River at Hondo reservoir site, New Mexico, for 1903.

Day.	June.	July.	Aug.	Sept.	Oct.	Nov.	Dec.
....		0.80	....	....	1.10	1.10	1.20
....		.75	....	....	.70	1.10	1.10
....		.60	....	....	.75	1.10	1.10
....					....	1.10	1.20
....					....	1.10	1.10
....					....	1.10	1.00
....	2.30				....	1.10	1.10
....	2.70				....	1.10	1.20
....	1.90				....	1.10	1.20
....	5.30				....	1.10	1.20
....	2.70				....	1.10	1.10
....	2.80				....	1.10	1.20
....	5.70		....		....	1.10	1.10
....	5.90	.60	....		....	1.10	.90
....	7.50	.55	3.60	....	1.20	1.10	.80
....	4.50	.50	1.05	....	1.20	1.10	.80
....	4.95	....	.70	....	1.20	1.10	.70
....	4.15	....	.85	....	1.20	1.00	.80
....	3.88	....	1.55	....	1.20	1.10	.70
....	3.65	2.70	1.15	....	1.20	1.20	.70
....	3.52	1.50	.80	....	1.20	1.10	.60
....	3.50	.80	.50	....	1.10	1.10	.60
....	3.20	.75	.50	....	1.20	1.00	.60
....	2.85	.40	....	....	1.20	1.10	.50
....	2.60	....		1.35	1.20	1.10	.60
....	2.48	....	2.10	1.05	1.10	1.10	.60
....	2.00	....	1.55	.95	1.10	1.10	.70
....	1.83	....	1.15	1.20	1.20	1.10	.70
....	1.45	....	.80	1.30	1.10	1.20	.70
....	1.00	....	.65	1.25	1.20	1.10	.70
....			.50	....	1.20	....	.70

PECOS RIVER AT SANTA ROSA, N. MEX.

ꞌ station was established May 5, 1903, by H. C. Hurd. It is
d at the bridge of the Chicago, Rock Island and Pacific Rail-

road. This is a five-span. deck girder bridge supported on tow
having masonry footings at the edges of the channel. It is 300 f
long and 75 feet above low water. The gage rod is a vertical 2 b
inch timber, graduated to feet and tenths with tacks. It is bol
to the masonry footing of the east tower. Daily readings are m
by L. M. Shely, a bank clerk. Discharge measurements are m
by wading when the stage of the river will permit and from a c
which will be placed later on, the bridge being too high and dan
ous to make measurements from during flood. The initial p
for soundings is the end of the girder at the east end of the bri
Both banks are high and can not overflow. The bed of the rive
solid rock overlain by quicksand to a depth of 2 or 3 feet.
cross section will be modified by the erosion of this quicksand in t
of flood. The current is never sluggish and becomes very swift c
ing floods. The channel is straight for one-fourth mile above
below the station. A shelf cut in the east abutment at an eleva
of 29.70 feet above the zero of the gage is the only bench mark.

Discharge measurements of Pecos River at Santa Rosa, N. Mex., in 1903.

Date.	Hydrographer.	Gage height.	Disch
		Feet.	Secon
May 7	H. C. Hurd	1.83	
August 17	Earl Marsh	1.78	

Mean daily gage height, in feet, of Pecos River at Santa Rosa, N. Mex., for .

Day.	May.	June.	July.	Aug.	Sept.	Oct.	Nov.	I
1		0.95	2.00	2.00	1.70	1.60	1.65	
2		.95	2.00	2.00	1.70	1.60	1.65	
3		.95	1.90	2.00	1.70	1.60	1.65	
4		.95	1.80	2.00	1.70	1.60	1.65	
5	1.45	1.25	1.50	2.00	1.70	1.60	1.65	
6	1.38	1.20	2.20	1.70	1.70	1.60	1.65	
7	1.35	1.60	2.20	1.40	1.70	1.60	1.60	
8	1.30	1.50	2.20	1.40	1.70	1.60	1.60	
9	1.30	6.00	2.20	1.40	1.70	1.60	1.60	
10	1.25	6.50	1.80	1.40	1.70	1.60	1.60	
11	1.30	3.00	1.80	1.50	1.70	1.60	1.60	
12	1.30	1.50	1.60	2.00	1.70	1.60	1.60	
13	1.35	2.80	1.90	2.00	1.70	1.60	1.60	
14	1.40	2.60	1.70	1.90	1.70	1.60	1.60	
15	1.40	2.80	1.70	2.00	1.70	1.60	1.40	
16	1.45	6.00	1.60	2.00	1.70	1.60	1.60	
17	1.45	3.80	1.70	1.70	1.70	1.60	1.60	
18	1.50	3.50	1.70	1.90	1.70	1.60	1.70	
19	1.45	2.80	1.70	2.20	1.70	1.60	1.65	
20	1.40	2.80	1.70	2.00	1.70	1.60	1.65	

Mean daily gage height, in feet, of Pecos River, etc.—Continued.

Day.	May.	June.	July.	Aug.	Sept.	Oct.	Nov.	Dec.
....	1.35	2.80	2.40	1.70	1.70	1.60	1.65	.65
....	1.35	2.50	2.00	1.70	1.70	1.60	1.65	.65
....	1.35	2.50	2.00	1.70	1.80	1.60	1.65	.65
....	1.20	2.60	2.00	1.70	1.70	1.60	1.65	.65
....	1.20	2.30	2.00	1.70	1.70	1.60	1.65	.65
....	1.20	2.20	2.00	1.70	1.70	1.60	1.65	.65
....	1.15	2....	2.00	1.70	1.70	1.65	1.65	.65
....	1.10	2.20	.00	1.70	1.70	1.65	1.65	.65
....	1.10	2.00	2.00	1.70	1.70	1.65	1.65	.65
....	.95	2.00	2.00	1.70	1.70	1.65	1.65	.65
....	.95	....	2.00	1.70	....	1.65	....	.65

RIO GRANDE NEAR LANGTRY, TEX.

station was established in April, 1900, by the International
) Boundary Commission. It is located one-half mile south of
' station, on the Southern Pacific Railway, and is about 440
elow El Paso, Tex., at the eastern end of the canyon section
Rio Grande, and a short distance to the west of the mouth of
River, one of the principal tributaries of the Rio Grande.

Discharge measurements of Rio Grande near Langtry, Tex., in 1903.

Date.	Hydrographer.	Gage height.	Discharge.
		Feet.	Second-feet.
1	J. D. Dillard	1.3	783
7	....do	1.2	714
12	....do	1.2	720
17	....do	1.2	719
23	....do	1.15	680
28	....do	1.1	658
2	....do	1.35	839
6	....do	1.3	795
16	....do	2.2	1,894
20	....do	1.9	1,396
25	....do	1.6	1,099
....	....do	2.0	1,519
....	....do	1.75	1,240
....	....do	1.5	980
....	....do	1.35	845
....	....do	1.2	724
....	....do	1.2	715
....	....do	1.25	724

Discharge measurements of Rio Grande near Langtry, Tex., in 1903—Contir

Date.	Hydrographer.	Gage height.	Disck
		Feet.	*Secon*
April 4	J. D. Dillard	1.1	
April 9	do	1.0	
April 14	do	.95	
April 19	do	1.4	
April 24	do	1.3	
April 29	do	2.6	
May 4	do	1.3	
May 9	do	1.65	
May 12	do	2.5	
May 15	do	2.1	
May 20	do	2.4	
May 25	do	1.95	
May 30	do	1.95	
June 2	do	8.7	
June 7	do	2.7	
June 11	do	8.9	
June 14	do	6.2	1
June 19	do	2.95	
June 24	do	2.8	
June 30	do	2.7	
July 5	do	3.0	
July 10	do	4.8	
July 14	do	4.2	
July 17	do	3.05	
July 25	do	2.0	
July 31	do	2.3	
August 3	do	1.65	
August 7	do	1.6	
August 11	do	1.6	
August 17	do	1.8	
August 22	do	3.1	
August 26	do	2.3	
August 30	do	2.5	
September 3	E. E. Winter	4.55	
September 7	do	3.6	
September 10	do	2.8	
September 14	do	2.3	
September 18	do	2.8	
September 22	do	2.7	
September 29	do	3.45	

ge measurements of Rio Grande near Langtry, Tex., in 1903—Continued.

Date.	Hydrographer.	Gage height.	Discharge.
		Feet.	Second-feet.
5	J. D. Dillard	5.3	8,836
10	do	3.0	8,047
15	do	2.25	1,946
19	do	2.0	1,607
23	do	1.9	1,500
27	do	1.7	1,280
31	do	1.5	1,089
)er 5	do	1.4	950
)er 9	do	1.35	900
)er 13	do	1.8	847
)er 18	do	1.15	743
)er 22	do	1.15	743
)er 26	do	1.1	722
)er 30	do	1.05	698
er 5	do	1.0	650
er 9	do	1.0	645
er 13	do	1.0	641
er 17	do	.95	626
er 21	do	.95	622
er 26	do	.95	622
er 31	do	.9	593

daily gage height, in feet, of Rio Grande near Langtry, Tex., for 1903.

Day.	Jan.	Feb.	Mar.	Apr.	May.	June.	July.	Aug.	Sept.	Oct.	Nov.	Dec.
	1.30	1.10	2.10	1.20	1.75	1.95	2.70	1.80	3.40	3.55	1.50	1.05
	1.30	1.35	1.95	1.20	1.65	3.45	2.80	1.80	3.65	4.70	1.45	1.05
	1.30	1.35	1.90	1.10	1.35	2.20	2.70	2.00	4.65	5.55	1.40	1.05
	1.30	1.30	1.90	1.10	1.30	2.10	2.80	2.75	4.25	5.95	1.40	1.05
	1.30	1.30	1.85	1.00	1.20	2.15	3.10	2.55	4.20	5.10	1.40	1.00
	1.25	1.30	1.75	1.00	1.45	3.00	3.50	1.85	3.80	5.00	1.35	1.00
	1.25	1.30	1.75	1.00	1.15	2.50	3.95	1.60	3.75	4.00	1.30	1.00
	1.20	1.30	1.70	1.00	1.10	2.60	4.30	1.60	4.10	3.85	1.30	1.00
	1.20	1.25	1.65	1.00	1.80	3.25	4.55	1.80	2.70	3.30	1.30	1.00
	1.20	1.20	1.60	1.00	2.05	2.90	4.80	1.60	2.80	2.95	1.30	1.00
	1.20	1.10	1.55	1.00	2.10	3.70	4.85	1.60	2.85	2.90	1.30	1.00
	1.20	1.10	1.50	1.00	3.00	8.95	4.65	1.55	2.70	2.70	1.30	1.00
	1.20	2.50	1.45	1.00	2.00	6.65	4.35	2.20	2.30	2.35	1.25	1.00
	1.20	2.50	1.40	1.00	2.00	6.05	4.10	2.40	2.20	2.30	1.25	1.00
	1.30	2.45	1.40	1.00	2.05	4.30	3.90	2.05	2.30	2.25	1.25	1.00
	1.25	2.25	1.40	1.00	2.70	3.30	3.30	1.90	2.65	2.10	1.25	1.00
	1.20	2.05	1.35	1.00	2.30	3.20	3.05	1.80	3.05	2.05	1.20	1.00
	1.15	2.00	1.30	1.20	2.10	3.00	2.90	1.80	2.80	2.00	1.20	1.00

Mean daily gage height, in feet, of Rio Grande, etc.—Continued.

Day.	Jan.	Feb.	Mar.	Apr.	May.	June.	July.	Aug.	Sept.	Oct.	Nov.	Dec
19	1.15	1.95	1.30	1.40	2.20	3.00	2.85	1.85	2.70	2.00	1.20	1.
20	1.15	1.90	1.30	1.20	2.30	2.90	2.70	2.10	2.75	1.90	1.10	1.
21	1.15	1.90	1.30	1.20	2.20	2.90	2.65	2.75	2.75	1.90	1.10	1.
22	1.15	1.90	1.20	1.20	2.20	2.85	2.65	3.10	2.55	1.90	1.10	1.
23	1.15	1.75	1.20	1.20	2.10	2.80	2.65	2.90	2.40	1.85	1.10	.
24	1.15	1.70	1.20	1.30	1.90	2.80	2.20	2.80	2.30	1.80	1.10	.
25	1.10	1.60	1.20	1.30	1.90	2.70	1.95	2.55	2.85	1.75	1.10	.
26	1.10	1.60	1.20	1.30	1.90	2.70	1.85	2.30	3.80	1.70	1.10	.
27	1.10	2.40	1.15	1.30	1.90	2.70	1.80	2.30	3.15	1.70	1.10	.
28	1.10	2.25	1.00	1.35	1.90	2.70	1.90	3.95	3.30	1.65	1.10	.
29	1.10	......	1.05	2.55	1.90	2.70	2.30	3.55	3.40	1.40	1.05	.
30	1.10	......	1.20	2.05	1.95	2.70	2.90	2.50	4.40	1.40	1.05	.
31	1.10	......	1.20	......	1.95	......	2.15	2.50	......	1.45	......	.

Estimated monthly discharge of Rio Grande near Langtry, Tex., for 1903.

Month.	Discharge in second-feet.			Total in acre-feet.
	Maximum.	Minimum.	Mean.	
January	780	660	711	43,69
February	2,340	660	1,277	70,93
March	1,700	575	960	59,03
April	2,320	530	738	43,91
May	2,920	650	1,507	92,68
June	18,000	1,510	3,993	237,60
July	6,600	1,350	3,383	207,98
August	4,750	1,070	1,983	121,90
September	5,860	1,720	2,922	173,87
October	10,790	960	3,134	192,67
November	1,040	700	811	48,27
December	700	595	642	39,50
The year	18,000	530	1,838	1,332,09

RIO GRANDE BELOW PRESIDIO, TEX.

This station was established April 8, 1900, by the International (Water) Boundary Commission. It is 6 miles below Presidio, also below the mouth of Conchos River, and about 215 miles below El Paso. It is at the western end of the canyon section of the Rio Grande. The discharge at this station minus the discharge at the station above Presidio, Tex., gives the discharge of the Conchos, except at rare intervals when some rain water enters the Rio Grande from the north.

charge measurements of Rio Grande below Presidio, Tex., in 1903.

Date.	Hydrographer.	Gage height.	Discharge.
		Feet.	*Second feet.*
2	Jas. P. Hague	4.9	375
5	do	4.8	314
8	do	4.8	306
11	do	4.7	298
15	do	4.7	312
17	do	4.7	303
20	do	4.7	302
23	do	4.7	317
26	do	5.2	513
29	do	5.25	497
¹ 1	do	5.2	481
¹ 4	do	5.0	438
¹ 7	do	7.65	2,622
¹ 9	do	7.55	2,506
¹ 11	do	6.9	1,398
¹ 13	do	6.9	1,352
¹ 16	do	6.65	1,057
¹ 18	do	6.2	911
¹ 20	do	5.85	721
¹ 23	do	6.9	1,382
¹ 25	do	6.9	1,381
¹ 27	do	6.6	1,141
	do	6.2	903
	do	5.95	750
	do	5.55	586
	do	5.4	532
	do	5.25	409
	do	5.15	379
	do	5.3	475
	do	5.0	342
	do	5.3	471
	do	5.0	347
	do	4.9	313
	do	4.6	225
	do	4.4	149
	do	4.3	105
	do	4.3	109
	do	5.8	628
	do	5.35	463
	do	5.3	457

Discharge measurements of Rio Grande below Presidio, Tex., in 1903—Contin

Date.	Hydrographer.	Gage height.	Disch
		Feet.	*Secon*
April 23	Jas. P. Hague	5.3	
April 26	....do	6.5	
April 29	....do	5.1	
May 3	....do	5.0	
May 5	....do	6.8	
May 7	....do	6.9	
May 9	....do	6.9	
May 11	....do	7.0	
May 13	....do	7.0	
May 16	....do	7.15	
May 18	....do	7.2	
May 20	....do	6.95	
May 22	....do	6.8	
May 25	....do	6.9	
May 27	....do	6.9	
June 2	....do	6.9	
June 4	....do	7.0	
June 6	....do	7.05	
June 9	....do	7.65	
June 12	....do	9.4	
June 14	....do	7.9	
June 16	....do	7.8	
June 18	....do	7.8	
June 20	....do	7.7	
June 22	....do	7.7	
June 24	....do	7.7	
June 26	....do	7.7	
June 28	....do	7.9	
June 30	....do	7.9	
July 2	....do	8.2	
July 4	....do	9.2	
July 6	....do	9.5	
July 8	....do	9.8	
July 10	....do	9.5	
July 13	....do	8.5	
July 16	....do	7.4	
July 19	....do	7.1	
July 22	....do	6.3	
July 25	....do	6.45	
July 28	....do	6.2	
July 30	....do	6.0	

ge mea urements of Rio Grande below Presidio, Tex., in 1903—Continued.

Date.	Hydrographer.	Gage height.	Discharge.
		Feet.	*Second-feet.*
7	E. E. Winter	6.5	972
10	do	7.1	1,381
13	do	6.3	773
16	do	6.9	1,196
19	do	8.55	3,937
22	do	7.5	3,132
25	do	7.3	1,385
28	do	8.45	3,925
30	do	8.0	2,624
ber 2	J. P. Hague	8.2	2,992
ber 4	do	8.0	2,245
ber 7	do	7.7	1,957
ber 9	do	7.7	1,981
ber 11	do	7.8	2,013
ber 14	do	8.3	3,189
ber 16	do	8.0	2,374
ber 18	do	7.7	2,210
ber 21	do	7.8	2,280
ber 23	do	8.7	4,043
ber 25	do	8.8	4,237
ber 27	do	8.75	3,983
ber 30	do	9.2	5,138
2	do	9.45	5,953
5	do	9.1	5,093
8	do	7.9	2,058
11	do	7.3	1,175
13	do	7.5	1,393
17	do	6.7	1,039
20	do	6.6	991
23	do	6.0	683
26	do	5.8	623
29	do	5.65	585
31	do	5.5	503
ber 4	do	5.5	502
ber 6	do	5.4	462
ber 9	do	5.4	437
ber 12	do	5.3	397
ber 14	do	5.2	385
ber 17	do	5.2	373
ber 20	do	5.0	335
ber 23	do	5.0	329

Discharge measurements of Rio Grande below Presidio, Tex., in 1903—Continued.

Date.	Hydrographer.	Gage height.	Discharge.
		Feet.	*Second-ft.*
November 26	J. P. Hague	5.0	334
November 29	do	5.0	317
December 2	do	4.9	293
December 5	do	4.8	275
December 8	do	4.8	271
December 12	do	4.8	271
December 16	do	4.8	259
December 19	do	4.8	241
December 23	do	4.75	237
December 26	do	4.6	209
December 28	do	4.6	201
December 31	do	4.6	216

Mean daily gage height, in feet, of Rio Grande below Presidio, Tex., for 1903.

Day.	Jan.	Feb.	Mar.	Apr.	May.	June.	July.	Aug.	Sept.	Oct.	Nov.	Dec.
1	4.90	5.20	6.45	4.75	6.00	7.10	8.65	8.95	9.25	10.45	5.80	4.90
2	4.90	5.00	6.25	4.65	5.05	7.05	8.55	8.85	8.30	9.55	5.80	4.90
3	4.90	5.00	6.15	4.50	4.75	6.90	8.60	5.75	8.05	9.45	5.80	4.90
4	4.80	4.95	5.95	4.45	6.50	6.95	9.05	5.75	8.00	9.25	5.80	4.90
5	4.80	4.90	5.85	4.40	6.75	8.30	9.35	6.45	8.10	9.15	5.40	4.90
6	4.80	4.90	5.75	4.40	7.05	8.50	9.45	6.15	8.10	8.80	5.40	4.90
7	4.80	6.65	5.55	4.40	6.90	7.35	9.70	6.30	7.75	8.40	5.40	4.90
8	4.80	7.80	5.50	4.30	6.90	7.00	9.75	5.60	7.50	7.95	5.40	4.90
9	4.80	7.55	5.45	4.30	6.85	7.60	9.80	6.60	7.80	7.90	5.40	4.90
10	4.70	7.80	5.40	4.30	7.15	7.70	9.55	7.10	7.40	7.75	5.35	4.90
11	4.70	6.95	5.30	4.80	7.10	7.75	9.45	6.70	7.75	7.30	5.30	4.90
12	4.70	6.90	5.25	5.20	7.00	9.45	9.20	6.30	7.55	7.05	5.30	4.90
13	4.70	6.90	5.20	6.25	7.00	8.40	8.70	6.30	7.05	7.40	5.25	4.90
14	4.70	6.90	5.20	5.95	6.95	7.95	8.35	6.65	7.60	7.20	5.20	4.90
15	4.70	6.70	5.20	5.60	6.95	7.90	7.95	6.60	8.10	6.95	5.20	4.90
16	4.70	6.65	5.15	5.35	7.15	7.80	7.55	6.90	7.90	6.95	5.20	4.90
17	4.70	6.35	5.10	5.35	7.15	7.60	7.50	7.30	7.85	6.80	5.20	4.90
18	4.70	6.25	5.05	5.30	7.20	7.85	7.30	8.65	7.70	6.70	5.15	4.90
19	4.70	6.05	5.15	5.55	7.05	7.70	7.05	8.40	7.55	6.60	5.10	4.90
20	4.70	5.85	5.15	5.35	6.95	7.70	7.20	8.30	7.55	6.60	5.05	4.90
21	4.70	5.80	5.10	5.50	6.90	7.70	6.80	7.75	7.85	6.35	5.00	4.90
22	4.70	7.10	5.00	5.40	6.85	7.75	6.40	7.80	11.35	6.25	5.00	4.90
23	4.70	6.90	4.95	5.20	6.85	7.70	6.30	7.20	8.70	6.05	5.00	4.90
24	4.70	6.90	5.15	5.00	6.90	7.75	6.35	8.85	8.45	6.00	5.00	4.90
25	5.00	6.90	5.30	4.90	6.90	7.75	6.45	7.00	8.80	5.90	5.00	4.90
26	5.30	6.75	5.15	6.90	6.90	7.75	6.55	8.45	8.70	5.80	5.00	4.90
27	5.40	6.70	5.10	6.00	6.95	7.80	6.30	7.60	8.75	5.70	5.00	4.90
28	5.30	6.60	5.00	5.30	6.95	7.85	6.05	8.05	8.95	5.70	5.00	4.90
29	5.25		5.00	5.10	7.00	7.90	6.10	8.15	10.35	5.65	5.00	4.90
30	5.25		4.90	5.00	7.00	7.90	6.05	8.40	9.40	5.60	5.00	4.90
31	5.20		4.90		7.00		6.00	10.25		5.50		4.90

ated monthly discharge of Rio Grande below Presidio, Tex., for 1903.

Month.	Discharge in second-feet.			Total in acre-feet.
	Maximum.	Minimum.	Mean.	
-----------------------	580	295	357	21,967
, -----------------------	2,560	400	1,153	64,026
-----------------------	1,050	315	489	30,069
-----------------------	1,270	105	405	24,109
-----------------------	1,720	270	1,305	80,251
-----------------------	5,700	1,340	2,530	150,526
-----------------------	7,200	720	3,184	192,714
-----------------------	8,100	530	2,048	125,911
er -----------------------	8,960	1,340	3,203	190,572
-----------------------	8,360	500	2,028	124,721
er -----------------------	520	320	391	23,236
r -----------------------	295	200	251	15,461
e year -----------------------	8,960	105	1,441	1,043,563

RIO GRANDE ABOVE PRESIDIO, TEX.

station was established April 4, 1900, by the International
) Boundary Commission. It is 7 miles above Presidio and
he mouth of Conchos River, one of the principal tributaries of
Grande, and is about 200 miles below El Paso. Its location
nough above the mouth of Conchos River to be free from the
of backwater from that stream.

scharge measurements of Rio Grande above Presidio, Tex., in 1903.

Date.	Hydrographer.	Gage height.	Discharge.
		Feet.	Second-feet.
0 -----------------	Jas. P. Hague ----------	1.5	74
3 -----------------	do	1.3	52
6 -----------------	do -----------------	1.85	134
9 -----------------	do -------- --- ---	1.7	105
1 -----------------	do -------- --	1.5	92
-----------------	do -- -- ----	1.0	23
-----------------	do -- --------	.9	16
-----------------	do ----------	4.0	711
-----------------	do -----------------	2.85	347
-------- -----	do ----------- ----	3.4	442
-----------------	do -- --------	2.9	363
-----------------	do -----------------	2.15	179

Discharge measurements of Rio Grande above Presidio, Tex., in 1903—Cont'd

Date.	Hydrographer.	Gage height.	Dischar
		Feet.	*Second-*
April 28	Jas. P. Hague	2.85	
April 30	....do	2.1	
May 3	....do	3.65	
May 6	....do	4.9	1,
May 8	....do	4.95	1,
May 10	....do	6.7	3
May 12	....do	5.0	1
May 14	....do	5.0	1
May 17	....do	5.0	1
May 19	....do	5.0	1
May 21	....do	5.0	1,
May 23	....do	5.0	1
May 26	....do	5.15	1
May 28	....do	5.35	1
June 3	....do	4.8	1.
June 5	....do	6.0	1,
June 8	....do	5.6	1,
June 10	....do	5.6	1,
June 13	....do	6.2	1,
June 15	....do	5.6	1,
June 17	....do	5.95	1.
June 19	....do	6.0	1,
June 21	....do	6.0	1,
June 23	....do	6.5	1,
June 25	....do	6.7	1.
June 27	....do	6.7	1,
June 29	....do	7.5	2,
July 1	....do	7.5	2,
July 3	....do	7.8	4,
July 5	....do	7.8	4.
July 7	....do	7.9	5,
July 9	....do	8.0	5,
July 11	....do	7.7	3,
July 14	....do	7.5	2.
July 17	....do	5.2	1,
July 20	....do	5.3	1,
July 23	....do	4.4	
July 26	....do	3.5	
July 29	....do	3.4	
July 31	....do	3.5	

ge measurements of Rio Grande above Presidio, Tex., in 1903—Cont'd:

Date.	Hydrographer.	Gage height.	Discharge.
		Feet.	Second-feet.
....	E. E. Winter	3.3	443
l	......do	3.8	403
5	......do	2.7	163
3	......do	2.6	157
l	......do	2.5	141
l	......do	2.0	48
7	......do	2.6	162
r 1	J. P. Hague	2.3	79
r 3................	......do	2.0	48
r 5................	......do	1.75	16
r 8................	......do	1.7	15
r 10................	......do	1.9	27
r 12................	......do	1.6	8
r 15................	......do	1.8	23
r 17................	......do	1.6	12
r 19................	......do	1.6	10
r 22....	......do	2.0	32
r 28................	......do	2.0	38
....	......do	2.1	49
l	......do	1.7	14

laily gage height, in feet, of Rio Grande above Presidio, Tex., for 1903.

Day.	Mar.	Apr.	May.	June.	July.	Aug.	Sept.	Oct.
....		1.35	1.95	5.45	7.50	3.40	2.25	2.15
....		1.15	2.05	5.00	7.60	3.20	2.30	2.00
....		1.00	2.95	4.85	7.80	3.10	2.00	2.00
....		.90	4.25	4.85	7.75	4.00	1.85	1.65
....		.90	4.80	6.00	7.80	4.25	1.75	.70
....		.85	5.30	5.95	7.90	3.00	1.70	......
....		.75	4.90	5.30	7.90	4.15	1.60	......
....		.50	5.00	5.35	8.00	3.30	1.75	......
....			5.05	6.15	8.00	3.30	1.75	......
....			6.05	5.55	7.95	3.70	1.90	......
....			5.30	6.20	7.75	3.50	2.05	......
....		3.40	5.00	7.15	7.70	3.05	1.60	......
....		4.15	5.05	6.80	7.65	2.95	2.85	......
....		3.60	5.00	5.45	7.40	2.80	2.35	......
....		3.15	5.05	5.70	6.95	2.70	1.90	......
....		2.85	5.20	5.85	6.35	2.65	1.75	......
....		3.10	5.20	5.95	5.40	2.65	1.60	......
....	1.10	3.10	5.00	6.00	5.20	2.50	1.00	......
....	1.90	3.25	5.00	6.05	5.45	2.50	1.55	......

Mean daily gage height, in feet, of Rio Grande, etc.—Continued.

Day.	Mar.	Apr.	May.	June.	July.	Aug.	Se
20	1.50	3.30	5.00	6.00	5.70	2.40	
21	1.40	3.20	5.00	6.05	4.60	2.50	..
22	1.40	2.85	5.00	6.15	4.40	2.20	2
23	1.30	2.35	5.00	6.60	4.40	2.05	2
24	2.30	2.30	5.00	6.70	4.40	2.00	3
25	2.30	2.20	5.05	6.70	4.40	2.00	2
26	1.90	5.00	5.15	6.70	3.45	2.00	2
27	1.85	4.15	5.15	6.70	3.00	2.50	2
28	1.70	2.85	5.25	7.15	3.45	2.15	2
29	1.70	2.35	5.30	7.50	3.45	2.00	2
30	1.60	2.00	5.40	7.50	3.55	1.90	3
31	1.50	.....	5.40	.....	3.50	1.20	..

NOTE.—There was no flow on dates where blanks occur.

Estimated monthly discharge of Rio Grande above Presidio, Tex., for

Month.	Discharge in second-feet.			
	Maximum.	Minimum.	Mean.	a
January	0	0	0	
February	0	0	0	
March	230	0	49	
April	1,060	0	280	
May	1,840	140	1,217	
June	2,620	1,120	1,734	
July	6,600	350	2,709	
August	755	20	262	
September	560	0	79	
October	60	0	5	
November	0	0	0	
December	0	0	0	
The year				

RIO GRANDE NEAR FORT HANCOCK, TEX.

This station was established by the International (Water) Bor
Commission March 27, 1900. It is 1½ miles southeast of For
cock, on the Southern Pacific Railway, in El Paso Valley,
about 55 miles below El Paso. It was washed out by the shif
the river channel and abandoned July 1, 1903. The following
urements were taken by E. E. Winter during the first six mo
1903.

Discharge measurements of Rio Grande near Fort Hancock, Tex., in 1903.

Date.	Hydrographer.	Gage height.	Discharge.
		Feet.	*Second-feet.*
March 14	E. E. Winter	5.3	362
March 23	do	4.6	174
March 26	do	4.2	170
March 30	do	3.2	68
April 2	do	2.2	2
April 9	do	7.75	1,094
April 13	do	5.6	399
April 16	do	6.3	768
April 18	do	6.3	789
April 21	do	5.2	292
April 25	do	5.45	409
April 28	do	5.4	360
April 30	do	6.35	680
May 2	do	7.8	1,363
May 5	do	8.0	1,720
May 8	do	8.0	1,687
May 11	do	8.05	1,788
May 16	do	7.9	1,967
May 19	do	7.9	1,985
May 21	do	8.2	2,372
May 23	do	8.3	2,699
May 26	do	8.65	2,985
May 29	do	8.8	3,510
June 2	do	7.6	1,646
June 5	do	7.6	1,669
June 8	do	7.9	1,856
June 11	do	8.25	2,430
June 13	do	8.6	4,088
June 16	do	9.4	4,479
June 18	do	9.7	5,185
June 20	do	10.1	6,092
June 23	do	12.5	10,285
June 25	do	13.1	11,188
June 27	do	13.2	11,147
June 30	do	12.7	11,849

Mean daily gage height, in feet, of Rio Grande near Fort Hancock, Tex., for 1903.

Day.	Mar.	Apr.	May.	June.	Day.	Mar.	Apr.	May.	June.
1...............	(a)	2.50	7.70	7.60	17...............	4.10	6.30	7.90	9.55
2...............	(a)	(a)	7.85	7.60	18...............	4.00	6.40	7.90	9.70
3...............	(a)	(a)	7.90	7.50	19...............	4.70	5.75	8.05	9.85
4...............	(a)	(a)	7.95	7.50	20...............	5.10	5.55	8.10	10.10
5...............	(a)	2.75	8.00	7.60	21...............	5.00	5.30	8.20	10.25
6...............	(a)	3.25	7.90	7.75	22...............	4.75	5.20	8.30	10.85
7...............	(a)	3.85	8.05	7.95	23...............	4.55	5.30	8.30	12.40
8...............	(a)	7.00	8.00	8.00	24...............	4.40	5.45	8.40	12.85
9...............	(a)	7.70	8.10	8.05	25...............	4.45	5.50	8.55	13.15
10...............	(a)	6.75	8.05	8.15	26...............	4.10	5.15	8.70	13.30
11...............	(a)	6.55	8.10	8.20	27...............	3.75	4.75	8.80	13.90
12...............	(a)	6.15	8.00	8.25	28...............	3.55	5.40	8.90	13.05
13...............	6.50	5.70	7.90	8.45	29...............	3.35	5.25	8.75	12.85
14...............	5.40	5.85	7.90	9.00	30...............	3.10	6.50	8.50	12.65
15...............	4.50	6.10	7.90	9.25	31...............	2.85	....	7.70	....
16...............	4.30	6.40	7.90	9.45					

a No flow.

Estimated monthly discharge of Rio Grande near Fort Hancock, Tex., in 1903.

Month.	Discharge in second-feet.			Total in acre-feet.
	Maximum.	Minimum.	Mean.	
January	0	0	0	0
February	0	0	0	0
March	840	0	133	8,132
April	1,080	0	436	25,964
May	3,760	1,180	2,182	134,142
June....	11,770	1,500	5,459	324,833

RIO GRANDE NEAR EL PASO, TEX.

This station was located at the pumping house of the smelter company, 3 miles north of El Paso, Tex. The bed of the stream here is composed of mud, constantly shifting and changing. May 1, 1897, the station was placed under the charge of W. W. Follett, consulting engineer, International (Water) Boundary Commission, and by him removed 1 mile farther up the river, to Courchesne's limekiln. The river heights were measured at the masonry pump foundation pier, 150 feet above the kiln. The top of the downstream chisel draft on this pier was assumed to be at a gage height of 15 feet, and the distance of the surface of the water below it was measured with a carefully graduated rod. This pier was torn down in October, 1902, and an inclined wooden gage established some 60 feet upstream. This is a 2

ch timber bolted to 1¼-inch steel bars set with cement in holes
in solid rock. The graduations are notches cut in the scant-
The bench mark is a ½-inch iron bolt set in solid rock at the
' the gage. Its elevation is 13 feet above the zero of the gage.
ft bank of the river is formed by the loose rock fill of the
on, Topeka and Santa Fe Railroad embankment, and will not
w. The right bank, however, is not so good, being made
and subject to overflow. The bottom of the river here has
oved unstable, scouring on a rise and filling on a falling river.
robably the best site for a station in the vicinity of El Paso,
r, as the entire river bed is constantly shifting for many miles
and below. On account of this shifting character of the
, the only accurate method of estimating the daily discharges
iking a large number of measurements.

ischarge measurements of Rio Grande near El Paso, Tex., in 1903.

Date.	Hydrographer.	Gage height.	Discharge.
		Feet.	*Second-feet.*
6	I. H. Huggett	5.0	26
8	do	4.6	3
16	do	5.2	33
19	do	5.0	24
27	do	5.25	39
y 9	do	5.0	25
y 23	E. E. Winter	5.7	106
y 25	do	5.4	69
y 27	do	5.2	26
	do	5.3	46
	do	5.9	139
	do	5.8	110
	do	5.4	57
1	do	8.7	1,363
3	do	7.6	651
7	do	6.9	423
9	do	7.8	747
1	do	7.3	554
4	do	7.0	375
7	do	6.2	203
9	do	5.8	156
1	do	5.5	110
	do	6.4	267
	do	7.8	786

Discharge measurements of Rio Grande near El Paso, Tex., in 1903—Contin

	Hydrographer.	Gage height.	Disch
		Feet.	Second
....	E. E. Winter............	9.8	
	.....do	8.45	
April 12............	.....do ....,.........	7.8	
April 15............	.....do	8.4	
April 17...	.....do	8.3	
April 20	.....do	7.45	
April 22............	.....do	7.9	
April 24............	.....do	7.95	
April 27............	.....do	7.8	
April 29...., ...	.....do	8.95	
May 1..............	.....do	9.85	
May 4..............	.....do	10.0	
May 7..............	.....do	10.2	
May 9..............	.....do	10.2	
May 12.............	.....do	10.45	
May 13.............	.....do	10.75	
May 15.............	.....do	10.9	
May 18.............	.....do	12.0	
May 20.............	.....do	12.4	
May 22.............	.....do	11.9	
May 25.............	.....do	11.7	
May 27.............	.....do	10.5	
May 30.............	.....do	9.1	
June 1	.....do	9.1	
June 4	.....do	9.0	
June 6	.....do	9.4	
June 9	.....do	10.4	
June 12	.....do	11.75	
June 15	.....do	13.3	
June 17	.....do	13.5	
June 19	...do	14.05	
June 22	...do	13.7	
June 24	.....do	13.65	
June 26	.....do	13.2	
June 29	.....do	12.75	
July 2.............	.....do	11.4	
July 5.............	.....do	10.8	
July 7.............	.....do	9.2	
July 10............	.....do	8.1	
July 12............	.....do	7.4	
July 14............	.....do	6.9	

ᵉ measurements of Rio Grande near El Paso, Tex., in 1903.—Continued.

Date.	Hydrographer.	Gage height.	Discharge.
		Feet.	Second-feet.
....	E. E. Winter............	6.4	942
....	.....do	6.8	1,097
....	.....do	6.0	561
....	.....do	5.85	497
....	.....do	6.45	1,107
....	.....do	5.8	472
....	.....do	5.4	401
....	.....do	5.1	239
3....................	.....do	4.6	156
6....................	T. M. Courchesne	4.7	216
8....................	.....do	4.4	120
11...................	.....do	4.4	128
14...................	.....do	3.8	30
ᵉʳ 29................	.....do	4.4	118
2	E. E. Winter	4.7	217
3	.....do	4.6	202
10	.....do	3.6	18
14	.....do	3.4	16
16	.....do	3.4	13
19	.....do	3.4	12
ᵉʳ 12	T. M. Courchesne	3.9	47
ᵉʳ 16	.....do	4.0	48
ᵉʳ 19	.....do	4.05	61
ᵉʳ 21	.....do	4.1	77
ᵃʳ 29	.....do	4.1	67
ᵉʳ 31	.....do	4.1	73

ᵗ daily gage height, in feet, of Rio Grande near El Paso, Tex., for 1903.

ᵃʸ.	Jan.	Feb.	Mar	Apr.	May.	June.	July.	Aug.	Sept.	Oct.	Nov.	Dec.
....	4.70	(ᵃ)	5.30	6.10	9.90	9.15	11.95	4.85	3.50	4.50	3.40	3.40
....	4.70	(ᵃ)	5.35	6.30	10.00	9.10	11.50	4.65	3.50	4.70	3.40	3.40
....	4.70	(ᵃ)	5.90	6.30	10.00	9.05	11.10	4.60	3.50	4.60	3.40	3.40
....	4.70	4.75	6.10	6.65	10.10	9.00	10.55	4.50	3.50	4.25	3.40	3.40
....	4.70	4.90	5.90	6.60	10.00	9.10	10.20	5.0	3.50	3.95	3.40	3.40
....	5.00	4.80	5.95	8.10	10.05	9.45	9.70	4.65	3.50	3.80	3.40	3.40
....	4.85	4.85	5.85	9.60	10.20	9.70	9.30	4.50	3.50	3.80	3.40	3.40
....	4.70	4.90	5.65	9.60	10.20	10.00	8.90	4.40	3.50	3.70	3.40	3.40
....	4.70	4.95	5.60	9.10	10.25	10.60	8.40	4.25	3.50	3.70	3.40	3.50
....	(ᵃ)	5.00	5.80	8.60	10.30	10.85	8.10	4.20	3.50	3.60	3.40	3.70
....	(ᵃ)	4.70	8.70	8.10	10.30	11.30	7.70	4.50	3.50	3.50	3.40	3.80
....	(ᵃ)	(ᵃ)	8.25	8.00	10.50	11.80	7.35	4.30	3.50	3.50	3.40	3.90

ᵃ No flow.

Mean daily gage height, in feet, of Rio Grande, etc.—Continued.

Day.	Jan.	Feb.	Mar.	Apr.	May.	June.	July.	Aug.	Sept.	Oct.	Nov.	D
13	(a)	(a)	7.60	8.10	10.75	12.90	7.14	4.05	3.60	3.40	3.40	
14	(a)	(a)	7.25	8.35	10.65	12.85	6.90	3.75	3.80	3.40	3.40	
15	4.80	(a)	6.80	8.60	11.00	12.90	7.10	3.60	3.50	3.40	3.40	
16	5.10	(a)	6.90	8.35	11.00	13.20	6.40	3.60	3.50	3.40	3.40	
17	5.20	(a)	7.15	8.85	11.75	13.65	6.45	3.60	3.50	3.40	3.40	
18	5.15	(a)	7.90	8.10	12.00	14.00	6.80	3.60	3.50	3.40	3.40	
19	5.00	5.00	7.90	7.65	12.15	14.00	6.40	3.60	3.50	3.40	3.40	
20	5.10	5.00	7.55	7.55	12.90	13.65	6.10	3.60	3.50	3.40	3.40	
21	4.80	5.10	7.85	7.75	12.05	14.05	5.95	3.60	3.50	3.40	3.40	
22	(a)	5.30	7.15	8.00	11.95	15.10	5.80	3.60	3.50	3.40	3.40	
23	(a)	5.70	7.15	8.10	12.10	13.70	5.95	3.60	3.50	3.40	3.40	
24	(a)	5.65	7.05	7.95	12.20	13.65	6.35	3.60	3.50	3.40	3.40	
25	(a)	5.65	6.50	7.30	11.80	13.40	6.30	3.60	3.60	3.40	3.40	
26	(a)	5.25	6.80	7.15	11.20	13.20	6.05	3.60	3.90	3.40	3.40	
27	5.00	5.20	6.20	7.80	10.45	13.30	5.85	3.60	4.00	3.40	3.40	
28	4.85	5.25	6.05	8.05	9.75	13.15	5.55	3.60	3.80	3.40	3.40	
29	4.70	......	5.85	9.05	9.80	12.70	5.35	3.60	4.50	3.40	3.40	
30	4.70	......	5.65	9.75	9.10	13.35	5.25	3.60	4.50	3.40	3.40	
31	4.90	......	5.60	......	9.10	......	5.05	3.60	......	3.40	......	

a No flow.

Estimated monthly discharge of Rio Grande near El Paso, Tex., for 1903

Month.	Discharge in second-feet.			Tota acre-
	Maximum.	Minimum.	Mean.	
January	35	0	10	
February	105	0	23	
March	1,360	40	368	2
April	2,050	200	831	4
May	5,340	2,100	3,312	20
June	18,070	1,700	9,863	58
July	10,930	215	2,573	15
August	280	10	70	
September	145	5	17	
October	220	5	33	
November	5	5	5	
December	75	5	40	
The year	18,070	0	1,429	1,08

RIO GRANDE NEAR SAN MARCIAL, N. MEX.

This station is located about one-half mile south of San Mai
N. Mex., at the bridge of the Atchison, Topeka and Santa Fe Raili
The original gage was established by Arthur P. Davis on Januar
1895. The observer was Bert Halseth, San Marcial, N. Mex., w

is about one-half mile distant. The gage was of hard pine
, 9 by 5 inches by 25 feet, anchored and bolted to the east end
iecond pier from the south. It was inclined and painted white.
stance between the footmarks was 1.6 feet. The 13-foot mark
vel with the extension of the pier, to which the gage was
ed. The 15-foot mark was level with the top of the capstone
ch the bridge truss rests. Measurements were made from the
ridge. On August 8, 1889, a station was established near San
l, and a measurement was made which gave a discharge of 19
-feet. Soon after this date, however, the river gage was
red and the locality was abandoned until 1895.
896 the inclined gage was carried away and a wire gage was put
place. The wire gage was attached to the guard rail of the
in the south span on the lower side. Bench mark No. 1 is the
the capstone on which the bridge truss rests, and is at an ele-
of 15 feet above gage datum; bench mark No. 2 is the top of
ension of the pier to which the old vertical gage was fastened,
at an elevation of 13 feet above gage datum. The wire gage
en abandoned and the gage heights are now measured with a
ted rod from the deck of the bridge, but using the old gage
. The channel is sandy and shifting. A number of bridge
nterfere with the current to a certain extent, but not with
served gage heights. They sometimes affect the discharge
ements.
: January 1, 1901, this station has been maintained under the
of the International (Water) Boundary Commission.

Discharge measurements of Rio Grande near San Marcial, N. Mex.

Date.	Hydrographer.	Gage height.	Discharge.
		Feet.	*Second-feet.*
3	J. R. Nisbet	5.9	79
6	do	5.6	57
10	do	6.9	258
13	do	6.85	349
16	do	6.6	310
20	do	6.8	351
23	do	6.7	305
26	do	6.6	254
30	do	7.0	416
y 3	do	6.9	343
y 6	do	7.0	428
y 9	do	6.9	368
y 12	do	7.1	384
y 15	do	7.2	306

Discharge measurements of Rio Grande near San Marcial, N. Mex.—Contɪ

Date.	Hydrographer.	Gage height.	Disc
		Feet.	*Secor*
February 18	J. R. Nisbet	7.8	
February 22	do	7.8	
February 26	do	7.4	
March 3	do	7.15	
March 6	do	7.2	
March 8	do	8.95	
March 10	do	7.8	
March 14	do	7.5	
March 17	do	7.7	
March 19	do	7.95	
March 24	do	7.3	
March 28	do	7.7	
April 3	do	7.8	
April 6	do	8.75	
April 11	do	8.5	
April 15	do	8.1	
April 18	do	8.3	
April 23	do	8.0	
April 27	do	8.8	
April 30	do	9.35	
May 4	do	9.7	
May 6	do	9.8	
May 10	do	10.25	
May 13	do	10.5	
May 15	do	10.4	
May 18	do	10.8	
May 20	do	10.7	
May 23	do	9.8	
May 26	do	8.7	
May 29	do	8.9	
May 31	do	8.8	
June 3	do	9.0	
June 6	do	9.9	
June 10	do	11.0	
June 12	do	12.4	
June 16	do	12.35	
June 20	do	12.5	
June 24	do	12.5	
June 28	do	10.5	
June 30	do	10.4	
July 2	do	9.8	

rge measurements of Rio Grande near San Marcial, N. Mex.—Continued.

Date.	Hydrographer.	Gage height.	Discharge.
		Feet.	*Second-feet.*
....	J. R. Nisbet..............	9.3	2,591
....	...do	8.7	1,919
....	....do	8.3	1,472
....	....do	8.1	1,017
....	....do	7.9	873
....	....do	8.3	555
....	....do	8.0	383
....	....do	7.8	825
....	....do	7.7	245
t 3	....do	7.5	75
t 7	....do	7.1	22
t 11	....do	6.5	16
t 15	....do	6.9	18
t 18	....do	6.6	9
t 22	....do	7.0	81
t 26	....do	7.95	151
t 29	....do	6.9	15
t 31	....do	6.6	7
iber 10	....do	6.7	26
iber 14	....do	6.5	9
iber 17	....do	6.6	10
iber 24	....do	6.8	33
iber 26	....do	8.15	284
iber 30	....do	6.6	19
r 4	....do	6.7	19
r 8	....do	6.6	13
iber 11	....do	6.7	28
iber 14	....do	6.8	100
iber 17	....do	7.0	73
iber 20	....do	7.2	72
iber 23	....do	7.3	149
iber 27	....do	7.5	229
iber 30	.... do	7.7	275
ber 4	....do	7.9	333
ber 9	.... do	7.9	308
ber 15	do	7.9	301
ber 19	do	7.9	302
ber 23	do	8.0	317
ber 26	do	7.9	310
ber 31	do	8.1	319

Mean daily gage height, in feet, of Rio Grande near San Marcial, N. Mex., for 1903.

Day.	Jan.	Feb.	Mar.	Apr.	May.	June.	July.	Aug.	Sept.	Oct.	Nov.	Dec.
1	6.25	7.00	7.30	7.90	9.35	8.85	10.00	7.70	(a)	7.50	(a)	7.90
2	6.20	6.80	7.35	7.80	9.35	8.90	9.90	7.55	(a)	7.45	(a)	7.90
3	5.85	6.90	7.20	8.00	9.35	9.00	9.70	7.55	(a)	6.90	(a)	7.90
4	5.90	6.80	7.20	9.45	9.55	9.40	9.55	7.50	(a)	6.70	(a)	7.90
5	5.85	6.85	7.15	9.90	9.70	9.65	9.35	7.40	(a)	6.60	(a)	7.90
6	5.85	6.95	7.20	8.65	9.80	9.85	9.25	7.20	(a)	6.60	6.60	7.80
7	5.85	7.10	7.05	8.35	9.70	9.95	8.90	7.20	(a)	6.60	6.60	7.90
8	6.90	6.80	8.80	8.00	9.70	10.30	8.75	7.15	(a)	6.60	6.60	7.90
9	6.75	6.85	8.50	8.00	9.95	10.95	8.50	6.85	6.45	6.50	6.60	7.90
10	6.80	6.95	7.80	7.95	10.15	10.80	8.30	6.65	6.70	6.50	6.60	7.85
11	6.85	6.95	7.40	8.25	10.30	10.75	8.20	6.50	6.85	(a)	6.70	8.00
12	6.60	7.10	7.45	8.50	10.30	12.25	8.30	6.80	6.65	(a)	6.70	7.95
13	6.70	7.00	7.30	8.35	10.45	11.95	8.25	6.70	6.60	(a)	6.80	7.90
14	6.60	7.15	7.50	8.50	10.55	11.70	8.25	6.70	6.55	(a)	6.80	8.00
15	6.55	7.20	7.90	8.10	10.45	11.70	8.25	7.05	6.50	(a)	6.90	7.95
16	6.65	7.05	7.90	7.95	10.45	12.30	8.10	6.80	6.40	(a)	6.90	7.80
17	6.60	6.95	7.75	8.10	10.45	12.45	8.00	6.75	6.60	(a)	7.00	7.90
18	6.00	7.25	7.75	8.30	10.80	12.60	7.90	6.60	6.50	(a)	7.10	7.60
19	6.70	7.20	7.90	8.15	10.70	11.95	7.80	6.70	(a)	(a)	7.	7.85
20	6.80	7.25	7.65	8.45	10.70	12.45	8.00	7.55	(a)	(a)	7.	7.90
21	6.70	7.30	7.70	8.50	10.30	12.40	8.40	7.15	(a)	(a)	7.	7.90
22	6.90	7.30	7.60	8.20	10.00	12.30	8.40	7.00	(a)	(a)	7.	7.90
23	6.70	7.30	7.50	8.05	9.85	12.20	8.10	7.00	(a)	(a)	7.	8.00
24	6.70	7.30	7.35	8.20	9.15	12.40	8.05	7.35	7.00	(a)	7.	8.00
25	6.75	7.20	7.30	8.30	8.65	11.60	8.00	7.10	6.55	(a)	7.	7.90
26	6.80	7.35	7.40	8.70	8.70	11.05	7.90	7.60	8.45	(a)	7.80	7.90
27	6.80	7.40	7.45	8.85	8.70	10.70	7.90	7.00	7.10	(a)	7.50	7.90
28	7.05	7.40	7.75	9.35	8.80	10.55	7.85	7.15	6.75	(a)	7.60	8.05
29	7.05	......	7.75	9.30	8.85	10.35	7.65	6.95	6.60	(a)	7.60	7.90
30	7.15	......	7.90	9.35	8.80	10.30	7.70	6.85	6.60	(a)	7.70	7.90
31	7.15	......	8.00	......	8.80	......	7.75	6.80	......	(a)	......	8.00

a No flow

Estimated monthly discharge of Rio Grande near San Marcial, N. Mex., for 1903.

[Drainage area, 28,057 square miles.]

Month.	Discharge in second-feet.			Total in acre-feet.	Run-off.	
	Maximum.	Minimum.	Mean.		Second-feet per square mile.	Depth in inches.
January	475	75	280	17,197	0.010	0.012
February	580	280	395	21,927	.014	.015
March	2,410	320	761	46,790	.027	.031
April	5,500	680	1,681	100,007	.060	.067
May	8,950	2,720	5,178	318,367	.184	.212
June	18,880	2,860	11,100	660,476	.395	.441
July	3,600	240	1,266	77,841	.045	.052
August	245	5	50	3,064	.002	.002
September	360	0	24	1,438	.001	.001
October	75	0	9	545	.0003	.0003
November	275	0	93	5,534	.003	.003
December	335	260	307	18,883	.011	.013
The year	18,880	0	1,762	1,272,069	.0627	.8493

RIO GRANDE NEAR RIO GRANDE, OR BUCKMAN, N. MEX.

Three miles below Embudo the river emerges into Espanola Valley, trough which it continues for a few miles and then enters White ock Canyon, flowing through that canyon for 30 miles. At the wer end of this canyon the river emerges into Albuquerque Valley, d so continues down to about Socorro. This valley averages from to 3 miles in width, and has been irrigated for a great many years y the Mexican settlers. Their primitive methods of irrigation are ery wasteful of the waters, so that the duty of water in this section, bout 17 acres per second-foot, is not as high as it might be. During le last few years, however, a number of important and modern irriation systems have been planned and built in the vicinity of Albuuerque. The gaging station, established February 3, 1895, is located bout one-fourth of a mile above Water Tank, a station on the Denver nd Rio Grande Railroad below Espanola. It was established by rthur P. Davis and P. E. Harroun on February 1, 1895.

The inclined portion of the gage reads from 1 to 10 feet. The vertical portion reads from 10 to 16 feet. The vertical section is fastened a large bowlder. The gage is read twice each day by A. L. Martiez, section foreman. Discharge measurements are made by means f a cable, car, and tagged wire, which are located about 15 feet bove the gage and about 200 feet above the Denver and Rio Grande lailroad bridge. The initial point for soundings is at the end of the able on the left bank, where it is fastened to two small trees. The hannel is straight for 150 feet above and for 300 feet below the station. The current is swift and there is but one channel at high and w water. The right bank is low and wooded. The left bank is igh, rocky, and wooded. Neither bank is subject to overflow. The able has a span of 220 feet. The bed of the stream is sandy and free rom vegetation, with a few bowlders near the left bank. The bench nark is the top of the bowlder to which the vertical gage is fastened. ts elevation is 17.82 feet above the zero of the gage.

Discharge measurements of Rio Grande near Rio Grande, N. Mex., in 1903.

Date.	Hydrographer.	Gage height.	Discharge.
		Feet.	*Second-feet.*
January 2	O. B. Powell	4.7	337
January 5	do	4.8	362
January 8	do	4.8	335
January 12	do	4.8	346
January 15	do	4.8	368
January 19	do	4.8	371
January 22	do	4.8	361
January 27	do	5.0	430

Discharge measurements of Rio Grande, etc.—Continued.

Date.	Hydrographer.	Gage height.	Discharg
		Feet.	*Second-fe*
January 31	O. B. Powell	4.9	3(
February 3	do	5.0	4:
February 6	do	4.8	3:
February 10	do	5.0	4:
February 14	do	5.1	4:
February 17	do	4.9	3:
February 24	do	5.4	6:
February 27	do	5.2	49
March 3	do	5.0	3:
March 6	do	7.9	2,6:
March 10	do	5.9	9:
March 12	do	7.4	1,9:
March 14	do	7.3	1,7:
March 17	do	6.3	1,09
March 20	do	5.9	84
March 23	do	6.0	8:
March 25	do	6.3	9:
March 27	do	6.4	1,2:
March 30	do	6.4	1,2:
April 2	do	11.4	9,2:
April 4	do	7.0	1.7:
April 6	do	6.5	1.3(
April 8	do	6.7	1.5:
April 10	do	7.4	2.2(
April 13	do	7.1	1.9:
April 15	do	6.9	1.7(
April 18	do	7.5	2.1:
April 21	do	7.0	1.7:
April 23	do	7.9	2.7:
April 25	do	8.9	3.7:
April 27	do	9.1	4.4:
April 29	do	9.1	4.6:
May 1	do	8.9	4.3(
May 4	do	9.6	5.6:
May 8	do	10.3	6.7:
May 13	do	11.0	8.5:
May 16	do	11.2	8.6:
May 19	do	10.8	6.8:
May 21	do	9.6	5.6(
May 23	do	8.8	4,5:

Discharg · measurements of Rio Grande, etc.—Continued.

Date.	Hydrographer.	Gage height.	Discharge.
		Feet.	*Second-feet.*
May 26	O. B. Powell	8.6	4,391
May 29	....do	8.7	4,545
May 31	....do	8.7	4,674
June 2	....do	9.6	6,469
June 4	....do	10.2	7,107
June 6	....do	10.5	7,789
June 9	....do	12.2	11,723
June 12	....do	12.0	9,644
June 16	....do	12.6	15,371
June 18	....do	12.6	15,492
June 20	....do	12.5	15,126
June 23	....do	11.4	13,863
June 25	....do	10.8	12,335
June 27	....do	10.5	11,485
July 2	....do	8.8	5,502
July 6	....do	7.1	3,281
July 9	....do	6.3	2,300
July 11	....do	6.8	2,637
July 14	....do	5.9	1,821
July 18	....do	6.2	1,994
July 22	....do	5.8	1,725
July 25	....do	5.1	781
July 28	....do	5.0	729
July 31	....do	4.7	647
August 22	F. Cogswell	4.1	376
September 16	....do	4.05	367
October 14	....do	4.0	364
November 12	....do	4.12	423

Mean daily gage height, in feet, of Rio Grande near Rio Grande, N. Mex., for 1903.

Day.	Jan.	Feb.	Mar.	Apr.	May.	June.	July.	Aug.	Sept.	Oct.	Nov.	Dec.
1	4.60	5.05	5.00	7.55	8.90	9.30	9.30	4.45	3.95	4.00	4.05	4.45
2	4.70	4.95	5.05	11.45	9.15	9.60	8.85	4.35	3.80	4.05	4.05	4.40
3	4.65	4.95	5.10	9.15	9.50	9.95	8.40	4.25	3.80	4.00	4.00	4.30
4	4.65	4.75	5.45	7.00	9.60	10.30	7.85	4.15	3.85	3.95	4.05	4.25
5	4.70	4.70	7.60	6.70	9.85	10.55	7.45	4.95	3.80	3.95	4.05	4.40
6	4.75	4.75	8.70	6.60	10.20	10.55	7.10	4.00	4.40	4.00	4.05	3.90
7	4.80	4.65	7.25	6.45	10.35	10.85	6.75	4.15	4.45	4.00	4.05	3.80
8	4.80	4.80	6.15	6.50	10.40	11.30	6.40	4.25	4.80	4.00	4.05	3.85
9	4.85	5.05	6.15	6.80	10.75	12.25	6.45	4.05	4.50	(0)	4.05	4.40

Mean daily gage height, in feet, of Rio Grande, etc.—Continued.

Day.	Jan.	Feb.	Mar.	Apr.	May.	June.	July.	Aug.	Sept.	Oct.	Nov.	Dec.
10	4.90	5.15	6.20	7.30	10.95	12.60	6.05	4.05	4.15	3.95	4.05	4.20
11	4.90	5.05	6.15	7.30	11.15	12.10	6.80	4.00	4.00	4.00	4.10	4.25
12	4.90	5.00	7.00	7.35	11.20	12.00	6.15	4.20	4.15	3.95	4.15	4.25
13	4.85	5.05	7.20	7.00	11.20	12.85	5.90	4.00	4.00	4.00	4.15	4.15
14	4.85	4.95	7.10	6.65	11.35	13.40	5.80	4.70	4.05	4.00	4.25	4.25
15	4.80	5.15	6.80	6.85	11.90	13.40	5.60	4.30	4.00	4.00	4.20	4.15
16	4.90	4.75	6.55	7.10	11.50	12.75	5.55	4.00	4.05	4.00	4.25	4.00
17	4.90	4.95	6.30	7.00	11.70	12.60	6.10	5.30	4.05	4.00	4.25	4.00
18	4.90	4.95	6.20	7.40	11.40	12.55	6.15	4.40	3.95	4.00	4.20	3.95
19	4.90	5.00	6.20	7.30	10.75	12.70	5.70	4.15	3.95	4.00	4.25	4.10
20	4.85	5.15	5.95	6.85	10.05	12.50	6.20	4.00	3.95	4.00	4.20	4.15
21	4.75	5.25	5.80	7.00	9.70	12.15	6.10	4.20	4.00	4.00	4.25	4.15
22	4.85	5.15	5.85	7.35	9.25	11.70	5.70	4.15	3.95	4.00	4.30	4.00
23	4.80	5.25	5.85	7.70	8.75	11.40	5.55	4.05	3.95	4.00	4.35	4.15
24	4.85	5.45	6.05	8.35	8.50	11.10	5.30	4.05	4.05	4.00	4.35	4.00
25	5.00	5.65	6.25	8.80	8.60	10.85	5.05	4.30	4.00	4.00	4.45	4.10
26	5.10	5.30	6.30	9.20	8.85	10.60	4.95	4.15	3.95	4.00	4.45	4.00
27	5.00	5.25	6.40	9.20	8.90	10.50	5.05	4.00	3.90	4.05	4.50	3.95
28	5.00	5.15	6.40	10.00	8.70	10.05	5.05	3.95	4.00	4.00	4.50	3.90
29	5.00	......	6.85	10.00	8.70	9.75	4.85	3.95	4.20	4.00	4.45	4.00
30	4.95	......	6.60	8.80	8.80	9.55	4.75	4.10	4.05	4.05	4.45	3.90
31	5.00	......	6.70	......	8.90	......	4.65	4.05	......	4.00	......	3.90

Estimated monthly discharge of Rio Grande near Rio Grande, N. Mex., for 1903.

Month.	Discharge in second-feet.			Total in acre-feet.
	Maximum.	Minimum.	Mean.	
January	470	290	376	23,127
February	810	290	445	24,724
March	3,750	380	1,223	75,193
April	9,450	1,260	2,896	172,324
May	10,900	4,290	6,613	406,612
June	19,300	5,870	11,923	709,468
July	7,260	630	2,225	136,780
August	995	340	432	26,563
September	645	300	375	22,314
October	370	340	355	21,82
November	520	355	423	25,170
December	500	300	384	23,61
The year	19,300	290	2,306	1,667,71

RIO GRANDE AT EMBUDO, N. MEX.

Embudo is a railroad station on the Denver and Rio Grande Railroad, in Rio Arriba County. The station is in a narrow canyon above the head of the valley in which Espanola and other towns are located, and in which the river Chama joins the Rio Grande. In November, 1888, an examination was made along the main stream for the purpose of selecting a point at which the total discharge of the Rio Grande entering New Mexico could be ascertained. Coming from Colorado southward, this is the first point at which the railroad reaches the river, and for this reason it was finally determined to establish a river station here. Measurements were begun at the rocky narrows about a mile above the railroad station, but in the spring of 1889 the observations were transferred to a point directly behind the railroad station, for convenience of the observer and consequent reduction of expense. The inclined gage, constructed at that time, is made of scantling, 4 by 4 inches, spiked to posts set firmly in the ground. It is on the right-hand side of the river, at a place where the slopes are very gentle, and therefore is of considerable length, in order to reach out to the low-water channel. It is graduated from about 7.30 feet, the low-water mark, up to 16 feet. The point 3.50 would correspond approximately to the deepest part of the section. The gage is about 75 feet above the cable and is in three parts, with different slopes. It is read twice each day by E. A. Kuhn, the station agent at Embudo, N. Mex. The cable is five-eighths-inch wire. It is fastened to a cedar tree on the left-hand side and to sand anchors on the right. Measurements were originally made from a boat held by traveling pulleys running on the cable. The boat being washed away by flood, later measurements were made from a box suspended from the cable.

The initial point for soundings is the deadman to which the cable is fastened on the right bank. The channel is straight for 300 feet above and below the station and the current is sluggish. Both banks are high and rocky, not liable to overflow and without trees. The width of the channel at the cable is 315 feet. The bed is composed of shifting sand, free from vegetation. There is but one channel at all stages. The current is too sluggish at low stages for accurate measurement at the cable, but a good velocity can be obtained by wading at a section 1,000 feet farther downstream.

Bench mark No. 1 is on a rock near the end of the cable, on the left-hand bank, marked "B. M." with white paint. It is 20.66 feet above zero of the gage. Bench mark No. 2 is on a rock about 100 feet above the cable, on the left bank of the river, and is marked "B. M." with white paint. It is 18.79 feet above datum. Bench mark No. 3 is a notch cut in the southeast corner of the station house, about 2 feet above the level of the platform, and is 30.48 feet above datum.

Discharge measurements of Rio Grande at Embudo, N. Mex., in 1903.

Date.	Hydrographer.	Gage height.	Discharge.
		Feet.	Second-feet.
January 3	O. B. Powell	7.5	333
January 6	do	7.4	310
January 9	do	7.4	307
January 13	do	7.6	406
January 16	do	7.4	330
January 20	do	7.4	353
January 23	do	7.4	306
January 26	do	7.4	345
January 29	do	7.4	376
February 2	do	7.3	291
February 5	do	7.2	270
February 9	do	7.4	335
February 13	do	7.6	430
February 16	do	7.4	333
February 25	do	7.6	433
February 28	do	7.4	331
March 2	do	7.6	437
March 5	do	8.3	787
March 9	do	8.2	704
March 11	do	8.0	592
March 13	do	8.2	707
March 16	do	8.4	696
March 18	do	8.4	739
March 21	do	8.3	743
March 24	do	8.0	559
March 26	do	8.6	896
March 28	do	8.4	779
March 31	do	8.6	976
April 5	do	8.5	911
April 7	do	8.0	589
April 9	do	8.4	739
April 11	do	8.4	745
April 14	do	8.3	679
April 16	do	8.3	693
April 20	do	8.4	729
April 22	do	8.5	853
April 24	do	8.7	919
April 28	do	9.8	2,061
April 30	do	9.5	1,789
May 2	do	9.5	1,815

Discharge measurements of Rio Grande at Embudo, N. Mex., in 1903—Cont'd.

Date.	Hydrographer.	Gage height.	Discharge.
		Feet.	*Second-feet.*
y 6	O. B. Powell	10.1	2,388
y 9	do	10.4	2,694
y 12	do	10.9	3,200
y 14	do	11.2	3,557
y 18	do	11.8	4,081
y 20	do	11.4	3,764
y 22	do	10.1	2,161
y 25	do	9.4	1,677
y 27	do	9.6	1,752
y 29	do	9.7	1,828
ne 1	do	10.2	2,580
ne 3	do	11.2	4,785
ne 5	do	11.7	5,048
ne 8	do	12.1	5,904
ne 10	do	12.8	6,945
ne 13	do	13.6	9,065
ne 15	do	15.0	12,653
ne 17	do	15.5	14,118
ne 19	do	15.8	15,858
ne 22	do	15.2	12,832
ne 24	do	14.5	10,290
ne 26	do	13.9	9,262
ne 29	do	13.0	7,464
ily 3	do	11.5	3,663
ily 7	do	10.1	1,860
uly 10	do	9.2	1,204
uly 13	do	9.1	1,146
uly 16	do	8.7	921
uly 20	do	9.1	1,089
uly 24	do	8.5	870
uly 27	do	8.3	754
uly 30	do	8.3	707
\ugust 21	F. Cogswell	7.6	274
eptember 17	do	7.7	378
)ctober 15	do	7.6	296
'ovember 11	do	7.78	435

Mean daily gage height, in feet, of Rio Grande at Embudo, N. Mex., for 190

Day.	Jan.	Feb.	Mar.	Apr.	May.	June.	July.	Aug.	Sept.	Oct.	Nov.	D
1	7.30	7.35	7.60	8.80	9.45	10.15	12.15	8.25	7.60	7.70	7.60	
2	7.35	7.30	7.50	9.15	9.50	10.55	11.90	8.20	7.60	7.65	7.60	
3	7.40	7.30	7.65	8.55	9.65	11.10	11.45	8.05	7.60	7.65	7.60	
4	7.45	7.30	7.95	8.35	9.75	11.60	11.00	7.90	7.60	7.60	7.60	
5	7.35	7.30	8.65	8.50	9.85	11.70	10.75	7.70	7.60	7.60	7.60	
6	7.35	7.45	9.80	8.20	10.00	11.80	10.45	7.65	7.60	7.60	7.60	
7	7.25	7.40	9.50	8.05	10.10	11.85	10.10	7.55	7.70	7.60	7.60	
8	7.30	7.45	8.20	8.10	10.25	12.10	9.70	7.50	7.60	7.60	7.60	
9	7.40	7.45	8.25	8.35	10.40	12.60	9.40	7.45	7.60	7.60	7.65	
10	7.40	7.45	7.95	8.40	10.60	12.85	9.25	7.40	7.60	7.60	7.65	
11	7.40	7.50	8.15	8.35	10.75	13.00	9.15	7.40	7.60	7.60	7.78	
12	7.35	7.50	8.45	8.10	10.90	13.10	9.10	7.25	7.65	7.60	7.80	
13	7.50	7.50	8.25	8.25	11.05	13.60	9.10	7.25	7.60	7.60	7.78	
14	7.35	7.50	8.35	8.30	11.15	14.10	9.00	7.30	7.60	7.60	7.75	
15	7.50	7.55	8.55	8.40	11.30	15.05	8.85	7.30	7.70	7.60	7.85	
16	7.35	7.40	8.45	8.25	11.60	15.30	8.70	7.30	7.70	7.60	7.88	
17	7.40	7.50	8.55	8.30	11.80	15.50	8.65	7.20	7.70	7.63	7.80	
18	7.30	7.50	8.50	8.45	11.75	15.60	8.75	7.20	7.70	7.60	7.83	
19	7.35	7.55	8.35	8.30	11.45	15.80	8.80	7.30	7.70	7.60	7.85	
20	7.35	7.60	8.35	8.40	11.10	15.70	8.80	7.55	7.70	7.60	7.80	
21	7.35	7.60	8.20	8.35	10.50	15.40	8.85	7.60	7.70	7.60	7.88	
22	7.25	7.60	8.25	8.45	10.10	15.15	8.90	7.80	7.70	7.60	7.95	
23	7.35	7.60	8.25	8.55	9.95	14.80	8.75	7.70	7.70	7.60	8.00	
24	7.35	7.65	8.80	8.75	9.60	14.50	8.45	7.95	7.70	7.60	8.00	
25	7.25	7.60	8.60	9.05	9.40	14.25	8.10	7.70	7.70	7.60	8.05	
26	7.40	7.60	8.50	9.35	9.40	13.90	8.15	7.70	7.70	7.60	8.08	
27	7.40	7.60	8.50	9.60	9.55	13.55	8.35	7.70	7.70	7.60	8.10	
28	7.40	7.50	8.40	9.75	9.70	13.30	8.45	7.75	7.70	7.60	8.05	
29	7.40	.....	8.35	9.65	9.70	12.90	8.50	7.70	7.70	7.60	8.00	
30	7.30	.....	8.30	9.55	9.80	12.60	8.35	7.65	7.70	7.60	8.00	
31	7.40	.....	8.40	.....	9.85	.....	8.25	7.60	.....	7.60	.....	

Estimated monthly discharge of Rio Grande at Embudo, N. Mex., for 1903

[Drainage area, 10,000 square miles.]

Month	Discharge in second-feet.				Run-off.	
	Maximum.	Minimum.	Mean.	Total in acre-feet.	Second-feet per square mile.	Depth inc
January	375	270	317	19,468	0.030	0
February	455	290	375	20,836	.034	
March	1,900	380	788	48,436	.078	
April	2,010	610	987	58,711	.098	
May	4,080	1,680	2,574	158,241	.250	
June	15,860	2,490	8,974	533,970	.890	
July	5,450	670	1,506	92,608	.150	
August	672	170	334	20,537	.030	
September	370	320	348	20,707	.030	
October	370	320	323	19,860	.030	
November	590	320	434	25,825	.040	
December	535	35	283	17,401	.030	
The year	15,860	35	1,437	1,036,600	.141	

RIO GRANDE NEAR CENICERO, COLO.

This station was located on June 28, 1899, by A. L. Fellows. It is at the State bridge across the Rio Grande, at a point near the Colorado-New Mexico State line and about 4 miles west of Eastdale, Colo. The station is favorably located for the purpose, the cross section being fairly uniform, the channel regular and not liable to overflow.

There are two gage rods, one for high water and the other for low water. The high-water rod is a 2 by 6 inch timber attached to the west side of the central downstream cylinder of the bridge. The low-water rod is a scale, marked in feet and tenths, on the perpendicular face of a large bowlder about a hundred yards below the bridge. The channel is in most respects an excellent one. The bed consists of bowlders and rock, and is subject to little change; the banks are high and are not subject to overflow. Gagings can be made at the bridge, but during low water they are usually made by wading. On June 22, 1900, both gages were referred to a bench mark consisting of a chiseled point marked "B. M." on the face of the lava bluff under the west end of the bridge, 7.42 feet above gage datum. The station is an extremely important one, giving, as it does, the discharge of the river at the Colorado State line, including practically all of the Colorado drainage. Roman Mondragon, who keeps a store at the west end of the bridge, has kept the records during the last year.

Discharge measurements of Rio Grande near Cenicero, Colo., in 1903.

Date.	Hydrographer.	Gage height.	Discharge.
		Feet.	*Second-feet.*
April 16	F. Cogswell	1.65	328
May 19	do	3.6	2,658
June 10	do	5.1	4,954
June 30	do	5.1	4,928
July 29	do	1.6	306
August 23	do	1.05	73
September 11	do	1.0	56
October 9	do	1.0	67
November 6	do	1.1	91

Mean daily gage height, in feet, of Rio Grande near Cenicero, Colo., for 19..

Day.	Jan.	Feb.	Mar.	Apr.	May.	June.	July.	Aug.	Sept.	Oct.	Nov.	D
1	0.80	0.80	0.80	1.80	1.90	3.40	4.80	1.20	1.10	1.10	1.00	
2	.80	.80	.80	1.80	2.00	3.80	4.50	1.10	1.00	1.10	1.00	
3	.80	.80	.80	1.80	2.80	4.10	4.30	1.20	1.00	1.10	1.00	
4	.80	.80	.80	1.80	2.90	4.50	4.00	1.00	1.00	1.10	1.10	
5	.80	.80	.80	1.70	3.00	4.50	3.80	.90	1.00	1.10	1.10	
6	.80	.80	.80	1.70	3.00	4.50	3.30	.90	1.00	1.00	1.10	
7	.80	.80	.80	1.70	3.00	4.50	2.80	.70	1.00	1.00	1.10	
8	.80	.80	.80	1.70	3.40	4.60	2.40	.90	1.00	1.00	1.10	
9	.80	.80	.80	1.60	3.40	5.00	2.30	.80	1.10	1.00	1.20	
10	.80	.80	.80	1.60	3.40	5.10	2.30	.70	1.10	1.00	1.40	
11	.80	.80	.80	1.60	3.50	5.20	2.20	.90	1.10	1.00	1.50	
12	.80	.80	.80	1.60	3.60	5.80	2.20	.70	1.20	1.00	1.60	
13	.80	.80	.80	1.60	3.70	6.20	2.10	.70	1.20	1.00	1.40	
14	.80	.80	.80	1.60	3.80	7.00	1.90	.90	1.20	1.00	1.40	
15	.80	.80	.80	1.50	3.80	7.20	1.80	.70	1.20	1.00	1.40	
16	.80	.80	.80	1.50	4.20	8.00	1.80	.80	1.20	1.00	1.50	
17	.80	.80	.80	1.50	4.40	8.00	1.70	.90	1.20	1.00	1.60	
18	.80	.80	.80	1.50	4.10	10.00	1.80	.80	1.20	1.00	1.50	
19	.80	.80	.80	1.50	3.70	7.80	2.10	.90	1.20	1.00	1.60	
20	.80	.80	.80	1.50	3.40	7.80	2.20	.80	1.20	1.00	1.60	
21	.80	.80	.80	1.50	2.90	7.70	1.90	.90	1.20	1.00	1.60	
22	.80	.80	.80	1.50	2.60	7.30	1.90	.90	1.20	1.00	1.60	
23	.80	.80	.80	1.50	2.50	7.20	1.90	.90	1.20	1.00	1.60	
24	.80	.80	.80	1.50	2.40	6.60	1.90	.90	1.10	1.00	1.60	
25	.80	.80	.80	1.50	2.20	6.20	1.70	1.10	1.10	1.00	1.60	
26	.80	.80	.80	1.50	2.40	6.00	1.80	1.00	1.10	1.00	1.60	
27	.80	.80	.80	1.60	2.80	5.80	1.80	1.10	1.10	1.00	1.60	
28	.80	.80	.90	1.60	2.60	5.50	2.00	1.10	1.10	1.00	1.60	
29	.80	......	.90	1.70	2.70	5.20	1.60	1.00	1.10	1.00	1.60	
30	.80	......	1.20	1.80	2.80	5.20	1.40	.90	1.10	1.00	1.60	
31	.80	......	1.40	......	3.00	......	1.30	1.00	......	1.00	......	

Rating table for Rio Grande near Cenicero, Colo.. from January 1, 1903,
December 31, 1903.

Gage height.	Discharge.	Gage height.	Discharge.	Gage height.	Discharge.	Gage height.	Dischar
Feet.	Second-feet.	Feet.	Second-feet.	Feet.	Second-feet.	Feet.	Second-f
0.7	14	1.7	376	3.4	2,385	6.0	6,38
.8	25	1.8	462	3.6	2,655	6.5	7,18
.9	40	1.9	556	3.8	2,930	7.0	7,98
1.0	60	2.0	656	4.0	3,225	7.5	8,78
1.1	84	2.2	874	4.2	3,530	8.0	9,58
1.2	112	2.4	1,110	4.4	3,840	8.5	10,38
1.3	144	2.6	1,355	4.6	4,150	9.0	11,18
1.4	184	2.8	1,605	4.8	4,460	9.5	11,98
1.5	236	3.0	1,865	5.0	4,780	10.0	12,78
1.6	300	3.2	2,125	5.5	5,580		

Curve not well defined between gage heights 1.7 and 5 feet, and very unce
between gage heights 5 and 10 feet.

nated monthly discharge of Rio Grande near Cenicero, Colo., for 1903.

[Drainage area, 7,695 square miles.]

Month.	Discharge in second-feet.			Total in acre-feet.	Run-off.	
	Maximum.	Minimum.	Mean.		Second-feet per square mile.	Depth in inches.
y	25	25	25	1,537	0.003	0.063
ry	25	25	25	1,388	.003	.068
....	184	25	34	2,091	.004	.005
....	462	236	314	18,684	.041	.046
....	3,840	556	2,012	123,713	.261	.301
....	12,780	2,385	6,375	379,339	.828	.924
....	4,460	144	1,178	72,432	.153	.176
....	112	14	47	2,890	.006	.007
ber	112	60	90	5,355	.012	.013
:	84	60	64	3,935	.008	.009
ber..........	300	60	213	12,674	.028	.031
)er	376	300	302	18,569	.039	.045
'he year	12,780	14	890	642,607	.116	1.563

CONEJOS RIVER NEAR MOGOTE, COLO.

ı station was established August 25, 1899, by A. L. Fellows.
age is located 500 feet below the highway bridge 4 miles from
e, Colo. It consists of a vertical pine board nailed to a stump of
on the right bank. It is read twice each day by Miss Josephine
Discharge measurements are made from the downstream side
highway bridge. The initial point for soundings is the inside
f the abutment on the right bank, downstream side. The chan-
straight above and below the station, and the banks, though
re not liable to overflow. The bed of the stream is composed
vel and cobblestones. There is but one channel, broken by the
ıl pier of the two-span bridge. No bench marks have been
ished.

Discharge measurements of Conejos River near Mogote, Colo., in 1903.

Date.	Hydrographer.	Gage height.	Discharge
		Feet.	Second-feet
April 17	F. Cogswell	1.9	18
May 18	do	3.3	1,10
June 9	do	3.7	1,7
July 1	do	3.4	1,3
July 30	do	1.8	2
August 24	do	1.5	:
September 12	do	1.4	
October 10	do	1.3	
November 7	do	1.0	

Mean daily gage height, in feet, of Conejos River near Mogote, Colo., for 190

Day.	Apr.	May.	June.	July.	Aug.	Sept.	Oc
1		2.90	a4.30	3.45	1.45	1.30	
2		3.00	4.35	3.15	1.70	1.30	
3		3.05	4.35	3.20	1.55	1.20	
4		3.40	3.65	3.20	1.40	1.30	
5		3.40	3.80	3.00	1.85	1.30	
6		3.40	3.75	2.90	1.40	1.40	
7		3.50	a4.30	3.00	1.40	a1.85	
8		3.50	4.00	2.50	1.50	1.80	
9		3.60	3.85	2.50	1.45	1.40	
10		3.60	3.65	a3.00	1.40	1.40	
11		3.40	3.75	3.05	1.35	1.35	
12		3.40	3.85	2.70	1.40	1.50	
13		3.50	3.85	2.65	1.40	1.40	
14		3.50	3.95	2.45	1.40	1.40	
15		3.65	3.90	2.70	1.40	1.30	
16		3.85	3.95	2.75	1.45	1.25	
17	1.95	3.85	a4.60	a3.05	1.70	1.35	
18	1.85	3.70	4.65	2.65	1.50	1.25	
19	1.90	3.15	4.45	2.45	1.35	1.25	
20	2.00	2.95	4.25	2.25	1.45	1.20	
21	1.95	3.05	4.40	2.50	1.45	1.30	
22	2.90	2.75	4.35	2.50	1.45	1.25	
23	2.55	2.80	4.05	2.35	1.40	1.20	
24	2.55	2.85	4.05	2.40	1.45	1.25	
25	2.90	2.85	3.75	2.45	1.45	1.40	
26	2.85	3.30	3.80	1.90	1.35	1.35	
27	2.65	3.35	3.85	1.85	1.30	1.40	
28	2.65	3.80	3.75	a2.45	1.30	1.40	
29	2.60	3.35	3.80	1.90	1.30	1.40	
30	2.70	3.55	3.75	1.85	1.30	1.40	
31		4.05		1.85	1.25		

a Heavy rain.

able for Conejos River near Magote, Colo., from April 17 to May 18, 1903.

Discharge.	Gage height.	Discharge.	Gage height.	Discharge.	Gage height.	Discharge.
Second-feet.	Feet.	Second-feet.	Feet.	Second-feet.	Feet.	Second-feet.
185	2.5	473	3.1	900	3.7	1,770
226	2.6	532	3.2	1,000	3.8	1,940
269	2.7	594	3.3	1,120	3.9	2,110
315	2.8	659	3.4	1,270		
364	2.9	730	3.5	1,432		
417	3.0	810	3.6	1,600		

table for Conejos River near Magote, Colo., from May 19 to December 31, 1903.

Discharge.	Gage height.	Discharge.	Gage height.	Discharge.	Gage height.	Discharge.
Second-feet.	Feet.	Second-feet.	Feet.	Second-feet.	Feet.	Second-feet.
58	2.0	360	3.0	840	4.0	2,280
83	2.1	396	3.1	910	4.1	2,450
109	2.2	433	3.2	1,000	4.2	2,620
137	2.3	472	3.3	1,120	4.3	2,790
167	2.4	514	3.4	1,270	4.4	2,960
197	2.5	558	3.5	1,432	4.5	3,130
227	2.6	605	3.6	1,600	4.6	3,300
258	2.7	655	3.7	1,770	4.7	3,470
291	2.8	710	3.8	1,940		
325	2.9	770	3.9	2,110		

1ated monthly discharge of Conejos River near Magote, Colo., for 1903.

[Drainage area, 282 square miles.]

Month.	Discharge in second-feet.			Total in acre-feet.	Run-off.	
	Maximum.	Minimum	Mean.		Second-feet per square mile.	Depth in inches.
'-30	730	165	436	12,107	1.55	0.81
	2,365	683	1,291	79,380	4.58	5.28
	3,385	1,685	2,323	138,228	8.24	9.20
	1,351	308	645	39,660	2.29	2.64
	258	123	173	10,637	.61	.70
1er	343	109	157	9,342	.56	.62
	182	109	138	8,485	.49	.56
he period				297,839		

RIO GRANDE NEAR DEL NORTE, COLO.

Measurements and observations were first begun in the vicinity of Del Norte in 1889 by George T. Quinby. The object of the measurements was to obtain the flow of the river before water was diverted for the agricultural region of San Luis Valley, and by a comparison of this with the figures obtained at Embudo to acquire data as to the effect of the numerous ditches taking out water between the two points. The river 25 miles above Del Norte flows out of the canyon at Wagon Wheel Gap. Little water, however, is diverted until the edge of the San Luis Valley is reached, the largest canal heading near the town of Del Norte. During freshets the river divides into a number of channels, making it difficult to obtain measurements near town. In order to avoid the expense of establishing a station during time of high water the first measurements—those about June 1—were made from several bridges crossing the numerous branches. The results were not wholly satisfactory, and on June 25 a station was established above the branches. Later a locality about 2 miles farther up was chosen.

The station is about 2 miles west of the town of Del Norte, above the main canal, taking water from the Rio Grande, and is above all the irrigating ditches of importance. The river flows in one channel, about 175 feet wide and of very regular section. The banks on each side are steep, and the water is reported never to overflow. The course of the stream is straight for several hundred yards both above and below the section. An inclined gage is set at an angle of about 30° to the horizontal on the right bank and is referred to bench marks. As noted on October 10, 1891, No. 1 is a large nail in the root of a tree 15 feet northwest of the end of the cable on right bank of river. Bench mark No. 2 is a large nail in the root of a tree 24 feet southwest of the end of the inclined gage. Both bench marks are 7.54 feet above the datum of the gage.

On June 16, 1900, the gage rod was connected with an iron bench mark of the United States Geological Survey set in the ground about 25 feet south of the rod, the zero of the rod being 9.25 feet below the bench mark.

Gagings were first made from a flatboat, 4 feet wide and 14 feet long, attached by rope and tackle to a ⅜-inch wire cable fastened to a large cottonwood tree on the left bank and to a sand anchor on the right bank. They are now made by means of a car which travels across the river along a steel cable, the distance being marked on a tag wire. The channel is excellent, the water, although falling rapidly, seldom scouring, and the bed therefore remaining practically the same from year to year. The bed of the channel is covered with small bowlders, and the sides, although not high, have never been known to overflow. The observer is J. S. Regan, who has kept the records regularly ever since the station was established.

Discharge measurements of Rio Grande near Del Norte, Colo., in 1903.

Date.	Hydrographer.	Gage height.	Discharge.
		Feet.	*Second-feet.*
April 18	F. Cogswell	1.85	569
May 20	do	3.65	2,150
June 11	do	5.7	5,195
July 2	do	4.45	3,198
July 31	do	2.05	699
September 14	do	1.9	592
October 12	do	1.52	385
November 9	do	1.4	260

Mean daily gage height, in feet, of Rio Grande near Del Norte, Colo., for 1903.

Day.	Apr.	May.	June.	July.	Aug.	Sept.	Oct.
1	1.74	2.90	5.76	4.80	1.98	1.66	1.62
2							
3	1.56	3.32	5.84	4.26	1.92	1.74	1.66
4							
5	1.48	3.84	5.36	3.58	1.86	1.70	1.70
6							
7	1.46	4.02	5.42	3.32	1.80	2.10	1.66
8							
9	1.74	4.36	5.48	3.46	1.80	2.12	1.64
10							
11	1.88	4.54	5.74	3.42	1.76	1.82	1.58
12							
13	1.58	4.96	6.06	3.14	1.74	1.96	1.52
14							
15	1.84	5.64	6.12	3.00	1.70	1.88	1.52
16							
17	1.92	5.16	6.20	3.30	1.80	1.82	1.50
18							
19	1.88	3.92	6.00	2.96	1.76	1.78	1.50
20							
21	2.40	3.78	5.74	2.74	1.70	1.70	1.48
22							
23	2.86	3.34	5.82	2.60	1.74	1.68	1.46
24							
25	3.08	3.46	5.54	2.42	1.92	1.62	1.44
26							
27	2.98	3.72	5.20	2.34	1.96	1.60	1.42
28							
29	2.74	3.80	5.12	2.20	1.74	1.60	1.40
30							
31		5.12		2.06	1.70		1.42

Rating table for Rio Grande near Del Norte, Colo., from January 1 to December 31, 1903.

Gage height.	Discharge.	Gage height.	Discharge.	Gage height.	Discharge.	Gage height.	Discharge.
Feet.	Second-feet.	Feet.	Second-feet.	Feet.	Second-feet.	Feet.	Second-feet.
1.4	269	2.4	965	3.4	1,880	4.8	3,720
1.5	328	2.5	1,045	3.5	1,985	5.0	4,045
1.6	389	2.6	1,125	3.6	2,095	5.2	4,375
1.7	454	2.7	1,210	3.7	2,210	5.4	4,705
1.8	522	2.8	1,300	3.8	2,325	5.6	5,035
1.9	592	2.9	1,390	3.9	2,445	5.8	5,365
2.0	665	3.0	1,485	4.0	2,570	6.0	5,695
2.1	740	3.1	1,580	4.2	2,835	6.2	6,025
2.2	815	3.2	1,675	4.4	3,115		
2.3	890	3.3	1,775	4.6	3,410		

Curve extended above 5.70 feet gage height.

Estimated monthly discharge of Rio Grande near Del Norte, Colo., for 1903.

[Drainage area, 1,400 square miles.]

Month.	Discharge in second-feet.				Run-off.	
	Maximum.	Minimum.	Mean.	Total in acre-feet.	Second-feet per square mile.	Depth in inches.
April [a]	1,561	304	748	44,509	0.53	0.5
May [a]	5,101	1,390	2,829	173,948	2.02	2.3
June [a]	6,025	4,243	5,189	308,767	3.71	4.1
July [a]	3,720	710	1,655	101,762	1.18	1.3
August [a]	650	454	526	32,342	.38	.4
September [a]	755	389	515	30,645	.37	.4
October [a]	454	269	349	21,459	.25	.2
The period	6,025	269				

[a] Gage readings every second day. Calculations from mean on assumption of full number of days in each month.

MISCELLANEOUS MEASUREMENTS IN THE RIO GRANDE DRAINAGE BASIN.

The following miscellaneous measurements were made in the Rio Grande drainage basin in 1903:

Miscellaneous measurements in Rio Grande drainage basin in 1903.

Date.	Stream	Locality	Discharge.
			Sec.-ft.
September 13	Rio Grande	Alamosa, Colo	
October 11	do	do	
November 8	do	do	

INDEX.

O

DEPARTMENT OF THE INTERIOR

UNITED STATES GEOLOGICAL SURVEY

CHARLES D. WALCOTT, Director

REPORT

OF

PROGRESS OF STREAM MEASUREMENTS

FOR

THE CALENDAR YEAR 1903

PREPARED UNDER THE DIRECTION OF F. H. NEWELL

BY

JOHN C. HOYT

PART IV.—Interior Basin, Pacific, and Hudson Bay Drainage

WASHINGTON

GOVERNMENT PRINTING OFFICE

1904

O

DEPARTMENT OF THE INTERIOR

UNITED STATES GEOLOGICAL SURVEY

CHARLES D. WALCOTT, Director

REPORT

OF

PROGRESS OF STREAM MEASUREMENTS

FOR

THE CALENDAR YEAR 1903

PREPARED UNDER THE DIRECTION OF F. H. NEWELL

BY

JOHN C. HOYT

PART IV.—Interior Basin, Pacific, and Hudson Bay Drainage

WASHINGTON

GOVERNMENT PRINTING OFFICE

1904

ε

CONTENTS.

CONTENTS.

4 CONTENTS.

ILLUSTRATION.

LETTER OF TRANSMITTAL.

DEPARTMENT OF THE INTERIOR,
UNITED STATES GEOLOGICAL SURVEY,
HYDROGRAPHIC BRANCH,
Washington, D. C., March 23, 1904.

ᴵR: I have the honor to transmit herewith Water-Supply Paper 100, which is Part IV of a series of four papers numbered 97 to ᵢ inclusive. These papers constitute the Report of Progress of ᵉam Measurements for the Calendar Year 1903. Parts I and II of ᵢ report contain the results of the data collected from the territory ᵗ of the Mississippi. Parts III and IV are devoted to the data ected in the territory west of the Mississippi.

ʰe work of assembling the original data on which this report is ed and the preparation of the same for publication has been done ler the immediate direction of John C. Hoyt, who has been assisted Frank H. Brundage, L. R. Stockman, R. H. Bolster, H. J. Saun-s, and W. A. Brothers. Acknowledgment is due these persons and ᵢ the various resident hydrographers and others as mentioned on following pages for the collection of the data herein presented.

Very respectfully,

F. H. NEWELL,
Chief Engineer.

lon. CHARLES D. WALCOTT,
Director United States Geological Survey.

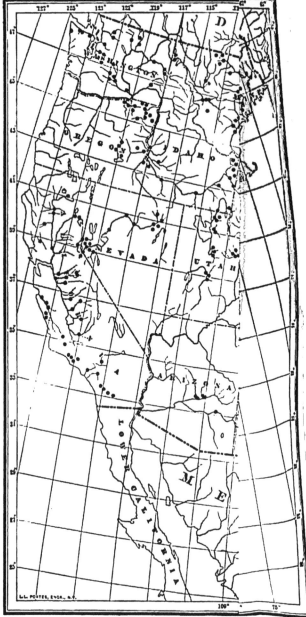

MAP OF THE U

Part III. Western Mississippi River and Western Gulf of M〈 Drainage.

Part IV. Interior Basin, Pacific, and Hudson Bay Drainage.

The territory covered by each paper is given in the subtitle. these larger drainages are, for convenience in arrangement, subdi〈 into smaller ones, under which the data are arranged, as far as p〈 cable, geographically.

These papers contain the data that have been collected at the re〈 gaging stations, the results of the computations based upon the o vations, and such other information as has been collected in the va drainage areas that may be of use in hydrographic studies, inclu〈 as far as available, a description of the drainage area and the str draining it.

For each regular station are given, as far as available, the fo〈 ing data:

1. Description of stations.
2. List of discharge measurements.
3. Gage-height table.
4. Rating table.
5. Table of estimated monthly and yearly discharges and run-〈

The descriptions of stations give, as far as possible, such ge〈 facts about the locality and equipment as would enable the read〈 find and use the station. They also contain, as far as possible, a plete history of all the changes that have occurred since the e lishment of the station that would be factors in using the collected.

The discharge-measurement table gives the results of the disc〈 measurements made during the year. This includes the date hydrographer's name, the gage height, and the discharge in se〈 feet.

The table of daily gage heights gives, for each day, the fluctua of the surface of the river as found from the mean of the gage ings taken on that day. At most of the stations the gage is re the morning and evening.

The rating table gives discharges in second-feet correspondi〈 each stage of the river as given by the gage heights. It depen〈 the general law that for streams of practically constant cross se the discharge is a function of the gage height and that like heights will have the same discharge. In its preparation the disc〈 measurements are plotted on cross-section paper to some conve scale, using gage heights as ordinates and discharges as absc Through these points a smooth curve is drawn, which is the for the table. From this curve are tabulated, on forms prepare the purpose, the discharges corresponding to each tenth of a fo〈

the gage. The first and second differences between the successive discharges are then taken. These are adjusted on the assumption that there is a gradual increase in the discharge as the gage height increases, and the discharge values in the table are then adjusted according to these revised differences. In preparing the rating table all available data are brought into use, including special conditions which might affect the discharge. For high waters above the stage covered by discharge measurements the general rule is to extend the curve by a line tangent to the curve. In case the river overflows its banks a per cent of the discharge is added, depending on the depth and velocity of the overflowed portion. For stages below that portion of the curve which is fixed by discharge measurements the curve has been extended, following the general form of the determined lower portion. Notes under each rating table indicate those portions based on actual observation and those that are estimated.

From the rating table and daily gage heights a table giving the daily discharge of the streams is prepared. From this the table of estimated monthly and yearly discharges and run-off is computed. This latter table gives in condensed form a summary of the results obtained from the observations made during the year at the station. In order to explain this table the following definitions are given:

The term "second-feet" (sec.-ft.) is an abbreviation for "cubic feet per second." It is the number of cubic feet of water flowing by the gaging station every second. The column headed "Maximum" gives the mean flow for the day when the mean gage height was the highest, and is the flow as given in the rating table for that mean gage height. As the gage height is the mean for the day, there might have been short periods when the water was higher and the corresponding discharge larger than given in this column. Likewise in the column of "Minimum" the quantity given is the mean flow for the day when the mean gage height was lowest. The column headed "Mean" is the average flow for each second during the month. Upon this the computations for the three remaining columns in the table are based.

An "acre-foot" is the quantity of water it would take to cover an acre to the depth of 1 foot, or it is 43,560 cubic feet of water. This quantity is used in making estimates for irrigation projects, and it is computed only for such streams as may be used for irrigation. The quantities in the column headed "Total in acre-feet," show the number of acres which would be covered 1 foot by the flow during the month had all the water been impounded.

The expression "second-feet per square mile" means the number of cubic feet of water flowing from every square mile of drainage area for each second.

"Depth in inches" means the depth of water in inches that would have covered the drainage area, uniformly distributed, if all the could have accumulated on the surface. This quantity is used for comparing run-off with rainfall, which quantity is also given in depth in inches.

It should be noticed that "acre-feet" and "depth in inches" represent the actual quantities of water which are produced during the periods in question, while "second-feet," on the contrary, is merely a rate of flow into which the element of time does not enter.

The results of stream measurements made during previous years the United States Geological Survey can be found in the following Survey publications:

1888. Tenth Annual Report, Part II.
1889. Eleventh Annual Report, Part II.
1890. Twelfth Annual Report, Part II.
1891. Thirteenth Annual Report, Part III.
1892. Fourteenth Annual Report, Part II.
1893. Bulletin No. 131.
1894. Bulletin No. 131, and Sixteenth Annual Report, Part II.
1895. Bulletin No. 140, and Seventeenth Annual Report, Part II.
1896. Water-Supply Paper No. 11; Eighteenth Annual Report, Part IV.
1897. Water-Supply Papers Nos. 15 and 16; Nineteeth Annual Report, Part IV.
1898. Water-Supply Papers Nos. 27 and 28; Twentieth Annual Report, Part IV.
1899. Water-Supply Papers Nos. 35 to 39, inclusive; Twenty-first Annual Report, Part IV.
1900. Water-Supply Papers Nos. 47 to 52, inclusive; Twenty-second Annual Report, Part IV.
1901. Water-Supply Papers Nos. 65, 66, and 75.
1902. Water-Supply Papers Nos. 82 to 85, inclusive.
1903. Water-Supply Papers Nos. 97 to 100, inclusive.

A limited number of these are for free distribution, and as long as the supply lasts the same may be obtained by application to the Director, United States Geological Survey. Aside from these, other copies are filed with the Superintendent of Public Documents, Washington, D. C., from whom they may be had at nominal cost. Copies of Government publications are, as a rule, furnished to the municipal public libraries in our large cities, where they may be consulted by those interested.

ACKNOWLEDGMENTS.

Most of the measurements presented in this paper have been obtained through local hydrographers. Acknowledgment is due to each of these persons, and thanks are extended to other persons and corporations who have assisted local hydrographers or have cooperated in any way, either by furnishing records of the height of water or by assisting in transportation.

Water-Supply and Irrigation Paper No. 100   Series P, Hydrographic Progress Reports, 27

DEPARTMENT OF THE INTERIOR

UNITED STATES GEOLOGICAL SURVEY

CHARLES D. WALCOTT, DIRECTOR

REPORT

OF

PROGRESS OF STREAM MEASUREMENTS

FOR

THE CALENDAR YEAR 1903

PREPARED UNDER THE DIRECTION OF F. H. NEWELL

BY

JOHN C. HOYT

PART IV.—Interior Basin, Pacific, and Hudson Bay Drainage

WASHINGTON
GOVERNMENT PRINTING OFFICE
1904

CONTENTS.

Numerous dam and reservoir sites and diversion points for canals were discovered, together with large tracts of fertile land capable of being irrigated. In November, 1902, extensive topographic, soil, and hydrographic surveys were begun in the Colorado River Valley by the reclamation service of the Geological Survey.

The Colorado River is formed by Grand and Green rivers, in the southeastern part of Utah. Of its tributaries Gila River rises in the western part of New Mexico and flows west into the Colorado at Yuma, Ariz., draining the southern half of Arizona. Salt River, its principal tributary, joins it about 15 miles west of Phoenix, Ariz. Rio Verde and Tonto Creek are tributaries of Salt River from the north. San Juan River, with its tributaries, drains southwestern Colorado, northwestern New Mexico, and northeastern Arizona. It takes a generally westward course, joining the Colorado north of the Utah-Arizona boundary. Animas, Los Pinos, and the Florida rivers are tributaries in southwestern Colorado. Grand River has its source on the Continental Divide, in the northern part of Colorado, and flows southwest to its junction with Green River. Gunnison River, its principal tributary, joins it from the south, a short distance south of Grand Junction, Colo. Dolores River rises in southwestern Colorado and flows northwest into the Grand, in eastern Utah. Green River rises in the Wind River Mountains in the western-central part of Wyoming, its main source being in the lofty peaks of the Continental Divide. The source of its tributaries is also among the higher snow-covered ranges, maintaining the volume of this stream late into the summer. The principal branches of White River rise in White River Plateau, a well-forested tract in the White River Forest Reserve. A number of lakes, among which are Oyster, Marvin, Traverse, and Deep lakes, furnish important reservoir sites, if such are ever needed. Duchesne River, with its tributaries, Uinta River, Lake Creek, and Whiterocks River, flows into the Green near the mouth of White River and near Ouray, Utah. The Ashley is a small tributary of the Green, in northeastern Utah. Yampa River rises in the eastern part of Routt County, Colo., flows in a general westerly direction through the entire county, and empties into Green River near the western boundary. The stream is somewhat peculiar in its character, the upper branches having considerable fall, and the water, flowing rapidly over shoals of gravel and rock, is easily taken out for utilization.

The following is a list of the stations maintained during 1903 in the Colorado River drainage basin:

    Colorado River at Yuma, Ariz
    Gila River at Yuma, Ariz.
    Imperial canal at canal heading near Yuma, Ariz.
    Colorado River at Bulls Head, Ariz.
    Verde River near McDowell, Ariz.

Salt River at McDowell, Ariz,
Salt River at reservoir site, Roosevelt, Ariz.
Tonto Creek at Roosevelt, Ariz.
Gila River at San Carlos, Ariz.
Animas River at Durango, Colo.
Animas River at Silverton, Colo.
Florida River near Durango, Colo.
Los Pinos River at Ignacio, Colo.
Dolores River at Dolores, Colo.
Gunnison River at Whitewater, Colo.
Uncompahgre River at Delta, Colo.
Uncompahgre River at Montrose, Colo.
Uncompahgre River at Colona, Colo.
Gunnison River near Cory, Colo.
Gunnison River (North Fork) at Hotchkiss, Colo.
Gunnison River near Cimarron, Colo.
Cimarron Creek at Cimarron, Colo.
Gunnison River at Iola, Colo.
Grand River near Palisades, Colo.
Grand River at Glenwood Springs, Colo.
White River at Meeker, Colo.
White River (South Fork) near Buford, Colo.
White River (North Fork) near Buford, Colo.
Marvine Creek near Buford, Colo.
Uinta River at Ouray School, Utah.
Duchesne River near Myton, Utah.
Lake Creek near Myton, Utah.
Strawberry Creek in Strawberry Valley, Utah.
Uinta River at Fort Duchesne, Utah.
Uinta River near Whiterocks, Utah.
Whiterocks River near Whiterocks, Utah.
Ashley Creek near Vernal, Utah.
Green River near Vernal, Utah.
Green River at Greenriver, Wyo.

COLORADO RIVER AT YUMA, ARIZ.

This station is located in the town of Yuma, Ariz., 1½ miles below
e mouth of Gila River and 10 miles by river above the Mexican
undary.

Records of the river height have been kept by the Southern Pacific
ilroad Company since April 1, 1878, on the gage which was estab-
hed by Arthur Brown, superintendent of the bridge and building
partment of the Southern Pacific Company, during the summer of
76. The lower section of the rod, reading from 10 to 22 feet, is
iled to the pile protection on the right bank above the Southern
cific Railroad bridge. The upper section, reading above 22 feet, is
stened on the lower side of the first bridge pier from the left bank.
his gage height, plus 100 feet, is the Southern Pacific elevation above
a level. At a later date the Southern Pacific Company established
vertical gage rod (the old rod still remaining), fastened to the pile

protection on the left bank of the river just below the railway bridge. This gage has been used continuously since it was established (date unknown), and is the one used by the United States Geological Survey at present. It corresponds in elevation with the old gage established in 1876. The gage is read twice each day by W. D. Smith, who is employed as local hydrographer for the stations in this vicinity.

Discharge measurements are made by means of a ¾-inch cable supported on masts. At low water measurements are made from a boat held in place by the cable. A car is used at flood stages. The initial point for soundings is the cable support on the south bank, about 20 feet from the water's edge at high water. The cable has a span of 650 feet. At low water the channel has a width of 325 feet. During floods a large part of the water flows through an old channel and does not pass under the cable. It is measured at the point where it passes under the railway trestle. The channel of the main river is straight for 600 feet above and 5,000 feet below the station. The current is swift and the gaging section is regular. The right bank is low, wooded, and liable to overflow. The left bank is not subject to overflow. The bed of the stream is composed of silt and sand and is very unstable. At low water a sand bar forms, which divides the channel into two parts. The bench mark is located on the first pier from the left bank. It is a standard bronze-cap United States Geological Survey bench mark and has an elevation of 137 feet above sea level, as determined by the topographic branch of the Geological Survey. Its elevation above the zero of the gage is 35.31 feet.

The observations at this station during 1903 have been made under the direction of S. G. Bennett, district hydrographer.

Discharge measurements of Colorado River at Yuma, Ariz., in 1903.

Date.	Hydrographer.	Gage height.	Discharge.
		Feet.	*Second-feet.*
January 2	R. M. Swain	17.40	3,423
January 3	do	17.30	3,503
January 5	W. D. Smith	17.00	3,133
January 7	do	16.90	2,826
January 9	do	16.90	2,722
January 10	do	16.90	2,818
January 12	do	16.80	2,694
January 14	do	16.95	2,905
January 19	do	17.00	2,939
January 21	do	17.00	2,900
January 23	do	17.10	3,078
January 24	do	17.20	3,176

urge measurements of Colorado River at Yuma, Ariz., in 1903—Continued.

Date.	Hydrographer.	Gage height.	Discharge.
		Feet.	*Second-feet.*
5	W. D. Smith	17.20	2,997
6	do	17.40	3,320
7	do	17.40	3,266
9	do	17.50	3,297
11	do	17.40	3,382
2	do	17.20	2,978
3	do	17.15	2,847
5	do	17.20	2,997
6	do	17.20	3,179
7	do	17.25	3,171
9	do	17.41	3,512
10	do	17.55	3,643
11	do	17.71	3,906
14	do	17.98	4,113
16	do	18.00	3,973
17	do	17.77	3,807
18	do	17.55	3,447
20	do	17.25	3,123
21	do	17.21	3,182
23	do	17.24	2,910
24	do	17.29	3,313
26	do	17.28	3,127
28	do	17.25	3,255
	do	17.50	3,500
	do	17.65	3,700
	do	17.69	3,695
	do	18.14	4,099
	do	19.01	5,791
	do	18.68	4,974
	do	18.48	4,695
	do	18.26	4,183
	do	18.23	4,097
	do	18.35	4,177
	do	19.84	7,146
	do	19.73	6,806
	do	19.68	6,284
	do	19.63	6,434
	do	20.54	9,408
	do	20.28	8,789
	do	20.05	8,682

Discharge measurements of Colorado River at Yuma, Ariz., in 1903—Continued.

Date.	Hydrographer.	Gage height.	Dischar:
		Feet.	*Second-fi*
March 28	W. D. Smith	19.84	7,6
March 30	do	19.58	6,5
March 31	do	20.20	8,2
April 2	do	20.27	8,9
April 8	do	21.55	20,3
April 9	do	22.85	32,(
April 10	do	22.75	25,
April 16	do	20.58	12.
April 18	do	20.40	11.
April 21	do	20.64	11,
April 23	do	20.55	10,
April 25	do	20.68	11,
April 27	do	20.90	12,
April 29	do	20.70	11,
May 2	do	21.60	17.
May 4	do	21.75	21
May 6	do	21.90	22
May 9	do	22.00	22
May 12	do	22.55	27
May 15	do	23.30	29
May 18	do	23.80	34
May 21	do	24.45	43
May 23	do	25.20	48
May 25	do	25.60	56
May 28	do	24.65	48
May 31	do	23.60	37
June 2	do	23.40	28
June 4	do	23.40	29
June 6	do	23.40	29
June 8	do	24.25	40
June 10	do	24.80	49
June 12	do	25.25	52
June 15	do	25.75	54
June 17	do	26.35	56
June 19	do	26.65	61
June 22	do	27.15	65
June 24	do	27.50	69
June 26	do	27.65	67
June 29	do	27.60	72
July 2	do	27.25	66

rge measurements of Colorado River at Yuma, Ariz., in 1903—Continued.

Date. .	Hydrographer.	Gage height.	Discharge.
		Feet.	*Second-feet.*
....	W. D. Smith..................	26.70	54,216
....	.....do	26.05	51,265
....	.....do	25.20	44,430
....	.....do	24.45	41,853
....	.....do	23.15	34,096
....	.....do	22.70	30,950
....	.....do	22.80	30,219
....	.....do	22.40	28,558
....	.....do	22.20	25,956
....	.....do	22.90	30,400
....	.....do	22.70	27,120
....	.....do	21.80	22,726
....	.....do	21.60	20,350
....	L. M. Barnes	20.80	13,325
....	.....do	20.50	11,233
....	.....do	19.75	8,170
....	.....do	19.83	9,193
r 2	W. D. Smith..................	19.20	5,759
r 4	.....do	19.00	5,501
r 10 ..:	.....do	19.40	5,942
r 12	.....do	19.35	6,309
r 15	.....do	19.10	5,138
r 17	.....do	19.50	7,399
r 19	.....do	19.50	7,714
r 21	.....do	19.30	6,943
r 23	.....do	19.60	8,331
r 25	.....do	19.60	7,991
r 29	.....do	19.70	7,649
....	.....do	20.40	9,352
....	.....do	20.70	11,493
....	.....do	21.00	15,806
....	.....do	20.35	9,852
)....................	.....do	19.90	7,906
2....................	.....do	20.10	8,403
4....................	.....do	20.40	9,916
8....................	.....do	20.00	8,044
9....................	.....do	19.80	6,745
2....................	.....do	19.70	6,763
5....................	.....do	19.70	6,396
7....................	.....do	19.70	6,365

Discharge measurements of Colorado River at Yuma, Ariz., in 1903—Continued.

Date.	Hydrographer.	Gage height.	Discharge.
		Feet.	*Second-feet.*
October 29.....................	W. D. Smith.................	19.75	6,128
October 31.....................	.....do	19.70	6,148
November 3	.....do	19.70	6,386
November 5	.....do	19.70	6,071
November 7	.....do	19.70	5,702
November 10	.....do	19.70	5,700
November 13	.....do	19.70	5,355
November 16	.....do	19.70	5,247
November 18	.....do	19.70	5,236
November 21	.....do	19.65	5,011
November 24	.....do	19.60	5,054
November 26	.....do	19.55	4,968
November 28	.....do	19.40	4,705
November 30	.....do	19.50	4,944
December 2.....................	.....do	19.70	5,348
December 5.....................	.....do	19.20	4,472
December 8.....................	.....do	19.42	4,68
December 10....................	.....do	19.60	5,05
December 12....................	.....do	19.65	5,02
December 14....................	.....do	19.50	4,90
December 16....................	.....do	19.36	4,71
December 18....................	.....do	19.20	4,3
December 20....................	.....do	18.85	3,8
December 22....................	.....do	18.55	3,6
December 25....................	.....do	18.35	3,2
December 28....................	.....do	18.40	3,4
December 30....................	.....do	18.90	3,8

Mean daily gage height, in feet, of Colorado River at Yuma, Ariz., for 1903.

Day.	Jan.	Feb.	Mar.	Apr.	May.	June.	July.	Aug.	Sept.	Oct.	Nov.	Dec
1.............	17.60	17.30	17.37	20.48	20.88	23.50	27.15	21.55	19.35	20.40	19.70	19.
2.............	17.40	17.25	17.53	20.30	21.55	23.40	27.25	21.50	19.25	20.35	19.70	19.
3.............	17.30	17.15	17.64	20.40	21.72	23.40	27.10	21.60	19.20	20.60	19.70	19.
4.............	17.20	17.10	17.70	21.05	21.75	23.40	26.90	21.45	19.05	20.70	19.70	19.
5.............	17.00	17.20	17.72	20.80	21.80	23.25	26.65	21.30	19.15	20.70	19.70	19.
6.............	16.90	17.20	18.02	20.98	21.90	23.40	26.35	21.10	19.50	21.00	19.70	19.
7.............	16.90	17.20	18.28	21.28	21.98	23.75	26.05	20.95	19.35	20.95	19.70	19.
8.............	16.90	17.28	18.92	21.50	21.95	21.25	25.75	20.85	19.30	20.40	19.70	19.
9.............	16.90	17.38	18.66	22.75	22.00	21.50	25.20	20.80	19.40	20.15	19.70	19.
10.............	16.90	17.52	18.48	22.65	22.15	21.80	24.75	20.65	19.40	19.95	19.70	19.
11.............	16.85	17.71	18.37	22.20	22.20	25.00	24.35	20.50	19.55	19.95	19.70	19.

Mean daily gage height, in feet, of Colorado River at Yuma, Ariz., for 1903—Continued.

Day.	Jan.	Feb.	Mar.	Apr.	May.	June.	July.	Aug.	Sept.	Oct.	Nov.	Dec.
12	16.80	17.82	18.22	21.92	22.50	25.25	23.65	20.45	19.40	20.20	19.70	19.65
13	16.80	17.90	18.22	21.42	22.75	25.55	23.25	20.15	19.15	20.40	19.70	19.60
14	16.95	17.95	18.32	21.05	23.05	25.65	22.90	20.10	19.05	20.35	19.70	19.55
15	17.00	18.04	19.85	20.78	23.30	25.75	22.70	20.10	19.05	20.10	19.70	19.50
16	17.10	18.00	19.80	20.45	23.40	26.20	22.75	20.05	19.05	19.95	19.70	19.40
17	17.10	17.76	19.52	20.50	23.50	26.35	22.80	19.80	19.40	19.95	19.70	19.30
18	17.05	17.54	19.75	20.40	23.80	26.50	22.60	19.80	19.95	19.85	19.70	19.20
19	17.00	17.36	19.88	20.55	23.98	26.65	22.40	19.80	19.50	19.80	19.65	19.05
20	17.00	17.25	19.65	20.65	24.20	26.80	22.40	19.85	19.30	19.80	19.65	18.85
21	17.00	17.21	19.62	20.60	24.45	26.95	22.30	19.80	19.30	19.80	19.65	18.65
22	17.00	17.22	20.58	20.60	24.90	27.15	22.15	19.70	19.55	19.70	19.65	18.55
23	17.10	17.24	20.52	20.55	25.20	27.25	22.15	19.65	19.60	19.70	19.60	18.50
24	17.20	17.27	20.30	20.50	25.45	27.50	22.90	19.45	19.60	19.70	19.60	18.40
25	17.35	17.30	20.10	20.68	25.60	27.65	23.00	19.25	19.60	19.70	19.60	18.35
26	17.40	17.27	20.02	20.88	25.45	27.65	22.90	19.20	19.70	19.70	19.60	18.30
27	17.40	17.24	20.02	20.90	25.10	27.70	22.60	19.20	19.75	19.70	19.45	18.35
28	17.45	17.26	19.88	20.80	24.65	27.60	22.30	19.20	19.70	19.75	19.40	18.45
29	17.50	......	19.65	20.70	24.25	27.60	21.85	19.20	19.70	19.80	19.35	18.75
30	17.50	......	19.60	20.72	23.85	27.50	21.75	19.25	20.10	19.80	19.50	18.90
31	17.40	......	20.15	......	23.60	......	21.60	19.40	......	19.70	......	18.90

Estimated monthly discharge of Colorado River at Yuma, Ariz., for 1903.[a]

[Drainage area, 225,049 square miles.]

Month.	Discharge in second-feet.			Total in acre-feet.	Run-off.	
	Maximum.	Minimum.	Mean.		Second-feet per square mile.	Depth in inches.
January	3,900	2,694	3,089	189,935	0.014	0.016
February	4,100	2,800	3,372	187,271	.015	.016
March	9,525	3,375	6,117	376,120	.027	.031
April	31,600	9,200	14,326	852,456	..064	.071
May	56,401	13,050	33,735	2,074,284	.150	.173
June	72,219	28,300	53,148	3,162,526	.236	.263
July	69,500	20,350	37,479	2,304,494	.166	.191
August	19,900	6,200	10,869	668,309	.048	.055
September	9,200	5,000	6,786	403,795	.030	.033
October	15,806	6,128	8,482	521,538	.038	.044
November	6,386	4,675	5,399	321,263	.024	.027
December	5,345	3,170	4,343	267,041	.019	.022
The year	72,219	2,694	15,595	11,329,032	.069	.942

[a] Computed by indirect method devised by W. B. Clapp. See article in Engineering News, April 21, 1904.

GILA RIVER AT YUMA AND GILA CITY, ARIZ.

Measurements of the discharge at the mouth of the river have been made, and a record of the periods during which the river was dry has been kept, by W. D. Smith, local hydrographer for this section. There is no gage at this point. The estimated monthly discharge made by Mr. Smith is based upon the measurements and estimates made by the hydrographer. The table of estimated monthly discharges must be considered as approximate. It is of value, however, for its maximum and minimum discharges as estimated by the local hydrographer based upon his measurements of discharge.

The observations at this station during 1903 have been made under the direction of S. G. Bennett, district hydrographer.

Discharge measurements of Gila River at Yuma, Ariz., in 1903.

Date.	Hydrographer.	Gage height.	Discharge.
		Feet.	*Second-feet.*
April 3	W. D. Smith		256
April 8	....do		1,981
April 10	....do		991
April 16	....do		314
April 20	....do		190
April 25	....do		62
May 7	....do		46
September 18	....do		81
October 1	....do		1,329
October 6	....do		522
October 12	....do		61

Estimated monthly discharge of Gila River at Yuma, Ariz., for 1903.

Month.	Discharge in second-feet.			Total for month in acre-feet.
	Maximum.	Minimum.	Mean.	
January			0	0
February			0	0
March			0	0
April	2,000	10	508	30,228
May a			13	799
June			0	0
July			0	0
August b			150	9,200
September c	1,400	0	123	7,319
October d	1,330	0	222	13,650
November			0	0
December			0	0

a May 1–10, discharge estimated at 40 second-feet; May 11–31, no flow.
b August 15 and 29–31, no flow; August 16–28, discharge estimated at 350 second-feet.
c September 1–10 and 22–29, no flow; September 11–21 and 31, discharge estimated at 207 second-feet.
d October 1–17, discharge estimated at 404 second feet; October 18–31, no flow.

Mean daily gage height, in feet, of Gila River at Gila City, Ariz., for 1903.

October 16 .... 2.15
October 17 .... 1.85
October 18 .... 1.70
October 19 to December 31, no flow.

IMPERIAL CANAL AT CALIFORNIA-MEXICO BOUNDARY LINE.

Imperial canal heads about 10 miles by river below Yuma, Ariz., on the California side. The station is located a half mile from the river and 600 feet below the wooden head gates. It was established October 24, 1903, by W. D. Smith. The vertical gage is located just above the boundary line on the right bank. It is read twice each day by J. S. Carter, the storekeeper. Discharge measurements are made by means of a boat and cable. The initial point for soundings is a charred post at the southeast corner of the corral about 150 feet west of the right bank. The channel is straight for 600 feet above and 300 feet below the cable and has a width of 70 feet. The velocity is moderate. There is but one channel at all stages, but when the gage at Yuma reads about 26 feet the river overflows into the canal below the gaging section. The bed of the canal is composed of silt and sand, free from vegetation, and is very unstable. The right bank is low and is liable to overflow. The left bank has an elevation of 6 feet above high water. Bench mark No. 1 is a standard iron bench-mark post of the United

June 5........................	....do ..................-...-..---		553
June 23........................	....do		605
July 8........................	....do	9. 7	449
July 21........................	....do	10. 7	722
August 10....................	L. M. Barnes	8. 8	426
September 11	W. D. Smith..................	8. 9	490
September 24	....do	10. 0	770
October 9.....................	....do	10. 6	820
October 24....................	....do	9. 5	644
November 6	....do	9. 2	634
November 20	....do	8. 9	456
December 7....................	....do	8. 7	526
December 21..................	....do	8. 45	459

Mean daily gage height, in feet, of Imperial canal at head, near Yuma, Ariz., for 1903.

Day.	Jan.	Feb.	Mar.	Apr.	May.	June.	July.	Aug.	Sept.	Oct.	Nov.	Dec.
....	8.5	8.2	8.3	10.6	10.6	......	11.0	10.0	9.00	10.50	9.40	9.02
....	8.3	8.0	8.3	10.7	10.7	......	10.9	9.7	8.95	10.60	9.40	9.50
....	8.1	8.0	8.4	10.8	11.3	......	10.9	......	8.90	10.75	9.30	9.02
....	8.0	7.9	8.5	11.4	11.7	(a)	10.5	9.7	8.80	11.15	9.30	8.78
....	7.8	8.0	8.4	11.3	(a)	9.0	10.4	8.0	8.80	11.10	9.30	8.70
....	7.7	8.0	8.5	11.5	(a)	9.0	10.0	9.2	9.00	11.00	9.20	8.68
....	7.7	8.1	9.4	11.5	(a)	9.3	9.8	8.1	9.10	11.80	9.10	8.70
....	7.6	8.1	9.6	11.6	(a)	......	9.7	9.0	8.95	10.80	9.05	8.80
....	7.5	8.2	9.4	12.0	(a)	......	9.4	9.3	8.90	10.70	9.10	8.82
....	7.5	8.5	9.2	11.7	(a)	......	9.2	9.0	9.00	10.30	9.10	8.82
....	7.5	8.6	9.1	10.8	(a)	......	10.3	8.2	9.00	10.00	9.00	9.00
ᵗ...............	7.4	8.7	9.0	10.7	(a)	9.7	10.0	8.3	9.20	10.00	9.00	9.00
3...............	7.4	8.8	8.9	10.4	(a)	......	10.3	8.7	9.50	10.80	9.00	8.95
ᵗ...............	7.4	8.8	8.9	10.4	(a)	......	9.2	10.2	8.90	11.00	9.00	9.00
.5...............	7.5	8.9	10.3	10.4	(a)	......	10.3	......	8.80	10.50	9.00	8.90
16...............	7.6	8.9	10.3	10.3	(a)	9.7	10.7	10.3	8.80	10.20	9.05	8.90
17...............	7.7	8.7	10.0	10.3	(a)	9.9	10.6	10.3	8.70	10.10	8.97	8.90
18...............	7.8	8.4	10.2	10.3	(a)	......	10.6	10.1	9.20	9.90	8.90	8.95
19...............	7.8	8.3	10.4	10.4	(a)	10.1	10.6	10.5	9.10	9.80	8.90	8.75
20...............	7.7	8.0	10.2	10.5	(a)	10.5	10.6	10.4	......	9.60	8.90	8.68
21...............	7.7	8.0	10.1	10.5	(a)	9.3	10.7	......	9.55	9.60	8.90	8.55
22...............	7.8	8.0	11.1	10.4	(a)	10.0	10.3	......	9.60	9.50	8.90	8.40
23...............	8.0	8.1	10.9	10.4	(a)	10.4	10.5	9.9	......	9.50	8.95	8.32
24...............	8.2	8.1	10.7	10.3	(a)	10.7	10.5	9.8	10.00	9.50	8.90	8.23
25...............	8.2	8.1	10.5	10.4	(a)	10.7	11.9	9.4	......	9.50	8.90	b8.20
26...............	8.5	8.1	10.4	10.6	(a)	11.0	11.7	9.4	9.90	9.50	8.88	b8.15
27...............	8.4	8.1	10.2	10.5	(a)	11.0	10.7	9.4	10.00	9.50	8.82	b8.17
28...............	8.4	8.2	9.9	10.4	(a)	......	10.7	9.2	8.85	9.50	8.90	b8.20
29...............	8.3	......	10.0	10.5	(a)	11.2	9.7	9.1	9.90	9.50	8.85	b8.40
30...............	8.3	......	10.5	10.5	(a)	11.2	10.0	9.5	10.10	9.50	9.02	b8.55
31...............	8.3	......	10.5	......	(a)	......	9.5	9.5	......	9.40	......	b8.55

a No record kept.   b Assumed from height of river at Yuma.

COLORADO RIVER AT BULLS HEAD, NEAR MOHAVE, ARIZ.

This station was established December 5, 1902, at a point of rocks known as the Bulls Head, by E. T. Perkins. The station is about 28 miles from Needles, Cal., and 12 miles from Mohave, Ariz. The gage is an inclined 1 by 5 inch board 24 feet long, spiked to a 4 by 6 inch timber. The timber is fastened to the rocks by iron braces. It is read once each day by T. M. Whedbee, hydrographic aid. Discharge measurements were made from a cable and car. The channel is straight for 300 feet above and below the station, and the current is swift. Both banks are high and rocky and are without trees. There is but one channel. The bed of the stream is composed of silt and sand and is shifting. The bench mark is a bronze tablet set in the rock on the Arizona side about 20 feet above low water and about 50 feet downstream from the bridge. Its elevation above the zero of the gage is 23.85 feet and above sea level is 530.7 feet. It is marked 531.

The observations at this station during 1903 have been made t
the direction of S. G. Bennett, district hydrographer.

Discharge measurements of Colorado River at Bulls Head, Arizona, in 1902 and

		Date.	Gage height.	Dis
1902.			*Feet.*	*Se*
		...	3. 15	
December 10		March 4	3. 20	
		...	3. 15	
December 17		...	3. 30	
		March 13 ...	4. 00	
December 24		March 14	3. 90	
December 29				
December 31	2, 913	March 18	4. 15	
January 3				
	3, 416			
	2, 978			
January 14	3, 415			
		April 1	5. 10	
January 20		April 3		
January 21	3, 700	April 6		
January 22	3, 776	April 10		
January 23	3. 10	3, 953	April 13	
January 24	4, 164	April 15	4. 95	
January 27	3, 651	April 17	5. 20	
January 29	3, 652	April 20	5. 20	
January 30	3, 455	April 22	5. 15	
January 31	3, 525	April 24	5. 30	
February 4	4, 299	April 30	6. 25	
February 7	4, 380	May 1	6. 80	
February 10	4, 924	May 4	7. 65	
February 19	3, 975	May 6	7. 55	
February 21	2. 60	3, 611	May 11	8. 40	
February 25	4, 582			

Mean daily gage height, in feet, of Colorado River at Bulls Head, Arizona, for 1903.

Day.	Jan.	Feb.	Mar.	Apr.	May.
....	2.02	....	3.60	5.10	6.85
1..	1.98	2.75	3.20	5.20	7.20
3..	2.20	2.90	3.40	5.30	7.55
4..	2.30	3.00	3.20	5.25	7.65
5..	2.40	3.09	3.20	6.65	7.70
6..	2.30	3.30	3.10	6.10	7.55
7..	2.20	3.35	3.15	8.00	7.60
8..	2.40	3.35	3.20	7.50	7.80
9..	2.70	3.38	3.80	....	8.10
10..	2.80	3.35	3.80	6.40	8.45
11..	2.50	3.15	3.95	6.00	8.50
12..	2.60	3.30	4.00	5.60	8.65
13..	2.50	2.50	3.85	5.30	8.90
14..	2.60	2.40	3.90	5.10	9.20
15..	2.60	2.40	4.15	4.95	9.50
16..	2.50	2.45	4.00	4.95	10.00
17..	2.65	2.60	4.00	5.20	10.25
18..	2.85	2.60	4.20	5.30	10.70
19..	2.98	2.65	4.55	5.30	....
20..	3.00	2.70	4.50	5.20	....
21..	3.00	2.60	4.55	....	....
22..	3.05	2.60	4.50	5.15	....
23..	3.10	3.00	4.45	....	....
24..	3.03	3.10	....	5.30	....
25..	2.96	3.20	3.90	5.30	....
26..	2.85	3.20	3.70	....	....
27..	2.74	3.00	4.00	5.25	....
28..	2.62	3.15	4.25	....	....
29..	2.60	....	4.40	5.75	....
30..	2.68	....	4.45	6.30	....
31..	2.70	....	4.65	....	....

Mean daily gage height, in feet, of Colorado River at Bulls Head, Arizona, from December 5 to 31, 1902.

Day.	Dec.	Day.	Dec.	Day.	Dec.	Day.	Dec.
5..............	3.00	12............	2.50	19..............	3.20	26..............	3.02
6..............	3.02	13............	2.45	20..............	3.50	27..............	2.82
7..............	2.90	14............	2.50	21..............	3.45	28..............	2.72
8..............	2.78	15............	2.64	22..............	3.40	29..............	2.48
9..............	2.72	16............	2.65	23..............	3.38	30..............	2.35
10..............	2.65	17............	2.95	24..............	3.20	31..............	2.18
11..............	....	18............	3.24	25..............	3.13		

VERDE RIVER AT M'DOWELL, NEAR LEHI, ARIZONA.

This station was established April 20, 1897, by J. B. Lippincott. It is located 30 miles northeast of Phoenix, 15 miles northeast of Mesa, 2⅛ miles above the Arizona canal diversion dam, and three-fourths of

mile above the mouth of the river. Three gages have been in use at
his station, as follows:

Gage No. 1 was established April 20, 1897, by J. B. Lippincott. It
consisted of a vertical rod attached to a large cottonwood tree on the
east bank about 60 feet below the cable. Readings were taken from
this gage until November 11, 1899, when the station was temporarily
abandoned. The bench mark is a point on a cats-claw (acacia) tree
about 100 feet southeast of the gage. Its elevation is 27.02 feet above
the zero of the gage.

Gage No. 2 was established in January, 1901, by H. G. Heisler, and
observations were resumed. Readings from this gage are used in the
1903 report. It is an inclined 2-inch by 4-inch timber fastened to the
rocks on the west bank about 500 feet above the cable, the zero of the
gage being 1,325.4 feet above sea level. Three bench marks have been
established for gage No. 2. First, a nail in a mesquite tree about 6
feet below the cable anchorage on the east bank; its elevation is 1,345.5
feet above sea level and 20.1 feet above the zero of the gage. Second,
a nail in the cable standard at the east bank; its elevation is 1,341.3
feet above sea level and 15.90 feet above the zero of the gage. Third,
a mark on rock at the gage; its elevation is 1,330.4 feet above sea level
and 5 feet above the zero of the gage.

On account of water piling up at gage No. 2 during flood, gage No.
3 was established May 16, 1904, by C. G. Williams. It is a vertical
1¼-inch by 6-inch rod spiked to a 2-inch by 6-inch timber fastened to
a willow tree on the east bank about one-half mile above the cable.
The zero of the gage is 1,339.26 feet above sea level. Two bench marks
have been established for gage No. 3. First, a nail in a large cottonwood tree on the top of the east bank near the gage; its elevation is
1,354.11 feet above sea level and 14.85 feet above the zero of the gage.
Second, a nail in the willow tree to which the gage is attached; its
elevation is 1,347.26 feet above sea level and 8 feet above the zero of
the gage.

Discharge measurements are made by means of a cable, car, and
tagged wire. At low water the channel is oblique to the gaging section and measurements are made by wading at a point 400 feet above
the cable. The channel is straight for a distance of 300 feet above
and below the station, and has a width at low water of 100 feet and at
high water of 450 feet. The current is swift. The right bank is
high, rocky, clean, and is not subject to overflow; the left bank is
low, clean, and is subject to overflow. The bed of the stream is composed of sand and is shifting.

Gage heights and discharge measurements at this station are taken
under the direction of C. G. Williams by W. Richins, who devotes
his whole time to the work.

ye measurements of Verde River at McDowell, near Lehi, Ariz., in 1903.

Date.	Hydrographer.	Gage height.	Discharge.
		Feet.	*Sec. feet.*
....	W. Richins	5.50	268
....	....do	5.48	245
....	....do	6.39	626
....	....do	4.05	734
....	....do	3.20	321
....	....do	2.57	107
....	....do	2.54	123
....	....do	2.48	104
....	....do	2.48	113
....	....do	2.48	113
....	....do	2.43	102
....	....do	2.37	88
....	....do	2.37	87
....	....do	2.33	82
....	....do	2.27	75
....	....do	3.77	408
....	....do	2.89	208
....	....do	3.40	387
....	....do	2.95	286
....	....do	2.65	184
....	....do	2.45	112
....	....do	2.35	99
....	....do	2.25	77
....	....do	2.14	57
....	....do	2.05	48
....	....do	2.01	41
....	....do	2.00	39
....	....do	2.02	42
....	....do	2.30	86
....	....do	4.23	873
....	....do	3.25	365
....	....do	3.77	548
....	....do	3.87	632
....	....do	3.15	323
....	....do	2.90	237
....	....do	2.75	185
....	....do	2.50	131
....	....do	2.85	214
....	....do	3.85	724
....	....do	5.15	1,448

Discharge measurements of Verde River at McDowell, near Lehi, Ariz., in 1903—Contin

Date.	Hydrographer.	Gage height.	Discha
		Feet.	*Sec. f*
August 15	W. Richins	3.88	
August 17	...do	3.50	
August 20	...do	2.95	
August 22	...do	3.00	
September 5	...do	2.70	
September 7	...do	6.35	
September 9	...do	4.55	
Do	...do	6.70	
September 11	...do	3.85	
September 15	...do	2.80	
September 17	...do	2.67	
September 22	...do	2.30	
September 26	...do	2.30	
September 29	...do	4.60	
October 1	...do	3.85	
October 3	...do	3.62	
October 7	...do	3.15	
October 10	...do	2.80	
October 13	...do	2.63	
October 17	...do	2.55	
October 20	...do	2.52	
October 24	...do	2.43	
October 27	...do	2.42	
October 31	...do	2.35	
November 4	...do	2.39	
November 7	...do	2.39	
November 10	...do	2.41	
November 14	...do	2.39	
November 17	...do	2.40	
November 21	...do	2.36	
November 24	...do	2.37	
November 28	...do	2.35	
December 1	...do	2.36	
December 4	...do	2.35	
December 8	...do	2.32	
December 11	...do	2.32	
December 15	...do	2.32	
December 18	...do	2.26	
December 22	...do	2.22	
December 28	...do	2.23	

Mean daily gage height, in feet, of Verde River at McDowell, near Lehi, Ariz., for

Day.	Jan.	Feb.	Mar.	Apr.	May.	June.	July.	Aug.	Sept.	Oct.	Nov.
1			5.90	8.10	2.80	2.41	2.07	2.70	2.45	4.03	2.42
2		5.54	5.80	19.00	2.75	2.42	2.05	2.60	2.72	3.77	2.38
3			5.75	16.40	2.75	2.38	2.05	2.53	2.85	3.68	2.38
4		5.75	5.70	15.10	2.70	2.37	2.06	2.46	2.85	3.86	2.39
5	5.60	5.75	5.70	11.30	2.70	2.38	......	2.39	2.65	3.45	2.39
6		5.60	5.70	9.40	2.70	2.37	2.01	2.34	2.50	3.22	2.39
7		5.58	6.90	6.70	2.65	......	2.01	2.61	5.35	3.13	2.38
8	5.50	5.65	7.10	5.00	2.65	2.35	2.00	2.70	5.00	2.97	2.40
9		5.65	7.05	4.85	2.70	2.32	1.99	4.40	5.30	2.87	2.41
10			7.00	4.05	2.70	2.29	2.00	3.70	4.45	2.78	2.41
11	5.50		6.15	3.85	2.65	2.27	2.00	2.95	3.73	2.73	2.41
12			6.05	3.80	2.60	3.12	......	5.10	3.33	2.68	2.40
13		5.90	6.05	3.75	2.70	2.80	2.00	3.92	3.10	2.64	2.40
14		5.48	7.15	3.75	2.70	2.60	2.02	3.45	2.91	2.62	2.39
15		5.48	5.85	3.60	2.60	3.25	2.37	3.80	2.77	2.59	2.39
16		5.48		7.00	3.60	2.50	3.33	2.39	4.22	2.71	2.56	2.39
17		5.48	6.70	3.55	2.50	3.05	3.85	3.43	2.65	2.55	2.40
18		5.60	6.70	3.40	2.50	2.90	4.22	3.17	2.55	2.55	2.39
19		5.46	5.70	6.75	3.20	2.50	2.74	3.70	3.12	2.48	2.52	2.38
20			5.70	6.40	3.30	2.56	2.64	3.47	2.93	2.41	2.51	2.36
21		5.45	5.75	6.25	3.20	2.57		3.14	3.22	2.35	2.49	2.36
22		5.45	5.75	6.15	3.20	2.56	2.50	3.11	2.95	2.30	2.46	......
23		5.48	5.70	6.00	3.15	2.53	2.43	3.68	......	2.40	2.44	2.37
24		5.46	5.75	6.00	3.20	2.49	2.40	3.52	2.77	2.39	2.42	2.37
25		5.46	6.20	6.00	3.15	2.47	2.32	3.84	2.73	2.33	2.41	2.36
26		5.46	6.15	6.35	3.10	2.47	2.33	3.62	2.68	2.32	2.42	2.36
27		5.46	6.10	9.50	3.10	2.45	2.25	3.33	2.62	2.57	2.42	2.35
28		5.95	9.50	3.00	2.48	......	3.10	2.58	4.00	2.	2.35
29			8.30	2.90	2.17	2.20	3.00	2.52	5.05	2.39	
30	5.44		7.90	2.80	2.47	2.12	2.87	2.50	1.07	2.36	2.36
31			7.80		2.45	......	2.78	2.40		2.35	

Estimated monthly discharge of Verde River at McDowell, near Lehi, Ariz., for 1903.

[Drainage area, 6,000 square miles.]

Month.	Discharge in second-feet.			Total in acre-feet.	Run-off.	
	Maximum.	Minimum.	Mean.		Second-feet per square mile.	Depth in inches.
January *a*			249	*b* 15,310	0.041	*b* 0.04
February *c*			362	*d* 20,104	.060	*d* .06
March 1–26 *e*			625	*f* 32,231	.104	*f* .10
April 8–30 *e*			496	*g* 22,627	.083	*g* .07
May	193	110	141	8,669	.024	.02
June	372	55	136	8,093	.023	.02
July	859	38	232	14,265	.039	.04
August	1,685	89	329	20,229	.055	.06
September	1,960	110	513	30,526	.085	.09
October	830	185	318	19,553	.053	.06
November	225	190	207	12,317	.035	.04
December	253	215	227	13,958	.038	.04
The period				217,882		

a January 1–4, 6–7, 9–10, 12–13, and 31 missing.
b Computed for 31 days.
c February 1, 3, and 10–12 missing.
d Computed for 28 days.
e Flood Mar. 27–Apr. 7, causing great change in stream bed.
f Computed for 26 days.
g Computed for 23 days.

NOTE.—Owing to the shifting character of the bed of the river the daily discharges were obtained from rating tables extending over short periods, by interpolation, and by Prof. Stout's method.

SALT RIVER AT M'DOWELL, NEAR LEHI, ARIZ.

This station was established April 20, 1897, by J. B. Lippincott It is located one-third mile above the junction of Salt and Verd rivers, 30 miles northeast of Phoenix, 15 miles northeast of Mesa, an 1½ miles above the Arizona canal diversion dam. There have bee three gages in use at this station, as follows:

Gage No. 1, set by J. B. Lippincott April 20, 1897, was a 2-inch by 6-inch timber, bolted to the rocks on the south bank of the rive about one-fourth mile above the cable from which discharge measure ments are made. This gage, which has since been removed, was use until November 30, 1899, when the station was temporarily abandoned The bench mark is a nail in a palo verde tree about 75 feet west o cable anchorage on the north bank. Its elevation is 17.33 feet abov the zero of the gage.

In 1901 observations were resumed, gage No. 2 being establishe by Mr. Appleby. It consists of a 2-inch by 6-inch timber fastened t a tree on the north bank of the river three-fourths mile above th cable. The zero of the gage is 1,323.59 feet above sea level, and its bench mark is a nail in a root of the willow tree to which the gage is fastened. Its elevation is 1,328.69 feet above sea level and 5.10 feet above the zero of the gage. On April 2, 1903, high water in Verde River backed up the water on gage No. 2, and changed the cross section by depositing sand.

Gage No. 3 was established May 19, 1903, by W. W. Schlecht. It consists of a 1-inch by 6-inch stadia rod spiked to a 2-inch by 4-inch timber and fastened to a tree on the south bank 1¼ miles above the cable. The water surface at this gage is about 15 feet higher than at the mouth of Verde River, and the zero of the gage is 1,336.27 feet above sea level. Three bench marks have been established for gage No. 3. First, a nail in a mesquite stump 200 feet east of Peters's corral. Its elevation is 1,363.2 feet above sea level and 26.93 feet above the zero of the gage. Second, a nail in a root of a mesquite tree on the top of the bank 50 feet northwest of the northwest corner of Peters's corral and about 75 feet from the gage. Its elevation is 1,356 feet above sea level and 19.73 feet above the zero of the gage. Third, a nail in the willow tree to which the gage is attached. Its elevation is 1,344.27 feet above sea level and 8 feet above the zero of the gage.

Discharge measurements are made by means of a cable and car. The south end of the cable is anchored to the rocks and the north end is run over an 8-inch by 8-inch standard 21 feet high. During low water discharge measurements are made, by wading, at a point about 1,000 feet upstream from the cable, where a tag wire has been placed. The initial point for soundings is 120 feet south of the standard under the cable at the north bank. The channel has a width at low water of 150 feet and at high water of about 700 feet. It is straight for about 500 feet above and below the station. The current is swift. The right bank is about 3½ feet high at the water's edge and rises with a gradual slope for 400 feet. It is clean and subject to overflow. The left bank rises vertically for about 5 feet, to a small bench from which the rocks rise to a considerable height. The bank is clean and is not subject to overflow. The bed of the stream is composed of sand and is shifting, and it is necessary to make a large number of measurements in order to obtain an accurate estimate of the discharge.

The 1903 rating tables do not apply to the gage heights from April 2, 1903, until May 19, 1903, when readings were begun on the new gage.

Gage heights and measurements at this station are taken under the direction of C. G. Williams by W. Richins, who devotes his whole time to the work.

Discharge measurements of Salt River at McDowell, near Lehi, Ariz., in 1903.

Date.	Hydrographer.	Gage height.	Disch
		Feet.	*Second*
January 8	W. Richins	1.51	
January 17	do	1.35	
March 25	do	1.64	
March 26	do	2.03	
April 10	do	4.06	1,
May 21	do	1.81	
May 23	do	1.75	
May 26	do	1.68	
May 28	do	1.63	
May 30	do	1.60	
June 2	do	1.56	
June 4	do	1.55	
June 6	do	1.54	
June 9	do	1.52	
June 11	do	1.50	
June 12	do	1.61	
June 16	do	1.94	
June 18	do	1.88	
June 20	do	1.80	
June 23	do	1.65	
June 25	do	1.60	
June 27	do	1.50	
June 30	do	1.41	
July 3	do	1.34	
July 7	do	1.23	
July 10	do	1.15	
July 14	do	1.16	
July 17	do	1.42	
July 18	do	1.38	
July 21	do	1.29	
July 23	do	1.45	
July 25	do	1.55	
July 28	do	1.36	
July 30	do	1.30	
August 1	do	1.35	
August 4	do	1.17	
August 8	do	2.29	
August 10	do	3.00	
August 12	do	1.68	
August 15	do	1.85	

Discharge measurements of Salt River at McDowell, near Lehi, Ariz., in 1903—Continued.

Date.	Hydrographer.	Gage height.	Discharge.
		Feet.	*Second-feet.*
August 17	W. Richins	1.75	319
August 20	do	1.88	424
August 22	do	1.66	339
September 4	do	1.65	311
September 7	do	1.65	220
September 9	do	2.28	871
September 11	do	1.70	294
September 15	do	1.58	235
September 17	do	1.53	211
September 22	do	1.45	172
September 26	do	1.48	200
September 29	do	2.84	868
October 1	do	1.95	450
October 3	do	1.75	348
October 7	do	1.66	275
October 10	do	1.60	241
October 13	do	1.52	220
October 17	do	1.48	200
October 20	do	1.47	194
October 24	do	1.46	184
October 27	do	1.46	185
October 31	do	1.46	175
November 4	do	1.47	192
November 7	do	1.47	191
November 10	do	1.48	200
November 14	do	1.49	198
November 17	do	1.49	199
November 21	do	1.49	192
November 24	do	1.51	205
November 28	do	1.51	202
December 1	do	1.50	193
December 4	do	1.50	196
December 8	do	1.50	200
December 11	do	1.50	203
December 15	do	1.50	198
December 18	do	1.49	189
December 22	do	1.48	191
December 28	do	1.48	196

Mean daily gage height, in feet, of Salt River at McDowell, near Lehi, Ariz., for 1903.

Day.	Jan.	Feb.	Mar.	Apr.	May.	June.	July.	Aug.	Sept.	Oct.	Nov.	Dec.
1			1.45	2.00	2.40	1.87	1.38	1.34	1.62	1.93	1.48	1.43
2		1.36	1.40	(a)	2.30	1.85	1.35	1.21	1.95	1.76	1.46	1.39
3			1.40	(a)	2.25	1.62	1.88	1.19	1.70	1.76	1.47	1.18
4		1.65	1.40	(a)	2.23	1.55	1.38	1.15	1.62	1.74	1.47	1.10
5	1.60	1.74	1.40	(a)	2.20	1.54		1.15	1.34	1.72	1.47	1.39
6		1.58	1.45	(a)	2.40	1.54	1.26	1.67	1.68	1.47		
7		1.75	1.45	3.10	2.30		1.23	1.14	1.65	1.68	1.47	1.38
8	1.51	1.70	1.60	3.20	2.15	1.58	1.30	2.14	2.18	1.62	1.48	1.33
9		1.70	1.75	3.65	2.15	1.51	1.16	2.19	2.52	1.62	1.48	1.6
10	1.50		2.00	4.04	2.10	1.50	1.14	2.40	1.72	1.59	1.48	1.6
11	1.48		2.10	4.10	2.00	1.50	1.14	2.15	1.67	1.87	1.49	1.18
12	1.46		1.95	4.00	1.95	1.52		2.01	1.68	1.54	1.48	1.52
13	1.43	2.02	1.85	3.95	1.90	1.61	1.15	2.13	1.62	1.52	1.48	
14	1.40		1.90	3.95	1.80	1.70	1.16	2.19	1.60	1.52	1.49	1.3
15	1.38	2.04	1.85	3.90	1.84	1.85	1.22	1.90	1.87	1.50	1.49	1.40
16	1.36	2.00	1.95	3.80	1.88	1.93	1.33	1.94	1.54	1.43	1.48	1.33
17	1.35	1.85	1.80	3.50	1.85	1.90	1.34	1.79	1.52	1.43	1.49	1.37
18		1.85	1.85	3.40	1.87	1.86	1.36	2.21	1.49	1.47	1.49	1.6
19	1.41	1.80	1.90	3.20	1.84	1.88	1.21	1.84	1.47	1.46	1.48	1.9
20		1.80	1.85	3.00	1.94	1.80	1.28	1.95	3.46	1.47	1.49	
21	1.36	1.80	1.80	2.90	1.81		1.58	1.66	1.45	1.46	1.49	1.8
22	1.34	1.45	1.75	2.80	1.77	1.66	1.70	1.71	1.43	1.48		1.8
23	1.33	1.45	1.70	2.70	1.74	1.64	1.47	1.56	1.48	1.48	1.51	1.6
24	1.28	1.45	1.65	2.70	1.70	1.62	1.83	1.52	1.47	1.65	1.50	1.6
25	1.28	1.45	1.60	2.65	1.69	1.58	1.56	1.44	1.61	1.46	1.50	1.8
26	1.29	1.50	2.05	2.65	1.68	1.55	1.44	1.62	1.45	1.46	1.50	1.8
27	1.29	1.53	2.05	2.60	1.67	1.49	1.39	1.72	1.54	1.46	1.51	1.6
28		1.50	2.10	2.55	1.63		1.35	1.67	1.66	1.46	1.50	1.67
29			2.05	2.55	1.61	1.42	1.31	1.53	2.82	1.45		1.8
30	1.28		2.00	2.50	1.60	1.41	1.29	1.60	2.23	1.49	1.51	1.6
31	1.26		2.00		1.57		1.26	1.67		1.46		1.8

a No record. Water backed upon gage from high water in Verde River.

Rating table for Salt River at McDowell, near Lehi, Ariz., from January 1, 1903, to April 1, 1903.[a]

Gage height.	Discharge.	Gage height.	Discharge.	Gage height.	Discharge.
Feet.	*Second-feet.*	*Feet.*	*Second-feet.*	*Feet.*	*Second-feet.*
1.30	161	1.60	267	1.90	402
1.40	193	1.70	310	2.00	450
1.50	228	1.80	355	2.10	500

a Table is approximate and applies only to gage No. 2.

Rating table for Salt River at McDowell, near Lehi, Ariz., from May 29, 1903, to December 31, 1903.[a]

Gage height.	Discharge.	Gage height.	Discharge.	Gage height.	Discharge.	Gage height.	Discharge.
Feet.	*Second-feet.*	*Feet.*	*Second-feet.*	*Feet.*	*Second-feet.*	*Feet.*	*Second-feet.*
1.20	95	1.80	365	2.40	775	3.00	1,195
1.30	129	1.90	425	2.50	845	3.10	1,265
1.40	164	2.00	495	2.60	915	3.20	1,335
1.50	201	2.10	565	2.70	985	3.30	1,405
1.60	247	2.20	635	2.80	1,055	3.40	1,475
1.70	305	2.30	705	2.90	1,125		

[a] Table applies only to gage No. 3.   Curve is well defined.

Estimated monthly discharge of Salt River at McDowell, near Lehi, Ariz., for 1903.

[Drainage area, 6,260 square miles.]

Month.	Discharge in second-feet.			Total in acre-feet.	Run-off.	
	Maximum.	Minimum.	Mean.		Second-feet per square mile.	Depth in inches.
January[a]			189	[b]11,621	0.030	[b]0.035
February[c]			304	[d]16,883	.049	[d].051
March	500	193	350	21,521	.056	.065
April						
May 19–31[b][g]			270	[e]6,962	.043	[e].021
June	446	168	262	15,590	.042	.047
July	305	75	141	8,669	.023	.027
August	1,475	71	365	22,443	.058	.067
September	1,069	157	312	18,565	.050	.056
October	446	182	231	14,204	.037	.043
November	205	186	196	11,663	.031	.035
December	209	186	197	12,113	.031	.036
The period				160,234		

[a] January 1–4, 6, 7, 9, 18, 20, 28, and 29 missing.
[b] Computed for 31 days.
[c] February 1, 3, 10, 11, 12, and 14 missing.
[d] Computed for 28 days.
[e] Computed for 13 days.
[g] May 19–28 estimated.

SALT RIVER AT RESERVOIR SITE, ROOSEVELT, ARIZ.

This station, established February 7, 1901, by H. G. Heisle
located at the town of Roosevelt, which is the United States Geolo
Survey construction camp for the Salt River dam and reservoir, a
about 12 miles west of Livingston. In previous reports this st
has been called Salt River at reservoir site below Tonto Creek,
Livingston, Ariz.

The gage rod and cable are at the upper end of the gorge, about ?
feet below the mouth of Tonto Creek and 1,500 feet above the dam
The gage is a vertical rod fastened to the rocks on the left bank o
river. Gagings are made from a traveling car suspended from a c

The observations at this station during 1903 have been made u
the direction of W. A. Farish, district hydrographer.

Discharge measurements of Salt River at reservoir site, Roosevelt, Ariz., in 190.

Date.	Hydrographer.	Gage height.	Disc
		Feet.	*Secon*
January 4	Osburn Richins..............	7.11	
January 12	.....do	7.08	
January 19	.....do	7.09	
January 26	.....do	7.08	
February 3	.....do	7.22	
February 6	.....do	7.30	
February 10	.....do	7.28	
February 13	.....do	7.51	
February 16	.....do	7.38	
February 28	.....do	7.20	
March 6	.....do	7.76	
March 11	.....do	7.50	
March 21	.....do	7.35	
March 25	.....do	7.30	
March 26	.....do	8.51	
April 3	.....do	8.85	
April 10	.....do	8.30	
May 11	.....do	7.40	
May 5	.....do	7.62	
May 20	.....do	7.30	
May 26	.....do	7.17	
May 30	.....do	7.10	
June 5	.....do	7.10	
June 11	.....do	7.05	
June 15	.....do	7.56	
June 18	.....do	7.48	
June 24.........	.....do	7.11	

Discharge measurements of Salt River at reservoir site, Roosevelt, Ariz., in 1903—Cont'd.

Date.	Hydrographer.	Gage height.	Discharge.
		Feet.	*Second-feet.*
June 30......................	Oeburn Richins..............	6.95	187
July 6........................	.....do	6.85	116
July 10.......................	.....do	6.83	98
July 14.......................	.....do	7.00	184
July 21.......................	.....do	7.47	466
July 24.......................	.....do	7.08	286
July 29.......................	.....do	6.93	135
August 4	.....do	6.83	89
August 7	.....do	9.25	*a* 2,024
August 7	.....do	6.89	101
August 8	.....do	7.75	532
August 10	.....do	7.70	516
August 12	.....do	8.21	861
August 17	.....do	7.75	621
August 25	.....do	7.54	455
August 28	.....do	7.30	258
September 1	.....do	7.51	461
September 7	.....do	7.66	515
September 8	.....do	9.62	*a* 4,016
September 14	.....do	7.21	248
September 21	.....do	7.10	209
September 28	.....do	7.37	350
September 28 (2 P. M.).......	.... .do	9.15	2,684
September 30	.....do	7.50	443
October 5.....................	.....do	7.30	362
October 9.....................	.....do	7.21	272
October 12....................	.....do	7.18	255
October 16....	.....do	7.19	231
October 24....................	.....do	7.18	223
October 26....................	.....do	7.18	243
October 31....................	.....do	7.18	211
November 2	.....do	7.17	207
November 9	.....do	7.18	214
November 16	.....do	7.18	211
November 23	.....do	7.18	214
November 30	.....do	7.17	207
December 7....................	.....do	7.17	210
December 14...................	.....do	7.18	208
December 21...................	.....do	7.18	208
December 31...................	.....do	7.18	212

a Float measurements.

Mean daily gage height, in feet, of Salt River at reservoir site, Roosevelt, Ariz., for 1901.

Day.	Jan.	Feb.	Mar.	Apr.	May.	June.	July.	Aug.	Sept.	Oct.	Nov.	Dec.
1	7.15	7.05	7.20	8.30	7.50	7.10	6.94	6.65	7.05	7.21	7.17	7.17
2	7.15	7.00	7.37	8.90	7.50	7.10	6.90	6.84	7.20	7.22	7.17	7.10
3	7.12	7.15	7.16	8.95	7.45	7.05	6.85	6.52	7.25	7.21	7.17	7.17
4	7.11	7.10	7.18	8.94	7.42	7.00	6.85	6.60	7.25	7.21	7.17	7.17
5	7.10	7.25	7.34	8.52	7.50	7.00	6.90	6.52	7.20	7.20	7.17	7.17
6	7.09	7.30	7.65	8.45	7.45	7.00	6.95	6.54	7.30	7.30	7.10	7.17
7	7.08	7.30	7.70	8.34	7.42	7.00	6.80	6.15	7.07	7.26	7.13	7.13
8	7.03	7.30	7.75	8.29	7.41	7.00	6.84	7.75	7.05	7.20	7.10	7.12
9	7.00	7.30	7.05	8.20	7.40	7.00	6.90	6.60	7.20	7.21	7.13	7.13
10	7.10	7.20	7.50	8.20	7.40	7.25	6.45	7.20	7.20	7.19	7.13	7.13
11	7.05	7.25	7.49	8.25	7.35	7.15	6.50	7.00	7.25	7.18	7.13	7.13
12	7.08	7.45	7.42	8.32	7.35	7.23	6.55	6.10	7.20	7.18	7.13	7.13
13	7.07	7.12	7.49	8.22	7.30	7.42	6.45	6.45	7.20	7.18	7.13	7.13
14	7.07	7.44	7.56	8.14	7.20	7.62	6.90	6.20	7.20	7.19	7.13	7.13
15	7.06	7.42	8.10	8.10	7.20	7.60	6.90	7.13	7.19	7.13	7.13	7.13
16	7.06	7.05	7.55	8.10	7.20	7.04	7.00	7.18	7.16	7.19	7.13	7.13
17	7.07	7.35	7.49	7.92	7.20	7.00	6.94	6.62	7.14	7.19	7.13	7.13
18	7.08	7.24	7.44	7.90	7.20	7.46	6.90	7.20	7.12	7.18	7.12	7.13
19	7.09	7.21	7.30	7.84	7.20	7.25	6.90	7.12	7.12	7.18	7.13	7.13
20	7.08	7.20	7.36	7.60	7.20	7.33	6.92	6.42	7.11	7.18	7.13	7.13
21	7.07	7.18	7.34	7.72	7.30	7.22	6.85	6.97	7.10	7.18	7.13	7.13
22	7.06	7.18	7.30	7.09	7.22	7.17	7.00	7.34	7.10	7.19	7.13	7.13
23	7.03	7.18	7.27	7.65	7.20	7.12	7.15	7.21	7.11	7.13	7.13	7.13
24	7.04	7.18	7.30	7.56	7.20	7.11	7.10	7.20	7.15	7.13	7.13	7.13
25	7.05	7.19	7.21	7.50	7.15	7.00	7.00	7.80	7.15	7.13	7.15	7.13
26	7.06	7.20	8.70	7.50	7.16	7.05	6.97	7.21	7.12	7.19	7.15	7.13
27	7.06	7.20	8.70	7.50	7.14	7.02	6.95	7.21	7.19	7.19	7.15	7.13
28	7.06	7.20	8.53	7.50	7.11	7.00	6.94	7.29	8.33	7.18	7.15	7.13
29	7.08	......	8.44	7.50	7.10	6.98	6.92	7.30	8.10	7.18	7.15	7.13
30	7.05	......	8.32	7.50	7.10	6.95	6.89	7.32	7.42	7.18	7.17	7.13
31	7.04	......	8.33	......	7.10	......	6.88	7.30	......	7.18	......	7.13

'ed monthly discharge of Salt River at reservoir site, Roosevelt, Ariz., for 1903.

[Drainage area, 5,756 square miles.]

onth.	Discharge in second-feet.			Total in acre-feet.	Run-off.	
	Maximum.	Minimum.	Mean.		Second-feet per square mile.	Depth in inches.
....	250	178	207	12,728	0.036	0.042
....	472	196	318	17,661	.055	.057
....	1,675	256	600	36,893	.104	.120
....	2,050	440	909	54,089	.158	.176
....	477	234	352	21,644	.061	.070
....	504	153	285	16,959	.050	.056
....	369	88	142	8,731	.025	.029
....	1,100	88	411	25,271	.071	.082
'r...........	955	197	316	18,803	.055	.061
....	319	233	253	15,556	.044	.051
'r	213	202	211	12,555	.037	.041
r	210	202	208	12,789	.036	.042
....	2,050	88	351	253,679	.061	.827

ıwing to the shifting character of the bed of the river the daily discharges were applied
al rating tables extending over short periods of time.

TONTO CREEK AT ROOSEVELT, ARIZ.

station, established April 1, 1901, by H. G. Heisler, is located
town of Roosevelt and about 12 miles west of Livingston.
ıtion has been called Tonto Creek near Livingston in previous

gage, a vertical rod fastened to a cliff of cemented gravel on
; bank, is about 3,500 feet above the mouth of the creek.
ements are made by wading with meter during low water and
ns of floats over a course of 25 feet during floods.
ıbservations at this station during 1903 have been made under
ıction of W. A. Farish, district hydrographer.

Discharge measurements of Tonto Creek at Roosevelt, Ariz., for 1903.

Date.	Hydrographer.	Gage height.	Discharge.
		Feet.	*Second-feet.*
January 4	O. Richins....................	3. 00	15
January 12	.....do	2. 96	9
February 3	.....do	3. 90	188
February 6	.....do	4. 29	197
February 10	.....do	4. 00	145
February 12 ....;....	.....do	4. 37	221
February 13	.....do	4. 55	279
February 16	.....do	4. 00	178
February 28	.....do	3. 70	41
March 6....................	.....do	4. 57	182
March 11....................	.....do	3. 70	47
March 21....................	.....do	3. 10	26
March 26....................	.....do	5. 61	899
April 3	.....do	4. 30	145
July 14....................	.....do	4. 00	55
July 21....................	.....do	4. 22	124
July 24....................	.....do	3. 96	82
August 7....................	.....do	3. 55	10
August 8	.....do	5. 10	590
August 13	.....do	5. 00	331
August 17	.....do	4. 29	177
September 7	.....do	4. 50	239
September 11	.....do	3. 35	9
September 28 ....do		5. 77	1, 134
September 30	.....do	3. 80	62

Mean daily gage height, in feet, of Tonto Creek at Roosevelt, Ariz., for 1903.

Day.	Jan.	Feb.	Mar.	Apr.	May.	June.	July.	Aug.	Sept.	Oct.	Nov.	Dec
1........................	3. 00	2. 90	3. 70	4. 30	3. 70	3. 70	3. 70	3. 30	3. 00	3. 47	3. 25	3. 25
2........................	3. 00	2. 91	3. 62	4. 30	3. 70	3. 70	3. 70	3. 30	3. 00	3. 45	3. 25	3. 25
3........................	3. 00	3. 50	3. 55	4. 30	3. 70	3. 70	3. 70	3. 30	3. 00	3. 40	3. 25	3. 25
4........................	3. 00	4. 00	3. 55	4. 00	3. 70	3. 70	3. 70	3. 30	3. 00	3. 35	3. 2b	3. 25
5........................	3. 00	4. 10	3. 58	3. 92	3. 70	3. 70	3. 70	3. 30	3. 00	3. 30	3. 25	3. 25
6........................	3. 00	4. 10	4. 58	3. 78	3. 70	3. 70	3. 70	3. 37	3. 00	3. 27	3. 25	3. 25
7........................	3. 00	4. 06	4. 30	3. 71	3. 70	3. 70	3. 70	3. 62	3. 25	3. 25	3. 25	3. 25
8........................	3. 00	4. 02	3. 90	3. 70	3. 70	3. 70	5. 00	5. 00	5. 40	3. 25	3. 25	3. 25
9........................	2. 96	4. 01	3. 86	3. 70	3. 70	3. 70	3. 70	5. 00	3. 88	3. 25	3. 25	3. 25
10........................	2. 96	4. 00	3. 77	3. 70	3. 70	3. 70	3. 70	3. 50	3. 78	3. 25	3. 25	3. 25
11........................	2. 96	4. 16	3. 69	3. 70	3. 70	5. 70	3. 70	3. 40	3. 32	3. 25	3. 25	3. 25
12........................	2. 96	4. 48	3. 58	3. 70	3. 70	3. 70	3. 70	3. 40	3. 15	3. 25	3. 25	3. 25
13........................	2. 96	4. 58	3. 44	3. 70	3. 70	3. 70	3. 70	4. 45	3. 10	3. 25	3. 25	3. 25

Mean daily gage height, in feet, of Tonto Creek at Roosevelt, Ariz., for 1903—Continued.

Day.	Jan.	Feb.	Mar.	Apr.	May.	June.	July.	Aug.	Sept.	Oct.	Nov.	Dec.
14	2.95	4.42	3.33	3.70	3.70	3.70	3.75	3.90	3.07	3.25	3.25	3.25
15	2.95	4.28	3.28	3.70	3.70	3.70	3.48	3.85	3.00	3.25	3.25	3.25
16	2.95	3.98	3.19	3.70	3.70	3.70	3.45	3.85	3.00	3.25	3.25	3.25
17	2.95	3.84	3.17	3.70	3.70	3.70	3.45	4.30	3.00	3.25	3.25	3.25
18	2.95	3.82	3.16	3.70	3.70	3.70	3.45	4.15	3.00	3.25	3.25	3.25
19	2.95	3.75	3.13	3.70	3.70	3.70	3.45	4.15	3.00	3.25	3.25	3.25
20	2.95	3.71	3.11	3.70	3.70	3.70	3.45	3.98	3.00	3.25	3.25	3.25
21	2.95	3.67	3.10	3.70	3.70	3.70	4.18	3.60	3.00	3.25	3.25	3.25
22	2.93	3.65	3.10	3.70	3.70	3.70	3.95	3.35	3.00	3.25	3.25	3.25
23	2.90	3.65	3.10	3.70	3.70	3.70	4.25	3.16	3.00	3.25	3.25	3.25
24	2.90	3.68	3.10	3.70	3.70	3.70	3.94	3.00	3.25	3.25	3.25	3.25
25	2.90	3.70	3.10	3.70	3.70	3.70	3.48	3.00	3.12	3.25	3.25	3.25
26	2.90	3.70	5.39	3.70	3.70	3.70	3.38	3.00	3.15	3.25	3.25	3.25
27	2.90	3.70	4.82	3.70	3.70	3.70	3.34	3.00	3.42	3.25	3.25	3.25
28	2.90	3.70	4.50	3.70	3.70	3.70	3.30	3.00	4.55	3.25	3.25	3.25
29	2.90	......	4.41	3.70	3.70	3.70	3.30	3.00	4.15	3.25	3.25	3.25
30	2.90	......	4.29	3.70	3.70	3.70	3.30	3.35	3.72	3.25	3.25	3.25
31	2.90	......	4.30	......	3.70	......	3.30	3.02	......	3.25	......	3.25

Estimated monthly discharge of Tonto Creek at Roosevelt, Ariz., for 1903.

[Drainage area, 1,030 square miles.]

Month.	Discharge in second-feet.			Total in acre-feet.	Run-off.	
	Maximum.	Minimum.	Mean.		Second-feet per square mile.	Depth in inches.
January	50	5	10	615	0.010	0.012
February	295	5	124	6,887	.120	.125
March	660	11	85	5,226	.083	.096
April	145	27	25	1,488	.024	.027
May	27	27	27	1,660	.026	.030
June	27	27	27	1,607	.026	.029
July	145	4	28	1,722	.027	.031
August	570	1	70	4,304	.068	.078
September	830	1	57	3,392	.055	.061
October	20	4	6	369	.006	.007
November	4	4	4	238	.004	.004
December	4	4	4	246	.004	.005
The year	830	1	39	27,754	.038	.505

NOTE.—Owing to the shifting character of the bed of the creek the daily discharges were estimated from the discharge measurements, taking into account the time interval between measurements and the change in section. A rating table was used from August 8 to December 31.

GILA RIVER AT SAN CARLOS, ARIZ.

The general character of the country through which the Gila River flows is a high and rolling plateau, with the river flowing through it in a deep canyon, and with practically no agricultural lands within its area. The river emerges from its upper canyon about 10 miles before it reaches the Arizona line, and thence flows through a valley of considerable width, known as Duncan Valley, until just before it receives the waters of San Francisco River.

Gila River is in a canyon for about 20 miles below the mouth of the San Francisco, or to within 10 miles of Solomonsville. At this point the hills separate, forming a large valley, which has been extensively settled and is now one of the finest irrigated portions of the Territory. This valley extends from a point 10 miles above Solomonsville to 6 miles below the mouth of San Carlos River, on the White Mountain Indian Reservation. At this latter place the mountains suddenly close in again and the river enters another canyon. Seven miles below the Indian agency at San Carlos the canyon narrows to a width of 100 feet, and at this point is located the San Carlos dam site, which was studied by the United States Geological Survey during 1899, in connection with the investigation of the water supply of Gila River. The results of this investigation are published in Water-Supply and Irrigation Paper No. 33, entitled "Storage of Water on Gila River, Arizona," by J. B. Lippincott. In connection with this investigation Cyrus C. Babb, on July 11, 1899, established a station on Gila River one-half mile south of the Indian agency at San Carlos and below the mouth of San Carlos Creek, where an inclined rod securely fastened to posts driven into the bank was erected. The original bench mark was a 20-penny nail in the base of a mesquite tree 5 inches in diameter, 85 feet west of the gage rod, at an elevation of 12.67 feet above gage datum. It was destroyed by a flood in the spring of 1903. A new bench mark was established May 9, 1903, by C. R. Olberg. It consists of a nail head in the west side of the post which supports the cable on the right bank of the river. Its elevation is 14.54 feet above the zero of the gage. Discharge measurements are made from a cable and car a short distance above the gage rod. The channel is straight for some distance above and below the station, and the water is comparatively swift. The right bank is high, but the left is low and liable to overflow. The bed of the stream is sandy and shifting.

The observations at this station during 1903 have been made under the direction of W. A. Farish, district hydrographer.

ean daily gage height, in feet, of Gila River at San Carlos, Ariz., for 1903.

Day.	Jan.	Feb.	Mar.	Apr.	May.	June.	July.	Aug.	Sept.	Oct.	Nov.	Dec.
....	2.20	1.80	1.65	1.95	1.40	1.30	(a)	(a)	1.90	2.80	1.30	1.40
....	2.20	1.80	1.65	1.90	1.40	1.30	(a)	(a)	1.85	2.05	1.30	1.40
....	2.20	1.80	1.65	1.85	1.40	1.30	(a)	(a)	1.75	2.00	1.30	1.40
....	2.20	1.80	1.65	1.80	1.40	1.30	(a)	(a)	1.68	1.90	1.30	1.40
....	2.10	1.80	1.65	1.90	1.40	1.30	(a)	(a)	1.55	1.90	1.30	1.40
....	2.10	1.80	1.65	2.10	1.40	1.30	(a)	(a)	2.05	1.78	1.30	1.40
....	2.10	1.80	1.60	2.10	1.40	1.30	(a)	(a)	1.75	1.68	1.30	1.40
....	2.10	1.80	1.60	2.05	1.40	1.30	(a)	2.80	1.70	1.62	1.30	1.40
....	2.00	1.80	1.60	1.90	1.40	1.30	(a)	3.30	1.70	1.58	1.30	1.40
....	2.00	1.80	1.60	1.80	1.40	1.30	(a)	2.85	1.60	1.50	1.30	1.40
....	2.00	1.80	1.55	1.75	1.40	2.50	(a)	2.50	1.60	1.48	1.30	1.38
....	2.00	1.80	1.55	1.70	1.40	2.50	(a)	3.07	1.50	1.45	1.40	1.35
.. ..	2.00	1.80	1.50	1.70	1.40	2.55	(a)	3.55	1.50	1.45	1.40	1.35
....	2.00	1.80	1.50	1.70	1.35	2.45	(a)	2.55	1.50	1.42	1.40	1.35
....	2.00	1.78	1.50	1.70	1.35	2.80	(a)	3.90	1.50	1.40	1.40	1.35
....	2.00	1.72	1.50	1.60	1.35	2.50	2.45	3.85	1.40	1.40	1.40	1.35
....	2.00	1.70	1.50	1.60	1.35	2.32	2.65	3.80	1.40	1.35	1.40	1.35
....	2.00	1.70	1.50	1.55	1.35	2.10	2.05	2.90	1.40	1.32	1.40	1.35
....	2.00	1.70	1.45	1.55	1.35	2.00	1.80	2.50	1.30	1.30	1.40	1.35
....	2.00	1.70	1.45	1.55	1.35	1.85	1.90	2.50	1.30	1.30	1.40	1.35
....	2.00	1.70	1.45	1.50	1.35	1.80	2.25	2.27	1.20	1.30	1.40	1.35
....	2.00	1.70	1.45	1.45	1.30	1.65	2.00	2.00	1.20	1.30	1.40	1.35
....	1.95	1.65	(a)	1.40	1.30	1.50	1.62	2.55	1.20	1.30	1.40	1.35
....	1.90	1.65	(a)	1.40	1.30	1.45	1.32	2.00	1.60	1.30	1.40	1.35
....	1.90	1.65	1.50	1.40	1.30	1.45	1.30	2.00	1.40	1.30	1.40	1.35
....	1.80	1.65	1.90	1.40	1.30	1.40	(a)	2.20	2.60	1.30	1.40	1.35
....	1.80	1.65	1.90	1.40	1.30	1.30	(a)	1.95	3.20	1.30	1.40	1.35
....	1.80	......	1.80	1.40	1.30	1.20	(a)	1.85	3.70	1.30	1.40	1.35
....	1.80	......	1.75	1.40	1.30	1.00	(a)	3.20	2.70	1.30	1.40	1.35
....	1.80	......	1.65	......	1.30	......	(a)	2.35	......	1.30	......	1.35

a No water at gage rod.

.... S. Janus.....................

.....do

2.4

1.30

July 12......................		....	----------	
July 18.....................	.....do	....		551
August 9....................	.....do	....	3.40	2,227
August 13...................	.....do	....	3.80	3,522
August 31...................	.....do	....	2.30	499
September 21	.....do	....	1.30	28
September 28	.....do	....	2.40	223
October 3...................	R. H. Ross	....	2.00	326
October 12..................	.....do	....	1.45	77
October 19..................	.....do	....	1.30	44
October 26.....	.....do	....	1.30	40
November 6	.....do	....	1.30	82
November 13	.....do	....	1.40	49
November 23	.....do	....	1.40	50
November 30	.....do	....	1.40	40
December 13.................	.....do	....	1.35	41
December 27.................	.....do	....	1.35	23

Estimated monthly discharge of Gila River at San Carlos, Ariz., for 1903.

[Drainage area, 13,455 square miles.]

Month.	Discharge in second-feet.			Total in acre-feet.	Run-off.	
	Maximum.	Minimum.	Mean.		Second-feet per square mile.	Depth in inches.
nuary	200	74	141	8,670	0.0105	0.0121
bruary	70	52	58	3,221	.0043	.0045
rch..............	96	0	37	2,275	.0028	.0032
ril..............	150	12	55	3,273	.0041	.0046
y................	12	1	4	246	.0003	.0003
ne	760	1	116	6,902	.0086	.0095
y................	620	0	52	3,197	.0039	.0045
gust	3,695	0	877	53,924	.0652	.0752
tember..........	3,170	20	281	16,721	.0209	.0232
ober	560	40	118	7,256	.0088	.0101
vember	82	40	57	3,392	.0042	.0047
cember	41	23	34	2,090	.0025	.0029
The year	3,695	0	152	111,167	.0113	.1548

NOTE.—Owing to the shifting character of the bed of the river the daily discharges were obtained m rating tables extending over brief periods of time, and by interpolation.

ANIMAS RIVER AT DURANGO, COLO.

The station was first established June 20, 1895, and has been main-ined during the greater part of each year since.

The original gage was located at the old wagon bridge, one-fourth ile west of the railroad bridge at Durango and about 200 feet above e Rio Grande Southern Railroad bridge. It was spiked to the est side of the south end of the middle pier of the wagon bridge. e head of a bolt at the east abutment of the railroad bridge is 17.24 et above this gage datum. During the early part of 1899 the old gon bridge was removed and a new one erected a short distance low. April 1, 1899, on the central pier of this bridge a new rod, hich consisted of a vertical piece of timber graduated to feet and nths, was fastened. The bench marks were three horizontal strips, posite the 10, 14, and 16.7 foot marks of the rod, respectively. wing to this change in location and height of the rod there is no parent relation between the rating tables before 1899 and the rating bles for 1899 and after.

The present gage was established June 20, 1901, by A. P. Davis, ssisted by F. Cogswell. It is located at the new wagon bridge one-ourth mile west of the railroad station at Durango, Colo. It is a ertical 2 by 6 inch timber 14 feet long. It is fastened to the masonry pier on the downstream side of the bridge, and is held in place by means of spikes driven into cracks in the masonry. It is read twice each day by C. G. Graden. On April 1–2, 1903, sand and gravel were washed down by Lightner Creek, which enters about 100 feet below the bridge, and were deposited around the gage rod. A new wire gage was established April 14, 1903, on the downstream side of the bridge in the span next the left bank. This gage was read until June 6, when it was stolen. In the meantime the sand and gravel had been washed away from the foot of the old gage, from which readings were taken for the remainder of the year. The wire gage read zero when the vertical gage read 5 feet. All 1903 readings have been reduced to the datum of the vertical gage. The 1903 rating table was prepared from measurements made after the deposit of sand and gravel had been scoured from around the vertical gage. It is therefore not applicable to gage heights taken previous to May 1, during which period the section at the gage was filled in. Discharge measurements are made from the bridge to which the gage is attached, except at low water, when they are made by wading. The initial point for soundings is the downstream edge of the right abutment. The channel is straight for 300 feet above and 400 feet below the station. The current is rough and has a rapid velocity. Both banks are high, rocky, and not liable to overflow. The bed of the stream is rocky and fairly permanent. Bench mark No. 1, established June 25, 1900, is a chiseled point at the southwest corner of the center pier of the highway bridge. Its elevation is 16.75 feet above the zero of the gage. Bench mark No. 2, established on the same date, is a similar point on the lower side of the left abutment. Its elevation is 16.84 feet above the zero of the gage. This relation was verified April 17, 1901.

The observations at this station during 1903 have been made under the direction of M. C. Hinderlider, district hydrographer.

Discharge measurements of Animas River at Durango, Colo., in 1903.

Date.	Hydrographer.	Gage height.	Discharge.
		Feet.	*Second-feet.*
April 14	F. Cogswell	7.90	484
May 15	do	10.70	4,208
June 6	do	10.05	3,766
June 27	do	10.20	4,051
July 27	do	8.05	1,231
August 17	do	7.20	455
September 8	do	7.85	1,008
October 5	do	7.05	319
November 2	do	6.75	195

Mean daily gage height, in feet, of Animas River at Durango, Colo., for 1903.

Day.	Mar.	Apr.	May.	June.	July.	Aug.	Sept.	Oct.
......	......	6.90	9.10	10.60	9.90	7.70	7.30	7.20
......	......	a 8.20	9.40	10.60	9.80	7.70	7.30	7.20
......	......	8.00	9.50	10.70	9.70	7.60	7.30	7.20
......	......	7.80	9.70	10.50	9.70	7.60	7.20	7.10
......	......	7.80	9.70	10.30	9.50	7.60	7.30	7.10
......	......	7.80	9.90	10.00	9.20	7.50	7.50	7.10
......	......	7.80	10.00	10.20	9.50	7.50	7.50	7.10
......	......	7.80	10.10	10.20	9.50	7.50	7.30	7.10
......	......	7.80	10.20	10.30	9.40	7.50	7.30	7.10
......	......	8.00	10.20	10.30	9.50	7.40	7.40	7.10
......	......	8.00	10.30	10.00	9.40	7.30	7.40	7.10
......	......	8.00	10.40	10.00	9.50	7.30	7.40	7.10
......	......	8.00	10.50	10.30	9.20	7.30	7.40	7.10
......	......	b 8.00	10.70	10.20	9.20	7.30	7.50	7.10
......	......	8.00	10.70	10.30	9.30	7.20	7.50	7.10
......	......	8.10	10.50	10.30	a 10.25	7.20	7.50	7.10
......	......	8.30	10.40	10.50	9.80	7.20	7.50	7.10
......	......	8.20	9.90	10.70	9.20	7.20	7.50	7.10
......	......	8.00	9.40	10.85	9.20	7.20	7.40	7.10
......	......	8.10	9.30	10.70	9.00	7.20	7.40	7.10
......	......	8.30	9.10	10.40	8.90	7.20	7.30	7.10
......	......	8.50	8.70	10.30	8.50	7.30	7.30	7.10
......	......	8.80	8.60	10.30	8.40	7.30	7.20	7.00
......	......	9.20	8.70	10.30	8.40	7.30	7.20	7.00
......	......	9.40	8.80	10.40	8.30	7.30	7.20	7.00
......	......	9.40	8.80	10.40	8.00	7.30	7.20	7.00
......	......	9.30	8.90	10.30	8.00	7.30	7.10	7.00
......	......	9.20	8.90	10.30	7.80	7.30	7.30	7.00
......	......	8.90	9.00	10.20	7.80	7.30	7.30	6.90
......	......	8.80	9.40	10.20	7.60	7.20	7.30	6.90
......	6.50	......	10.00	......	7.70	7.20	......	6.90

a Heavy rain.   b New gage; 5.00 on new gage = 10.00 on old gage.

Rating table for Animas River at Durango, Colo., from May 1 to December 31, 1903.

Gage height.	Discharge.	Gage height.	Discharge.	Gage height.	Discharge.	Gage height.	Discharge.
Feet.	*Second-feet.*	*Feet.*	*Second-feet.*	*Feet.*	*Second-feet.*	*Feet.*	*Second-feet.*
6.5	122	7.5	672	8.5	1,810	9.5	3,060
6.6	150	7.6	761	8.6	1,935	9.6	3,185
6.7	179	7.7	854	8.7	2,060	9.7	3,310
6.8	213	7.8	954	8.8	2,185	9.8	3,435
6.9	257	7.9	1,065	8.9	2,310	9.9	3,560
7.0	306	8.0	1,185	9.0	2,435	10.0	3,685
7.1	362	8.1	1,310	9.1	2,560	10.2	3,935
7.2	429	8.2	1,435	9.2	2,685	10.4	4,185
7.3	504	8.3	1,560	9.3	2,810	10.6	4,435
7.4	586	8.4	1,685	9.4	2,935	10.8	4,685

Curve well defined except between 8.10 and 10 feet gage height.

Estimated monthly discharge of Animas River at Durango, Colo., for 1903.

[Drainage area, 812 square miles.]

Month.	Discharge in second-feet.			Total in acre-feet.	Run-off.	
	Maximum.	Minimum.	Mean.		Second-feet per square mile.	Depth inches
May	4,560	1,935	3,241	199,281	3.99	4.
June	4,745	3,685	4,129	245,692	5.09	5.
July	3,997	761	2,446	150,399	3.01	3.
August	854	429	554	34,064	.68	.
September..........	672	362	542	32,251	.66	.
October	429	257	347	21,336	.43	.
The period ...	....	....	....	683,023	....	.....

ANIMAS RIVER AT SILVERTON, COLO.

This station was established June 4, 1903, by F. Cogswell. It
located at the East Fourteenth Street bridge, one-fourth mile north
the Denver and Rio Grande Railroad station. The gage is a 1-in
vertical board spiked to the downstream side of the rock-filled cr
pier near the left bank. The lower end of the rod reads 2 fee
During 1902 it was read twice each day by Andy Coyle. Dischar
measurements are made from the bridge to which the gage is attacho
The initial point for soundings is the end of the bridge on the le
bank. The channel is straight for 500 feet above and below t
station. The current is swift. There is one channel at low wat
and three at high water. At a gage reading of 4 feet the gaging st
tion has a width of 53 feet. Both banks are low and liable to ove
flow. Cement Creek enters Animas River about 800 feet below t
gage on the left side. Red Mountain Creek is tributary to Anin
River on the left side about three-fourths of a mile below the gagi
station. No bench marks have been established, as the station v
abandoned October 31, 1903.

The observations at this station during 1903 have been made und
the direction of M. C. Hinderlider, district hydrographer.

Discharge measurements of Animas River at Silverton, Colo., in 1903.

Date.	Hydrographer.	Gage height.	Discharge.
		Feet.	*Second-feet.*
June 4	F. Cogswell	3.90	527
June 25	do	4.35	847
July 24	do	3.30	216
August 15	do	2.90	65
September 9	do	3.10	98
October 7	do	2.80	44
November 4	do	2.70	51

Mean daily gage height, in feet, of Animas River at Silverton, Colo., for 1903.

Day.	June.	July.	Aug.	Sept.	Oct.	Day.	June.	July.	Aug.	Sept.	Oct.
1		4.30	3.15	2.80	2.90	17	4.55	3.85	2.95	2.90	2.80
2		4.15	3.05	2.85	2.90	18	4.45	3.60	2.90	2.95	2.80
3		3.85	3.15	2.90	2.90	19	4.45	3.50	2.90	2.90	2.85
4	4.00	3.80	3.05	2.85	2.90	20	4.35	3.50	2.90	2.90	2.80
5	4.05	3.80	3.05	2.90	2.90	21	4.50	3.45	2.90	2.90	2.75
6	4.15	3.65	3.05	3.00	2.75	22	4.30	3.40	2.90	2.90	2.75
7	4.30	4.10	3.00	3.15	2.70	23	4.50	3.40	2.90	2.90	2.70
8	4.05	4.00	3.00	3.10	2.85	24	4.50	3.40	3.05	2.90	2.70
9	3.95	3.90	3.00	3.05	2.80	25	4.45	3.30	3.00	2.90	2.70
10	3.95	3.90	3.00	3.00	2.80	26	4.40	3.30	3.00	2.90	2.70
11	3.95	3.85	3.00	3.00	2.80	27	4.35	3.30	2.90	2.90	2.70
12	4.05	3.75	3.00	3.05	2.80	28	4.40	3.25	2.90	2.90	2.60
13	4.20	3.60	2.95	3.00	2.80	29	4.45	3.25	2.90	2.90	2.65
14	4.25	3.60	3.00	3.00	2.80	30	4.35	3.20	2.85	2.90	2.75
15	ᵃ4.40	3.55	3.00	3.00	2.80	31		3.15	2.80		2.70
16	4.15	3.50	2.95	2.90	2.80						

ᵃ Rain.

Rating table for Animas River at Silverton, Colo., from June 4 to December 31, 1903.

Gage height.	Discharge.	Gage height.	Discharge.	Gage height.	Discharge.	Gage height.	Discharge.
Feet.	*Second-feet.*	*Feet.*	*Second-feet.*	*Feet.*	*Second-feet.*	*Feet.*	*Second-feet.*
2.6	15	3.2	159	3.8	460	4.4	883
2.7	27	3.3	201	3.9	528	4.5	954
2.8	43	3.4	246	4.0	599	4.6	1,025
2.9	64	3.5	295	4.1	670		
3.0	90	3.6	346	4.2	741		
3.1	122	3.7	400	4.3	812		

Estimated monthly discharge of Animas River at Silverton, Colo., for 1903.

[Drainage area, 66 square miles.]

Month.	Discharge in second-feet.			Total in acre-feet.	Run-off.	
	Maximum.	Minimum.	Mean.		Second-feet per square mile.	Depth in inches.
June 4–30..........	990	564	795	42,575	12.06	12.14
July	812	140	382	23,488	5.79	6.67
August	140	43	84	5,165	1.27	1.46
September..........	140	43	75	4,463	1.14	1.27
October	64	15	41	2,521	.62	.71
The period ...	....	....	....	78,212	....	....

FLORIDA RIVER NEAR DURANGO, COLO.

During 1899 work was being done on a large storage project near the head of the river, and on this account a gaging station was established by A. L. Fellows, May 19, 1899.

Information derived at this point is of importance, as it is desired that the excess water of this stream shall be made available by means of storage reservoirs in the upper part of this drainage basin. Much valuable land can be irrigated if this water is properly conserved. The gage is located at a wagon bridge at Stewart's ranch, which is about 6½ miles east of Durango, and is reached by driving. The gage rod consists of a vertical 4-inch strip, 7½ feet long, graduated to feet and tenths and spiked to the east abutment of the bridge. The station was discontinued during 1900, but daily observations were resumed April 1, 1901. It is read twice each day by Mrs. Annie Stewart. Discharge measurements are made from the downstream side of the single-span highway bridge to which the gage is attached. The bridge makes an angle of 75° with the direction of the current, and is marked on the downstream side at intervals of 5 feet. The initial point for soundings is the inside edge of the abutment on the right bank, downstream side of the bridge. The channel is straight for 200 feet above and below the station. The right bank is high and will overflow only at very high water. The left bank is low and subject to overflow. Both banks are covered with brush. The bed of the stream is composed of gravel and bowlders, and is probably permanent. The bench mark established and verified April 15, 1901, consists of a spike in an 8-inch cottonwood stump 50 feet south of the gage rod. Its elevation is 5.86 feet above the zero of the gage.

The observations at this station during 1903 have been made under the direction of M. C. Hinderlider, district hydrographer.

Discharge measurements of Florida River near Durango, Colo., in 1903.

Date.	Hydrographer.	Gage height.	Discharge.
		Feet.	Second-feet.
May 15	F. Cogswell	3.80	844
June 6	do	3.40	716
June 27	do	3.25	677
July 27	do	1.60	95
August 17	do	1.00	25
September 8	do	2.05	182
October 5	do	1.20	33
November 2	do	.70	6

Mean daily gage height, in feet, of Florida River near Durango, Colo., for 1903.

Day.	Apr.	May.	June.	July.	Aug.	Sept.	Oct.
1	2.10	2.60	3.75	2.70	1.00	0.80	1.10
2	2.00	2.70	3.85	2.50	.96	.96	1.15
3	1.35	2.75	3.40	2.40	.95	1.10	1.10
4	1.30	2.80	3.25	2.10	.90	1.00	1.15
5	1.50	2.85	3.10	2.00	.85	.95	1.20
6	1.40	2.90	3.40	1.75	.90	1.45	1.15
7	1.45	3.05	3.40	1.75	.85	1.95	1.10
8	1.65	3.05	3.40	1.95	.75	2.00	1.10
9	1.90	3.15	3.40	1.85	.75	1.70	1.10
10	1.95	3.20	3.25	1.85	.75	1.50	1.05
11	1.90	3.30	3.35	1.70	.70	1.40	1.00
12	1.80	3.45	3.50	1.65	.80	1.60	1.00
13	1.75	3.60	3.65	1.50	.75	1.50	.90
14	1.85	3.70	3.80	1.50	.70	1.45	.90
15	1.90	3.70	3.65	1.45	.85	1.40	.90
16	1.85	3.50	3.65	1.85	.85	1.30	.90
17	1.85	3.25	3.75	a 2.05	.95	1.30	1.15
18	1.85	2.95	3.90	2.10	.90	1.20	1.40
19	1.85	2.60	3.60	1.85	.90	1.20	1.35
20	2.00	2.20	3.60	1.80	.95	1.15	1.30
21	2.10	2.15	3.50	1.80	.90	1.10	1.25
22	2.25	2.15	3.50	1.65	.85	1.10	1.20
23	2.45	2.15	3.60	1.55	1.00	1.10	1.00
24	2.50	2.15	3.60	1.40	1.10	1.10	.90
25	2.60	2.15	3.50	1.25	1.05	1.10	.90
26	2.65	2.25	3.40	1.15	1.05	1.00	.80
27	2.60	2.45	3.35	1.55	1.05	1.05	.80
28	2.55	2.35	3.15	1.45	.95	1.05	.70
29	2.45	2.30	3.00	1.25	.90	1.15	.70
30	2.45	2.90	2.90	1.15	.85	1.10	.60
31		3.30		1.05	.70		.60

a Rain.

Rating table for Florida River near Durango, Colo., from January 1 to December 31, 1903.

Gage height.	Discharge.	Gage height.	Discharge.	Gage height.	Discharge.	Gage height.	Discharge.
Feet.	*Second-feet.*	*Feet.*	*Second-feet.*	*Feet.*	*Second-feet.*	*Feet.*	*Second-feet.*
0.6	5	1.5	79	2.4	269	3.3	645
.7	6	1.6	95	2.5	300	3.4	697
.8	9	1.7	112	2.6	336	3.5	751
.9	14	1.8	129	2.7	375	3.6	809
1.0	21	1.9	148	2.8	417	3.7	872
1.1	29	2.0	169	2.9	460	3.8	941
1.2	39	2.1	191	3.0	504	3.9	1,012
1.3	51	2.2	215	3.1	549		
1.4	64	2.3	241	3.2	596		

Curve fairly well defined.

Estimated monthly discharge of Florida River near Durango, Colo., for 1903.

[Drainage area, 136 square miles.]

Month.	Discharge in second-feet.			Total in acre-feet.	Run-off.	
	Maximum.	Minimum.	Mean.		Second-feet per square mile.	Depth in inches.
April............	356	51	181	10,770	1.33	1.4
May	872	203	455	27,977	3.34	3.8
June	1,012	460	745	44,331	5.48	6.1
July	375	25	127	7,809	.93	1.0
August	29	6	15	922	.11	.1
September........	169	9	52	3,094	.38	.4
October	64	5	25	1,537	.18	.2
The period....				96,440		

LOS PINOS RIVER AT IGNACIO, COLO.

The station was established April 22, 1899, by A. L. Fellows, at the request of the Commissioner of Indian Affairs, for the purpose of ascertaining the quantity of water available for irrigation along the stream. It is located at the wagon bridge at Ignacio, the subagency of the Southern Ute Indian Reservation, 2 miles north of the Denver and Rio Grande Railroad station. The rod is a vertical 2 by 4 inch timber, 10 feet long, spiked to the bridge, the marks being strips of brass securely nailed to the post.

The gage is read twice each day by John Wesch, the clerk at the Indian agency. Discharge measurements are made from the downstream side of the 4-span highway bridge, to which the gage is attached. The initial point for soundings is the end of the right abutment at the downstream side of the bridge. The channel is straight for a considerable distance above and below the station, and is 114 feet wide between bridge abutments. The current is swift. The banks are not high, but are not liable to overflow. The bed of the stream is composed of gravel, free from vegetation, and fairly permanent. Bench mark No. 1 is the 8-foot mark on the gage rod, which is level with the top of the lower end of a 6 by 8 inch timber protruding downstream from the pier on the right-hand side. Bench mark No. 2, placed April 7, is an iron bench-mark post set 30 feet northwest from the northwest corner of the bridge, its top being 7.64 feet above the zero of the gage.

The observations at this station during 1903 have been made under the direction of M. C. Hinderlider, district hydrographer.

Discharge measurements of Los Pinos River at Ignacio, Colo., in 1903.

Date.	Hydrographer.	Gage height.	Discharge.
		Feet.	*Second-feet.*
April 13	F. Cogswell	3.20	385
May 16	....do	5.20	2,239
June 5	....do	4.90	2,059
June 26	....do	5.10	2,245
July 25	....do	3.10	426
August 18	....do	2.20	87
September 7	....do	3.95	1,079
October 6	....do	2.60	177
November 3	....do	2.30	93

Mean daily gage height, in feet, of Los Pinos River at Ignacio, Colo., for 1903.

Day.	Apr.	May.	June.	July.	Aug.	Sept.	Oct.
1		4.00	5.65	4.60	2.55	2.55	12
2		4.05	5.70	4.55	2.55	2.55	1?
3		4.25	5.80	4.55	2.55	2.80	12
4		4.40	5.80	4.20	2.55	2.70	12
5		4.45	4.90	3.95	2.50	2.65	16
6		4.65	5.20	3.85	2.50	2.65	14
7		5.05	5.20	2.80	2.50	2.60	14
8		5.15	5.20	3.85	2.45	2.65	14
9		5.20	5.02	2.70	2.40	2.50	14
10		5.20	4.95	2.70	2.40	2.15	14
11		5.90	6.15	3.60	2.40	2.60	14
12		5.50	5.20	2.55	2.40	2.10	14
13	3.25	5.90	5.15	3.45	2.40	2.65	24
14	3.90	5.45	5.20	3.45	2.35	3.05	14
15	3.25	5.55	5.15	2.45	2.85	2.90	12
16	3.90	5.25	5.15	3.75	2.35	2.70	14
17	3.40	4.95	5.40	3.65	2.30	2.80	14
18	3.30	4.85	5.60	3.75	2.30	2.80	14
19	3.85	4.20	5.40	3.50	2.25	2.75	14
20	3.33	4.05	5.20	(a)	2.20	2.70	14
21	3.40	4.05	5.25	....	2.20	2.70	14
22	3.50	3.65	5.15	....	2.40	2.70	14
23	3.73	3.85	5.40	....	2.50	2.65	14
24	3.95	3.75	5.15	....	2.60	2.65	14
25	4.20	3.75	5.15	(a)	2.55	2.60	14
26	4.20	3.85	5.10	....	3.05	2.60	14
27	4.10	3.90	5.05	....	3.20	2.65	14
28	4.00	3.85	4.65	....	2.95	2.60	14
29	3.85	3.95	4.70	....	2.80	2.65	14
30	3.80	4.55	4.65	....	2.65	2.70	14
31		5.35	....	(a)	2.55	....	14

a Observer absent.

Rating table for Los Pinos River at Ignacio, Colo., for 1903.

Gage height.	Discharge.	Gage height.	Discharge.	Gage height.	Discharge.	Gage height.	Discharge.
Feet.	*Second-feet.*	*Feet.*	*Second-feet.*	*Feet.*	*Second-feet.*	*Feet.*	*Second-feet.*
2.2	80	3.1	372	4.0	1,180	4.9	2,008
2.3	98	3.2	444	4.1	1,272	5.0	2,100
2.4	118	3.3	536	4.2	1,364	5.1	2,192
2.5	142	3.4	628	4.3	1,456	5.2	2,284
2.6	170	3.5	720	4.4	1,548	5.3	2,376
2.7	203	3.6	812	4.5	1,640	5.4	2,468
2.8	239	3.7	904	4.6	1,732	5.5	2,560
2.9	279	3.8	996	4.7	1,824	5.6	2,652
3.0	322	3.9	1,088	4.8	1,916	5.7	2,744

Estimated monthly discharge of Los Pinos River at Ignacio, Colo., for 1903.

[Drainage area, 450 square miles.]

Month.	Discharge in second-feet.			Total in acre-feet.	Run-off.	
	Maximum.	Minimum.	Mean.		Second-feet per square mile.	Depth in inches.
April 13–30	1,364	489	840	29,990	1.87	1.25
May	2,606	950	1,685	103,607	3.74	4.31
June	2,744	1,778	2,267	134,896	5.04	5.62
July 1–19 a	1,732	674	1,030	38,817	2.29	1.62
August	444	80	163	10,022	.36	.42
September...........	996	156	289	17,197	.64	.71
October	203	118	148	9,100	.33	.38
The period....	2,744	80	....	343,629	....	....

a July 20–31, observer absent.

DOLORES RIVER AT DOLORES, COLO.

This station was established June 25, 1895, by A. P. Davis, assisted by F. Cogswell. It is located at the footbridge, one-half mile east of Dolores, Colo. The gage is a vertical 2 by 6 inch board spiked and bolted to the left abutment of the bridge. It is read twice each day by Mrs. Mary D. Smith. Discharge measurements are made from the downstream side of the single-span footbridge to which the gage is attached. The initial point for soundings is the left abutment on the downstream side. The channel is straight for a considerable distance above and below the station. Both banks are high and are not liable to overflow. The bed of the stream is composed of gravel and is not subject to much change. The bench mark is a nail in the base of a cottonwood tree 18 feet southwest of the gage. Its elevation is 15.57 feet above the zero of the gage. This bench mark was verified April 15, 1901.

The observations at this station during 1903 have been made under the direction of M. C. Hinderlider, district hydrographer.

Discharge measurements of Dolores River at Dolores, Colo., in 1903.

Date.	Hydrographer.	Gage height.	Discharge.
		Feet.	Second-feet.
April 11	F. Cogswell	3.40	292
May 13	do	5.30	2,362
June 2	do	5.35	2,461
June 23	do	5.10	2,106
July 22	do	3.50	431
August 13	do	2.90	125
September 4	do	2.95	124
October 2	do	2.85	103
October 30	do	2.70	53

Mean daily gage height, in feet, of Dolores River at Dolores, Colo., for 1903.

Day.	Apr.	May.	June.	July.	Aug.	Sept.	Oct.
1	3.55	4.60	5.25	4.75	3.10	2.90	2.80
2	3.40	4.70	5.40	4.75	3.10	2.80	2.90
3	3.20	4.90	5.20	4.70	3.00	2.85	2.80
4	3.10	5.05	5.15	4.20	3.00	3.00	2.80
5	3.10	5.10	5.15	4.10	3.00	3.00	2.85
6	3.10	5.10	4.85	3.90	3.00	3.10	2.80
7	3.10	5.15	5.10	3.80	3.00	3.65	2.80
8	3.30	5.20	5.25	3.95	3.00	3.35	2.80
9	a3.55	5.30	5.25	4.00	3.00	3.15	2.80
10	3.55	5.20	5.40	3.95	3.00	3.10	2.80
11	3.50	5.25	5.40	3.75	3.00	3.00	2.80
12	3.40	5.30	5.45	3.70	3.00	3.10	2.80
13	3.25	5.40	5.60	3.60	2.95	3.10	2.80
14	3.45	5.50	5.65	3.60	2.90	3.05	2.80
15	3.50	5.35	5.40	3.55	2.90	3.00	2.80
16	3.50	5.30	6.35	3.95	2.90	3.00	2.80
17	3.60	4.50	5.55	4.00	2.90	3.00	2.70
18	3.50	4.30	5.60	3.85	2.90	2.95	2.70
19	3.40	4.30	5.35	3.55	2.90	2.90	2.70
20	3.50	4.35	5.10	3.50	2.90	2.90	2.70
21	3.70	4.45	5.10	3.50	2.90	2.80	2.70
22	4.00	4.25	5.10	3.50	2.90	2.80	2.70
23	4.10	4.15	5.05	3.40	2.90	2.80	2.70
24	4.30	4.15	5.10	3.40	2.90	2.90	2.70
25	4.45	4.25	5.05	3.30	2.90	2.90	2.70
26	4.60	4.55	5.00	3.25	2.90	2.85	2.70
27	4.60	4.45	5.00	3.40	2.95	2.80	2.70
28	4.45	4.35	4.85	3.30	2.90	2.80	2.70
29	4.40	4.50	4.80	3.20	2.90	2.80	2.70
30	4.25	4.80	4.85	3.20	2.90	2.80	2.70
31		5.15		3.10	2.90		2.65

a Rain.

Rating table for Dolores River at Dolores, Colo., for 1903.

Discharge.	Gage height.	Discharge.	Gage height.	Discharge.	Gage height.	Discharge.
Second-feet.	*Feet.*	*Second-feet.*	*Feet.*	*Second-feet.*	*Feet.*	*Second-feet.*
35	3.4	376	4.2	1,046	5.0	1,980
53	3.5	444	4.3	1,148	5.1	2,110
80	3.6	518	4.4	1,254	5.2	2,240
116	3.7	596	4.5	1,364	5.3	2,370
158	3.8	678	4.6	1,478	5.4	2,500
206	3.9	764	4.7	1,598	5.5	2,630
258	4.0	854	4.8	1,722	5.6	2,760
314	4.1	948	4.9	1,850	5.7	2,890

stimated monthly discharge of Dolores River at Dolores, Colo., for 1903.

[Drainage area, 524 square miles.]

nth.	Discharge in second-feet.			Total in acre-feet.	Run-off.	
	Maximum.	Minimum.	Mean.		Second-feet per square mile.	Depth in inches.
....	1,478	206	629	37,428	1.20	1.34
...,......	2,630	997	1,754	107,849	3.35	3.86
....	2,825	1,722	2,255	134,182	4.30	4.80
....	1,660	206	662	40,705	1.26	1.45
....	206	116	137	8,424	.26	.30
r..........	557	80	155	9,223	.30	.33
....	116	43	71	4,366	.14	.16
ə period ...	....	....	....	342,177	....	....

MISCELLANEOUS DISCHARGE MEASUREMENTS IN THE SAN JUAN RIV DRAINAGE BASIN, COLORADO.

The following miscellaneous discharge measurements were made the San Juan drainage basin in 1903:

Miscellaneous discharge measurements in the San Juan drainage basin, 1903

Date.	Stream.	Locality.	Gage height.	D cha
			Feet.	*Sec.*
June 4	Cement Creek	Silverton, Colo		1
25	....do	....do	2.10	1
July 24	....do	....do	1.60	
Aug. 15	....do	....do	1.45	
Sept. 5	....do	....do		
9	....do	....do	1.40	
Oct. 7	....do	....do	1.70	
Nov. 4	....do	....do	1.60	
Apr. 14	Lightner Creek	Durango, Colo		
May 15	....do	....do		
June 6	....do	....do		
27	....do	....do	1.80	
July 27	....do	....do	1.00	
Aug. 17	....do	....do		
Sept. 8	....do	....do		
Oct. 5	....do	....do		
Nov. 2	....do	....do		
June 4	Mineral Creek	Silverton, Colo		2
25	....do	....do	3.00	2
July 24	....do	....do	2.30	2
Aug. 15	....do	....do	1.90	
Sept. 9	....do	....do	2.10	2
Oct. 7	....do	....do	1.70	
Nov. 4	....do	....do	1.60	
Sept. 5	....do	....do		
5	Red Mountain Creek	Road bridge 2 miles above Ironton, Colo.		
	Animas River	At bridge one-half mile below How-ardsville.		

GUNNISON RIVER AT WHITEWATER. COLO.

This station was established by J. E. Field, April 10, 1902, at a wagon bridge constructed by the State of Colorado at a point ab half a mile above the railroad station at Whitewater, on the Den and Rio Grande Railroad. During 1895, 1897, and 1901 incompl series of gage heights were obtained at this point, but the stat was not regularly established until 1902. It was intended that station should take the place of the one formerly maintained on Gunnison at Grand Junction. The latter station was abandoned account of inaccuracies that could not be overcome. They w mainly due to the fact that high stages of water in Grand Ri

the gage rod in the Gunnison, and that the stream bed was
lled with great bowlders, making accurate gaging impossible.
e gage is attached to the guard rail on the downstream side of
ge in the middle of the right span. The length of the wire
.e end of the weight to the marker is 25.1 feet. It is read
ich day by James Page. Discharge measurements are made
e downstream side of the two-span highway bridge to which
? is attached. The initial point for soundings is the end of the
on the right bank. The channel is straight for several hun-
?t above and below the station. It is broken by the center
the bridge. The right bank is high, and not liable to over-
The left bank will overflow at very high stages. The bed of
am is composed of bowlders and sand, with stone riprap on
of the channel. There is a deposit of mud and silt at low
No bench marks have been established.
ibservations at this station during 1903 have been made under
ction of M. C. Hinderlider, district hydrographer.

scharge measurements of Gunnison River at Whitewater, Colo., in 1903.

Date.	Hydrographer.	Gage height.	Discharge.
		Feet.	*Second-feet.*
....	F. Cogswell....................	4. 90	1, 653
....	....do	9. 70	10, 164
....	....do	7. 95	6, 197
....	....do	11. 35	15, 111
....	....do	7. 30	4, 841
....	....do	4. 05	1, 006
r 1...................	....do	3. 90	872
r 29..................	....do	4. 00	864
4.....................	....do	3. 85	723
r 23	....do	4. 15	889

Mean daily gage height, in feet, of Gunnison River at Whitewater, Colo., for 1903.

Day.	Apr.	May.	June.	July.	Aug.	Sept.	Oct.	Nov.	Dec.
1	3.70	7.50	9.15	9.15	7.05	3.95	3.95	3.90	4.0
2	3.80	7.60	9.70	9.05	6.70	3.85	3.95	3.90	4.0
3	3.90	7.85	10.05	8.90	6.15	3.80	3.80	3.90	10
4	4.00	8.05	10.30	8.65	5.55	3.95	3.80	4.00	10
5	4.00	8.45	10.10	8.40	5.10	4.10	3.90	3.95	10
6	4.10	8.80	9.75	7.60	4.70	4.00	4.00	3.90	10
7	4.25	9.05	9.80	7.15	4.15	4.20	4.10	3.80	10
8	4.20	9.10	10.10	6.95	3.90	4.50	4.10	3.80	10
9	4.30	9.45	10.40	6.85	4.05	4.45	4.20	3.80	10
10	4.85	9.65	10.55	6.65	4.05	4.65	4.30	3.80	10
11	4.75	9.85	10.75	6.95	4.00	4.95	4.20	3.80	10
12	5.00	9.75	a10.85	7.00	4.00	5.30	4.20	3.90	10
13	4.90	9.75	11.50	6.95	4.00	5.35	4.10	3.90	10
14	4.65	10.10	12.05	6.80	4.00	5.10	4.00	4.00	10
15	4.70	10.60	12.05	6.80	4.00	5.00	4.00	3.90	10
16	4.75	10.75	12.10	6.90	4.00	4.90	4.00	3.90	10
17	4.90	11.00	11.95	7.10	4.00	4.85	3.95	3.80	10
18	4.90	10.50	11.75	7.40	4.00	4.80	4.00	3.80	10
19	5.00	9.45	11.45	7.05	4.15	4.70	4.10	3.80	10
20	5.10	8.85	11.00	6.95	4.30	4.70	4.10	3.70	10
21	5.30	8.25	10.80	6.80	4.20	4.60	4.00	3.65	10
22	5.50	8.10	10.50	6.65	4.25	4.60	4.00	3.70	10
23	5.50	8.00	10.35	6.85	4.35	4.50	3.90	3.90	10
24	6.35	7.70	10.10	6.95	4.15	4.40	3.80	4.15	10
25	6.85	7.40	9.80	7.25	3.95	4.30	3.80	4.20	10
26	7.45	7.50	9.70	b7.85	3.80	4.20	3.90	4.30	3.8
27	7.80	7.80	9.60	7.60	3.95	4.20	4.00	4.40	3.8
28	7.80	7.80	9.55	7.55	4.20	4.10	4.00	4.30	3.9
29	7.70	7.90	9.15	7.50	4.05	4.05	4.00	4.20	3.8
30	7.50	8.00	9.30	7.40	4.00	4.00	3.90	4.05	3.7
31		8.55		7.30	4.00		3.90		3.7

a Heavy rain.　　　　b Waterspout 10 miles east.

Rating table for Gunnison River at Whitewater, Colo., from April 1 to December 31, 1903

Gage height.	Discharge.	Gage height.	Discharge.	Gage height.	Discharge.	Gage height.	Discharge.
Feet.	*Second-feet.*	*Feet.*	*Second-feet.*	*Feet.*	*Second-feet.*	*Feet.*	*Second-feet.*
3.6	600	4.6	1,380	6.2	3,190	8.5	7,240
3.7	670	4.7	1,470	6.4	3,480	9.0	8,400
3.8	740	4.8	1,560	6.6	3,780	9.5	9,640
3.9	810	4.9	1,660	6.8	4,080	10.0	10,970
4.0	890	5.0	1,760	7.0	4,390	10.5	12,420
4.1	970	5.2	1,970	7.2	4,710	11.0	13,980
4.2	1,050	5.4	2,185	7.4	5,040	11.5	15,635
4.3	1,130	5.6	2,415	7.6	5,400	12.0	17,430
4.4	1,210	5.8	2,665	7.8	5,790	12.1	17,810
4.5	1,290	6.0	2,925	8.0	6,190		

Estimated monthly discharge of Gunnison River at Whitewater, Colo., for 1903.

[Drainage area, 7,868 square miles.]

Month.	Discharge in second-feet.			Total in acre-feet.	Run-off.	
	Maximum.	Minimum.	Mean.		Second-feet per square mile.	Depth in inches.
April	5,790	670	2,263	134,658	0.288	0.32
May	13,980	5,040	8,163	501,923	1.037	1.20
June	17,810	8,768	12,545	746,480	1.594	1.77
July	8,768	3,855	5,132	315,554	.652	.75
August	4,470	740	1,312	80,672	.167	.20
September	2,130	740	1,284	76,403	.163	.18
October	1,130	740	890	54,724	.113	.13
November	1,210	635	844	50,221	.107	.12
December	970	670	810	49,805	.103	.12
The period				2,010,440		

UNCOMPAHGRE RIVER AT DELTA, COLO.

This station was originally established April 29, 1903, by W. N. Sammis, assisted by F. Cogswell. The original station was located at a highway bridge one-fourth mile above the Denver and Rio Grande Railroad bridge. The discharge at this point did not include the mill-ditch waste. On November 17 the station was reestablished at the Denver and Rio Grande Railroad bridge one-fourth mile northwest from the Denver and Rio Grande Railroad station. The new location was selected for the reason that the discharge of the river at this point includes the mill-ditch waste. The new gage consists of a 2 by 4 inch vertical rod nailed to a 12-inch pile on the downstream side of the railroad bridge. It reads from 1 foot to 6 feet. During 1903 gage readings have been made twice each day by Michael O'Rourke. Up to November 17, 1903, discharge measurements were made at the highway bridge to which the old gage was attached, or by wading near the old gage. Discharge measurements are now made from the railroad bridge at which the new gage is located. The initial point for soundings is a chisel mark on the east side of the bridge. The channel is straight for 100 feet above and below the railroad bridge. The right bank is high. The left bank is low and liable to overflow. The bed of the stream is muddy and shifting. All water passes beneath the bridge at flood stages. The bench mark is a cross on the top of the lower cord of the bridge over gage rod. Its elevation is 8.75 feet above the zero of the gage at the railroad bridge.

The observations at this station during 1903 have been made under the direction of M. C. Hinderlider, district hydrographer.

4		3.70			2.90	2.90	2.75	2.60	2
5		3.95			2.95	2.85	2.70	2.00	2
6		4.05			2.95	2.90	2.70	2.00	
7		4.05			2.95	3.55	2.63	2.00	2
8		3.95			2.95	4.60	2.55	2.55	2
9		4.20			2.95	4.30	2.65	2.55	2
10		4.25			2.95	3.15	2.65	2.55	2
11					2.95			2.00	2
12		4.00	6.00	4.30	2.95	4.00	2.65	2.65	2
13		4.15	5.65	4.15	3.00	3.65	2.65	2.75	2
14		4.25	6.30	4.00	3.00	3.35	2.65	2.85	1
15		4.40	6.85	4.00	3.02	3.12	2.65	2.70	1
16		4.55	6.65	4.85	3.05	3.10	2.65	2.85	1
17		4.75	6.55	4.95	2.98	3.10	2.65	2.30	1
18		4.65	6.35	4.75	2.98	3.10	2.65	2.35	1
19		4.00	6.30	4.50	2.95	3.10	2.63	2.40	1
20		3.65	6.20	4.28	2.98	3.05	2.60	2.75	1
21		3.30	5.95	4.10	2.98	3.05	2.60	2.88	1
22		3.15	5.70	3.98	2.95	3.05	2.60	2.83	1
23		3.15	5.60	3.95	2.95	3.05	2.60	2.80	1
24		3.00	5.50	3.95	3.05	3.00	2.60	2.75	1
25		3.00	5.45	3.80	3.05	2.95	2.60	2.75	1
26		2.85	5.40	3.80	3.05	2.85	2.60	2.70	1
27		2.65	5.35	3.85	3.00	2.75	2.60	2.70	1
28		2.50	5.25	3.60	2.95	2.75	2.60	2.30	1
29	8.90	2.45	5.40	3.40	2.95	2.68	2.60	2.70	1
30	8.20	2.60	5.40	3.20	2.95	2.73	2.60	2.70	1
31		2.70	....	3.10	2.95	....	2.60	....	1

Rating table for Uncompahgre River at Delta, Colo., for May, 1903.

Discharge.	Gage height.	Discharge.	Gage height.	Discharge.	Gage height.	Discharge.
Second-feet.	*Feet.*	*Second-feet.*	*Feet.*	*Second-feet.*	*Feet.*	*Second-feet.*
13	3.1	37	3.8	136	4.5	390
15	3.2	44	3.9	165	4.6	440
17	3.3	53	4.0	197	4.7	495
19	3.4	64	4.1	231	4.8	555
22	3.5	77	4.2	267	4.9	620
26	3.6	93	4.3	305		
31	3.7	112	4.4	345		

ble for Uncompahgre River at Delta, Colo., from June 1 to November 16, 1903.

Discharge.	Gage height.	Discharge.	Gage height.	Discharge.	Gage height.	Discharge.
Second-feet.	*Feet.*	*Second-feet.*	*Feet.*	*Second-feet.*	*Feet.*	*Second-feet.*
1	3.7	113	4.8	555	5.9	1,368
2	3.8	136	4.9	620	6.0	1,451
5	3.9	165	5.0	685	6.1	1,535
9	4.0	197	5.1	750	6.2	1,621
15	4.1	231	5.2	818	6.3	1,709
23	4.2	267	5.3	889	6.4	1,800
33	4.3	305	5.4	963	6.5	1,892
45	4.4	345	5.5	1,041	6.6	1,985
59	4.5	390	5.6	1,122	6.7	2,078
75	4.6	440	5.7	1,204	6.8	2,172
93	4.7	495	5.8	1,286	6.9	2,266

d raised in fall by deposition of silt.

mated monthly discharge of Uncompahgre River at Delta, Colo., for 1903.

[Drainage area, 1,130 square miles.]

onth.	Discharge in second-feet.			Total in acre-feet.	Run-off.	
	Maximum.	Minimum.	Mean.		Second-feet per square mile.	Depth in inches.
....	525	14	158	9,715	0.140	0.16
....	2,219	52	937	55,755	.829	.93
....	854	23	309	19,000	.273	.31
....	19	12	14	861	.012	.01
r..........	440	2	51	3,035	.045	.04
....	11	1	2	123	.002	.00
r..........	123	0	40	2,380	.035	.04
•	150	90	118	7,256	.104	.12
e period				98,125	....	

UNCOMPAHGRE RIVER AT MONTROSE, COLO.

This station was established April 22, 1903, by A. L. Fellov
assisted by W. N. Sammis. It is located at the highway bridge w
of Montrose and one-fourth mile west of the Denver and Rio Grav
Railroad. The gage is a vertical 2 by 4 inch timber fastened to
wing dam 20 feet above the bridge. It is read twice each day
Thomas Allen. Discharge measurements are made from the high
bridge and its approaches on both banks. The initial point for sou
ings is at the lower cylinder of the bridge on the right bank. '
channel is straight for 300 feet above and below the station. Thei
but one channel at all stages, broken by two bridge piers at high wa
The right bank is high and composed of earth. The left bank is
and composed of gravel. Both banks are subject to overflow.
bed of the stream is composed of gravel. The bench mark consist
two nails driven into a blaze on the root of a cottonwood tree.
tree is 50 feet east of the rod. The elevation of the bench mai
9.965 feet above the zero of of the gage.

The observations at this station during 1903 have been made ui
the direction of M. C. Hinderlider, district hydrographer.

Discharge measurements of Uncompahgre River at Montrose, Colo., in 1903.

Date.	Hydrographer.	Gage height.	Disch
		Feet.	Secon
April 22	W. N. Sammis	3. 40	
April 24	do	4. 10	
May 4	do	4. 55	
May 11	do	4. 45	
May 21	do	3. 70	
June 3	do	4. 45	
June 10	do	6. 25	1
June 13	do	7. 25	2
July 14	do	4. 50	
August 7	do	3. 10	
August 21	do	2. 40	
October 19	do	2. 70	
October 21	E. C. Murphy	2. 68	
November 13	W. N. Sammis	2. 90	

laily gage height, in feet, of Uncompahgre River at Montrose, Colo., for 1903.

Day.	Apr.	May.	June.	July.	Aug.	Sept.	Oct.	Nov.	Dec.
....		3.90	4.75	5.60	2.85	2.20	3.00	2.75	3.10
....		4.00	4.90	5.50	2.85	2.15	3.05	2.60	3.15
....		4.35	4.60	5.15	2.65	3.00	3.15	2.60	3.15
....		4.45	4.70	4.50	2.80	3.20	3.20	2.60	2.95
....		4.65	4.70	4.50	2.85	2.40	3.05	2.55	2.65
....		4.60	5.10	4.50	2.95	a 3.00	2.85	2.65	2.85
....		4.60	5.25	4.70	3.00	4.20	2.60	2.73	2.95
....		4.55	5.50	4.85	2.95	3.95	2.45	2.70	3.10
....		4.75	5.50	4.40	2.90	3.60	2.65	2.50	2.90
....		4.45	6.10	4.95	2.80	3.40	2.70	2.40	2.90
....		4.35	5.85	4.90	2.60	a 3.45	2.55	2.40	2.95
....		4.35	6.50	4.75	2.75	3.95	2.55	2.40	3.10
....		4.60	7.15	5.00	2.50	3.55	2.50	2.30	3.05
....		5.00	6.90	4.35	2.55	3.40	2.55	2.30	3.15
....		5.25	6.60	5.65	2.65	3.40	2.55	2.50	3.15
....		5.40	6.45	5.25	2.60	3.35	2.55	2.55	3.40
....		5.50	6.40	5.75	2.45	3.30	2.30	2.55	3.35
....		4.70	6.45	5.20	2.55	3.15	2.35	2.35	3.20
....		4.10	6.20	4.75	2.30	3.45	2.10	2.50	3.10
....		3.80	6.40	4.50	2.45	3.20	2.30	3.00	3.05
....		3.70	6.10	4.35	2.25	3.00	2.40	3.10	3.20
....	3.68	3.60	6.30	4.50	2.35	3.05	2.30	2.90	2.95
....	3.90	3.55	6.25	4.35	2.45	2.85	2.30	2.85	2.85
....	4.30	3.45	6.00	4.35	2.75	2.90	2.30	2.85	2.75
....	4.30	3.45	6.00	4.10	2.55	2.90	2.35	2.85	2.85
....	4.55	3.60	5.85	4.20	2.35	2.85	2.30	3.00	2.90
....	4.20	3.40	5.85	4.05	2.40	2.70	2.40	2.90	2.90
....	4.40	3.35	6.00	4.25	2.30	2.80	2.40	2.90	2.80
....	3.81	3.30	6.10	3.70	2.40	a 3.10	2.35	2.85	2.80
....	3.65	3.35	5.95	3.10	2.35	2.95	2.40	3.00	2.80
....		3.60	....	2.80	2.20	....	2.55	....	2.90

a Rain.

able for Uncompahgre River at Montrose, Colo., from April 22 to May 18, 1903.

Discharge.	Gage height.	Discharge.	Gage height.	Discharge.	Gage height.	Discharge.
Second-feet.	Feet.	Second-feet.	Feet.	Second-feet.	Feet.	Second-feet.
130	4.2	281	4.8	462	5.4	668
152	4.3	309	4.9	495	5.5	706
176	4.4	338	5.0	529	5.6	744
201	4.5	368	5.1	563	5.7	782
227	4.6	399	5.2	597		
254	4.7	430	5.3	632		

Rating table for Uncompahgre River at Montrose, Colo., from May 19 to December 31, 1903

Gage height.	Discharge.	Gage height.	Discharge.	Gage height.	Discharge.	Gage height.	Discharge.
Feet.	*Second-feet.*	*Feet.*	*Second-feet.*	*Feet.*	*Second-feet.*	*Feet.*	*Second-feet.*
2.1	7	3.1	84	4.1	306	5.4	686
2.2	8	3.2	103	4.2	332	5.6	754
2.3	9	3.3	122	4.3	358	5.8	828
2.4	10	3.4	143	4.4	385	6.0	912
2.5	12	3.5	165	4.5	412	6.2	1,004
2.6	15	3.6	187	4.6	440	6.4	1,110
2.7	20	3.7	210	4.7	470	6.6	1,250
2.8	30	3.8	234	4.8	500	6.8	1,436
2.9	46	3.9	258	5.0	560	7.0	1,712
3.0	65	4.0	282	5.2	622	7.2	2,200

Estimated monthly discharge of Uncompahgre River at Montrose, Colo., for 1903.

[Drainage area, 565 square miles.]

Month.	Discharge in second-feet.			Total in acre-feet.	Run-off.	
	Maximum.	Minimum.	Mean.		Second-feet per square mile.	Depth in inches.
April 22–30	383	141	254	4,534	0.45	0.1
May	706	122	317	19,492	.56	.6
June	2,060	440	925	55,041	1.64	1.8
July	809	30	445	27,362	.79	.9
August	65	8	22	1,353	.04	.0
September	332	8	106	6,307	.19	.2
October	103	7	24	1,476	.04	.0
November	84	9	28	1,666	.05	.0
December	143	18	65	3,997	.12	..
The period				121,228		

UNCOMPAHGRE RIVER AT COLONA, COLO.

This station was originally established April 23, 1903, by W. ?
Sammis, assisted by F. Cogswell. It was located at a highway bridge
one-half mile northeast of Colona, Colo. The gage consisted of a ve
tical 2 by 4 inch timber spiked to the log abutment on the downstrea
side near the left bank. Gage readings were taken twice each day l
Chester Olsen, from April 26 to August 10, 1903. Discharge mea
urements were made from the two-span bridge to which the gage w
attached. The initial point for soundings was the face of the log abu
ment on the left bank, at the downstream side. The channel is straig

For 100 feet above and 200 feet below the bridge, and the current has a well-distributed velocity. Both banks are wooded and will overflow. The bench mark is a nail in a cottonwood tree 6 inches in diameter and 50 feet northwest of the gage. It is marked "B. M. U. S. G. S." Its elevation is 9.65 feet above the zero of the gage. On August 2, 1903, a dam was put in below the station for the purpose of diverting the water for irrigation. This backed up the water on the gage, and the station was discontinued August 10. It was reestablished on the same date by W. N. Sammis, assisted by I. W. McConnell, at Sam S. Kettle's bridge, about 1 mile south of Colona, Colo. The new gage is a vertical 2 by 4 inch timber nailed to the center pier of the two-span bridge, on the upstream side. It is read twice each day by Sam S. Kettle. Discharge measurements are made from the two-span wooden highway bridge at which the gage is located. The initial point for soundings is the face of the right abutment, on the downstream side of the bridge. The channel is straight for 100 feet above and for 50 feet below the station. The right bank is low and overflows at high water for a distance of 130 feet. The left bank is high and rocky and will not overflow. The bed of the stream is composed of gravel, free from vegetation, and is shifting. There are two channels at low and at high water. The bridge has a total span of about 60 feet between abutments. The bench mark is a nail driven in a blaze on a cottonwood tree. The tree is about 4 inches in diameter, and is about 75 feet northeast of the gage. The elevation of the bench mark is 4.49 feet above the zero of the gage.

The observations at this station during 1903 have been made under the direction of M. C. Hinderlider, district hydrographer.

Discharge measurements of Uncompahgre River at Colona, Colo., in 1903.

Date.	Hydrographer.	Gage height.	Discharge.
		Feet.	*Second-feet.*
April 25	W. N. Sammis	3.25	387
May 4	...do	3.40	480
May 11	...do	3.50	554
May 16	...do	3.85	901
May 23	...do	3.00	337
June 4	...do	3.70	764
June 11	...do	3.90	1,110
August 3	...do	2.60	284
August 10	...do	2.70	191
September 4	...do	2.60	220
Do	...do	2.70	206
October 20	...do	2.50	106
November 10	...do	2.50	99

Mean daily gage height, in feet, of Uncompahgre River at Colona, Colo., for 1903.

Day.	Apr.	May.	June.	July.	Aug.	Sept.	Oct.	Nov.	Dec.
1		3.15	3.80	2.90	2.20	2.50	2.60	2.50	2.50
2		3.30	3.60	2.85	2.20	2.50	2.60	2.50	2.50
3		3.30	3.75	2.75	2.20	a3.20	2.60	2.50	2.50
4		3.45	3.70	2.55	2.20	2.70	2.60	2.50	2.50
5		3.45	3.85	2.45	2.15	2.80	2.60	2.50	2.85
6		3.55	3.90	2.40	2.15	2.80	2.60	2.50	2.70
7		3.60	4.00	2.65	2.15	a3.65	2.60	2.50	2.75
8		3.55	4.00	2.65	2.10	3.00	2.60	2.50	2.75
9		3.60	4.00	2.70	2.10	2.80	2.60	2.50	2.80
10		3.60	4.00	2.65	b2.10	2.80	2.60	2.50	2.80
11		3.55	3.95	2.70	2.70	2.80	2.60	2.50	2.80
12		3.75	c4.25	2.65	2.00	2.90	2.60	2.50	2.80
13		3.80	c4.35	2.55	2.60	2.80	2.60	2.50	2.90
14		3.90	3.85	2.55	2.60	2.80	2.60	2.50	3.00
15		c4.05	3.95	c2.55	2.50	2.80	2.60	2.50	3.00
16		4.05	3.85	c2.75	2.50	2.80	2.55	2.50	3.00
17		3.90	3.85	c2.85	2.50	2.80	2.50	2.50	3.00
18		3.60	3.95	2.60	2.50	2.80	2.50	2.40	3.00
19		3.30	3.75	2.50	2.50	2.80	2.50	2.40	3.00
20		3.30	3.65	2.50	2.50	2.80	2.50	2.40	3.00
21		3.20	3.65	2.50	3.50	2.70	2.50	2.50	3.00
22		3.15	3.75	c2.80	2.50	2.70	2.50	2.50	2.90
23		3.00	3.55	c2.50	2.50	2.70	2.50	2.50	2.90
24		3.00	3.25	2.45	c2.70	2.70	2.50	2.50	2.90
25		3.00	3.25	2.45	2.60	2.70	2.50	2.50	3.00
26	3.30	3.00	3.25	2.45	2.60	2.70	2.50	2.50	3.30
27	3.30	3.00	3.35	2.35	2.60	2.60	2.50	2.50	3.00
28	3.35	2.95	3.25	2.35	2.60	2.60	2.50	2.50	3.00
29	3.30	3.10	3.25	2.30	2.60	2.60	2.50	2.50	3.00
30	3.15	3.30	3.20	2.30	2.60	2.60	2.50	2.50	3.30
31		3.85		2.30	2.60		2.50		3.00

a Heavy rain.
b Old station abandoned on account of unfavorable conditions. New station established about 1 mile upstream from old station. Observations at new station commenced on August 10.
c Rain.

Rating table for Uncompahgre River at Colona, Colo., from April 26 to June 30, 1903.

Gage height.	Discharge.	Gage height.	Discharge.	Gage height.	Discharge.	Gage height.	Discharge.
Feet.	Second-feet.	Feet.	Second-feet.	Feet.	Second-feet.	Feet.	Second-feet.
2.9	301	3.3	429	3.7	727	4.1	1,280
3.0	323	3.4	480	3.8	860	4.2	1,420
3.1	351	3.5	542	3.9	1,000	4.3	1,560
3.2	386	3.6	622	4.0	1,140	4.4	1,700

Table for station one-half mile northeast of Colona. This station was abandoned August 10, 1903, owing to water backing up on gage rod from dam below. Discharge for remainder of year were taken at Kettles Crossing, 1 mile above this station.

Rating table for Uncompahgre River at Colona, Colo., from July 1 to August 10, 1903.

Gage height.	Discharge.	Gage height.	Discharge.	Gage height.	Discharge.	Gage height.	Discharge.
Feet.	*Second-feet.*	*Feet.*	*Second-feet.*	*Feet.*	*Second-feet.*	*Feet.*	*Second-feet.*
2.0	190	2.4	231	2.8	283	3.2	386
2.1	199	2.5	243	2.9	301		
2.2	209	2.6	255	3.0	323		
2.3	220	2.7	268	3.1	351		

Discharges from July 1 to August 10, inclusive, are approximate. They depend upon a single measurement the gage height of which is somewhat approximate.

Rating table for Uncompahgre River at Kettles Crossing, near Colona, Colo., from August 11 to October 31, 1903.

Gage height.	Discharge.	Gage height.	Discharge.	Gage height.	Discharge.	Gage height.	Discharge.
Feet.	*Second-feet.*	*Feet.*	*Second-feet.*	*Feet.*	*Second-feet.*	*Feet.*	*Second-feet.*
2.2	30	2.6	149	3.0	341	3.4	533
2.3	46	2.7	197	3.1	389	3.5	581
2.4	69	2.8	245	3.2	437	3.6	629
2.5	102	2.9	293	3.3	485	3.7	677

Measurements were made between 2.50 and 2.70 feet gage height. Above and below these limits curve and table are approximate.

Estimated monthly discharge of Uncompahgre River at Colona, Colo., for 1903.

[Drainage area, 433 square miles.]

Month.	Discharge in second-feet.				Run-off.	
	Maximum.	Minimum.	Mean.	Total in acre-feet.	Second-feet per square mile.	Depth in inches.
April (5 days)	454	368	422	4,185	0.974	0.18
May	1,210	312	559	34,372	1.291	1.49
June	1,630	386	834	49,626	1.926	2.15
July	301	220	252	15,495	.582	.67
August	209	102	156	9,592	.360	.42
September	653	102	236	14,043	.545	.60
October	149	102	126	7,747	.291	.33
November	102	69	99	5,891	.229	.26
December	485	102	285	17,524	.658	.76
The period				158,475		

Mean daily gage height, in feet, of Uncompahgre River at Colona, Colo., for 1903.

Day.	Apr.	May.	June.	July.	Aug.	Sept.	Oct.	Nov.	De?
1		3.15	3.80	2.90	2.20	2.50	2.60	2.50	2.?
2		3.20	3.80	2.85	2.20	2.50	2.60	2.50	2.?
3		3.30	3.75	2.75	2.20	a 3.20	2.60	2.50	2.?
4		3.45	3.70	2.55	2.20	2.70	2.60	2.50	2.?
5		3.45	3.85	2.45	2.15	2.70	2.60	2.50	2.?
6		3.55	3.90	2.40	2.15	2.80	2.60	2.50	2.?
7		3.60	4.00	2.65	2.15	a 2.85	2.60	2.50	2.?
8		3.85	4.00	2.65	2.10	3.00	2.60	2.85	2.?
9		3.60	4.00	2.70	2.10	2.80	2.60	2.50	2.?
10		3.60	4.00	2.65	b 2.10	2.80	2.60	2.50	2.?
11		3.55	3.95	2.70	2.70	2.80	2.60	2.50	2.?
12		3.75	c 4.25	2.65	2.00	2.90	2.60	2.50	2.?
13		3.80	c 4.35	2.55	2.60	2.80	2.60	2.50	2.?
14		3.90	3.85	2.55	2.60	2.80	2.60	2.50	2.?
15		c 4.05	3.95	c 2.55	2.50	2.80	2.60	2.50	2.?
16		4.05	3.85	c 2.75	2.50	2.80	2.55	2.50	2.?
17		3.90	3.85	c 2.85	2.50	2.80	2.50	2.50	2.?
18		3.90	3.95	2.60	2.50	2.70	2.50	2.40	2.?
19		3.90	3.75	2.50	2.50	2.80	2.50	2.40	2.?
20		3.80	3.65	2.50	2.50	2.80	2.50	2.40	2.?
21		3.20	3.65	2.50	2.50	2.70	2.50	2.50	2.?
22		3.15	3.75	c 2.80	2.50	2.70	2.50	2.50	2.?
23		3.00	3.55	c 2.50	2.50	2.70	2.50	2.50	2.?
24		3.00	3.25	2.45	c 2.70	2.70	2.50	2.50	2.?
25		3.00	3.25	2.45	2.60	2.70	2.50	2.50	2.?
26	3.30	3.00	3.25	2.45	2.60	2.70	2.50	2.50	2.?
27	3.30	3.00	3.35	2.35	2.60	2.60	2.50	2.50	2.?
28	3.35	2.95	3.25	2.35	2.60	2.60	2.50	2.50	2.?
29	3.30	3.10	3.25	2.30	2.60	2.60	2.50	2.50	2.?
30	3.15	3.30	3.20	2.30	2.60	2.60	2.50	2.50	2.?
31		3.35		2.30	2.60		2.50		2.?

a Heavy rain.

b Old station abandoned on account of unfavorable conditions. New station established about 1 mile upstream from old station. Observations at new station commenced on August 10.

c Rain.

Rating table for Uncompahgre River at Colona, Colo., from April 26 to June 30, 1903.

Gage height.	Discharge.	Gage height.	Discharge.	Gage height.	Discharge.	Gage height.	Discharge.
Feet.	Second-feet.	Feet.	Second-feet.	Feet.	Second-feet.	Feet.	Second-feet.
2.9	301	3.3	429	3.7	727	4.1	1,280
3.0	323	3.4	480	3.8	860	4.2	1,420
3.1	351	3.5	542	3.9	1,000	4.3	1,560
3.2	386	3.6	622	4.0	1,140	4.4	1,700

Table for station one-half mile northeast of Colona. This station was abandoned August 10, 1903, owing to water backing up on gage rod from dam below. Discharges for remainder of year were taken at Kettles Crossing, 1 mile above this station.

table for Uncompahgre River at Colona, Colo., from July 1 to August 10, 1903.

Discharge.	Gage height.	Discharge.	Gage height.	Discharge.	Gage height.	Discharge.
Second-feet.	*Feet.*	*Second-feet.*	*Feet.*	*Second-feet.*	*Feet.*	*Second-feet.*
190	2. 4	231	2. 8	283	3. 2	386
199	2. 5	243	2. 9	301		
209	2. 6	255	3. 0	323		
220	2. 7	268	3. 1	351		

ges from July 1 to August 10, inclusive, are approximate. They depend
ngle measurement the gage height of which is somewhat approximate.

*le for Uncompahgre River at Kettles Crossing, near Colona, Colo., from August
11 to October 31, 1903.*

Discharge.	Gage height.	Discharge.	Gage height.	Discharge.	Gage height.	Discharge.
Second-feet.	*Feet.*	*Second-feet.*	*Feet.*	*Second-feet.*	*Feet.*	*Second-feet.*
30	2. 6	149	3. 0	341	3. 4	533
46	2. 7	197	3. 1	389	3. 5	581
69	2. 8	245	3. 2	437	3. 6	629
102	2. 9	293	3. 3	485	3. 7	677

ements were made between 2.50 and 2.70 feet gage height. Above and below
its curve and table are approximate.

mated monthly discharge of Uncompahgre River at Colona, Colo., for 1903.

[Drainage area, 433 square miles.]

onth.	Discharge in second-feet.				Run-off.	
	Maximum.	Minimum.	Mean.	Total in acre-feet.	Second-feet per square mile.	Depth in inches.
lays)	454	368	422	4, 185	0. 974	0. 18
....	1, 210	312	559	34, 372	1. 291	1. 49
....	1, 630	386	834	49, 626	1. 926	2. 15
....	301	220	252	15, 495	. 582	. 67
....	209	102	156	9, 592	. 360	. 42
r..........	653	102	236	14, 043	. 545	. 60
....	149	102	126	7, 747	. 291	. 33
r..........	102	69	99	5, 891	. 229	. 26
r	485	102	285	17, 524	. 658	. 76
e period......				158, 475		

GUNNISON RIVER NEAR CORY, COLO.   -

This station was established April 30, 1903, by W. N. Sam̄
assisted by F. Cogswell.   It is located at the highway bridge
the road between Delta and Cory, Colo., and is about 6 miles ea
Delta, Colo.   It is about 1 mile above the mouth of Surface Cr
The wire gage is located on the downstream side of the right s
near the right bank.   The length of the chain from the end of
weight to the marker is 23 feet.   The gage is read twice each da
R. Shea.   The gage scale is painted on the hand rail of the bri
Discharge measurements are made from the 2-span highway br
and its approach.   The initial point for soundings is a paint mar
the east approach to the bridge.   The channel is straight for 300
above and 500 feet below the station.   The right bank is high, wo
and will not overflow.   The left bank is low, but will not over
There is but one channel at all stages.   The bed of the stream is
posed of gravel.   The bench mark is a cross inclosed in a circ
the downstream side of the right abutment.   Its elevation is
feet above gage datum.

The observations at this station during 1903 have been made t
the direction of M. C. Hinderlider, district hydrographer.

Discharge measurements of Gunnison River near Cory, Colo., in 1903.

Date.	Hydrographer.	Gage height.	Disc
		Feet.	*Seco*
April 30	W. N. Sammis................	8. 00	
May 19......................	.....do	9. 45	
May 28......................	.....do	8. 45	
June 16	.....do	11. 30	1
July 8.......................	.....do	8. 60	
July 18......................	.....do	8. 15	
August 12	.....do	5. 95	
September 9	.....do	6. 45	
October 22..................	E. C. Murphy	5. 68	
November 17	W. N. Sammis................	5. 70	

in daily gage height, in feet, of Gunnison River near Cory, Colo., for 1903.

Day.	Apr.	May.	June.	July.	Aug.	Sept.	Oct.	Nov.	Dec.
....		8.00	9.95	9.85	6.50	5.75	5.80	5.65	5.45
....		8.15	10.30	9.15	6.40	5.70	5.80	5.65	5.65
....		8.45	10.45	8.90	6.30	5.70	5.65	5.65	5.55
....		8.75	10.55	8.60	6.20	5.90	5.70	5.65	5.30
....		9.10	10.30	8.20	6.20	5.85	5.70	5.65	5.10
....		9.40	10.10	8.10	6.10	5.70	5.70	5.65	5.25
....		9.50	10.20	8.10	6.10	6.80	5.70	5.65	5.20
....		9.45	10.50	8.55	6.00	6.65	5.70	5.65	5.25
....		9.80	10.60	8.35	6.00	6.40	5.70	5.65	5.30
....		10.00	10.65	8.30	6.00	6.15	5.70	5.65	5.40
....		10.00	10.55	8.40	5.90	6.05	5.70	5.45	5.45
....		9.95	10.45	8.25	5.90	6.25	5.70	5.35	5.45
....		10.00	a11.30	8.00	5.90	6.20	5.70	5.50	5.45
....		10.35	11.60	7.90	5.90	6.10	5.70	5.75	5.55
....		10.60	11.45	7.80	5.90	6.10	5.70	5.70	5.45
....		10.80	11.40	8.05	5.90	6.10	5.70	5.70	5.45
....		10.80	11.15	8.20	5.90	6.00	5.70	5.70	5.30
....		10.05	11.30	8.10	5.90	6.00	5.70	5.40	5.30
....		9.35	11.15	7.80	5.90	6.00	5.70	5.25	5.35
....		9.00	10.85	7.60	5.85	5.95	5.60	5.50	5.50
....		8.75	10.60	7.45	5.80	5.90	5.60	5.55	5.50
....		8.65	10.40	7.45	5.80	5.90	5.60	5.75	5.55
....		8.45	10.30	7.35	5.80	5.90	5.60	5.65	5.55
....		8.25	10.15	7.30	5.80	5.80	5.60	5.75	5.35
....		8.20	10.10	a7.45	6.00	5.80	5.60	5.70	5.45
....		8.30	9.90	7.15	6.00	5.80	5.50	5.60	5.45
....		8.40	9.80	7.10	6.15	5.80	5.50	5.45	5.35
....		8.40	9.80	7.00	6.05	5.75	5.50	5.45	5.45
....		8.45	9.80	6.85	5.95	5.70	5.50	5.45	5.45
............t............	7.80	8.70	9.65	6.65	5.80	5.75	5.50	5.45	5.45
....		9.40	....	6.60	5.80	....	5.50	....	5.45

a Heavy rain.

table for Gunnison River near Cory, Colo., from April 30 to December 31, 1903.

e. t.	Discharge.	Gage height.	Discharge.	Gage height.	Discharge.	Gage height.	Discharge.
t.	Second-feet.	Feet.	Second-feet.	Feet.	Second-feet.	Feet.	Second-feet.
1	450	6.1	1,155	7.2	2,455	9.2	7,055
2	490	6.2	1,255	7.4	2,760	9.4	7,780
3	540	6.3	1,355	7.6	3,095	9.6	8,565
4	600	6.4	1,460	7.8	3,450	9.8	9,395
5	660	6.5	1,565	8.0	3,830	10.0	10,290
6	730	6.6	1,675	8.2	4,250	10.5	12,740
7	800	6.7	1,790	8.4	4,700	11.0	15,430
.8	880	6.8	1,910	8.6	5,205	11.5	18,130
.9	970	6.9	2,035	8.8	5,770	11.7	19,210
.0	1,060	7.0	2,170	9.0	6,385		

Estimated monthly discharge of Gunnison River near Cory, Colo., for 1903.

[Drainage area, 5,223 square miles.]

Month.	Discharge in second-feet.			Total in acre-feet.	Run-off	
	Maximum.	Minimum.	Mean.		Second-feet per square mile.	Depth in inches
May...............	14,350	3,830	7,554	464,477	1.444	
June	18,670	8,770	12,903	767,782	2.465	
July...............	7,595	1,675	3,774	232,054	.721	
August	1,565	880	1,057	64,992	.202	
September	1,910	800	1,056	62,836	.202	
October...........	880	660	765	47,038	.146	
November	840	515	723	43,021	.138	
December.........	765	450	609	37,446	.116	
The period.......				1,719,646		

NORTH FORK OF GUNNISON RIVER AT HOTCHKISS, COLO.

This station was established May 2, 1903, by W. N. Sammis. located at the highway bridge, one-half mile east of Hotchkiss. The gage is a vertical 2 by 4 inch rod, 3 feet long, nailed to the stream side of the log abutment on the right bank. Another 3 section is nailed to the middle pier on the downstream side. gage is read twice each day by Frank Visner. Discharge mea ments are made from the two-span highway bridge and its appr The initial point for soundings is the end of the hand rail on the d stream side at the left bank. The channel is straight for 100 above and below the station, and the current is swift. The right is low and liable to overflow. The left bank is high and is not to overflow. The bed of the stream is rocky. The bench mar bolt on the upstream side of the southwest pier on the right bank is marked "B. M. U. S. G. S." Its elevation is 8.58 feet abov zero of the gage.

The observations at this station during 1903 have been made u the direction of M. C. Hinderlider, district hydrographer.

ge measurements of North Fork of Gunnison River at Hotchkiss, Colo., in 1903.

Date.	Hydrographer.	Gage height.	Discharge.
		Feet.	*Second-feet.*
....	W. N. Sammis	4.30	1,815
.....................do		4.30	2,478
.....................do		4.15	2,096
.....................do		5.25	3,501
.....................do		3.40	761
.....................do		3.20	892
? ....do		1.90	87
r 9 ....do		2.10	129

y gage height, in feet, of North Fork of Gunnison River at Hotchkiss, Colo., for 1903.

Day.	May.	June.	July.	Aug.	Sept.	Oct.	Nov.	Dec.
....		(a)	4.30	2.15	1.65	2.10	2.00	2.00
....		(a)	4.25	2.10	1.65	2.10	2.00	2.00
....	4.20	(a)	4.00	2.05	1.60	2.10	2.00	2.00
....	4.85	(a)	3.90	2.00	1.60	2.10	2.00	2.10
....	4.95	5.10	(a)	2.00	1.60	2.10	2.00	2.10
....	5.05	5.25	(a)	1.95	1.60	2.10	2.00	2.10
....	5.15	5.25	3.60	1.90	b2.75	2.10	2.00	2.10
....	5.20	5.30	3.65	1.90	2.70	2.10	1.90	2.20
....	5.35	5.35	3.55	1.90	2.60	2.10	1.90	2.20
....	5.35	5.30	3.45	1.90	2.40	2.10	1.90	2.20
....	5.45	5.35	3.45	1.90	2.30	2.00	1.90	2.20
....	5.30	5.45	3.40	1.85	2.20	2.00	1.90	2.20
....	5.40	b6.10	3.25	1.90	2.20	2.00	1.90	2.20
....	5.60	5.80	3.25	1.85	2.20	2.00	1.90	2.20
....	5.65	5.90	3.30	1.80	2.20	2.00	1.90	2.10
....	5.65	5.90	(a)	1.80	2.10	2.00	1.90	2.20
....	5.65	5.80	3.20	1.80	2.10	2.00	1.90	2.20
....	5.00	5.80	3.15	1.70	2.10	1.90	1.90	2.20
....	4.50	5.60	2.90	1.70	2.10	1.90	1.90	2.20
....	4.30	5.55	2.85	1.60	2.10	1.90	1.90	2.40
....	4.25	5.25	2.75	1.60	2.10	1.90	1.90	2.40
....	4.15	5.25	2.80	1.60	2.10	1.90	1.90	2.60
....	4.00	5.15	2.75	1.60	2.10	1.90	1.90	2.60
....	4.10	5.15	2.65	1.60	2.00	1.90	2.00	2.60
....	4.10	5.10	2.60	1.60	2.00	1.90	2.00	2.60
....	4.10	4.85	(c)	1.60	2.00	1.90	2.00	2.85
....	4.15	4.75	(c)	1.70	2.00	1.90	2.00	2.90
....	4.10	4.90	2.50	1.65	2.10	1.90	2.00	2.90
....	4.20	4.80	2.35	1.60	2.10	1.90	2.00	2.90
....	4.70	4.65	2.30	1.60	2.10	1.90	2.00	2.90
....	(a)	....	2.20	1.60	....	1.90	....	2.90

a Gage broken. b Heavy rain. c Observer absent.

Rating table for North Fork of Gunnison River at Hotchkiss, Colo., from May 3 to December 31, 1903.

Gage height.	Discharge.	Gage height.	Discharge.	Gage height.	Discharge.	Gage height.	Discharge.
Feet.	*Second-feet.*	*Feet.*	*Second-feet.*	*Feet.*	*Second-feet.*	*Feet.*	*Second-feet.*
1.6	45	2.6	306	3.6	940	4.6	2,305
1.7	55	2.7	348	3.7	1,045	4.7	2,475
1.8	70	2.8	394	3.8	1,155	4.8	2,650
1.9	88	2.9	443	3.9	1,270	4.9	2,830
2.0	108	3.0	495	4.0	1,395	5.0	3,020
2.1	133	3.1	550	4.1	1,530	5.2	3,430
2.2	162	3.2	610	4.2	1,675	5.4	3,890
2.3	195	3.3	680	4.3	1,825	5.6	4,400
2.4	230	3.4	760	4.4	1,980	5.8	4,950
2.5	267	3.5	845	4.5	2,140	6.0	5,530

Estimated monthly discharge of North Fork of Gunnison River at Hotchkiss, Colo., for 1903.

[Drainage area, 850 square miles.]

Month.	Discharge in second-feet.			Total in acre-feet.	Run-off.	
	Maximum.	Minimum.	Mean.		Second-feet per square mile.	Depth in inches.
May (28 days)	4,535	1,395	2,783	154,560	3.274	3.41
June (26 days)	5,820	2,390	3,812	196,585	4.484	4.34
July (26 days)	1,825	162	694	35,790	.816	.79
August	148	45	73	4,489	.086	.10
September..........	371	45	142	8,450	.167	.19
October	133	88	107	6,579	.126	.15
November	108	88	97	5,772	.114	.12
December	443	108	229	14,081	.269	.31
The period				426,306		

GUNNISON RIVER NEAR CIMARRON, COLO.

This station was established September 18, 1903, by W. N. Sammis. It is located just above the Denver and Rio Grande Railroad bridge about 1 mile from Cimarron, Colo., and about 1,000 feet above the mouth of Cimarron River. The chain gage is located on the railroad bridge near the center of the stream and about 100 feet below the gaging section. It is read twice each day by J. J. McNamara. Discharge measurements are made by means of a cable and car. The righ

the cable is directly under the end of the bridge on the right The initial point for soundings is about 20 feet from the cable t on the right bank. Distances are marked on a tagged wire he cable. The channel is straight for 1,000 feet above and 300 low the station. The current has a high velocity at the gaging , increasing about 200 feet below. Both banks are high and nd will not overflow. The bed of the stream is composed of bowlders and sand, and is not subject to change. The bench s on a sharp-pointed granite bowlder, at a point 5 feet from the point for soundings under the cable. It is 20 feet north of the anite pier from the north end of the bridge, and is directly he east side of the railroad bridge. Its elevation is 10.68 feet rage datum.

observations at this station during 1903 have been made under ection of M. C. Hinderlider, district hydrographer.

ischarge measurements of Gunnison River near Cimarron, Colo., in 1903.

Date.	Hydrographer.	Gage height.	Discharge.
		Feet.	Second-feet.
er 18	W. N. Sammis	4.90	861
17	do	4.30	576
21	E. C. Murphy	4.29	557
er 25	F. Cogswell	4.10	481

daily gage height, in feet, of Gunnison River near Cimarron, Colo., for 1903.

Day.	Sept.	Oct.	Nov.	Dec.	Day.	Sept.	Oct.	Nov.	Dec.
		4.50	4.10	3.90	17		4.40	3.40	3.70
		4.50	4.00	3.90	18		4.30	3.40	3.70
		4.50	4.00	3.95	19	4.80	4.40	3.40	3.70
		4.50	4.00	3.15	20	4.70	4.40	4.00	3.70
		4.50	4.10	3.80	21	4.60	4.30	4.20	3.70
		4.50	4.10	3.60	22	4.50	4.30	4.20	
		4.50	4.10	3.80	23	4.50	4.30	4.10	
		4.40	4.00	3.80	24	4.50	4.30	4.10	
		4.40	4.00	3.80	25	4.50	4.20	4.10	
		4.40	4.00	3.75	26	4.50	4.20	4.10	
		4.40	3.80	3.75	27	4.50	4.10	3.90	
		4.40	4.35	3.75	28	4.50	4.10	3.80	
		4.10	4.30	3.70	29	4.40	4.00	3.80	
		4.40	4.30	3.70	30	4.40	4.00	3.80	
		4.40	4.30	3.70	31		4.10		

Rating table for Gunnison River near Cimarron, Colo., from September 19 to December 31, 1903.

Gage height.	Discharge.	Gage height.	Discharge.	Gage height.	Discharge.	Gage height.	Discharge.
Feet.	*Second-feet.*	*Feet.*	*Second-feet.*	*Feet.*	*Second-feet.*	*Feet.*	*Second-feet.*
3.1	187	4.0	448	4.9	855	5.8	1,350
3.2	212	4.1	483	5.0	910	5.9	1,405
3.3	238	4.2	521	5.1	965	6.0	1,460
3.4	265	4.3	561	5.2	1,020	6.1	1,515
3.5	293	4.4	605	5.3	1,075	6.2	1,570
3.6	322	4.5	652	5.4	1,130	6.3	1,625
3.7	352	4.6	700	5.5	1,185	6.4	1,680
3.8	383	4.7	750	5.6	1,240	6.5	1,735
3.9	415	4.8	802	5.7	1,296	6.6	1,790

Estimated monthly discharge of Gunnison River near Cimarron, Colo., for 1903.

[Drainage area, 3,844 square miles.]

Month.	Discharge in second-feet.			Total in acre-feet.	Run-off.	
	Maximum.	Minimum.	Mean.		Second-feet per square mile.	Depth in inches.
September (12 days).	802	605	669	15,923	0.174	0.076
October	652	448	583	35,847	.152	.170
November	583	265	451	26,836	.117	.131
December (21 days).	432	322	374	15,578	.097	.070
The period	....			94,184	....	

CIMARRON RIVER AT CIMARRON, COLO.

This station was established April 28, 1903, by W. N. Sammis, assisted by F. Cogswell. It is located at the footbridge 1,200 feet south of the Denver and Rio Grande Railroad station. The gage is a vertical 2 by 4 inch rod, fastened to the left abutment on the down-stream side. It is read twice each day by G. L. Linscott. Discharge measurements are made from the footbridge, which is supported by two log abutments and one log pier in the center of the stream. The initial point for soundings is the left abutment on the downstream side. The channel is straight for 100 feet above and and 250 feet below the station. Both banks are high, wooded, and have not been known to overflow for the past eight years. At low water there is but one channel, and all the water passes beneath the left span. At high water there are two channels, separated by the center pier of the

. The bridge has a span between abutments of about 60 feet.
nch mark consists of two nails driven into the root of a cotton-
tree 50 feet west of the gage. Its elevation is 7.82 feet above
·o of the gage.
observations at this station during 1903 have been made under
·ection of M. C. Hinderlider, district hydrographer.

Discharge measurements of Cimarron River at Cimarron, Colo., in 1903.

Date.	Hydrographer.	Gage height.	Discharge.
		Feet.	*Second-feet.*
....	W. N. Sammis...............	2.30	238
....	......do	2.70	363
....	......do	3.40	686
....	......do	2.80	419
....	......do	3.40	711
....	......do	3.35	705
....	......do	2.80	391
....	......do	2.50	314
....	......do	2.30	247
5	......do	1.90	131
)er 18	......do	1.80	106
18.......................	......do	1.70	68
20....................	E. C. Murphy	1.66	44
)er 24	W. N. Sammis...............	1.69	55

in daily gage height, in feet, of Cimarron River at Cimarron, Colo., for 1903.

Day.	Apr.	May.	June.	July.	Aug.	Sept.	Oct.	Nov.	Dec.
....		2.44	3.30	(a)	2.10	1.65	1.80	1.70	(b)
....		2.50	3.30	(a)	2.05	1.65	1.75	1.70	....
....		2.50	3.31	(a)	2.00	c 2.20	1.70	1.70	....
....		2.43	3.30	(a)	2.00	1.70	1.70	1.70	....
....		2.50	d 3.30	(a)	1.95	1.75	1.65	1.70	....
....		2.35	3.30	(a)	1.90	1.90	1.60	1.70	....
....		2.68	3.40	(a)	1.95	c 2.50	1.60	1.70	....
....		2.33	3.40	(a)	1.90	2.10	1.60	1.70	....
....		2.33	(a)	(a)	1.85	2.00	1.60	1.70	....
....		2.67	(a)	(a)	1.85	1.90	1.60	(e)	....
....		2.69	(a)	(a)	1.80	1.90	1.70	(e)	....
....		3.00	(a)	(a)	1.80	1.95	1.70	1.90	....
....		3.32	(a)	(a)	1.85	1.90	1.70	1.80	....
....		3.17	(a)	(a)	1.80	1.90	1.70	1.73	....
....		3.17	(a)	(a)	1.80	1.90	1.70	1.70	....

:ords.
n December 1 to 21, inclusive.
y rain on headwaters.
:ds from June 8 to July 28, inclusive, practically worthless. See letter M. C. Hinderlider,

Mean daily gage height, in feet, of Cimarron River at Cimarron, Colo., for 1903—Cont'd.

Day.	Apr.	May.	June.	July.	Aug.	Sept.	Oct.	Nov.	Dec.
16................	....	3.17	(a)	(a)	1.85	1.85	1.70	1.70	....
17................	....	3.40	(a)	(a)	1.80	1.85	1.70	(b)	....
18................	....	3.32	(a)	(a)	1.80	1.85	1.70	(b)	....
19................	....	3.32	(a)	(a)	1.75	1.85	1.70	(b)	....
20................	....	3.43	(a)	(a)	1.80	1.85	1.70	(b)	....
21................	....	2.79	(a)	(a)	1.75	1.75	1.70	(b)	....
22................	....	2.25	(a)	(a)	1.70	1.80	1.70	(b)	....
23................	....	2.38	(a)	(a)	1.70	1.80	1.70	(b)	....
24................	....	2.50	(a)	(a)	1.80	1.70	1.70	1.80	....
25................	....	2.43	(a)	(a)	1.80	1.65	1.70	1.75	....
26................	....	2.34	(a)	(a)	1.80	1.65	1.70	1.70	....
27................	....	2.34	(a)	(a)	1.80	1.60	1.70	1.80	....
28................	2.33	2.35	(a)	(a)	1.75	1.65	1.70	1.80	....
29................	2.40	2.42	(a)	2.30	1.75	1.75	1.70	(b)	....
30................	2.40	2.34	(a)	2.20	1.75	1.70	1.70	(b)	....
31................	....	3.16	(a)	2.20	1.65	....	1.70	....	....

a No records.   b Ice.

Rating table for Cimarron River at Cimarron, Colo., from April 28 to December 31, 1903.

Gage height.	Discharge.	Gage height.	Discharge.	Gage height.	Discharge.	Gage height.	Discharge.
Feet.	*Second-feet.*	*Feet.*	*Second-feet.*	*Feet.*	*Second-feet.*	*Feet.*	*Second-feet.*
1.6	32	2.1	187	2.6	342	3.1	536
1.7	63	2.2	218	2.7	373	3.2	588
1.8	94	2.3	249	2.8	407	3.3	646
1.9	125	2.4	280	2.9	446	3.4	709
2.0	156	2.5	311	3.0	489	3.5	777

Estimated monthly discharge of Cimarron River at Cimarron, Colo., for 1903.

[Drainage area, 210 square miles.]

Month.	Discharge in second feet.			Total in acre-feet.	Run-off.	
	Maximum.	Minimum.	Mean.		Second-feet per square mile.	Depth in inches.
April (3 days)	280	258	273	1,624	1.30	0.14
May	729	234	401	24,657	1.91	2.20
June (8 days)	709	646	662	10,505	3.15	.94
July (3 days).......	249	218	228	1,357	1.09	.13
August	187	48	104	6,395	.50	.50
September..........	311	32	106	6,307	.50	.5
October	94	32	59	3,628	.28	.3
November (19 days).	125	63	74	2,789	.35	.2
The period	....	....	....	57,262	....	....

GUNNISON RIVER AT IOLA, COLO.

station was established April 1, 1900, by A. L. Fellows, for the
? of determining the amount of water available for the irriga-
oject of Uncompahgre Valley. Owing to the fact that a num-
streams enter the Gunnison below the station, the results
d do not show the total amount of water available. The gage
tical 2 by 4 inch timber 8 feet long, spiked to the lower side of
t pier from the left bank at a point 141 feet from the initial
or soundings. The gage is read once each day by C. A. Green,
stmaster at Iola. Discharge measurements are made from the
ide of the highway bridge, to which the gage is attached. At
ter they are made by wading. The bridge is located one-half
ortheast of the railroad station at Iola. The initial point for
ngs is the edge of the right abutment on the downstream side.
annel is straight for 1,000 feet above and for 500 feet below the
. It is broken by three bridge piers. The current is swift.
anks are low, but are not liable to overflow. The bed of the
is composed of gravel and small bowlders and is permanent.
·nch mark is a spike in the base of a stump of a post 40 feet
of the south end of the bridge, on the west side of the road.
·ation is 7.43 feet above the zero of the gage. This relation was
l April 11, 1901.
observations at this station during 1903 have been made under
ection of M. C. Hinderlider, district hydrographer.

Discharge measurements of Gunnison River at Iola, Colo., in 1903.

Date.	Hydrographer.	Gage height.	Discharge.
		Feet.	*Second-feet.*
....	F. Cogswell..................	3. 10	1,435
....	.....do	3. 95	2,750
....	.....do	4. 10	2,933
....	.....do	5. 10	4,642
....	.....do	3. 30	1,467
l2	.....do	2. 40	544
ier 3	.....do	2. 30	467
l...................	.....do	2. 30	458
23.................	.....do	2. 22	371

Mean daily gage height, in feet, of Gunnison River at Iola, Colo., for 1903.

Day.	Apr.	May.	June.	July.	Aug.	Sept.	Oct.	Nov.	D
1	3.50	2.90	4.00	4.40	3.00	2.50	2.30	2.30	
2	3.50	3.00	4.20	4.30	3.00	2.40	2.40	2.30	
3	3.60	2.90	4.50	4.20	3.00	2.30	2.40	2.30	
4	3.60	3.10	4.50	3.90	3.00	2.40	2.30	2.30	
5	3.40	3.20	4.40	3.70	3.00	2.40	2.40	2.30	
6	3.20	3.30	4.30	3.60	3.10	2.40	2.40	2.30	
7	3.00	3.40	4.50	3.50	2.90	2.40	2.30	2.30	
8	2.80	3.50	4.60	3.80	2.70	2.70	2.40	2.30	
9	2.60	3.60	4.70	3.70	2.60	2.60	2.30	2.30	
10	2.50	3.80	4.80	3.70	2.50	2.50	2.40	2.40	
11	2.40	3.70	4.90	3.80	2.40	2.50	2.40	2.40	
12	2.40	3.90	5.00	3.80	2.40	2.60	2.40	2.40	
13	2.40	3.90	5.20	3.70	2.30	2.50	2.40	2.40	
14	2.60	3.90	5.40	3.70	2.20	2.40	2.30	2.40	
15	2.50	4.00	5.20	3.70	2.00	2.70	2.30	2.40	
16	2.40	4.10	5.00	3.80	2.10	2.60	2.30	2.30	
17	2.40	4.50	5.50	3.80	2.20	2.50	2.30	2.30	
18	2.50	4.00	6.00	3.70	2.20	2.50	2.30	2.30	
19	2.30	3.80	5.80	3.50	2.20	2.40	2.30	2.30	
20	2.40	3.50	5.50	3.40	2.30	2.40	2.30	2.30	
21	2.40	3.40	4.90	3.40	2.30	2.40	2.30	2.30	
22	2.50	3.30	4.80	3.40	2.40	2.40	2.30	2.30	
23	2.60	3.20	4.70	3.40	2.40	2.40	2.30	2.30	
24	2.70	3.20	4.60	3.30	2.40	2.40	2.30	2.30	
25	2.80	3.10	4.50	3.20	2.40	2.30	2.30	2.30	
26	3.00	3.00	4.40	3.10	2.40	2.30	2.30	2.30	
27	3.10	3.20	4.40	3.10	2.50	2.30	2.30	2.30	
28	3.20	3.40	4.40	3.00	2.50	2.30	2.30	2.30	
29	3.00	3.60	4.50	3.00	2.40	2.30	2.30	2.30	
30	2.90	3.80	4.50	3.00	2.40	2.40	2.40	2.30	
31		3.80		3.00	2.50		2.30		

a Ice.

Rating table for Gunnison River at Iola, Colo., from April 1 to December 31, 190 .

Gage height.	Discharge.	Gage height.	Discharge.	Gage height.	Discharge.	Gage height.	Discharge
Feet.	*Second-feet.*	*Feet.*	*Second-feet.*	*Feet.*	*Second-feet.*	*Feet.*	*Second-fe*
2.0	250	3.0	1,185	4.0	2,789	5.0	4,459
2.1	314	3.1	1,321	4.1	2,956	5.1	4,620
2.2	382	3.2	1,467	4.2	3,123	5.2	4,793
2.3	458	3.3	1,623	4.3	3,290	5.3	4,960
2.4	544	3.4	1,787	4.4	3,457	5.4	5,127
2.5	640	3.5	1,954	4.5	3,624	5.5	5,294
2.6	740	3.6	2,121	4.6	3,791	5.6	5,461
2.7	840	3.7	2,288	4.7	3,958	5.7	5,628
2.8	945	3.8	2,455	4.8	4,125	5.8	5,795
2.9	1,060	3.9	2,622	4.9	4,292	6.0	6,129

Below 2.20 feet gage height and above 5.10 feet gage height the table is not determined.

Estimated monthly discharge of Gunnison River at Iola, Colo., for 1903.

[Drainage area, 2,298 square miles.]

Month.	Discharge in second-feet.			Total in acre-feet.	Run-off.	
	Maximum.	Minimum.	Mean.		Second-feet per square mile.	Depth in inches.
il...............	2,121	458	1,019	60,635	0.44	0.49
....	3,624	1,060	1,996	122,729	.87	1.00
?...............	6,129	2,789	4,108	244,443	1.79	2.00
'...............	3,457	1,185	2,078	127,771	.90	1.04
ust	1,321	250	671	41,258	.29	.33
ember..........	840	458	585	34,810	.25	.28
)ber	544	458	486	29,883	.21	.24
ember..........	544	458	475	28,264	.21	.23
ember 1–23......	544	458	465	21,214	.20	.17
The period....	....	....	....	711,007	....	....

GRAND RIVER NEAR PALISADES, COLO.

This station was established April 9, 1902, by John E. Field. It is cated at the iron highway bridge at the point where the river enters nnd Valley, about 2 miles above Palisades. The station is above all rigating ditches supplying water to Grand Valley, with the excep-)n of one pumping plant, which takes about 20 second-feet from the rer one-fourth mile above the station. The wire gage is located on e downstream side near the center of the bridge. The length of the ire, from the end of the weight to the end of the wire, is 35.85 feet. 1e gage is read twice each day by S. L. Purdy. Discharge measure- ents are made from the lower side of the single-span bridge to which e gage is attached. The initial point for soundings is the end of the idge on the left bank. The channel is straight for 600 feet above d below the station. The right bank is high and rocky and will not 'erflow. The left bank is somewhat lower and will overflow at tremely high water. Above the banks the mountains rise rather)ruptly on either side. The bed of the stream is composed of large)wlders, which are covered at low water with a deposit of mud. uring high water this deposit is carried away, and at low water a ew deposit forms. No bench marks have been established.

The observations at this station during 1903 have been made under 1e direction of M. C. Hinderlider, district hydrographer.

Discharge measurements of Grand River near Palisades, Colo., in 1903.

Date.	Hydrographer.	Gage height.	Discha
		Feet.	Second
April 8	F. Cogswell	12. 15	1
May 9	....do	15. 70	ᴻ
May 29	....do	15. 80	ᴻ
June 18	....do	20. 00	23
July 17	....do	16. 40	10
August 29	....do	12. 80	1
September 26	....do	12. 85	2,
November 21	....do	12. 20	1,

Mean daily gage height, in feet, of Grand River near Palisades, Colo., for 1903.

Day.	Apr.	May.	June.	July.	Aug.	Sept.	(
1	12.35	14.20	17.35	18.15	13.85	12.60	
2	13.00	14.15	18.00	17.80	13.65	12.60	
3	12.60	14.05	18.20	17.40	13.55	12.55	
4	12.40	14.15	18.45	17.00	13.35	12.55	
5	12.40	14.35	18.55	16.50	13.30	12.45	
6	12.33	14.65	18.65	16.15	13.15	12.50	1
7	12.20	15.10	18.00	16.00	13.10	12.70	1
8	12.15	15.40	18.95	16.35	13.05	13.20	1
9	12.18	15.70	19.10	16.40	13.00	13.20	1
10	12.25	16.05	19.15	16.45	12.95	13.20	1
11	12.33	16.35	19.15	16.40	12.90	13.10	1
12	12.65	16.40	18.65	16.25	12.80	13.30	1
13	12.65	16.35	19.25	16.05	12.80	13.25	1
14	12.55	16.75	19.60	15.90	12.75	13.60	1
15	12.43	17.35	19.60	16.05	12.70	13.40	1
16	12.40	17.75	19.65	16.10	12.70	13.30	1
17	12.40	18.25	19.75	16.40	12.70	13.20	1
18	12.48	18.00	20.05	16.55	12.60	13.20	1
19	12.50	17.45	20.05	16.05	12.60	13.20	1
20	12.50	16.85	20.05	15.55	12.60	13.05	1
21	12.50	16.25	19.80	15.35	12.50	13.00	1
22	12.50	16.00	19.50	15.10	12.50	12.90	1
23	12.85	15.75	19.10	15.15	12.50	12.90	1
24	13.15	15.55	18.90	15.05	12.55	12.85	1
25	13.80	15.35	18.60	15.00	12.75	12.80	1
26	14.15	15.40	18.35	14.85	13.00	12.80	1
27	14.50	15.65	18.70	14.95	13.00	12.70	1
28	14.50	15.75	18.10	14.35	12.90	12.65	1
29	14.50	15.85	18.40	14.15	12.80	12.60	1
30	14.45	16.05	18.40	14.20	12.65	12.75	1
31		16.50		14.05	12.60		1

table for Grand River near Palisades, Colo., from April 1 to December 31, 1903.

e. ht.	Discharge.	Gage height.	Discharge.	Gage height.	Discharge.	Gage height.	Discharge.
t.	Second-feet.	Feet.	Second-feet.	Feet.	Second-feet.	Feet.	Second-feet.
1	1,350	13.1	2,540	14.2	4,700	16.5	10,440
2	1,440	13.2	2,710	14.4	5,150	17.0	11,890
3	1,530	13.3	2,890	14.6	5,610	17.5	13,420
4	1,630	13.4	3,080	14.8	6,070	18.0	15,040
5	1,730	13.5	3,270	15.0	6,540	18.5	16,750
6	1,840	13.6	3,470	15.2	7,030	19.0	18,580
7	1,960	13.7	3,670	15.4	7,530	19.5	20,500
8	2,090	13.8	3,870	15.6	8,040	20.0	22,500
9	2,230	13.9	4,070	15.8	8,560	20.1	22,900
0	2,380	14.0	4,270	16.0	9,080	....	....

Estimated monthly discharge of Grand River near Palisades, Colo., for 1903.

[Drainage area, 8,546 square miles.]

Month.	Discharge in second-feet.			Total in acre-feet.	Run-off.	
	Maximum.	Minimum.	Mean.		Second-feet per square mile.	Depth in inches.
....	5,380	1,395	2,402	142,929	0.28	0.31
....	15,880	4,375	9,033	555,417	1.06	1.22
....	22,700	12,955	18,388	1,094,162	2.15	2.40
....	15,540	4,375	8,849	544,104	1.03	1.19
....	3,970	1,730	2,282	140,315	.27	.31
ber.........	3,470	1,680	2,331	138,704	.27	.30
r	2,985	1,730	2,097	128,940	.25	.29
The period ...	....	....	....	2,744,571	....	....

GRAND RIVER AT GLENWOOD SPRINGS, COLO.

s station was first located May 12, 1899, at the request of the
er and Rio Grande Railway Company, at the railroad bridge, a
er of a mile west of the depot and just above the mouth of
ng Fork. A wire gage was used. At the beginning of 1900,
er, a new gage rod was located near the electric-light works.
gage consisted of a light vertical staff, to the lower end of which
ttached a wooden float which rested on the surface of the water,
ing in a well or box made of 6-inch boards. The bottom of this
onnected with a small wooden flume extending out into the river,
allowing the water in the well to assume the level of the river

surface. The vertical staff attached to the float was graduated to feet and tenths. In July, 1902, this gage, being in bad repair, was replaced with an automatic water register, the site being the same as that of the old float gage. This register never worked satisfactorily, and was later replaced with a float gage, using the same well and intake flume as was previously used. The present gage consists of a metal float and counterweight connected with a pliable wire passing over pulleys so arranged that a rise of 1 foot in the river registers but one-half the amount on the rod. By this arrangement the large rise and fall of the river is readily accommodated by the length of gage, which would be impossible with a direct-reading gage. The gage rod proper consists of a light pine rod, 1 by 1 inch by 5 feet long, securely fastened to the small house in which the apparatus is located. A pointer attached to a counterweight slides along the gage rod and registers the rise and fall of the river. This gage has given entire satisfaction. The gage and equipment are inclosed within a small house, which is kept locked. The observer, who is the engineer of the electric-light works, is located immediately across the street from the gage. Readings are taken in the morning and evening during the entire year. Measurements are made from the single-span steel wagon bridge crossing the river between the town and the hotel. The channel is good, being composed of gravel and rock, and does not change much; the banks are high and not subject to overflow. Current medium to swift.

The observations at this station during 1903 have been made under the direction of M. C. Hinderlider, district hydrographer.

Discharge measurements of Grand River at Glenwood Springs, Colo., in 1903.

Date.	Hydrographer.	Gage height.	Discharge.
		Feet.	*Second-feet.*
February 7	M. C. Hinderlider	3. 09	527
April 7	F. Cogswell...................	3. 70	900
May 8.......................	....do	6. 25	4, 764
May 28......................	....do	6. 55	5, 549
June 17......................	....do	9. 20	16, 688
July 16......................	....do	6. 80	5, 824
August 8.....................	....do	4. 45	1, 633
August 31....................	....do	4. 05	1, 252
September 28	....do	3. 95	1, 215
October 26[a]	....do	3. 95	1, 100
November 20	....do	3. 33	764

[a] New railroad bridge building across river 1,000 feet below station; flow affected by false work in channel.

Mean daily gage height, in feet, of Grand River at Glenwood Springs, Colo., for 1903.

Day.	Jan.	Feb.	Mar.	Apr.	May.	June.	July.	Aug.	Sept.	Oct.	Nov.	Dec.
....	(a)	3.08	3.15	3.73	5.30	7.50	7.88	5.05	3.98	4.16	3.83	3.53
....	(a)	3.05	3.15	3.85	5.35	7.80	7.70	4.95	3.93	4.25	3.75	3.60
....	(a)	3.00	3.23	3.80	5.30	8.05	7.48	4.85	3.90	4.30	3.75	3.50
....	(a)	2.98	3.25	3.95	5.30	8.40	7.20	4.75	3.90	4.28	3.75	3.23
....	(a)	2.95	3.30	3.85	5.48	8.45	6.08	4.68	3.85	4.30	3.83	3.03
....	(a)	3.05	3.35	3.70	5.70	8.48	6.58	4.63	3.88	4.23	3.80	3.00
....	(a)	3.13	3.25	3.68	6.08	8.58	6.45	4.58	4.00	4.20	3.73	2.95
....	(a)	3.13	3.23	3.75	6.25	8.75	6.68	4.48	4.30	4.20	3.75	2.95
....	(a)	3.15	3.33	3.75	6.55	8.85	6.78	4.38	4.45	4.18	3.75	2.98
....	(a)	3.13	3.35	3.85	6.83	8.83	6.75	4.35	4.35	4.13	3.73	3.05
....	(a)	3.15	3.38	4.08	6.95	8.60	6.68	4.30	4.28	4.10	3.45	3.03
....	(a)	3.15	3.48	4.20	6.85	8.35	6.55	4.28	4.28	4.	3.25	3.13
....	2.95	3.13	3.55	4.03	6.78	8.65	6.43	4.25	4.30	4.15	3.38	3.18
....	2.95	3.13	3.60	3.93	7.05	8.90	6.45	4.25	4.40	4.18	3.53	3.18
....	2.98	3.10	3.70	3.90	7.55	8.83	6.55	4.20	4.35	4.20	3.78	3.15
....	2.90	3.08	3.75	3.95	7.90	8.80	6.75	4.20	4.35	4.20	3.85	2.90
....	2.90	3.08	3.63	4.03	8.18	9.	6.75	4.20	4.33	4.18	3.68	2.98
....	2.93	3.15	3.70	4.08	8.05	9.40	6.83	4.20	4.33	4.15	3.33	2.93
....	2.93	3.15	3.50	4.10	7.58	9.38	6.45	4.20	4.28	4.18	3.23	3.08
....	2.95	3.13	3.38	4.10	7.10	9.23	6.13	4.15	4.25	4.13	3.28	3.00
....	2.95	3.18	3.35	4.08	6.78	8.88	5.95	4.10	4.25	4.13	3.45	3.05
....	2.95	3.18	3.45	4.33	6.68	8.63	5.83	4.10	4.18	4.08	3.63	3.15
....	2.95	3.18	3.43	4.63	6.43	8.53	5.90	4.10	4.13	4.10	3.70	3.18
....	2.98	3.20	3.43	4.95	6.23	8.40	5.80	4.10	4.10	4.08	3.78	3.05
....	3.00	3.18	3.48	5.33	6.13	8.15	5.78	b4.40	4.08	4.03	3.75	2.95
....	3.00	3.25	3.55	5.58	6.30	7.95	5.73	4.58	4.05	4.00	3.70	3.03
....	3.00	3.25	3.55	5.80	6.50	7.83	5.68	4.43	3.95	3.95	3.65	2.85
....	3.05	3.25	3.65	5.80	6.53	7.83	5.55	4.38	3.95	3.95	3.53	2.90
....	3.03	......	3.68	5.83	6.60	8.05	5.40	4.28	3.95	3.88	3.53	2.85
....	2.90	......	3.70	5.70	6.70	8.08	5.25	4.15	3.98	3.85	3.48	2.82
....	2.90	......	3.70	......	7.00	......	5.15	4.03	......	3.85	......	2.88

a No readings, on account of wire on gage being broken.   b Heavy rain.

Rating table for Grand River at Glenwood Springs, Colo., for 1903.

Gage height.	Discharge.	Gage height.	Discharge.	Gage height.	Discharge.	Gage height.	Discharge.
Feet.	Second-feet.	Feet.	Second-feet.	Feet.	Second-feet.	Feet.	Second-feet.
2.8	380	3.8	990	5.6	3,400	7.6	8,620
2.9	430	3.9	1,080	5.8	3,790	7.8	9,380
3.0	480	4.0	1,180	6.0	4,210	8.0	10,210
3.1	530	4.2	1,400	6.2	4,640	8.2	11,080
3.2	580	4.4	1,630	6.4	5,090	8.4	12,020
3.3	630	4.6	1,870	6.6	5,580	8.6	13,040
3.4	690	4.8	2,140	6.8	6,110	8.8	14,150
3.5	760	5.0	2,440	7.0	6,670	9.0	15,390
3.6	830	5.2	2,740	7.2	7,260	9.2	16,650
3.7	910	5.4	3,050	7.4	7,910	9.4	17,910

Estimated monthly discharge of Grand River at Glenwood Springs, Colo., for 1903.

[Drainage area, 5,838 square miles.]

Month.	Discharge in second-feet.			Total in acre-feet.	Run-off.	
	Maximum.	Minimum.	Mean.		Second-feet per square mile.	Depth inch
January a............	505	430	458	28, 161	0. 078	0.
February............	605	455	544	30, 212	. 093	.
March..............	950	555	740	45, 501	. 127	.
April..............	3, 853	894	1, 685	100, 264	. 289	.
May	10, 992	2, 890	5, 803	356, 813	. 994	1.
June	17, 910	8, 260	12, 804	761, 891	2. 193	2.
July	9, 708	2, 665	5, 222	321, 088	. 894	1.
August............	2, 515	1, 213	1, 613	99, 180	. 276	.
September..........	1, 690	1, 035	1, 342	79, 855	. 230	.
October............	1, 510	1, 035	1, 313	80, 733	. 225	.
November..........	1, 035	595	857	50, 995	. 147	.
December	830	395	518	31, 851	. 089	.
The year	17, 910	395	2, 742	1, 986, 544	. 470	6.

a January 1 to 12 estimated.

MISCELLANEOUS MEASUREMENTS IN THE GRAND RIVER DRAINAGE BASIN, COLORADO.

The following miscellaneous measurements were made by W. Sammis, F. Cogswell, and W. P. Edwards in 1903:

Miscellaneous measurements in Grand River drainage basin in 1903.

Date.	Stream.	Locality.	Gage height.	Di char
			Feet.	Secon
May 12.....	Cushman ditch	Near Montrose, Colo...............		
22.....	......do	3 miles south of Montrose, Colo......		
June 23.....	......do	4 miles south of Montrose, Colo......		
24.....	......do	13 miles west of Montrose, Colo......		
May 12.....	Dry Creek	Montrose Canal flume		
22.....	......do	10 miles west of Montrose, Colo......		
June 24.....	......do	13 miles west of Montrose, Colo......		
May 20.....	Ironstone ditch	10 miles west of Montrose, Colo......		
30.....	......do	......do............		
20.....	Leroux Creek	Hotchkiss, Colo...............		
29.....	......do	......do............		
June 4.....	Loutsenhizer ditch	5 miles south of Montrose, Colo......		
25.....	......do	3 miles south of Montrose, Colo......		
Apr. 29.....	Mill ditch	Delta, Colo............		
May 7.....	......do	......do............		
18.....	......do	......do............		
27.....	......do	......do............		
July 0..	do	......do............		

Miscellaneous measurements in Grand River drainage basin in 1903—Continued.

Date.	Stream.	Locality.	Gage height.	Discharge.
			Feet.	*Second-ft.*
ʙ. 23.....	Mill ditch	Delta, Colo....................		.9
ᴍʳ. 8.....	Roaring Forks..............	Glenwood Springs, Colo	....	415
ᴊ 8........	....do.........	....do.........		2,143
28........	....do.........	....do.........		2,268
ᴍᴀᴇ 17.....	....do.........	....do.........	6.80	8,052
ᴋy 16......	....do.........	....do.........	5.10	3,448
ᴇ. 7.......	....do.........	....do.........	3.90	1,129
30........	....do.........	....do.........	3.20	655
ᴍᴛ. 27.....	....do.........	....do.........	3.25	660
ᴛ. 26.....	....do.........	....do.........	3.10	547
ᴠ. 19.....	....do.........	....do.........	3.00	519
ꜰ. 11.....	Lost Canyon Creek........	Dolores, Colo	....	110
ᴊ 13.....	....do.........	....do.........		183
ᴀᴇ 3.....	....do.........	....do.........		113
23.....	....do.........	....do.........	2.30	48
ᴊ 22.....	....do.........	....do.........	1.70	.4
ᴇ. 13.....	....do.........	....do.........		0
ᴏᴛ. 4.....	....do.........	....do.........		0
ᴛ. 2.....	....do.........	....do.........		0
30.....	....do.........	....do.........		0
ᴊʳ. 25.....	Montrose canal..........	Montrose, Colo..........		166
ᴀy 4.....	....do.........	In flume	2.00	186
11.....	....do.........	Near head-gate		225
16.....	....do.........	At head-gate...........	2.40	200
23.....	....do.........	Flume 6 miles south of Montrose, Colo.	1.00	87
ᴍᴇ 4.....	....do.........	7 miles south of Montrose, Colo.......	2.40	241
11.....	....do.........	6 miles south of Montrose, Colo.......	2.70	212
26.....	....do.........	....do.........		185
ᴜʏ 10.....	....do.........	At head gate...........		216
13.....	....do.........	6 miles south of Montrose, Colo.......		220
ᴜɢ. 3.....	....do.........	6 miles south of Montrose, Colo., at head gate.	1.60	112
ᴘᴛ. 4.....	....do.........	....do.........	1.10	65
ᴀʏ 13.....	Selig ditch.............	....		22
ᴜʏ 14.....	....do.........	3 miles west of Montrose, Colo......		3
ᴀʏ 12.....	Spring Creek..........	4 miles from Montrose, Colo		134
22.....	....do.........	3 miles south of Montrose, Colo.....		30
ᴜɴᴇ 23.....	....do.........	4 miles south of Montrose, Colo......		1.8
ᴘʀ. 28.....	Squaw Creek	Cimarron, Colo...........		28
ᴀy 6.....	....do.........	....do.........		8
15.....	....do.........	....do.........		1
ᴜɴᴇ 9.....	....do.........	....do.........		32
ᴀy 12.....	Supply ditch	Water taken from Spring Creek into Montrose canal.	....	19
22.....	....do.........	....do.........		24
ᴊᴜɴᴇ 23.....	....do.........	....do.........		4
May 12.....	....do.........	Water taken from Dry Creek into Montrose canal.	....	35
22.....	....do.........	....do.........		7
June 24.....	....do.........	....do.........		1.2
May 13.....	Midland ditch	At head-gate...........		21
26.....	Cedar Creek.............	10 miles east of Montrose, Colo........		6
29.....	Gunnison River	State bridge near Cory, Colo...........		5,717
June 4.....	City ditch	2 miles south of Montrose, Colo.		9
26.....	Highline ditch	8 miles south of Montrose, Colo.........		45

Miscellaneous measurements in Grand River drainage basin in 1903—Continued.

Date.	Stream.	Locality.	Gage height.	Dis-charge
			Feet.	*Sec.-ft*
Sept. 25.....	Ten Mile Creek	At Uneva Lake, Colo..............	1.30	
Oct. 27.....	.....do...............	.....do...............	1.05	
Sept. 8.....	Beaver Creek.............	3 miles above junction with Dallas...		
Aug. 1.....	Big Blue Creek............	....		
Sept. 29.....	.....do.........	....		
28.....	East Fork Big Blue Creek..	Above junction with West Fork....		
Aug. 18.....	Bush Creek............	6 miles above junction with Slate River.		
21.....	Carbon Creek	Above junction with Ohio Creek		
1.....	Cebolla River	Road bridge near Du Bois		
Sept. 1.....	.....do...........	3 miles above junction of Spring Creek.		
2.....	.....do.........	Hot Springs bridge		
18.....	Cimarron River	Head-gate Cimarron ditch....		
20.....	.....do.........	Jackson, East Fork..............		
20.....	.....do.........	Jackson, Middle Fork..........		
29.....	.....do.........	Cimarron, Colo	1.70	
Aug. 26.....	Cochetopa Creek	Above junction with Tomichi Creek ..		
28.....	.....do.........	Above junction of Los Pinos Creek....		
18.....	Copper Creek	1/4 mile below Gothic		
Sept. 9.....	Cow Creek	Upper Ford, 9 miles above Uncompahgre		
9.....	.....do.........	Junction with Uncompahgre......		
Aug. 4.....	Crystal Creek..........	Junction, 5 miles above Spring		
Sept. 8.....	Dallas Creek.........	Main fork 3 miles above Uncompahgre River ...		
8.....	.....do.........	Road between Hager and Ridgway .		79
Aug. 4.....	East Creek...........	At Almont, above junction......		576
14.....	.....do.........	Oversteg, below junction with Slate River		190
18.....	East River............	Bridge, 4 miles above junction with Slate River....		117
18.....	.....do.........	1/4 mile below Gothic		36
Sept. 2.....	Henson Creek	3 miles above Capitol City		19
2.....	.....do.........	1/4 mile above Lake City		56
Aug. 12.....	Illinois Creek	Main Tincup road in Taylor Park		9
1.....	Lake Fork River	Lake Fork Road		277
Sept. 2.....	.....do.........	Bridge, 2 miles above Gateview		166
4.....	.....do.........	Lake shore, 4 miles above Lake City .		56
4.....	.....do.........	1 mile above Sherman........		24
Aug. 1.....	Little Blue Creek..........	Lake Fork road...........		8
Sept. 17.....	Little Cimarron	Above junction with Cimarron......		17
29.....	.....do.........	.....do....		19
22.....	.....do.........	East of Jackson		17
Aug. 24.....	Little Tomichi Creek......	Road bridge between Urianita and Sargent.		31
5.....	Lottys Creek.............	Union Park, above junction Taylor River.		9
5.....	.....do.........	Flume, exit from Union Park.......		27
28.....	Los Pinos.............	Above junction with Cochetopo		8
Sept. 1.....	Mineral Creek..........	Junction with Cebolla		10
Aug. 21.....	Ohio Creek	Road 1/4 mile below Baldwin's mine...		42
12.....	Pie Plant Creek	3 miles west of Illinois Creek		10
1.....	Pine Creek	Lake Fork road.........		10
22.....	Quartz Creek	Wagon road at Parlin........		25
22.....	.....do.........	At Pitkin		17
15.....	Slate River	Above junction with East River......		70

Miscellaneous measurements in Grand River drainage basin in 1903—Continued.

Date.	Stream.	Locality.	Gage height.	Discharge.
			Feet.	Second-ft.
Aug. 18.....	Slate River...................	¼ mile below Anthracite..................		20
4.....	Spring Creek...............	8 miles above Almont and above junction.		55
31..... .....do........		5 miles above Cathedral..................		13
4.....	Taylor River...............	Bridge 7 miles above Spring Creek		366
12..... .....do........		Canyon entrance from Taylor Park.........		141
14..... .....do........		1½ miles below Dorchester post-office, in Taylor Park.		37
13.....	Texas Creek...............	Main Tincup road in Taylor Park		36
22.....	Tomichie Creek............	Bridge, Haverly siding, 2½ miles from Gunnison.		71
24..... .....do........		1 mile below Sargent....................		78
Sept. 7.....	Uncompahgre River	Road bridge at Ridgway..................		255
8.....	West Dallas Creek..........	2 miles above junction with Dallas............		17
Aug. 1.....	Willow Creek...............	Lake Fork road		4
3..... .....do........		Road between Iola and Gunnison		10
12..... .....do........		From Tincup into Taylor Park, 1 mile above junction.		35

Mean daily gage height, in feet, of North Tenmile Creek, at Unera Lake, Colo., for 1903.

Day.	Sept.	Oct.	Nov.	Day.	Sept.	Oct.	Nov.	Day.	Sept.	Oct.	Nov.
1..............	.....	2.10	1.08	12..............		1.15	.....	23..............			1.10	.....
2..............	.....	2.00	1.08	13..............		1.13	.....	24..............			1.10	.....
3..............	.....	2.15	1.08	14..............		1.13	.....	25..............			1.10	.....
4..............	.....	2.40	1.08	15..............		1.10	.....	26..............			1.10	.....
5..............	.....	1.75	1.08	16..............		1.15	.....	27..............		1.10	1.07	.....
6..............	.....	1.55	1.08	17..............		1.23	.....	28..............		1.15	1.08	.....
7..............	.....	1.35	1.08	18..............		1.20	.....	29..............		2.15	1.08	.....
8..............	.....	1.25		19..............		1.15	.....	30..............		2.40	1.13	.....
9..............	.....	1.25		20..............		1.10	.....	31..............			1.08	.....
10..............		1.15	.....	21..............		1.10	.....				
11..............		1.10	.....	22..............		1.10	.....				

WHITE RIVER AT MEEKER, COLO.

This station was established by A. L. Fellows, May 24, 1901, about one-half mile above the town of Meeker, at a point where a wagon bridge crosses the stream on the ranch of L. F. Van Cleave. The gage rod consists of a vertical 2 by 4 inch timber nailed to the left abutment of the bridge on the downstream side. The bridge is marked every 5 feet. The initial point for soundings is at the rod at the left or south end of the bridge. The channel is straight for 500 feet above and below the station. The current is swift. There is but one channel at all stages. The banks are so high that they are not liable to overflow; the channel is of rock and gravel and seems permanent in its nature. Measurements are made from the lower side of the wagon bridge. The observer is L. F. Van Cleave, who reads the rod twice each day. Bench mark No. 1 is the top of a bolt in a truss immediately above the gage rod; elevation above zero of gage, 10.83 feet.

Bench mark No. 2 is a spike driven in the left post of the gate ne bottom; elevation above zero of gage, 9.29 feet.   Bench mark No. is a nail on south side of the post directly east of bench mark No. elevation above zero of gage, 9.14 feet.

The observations at this station during 1903 have been made und the direction of M. C. Hinderlider, district hydrographer.

Discharge measurements of White River at Meeker, Colo., in 1903.

Date.	Hydrographer.	Gage height.	Dischar
		Feet.	*Second-f*
July 22......................	R. S. Stockton	4.08	
July 22......................	McDermith & Stockton.......	4.08	
August 1	R. S. Stockton	3.80	
August 21	O. McDermith...............	3.60	
September 1	McDermith & Stockton.......	3.77	
September 16	.....do	4.05	

Mean daily gage height, in feet, of White River at Meeker, Colo., for 1903.

Day.	Apr.	May.	June.	July.	Aug.	Sept.	Oc
1...	3.75	4.20	5.50	4.95	3.80	3.75	
2...	3.90	4.15	5.60	4.85	3.70	3.70	
3...	3.75	4.15	5.65	4.70	3.70	3.70	
4...	3.70	4.25	5.70	4.65	3.70	3.70	
5...	3.70	4.45	5.35	4.55	3.65	3.70	
6...	3.65	4.55	5.65	4.50	3.60	3.65	
7...	3.65	4.75	5.65	4.50	3.60	a 4.05	
8...	3.70	4.65	5.60	4.45	3.60	3.90	
9...	3.75	4.90	5.60	4.40	3.60	3.80	
10...	3.80	5.00	5.60	4.40	3.60	3.80	
11...	3.80	5.00	5.55	4.30	3.50	3.80	
12...	3.75	5.10	5.35	4.25	3.60	3.95	
13...	3.65	5.20	5.75	4.20	3.60	4.00	
14...	3.75	5.40	5.85	4.20	3.60	4.30	
15...	3.75	5.50	5.85	4.20	3.60	4.10	
16...	3.70	5.55	5.75	4.30	3.60	4.00	
17...	3.75	5.75	5.75	4.35	3.60	3.95	
18...	3.75	5.35	5.85	4.15	3.60	3.95	
19...	3.75	5.20	5.75	4.05	3.60	3.90	
20...	3.75	4.95	5.80	4.10	3.55	3.95	
21...	3.80	5.05	5.60	4.05	3.55	3.95	
22...	3.85	4.90	5.65	4.00	3.55	3.95	
23...	3.90	4.75	5.65	3.95	3.65	3.95	
24...	3.95	4.75	5.50	3.90	3.80	3.90	
25...	4.10	4.75	5.40	3.90	3.80	3.90	
26...	4.20	4.70	5.30	3.90	3.70	3.90	
27...	4.20	4.75	5.20	3.95	3.70	3.85	
28...	4.20	4.85	5.20	3.90	3.70	3.85	
29...	4.30	4.85	5.15	3.90	3.70	3.85	
30...	4.10	5.00	5.05	3.80	3.75	4.00	
31...		5.35		3.80	3.75		

a Rain.

Rating table for White River at Meeker, Colo., from April 1 to December 31, 1903.

Gage height.	Discharge.	Gage height.	Discharge.	Gage height.	Discharge.	Gage height.	Discharge.
Feet.	*Second-feet.*	*Feet.*	*Second-feet.*	*Feet.*	*Second-feet.*	*Feet.*	*Second-feet.*
3.5	321	4.1	625	4.7	1,417	5.3	2,209
3.6	340	4.2	757	4.8	1,549	5.4	2,341
3.7	362	4.3	889	4.9	1,681	5.5	2,473
3.8	392	4.4	1,021	5.0	1,813	5.6	2,605
3.9	434	4.5	1,153	5.1	1,945	5.7	2,737
4.0	510	4.6	1,285	5.2	2,077	5.8	2,869

Table well defined from 3.50 to 4.10 feet gage height, the remainder is based on a high-water measurement made in 1901 at a gage height of 5.60 feet.

Estimated monthly discharge of White River at Meeker, Colo., for 1903.

[Drainage area, 634 square miles.]

Month.	Discharge in second-feet.			Total in acre-feet.	Run-off.	
	Maximum.	Minimum.	Mean.		Second-feet per square mile.	Depth in inches.
April...............	889	351	458	27,253	0.72	0.80
May................	2,803	691	1,675	102,992	2.64	3.04
June	2,935	1,879	2,563	152,509	4.04	4.51
July...............	1,747	392	822	50,543	1.30	1.50
August	392	321	350	21,521	.55	.63
September	889	362	458	27,253	.72	.80
October	568	392	450	27,669	.71	.82
The period	2,935	321	....	409,740	....	....

SOUTH FORK OF WHITE RIVER NEAR BUFORD, COLO.

This station was established July 25, 1903, by Robert S. Stockton. It is located at the county bridge at the lower end of a section of the river known as the "Stillwater." It is about 7 miles from Buford, Colo., the nearest post-office, and is about 30 miles from Meeker, Colo. The elevation above sea level as determined by aneroid barometer is 7,400 feet. The gage is a 2 by 4 inch vertical pine timber 10 feet long. It is spiked to the upper side of the middle pier. It is read twice each day by George Hazard. Discharge measurements are made from the bridge at high water and by wading at low water. The initial point for soundings is the edge of the abutment at the west end of the bridge. The bridge makes an angle of 22° with the normal to the stream. This is taken into account in making measurements. The channel is straight for 50 feet above the station and for 300 feet below. The current is swift. Both banks are high, covered with grass and sagebrush, and are not subject to overflow. The bed of the stream below the Stillwater and at the gaging station is covered with bowlders, some of which are 2 or 3 feet in diameter. There is but one channel, broken by the middle pier of the bridge. Bench mark No. 1 is the first bolt on the bridge east of the gage. Its elevation is 11.95 feet above the zero of the gage. Bench mark No. 2 is a spike in one of the logs of the abutment at the southwest corner of the bridge. Its elevation is 8.57 feet above the zero of the gage. Bench mark No. 3 is the corner of a large rock 30 feet west of the bridge and north of the road. Its elevation is 13.14 feet above the zero of the gage. Bench mark No. 4 is the top of a pyramid-shaped rock on the east side of the river below the bridge, 41 feet distant from the northeast bolt on the bridge tie. The rock projects 8 inches above the ground. The elevation of the bench mark is 7.02 feet above the zero of the gage. The bridge has a span between abutments of 55 feet.

The observations at this station during 1903 have been made under the direction of M. C. Hinderlider, district hydrographer.

Discharge measurements of South Fork of White River near Buford, Colo., in 1903.

Date.	Hydrographer.	Gage height.	Discharge.
		Feet.	*Second-feet.*
July 25	R. S. Stockton	2.80	290
September 18	O. McDermith	2.63	240
August 27	O. McDermith and R. S. Stockton.	2.52	207

Mean daily gage height, in feet, of South Fork of White River near Buford, Colo., for 1903.

Day.	July.	Aug.	Sept.	Oct.	Day.	July.	Aug.	Sept.	Oct.
1		2.70	2.50	2.63	17		2.50	2.60	2.65
2		2.65	2.48	2.65	18		2.50	2.63	(a)
3		2.60	2.50	2.73	19		2.50	2.60	(a)
4		2.60	2.45	2.65	20		2.50	2.60	(a)
5		2.60	2.50	2.68	21		2.50	2.60	(a)
6		2.60	2.55	2.68	22		2.50	2.65	(a)
7		2.60	2.65	2.65	23		2.50	2.65	2.60
8		2.60	2.60	2.65	24		2.60	2.63	2.60
9		2.60	2.50	2.65	25		2.50	2.60	2.58
10		2.60	2.70	2.65	26	2.80	2.50	2.60	2.55
11		2.55	2.55	2.73	27	2.80	2.50	2.60	2.55
12		2.60	2.60	2.70	28	2.75	2.50	2.60	2.55
13		2.55	2.75	2.70	29	2.70	2.50	2.63	2.55
14		2.50	2.70	2.70	30	2.70	2.50	2.63	2.55
15		2.50	2.63	2.70	31	2.70	2.50		2.55
16		2.55	2.55	2.65					

a Observer absent.

thing table for South Fork of White River near Buford, Colo., from July 26 to December 31, 1903.

Gage height.	Discharge.	Gage height.	Discharge.	Gage height.	Discharge.	Gage height.	Discharge.
Feet.	*Second-feet.*	*Feet.*	*Second-feet.*	*Feet.*	*Second-feet.*	*Feet.*	*Second-feet.*
2. 45	186	2. 55	215	2. 65	245	2. 75	275
2. 50	200	2. 60	230	2. 70	260	2. 80	290

Straight-line curve defined by three points.

timated monthly discharge of South Fork of White River near Buford, Colo., for 1903.

[Drainage area, 148 square miles.]

Month.	Discharge in second-feet.			Total in acre-feet.	Run-off.	
	Maximum.	Minimum.	Mean.		Second-feet per square mile.	Depth in inches.
July 26–31..........	290	260	273	3, 249	1. 85	0. 34
August	260	200	215	13, 220	1. 45	1. 67
September..........	275	186	226	13, 448	1. 53	1. 71
October a	269	215	240	14, 757	1. 62	1. 87
The period ...	....	....	....	44, 674	....	....

a October 18 to 22 estimated.

NORTH FORK OF WHITE RIVER NEAR BUFORD, COLO.

This station was established July 28, 1903, by Robert S. Stockton. It is located at the county bridge at Rawson's ranch, below the mouth of Marvine Creek, 7 miles from Buford, the nearest post-office, and 32 miles from Meeker, Colo. The gage is a 2 by 4 inch vertical pine timber, spiked to the lower side of the first pier from the south end of the bridge. It reads from 1 to 9 feet. It is read twice each day by H. N. Rawson. Discharge measurements are made from the three-span highway bridge to which the gage is attached. The bridge has a total span of 85 feet. The initial point for soundings is the edge of the abutment at the south end of the bridge. The channel is straight for 200 feet above and 300 feet below the station, and the current is swift. Both banks are high and are not liable to overflow. The right bank is covered with trees and brush. The bed of the stream is rocky and free from vegetation. There is but one channel at all stages, broken by the two narrow bridge piers. Bench mark No. 1 is the top of a bolt on the bridge nearest the gage. Its elevation is 11.51 feet above the zero of the gage. Bench mark No. 2 is a spike in the trunk of a large cottonwood tree at the southeast corner of the bridge. Its elevation is

7.46 feet above the zero of the gage. Bench mark No. 3 is a spike in one of the abutment logs at the end of the upper sill at the southwest corner of the bridge. Its elevation is 7.88 feet above the zero of the gage..

The observations at this station during 1903 have been made under the direction of M. C. Hinderlider, district hydrographer.

Discharge measurements of North Fork of White River near Buford, Colo., in 1903.

Date.	Hydrographer.	Gage height.	Discharge
		Feet.	Second-feet
July 28......................	R. S. Stockton	2.10	343
September 19	O. McDermith	2.00	291
August 23	O. M. McDermith and Stockton.	1.95	281

Mean daily gage height, in feet, of North Fork of White River near Buford, Colo., for 1903.

Day.	July.	Aug.	Sept.	Oct.	Nov.	Day.	July.	Aug.	Sept.	Oct.	Nov.
1..................	......	2.00	1.90	2.03	1.90	17..................	......	2.00	1.98	1.93	......
2..................	......	2.00	1.90	2.05	1.88	18..................	......	2.00	1.98	1.98	......
3..................	......	2.00	1.90	2.03	1.88	19..................	......	1.95	1.95	1.98	......
4..................	......	2.00	1.90	1.98	1.90	20..................	......	1.95	1.98	1.98	......
5..................	......	2.00	1.98	2.00	......	21..................	......	1.90	1.98	1.98	......
6..................	......	2.00	2.15	1.98	......	22..................	......	1.93	2.00	1.98	......
7..................	......	2.00	2.00	1.98	......	23..................	......	1.98	2.00	1.93	......
8..................	......	2.00	1.98	1.98	......	24..................	......	2.00	1.98	1.93	......
9..................	......	2.00	1.95	1.98	......	25..................	......	1.95	1.98	1.93	......
10..................	......	2.00	1.90	2.00	......	26..................	......	1.95	1.95	1.93	......
11..................	......	2.00	2.00	2.03	......	27..................	......	1.95	1.98	1.93	......
12..................	......	2.00	1.95	2.00	......	28..................	......	1.95	1.98	1.93	......
13..................	......	2.00	2.08	1.98	......	29..................	2.10	1.95	2.00	1.93	......
14..................	......	2.00	2.03	1.98	......	30..................	2.10	1.95	2.03	1.90	......
15..................	......	2.00	2.00	1.98	......	31..................	2.00	1.90	......	1.90	......
16..................	......	2.00	1.98	1.93	......						

Rating table for North Fork of White River near Buford, Colo., from July 29 to December 31, 1903.

Gage height.	Discharge.	Gage height.	Discharge.	Gage height.	Discharge.	Gage height.	Discharge.
Feet.	Second-feet.	Feet.	Second-feet.	Feet.	Second-feet.	Feet.	Second-feet.
1.85	168	1.95	242	2.05	317	2.15	392
1.90	205	2.00	280	2.10	355	2.20	430

Straight-line curve defined by three points.

imated monthly discharge of North Fork of White River near Buford, Colo., for 1903.

[Drainage area, 181 square miles.]

Month.	Discharge in second-feet.			Total in acre-feet.	Run-off.	
	Maximum.	Minimum.	Mean.		Second-feet per square mile.	Depth in inches.
y 29–31..........	355	280	330	1,964	1.82	0.20
gust	280	205	263	16,171	1.45	1.67
ptember..........	392	205	263	15,650	1.45	1.62
tober	340	205	252	15,495	1.39	1.60
The period....	....	....	....	49,280	....	....

MARVINE CREEK NEAR BUFORD, COLO.

This station is located at a point where the stream is crossed by a large aspen log. The nearest post-office is Buford, Colo., about 10 miles distant. Meeker, Colo., is about 35 miles distant. The elevation of this point, as determined by an aneroid barometer, is 7,550 feet above sea level. The gage is a vertical 2 by 4 inch pine timber 5 feet long, fastened to the lower side of ᴊe foot log, which is used as a bridge. Discharge measurements are made from the log which spans the stream at the gage. The channel is 30 feet wide at this point. The initial point for soundings is at the gage rod. The channel is straight for 100 feet above and below the station, and the current is swift. The right bank is sloping and will overflow for 10 or 15 feet at high water. The left bank is steep and will not overflow. Both banks are covered with thick brush. The bed of the stream at the station is covered with bowlders and is free from vegetation. The channel is divided into two parts by a large sunken log, which supports the middle of the foot bridge. Bench mark No. 1 is the top of a rock 22 feet west of the gage rod. Its elevation is 4.62 feet above the zero of the gage. Bench mark No. 2 is the top of a triangular-shaped rock between two spruce trees on the west bank. Its elevation is 6.01 feet above the zero of the gage. Bench mark No. 3 is the top of a large rock 30 feet northwest of the gage. Its elevation is 4.61 feet above the zero of the gage.

The observations at this station during 1903 have been made under the direction of M. C. Hinderlider, district hydrographer.

Discharge measurements of Marvine Creek near Buford, Colo., in 1903.

Date.	Hydrographer.	Gage height.	Discharge.
		Feet.	*Second-feet.*
July 27......................	R. S. Stockton	2.05	137
August 26	O. McDermith and Stockton...	1.94	111
September 19	O. McDermith	1.91	100

Mean daily gage height, in feet, of Marvine Creek near Buford, Colo., for 1903.

Day.	July.	Aug.	Sept.	Oct.	Day.	July.	Aug.	Sept.	Oct.
1....................	......	2.05	1.93	1.93	17....................	......	2.00	1.93	......
2....................		2.00	1.90	1.95	18....................		2.00	1.93	......
3....................		2.00	1.90	1.95	19....................		2.00	1.93	......
4....................		2.00	1.90	1.95	20....................		2.00	1.93	......
5....................		2.00	1.90	1.93	21....................		2.00	1.93	......
6....................		2.00	1.95	1.93	22....................		1.95	1.90	......
7....................		2.00	1.95	1.90	23....................		1.95	1.90	......
8....................		2.00	1.95	1.90	24....................		2.00	1.90	......
9....................		2.00	1.93	1.90	25....................		2.00	1.90	......
10....................		2.00	1.90	1.90	26....................		1.95	1.90	......
11....................		2.00	1.93	1.90	27....................		1.95	1.90	......
12....................		2.00	1.95	1.93	28....................	2.10	1.95	1.90	......
13....................		2.00	2.00	1.90	29....................	2.05	1.95	1.93	......
14....................		2.00	2.00	......	30....................	2.10	1.95	1.93	......
15....................		2.00	1.95	......	31....................	2.10	1.95	......	......
16....................		2.00	1.95	......					

Rating table for Marvine Creek near Buford, Colo., from July 28 to December 31, 1903.

Gage height.	Discharge.	Gage height.	Discharge.	Gage height.	Discharge.	Gage height.	Discharge.
Feet.	*Second-feet.*	*Feet.*	*Second-feet.*	*Feet.*	*Second-feet.*	*Feet.*	*Second-feet.*
1.90	98	1.95	112	2.00	125	2.05	139
						2.10	152

Straight-line curve defined by three points.

Estimated monthly discharge of Marvine Creek near Buford, Colo., for 1903.
[Drainage area, 50 square miles.]

Month.	Discharge in second-feet.			Total in acre-feet.	Run-off.	
	Maximum.	Minimum.	Mean.		Second-feet per square mile.	Depth in inches.
July 28–31..........	152	139	149	1,182	2.98	0.44
August	139	112	122	7,501	2.44	2.81
September..........	125	98	106	6,307	2.12	2.37
October 1–13........	120	98	105	2,707	2.10	1.01
The period ...	....	....	....	17,697	....	....

The following miscellaneous measurements were made by Ora
cDermith, L. C. Hill, H. C. Berry, and R. S. Stockton in the Green
ᵢver basin in 1903:

Miscellaneous measurements in the Green River basin in 1903.

Date.	Stream.	Locality.	Discharge.
			Second-feet.
ₚt. 10	Elk River	Trull, Colo	166
9	Elkhead Creek	Craig, Colo	2. 6
11	Fish Creek	Steamboat Springs, Colo.......	7
9	Fortification Creek	Craig, Colo	.7
ne 24	Roaring Fork of Yampa River.	Lower Stillwater, Colo	181
ₓt. 10	Trout Creek	Pool, Colo.....................	13
ly 27	Ute Creek	Near Buford, Colo.............	22
ₓt. 2	White River...........	Wilburs Bridge, Colo..........	347
3	.....do	White River City, Colo........	364
13	Williams Fork River ...	Hamilton, Colo	100
9	Yampa River	Craig, Colo	348
11	.....do	Steamboat Springs, Colo.......	129

UINTA RIVER AT OURAY SCHOOL, UTAH.

This station was established November 8, 1899, by C. C. Babb,
ᵢsted by C. T. Prall. It is located at the highway bridge 5 miles
low the station at Fort Duchesne. The gage is a vertical 1½ by 5
·h board, 9 feet long, nailed to the east side of the north crib of the
ᵢdge. It is read twice each day by O. M. Waddell, the superintend-
ₜof the Indian school. Discharge measurements are made at high
ᵧes from the bridge and at ordinary stages by wading about 300 feet
low. The initial point for soundings for the section at the bridge is
₂ zero mark on the bridge railing. The initial point for the wading
ₜion is the first tag from the post on the right bank to which the wire
fastened. The channel is curved above the bridge and is straight
ᵣ 600 feet below. The right bank is high, is composed of gravel, and
not subject to overflow. The left bank is low and will overflow at
ᵧh stages. The bed of the stream is rocky and is filled in with sedi-
ₑnt during a part of the year. The central pier of the bridge divides
₃ bridge section into two parts. Bench mark No. 1 is the center one
the line of nails driven in a cottonwood tree at the northwest corner
the bridge. Its elevation is 7.48 feet above the zero of the gage.

Bench mark No. 2 is a nail in the flagstaff in the school grounds. elevation is 22.64 feet above the zero of the gage and 4,760 feet ι sea level.

· The observations at this station during 1903 have been made ι the direction of H. S. Reed, district hydrographer.

Discharge measurements of Uinta River at Ouray School, Utah, in 1903.

Date.	Hydrographer.	Gage height.	Dis
		Feet.	*Seco*
March 29	H. S. Reed	1.06	
April 4	do	.63	
April 17	do	.41	
April 23	do	.46	
May 1	do	.40	
May 7	do	.80	
May 14	do	2.80	
May 21	do	1.90	
May 28	do	1.63	
June 4	do	3.90	
June 19	do	3.00	
June 26	do	2.18	
July 3	do	1.70	
July 9	do	1.57	
July 16	do	1.65	
July 23	do	1.40	
July 30	do	.88	
August 6	do	.47	
August 14	do	.23	
August 20	do	.18	
August 27	do	.30	
September 3	do	.30	
September 11	do	.50	
September 18	do	.66	
September 26	do	.43	
October 2	do	1.13	
October 7	do	.70	
October 20	do	.62	
October 29	do	.60	
November 13	do	.80	
November 20	do	.84	
November 27	do	.55	
December 4	do	.40	

Mean daily gage height, in feet, of Uinta River at Ouray School, Utah, for 1903.

Day.	Mar.	Apr.	May.	June.	July.	Aug.	Sept.	Oct.	Nov.	Dec.
1		0.98	.43	3.08	1.82	0.66	0.18	0.65	0.56	0.61
2		1.18	.40	3.65	1.73	.66	.21	1.07	.57	.53
3		.72	.50	3.95	1.70	.65	.28	1.48	.60	.62
4		.61	.54	3.85	1.62	.53	.25	.92	.60	.52
5		.65	.60	3.68	1.56	.50	.29	.78	.61	.60
6		.53	.72	3.72	1.42	.45	.35	.75	.63	
7		.50	.88	3.85	1.50	.38	1.00	.69	.60	.30
8		.55	1.01	4.13	1.85	.30	.65	.65	.61	.32
9		.54	1.06	4.15	1.60	.27	.59	.68	.55	.64
10		.58	1.22	3.80	1.41	.26	.56	.75	.43	.62
11		.59	1.36	3.58	1.30	.28	.52	.72	.42	.75
12		.48	1.53	3.50	1.25	.25	.78	.70	.45	.84
13		.43	1.87	3.53	1.19	.23	.76	.71	.75	
14		.37	2.56	3.78	1.21	.22	.75	.68	.72	
15		.43	2.98	3.45	1.20	.22	.82	.67	.74	
16		.45	3.13	3.18	1.17	.22	.86	.70	.70	
17		.43	2.95	3.05	2.15	.22	.72	.67	.37	
18		.44	2.46	3.03	2.35	.20	.69	.65	.34	
19		.42	2.20	2.83	1.45	.18	.65	.64	.60	
20		.40	2.01	2.73	1.25	.18	.57	.64	.68	
21		.36	1.93	2.60	1.18	.18	.52	.63	.88	
22		.41	1.90	2.56	1.40	.18	.50	.64	.85	
23		.48	1.75	2.45	1.48	.24	.48	.64	.82	
24		.52	1.67	2.34	1.27	.35	.47	.60	.77	
25		.57	1.58	2.25	1.28	.48	.46	.62	.53	
26		.58	1.50	2.08	1.25	.30	.44	.63	.52	
27		.60	1.50	2.06	(a)	.30	.44	.62	.53	
28		.60	1.58	2.03	(a)	.32	.45	.61	.52	
29	1.08	.54	1.57	2.00	1.00	.33	.52	.60	.55	
30	1.28	.50	1.90	1.94	.88	.28	.55	.58	.60	
31	1.38		2.43		.75	.25		.57		

a No record.

Rating table for Uinta River at Ouray School, Utah, for 1903.

Gage height.	Discharge.	Gage height.	Discharge.	Gage height.	Discharge.	Gage height.	Discharge.
Feet.	*Second-feet.*	*Feet.*	*Second-feet.*	*Feet.*	*Second-feet.*	*Feet*	*Second-feet.*
0.2	58	1.2	248	2.2	620	3.2	1,450
.3	72	1.3	273	2.3	680	3.3	1,570
.4	88	1.4	300	2.4	740	3.4	1,700
.5	105	1.5	328	2.5	810	3.5	1,840
.6	123	1.6	358	2.6	885	3.6	1,980
.7	141	1.7	390	2.7	965	3.7	2,120
.8	160	1.8	425	2.8	1,055	3.8	2,260
.9	180	1.9	467	2.9	1,145	3.9	2,400
1.0	202	2.0	515	3.0	1,240	4.0	2,540
1.1	224	2.1	565	3.1	1,340	4.2	2,820

Table well defined.

Estimated monthly discharge of Uinta River at Ouray School, Utah, for 1903.

[Drainage area, 967 square miles.]

Month.	Discharge in second-feet.			Total in acre-feet.	Run-off.	
	Maximum.	Minimum.	Mean.		Second-feet per square mile.	Depth in inches.
April	248	80	115	6,843	0.12	0.13
May	1,395	88	447	27,485	.46	.53
June	2,750	490	1,498	89,137	1.55	1.73
July [a]	710	150	313	19,246	.32	.37
August	132	58	78	4,796	.08	.09
September	202	58	114	6,783	.12	.13
October.	328	114	144	8,854	.15	.17
November	180	80	124	7,379	.13	.15
December 1–12 [b]	....	....	117	2,785	.12	.05

[a] July 27 and 28 interpolated.   [b] December 6 interpolated.

DUCHESNE RIVER NEAR MYTON, UTAH.

This station was established October 26, 1899, by C. C. Babb, assisted by C. T. Prall. It is located at the highway bridge, on the road from Fort Duchesne to Price, Utah, 14 miles from Fort Duchesne. It is 3 miles below the mouth of Lake Creek. The gage is a vertical 2 by 5 inch timber, 12 feet long, nailed to the south side of the west abutment. It is read twice each day by H. Calvert, the storekeeper at Myton, Utah. Discharge measurements are made at all stages from the two-span highway bridge to which the gage is attached. The initial point for soundings is the extreme east end of the bridge stringer. The channel is straight for 100 feet above and for 500 feet below the bridge. The current is sluggish at ordinary stages. The right bank is high, without trees, and will not overflow. The left bank is lower than the right, is covered with underbrush and trees, and will overflow at extreme flood stages. The bed of the stream is sandy and somewhat shifting. The channel is divided into two parts by the center pier of the bridge. The bench mark is a nail in the northwest corner of the store. Its elevation is 15.72 feet above the zero of the gage.

The observations at this station during 1903 have been made under the direction of H. S. Reed, district hydrographer.

Discharge measurements of Duchesne River near Myton, Utah, in 1903.

Date.	Hydrographer.	Gage height.	Discharge.
		Feet.	*Second-feet.*
....	H. S. Reed	5.37	328
....	....do	5.36	351
....	....do	5.64	530
....	....do	6.03	906
....	....do	6.43	1,164
....	....do	7.30	2,030
....	....do	6.75	1,368
....	....do	6.48	1,183
....	....do	8.35	4,138
....	....do	8.40	4,202
....	....do	7.98	3,431
....	....do	7.00	1,911
....	....do	6.52	1,440
....	....do	6.08	943
....	....do	6.20	1,029
....	....do	5.95	833
....	....do	5.60	563
....	....do	5.38	400
....	....do	5.30	343
....	....do	5.18	288
....	....do	5.18	291
r 5	....do	5.10	284
r 20	....do	5.19	283
....	....do	5.52	520
....	....do	5.30	398
1....................	....do	5.25	348
1....................	....do	5.25	348
)....................	....do	5.21	304
· 28	....do	5.38	412

daily gage height, in feet, of Duchesne River near Myton, Utah, for 1903.

Day.	Apr.	May.	June.	July.	Aug.	Sept.	Oct.	Nov.	Dec.
....	....	5.86	7.15	6.62	5.56	5.12	5.61	5.20	5.47
....	....	6.04	7.48	6.50	5.52	5.10	5.67	5.20	5.45
....	....	6.14	....	6.42	5.50	5.10	5.55	5.20	5.39
....	....	6.11	....	6.36	5.48	5.10	5.41	5.23	5.31
....	5.40	6.15	8.38	6.31	5.45	5.10	5.37	5.19	5.27
....	5.36	6.23	8.34	6.25	5.41	5.51	5.35	5.18	....
....	5.30	6.30	8.36	6.23	5.33	5.45	5.31	5.19	....
....	5.35	6.40	8.51	6.21	5.30	5.31	5.31	5.20	....
....	5.36	6.40	8.70	6.16	5.29	5.28	5.35	5.20	....
....	5.40	6.48	8.60	6.09	5.28	5.25	5.45	5.19	....

Mean daily gage height, in feet, of Duchesne River near Myton, Utah, for 1903—Continued.

Day.	Apr.	May.	June.	July.	Aug.	Sept.	Oct.	Nov.	Dec.
11	5.41	6.51	8.42	6.01	5.28	5.20	5.31	5.18	
12	5.42	6.55	8.42	5.96	5.28	5.20	5.30	5.16	
13	5.35	6.71	8.35	5.91	5.28	5.21	5.31	5.21	
14	5.30	6.97	8.45	5.90	5.29	5.25	5.32	5.31	
15	5.30	7.21	8.26	5.85	5.30	5.29	5.31	5.30	
16	5.31	7.36	8.05	5.85	5.28	5.29	5.29	5.77	
17	5.31	7.47	8.07	6.34	5.25	5.25	5.29	5.20	
18	5.83	7.17	8.05	6.05	5.21	5.23	5.28	5.30	
19	5.32	6.96	7.97	5.91	5.20	5.21	5.26	5.24	
20	5.31	6.85	7.87	5.81	5.19	5.20	5.27	5.34	
21	5.34	6.78	7.80	5.80	5.19	5.18	5.26	5.42	
22	5.36	6.73	7.70	5.81	5.19	5.17	5.26	5.40	
23	6.45	6.64	7.51	6.01	5.45	5.15	5.26	5.40	
24	5.60	6.55	7.32	5.95	5.20	5.11	5.26	5.35	
25	5.87	6.50	7.12	5.97	5.24	5.10	5.26	5.38	
26	5.99	6.49	7.02	5.96	5.19	5.10	5.24	5.31	
27	6.04	6.18	6.97	6.10	5.18	5.10	5.22	5.30	
28	6.15	6.50	6.92	5.86	5.17	5.11	5.21	5.30	
29	6.05	6.49	6.80	5.76	5.16	5.21	5.20	5.28	
30	5.95	6.50	6.72	5.69	5.16	5.45	5.20	5.30	
31		6.71		5.61	5.13		5.20		

Rating table for Duchesne River near Myton, Utah, from April to 1 to June 2, 1903.

Gage height.	Discharge.	Gage height.	Discharge.	Gage height.	Discharge.	Gage height.	Discharge.
Feet.	Second-feet.	Feet.	Second-feet.	Feet.	Second-feet.	Feet.	Second-feet.
5.3	320	5.7	560	6.2	940	7.0	1,760
5.4	380	5.8	630	6.4	1,120	7.2	2,000
5.5	440	5.9	700	6.6	1,320	7.4	2,240
5.6	500	6.0	780	6.8	1,520		

Rating table for Duchesne River near Myton, Utah, from June 5 to December 31, 1903.

Gage height.	Discharge.	Gage height.	Discharge.	Gage height.	Discharge.	Gage height.	Discharge.
Feet.	Second-feet.	Feet.	Second-feet.	Feet.	Second-feet.	Feet.	Second-feet.
5.1	275	5.7	640	6.6	1,460	7.8	3,155
5.2	319	5.8	710	6.8	1,705	8.0	3,490
5.3	371	5.9	790	7.0	1,970	8.2	3,850
5.4	431	6.0	870	7.2	2,250	8.4	4,210
5.5	500	6.2	1,040	7.4	2,530	8.6	4,570
5.6	570	6.4	1,240	7.6	2,835		

Estimated monthly discharge of Duchesne River near Myton, Utah, for 1903.

[Drainage area, 2,746 square miles.]

Month.	Discharge in second-feet.			Total in acre-feet.	Run-off.	
	Maximum.	Minimum.	Mean.		Second-feet per square mile.	Depth in inches.
1 5–30	....	....	456	23,516	0.17	0.16
....	2,300	665	1,332	81,902	.49	.56
: (28 days)a.....	....	....	3,248	180,385	1.18	1.23
....	1,460	570	912	56,077	.33	.38
ust	535	296	375	23,058	.14	.16
ember...:......	500	275	329	19,577	.12	.13
iber	605	319	383	23,550	.14	.16
ember	431	296	353	21,005	.13	.15
ember 1–5	....	....	415	4,116	.15	.03

a June 3 and 4 missing.

LAKE CREEK NEAR MYTON, UTAH.

'his station was established July 3, 1900, by C. T. Prall. It is
ited at the wagon bridge one-half mile above the mouth of the
ek. It is 3 miles from the gaging station on Duchesne River at
ce Road bridge near Myton and is 17 miles southwest of Fort
hesne. The gage is a 1 by 4 inch vertical board, 9 feet long,
ed to the down-stream side of the west abutment. During 1903
discharge measurements were made from the bridge. There is a
ion 400 feet below the bridge at which measurements may be made
vading. The initial point for soundings is a point marked on the
lge floor at the edge of the east abutment on the right bank. The
inel is straight for 75 feet above and for 200 feet below the station.
current is never swift except during flood stages. Both banks
high and are not subject to overflow. The bed of the stream is
posed of cobblestones and there is but one channel at all stages.
ch mark No. 1 is a nail in the bridge abutment opposite the 4.5-foot
k on the gage rod. Bench mark No. 2 is a nail in the bridge
ight directly over the gage rod. Its elevation is 10.59 feet above
zero of the gage.
'he observations at this station during 1903 have been made under
direction of H. S. Reed, district hydrographer.

Discharge measurements of Lake Creek near Myton, Utah, in 1903.

Date.	Hydrographer.	Gage height.	Discha
		Feet.	*Second-*
April 3	H. S. Reed	1.93	
April 9	....do	1.98	
April 18	....do	1.90	
April 24	....do	2.03	
May 2	....do	2.11	
May 8	....do	2.42	
May 16	....do	3.60	
May 22	....do	2.72	
May 29	....do	2.55	
June 5	....do	4.70	1
June 13	....do	5.00	1
June 20	....do	4.55	1
June 27	....do	3.50	
July 3	....do	3.05	
July 10	....do	2.70	
July 17	....do	2.75	
July 24	....do	2.60	
July 31	....do	2.35	
August 7	....do	2.15	
August 15	....do	2.12	
August 22	....do	2.05	
August 28	....do	2.05	
September 5	....do	2.00	
September 21	....do	2.03	
October 9	....do	2.15	
October 21	....do	2.08	
Do	....do	2.08	
October 30	....do	2.05	
November 28	....do	2.18	

Mean daily gage height, in feet, of Lake Creek near Myton, Utah, for 1903.

Day.	Apr.	May.	June.	July.	Aug.	Sept.	Oct.	Nov.
1								
2		2.11		3.05				
3	1.93							
4								
5			4.70			2.00		
6								
7					2.15			
8		2.42						
9	1.98						2.15	
10				2.70				
11								
12								
13			5.00					
14								
15					2.12			
16		3.60						
17				2.75				
18	1.90							
19								
20				4.55				
21						2.03	2.08	
22		2.72			2.05			
23								
24	2.03			2.60				
25								
26								
27				3.50				
28					2.05			2.18
29		2.55						
30							2.05	
31				2.35				

Estimated monthly discharge of Lake Creek [a] near Myton, Utah, for 1903.

[Drainage area, 475 square miles.]

Month.	Discharge in second-feet.				Run-off.	
	Maximum.	Minimum.	Mean.	Total in acre-feet.	Second-feet per square mile.	Depth in inches.
April	139	82	108	6,426	0.23	0.26
May	966	139	459	28,223	.97	1.12
June	1,780	687	1,341	79,795	2.82	3.15
July	687	241	411	25,271	.87	1.00
August	205	124	157	9,654	.33	.38
September	136	110	123	7,319	.26	.29
October	149	122	135	8,301	.28	.32
November	152	137	139	8,271	.29	.32

[a] Gage heights were taken at this station only when discharge measurements were taken, and daily discharges were obtained from the discharge measurements by interpolation; therefore the above data are approximate to that extent.

STRAWBERRY RIVER IN STRAWBERRY VALLEY, UTAH.

This station was established May 12, 1903, by C. Tanner. The station is located just below the junction of Big and Little Strawberry rivers. The observer is paid by residents of Utah County, who wish to determine the discharge of the river in view of its proposed diversion into the Spanish Fork. The gage is a vertical hard-pine stick driven into the bed of the river 100 feet below the junction of Big and Little Strawberry rivers and 300 feet above the gaging section. Discharge measurements are made by wading. The gage is read once each day by Frank Thomas. The initial point for soundings is a willow stick, driven into the north bank, to which the tagged line is fastened when measurements are made. The channel is straight for 100 feet above and 40 feet below the station. The current is swift. The right bank is high and rocky and is covered with a growth of pine and aspen. The left bank is low and is covered with a dense growth of willows. At the gaging section the south side of the channel is rough and covered with large angular bowlders. The north side of the channel is smooth. The bed is permanent and free from vegetation. No bench marks have been established, but a second gage rod has been established at the measuring section. The relation between the two gages has been determined by simultaneous readings.

The observations at this station during 1903 have been made under the direction of G. L. Swendsen, district hydrographer.

Discharge measurements of Strawberry River in Strawberry Valley, Utah, in 1903.

Date.	Hydrographer.	Gage height.	Discharge.
		Feet.	*Second-feet.*
May 12	C. Tanner	2.10	231
June 18	T. C. Callister	1.48	133
July 22	C. Tanner	.99	34
October 24	do	.78	26

ꜱᴇᴀɴ daily gage height, in feet, of Strawberry River in Strawberry Valley, Utah, for 1903.

Day.	May.	June.	July.	Aug.	Day.	May.	June.	July.	Aug.
....	......	......	1.02	0.82	17	2.22	1.57	0.87	0.77
....	......	2.42	1.02	.82	18	2.10	1.42	.95	.77
....	......	2.52	1.02	.82	19	2.00	1.40	.95	.77
....	......	2.53	1.00	.80	20	1.97	1.33	.90	.77
....	......	2.42	.97	.80	21	2.00	1.30	.90	.80
....	......	2.48	.97	.80	22	2.07	1.23	.90	.80
....	......	2.45	.97	.80	23	1.83	1.20	1.05	.80
....	......	2.15	.95	.80	24	2.01	1.15	1.00	.80
....	......	2.07	.95	.77	25	1.93	1.15	.95	.80
....	......	2.01	.95	.77	26	1.85	1.12	.92	......
....	......	1.91	.90	.77	27	1.87	1.07	.90	......
....	2.10	1.83	.90	.77	28	2.00	1.05	.90	......
....	2.22	1.78	.90	.77	29	1.97	1.05	.87	......
....	2.36	1.73	.90	.77	30	......	1.02	.85	......
....	2.22	1.65	.87	.77	31	......	......	......	......
....	2.36	1.60	.87	.77					

Rating table for Strawberry River in Strawberry Valley, Utah, for 1903.

Gage height.	Discharge.	Gage height.	Discharge.	Gage height.	Discharge.	Gage height.	Discharge.
Feet.	*Second-feet.*	*Feet.*	*Second-feet.*	*Feet.*	*Second-feet.*	*Feet.*	*Second-feet.*
0.7	15	1.2	90	1.7	165	2.2	245
.8	30	1.3	105	1.8	181	2.3	261
.9	45	1.4	120	1.9	197	2.4	277
1.0	60	1.5	135	2.0	213	2.5	293
1.1	75	1.6	150	2.1	229	2.6	309

ꜱᴛimated monthly discharge of Strawberry River in Strawberry Valley, Utah, for 1903.

Month.	Discharge in second-feet.			Total in acre-feet.
	Maximum.	Minimum.	Mean.	
May 12–31 [a]	271	186	225	8,926
June [b]	298	63	166	9,878
July [c]	63	35	50	3,074
August 1–25	33	25	28	1,388

[a] May 30 and 31 interpolated.   [b] June 1 interpolated.   [c] July 31 interpolated.

UINTA RIVER AT FORT DUCHESNE, UTAH.

This station was established September 14, 1899, by C. C. Babb, assisted by C. T. Prall. It is located at the highway bridge at the military post. The gage is a vertical rod nailed to the south end of the east crib of the bridge. It is read twice each day by A. J.

McDonald, quartermaster-sergeant at the post. Discharge measurements are made at high water from the bridge, and at ordinary stages by wading at a point 200 feet below. The initial point for soundings is a zero marked on the west end of the bridge stringer on the downstream side. A tagged wire is stretched just below the bridge. The channel is curved both above and below the station and makes a half circle at the gaging section. The current is sluggish near the left bank, but is swift near the right bank. Both banks are low and subject to overflow. The right bank is covered with a heavy undergrowth. The bed of the stream is rocky, though at times the section is filled in with sediment brought down by Deep Creek during floods. The bench mark is a nail in the southeast crib in the first header above the bottom. Its elevation is 2.93 feet above the zero of the gage.

The observations at this station during 1903 have been made under the direction of H. S. Reed, district hydrographer.

Discharge measurements of Unita River at Fort Duchesne, Utah, in 1903.

Date.	Hydrographer.	Gage height.	Discharge.
		Feet.	*Second-feet.*
January 2	H. S. Reed	....	65
January 12	....do	....	66
April 2	....do	2.88	176
April 8	....do	2.65	109
April 16	....do	2.62	98
April 22	....do	2.63	105
April 29	....do	2.67	123
May 6	....do	2.85	182
May 13	....do	3.47	614
May 17	....do	3.95	1,157
May 20	....do	3.45	558
May 27	....do	3.15	348
June 4	....do	4.30	2,476
June 9	....do	4.40	3,019
June 18	....do	4.10	1,455
June 25	....do	3.57	635
July 1	....do	3.42	523
July 9	....do	3.23	406
July 15	....do	3.05	281
July 22	....do	3.10	328
July 28	....do	2.97	240
August 4	....do	2.72	126
August 14	....do	2.60	84
August 20	....do	2.53	72
August 26	....do	2.60	82
September 2	....do	2.48	70

re measurements of Unita River at Fort Duchesne, Utah, in 1903—Continued.

Date.	Hydrographer.	Gage height.	Discharge.
		Feet.	Second-feet.
r 10	H. S. Reed	2.65	107
r 17	do	2.75	158
r 28	do	2.60	100
	do	2.93	231
	do	2.80	162
9	do	2.73	125
4	do	2.69	125
4	do	2.69	88
4	do	2.69	92
3	do	2.66	107
·12	do	2.75	134
·19	do	2.69	100
·25	do	2.75	149
7	do	2.65	64

daily gage height, in feet, of Uinta River at Fort Duchesne, Utah, for 1903.

Day.	Mar.	Apr.	May.	June.	July.	Aug.	Sept.	Oct.	Nov.	Dec.
		2.96	2.68	4.00	3.32	2.80	2.50	2.80	2.68	2.70
		2.90	2.66	4.27	3.35	2.80	2.50	2.90	2.65	2.70
		2.75	2.67	4.32	3.32	2.80	2.50	2.92	2.69	2.65
		2.74	2.73	4.29	3.27	2.72	2.50	2.87	2.69	2.60
		2.71	2.78	4.24	3.21	2.70	2.55	2.86	2.70	2.53
		2.62	2.82	4.23	3.17	2.69	3.00	2.87	2.70	2.50
		2.67	2.91	4.19	3.35	2.66	2.88	2.85	2.70	2.42
		2.68	2.96	4.30	3.29	2.63	2.79	2.80	2.70	2.40
		2.71	3.00	4.37	3.21	2.61	2.70	2.73	2.70	2.35
		2.74	3.07	4.24	3.12	2.60	2.64	2.70	2.67	2.30
		2.77	3.11	4.20	3.08	3.60	2.66	2.76	2.67	2.30
		2.74	3.18	4.20	3.07	2.60	2.82	2.78	2.75	2.30
		2.62	3.44	4.15	3.07	2.62	2.67	2.78	2.75	......
		2.62	3.88	4.12	3.08	2.63	2.92	2.78	2.79	......
		2.65	4.04	4.10	3.15	2.61	2.87	2.78	2.75	......
		2.65	4.07	4.07	3.10	2.59	2.82	2.78	2.75	......
		2.65	3.90	4.06	3.38	2.58	2.77	2.78	2.70	......
		2.61	3.67	4.05	3.14	2.58	2.72	2.74	2.65	......
		2.60	3.46	3.96	3.08	2.56	2.65	2.72	2.70	......
		2.60	3.41	3.92	3.04	2.55	2.63	2.74	2.75	......
		2.61	3.40	3.90	3.07	2.51	2.62	2.73	2.90	......
		2.64	3.39	3.86	3.15	2.57	2.61	2.75	2.90	......
		2.67	3.26	3.80	3.12	2.64	2.60	2.73	2.81	......
		2.71	3.20	3.77	3.09	2.62	2.60	2.71	2.80	......
		2.74	3.11	3.62	3.02	2.62	2.60	2.70	2.77	......
		2.77	3.11	3.66	3.02	2.62	2.58	2.70	2.70	......
		2.75	3.16	3.55	3.02	2.62	2.59	2.70	2.70	......
		2.72	3.11	3.49	2.97	2.59	2.60	2.70	2.70	......
	2.72	2.70	3.14	3.48	2.91	2.57	2.60	2.68	2.70	......
	2.90	2.69	3.11	3.46	2.87	2.55	2.03	2.68	2.70	......
	2.95		3.60		2.82	2.52		2.68		......

Rating table for Uinta River at Fort Duchesne, Utah, for 1903.

Gage height.	Discharge.	Gage height.	Discharge.	Gage height.	Discharge.	Gage height.	Discharge.
Feet.	*Second-feet.*	*Feet.*	*Second-feet.*	*Feet.*	*Second-feet.*	*Feet.*	*Second-feet.*
2.3	43	2.9	205	3.5	600	4.1	1,450
2.4	53	3.0	259	3.6	685	4.2	1,800
2.5	70	3.1	320	3.7	790	4.3	2,400
2.6	94	3.2	385	3.8	915	4.4	3,060
2.7	123	3.3	453	3.9	1,060		
2.8	159	3.4	524	4.0	1,230		

Table well defined.

Estimated monthly discharge of Uinta River at Fort Duchesne, Utah, for 1903.

[Drainage area, 672 square miles.]

Month.	Discharge in second-feet.			Total in acre-feet.	Run-off.	
	Maximum.	Minimum.	Mean.		Second-feet per square mile.	Depth in inches.
March 29–31			186	1,107	0.28	
April	259	94	125	7,438	.19	
May	1,330	108	461	28,346	.69	
June	2,730	561	1,440	85,686	2.14	
July	524	159	343	21,090	.51	
August	159	70	102	6,272	.15	
September	259	70	121	7,200	.18	
October	205	123	149	9,162	.22	
November	205	108	133	7,914	.20	
December 1–12			73	1,738	.11	

UINTA RIVER NEAR WHITEROCKS, UTAH.

This station was established September 16, 1899, by C. C. B
assisted by C. T. Prall. It is located at the point where the r
emerges from its canyon, about 10 miles northwest of the In
agency at Whiterocks, Utah. The station is on the road to
Government sawmill, and is three-fourths of a mile above the bri
It is 600 feet below the mouth of Pole Creek. The gage is an incl
2 by 4 inch timber, 12 feet long, bolted to two trees on the left b
Discharge measurements are made at flood stages by means of a c
and car. At ordinary stages they are made by wading. The in
point for soundings is the first tag on the barbed wire from the
cable support. The channel is straight for 600 feet above and be
the station. The current is swift, and at high stages difficulty is e:
rienced in keeping the meter in position on account of the high ve

rough bed. The right bank is high and rocky, with a few
It probably will not overflow. The left bank is lower than
t, is covered in places with a growth of willows, and will over-
flood stages. The bed of the stream is rough and rocky and
ed with large bowlders. There are two channels at all stages.
ch mark is a nail on an aspen tree 125 feet north of the gage.
ition is 8.93 feet above the zero of the gage.
bservations at this station during 1903 have been made under
ction of H. S. Reed, district hydrographer.

Discharge measurements of Uinta River near Whiterocks, Utah, in 1903.

Date.	Hydrographer.	Gage height.	Discharge.
		Feet.	*Second-feet.*
....	H. S. Reed	1.03	a 103
1	....do	(b)	71
7	....do	(b)	82
2	....do	(b)	104
10	....do	(b)	81
....	....do	(b)	113
....	....do	.75	111
....	....do	.95	165
....	....do	.80	134
....	....do	.85	125
....	....do	.85	140
....	....do	1.05	186
....	....do	1.08	194
....	....do	1.78	518
....	....do	1.35	301
....	....do	2.70	1,255
....	....do	2.45	1,087
....	....do	2.20	859
....	....do	2.00	733
....	....do	1.55	433
....	....do	1.47	349
....	....do	1.15	204
....	....do	1.10	232
....	....do	1.07	178
r 8	....do	1.20	240
r 15	....do	1.13	220
r 29	....do	1.07	182
4.....................	....do	1.05	189
6.....................	....do	1.00	162
1.....................	....do	.95	144

a Gage height *inaccurate on account of ice.*   b No record; ice.

pae...:....
&..........
....
....

12.........................

15.........................
16.........................

25.........................

31.........................

Estimated monthly discharge of Uinta River near Whiterocks, Utah, for 1903. [a]

Month.	Discharge in second-feet.			Total in second-feet.	Run-off.
	Maximum.	Minimum.	Mean.		Second-feet per square mile.
January	103	71	86	5, 288	0. 39
February............	104	81	95	5, 276	. 44
March...............	165	97	122	7, 501	. 56
April...............	190	125	148	8, 807	. 68
May	778	190	430	26, 440	1. 97
June	1, 255	583	894	53, 197	4. 10
July	583	277	382	23, 488	1. 75
August	276	178	232	14, 265	1. 06
September..........	240	182	209	12, 436	. 96
October	189	153	176	10, 822	. 81
November...........	153	153	153	9, 104	. 70
December	145	144	145	8, 916	. 67
The year	1, 255	71	256	185, 540	1. 17

[a] Gage heights were taken at this station only when discharge measurements were taken; daily discharges were obtained from the discharge measurements by interpolation; therefor above data is approximate to that extent.

WHITEROCKS RIVER NEAR WHITEROCKS, UTAH.

station was established September 15, 1899, by C. C. Babb,
l by C. T. Prall.   It is located at the mouth of the canyon at
t of the "dug way" leading to the river bottom from the plateau
It is 10 miles above the Indian agency at Whiterocks, which
iearest settlement.   The gage is an inclined 2 by 4 inch timber,
long, bolted to the triple trunk of a tree on the left bank, 200
low the gaging section.   Discharge measurements are made at
ages by means of a cable and car located 200 feet above the
od.   At ordinary stages measurements are made by wading.
tial point for soundings is the tree to which the cable is fastened
right bank.   The channel is straight for 150 feet above and for
t below the cable.   The current is swift.   Both banks are of
n height and are covered with a thick growth of trees and
rush.   The bed of the stream is rough and rocky and is cov-
ith large bowlders.   The bed is permanent, but it is hard to get
e soundings on account of its roughness.   The bench mark is
n a burnt aspen tree 50 feet east of the gage.   Its elevation is
eet above the zero of the gage.
observations at this station during 1903 have been made under
ection of H. S. Reed, district hydrographer.

scharge measurements of Whiterocks River near Whiterocks, Utah, in 1903.

Date.	Hydrographer.	Gage height.	Discharge.
		Feet.	Second-feet.
7	H. S. Reed	0. 80	33
22	.....do	1. 65	29
28	.....do	. 75	39
y 3	.....do	. 75	41
y 11	.....do	. 70	37
1	.....do	. 80	48
....	.....do	. 75	45
....	.....do	. 80	49
....	.....do	. 80	52
....	.....do	1. 00	82
....	.....do	1. 25	110
....	.....do	1. 78	275
....	.....do	1. 50	169
....	.....do	2. 95	1, 146
....	.....do	2. 50	783
....	.....do	2. 30	633
....	.....do	1. 85	368
....	.....do	1. 75	280

Discharge measurements of Whiterocks River near Whiterocks, Utah, in 1903—Continued.

Date.	Hydrographer.	Gage height.	Discharge.
		Feet.	*Second-ft.*
July 21	H. S. Reed	1.30	134
August 13	...do	1.00	83
August 19	...do	1.00	89
September 1	...do	.95	78
September 9	...do	1.05	105
September 16	...do	1.00	93
September 30	...do	.95	77
October 15	...do	.95	79
October 27	...do	.90	74
December 2	...do	.85	61

Mean daily gage height, in feet, of Whiterocks River near Whiterocks, Utah, for 1903.

Day.	Jan.	Feb.	Mar.	Apr.	May.	June.	July.	Aug.	Sept.	Oct.	Nov.	Dec.
1									4.85			
2												0.85
3		0.75				2.95						
5					1.35							
6							1.75					
7	0.80			0.75								
9									1.05			
11		.70										
12						2.50						
13								1.00				
15				.80						0.95		
16						2.30			1.00			
17						2.30						
19					1.78			1.00				
21				.80			1.30					
22	1.65											
24						1.85						
26					1.50							
27										.90		
28	.75		1.00									
30									.95			
31			0.80									

uted monthly discharge of Whiterocks River near Whiterocks, Utah, for 1903. [a]

[Drainage area, 114 square miles.]

fonth.	Discharge in second-feet.			Total in acre-feet.	Run-off.	
	Maximum.	Minimum.	Mean.		Second-feet per square mile.	Depth in inches.
....	40	29	34	2,091	0.30	0.35
....	42	37	41	2,277	.36	.37
....	48	42	43	2,644	.38	.44
....	96	45	56	3,332	.49	.55
....	655	96	260	15,987	2.28	2.63
....	1,146	324	658	39,154	5.77	6.44
....	324	108	198	12,175	1.74	2.01
....	108	82	93	5,718	.82	.95
er...........	105	76	89	5,296	.78	.87
....	79	68	76	4,673	.67	.77
ɔr	68	67	67	3,987	.59	.66
ir [b].........	67	59	60	3,689	.53	.61
ie year......	1,146	29	140	101,023	1.23	16.65

eights were taken at this station only when discharge measurements were taken, and daily
were obtained from the discharge measurements by interpolation; therefore the above
pproximate to that extent.
rge for December estimated.

ASHLEY CREEK NEAR VERNAL, UTAH.

stream drains an area directly east of the Uinta basin. The
was established March 15, 1900, by C. T. Prall. The river
s from its canyon about 7½ miles north of Vernal, Utah. The
is located near this point, at the highway bridge leading by the
the gage reader. The station is one-half mile below the mouth
Fork and is above the series of canals by which Vernal Valley
ated. The gage is a vertical 1 by 5 inch board, 10 feet long,
1 to overhanging trees on the right bank just above the bridge.
id twice each day by E. Marett. Discharge measurements are
: high water from the bridge and at ordinary stages by wading.
tial point for soundings is on the right bank. The channel is
t for 200 feet above and below the station. The current is
The right bank is high, not liable to overflow, and covered
ees. The left bank is low, covered with underbrush, and is
to overflow at flood stages, at which time there are two or three
s. The bed of the stream is rough and rocky, and is not subject
ge. The bench mark is a large nail, about which are driven
naller nails, in a stump 50 feet west of the bridge. Its eleva-
7.21 feet above the zero of the gage.

H. S. Reed ....

April 12 .....do ....

....

.....do ....

Mean daily gage height, in feet, of Ashley Creek near Vernal, Utah, for 1903.

	Jan.	Feb.	Mar.	Apr.	May.	June.	July.	Aug.	Sept.	Oct.	Nov.
1..........................	0.60	0.60	0.55	0.75	1.85	3.80	2.45	2.00	1.70	1.80	1.70
2..........................	.60	.60	.55	.63	1.70	3.95	2.40	2.00	1.70	1.90	1.70
3..........................	.60	.80	.55	.60	1.95	4.40	2.40	2.00	1.70	1.95	1.60
4..........................	.60	.60	.55	.60	2.05	4.10	2.40	1.90	1.70	1.80	1.60
5..........................	.60	.60	.50	.75	2.25	3.95	2.40	1.90	1.70	1.80	1.60
6..........................	.60	.60	.50	.63	2.40	4.10	2.40	1.90	2.00	1.80	1.60
7..........................			.50	.60	2.35	4.65	2.60	1.90	1.85	1.80	1.60
8..........................	.60	.60	.50	.60	2.35	4.60	2.40	1.90	1.70	1.80	1.60
9..........................	.60	.60	.50	.60	2.35	a4.50	2.30	1.90	1.70	1.80	1.60
10..........................	.60	.60	.50	.65	2.45	3.90	2.30	1.90	1.70	1.80	1.60
11..........................	.60	.60	.50	.60	2.60	3.90	2.30	1.90	1.70	1.80	1.60
12..........................	.60	.60	.50	.60	2.70	4.05	2.30	1.90	1.70	1.70	1.60
13..........................	.60	.60	.50	.60	3.05	4.05	2.20	1.90	1.70	1.70	1.60
14..........................	.60	.60	.50	.60	3.60	3.80	2.20	1.90	1.70	1.70	1.60
15..........................	.60	b2.60	.50	.60	3.75	3.85	2.20	1.90	1.70	1.70	1.60
16..........................	.60	b2.60	.50	.60	3.75	3.70	2.40	1.90	1.80	1.70	1.60
17..........................	.60	b2.00	.50	.60	3.05	3.55	2.40	1.80	1.80	1.70	1.60
18..........................	.60	b2.60	.50	.60	2.90	3.45	2.35	1.80	1.80	1.70	1.60
19..........................	.60	.60	.50	.60	2.70	3.35	2.30	1.80	1.80	1.70	1.60
20..........................	.60	.60	.50	.60	2.60	3.20	2.20	1.80	1.80	1.70	1.60
21..........................	.60	.60	.50	.60	2.50	3.05	2.20	1.70	1.70	1.70	1.60
22..........................	.60	.60	.50	.63	2.30	2.95	2.20	1.70	1.70	1.70	1.60
23..........................	.60	.60	.55	.65	2.15	2.85	2.20	1.70	1.70	1.70	1.50
24..........................	.60	.60	.65	.95	2.15	2.75	2.20	1.70	1.70	1.70	1.50
25..........................	.60	.60	.65	1.35	2.20	2.65	2.20	1.70	1.60	1.70	1.50
26..........................	.60	.65	.65	1.80	2.25	2.65	2.25	1.70	1.60	1.70	1.50
27..........................	.60	.60	.65	1.80	2.30	2.60	2.20	1.70	1.60	1.70	1.50
28..........................	.60	.60	.65	1.70	2.35	2.55	2.10	1.70	1.60	1.70	1.50
29..........................	.60	......	.65	1.55	2.60	2.50	2.10	1.70	1.60	1.70	1.50
30..........................	.60	......	.80	1.75	2.80	2.50	2.00	1.70	1.60	1.70	1.50
31..........................	.60	......	.75	......	3.65	......	2.00	1.70	......	1.70	......

a New rod.　　　　　——　　　　b Back water caused by ice.

Rating table for Ashley Creek near Vernal, Utah, from January 1 to June 8, 1903.

Gage height.	Discharge.	Gage height.	Discharge.	Gage height.	Discharge.	Gage height.	Discharge.
Feet.	*Second-feet.*	*Feet.*	*Second-feet.*	*Feet.*	*Second-feet.*	*Feet.*	*Second-feet.*
0.5	37	1.5	230	2.5	538	3.5	1,065
.6	45	1.6	258	2.6	570	3.6	1,145
.7	55	1.7	286	2.7	605	3.8	1,310
.8	69	1.8	315	2.8	645	4.0	1,480
.9	85	1.9	346	2.9	690	4.2	1,660
1.0	106	2.0	378	3.0	740	4.4	1,840
1.1	128	2.1	410	3.1	795	4.6	2,020
1.2	152	2.2	442	3.2	855		
1.3	176	2.3	474	3.3	920		
1.4	202	2.4	506	3.4	990		

Estimated monthly discharge of Ashley Creek near Vernal, Utah, for 1903.

[Drainage area, 250 square miles.]

Month.	Discharge in second-feet.			Total in acre-feet.	Run-off.	
	Maximum.	Minimum.	Mean.		Second-feet per square mile.	Depth in inches.
January	45	45	45	2,767	0.18	0.21
February	45	45	45	2,499	.18	.19
March	69	37	42	2,582	.17	.20
April	315	45	94	5,593	.38	.42
May	1,267	286	602	37,016	2.41	2.78
June 1 to 8	2,065	1,310	1,656	26,277	6.62	1.97

GREEN RIVER NEAR VERNAL, UTAH.

This station was established November 7, 1903, by H. S. Reed. It is located about 300 feet below Billings Ferry and about 15 miles from Vernal, Utah. It is 1½ miles below the mouth of Brush Creek and 3 miles above the mouth of Ashley Creek. The gage is a vertical 2 by 5 inch timber 10 feet long, braced to a cottonwood tree about 10 feet from the edge of the river. It is read twice each day by Victor Billings. Discharge measurements are made from the ferryboat. The initial point for soundings is the post on the right bank to which the ferry cable is attached. The channel is straight for 1,000 feet above and below the station. The right bank is high, is composed of gravel, and will not overflow. The left bank is low and sandy, and covered with underbrush. The bed of the stream is sandy and shifting. There is but one channel at all stages. Bench mark No. 1 is a

40-penny spike driven into the cottonwood tree to which the gage is attached. Its elevation is 10.66 feet above the zero of the gage. Bench mark No. 2 is a 40-penny spike driven into the southwest corner of Mr. Billings's grain house about 3 feet above the ground. Its elevation is 25.67 feet above the zero of the gage.

The observations at this station during 1903 have been made under the direction of H. S. Reed, district hydrographer.

The following discharge measurement was made by H. S. Reed in 1903:

November 10: Gage height, 2.45 feet; discharge, 1,453 second-feet.

Mean daily gage height, in feet, of Green River near Vernal, Utah, for 1903.

Day.	Nov.	Dec.	Day.	Nov.	Dec.	Day.	Nov.	Dec.
1		2.50	12	2.45	2.66	23	2.05	2.63
2		2.49	13	2.31	2.66	24	2.13	2.60
3		2.49	14	2.36	2.70	25	2.17	2.58
4		2.39	15	2.40	2.56	26	2.21	2.56
5		2.43	16	2.45	2.56	27	2.33	2.53
6		2.46	17	2.30	2.60	28	2.36	2.50
7	2.50	3.33	18	2.21	2.43	29	2.38	2.55
8	2.51	3.05	19	2.15	2.60	30	2.53	2.55
9	2.50	3.15	20	1.98	2.65	31		2.55
10	2.50	3.08	21	1.78	2.57			
11	2.47	2.70	22	1.95	2.59			

GREEN RIVER AT GREENRIVER, WYO.

The gaging station at Greenriver, Wyo., is at the crossing of the Union Pacific Railroad. It was established May 2, 1895, at the pump house of the Union Pacific Railroad Company. The rod is fastened to a pile near the east end of the bridge. The bench mark consists of a cross on the third step from the bottom of the south end of the east abutment. As the section under the railroad bridge is poor, discharge measurements are made from the iron highway bridge about one-half mile below. The station was temporarily discontinued during the latter part of 1900 and for part of 1901.

No discharge measurements were made in 1903.

The original elevation of the bench mark above the zero of the gage was 12.48 feet. During the winter of 1903-4 the gage rod was raised from its original position. The present elevation of the bench mark is 12.46 feet above the zero of the gage.

The observations at this station during 1903 have been made under the direction of A. J. Parshall, district hydrographer.

n daily gage height, in feet, of Green River at Greenriver, Wyo., for 1903.

Day.	Apr.	May.	June.	July.	Aug.	Sept.	Oct.
....	0.75	1.65	2.03	4.18	2.13	1.25	1.00
....	1.10	1.53	2.35	4.15	2.03	1.20	1.02
....	1.57	1.52	2.85	4.05	2.00	1.15	1.10
....	1.70	1.48	3.33	3.85	2.00	1.00	1.10
....	1.55	1.42	3.72	3.65	1.98	1.00	1.15
....	1.15	1.50	4.15	3.55	1.90	.97	1.18
....	.95	1.45	4.35	3.43	1.87	.95	1.20
....	1.15	1.47	4.52	3.25	1.83	1.10	1.20
....	1.40	1.45	4.70	3.10	1.68	1.33	1.23
....	1.57	1.55	4.97	3.03	1.65	1.65	1.25
....	1.40	1.57	5.17	2.93	1.63	1.90	1.30
....	1.20	1.57	5.30	2.88	1.60	2.05	1.25
....	1.25	1.53	5.35	2.78	1.60	2.65	1.25
....	1.22	1.55	5.35	2.72	1.55	2.70	1.25
....	1.10	1.60	5.35	2.70	1.50	2.75	1.20
....	1.20	1.65	5.40	2.70	1.48	2.75	1.20
....	1.15	2.02	5.40	2.70	1.43	2.60	1.20
....	1.27	2.30	5.32	2.75	1.37	2.60	1.15
....	1.22	2.40	5.35	2.78	1.35	2.05	1.15
....	1.25	2.32	5.37	2.73	1.30	1.47	1.15
....	1.22	2.25	5.22	2.68	1.28	1.37	1.15
....	1.20	2.17	5.12	2.53	1.25	1.35	1.15
....	1.30	2.23	4.87	2.50	1.25	1.35	1.13
....	1.37	2.37	4.65	2.50	1.30	1.18	1.10
....	1.47	2.33	4.45	2.53	1.30	1.15	1.10
....	1.50	2.20	4.25	2.55	1.32	1.02	1.10
....	1.55	2.25	4.07	2.48	1.35	1.00	1.10
....	1.60	2.17	4.08	2.40	1.40	1.00	1.10
....	1.72	2.03	4.10	2.33	1.35	1.00	1.10
....	1.82	2.00	4.15	2.28	1.32	1.00	1.10
....		2.02	....	2.28	1.25	....	1.10

SCELLANEOUS MEASUREMENTS IN COLORADO RIVER BASIN.

ollowing miscellaneous measurements were made in the Colover basin in 1903:

Miscellaneous discharge measurements in Colorado River basin.

Hydrographer.	Stream.	Locality.	Gage height.	Discharge.
			Feet.	*Sec.-ft.*
W. D. Smith	Gila River........	At mouth...	....	236
.....do	....do	.....do	....	1,981
.....do	....do	.....do	....	991
.....do	....do	.....do	....	314
.....do	....do	.....do	....	190
.....do	....do	.....do	....	62
.....do	....do	.....do	....	46

Miscellaneous discharge measurements in Colorado River basin—Continued.

Date.	Hydrographer.	Stream.	Locality.	Gage height.	Discharge.
				Feet.	*Sec.ft.*
Sept. 18	W. D. Smith........	Gila River.......	At mouth........	....	81
Oct. 1	.....do	.....do	.....do	....	1,329
6	.....do	.....do	.....do	....	522
12	.....do	.....do	.....do	....	61
Mar. 4	.....do	Colorado Valley Pumping and Irrigation Co.'s canal.	At heading.	....	20
May 9	.....do	.....do	.....do	....	51
June 17	.....do	.....do	.....do	....	46
19	.....do	.....do	.....do	....	44
30	.....do	.....do	.....do	....	48
July 1	.....do	.....do	.....do	....	46
18	.....do	.....do	.....do	....	42
29	.....do	.....do	.....do	....	50
Aug. 8	L. M. Barnes	.....do	.....do	....	51
Sept. 14	W. D. Smith........	.....do	.....do	....	42
Oct. 19	.....do	.....do	.....do	....	33
Nov. 17	.....do	.....do	.....do	....	34
Dec. 18	.....do	.....do	.....do	....	31
May 20	.....do	Ludy canal	.....do	....	60
June 5	.....do	.....do	.....do	....	79
23	.....do	.....do	.....do	....	38
July 21	.....do	.....do	.....do	....	71
Sept. 11	.....do	.....do	.....do	....	63
Oct. 9	.....do	.....do	.....do	....	25
24	.....do	.....do	.....do	....	66
Nov. 20	.....do	.....do	.....do	....	24
Dec. 7	.....do	.....do	.....do	....	37
21	.....do	.....do	.....do	....	14
June 5	.....do	Farmers' canal...	.....do	....	25
23	.....do	.....do	.....do	....	38
20	.....do	Rose flume a	At Rose's ranch.	0.96	10
20	.....do	.....do	.....do	.91	8.1
20	.....do	.....do	.....do	.86	8
20	.....do	.....do	.....do	.97	10
Oct. 1	.....do	.....do	.....do	.685	7
1	.....do	.....do	.....do	.68	7

a These measurements were made to rate flume for ascertaining duty of water on 12-acre alfalfa field

Tributaries of Green River in Newfork drainage basin, Wyoming.

Date.	Hydrographer.	Stream.	Locality.	Gage height.	Discharge.
				Feet.	*Sec.-ft.*
ept. 10	A. J. Parshall......	Pine Creek	1 mile below Fremont Lake.	....	73
18	Rex. G. Schnitger..	Bowlder Creek...	1 mile below Bowlder Lake.	....	40
18	.....do	Fall Creek......	Near Fisher's ranch.	....	10
18	.....do	Pole Creek......	Below Half Moon Lake.	....	42
20	.....do	East Fork	At Gilligan's ranch.	....	26
21	.....do	Newfork.........	At Newfork.	....	128
22	.....do	.....do	Below the mouth of East Fork.	....	184
9	A. J. Parshall......	Big Sandy	At Ten Trees	....	51

Creeks, canals, and ditches in northeastern Utah.

26	H. S. Reed	Canal No. 1......	Head-gate ..	0.90	7
27	.....do	.....do	....do	1.00	14
13	.....do	.....do	....do	1.20	17
26	.....do	Bench ditch	....do	.57	48
13	.....do	.....do	....do	.25	16
26	.....do	Pole Creek......	Just above mouth.	....	11
9	.....do	Dry Gulch Creek.	At mouth......	....	12
9	.....do	Government canal below army post.	Below head-gate.	....	13
9	.....do	Deep Creek......	At mouth......	....	1.2
27	.....do	.....do	....do	....	9.2
30	.....do	Farm Creek......	Road crossing at mouth.	....	5.5
27	.....do	.....do	....do	....	1.8
4	.....do	.....do	....do	....	13
25	.....do	.....do	....do	....	16
20	.....do	.....do	....do	....	2.6
12	.....do	Upper Ashley canal.	Below bridge crossing.	....	14
12	.....do	Island ditch	Below weir	....	1.5
12	.....do	Dodd ditch	....do	....	2.8

Creeks, canals, and ditches in northeastern Utah—Continued.

Date.	Hydrographer.	Stream.	Locality.	Gage height.	Discharge.
				Feet.	*Sec.-ft.*
Apr. 12	H. S. Reed	Rock Point ditch .	Below weir .	....	2.4
12	.....do	Stemaker ditch ..	......do		1.5
12	.....do	Ashley Central ditch.	Above measuring weir.	....	14
July 20	.....do	Uinta River	½ mile below regular station.	....	314
Dec. 1	.....do	.....do	......do		134

INTERIOR BASIN.

Under this head is comprised a large extent of arid country which includes nearly the whole of Nevada, the northern and western parts of Utah, and small portions of California and Idaho. Having no outlet to the sea, the entire drainage of this vast basin is lost mainly through evaporation from the numerous lakes and sinks in which the waters of the rivers collect.

The largest of the lakes is Great Salt Lake, which receives the waters of that portion of the basin lying in northern Utah and Idaho. Southwestern Utah drains into Sevier Lake. Other important lakes are Humboldt, Pyramid, Winnemucca, Carson, and Walker, all in western Nevada. These lakes receive the basin drainage from Nevada and California. The principal rivers in this section are as follows:

Bear River has its source on the northern slope of the Uinta Mountains in the northeastern part of Utah. After a circuitous course, in which it leaves Utah and enters Wyoming, reenters Utah, appears again in Wyoming, visits Idaho, and reenters Utah, it finally discharges its waters into Great Salt Lake. Considerable irrigation is practiced on certain portions of the river. Logan River empties into Bear River in Cache Valley, Utah.

Weber River rises in the high country east of the Wasatch Mountains. Passing through that range, it appears in the plains region in the vicinity of Ogden, where, after receiving the waters of Ogden River, it discharges into Great Salt Lake. There are a number of good reservoir sites on its upper tributaries, some of which have been utilized within the last year.

Provo River rises on the western slope of the Uinta Mountains and after receiving a number of tributaries enters what is known as Heber Valley, where considerable irrigation is practiced. After crossing this valley it passes through the Wasatch Mountains in a picturesque canyon, and finally enters Utah Valley, where its summer flow is comately diverted for irrigation purposes. Its flood waters discharge into Utah Lake.

Sevier River drains a large area in the southwestern part of Utah. It flows northerly until it enters Juab County, then makes a short bend, and flows southwest into Sevier Lake. San Pitch River joins Sevier River near Gunnison, Utah.

Humboldt River rises in the extreme northeastern part of Nevada and flows in a general westerly and southerly direction, finally entering Humboldt Lake, whence its waters find their way into the Humboldt and Carson Sinks. The general direction of the mountain ranges of this basin is north and south, crossed at nearly right angles by the main Humboldt River. The tributaries flow in the general direction of the mountain ranges, and drain either to the north or to the south. During low stages the water of the river is almost wholly diverted. For the future development of the country recourse must be had to the construction of storage reservoirs. Of its tributaries the North Fork enters it west of Peko, Nev., and the South Fork enters it about 9 miles below Elko, Nev.

Pine Creek is a tributary from the south and joins it near Palisade, Nev. Marys River is one of the headwater tributaries of the Humboldt.

Walker River is formed by two branches which have their sources across the Nevada-California boundary, in California. It flows north and then takes a sharp bend to the southeast, emptying into Walker Lake.

Carson River has its source on the slopes of the Sierra Nevada, in eastern California, and flows northward into the State of Nevada. East and West forks unite near Genoa, Nev., in Carson Valley. At Empire, 3 miles east of Carson, after having traversed the upper Carson Valley, it turns to the northeast and enters a deep canyon, through which it flows for several miles, emerging into a second smaller valley a short distance above the town of Dayton. After leaving this valley it passes through two other shorter canyons and through one rather large valley before entering lower Carson Valley, or Carson Sink Valley, as it is also known, and discharging its waters into the Carson Sink.

The Truckee has its source on the slopes of the Sierra Nevada, in eastern California, and flows northward, entering Lake Tahoe, which is at an elevation of 6,225 feet and is the largest body of fresh water in the United States at this considerable altitude. The area of the lake is 193 square miles. Its outlet is at Tahoe, Cal., from which point Truckee River has a general northward and eastward course, receiving several important tributaries which contribute to its flow. It drains into Pyramid and Winnemucca lakes, which have no outlets. Donner and Prosser creeks are tributaries of Truckee River. Independence Creek discharges into the Little Truckee, a main branch of Truckee River, entering it at Boca, Cal. They drain areas of 31, 56, and 8.5 square miles, respectively, lying to the northwest of Lake Tahoe, in California.

Susan River has its source in the Sierra Nevada in northeastern California, and flows eastward, discharging into Honey Lake—one of the land-locked lakes of the Great Basin—of which it is the principal feeder. A considerable area of land is irrigated from the waters of the river below the gaging station, and during the last ten or twelve years several projects have been started for irrigating other extensive areas by the storage of its waters both above and below the town of Susanville.

The following is a list of the stations maintained during 1903 in the interior drainage basin:

Bear River near Collinston, Utah.
Bear River near Preston, Idaho.
Bear River at Dingle, Idaho.
Logan River near Logan, Utah.
Weber River near Uinta, Utah.
American Fork River near American Fork, Utah.
Provo River at San Pedro, Los Angeles and Salt Lake Railroad bridge, Provo, Utah.
Provo River at mouth of canyon near Provo, Utah.
Spanish Fork River near Spanish Fork, Utah.
Sevier River near Gunnison, Utah.
San Pitch River near Gunnison, Utah.
Humboldt River near Oreana, Nev.
Humboldt River near Golconda, Nev.
Humboldt River at Palisade, Nev.
Pine Creek near Palisade, Nev.
Humboldt River (South Fork) near Elko, Nev.
Humboldt River (North Fork) near Elburz, Nev.
Marys River near Deeth, Nev.
Walker River near Wabuska, Nev.
Walker River (East Fork) near Yerington, Nev.
Walker River (West Fork) near Coleville, Cal.
Carson River near Empire, Nev.
Carson River (East Fork) near Gardnerville, Nev.
Carson River (West Fork) at Woodfords, Cal.
Truckee River at Tahoe, Cal.
Truckee River at Vista, Nev.
Truckee River at Pyramid Lake Indian Agency, near Wadsworth, Nev.
Lake Minnemucca Inlet north of Pyramid Lake Indian Agency, near Wadsworth, Nev.
Truckee River at Nevada-California State line, near Mystic, Cal.
Little Truckee River near Pine station, Cal.
Independence Creek below Independence Lake, California.
Prosser Creek near Boca, Cal.
Prosser Creek near Hobart Mills, Cal.
Donner Creek near Truckee, Cal.
Susan River near Susansville, Cal.

Owens River near Round Valley, Cal.
Rock Creek near Round Valley, Cal.
Pine Creek near Round Valley, Cal.
Owens River canal near Bishop, Cal.
Bishop Creek canal near Bishop, Cal.
Farmers canal near Bishop, Cal.
McNally canal near Bishop, Cal.
George Collins's canal near Bishop, Cal.
Bishop Creek near Bishop, Cal.
Rawson canal near Bishop, Cal.
A. O. Collins's canal near Bishop, Cal.
Dell canal near Bishop, Cal.
Big Pine and Owens River canal near Bishop, Cal.
Sanger canal at Alvord, Cal.
East Side canal near Citrus, Cal.
Stevens canal near Citrus, Cal.
Owens River near Independence, Cal.

BEAR RIVER NEAR COLLINSTON, UTAH.

This station was established July 1, 1889, by Samuel Fortier. It is
ated about 4 miles from the railroad station at Collinston, 2 miles
t of Fielding, Utah, and below the headworks of the Bear River
al. The gage is a vertical iron rod graduated to tenths of a foot.
ere is also an iron gage fastened in a hole drilled in bed rock for
·-water readings. The gage is read by D. A. Cannon. Discharge
asurements are now made one-fourth mile above the gage from the
dge carrying the water-supply pipe of the Utah Light and Power
npany. The section is only fair for high-water measurements, but
d for low water. The right bank is high and sloping. The left
k is a steep cliff about 100 feet high composed of gravel. The bed
the stream is solid rock covered with small stones. The bench
rk is a nail in an oak post 20 feet west of the gage and 20 feet north
the cable at which discharge measurements were formerly taken.
elevation is 7.35 feet above gage datum.
The observations at this station during 1903 have been made under
· direction of G. L. Swendsen, district hydrographer.

Discharge measurements of Bear River near Collinston, Utah, in 1903.

Date.	Hydrographer.	Gage height.	Discharge.
		Feet.	*Second-feet.*
nuary 5	G. L. Swendsen	0.50	349
pril 6	do	2.70	2,332
ay 30	do	3.20	2,863
ly 1	W. W. McLaughlin	1.71	1,366
ugust 18	do	·1.20	31
·tober 24	do	1.20	903
ovember 19	C. Tanner	1.05	722

Mean daily gage height, in feet, of Bear River near Collinston, Utah, for 1903.

Day.	Jan.	Feb.	Mar.	Apr.	May.	June.	July.	Aug.	Sept.	Oct.	Nov.	Dec.
1	0.30	1.70	1.30	2.40	3.10	2.80	1.85	-0.60	-0.85	1.00	1.20	
2	.40	1.60	1.30	2.70	3.00	2.95	1.75	-.65	-.85	1.10	1.30	
3	.50	1.50	1.55	2.20	3.00	3.10	1.65	-.60	-.80	1.20	1.25	
4	.50	1.50	1.35	2.80	2.90	3.25	1.30	-.65	-.80	1.25	1.5	
5	.40	1.40	1.30	2.70	2.90	3.30	1.40	-.70	-.80	1.25	1.25	
6	.40	1.40	1.30	2.60	2.80	3.45	1.30	-.75	-.75	1.30	1.30	
7	.40	1.40	1.30	2.60	2.80	3.45	1.30	-.80	-.70	1.30	1.35	
8	.45	1.30	1.30	2.70	2.90	3.50	1.10	-.90	-.60	1.30	1.30	
9	.40	1.30	1.35	2.80	2.90	3.45	.90	-1.00	-.50	1.30	1.35	
10	.40	1.30	1.40	2.80	2.90	3.45	.80	-1.05	-.40	1.30	1.45	
11	.40	1.40	1.40	3.00	2.90	3.55	.70	-1.10	-.30	1.30	1.45	
12	.40	1.40	1.50	3.20	2.80	3.55	.65	-1.10	-.20	1.30	1.50	
13	.35	1.30	1.70	3.00	2.70	3.45	.60	-1.15	.10	1.25	1.55	
14	.30	1.40	2.20	3.00	2.65	3.30	.50	-1.15	.00	1.25	1.60	
15	.30	1.30	3.70	2.80	2.60	3.20	.40	-1.15	.10	1.25	1.70	
16	.25	1.35	4.20	2.90	2.60	3.15	.30	-1.15	.20	1.25	1.70	
17	.25	1.35	4.20	2.90	2.60	3.00	.20	-1.15	.30	1.25	1.65	
18	.30	1.40	3.60	2.80	2.60	2.90	.10	-1.20	.40	1.25	1.60	
19	.30	1.40	3.00	2.80	2.60	2.80	.00	-1.20	.50	1.25	1.60	
20	.40	1.35	2.40	2.70	2.60	2.70	-.10	-1.15	.55	1.25	1.40	
21	.40	1.35	2.10	2.70	2.60	2.60	-.15	-1.15	.60	1.20	1.40	
22	.70	1.30	2.20	2.60	2.75	2.50	-.20	-1.15	.65	1.25	1.45	
23	2.00	1.30	2.15	2.70	2.90	2.45	-.25	-1.15	.70	1.25		
24	2.20	1.35	2.20	2.80	3.00	2.45	-.30	-1.10	.75	1.25	1.60	
25	3.00	1.35	2.30	2.90	3.00	2.40	-.30	-1.05	.80	1.20		
26	2.80	1.35	2.35	3.00	2.90	2.35	-.30	-1.00	.80	1.30	1.60	
27	2.80	1.40	2.25	3.10	2.80	2.25	-.30	-1.00	.80	1.20		
28	2.70	1.40	2.30	3.20	2.80	2.20	-.30	-.95	.85	1.20	1.50	
29	1.80		2.30	3.30	2.80	2.10	-.35	-.90	.95	1.20	1.50	
30	1.80		2.30	3.20	2.75	2.00	-.40	-.90	.95	1.20	1.50	
31	1.80		2.40		2.80		-.50	-.90		1.20		

Rating table for Bear River near Collinston, Utah, for 1903.

Gage height.	Discharge.	Gage height.	**Discharge.**	Gage height.	Discharge.	Gage height.	Discharge.
Feet.	Second-feet.	Feet.	Second-feet.	Feet.	Second-feet.	Feet.	Second-feet.
-1.2	31	-0.1	110	1.0	700	2.2	1,850
-1.1	34	+.0	135	1.1	785	2.4	2,050
-1.0	37	.1	167	1.2	870	2.6	2,250
-.9	41	.2	206	1.3	960	2.8	2,450
-.8	45	.3	250	1.4	1,055	3.0	2,650
-.7	49	.4	298	1.5	1,150	3.2	2,850
-.6	54	.5	351	1.6	1,250	3.4	3,050
-.5	60	.6	410	1.7	1,350	3.6	3,250
-.4	67	.7	475	1.8	1,450	3.8	3,450
-.3	76	.8	545	1.9	1,550	4.0	3,650
-.2	90	.9	620	2.0	1,650	4.2	3,850

Table well defined.

Estimated monthly discharge of Bear River near Collinston, Utah, for 1903.

[Drainage area, 6,000 square miles.]

Month.	Discharge in second-feet.			Total in acre-feet.	Run-off.	
	Maximum.	Minimum.	Mean.		Second-feet per square mile.	Depth in inches.
January	2,650	228	781	48,022	0.13	0.15
February	1,350	960	1,044	57,981	.17	.18
March	3,850	960	1,782	109,571	.30	.35
April	2,950	1,850	2,480	147,570	.41	.46
May	3,350	2,250	2,474	152,120	.41	.47
June	3,200	1,650	2,570	152,926	.43	.48
July	1,500	60	448	27,546	.07	.09
August	57	31	38	2,337	.01	.01
September	660	43	264	15,709	.04	.05
October	960	700	902	55,462	.15	.17
November *a*	1,350	870	1,115	66,347	.19	.21
December *b*	1,150	785	924	56,815	.15	.17
The year	3,850	31	1,235	892,406	.20	2.79

a November 23, 25, and 27 interpolated.   *b* December 1, 3, 7, 9, 13, 15, and 17 interpolated. .

BEAR RIVER NEAR PRESTON, IDAHO.

This station, established October 11, 1889, is about 6 miles from Preston, Idaho, and 10 miles north of the Utah-Idaho boundary line. The station is of considerable importance from the fact that its location is near the Utah-Idaho line, and the measurements there will indicate the volume of water that passes from Idaho into Utah. During 1901 a large canal was completed, appropriating the waters of the Bear about 8 miles below Soda Springs in sufficient quantity to irrigate about 35,000 acres of very fine land in that locality. The original gage consisted of a vertical board nailed to a pile of the highway bridge. This was carried away June 30, 1899, but was replaced on August 4 by a wire gage. It is read by J. A. Nelson. Discharge measurements are made by means of a cable and car. The initial point for soundings is the post over which the cable passes on the right bank. The channel is straight for 250 feet above and below the station. Both banks are high, clean, and not liable to overflow. The bed of the stream is smooth, and there is but one channel at low water. The bench mark for the readings of 1903 is that of the old gage, which is a nail in the southeast corner of a house near the gage, about 1.5 feet from the ground and 10.95 feet above gage datum. Because of some difficulty in maintaining the wire gage, a temporary gage was nailed

to the pile of the highway bridge above the station. The readings
this gage have been reduced to the same datum as the wire gage. (
October 31, 1903, a new gage was nailed to a pile on the highw
bridge above the station. Its elevation is 10.80 feet below the (
bench mark; therefore the gage heights from November 1 to Dece
31, 1903, have been increased 0.15 foot.

The observations at this station during 1903 have been made und
the direction of G. L. Swendsen, district hydrographer.

Discharge measurements of Bear River near Preston, Idaho, in 1903.

Date.	Hydrographer.	Gage height.	Dischar
		Feet.	*Second-f*
March 16	G. L. Swendsen	2.40	
October 31	W. W. McLaughlin	1.55	

Mean daily gage height, in feet, of Bear River near Preston, Idaho, for 1903.

Day.	Jan.	Feb.	Mar.	Apr.	May.	June.	Nov.	De
1	1.60	1.60	1.60	2.55	2.95	2.90	1.67	
2	1.60	1.60	1.60	2.70	2.95	2.95	1.67	
3	1.60	1.60	1.60	2.50	2.85	2.95	1.67	
4	1.60	1.60	1.60	2.60	2.85	3.05	1.67	
5	1.60	1.60	1.60	2.72	2.80	3.12	1.67	
6	1.60	1.60	1.65	2.72	2.80	3.20	1.68	
7	1.60	1.60	1.70	2.93	2.80	....	1.68	
8	1.60	1.60	1.70	2.97	2.80	....	1.68	
9	1.60	1.60	1.65	3.05	2.80	....	1.68	
10	1.60	1.60	1.70	3.22	2.80	....	1.68	
11	1.60	1.60	1.75	3.30	2.80	....	1.68	
12	1.60	1.60	2.30	3.70	2.80	....	1.68	
13	1.60	1.60	2.30	3.36	2.80	....	1.69	
14	1.60	1.60	2.45	3.20	2.80	....	1.70	
15	1.60	1.60	2.40	3.10	2.75	....	1.70	
16	1.60	1.60	2.38	3.00	2.70	....	1.69	
17	1.60	1.60	2.32	3.00	2.70	....	1.68	
18	1.60	1.60	3.00	2.95	2.80	....	1.69	
19	1.60	1.60	2.80	2.90	2.90	....	1.69	
20	1.60	1.60	2.70	2.90	2.92	....	1.69	
21	1.60	1.60	2.70	2.87	2.95	....	1.69	
22	1.60	1.60	1.70	2.85	2.95	....	1.69	
23	1.60	1.60	1.75	2.80	2.95	....	1.69	
24	1.60	1.60	1.80	2.80	2.95	....	1.69	
25	1.60	1.60	1.95	2.80	2.95	....	1.69	
26	1.60	1.60	2.00	2.80	2.95	....	1.69	
27	1.60	1.60	2.00	2.90	2.95	....	1.69	...
28	1.60	1.60	2.30	2.95	2.95	....	1.69	...
29	1.60	....	2.30	2.95	2.80	....	1.69	...
30	1.60	....	2.22	2.98	2.80	....	1.69	...
31	1.60	....	2.40	....	2.80	....		...

BEAR RIVER AT DINGLE, IDAHO.

This station was established May 9, 1903, by G. L. Swendsen. It
s located in the cut made by the Oregon Short Line Railroad, one-
fourth mile east of the Dingle railroad station. The station is 250
feet south of the railroad track. The gage is a vertical wooden rod,
bolted to an iron rod which is driven into the bed of the stream. It
s read once each day by M. K. Hopkins. Discharge measurements
are made by means of a cable and car. The cable has a total span of
150 feet. The initial point for soundings is the zero mark on the
cable at the north bank. The channel is straight for 300 feet above
and below the station. The velocity is moderate at ordinary stages
and is well distributed. Both banks are high, not liable to overflow,
and cleared of everything except some small brush. The bed of the
stream is composed of gravel, and there is but one channel at all
stages. The bench mark is a heavy hub 25 feet north of the gage on
the north bank. Its elevation is 14.13 feet above the zero of the gage.
The zero of the gage is 56.5 feet above the datum of the survey for
the Bear Lake reservoir project made during the summer of 1903.
The zero of the gage is 6,015.3 feet above the datum of the Oregon
Short Line Railroad levels.

The observations at this station during 1903 have been made under
the direction of G. L. Swendsen, district hydrographer.

Discharge measurements of Bear River at Dingle, Idaho, in 1903.

Date.	Hydrographer.	Gage height.	Discharge.
		Feet.	*Second-feet.*
April 27	Geo. L. Swendsen	....	976
May 9	W. G. Swendsen	4.60	581
June 20	.....do	5.95	1,333
September 3	W. P. Hardesty	3.52	135
October 7	.....do	3.66	168

Mean daily gage height, in feet, of Bear River at Dingle, Idaho, for 1903.

Day.	May.	June.	July.	Aug.	Sept.	Oct.	Nov.	De
1		5.00	5.00	3.90	3.50	3.60	3.60	
2		5.00	4.90	3.80	3.50	3.00	3.60	
3		5.20	4.70	3.90	3.50	3.70	3.60	
4		5.30	4.70	3.90	3.50	3.70	3.60	
5		5.50	4.60	3.90	3.50	3.75	3.60	
6		5.80	4.00	3.90	3.50	3.70	3.60	
7		5.90	4.60	3.80	3.60	3.70	3.70	
8		6.00	4.60	3.80	3.50	3.70	3.70	
9		6.00	4.50	3.80	3.50	3.70	3.70	
10	4.65	6.00	4.40	3.80	3.50	3.70	3.60	
11	4.60	6.10	4.50	3.80	3.50	3.00	3.50	
12	4.40	6.10	4.50	3.80	3.50	3.60	3.60	
13	4.40	5.80	4.50	3.80	3.60	3.60	3.60	
14	4.50	5.70	4.50	3.80	3.60	3.60	3.80	
15	4.60	5.60	4.40	3.80	3.60	3.60	3.70	
16	4.70	6.00	4.40	3.70	3.60	3.60	3.90	
17	4.90	5.90	4.30	3.70	3.60	3.60	3.50	
18	5.00	5.90	4.30	3.70	3.60	3.60	3.70	
19	4.90	5.90	4.30	3.70	3.60	3.60		
20	5.20	6.10	4.30	3.70	3.60	3.60	3.90	
21	5.00	5.90	4.20	3.70	3.60	3.60	3.80	
22	5.00	5.90	4.20	3.70	3.60	3.60	3.90	
23	4.90	5.80	4.10	3.70	3.60	3.60	3.80	
24	4.80	5.70	4.10	3.60	3.60	3.60	3.80	...
25	4.90	5.70	4.00	3.60	3.60	3.60	3.70	
26	4.90	5.50	4.00	3.60	3.60	3.60	4.50	
27	4.90	5.40	4.00	3.60	3.60	3.60	4.00	
28	4.90	5.40	4.00	3.60	3.60	3.60	3.80	....
29	4.90	5.30	4.00	3.60	3.60	3.60	4.40	
30	4.90	5.20	3.90	3.60	3.60	3.60	3.90	
31	5.00		3.90	3.50		3.60		....

Rating table for Bear River at Dingle, Idaho, from May 10 to December 31, 1903.

Gage height.	Discharge.	Gage height.	Discharge.	Gage height.	Discharge.	Gage height.	Discharge
Feet.	*Second-feet.*	*Feet.*	*Second-feet.*	*Feet.*	*Second-feet.*	*Feet.*	*Second-feet*
3.2	82	4.0	279	4.8	695	5.6	1,135
3.3	96	4.1	322	4.9	750	5.7	1,190
3.4	112	4.2	369	5.0	805	5.8	1,245
3.5	131	4.3	421	5.1	860	5.9	1,300
3.6	153	4.4	475	5.2	915	6.0	1,355
3.7	178	4.5	530	5.3	970	6.1	1,410
3.8	207	4.6	585	5.4	1,025		
3.9	241	4.7	640	5.5	1,080		

Table well defined. Rating table does not take into account canals taken immediately above this station.

Estimated monthly discharge of Bear River at Dingle, Idaho, for 1903.

Month.	Discharge in second-feet.			Total in acre-feet.
	Maximum.	Minimum.	Mean.	
ay 10–31	915	475	704	30,720
me	1,410	805	1,179	70,155
aly	805	241	458	28,161
ugust	241	131	191	11,744
eptember	153	131	145	8,628
)ctober	178	153	159	9,776
ƒovember	530	131	205	12,198
December ͣ	860	82	218	13,404

ͣ December 24 and 27–31 estimated.

LOGAN RIVER NEAR LOGAN, UTAH.

The station on Logan River was established June 1, 1896, about 2 miles east of Logan, Utah. The original equipment consisted of a table and car and two gages. One gage was a vertical iron post set in the bed of the stream, and the other a vertical wooden gage near the above. The bench mark for both gages is a stone marked "B. M., 35 ft." northwest of the wooden gage. Its elevation is 14.01 feet above gage datum.

The observations are taken by the Hercules Power Company free of charge, but the local conditions at the station became unsatisfactory, and readings were discontinued July 18, 1903.

However, local interest in the station was such that plans were made for a reestablishment of the station at a more suitable point. The new station will be ready for use early in 1904.

The observations at this station during 1903 have been made under the direction of G. L. Swendsen, district hydrographer.

Discharge measurements of Logan River near Logan, Utah, in 1903.

Date.	Hydrographer.	Gage height.	Discharge.
		Feet.	*Second-feet.*
January 17	G. L. Swendsen	2.63	164
March 19	do	2.61	159
April 8	do	2.63	165

Mean daily gage height, in feet, of Logan River near Logan, Utah, for 1903.

Day.	Jan.	Feb.	Mar.	Apr.	May.	June.	July.
1	2.68	2.54	2.54	2.75	3.38	3.75	1.57
2	2.75	2.66	2.54	2.75	3.00	3.87	1.6
3	2.66	2.57	2.62	2.69	3.62	3.67	1.5
4	2.66	2.66	2.62	2.71	3.56	....	1.5
5	2.66	2.52	2.62	2.62	3.58	....	1.5
6	2.60	2.54	2.62	2.69	3.54	4.34	1.5
7	2.62	2.52	2.66	2.69	3.60	....	1.5
8	2.60	2.64	2.54	2.62	3.54	....	1.5
9	2.60	2.63	2.66	....	3.00	....	1.5
10	2.60	2.63	2.52	2.62	....	....	1.6
11	2.62	2.63	2.62	2.94	3.34	4.42	1.5
12	3.56	2.66	2.62	2.75	3.44	4.50	3.54
13	3.58	2.62	2.66	2.75	3.52	4.42	1.54
14	3.54	2.52	2.64	2.75	3.62	4.34	3.54
15	3.60	2.54	2.64	2.74	3.75	4.35	3.5
16	2.58	2.62	2.64	2.86	3.67	....	1.5
17	2.60	2.62	2.64	2.87	....	4.35	1.5
18	2.63	2.52	2.64	2.84	3.50	4.00	1.50
19	2.64	2.62	2.62	2.79	3.50	3.70	....
20	2.64	2.66	2.62	2.84	3.44	3.70	....
21	2.64	2.86	2.62	2.87	3.44	3.90	....
22	2.63	2.66	....	....	3.50	3.87	....
23	2.64	2.86	2.62	2.91	3.50	3.85	....
24	2.50	2.86	2.62	3.29	3.37	3.79	....
25	....	2.60	2.64	3.21	3.36	3.75	....
26	2.64	2.62	2.64	3.28	3.37	3.88	....
27	2.50	2.51	2.64	3.26	....	3.82	....
28	2.50	2.66	2.62	3.38	3.44	3.75	....
29	2.64	....	....	3.35	3.44	3.75	....
30	2.54	....	2.61	3.35	3.54	3.67	....
31	2.64	....	2.72	....	3.53	....	....

WEBER RIVER NEAR UINTA, UTAH.

The gaging station on Weber River, established in October, 1899, is located in the canyon 5 miles east of Uinta, on the Union Pacific Railroad, immediately above the narrows known as Devils Gate. The gage is vertical and is supported from above by a projecting timber placed out of reach of high water. It is read by Hugh McQueen. The bench mark consists of a spike driven into the first telegraph pole in the canyon above the gage, at an elevation of 17.44 feet above gage datum. The equipment consisted of a cable and car, but these were condemned early in 1903 and are not now in use. Observations were continued until July 11, 1903, when the State engineer began a systematic study of the Weber River water supply.

The observations at this station during 1903 have been made under the direction of G. L. Swendsen, district hydrographer.

Discharge measurements of Weber River near Uinta, Utah, in 1903.

Date.	Hydrographer.	Gage height.	Discharge.
		Feet.	*Second-feet.*
nuary 12	G. L. Swendsen..............	1. 41	124
pril 13.....................	.....do	2. 50	621
ay 18.....................	.....do	3. 40	677

Mean daily gage height, in feet, of Weber River near Uinta, Utah, for 1903.

Day.	Mar.	Apr.	May.	June.	July.	Day.	Mar.	Apr.	May.	June.	July.
....	.....	4.10	2.80	3.70	2.40	17................	.....	2.60	3.40	3.55	......
....	.....	3.80	2.90	3.90	2.30	18................	.....	2.50	3.40	3.45	......
....	.....	3.10	3.00	4.10	2.20	19................	.....	2.60	3.30	3.45	......
....	.....	2.70	3.10	4.20	2.20	20................	.....	2.60	3.10	3.35	......
....	.....	2.60	3.20	4.20	2.20	21................	.....	2.70	3.20	3.20	......
....	.....	2.50	3.30	4.10	2.10	22................	.....	2.80	3.20	3.10	......
....	.....	2.50	3.30	4.05	2.00	23................	.....	2.90	3.10	3.00	......
....	.....	2.60	3.30	4.05	2.00	24................	.....	3.00	3.20	2.90	......
....	.....	3.10	3.20	4.05	1.90	25................	.....	3.10	3.20	2.75	......
....	.....	2.90	3.20	4.05	1.90	26................	.....	3.20	3.20	2.65	......
....	.....	2.80	3.10	3.65	1.80	27................	.....	3.40	3.40	2.55	......
....	.....	2.60	3.00	3.80	......	28................	.....	3.30	3.40	2.50	......
....	.....	2.50	3.10	3.65	......	29................	.....	3.10	3.40	2.50	......
....	.....	2.50	3.30	3.55	......	30................	3.50	2.90	3.40	2.40	......
....	.....	2.60	3.30	3.55	......	31................	4.30	......	3.50	......	......
....	.....	2.70	3.30	3.45	......						

AMERICAN FORK RIVER NEAR AMERICAN FORK, UTAH.

This station was established May 21, 1900, by C. C. Babb. It is
:ated about 6 miles northeast of the town of American Fork, Utah,
feet north of the county road, and 200 feet southwest of the power
use of the electric company. Measurements are made over a
arp, crested, rectangular weir. A nail driven into the weir structure,
st south of the south opening and level with the crest of the weir,
rves to determine gage heights. The observer is Peter Anderson.
uring floods considerable gravel runs in the stream and is depos-
:d just above the weir, making measurement at such times erroneous.
'hen this difficulty becomes pronounced, several planks in the dam
:e raised and the gravel is sluiced out.

The observations at this station during 1903 have been made under
ie direction of G. L. Swendsen, district hydrographer.

Discharge measurements of American Fork River near American Fork, Utah, in 1903.

Date.	Hydrographer.	Gage height.	Discharge.
		Feet.	*Second-ft.*
April 6......................	C. Tanner	0.49	*3
December 1.................	.....do.......................	.34	*2

* Weir measurement.

Mean daily gage height, in feet, of American Fork River near American Fork, Utah, for 1903.

Day.	Apr.	May.	June.	July.	Aug.	Sept.	Oct.	Nov.	Dec
1.........................	......	0.68	1.30	0.94	0.45	0.38	0.46	0.35	0.1
2.........................	......	.68	1.44	.90	.45	.33	.44	.35	.1
3.........................	......	.77	1.43	......	.45	.33	.43	.35	.1
4.........................	......	.82	1.45	......	.44	.33	.41	.35	.1
5.........................	......	.88	1.50	.76	.44	.33	.41	.35	.
6.........................	0.49	.92	1.56	.79	.42	.33	.42	.34	.
7.........................	.35	.96	1.50	.76	.41	.33	.40	.34	.
8.........................	.33	.92	1.50	.74	.41	.33	.40	.34	.
9.........................	.39	.95	1.68	.72	.41	.38	.40	.34	.
10........................	.43	.98	1.52	.71	.41	.33	.40	.33	
11........................	.42	.99	1.48	.70	.42	.33	.42	.33	
12........................	.40	1.11	1.50	.70	.40	.33	.42	.36	
13........................	.43	1.25	1.44	.70	.40	.33	.41	.33	
14........................	.44	1.33	1.48	.68	.39	.32	.40	.37	
15........................	.40	1.26	1.37	.66	.39	.32	.40	.36	
16........................	.41	1.25	1.48	.66	.38	.32	.39	.35	
17........................	.42	1.07	1.50	.64	.38	.32	.39	.33	
18........................	.44	.95	1.52	.64	.37	.32	.39	.32	
19........................	.40	.84	1.41	.61	.37	.32	.39	.34	
20........................	.42	.78	1.31	.60	.37	......	.38	.34	
21........................	.41	.78	......	.58	.37	......	.38	.34	
22........................	.50	.76	......	.56	.36	......	.38	.34	
23........................	.60	.72	......	.55	.36	......	.37	.34	
24........................	.75	.74	......	.52	.36	......	.37	.35	
25........................	.81	.71	......	.52	.36	......	.37	.34	
26........................	.87	.74	1.02	.50	.35	......	.37	.33	
27........................	.80	.81	1.04	.49	.35	......	.36	.32	
28........................	.70	.84	1.05	.48	.34	......	.36	.31	.
29........................	.61	.85	1.02	.48	.35	......	.36	.31	...
30........................	.59	1.01	.98	.47	.33	......	.35	.34	...
31........................	......	1.16	......	.45	.33	......	.35	......	...

PROVO RIVER AT S. P., L. A. & S. L. R. R. BRIDGE, PROVO, UTAH.

This station was established May 24, 1903, by C. Tanner. It
located at the San Pedro, Los Angeles and Salt Lake Railroad bridg
one-half mile from Provo, Utah. The Rio Grande Western Railw
bridge is about 300 feet east of the station. The gage is a vertic
rod fastened to a cottonwood post, which is set in the bed of the riv

near the left, or south, bank. The top of the post is spiked to a stringer of the bridge. The gage is read once each day by Fred Thomson. Discharge measurements at high water are made from the railroad bridge to which the gage is attached. At low stages measurements are made by wading. The initial point for soundings is the north face of the left, or south, abutment. The channel is straight for 175 feet above and for 300 feet below the station. The current is swift. There are two channels at low water and one channel at high water. Both banks have an elevation of about 8 feet above the river bed, and will overflow only at extreme flood stages. The bed of the stream is composed of gravel and bowlders and is somewhat shifting. The bench mark is a cross chiseled on the top of the south, or left, bridge abutment near the northwest corner. The letters "B. M." are chiseled near the bench mark. Its elevation is 6.66 feet above the zero of the gage.

The observations at this station during 1903 have been made under the direction of G. L. Swendsen, district hydrographer.

Discharge measurements of Provo River at S. P., L. A. and S. L. R. R. bridge, Provo, Utah, in 1903.

Date.	Hydrographer.	Gage height.	Discharge.
		Feet.	*Second-feet.*
May 24	C. Tanner	0.92	180
June 6	....do	1.82	738
October 31	....do	.20	10
November 23	....do	.88	130

Mean daily gage height, in feet, of Provo River at S. P., L. A. and S. L. R. R. bridge, Provo, Utah, for 1903.

Day.	May.	June.	Nov.	Dec.	Day.	May.	June.	Nov.	Dec.
1		1.06		0.89	17		1.40	(a)	
2		1.42			18		1.88	0.76	0.97
3		1.68			19		1.28		
4		1.74			20		1.20		
5		1.84			21		1.10		
6		1.86		.95	22		1.08	.88	
7		1.78			23		.90		.94
8		1.80		.86	24	0.92	.74		
9		1.92			25	.82	.44	.90	
10		1.72			26	.68	.12		
11		1.74			27	.70	.06		
12		1.68		.90	28	.68	(a)		
13		1.66			29	.68			.90
14		1.68			30	.66			
15		1.64			31	.82			
16		1.44		.90					

a No discharge from June 27 to November 18, 1903.

Rating table for Provo River at S. P., L. A. and S. L. R. R. bridge, Provo, Utah, for 19??

Gage height.	Discharge.	Gage height.	Discharge.	Gage height.		Gage height.	Discharge.
Feet.	*Second-feet.*	*Feet.*	*Second-feet.*	*Feet.*	*Second-feet.*	*Feet.*	*Second-feet.*
0.0	0	0.5	55	1.0	200	1.5	465
.1	3	.6	76	1.1	242	1.6	540
.2	10	.7	101	1.2	289	1.7	620
.3	22	.8	130	1.3	342	1.8	710
.4	37	.9	163	1.4	400	1.9	810

Estimated monthly discharge of Provo River at S. P., L. A. and S. L. R. R. bridge, Provo, Utah, for 1903.

Month.	Discharge in second-feet.			Total in acre-feet
	Maximum.	Minimum.	Mean.	
May 24–31..........................	....	....	116	1,?
June 1–27.............................	832	2	443	23,?
November 18, 22, 25..................	....	....	146	....
December (8 days)....................	189	150	168	a 10,

a The flow was so uniform that the mean for 8 days (168) was assumed to hold for the entire m?
and the acre-feet were computed for the whole month.

PROVO RIVER AT MOUTH OF CANYON, NEAR PROVO, UTAH.

This station was established July 27, 1889, by Samuel Fortier.
is located about 6 miles north of Provo and above the head of m
of the irrigation canals of Utah Valley. The diversion works of
company which develop power at the mouth of Provo Canyon
electric transmission to the mines west of Provo are located abo?
miles above the station. The Heber branch of the Denver and
Grande Railway passes within 20 feet of the new gage rod. The
tion is one-eighth mile above the county bridge. The old gage i?
inclined rod set on the left bank. The new gage is a vertical as
stick set on the right bank 90 feet northeast of the old gage. ?
gages read 4.80 feet when the new gage was installed, but the zer?
the new gage is 0.10 foot above the zero of the old gage. The g
is read twice each day by L. T. Walter. Discharge measurem?
are made by means of a cable and car located at the gage. Instea
using a tagged wire, the cable has been marked with white paint e?
3 feet. The initial point for soundings is the first white mark on
cable south of the vertical post which supports the cable on the r
bank. The channel is straight for 200 feet above and for 100
below the station. The right bank is steep and rocky and ha?

:levation of about 10 feet above the zero of the gage. The left bank
s sloping and has an elevation of about 7 feet above the zero of the
;age. It is liable to overflow at flood stages. The bed of the stream
s composed of bowlders and is permanent. The current is swift and
it flood stages has a velocity that is too high for accurate measurement.
There is one channel up to the point where the river overflows the
eft bank. Above this stage there are two channels. The bench
mark is a cross chiseled in a limestone rock about 1 foot square, which
bears south 15° E. and is 100 feet distant from the old gage. Its
elevation is 6.98 feet above the zero of the old (inclined) gage. The
letters "B. M." are chiseled in the rock.

The observations of this station during 1903 have been made under
the direction of G. L. Swendsen, district hydrographer.

Discharge measurements of Provo River at mouth of canyon, near Provo, Utah, in 1903.

Date.	Hydrographer.	Gage height.	Discharge.
		Feet.	*Second-feet.*
January 19	G. L. Swendsen..............	4. 02	182
April 4.....................	C. Tanner	4. 66	381
May 27.....................	......do	4. 75	393
June 5.....................	......do	6. 10	1, 197
July 9.....................	......do	4. 28	220
August 3..................	......do	4. 20	193
September 20	......do	4. 20	182
October 21................	......do	4. 40	215

Mean daily gage height, in feet, of Provo River at mouth of canyon, near Provo, Utah, for 1903.

Day.	June.	July.	Aug.	Sept.	Oct.	Nov.	Dec.
1....	....	4. 35	4. 25	4. 15	4. 35	4. 40	4. 55
2....	....	4. 35	....	4. 15	4. 40	4. 40	4. 55
3....	....	4. 35	4. 20	4. 15	4. 40	4. 40	4. 55
4....	....	4. 30	4. 20	4. 10	....	4. 40	4. 55
5....	....	....	4. 20	4. 10	4. 40	4. 40	1. 55
6....	....	4. 30	4. 20	....	4. 40	4. 40	....
7....	6. 10	4. 30	4. 20	4. 10	4. 40	4. 40	4. 55
8....	5. 90	4. 30	4. 20	4. 15	4. 35	....	4. 55
9....	5. 97	4. 30	....	4. 15	4. 35	4. 45	4. 55
10....	5. 90	4. 25	4. 20	4. 20	4. 35	4. 45	4. 55
11....	5. 73	4. 30	4. 20	4. 30	4. 35	4. 55	4. 55
12....	5. 77	....	4. 20	4. 30	4. 35	4. 55	4. 55
13....	5. 67	4. 30	4. 20	....	4. 35	4. 80	....
14....	....	4. 30	4. 25	4. 30	4. 35	4. 70	4. 50
15....	5. 60	4. 30	4. 20	4. 30	4. 40	....	4. 50
16....	5. 47	4. 30	....	4. 25	4. 40	4. 65	4. 45
17....	5. 42	4. 30	4. 20	4. 25	4. 40	4. 65	4. 45
18....	5. 37	4. 30	4. 20	4. 20	....	4. 65	4. 40

Mean daily gage height, in feet, of Provo River at mouth of canyon, near Provo, U
1903—Continued.

Day.	June.	July.	Aug.	Sept.	Oct.	Nov.
19...	5.20	....	4.20	4.20	4.40	4.60
20...	5.22	4.30	4.20	....	4.40	4.60
21...	....	4.30	4.20	4.20	4.40	4.55
22...	5.05	4.30	4.20	4.25	4.40	....
23...	5.00	4.30	....	4.25	4.35	4.55
24...	4.85	4.30	4.15	4.25	4.35	4.55
25...	4.70	4.30	4.15	4.25	....	4.55
26...	4.60	....	4.15	4.25	4.35	4.55
27...	4.55	4.30	4.15	....	4.35	4.55
28...	....	4.30	4.20	4.25	4.35	4.55
29...	4.50	4.27	4.20	4.30	4.40	....
30...	4.35	4.25	....	4.30	4.40	4.55
31...	....	4.25	4.10	....	4.40	....

Rating table for Provo River at mouth of canyon, near Provo, Utah, for 190

Gage height.	Discharge.	Gage height.	Discharge.	Gage height.	Discharge.	Gage height.	Disc
Feet.	*Second-feet.*	*Feet.*	*Second-feet.*	*Feet.*	*Second-feet.*	*Feet.*	*Secon*
4.0	155	4.6	332	5.2	620	5.8	
4.1	168	4.7	375	5.3	670	5.9	1,
4.2	188	4.8	420	5.4	730	6.0	1,
4.3	218	4.9	465	5.5	790	6.1	1,
4.4	253	5.0	515	5.6	850		
4.5	291	5.1	565	5.7	915		

Table uncertain owing to change of gage and an obstruction in the river du
latter part of the year.

Estimated monthly discharge of Provo River at mouth of canyon, near Provo, Utah,

Month.	Discharge in second-feet.			1
	Maximum.	Minimum.	Mean.	a
June 7–30	1,195	235	685	
July	235	203	218	
August.................................	203	168	187	
September..............................	218	168	195	
October	253	235	245	
November..............................	420	253	306	
December	311	253	292	
The period	....	....	....	

Sunday discharges estimated.

SPANISH FORK RIVER NEAR SPANISH FORK, UTAH.

This station was originally established May 23, 1900, by C. C. Babb.
It was reestablished March 26, 1903, by C. Tanner. It is located 600
feet above the dam of the East Bench Irrigation Company and 5 miles
southeast of Spanish Fork, Utah. The Rio Grande Western Railroad
track is about 300 feet northeast of the gage. The gage is a 1 by
4 inch redwood stick 4½ feet long. It is driven firmly in the bed of
the river and is well braced. It is graduated to feet, tenths, and fiftieths, and located near the right bank about 1 foot from the water's
edge. It is read once each day by Levi Thorpe. Discharge measurements are made by wading. The initial point for soundings is a peeled
willow stake on the right bank 2 feet northeast of the gage. The
channel is straight for 150 feet above and below the station and the
current is swift. The right bank is about 4 feet above the bed of the
stream. It is covered with a dense growth of willows and is not liable
to overflow. The left bank is about 5½ feet above the bed of the
stream. It is also covered with willows. The bed of the stream is
composed of coarse gravel and sand, somewhat shifting. The change
in the cross section takes place principally in the right half. There is
but one channel at all stages. The bench mark is a cross on a limestone rock, which bears south 36° east and is 29 feet distant from the
gage. On the south face of the rock are the letters "U. S. G. S." in
black paint. Its elevation is 7.16 feet above the zero of the gage.

The observations at this station during 1903 have been made under
the direction of G. L. Swendsen, district hydrogragher.

Discharge measurements of Spanish Fork River near Spanish Fork, Utah, for 1903.

Date.	Hydrographer.	Gage height.	Discharge.
		Feet.	*Second-feet.*
March 26	C. Tanner	1.61	124
April 27	.....do	1.88	172
May 28	.....do	2.42	268
June 20	C. Callister	1.88	145
July 29	C. Tanner	1.50	72
September 22	.....do	1.41	53
October 26	.....do	1.39	51
December 15	.....do	1.38	53

Mean daily gage height, in feet, of Spanish Fork River near Spanish Fork, Utah, for 1903.

Day.	Apr.	May.	June.	July.	Aug.	Sept.	Oct.	Nov.	Dec.
1.........	......	1.72	2.54	1.68	1.46	1.38	1.48	1.38	1.50
2.........	......	1.76	2.58	1.66	1.46	1.36	1.56	1.38	1.48
3.........	......	1.80	2.66	1.64	1.44	1.36	1.54	1.35	1.50
4.........	......	1.88	2.56	1.64	1.44	1.36	1.44	1.34	1.50
5.........	......	1.92	2.90	1.68	1.44	1.32	1.44	1.46	1.52
6.........	1.40	1.97	2.76	1.66	1.42	1.36	1.45	1.40	1.50
7.........	1.42	1.98	2.70	1.66	1.42	1.38	1.44	1.42	1.52
8.........	1.44	2.00	2.66	1.64	1.42	1.36	1.44	1.48	1.52
9.........	1.50	2.10	2.54	1.60	1.36	1.36	1.42	1.44	1.52
10.........	1.54	2.10	2.52	1.60	1.34	1.36	1.42	1.42	1.52
11.........	1.54	2.16	2.44	1.58	1.40	1.36	1.42	1.30	1.50
12.........	1.50	2.16	2.42	1.54	1.34	1.38	1.42	1.30	1.50
13.........	1.48	2.20	2.48	1.50	1.34	1.42	1.40	1.34	1.52
14.........	1.46	2.32	2.20	1.48	1.34	1.42	1.40	1.34	1.50
15.........	1.50	2.30	2.20	1.48	1.34	1.42	1.40	1.46	1.50
16.........	1.48	2.36	2.14	1.48	1.34	1.42	1.40	1.46	1.48
17.........	1.50	2.50	2.10	1.58	1.32	1.42	1.38	1.30	1.50
18.........	1.48	2.34	2.00	1.52	1.32	1.42	1.38	1.32	1.50
19.........	1.50	2.30	1.90	1.50	1.32	1.42	1.38	1.32	1.50
20.........	1.50	2.14	1.84	1.48	1.36	1.40	1.38	1.42	1.54
21.........	1.50	2.28	1.84	1.48	1.44	1.40	1.38	1.42	1.52
22.........	1.56	2.22	1.80	1.48	1.36	1.40	1.38	1.44	1.52
23.........	1.64	2.12	1.82	1.50	1.40	1.38	1.38	1.44	1.52
24.........	1.72	2.30	1.76	1.54	1.38	1.36	1.38	1.44	1.52
25.........	1.90	2.24	1.78	1.48	1.34	1.36	1.38	1.44	1.50
26.........	2.00	2.28	1.74	1.58	1.34	1.36	1.38	1.41	1.50
27.........	1.98	2.36	1.72	1.50	1.34	1.36	1.40	1.36	1.52
28.........	1.90	2.48	1.72	1.40	1.34	1.38	1.38	1.34	1.54
29.........	1.78	2.50	1.70	1.48	1.38	1.38	1.38	1.32	1.54
30.........	1.69	2.56	1.70	1.46	1.38	1.46	1.38	1.34	1.54
31.........	......	2.70	......	1.46	1.36	......	1.32	......	1.52

Rating table for Spanish Fork River near Spanish Fork, Utah, from April 6 to June 8, 1903.

Gage height.	Discharge.	Gage height.	Discharge.	Gage height.	Discharge.	Gage height.	Discharge.
Feet.	*Second-feet.*	*Feet.*	*Second-feet.*	*Feet.*	*Second-feet.*	*Feet.*	*Second-feet.*
1.4	89	1.9	175	2.4	265	2.8	338
1.5	105	2.0	193	2.5	283	2.9	358
1.6	121	2.1	211	2.6	301	3.0	378
1.7	139	2.2	229	2.7	319	3.1	398
1.8	157	2.3	247				

ible for Spanish Fork River near Spanish Fork, Utah, from June 4 to December 31, 1903.

t.	Discharge.	Gage height.	Discharge.	Gage height.	Discharge.	Gage height.	Discharge.
	Second-feet.	*Feet.*	*Second-feet.*	*Feet.*	*Second-feet.*	*Feet.*	*Second-feet.*
	23	1.7	107	2.2	207	2.7	314
	36	1.8	127	2.3	227	2.8	336
	53	1.9	147	2.4	248	2.9	358
	71	2.0	167	2.5	270	3.0	380
	89	2.1	187	2.6	292		

d monthly discharge of Spanish Fork River near Spanish Fork, Utah, for 1903.

Month	Discharge in second-feet.			Total in acre-feet.
	Maximum.	Minimum.	Mean.	
30	193	89	122	6,049
....	319	143	227	13,958
....	388	107	218	12,972
....	103	53	79	4,858
....	64	39	48	2,951
)er...............................	85	46	52	3,094
....	96	39	55	3,382
er	78	28	50	2,975
er	85	28	50	3,074
iod	....		....	53,313

SEVIER RIVER NEAR GUNNISON, UTAH.

station, established by Caleb Tanner on June 29, 1900, is at the bridge which crosses the stream 4 miles west of the town of
ion on the road to West View precinct. The gage is a vertical
inch redwood timber painted white and nailed to a bridge pile
ght bank of the stream. It is graduated by means of black
o feet, tenths, and fiftieths. Discharge measurements are made
nary stages from the bridge to which the gage is attached. At
ter they are made by wading about 200 feet below the bridge.
itial point for soundings, when measurements are made from
dge, is the bridge pile to which the gage is fastened on the right
f the stream. When measurements are made by wading the
point for soundings is a blazed willow peg driven into the right
of the stream. On the left bank directly opposite the initial
s a 1 by 3 inch wooden peg standing 5 inches above the ground.
iannel is straight for 300 feet above and below the station.
anks are about 10 feet above the bed of the stream, are covered

with salt grass, and are not liable to overflow. The bed of the stream is composed of sand and gravel, free from vegetation, and is permanent. There is one channel at all stages. At low water the low velocity at the bridge makes the measurement inaccurate because of the bridge piles and floating débris. This is overcome by using the section below. The bench mark is the top of a post at the southeast corner of the bridge marked in pencil "U. S. G. S. gage B. M." Its elevation is 13.23 feet above gage datum.

The observations at this station during 1903 have been made under the direction of G. L. Swendsen, district hydrographer.

Discharge measurements of Sevier River near Gunnison, Utah, in 1903.

Date.	Hydrographer.	Gage height.	Discharge.
		Feet.	*Second-feet.*
January 25	O. Tanner	1.80	195
March 14	....do	2.50	380
April 11	....do	1.22	91
June 26	....do	.78	40
July 30	....do	.70	33
September 7	....do	.61	25
November 11	....do	.85	61

Mean daily gage height, in feet, of Sevier River near Gunnison, Utah, for 1903.

Day.	Jan.	Feb.	Mar.	Apr.	May.	June.	July.	Aug.	Sept.	Oct.	Nov.	Dec.
1	1.70	......	2.15	1.60	0.30	1.50	0.70	0.68	0.62	0.70	0.74	0.74
2	1.70	......	2.15	1.62	.30	1.50	.68	.70	.64	.70	.80	(a)
3	1.70	......	2.15	1.65	.50	1.52	.72	.62	.62	.70	.80	(a)
4	1.80	......	2.15	1.67	.45	1.55	.78	.66	.62	.70	.82	(a)
5	1.78	......	2.15	1.50	.60	1.75	.70	.70	.62	.72	.84	(a)
6	1.80	......	2.15	1.50	.60	1.75	.80	.70	.60	.72	.84	(a)
7	1.80	......	2.10	1.32	.72	1.45	.74	.62	.62	.74	.84	(a)
8	1.78	......	2.10	1.40	.72	1.35	.68	.62	.62	.74	.84	(a)
9	1.80	......	2.10	1.25	.77	1.60	.70	.62	.62	.76	.84	(a)
10	1.80	......	2.10	1.20	.80	1.70	.70	.62	.62	.76	.84	1.00
11	1.80	......	2.00	1.22	.90	1.70	.72	.62	.64	.74	.84	1.00
12	1.80	......	2.00	1.20	.95	1.55	.72	.68	.64	.76	.86	1.00
13	1.85	......	2.00	1.10	1.05	1.55	.68	.64	.64	.78	.90	1.04
14	1.85	......	2.05	1.10	1.05	1.55	.70	.64	.64	.76	.90	1.05
15	1.80	......	2.40	1.00	1.20	1.65	.70	.64	.64	.76	.94	(a)
16	1.85	......	2.45	.92	1.20	1.80	.72	.62	.66	.76	.96	(a)
17	1.80	......	2.45	.92	1.07	1.90	.72	.60	.66	.74	.96	(a)
18	1.85	......	2.10	.90	1.07	1.70	.74	.60	.66	.74	.96	(a)
19	1.85	......	2.10	.90	1.20	1.40	.72	.60	.66	.72	1.00	1.04
20	1.85	......	2.10	.87	1.20	1.40	.78	.60	.68	.72	1.04	2.
21	1.85	(b)	1.40	.77	1.40	1.40	.70	.60	.70	.70	1.10	2.
22	1.85	2.20	1.40	.72	1.55	1.40	.70	.60	.66	.70	1.02	2.
23	1.82	2.20	1.50	.72	1.60	1.22	.70	.60	.64	.70	1.00	(a)
24	1.62	2.20	1.50	.72	1.60	.95	.70	.60	.64	.72	1.00	(a)

a Ice.   b River frozen from January 28 to February 21, inclusive.

n daily gage height, in feet, of Sevier River near Gunnison, Utah, for 1903—Continued.

Day.	Jan.	Feb.	Mar.	Apr.	May.	June.	July.	Aug.	Sept.	Oct.	Nov.	Dec.
....	1.82	2.15	1.85	0.78	1.62	0.82	0.70	0.60	0.64	0.72	0.98	(a)
....	1.87	2.15	1.80	.72	1.50	.62	.90	.62	.64	.70	.96	(a)
....	1.87	2.15	1.40	.72	1.50	.62	.84	.60	.64	.72	.96	2.00
....	(?)	2.15	1.57	.70	1.52	.74	.74	.60	.66	.72	.94	2.00
....			1.57	.70	1.40	.76	.70	.62	.70	.72	.92	1.90
....			1.40	.70	1.20	.76	.70	.62	.70	.72	.94	(a)
....			1.62		1.22		.68	.62		.72		(a)

a Ice.　　　　b River frozen from January 28 to February 21, inclusive.

Rating table for Sevier River near Gunnison, Utah, for 1903.

Gage height.	Discharge.	Gage height.	Discharge.	Gage height.	Discharge.	Gage height.	Discharge.
Feet.	*Second-feet.*	*Feet.*	*Second-feet.*	*Feet.*	*Second-feet.*	*Feet.*	*Second-feet.*
0.3	8	0.9	47	1.5	135	2.1	268
.4	12	1.0	58	1.6	154	2.2	296
.5	16	1.1	71	1.7	174	2.3	324
.6	22	1.2	85	1.8	194	2.4	352
.7	29	1.3	101	1.9	216	2.5	380
.8	37	1.4	117	2.0	241		

Table well determined.

Estimated monthly discharge of Sevier River near Gunnison, Utah, for 1903.

[Drainage area, 3,966 square miles.]

Month.	Discharge in second-feet.			Total in acre-feet.	Run-off.	
	Maximum.	Minimum.	Mean.		Second-feet per square mile.	Depth in inches.
nuary 1–27 a	209	158	195	10,443	0.049	0.049
bruary 22–28 a	296	282	288	3,999	.072	.019
irch...............	366	101	221	13,589	.055	.063
ril................	168	29	74	4,403	.019	.021
ly.................	158	8	76	4,673	.019	.022
ne	216	23	121	7,200	.030	.033
ly.................	47	27	31	1,906	.008	.009
gust	29	22	24	1,476	.006	.007
)tember...........	29	22	25	1,488	.006	.007
tober	35	29	31	1,906	.008	.009
vember	71	32	48	2,856	.012	.013
cember (12 days) a.	....	....	167	3,975	.042	.019
The period....	366	8	108	57,914	.027	.271

a River frozen from January 28 to February 21, and December 2–9, 15–19, 23–26, and 30–31.

SAN PITCH RIVER NEAR GUNNISON, UTAH.

This station was established June 30, 1900, by Caleb Tanner. It is located 4 miles northeast of Gunnison and one-eight mile west of the Rio Grande and Western Railroad. The station is just west of the second farmhouse along the railroad track north from Gunnison. The gage consists of a strongly braced vertical timber driven firmly into the bed of the stream near the left bank, about 100 feet above the ford. It is painted white and graduated by means of black paint to feet, tenths, and fiftieths. Discharge measurements are made by wading directly under a wire stretched across the river at the station. The initial point for soundings is the west side of a post about 3 feet high, on the left bank, to which the tagged line is attached. The channel is straight for 100 feet above and for 200 feet below the station. The current is swift. Both banks are about 5 feet above the bed of the stream, composed of clay, and covered with grass. The bed of the stream is composed of sand and gravel. It is clean and fairly permanent. There is but one channel at all stages. The bed shifts somewhat during floods, but except for this the conditions are good for accuracy. The bench mark is the top of a cedar post, 1 foot in diameter, set firmly in the ground, 40 feet west of the gage rod. Its elevation is 5.96 feet above the gage datum.

The observations at this station during 1903 have been made under the direction of G. L. Swendsen, district hydrographer.

Discharge measurements of San Pitch River near Gunnison, Utah, in 1903.

Date.	Hydrographer.	Gage height.	Discharge.
		Feet.	*Second-feet.*
January 24	Caleb Tanner	1.72	1
March 15	....do	1.76	1
April 12	....do	1.76	1
June 27	....do	2.18	6
July 31	....do	2.18	6
November 11	....do	1.73	1

Mean daily gage height, in feet, of San Pitch River near Gunnison, Utah, for 1903.

Day.	Jan.	Feb.	Mar.	Apr.	May.	June.	July.	Aug.	Sept.	Oct.	Nov.	Dec.
1	1.78	1.71	1.66	1.92	2.28	2.44	2.18	2.18	1.86	1.88	1.76	1.80
2	1.70	1.70	1.70	1.92	2.34	2.40	2.16	2.16	1.88	2.00	1.74	1.78
3	1.70	1.70	1.70	1.86	2.68	2.40	2.20	2.14	1.88	1.96	1.76	1.80
4	1.70	1.62	1.72	1.84	2.36	2.42	2.20	2.14	1.88	1.90	1.76	1.74
5	1.70	1.68	1.72	1.76	2.34	2.52	2.20	2.00	1.86	1.90	1.76	1.70
6	1.70	1.64	1.70	1.76	2.34	2.52	2.20	1.92	1.84	1.90	1.76	1.76
7	1.70	1.68	1.66	1.76	2.42	2.40	2.18	1.94	1.86	1.90	1.76	1.76
8	1.70	1.64	1.70	1.78	2.44	2.36	2.18	1.90	1.86	1.90	1.76	1.74
9	1.70	1.70	1.72	1.80	2.44	2.34	2.20	1.90	1.88	1.90	1.76	1.76
10	1.70	1.68	1.70	1.82	2.66	2.52	2.22	1.90	1.88	1.88	1.78	1.76
11	1.70	1.71	1.70	1.82	2.66	2.56	2.22	1.90	1.86	1.86	1.78	1.76
12	1.70	1.71	1.70	1.88	2.68	2.54	2.22	1.88	1.86	1.80	1.78	1.76
13	1.70	1.68	1.74	1.72	2.68	2.54	2.16	1.86	1.88	1.80	1.76	1.78
14	1.70	1.62	1.74	1.76	2.66	2.54	2.18	1.88	1.88	1.80	1.76	1.78
15	1.70	1.62	1.76	1.78	2.68	2.52	2.14	1.88	1.86	1.80	1.73	1.78
16	1.70	1.62	1.76	1.78	2.68	2.40	2.18	1.90	1.88	1.86	1.73	1.78
17	1.70	1.64	1.74	1.78	2.68	2.40	2.16	1.90	1.88	1.88	1.72	1.78
18	1.70	1.68	1.76	1.80	2.48	2.40	2.16	1.90	1.88	1.80	1.72	1.78
19	1.70	1.62	1.74	1.76	2.48	2.40	2.18	1.88	1.88	1.80	1.78	1.76
20	1.70	1.68	1.74	1.76	2.56	2.30	2.18	1.88	1.88	1.80	1.80	1.78
21	1.70	1.68	1.70	1.78	2.58	2.30	2.18	1.88	1.88	1.82	1.80	1.78
22	1.70	a1.70	1.74	1.78	2.58	2.34	2.16	1.88	1.88	1.82	1.78	1.76
23	1.70	1.70	1.76	1.78	2.58	2.34	2.18	1.86	1.88	1.82	1.78	1.78
24	1.72	1.72	1.76	1.80	2.48	2.20	2.20	1.88	1.86	1.82	1.80	1.78
25	1.74	1.72	1.74	2.28	2.52	2.20	2.18	1.88	84	1.80	1.80	1.78
26	1.72	1.72	1.76	2.28	2.52	2.20	2.20	1.88	84	1.78	1.78	1.68
27	1.74	1.68	1.76	2.28	2.52	2.18	2.20	1.86	84	1.78	1.76	1.78
28	(a)	1.72	1.76	2.26	2.52	2.18	2.20	1.86	1.84	1.78	1.78	1.78
29	(a)		1.76	2.28	2.52	2.20	2.18	1.88	1.84	1.78	1.80	1.80
30	(a)		1.78	2.28	2.56	2.18	2.18	1.88	84	1.78	1.80	1.80
31	(a)		1.80		2.72		2.18	1.86		1.78		1.80

a River frozen from January 28 to February 22.

Rating table for San Pitch River near Gunnison, Utah, for 1903.

Gage height.	Discharge.	Gage height.	Discharge.	Gage height.	Discharge.	Gage height.	Discharge.
Feet.	*Second-feet.*	*Feet.*	*Second-feet.*	*Feet.*	*Second-feet.*	*Feet.*	*Second-feet.*
1.7	9	2.0	39	2.3	82	2.6	136
1.8	18	2.1	52	2.4	100	2.7	154
1.9	28	2.2	66	2.5	118	2.8	172

Estimated monthly discharge of San Pitch River near Gunnison, Utah, for 1903.

[Drainage area, 836 square miles.]

Month.	Discharge in second-feet.			Total in acre-feet.	Run-off.	
	Maximum.	Minimum.	Mean.		Second-feet per square mile.	Depth in inches.
January a............	13	9	10	615	0.012	0.014
February a..........	11	3	6	333	.007	.007
March..............	18	5	12	738	.014	.016
April...............	79	11	30	1,785	.036	.040
May	158	79	123	7,563	.147	.169
June	129	63	95	5,653	.114	.127
July	69	60	64	3,935	.077	.089
August.............	63	24	31	1,906	.037	.043
September..........	26	22	25	1,488	.030	.033
October	39	16	22	1,353	.026	.030
November	18	11	15	893	.018	.020
December ... .:....	18	7	15	922	.018	.021
The year	158	3	37	27,184	.045	.609

a Ice, January 28 to February 22.

HUMBOLDT RIVER NEAR OREANA, NEV.

On the lower reaches of this river measurements have been made for a number of years near Oreana, and the results show the amount of water available for storage at the possible reservoir sites in the vicinity of Humboldt station and also for the six canal systems now in operation below Oreana. The station established by L. H. Taylor January 27, 1896, was located at the old Oreana highway bridge, about 12 miles northeast of Lovelocks, Nev. The bridge abutment to which the gage was fastened was undermined and fell May 26, 1897. A temporary gage was used until September 8, 1897, when a new incline one was placed on the left bank of the river about a mile and a half above the site of the old gage and opposite the Central Pacific Railroad section house.

This gage was washed out. The present gage was established November 29, 1902, by E. C. Murphy. It is vertical, in two sections and is spiked to piles at the site of the old dam. The datum is the same as that of the old gage. The gage is read once each day by John Hart. Discharge measurements are made at high water by means of a cable and car located at the gage. At low water measurements are made by wading a short distance below the station. The channel is straight for 300 feet above and for 200 feet below the station. The current is moderate. The right bank is high and will not overflow. The left bank will overflow only at extreme high

━━ter. There is but one channel at all stages. The bed of the stream
━━andy and shifting. The bench mark consists of 4 nails driven
━━to the pile to which the upper section of the gage is fastened. Its
━━vation is 5 feet above the zero of the gage.

━━The observations at this station during 1903 have been made under
━━e direction of A. E. Chandler, district hydrographer.

Discharge measurements of Humboldt River near Oreana, Nev., in 1903.

Date.	Hydrographer.	Gage height.	Discharge.
		Feet.	*Second-feet.*
━━rch 25	D. W. Hays	1.02	191
━━y 28	W. A. Wolf..................	.98	145
━━ne 21	.....do	.73	109
━━ly 8	.....do	2.45	479
━━uly 18	.....do	1.95	314
━━ugust 14......................	.....do	.62	82
━━ovember 19	A. E. Chandler	− .10	20

Mean daily gage height, in feet, of Humboldt River near Oreana, Nev., for 1903.

Day.	Jan.	Feb.	Mar.	Apr.	May.	June.	July.	Aug.	Sept.	Oct.	Nov.	Dec.
1..............	0.20	0.60	1.20	1.30	2.10	1.10	2.60	1.00	0.10	−0.10	−0.10	−0.10
2..............	.20	.60	1.30	1.60	1.90	1.10	2.50	1.00	.10	− .20	− .10	− .10
3..............	.20	.70	1.20	1.30	1.70	1.10	2.40	.90	.10	− .20	− .10	− .10
4..............	.20	.80	1.10	1.30	1.50	1.10	2.20	.90	.10	− .20	− .10	− .10
5..............	.30	.80	1.00	1.40	1.30	1.10	2.20	.90	.10	− .20	− .10	− .10
6..............	.30	.90	.90	1.40	1.10	1.10	2.20	.80	− .10	− .20	− .10	− .10
7..............	.40	.90	.80	1.50	1.10	1.00	2.20	.80	− .10	− .10	− .10	− .10
8..............	.50	.90	.70	1.60	1.10	.90	2.20	.70	− .10	.10	− .10	− .10
9..............	.50	.90	.70	1.70	1.10	.80	2.10	.70	− .10	− .10	− .10	− .10
10..............	.50	.90	.60	1.80	1.10	.80	2.10	.60	− .10	− .10	·· .10	− .10
11..............	.50	.90	.60	1.90	1.10	.70	2.00	.60	− .10	− .10	− .10	− .10
12..............	.50	.90	.60	2.00	1.10	.70	2.00	.60	− .10	− .10	− .10	− .10
13..............	.50	.90	.50	2.10	1.10	.60	2.00	.50	− .10	− .10	− .10	− .10
14..............	.50	.90	.50	2.20	1.20	.60	2.00	.50	− .10	− .10	− .10	− .10
15..............	.50	.90	.50	2.30	1.20	.60	1.90	.50	·· .10	· .10	− .10	− .10
16..............	.50	.90	.50	2.40	1.20	.50	1.90	.40	− .10	− .10	− .20	− .10
17..............	.50	.90	.60	2.40	1.20	.50	1.90	.40	− .10	− .20	− .20	− .10
18..............	.50	.90	.60	2.40	1.10	.60	1.90	.40	− .10	− .20	− .20	− .10
19..............	.50	.90	.70	2.40	1.10	.70	1.80	.40	− .10	− .30	− .20	− .10
20..............	.50	.80	.80	2.40	1.10	.70	1.80	.30	− .10	− .30	− .20	− .20
21..............	.40	.80	.90	2.40	1.00	.90	1.70	.30	− .10	− .30	− .10	− .20
22..............	.40	.80	.90	2.30	1.00	1.10	1.70	.30	− .10	− .30	− .10	− .20
23..............	.40	.90	.90	2.30	1.00	1.40	1.60	.30	− .10	− .30	− .10	− .20
24..............	.30	.90	.90	2.30	1.00	1.40	1.50	.20	− .10	·· .30	− .10	− .20
25..............	.30	1.00	1.00	2.30	.90	1.50	1.40	.20	− .10	− .20	− .10	− .20
26..............	.30	1.00	1.00	2.30	.90	1.50	1.40	.20	− .10	− .20	− .10	− .20
27..............	.70	1.10	1.10	2.30	.90	1.50	1.40	.20	− .10	− .20	− .10	− .10
28..............	.70	1.20	1.10	2.20	.90	1.50	1.30	.20	− .10	− .20	− .10	− .10
29..............	.70		1.10	2.20	1.00	1.50	1.30	.10	− .10	− .10	− .10	− .10
30..............	.60		1.20	2.20	1.02	2.70	1.20	.10	− .10	− .10	− .10	+ .10
31..............	.60		1.20		1.02		1.10	.10		− .10		.10

Rating table for Humboldt River near Oreana, Nev., for 1905.

Gage height.	Discharge.	Gage height.	Discharge.	Gage height.	Discharge.	Gage height.	Discharge.
Feet.	Second-feet.	Feet.	Second-feet.	Feet.	Second-feet.	Feet.	Second-feet.
−0.3	8	0.5	77	1.3	187	2.1	
− .2	13	.6	89	1.4	203	2.2	
− .1	20	.7	101	1.5	220	2.3	
.0	28	.8	114	1.6	238	2.4	
.1	37	.9	128	1.7	256	2.5	
.2	46	1.0	142	1.8	279	2.6	
.3	56	1.1	156	1.9	302	2.7	
.4	66	1.2	171	2.0	328		

Table fairly well determined.

Estimated monthly discharge of Humboldt River near Oreana, Nev., for 1905.

[Drainage area, 13,800 square miles.]

Month.	Discharge in second-feet.			Total in acre-feet.	Run-off.	
	Maximum.	Minimum.	Mean.		Second-feet per square mile.	Depth in inches.
January	101	46	71	4,366	0.005	0.006
February	171	89	125	6,942	.009	.009
March...............	187	77	124	7,624	.009	.010
April................	458	187	349	20,767	.025	.028
May	358	128	169	10,391	.012	.014
June	580	77	157	9,342	.011	.012
July	537	156	307	18,877	.022	.025
August	142	37	78	4,796	.006	.007
September..........	37	20	23	1,369	.002	.002
October	20	8	15	922	.001	.001
November	20	13	19	1,131	.001	.001
December	37	13	20	1,230	.001	.001
The year	580	8	121	87,757	.009	.116

HUMBOLDT RIVER NEAR GOLCONDA, NEV.

he gaging station near Golconda is located near the great northern
l of Humboldt River and below the central valley. It is about 12
s above the mouth of Little Humboldt River. The station was
blished by L. H. Taylor, October 24, 1894, and has been maintained
inuously since that time. It is located 1¼ miles north of the town.
new inclined gage, installed November 28, 1902, by E. C. Murphy,
stened to the left bank by 4 by 4 inch stakes. The zeros of the
and new gages are at the same elevation. The bench mark is a 4
: inch timber driven 4 feet north of the cable post on the left bank,
n elevation of 10.75 feet above the zero of the gage. Measure-
ts are made from a cable and suspended car. The banks are
erately high, but liable to overflow at extreme high water. The
of the stream is of gravel and sand and is somewhat shifting.
he channel is straight for 300 feet above and for 100 feet below the
ion. There is but one channel at all stages. At the cable the cur-
: is sluggish at low stages, but measurements can be made above or
w by wading. ' The gage is read once each day by Irene Lyng.
he observations at this station during 1903 have been made under
direction of A. E. Chandler, district hydrographer.

Discharge measurements of Humboldt River near Golconda, Nev., in 1903.

Date.	Hydrographer.	Gage height.	Discharge.
		Feet.	*Second-feet.*
ch 24	D. W. Hays	3. 80	234 ·
e 6	W. A. Wolf	4. 22	331
e 11	D. W. Hays	4. 46	315
y 20	W. A. Wolf	3. 36	221
гust 12	do	. 85	29. 6
ember 6	A. E. Chandler	- . 23	1. 5

Mean daily gage height, in feet, of Humboldt River near Golconda, Nev., for 1903.

Day.	Jan.	Feb.	Mar.	Apr.	May.	June.	July.	Aug.	Sept.	Oct.	Nov.	Dec.
1	0.80	2.40	3.10	4.30	5.00	4.50	6.30	2.30	0.60	0.20	−0.20	−0.6
2	.90	2.50	3.20	4.40	5.00	4.40	6.30	2.30	.50	.20	−.20	−.6
3	.90	2.60	3.30	4.50	4.80	4.40	6.10	2.00	.50	.20	−.20	−.6
4	.90	3.00	3.30	4.60	4.70	4.40	6.10	1.70	.50	.30	−.20	−.6
5	1.00	3.60	3.20	4.70	4.60	4.30	5.80	1.50	.50	.20	−.30	−.6
6	1.00	3.00	3.00	4.80	4.50	4.30	5.60	1.30	.40	.30	−.30	−.3
7	1.10	3.00	2.90	4.90	4.40	4.20	5.30	1.30	.40	.30	−.20	−.5
8	1.10	3.10	3.00	5.00	4.20	4.20	5.00	1.20	.40	.30	−.30	−.5
9	1.10	3.10	3.20	5.40	4.10	4.10	4.70	1.20	.80	.30	−.40	−.5
10	1.10	3.20	3.50	5.70	4.10	4.10	4.40	1.20	.80	.30	−.40	−.5
11	1.10	3.20	3.50	6.00	4.00	4.10	4.30	1.00	.80	.30	−.40	−.5
12	1.20	3.20	3.60	6.00	4.00	4.00	4.00	1.00	.80	.30	−.10	−.5
13	1.20	3.20	3.70	6.00	3.90	4.00	4.00	.90	.80	.20	−.10	−.5
14	1.20	3.20	3.90	6.00	3.80	4.30	3.90	.70	.80	.20	−.50	−.6
15	1.30	3.20	3.90	5.90	3.80	4.50	3.90	.70	.80	.20	−.40	−.6
16	1.30	3.20	3.90	5.90	3.80	5.00	3.70	.70	.80	.20	−.30	−.6
17	1.30	3.10	3.90	5.80	3.70	5.00	3.70	.70	.80	.30	−.30	−.6
18	1.30	3.00	3.90	5.80	3.60	5.10	3.70	.70	.80	.20	−.20	−.6
19	1.30	3.00	3.90	5.70	3.50	5.20	3.60	.70	.80	.30	−.20	−.5
20	1.30	3.00	4.00	5.60	3.50	5.20	3.40	.70	.80	.30	−.20	−.6
21	1.40	3.00	4.00	5.50	3.50	5.40	3.30	.70	.80	.20	−.20	−.6
22	1.40	3.00	4.00	5.40	3.50	5.50	3.30	.70	.80	.20	−.30	−.6
23	1.50	3.00	4.00	5.30	3.50	5.80	3.20	.70	.80	.20	−.30	−.6
24	1.60	3.00	4.10	5.30	3.60	6.00	3.00	.70	.80	.20	−.40	−.6
25	1.80	3.00	4.10	5.30	3.80	6.10	3.00	.70	.80	.20	−.40	−.6
26	1.90	3.00	4.10	5.30	3.90	6.20	2.70	.70	.80	.20	−.40	−.6
27	2.00	3.00	4.10	5.30	4.00	6.30	2.60	.60	.30	.20	−.50	−.6
28	2.00	3.00	4.10	5.20	4.30	6.30	2.40	.60	.30	.20	−.50	−.6
29	2.10	......	4.20	5.10	4.50	6.30	2.40	.60	.30	.10	−.50	−.6
30	2.20	......	4.20	5.00	4.50	6.30	2.30	.60	.30	.10	−.50	−.6
31	2.30	......	4.30	......	4.50	......	2.30	.60	......	.10	......	−.5

Rating table for Humboldt River near Golconda, Nev., for 1903.

Gage height.	Discharge.	Gage height.	Discharge.	Gage height.	Discharge.	Gage height.	Discharge.
Feet.	Second-feet.	Feet.	Second-feet.	Feet.	Second-feet.	Feet.	Second-feet.
−0.8	0.5	0.3	10	2.0	86	4.2	291
−.7	.5	.4	13	2.2	99	4.4	318
−.6	1.0	.5	16	2.4	113	4.6	349
−.5	1.0	.6	19	2.6	129	4.8	383
−.4	1.5	.7	22	2.8	145	5.0	420
−.3	1.5	.8	26	3.0	163	5.2	460
−.2	2.0	1.0	34	3.2	182	5.4	503
−.1	2.5	1.2	43	3.4	202	5.6	550
.0	3.0	1.4	53	3.6	222	5.8	600
.1	5.0	1.6	63	3.8	244	6.0	650
.2	7.0	1.8	74	4.0	266	6.2	710

Table fairly well determined. Curve extended above 4.50 feet gage height. Below −0.2 feet gage height discharge is estimated.

nated monthly discharge of Humboldt River near Golconda, Nev., for 1903.

[Drainage area, 10,780 square miles.]

[onth.	Discharge in second-feet.			Total in acre-feet.	Run-off.	
	Maximum.	Minimum.	Mean.		Second-feet per square mile.	Depth in inches.
....	106	26	53	3,259	0.0049	0.0056
....	182	113	164	9,108	.0152	.0158
....	304	154	236	14,511	.0219	.0252
....	650	304	496	29,514	.0460	.0513
....	420	212	283	17,401	.0263	.0303
....	740	266	445	26,479	.0413	.0461
....	740	106	311	19,123	.0289	.0333
....	106	19	37	2,275	.0034	.0039
er..........	19	10	11	655	.0010	.0010
....	10	5	8	492	.0007	.0008
er..........	2.5	1.0	1.6	95	.0001	.0001
er	2.5	.5	1.1	68	.0001	.0001
he year	740	.5	171	122,980	.0158	.2135

HUMBOLDT RIVER AT PALISADE, NEV.

station was established November 27, 1902, by E. C. Murphy.
ge is a vertical 1 by 4 inch board spiked to the right abutment
single-span highway bridge one-fourth mile from the hotel at
le. It is read once each day by T. H. Jewell, the hotel keeper.
is a railroad bridge about 500 feet below. Discharge measure-
are made by means of a cable and car about one-fourth mile
the gage. At very low stages, when the current becomes slug-
the cable, measurements are made by wading a short distance

The initial point for soundings is the zero on the tagged wire.
annel is straight for 200 feet above and for 300 feet below the
. The right bank is low and liable to overflow. The left bank
. The bed of the stream is composed of gravel and sand and is
nent. There is but one channel at all stages. The bench mark
s of a spike and three nails driven into the bridge abutment to
the gage is fastened. Its elevation is 7 feet above the zero of
ge.

observations at this station during 1903 have been made under
ection of A. E. Chandler, district hydrographer.

Discharge measurements of Humboldt River at Palisade, Nev., in 1903.

Date.	Hydrographer.	Gage height.	Discharge.
		Feet.	*Second-feet.*
March 21	D. W. Hays	2.85	403
June 27	W. A. Wolf	3.91	818
July 14	A. E. Chandler	2.10	176
August 10	W. A. Wolf	1.22	35

Mean daily gage height, in feet, of Humboldt River at Palisade, Nev., for 1903.

Day.	Jan.	Feb.	Mar.	Apr.	May.	June.	July.	Aug.	Sept.	Oct.	Nov.	Dec.
1	1.60	1.85	1.55	5.50	4.00	4.00	3.70	1.50	1.05	1.20	1.20	1.50
2	1.50	1.80	1.60	5.30	3.95	4.10	3.55	1.50	1.05	1.15	1.20	1.50
3	1.55	1.80	1.70	5.50	4.00	4.20	3.30	1.45	1.05	1.15	1.20	1.30
4	1.55	1.90	1.70	5.70	3.90	4.40	3.20	1.45	1.05	1.15	1.20	1.50
5	1.50	1.85	1.70	5.40	3.90	4.50	3.00	1.45	1.05	1.15	1.20	1.50
6	1.50	1.60	1.70	5.00	3.80	5.00	2.95	1.45	1.05	1.15	1.30	1.30
7	1.40	1.65	1.75	4.60	3.80	5.10	2.90	1.40	1.05	1.15	1.30	1.30
8	1.50	1.65	1.60	4.35	3.80	5.20	2.80	1.35	1.05	1.20	1.30	1.30
9	1.55	1.65	1.65	4.30	3.70	5.40	2.70	1.30	1.05	1.20	1.30	1.30
10	1.55	1.70	1.70	4.30	3.70	5.50	2.60	1.20	1.10	1.20	1.25	1.15
11	1.55	1.70	1.90	4.15	3.70	5.60	2.50	1.20	1.10	1.20	1.35	1.15
12	1.50	1.65	2.30	4.00	3.70	5.90	2.30	1.20	1.10	1.20	1.40	1.45
13	1.45	1.60	2.50	4.00	3.70	6.00	2.20	1.15	1.10	1.20	1.40	1.45
14	1.50	1.50	2.50	4.10	3.70	5.90	2.10	1.15	1.10	1.25	1.40	1.45
15	1.50	1.50	2.55	4.20	3.70	5.85	2.05	1.15	1.10	1.25	1.50	1.45
16	1.60	1.55	2.60	4.10	3.75	5.80	2.00	1.15	1.10	1.25	1.50	1.45
17	1.50	1.65	2.50	3.90	3.75	5.60	1.90	1.15	1.10	1.30	1.50	1.50
18	1.55	1.70	2.60	3.80	3.75	5.40	1.90	1.10	1.15	1.30	1.50	1.50
19	1.65	1.65	2.60	3.80	3.75	5.20	1.90	1.10	1.15	1.30	1.50	1.50
20	1.50	1.60	2.70	3.75	3.75	5.15	1.90	1.10	1.15	1.30	1.50	1.50
21	1.55	1.55	2.85	3.75	3.80	5.05	1.90	1.05	1.15	1.30	1.50	1.50
22	1.55	1.55	2.85	3.70	3.80	4.80	1.90	1.05	1.10	1.30	1.50	1.55
23	1.65	1.60	2.85	3.80	3.80	4.50	1.85	1.05	1.10	1.30	1.50	1.55
24	1.75	1.60	2.90	3.75	3.80	4.30	1.85	1.05	1.10	1.30	1.50	1.55
25	2.00	1.60	3.15	3.75	3.80	4.25	1.85	1.05	1.10	1.30	1.50	1.60
26	1.80	1.60	3.30	3.70	3.75	4.20	1.85	1.05	1.10	1.30	1.50	1.60
27	1.80	1.55	3.45	3.75	3.75	4.10	1.80	1.05	1.10	1.30	1.50	1.60
28	1.80	1.55	3.70	3.85	3.70	4.00	1.80	1.05	1.10	1.30	1.50	1.60
29	1.65	......	4.40	3.80	3.70	3.90	1.75	1.05	1.10	1.30	1.50	1.60
30	1.70	......	5.00	3.90	3.70	3.80	1.70	1.05	1.10	1.30	1.50	1.60
31	1.80	......	5.70	......	3.70	......	1.65	1.05	......	1.30	......	1.60

ing table for Humboldt River at Palisade, Nev., from November 27, 1902, to December 31, 1903.

Gage height.	Discharge.	Gage height.	Discharge.	Gage height.	Discharge.	Gage height.	Discharge.
Feet.	*Second-feet.*	*Feet.*	*Second-feet.*	*Feet.*	*Second-feet.*	*Feet.*	*Second-feet.*
1.0	18	2.0	153	3.0	460	4.2	940
1.1	24	2.1	176	3.1	500	4.4	1,020
1.2	32	2.2	200	3.2	540	4.6	1,100
1.3	41	2.3	224	3.3	580	4.8	1,180
1.4	52	2.4	251	3.4	620	5.0	1,260
1.5	64	2.5	280	3.5	660	5.2	1,340
1.6	79	2.6	310	3.6	700	5.4	1,420
1.7	95	2.7	343	3.7	740	5.6	1,500
1.8	113	2.8	380	3.8	780	5.8	1,580
1.9	132	2.9	420	4.0	860	6.0	1,660

Table well determined; extended above 3.90 feet gage height.

Estimated monthly discharge of Humboldt River at Palisade, Nev., in 1902 and 1903.

[Drainage area, 5,014 square miles.]

Month.	Discharge in second-feet.			Total in acre-feet.	Run-off.	
	Maximum.	Minimum.	Mean.		Second-feet per square mile.	Depth in inches.
1902.						
vember 27–30....	58	46	50	397	.010	.001
cember..........	79	41	64	3,935	.013	.015
1903.						
luary	153	52	80	4,919	.016	.018
bruary	132	71	87	4,832	.017	.018
rch..............	1,540	71	375	23,058	.075	.086
ril..............	1,540	740	960	57,124	.191	.213
y................	860	740	771	47,407	.154	.178
ie	1,660	780	1,216	72,357	.243	.271
y................	740	87	254	15,618	.051	.059
gust	64	21	34	2,091	.007	.008
ptember	28	21	24	1,428	.005	.006
tober............	41	24	36	2,214	.007	.008
ovember	64	41	55	3,273	.011	.012
cember..........	79	58	67	4,120	.013	.015
The year.....	1,660	21	330	238,441	.066	.892

PINE CREEK NEAR PALISADE, NEV.

This station was established November 27, 1902, by E. C. Murphy. It is located at the Eureka and Palisade Railroad bridge, 1 mile southwest of Palisade, Nev. The gage is a 1 by 4 inch vertical board spiked to the right abutment of the railroad bridge on the downstream side. It is read once each day by T. H. Jewell. Discharge measurements are made from the upstream side of the single-span railroad bridge at which the gage is located. The initial point for soundings is the right abutment. The channel is straight for 40 feet above and for 200 feet below the station. The current has a moderate velocity. The right bank is high and will not overflow. The left bank will overflow at extreme high water. There is but one channel at all stages, but during floods the entire flat on the left bank is under water. The bed of the stream is composed of gravel and sand and is shifting. The bench mark consists of a spike and three nails driven into the abutment near the gage. Its elevation is 7 feet above the zero of the gage.

The observations at this station during 1903 have been made under the direction of A. E. Chandler, district hydrographer.

Discharge measurements of Pine Creek near Palisade, Nev., in 1903.

Date.	Hydrographer.	Gage height.	Discharge
		Feet.	*Second-feet.*
March 21	D. W. Hays	1.75	25.
June 10	....do	1.50	1.
November 7	A. E. Chandler	1.80	5.

Mean daily gage height, in feet, of Pine Creek near Palisade, Nev., for 1903.

Day.	Jan.	Feb.	Mar.	Apr.	May.	June.	July.	Aug.	Sept.	Oct.	Nov.	Dec.
1	1.75	1.65	1.30	2.60	1.40	1.35	1.40	1.50	1.30	1.70	1.85	1.8
2	1.75	1.70	1.60	2.50	1.40	1.40	1.40	1.45	1.30	1.75	1.85	1.8
3	1.80	1.75	1.35	2.50	1.35	1.40	1.40	1.40	1.35	1.75	1.85	1.8
4	1.35	1.75	1.30	2.60	1.30	1.45	1.40	1.40	1.35	1.75	1.85	1.8
5	1.30	1.70	1.30	2.20	1.30	1.45	1.40	1.35	1.40	1.75	1.85	1.8
6	1.60	1.70	1.30	2.00	1.30	1.40	1.40	1.35	1.40	1.75	1.85	2.0
7	1.55	1.75	1.60	1.95	1.30	1.40	1.45	1.30	1.45	1.75	1.85	2.0
8	1.40	1.85	1.40	1.85	1.30	1.40	1.45	1.30	1.45	1.75	1.85	1.
9	1.30	2.05	1.25	1.85	1.30	1.45	1.40	1.30	1.50	1.80	1.85	1.
10	1.30	1.90	1.35	1.90	1.30	1.50	1.40	1.30	1.50	1.80	1.85	1.
11	1.75	1.85	1.65	1.95	1.30	1.55	1.40	1.30	1.55	1.80	1.85	1.
12	1.80	1.85	1.70	2.00	1.30	1.55	1.45	1.30	1.60	1.80	1.90	1.
13	1.90	1.80	1.95	2.10	1.40	1.55	1.50	1.30	1.60	1.85	1.90	1.
14	1.85	1.75	2.40	2.00	1.40	1.55	1.60	1.30	1.60	1.85	1.90	1
15	1.75	1.70	2.25	1.95	1.40	1.55	1.60	1.35	1.60	1.85	1.90	1
16	1.70	1.70	1.90	1.85	1.40	1.55	1.60	1.35	1.60	2.00	1.90	1
17	1.60	1.75	1.85	1.90	1.40	1.50	1.60	1.35	1.60	2.00	1.90	1

Mean daily gage height, in feet, of Pine Creek near Palisade, Nev., for 1903—Continued.

Day.	Jan.	Feb.	Mar.	Apr.	May.	June.	July.	Aug.	Sept.	Oct.	Nov.	Dec.
18	1.80	1.80	1.85	1.90	1.40	1.50	1.60	1.35	1.60	.95	1.90	1.85
19	1.75	1.65	1.75	1.80	1.45	1.55	1.60	1.35	1.60	.95	1.85	1.90
20	1.60	1.90	1.75	1.80	1.50	1.55	1.60	1.35	1.60	1.90	1.85	1.90
21	1.40	1.90	1.75	1.75	1.50	1.50	1.60	1.30	1.60	1.85	1.85	1.90
22	1.35	1.50	1.75	1.70	1.50	1.50	1.55	1.30	1.55	1.85	1.85	1.95
23	1.40	1.30	1.80	1.70	1.50	1.45	1.55	1.30	1.55	1.80	1.85	1.95
24	1.45	1.30	1.85	1.65	1.45	1.45	1.55	1.30	1.60	1.80	1.85	2.00
25	1.45	1.30	1.90	1.60	1.40	1.40	1.55	1.30	1.60	1.80	1.85	2.05
26	1.45	1.30	2.00	1.65	1.40	1.40	1.55	1.30	1.60	1.80	1.80	2.15
27	1.50	1.55	2.05	1.40	1.40	1.40	1.55	1.30	1.60	1.80	1.80	2.20
28	1.40	1.30	2.10	1.40	1.40	1.40	1.55	1.30	1.65	1.85	1.80	2.20
29	1.45		2.30	1.40	1.40	1.40	1.55	1.30	1.65	1.85	1.80	2.20
30	1.50		2.40	1.40	1.40	1.40	1.55	1.30	1.70	1.90	1.80	2.20
31	1.60		2.40		1.40		1.55	1.30		1.90		2.20

SOUTH FORK OF HUMBOLDT RIVER NEAR ELKO, NEV.

The station, established August 29, 1896, by L. H. Taylor, is located 10 miles southwest of the town of Elko and about 6 miles above the junction of the South Fork with the main stream. The gage is inclined and spiked to posts driven firmly into the right bank. A new inclined gage was installed by E. C. Murphy November 22, 1902. It is at the site of the old one, the 4-foot marks of the old and new gages coinciding. The bench mark is a 4 by 4 inch timber driven 4 feet south of the gage. It is 6.29 feet above gage datum. The measurements are made from a cable and suspended car at a point 1 mile above the gage, the latter being placed near the home of the observer, for his convenience. At the point of measurement the banks are high, and the channel is straight for some distance above and below the station. The bed of the stream is composed of gravel and is stable. There is a good site for a reservoir a short distance above the station.

The observations at this station during 1903 have been made under the direction of A. E. Chandler, district hydrographer.

Discharge measurements of South Fork of Humboldt River near Elko, Nev., in 1903.

Date.	Hydrographer.	Gage height.	Discharge.
		Feet.	*Second-feet.*
March 23	D. W. Hays	1.30	68
June 9	do	4.90	1,063
June 26	do	2.52	428
August 9	W. A. Wolf	.20	5.3
November 9	A. E. Chandler	.28	11.4

Mean daily gage height, in feet, of South Fork of Humboldt River near Elko, Nev., for 1903.

Day.	Jan.	Feb.	Mar.	Apr.	May.	June.	July.	Aug.	Sept.	Oct.	Nov.	Dec.
1	1.30	1.60	1.90	2.50	1.70	3.20	2.30	0.70	0.05	0.10	0.25	0.20
2	1.30	1.60	1.90	2.30	1.65	3.20	2.00	.70	.05	.10	.25	.20
3	1.90	1.60	1.90	2.30	1.70	4.90	2.00	.70	.00	.15	.25	.10
4	1.30	1.60	1.90	2.30	1.70	4.90	1.50	.70	.00	.15	.25	.20
5	1.38	1.65	1.90	1.00	1.90	4.90	1.30	.70	.00	.15	.25	.20
6	1.30	1.60	1.90	1.60	1.80	4.90	1.30	.70	.00	.15	.25	.20
7	1.30	1.60	1.90	1.60	1.90	4.70	1.30	.70	.00	.20	.25	.20
8	1.30	1.60	1.90	1.60	2.00	4.70	1.30	.70	.00	.20	.25	.20
9	1.30	1.60	2.00	1.60	2.00	4.70	1.20	.20	.00	.20	.30	.20
10	1.30	1.60	2.50	1.70	1.90	4.70	1.20	.20	.00	.20	.30	.20
11	1.30	1.60	3.00	1.70	1.90	5.00	1.20	.20	.00	.20	.30	.20
12	1.40	1.60	3.00	1.70	1.90	5.00	1.15	.20	.00	.20	.30	.20
13	1.40	1.60	3.00	1.70	2.00	5.00	1.15	.20	.00	.20	.30	.20
14	1.40	1.60	3.00	1.70	2.70	4.30	1.10	.20	.00	.20	.30	.20
15	1.40	1.60	2.50	1.70	2.70	4.20	1.10	.20	.00	.20	.30	.20
16	1.40	1.60	2.00	1.70	2.70	4.30	1.00	.20	.00	.20	.30	.20
17	1.40	1.65	1.80	1.70	2.05	4.00	1.00	.20	.00	.20	.30	.20
18	1.40	1.65	1.50	1.60	2.05	4.00	1.00	.25	.00	.20	.30	.20
19	1.40	1.70	1.50	1.60	2.05	4.00	1.00	.20	.00	.20	.30	.20
20	1.40	1.70	1.50	1.60	2.05	3.70	.95	.20	.00	.20	.30	.20
21	1.40	1.70	1.50	1.60	2.10	3.00	.95	.20	.00	.20	.30	.20
22	1.40	1.80	1.30	1.60	2.10	3.00	.90	.20	.00	.20	.30	.20
23	1.40	1.80	1.30	1.65	2.10	3.00	.90	.20	.00	.20	.30	.20
24	1.40	1.80	1.30	1.65	2.15	2.50	.90	.20	.00	.20	.30	.20
25	1.40	1.90	1.30	1.70	2.15	2.50	.85	.20	.00	.20	.30	.20
26	1.60	1.80	1.30	1.80	2.15	2.50	.80	.10	.00	.20	.30	.20
27	1.70	1.80	1.35	1.80	2.15	2.50	.80	.10	.00	.20	.30	.20
28	1.70	1.80	1.35	2.00	3.20	2.40	.75	.10	.05	.20	.30	.20
29	1.70	......	2.50	1.70	3.20	2.40	.75	.10	.05	.20	.30	.20
30	1.70	......	2.50	1.70	3.20	2.30	.70	.10	.05	.20	.30	.20
31	1.70	......	2.50	......	3.20	......	.70	.05	......	.20	......	.20

Rating table for South Fork of Humboldt River near Elko, Nevada, for 1903.

Gage height.	Discharge.	Gage height.	Discharge.	Gage height.	Discharge.	Gage height.	Discharge.
Feet.	*Second-feet.*	*Feet.*	*Second-feet.*	*Feet.*	*Second-feet.*	*Feet.*	*Second-feet.*
0.0	2	1.0	74	2.2	327	4.2	922
.1	3	1.1	89	2.4	382	4.4	983
.2	5	1.2	105	2.6	439	4.6	1,045
.3	8	1.3	122	2.8	499	4.8	1,106
.4	12	1.4	141	3.0	559	5.0	1,168
.5	17	1.5	161	3.2	619	· 5.2	1,230
.6	24	1.6	182	3.4	679	5.4	1,292
.7	35	1.7	204	3.6	740	5.6	1,354
.8	47	1.8	227	3.8	801	5.8	1,416
.9	60	2.0	276	4.0	862	6.0	1,478

ed monthly discharge of South Fork of Humboldt River near Elko, Nev., for 1903.

[Drainage area, 1,150 square miles.]

Month.	Discharge in second-feet.			Total in acre-feet.	Run-off.	
	Maximum.	Minimum.	Mean.		Second-feet per square mile.	Depth in inches.
....	204	122	145	8,916	0.126	0.145
?	227	182	196	10,885	.170	.177
....	559	122	276	16,970	.240	.277
....	410	182	222	13,210	.192	.214
....	619	193	332	20,414	.289	.333
....	1,168	354	808	48,079	.702	.783
....	354	35	102	6,272	.088	.101
....	35	3	12	738	.010	.012
er............	3	2	2	119	.002	.002
....	5	3	5	307	.004	.005
er............	8	6	7	417	.006	.007
er	8	8	8	492	.007	.008
ıe year ,....	1,168	2	176	126,819	.153	2.064

NORTH FORK OF HUMBOLDT RIVER NEAR ELBURZ, NEV.

station was established October 10, 1902, by E. C. Murphy.
ated 150 feet below the Southern Pacific Railroad bridge and
one-fourth mile above the junction of the North Fork with
ldt River. It is 2 miles west of the Southern Pacific Railroad
at Elburz. The nearest post-office is Halleck, Nev. The gage
clined 4 by 4 inch timber fastened to the left bank just above
le. It is read once each day by A. R. Blevins. Discharge
ements are made by means of a cable and car. The initial point
ndings is the zero on the tagged wire. The channel is straight
) feet above and below the station. There is but one channel
ages and the current is moderate. Both banks are high. The
ık will overflow at extreme flood stages. The bed of the stream
ıosed of gravel and silt and changes slightly. The bench mark
y 4 inch timber driven in the ground on the left bank about 20
ıstream from the gage. Its elevation is 6.99 feet above the
the gage.
ɔbservations at this station during 1903 have been made under
ection of A. E. Chandler, district hydrographer.

Discharge measurements of North Fork of Humboldt River near Elburz, Nev., in 1903.

Date.	Hydrographer.	Gage height.	Discharge.
		Feet.	*Second-feet.*
March 22	D. W. Hays	2.84	?
June 8	do	4.08	1?
June 27	W. A. Wolf	2.86	?

Mean daily gage height, in feet, of North Fork of Humboldt River near Elburz, Nev., for 1903.

Day.	Jan.	Feb.	Mar.	Apr.	May.	June.	July.	Aug.	Sept.	Oct.	Nov.	Dec.
1	2.80	3.60	3.50	6.10	4.10	3.40	2.80	2.20	2.00	2.30	2.60	2.
2	2.80	3.50	3.30	5.80	4.00	3.40	2.70	2.20	2.00	2.40	2.60	2.
3	2.80	3.50	3.30	5.20	4.00	3.50	2.70	2.20	2.00	2.40	2.60	2.
4	2.70	3.40	3.30	4.40	3.90	3.60	2.60	2.10	2.00	2.40	2.60	4.
5	2.70	3.20	3.40	4.70	3.90	3.80	2.60	2.10	2.00	2.40	2.60	4.
6	2.70	3.20	3.30	4.40	3.80	4.00	2.60	2.10	2.00	2.40	2.60	2.
7	2.60	3.30	3.40	4.00	3.80	4.10	2.60	2.00	2.00	2.40	2.60	2.
8	2.80	3.30	3.30	4.10	3.70	4.20	2.60	2.00	2.00	2.30	2.60	2.
9	2.70	3.30	3.40	4.10	3.60	4.40	2.50	2.00	2.00	2.30	2.60	2.
10	2.60	3.20	3.30	4.30	3.60	4.60	2.50	2.00	2.00	2.30	2.60	2.
11	2.50	3.20	3.40	4.00	3.50	4.80	2.50	2.00	2.00	2.30	2.60	2.
12	2.50	3.30	3.40	4.80	3.50	4.90	2.50	2.00	2.00	2.30	2.60	2.
13	2.50	3.30	3.50	4.40	3.50	5.00	2.50	2.00	2.00	2.50	2.60	2.
14	2.50	3.30	3.50	4.00	3.40	4.80	2.50	2.00	2.10	2.50	2.60	2.
15	2.50	3.30	3.50	4.00	3.40	4.60	2.40	2.00	2.10	2.50	2.50	2.
16	2.50	3.30	3.40	3.90	3.40	4.30	2.40	2.00	2.20	2.50	2.50	2.
17	2.50	3.30	3.40	3.90	3.40	4.00	2.40	2.00	2.20	2.50	2.50	2.
18	2.60	3.30	3.40	4.00	3.40	3.80	2.40	2.00	2.30	2.50	2.50	2.
19	2.70	3.30	3.30	4.00	3.50	3.60	2.40	2.00	2.30	2.50	2.50	2.
20	2.70	3.30	3.20	3.90	3.60	3.40	2.40	2.00	2.30	2.50	2.55	(a
21	2.80	3.30	3.00	3.90	3.50	3.30	2.40	2.00	2.30	2.50	2.55	(a
22	2.70	3.30	3.00	4.00	3.50	3.20	2.40	2.00	2.30	2.50	2.60	(a
23	2.80	3.30	2.90	4.10	3.50	3.20	2.40	2.00	2.30	2.50	2.60	(a
24	2.80	3.30	3.10	4.00	3.50	3.10	2.30	2.00	2.30	2.50	2.60	(a
25	2.80	3.30	3.70	3.90	3.50	3.00	2.30	2.00	2.30	2.50	2.60	(a
26	2.90	3.30	4.40	3.90	3.60	3.00	2.30	2.00	2.30	2.50	2.60	(a
27	3.30	3.30	5.00	4.00	3.60	2.90	2.30	2.00	2.30	2.50	2.70	(
28	3.60	3.30	5.30	4.10	3.50	2.90	2.30	2.00	2.30	2.50	2.70	
29	3.60	......	5.40	4.20	3.50	2.80	2.20	2.00	2.30	2.50	2.70	
30	3.60	......	5.50	4.10	3.60	2.80	2.20	2.00	2.30	2.60	2.60	
31	3.50	......	6.00	......	3.50	......	2.20	2.00	......	2.60	......	

a No flow; solid ice.

MARYS RIVER NEAR DEETH, NEV.

This station was established November 24, 1902, by E. C. Murphy, and was discontinued July 14, 1903. It is located at the wagon bridge about 500 feet from Bradley's home ranch, which is known as Malovista. It is about 20 miles upstream from the mouth of the river or from Deeth, Nev. The gage is a vertical 4 by 4 inch timber spiked to the upstream side of the middle pier of the bridge. Up to the time the station was discontinued it was read twice each day by George Murry. Discharge measurements were made from the bridge to which the gage was attached. The initial point for soundings was taken at the end of the bridge on the right bank. The channel is straight for 50 feet above and for 100 feet below the station. The right bank is low and liable to overflow. The left bank is high. There is but one channel at all stages, broken by the middle pier of the 2-span bridge. The bed of the stream is composed of silt and clay, and changes slightly. The current has a moderate velocity. The bench mark is a 4 by 4 inch timber driven into the ground on the left bank just above the bridge and 35 feet from the gage. Its elevation is 9.28 feet above the zero of the gage. During September and October the discharge at this point becomes very small.

The observations at this station during 1903 have been made under the direction of A. E. Chandler, district hydrographer.

Discharge measurements of Marys River near Deeth, Nev., in 1903.

Date.	Hydrographer.	Gage height.	Discharge.
		Feet.	*Second-feet.*
March 17	D. W. Hays	2.80	37
June 6	A. E. Chandler	4.87	262
June 25	W. A. Wolf	3.30	75
August 6	......do	2.08	4

Mean daily gage height, in feet, of Marys River near Deeth, Nev., for 1903.

Day.	Jan.	Feb.	Mar.	Apr.	May.	June.	July.
1	2.45	3.05	3.50	5.10	......	4.45	4.6
2	2.45	2.80	3.50	4.70	......	4.30	4.5
3	2.45	3.00	3.50	4.30	4.60	5.00	2.5
4	2.55	3.00	3.40	4.20	4.60	5.10	2.6
5	2.45	2.75	3.50	4.10	4.70	5.65	2.8
6	2.40	2.80	3.55	4.10	4.60	5.10	2.8
7	2.40	2.80	3.55	4.00	4.60	5.40	1.8
8	2.45	2.70	2.85	3.90	4.60	5.25	1.6
9	2.40	2.80	2.85	3.80	4.60	4.80	1.5
10	2.65	2.90	2.90	4.10	4.60	4.85	1.6
11	2.40	2.90	2.50	4.90	4.60	4.65	1.6
12	2.45	3.40	2.55	4.60	4.60	4.60	1.6
13	2.40	2.40	2.65	4.30	4.60	4.50	1.5
14	2.45	3.46	2.65	4.20	4.65	4.40	1.5
15	2.45	3.40	......	4.20	5.00	4.30	
16	2.45	3.00	......	4.10	5.00	4.20	
17	2.40	5.00	2.80	4.10	4.85	4.10	
18	2.40	3.00	2.90	4.10	4.70	4.00	
19	2.40	3.00	2.70	4.30	4.45	3.80	
20	2.45	3.20	2.60	4.50	4.35	3.80	
21	2.45	3.00	2.90	4.60	4.30	3.70	
22	2.45	3.00	3.20	4.70	4.25	3.60	
23	2.40	3.00	3.10	4.90	4.20	3.40	
24	2.50	3.25	3.10	5.00	4.20	3.40	
25	3.65	3.55	3.20	5.10	4.10	3.30	
26	3.50	3.05	3.40	5.20	4.10	3.30	
27	2.95	3.25	4.00	5.30	4.10	3.00	
28	2.70	3.40	4.30	5.30	4.20	2.90	
29	3.10	......	4.70	5.40	4.25	2.90	
30	3.05	......	5.00	5.00	4.30	2.90	
31	3.05	......	5.30	......	4.40	......	

Rating table for Marys River near Deeth, Nev., from November 24, 1902, to July 14, 1903.

Gage height.	Discharge.	Gage height.	Discharge.	Gage height.	Discharge.	Gage height.	Discharge.
Feet.	Second-feet.	Feet.	Second-feet.	Feet.	Second-feet.	Feet.	Second-feet.
2.1	5	3.0	51	3.9	135	4.8	252
2.2	8	3.1	58	4.0	146	4.9	267
2.3	12	3.2	66	4.1	158	5.0	282
2.4	16	3.3	75	4.2	171	5.1	297
2.5	21	3.4	84	4.3	184	5.2	312
2.6	26	3.5	93	4.4	197	5.3	327
2.7	32	3.6	103	4.5	210	5.4	342
2.8	38	3.7	113	4.6	223	5.5	357
2.9	44	3.8	124	4.7	237		

Table is extended above 4.87 feet gage height. Table fairly well determined.

nated monthly discharge of Marys River near Deeth, Nev., in 1902 and 1903.

Month.	Discharge in second-feet.			Total in acre-feet.
	Maximum.	Minimum.	Mean.	
1902.				
er 24–30	21	10	15	208
·r	.23	12	16	984
1903.				
....	93	16	27	1,660
·.............................	98	32	57	3,166
....	327	21	85	5,226
....	342	124	218	12,972
....	282	158	211	12,974
....	349	44	177	10,532
·.............................	41	8	22	611
·e period				47,141

WALKER RIVER NEAR WABUSKA, NEV.

station was established July 22, 1902, by L. H. Taylor. It uipped December 12, 1902, by E. C. Murphy. It is located :00 feet above the Carson and Colorado Railroad bridge near :tion house at Clever station. It is about 2¼ miles east of ka. The original gage was washed out December 30, 1903. gage, consisting of a 1 by 4 inch board fastened to a vertical 4 ich timber, was installed January 17, 1904. It is on the left 5 feet below the point at which the old gage was located and t south of the section house. The datum is the same as that of rinal gage. The gage is read twice each day by Charles Nelson. ·ge measurements are made by means of a cable and car 35 feet :he gage. The initial point for soundings is the zero on the wire. The channel is straight for 150 feet above the station ghtly curved for the same distance below. Both banks are d not liable to overflow. The bed of the stream is sandy and ;. There is but one channel at all stages. The current has a te velocity and the discharge is small from August to October. ich mark consist of four nails driven into the northwest corner pump platform 35 feet east of the gage. Its elevation is 6.94 ·ve the zero of the gage.

·bservations at this station during 1903 have been made under :ction of A. E. Chandler, district hydrographer.

10	1.55	1.90	1.75	1.85	2.30	3.60	1.90	.30	.20	.15	.40	2.20
11	1.55	1.90	1.75	1.90	2.40	3.70	1.85	.30	.20	.15	.40	1.75
12	1.50	1.75	1.75	1.90	2.70	3.70	1.80	.25	.20	.15	.40	1.75
13	1.50	1.45	1.75	1.90	2.80	3.70	1.70	.25	.20	.15	.42	1.85
14	1.40	1.40	1.75	1.80	3.00	3.60	1.70	.25	.20	.15	.45	1.85
15	1.40	1.45	1.75	1.75	3.10	3.60	1.60	.20	.20	.15	.45	1.90
16	1.45	1.50	1.85	1.70	2.90	3.60	1.50	.20	.15	.15	.60	1.20
17	1.45	1.00	1.90	1.70	2.80	3.40	1.40	.20	.15	.20	.78	1.25
18	1.50	1.70	1.80	1.60	2.80	3.10	1.35	.20	.15	.20	.80	1.25
19	1.60	1.70	1.70	1.50	2.60	2.90	1.25	.25	.15	.20	.75	1.20
20	1.70	1.80	1.70	1.50	2.50	2.95	1.15	.25	.15	.20	.75	1.20
21	1.70	1.80	1.65	1.50	2.20	3.00	1.00	.25	.15	.20	.75	1.25
22	1.60	1.80	1.65	1.50	2.10	3.20	1.00	.25	.15	.20	.85	1.25
23	1.60	1.90	1.75	1.50	2.10	3.20	.90	.25	.15	.20	.90	1.25
24	1.60	1.90	1.90	1.50	2.10	3.20	.80	.25	.15	.20	.90	1.20
25	1.60	1.80	2.10	1.50	2.10	3.10	.80	.25	.15	.20	.95	1.2
26	1.60	1.80	2.10	1.60	2.00	3.00	.70	.25	.15	.20	.95	1.2
27	1.65	1.80	2.10	1.60	2.00	3.10	.60	.25	.15	.20	.95	1.2
28	1.65	1.80	2.00	1.70	2.00	3.20	.50	.25	.15	.20	1.05	1.2
29	1.70	......	1.90	1.70	1.90	3.20	.50	.25	.15	.20	1.20	2.2
30	1.70	......	1.90	1.60	1.90	3.20	.40	.20	.15	.20	1.20	2.2
31	1.80	......	2.00	......	2.20	......	.40	.20	......	.20	......	......

uting table for Walker River near Wabuska, Nev., from July 22, 1902, to December 31, 1903.

Gage height.	Discharge.	Gage height.	Discharge.	Gage height.	Discharge.	Gage height.	Discharge.
Feet.	*Second-feet.*	*Feet.*	*Second-feet.*	*Feet.*	*Second-feet.*	*Feet.*	*Second-feet.*
0.2	1	1.1	59	2.0	224	2.9	566
.3	2	1.2	73	2.1	251	3.0	614
.4	3	1.3	89	2.2	282	3.1	664
.5	5	1.4	105	2.3	316	3.2	714
.6	9	1.5	121	2.4	352	3.3	764
.7	16	1.6	139	2.5	390	3.4	814
.8	25	1.7	158	2.6	430	3.5	864
.9	35	1.8	178	2.7	474	3.6	914
1.0	47	1.9	200	2.8	520	3.7	964

Table is well determined.

Estimated monthly discharge of Walker River near Wabuska, Nev., for 1902 and 1903.

[Drainage area, 2,420 square miles.]

Month.	Discharge in second-feet.			Total in acre-feet.	Run-off.	
	Maximum.	Minimum.	Mean.		Second-feet per square mile.	Depth in inches.
1902.						
ly 22–31	47	9	18	357	0.0074	0.0028
igust	9	2	3	184	.0012	.0014
ptember	1	1	1	59	.0004	.0004
tober	9	1	2	123	.0008	.0009
vember	105	7	53	3,154	.0219	.0244
cember	200	89	148	9,100	.0612	.0706
1903.						
iuary	178	105	137	8,424	.0566	.0652
iruary	200	105	159	8,830	.0657	.0684
rch	251	148	190	11,683	.0785	.0905
il	352	121	184	10,949	.0760	.0848
y	664	139	315	19,369	.1302	.1501
ie	964	282	727	43,259	.3004	.3351
y	664	3	195	11,990	.0806	.0929
gust	3	1	2	123	.0008	.0009
tember	1	1	1	59	.0004	.0004
ober	1	1	1	61	.0004	.0005
vember	73	1	19	1,131	.0079	.0088
cember a	334	73	114	7,010	.0471	.0543
The year	964	1	170	122,888	.0704	.9519

a December 31, 1903, estimated.

EAST FORK OF WALKER RIVER NEAR YERINGTON, NEV.

This station was established October 6, 1902, by E. C. Murphy. It is located at Ross's ranch, about 10 miles southeast of Yerington, Nev. The station is just above the point where the road crosses the river. The gage is a vertical 4 by 4 inch timber fastened to the right bank just below the cable. It is read once each day by I. A. Strosnider, the ranch owner. Discharge measurements are made by means of a cable and car. The initial point for soundings is the zero of the tagged wire. The channel is straight for 200 feet above and for 100 feet below the station. At low water the sand bar in the middle of the stream divides the channel into two parts. Both banks are low and are liable to overflow. The bed of the stream is composed of sand and clay and is liable to shift. The current has a moderate velocity. The bench mark consists of 3 nails and a spike in a stump 6 inches in diameter, 15 feet east of the gage. Its elevation is 5.85 feet above the zero of the gage.

The observations at this station during 1903 have been made under the direction of A. E. Chandler, district hydrographer.

Discharge measurements of East Fork of Walker River near Yerington, Nev., in 1903.

Date.	Hydrographer.	Gage height.	Discharge.
		Feet.	*Second-ft.*
April 10	D. W. Hays	1.94	125
April 25	I. W. Huffaker	1.80	122
May 8	do	1.37	54
May 21	do	1.78	124
June 13	do	3.09	434
June 30	do	2.70	292
July 20	do	1.90	124
July 23	do	1.70	83
August 9	do	1.09	18

Mean daily gage height, in feet, of East Fork of Walker River near Yerington, Nev., for 1903.

Day.	Jan.	Feb.	Mar.	Apr.	May.	June.	July.	Aug.	Sept.	Oct.	Nov.	Dec.
1	1.80	1.70	1.70	2.20	1.60	1.60	2.80	1.20	1.00	1.20	1.60	1.80
2	1.80	1.70	1.70	2.10	1.50	1.60	2.75	1.20	1.00	1.20	1.60	1.80
3	1.80	1.70	1.70	2.00	1.50	1.70	2.65	1.20	1.00	1.20	1.60	1.80
4	1.80	1.70	1.80	1.90	1.50	1.70	2.55	1.20	1.00	1.20	1.60	1.80
5	1.80	1.80	1.80	1.90	1.50	1.80	2.50	1.10	1.10	1.30	1.60	1.80
6	1.80	1.90	1.80	1.90	1.40	2.15	2.45	1.10	1.10	1.30	1.60	1.70
7	1.80	2.00	1.80	1.90	1.40	2.35	2.45	1.10	1.10	1.30	1.60	1.70
8	1.80	2.00	1.80	2.00	1.40	2.50	2.40	1.00	1.10	1.30	1.60	1.70
9	1.80	1.80	1.80	2.00	1.45	2.70	2.30	1.00	1.20	1.30	1.60	1.70

Mean daily gage height, in feet, of East Fork of Walker River near Yerington, Nev., for 1903—Continued.

Day.	Jan.	Feb.	Mar.	Apr.	May.	June.	July.	Aug.	Sept.	Oct.	Nov.	Dec.
10	1.80	1.80	1.90	2.00	.....	2.75	2.20	1.00	1.20	1.30	1.60	1.70
11	1.80	1.70	2.00	2.00	.....	2.85	2.10	1.00	1.20	1.30	1.65	1.70
12	1.80	1.60	2.00	1.90	.....	3.10	2.00	.90	1.20	1.30	1.70	1.70
13	1.80	1.60	2.10	1.80	.....	3.10	1.95	.90	1.20	1.40	1.75	1.70
14	1.80	1.60	2.20	1.80	.....	3.00	1.90	.80	1.20	1.40	1.80	1.70
15	1.80	1.65	2.20	1.70	.....	2.85	2.00	.80	1.20	1.50	1.80	1.70
16	1.80	1.65	2.30	1.70	.....	2.80	2.00	.80	1.20	1.50	1.80	1.70
17	1.80	1.60	2.30	1.70	.....	2.70	2.10	.80	1.20	1.60	1.80	1.70
18	1.80	1.60	2.40	1.70	.....	2.60	2.	.80	1.20	1.60	1.80	1.70
19	1.80	1.60	2.50	1.70	.....	2.50	2.00	.65	1.20	1.60	1.80	1.70
20	1.80	1.60	2.60	1.75	.....	2.40	2.00	.65	1.20	1.60	1.80	1.70
21	1.80	1.60	2.60	1.80	.....	2.30	1.95	.65	1.20	1.60	1.80	1.70
22	1.80	1.60	2.70	1.85	.....	2.40	.90	.80	1.20	1.60	1.80	1.70
23	1.80	1.65	2.80	1.90	.....	2.40	.90	.80	1.20	1.60	1.80	1.70
24	1.80	1.70	2.90	1.90	.....	2.50	1.80	.80	1.20	1.60	1.80	1.70
25	1.80	1.70	3.00	1.90	.....	2.60	1.75	.85	1.20	1.60	1.80	1.70
26	1.80	1.70	2.90	1.90	.....	2.70	1.70	.90	1.20	1.60	1.80	1.70
27	1.80	1.70	2.80	1.80	.....	2.70	1.60	.90	1.20	1.60	1.80	1.70
28	1.80	1.70	2.75	1.75	.....	2.80	1.50	.90	1.20	1.60	1.80	1.70
29	1.80	.....	2.65	1.70	.....	2.90	1.40	.95	1.20	1.60	1.80	1.70
30	1.70	.....	2.45	1.70	.....	2.90	1.40	1.00	1.20	1.60	1.80	1.60
31	1.70	.....	2.30	.....	1.50	.....	1.30	1.00	.....	1.60	.....	1.60

Rating table for East Fork of Walker River near Yerington, Nev., from October 6, 1902, to December 31, 1903.

Gage height.	Discharge.	Gage height.	Discharge.	Gage height.	Discharge.	Gage height.	Discharge.
Feet.	*Second-feet.*	*Feet.*	*Second-feet.*	*Feet.*	*Second-feet.*	*Feet.*	*Second-feet.*
0.7	3	1.4	61	2.1	164	2.8	322
.8	10	1.5	73	2.2	181	2.9	355
.9	17	1.6	86	2.3	199	3.0	393
1.0	25	1.7	100	2.4	219	3.1	440
1.1	33	1.8	115	2.5	241		
1 2	42	1.9	131	2.6	266		
1.3	51	2.0	147	2.7	292		

Table is extended below 1.10 feet gage height and is not well determined in lower part.

Estimated monthly discharge of East Fork of Walker River near Yerington, Nev.

[Drainage area, 1,100 square miles.]

| Month. | Discharge in second-feet. | | | Total in acre-feet. | Run-off. | |
	Maximum.	Minimum.	Mean.		Second-feet per square mile.	Depth inche
1902.						
October 6–31........	100	51	84	4,332	0.076	0.0
November	115	100	113	6,724	.102	.1
December	164	115	126	7,747	.114	.1
1903.						
January	115	100	114	7,010	.103	.1
February	147	86	101	5,609	.092	.1
March..............	393	100	207	12,728	.188	.
April..............	181	100	121	7,200	.110	.
May 1–9 and 31	....	....	70	1,388	.063	.
June	440	86	250	15,412	.235	.
July	322	51	162	9,961	.147	.
August	42	0	20	1,230	.018	.
September..........	42	25	39	2,321	.035	.
October	86	42	69	4,243	.063	.
November	115	86	104	6,188	.094	.
December	115	86	102	6,272	.092	:
The period ...	....	....	....	79,562	....	....

WEST FORK OF WALKER RIVER NEAR COLEVILLE, CAL.

This station was established October 5, 1902, by E. C. Murp
It is located about 4 miles southwest of Coleville, Cal. The g
is a 4 by 4 inch vertical timber on the left bank, about one-
mile above the cable and 300 feet from the observer's house. I
read once each day by J. S. Trumble. Discharge measurements
made by means of a cable and car near the mouth of the canyon al
1 mile east of the point where the main road from Topaz to Bri
port crosses Lost Canyon Creek. The cable is located about 600
from the road. The initial point for soundings is the zero on
tagged wire. The channel is straight for 150 feet above and below
station. The current is swift. The right bank is low and rocky
is liable to overflow; the left bank is high and rocky and will
overflow. The bed of the stream is rocky and uneven. There is
one channel at all stages. The bench mark is a spike driven into
tree to which the gage is fastened. Its elevation is 6 feet above
zero of the gage.

The observations at this station during 1903 have been made under the direction of A. E. Chandler, district hydrographer.

Discharge measurements of West Fork of Walker River near Coleville, Cal., in 1903.

Date.	Hydrographer.	Gage height.	Discharge.
		Feet.	*Second-feet.*
April 8	D. W. Hays	2.00	225
April 30	I. W. Huffaker	2.65	476
June 8	do	4.00	1,887
July 14	do	2.70	451
July 28	do	2.20	260
August 7	do	1.85	184
September 29	W. A. Wolf	1.08	65

Mean daily gage height, in feet, of West Fork of Walker River near Coleville, Cal., for 1903.

Day.	Jan.	Feb.	Mar.	Apr.	May.	June.	July.	Aug.	Sept.	Oct.	Nov.	Dec.
1	1.20	1.40	1.20	1.60	2.70	4.10	3.50	2.00	1.30	1.10	1.10	1.20
2	1.40	1.30	1.30	1.70	2.80	4.00	3.30	2.00	1.30	1.10	1.10	1.10
3	1.20	1.20	1.80	1.70	3.00	3.90	3.00	1.90	1.30	1.10	1.10	1.10
4	1.10	1.10	1.20	1.60	3.10	3.80	3.10	1.90	1.30	1.10	1.10	1.10
5	1.30	1.00	1.20	1.50	3.20	3.90	3.00	1.80	1.30	1.10	1.10	1.10
6	1.30	1.20	1.30	1.50	3.20	4.00	3.00	1.80	1.30	1.10	1.10	1.10
7	1.20	1.30	1.10	1.70	3.40	3.90	2.90	1.70	1.20	1.10	1.10	1.10
8	1.10	1.40	1.20	1.90	3.40	1.00	2.50	1.70	1.20	1.10	1.10	1.10
9	1.00	1.30	1.20	2.20	3.30	4.10	2.50	1.90	1.20	1.20	1.10	1.10
10	1.00	1.20	1.30	2.30	3.50	4.00	2.50	1.80	1.20	1.10	1.10	1.10
11	1.00	1.20	1.10	2.30	3.70	3.90	2.50	1.80	1.20	1.10	1.10	1.10
12	1.00	1.30	1.10	1.90	3.80	3.80	2.40	1.80	1.20	1.10	1.10	1.20
13	1.00	1.20	1.20	2.00	4.00	3.90	2.30	1.70	1.20	1.10	1.20	1.20
14	1.00	1.30	1.30	1.60	3.60	3.80	2.30	1.70	1.20	1.10	1.10	1.15
15	1.00	1.30	1.30	1.70	3.50	3.70	2.20	1.70	1.20	1.10	1.20	1.10
16	1.00	1.20	1.40	1.80	3.40	3.70	2.30	1.60	1.20	1.10	1.20	1.10
17	1.10	1.30	1.30	1.80	3.30	3.30	2.30	1.60	1.20	1.10	1.10	1.10
18	1.00	1.30	1.30	1.90	3.10	3.20	2.30	1.60	1.20	1.10	1.20	1.10
19	1.00	1.20	1.20	1.80	2.90	3.50	2.20	1.60	1.20	1.10	1.20	1.10
20	1.10	1.20	1.30	1.80	2.80	3.60	2.30	1.60	1.20	1.10	1.20	1.10
21	1.00	1.10	1.40	1.90	2.50	3.60	2.40	1.60	1.20	1.10	1.20	1.10
22	1.00	1.20	1.30	2.10	2.50	3.60	2.40	1.60	1.20	1.10	1.20	1.10
23	1.20	1.20	1.30	2.50	2.50	3.50	2.40	1.10	1.20	1.10	1.30	1.10
24	1.00	1.30	1.30	2.70	2.40	3.60	2.20	1.50	1.10	1.10	1.40	1.10
25	1.10	1.30	1.40	2.50	2.10	3.70	2.10	1.50	1.10		1.40	1.10
26	1.20	1.20	1.60	2.70	2.40	3.70	2.10	40	1.10	10	1.40	1.10
27	1.40	1.20	1.60	2.50	2.90	3.70	2.10	1.40	1.20		1.40	1.10
28	1.10	1.30	2.30	2.60	3.20	3.60	2.10	1.40	1.10		1.40	1.10
29	1.00		2.20	2.70	3.40	3.60	2.10	1.40	1.10		1.30	1.10
30	1.00		2.10	2.50	3.90	3.50	2.00	1.40	1.10		1.30	1.10
31	1.30		1.70		4.00		1.90	1.30		10		1.10

Rating table for West Fork of Walker River near Coleville, Cal., from October 4, 1902, to December 31, 1903.

Gage height.	Discharge.	Gage height.	Discharge.	Gage height.	Discharge.	Gage height.	Discharge.
Feet.	*Second feet.*	*Feet.*	*Second feet.*	*Feet.*	*Second feet.*	*Feet.*	*Second feet.*
1.0	60	1.8	170	2.6	426	3.4	1,025
1.1	67	1.9	192	2.7	476	3.5	1,100
1.2	75	2.0	215	2.8	530	3.6	1,305
1.3	85	2.1	240	2.9	590	3.7	1,450
1.4	98	2.2	269	3.0	655	3.8	1,595
1.5	113	2.3	302	3.1	730	3.9	1,740
1.6	130	2.4	340	3.2	815	4.0	1,885
1.7	149	2.5	381	3.3	910	4.1	2,030

Curve is not very well determined above 3 feet gage height.

Estimated monthly discharge of West Fork of Walker River near Coleville, Cal., for 1902 and 1903.

[Drainage area 306 square miles.]

Month.	Discharge in second-feet.			Total in acre-feet.	Run-off.	
	Maximum.	Minimum.	Mean.		Second-feet per square mile.	Depth in inches.
1902.						
October 5-31	85	60	67	3,588	0.22	0.22
November	113	67	83	4,939	.27	.30
December	98	60	75	4,612	.25	.29
1903.						
January	98	60	69	4,243	.23	.27
February	98	60	79	4,387	.26	.27
March	302	67	105	6,456	.34	.39
April	476	113	246	14,638	.80	.89
May	1,885	340	888	54,601	2.90	3.34
June	2,030	815	1,512	89,970	4.94	5.51
July	1,160	192	402	24,718	1.31	1.51
August	215	85	143	8,793	.47	.54
September	85	67	75	4,463	.25	.28
October	75	67	67	4,120	.22	.25
November	98	67	77	4,582	.25	.28
December	75	67	67	4,120	.22	.25
The year	2,030	60	311	225,091	1.02	13.78

CARSON RIVER NEAR EMPIRE, NEV.

This station was established October 21, 1900, by L. H. Taylor. e original gage was located upstream from the site of the present tion. On February 18, 1901, the erosion of a bar in the channel)ve the gage caused a division of the stream into two channels. On s account, the present gage was installed on March 13, 1901. It is ated about three-fourths mile east of Brunswick Mill and 2¼ miles t of Empire, Nev. The gage is an inclined 4 by 4 inch timber spiked a cottonwood stump on the left bank. The gage datum was lowered feet August 11, 1903, to enable readings to be made at extreme low ter. It is read once each day by David Lloyd. Discharge measure- nts are made by means of a cable and car just below the gage. The tial point for soundings is the zero of the tagged wire on the left ak. The channel is straight for 100 feet above and below the station. th banks are high, rocky, and will not overflow. The bed of the eam is composed of solid rock, gravel, and cobble stones and is t likely to shift. At flood stages the stream is too deep and swift obtain the best results. At low water the stream is sluggish at the)le, and measurements are made by wading, one-fourth mile below. nch mark No. 1 is the top of a bowlder on the left bank 2 feet west the gage. Its elevation is 7.10 feet above the zero of the gage as iginally established and 8.40 feet above the zero of the gage in its 'esent position. Bench mark No. 2 is the top of a bowlder on the left mk 10 feet north of the gage. Its elevation is 8.38 feet above the :ro of the gage, as originally established, and 9.68 feet above the zero [the gage in its present position.

The observations at this station during 1903 have been made under le direction of A. E. Chandler, district hydrographer.

Discharge measurements of Carson River near Empire, Nev., in 1903.

Date.	Hydrographer.	Gage height.	Discharge.
		Feet.	*Second-feet.*
pril 6	D. W. Hays	2.62	646
pril 12	do	2.50	573
pril 24	A. H. Schadler	2.65	671
ay 7	do	3.75	1,564
me 4	do	4.10	1,786
me 19	do	3.10	955
ıly 20	do	1.00	110
ugust 11	W. B. Harrington	.05	13
eptember 3	do	.05	20

Mean daily gage height, in feet, of Carson River near Empire, Nev., for 1903.

Day.	Jan.	Feb.	Mar.	Apr.	May.	June.	July.	Aug.	Sept.	Oct.	Nov.	Dec.
1	1.40	2.00	2.00	3.40	3.00	4.00	3.00	0.50	0.00	-0.10	0.00	1.4
2	1.40	2.00	1.90	3.30	3.10	4.10	2.90	.50	.00	.00	.90	1.5
3	1.40	2.00	1.80	3.00	3.30	4.20	2.90	.40	.00	.10	.90	1.5
4	1.40	1.90	1.80	2.90	3.50	4.10	2.80	.30	-.10	.20	.90	1.5
5	1.40	1.90	1.80	2.80	3.60	4.00	2.60	.30	-.10	.20	.90	1.5
6	1.40	1.90	1.80	2.70	3.70	3.90	2.40	.30	-.10	.30	.90	1.5
7	1.40	1.80	1.80	2.60	3.80	4.00	2.20	.20	.00	.30	1.00	1.5
8	1.40	1.80	1.80	2.50	3.90	4.10	2.20	.20	-.20	.00	1.00	1.4
9	1.40	1.80	1.80	2.70	3.90	4.10	2.00	.20	-.10	.20	1.00	1.4
10	1.40	2.20	1.70	2.80	3.80	4.20	1.90	.20	-.10	.30	1.10	1.4
11	1.50	2.40	1.70	2.60	4.00	4.20	1.80	.10	-.10	.30	1.20	1.4
12	1.50	2.50	1.60	2.50	4.10	4.20	1.70	.10	-.10	.40	1.20	1.4
13	1.60	2.30	1.60	2.30	4.20	4.20	1.50	.10	-.10	.30	1.30	1.4
14	1.60	2.20	1.60	2.40	4.30	3.90	1.40	.10	.00	.50	1.50	1.4
15	1.70	2.10	1.60	2.30	4.40	3.80	1.30	.10	.10	.50	1.60	1.4
16	1.60	2.00	1.50	2.30	4.20	3.70	1.30	-.10	.10	.60	1.60	1.4
17	1.60	1.90	1.50	2.30	4.00	3.40	1.30	-.10	.10	.60	1.70	1.4
18	1.50	1.90	1.50	2.20	3.70	3.20	1.20	-.10	.70	.70	1.80	1.4
19	1.50	1.90	1.50	2.20	3.50	3.10	1.10	-.10	.10	.70	1.80	1.4
20	1.60	1.70	1.50	2.20	3.30	3.00	1.00	.10	.10	.80	1.90	1.4
21	1.60	1.60	1.50	2.30	3.20	3.10	.90	.10	.10	.70	2.00	1.4
22	1.50	1.70	1.50	2.40	3.00	3.20	.80	.10	.10	.70	1.90	1.4
23	1.40	1.80	1.50	2.40	3.00	3.30	.70	.10	.20	.70	1.90	1.4
24	1.40	1.90	1.50	2.60	3.00	3.20	.60	.10	-.10	.70	1.90	1.4
25	1.50	2.00	1.50	2.80	2.90	3.00	.60	.10	.00	.70	1.80	1.4
26	1.90	2.10	1.70	3.10	2.80	3.10	.60	.10	-.20	.70	1.80	1.4
27	2.30	2.10	1.90	3.00	2.90	3.30	.60	.10	.00	.80	1.70	1.4
28	2.70	2.10	2.20	2.90	3.00	3.20	.60	.00	.20	.80	1.70	1.5
29	2.20		2.40	2.80	3.00	3.00	.60	.00	-.20	.80	1.70	1.5
30	2.10		2.40	2.80	3.20	3.00	.60	-.10	-.10	.90	1.60	1.5
31	2.00		2.80		3.80		.60	.00		.90		1.50

a Gage readings from August 11 to December 31 are referred to old gage.

Rating table for Carson River near Empire, Nev., for 1903.

Gage height.	Discharge.	Gage height.	Discharge.	Gage height.	Discharge.	Gage height.	Discharge.
Feet.	Second-feet.	Feet.	Second-feet.	Feet.	Second-feet.	Feet.	Second-feet.
-0.2	12	0.8	79	1.8	312	2.8	748
-.1	13	.9	94	1.9	346	2.9	818
.0	14	1.0	110	2.0	382	3.0	892
.1	17	1.1	128	2.1	420	3.2	1,048
.2	21	1.2	148	2.2	458	3.4	1,215
.3	27	1.3	170	2.3	498	3.6	1,385
.4	35	1.4	193	2.4	540	3.8	1,555
.5	44	1.5	218	2.5	584	4.0	1,725
.6	54	1.6	248	2.6	632	4.2	1,895
.7	66	1.7	280	2.7	686	4.4	2,065

Table is not accurately determined in the lower part.

Estimated monthly discharge of Carson River near Empire, Nev., for 1903.

[Drainage area, 988 square miles.]

Month.	Discharge in second-feet.			Total in acre-feet.	Run-off.	
	Maximum.	Minimum.	Mean.		Second-feet per square mile.	Depth in inches.
January	686	193	264	16,233	0.27	0.31
February	584	248	378	20,993	.38	.40
March..............	748	218	308	18,938	.31	.36
April..............	1,215	458	681	40,522	.69	.77
May................	2,065	748	1,321	81,225	1.34	1.54
June	1,895	892	1,411	83,960	1.43	1.59
July	892	54	279	17,155	.28	.32
August	44	13	20	1,230	.02	.02
September..........	21	12	15	893	.02	.02
October	94	13	49	3,013	.05	.06
November	382	94	221	13,150	.22	.25
December	248	170	203	12,482	.20	.24
The year	2,065	12	429	309,794	.43	5.88

EAST FORK OF CARSON RIVER NEAR GARDNERVILLE, NEV.

The gaging station was established by L. H. Taylor on October 17, 1900, at the place where measurements were made in the years 1890, 1891, and 1892. It is located about 5 miles southeast of Gardnerville, Nev., and one-half mile southwest of Rodenbah's ranch. The old gage was an inclined timber securely fastened to posts set in the right bank of the stream. The old bench mark was on a basalt rock in the edge of the stream, 20 feet from the gage, at an elevation of 6.3 feet above gage datum.

On August 2, 1901, a loose-rock dam was raised a short distance below the gaging station, which affected the velocity at the latter point. The dam was partly washed out by a freshet on December 4, 1901. A new gage was established on March 10, 1901, a short distance downstream from the original one, which had been destroyed. It consists of a vertical timber driven into the stream bed at the right bank and spiked to a cottonwood tree. On October 3, 1902, a new inclined gage was installed by E. C. Murphy at a point on the left bank of the river 600 feet above the cable.

The gage is read twice each day by Miss Susie Rodenbah. Discharge measurements are made by means of a cable and car located at the vertical gage about 400 feet above the bridge. The initial point for soundings is the zero of the tagged wire on the right bank. The

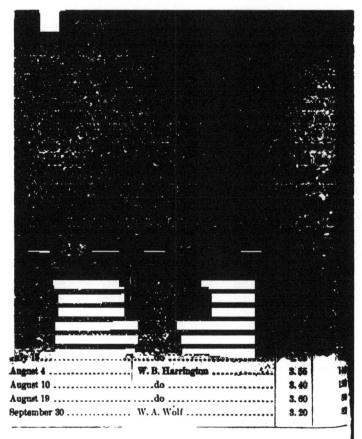

August 4	W. B. Harrington	3.55	1...
August 10	do	3.40	1...
August 19	do	3.60	...
September 30	W. A. Wolf	3.20	...

Mean daily gage height, in feet, of East Fork of Carson River near Gardnerville, Nev., for 1903.

Day.	Jan.	Feb.	Mar.	Apr.	May.	June.	July.	Aug.	Sept.	Oct.	Nov.	Dec.
1	2.50	2.65	2.70	4.35	5.80	6.65	4.90	3.45	3.82	2.60	3.20	3.00
2	2.45	2.60	2.75	4.40	5.90	6.75	4.75	3.50	3.82	2.60	3.20	3.00
3	2.50	2.50	2.80	4.50	5.90	6.50	4.75	3.65	3.42	2.60	3.20	3.15
4	2.55	2.50	2.75	4.55	5.85	5.85	4.65	3.56	3.42	2.70	3.25	3.15
5	2.55	2.55	2.75	4.65	5.80	6.10	4.55	3.45	3.25	3.05	3.35	3.25
6	2.60	2.50	2.80	4.75	5.85	6.45	4.50	3.50	3.25	3.30	3.20	3.20
7	2.60	2.50	2.80	4.85	5.95	6.45	4.45	3.45	3.15	3.20	3.20	3.16
8	2.60	2.50	2.85	4.85	6.00	6.35	4.40	3.45	3.05	3.20	3.20	3.10
9	2.50	2.50	2.90	4.95	6.10	6.25	4.45	3.45	3.00	3.25	3.30	3.05
10	2.55	2.40	2.90	4.95	6.35	6.15	4.45	3.40	3.00	3.35	3.35	3.00
11	2.40	2.40	2.95	4.85	6.45	6.25	4.35	3.35	3.00	3.40	3.45	3.00
12	2.45	2.30	2.95	4.70	6.70	6.15	4.30	3.45	3.00	3.40	3.85	3.05
13	2.45	2.30	2.95	4.70	6.95	6.05	4.28	3.45	3.05	3.35	4.90	3.05
14	2.40	2.25	3.00	4.75	7.05	5.70	4.05	3.35	3.05	3.80	4.75	3.00
15	2.40	2.35	3.15	4.75	6.95	5.35	4.00	3.30	3.05	3.22	4.80	2.95

m daily gage height, in feet, of East Fork of Carson River near Gardnerville, Nev.,
for 1903—Continued.

Day.	Jan.	Feb.	Mar.	Apr.	May.	June.	July.	Aug.	Sept.	Oct.	Nov.	Dec.
....	2.40	2.40	3.20	4.70	6.85	5.20	3.90	3.35	3.00	3.35	4.35	2.85
....	2.40	2.40	3.35	4.65	6.75	5.20	3.90	3.60	3.00	3.35	4.30	2.85
....	2.40	2.35	3.40	4.75	6.35	5.25	3.85	3.65	3.00	3.38	4.15	2.85
....	2.40	2.35	3.45	4.75	6.10	5.35	3.90	3.45	3.00	3.40	4.05	2.80
....	2.40	2.40	3.55	4.85	6.85	5.45	3.80	3.35	2.95	3.40	3.95	2.65
....	2.30	2.30	3.55	4.95	5.70	5.75	3.80	3.60	2.90	3.40	3.90	2.60
....	2.40	2.40	3.60	5.05	5.25	5.95	3.90	3.85	2.90	3.40	3.90	2.50
....	2.40	2.40	3.70	5.15	4.90	6.00	3.80	3.65	2.90	3.40	3.85	2.50
....	2.30	2.45	3.80	5.25	4.50	5.90	3.80	3.65	2.85	3.40	3.80	2.45
....	2.45	2.45	3.75	5.35	4.60	5.75	3.70	3.65	2.85	3.38	3.80	2.45
....	2.45	2.50	3.65	5.45	4.65	5.65	3.60	3.65	2.85	3.35	3.80	2.40
....	2.50	2.55	3.85	5.55	4.80	5.50	3.55	3.72	2.85	3.35	3.75	2.36
....	2.55	2.60	4.05	5.60	5.20	5.45	3.50	3.68	2.90	3.20	3.75	2.35
....	2.60	......	4.15	5.65	5.80	5.15	3.45	3.62	2.80	3.20	3.55	2.40
....	2.60	......	4.25	5.75	6.20	4.95	3.50	3.62	2.80	3.20	3.50	2.40
....	2.50	......	4.30	......	6.50	......	3.42	3.60	......	3.20	......	2.40

Rating table for East Fork of Carson River near Gardnerville, Nev., for 1903.

Gage height.	Discharge.	Gage height.	Discharge.	Gage height.	Discharge.	Gage height.	Discharge.
Feet.	*Second-feet.*	*Feet.*	*Second-feet.*	*Feet.*	*Second-feet.*	*Feet.*	*Second-feet.*
2.2	21	3.2	82	4.2	375	5.4	1,285
2.3	25	3.3	97	4.3	430	5.6	1,475
2.4	29	3.4	115	4.4	485	5.8	1,665
2.5	34	3.5	135	4.5	550	6.0	1,855
2.6	39	3.6	155	4.6	620	6.2	2,045
2.7	45	3.7	180	4.7	690	6.4	2,235
2.8	51	3.8	210	4.8	760	6.6	2,425
2.9	58	3.9	240	4.9	840	6.8	2,615
3.0	65	4.0	280	5.0	920	7.0	2,805
3.1	72	4.1	325	5.2	1,100		

Table extended above 6.15 feet gage height and below 3.20 feet gage height. Below
1.20 feet gage height curve is only approximate.

Estimated monthly discharge of East Fork of Carson River near Gardnerville, Nev., for 190

[Drainage area, 381 square miles.]

Month.	Discharge in second-feet.			Total in acre feet.	Run-off.	
	Maximum.	Minimum.	Mean.		Second-feet per square mile.	Depth inch
January	39	25	32	1,968	0.08	
February	42	21	31	1,722	08	
March	430	45	135	8,301	.36	
April	1,617	457	890	52,959	2.34	
May	2,852	550	1,794	110,309	4.71	
June	2,567	880	1,714	101,990	4.50	
July	840	115	356	21,889	.93	
August	225	97	143	8,793	.38	
September	135	51	73	4,344	.19	
October	115	51	93	5,718	.24	
November	725	82	229	13,626	.60	
December	115	27	57	3,505	.15	
The year	2,852	21	462	335,124	1.21	1

WEST-FORK OF CARSON RIVER AT WOODFORDS, CAL.

This station was established October 18, 1900, by L. H. Taylor. is located about three-fourths mile above the post-office at Woodfc and 200 feet from the main road between Woodfords and Blue Lal Cal. The gage is a vertical board nailed to a cottonwood tree on left bank. It is read once each day by Miss Bernice Merrill. 1 charge measurements are made by means of a cable and car just be the gage. The initial point for soundings is the zero of the tag wire. The channel is straight for 100 feet above and below the ca and there is but one channel at all stages. Both banks are high rocky and are not liable to overflow. The bed of the stream is ro and uneven and is not liable to change. At high water the curt velocity is too high for accurate measurement.

The observations at this station during 1903 have been made un the direction of A. E. Chandler, district hydrographer.

Discharge measurements of West Fork of Carson River at Woodfords, Cal., in 1903.

Date.	Hydrographer.	Gage height.	Discharge.
		Feet.	*Second-feet.*
April 7	D. W. Hays	3.06	97
April 22	A. H. Schadler	3.80	183
June 24	do	4.20	247
July 10	do	3.30	128
July 17	do	3.05	107
August 4	W. B. Harrington	2.60	48
September 9	do	2.20	37

Mean daily gage height, in feet, of West Fork of Carson River at Woodfords, Cal., for 1903.

Day.	Jan.	Feb.	Mar.	Apr.	May.	June.	July.	Aug.	Sept.	Oct.	Nov.	Dec.
1	2.40	2.90	2.70	3.40	4.90	4.90	3.70	2.70	2.20	2.20	2.25	2.70
2	2.40	2.80	2.70	3.50	4.90	4.90	3.60	2.70	2.20	2.20	2.25	2.65
3	2.40	2.80	2.70	3.40	5.00	5.00	3.50	2.60	2.20	2.25	2.25	2.60
4	2.40	2.80	2.70	3.30	5.00	5.20	3.50	......	2.20	2.25	2.30	2.60
5	2.40	2.70	2.80	3.20	5.10	5.30	3.50	2.50	2.20	2.25	2.25	2.60
6	2.30	2.50	2.90	3.00	5.20	5.30	3.50	2.50	2.20	2.20	2.30	2.70
7	2.30	2.50	2.90	2.90	5.30	5.30	3.50	2.50	2.20	2.30	2.25	2.75
8	2.30	2.60	2.90	2.90	5.50	5.30	......	......	2.20	2.30	2.20	2.75
9	2.30	2.70	2.80	3.10	5.50	5.20	......	2.50	2.20	2.25	2.20	2.80
10	2.40	2.70	2.70	3.20	5.60	5.10	3.30	2.10	2.20	2.30	2.20	2.80
11	2.40	2.70	2.60	3.20	5.60	5.10	3.20	2.40	2.20	2.20	2.35	2.80
12	2.40	2.80	2.60	3.30	5.60	5.10	3.20	2.40	2.20	2.20	2.50	2.80
13	2.30	2.60	2.50	3.40	5.60	5.00	3.20	2.40	2.10	2.20	2.90	2.75
14	2.30	2.50	2.50	3.50	5.50	5.00	3.10	......	2.15	2.20	3.05	2.75
15	2.30	2.50	2.40	3.40	5.30	4.90	3.00	2.10	2.15	2.20	3.05	2.75
16	2.30	2.50	2.40	3.40	5.00	4.80	3.00	2.10	2.15	2.20	3.00	2.75
17	2.30	2.50	2.50	3.30	5.00	4.80	3.00	2.30	2.20	2.20	3.00	2.75
18	2.30	2.60	2.60	3.30	4.80	4.80	2.90	......	2.20	2.20	3.00	2.75
19	2.30	2.60	2.60	3.30	4.70	4.80	......	2.30	2.20	2.20	2.95	2.75
20	2.30	2.70	2.60	3.50	4.50	4.70	2.90	2.10	2.20	2.20	2.95	2.80
21	2.30	2.70	2.60	3.60	4.40	4.60	......	2.30	2.20	2.20	2.90	2.85
22	2.30	2.70	2.60	3.80	4.40	4.50	2.90	2.30	2.20	2.20	2.90	2.85
23	2.40	2.80	2.60	4.10	4.40	4.40	......	2.30	2.20	2.20	2.90	2.75
24	2.60	2.70	2.60	4.30	4.40	4.10	......	2.30	2.25	2.20	2.80	2.75
25	2.50	2.70	2.60	4.50	4.50	4.30	2.90	2.30	2.25	2.20	2.80	2.70
26	2.70	2.70	2.60	4.60	4.60	4.20	2.80	2.30	2.25	2.20	2.80	2.65
27	3.00	2.70	2.60	4.50	4.70	4.20	......	2.30	2.25	2.15	2.80	2.65
28	3.00	2.70	2.70	4.60	4.70	4.10	2.80	2.30	2.25	2.20	2.75	2.65
29	3.00	......	2.80	4.60	4.80	4.00	2.80	2.20	2.25	2.20	2.70	2.50
30	2.90	......	3.00	4.70	4.80	3.90	2.70	2.20	2.25	2.25	2.70	2.15
31	3.00	......	3.20	......	4.80	......	......	2.20	......	2.25	......	2.45

Rating table for West Fork of Carson River at Woodfords, Cal., for 1903.

Gage height.	Discharge.	Gage height.	Discharge.	Gage height.	Discharge.	Gage height.	Discharge.
Feet.	*Second-feet.*	*Feet.*	*Second-feet.*	*Feet.*	*Second-feet.*	*Feet.*	*Second-feet.*
2.0	20	2.9	83	3.8	188	4.7	340
2.1	25	3.0	93	3.9	202	4.8	358
2.2	30	3.1	103	4.0	217	4.9	375
2.3	36	3.2	114	4.1	233	5.0	394
2.4	42	3.3	125	4.2	250	5.1	412
2.5	49	3.4	137	4.3	268	5.2	430
2.6	57	3.5	149	4.4	286	5.3	448
2.7	65	3.6	161	4.5	304	5.4	466
2.8	74	3.7	174	4.6	322	5.5	484

Table extended beyond 4.20 feet gage height; well defined below that point.

Estimated monthly discharge of West Fork of Carson River at Woodfords, Cal., for 1903.

[Drainage area, 70 square miles.]

Month.	Discharge in second-feet.			Total in acre-feet.	Run-off.	
	Maximum.	Minimum.	Mean.		Second-feet per square mile.	Depth in inches.
January	93	36	49	3,013	0.70	0.81
February	83	49	63	3,499	.90	.94
March	114	42	64	3,935	.91	1.05
April	340	83	174	10,354	2.49	2.78
May	502	286	389	23,919	5.56	6.41
June	448	202	353	21,005	5.04	5.62
July [a]	174	65	106	6,518	1.51	1.74
August [b]	65	30	42	2,582	.60	.69
September [c]	33	25	30	1,785	.43	.48
October	36	27	31	1,906	.44	.51
November	98	30	63	3,749	.90	1.00
December	78	45	66	4,058	.94	1.08
The year	502	25	119	86,323	1.70	23.11

[a] Values for 7th, 8th, 9th, 19th, 21st, 23d, 24th, 27th, and 31st are interpolated.
[b] Values for 4th, 8th, 14th, and 18th are interpolated.
[c] Value for 6th is interpolated.

TRUCKEE RIVER AT TAHOE, CAL.

ckee River, the natural outlet of Lake Tahoe, leaves the lake at
ty of Tahoe. About 500 feet from the lake there is a timber
cross the river, which has been maintained for more than twenty
for the purpose of controlling the discharge from the lake.
17, 1900, a gage was placed in the stream for the purpose of
ling the height of the water in the river. The gage is a vertical
r driven into the stream bed at the left bank about 300 feet below
im, and is spiked to the root of a cottonwood tree growing on
ink. On November 18, 1902, a new gage was established by
Murphy in the exact position of the old gage. It is a 4 by 4
ertical timber. The elevation of its zero is the same as that of
d gage. The bench mark is cut in the side of the tree and is
above gage datum. The measurements are made from a cable
spended car about one-fourth mile below the gage, which was
l as near the city of Tahoe as possible for the convenience of the
er. At the point of measurement the right bank is low and is
t to overflow at very high stages of the stream. The left bank
ier high. The channel is nearly straight for 300 feet above and
the station, and the bed of the river, which is of gravel and
sand, is smooth and stable. The current has a moderate veloc-
The purpose of the station is to ascertain the actual overflow
Lake Tahoe with a view to determining its real value as a storage
oir.
observations at this station during 1903 have been made under
rection of A. E. Chandler, district hydrographer.

Discharge measurements of Truckee River at Tahoe, Cal., in 1903.

Date.	Hydrographer.	Gage height.	Discharge.
		Feet.	*Second-feet.*
).....................	G. B. Lorenz.................	0.10	13
3.....................	do	.79	116
).....................	do	1.50	235
).....................	do	1.90	312

Mean daily gage height, in feet, of Truckee River at Tahoe, Cal., for 1903.

Day.	Jan.	Feb.	Mar.	Apr.	May.	June.	July.	Aug.	Sept.	Oct.	Nov.	Dec.
1	1.65	0.80	1.35	0.10	0.10	0.10	0.70	1.70	1.95	1.85	1.90	
2	1.65	.80	1.35	.10	.10	.10	.70	1.70	1.95	1.85	2.10	
3	1.65	.40	1.35	.10	.10	.10	.70	1.70	2.05	1.80	2.10	
4	1.65	.40	1.35	.10	.10	.10	.70	1.70	2.05	1.80	2.10	
5	1.65	.80	1.35	.10	.10	.10	.70	1.70	2.05	1.80	1.90	
6	1.65	.80	1.40	.10	.10	.10	.70	1.70	2.05	1.80	1.90	
7	1.65	.80	1.40	.10	.10	.10	.90	1.70	2.05	1.80	1.90	
8	1.65	.40	1.40	.10	.10	.10	1.00	1.70	2.05	1.80	1.90	
9	1.65	.40	1.40	.10		.10	1.20	1.70	2.05	1.80	1.90	
10	1.65	.40	1.40	.10		.10	1.20	1.70	2.05	1.80	1.90	
11	1.65	.40	1.40	.10	.10	.10	1.20	1.70	2.05	1.85		
12	1.65	.40	1.40	.10	.10	.10	1.40	1.70	2.05	2.00		
13	1.60	.40	1.40	.10	.10	.10	1.50	1.70	2.05	2.00		
14	1.60	.40	1.40	.10	.10	.10	1.40	1.80	2.05	2.00		
15	1.60	.90	1.40	.10	.10	.10	1.50	1.80	1.95	2.00		
16	1.60	.90	1.40	.10	.10	.10	1.50	1.80	1.95	2.00		
17	1.57	.90	1.40	.10	.10	.10	1.50	1.80	1.95	1.95	.95	
18	1.56	.90	1.40	.10	.10	.10	1.50	1.80	1.95	1.95	.95	
19	1.55	.90	1.40	.10	.10	.10	1.50	1.95		1.95		
20	1.54	.90	1.40	.10	.10	.10	1.50	1.95		1.95		
21		.90	1.40	.10	.10	.10	1.50	1.95	1.90	1.95	.90	
22	1.56	.90	1.40	.10	.10	.70	1.60	1.95	1.90	1.95	.90	
23	1.58	.90	1.40	.10	.10	.70	1.50	1.95	1.90	1.95	.90	
24	1.58	.90	1.40	.10	.10	.70	1.50	1.95	1.90	1.95	.90	
25	1.58	1.00	.35	.10	.10	.70	1.50	1.95	1.90	1.95	.90	
26	1.58	1.00	1.10	.10	.10	.70	1.50	1.95	1.90	1.95	.90	
27	1.58	1.35	.70	.10	.10	.70	1.60	1.95	1.90	1.95	1.20	
28	1.58		.10	.10	.10	.70	1.60	1.95	1.85	1.95	1.20	
30	.35		.10	.10	.10	.70	1.70	1.95	1.85	1.95	1.20	
31	.35		.10		.10		1.70	1.95		1.95		

Rating table for Truckee River at Tahoe, Cal., for 1903.

Gage height.	Discharge.	Gage height.	Discharge.	Gage height.	Discharge.	Gage height.	Discharge.
Feet.	Second-feet.	Feet.	Second-feet.	Feet.	Second-feet.	Feet.	Second-feet.
0.1	13	0.7	93	1.3	201	1.9	372
.2	25	.8	109	1.4	225	2.0	407
.3	38	.9	125	1.5	251	2.1	445
.4	51	1.0	142	1.6	279	2.2	486
.5	64	1.1	160	1.7	308	2.3	531
.6	78	1.2	179	1.8	339	2.4	580

Above 1.20 feet gage height the table is the same as that of 1902.

Estimated monthly discharge of Truckee River at Tahoe, Cal., for 1903.

[Drainage area, 519 square miles.]

onth.	Discharge in second-feet.			Total in acre-feet.	Run-off.	
	Maximum.	Minimum.	Mean.		Second-feet per square mile.	Depth in inches.
....	293	44	266	16,356	0.51	0.59
....	213	51	103	5,720	.20	.21
....	225	13	190	11,683	.37	.43
......	13	13	13	774	.03	.03
....	13	13	13	799	.03	.03
....	93	13	38	2,261	.07	.08
....	308	93	205	12,605	.39	.45
....	389	308	344	21,152	.66	.76
....	426	355	398	23,683	.77	.86
....	407	339	374	22,966	.72	.83
r........	445	125	255	15,174	.49	.55
....	372	179	260	15,987	.50	.58
e year	445	13	205	149,190	.40	5.40

TRUCKEE RIVER AT VISTA, NEV.

station was originally established August 18, 1899, by L. H.
It is located 7 miles east of Reno, Nev., and one-fourth
m the Southern Pacific Railroad station at Vista, Nev. On
er 12, 1902, a new gage was installed on the left bank, 150 feet
he railroad bridge, by E. C. Murphy. Its zero has the same
n as that of the original gage. On April 3, 1903, a new gage
ablished by D. W. Hays. It is located on the left bank and is
al 4 by 4 inch timber established on the same datum as the
s gages. On June 23, 1903, this gage was torn out and moved
arther into the river, in order to establish it in deeper water.
o of the gage was lowered 2.72 feet. The gage is read once
y by M. Tuomey. Discharge measurements are made by
f a cable and car below the railroad bridge. The initial point
dings is the zero of the tagged wire. The channel is straight
feet above and 400 feet below the station. Both banks are
The left bank is liable to overflow only at extreme high stages.
s but one channel at all stages. The bench mark is the head
t set in the concrete on the upstream side of the left abutment
Southern Pacific Railroad bridge 300 feet from the gage. Its
n is 19.74 feet above the zero of the gage.

Day.	Jan.	Feb.	Mar.	Apr.	May.	June.	July.	Aug.	Sept.	Oct.	Nov.	Dec.
1		3.40	3.40	5.00	5.60	5.30	3.40	2.00	2.40	2.50	3.00	3.20
2		3.50	3.50	4.10	5.90	5.30	3.30	2.00	2.30	2.50	3.00	3.20
3		3.50	3.60	5.20	6.00	4.90	3.10	2.10	2.30	2.50	3.00	3.10
4		3.50	3.60	5.10	6.10	4.80	3.00	2.10	2.30	2.60	3.10	3.00
5		3.40	3.50	5.00	6.50	4.70	2.80	2.10	2.30	2.70	3.20	3.10
6		3.40	3.40	4.40	6.50	4.80	2.60	2.10	2.40	2.70	3.10	3.10
7	3.20	3.40	3.30	4.50	6.50	4.80	2.50	2.20	2.40	2.70	3.20	3.10
8	3.20	3.40	3.40	5.00	6.00	4.80	2.50	2.10	2.50	2.70	3.00	3.10
9	3.20	3.30	3.40	5.50	6.00	4.60	2.40	2.10	2.50	2.80	3.10	3.10
10	3.20	3.30	3.40	6.00	6.00	4.80	2.80	2.10	2.60	2.80	3.10	3.10
11	3.20	3.20	3.40	5.10	6.30	4.80	2.40	2.00	2.70	2.90	3.10	3.10
12	3.20	3.20	3.40	4.90	6.30	4.60	2.20	2.00	2.70	2.80	3.10	3.00
13	3.20	3.10	3.40	4.80	6.30	4.60	2.10	2.10	2.50	2.80	5.00	25.30
14	3.10	3.00	3.40	4.80	6.10	4.40	2.00	2.10	2.60	2.90	5.50	25.30
15	3.10	3.00	3.40	4.00	5.90	4.10	2.10	2.10	2.70	2.90	4.60	25.20
16	3.10	3.00	3.30	4.60	5.80	3.60	2.10	2.00	2.60	2.90	3.60	25.20
17	3.00	3.00	3.30	4.00	5.40	3.60	2.10	2.10	2.60	3.00	3.30	25.20
18	3.00	3.00	3.20	4.60	5.00	3.80	2.10	2.10	2.60	3.00	3.20	25.10
19	3.00	3.00	3.20	4.00	4.00	3.40	2.10	2.00	2.50	3.00	3.10	25.00
20	3.00	3.00	3.20	4.70	4.50	3.40	2.10	2.30	2.50	3.10	4.10	25.10
21	3.00	3.00	3.20	5.00	4.40	3.40	2.10	2.30	2.50	3.00	5.50	25.10
22	3.10	3.00	3.40	5.40	4.30	3.40	2.10	2.20	2.40	3.10	5.10	25.00
23	3.10	3.20	3.60	6.00	4.20	3.40	2.20	2.20	2.40	3.10	4.80	25.00
24	3.20	3.40	3.60	5.70	4.00	3.60	2.10	2.30	2.40	3.10	4.30	25.00
25	4.10	3.40	3.70	5.90	4.00	3.50	2.10	2.30	2.40	3.10	3.40	25.00
26	3.80	3.90	3.70	5.60	4.00	3.30	2.10	2.40	2.40	3.10	3.70	25.00
27	3.90	3.30	3.80	5.40	4.00	3.30	2.10	2.50	2.40	3.10	3.70	25.10

daily gage height, in feet, of Truckee River at Vista, Nev., for 1903—Continued.

Day.	Jan.	Feb.	Mar.	Apr.	May.	June.	July.	Aug.	Sept.	Oct.	Nov.	Dec.
....	4.00	3.40	5.60	5.20	4.00	3.00	2.00	2.20	2.50	3.10	3.50	3.10
....	4.10	....	5.40	5.20	4.00	3.30	2.00	2.20	2.50	3.10	3.50	3.10
....	3.80	....	7.60	5.40	4.30	3.60	2.00	2.20	2.50	3.00	3.10	3.10
....	3.20	....	8.00	....	5.00	....	2.00	2.30	....	3.00	....	3.10

June 23, 1903, gage was lowered 2.72 feet. All gage heights refer to original ·
.

Rating table for Truckee River at Vista, Nev., for 1903.

Gage height.	Discharge.	Gage height.	Discharge.	Gage height.	Discharge.	Gage height.	Discharge.
Feet.	Second-feet.	Feet.	Second-feet.	Feet.	Second-feet.	Feet.	Second-feet.
2.0	130	3.0	530	4.0	955	6.0	2,350
2.1	170	3.1	570	4.2	1,055	6.2	2,610
2.2	210	3.2	610	4.4	1,155	6.4	2,900
2.3	250	3.3	650	4.6	1,255	6.6	3,215
2.4	290	3.4	690	4.8	1,355	6.8	3,550
2.5	330	3.5	730	5.0	1,455	7.0	3,900
2.6	370	3.6	775	5.2	1,575	7.2	4,250
2.7	410	3.7	820	5.4	1,735	7.4	4,600
2.8	450	3.8	865	5.6	1,920	7.6	4,950
2.9	490	3.9	910	5.8	2,125	8.0	5,650

ble fairly well defined. Curve extended above 5.85 feet gage height.

Estimated monthly discharge of Truckee River at Vista, Nev., for 1903.

[Drainage area, 1,519 square miles.]

Month.	Discharge in second-feet.				Run-off.	
	Maximum.	Minimum.	Mean.	Total in acre-feet.	Second-feet per square mile.	Depth in inches.
ary	1,005	530	653	40,151	0.43	0.50
uary	730	530	624	34,655	.41	.43
·h	5,650	610	1,077	66,222	.71	.82
l	2,350	1,155	1,537	94,458	1.01	1.13
....	3,055	955	1,852	113,875	1.22	1.41
....	1,650	530	1,020	60,694	.67	.75
....	690	130	259	15,925	.17	.20
ıst	330	130	192	11,806	.13	.15
·mber	410	250	321	19,101	.21	.23
ber	570	330	486	29,883	.32	.37
·mber	2,125	530	845	50,281	.56	.62
mber	650	530	569	34,986	.37	.43
The year	5,650	130	786	569,037	.52	7.04

nuary 1 to 6 estimated.

TRUCKEE RIVER AT PYRAMID LAKE INDIAN AGENCY, NEAR WADS-
WORTH, NEV.

This station was established November 6, 1902, by E. C. Murphy.
It is located one-fourth mile west of the school at the Indian Agency
and 18 miles north of Wadsworth, Nev. The gage is a 4 by 4 inch
vertical timber spiked to a cottonwood tree on the right bank. It is
read twice each day by John B. Woods. Discharge measurements are
made by means of a cable and car about 200 feet below the gage. The
initial point for soundings is the zero of the tagged wire. The channel is straight for 200 feet above and below the cable and the current
is swift. The right bank is low, and is liable to overflow beyond the
cable support at very high water. The left bank is high and will not
overflow. The bed of the stream is composed of sand and gravel and
is permanent. There is but one channel at all stages. The bench
mark consists of 3 nails driven into the root of a cottonwood tree 18
feet north of the gage. Its elevation is 12.07 feet above the zero of
the gage.

The observations at this station during 1903 have been made under
the direction of A. E. Chandler, district hydrographer.

Discharge measurements of Truckee River at Pyramid Lake Indian Agency, Nev., in 1903.

Date.	Hydrographer.	Gage height.	Discharge
		Feet.	Second-feet.
March 15	D. W. Hays	4.40	608
May 9	H. B. Jameson	6.51	2,028
June 15	do	5.30	1,054
July 16	do	3.10	202
August 21	do	2.90	164

*Mean daily gage height, in feet, of Truckee River at Pyramid Lake Indian Agency, for
1903.*

Day.	Jan.	Feb.	Mar.	Apr.	May.	June.	Oct.	Nov.	Dec.
1	4.20		4.40	7.62	6.30	6.21	3.50	3.70	3.84
2	4.19		4.40	6.88	6.29	6.20	3.54	3.70	3.80
3	4.16	4.30	4.50	6.31	6.38	6.09	3.52	3.70	3.80
4	4.26	4.22	4.46	6.30	6.46	5.95	3.51	3.70	3.76
5	4.14	4.11	4.41	6.15	6.67	5.70	3.52	3.70	3.78
6	4.11	4.06	4.41	5.65	6.91	5.83	3.54	3.70	3.74
7	4.19	3.97	4.34	5.52	6.75	5.75	3.54	3.70	3.70
8	4.26	4.07	4.35	5.60	6.80	5.75	3.54	3.70	3.67
9	4.24	4.01	4.33	6.10	6.70	5.75	3.57	3.68	3.66
10	4.21	4.14	4.34	6.45	6.50	5.80	3.58	3.70	3.67
11	4.04	4.11	4.32	6.45		5.84	3.59	3.70	3.66
12	4.05	4.10	4.35	6.15	6.80	5.86	3.62		3.67
13	4.16	3.98	4.40	5.95	6.77		3.68	4.23	3.65
14	4.13	4.01	4.41	5.86	6.80	5.46	3.62	4.84	3.64

y gage height, in feet, of Truckee River at Pyramid Lake Indian Agency, Nev.,
for 1903—Continued.

Day.	Jan.	Feb.	Mar.	Apr.	May.	June.	Oct.	Nov.	Dec.
....	4.15	4.05	4.37	5.52	6.67	5.36	3.63	5.18	3.63
....	4.13	3.83	4.33	5.49	6.55	5.11	3.66	4.26	3.61
....	4.08	3.77	4.31	5.50	6.39	4.75	3.68	3.92	3.61
....	4.09	3.79	4.24	5.50	6.04	4.37	3.68	3.80	3.64
....	4.13	3.98	4.17	5.49	5.61	4.26	3.68	3.78	3.60
....		4.06	4.24	5.50	5.44	....	3.68	3.75	3.60
....		4.05	4.22	5.50	5.34	....	3.69	4.24	3.61
....		4.06	4.14	5.69	5.33	....	3.70	5.42	3.61
....	4.12	4.05	4.29	5.97	5.30	....	3.68	4.75	3.56
....	4.22	4.14	4.43	6.20	5.20	....	3.69	4.46	3.58
....	4.79	4.21	4.51	6.45	4.99	....	3.70	4.42	3.51
....	4.99	4.22	4.60	6.85	5.02	....	3.70	4.12	3.55
....	5.05	4.20	4.65	6.60	4.92	....	3.74	4.08	3.70
....	4.94	4.20	4.81	6.13	5.00	....	3.69	4.00	3.71
....	4.94	....	5.95	5.93	5.40	....	3.68	3.92	3.57
....	4.97	....	6.70	5.90	6.05	....	3.68	3.93	3.55
....	4.83	....	8.30	....	6.18	....	3.65	....	3.55

ble for Truckee River at Pyramid Lake Indian Agency, Nev., from November 6,
1902, to December 31, 1903.

Discharge.	Gage height.	Discharge.	Gage height.	Discharge.	Gage height.	Discharge.
Second-feet.	Feet.	Second-feet.	Feet.	Second-feet.	Feet.	Second-feet.
115	3.5	297	4.5	648	6.2	·1,795
125	3.6	323	4.6	692	6.4	1,955
137	3.7	350	4.7	740	6.6	2,115
150	3.8	380	4.8	790	6.8	2,275
165	3.9	412	5.0	900	7.0	2,435
182	4.0	448	5.2	1,020	7.6	2,915
202	4.1	484	5.4	1,155	8.0	3,235
224	4.2	524	5.6	1,315	8.3	3,475
247	4.3	564	5.8	1,475		
271	4.4	604	6.0	1,635		

fairly well determined. Curve is extended below 2.90 feet gage height.

Estimated monthly discharge of Truckee River at Pyramid Lake Indian Agency, Nev., for 1903.

[Drainage area, 2,130 square miles.]

Month.	Discharge in second-feet.			Total in acre-feet.	Run-off.	
	Maximum.	Minimum.	Mean.		Second-feet per square mile.	Depth in inches.
1902.						
November 7-30	648	504	555	26,420	0.26	0.3
December	716	466	533	32,773	.25	.9
1903.						
January	930	466	586	36,082	.28	.33
February	815	364	487	27,047	.23	.24
March.............	3,475	504	775	47,653	.36	.42
April.............	2,915	1,235	1,645	97,884	.77	.86
May	2,355	840	1,701	104,590	.80	.92
June 1-19.........	....	....	1,323	49,856	.62	.44
October	564	297	332	20,414	.16	.18
November	1,155	350	480	29,097	.23	.26
December	396	297	337	20,721	.16	.18

LAKE WINNEMUCCA INLET NORTH OF PYRAMID LAKE INDIAN AGENCY, NEV.

This station was established November 7, 1902, by E. C. Murphy. It is located 3 miles north of Pyramid Lake Indian School and 21 miles north of Wadsworth, Nev. The gage is a 4 by 4 inch vertical timber spiked to a willow tree on the right bank. It is read three times each week by John B. Woods. The discharge measurements are made by means of a cable and car near the gage. The initial point for soundings is the zero of the tagged wire. The channel is straight for 150 feet above and below the cable and the current is moderate. Both banks are high and not subject to overflow. The bed is composed of gravel and silt and may change slightly. There is but one channel at all stages. The bench mark consists of two spikes driven into the willow tree to which the gage is attached. Its elevation is 6 feet above the zero of the gage.

The observations at this station during 1903 have been made under the direction of A. E. Chandler, district hydrographer.

Discharge measurements of Lake Winnemucca Inlet north of Pyramid Lake Indian Agency, Nev., in 1903.

Date.	Hydrographer.	Gage height.	Discharge.
		Feet.	*Second-feet.*
March 15	D. W. Hays...................	1.62	60
May 10	H. B. Jameson..............	4.60	318
June 15	....do	2.60	139
August 21	....do	.30	1

Mean daily gage height, in feet, of Lake Winnemucca Inlet north of Pyramid Lake Indian Agency, Nev., for 1903.

Day.	Jan.	Feb.	Apr.	May.	June.	Oct.	Nov.	Dec.
1		1.50			3.10		1.55	1.88
2								
3		1.50		4.80			1.70	
4	1.60			4.90	3.70	1.90		1.77
5		1.60						
6	1.70					1.80	1.90	1.60
7				5.30	3.70			
8	1.50				3.70		1.88	1.70
9						1.50		
10				4.20			1.90	
11	1.70			4.10	3.60	1.55		1.80
12								
13	1.60					1.66	1.77	1.77
14				4.10	2.80			
15	1.50				2.70		4.77	1.70
16						1.77		
17				3.80			1.70	
18	1.50			3.70	2.00	1.80		1.80
19			3.60					
20	1.50					1.88	5.90	1.80
21			2.90	2.50	1.70			
22	1.50				1.50		2.60	1.70
23						1.70		
24			3.50	2.20			1.80	
25	1.50			2.10	1.50	1.70		1.70
26			5.10					
27	1.70					1.68	1.99	
28			4.80	2.10				
29	1.60						1.99	
30			4.80			1.60		
31				2.30				

Estimated monthly discharge of Lake Winnemucca Inlet north of Pyramid Lake Indian Agency, Nev., for 1903.

Gage height.	Discharge.	Gage height.	Discharge.	Gage height.	Discharge.	Gage height.	Discharge.
Feet.	*Second-feet.*	*Feet.*	*Second-feet.*	*Feet.*	*Second-feet.*	*Feet.*	*Second-feet.*
0.3	2	1.3	40	2.3	112	4.2	283
.4	3	1.4	46	2.4	121	4.4	301
.5	6	1.5	52	2.6	139	4.6	319
.6	10	1.6	59	2.8	157	4.8	337
.7	14	1.7	66	3.0	175	5.0	355
.8	18	1.8	73	3.2	193	5.2	373
.9	22	1.9	80	3.4	211	5.4	391
1.0	26	2.0	87	3.6	229	5.6	409
1.1	30	2.1	95	3.8	247	5.8	427
1.2	35	2.2	103	4.0	265	6.0	445

Table well defined. Curve extended above 4.60 feet gage height.

Rating table for Lake Winnemucca Inlet north of Pyramid Lake Indian Agency, Nev., for 1903.

Month.	Mean discharge in second-feet.	Total in acre-feet.
January, 12 days	57	a 3,505
February, 3 days	54	b 536
April, 6 days	276	c 6,561
May, 13 days..	224	a 13,773
June, 11 days	154	a 9,164
October, 12 days	67	a 4,120
November, 13 days	127	a 7,557
December, 11 days	69	a 4,243

a Acre-feet computed for entire month.   b Acre-feet computed from February 1-5.
c Acre-feet computed from April 19-30.

TRUCKEE RIVER AT NEVADA-CALIFORNIA STATE LINE, NEAR MYSTIC, CAL.

This station was originally established September 7, 1899, by L. H.
Taylor. It is located at the State line, 17 miles west of Reno, Nev.
The original gage was vertical, driven into the bed of the river, and
wired to a granite bowlder. On November 11, 1902, a new gage was
established by E. C. Murphy. It is located on the right bank, 40
feet below the point at which the old gage was located, and consist
of 2 sections of 4 by 4 inch timber. The upper section is vertical, and
is spiked to a cottonwood tree. The lower section is inclined, and
immediately under the vertical section. The gage datum is the same
as that of the old gage. The gage is read once each day by H. E.
Dickinson. Discharge measurements are made by means of a cable

car 2¼ miles below the gage at Linham Siding, and 100 feet below lge No. 2. The initial point for soundings is the zero of the tagged ¦. The channel is straight for 60 feet above and for 75 feet below cable. The current is swift. Both banks are high and rocky, and not liable to overflow. The bed of the stream is composed of lders and cobblestones. The bench mark consists of 2 spikes en into the root of the cottonwood tree to which the vertical sec- of the gage is attached. Its elevation is 5.99 feet above the zero he gage.

he observations at this station during 1903 have been made under direction of A. E. Chandler, district hydrographer.

harge measurements of Truckee River at Nevada-California State line, near Mystic, Cal., in 1903.

Date.	Hydrographer.	Gage height.	Discharge.
		Feet.	*Second-feet.*
'1	H. B. Jameson	3.95	1,915
'26	......do	3.00	945
e 25	......do	2.90	1,070
'18	......do	2.10	457
ust 27	......do	2.10	477

n daily gage height, in feet, of Truckee River at Nevada-California State line, near Mystic, Cal., for 1903.

Day.	Jan.	Feb.	Mar.	Apr.	May.	June.	July.	Aug.	Sept.	Oct.	Nov.	Dec.
	2.10	1.57	1.65	4.10	4.00	3.95	2.68	2.05	2.15	1.95	2.20	2.00
	2.00	2.25	1.80	3.10	4.10	3.90	2.60	2.02	2.02	2.00	2.10	1.95
	1.95	2.00	1.80	3.10	4.00	3.62	2.40	2.08	1.98	2.10	1.90	2.00
	2.05	2.20	1.90	3.10	4.10	3.55	2.22	2.00	2.15	2.12	2.15	.....
	2.25	2.10	2.00	2.80	4.22	3.55	2.15	2.00	2.08	2.10	2.30	1.50
	2.25	1.90	1.80	2.80	4.50	3.50	2.15	2.02	2.10	1.95	2.20	1.50
	2.15	2.00	1.80	2.75	4.35	3.47	2.10	2.00	2.00	2.00	2.00	1.60
	2.25	2.05	2.10	3.30	4.17	3.52	2.00	2.00	2.15	1.95	2.00	1.90
	2.25	2.25	1.90	3.52	4.25	3.47	2.00	2.00	2.10	1.95	1.90	2.02
	2.00	1.80	2.00	3.80	4.15	3.52	2.20	2.00	2.12	2.05	1.95	1.90
	2.10	1.95	2.00	3.30	4.15	3.50	2.08	2.00	2.12	2.15	2.20	1.90
	1.90	2.10	2.00	3.25	4.30	3.37	2.00	2.00	2.08	2.05	3.10	1.90
	1.90	2.25	1.85	3.00	4.40	3.37	2.00	2.00	2.10	2.05	4.10	2.05
	1.90	2.20	2.00	3.00	4.22	3.20	2.10	2.00	2.10	2.10	4.05	1.90
	1.60	2.05	2.00	3.25	4.07	2.86	2.00	2.00	2.20	2.12	3.65	1.90
	1.85	2.00	1.95	3.05	4.12	2.75	2.00	2.00	2.08	2.15	2.40	2.00
	1.90	2.20	1.95	3.00	3.72	2.72	2.00	2.00	2.20	2.05	2.00	1.95
	1.50	2.20	1.95	3.00	3.50	2.75	2.05	2.00	2.05	2.20	1.95	2.02
	1.75	2.00	1.90	3.00	3.00	2.80	1.98	2.00	2.00	2.30	2.10	1.95
	1.95	2.00	2.00	3.00	3.10	2.70	2.00	2.10	2.00	2.17	2.95	1.95
	1.90	2.10	2.10	3.20	3.12	2.72	2.02	2.08	2.05	2.10	4.80	2.00
	1.95	2.10	2.10	3.30	3.15	2.82	2.25	2.10	2.00	2.00	3.80	1.80
	1.87	2.00	2.15	3.50	3.05	2.70	2.00	2.10	2.00	2.18	3.30	1.80
	2.55	1.75	2.10	3.80	2.87	2.48	2.00	2.10	2.02	2.25	3.15	1.75
	2.90	1.55	2.15	4.10	2.87	2.78	2.00	2.02	2.00	2.22	2.55	1.80

Mean daily gage height, in feet, of Truckee River at Nevada-California State line, near Mystic, Cal., for 1903—Continued.

Day.	Jan.	Feb.	Mar.	Apr.	May.	June.	July.	Aug.	Sept.	Oct.	Nov.	Dec.
26	2.30	1.45	2.15	4.00	2.97	2.55	2.00	2.02	1.90	2.18	2.30	1.70
27	2.65	1.65	2.30	3.67	3.00	2.82	2.00	2.15	1.90	2.10	2.30	1.70
28	2.30	1.70	3.85	3.47	3.05	2.60	2.00	2.30	1.90	2.00	2.30	1.60
29	2.55		4.05	3.42	3.30	2.80	2.00	2.20	2.00	2.05	2.10	1.60
30	2.25		5.15	3.80	3.62	2.75	2.00	2.15	2.00	2.15	2.00	1.70
31	1.90		4.75		2.97		2.00	2.25		2.10		1.50

Rating table for Truckee River at Nevada-California State line, near Mystic, Cal., for 1903.

Gage height.	Discharge.	Gage height.	Discharge.	Gage height.	Discharge.	Gage height.	Discharge.
Feet.	*Second-feet.*	*Feet.*	*Second-feet.*	*Feet.*	*Second-feet.*	*Feet.*	*Second-feet.*
1.3	247	2.3	613	3.3	1,268	4.3	2,173
1.4	268	2.4	671	3.4	1,345	4.4	2,280
1.5	292	2.5	732	3.5	1,425	4.5	2,391
1.6	319	2.6	795	3.6	1,508	4.6	2,505
1.7	349	2.7	859	3.7	1,594	4.8	2,745
1.8	383	2.8	924	3.8	1,683	5.0	3,002
1.9	421	2.9	989	3.9	1,775	5.2	3,283
2.0	463	3.0	1,055	4.0	1,870	5.4	3,506
2.1	509	3.1	1,123	4.1	1,968	5.6	3,953
2.2	559	3.2	1,194	4.2	2,069	5.8	4,370

Estimated monthly discharge of Truckee River at Nevada-California State line, near Mystic Cal., for 1903.

[Drainage area, 955 square miles.]

Month.	Discharge in second-feet.			Total in acre-feet.	Run-off.	
	Maximum.	Minimum.	Mean.		Second-feet per square mile.	Depth in inches.
January	989	292	522	32,097	0.55	0.63
February	586	280	463	25,714	.49	.51
March	3,211	334	686	42,180	.72	.83
April	1,968	891	1,301	77,415	1.36	1.52
May	2,391	969	1,658	101,946	1.74	2.01
June	1,822	720	1,148	68,311	1.20	1.34
July	846	455	513	31,543	.54	.62
August	613	463	490	30,129	.51	.59
September	559	421	489	29,098	.51	.57
October	613	442	507	31,174	.53	.61
November	2,745	421	855	50,876	.90	1.00
December [a]	486	292	403	24,780	.42	.48
The year	3,211	280	753	545,263	.79	10.71

[a] December 4, estimated.

LITTLE TRUCKEE RIVER NEAR PINE STATION, CAL.

The station was established June 25, 1903, by George B. Lorenz. ıe cable station is located about one-fourth of a mile upstream from ˙uhn's mill, known also as Pine˙ station. The dam at this point cks the water up for about 300 yards. The road which follows the ttle Truckee north from Boca crosses the river at a ford about 150 et below the cable and the same distance above the gage rod. Pine ation, on the Boca and Loyalton Railroad, about one-fourth of a ıle distant, is the nearest railroad point. The gage rod is located out 300 feet below the cable and consists of a vertical 2 by 6 inch ard graduated to feet and tenths from zero to 4 feet. The foot arks consist of 2-inch brass numbers and the tenths are marked with pper staples. The gage is read twice each day by W. R. Noyes. ıscharge measurements are made from a wooden car running on a inch steel cable which is stretched across the river about 5¼ feet ıove the high-water level. A tag wire is stretched just above the ˙ble. The initial point for soundings is the post which supports the ˙ble on the left or north bank. Above the station the channel is raight for 100 feet at high water and for 30 feet at low water. elow the station the channel is straight for 200 feet at high and low ˙ater. The current is swift at all stages. The right or south bank � high, steep, and rocky, and is not subject to overflow. It supports ome underbrush and a few trees. The left or north bank is com- ˙osed of gravel for 40 feet back from the high-water channel. It is ow and may be overflowed. The bed of the stream consists of coarse ˙ravel and is free from vegetation. There is but one channel at all ˙tages.

The bench mark is the head of a large nail driven into a 2 by 4 inch ˙imber set in the ground in a clump of small willows, 8 feet from the ˙age. The willows surrounding the bench mark are blazed and marked "B. M. U. S. G. S." Its elevation is 4.83 feet above the zero of the ˙age.

The observations at this station during 1903 have been made under ˙he direction of A. E. Chandler, district hydrographer.

Discharge measurements of Little Truckee River near Pine station, Cal., in 1903.

Date.	·	Hydrographer.	Gage height.	Discharge.
			Feet.	*Second-feet.*
July 17		G. B. Lorenz	1.20	50
July 30		do	1.00	29
August 14		do	.84	19
October 9		do	.80	20

Mean daily gage height, in feet, of Little Truckee River near Pine station, Cal., for 1903.

Day.	June.	July.	Aug.	Sept.	Oct.	Nov.	Dec.
1		2.15	1.00	0.79		0.98	1.6
2		2.15	.99	.79		.92	1.4
3		2.10	.95	.76		.92	1.4
4		2.10	.95	.76		.96	1.3
5		2.10	.95	.74		1.06	1.3
6		2.10	.95	.76		1.03	1.3
7		2.00	.95	.78		1.03	1.3
8		1.60	.94	.79		1.03	1.3
9		1.59	.89	.78		.94	1.3
10		1.55	.94	.79		1.00	1.3
11		1.48	.91	.79		1.02	1.3
12		1.42	.91	.79		1.42	1.7
13		1.45	.92	.76		2.18	1.3
14		1.42	.91	.82		2.75	1.3
15		1.41	.85	.86		2.25	1.3
16		1.38	.84	.82		1.82	1.3
17		1.25	.82	.81		1.09	1.7
18		1.25	.79	.80		1.56	1.3
19		1.18	.79	.80		1.50	1.3
20		1.15	.81	.80		1.86	1.3
21		1.05	.81			2.54	1.3
22		1.05	.82			2.58	1.3
23		1.05	.79			2.45	1.3
24		1.10	.79			2.35	1.3
25	2.06	1.05	.81		1.20	1.99	1.3
26	2.05	1.09	.79		1.20	1.82	1.3
27	2.15	1.08	.79		1.18	1.68	1.3
28	2.10	1.02	.81		1.12	1.64	1.3
29	2.10	1.04	.79		1.00	1.56	1.3
30	2.10	1.02	.79		.98	1.52	1.2
31		1.02	.81		.95		1.3

Rating table for Little Truckee River near Pine station, Cal., from June 25 to December 31, 1903.

Gage height.	Discharge.	Gage height.	Discharge.	Gage height.	Discharge.	Gage height.	Discharge.
Feet.	*Second-feet.*	*Feet.*	*Second-feet.*	*Feet.*	*Second-feet.*	*Feet.*	*Second-feet.*
0.7	14	1.3	64	1.9	160	2.5	256
.8	18	1.4	80	2.0	176	2.6	272
.9	23	1.5	96	2.1	192	2.7	288
1.0	29	1.6	112	2.2	208	2.8	304
1.1	38	1.7	128	2.3	224		
1.2	50	1.8	144	2.4	240		

Table only approximate beyond 1.20 feet gage height.   Not well defined below 1 foot gage height.

Estimated monthly discharge of Little Truckee River near Pine station, Cal.

[Drainage area, 166 square miles.]

Month.	Discharge in second-feet.			Total in acre-feet.	Run-off.	
	Maximum.	Minimum.	Mean.		Second-feet per square mile.	Depth in inches.
June 25–30	200	184	191	2,273	1.15	0.26
July	200	31	89	5,472	.54	.62
August	29	18	22	1,353	.13	.15
September 1–20	....	....	18	714	.11	.08
October 25–31	....	....	39	541	.23	.06
November	296	25	117	6,962	.70	.78
December	94	45	63	3,874	.38	.44

INDEPENDENCE CREEK BELOW INDEPENDENCE LAKE, CALIFORNIA.

This station was established October 24, 1902, by E. C. Murphy. It is located about one-eighth of a mile below the dam at the lower end of Independence Lake, California. The gage is a vertical 4 by 4 inch timber, and is located about 200 feet upstream from the cable and car from which the measurements are made. The gage is read twice daily by C. J. Gepper. The channel is straight for 50 feet above and below the station. The banks are low, but not subject to overflow. The bed of the stream is permanent, being composed of gravel and clay. The current is moderately rapid, and the discharge is controlled by the dam at the end of the lake. The bench mark is three nails driven in the root of a pine tree 55 feet northwest of the gage. Its elevation is 5.82 feet above the zero of the gage.

The observations at this station during 1903 have been made under the direction of A. E. Chandler, district hydrographer.

Discharge measurements of Independence Creek below Independence Lake, California, in 1902 and 1903.

Date.	Hydrographer.	Gage height.	Discharge.
1902.		*Feet.*	*Second-fect.*
October 24	E. C. Murphy	1.23	0.3
1903.			
June 21	G. B. Lorenz	1.90	15.2
July 19	......do	1.60	7.3
August 6	......do	1.50	4.7
August 27	......do	1.40	1.0

Mean daily gage height, in feet, of Independence Creek below Independence Lake, California, for 1902.

Day.	Oct.	Nov.	Dec.	Day.	Oct.	Nov.	Dec.	Day.	Oct.	Nov.	Dec.
1...............		1.33	2.20	12...............		1.93	1.60	23...............		1.93	1.60
2...............		1.33	2.20	13...............		1.73	1.60	24...............	1.23	1.93	1.55
3...............		1.33	2.00	14...............			1.60	25...............	1.23	1.93	1.55
4...............		1.23	1.95	15...............			1.50	26...............	1.23	1.93	1.50
5...............		1.23	1.90	16...............		1.93	1.50	27...............	1.13	1.93	1.50
6...............		1.13	1.90	17...............		1.93	1.50	28...............	1.23	1.93	1.75
7...............		1.23	1.80	18...............		1.93	1.50	29...............	1.23	1.93	1.55
8...............		1.33	1.80	19...............		1.93	1.50	30...............	1.23	1.90	1.53
9...............		1.18	1.80	20...............		1.93	1.50	31...............	1.33		1.50
10...............		1.93	1.60	21...............		1.93	1.50				
11...............		1.98	1.60	22...............		1.93	1.60				

Mean daily gage height, in feet, of Independence Creek below Independence Lake, California, for 1903.

Day.	Jan.	Feb.	Mar.	Apr.	May.	June.	July.	Aug.	Sept.	Oct.	Nov.	Dec.
1...............	1.70	2.45	2.00	2.00	2.80		2.98			1.38	1.44	1.90
2...............	1.70	2.50	2.00	2.00	2.80		2.88	1.51		1.38	1.43	1.90
3...............	1.75	2.50	2.00	2.00	2.80		2.75	1.51		1.37	1.42	1.90
4...............	1.85	2.50	2.00	2.00	2.80		2.56	1.50		1.35	1.50	1.90
5...............	1.90	2.50	2.10	2.00	2.80		2.45	1.50		1.40	1.50	1.90
6...............	1.95	2.50	2.10	2.00	2.95		2.25	1.50		1.39	1.50	1.90
7...............	2.00	2.50	2.10	2.00	3.00		2.00	1.50		1.39	1.50	1.90
8...............	2.10	2.50	2.10	2.00	3.00		1.75	1.50		1.39	1.49	1.88
9...............	2.20	2.50	2.10	1.90	3.00		1.58	1.50		1.38	1.48	1.86
10...............	2.20	2.50	2.10	1.90	3.00		1.50	1.50		1.38	1.48	1.85
11...............	2.20	2.50	2.10	1.90	3.00		1.50	1.50		1.37	1.48	1.85
12...............	1.95	2.50	2.10	1.90	3.00		1.50	1.50	1.35	1.39	1.63	1.84
13...............	1.80	2.50	2.10	1.90	3.10		1.50	1.50	1.35	1.41	2.96	1.83
14...............	1.60	2.50	2.20	1.90	3.05		1.50	1.50	1.34	1.41		1.82
15...............	1.60	2.40	2.20	2.00	3.05	3.08	1.61	1.50	1.34	1.41	2.91	1.82
16...............	1.65	2.30	2.20	2.00	3.05	2.85	1.56	1.50	1.34	1.41	2.50	1.85
17...............	1.75	2.25	2.70	2.00	2.85	2.70	1.60	1.45	1.34	1.46	2.21	1.85
18...............	1.80	2.20	2.10	2.10	2.80	1.90	1.60	1.45	1.33	1.50	2.12	1.85
19...............	1.80	2.20	2.00	2.10	2.75	1.70	1.60	1.45	1.32	1.50	1.98	1.85
20...............	1.90	2.20	2.00	2.10	2.75	1.78	1.60	1.45	1.31	1.50	2.25	1.84
21...............	1.90	2.10	2.00	2.10	2.65	1.90	1.60	1.45	1.35	1.50	2.65	1.83
22...............	1.90	2.10	2.00	2.10	2.60	1.92	1.54	1.42	1.39	1.50	2.70	1.82
23...............	1.90	2.10	2.00	2.10	2.60	2.01	1.50		1.39	1.50	2.80	1.57
24...............	2.15	2.10	2.00	2.15	2.50	2.12	1.50		1.39	1.50	2.85	1.50
25...............	2.20	2.10	2.00	2.25	2.50	2.18			1.39	2.25	2.42	1.50
26...............	2.20	2.00	2.00	2.35	2.55	2.29			1.39	2.20	2.26	1.50
27...............	2.20	2.00	2.00	2.45	2.50	2.38			1.39	1.89	2.10	1.50
28...............	2.30	2.00	2.00	2.70	2.50	2.40			1.39	1.60	1.95	1.50
29...............	2.35		2.00	2.80	2.50	2.48			1.39	1.52	1.90	1.50
30...............	2.40		2.00	2.80	2.50	2.95			1.39	1.45	1.90	1.50
31...............	2.40		2.00							1.45		1.50

Rating table for Independence Creek below Independence Lake, California, from October 24, 1902, to December 31, 1903.

Gage height.	Discharge.	Gage height.	Discharge.	Gage height.	Discharge.	Gage height.	Discharge.
Feet.	*Second-feet.*	*Feet.*	*Second-feet.*	*Feet.*	*Second-feet.*	*Feet.*	*Second-feet.*
1.1	0.1	1.6	6.9	2.1	21	2.6	36
1.2	.3	1.7	9.5	2.2	24	2.7	39
1.3	.8	1.8	12.3	2.3	27	2.8	42
1.4	2.2	1.9	15.2	2.4	30	2.9	45
1.5	4.5	2.0	18.0	2.5	33	3.0	48

Remarks: Curve fairly well defined.   Extended above gage height 1.90 feet.

Estimated monthly discharge of Independence Creek below Independence Lake, California, for 1902 and 1903.

[Drainage area 8.5 square miles.]

Month.	Discharge in second-feet.			Total in acre-feet.	Run-off.	
	Maximum.	Minimum.	Mean.		Second-feet per square mile.	Depth in inches.
1902.						
October, 24–31	1.4	0.2	0.5	8	0.06	0.02
November, 28 days..	....	....	11	611	1.29	1.34
December	24	4.5	10	615	1.18	1.36
1903.						
January	30	6.9	18	1,107	2.12	2.44
February	33	18	28	1,555	3.29	3.43
March..............	24	18	20	1,230	2.35	2.71
April..............	42	15.2	22	1,309	2.59	2.89
May, 1–30	....	....	42	2,499	4.94	5.51
June, 15–30.........	....	....	27	857	3.18	1.89
July, 1–24	....	....	15	714	1.76	1.57
August, 2–22	....	....	4	167	.47	.37
September, 12–30 ...	....	....	1.6	60	.19	.13
October	26	1.4	5	307	.59	.68
November	47	2.7	20	1,190	2.35	2.62
December	15	4.5	11	676	1.29	1.49

PROSSER CREEK NEAR BOCA, CAL.

This station was established October 23, 1902, by E. C. Murphy. It is located about 2 miles from Boca, Cal., and about 500 feet below the dam of the Prosser Creek Ice Company, about one-eighth of a mile above the mouth of the creek, at the footbridge between two ice houses. The gage, which is a vertical staff graduated to feet and tenths, is located about 10 feet below the bridge. It is read twice daily by C. Lindsley. The channel is straight both above and below the station. The banks are high. The bed of the stream is permanent and is composed of sandy gravel and stone. The bench mark is a spike driven into the timber to which the gage is fastened. Its elevation is 6 feet above the zero of the gage. This station was discontinued June 27, 1903, at which time the station near Hobart Mills was established.

The observations at this station during 1903 have been made under the direction of A. E. Chandler, district hydrographer.

Mean daily gage height, in feet, of Prosser Creek near Boca, Cal., for 1903.

Day.	Jan.	Feb.	Apr.	May.	June.	Day.	Jan.	Feb.	Apr.	May.	June
1	0.55	0.65	....	2.32	2.28	17	0.50	....	....	1.40	....
2	.60	.50	....	2.35	1.98	18	.60	....	....	1.30	....
3	.65	.55	....	2.35	1.75	19	.63	....	....	1.18	....
4	.55	.55	....	2.35	1.68	20	.65	....	....	1.18	....
5	.47	.50	....	2.40	1.92	21	.78	....	1.30	1.15	....
6	.72	.55	....	2.55	1.95	22	.80	....	1.30	1.10	....
7	.96	.55	....	2.62	....	23	1.65	....	2.07	1.10	....
8	.82	.60	....	2.62	....	24	2.15	....	2.25	1.10	....
9	.75	.60	....	2.62	....	25	2.15	....	2.30	1.10	....
10	.87	.55	....	2.50	....	26	1.10	....	2.35	1.18	....
11	....	.55	....	2.60	....	27	1.05	....	2.05	1.10	....
12	....	....	....	2.58	....	28	.90	....	2.04	1.15	....
13	....	....	....	2.50	....	29	.85	....	2.07	1.20	....
14	.55	....	....	2.55	....	30	.80	....	2.11	2.15	....
15	.55	....	....	2.42	....	31	.85	....	....	2.30	....
16	.65	....	....	2.40	....						

PROSSER CREEK NEAR HOBART MILLS, CAL.

This station was established June 27, 1903, by George B. Lorenz. It is located about 150 feet below the wagon bridge and the road house on the Boca and Truckee highway, 4 miles from Truckee and 3 miles from Hobart Mills. Alder Creek joins Prosser Creek about 200 feet above the station. The nearest dam is located at the ice pond, about 2 miles below the station. The gage consists of a 2 by 4 inch vertical timber graduated to feet and tenths by means of brass figures and copper staples, and reading from zero to 4 feet gage height. It is located on the left, or north, bank 50 feet upstream from the cable. Readings are taken twice each day by Robert Dorian. Discharge measurements are made from a wooden car running on a ⅜-inch steel cable, which is stretched across the river about 5 feet above high-water level. A tag wire is stretched just above the cable. The channel is straight for 150 feet above and 200 feet below the cable. The creek makes a bend at the highway bridge. Both banks are high and not subject to overflow. The right bank is wooded; the left is cleared, except for a few willows. The bed is composed of coarse gravel and is free from vegetation. The water flows in one channel at all stages. The bench mark is on the northeast pier of the steel wagon bridge and is the top of the steel casing of the concrete pier above the arrow mark. Its elevation is 9.87 feet above the zero of the gage. The drainage area is 48 square miles.

The observations at this station during 1903 have been made under the direction of A. E. Chandler, district hydrographer.

Discharge measurements of Prosser Creek near Hobart Mills, Cal., in 1903.

Date.	Hydrographer.	Gage height.	Discharge.
		Feet.	*Second-feet.*
July 18	G. B. Lorenz	2.5	41
July 31	do	2.3	28
August 6	do	2.3	20
August 14	do	2.18	15
August 27	do	2.10	10

Mean daily gage height, in feet, of Prosser Creek near Hobart Mills, Cal., for 1903.

Day.	June.	July.	Aug.	Sept.	Oct.	Nov.	Dec.
1		2.75	2.35	2.15	2.15		2.65
2		2.72	2.35	2.12	2.15		2.65
3		2.72	2.30	2.12	2.18		2.55
4		2.62	2.30	2.12	2.18		2.75
5		2.65	2.30	2.12	2.18		2.95
6		2.55	2.30	2.12	2.15		2.85
7		2.55	2.30	2.12	2.15		2.80
8		2.55	2.27	2.12	2.15		2.65
9		2.50	2.25	2.12	2.18		2.60
10		2.48	2.25	2.12	2.30		2.75
11		2.50	2.25	2.12			2.65
12		2.55	2.22	2.12		3.40	2.70
13		2.48	2.22	2.12		3.65	2.55
14		2.45	2.22	2.12		3.70	2.75
15		2.48	2.22	2.12		3.45	2.65
16		2.48	2.22	2.12		3.20	2.55
17		2.50	2.18	2.12		2.60	2.60
18		2.48	2.18	2.12		2.50	2.55
19		2.42	2.18	2.12		2.55	2.55
20		2.45	2.18	2.18		2.95	2.55
21		2.45	2.18	2.18		3.85	2.65
22		2.42	2.18	2.18		3.65	2.70
23		2.40	2.18	2.18		3.30	2.90
24		2.38	2.20	2.18		2.90	3.00
25		2.38	2.18	2.18		2.75	2.90
26		2.35	2.18	2.18		2.70	2.85
27		2.35	2.18	2.12		2.60	2.85
28	2.75	2.35	2.18	2.12		2.65	2.80
29	2.75	2.35	2.18	2.12		2.65	2.75
30	2.95	2.32	2.18	2.12		2.70	2.95
31	2.78	2.35	2.18				2.90

DONNER CREEK NEAR TRUCKEE, CAL.

This station was established October 23, 1902, by E. C. Murphy. It is located about 150 feet below the dam of the Donner Creek Ice Company and 1¼ miles west of Truckee, Cal. The gage is a 4 by 4 inch vertical timber located on the left bank. It is read twice each day by F. R. Williams. Discharge measurements are made by means of a cable and car 50 feet downstream from the gage. The channel is straight for 150 feet above and below the cable. Both banks are high and will not overflow. The bed of the stream is composed of gravel, permanent and free from vegetation. There is but one channel at all stages and the current is swift. The bench mark consists of 4 nails driven into the root of a pine tree 12 inches in diameter and 35 feet north of the gage. Its elevation is 6.30 feet above the zero of the gage.

The observations at this station during 1903 have been made under the direction of A. E. Chandler, district hydrographer.

Discharge measurements of Donner Creek near Truckee, Cal., in 1903.

Date.	Hydrographer.	Gage height.	Discharge
		Feet.	Second-feet.
June 21	G. B. Lorenz	2.80	108
July 20	......do	1.80	11
July 31	......do	1.60	5
August 31	......do	1.40	1

Mean daily gage height, in feet, of Donner Creek near Truckee, Cal., for 1903.

Day.	Jan.	Feb.	Mar.	Apr.	May.	June.	July.	Aug.	Sept.	Oct.	Nov.	Dec.
1	2.30	2.50	2.40	3.35	3.65	4.00	2.40	1.65	1.45	1.40	1.45	2.45
2	2.35	2.50	2.40	3.30	3.75	3.80	2.40	1.65	1.45	1.40	1.45	2.45
3	2.30	2.50	2.40	3.10	3.80	3.70	2.40	1.60	1.40	2.10	1.45	2.45
4	2.30	2.50	2.40	3.10	4.05	3.60	2.40	1.60	1.40	1.65	1.48	2.40
5	2.30	2.50	2.40	3.15	4.00	3.50	2.40	1.40	1.45	1.50	1.9	
6	2.30	2.50	2.40	3.00	4.15	3.50	2.30	1.55	1.40	1.40	1.40	1.7
7	2.30	2.50	2.40	3.05	4.05	3.50	2.30	1.60	1.40	1.40	1.58	1.9
8	2.30	2.50	2.40	3.10	4.05	3.40	2.30	1.60	1.40	1.40	2.10	1.9
9	2.30	2.50	2.40	3.30	4.20	3.40	2.15	1.60	1.40	1.45	2.10	1.9
10	2.40	2.50	2.40	3.30	4.20	3.35	2.10	1.60	1.40	1.50	2.00	1.9
11	2.40	2.40	2.40	3.20	4.15	3.35	2.10	1.50	1.40	1.50	2.00	1.8
12	2.35	2.40	2.40	3.05	4.30	3.30	1.90	1.90	1.40	1.45	2.90	1.8
13	2.35	2.40	2.40	3.00	4.35	3.30	1.90	1.80	1.40	1.45	3.20	1.8
14	2.30	2.40	2.40	3.10	3.95	3.10	1.90	1.80	1.40	1.45	3.10	1.8
15	2.25	2.40	2.40	3.00	3.60	2.95	1.90	1.40	1.40	1.50	2.90	1.8
16	2.25	2.40	2.40	2.90	4.10	3.15	1.90	1.40	1.40	1.50	2.70	1.8
17	2.20	2.40	2.40	2.90	3.95	2.90	1.90	1.40	1.40	1.55	2.75	1.8
18	2.20	2.40	2.40	2.90	3.80	3.00	1.85	1.40	1.40	1.50	2.90	1.7

m daily gage height, in feet, of Donner Creek near Truckee, Cal., for 1903—Continued.

Day.	Jan.	Feb.	Mar.	Apr.	May.	June.	July.	Aug.	Sept.	Oct.	Nov.	Dec.
....	2.15	2.40	2.40	2.90	3.40	2.85	1.80	1.40	1.40	1.50	2.80	1.70
....	2.20	2.40	2.40	2.90	3.35	2.90	1.80	1.30	1.40	1.50	3.00	1.70.
....	2.25	2.40	2.50	3.00	3.40	2.90	1.80	1.30	1.40	1.45	3.90	1.70
....	2.20	2.40	2.80	3.15	3.70	2.80	1.80	1.30	1.40	1.45	3.55	1.70
....	2.50	2.40	2.80	3.20	3.45	2.70	1.80	1.30	1.40	1.45	3.45	1.70
....	2.70	2.40	2.80	3.35	3.50	2.75	1.70	1.35	1.40	1.45	3.25	1.70
....	2.80	2.40	2.75	3.60	3.50	2.95	1.70	1.35	1.40	1.52	3.05	1.70
....	2.50	2.40	2.75	3.60	3.50	2.75	1.65	1.40	1.40	1.45	3.00	1.70
....	2.50	2.40	2.40	3.50	3.45	2.60	1.65	1.40	1.40	1.45	2.90	1.70
....	2.70	2.40	3.00	3.50	3.45	2.45	1.65	1.40	1.40	1.45	2.85	1.70
....	2.70	......	3.00	3.00	3.65	2.45	1.65	1.40	1.40	1.45	2.25	1.70
....	2.60	......	4.10	3.00	3.85	2.40	1.65	1.40	1.40	1.45	2.70	1.70
....	2.50	......	3.55	......	4.05	......	1.70	1.40	......	1.45	......	1.70

ting table for Donner Creek near Truckee, Cal., from October 23, 1902, to December 31, 1903.

Gage height.	Discharge.	Gage height.	Discharge.	Gage height.	Discharge.	Gage height.	Discharge.
Feet.	*Second-feet.*	*Feet.*	*Second-feet.*	*Feet.*	*Second-feet.*	*Feet.*	*Second-feet.*
1.3	1	2.1	28	2.9	120	3.7	248
1.4	1	2.2	36	3.0	136	3.8	264
1.5	3	2.3	46	3.1	152	3.9	280
1.6	5	2.4	56	3.2	168	4.0	296
1.7	7	2.5	67	3.3	184	4.1	312
1.8	11	2.6	79	3.4	200	4.2	328
1.9	16	2.7	92	3.5	216	4.3	344
2.0	22	2.8	106	3.6	232	4.4	360

Table not well defined.   Curve extended above 2.80 feet gage height.

Estimated monthly discharge of Donner Creek near Truckee, Cal., for 1903.

[Drainage area, 30 square miles.]

Month.	Discharge in second-feet.			Total in acre-feet.	Run-off.	
	Maximum.	Minimum.	Mean.		Second-feet per square mile.	Depth in inches
1902.						
October 23–31	11	5	7	125	0.23	0.10
November	36	5	14	833	.47	.52
December	51	9	17	1,045	.57	.66
1903.						
January	106	32	54	3,320	1.80	2.08
February	67	56	60	3,332	2.00	2.08
March	312	56	83	5,103	2.77	3.19
April	232	120	166	9,878	5.53	6.17
May	352	192	267	16,417	8.90	10.26
June	296	56	156	9,283	5.20	5.80
July	56	6	23	1,414	.77	.89
August	16	1	3	184	.10	.12
September	2	1	1	60	.03	.03
October	28	1	3	184	.10	.12
November	280	1	92	5,474	3.07	3.43
December	61	7	16	984	.53	.61
The year	352	1	77	55,633	2.57	34.73

SUSAN RIVER NEAR SUSANVILLE, CAL.

This station was established June 3, 1900, by L. H. Taylor, at which time a temporary gage was installed on the right bank. It is located about three-fourths mile southwest of Susanville, at the electric-light plant. A short distance above the station a small irrigating ditch, known as the "Masten ditch," is taken out on the right bank. There is a flume near its head in which a gage has been placed. On December 20, 1903, the station at the electric-light plant was reestablished by H. E. Green. A cable was installed for high-water measurements. The initial point for soundings is a post in the fence in line with the cable. The cable support on the left bank is 34.8 feet from the initial point for soundings. The channel is straight for 150 feet above and for 250 feet below the cable. The current is swift. There is a riffle immediately above the cable. The right bank is high and is composed of clay covered with vegetation. It is not liable to overflow. The left bank is low, liable to overflow, and is covered with a sparse growth of willows.

'he bed of the stream is composed of gravel and cobblestones and is
ermanent. Bench mark No. 1 is a nail in the fence post which is used
ᵴ the initial point for soundings. Its elevation is 11.35 feet above
he zero of the gage. Bench mark No. 2 is a nail in the cable post.
ts elevation is 9 feet above the zero of the gage. Bench mark No.
is a nail in the cottonwood tree to which the cable is attached. Its
levation is 10 feet above the zero of the gage. The gage is read once
ach day by James Branham.

The observations at this station during 1903 have been made under
he direction of A. E. Chandler, district hydrographer.

The following discharge measurement was made by G. B. Lorenz in
.903:

> July 9: Gage height, 190 feet; discharge, 7.60 second-feet.

The Masten ditch diverts water from Susan River above this station.
On July 9 it was discharging 7.1 second-feet.

Mean daily gage height, in feet, of Susan River near Susanville, Cal., for 1903.

Day.	Jan.	Feb.	Mar.	Apr.	May.	June.	July.	Aug.	Sept.	Oct.	Nov.	Dec.
1	3.70		2.40	5.30	4.75	3.20	2.10	1.60	1.70	1.85	2.05	3.02
2	3.55		2.40	4.75	4.75	3.20	2.05	1.60	1.70	1.85	2.05	2.92
3	3.50		2.40	4.50	4.75	3.15	2.05	1.60	1.70	1.90	2.05	2.82
4	3.40		2.40	4.45	4.75	3.15	2.00	1.60	1.75	1.90	2.40	2.72
5	3.30		2.40	4.30	4.67	3.10	2.00	1.60	1.75	1.90	2.40	2.62
6	3.30		2.40	4.15	4.65	3.10	1.95	1.65	1.75	1.90	2.20	2.50
7	3.30		2.40	4.15	4.55	3.05	1.95	1.65	1.75	1.95	2.10	2.45
8	2.90		2.40	4.30	4.47	3.05	1.90	1.65	1.75	1.95	2.10	2.45
9	2.80		2.45	4.45	4.00	3.00	1.90	1.65	1.80	2.00	2.10	2.42
10	2.70		2.45	4.60	4.17	3.00	1.85	1.65	1.80	2.05	2.10	2.40
11	2.60		2.60	4.45	4.35	2.95	1.85	1.65	1.80	2.05	2.10	2.38
12	2.60		3.00	4.35	4.30	2.95	1.85	1.65	1.80	2.05	3.55	2.32
13	2.50		3.15	4.35	4.27	2.87	1.85	1.65	1.85	2.05	3.10	2.35
14	2.50		3.25	4.25	4.20	2.80	1.75	1.65	1.85	2.05	5.60	2.35
15	2.40		3.20	4.22	4.12	2.72	1.75	1.65	1.85	2.05	3.30	2.35
16	2.40		3.15	4.20	4.08	2.62	1.75	1.65	1.85	2.05	2.90	2.60
17	2.40		3.07	4.20	3.97	2.52	1.75	1.65	1.85	2.05	2.58	2.65
18	2.40		2.92	4.20	3.92	2.48	1.75	1.65	1.85	2.05	2.55	2.55
19	2.40		2.87	4.20	3.87	2.45	1.75	1.65	1.85	2.05	2.50	2.50
20	2.40		2.90	4.30	3.77	2.42	1.70	1.65	1.85	2.05	4.10	2.45
21	2.45		3.10	4.32	3.80	2.40	1.70	1.65	1.85	2.05	4.95	2.45
22	2.52		3.20	4.50	3.65	2.35	1.65	1.65	1.85	2.05	4.35	2.45
23	2.85		3.60	4.62	3.57	2.30	1.65	1.65	1.85	2.05	4.25	2.40
24	3.95		3.50	4.72	3.52	2.25	1.65	1.65	1.85	2.05	3.95	2.40
25	4.55		4.00	4.80	3.45	2.25	1.65	1.70	1.85	2.05	3.75	2.35
26	3.65		4.18	4.95	3.40	2.20	1.65	1.70	1.85	2.05	3.55	2.35
27	3.05		4.10	4.77	3.35	2.15	1.65	1.70	1.85	2.05	3.35	2.35
28	2.80		4.85	4.70	3.32	2.15	1.65	1.70	1.85	2.05	3.30	2.35
29	2.70		5.00	4.67	3.30	2.15	1.65	1.70	1.85	2.05	3.22	2.30
30	2.60		8.00	4.67	3.25	2.10	1.60	1.70	1.85	2.05	3.12	2.30
31	2.50		6.00	..	3.25		1.60	1.70		2.05		2.30

Gage read from surface of ice during February.

OWENS RIVER DRAINAGE BASIN.

During the summer and autumn of 1903 an investigation of the flow of the waters which drain into Owens River was begun by J. C. Clausen, assisted by R. S. Hawley. Twenty gaging stations were established on streams and canals in this section. The results of the data collected at these stations are shown in the following pages.

OWENS RIVER NEAR ROUND VALLEY, CAL.

This station was established by J. C. Clausen, assisted by R. S. Hawley, on August 3, 1903. It is located at the footbridge, 700 feet above the junction of Owens River and Rock Creek. The river at this point cuts through a lava deposit about 100 feet thick and forms a gorge which is about 250 feet wide at the top. The gage is a 1 by 4 inch vertical rod, fastened to the concrete bridge abutment on the left bank. It is read once each day by Roscoe Jones. Discharge measurements are made from the single-span footbridge to which the gage is attached. The bridge is 37 feet long and has a clear span of 35 feet. The initial point for soundings is the anchor bolt of the right abutment. The channel is straight for 175 feet above and for 250 feet below the station. The current is swift at all stages. Both banks are high and rocky and are not liable to overflow. The bed of the stream is composed of rock and lava bowlders and is not subject to much change. The bench mark is a bolt set in a lava bowlder 97.4 feet north of the right abutment.

The observations at this station during 1903 have been made under the direction of S. G. Bennett, district hydrographer.

Discharge measurements of Owens River near Round Valley, Cal., in 1903.

Date.	Hydrographer.	Gage height.	Discharge.
		Feet.	*Second-feet.*
August 3	R. S. Hawley	1.80	161
August 9	...do	1.85	160
August 18	...do	1.75	159
August 22	...do	1.75	160
September 1	...do	1.74	159
September 12	...do	1.80	175
October 8	...do	1.85	185
October 25	...do	1.82	173
November 24	...do	1.88	196
December 29	...do	1.70	151

Mean daily gage height, in feet, of Owens River near Round Valley, Cal., for 1903.

Day.	Aug.	Sept.	Oct.	Nov.	Dec.	Day.	Aug.	Sept.	Oct.	Nov.	Dec.
....	....	1.75	1.77	1.80	1.80	17..................	1.80	1.70	1.80	1.70	1.80
....	....	1.75	1.80	1.80	1.80	18..................	1.75	1.80	1.80	1.70	1.80
....	....	1.75	1.80	1.80	1.80	19..................	1.75	1.80	1.80	1.70	1.80
....	1.80	1.73	1.83	1.80	1.70	20..................	1.75	1.80	1.80	1.70	1.80
....	1.80	1.70	1.80	1.80	1.70	21..................	1.68	1.80	1.80	1.70	1.80
....	1.80	1.72	1.80	1.75	1.70	22..................	1.72	1.80	1.80	1.70	1.80
....	1.80	1.73	1.80	1.70	1.70	23..................	1.80	1.80	1.80	1.80	1.80
....	1.80	1.75	1.82	1.70	1.70	24..................	1.75	1.78	1.80	1.90	1.80
....	1.85	1.77	1.80	1.70	1.70	25..................	1.80	1.77	1.80	1.90	1.80
....	1.80	1.80	1.83	1.70	1.70	26..................	1.80	1.80	1.80	1.80	1.80
....	1.80	1.80	1.80	1.70	1.70	27..................	1.78	1.80	1.80	1.80	1.70
....	1.90	1.80	1.80	1.70	1.70	28..................	1.72	1.80	1.80	1.80	1.70
....	1.90	1.80	1.80	1.70	1.70	29..................	1.70	1.80	1.80	1.80	1.70
....	1.90	1.80	1.82	1.70	1.70	30..................	1.70	1.78	1.80	1.80	1.70
....	1.80	1.80	1.80	1.70	1.70	31..................	1.70	......	1.70	......	1.70
....	1.80	1.80	1.80	1.70	1.80						

ROCK CREEK NEAR ROUND VALLEY, CAL.

This station was established August 3, 1903, by J. C. Clausen and L. S. Hawley. It is located at the wagon bridge on the road from Long Valley to Bishop, 3,500 feet above the mouth of the creek. The gage is a 1 by 4 inch vertical rod fastened to the left end of the bridge. It is read once each day by Roscoe Jones. Discharge measurements are made from a footbridge which has a span of 18 feet. The initial point for soundings is on right bank of stream. The distances across stream are marked on the footbridge at intervals of 2 feet. The channel is straight for 50 feet above and for 40 feet below the footbridge. The current is swift. Both banks are high and rocky and are not liable to overflow. The bed of the stream is composed of gravel and is not subject to much change. The bench mark is a point marked on lava rock 15 feet east of the left end of the footbridge. The observations at this station during 1903 have been made under the direction of S. G. Bennett, district hydrographer.

Discharge measurements of Rock Creek near Round Valley, Cal., in 1903.

Date.	Hydrographer.	Gage height.	Discharge.
		Feet.	*Second-feet.*
August 3..................	R. S. Hawley	1.55	20.0
August 8..................	.....do	1.62	21.0
August 18..................	.....do	1.55	19.5
September 1	.....do	1.42	16.8
September 10	.....do	1.49	17.5
November 24..................	.....do	1.51	18.3
December 29..................	.....do	1.45	18.1

Mean daily gage height, in feet, of Rock Creek near Round Valley, Cal., for 1903.

Day.	Aug.	Sept.	Oct.	Nov.	Dec.	Day.	Aug.	Sept.	Oct.	Nor.	Dec.
1		1.40	1.35	1.40	1.50	17	1.60	b1.00	1.30	1.50	1.5
2		1.45	1.30	1.40	1.50	18	1.55	1.50	1.30	1.50	1.5
3		1.40	1.30	1.40	1.50	19	1.50	1.50	1.30	1.50	1.5
4	a1.55	1.40	1.30	1.40	1.50	20	1.50	1.50	1.30	1.50	1.5
5	1.50	1.40	1.35	1.40	1.50	21	1.45	1.50	1.30	1.50	1.5
6	1.60	1.45	1.30	1.45	1.55	22	1.50	1.55	1.30	1.50	1.5
7	1.60	1.40	1.30	1.50	1.55	23	1.60	1.50	1.30	1.50	1.5
8	1.60	1.45	1.30	1.50	1.55	24	1.60	1.50	1.30	1.40	1.55
9	1.55	1.45	1.30	1.50	1.55	25	1.50	1.55	1.30	1.50	1.5
10	1.60	1.50	1.35	1.40	1.55	26	1.50	1.50	1.30	1.50	1.5
11	1.60	1.55	1.30	1.40	1.60	27	1.50	1.50	1.30	1.50	1.6
12	1.70	1.60	1.30	1.40	1.60	28	1.45	1.40	1.30	1.50	1.6
13	1.70	1.60	1.30	1.40	1.60	29	1.40	1.30	1.30	1.50	1.6
14	1.60	1.55	1.30	1.40	1.60	30	1.40	1.30	1.40	1.50	1.5
15	1.50	1.50	1.30	1.40	1.60	31	1.40		1.40		1.5
16	1.60	1.40	1.30	1.45	1.60						

a Given as 1.10, probably is 1.55.　　　　b Probably is 1.50.

PINE CREEK NEAR ROUND VALLEY, CAL.

This station was established August 3, 1903, by J. C. Clausen and R. S. Hawley. It is located 150 feet below the wagon bridge on the road from Bishop to Long Valley and 100 feet above the mouth of the creek. The gage is a 1 by 3 inch vertical rod fastened in the rocks near the right bank. It is read once each day by Roscoe Jones. Discharge measurements are made by wading. The initial point for soundings is a stake on the right bank of the stream.

The channel is straight for about 50 feet above and for about 100 feet below the station. The current has a velocity of about 1 foot per second at ordinary stages. Both banks are high and rocky and are not liable to overflow. The bed of the stream is rocky and permanent. The bench mark is the one at the Rock Creek station. It is a point marked on a lava rock 15 feet east of the left end of the footbridge.

The observations at this station during 1903 have been made under the direction of S. G. Bennett, district hydrographer.

ischarge measurements of Pine Creek near Round Valley, Cal., in 1903.

Date.	Hydrographer.	Gage height.	Discharge.
		Feet.	*Second-feet.*
....	R. S. Hawley	1.90	15.0
....	....do	1.87	10.4
l...................	....do	1.73	5.7
r 1................	....do	1.71	5.9
r 10................	....do	1.71	6.3
r 24................	....do	1.89	10.1

daily gage height, in feet, of Pine Creek near Round Valley, Cal., for 1903.

.	Aug.	Sept.	Oct.	Nov.	Dec.	Day.	Aug.	Sept.	Oct.	Nov.	Dec.
....		1.75	1.80	1.90	1.85	17............	1.80	1.70	1.90	1.90	1.90
....		1.75	1.83	1.90	1.85	18............	1.73	1.70	1.90	1.90	1.90
....		1.72	1.80	1.90	1.85	19............	1.70	1.70	1.93	1.90	1.90
....	1.90	1.70	1.80	1.90	1.90	20............	1.70	1.70	1.92	1.90	1.90
....	1.90	1.70	1.80	1.90	1.90	21............	1.70	1.70	1.90	1.80	1.90
....	1.90	1.72	1.80	1.85	1.95	22............	1.72	1.70	1.95	1.80	1.85
....	1.90	1.73	1.80	1.80	1.95	23............	1.80	1.73	1.90	1.85	1.85
....	1.90	1.70	1.82	1.90	1.95	24............	1.75	1.72	1.90	1.90	1.80
....	1.80	1.72	1.80	1.90	1.95	25............	1.80	1.70	1.90	1.90	1.85
....	1.80	1.75	1.80	1.90	1.95	26............	1.80	1.70	1.90	1.90	1.85
....	1.80	1.70	1.83	1.90	1.90	27............	1.75	1.70	1.90	1.85	1.90
....	1.90	1.70	1.80	1.90	1.90	28............	1.73	1.75	1.90	1.85	1.90
....	1.90	1.70	1.80	1.90	1.90	29............	1.70	1.80	1.90	1.85	1.90
....	1.80	1.70	1.82	1.90	1.90	30............	1.70	1.80	1.90	1.80	1.90
....	1.80	1.70	1.90	1.90	1.90	31............	1.70	......	1.90	......	1.90
....	1.80	1.70	1.80	1.90	1.90						

OWENS RIVER CANAL NEAR BISHOP, CAL.

station was established August 5, 1903, by J. C. Clausen and
awley. It is located at the footbridge near the quarter-section
ich divides the north half of sec. 27, T. 6 S., R. 32 E., of the
Diablo meridian. The gage is a vertical rod nailed to the
It is read once each day by A. S. Kinsley, the ditch tender.
ge measurements are made from the footbridge to which the
attached. The initial point for soundings is on right bank of
The distances across stream are marked on the bridge at
s of 2 feet. The channel is straight for 300 feet above and for
below the station. The current is sluggish. The right bank
and rocky and will not overflow. The left bank is low and
rflow. The bed of the stream is composed of gravel and is
ent.

The observations at this station during 1903 have been made under the direction of S. G. Bennett, district hydrographer.

Discharge measurements of Owens River canal near Bishop, Cal., in 1903.

Date.	Hydrographer.	Gage height.	Discharge.
		Feet.	*Second-feet.*
July 30	R. S. Hawley		39.0
August 5	....do	2.50	21.8
August 15	....do	2.69	28.6
August 26	....do	2.82	36.5
September 7	....do	2.89	33.1
September 15	....do	3.02	39.5
October 23	....do	2.57	23.2

Mean daily gage height, in feet, of Owens River canal near Bishop, Cal., for 1903.

Day.	Aug.	Sept.	Oct.	Day.	Aug.	Sept.	Oct.	Day.	Aug.	Sept.	Oct.
1		2.90	3.10	12	2.80	3.00		23	2.85	3.00	
2		2.90	3.10	13	2.80	3.05		24	2.85	3.00	
3		2.90	3.10	14	2.80	3.05		25	2.80	3.00	
4		2.95	3.10	15	(b)	3.05		26	2.80	3.05	
5	a2.50	2.95	3.10	16	(b)	3.05		27	2.85	3.05	
6	2.50	(b)	3.10	17	2.75	3.05		28	2.85	3.05	
7	2.50	2.95	3.10	18	2.80	3.05		29	2.90	(b)	
8	(b)	2.90	3.10	19	(b)	3.10		30	2.90	2.10	
9	2.70	2.90	3.10	20	2.80	3.10		31	2.90		
10	2.75	2.95	2.70	21	(b)	(b)					
11	(b)	3.00		22	2.85	3.00					

a Lowest water known in twenty-four years.   b No record.

On October 10 the directors of the canal ordered part of the water turned off. Irrigation was practically over for season. From October 10 no daily record was kept, but beginning October 11 the gage read about 1.60 until October 24; the water was then all turned out.

BISHOP CREEK CANAL NEAR BISHOP, CAL.

This station was established August 5, 1903, by J. C. Clausen, assisted by R. S. Hawley. It is located at the footbridge below the waste gate near the house of the observer, A. Fitzgerald. It is 3¼ miles northwest of Bishop, Cal. The gage is a 1 by 3 inch vertical rod fastened to the bridge anchor. Discharge measurements are made from the footbridge at which the gage is located. The initial point for soundings is on right bank; the initial point and points at which soundings are made are marked on footbridge. The channel is straight for 50 feet above and for 100 feet below the station. The current is swift. The right bank is high and the left bank is low. Neither bank is liable to overflow. The bed of the stream is composed of sand and gravel and is fairly permanent. The gaging stations on anals taking water from Owens River may be considered as temporary. The gage rods in most instances are securely fastened to the ootbridges and are not likely to change. No permanent bench marks vere established.

The observations at this station during 1903 have been made under he direction of S. G. Bennett, district hydrographer.

Discharge measurements of Bishop Creek canal near Bishop, Cal., in 1903.

Date.	Hydrographer.	Gage height.	Discharge.
		Feet.	*Second-feet.*
uly 30	R. S. Hawley		101
August 5	do	3.0	72
August 15	do	2.89	61
August 26	do	3.25	84
September 5	do	3.17	81
September 15	do	3.31	82
October 23	do	1.97	27

Mean daily gage height, in feet, of Bishop Creek canal near Bishop, Cal., for 1903.

Day.	Aug.	Sept.	Oct.	Day.	Aug.	Sept.	Oct.	Day.	Aug.	Sept.	Oct.
1				12				23		3.30	
2		3.15		13				24		3.40	
3		3.15		14	3.10			25		3.40	
4				15	2.90	3.30		26	3.25	3.40	
5	3.00	3.15		16		3.30		27		3.40	
6	3.10		2.00	17		3.25		28		3.40	
7				18	3.10	3.25		29		(a)	
8	3.10	3.30		19	3.00	3.30		30			
9				20		3.30		31			
10	3.10	3.30		21		3.30					
11	3.10			22	3.10	3.30					

a No record until October 6; but not much change in the canal.

NOTE.—On October 6 the head-gate was shut down, and until October 29, when the water was turned out, the gage read about 2.0 feet.

FARMERS CANAL NEAR BISHOP, CAL.

This station was established August 6, 1903, by J. C. Clausen and R. S. Hawley. It is located at the footbridge near the house of the observer, Robert Love, and 3 miles north of Bishop, Cal. The gage is a 1 by 3 inch vertical rod fastened to the bridge pier. Discharge measurements are made from the footbridge to which the gage is attached. The initial point for soundings is on footbridge near the right bank. The distances across stream are marked every 2 feet on the footbridge.

The channel is straight for 300 feet above and for 50 feet below the station. The current is sluggish. Both banks are low and are liable to overflow. The bed of the stream is sandy and shifting.

The observations at this station during 1903 have been made under the direction of S. G. Bennett, district hydrographer.

Discharge measurements of Farmers canal near Bishop, Cal., in 1903.

Date.	Hydrographer.	Gage height.	Discharge.
		Feet.	*Second-feet.*
July 31	R. S. Hawley		12.9
August 6	do	2.4	7.0
August 17	do	2.7	14.7
August 25	do	2.5	10.7
September 3	do	2.42	9.7
September 14	do	2.58	10.4
October 23	do	2.55	16.3
November 27	do	1.54	16.0

Mean daily gage height, in feet, of Farmers canal near Bishop, Cal., for 1903.

Day.	Aug.	Sept.	Oct.	Nov.	Dec.	Day.	Aug.	Sept.	Oct.	Nov.	Dec.
1		2.40	2.95	2.55	2.50	17	2.70	*2.50	(2.55)	2.50
2		2.40	2.95			18	2.68	2.67		(2.55)	
3		2.43	2.95	2.55	2.50	19		2.80	2.60	(2.55)	2.45
4		2.40	3.00			20	2.50	2.82	2.50	(2.55)	
5		2.45	3.08	2.55	2.50	21		2.85	2.50	2.55	2.50
6		2.45				22	2.50	2.83	2.50	2.55	
7	2.40	2.50	3.05	2.55	2.50	23	2.50		2.55	(2.55)	2.45
8	2.33	2.58		2.55		24	2.55	2.87		(2.55)	
9	2.40	2.67	2.82	(2.55)	2.50	25	2.50			(2.55)	2.50
10	2.60	2.65		(2.55)		26	2.50	2.90	2.53	(2.55)	
11	2.60		2.80	(2.55)	2.50	27	2.55	2.93		(2.55)	2.50
12	2.67	2.68		(2.55)		28	2.56	2.90	2.55	
13	2.75	2.57	2.80	(2.55)	2.50	29			2.55		2.50
14	2.60		2.55		30	2.45	2.92			2.50
15	2.60		2.55	2.48	31					2.50
16	2.70	2.63	2.90	(2.55)							

a The rock dam below the gage rod was partly removed.

M'NALLY CANAL NEAR BISHOP, CAL.

; station was established July 31, 1903, by J. C. Clausen and R.
wley. It is located at the head of the canal, 3¾ miles north of
), Cal. The gage is a 1 by 3 inch vertical board fastened to the
orks of the canal. Discharge measurements are made from a
idge. The initial point for soundings is marked on the foot-
; near the right bank. The channel is straight for 50 feet below
ge. The current is swift. Both banks are high and will not
w. The bed of the stream is rocky and permanent.
gage reader could be obtained for this station, but since the gage
istalled there has been little variation in the amount of water
g in the canal.
observations at this station during 1903 have been made under
rection of S. G. Bennett, district hydrographer.

Discharge measurements of McNally canal near Bishop, Cal., in 1903.

Date.	Hydrographer.	Gage height.	Discharge.
		Feet.	*Second-feet.*
....	R. S. Hawley	1.4	66
6	.....do....................	1.4	66
17	.....do	1.36	63
25	.....do	1.25	54
ber 4.................	.....do	1.25	52
ber 14.....	.....do	1.26	52
ber 21...............	.....do	1.10	40

GEORGE COLLINS'S CANAL NEAR BISHOP, CAL.

i station was established August 17, 1903, by R. S. Hawley. It
ted at the footbridge 3 miles east and one-half mile north of
), Cal. The gage is a vertical rod fastened to the bridge near
use of the observer, Arthur Wines. Discharge measurements
ide from the bridge. The initial point for soundings is marked
footbridge near the right bank of the canal. The channel is
it for 75 feet above and for 50 feet below the station. The
t is sluggish. The right bank is low and the left bank is high.
r bank is liable to overflow. The bed of the stream is composed
l and is fairly permanent.
observations at this station during 1903 have been made under
rection of S. G. Bennett, district hydrographer.

Discharge measurements of George Collins's canal near Bishop, Cal., in 1903.

Date.	Hydrographer.	Gage height.	Discharge.
		Feet.	*Second-feet.*
July 31	R. S. Hawley		4.9
August 11	do		5.6
August 17	do	1.00	6.3
August 27	do	1.01	6.0
September 7	do	.94	5.5
September 14	do	.98	5.5
October 23	do	.60	2.8

Mean daily gage height, in feet, of George Collins's canal near Bishop, Cal., for 1903.

Day.	Aug.	Sept.	Oct.	Nov.	Dec.	Day.	Aug.	Sept.	Oct.	Nov.	Dec.
1		1.03	0.95	0.75	0.75	17		0.98	0.95	0.80	b0.50	0.65
2		.98	.95	.75	.75	18		.92	.97	.80	.50	.60
3		.97	.95	.75	.75	19		.93	.95	.80	.50	.60
4		.98	.95	.75	.75	20		.92	1.35	.80	.50	.60
5		.97	.95	.75	.70	21		.70	1.35	.85	.55	(d)
6		.98	.95	.70	.65	22		.75	1.35	.85	.55	
7		.97	.95	.70	.65	23		.70	1.35	c.75	.55	
8		.95	.95	.70	.65	24		.90	1.35	.75	.55	
9		.95	.95	.70	.65	25		.90	1.35	.75	.55	
10		.95	.95	.70	.65	26		.90	1.35	.75	.55	
11		.95	.95	.75	.65	27		.97	1.35	.75	.62	
12		.92	a.80	.75	.65	28		1.03	1.20	.75	.70	
13		.93	.80	.75	.63	29		1.02	1.20	.75	.70	
14		.95	.80	.75	.62	30		1.03	1.20	.75	.85	
15		.95	.80	.75	.63	31		1.04		.75		
16	1.00	.95	.80	.75	.62						

a Falling slowly; no storms.　　*b* Falling fast.　　*c* Falling slightly.　　*d* Canal turned off.

BISHOP CREEK NEAR BISHOP, CAL.

This station was established August 10, 1903, by J. C. Clausen and R. S. Hawley. It is located at the wagon bridge on the Bishop road, about 4½ miles from Bishop and about 2 miles from the point where the creek leaves the canyon. North Hillside canal, South Hillside canal, and Power's canal are taken out above the station. The gage is a 1 by 3 inch vertical rod fastened in the rocks and braced to the right bank just above the wagon bridge. It is read once each day by A. S. Kilpatrick. Discharge measurements are made from the wagon bridge. The initial point for soundings is marked on the bridge near the right bank of the stream. The channel is straight for 100 feet above and for 50 feet below the station. The current is swift. Both banks are high and rocky and are not liable to overflow. The bed of the stream is ad rocky and is permanent. The bench mark is a large flat owlder on the right bank, 40 feet above the bridge. eervations at this station during 1903 have been made under tion of S. G. Bennett, district hydrographer.

Discharge measurements of Bishop Creek near Bishop, Cal., in 1903.

Date.	Hydrographer.	Gage height.	Discharge.
		Feet.	*Second-feet.*
....	R. S. Hawley	2.45	121
....	.....do	2.17	95
....	.....do	2.20	96
·3	.....do	1.99	87
·7	.....do	1.80	52
·15	.....do	1.65	40
·21	.....do	1.60	38
l...................	.....do	1.39	27
20	.....do	1.41	26
26	.....do	1.35	25

ɪn daily gage height, in feet, of Bishop Creek near Bishop, Cal., for 1903.

	Aug.	Sept.	Oct.	Nov.	Dec.	Day.	Aug.	Sept.	Oct.	Nov.	Dec.
....		2.00	1.60	1.45	1.30	17..............	2.30	1.60	1.70	1.40	1.45
....		1.95	1.60	1.45	1.30	18..............	2.20	1.60	1.65	1.40	1.45
....		1.90	1.70	1.45	1.30	19..............	2.20	1.60	1.50	1.40	1.45
....		2.00	1.70	1.45	1.35	20..............	2.50	1.60	1.50	1.40	1.45
....		2.00	1.70	1.45	1.35	21..............	2.20	1.60	1.45	1.40	1.45
....		2.00	1.70	1.45	1.35	22..............	2.20	1.70	1.45	1.40	1.45
....		2.00	1.70	1.45	1.40	23..............	2.25	1.60	1.45	1.40	1.45
....		1.90	1.70	1.45	1.40	24..............	2.20	1.60	1.45	1.40	1.45
....		1.90	1.70	1.45	1.40	25..............	2.15	1.60	1.45	1.40	1.45
....		1.80	1.70	1.45	1.40	26..............	2.15	1.65	1.45	1.40	1.45
....	2.50	1.80	1.70	1.45	1.40	27..............	2.10	1.60	1.45	1.40	1.40
....	2.50	1.75	1.70	1.45	1.40	28..............	2.05	1.60	1.45	1.40	1.40
....	2.50	1.75	1.70	1.45	1.40	29..............	2.05	1.70	1.45	1.40	1.40
....	2.40	1.70	1.70	1.45	1.40	30..............	2.05	1.65	1.45	1.40	1.40
....	2.30	1.70	1.65	1.45	1.40	31..............	2.00	......	1.45	......	1.40
....	2.30	1.65	1.65	1.45	1.45						

RAWSON CANAL NEAR BISHOP, CAL.

tation was established August 7, 1903, by R. S. Hawley. It ɪ at the county bridge 2¼ miles east of Bishop, Cal. The a vertical rod, fastened to the bridge from which discharge ments are made. The gage is read once each day by Paul ld. The initial point for soundings is marked on the bridge near the right bank of the canal. The channel is straight for 100 feet above and below the station, and the current is swift. The right bank is high, and the left bank is low. Neither bank is liable to overflow. The bed of the stream is composed of gravel and is permanent.

The observations at this station during 1903 have been made under the direction of S. G. Bennett, district hydrographer.

Discharge measurements of Rawson canal near Bishop, Cal., in 1903.

Date.	Hydrographer.	Gage height.	Discharge.
		Feet.	*Second-feet.*
July 31	R. S. Hawley		17.0
August 7	do	1.7	17.6
August 11	do	1.62	13.6
August 17	do	1.85	20.3
August 27	do	1.73	17.1
September 7	do	1.77	18.2
September 14	do	1.81	17.1
October 24	do	1.73	15.4
November 27	do	1.6	13.0

Mean daily gage height, in feet, of Rawson canal near Bishop, Cal., for 1903.

Day.	Aug.	Sept.	Oct.	Nov.	Dec.	Day.	Aug.	Sept.	Oct.	Nov.	Dec.
1		1.74	2.00	1.59	1.65	17	1.85	1.93	1.65	1.92	1.45
2		1.72	2.00	1.63	1.64	18	1.82	1.94	1.65	1.92	1.10
3		1.76	1.96	1.60	1.70	19	1.81	1.99	1.80	1.93	1.10
4		1.73	1.93	1.60	1.71	20	1.86	2.10	1.75	1.90	1.10
5		1.74	1.93	1.60	1.70	21	1.84	2.10	1.70	1.90	1.10
6		1.80	1.94	1.63	1.35	22	1.82	1.94	1.65	1.90	1.20
7		1.79	1.91	1.63	1.35	23	1.81	1.92	1.70	1.91	1.10
8		1.80	1.92	1.63	1.43	24	1.80	1.93	1.73	1.90	1.10
9		1.76	2.01	1.65	1.35	25	1.75	1.94	1.75	1.90	1.10
10		1.78	2.01	1.66	1.30	26	1.74	2.04	1.76	1.90	1.30
11	1.60	1.77	1.80	1.65	1.30	27	1.74	2.00	1.77	1.53	1.30
12	1.80	1.85	1.75	1.61	1.30	28	1.76	2.00	1.63	1.50	1.30
13	1.90	1.90	1.75	1.63	1.39	29	1.73	2.01	1.59	1.51	1.30
14	1.90	1.72	1.70	1.60	1.30	30	1.74	2.00	1.59	1.65	1.30
15	1.85	1.80	1.70	1.70	1.42	31	1.72		1.50		1.30
16	1.80	1.85	1.70	1.80	1.44						

A. O. COLLINS'S CANAL NEAR BISHOP, CAL.

This station was established August 7, 1903, by R. S. Hawley. It is located at the county bridge 3 miles east of Bishop, Cal. The gage is a vertical rod fastened to the right bank, just above the bridge from which discharge measurements are made. It is read once each day by

cDonald. The initial point for soundings is marked on the
near the right bank of the canal. The channel is straight for
; above and for 50 feet below the station. The current is
. Both banks are high and are not liable to overflow. The
he stream is sandy and shifting.
bservations at this station during 1903 have been made under
:tion of S. G. Bennett, district hydrographer.

charge measurements of A. O. Collins's canal near Bishop, Cal., in 1903.

Date.	Hydrographer.	Gage height.	Discharge.
		Feet.	*Second-feet.*
-------------------	R. S. Hawley-------------------	-------	12. 1
-------------------	-----do ---------------------	2. 2	8. 3
-------------------	-----do ------------------	2. 0	3. 9
-------------------	-----do ---------------------	1. 82	2. 6
-------------------	-----do ----------------------	2. 1	8. 8
r 7 -----------------	-----do ---------------------	2. 1	7. 9
r 14 -----------------	-----do ---------------------	2. 1	7. 6
3-----------------------	-----do ---------------------	1. 65	1. 2

rily gage height, in feet, of A. O. Collins's canal near Bishop, Cal., for 1903.

	Aug.	Sept.	Oct.	Nov.	Dec.	Day.	Aug.	Sept.	Oct.	Nov.	Dec.
....		2. 06	2. 40	1. 99	1. 80	17.................	1. 82	2. 14	1. 65	1. 80	1. 70
....		2. 10	2. 00	1. 99	1. 52	18.................	1. 74	2. 10	1. 65	1. 81	2. 10
....		2. 10	1. 65	2. 00	1. 60	19.................	2. 20	2. 07	1. 65	1. 82	1. 70
....		2. 10	1. 65	2. 00	1. 65	20.................	2. 12	2. 09	1. 65	1. 80	1. 72
....		2. 01	2. 10	2. 00	1. 60	21.................	2. 10	2. 07	1. 65	1. 80	1. 70
....		2. 01	2. 30	2. 40	1. 90	22..................	2. 04	2. 20	1. 65	1. 83	1. 73
....		2. 01	2. 40	2. 00	1. 85	23.................	2. 10	2. 30	1. 65	1. 60	1. 70
....		2. 01	2. 40	2. 10	1. 90	24.................	2. 10	2. 40	1. 65	1. 62	1. 72
....		2. 01	2. 00	2. 00	1. 90	25.................	2. 03	2. 50	2. 40	1. 53	1. 70
....		2. 02	1. 65	2. 20	1. 82	26.................	2. 03	2. 40	2. 50	1. 51	1. 70
....	2. 00	2. 02	1. 65	2. 10	1. 80	27.................	2. 10	2. 70	2. 40	1. 50	1. 70
....	1. 80	2. 03	1. 65	2. 40	1. 80	28......;.........	2. 01	2. 40	2. 40	1. 65	1. 70
....	1. 80	2. 07	1. 65	2. 30	1. 60	29.................	2. 10	2. 04	2. 40	1. 51	1. 70
....	1. 80	2. 02	1. 65	1. 50	1. 72	30.................	2. 00	2. 40	2. 06	1. 65	1. 70
....	1. 80	2. 03	1. 65	1. 50	1. 72	31.................	2. 04	....	2. 09	....	1. 70
....	1. 70	2. 04	1. 65	1. 66	1. 90						

DELL CANAL NEAR BISHOP, CAL.

tation was established August 24, 1903, by R. S. Hawley. It
d at a flume 3 miles from the head-gate at a point where the
)sses a slough in Sanders's field. The gage is a vertical rod
to the flume. Discharge measurements are made from a bridge

across the flume. The observer is F. Sanders. The initial point for soundings is the edge of the flume at the right side. The channel is straight for 150 feet above and for 200 feet below the station. The current is sluggish at all times.

The observations at this station during 1903 have been made under the direction of S. G. Bennett, district hydrographer.

Discharge measurements of Dell canal near Bishop, Cal., in 1903.

Date.	Hydrographer.	Gage height.	Discharge.
		Feet.	*Second-feet.*
August 11	R. S. Hawley	....	24.2
August 20	.....do	....	11.0
August 24	.....do	....	10.0
Do	.....do	0.85	8.0
September 4	.....do	.78	7.0
September 22	.....do	1.47	19.6
October 7....................	.....do	1.52	19.0

Mean daily gage height, in feet, of Dell canal near Bishop, Cal., for 1903.

Day.	Aug.	Sept.	Oct.	Nov.	Day.	Aug.	Sept.	Oct.	Nov.
1....................		0.79	0.80	0.90	17		0.80	0.84	1.08
2....................		.80	.80	.90	18		.78	......	1.11
3....................		.80	.81	.90	19		.78	.84	1.12
4....................		.79	......	.91	20		......	.85	1.14
5....................		.79	.80	.91	21		.79	.86	1.13
6....................			.80	.92	22		.80	.86	1.12
7....................		.78	.81	.92	23	0.85	.81	.86	1.11
8....................		.78	.81	......	24	.85	.81	.87	1.12
9....................		.79	.80	.95	25	.86	.80	.88	1.15
10....................		.80	.80	.94	26	.86	.81	.88	1.16
11....................		.80	......	.95	27			.89	1.12
12....................		.80	.82	.95	28	.85	.78	.81	(a)
13....................		.80	.82	1.01	29	.85	.78	.89	......
14....................		.79	.83	1.02	30		.79	.89	......
15....................		.79	.83	......	31			.89	......
16....................		.80	.83	1.06					

a No record obtained from Nov. 27, 1903, to Feb. 1, 1904, when the water was turned out.

BIG PINE AND OWENS RIVER CANAL NEAR BISHOP, CAL.

This station was established by J. C. Clausen and R. S. Hawley, August 4, 1903. It is located at a footbridge near the house of the observer, William Oliver. It is 7½ miles south and 3 miles east of Bishop, Cal. The gage is a vertical rod securely nailed to the footbridge. Discharge measurements are made from the footbridge. The initial point for soundings is marked on the footbridge near the right bank of the canal. The channel is straight for 600 feet above and for 300 feet below the station. The current is sluggish. Both banks are high and are not liable to overflow. The bed of the stream is sandy and somewhat shifting.

The observations at this station during 1903 have been made under the direction of S. G. Bennett, district hydrographer.

Discharge measurements of Big Pine and Owens River canal near Bishop, Cal., in 1903.

Date.	Hydrographer.	Gage height.	Discharge.
		Feet.	*Second-feet.*
August 4....................	R. S. Hawley.................	1.60	19.0
August 11...................	.....do	1.55	16.3
August 20...................	.....do	1.50	17.3
September 14	.....do	.90	1.3
September 22	.....do	1.15	4.3
October 27..................	.....do	2.29	46.4

Mean daily gage height, in feet, of Big Pine and Owens River canal near Bishop, Cal., for 1903.

Day.	Aug.	Sept.	Oct.	Nov.	Dec.	Day.	Aug.	Sept.	Oct.	Nov.	Dec.
1....................		0.90	1.20	2.53	2.35	17..................	1.60	1.05	2.45	2.55	1.50
2....................		.90	1.25	2.58	2.40	18..................	1.60	1.10	2.42	2.50	1.52
3....................		.90	1.35	2.53	2.40	19..................	1.60	1.15	2.43	2.50	1.58
4....................		.90	1.35	2.53	2.40	20..................	1.60	1.15	2.45	2.50	1.47
5....................	1.55	.90	1.50	2.53	2.30	21..................	1.40	1.12	2.47	2.53	1.48
6....................	1.50	.90	1.80	2.53	2.30	22..................	1.40	1.10	2.46	2.53	1.53
7....................	1.63	.90	1.90	2.58	2.30	23..................	1.40	1.10	2.30	2.53	1.53
8....................	1.62	.90	1.95	2.57	2.30	24..................	1.40	1.10	2.30	2.43	1.47
9....................	1.80	.90	1.95	2.57	2.35	25..................	1.35	1.10	2.33	2.41	1.47
10....................	1.70	.97	2.03	2.57	2.35	26..................	1.35	1.10	2.32	2.35	1.50
11....................	1.53	.95	2.22	2.62	2.35	27..................	1.75	1.12	2.28	2.35	1.50
12....................	1.55	1.00	2.25	2.57	2.35	28..................	1.75	1.15	2.27	2.35	1.50
13....................	1.40	1.00	2.25	2.57	1.50	29..................	1.75	1.18	2.25	2.38	1.50
14....................	1.40	1.05	2.27	2.60	1.50	30..................	1.30	1.22	2.45	2.40	1.50
15....................	1.40	1.05	2.40	2.60	1.50	31..................	.90		2.45		1.50
16....................	1.60	1.05	2.42	2.55	1.50						

SANGER CANAL AT ALVORD, CAL.

This station was established August 4, 1903, by J. C. Clausen and R. S. Hawley. It is located at the county-road bridge, one-fourth mile east of the Southern Pacific Railroad station at Alvord, Cal. The gage is a vertical rod, fastened to the bridge from which discharge measurements are made. It is read once each day by John Hale. The initial point for soundings is marked on the bridge near the left bank of the canal. The channel is straight for 300 feet above and for

100 feet below the station. The current is sluggish. Both banks are low and liable to overflow. The bed of the stream is shifting.

The observations at this station during 1903 have been made under the direction of S. G. Bennett, district hydrographer.

Discharge measurements of Sanger canal at Alvord, Cal., in 1903.

Date.	Hydrographer.	Gage height.	Discharge.
		Feet.	*Second-feet.*
August 4	R. S. Hawley....................	2.00	3.1
August 13	....do	2.60	10.0
September 8	....do	2.45	11.3
September 22	....do	2.35	10.1
October 27	....do	2.70	14.5
December 4	....do	2.90	13.4

Mean daily gage height, in feet, of Sanger canal at Alvord, Cal., for 1903.

Day.	Sept.	Oct.	Nov.	Dec.	Day.	Sept.	Oct.	Nov.	Dec.
1		2.50	2.60	2.45	17		2.50	2.30	2.20
2			2.60		18		2.50		
3		2.50	2.60	2.40	19		2.50	2.40	2.20
4		2.50	2.55		20		2.50		2.20
5			2.50	2.40	21			2.50	
6		2.50	2.50	2.40	22		2.60	2.50	2.20
7			2.50		23				
8		2.50	2.40	2.30	24		2.60	2.50	2.20
9			2.40		25		2.60		
10		2.50	4.40	2.30	26			2.50	2.20
11		2.20	2.40		27	2.60	2.50		2.20
12			2.40	2.30	28			2.50	2.20
13		2.50	2.30	2.30	29	2.60	2.50	2.40	2.20
14			2.30		30	2.50		2.40	2.20
15		2.50	2.30	2.30	31		2.40		2.20
16									

EAST SIDE CANAL NEAR CITRUS, CAL.

This station was established August 27, 1903, by R. S. Hawley. It is located at the head-gate of the canal. The gage is a vertical rod fastened to the head-gate at which discharge measurements are made. The gage is read once each day by J. Vaughn. At low stages the meter measurements are made by wading at a point below head-gate. High-stage measurements are made from Southern Pacific Railroad bridge one-half mile below head-gate.

The observations at this station during 1903 have been made under the direction of S. G. Bennett, district hydrographer.

Discharge measurements of East Side canal near Citrus, Cal., in 1903.

Date.	Hydrographer.	Gage height.	Discharge.
		Feet.	*Second feet.*
August 13	R. S. Hawley	1.95	32
August 27	....do	2.00	44
September 16	....do	1.98	36
October 30	....do	2.95	80
December 4	....do	3.23	96

Mean daily gage height, in feet, of East Side canal near Citrus, Cal., for 1903.

Day.	Aug.	Sept.	Oct.	Nov.	Dec.	Day.	Aug.	Sept.	Oct.	Nov.	Dec.
						17					
		2.00	2.00	3.00	3.20	18		2.00		3.00	2.00
						19			2.70		
		2.00		3.00	3.20	20				3.00	
			2.20			21		2.00	2.70		2.00
				3.00	(a)	22					
		2.00	2.30		(a)	23		2.00	3.10		2.40
		2.00	2.30	3.00	2.35	24					
						25		2.00	3.05		2.40
		2.00		3.00	2.40	26			3.10		
			2.40			27	2.00		3.15		
			3.00			28	2.00	2.10	3.00		1.80
		1.90	2.50		2.40	29					
						30		2.10	3.00	3.15	
		2.00	2.60	3.00	2.40	31	2.00				

a Head-gate shut down.

STEVENS CANAL NEAR CITRUS, CAL.

This station was established August 27, 1903, by R. S. Hawley. It is located at the waste gate of the canal, 3½ miles north of Citrus, Cal. The gage is a vertical rod fastened to the waste gate. It is read once each day by J. Vaughn. Discharge measurements are made by wading. The initial point for soundings is on the right bank of the canal. The channel is straight for 300 feet above and for 200 feet below the station. The current is sluggish. Both banks are high and are not liable to overflow. The bed of the stream is composed of gravel and earth and is fairly permanent.

The observations at this station during 1903 have been made under the direction of S. G. Bennett, district hydrographer.

Discharge measurements of Stevens canal near Citrus, Cal., in 1903.

Date.	Hydrographer.	Gage height.	Discharge.
		Feet.	*Second-feet.*
August 13	R. S. Hawley		17.7
August 27	.....do	2.3	23.0
September 16	.....do	2.4	21.3

Mean daily gage height, in feet, of Stevens canal near Citrus, Cal., for 1903.

Day.	Aug.	Sept.	Oct.	Day.	Aug.	Sept.	Oct.	Day.	Aug.	Sept.	Oct.
1				12				22			
2		2.30	2.50	13				23		2.00	
3				14		2.60		24			
4		2.30		15				25		2.00	
5			2.50	16		2.40		26			
6				17				27	2.30		
7		2.35	2.50	18		2.00		28	2.30	2.30	
8				19				29			
9		2.35		20				30		2.35	
10				21		2.40		31	2.30		
11		2.30									

On November 7 the water was turned out.

OWENS RIVER NEAR CITRUS, CAL.

This station was established October 30, 1903, by R. S. Hawley. It is located at the county bridge 4 miles east of Independence, Cal., and 1 mile from the Southern Pacific Railroad station at Citrus, Cal. The station at this point shows the amount of waste water which is discharged into Owens Lake. The gage is a 1 by 4 inch vertical rod nailed to a pile on the upstream side of the middle pier of the bridge. It is read twice each day by Milton Levy. Discharge measurements are made from the county bridge at which the gage is located. The initial point for soundings is the end of the rail at the right end of the bridge. The channel is straight for 200 feet above and for 300 feet below the station. The current has a mean velocity of about 2 feet at ordinary stages. Both banks are high and are not liable to overflow. The bed of the stream is sandy and is liable to shift. The bench mark is a copper tack in the floor beam directly over the gage rod. Its elevation is 1.50 feet above the 12-foot mark of the gage.

The observations at this station during 1903 have been made under the direction of S. G. Bennett, district hydrographer.

Discharge measurements of Owens River near Citrus, Cal., in 1903.

Date.	Hydrographer.	Gage height.	Discharge.
		Feet.	*Second-feet.*
....	J. C. Clausen	....	534
28	R. S. Hawley	3.7	8.6
r 30.................	.....do	4.4	195
ᵍer 4.................	.....do	5.0	215

Mean daily gage height, in feet, of Owens River near Citrus, Cal., for 1903.

'.	Oct.	Nov.	Dec.	Day.	Oct.	Nov.	Dec.	Day.	Oct.	Nov.	Dec.
....	......	4.48	4.95	12...............	......	4.83	5.01	23.............	....	4.90	5.30
....	......	4.45	5.00	13...............	......	4.85	5.01	24.............	....	4.95	5.20
....	......	4.65	4.95	14...............	......	4.85	5.20	25.............	....	5.00	5.15
....	......	4.70	5.00	15...............	......	4.87	5.20	26.............	....	5.00	5.15
....	......	4.75	5.00	16...............	......	4.85	5.00	27.............	....	5.00	5.20
....	......	4.70	5.00	17...............	......	4.85	5.20	28.............	....	5.00	5.20
....	......	4.60	5.00	18...............	......	4.85	5.20	29.............	....	5.00	5.35
....	......	4.65	5.03	19...............	......	4.85	5.20	30.............	4.40	5.00	5.35
....	......	4.70	5.03	20...............	......	4.88	5.20	31.............	4.47	......	5.35
....	......	4.77	5.01	21...............	......	4.95	5.25				
....	......	4.80	5.00	22...............	......	4.97	5.30				

MISCELLANEOUS MEASUREMENTS IN INTERIOR BASIN.

The following miscellaneous measurements were made in the interior basin in 1903:

Miscellaneous measurements in Utah in 1903.

Date.	Hydrographer.	Stream.	Location.	Discharge.
				Sec.-feet.
Apr. 7	O. Tanner	Hobble Creek	Above canals	22
Aug. 14	....do	....do	....do	14
Nov. 4	....do	....do	....do	13
Dec. 5	....do	....do	....do	7
Dec. 16	....do	....do	....do	13
May 2	....do	Peteetneet Creek		24
Aug. 14	....do	....do	At weir	6
Nov. 4	....do	....do	....do	2
Dec. 8	....do	....do	....do	2
May 2	....do	Santaquin Creek		27
Aug. 13	....do	....do		7
Nov. 3	....do	....do		6
Dec. 2	....do	Battle Creek		3
Do...	....do	Grove Creek		1
Dec. 3	....do	Dry Creek		7
Dec. 1	....do	Fort Canyon Creek		3
Dec. 9	....do	Currant Creek		19
Dec. 5	....do	Little Spring Creek		5
Dec. 10	....do	Spanish Fork River	At lake shore (gage height, 2.99).	46
Apr. 11	....do	Salina Creek	Near Salina	39
Jan. 25	....do	....do	....do	14

Measurements of tributaries of Bear Lake, Idaho, in 1903.

Oct. 6	W. P. Hardesty	Big Creek	Above junction with Spring Creek.	11
Do...	....do	Spring Creek		14
Do...	....do	Bloomington Creek		9
Sept. 27	....do	Paris Creek		31

Miscellaneous measurements in Owens River drainage, California, in 1903.

Date.	Hydrographer.	Stream.	Location.	Discharge.
				Sec.-feet.
ㄱy 30	R. S. Hawley	Hillside canal......	Bishop	3.8
g. 5	.....do	.....do	.....do	1.9
g. 15	.....do	.....do	.....do	2.3
ㄱt. 7	.....do	.....do	.....do	2.7
g. 5	.....do	Loves ditch.........	.....do	.9
g. 15	.....do	.....do	.....do	.5
ㄱt. 5	.....do	.....do	.....do	1.5
ㄱt. 15	.....do	.....do	.....do	2.4
ㄴ 15	.....do	.....do	.....do	2.9
ㄴ 10	.....do	McGee Creek	.....do	2.6
Do ...	.....do	Birch Creek	.....do	5.0
ㄴ 12	.....do	Coyote Creek	.....do	1.6
Do ...	.....do	Mill Creek.........	Bigpine	7.0
g. 14	.....do	Big Pine Creek	.....do	112
ㄴ 12	.....do	.....do	.....do	16.4
ㄴ 13	.....do	Birch Creek	Fish Springs........	3.5
ㄱt. 21	.....do	Fish Slough ditch ..	.....do	11.2
ㄴ 13	.....do	Tinemaha Creek	.....do	3.6
Do ...	.....do	Taboosa Creek	Black Rock..........	4.8
Do ...	.....do	Clear Creek........	.....do	1.7
Do ...	.....do	Goodale Creek	.....do	6.6
Do ...	.....do	Division Creek	Near Independence ..	2.5
Do ...	.....do	Oak Creek..........	.....do	10.7
g. 13	.....do	Independence Creek	Independence........	8.6
ㄱt. 17	.....do	.....do	.....do	4.5
ㄴ 14	.....do	.....do	.....do	4.3
Do ...	.....do	Shepherd Creek....	Near Independence ..	2.1
Do ...	.....do	Moffitt Creek	.....do	.3
Do ...	.....do	Georges Creek.....	.....do	.8
ㄴ 15	.....do	Lonepine Creek....	At Lonepine.........	3.3
Do ...	.....do	Tuttle Creek.......	Near Lonepine.......	2.3
ㄴ 31	.....do	Cottonwook Creek .	Near Olancho........	7.6
Do ...	.....do	Ash Creek.........	.....do	1.2

Miscellaneous stream measurements in Humboldt River basin, Nevada, in 1903.

Date.	Hydrographer.	Stream.	Location.	Discharge.
				Sec.-ft.
May 30	W. A. Wolf	Humboldt River	Roger ranch	43
June 17	....do	....do	....do	44
July 1	....do	....do	....do	124
July 10	....do	....do	....do	112
July 31	....do	Cottonwood Creek	Above all diversions	4.1
Aug. 2	....do	Martin Creek	....do	8.0
Aug. 3	....do	Little Humboldt River.	Cathcart lane	1.4

Miscellaneous stream measurements in Diamond Valley, Carson River basin, California, in 1903.

Date.	Hydrographer.	Creek.	Location.	Discharge.
				Sec.-ft.
Apr. 29	A. H. Schadler	Harveys	Above all points of diversion.	2.4
May 21	....do	....do	....do	2.2
June 24	....do	....do	....do	2.4
July 18	....do	....do	....do	.7
July 23	W. B. Harrington	....do	....do	1.3
Aug. 5	....do	....do	....do	1.8
Sept. 9	....do	....do	....do	.4
July 23	....do	Hawkins	....do	.8
Aug. 5	....do	....do	....do	1.0
Sept. 9	....do	....do	....do	.6

Measurements of creeks on west side of Carson Valley, California, in 1903.

Aug. 6	W. B. Harrington	Fredericksburg	Above all diversions..	4.2
Sept. 9	....do	....do	....do	2.9
Apr. 28	A. H. Schadler	Fairview	....do	2.8
May 20	....do	....do	....do	4.2
July 9	....do	....do	....do	3.3
July 24	W. B. Harrington	....do	....do	3.3
Aug. 5	....do	....do	....do	2.2
Sept. 9	....do	....do	....do	3.1

Measurements of creeks on west side of Carson Valley, Nevada, in 1903.

ate.	Hydrographer.	Creek.	Location.	Discharge.
				Sec.-ft.
r. 28	A. H. Schadler	Wehrman or Heitman springs.	Above all diversions..	0.2
y 20	.....do	.....do	.....do	.6
ly 9	.....do	.....do	.....do	.7
r. 28	.....do	Garaveti No. 1	.....do	2.4
y 20	.....do	.....do	.....do	0.0
ly 9	.....do	.....do	.....do	0.0
r. 28	.....do	Garaveti No. 2	.....do	.2
y 20	.....do	.....do	.....do	2.6
ly 9	.....do	.....do	.....do	2.5
r. 28	.....do	Monfrena Springs (Kerry).	.....do	Trace.
y 20	.....do	.....do	.....do	.1
y 9	.....do	.....do	.....do	.5
g. 6	W. B. Harrington ..	.....do	.....do	1.4
t. 9	.....do	.....do	.....do	.8
r. 28	A. H. Schadler	Scossa ranch	.....do	.3
y 20	.....do	.....do	.....do	.2
y 9	.....do	.....do	.....do	0.0
g. 6	W. B. Harrington ..	.....do	.....do	.1
t. 9	.....do	.....do	.....do	.1
g. 6	.....do	Palmers	.....do	.2
t. 9	.....do	.....do	.....do	0.0
r. 28	A. H. Schadler	South Sheridan (Barbers).	.....do	1.2
y 20	.....do	.....do	.....do	2.7
1e 23	.....do	.....do	.....do	.7
y 9	.....do	.....do	.....do	1.0
g. 6	W. B. Harrington ..	.....do	.....do	2.9
t. 9	.....do	.....do	.....do	2.0
r. 28	A. H. Schadler	Sheridan	.....do	1.6
y 20	.....do	.....do	.....do	3.1
1e 23	.....do	.....do	.....do	2.3
ly 9	.....do	.....do	.....do	1.2
g. 6	W. B. Harrington ..	.....do	.....do	1.9
t. 9	.....do	.....do	.....do	1.8
r. 28	A. H. Schadler	Parks	.....do	2.4
y 20	.....do	.....do	.....do	2.5
1e 23	.....do	.....do	.....do	2.1
ly 9	.....do	.....do	.....do	2.1
g. 6	W. B. Harrington ..	.....do	.....do	.6

Measurements of creeks on west side of Carson Valley, Nevada, in 1903—Continued.

Date.	Hydrographer.	Creek.	Location.	Discharge.
				Sec.-ft.
Sept. 10	W. B. Harrington ..	Parks	Above all diversions..	0.5
Apr. 28	A. H. Schadler.....	Motts Canyon......	......do	1.9
May 20	.....do	......do	......do	1.9
June 22	.....do	......do	......do	2.2
July 9	.....do	......do	......do	2.3
Aug. 6	W. B. Harrington..	......do	......do	1.1
Sept. 10	.....do	......do	......do	2.4
Apr. 28	A. H. Schadler.....	Kingsbury...........	......do	3.0
May 20	.....do	......do	......do	2.5
June 22	.....do	......do	......do	2.3
July 9	.....do	......do	......do	.7
Aug. 6	W. B. Harrington..	......do	......do	1.2
Sept. 10	.....do	......do	......do	2.4
Apr. 27	A. H. Schadler.....	Genoa................	......do	2.3
May 19	.....do	......do	......do	1.8
June 22	.....do	......do	......do	1.2
July 8	.....do	......do	......do	1.1
Aug. 6	W. B. Harrington..	......do	......do	.5
Sept. 10	.....do	......do	......do	1.4
Aug. 6	.....do	Childs Canyon......	......do	0.0
Apr. 27	A. H. Schadler.....	Sierra	......do	4.0
May 19	.....do	.....do	......do	3.6
June 22	.....do	.....do	......do	.7
July 8	.....do	.....do	......do	2.4
Aug. 7	W. B. Harrington..	.....do	......do	1.4
Sept. 10	.....do	.....do	......do	.8

Miscellaneous stream measurements on west side of Jacks Valley, Nevada, in 1903.

Apr. 27	A. H. Schadler.....	Jacks Valley No. 1 .	Above all diversions..	2.5
May 19	.....do	......do	......do	2.7
July 8	.....do	......do	......do	1.5
Aug. 7	W. B. Harrington..	......do	......do	1.3
Sept. 10	.....do	......do	......do	1.2
Apr. 27	A. H. Schadler.....	Jacks Valley No. 2 .	......do	1.7
July 8	.....do	......do	......do	0.1
Aug. 7	W. B. Harrington..	......do	......do	0.0
Sept. 10	.....do	......do	......do	0.0

Miscellaneous stream measurements on west side of Eagle Valley, Nevada, in 1903.

Date.	Hydrographer.	Creek.	Location.	Discharge.
				Sec.-ft.
r. 27	A. H. Schadler.......	Clear...............	Above all diversions..	11.2
y 18	.....do	.....do	.....do	9.5
ie 20	.....do	.....do	.....do	3.4
ly 8	.....do	.....do	.....do	2.1
g. 7	W. B. Harrington..	.....do	.....do	2.2
it. 10	.....do	.....do	.....do	2.6
ir. 25	A. H. Schadler.....	Kings Canyon	.....do	1.5
iy 15	.....do	.....do	.....do	1.2
ne 18	.....do	.....do	.....do	2.8
ly 7	W. B. Harrington..	.....do	.....do	2.8
ly 25	.....do	.....do	.....do	2.8
ig. 11	.....do	.....do	.....do	2.0
ir. 25	A. H. Schadler.....	Ash Canyon	.....do	4.5
iy 15	.....do	.....do	.....do	5.4
ne 18	.....do	.....do	.....do	5.3
ly 7	.....do	.....do	.....do	3.5
ly 25	W. B. Harrington..	.....do	.....do	3.3
ig. 2	.....do	.....do	.....do	2.0
ig. 8	.....do	.....do	.....do	2.8
ig. 12	.....do	.....do	.....do	2.8

scellaneous stream measurements in Walker River basin, California and Nevada, in 1903.

Date.	Hydrographer.	Stream.	Location.	Discharge.
				Sec.-ft.
iy 4	I. W. Huffaker	West Walker River.	Hoyes Bridge	662
ne 23	.....do	.....do	.....do	930
ly 24	.....do	.....do	Lower end of Smiths Valley.	101
ly 25	.....do	.....do	Lower end of Antelope Valley.	134
ly 28	.....do	.....do	Hoyes Bridge	132
ly 29	.....do	.....do	Below Swauger ditch .	66
Do ...	.....do	.....do	Topaz	71
g. 6	.....do	.....do	Above Little Walker..	108
g. 8	.....do	.....do	Hoyes Bridge	53
Do ...	.....do	.....do	Below Swauger ditch .	0.0
Do ...	.....do	.....do	Lower end of Smiths Valley.	33

Miscellaneous stream measurements in Walker River basin, California and Nevada, 1903—Continued.

Date.	Hydrographer.	Stream.	Location.	D cha
				Se
June 5	I. W. Huffaker......	East Walker River.	Lower end big meadow	2
July 9	.....do	....do	....do	1
Aug. 5	.....do	....do	....do	
Apr. 29	.....do	Dogtown Creek or Virginia Creek.	Above ditches........	
May 30	.....do ....,	....do	....do	
July 7	.....do	....do	....do	
Aug. 3	.....do	....do	....do	
Apr. 29	.....do	Green Creek............	....do	
May 30	.....do	....do	....do	
June 3	.....do	....do	Bridgeport............	
July 7	.....do	....do	Above ditches	
Aug. 3	.....do	....do	....do	
Apr. 29	.....do	Somers Creek........	....do	
May 30	.....do	....do	....do	
July 7	.....do	....do	....do	
Aug. 3	.....do	....do	Lower end of meadow.	
May 28	.....do	Robinson Creek....	Above Twin Lakes...	
June 6	.....do	....do	.....do	1
July 6	.....do	....do	.....do	
July 11	.....do	....do	Below Twin Lakes ...	1
Aug. 2	.....do	....do	Above Twin Lakes...	
Aug. 4	.....do	....do	Below Twin Lakes ...	
Apr. 28	.....do	Buckeye Creek......	Above ditches	
May 29	.....do	....do	.....do	1
July 8	.....do	....do	.....do	1
Aug. 4	.....do	....do	.....do	
Apr. 30	.....do	Swauger Creek......	Worthley cabin	
June 7	.....do	....do	.....do	
July 13	.....do	....do	.....do	
Aug. 6	.....do	....do	.....do	
Aug. 7	.....do	Little Walker River.		
May 1	.....do	Lost Canyon Creek.	Above ditches	
May 2	.....do	....do	Below sawmill bridge.	
June 9	.....do	....do	Below Little Antelope ditches.	
July 14	.....do	....do	.....do	
May 27	.....do	Sweetwater Creek..	Below ditches........	
July 2	.....do	....do	Entire creek	
Do ...	.....do	....do	Above Roach ranch ..	

Miscellaneous stream measurements in Walker River basin, California and Nevada, in 1903—Continued.

Date.	Hydrographer.	Stream.	Location.	Discharge.
				Sec.-ft.
July 3	I. W. Huffaker.....	Sweetwater Creek..	Above Conway ditches	8.2
Do...	.....do	.....do	Above Compton ditches.	10.2
Do...	.....do	.....do	Below ditches........	1.3
May 27	.....do	Fryingpan Creek...	Above ditches	2.8
May 6	.....do	Desert Creek........	.....do	15.1
June 23	.....do	.....do	.....do	30
July 27	.....do	.....do	.....do	7.6
July 24	.....do	Walker River......	Below forks	68
Aug. 9	.....do	.....do	.....do	15.0
July 22	.....do	.....do	At Lower Mason Valley Bridge.	45.0

Miscellaneous stream measurements in Washoe Valley, Nevada, in 1903.

Aug. 12	H. Jameson........	Simmonds.........	Above all diversions..	0.3
July 23	.....do	Musdrove	.....do	.5
Aug. 19	.....do	.....do	.....do	.5
July 23	.....do	Lewers	.....do	.5
Do...	.....do	Franktown	.....do	5.1
Aug. 19	.....do	.....do	.....do	3.6
July 22	.....do	Ophir	.....do	12.5
Aug. 19	.....do	.....do	.....do	5.2
July 23	.....do	Browns............	.....do	1.2
Aug. 18	.....do	.....do	.....do	.4
July 22	.....do	Galena	.....do	13.9

Miscellaneous stream measurements in Truckee Meadows, Nevada, in 1903.

July 23	H. Jameson........	Steamboat	Above all diversions..	20
Do...	.....do	.....do	.....do	22
July 22	.....do	Thomas	.....do	2.2
Aug. 3	.....do	Evans	.....do	1.2
June 20	.....do	Whites	.....do	11.6
July 22	.....do	.....do	.....do	5.7
June 4	.....do	Hunter.............	.....do	18.5
July 26	.....do	.....do	Above Steamboat ditch	2.3
Aug. 28	.....do	.....do	Above all diversions..	7.3
May 2	.....do	Dog	.....do	25

Miscellaneous measurements of streams flowing into Lake Tahoe, California and Nevada, in 1903.

Date.			Location.	Discharge
				Sec.ft
Aug. 1	G. B. Lorenz	Ward Creek	Near Lake Tahoe......	5.1
Do ...	.....do	Blackwood Creek ..	.....do	7.3
Do ...	.....do	Madden Creek.....	.....do	.3
Do ...	.....do	Unnamed creek....	.....do	Trace.
Do ...	.....do	McKinney Creek...	.....do	.4
Do ...	.....do	General Creek....	.....do	1.3
Do ...	.....do	Meiggs Creek	.....do	2.7
Do ...	.....do	Lonely Gulch Creek	.....do	1.0
Aug. 2	.....do	Frosts Creek.......	.....do	2.3
Do ...	.....do	Unnamed creek....	.....do	.9
Do ...	.....do	.....do	.....do	1.0
Do ...	.....do	Emerald Creek	.....do	5.4
Do ...	.....do	Cascade	.....do	4.9
Do ...	.....do	Trout	.....do	29
Do ...	.....do	Little Truckee	.....do	61
Aug. 3	.....do	Zephyr Cove.......	.....do	Trace.
Do ...	.....do	Unnamed	.....do	1.0
Do ...	.....do	Unnamed (South Cave Rock).	.....do	.1
Do ...	.....do	Unnamed (North Glenbrook).	.....do	1.1
Do ...	.....do	Unnamed (at Glenbrook).	.....do	.6
Aug. 4	.....do	Unnamed...........	.....do	Trace.
Do ...	.....do	.....do	.....do	1.2
Do ...	.....do	Unnamed (at Breakwater).	.....do	.5
Do ...	.....do	Unnamed (south of Incline).	.....do	.5
Do ...	.....do	Incline	.....do	1.3
Do ...	.....do	Unnamed (West Incline).	.....do	1.8
Do ...	.....do	.....do	.....do	.9
Do ...	.....do	.....do	.....do	1.0
Do ...	.....do	.....do	.....do	1.0
Aug. 5	.....do	Unnamed (west of Brockway).	.....do	.8
Do ...	.....do	Unnamed (east of Woodward).	.....do	Trace.

urements of ditches on south side of Truckee Meadows, Nevada, in 1903.

Hydrographer.	Ditch.	Location.	Discharge.
			Sec.-ft.
H. Jameson	Steamboat canal	Near head-gate	0.0
....do	....do	....do	0.0
....do	....do	....do	33
....do	....do	....do	64
....do	....do	....do	67
....do	Mayberry or Holcolm	....do	0.0
....do	....do	....do	0.0
....do	....do	....do	24
....do	....do	....do	41
....do	....do	....do	47
....do	Truckee Meadows or Lake.	....do	0.0
....do	....do	....do	0.0
....do	....do	....do	29
....do	....do	....do	38
....do	....do	....do	39
....do	....do	....do	31
....do	South Side Irrigation Co., or Wheeler.	....do	6.9
....do	....do	....do	25
....do	....do	....do	52
....do	....do	....do	52
....do	Indian Flat, or Fray	....do	1.0
....do	....do	....do	0.0
....do	....do	....do	12.5
....do	....do	....do	16.8
....do	....do	....do	16.8
....do	....do	....do	14.6
....do	Haydon	....do	3.4
....do	....do	....do	1.5
....do	Cochran	....do	20
....do	....do	....do	22
....do	....do	....do	33.
....do	....do	....do	22
....do	....do	....do	31
....do	....do	....do	28
....do	....do	....do	27
....do	Scott ranch	....do	.5
....do	....do	....do	11.6
....do	....do	....do	36
....do	....do	....do	36

Measurements of ditches on south side of Truckee Meadows, Nevada, in 1903—Continued

Date.	Hydrographer.	Ditch.	Location.	Discharge.
				Sec.-ft.
June 1	H. Jameson	Scott ranch	Near head-gate...	36
July 17	.....do	.....do	.....do	28
Apr. 22	.....do	Abbey	.....do	8.
May 25	.....do	.....do	.....do	11 .
June 5	.....do	.....do	.....do	17 -
June 26	.....do	.....do	.....do	14 -
July 17	.....do	.....do	.....do	10 -
Apr. 29	.....do	Wilson	.....do	5 -
May 25	.....do	.....do	.....do	4 -
June 6	.....do	.....do	.....do	5 -
June 26	.....do	.....do	.....do	5 -
July 17	.....do	.....do	.....do	3 -
Aug. 1	.....do	.....do	.....do	6 -
Apr. 18	.....do	Pioneer.	.....do	14 -
Apr. 29	.....do	.....do	.....do	22
May 25	.....do	.....do	.....do	30
June 6	.....do	.....do	.....do	43
June 26	.....do	.....do	.....do	36
July 17	.....do	.....do	.....do	15 -

Measurements of ditches on north side of Truckee Meadows, Nevada, in 1903.

Date.	Hydrographer.	Ditch.	Location.	Discharge.
Apr. 17	H. Jameson........	Merrill	Near head-gate...	0 . 0
May 2	.....do	.....do	.....do	13 . 6
May 21	.....do	.....do	.....do	13 . 6
June 19	.....do	.....do	.....do	23
July 20	.....do	.....do	.....do	14 . 0
Apr. 17	.....do	Kruger or Kates.........	.....do	0 . 0
May 2	.....do	.....do	.....do	2 . 0
June 19	.....do	.....do	.....do	11 . 3
July 20	.....do	.....do	.....do	11 . 8
Apr. 15	.....do	Highland.	.....do	0 . 0
Apr. 17	.....do	.....do	.....do	14 . 7
Apr. 27	.....do	.....do	.....do	18 . 1
May 22	.....do	.....do	.....do	25
June 19	.....do	.....do	.....do	35
July 20	.....do	.....do	.....do	31
Apr. 27	.....do	Hogan	.....do	5 . 2
May 20	.....do	.....do	.....do	5 . 2
June 19	.....do	.....do	.....do	8 . 7

nmeasurements of ditches on north side of Truckee Meadows, Nevada, in 1903—Continued.

Date.	Hydrographer.	Ditch.	Location.	Dis-charge.
				Sec.-ft.
ıy 2	H. Jameson........	Mayberry & Carr	Near head-gate...	3.3
Do...	.....do	Hunter..............	..do	1.2
ıy 20	.....do	Mayberry, north side ..	.....do	2.1
ıy 26	.....do	.....do	.....do	2.1
r. 15⎱ r. 21⎰	.....do	Orr	.....do	0.0
r. 23	.....do	.....do	.....do	35
y 14	.....do	.....do	.....do	84
y 2	.....do	.....do	.....do	84
y 27	.....do	.....do	.....do	84
ıe 1	.....do	.....do	.....do	90
r. 23	.....do	Countryman........	.....do	0.0
y 20	.....do	.....do	.....do	1.0
ıe 1	.....do	.....do	.....do	4.9
ıe 4	.....do	.....do	.....do	4.9
r. 23	.....do	Chrisholm or Chism....	.....do	1.7
ıy 20	.....do	.....do	.....do	1.0
ne 1	.....do	.....do	.....do	2.5
ne 4	.....do	.....do	.....do	2.5
pr. 15	.....do	Sullivan.............	.....do	26
pr. 22	.....do	.....do	.....do	37
ıne 1	.....do	.....do	.....do	54
me 22	.....do	.....do	.....do	34
ly 21	.....do	.....do	.....do	20
r. 15	.....do	English Mill or Auburn	.....do	0.0
r. 21	.....do	.....do	.....do	13.9
ıy 14	.....do	.....do	.....do	26
y 27	.....do	.....do	.....do	20
ıe 17	.....do	English Mill........	.....do	24
r. 20	.....do	Perry or Columbo......	.....do	2.7
ıe 26	.....do	.....do	.....do	4.6
ʼ. 18	.....do	North Truckee Irrigation	.....do	0.0
ʼ. 20	.....do	.....do	.....do	0.0
r 14	.....do	.....do	.....do	44
r 27	.....do	.....do	.....do	29
e 17	.....do	.....do	.....do	23
e 26	.....do	.....do	.....do	23
r 17	.....do	.....do	.....do	25
ʼ. 1	.....do	.....do	.....do	27
. 18	.....do	Sessions	.....do	2.0

Measurements of ditches on north side of Truckee Meadows, Nevada, in 1903—Continued.

Date.	Hydrographer.	Ditch.	Location.	Discharge.
				Sec.-ft.
Apr. 20	H. Jameson.........	Sessions	Near head-gate...	2.3
May 14	....do	....do	....do	20
May 27	....do	....do	....do	13.1
June 17	....do	....do	....do	16.5
June 26	....do	....do	....do	13.8
July 20	....do	....do	....do	8.6
Aug. 1	....do	....do	....do	8.4
Apr. 18	....do	Mitchell & Carmac.....	....do	13.5
Apr. 28	....do	....do	....do	22
May 18	....do	....do	....do	15.8
June 2	....do	....do	....do	23
June 17	....do	....do	....do	19.1
June 26	....do	....do	....do	19.1
Apr. 18	....do	Glendale	....do	15.6
Apr. 28	....do	....do	....do	16.8
May 18	....do	....do	....do	12.2
June 2	....do	....do	....do	12.9
June 17	....do	....do	....do	7.6
June 26	....do	....do	....do	14.4
July 17	....do	....do	....do	14.5

Measurements of ditches on Truckee River below Truckee Cox Meadows, Nevada, in 1903.

May 7	H. Jameson.........	Largnasino	Near head-gate...	18.1
May 12	....do	Sheep Ranch	....do	9.2
May 11	....do	O'Brien	....do	8.2
May 8	....do	Wadsworth Irrigation Co	....do	17.1
May 11	....do	Proctor..................	....do	7.8
Do...	....do	Olinghouse	....do	3.1
Do...	....do	Fellnagle	....do	7.9
May 7	....do	Hill.....................	....do	2.4
May 10	....do	Indian Agency..........	....do	26

Measurements of ditches from Buckeye Creek, California, in 1903.

Apr. 28	I. W. Huffaker	Reason Barnes.........	Near head-gate...	1.8
May 29	....do	....do	....do	20
July 8	....do	....do	....do	13.7
Apr. 28	....do	Rickey	....do	16.4
May 29	....do	....do	....do	35

Measurements of ditches from Buckeye Creek, California, in 1903—Continued.

ate.	Hydrographer.	Ditch.	Location.	Discharge.
				Sec.-ft.
8	I. W. Huffaker	Rickey	Near head-gate...	22
28	.....do	Severe..................	.....do	7.0
2	.....do	.....do	.....do	34
8	.....do	.....do	.....do	28
2	.....do	Elliott.................	.....do	11.6
9	.....do	.....do	.....do	7.2
2	.....do	Elliott, Day & Simpson.	.....do	4.8
o	.....do	Whitney Day............	.....do	5.6
		UPPER NORTH CHANNEL.		
2	I. W. Huffaker	Elliott 1	Near head-gate...	5.1
o	.....do	Elliott 2	.....do	.7
o	.....do	Elliott 3	.....do	1.6
o	.....do	Elliott 4	.....do	11.6
		LOWER NORTH CHANNEL.		
2	I. W. Huffaker	Elliott 1	Near head-gate...	.9
o	.....do	Elliott 2	.....do	9.4
o	.....do	Elliott 3	.....do	2.9
o	.....do	Day & Simpson	.....do	2.2
		LOWER SOUTH CHANNEL.		
2	I. W. Huffaker	Elliott, Day & Simpson.	Near head-gate...	10.1
9	.....do	.....do	.....do	4.2

Measurements of Robinson Creek ditches, Nevada, in 1903.

	Hydrographer.	Ditch.	Location.	Discharge.
28	I. W. Huffaker	Huniwel, New	Near head-gate...	0.0
27	.....do	.....do	.....do	0.0
6	.....do	.....do	.....do	4.4
28	.....do	Huniwel, Old...........	.....do	13.2
4	.....do	.....do	.....do	93
8	.....do	.....do	.....do	71
4	.....do	Huniwel and Elliott....	.....do	31
8	.....do	.....do	.....do	30
4	.....do	Rickey	.....do	7.9
8	.....do	.....do	.....do	7.9
4	.....do	Huniwel	.....do	7.3
8	.....do	.....do	.....do	7.5
29	.....do	Kirkwood, Hays, Brandon.	.....do	12.4

Measurements of Robinson Creek ditches, Nevada, in 1903—Continued.

Date.	Hydrographer.	Ditch.	Location.	Dis-charge.
				Sec.-ft.
July 6	I. W. Huffaker.....	Severe flume, lateral of Severe (Buckeye).	Near head-gate...	1.1
May 29	.....do	Open Chain	.....do	11.1
Do...	.....do	McCullough	.....do	10.1
Do...	.....do	Huntoon	.....do	14.1
July 9	.....do	.....do	.....do	11.1
June 2	.....do	Kelso	.....do	4.8
July 9	.....do	.....do	.....do	5.5
June 2	.....do	Elliott Severe......	.....do	17.7
July 9	.....do	.....do	.....do	10.4
Apr. 28	.....do	Elliott 2...........	.....do	2.4
May 28	.....do	Elliott 1...........	.....do	9.0
July 6	.....do	Elliott No. 1.......	.....do	10.9
June 2	.....do	Elliott No. 2.......	.....do	1.6

Measurements of Green Creek ditches, California, in 1903.

Date.	Hydrographer.	Ditch.	Location.	Dis-charge.
July 7	I. W. Huffaker	Dynamo..............	Near head-gate...	5.3
June 3	.....do	.....do	.....do	0.1
Do...	.....do	Shell...............	.....do	29
July 7	.....do	.....do	.....do	0.
June 3	.....do	Green Creek No. 1 ..	.....do	13.
July 7	.....do	.....do	.....do	0.
June 3	.....do	Green Creek No. 2 ..	.....do	38
July 7	.....do	.....do	.....do	0.0
June 4	.....do	Lower Green Creek ...	.....do	22
July 11	.....do	.....do	.....do	12.4
June 4	.....do	.....do	Near head-gate...	22
July 11	.....do	.....do	.....do	17.2

Measurements of Dogtown Creek ditches, California, in 1903.

Date.	Hydrographer.	Ditch.	Location.	Dis-charge.
July 7	I. W. Huffaker	Osborne & Kinney	Near head-gate...	3.8
June 3	.....do	Bryant	.....do	0.0
July 11	.....do	.....do	.....do	9.5
June 3	.....do	Dogtown No. 1.........	.....do	24
July 7	.....do	.....do	.....do	0.0
June 3	.....do	Dogtown No. 2.........	.....do	0.0

asurements of East Walker River ditches, California, in 1903.

Hydrographer.	Ditch.	Location.	Discharge.
			Sec.-ft.
V. Huffaker.....	Williams	Near head-gate...	5.8
.do	Stewart..................	.....do	1.4

Measurements of East Walker River ditches, Nevada, in 1903.

Hydrographer.	Ditch.	Location.	Discharge.
V. Huffaker.....	Fulston, lower east side .	Near head-gate...	8.7
.do	Fulston, lower east side A.	.....do	18.7
.do	Fulston, lower east side B.	.....do	1.2
.do ...:.......	Fulston, upper west side.	.....do	1.9
.do ....'.	.....do	.....do	3.4
.do	.....do	.....do	3.9
.do	Fridell, west side.......	.....do	3.5
.do	Fridell, east side........	.....do	3.7
.do	Peterson.................	.....do	0.0
.do	Conway, upper east.....	.....do	12.2
.do	Conway, upper west	.....do	8.7
.do	Conway, lower east.....	.....do	0.0
.do	Conway, lower west	.....do	0.0
.do	Boerlin..................	.....do	2.7
.do	Grullie, upper east	.....do	0.0
.do	Grullie, lower east......	.....do	0.0
.do	Morgan, upper west	.....do	4
.do	Morgan, upper east.....	.....do	3.2
.do	Morgan, lower west.....	.....do	5.9
.do	Morgan, lower east	.....do	0.0
.do	Poli, east	.....do	.8
.do	Poli, west..............	.....do	2.4
.do	Webster, Wheeler, west.	.....do	4.5
.do	Webster, Wheeler, east .	.....do	5.9
.do	Webster, Webster Bros..	.....do	14.9
.do	Latipee, east..........	.....do	6
.do	Latipee, upper west	.....do	14.1
.do	Latipee, lower west	.....do	5.1
.do	Gignoux, west..........	.....do	6.7
.do	Gignoux, east..........	.....do	1.1
.do	Glenn, west	.....do	7.9
.do	Charlesbois.............	.....do	8.2

Measurements of Sweetwater Creek ditches, Nevada, in 1903.

Date.	Hydrographer.	Ditch.	Location.	
lay 27	I. W. Huffaker......	Yparraguerre............	Near head-gate...	
uly 2	.....do	.....do	.....do	
lay 27	.....do	Roach................	.....do	
uly 2	.....do	.....do	.....do	
lay 27	.....do	Compton & Conway......	.....do	
uly 3	.....do	.....do	.....do	

Measurements of Bodie Creek ditches, Nevada, in 1903.

lay 26	I. W. Huffaker	Ravanelle, south	Near head-gate...	
Do ...	.....do	Ravanelle, north	.....do	

Measurements of Antelope Valley ditches, California, in 1903.

May	1	I. W. Huffaker	Taylor	Near head-gate...	
June	8	.....do	.....do	.....do	
May	1	.....do	.....do	.....do	
June	9	.....do	Hardy	.....do	
July	16	.....do	.'.do	.....do	
May	1	.....do	West Goodnow	.....do	
June	9	.....do	.....do	.....do	
July	16	.....do	.....do	.....do	
May	1	.....do	Chichester No. 1	.....do	
June	9	.....do	.....do	.....do	
July	16	.....do	.....do	.....do	
May	1	.....do	Chichester No. 2	.....do	
June	9	.....do	.....do	.....do	
July	16	.....do	.....do	.....do	
May	1	.....do	Coleville No. 2........	.....do	
June	9	.....do	.....do	.....do	
July	16	.....do	.....do	.....do	
May	1	.....do	Tunnel and Alkali	.....do	
June	9	.....do	.....do	.....do	
July	16	.....do	.....do	.....do	
May	1	.....do	Swauger............	.....do	
June	9	.....do	.....do	.....do	
July	16	.....do	.....do	.....do	
May	2	.....do	Wiley	.....do	
June	9	.....do	.....do	.....do	

ate.	Hydrographer.	Ditch.	Location.	Discharge.
				Sec.-ft.
4	I. W. Huffaker.....	Carney................	Near head-gate...	0.0
11	.....do	....do	.....do	7.2
15	.....do	.....do	.....do	6.3
4	.....do	Kirman & Rickey	.....do	53
11	.....do	.....do	.....do	42
15	.....do	.....do	.....do	34
4	.....do	Larsen	.....do	0.0
11	.....do	.....do	.....do	23
15	.....do	.....do	.....do	.22
4	.....do	Gullickson	.....do	0.0
11	.....do	.....do	.....do	11.1
15	.....do	.....do	.....do	.4
4	.....do	W. W. Co. or Slough....	.....do	34
11	.....do	.....do	.....do	75
15	.....do	.....do	.....do	60
4	.....do	Eggleston	.....do	0.0
16	.....do	.....do	.....do	0.0
4	.....do	Powell	.....do	0.0
11	.....do	.....do	.....do	9.7
16	.....do	.....do	.....do	4.1
4	.....do	Goodnow No. 1	.....do	0.0
11	.....do	.....do	.....do	7.2
16	.....do	.....do	.....do	0.0
4	.....do	Goodnow No. 2	.....do	0.0
11	.....do	.....do	.....do	12.6
16	.....do	.....do	.....do	0.0
2	.....do	Lancaster.............	.....do	0.0
11	.....do	.....do	.....do	15.8
10...	.....do	Pitts & Radley........	.....do	7.7
17	.....do	.....do	.....do	0.0

Measurements of Lost Canyon Creek ditches, California, in 1903.

2	I. W. Huffaker.....	Allard & Rickey	Near head-gate...	0.0
14	.....do	.....do	.....do	5.1
2	.....do	McKay & Rickey	.....do	1.5
14	.....do	.....do	.....do	3.2
9	.....do	Shields...............	.....do	2.2

May 5	......do	Rodger	..........do	
June 22	......do	......do	..........do	2?
July 25	......do	......do	..........do	15
May 5	......do	West Walker Co	..........do	2?
June 22	......do	......do	..........do	21
July 25	......do	......do	..........do	10.?
May 5	......do	Smith	..........do	5
June 22	......do	......do	..........do	4.6
July 25	......do	......do	..........do	5.5
May 6	......do	Simpson Colony	..........do	4?
June 22	......do	......do	..........do	47
July 25	......do	......do	..........do	39
May 6	......do	Turner	..........do	1.3
June 22	......do	......do	..........do	1.4
July 25	......do	......do	..........do	.7

nents of Antelope Valley ditches, California, in 1903—Continued.

Hydrographer.	Ditch.	Location.	Discharge.
			Sec.-ft.
. Huffaker......	Carney	Near head-gate....	0.0
do	.....do	.....do	7.2
do	.....do	.....do	6.3
do	Kirman & Rickey	.....do	53
do	.....do	.....do	42
do	.....do	.....do	34
do	Larsen	.....do	0.0
do	.....do	.....do	23
do	.....do	.....do	22
do	Gullickson	.....do	0.0
do	.....do	.....do	11.1
do	.....do	.....do	.4
do	W. W. Co. or Slough....	.....do	34
do	.....do	.....do	75
do	.....do	.....do	60
do	Eggleston	.....do	0.0
do	.....do	.....do	0.0
do	Powell	.....do	0.0
do	.....do	.....do	9.7
do	.....do	.....do	4.1
do	Goodnow No. 1	.....do	0.0
do	.....do	.....do	7.2
do	.....do	.....do	0.0
do	Goodnow No. 2	.....do	0.0
do	.....do	.....do	12.6
do	.....do	.....do	0.0
do	Lancaster.............	.....do	0.0
do	.....do	.....do	15.8
do	Pitts & Radley........	.....do	7.7
do	.....do	.....do	0.0

surements of Lost Canyon Creek ditches, California, in 1903.

. Huffaker......	Allard & Rickey	Near head-gate....	0.0
do	.....do	.....do	5.1
do	McKay & Rickey	.....do	1.5
do	.....do	.....do	3.2
do	Shields................	.....do	2.2

Measurements of East Fork of Walker River ditches, Mason Valley, Nevada, in 190?—
Continued.

Date.	Hydrographer.	Ditch.	Location.	N...
				Sec.
June 16	I. W. Huffaker......	Bewley...................	Near head-gate...	
May 19	.....do	Mickey................	....do	
June 16	.....do	....do	....do	
July 20	.....do	....do	....do	
May 19	.....do	Fox..................	....do	
June 16	.....do	....do	....do	
July 20	.....do	....do	....do	
May 19	.....do	Sweetman...............	....do	
June 16	.....do	....do	....do	
July 20	.....do	....do	....do	

Measurements of Walker River ditches, Mason Valley, Nevada, in 1903.

May 15	I. W. Huffaker......	Spragg Woodcock	Near head-gate...	
June 17	.....do	....do	....do	
July 21	.....do	....do	....do	
May 15	.....do	McCloud	....do	
June 17	.....do	....do	....do	
July 21	.....do	....do	....do	
May 15	.....do	Nichols................	....do	
June 17	.....do	....do	....do	
May 15	.....do	Campbell...............	....do	
June 17	.....do	....do	....do	
May 15	.....do	Merritt................	....do	
June 17	.....do	....do	....do	
May 15	.....do	Sprague, Allcorn, Bewley & Kimball.	....do	
June 17	.....do	....do	....do	
May 16	.....do	Gold Hill..............	....do	
June 15	.....do	....do	....do	
May 16	.....do	Johnson Lane	....do	
June 15	.....do	....do	....do	
May 16	.....do	Joggles................	....do	
June 15	.....do	....do	....do	
May 16	.....do	Cleaver................	....do	
June 15	.....do	....do	....do	

Measurements of West Fork of Walker River, Mason Valley, Nevada, in 1903.

Date.	Hydrographer.	Ditch.	Location.	Discharge.
				Sec.-ft.
ly 20	I. W. Huffaker	Wilson tunnel	Near head-gate ..	19.4
ne 20	.....do	.....do	.....do	30
ly 23	.....do	.....do	.....do	24
ly 20	.....do	Wilson mill............	.....do	8
ne 20	.....do	.....do	.....do	4.2
ly 23	.....do	.....do	.....do	4.9
ly 20	.....do	Wilson Palmer.........	.....do	2.6
me 20	.....do	.....do	.....do	11.2
ly 23	.....do	.....do	.....do	7.4
ly 20	.....do	Lee & Sanders	.....do	14
me 20	.....do	.....do	.....do	12
ly 24	.....do	.....do	.....do	13.5
ly 20	.....do	Wheeler...............	.....do	5.1
me 20	.....do	.....do	.....do	7.4
ly 23	.....do	.....do	.....do	5.2
ly 20	.....do	Alkali.................	.....do	57
ne 20	.....do	.....do	.....do	71
ly 24	.....do	.....do	.....do	13.3

Measurements of East Fork of Walker River ditches, Mason Valley, Nevada, in 1903.

Date.	Hydrographer.	Ditch.	Location.	Discharge.
ly 21	I. W. Huffaker	Ross	Near head-gate ..	8.5
ne 30	.....do	.....do	.....do	5.9
ly 21	.....do	Ross No. 2.............	.....do	0.0
ne 30	.....do	.....do	.....do	1.9
ly 19	.....do	Greenwood	.....do	46
ne 16	.....do	.....do	.....do	41
ly 20	.....do	.....do	.....do	44
ly 19	.....do	Nelson	.....do	0.0
ne 16	.....do	.....do	.....do	8.2
ly 20	.....do	.....do	.....do	1.0
ly 19	.....do	Daniels................	.....do	18.1
ne 16	.....do	.....do	.....do	20
ly 20	.....do	.....do	.....do	17
ly 19	.....do	Baker	.....do	8.6
me 16	.....do	.....do	.....do	31
me 16	.....do	.....do	.....do	17.8
ly 20	.....do	.....do	.....do	9.2
ly 19	.....do	Bewley................	.....do	1.1

Measurements of East Fork of Walker River ditches, Mason Valley, Nevada, in 1903—Continued.

Date.	Hydrographer.	Ditch.	Location.	Dis- charge
				Sec.
June 16	I. W. Huffaker......	Bewley..................	Near head-gate ..	
May 19	.....do	Mickey..........	.....do	3
June 16	.....do	....do	.....do	3
July 20	.....do	....do	.....do	2
May 19	.....do	Fox.....................	.....do	3
June 16	.....do	....do	.....do	1
July 20	.....do	....do	.....do	2
May 19	.....do	Sweetman	.....do	
June 16	.....do	....do ....:......	.....do	
July 20	.....do	....do	.....do	

Measurements of Walker River ditches, Mason Valley, Nevada, in 1903.

Date.	Hydrographer.	Ditch.	Location.	Dis- charge
May 15	I. W. Huffaker......	Spragg Woodcock	Near head-gate...	
June 17	.....do	....do	.....do	
July 21	....do	....do	.....do	
May 15	.....do	McCloud	.....do	
June 17	.....do	....do	.....do	
July 21	.....do	....do	.....do	
May 15	.....do	Nichols.............	.....do	
June 17	.....do	....do	.....do	
May 15	.....do	Campbell.............	.....do	
June 17	.....do	.....do	.....do	
May 15	.....do	Merritt.............	.....do	
June 17	.....do	....do	.....do	
May 15	.....do	Sprague, Allcorn, Bewley & Kimball.	.....do	
June 17	.....do	....do	.....do	1
May 16	.....do	Gold Hill.............	.....do	
June 15	.....do	.....do	.....do	1
May 16	.....do	Johnson Lane	.....do	
June 15	.....do	.....do	.....do	
May 16	.....do	Joggles	.....do	
June 15	.....do	....do	.....do	
May 16	.....do	Cleaver.............	.....do	
June 15	.....do	....do	.....do	

surements of Humboldt River ditches, Lovelock Valley, Nevada, in 1903—Continued.

Date.	Hydrographer.	Ditch.	Location.	Gage.	Discharge.
					Sec. ft.
y 9	W. A. Wolf....	Union canal	Near head-gate.....	1.90	47
ly 13	.....do	.....do	.....do	2.62	72
ly 15	.....do	.....do	.....do	1.85	41
g. 15	.....do	.....do	.....do	.40	9.0
y 29	.....do	Union Canal Fuse..	.....do	......	2.6
y 26	.....do	Marzen Slough.....	¼ mile south of rail-road.	......	8.2
e 18	.....do	.....do	.....do	......	7.5
r 2	.....do	.....do	.....do	......	1.7
. 18	.....do	.....do	.....do	......	1.7
' 26	.....do	Marzen Slough drain.	Near head	......	5.9
' 2	.....do	.....do	.....do	......	.8
' 16	.....do	Sommers	Near head-gate	......	10.7
. 18	.....do	.....do .	.....do	......	0.0
r 21	.....do	Kewley Slough	.....do	......	2.9
e 17	.....do	.....do	.....do	......	0.0
y 25	.....do	.....do	.....do	......	5.0
y 1	.....do	.....do	.....do	......	8.7
g. 18	.....do	.....do	.....do	......	0.0
y 25	.....do	Reservoir...........	.....do	......	2.3
ne 17	.....do	.....do	.....do	......	0.0
ly 1	.....do	.....do	.....do	......	8.9
ly 10	.....do	.....do	.....do	......	19.0
g. 18	.....do	.....do	.....do	......	.5
y 21	.....do	Drain at camp No. 2.	At camp No. 2	......	3.4
y 25	.....do	.....do	.....do	......	6.8
ne 17	.....do	.....do	.....do	......	0.0
y 1	.....do	.....do	.....do	......	2.8
y 10	.....do	.....do	.....do	......	0.0
g. 18	.....do	.....do	.....do	......	0.0
y 10	.....do	Reservoir and Kewley Slough.	Below junction.....	......	30
y 16	.....do	do	.....do	......	43

Measurements of Humboldt River ditches between Battle Mountain and Golconda, Nev., in 1903.

Date.	Hydrographer.	Ditch.	Location.	Gage.	
June 8	W. A. Wolf	McIntyre	Near head-gate		
July 22	...do	do	do		
June 7	...do	Nelson, north side	do		
Do	...do	Nelson, south side	do		
June 10	...do	Pinson No. 1	do		
Do	...do	Pinson No. 2	do		

Measurements of Humboldt River ditches below Golconda, Nev., in 1903.

June 10	W. A. Wolf	Humboldt canal	Near head-gate		
Do	...do	Pole Creek	do		
June 5	...do	Button's upper	do		
Do	...do	Pioneer Extension	do		
June 11	...do	Stauffer No. 1	do		
Do	...do	Stauffer No. 2	do		
June 12	...do	Trousdale No. 1	do		
Do	...do	Trousdale No. 2	do		
Do	...do	Trousdale Main	do		28
Do	...do	Stauffer No. 3	do		3.6
Do	...do	Stauffer No. 4	do		3.1

Measurements of East Carson ditches, Nevada, in 1903.

Date.	Hydrographer.	Ditch.	Location.	Discharge.
				Sec.-ft.
May 11	A. H. Schadler	Heitman	Near head-gate	0.0
May 28	...do	do	do	1.2
June 29	...do	do	do	8.4
July 14	...do	do	do	14.6
May 2	...do	Rodenbah	do	0.0
May 13	...do	do	do	1.4
May 28	...do	do	do	.8
June 29	...do	do	do	1.4
July 14	...do	do	do	1.9
May 2	...do	High flier	do	2.9
May 28	...do	do	do	.6
June 24	...do	do	do	.2
July 14	...do	do	do	.6

Measurements of Humboldt River ditches, Lovelock Valley, Nevada, in 1903—Continued.

Date.	Hydrographer.	Ditch.	Location.	Gage.	Dis-charge.
					Sec. ft.
July 9	W. A. Wolf....	Union canal	Near head-gate.....	1.90	47
July 13	.....do	.....do	.....do	2.62	72
July 15	.....do	.....do	.....do	1.85	41
Aug. 15	.....do	.....do	.....do	.40	9.0
May 29	.....do	Union Canal Fuss..	.....do		2.6
May 26	.....do	Marzen Slough.....	¼ mile south of rail-road.		8.2
June 18	.....do	.....do	.....do		7.5
July 2	.....do	.....do	.....do		1.7
Aug. 18	.....do	.....do	.....do		1.7
May 26	.....do	Marzen Slough drain.	Near head		5.9
July 2	.....do	.....do	.....do		.8
July 16	.....do	Sommers	Near head-gate		10.7
Aug. 18	.....do	.....do	.....do		0.0
May 21	.....do	Kewley Slough	.....do		2.9
June 17	.....do	.....do	.....do		0.0
May 25	.....do	.....do	.....do		5.0
July 1	.....do	.....do	.....do		8.7
Aug. 18	.....do	.....do	.....do		0.0
May 25	.....do	Reservoir...........	.....do		2.3
June 17	.....do	.....do	.....do		0.0
July 1	.....do	.....do	.....do		8.9
July 10	.....do	.....do	.....do		19.0
Aug. 18	.....do	.....do	.....do		.5
May 21	.....do	Drain at camp No. 2.	At camp No. 2		3.4
May 25	.....do	.....do	.....do		6.8
June 17	.....do	.....do	.....do		0.0
July 1	.....do	.....do	.....do		2.8
July 10	.....do	.....do	.....do		0.0
Aug. 18	.....do	.....do	.....do		0.0
July 10	.....do	Reservoir and Kewley Slough.	Below junction.....		30
July 16	.....do	do	.....do		43

Measurements of East Carson ditches, Nevada, in 1903—Continued.

Date.	Hydrographer.	Ditch.	Location.	
July 14	A. H. Schadler	I. M. Christenson	Near head-gate	
May 11	....do	Nelson	....do	
May 28	....do	....do	....do	
July 1	....do	....do	....do	
July 14	....do	....do	....do	
May 11	....do	Wehrman	....do	
May 28	....do	....do	....do	
July 1	....do	....do	....do	
July 14	....do	....do	....do	
May 11	....do	Lang & Sill	....do	
May 27	....do	....do	....do	
July 1	....do	....do	....do	
July 13	....do	....do	....do	
May 11	....do	Jensen, Heise & Co.	....do	
May 27	....do	....do	....do	
July 1	....do	....do	....do	
July 13	....do	....do	....do	
May 12	....do	Company	....do	
June 3	....do	....do	....do	
June 29	....do	....do	....do	
July 14	....do	....do	....do	
May 11	....do	Christenson Co.	....do	
May 27	....do	....do	....do	
July 1	....do	....do	....do	
July 14	....do	....do	....do	
May 12	....do	Dangberg & Hussman	....do	
June 3	....do	....do	....do	
June 29	....do	....do	....do	
July 15	....do	....do	....do	
May 12	....do	Hussman	....do	
June 3	....do	....do	....do	
June 29	....do	....do	....do	
May 12	....do	Hussman, Helewinkle & Lampe.	....do	
June 3	....do	....do	....do	
June 29	....do	....do	....do	
May 4	....do	Lampe Slough	....do	
May 2	....do	....do	....do	
May 12	....do	....do	....do	
June 30	....do	....do	....do	

Measurements of East Carson ditches, Nevada, in 1903—Continued.

ate.	Hydrographer.	Ditch.	Location.	Discharge.
				Sec.-ft.
y 15	A. H. Schadler	Lampe Slough	Near head-gate...	0. 6
y 4	.....do	Emigrant..............	.....do	1. 5
1e 2	.....do	.....do	.....do	30
1e 30	.....do	.....do	.....do	1. 3
ly 15	.....do	.....do	.....do	5. 3
y 4	.....do	Banning & McFanning..	.....do	14. 5
1e 2	.....do	.....do	.....do	21
1e 30	.....do .....:..	.....do	.....do	14. 5
y 15	.....do	.....do	.....do	7. 8
1e 2	.....do	Jepsen, Martin & Ezell	.....do	9. 1
y 1	.....do	.....do	.....do	8. 2
y 15	.....do	.....do	.....do	0. 0
1e 2	.....do	Springmeyer & Cohn...	.....do	7
y 1	.....do	.....do	.....do	3. 6
y 15	.....do	.....do	.....do	0. 0
y 12	.....do	Springmeyer, Jepsen...	.....do	8. 8
1e 2	.....do	.....do	.....do	10. 8
1e 30	.....do	.....do	.....do	2
y 15	.....do	.....do	.....do	4. 6
y 12	.....do	Yori	.....do	0. 0
1e 2	.....do	.....do	.....do	1. 6
1e 30	.....do	.....do	.....do	5. 3
y 15	.....do	.....do	.....do	4. 9
y 27	.....do	Thompson & Christenson.	.....do	1. 8
1e 30	.....do	.....do	.....do	.5
y 11	.....do	Heitmen Co	.....do	14. 5
y 27	.....do	.....do	.....do	21
y 1	.....do	.....do	.....do	14. 1
y 14	.....do	.....do	.....do	14. 6
y 27	.....do	Henningsen Co	.....do	16. 7
y 1	.....do	.....do	.....do	20
y 14	.....do	.....do	.....do	12. 7
y 12	.....do	Stoddicke	.....do	0. 0
1e 3	.....do	.....do	.....do	8. 8
y 1	.....do	.....do	.....do	7. 2
y 13	.....do	.....do	.....do	4. 8
y 12	.....do	Hellwinkle	.....do	0. 0
1e 3	.....do	.....do	.....do	7. 3
ly 1	.....do	.....do	.....do	.3

Measurements of East Carson ditches, Nevada, in 1903—Continued.

Date.	Hydrographer.	Ditch.	Location.	Discharge.
				Sec.-ft.
July 14	A. H. Schadler......	L. M. Christenson........	Near head-gate...	4.1
May 11	.....do	Nelson	.....do	
May 28	.....do	......do	.....do	
July 1	.....do	......do	.....do	
July 14	.....do	......do	.....do	
May 11	.....do	Wehrman	.....do	
May 28	.....do	......do	.....do	
July 1	.....do	......do	.....do	
July 14	.....do	......do	.....do	
May 11	.....do	Lang & Sill	.....do	
May 27	.....do	......do	.....do	
July 1	.....do	......do	.....do	
July 13	.....do	......do	.....do	
May 11	.....do	Jensen, Heise & Co......	.....do	
May 27	.....do	......do	.....do	
July 1	.....do	......do	.....do	
July 13	.....do	......do	.....do	
May 12	.....do	Company...............	.....do	
June 3	.....do	......do	.....do	28
June 29	.....do	......do	.....do	33
July 14	.....do	......do	.....do	28
May 11	.....do	Christenson Co.........	.....do	11.5
May 27	.....do	.....do	.....do	1.6
July 1	.....do	.....do	.....do	2.9
July 14	.....do	.....do	.....do	17.1
May 12	.....do	Dangberg & Hussman ..	.....do	35
June 3	.....do	.....do	.....do	11.1
June 29	.....do	.....do	.....do	12.3
July 15	.....do	.....do	.....do	5.5
May 12	.....do	Hussman	.....do	.7
June 3	.....do	.....do	.....do	1
June 29	.....do	.....do	.....do	0.0
May 12	.....do	Hussman, Helewinkle & Lampe.	.....do	6.4
June 3	.....do	.....do	.....do	2.1
June 29	.....do	.....do	.....do	2.8
May 4	.....do	Lampe Slough	.....do	2.4
May 2	.....do	.....do	.....do	0.0
May 12	.....do	.....do	.....do	0.0
June 30	.....do	.....do	.....do	.6

Measurements of West Carson ditches, California, in 1903—Continued.

Hydrographer.	Ditch.	Location.	Dis-charge.
			Sec.-ft.
A. H. Schadler.....	Henningson North	Near head-gate...	1.3
.....do	.....do	.....do	3.7
.....do	.....do	.....do	2.3
.....do	.....do	.....do	6.9
.....do	.....do	.....do	4.5
.....do	Heimsoth New...........	.....do	1.5
.....do	.....do	.....do	.6
.....do	.....do	.....do	1.9
.....do	.....do	.....do	1.4
.....do	Heimsoth Old	.....do	4.1
.....do	.....do	.....do	2.3
.....do	.....do	.....do	3
.....do	.....do	.....do	2.9
.....do	Stuart No. 1	.....do	.5
.....do	.....do	.....do	.3
.....do	.....do	.....do	1.1
.....do	.....do	.....do	.7
.....do	J. Scossa	.....do	4.6
.....do	.....do	.....do	0.0
.....do	.....do	.....do	6.3
.....de	.....do	.....do	8.9
.....do	Stuart No. 2	.....do	1.6
.....do	.....do	.....do	.5
.....do	.....do	.....do	1.9
.....do	.....do	.....do	.7
.....do	Stuart flume	.....do	.6
.....do	.....do	.....do	0.0
.....do	.....do	.....do	2.8
.....do	.....do	.....do	2.5
.....do	Fredericksburg	.....do	20
.....do	.....do	.....do	27
.....do	.....do	.....do	2.5
.....do	.....do	.....do	29
.....do	.....do	.....do	14
.....do	Chambers	.....do	5.8
.....do	.....do	.....do	7.5
.....do	.....do	.....do	1.2
.....do	.....do	.....do	6.1
.....do	.....do	.....do	6
.....do	Jarvis No. 1.............	.....do	3

May	1	A. H. Schadler	Deluchi No. 2	Near head-gate ...
May	23	.....do	.....do	.....do ....
June	26	.....do	.....do	.....do ....
July	11	.....do	.....do	.....do ....
May	1	.....do	Deluchi No. 3	.....do ....
May	23	.....do	.....do	.....do ....
June	26	.....do	.....do	.....do ....
July	11	.....do	.....do	.....do ....
May	1	.....do	Wilkerson, Fay, Berry & Thran.	.....do ....
May	23	.....do	.....do	.....do ....
July	11	.....do	.....do	.....do ....
June	26	.....do	.....do	.....do ....
May	1	.....do	F. Dressler	.....do ....
May	23	.....do	.....do	.....do ....
June	26	.....do	.....do	.....do ....
July	11	.....do	.....do	.....do ....
May	1	.....do	Wyatt	.....do ....
May	23	.....do	.....do	.....do ....
June	26	.....do	.....do	.....do ....
July	11	.....do	.....do	.....do ....
May	8	.....do	Wyatt, Dressler	.....do ....

West Carson ditches, Nevada, in 1903—Continued.

Hydrographer.	Ditch.	Location.	Dis-charge.
			Sec.-ft.
A. H. Schadler.....	Wyatt, Dressler........	Near head-gate...	0.0
....do	....do	.....do	2.0
.....do	....do	.....do	0.0
.....do	Jones Co	.....do	17.8
.....do	....do	.....do	9.0
.....do	....do	.....do	12.6
.....do	....do	.....do	5.3
.....do	New Saddlemeyer......	......do	6.9
.....do	....do	.....do	3.6
.....do	....do	.....do	3.4
.....do	....do	.....do	.7
.....do	Tucke Co.............	.....do	8.9
.....do	....do	.....do	.9
.....do	....do	.....do	2.8
.....do	....do	.....do	.4
.....do	Bart Cary	.....do	11.4
.....do	....do	.....do	3.7
.....do	....do	.....do	4.5
.....do	....do	.....do	.9
.....do	Parsons	.....do	2.6
.....do	....do	.....do	0.0
.....do	....do	.....do	1.4
.....do	Winkleman & Brockless.	.....do	.4
.....do	....do	.....do	3.8
.....do	....do	.....do	0.0
.....do	....do	.....do	0.0
.....do	Park dam No. 1........	.....do	8.0
.....do	....do	.....do	0.0
.....do	....do	.....do	0.0
.....do	....do	.....do	6.1
.....do	Park dam No. 2........	.....do	17.2
.....do	....do	.....do	3.8
.....do	....do	.....do	14.0
.....do	....do	.....do	0.0
.....do	Park dam No. 3........	.....do	0.0
.....do	....do	.....do	0.0
.....do	....do	.....do	0.0
.....do	Hickey dam No. 1	.....do	.9
.....do	....do	.....do	0.0
.....do	....do	.....do	.1

West Carson ditches, Nevada, in 1903—Continued.

Date.	Hydrographer.	Ditch.	Location.	Dis-charge.
May 27	A. H. Schadler	Hickey dam No. 2	Near head-gate	
June 30	....do	Hickey dam No. 3	....do	
May 27	....do	Hickey dam No. 4	....do	
June 30	....do	....do	....do	
May 8	....do	Jones dam No. 1	....do	
May 25	....do	....do	....do	
June 27	....do	....do	....do	
May 8	....do	Jones dam No. 2	....do	
May 25	....do	....do	....do	
June 27	....do	....do	....do	
July 13	....do	....do	....do	
May 8	....do	Jones dam No. 3	....do	
May 25	....do	....do	....do	
June 27	....do	....do	....do	
July 13	....do	....do	....do	
May 9	....do	Squires	....do	
May 25	....do	....do	....do	
June 27	....do	....do	....do	
July 13	....do	....do	....do	
May 9	....do	Winkleman	....do	
May 25	....do	....do	....do	
June 27	....do	....do	....do	
July 13	....do	....do	....do	
May 9	....do	Rabe	....do	
May 27	....do	....do	....do	
June 30	....do	....do	....do	
May 27	....do	Thompson & Christenson	....do	
June 30	....do	....do	....do	
May 27	....do	Johnson No. 1	....do	
June 30	....do	....do	....do	
Do	....do	Johnson middle	....do	
May 27	....do	Johnson dam	....do	
June 30	....do	....do	....do	

ments of ditches between Dayton and Carson Sink, Nevada, in 1903.

Hydrographer.	Ditch.	Location.	Discharge.
			Sec.-ft.
B. Harrington..	Douglas mill............	Near head-gate...	26
.do	.....do	.....do	28
H. Schadler.....	Tallier	.....do	1.8
B. Harrington..	Ophir	.....do	6.9
.do	.....do	.....do	3.6
H. Schadler.....	Jerusalem (Dayton Co.).	.....do	9.6
B. Harrington..	.....do	.....do	7.8
.do	.....do	.....do	1.5
H. Schadler.....	Fish (Cassinelli)	.....do	2.2
.do	Venturi (Baroni?)	.....do	1.8
B. Harrington..	Venturi No. 1...........	.....do	0.0
.do	.....do	.....do	0.0
.do	Venturi No. 2	.....do	1.2
.do	.....do	.....do	0.0
.do	Rock Point mill.........	.....do	37
.do	.....do	.....do	10.8
.do	Cordelli	.....do	0.0
.do	.....do	.....do	7.9
.do	Gotelli	.....do	16.8
.do	.....do	.....do	7.2
.do	Colishi	.....do	4.7
.do	.....do	.....do	2.7
H. Schadler.....	Sutro Co	.....do	10.3
B. Harrington..	.....do	.....do	2.8
.do	.....do	.....do	0.0
.do	Depoali & Co	.....do	1.3
.do	.....do	.....do	0.0
H. Schadler.....	Cook & Co...............	.....do	6.8
.do	Newman & Howard........	.....do	12.7
B. Harrington..	.....do	.....do	10.1
.do	.....do	.....do	7.3
H. Schadler	Buckland ranch	.....do	11.6
B. Harrington..	.....do	.....do	4.0
.do	.....do	.....do	2.3
H. Schadler	Buckland, east..........	.....do	17.2
.do	Newlands, east..........	.....do	14.5
B. Harrington..	.....do	.....do	0.0
.do	.....do	.....do	7.0
H. Schadler	Newlands, west	.....do	22

West Carson ditches, Nevada, in 1903—Continued.

Date.	Hydrographer.	Ditch.	Location.	Discharge.
				Sec.-ft.
May 27	A. H. Schadler.....	Hickey dam No. 2	Near head-gate...	2.(
June 30	.....do	Hickey dam No. 3	.....do	0.0
May 27	.....do	Hickey dam No. 4	.....do	5.7
June 30	.....do	.....do	.....do	4.2
May 8	.....do	Jones dam No. 1	.....do	1.1
May 25	.....do	.....do	.....do	0.4
June 27	.....do	.....do	.....do	1.9
May 8	.....do	Jones dam No. 2	.....do	3.2
May 25	.....do	.....do	.....do	1.4
June 27	.....do	.....do	.....do	0.0
July 13	.....do	.....do	.....do	.9
May 8	.....do	Jones dam No. 3	.....do	3.7
May 25	.....do	.....do	.....do	1.1
June 27	.....do	.....do	.....do	1.2
July 13	.....do	.....do	.....do	1.5
May 9	.....do	Squires	.....do	4.5
May 25	.....do	.....do	.....do	.1
June 27	.....do	.....do	.....do	1.0
July 13	.....do	.....do	.....do	4.1
May 9	.....do	Winkleman	do	2.2
May 25	.....do	.....do	.....do	.5
June 27	.....do	.....do	.....do	7.6
July 13	.....do	.....do	.....do	3.4
May 9	.....do	Rabe............	.....do	0.0
May 27	.....do	.....do............	.....do	
June 30	.....do	.....do............	.....do	0
May 27	.....do	Thompson & Christenson	.....do	
June 30	.....do	.....do............	.....do	
May 27	.....do	Johnson No. 1	.....do	
June 30	.....do	.....do............	.....do	
Do ...	.....do	Johnson middle	.....do	
May 27	.....do	Johnson dam	.....do	
June 30	.....do	.....do............	.....do	

nts of ditches between Dayton and Carson Sink, Nevada, in 1903.

rdrographer.	Ditch.	Location.	Dis-charge.
			Sec.-ft.
. Harrington..	Douglas mill............	Near head-gate...	26
ol	.....do	.....do	28
. Schadler.....	Tallier	.....do	1.8
. Harrington..	Ophir	.....do	6.9
.o	.....do	.....do	3.6
. Schadler.....	Jerusalem (Dayton Co.).	.....do	9.6
. Harrington..	.....do	.....do	7.8
.o	.....do	.....do	1.5
. Schadler.....	Fish (Cassinelli)	.....do	2.2
:o	Venturi (Baroni?)	.....do	1.8
. Harrington..	Venturi No. 1...........	.....do	0.0
lo	.....do	.....do	0.0
:o	Venturi No. 2	.....do	1.2
lo	.....do	.....do	0.0
io	Rock Point mill.........	.....do	37
io	.....do	.....do	10.8
lo	Cordelli	.....do	0.0
lo	.....do	.....do	7.9
lo	Gotelli	.....do	16.8
io	.....do	.....do	7.2
lo	Colishi	.....do	4.7
lo	.....do	.....do	2.7
. Schadler.....	Sutro Co	.....do	10.3
. Harrington..	.....do	.....do	2.8
lo	.....do	.....do	0.0
lo	Depoali & Co	.....do	1.3
lo	.....do	.....do	0.0
. Schadler.....	Cook & Co.	.....do	6.8
lo	Newman & Howard	.....do	12.7
i. Harrington..	.....do	.....do	10.1
lo	.....do	.....do	7.3
. Schadler	Buckland ranch	.....do	11.6
i. Harrington..	.....do	.....do	4.0
lo	.....do	.....do	2.3
. Schadler	Buckland, east.........	.....do	17.2
lo	Newlands, east.........	.....do	14.5
i. Harrington..	.....do	.....do	0.0
lo	.....do	.....do	7.0
. Schadler	Newlands, west	.....do	22

04——17

Measurements of Carson Sink ditches, Nevada, in 1903.

Date.	Hydrographer.	Ditch.	Location.	Discharge.
		MAIN CARSON RIVER.		*Sec.-ft.*
June 6	A. H. Schadler	Kaiser's	Near head-gate...	3.9
July 2	H. Jameson	......do	......do	6.6
July 7	.....do	......do	......do	2.6
June 7	A. H. Schadler	Leets	......do	4.0
July 2	H. Jameson	......do	......do	3.4
July 7	.....do	......do	......do	8.5
June 7	A. H. Schadler	Thelands	......do	2.0
July 2	H. Jameson	......do	......do	8.9
July 13	.....do	......do	......do	4.6
June 11	A. H. Schadler	Allen & Thomas.....	......do	9.6
July 3	H. Jameson	......do	......do	6.5
July 9	.....do	......do	......do	10.1
June 10	A. H. Schadler	Johnson Bros	......do	2.8
June 11	.....do	......do	......do	2.5
July 3	H. Jameson	......do	......do	3.7
		CARSON RIVER.		
June 11	A. H. Schadler	Allen	Near head-gate...	4.6
July 12	H. Jameson	......do	......do	0.0
June 11	A. H. Schadler	Furgeson	......do	4.3
July 12	H. Jameson	......do	......do	2.9
July 9	.....do	......do	......do	3.2
June 11	A. H. Schadler	Lee & Wightman.....	......do	15.1
July 9	H. Jameson	......do	......do	11.3
July 12	.....do	......do	......do	13.3
June 11	A. H. Schadler	Springer	......do	12.3
July 9	H. Jameson	......do	......do	2.4
July 12	.....do	......do	......do	6.1
July 9	.....do	Hardy & McCullough...	......do	3.4
July 12	.....do	......do	......do	4.7
June 11	A. H. Schadler.....	Cushman	......do	5.3
July 12	H. Jameson........	......do	......do	0.0
June 12	.....do	Cushman & Wightman..	......do	6.9
June 11	A. H. Schadler.....	Dillan	......do	0.5
July 9	H. Jameson........	......do	......do	6.8
July 12	.....do	......do	......do	14.0

Measurements of Carson Sink ditches, Nevada, in 1903—Continued.

ate.	Hydrographer.	Ditch.	Location.	Dis-charge.
		OLD RIVER.		*Sec.-ft.*
ə 11	A. H. Schadler.....	Old River bridge	Near head-gate ..	8.4
· 3	H. Jameson............	.....do	.....do	7.5
· 10	.....do	.....do	.....do	5.3
ə 8	A. H. Schadler.....	Ernsts...............	.....do	26
· 3	H. Jameson............	.....do	.....do	12.7
· 9	.....do	.....do	.....do	6.9
· 5	.....do	Williams	.....do	2.9
ə 10	A. H. Schadler.....	Indian	.....do	24
· 5	H. Jameson............	.....do	.....do	7.8
· 6	.....do	.....do	.....do	8.3
		NEW RIVER.		
e 11	A. H. Schadler	Summer's Slough	Near head-gate...	2.4
· 10	.....do	.....do	.....do	12.1
ə 11	.....do	Allen	.....do	11.4
)o...	.....do	Allen & Smart	.....do	4.7
e 10	.....do	New River, west	.....do	42
· 3	H. Jameson	.....do	.....do	4.4
· 10	.....do	.....do	.....do	30
e 13	A. H. Schadler	New River, east........	.....do	2.8
e 10	.....do	Browns...............	.....do	10.1
ſ 3	H. Jameson	.....do	.....do	17.3
ſ 10	.....do	.....do	.....do	8.4
e 10	.....do	Webb-Small	.....do	21
ſ 1	.....do	.....do	.....do	9.3
ſ 6	.....do	.....do	.....do	3.7
ſ 10	.....do	.....do	.....do	15.3
e 10	A. H. Schadler	Harriman	.....do	1.0
e 11	.....do	Lydell	.....do	16.1
Do...	.....do	Cape	.....do	32
e 12	.....do	Grimes...................	.....do	10.4
ſ 9	H. Jameson ....do		.....do	6.9

Measurements of Carson Sink ditches, Nevada, in 1903—Continued.

Date.	Hydrographer.	Ditch.	Location.	
		STILLWATER SLOUGH.		*Sec. ft.*
June 9	A. H. Schadler	Dyer	Near head-gate..	4.0
July 6	H. Jameson	....do	....do	6.0
July 11	.....do	....do	....do	0.6
June 9	A. H. Schadler	Brannan	....do	11.2
July 6	H. Jameson	....do	....do	10.1
July 11	.....do	....do	....do	10.2
June 8	A. H. Schadler	Cirac...............	....do	1.0
July 5	H. Jameson	....do	....do	1.6
July 11	.....do	....do	....do	2.2
June 8	A. H. Schadler	Doolittle	....do	29
July 5	H. Jameson	....do	....do	12.6
July 11	....do	....do	....do	8.6
July 5	.....do	Freeman	....do	74
June 8	A. H. Schadler	Kents................	....do	30
July 8	H. Jameson	.....do	....do	8
July 11	.....do	.....do	....do	0.8

Measurements of East Carson ditches, California, in 1903.

Date.	Hydrographer.	Ditch.	Location.	
June 24	A. H. Schadler	Grover Hot Springs	Near head-gate...	3.0
Do ...	.....do	Markleeville.............	....do	7.1
Do ...	.....do	Creek above Markleeville.	....do	.4
June 25	.....do	Mayo	....do	2.9

SAN FRANCISCO BAY DRAINAGE BASIN.

acramento River, rising in northern California and flowing south,
San Joaquin River, rising in the southern Sierras and flowing
theast, drain the western slope of the Sierra Nevada, traverse what
ften called the Valley of California, and meet near Suisun Bay,
lly discharging their waters into the Pacific Ocean through San
ncisco Bay.

acramento River derives its water supply largely from Mount
sta and the surrounding high ranges in the extreme northern por-
1 of California. The stream does not have the same regular annual
tuations that characterize the rivers discharging from the higher
rra Nevada, since a large part of its basin is not at an elevation
ficient to cause the winter snows to remain unmelted until the sum-
r months. The greatest floods of this basin usually occur in Janu-
' and February, when the snow is accompanied or followed by rain.
San Joaquin River is divided into two distinct parts. The valley
rtion forms the central drainage line of San Joaquin Valley, and
ring the spring is navigable for a hundred or more miles. Stanis-
s, Tuolumne, and King rivers are the largest affluents of this por-
n of the stream. Its valley is fertile and almost destitute of timber.
e mountainous portion of the stream drains the western slopes of
 Sierra Nevada between Yosemite National Park and Mount God-
rd, the crest of its divide reaching on the north an elevation of
,000 feet in Mount Lyell, and an elevation of 14,000 feet in Mount
ddard. The resulting steep grades of this river offer exceptional
portunities for water-power developments, and the high elevations
the basin insure a well-sustained summer flow from perpetual snow
ks.

The following streams are tributary to either Sacramento or San
quin rivers:

ache Creek is the outlet of Clear Lake, in Lake County, Cal.
wing southeasterly, its flood waters find their way into Sacramento
er between the mouths of Feather and American rivers. In 1889
ar Lake was segregated as a reservoir site, as described in the
irteenth Annual Report, part 3, pages 405-409. During 1900 a
lrographic examination of the entire basin of Cache Creek was
de by A. E. Chandler, whose detailed report has been published as
iter-Supply Paper No. 45.

Feather River is the second largest tributary of Sacramento River,
 River being the first. Its basin line follows the crest of the
rra Nevada for about 130 miles. The rainfall in this basin is large;
 mean for nineteen years at Mumfords Hill, Plumas County, is

71.64 inches; the rainfall for the year 1889-90 at this point reached 138.85 inches. The water collected by the river when rains are general sometimes causes tremendous freshets, usually of short duration. The river has at such times overtopped its right bank and overflowed the plain lands to the north of Sutter Buttes. This occurred even before the great reduction of waterway below the mouth of Yuba River. The channel of the river below its junction with the Yuba has become the repository of so much mining débris that its bed has become nearly filled. Its bottom is almost at the heights of its former banks, and levees have been built to prevent overflowing. Only a comparatively small portion of the summer flow of this stream is used for irrigation, though the possibilities of irrigation and power development are great.

There are some excellent reservoir sites on the upper tributaries of the North Fork of the Feather. This stream is fed in part by large springs, one of which, in Big Meadows, was flowing 109 cubic feet per second in September, 1902, and another 64 cubic feet per second.

Stony Creek drains 760 square miles of the eastern slopes of the Coast Range. After reaching the Sacramento Valley it flows north for a number of miles, contrary to the general drainage, and then turns east and enters Sacramento River below Vina, Cal. A large portion of the basin near the heads of the stream is heavily covered with commercial timber. There are a number of good reservoir sites on this stream and its tributaries.

Tuolumne River rises on the western slope of the Sierra Nevada in California and drains the country between Stanislaus River on the north and Merced River on the south. The northern half of Yosemite National Park includes a portion of the drainage basin of this stream. The river is fed largely from small mountain lakes occurring high in the drainage basin, where the snow remains on the mountain slopes throughout the year, thus insuring a large run-off. The stream has a heavy fall, and the opportunities for power development are numerous. There are also a number of reservoir sites in the basin where flood water could be stored for use during the irrigation season. The Tuolumne is an important tributary of the San Joaquin.

Merced River above Merced Falls drains approximately 1,090 square miles of the western slopes of the Sierra Nevada. There are included in the eastern portion of its drainage area a large number of big peaks, the highest, Mount Lyell, reaching an elevation of 13,042 feet Its basin lies south of that of the Tuolumne, the courses of the two streams being nearly parallel.

King River rises on the western slope of the Sierra Nevada, in Fresno County, Cal. The waters coming from the high catchment basin are probably of greater value for irrigation purposes than those of any other stream in central California, being used for raising grapes and deciduous fruits in the neighborhood of Fresno, Selma, and Hanford. The summer flow of the river is now entirely diverted, and during the dry season of the last few years the scarcity of water has caused many hardships. In the spring there is a large surplus, due to the melting of snows, which, if stored in suitable reservoirs, would bring larger areas under cultivation. The river has a relatively gentle grade, affording little opportunity for power development.

Tule River drains a portion of the western slope of the Sierra Nevada. Its basin has somewhat less run-off than that of Kaweah River, which joins it on the north, and is much less elevated and snow covered than King River basin. The water of this stream is all appropriated during the irrigation season, and a portion is used in irrigating valuable orange lands in the vicinity of Portersville, Cal.

Kern River flows from the southern end of the Sierra Nevada, being formed by two large tributaries, known as North and South forks. These have a general southerly and parallel course and unite a short distance below the town of Kernville. The run-off from the drainage basin as a whole is notably less than that from the catchment areas to the north. This is probably due to the fact that a portion of the basin is to the east of the high crest, and is sheltered by the mountain mass from the rain-bearing winds. The waters of Kern River are almost completely used for irrigation by the large canals in the southern end of the San Joaquin Valley. The greater part of the land is included in large holdings owned by the Kern County Land Company or by the Miller and Lux estate. The winter waters are in part stored by the Miller and Lux estate in Buena Vista Lake, into which the river naturally discharges. The waters of this lake are controlled by a system of levees, so that they can be used during the following summer to irrigate lands lying to the northwest. This lake is very broad and shallow and there is great loss by evaporation, so that as a matter of economy it would be desirable to hold this water in the upper mountain valleys. This would afford also a large supply for water power.

The fall of Kern River is sufficiently large for the development of a considerable amount of water power.

The following is a list of the stations in the San Francisco bay drainage basin:

Cache Creek near Yolo, Cal.
Cache Creek at Lower Lake, Cal.
Yuba River near Smartsville, Cal.
Feather River at Oroville, Cal.
Stony Creek near Fruto, Cal.
McCloud River near Gregory, Cal.
Sacramento River near Red Bluff, Cal.
Mokelumne River at Electra. Cal.
Stanislaus River at Knights Ferry, Cal.
Tuolumne River at Lagrange, Cal.
Turlock canal at Lagrange, Cal.
Modesto canal at Lagrange, Cal.
Merced River above Merced Falls, Cal.
King River at Kingsburg, Cal.
King River near Sanger, Cal.
Kaweah River below Threerivers, Cal.
Tule River near Portersville, Cal.
Kern River near Bakersfield, Cal.
San Joaquin River at Herndon, Cal.

CACHE CREEK NEAR YOLO, CAL.

This station was established January 1, 1903, by S. G. Bennett. It is located at the wagon bridge on the road from Woodlands to Yolo about 1,000 feet above the Southern Pacific Railway bridge. The gage is a 2 by 12 inch vertical plank nailed to the upstream side of the right abutment. It is read twice each day by John Woodard. Discharge measurements are made from the downstream side of the bridge, to which the gage is attached. The initial point for soundings is the end of the bridge, on the right bank. The channel is straight for 1,000 feet above and below the station. The current is swift at ordinary and high stages. The banks are steep and wooded and their height has been increased by levees. They are said to overflow at extreme high water. The bed of the stream is composed of earth and gravel, with a little sand. It is subject to some change.

The observations at this station during 1903 have been made under the direction of J. B. Lippincott, supervising engineer.

Discharge measurements of Cache Creek near Yolo, Cal., in 1903.

Date.	Hydrographer.	Gage height.	Discharge.
		Feet.	*Second-feet.*
January 25	S. G. Bennett	4.80	1,517
May 13	do	3.05	675
August 3	do	1.00	23

Mean daily gage height, in feet, of Cache Creek near Yolo, Cal., for 1903.

Day.	Jan.	Feb.	Mar.	Apr.	May.	June.	July.	Aug.	Sept.	Oct.	Nov.	Dec.
....	2.90	6.00	3.90	5.70	3.40	2.60	1.70	0.90	0.	0.	0.	2.10
....	2.80	5.40	3.90	5.20	3.40	2.50	1.70	.90	......	......	......	2.00
....	2.80	5.10	3.90	5.00	3.40	2.50	1.70	.80	......	......	......	1.90
....	2.80	4.70	3.90	4.70	3.30	2.50	1.60	.80	......	......	......	1.90
....	2.80	4.40	3.90	4.70	3.30	2.50	1.60	.80	......	......	......	1.80
....	2.80	4.40	3.90	4.60	3.30	2.40	1.60	.80	......	......	......	1.80
....	2.70	4.30	3.80	4.50	3.30	2.40	1.50	.70	......	......	......	1.80
....	2.70	10.20	3.80	4.40	3.30	2.40	1.50	.70	......	......	......	1.70
....	2.70	5.80	3.80	4.30	3.20	2.40	1.50	(a)	......	......	......	1.70
....	2.70	5.20	3.80	4.30	3.20	2.30	1.40	......	......	......	......	1.60
....	2.60	4.90	3.80	4.10	3.10	2.30	1.40	......	......	......	......	1.60
....	2.60	4.80	3.80	4.00	3.10	2.30	1.40	......	......	......	......	1.50
....	2.60	4.70	4.80	4.00	3.10	2.30	1.40	......	......	......	......	1.50
....	2.60	4.50	7.70	3.90	3.10	2.20	1.30	......	......	......	......	1.50
....	2.60	4.50	5.80	3.90	3.00	2.30	1.30	......	......	......	......	1.50
....	2.60	4.40	5.20	3.80	3.00	2.30	1.30	......	......	......	......	5.40
....	2.50	4.40	5.00	3.80	3.00	2.30	1.30	......	......	......	......	3.80
....	2.50	4.30	4.80	3.80	2.90	2.30	1.20	......	......	......	......	3.30
....	2.50	4.30	4.70	3.70	2.90	2.20	1.20	......	......	......	(a)	3.00
....	2.50	4.30	4.50	3.70	2.90	2.20	1.20	......	......	......	4.00	2.80
....	2.50	4.20	4.40	3.60	2.90	2.10	1.20	......	......	......	9.30	2.80
....	2.50	4.10	4.40	3.60	2.80	2.10	1.20	......	......	......	5.60	2.60
....	2.50	4.10	4.30	3.60	2.80	2.00	1.10	......	......	......	4.30	2.60
....	3.70	4.10	4.20	3.50	2.80	2.00	1.10	......	......	......	3.50	2.40
....	5.20	4.00	4.10	3.50	2.70	1.90	1.10	......	......	......	3.30	2.40
....	5.50	4.00	4.00	3.50	2.70	1.90	1.10	......	......	......	2.90	2.20
....	8.00	4.00	4.00	3.50	2.70	1.80	1.00	......	......	......	2.50	2.20
....	10.20	4.00	4.30	3.50	2.70	1.80	1.00	......	......	......	2.30	2.20
....	6.00	......	5.90	3.50	2.70	1.80	1.00	...+...	......	......	2.20	2.20
....	5.70	......	5.10	3.40	2.60	1.80	.90	......	......	......	2.20	2.10
....	8.10	......	5.30	......	2.60	......	.90	......	......	......	......	2.10

a August 9 to November 9, 1903, inclusive, no discharge. Gage height 0.6 feet; water in pools.

Rating table for Cache Creek near Yolo, Cal., for 1903.

Gage height.	Discharge.	Gage height.	Discharge.	Gage height.	Discharge.	Gage height.	Discharge.
Feet.	*Second-feet.*	*Feet.*	*Second-feet.*	*Feet.*	*Second-feet.*	*Feet.*	*Second-feet.*
0.7	4	2.0	210	4.0	1,000	6.0	2,300
.8	9	2.2	270	4.2	1,120	6.5	2,675
.9	15	2.4	330	4.4	1,240	7.0	3,060
1.0	22	2.6	395	4.6	1,365	7.5	3,460
1.1	31	2.8	465	4.8	1,495	8.0	3,885
1.2	43	3.0	535	5.0	1,625	8.5	4,320
1.3	55	3.2	615	5.2	1,755	9.0	4,770
1.4	70	3.4	705	5.4	1,885	9.5	5,220
1.6	110	3.6	800	5.6	2,020	10.0	5,695
1.8	160	3.8	900	5.8	2,160	10.5	6,190

Estimated monthly discharge of Cache Creek near Yolo, Cal., for 1903.

[Drainage area, 1,250 square miles.]

Month	Discharge in second-feet.			Total in acre-feet.	Run-off.	
	Maximum.	Minimum.	Mean.		Second-feet per square mile.	Depth in inches.
January.............	5,890	360	1,036	63,701	0.81	0.93
February...........	5,890	1,000	1,506	83,639	1.18	1.2
March.............	3,630	900	1,315	80,856	1.03	1.19
April.............	2,090	705	1,052	62,598	.82	.9
May	705	395	545	33,511	.43	.5
June	395	160	276	16,423	.22	.25
July	135	15	63	3,874	.05	.06
August a...........	15	0	2	123	.00	.00
September.........	0	0	0	0	.00	.00
October	0	0	0	0	.00	.00
November a........	5,040	0	412	24,516	.32	.36
December	1,885	90	325	19,983	.25	.28
The year	5,890	0	544	389,224	.43	5.73

a Creek dry from August 9 to November 19.

CACHE CREEK AT LOWER LAKE, CAL.

This station was established January 1, 1900, by S. G. Bennett. The original gage was located at the wagon bridge, from which discharge measurements were made. On March 26, 1903, a cable was installed 300 feet above the bridge, and a new gage was established 100 feet above the cable. The present gage is a vertical 1 by 3 inch plank nailed to a 6 by 6 inch timber driven into the bed of the river and fastened to a large willow tree on the left bank. On March 26, when the new gage was put in place, the reading was 5.7 feet. The old gage read 4.4 feet on the same date. The gage is read once each day by Mrs. J. R. Anderson. The initial point for soundings is a small tree in line with the cable on the left bank, 28 feet from the tree to which the cable is attached. The channel is straight for 150 feet above and for 300 feet below the station. The current has a moderate velocity at ordinary stages. The right bank is low and will overflow at a gage height of about 10 feet. It is covered with a thick growth of willow and oak trees for 100 feet back from the water's edge. The left bank is high and rocky and is not liable to overflow. The bed of the stream is composed of firm gravel and changes only slightly Gravel is sometimes washed in from Siegler Creek, 300 feet below the cable. The bench mark is a nail in the root of the oak tree to which

stened on the left bank. Its elevation is 8.32 feet above the gage.

ervations at this station during 1903 have been made under on of J. B. Lippincott, supervising engineer.

charge measurements of Cache Creek at Lower Lake, Cal., in 1903.

Date.	Hydrographer.	Gage height.	Discharge.
		Feet.	*Second-feet.*
....	J. R. Anderson	2.90	344
....	....do	3.50	575
....	....do	4.00	747
....	....do	4.40	830
....	....do	4.30	847
....	....do	4.20	775
....	....do	4.20	776
....	....do	4.50	831
....	....do	a 4.40	837
....	S. G. Bennett...............	b 5.70	752
....	J. R. Anderson	5.90	808
....	....do	5.80	795
....	....do	5.60	732
....	....do	5.20	608
....	....do	5.00	562
....	....do	4.80	501
....	....do	4.60	429
....	....do	4.40	409
....	....do	4.20	363
....	....do	4.00	333
....	....do	3.90	294
....	....do	3.70	246
....	....do	3.50	212
....	....do	3.40	188
....	....do	3.30	171
....	....do	3.10	137
....	....do	3.00	124
....	....do	2.90	106
....	....do	2.80	86
....	....do	2.70	68
2	....do	2.60	75
9	....do	2.45	39
6	....do	2.40	32
....	....do	2.35	29
....	....do	2.35	25
....	....do	2.30	23

a Old gage. b New gage = 4.40 old.

7	2.9				u..	4.0						
8	2.9				5.1	4.3						
9	2.9				5.1	4.3						
10	2.9				5.1							
11	2.9				5.0							
12												
13	2.9	4.4	5.2	5.7	5.0	4.2	3.6	3.0	2.6	2.3		2.1
14	2.9	4.4	4.5	5.7	4.9	4.1	3.5	3.0	2.6	2.3		2.2
15	2.9	4.4	4.4	5.6	5.0	4.1	3.5	3.0	2.5	2.3		2.2
16	2.9	4.3	4.4	5.6	5.0	4.1	3.5	3.0	2.5	2.3		2.4
17	2.9	4.3	4.4	5.6	4.8	4.1	3.5	3.0	2.5	2.3	2.3	2.4
18	2.9	4.3	4.4	5.6	4.8	4.0	3.5	2.9	2.5	2.3	2.3	2.4
19	2.9	4.3	4.4	5.5	4.8	4.0	3.5	2.9	2.4	2.3	2.4	2.4
20	2.9	4.3	4.4	5.5	4.8	4.0	3.4	2.9	2.5	2.3	2.6	2.4
21	2.9	4.3	4.4	5.5	4.8	4.0	3.4	2.9	2.4	2.3	2.5	2.5
22	2.9	4.3	4.4	5.5	4.7	4.0	3.4	2.9	2.4	2.3	2.9	2.5
23	3.0	4.3	4.4	5.5	4.7	4.0	3.4	2.9	2.4	2.3	2.9	2.6
24	3.0	4.3	4.4	5.4	4.7	4.0	3.3	2.8	2.4	2.3	3.0	2.6
25	3.2	4.3	4.4	5.5	4.6	4.0	3.3	2.8	2.4	2.3	3.0	2.5
26	3.2	4.2	a 5.7	5.4	4.6	3.9	3.3	2.8	2.4	2.2	3.1	2.6
27	3.7	4.2	5.6	5.4	4.6	3.9	3.3	2.8	2.4	2.3	3.1	2.5
28	3.5	4.2	6.0	5.4	4.6	3.9	3.3	2.8	2.4	2.2	3.2	4.5
29	3.5		5.8	5.4	4.6	3.9	3.3	2.8	2.4	2.2	3.2	2.6
30	3.9		5.8	5.4	4.5	3.9	3.3	2.8	2.4	2.2	3.2	2.3
31	3.8		5.9		4.5		3.2	2.7		2.2		2.6

a New gage; old gage 4.4.

ting table for Cache Creek at Lower Lake, Cal., from January 1, to March 25, 1903.

Gage height.	Discharge.	Gage height.	Discharge.	Gage height.	Discharge.	Gage height.	Discharge.
Feet.	*Second-feet.*	*Feet.*	*Second-feet.*	*Feet.*	*Second-feet.*	*Feet.*	*Second-feet.*
2.8	315	3.3	440	3.8	580	4.3	730
2.9	340	3.4	465	3.9	610	4.4	760
3.0	365	3.5	490	4.0	640	4.5	790
3.1	390	3.6	520	4.1	670	4.6	820
3.2	415	3.7	550	4.2	700		

ting table for Cache Creek at Lower Lake, Cal., from March 26 to December 31, 1903

Gage height.	Discharge.	Gage height.	Discharge.	Gage height.	Discharge.	Gage height.	Discharge.
Feet.	*Second-feet.*	*Feet.*	*Second-feet.*	*Feet.*	*Second-feet.*	*Feet.*	*Second-feet.*
2.2	15	3.2	147	4.2	350	5.2	610
2.3	20	3.3	165	4.3	375	5.3	640
2.4	32	3.4	185	4.4	400	5.4	670
2.5	45	3.5	205	4.5	425	5.5	700
2.6	58	3.6	225	4.6	450	5.6	730
2.7	72	3.7	245	4.7	475	5.7	760
2.8	86	3.8	265	4.8	500	5.8	790
2.9	100	3.9	285	4.9	525	5.9	820
3.0	115	4.0	305	5.0	550	6.0	850
3.1	130	4.1	325	5.1	580		

Estimated monthly discharge of Cache Creek at Lower Lake, Cal., for 1903.

[Drainage area, 500 square miles.]

Month.	Discharge in second-feet.			Total in acre-feet.	Run-off.	
	Maximum.	Minimum.	Mean.		Second-feet per square mile.	Depth in inches.
January	610	315	379	23,304	0.76	0.88
February	820	580	712	39,542	1.42	1.48
March	850	360	724	44,517	1.45	1.67
April	820	670	740	44,033	1.48	1.65
May	670	425	533	32,773	1.07	1.23
June	425	285	336	19,993	.67	.75
July	285	147	208	12,789	.42	.48
August	147	72	111	6,825	.22	.25
September	72	32	50	2,975	.10	.11
October	32	15	22	1,353	.04	.05
November	147	15	54	3,213	.11	.12
December	225	130	168	10,330	.34	.39
The year	850	15	336	241,647	.67	9.06

YUBA RIVER NEAR SMARTSVILLE, CAL.

This station was established June 2, 1903, by W. H. Stearns. It is located at what is called "The Narrows," 1 mile from Smartsville, Cal., 18 miles from the Southern Pacific Railway station at Wheatland, Cal., and the same distance from Marysville, Cal. There are daily stages from each point. The gage is a vertical rod bolted to the rock cliff on the left bank of the river; it is graduated to feet and tenths. The observer, J. McKeel, reads the gage once each day. Discharge measurements are made from a car suspended from a three-fourths-inch galvanized-iron cable. An auxiliary cable is stretched parallel to and 100 feet upstream from the main cable. The initial point for soundings is on the left bank at the eyebolt to which the cable is fastened. The channel is straight for 200 feet above and 300 feet below the cable, and the current is swift at all stages. In the 150 feet above the cable the stream has a fall of 0.2 foot and of 0.9 foot in the 200 feet below. Both banks are high and rocky and are not subject to overflow. The bed of the stream is composed of gravel and sand—tailings from hydraulic mining—and is constantly shifting. The bench mark is a cross in a projecting point of rock under the cable on the right bank of the river 335 feet from the initial point for soundings. It is 38.65 feet above the zero of the gage. The discharge measurements at this station will be approximate only, but this is the best place that can be found on the main river.

The observations at this station during 1903 have been made under the direction of J. B. Lippincott, supervising engineer.

Discharge measurements of Yuba River near Smartsville, Cal., in 1903.

Date.	Hydrographer.	Gage height.	Discharge.
		Feet.	*Second-feet.*
November 4	S. G. Bennett	3.8	570
November 17	J. R. McKeel	6.8	3,196
November 30	do	6.2	2,452
December 4	do	5.5	1,482
December 28	do	5.6	1,593

Mean daily gage height, in feet, of Yuba River near Smartsville, Cal., for 1903.

Day.	June.	July.	Aug.	Sept.	Oct.	Nov.	Dec.
....	(8.2)	5.2	3.8	3.5	3.6	3.6	6.1
....	8.2	5.1	3.8	3.5	3.6	3.6	6.0
....	7.8	5.0	3.8	3.5	3.6	3.6	5.9
....	7.5	5.0	3.8	3.5	3.6	3.8	5.9
....	7.5	(4.9)	3.8	3.5	3.6	3.9	5.8
....	7.7	(4.9)	3.7	3.5	3.6	3.8	5.8
....	7.8	(4.9)	3.7	3.5	3.6	4.1	5.7
....	7.5	(4.8)	3.7	3.5	3.6	4.0	5.7
....	7.3	(4.8)	3.7	3.4	3.6	3.9	5.6
....	7.4	(4.8)	3.7	3.5	4.2	3.8	5.5
....	7.0	(4.8)	3.7	3.5	4.7	3.8	5.5
....	6.8	(4.7)	3.7	3.5	4.0	8.3	5.4
....	6.7	(4.7)	3.7	3.5	3.9	8.6	5.5
....	6.6	(4.6)	3.6	3.5	3.8	10.4	5.5
....	6.3	(4.6)	3.6	3.5	3.7	10.0	5.5
....	6.2	(4.5)	3.6	3.5	3.6	7.4	5.8
....	5.9	(4.5)	3.6	3.5	3.6	6.8	7.5
....	5.7	(4.4)	3.6	3.5	3.6	6.5	6.9
....	5.9	(4.4)	3.6	3.5	3.6	6.5	6.4
....	5.7	(4.3)	3.5	3.5	3.6	11.0	6.5
....	5.5	(4.2)	3.5	3.5	3.7	16.0	6.3
....	5.5	(4.2)	3.5	3.5	3.7	11.0	6.1
....	5.5	(4.1)	3.5	3.5	3.7	9.0	5.9
....	5.3	(4.1)	3.5	3.5	3.7	8.3	5.7
....	5.2	(4.0)	3.5	3.5	3.7	7.5	5.7
....	5.2	(4.0)	3.5	3.5	3.7	7.0	5.6
....	5.1	(3.9)	3.5	3.5	3.7	6.7	5.6
....	5.1	3.9	3.5	3.5	3.6	6.5	5.6
....	5.2	3.9	3.5	3.5	3.6	6.3	5.5
....	5.3	3.9	3.5	3.5	3.6	6.2	5.5
....		3.8	3.5		3.6		5.7

On July 5 water fell below gage. Estimate of gage heights from July 5 to 27, clusive, made from statement of observer, Mr. McKeel.

Rating table for Yuba River near Smartsville, Cal., for 1903.

Gage height.	Discharge.	Gage height.	Discharge.	Gage height.	Discharge.	Gage height.	Discharge.
Feet.	*Second-feet.*	*Feet.*	*Second-feet.*	*Feet.*	*Second-feet.*	*Feet.*	*Second-feet.*
3.4	460	4.8	1,060	6.8	3,200	11.5	15,150
3.5	480	5.0	1,200	7.0	3,530	12.0	17,000
3.6	510	5.2	1,350	7.5	4,430	12.5	19,100
3.7	540	5.4	1,510	8.0	5,400	13.0	21,200
3.8	570	5.6	1,690	8.5	6,400	13.5	23,400
3.9	610	5.8	1,890	9.0	7,500	14.0	25,600
4.0	650	6.0	2,110	9.5	8,800	14.5	27,800
4.2	730	6.2	2,350	10.0	10,200	15.0	30,000
4.4	830	6.4	2,610	10.5	11,700	15.5	32,300
4.6	940	6.6	2,890	11.0	13,300	16.0	34,600

Estimated monthly discharge of Yuba River near Smartsville, Cal., for 1903.

[Drainage area, 1,220 square miles.]

Month.	Discharge in second-feet.			Total in acre-feet.	Run-off.	
	Maximum.	Minimum.	Mean.		Second-feet per square mile.	Depth in inches
June	5,800	1,270	2,911	173,217	2.39	2.6
July	1,350	570	899	55,277	.74	.8
August	570	480	516	31,728	.42	.4
September..........	480	460	479	28,502	.39	.4
October	1,000	510	550	33,818	.45	.5
November..........	34,600	510	4,893	291,154	4.01	4.6
December	4,430	1,510	2,009	123,529	1.65	1.90
The period	34,600	460	1,751	737,225	1.44	11.2

FEATHER RIVER AT OROVILLE, CAL.

At Oroville, where Feather River breaks from the foothills on the western slope of the Sierra Nevada into Sacramento Valley, it has a drainage area of 3,350 square miles. This station was established January 1, 1902, by S. G. Bennett. It is located at the northeast edge of the town of Oroville, Cal. Observations of the daily river height were begun January 1, 1902, using the rod of the Weather Bureau, which has been in place for a number of years. This is a 2-inch vertical iron pipe. Readings on this rod have been taken and reported by the Weather Bureau during floods when there was danger of overflow on the lower Feather and Sacramento rivers. To avoid negative readings the height as read on the gage has been increased by 2 feet. The gage is read twice each day by D. G. Page. Discharge measurements are made by means of a cable and a boat, located 500 feet above the gage. Flood measurements are made by means of floats. The channel is straight for 300 feet above and for 100 feet below the station. It has a width of about 200 feet at ordinary stages and about 700 feet at flood stages. The bed of the stream is composed mainly of rock and is rough and permanent. The current is swift at ordinary stages. The minimum midsummer discharge, as far as known, was in 1900, when the stream was flowing 1,123 second-feet.

The observations at this station during 1903 have been made under the direction of J. B. Lippincott, supervising engineer.

Discharge measurements of Feather River at Oroville, Cal., in 1903.

Date.	Hydrographer.	Gage height.	Discharge.
		Feet.	*Second-feet.*
7	S. G. Bennett	12. 20	31, 256
....do		8. 07	12, 001
· 3 ....do		1. 48	1, 289
19.................. .....do		4. 80	5, 759

m daily gage height, in feet, of Feather River at Oroville, Cal., for 1903.

ay.	Jan.	Feb.	Mar.	Apr.	May.	June.	July.	Aug.	Sept.	Oct.	Nov.	Dec.
....	4.0	6.9	4.0	14.6	9.0	5.7	2.9	1.9	1.5	1.6	1.5	4.5
....	3.8	6.3	4.0	13.5	9.0	5.6	2.8	1.9	1.5	1.6	1.5	4.4
....	3.5	6.9	5.0	12.8	8.9	5.4	2.7	1.9	1.5	1.6	1.5	4.4
....	3.5	6.5	4.5	11.5	8.8	5.1	2.6	1.9	1.5	1.6	1.6	4.3
....	3.6	6.0	4.5	10.8	8.7	5.0	2.5	1.9	1.5	1.6	2.1	4.3
....	3.8	5.0	4.5	10.2	8.7	4.9	2.5	1.8	1.4	1.6	1.9	4.1
....	3.8	5.2	4.5	10.2	8.6	4.8	2.5	1.8	1.4	1.6	1.9	3.7
....	3.9	5.5	4.5	10.2	8.4	4.8	2.4	1.8	1.4	1.6	5.2	3.6
....	4.0	5.0	4.5	10.2	8.2	4.6	2.5	1.8	1.4	1.8	7.3	3.6
....	4.0	4.8	4.5	10.2	8.1	4.6	2.5	1.8	1.4	2.5	9.9	3.4
....	3.9	4.8	4.5	10.2	8.1	4.4	2.4	1.8	1.4	2.4	11.5	3.4
....	3.7	4.8	4.5	10.2	7.9	4.4	2.3	1.7	1.4	2.1	13.8	3.2
....	3.5	4.6	6.3	10.2	7.8	4.3	2.3	1.7	1.4	1.9	17.4	3.2
....	3.5	4.6	8.2	10.2	7.7	3.9	2.3	1.7	1.4	1.8	19.2	3.0
....	3.3	4.3	7.2	10.2	7.4	3.9	2.2	1.7	1.4	1.6	16.8	3.2
....	3.3	4.3	7.1	10.2	7.2	3.9	2.2	1.7	1.4	1.6	12.3	3.6
....	3.3	4.0	6.0	10.1	7.0	3.8	2.2	1.7	1.5	1.6	6.7	7.1
....	3.1	4.0	6.0	10.0	6.7	3.7	2.2	1.6	1.5	1.6	5.0	5.2
....	3.1	4.0	5.8	9.9	6.5	3.6	2.2	1.6	1.5	1.6	6.1	4.8
....	3.0	3.8	5.5	9.8	6.1	3.5	2.1	1.6	1.5	1.6	14.9	4.6
....	3.0	3.8	5.5	9.8	6.1	3.4	2.1	1.6	1.4	1.6	19.1	4.4
....	6.2	3.8	5.4	9.7	6.0	3.3	2.1	1.6	1.4	1.5	14.0	4.3
....	7.0	4.0	6.0	9.7	6.0	3.2	2.1	1.6	1.4	1.5	11.3	4.0
....	9.1	4.0	6.0	9.6	5.8	3.2	2.1	1.6	1.4	1.5	8.5	3.9
....	12.5	4.0	5.5	9.6	5.5	3.0	2.0	1.6	1.4	1.5	6.0	3.8
....	11.3	4.0	6.2	9.5	5.5	3.0	2.0	1.6	1.4	1.5	4.5	3.6
....	12.2	4.0	7.7	9.4	5.5	2.9	2.0	1.6	1.4	1.5	4.0	3.6
....	9.3	4.0	9.1	9.3	5.4	3.0	2.0	1.6	1.4	1.5	4.8	3.4
....	7.8	......	12.0	9.2	5.5	3.0	2.0	1.6	1.4	1.5	4.6	3.4
....	7.9	......	20.3	9.1	5.6	2.9	1.9	1.6	1.5	1.5	4.5	3.2
....	7.2	......	16.1	......	5.7	......	1.9	1.6	.....	1.5	......	3.5

Rating table for Feather River at Oroville, Cal., for 1903.

Gage height.	Discharge.	Gage height.	Discharge.	Gage height.	Discharge.	Gage height.	Discharge
Feet.	Second-feet.	Feet.	Second-feet.	Feet.	Second-feet.	Feet.	Second-feet.
1.4	1,200	2.8	2,680	6.0	7,500	11.0	21,100
1.5	1,300	3.0	2,900	6.5	8,450	12.0	25,800
1.6	1,400	3.2	3,140	7.0	9,450	13.0	31,500
1.7	1,500	3.4	3,380	7.5	10,550	14.0	38,100
1.8	1,600	3.6	3,640	8.0	11,750	15.0	45,700
1.9	1,700	3.8	3,920	8.5	13,000	16.0	53,900
2.0	1,800	4.0	4,200	9.0	14,300	17.0	62,800
2.2	2,020	4.5	4,950	9.5	15,700	18.0	72,100
2.4	2,240	5.0	5,750	10.0	17,300	19.0	81,800
2.6	2,460	5.5	6,600	10.5	19,100	20.0	92,000

Estimated monthly discharge of Feather River at Oroville, Cal., for 1903.

[Drainage area, 3,350 square miles.]

Month.	Discharge in second-feet.			Total in acre-feet.	Run-off.	
	Maximum.	Minimum.	Mean.		Second-feet per square mile.	Depth inches
January	28,500	2,900	7,517	462,202	2.24	
February	9,250	3,920	5,431	301,623	1.62	
March	95,150	4,200	11,973	736,191	3.57	
April	42,460	14,580	19,165	1,140,396	5.72	
May	14,300	6,430	10,019	616,044	2.99	
June	6,960	2,790	4,329	257,593	1.29	
July	2,790	1,700	2,102	129,247	.63	
August	1,700	1,400	1,506	92,600	.45	
September	1,300	1,200	1,233	73,369	.37	
October	2,350	1,300	1,464	89,833	.44	
November	83,800	1,300	19,934	1,186,155	5.95	
December	9,670	2,900	4,213	259,047	1.26	
The year	95,150	1,200	7,407	5,344,300	2.11	2

STONY CREEK NEAR FRUTO, CAL.

On January 30, 1901, a gaging station was established by Burt (
at Julian's ranch, 6 miles northwest of the town of Fruto, Cal.
determine the amount of water available for storage. The obse
is Mrs. Lee Julian.

ɔbservations at this station during 1903 have been made under ɔction of J. B. Lippincott, supervising engineer.

Discharge measurements of Stony Creek near Fruto, Cal., in 1903.

Date.	Hydrographer.	Gage height.	Discharge.
		Feet.	*Second-feet.*
30	S. G. Bennett..................	7.65	3,374
....	.....do	4.15	322
....	.....do	3.10	10
30.....................	.....do	3.20	13

Mean daily gage height, in feet, of Stony Creek near Fruto, Cal., for 1903.

Day.	Jan.	Feb.	Mar.	Apr.	May.	June.	July.	Aug.	Sept.	Oct.	Nov.	Dec.
....	5.50	6.10	5.30	6.60	4.70	3.50	3.10	3.10	3.00	3.10	3.20	4.10
....	5.50	5.90	5.20	6.20	4.70	3.40	3.10	3.10	3.00	3.10	3.20	4.10
....	5.40	5.60	5.20	6.20	4.70	3.40	3.10	3.10	3.00	3.10	3.20	4.10
....	5.40	5.50	5.20	6.10	4.60	3.30	3.10	3.10	3.00	3.20	3.20	4.00
....	5.30	5.50	5.40	5.90	4.60	3.20	3.10	3.10	3.00	3.20	3.20	4.00
....	5.30	5.50	5.40	5.70	4.60	3.20	3.10	3.10	3.00	3.20	3.20	3.90
....	5.30	6.50	5.30	5.70	4.60	3.20	3.10	3.10	3.00	3.20	3.20	3.90
....	5.20	6.30	5.20	5.60	4.50	3.20	3.10	3.10	3.00	3.20	3.20	3.90
....	5.20	5.80	5.20	5.50	4.50	3.20	3.00	3.10	3.00	3.20	3.20	3.90
....	5.10	5.80	5.20	5.40	4.50	3.20	3.00	3.10	3.00	3.20	3.20	3.90
....	5.10	5.90	5.10	5.30	4.50	3.10	3.00	3.10	3.00	3.20	3.20	3.90
....	5.10	5.60	5.10	5.20	4.40	3.10	3.00	3.10	3.00	3.20	4.30	3.90
....	5.10	5.50	7.00	5.20	4.40	3.10	3.00	3.10	3.00	3.20	4.80	3.90
....	5.00	5.50	7.10	5.20	4.30	3.10	3.00	3.10	3.00	3.20	6.90	4.00
....	5.00	5.50	7.60	5.20	4.20	3.10	3.00	3.10	3.00	3.20	5.30	4.00
....	5.00	5.40	7.30	5.10	4.20	3.10	3.00	3.10	3.00	3.20	4.40	7.40
....	5.00	5.40	6.70	5.10	4.20	3.10	2.90	3.10	3.00	3.20	4.10	6.90
....	5.00	5.30	6.30	5.20	4.20	3.10	2.90	3.10	3.00	3.20	3.70	5.80
....	5.00	5.30	6.20	5.20	4.10	3.10	2.90	3.10	3.00	3.20	4.90	4.80
....	4.90	5.30	6.10	5.20	4.10	3.10	2.90	3.10	3.00	3.30	6.90	4.80
....	4.90	5.20	6.00	5.10	4.10	3.10	2.90	3.10	3.00	3.30	8.10	4.70
....	5.70	5.20	5.80	5.10	4.10	3.10	2.90	3.10	3.00	3.30	6.80	4.70
....	6.20	5.80	5.80	5.00	4.00	3.10	2.90	3.10	3.00	3.30	5.80	4.60
....	7.10	5.60	6.00	5.00	4.00	3.10	2.90	3.10	3.10	3.30	5.30	4.50
....	8.60	5.50	6.00	4.90	3.90	3.10	2.90	3.10	3.10	3.20	4.90	4.50
....	8.00	5.40	5.90	4.90	3.80	3.10	3.10	3.10	3.10	3.20	4.60	4.30
....	7.80	5.40	6.	4.80	3.70	3.10	3.10	3.10	3.10	3.20	4.40	4.30
....	7.30	5.30	9.10	4.80	3.70	3.10	3.10	3.10	3.10	3.20	4.30	4.30
....	6.60		7.20	4.80	3.60	3.10	3.10	3.10	3.10	3.20	4.30	4.20
....	7.90		7.40	4.70	3.60	3.	3.10	3.00	3.10	3.20	4.20	4.20
....	6.70	......	7.00	......	3.50		3.10	3.00	......	3.20	......	4.20

Rating table for Stony Creek near Fruto, Cal., for 1903.

Gage height.	Discharge.	Gage height.	Discharge.	Gage height.	Discharge.	Gage height.	Discharge.
Feet.	*Second-feet.*	*Feet.*	*Second-feet.*	*Feet.*	*Second-feet.*	*Feet.*	*Second-feet.*
2.9	4	3.9	170	5.8	1,250	7.8	3,608
3.0	7	4.0	200	6.0	1,430	8.0	3,900
3.1	10	4.2	260	6.2	1,625	8.2	4,205
3.2	15	4.4	330	6.4	1,835	8.4	4,530
3.3	26	4.6	410	6.6	2,050	8.6	4,840
3.4	43	4.8	500	6.8	2,275	8.8	5,170
3.5	63	5.0	620	7.0	2,510	9.0	5,510
3.6	88	5.2	765	7.2	2,765	9.2	5,860
3.7	113	5.4	920	7.4	3,035	9.4	6,220
3.8	140	5.6	1,080	7.6	3,315	9.6	6,600

Estimated monthly discharge of Stony Creek near Fruto, Cal., for 1903.

[Drainage area, 760 square miles.]

Month.	Discharge in second-feet.			Total in acre-feet.	Run-off.	
	Maximum.	Minimum.	Mean.		Second-feet per square mile.	Depth in inches.
January	4,840	560	1,423	87,497	1.87	2.16
February	1,940	765	1,087	60,369	1.43	1.49
March...............	5,685	690	1,663	102,254	2.19	2.52
April...............	2,050	450	892	53,078	1.17	1.31
May.................	450	63	276	16,971	.36	.42
June	63	10	16	952	.02	.02
July	10	4	7	430	.01	.01
August	10	7	10	615	.01	.01
September..........	10	7	8	476	.01	.01
October	26	10	16	984	.02	.02
November..........	4,050	15	613	36,476	.81	.90
December	3,105	170	462	28,407	.61	.70
The year	5,685	4	539	388,509	.71	9.57

M'CLOUD RIVER NEAR GREGORY, CAL.

This station was established March 23, 1902, at the request of the consulting engineer of the McCloud River Electric Company, with the understanding that they would maintain the station. The watershed of McCloud River includes the southern and eastern slopes

Mount Shasta and is heavily timbered. The river is fed by the
nerous springs and its minimum flow is about 1,100 second-feet.
he gage rod is a 2 by 3 inch timber nailed and wired to a tree on
right bank of the stream at Johns Camp, 14 miles east of Gregory.
asurements are made from a car suspended from a wire cable.
ire is an auxiliary cable 50 feet upstream from the main cable.
he initial point of sounding is at the gage rod on the right bank
the stream. The channel is straight for 300 feet above and for 600
t below the gaging station. The current is swift at all stages.
: banks are high and wooded and not liable to overflow. The
l is composed of limestone on the sides, with some large river
vel and bowlders in the center of the channel. Sufficient meter
isurements have not been taken to construct a rating curve.
he observations at this station during 1903 have been made under
direction of J. B. Lippincott, supervising engineer.

Discharge measurements of McCloud River near Gregory, Cal., in 1902-3.

Date.	Hydrographer.	Gage height.	Discharge.
1902.		*Feet.*	*Second-feet.*
ember 23	S. G. Bennett...............	1.45	1,272
ember 9	J. D. Schuyler...............	6.00	9,071
1903.			
31.....................	S. G. Bennett...............	1.50	1,449

Iean daily gage height, in feet, of McCloud River near Gregory, Cal., for 1902.

Day.	Sept.	Oct.	Nov.	Dec.	Day.	Sept.	Oct.	Nov.	Dec.
....		1.5	1.5	1.9	17	1.5	3.1	2.6	
....		1.5	1.5	1.9	18	1.4	3.7	2.5	
....		1.5	1.4	1.8	19	1.4	3.6	2.4	
....		1.5	1.4	1.8	20	1.4	2.9	2.3	
....		1.4	1.4	1.9	21	1.5	2.7	2.2	
....		1.4	1.5	1.9	22	1.6	2.5	2.3	
....		1.4	1.6	2.9	23	1.5	1.7	2.4	2.3
....		1.3	3.8	3.4	24	1.5	2.1	2.3	2.4
....		1.3	7.5	3.4	25	1.5	2.5	2.2	2.5
....		1.4	6.4	4.6	26	1.5	1.8	2.1	3.2
....		1.4	3.7	4.9	27	1.5	1.6	2.0	3.1
....		1.4	3.2	4.3	28	1.5	1.6	2.0	2.9
....		1.4	3.1	3.8	29	1.5	1.6	1.9	2.7
....		1.5	3.0	3.2	30	1.4	1.5	1.9	2.5
....		1.5	2.8	2.9	31	1.5	......	2.4	
....		1.5	2.8	2.8					

Mean daily gage height, in feet, of McCloud River near Gregory, Cal., for 1901.

Day.	Jan.	Feb.	Mar.	Apr.	May.	June.	July.	Aug.	Sept.	Oct.	Nov.	Dec.
1	2.4	2.9	2.4	4.9	2.6	2.0	1.7	1.52	1.45	1.45	1.9	2.9
2	2.3	2.7	2.4	4.4	2.5	1.9	1.6	1.52	1.45	1.45	1.9	1.7
3	2.2	2.6	2.6	4.0	2.5	2.0	1.7	1.52	1.45	1.45	1.9	1.6
4	2.1	2.5	2.6	3.8	2.5	1.9	1.6	1.52	1.45	1.30	1.9	1.7
5	2.1	2.5	2.5	3.5	2.6	1.9	1.7	1.52	1.45	1.45	1.9	1.9
6	2.1	2.4	2.5	3.4	2.6	1.9	1.6	1.52	1.45	1.40	1.9	1.9
7	2.0	2.5	2.4	3.2	2.5	1.9	1.7	1.48	1.45	1.40	1.9	1.7
8	2.0	2.3	2.5	3.2	2.4	1.9	1.6	1.48	1.45	1.40	1.9	1.7
9	2.0	2.3	2.4	3.2	2.4	1.8	1.6	1.48	1.45	1.40	1.9	1.7
10	2.0	2.2	2.4	3.1	2.4	1.9	1.6	1.48	1.45	1.40	1.90	1.7
11	1.9	2.3	2.7	3.0	2.4	1.8	1.5	1.48	1.45	1.40	1.9	1.9
12	1.9	2.4	4.0	2.9	2.4	1.9	1.6	1.48	1.45	1.35	1.80	1.9
13	1.9	2.3	4.1	2.9	2.4	1.5	1.5	1.44	1.45	1.65	1.95	1.7
14	1.9	2.3	4.6	2.8	2.3	1.8	1.6	1.48	1.45	1.85	6.05	1.4
15	1.8	2.2	4.3	2.8	2.3	1.5	1.5	1.45	1.45	1.35	3.75	1.9
16	1.8	2.2	3.8	2.7	2.2	1.5	1.5	1.45	1.45	1.65	1.90	1.5
17	1.8	2.1	3.5	2.7	2.2	1.6	1.5	1.45	1.45	1.45	1.62	1.5
18	1.8	2.1	3.0	2.6	2.2	1.5	1.5	1.45	1.45	1.45	1.80	1.8
19	1.8	2.1	2.9	2.6	2.1	1.5	1.5	1.45	1.45	1.45	2.8	1.9
20	1.8	2.1	2.8	2.6	2.1	1.8	1.5	1.45	1.45	1.45	6.80	1.6
21	2.1	2.1	2.7	2.6	2.1	1.7	1.5	1.45	1.45	1.40	8.10	1.6
22	3.0	2.2	2.6	2.6	2.1	1.7	1.5	1.45	1.45	1.40	7.35	1.5
23	3.5	2.7	2.6	2.7	2.1	1.7	1.5	1.45	1.45	1.40	6.20	1.8
24	6.3	2.7	2.7	2.7	2.0	1.7	1.5	1.45	1.40	1.40	4.95	1.8
25	7.7	2.7	3.1	2.7	2.0	1.7	1.5	1.45	1.40	1.40	4.60	1.8
26	6.5	2.6	3.5	2.7	2.0	1.7	1.5	1.45	1.40	1.40	4.10	1.6
27	4.7	2.5	3.8	2.7	2.0	1.7	1.5	1.45	1.40	1.40	3.70	2.0
28	4.0	2.5	6.0	2.6	2.0	1.7	1.5	1.45	1.40	1.40	2.65	2.2
29	3.5	......	7.2	2.6	2.0	1.7	1.5	1.45	1.40	1.40	2.55	2.2
30	3.3	......	6.5	2.6	2.0	1.7	1.5	1.45	1.43	1.40	2.50	2.20
31	3.1	......	5.8	......	1.9	......	1.5	1.45	......	1.40	......	2.12

SACRAMENTO RIVER NEAR RED BLUFF, CAL.

The gaging station at Jellys Ferry, which is located 12 miles above the town of Red Bluff, was established April 30, 1895. The right bank of the river is high, but the left bank is liable to overflow when the river rises above the 25-foot mark. The river has been known to reach the 35-foot mark. Because of the liability to overflow, it was deemed advisable to select a new gaging station where the water at flood stage would be more confined. A point in Iron Canyon, where the river had been gaged by the State engineering department in 1879 and by the commissioner of public works in 1893–94, was chosen as a new gaging station. The river-stage rod used by the commissioner of public works was still in place and has been used in making river-height observations since January 28, 1902, the date upon which the observations were begun. The river at this point in lower portion of Iron Canyon, 4 miles above Red Bluff, has a direct course for 2 or 3 miles. The width between banks at low water is about 500 feet. The depth

it low stages averages 6 feet with a maximum depth of 9 feet.
s are steep and firm. The river flows in a bed of coarse·
d cobbles with here and there a small bowlder. The bed
ıva. Discharge measurements are made from a cable 800
, which is anchored in the lava rock that forms the wall of
n. The observer at Iron Canyon was F. A. Wilcox.
servations at this station during 1903 have been made under
ion of J. B. Lippincott, supervising engineer.

arge measurements of Sacramento River near Red Bluff, Cal., in 1903.

Date.	Hydrographer.	Gage height.	Discharge.
		Feet.	*Second-feet.*
....	S. G. Bennett..................	10. 90	45, 654
....	.....do	2. 95	10, 034
....	.....do	1. 05	4, 625
....	.....do	1. 07	4, 501
5....................	.....do	2. 20	7, 461

ly gage height, in feet, of Sacramento River near Red Bluff, Cal., for 1903.

r.	Jan.	Feb.	Mar.	Apr.	May.	June.	July.	Aug.	Sept.	Oct.	Nov.	Dec. •
....	4.3	7.4	(4.8)	10.0	4.0	2.4	1.5	1.05	.95	1.1	1.1	3.5
....	3.9	6.5	(4.8)	8.7	4.0	2.4	1.5	1.05	.95	1.1	1.1	3.4
....	3.6	5.8	5.1	7.7	4.0	2.3	1.5	1.05	.95	1.1	1.1	3.2
....	3.5	5.3	(5.7)	7.4	4.0	2.2	1.4	1.05	.95	1.1	1.1	3.1
....	3.4	5.0	6.3	6.8	4.0	2.1	1.4	1.05	.95	1.1	1.4	3.0
....	3.3	4.7	6.3	6.4	4.0	2.0	1.4	1.05	.95	1.1	1.5	2.7
....	3.2	4.7	6.3	6.1	3.9	2.0	1.4	1.05	.90	1.1	1.6	2.6
....	3.1	5.4	8.0	5.9	3.8	2.0	1.4	1.05	.90	1.1	1.5	2.5
....	3.0	5.1	6.3	5.9	3.6	2.0	1.4	1.05	.90	1.1	1.4	2.5
....	2.9	5.3	5.6	5.8	3.6	2.0	1.3	1.00	.90	1.8	1.3	2.4
... .	2.8	6.0	5.3	5.4	3.5	2.0	1.3	1.00	.90	2.1	1.3	2.4
....	2.8	6.0	8.8	5.2	3.4	1.9	1.3	1.00	.90	1.1	1.3	2.3
....	2.7	5.2	11.0	5.0	3.3	1.9	1.3	.96	.90	1.1	3.3	2.3
....	2.7	4.8	15.3	4.9	3.2	1.9	1.3	.95	.90	1.1	6.8	2.3
....	2.7	4.4	13.3	4.8	3.1	1.9	1.3	.95	.90	1.1	4.3	2.3
....	2.5	4.2	11.0	4.6	3.0	1.9	1.2	.95	.90	1.1	3.8	7.5
....	2.5	4.1	9.3	4.5	3.0	1.8	1.2	.95	.90	1.1	3.6	11.7
....	2.4	3.9	8.0	4.9	3.9	1.7	1.2	.95	.90	1.1	3.6	4.8
....	2.4	3.8	7.1	4.5	3.9	1.7	1.2	.95	.90	1.1	4.2	4.7
....	2.4	3.8	6.3	4.3	2.8	1.7	1.2	.95	.90	1.1	12.0	7.1
....	2.8	3.8	6.1	4.3	2.7	1.7	1.2	.95	.90	1.1	19.0	4.9
....	8.5	3.9	5.8	4.3	2.6	1.7	1.2	.95	.90	1.1	21.5	4.7
....	13.9	5.1	5.6	4.3	2.6	1.7	1.2	.95	.90	1.1	14.0	4.1
....	13.5	5.4	5.6	4.3	2.6	1.7	1.1	.95	.90	1.1	11.3	3.9
....	22.8	5.3	6.3	4.3	2.6	1.6	1.2	.95	.90	1.1	9.0	3.5
....	(18.3)	5.1	7.4	4.3	2.6	1.5	1.2	.95	.90	1.1	6.7	3.3
....	14.0	4.9	8.0	4.2	2.6	1.5	1.1	.95	.90	1.1	4.9	3.2
....	11.0	4.8	13.6	4.1	2.5	1.5	1.1	.95	.90	1.1	4.0	3.1
....	8.7	......	15.7	4.0	2.5	1.5	1.1	.95	1.00	1.1	4.0	2.9
....	8.8	......	12.0	4.0	2.4	1.5	1.1	.95	1.00	1.1	4.0	2.7
....	8.3	......	11.4	....	2.4	......	1.1	.95	......	1.1	......	3.1

Rating table for Sacramento River near Red Bluff, Cal., for 1903.

Gage height.	Discharge.	Gage height.	Discharge.	Gage height.	Discharge.	Gage height.	Discharge.
Feet.	*Second-feet.*	*Feet.*	*Second-feet.*	*Feet.*	*Second-feet.*	*Feet.*	*Second-feet.*
1.0	5,000	2.2	7,840	5.0	17,100	15.0	68,200
1.1	5,220	2.4	8,380	5.5	19,100	16.0	75,100
1.2	5,440	2.6	8,950	6.0	21,100	17.0	82,200
1.3	5,660	2.8	9,550	7.0	25,300	18.0	89,700
1.4	5,880	3.0	10,150	8.0	29,700	19.0	97,600
1.5	6,100	3.2	10,790	9.0	34,300	20.0	105,900
1.6	6,340	3.4	11,430	10.0	39,100	21.0	114,600
1.7	6,580	3.6	12,090	11.0	44,200	22.0	123,700
1.8	6,820	3.8	12,770	12.0	49,700	23.0	133,200
1.9	7,060	4.0	13,450	13.0	55,600	24.0	143,100
2.0	7,300	4.5	15,200	14.0	61,700	25.0	153,400

Estimated monthly discharge of Sacramento River near Red Bluff, Cal., for 1903.

[Drainage area, 9,286 square miles.]

Month.	Discharge in second-feet.			Total in acre-feet.	Run-off.	
	Maximum.	Minimum.	Mean.		Second-feet per square mile.	Depth in inches.
January	131,300	8,380	25,561	1,571,685	2.75	3.17
February	27,060	12,770	17,186	954,462	1.85	1.93
March..............	73,000	16,340	31,555	1,940,242	3.39	3.91
April..............	39,100	13,450	18,791	1,118,142	2.02	2.25
May	13,450	8,380	10,948	673,166	1.18	1.36
June	8,380	6,100	6,970	414,744	.75	.84
July	6,100	5,220	5,589	343,654	.60	.69
August	5,110	4,890	4,965	305,286	.53	.61
September..........	5,000	4,780	4,813	286,393	.52	.58
October	7,570	5,220	5,347	328,774	.58	.67
November	119,100	5,220	21,955	1,306,413	2.36	2.63
December	48,020	8,110	13,101	805,549	1.41	1.63
The year	131,300	4,780	13,898	10,048,510	1.50	20.27

MOKELUMNE RIVER AT ELECTRA, CAL.

This station was established January 1, 1900, by Burr Bassell. It is located 3 miles above the wagon bridge, on the Mokelumne Hill and Jackson road. It is one-half mile below the Standard Electric Company's power house and the post-office at Electra, Cal. The gage is a large inclined timber bolted to a tree on the left bank. It is read twice each day by H. F. Vogt. Discharge measurements are made by

)f a cable and car 200 feet below the gage. The initial point
ndings is the anchor sheave to which the cable is fastened on
ht bank. The channel is straight for 300 feet above and for
t below the station. The current is swift at high stages, but
re cross currents at extreme low water. Both banks are high
; not liable to overflow. The lower part of the right bank is
ed of hard gravel. The upper part of the right bank and the
eft bank are composed of solid rock. The bed of the stream
osed of rock and gravel and is fairly permanent.
)bservations at this station during 1903 have been made under
:ction of J. B. Lippincott, supervising engineer.

Discharge measurements of Mokelumne River at Electra, Cal., in 1903.

Date.	Hydrographer.	Gage height.	Discharge.
		Feet.	*Second-feet.*
....	S. G. Bennett..............	8.10	4,767
....do ....,........		6.45	2,039

m daily gage height in feet of Mokelumne River at Electra, Cal., for 1903.

Day.	May.	June.	July.	Aug.	Sept.	Oct.	Nov.	Dec.
....		8.4	5.3	4.2	4.3	4.3	4.3	4.8
....		8.7	5.2	4.1	4.3	4.3	4.3	4.6
....		8.5	4.7	4.2	4.3	4.2	4.4	4.6
....		8.1	4.9	4.3	4.4	4.3	4.3	4.4
....		7.5	4.7	4.3	4.3	4.4	4.3	4.4
....		8.0	4.6	4.3	4.2	4.3	4.3	4.3
....		7.9	4.7	4.2	4.3	4.3	4.3	4.3
....		8.0	4.8	4.2	4.4	4.3	4.4	4.2
....		7.6	4.7	4.3	4.4	4.3	4.5	4.2
....		7.2	4.7	4.2	4.3	4.3	4.4	4.2
....	8.2	4.7	4.5	4.2	4.3	4.3	4.5	4.3
....		6.8	4.4	4.2	4.4	4.3	4.7	4.3
....		7.0	4.6	4.2	4.3	4.2	8.4	4.2
....		7.1	4.6	4.2	4.3	4.3	6.7	4.1
....		7.0	4.7	4.3	4.3	4.2	5.8	4.1
....		5.6	4.4	4.1	4.4	4.1	5.4	4.1
....	5.6	5.7	4.4	4.2	4.3	4.2	4.7	4.1
....	5.6	6.3	4.4	4.3	4.3	4.2	4.5	4.3
....	4.8	6.8	4.2	4.2	4.4	4.2	4.6	4.3
....	5.1	6.6	4.3	4.2	4.4	4.3	5.9	4.2
....	4.2	6.5	4.3	4.3	4.4	4.3	6.2	4.3
....	4.5	6.5	4.3	4.3	4.3	4.3	5.7	4.2
....	3.9	6.1	4.3	4.2	4.3	4.3	5.5	4.2
....	3.9	6.1	4.2	4.2	4.3	4.2	5.7	4.3
....	4.5	6.3	4.3	4.3	4.4	4.2	5.7	4.1
....	4.9	6.1	4.2	4.3	4.3	4.3	5.4	4.1
....	6.1	6.1	4.2	4.3	4.4	4.3	5.1	4.2
....	7.5	5.5	4.4	4.3	4.3	4.2	4.9	4.2
....	7.8	5.5	4.2	4.3	4.4	4.3	4.9	4.1
....	8.5	5.3	4.2	4.4	4.4	4.3	4.7	4.2
....	9.8	....	4.3	4.4	....	4.3	....	4.6

Rating table for Mokelumne River at Electra, Cal., for 1903.

Gage height.	Discharge.	Gage height.	Discharge.	Gage height.	Discharge.	Gage height.	Discharge.
Feet.	Second-feet.	Feet.	Second-feet.	Feet.	Second-feet.	Feet.	Second-feet.
				6.2	1,800	8.2	4,480
				6.4	2,000	8.4	4,820
				6.6	2,240	8.6	5,100
				6.8	2,480	8.8	5,500
			1,100	7.0	2,730	9.0	5,870
				7.2	2,990	9.2	6,280
				7.4	3,260	9.4	6,700
				7.6	3,550	9.6	7,140
				7.8	3,850	9.8	7,610
				8.0	4,160	10.0	8,100

Estimated monthly discharge of Mokelumne River at Electra, Cal., for 1903.

[Drainage area, 587 square miles.]

Month.	Discharge in second-feet.			Total in acre-feet.	Run-off.	
	Maximum.	Minimum.	Mean.		Second-feet per square mile.	Depth in inches.
June	5,330	470	2,626	155,962		5.43
July	940	140	344	21,152		
August	260	90	170	10,453		
September	260	140	222	13,210		.44
October	260	90	181	11,129		.39
November	4,820	200	855	50,876		1.77
December	540	90	188	11,560		.40
The period	5,330	90	654	274,043		9.36

STANISLAUS RIVER AT KNIGHTS FERRY, CAL.

his station was established May 19, 1903, by W. H. Stearns. The
ion is located 200 feet from the post-office at Knights Ferry. There
a island 800 feet above the gaging station and a dam on each chan-
at the head of the island. The Stanislaus Milling and Power
npany's power house is on the right bank of the river below one of
ie dams and about 1,000 feet above the gaging station. Ordinary
low water stages are read on a 2-inch iron pipe driven into the bed
he stream. For high stages the gage is a 1 by 6 inch plank nailed
, 6 by 6 inch post on the right bank of the river. The graduations
both gages are to feet and tenths. The gage height is read twice
h day by J. F. Cannon. Discharge measurements are made from
ar suspended from a ¾-inch galvanized-iron cable. Above and
allel with the main cable is another smaller one, from which a stay
i can be run to the meter to hold it, in flood measurements, against
swift current. The initial point for soundings is the eyebolt, to
ich the cable is fastened on the right bank. The channel is straight
500 feet above and below the cable. At ordinary and high stages
stream has a fall of 0.47 foot in the 500 feet above the cable, and
0.68 foot in the 500 feet below. Both banks are composed of
iented gravel and are high. The left bank is not subject to over-
v. In extreme floods the right bank has been known to be over-
ved, flooding the yards and houses next the river. The bed is of
ivel and is subject to some change from the addition of material
ich is washed down from the island above. The bench mark is a
ke driven into a 6 by 6 inch redwood post near the upper section of
gage. Its elevation is 12 feet above the zero of the gage.
The observations at this station during 1903 have been made under
direction of J. B. Lippincott, supervising engineer.

Discharge measurements of Stanislaus River at Knights Ferry, Cal., in 1903.

Date.	Hydrographer.	Gage height.	Discharge.
		Feet.	*Second-feet.*
ie 13	S. G. Bennett	10.05	3,704
ie 15	do	9.60	2,854
itember 22	do	5.60	32
vember 26	do	7.15	648

Mean daily gage height, in feet, of Stanislaus River at Knights Ferry, Cal., for 1903.

Day.	May.	June.	July.	Aug.	Sept.	Oct.	Nov.	Dec.
1		11.4	8.4	6.4	5.7	5.8	5.7	4.9
2		11.4	8.3	6.5	5.7	5.8	5.7	4.7
3		10.9	8.1	6.4	5.7	5.8	5.7	4.6
4		10.5	8.1	6.3	5.7	5.8	5.7	4.6
5		10.4	8.0	6.4	5.6	5.8	5.7	4.5
6		10.8	7.7	6.4	5.8	5.8	5.7	4.4
7		10.9	7.7	6.3	5.7	5.7	5.7	4.4
8		10.6	7.2	6.3	5.6	5.7	5.7	4.3
9		10.5	7.1	6.2	5.6	5.7	5.7	4.3
10		10.1	7.1	6.2	5.5	5.7	5.7	4.6
11		10.1	7.0	6.1	5.6	5.7	5.7	4.6
12		9.9	6.9	6.1	5.6	5.7	7.3	4.3
13		9.8	6.9	6.1	5.5	5.7	8.3	4.6
14		9.8	6.9	6.0	5.6	5.7	8.5	4.3
15		9.6	6.9	6.0	5.5	5.7	8.7	4.5
16		9.1	6.9	5.9	5.5	5.7	7.6	4.3
17		9.1	7.0	5.9	5.5	5.6	6.9	4.4
18		8.8	6.9	5.9	5.6	5.6	6.9	4.4
19	9.6	9.3	6.8	5.9	5.5	5.6	6.6	4.3
20	9.6	9.0	6.9	5.9	5.6	5.7	7.3	4.4
21	9.5	9.4	6.7	5.9	5.6	5.7	8.1	4.4
22	9.3	9.1	6.6	5.9	5.6	5.7	8.1	4.3
23	9.1	9.0	6.6	5.8	5.6	5.7	7.7	4.3
24	9.0	8.9	6.6	5.8	5.6	5.7	7.6	4.3
25	8.9	9.1	6.5	5.8	5.6	5.7	7.4	4.3
26	9.1	9.0	6.5	5.8	5.6	5.7	7.3	4.3
27	9.2	8.9	6.5	5.8	5.5	5.7	7.0	4.3
28	9.4	9.0	6.5	5.8	5.5	5.7	7.0	4.1
29	10.2	8.6	6.5	5.8	5.5	5.7	6.9	4.1
30	11.0	8.6	6.5	5.7	5.5	5.7	6.8	4.1
31	11.7	....	6.4	5.7	....	5.7	....	4.1

Rating table for Stanislaus River at Knights Ferry, Cal., from May 19 to December 31, 1903.

Gage height.	Discharge.	Gage height.	Discharge.	Gage height.	Discharge.	Gage height.	Discharge.
Feet.	*Second-feet.*	*Feet.*	*Second-feet.*	*Feet.*	*Second-feet.*	*Feet.*	*Second-feet.*
5.5	25	6.5	300	8.0	1,310	10.0	3,750
5.6	35	6.6	340	8.2	1,500	10.2	4,110
5.7	50	6.7	390	8.4	1,700	10.4	4,510
5.8	75	6.8	440	8.6	1,910	10.6	4,930
5.9	100	6.9	495	8.8	2,140	10.8	5,400
6.0	125	7.0	550	9.0	2,400	11.0	5,920
6.1	155	7.2	670	9.2	2,660	11.2	6,500
6.2	185	7.4	810	9.4	2,920	11.4	7,130
6.3	220	7.6	960	9.6	3,180	11.6	7,830
6.4	260	7.8	1,130	9.8	3,460		

Estimated monthly discharge of Stanislaus River at Knights Ferry, Cal., for 1903.

[Drainage area, 935 square miles.]

Month.	Discharge in second-feet.			Total in acre-feet.	Run-off.	
	Maximum.	Minimum.	Mean.		Second-feet per square mile.	Depth in inches.
May	8,210	2,270	3,519	216,375	3.76	4.33
June	7,130	1,910	3,576	212,787	3.82	4.26
July	1,700	260	840	39,352	.68	.78
August	1,300	50	142	8,731	.15	.17
September	50	25	32	1,904	.03	.03
October	50	25	44	2,705	.05	.06
November	2,530	50	638	37,964	.68	.76
December	440	220	267	16,417	.29	.33
The period ...	8,210	25	1,107	536,235	1.18	10.72

TUOLUMNE RIVER AT LAGRANGE, CAL.

This station was established August 29, 1895, by J. B. Lippincott. It is located at the wagon bridge in the town of Lagrange, Cal. It is below the high dam of the Turlock and Modesto irrigation districts and also below the head of the canal of the Lagrange Ditch and Hydraulic Mining Company, which diverts water from the left bank of the river 15 miles above the Lagrange dam. During 1903 this canal was being repaired, and for the greater part of the year carried no water; for the remainder of the year it carried only enough water to keep the flumes wet. The gage is a vertical timber fastened to the right abutment. It is read twice each day by Miss Annie P. McGinn. Discharge measurements are made from the bridge to which the gage is attached. The initial point for soundings is at the south end of the trestle approach to bridge on left bank of river.

The channel is straight for 400 feet above and for 600 feet below the station. It is broken by two iron piers and has a width at high water of about 575 feet. During the season of low water all the water is taken out by the canals above the station. The bed of the stream is composed of gravel and is fairly permanent. The current is swift. Both banks are high and are not subject to overflow.

The observations at this station during 1903 have been made under the direction of J. B. Lippincott, supervising engineer.

Discharge measurements of Tuolumne River at Lagrange, Cal., in 1903.

Date.	Hydrographer.	Gage height.	Discharge.
		Feet.	*Second-ft.*
February 7	S. G. Bennett	5.50	1,20
April 7	do	6.75	4,13
June 16	do	6.85	4,38
November 27	do	5.05	99

Mean daily gage height, in feet, of Tuolumne River at Lagrange, Cal., for 1903.

Day.	Jan.	Feb.	Mar.	Apr.	May.	June.	July.	Aug.	Sept.	Oct.	Nov.	Dec.
1	4.8	6.7	5.4	11.1	7.5	9.0	6.7	3.3	3.2	3.1	3.7	4.7
2	4.7	6.2	5.4	10.1	7.8	9.1	6.5	3.4	3.2	3.1	3.7	4.7
3	4.7	6.1	5.5	7.9	7.8	8.8	6.3	3.3	3.2	3.1	3.7	4.7
4	4.7	5.8	5.6	7.5	8.0	8.6	5.9	3.4	3.2	3.4	3.7	4.6
5	4.6	5.7	6.2	7.2	8.2	8.4	5.9	3.3	3.2	3.5	3.7	4.6
6	4.6	5.6	5.8	7.0	8.2	8.4	5.7	3.3	3.2	3.6	3.7	4.5
7	4.6	5.6	5.7	6.7	8.5	8.6	5.6	3.3	3.2	3.7	3.7	4.4
8	4.6	5.7	5.7	6.9	8.4	8.5	5.5	3.4	3.2	3.8	3.85	4.4
9	4.6	5.6	5.8	7.1	8.6	8.5	5.5	3.4	3.2	3.7	3.86	4.4
10	4.5	5.6	5.7	7.3	8.9	8.3	5.2	3.3	3.2	3.7	3.90	4.3
11	4.5	5.7	5.6	7.0	9.0	8.0	5.0	3.3	3.2	3.7	3.90	4.3
12	4.5	5.5	5.7	6.7	9.0	7.9	4.9	3.3	3.2	3.7	3.85	4.4
13	4.4	5.5	5.7	6.6	9.4	7.7	4.8	3.3	3.2	3.7	3.80	4.1
14	4.3	5.4	5.5	6.5	9.2	7.6	4.8	3.3	3.2	3.7	6.15	4.1
15	4.4	5.5	6.3	6.5	9.0	7.5	4.7	3.3	3.2	3.7	6.3	4.0
16	4.3	5.4	6.2	6.3	8.8	7.2	4.6	3.3	3.2	3.6	5.8	4.0
17	4.3	5.4	6.1	6.3	8.7	6.8	4.6	3.3	3.2	3.6	5.4	4.0
18	4.3	5.4	6.0	6.2	7.7	7.0	4.5	3.3	3.2	3.6	5.0	4.1
19	4.4	5.4	6.0	6.1	7.2	7.3	4.4	3.3	3.2	3.7	4.9	4.2
20	4.3	5.5	5.9	6.2	6.9	7.3	4.6	3.3	3.2	3.7	5.5	4.3
21	4.4	5.5	6.0	6.3	6.6	7.5	4.6	3.2	3.2	3.7	6.5	4.2
22	4.4	5.4	5.8	6.6	6.4	7.5	4.7	3.2	3.2	3.7	6.2	4.2
23	4.5	5.5	5.8	7.0	6.3	7.4	4.2	3.2	3.2	3.6	5.9	4.2
24	5.2	5.6	5.8	7.6	6.2	7.3	3.7	3.3	3.2	3.7	5.7	4.3
25	6.0	5.4	6.0	7.8	6.0	7.2	3.6	3.2	3.2	3.7	5.4	4.3
26	7.3	5.8	6.1	7.8	5.8	7.1	3.5	3.2	3.2	3.7	5.2	4.4
27	9.6	5.3	6.3	7.6	6.2	7.2	3.6	3.2	3.1	3.7	5.0	4.32
28	10.4	5.2	7.2	7.6	7.3	7.0	3.8	3.2	3.1	3.7	4.9	4.40
29	7.1	......	7.6	7.3	7.8	6.9	3.4	3.2	3.1	3.7	4.8	4.38
30	6.7	......	8.5	7.2	8.6	6.8	3.4	3.2	3.1	3.7	4.8	4.36
31	6.7	......	10.2	......	9.0	......	3.3	3.2	......	3.7	......	4.75

Rating table for Tuolumne River at Lagrange, Cal., for 1903.

Discharge.	Gage height.	Discharge.	Gage height.	Discharge.	Gage height.	Discharge.
Second-feet.	*Feet.*	*Second-feet.*	*Feet.*	*Second-feet.*	*Feet.*	*Second-feet.*
7	4.8	560	6.8	4,230	. 8.8	11,100
15	5.0	750	7.0	4,800	9.0	11,860
25	5.2	1,000	7.2	5,410	9.2	12,620
35	5.4	1,310	7.4	6,060	9.4	13,400
55	5.6	1,640	7.6	6,740	9.6	14,180
80	5.8	1,980	7.8	7,440	9.8	14,960
115	6.0	2,340	8.0	8,150	10.0	15,750
200	6.2	2,760	8.2	8,870	10.5	17,750
300	6.4	3,200	8.4	9,600	11.0	19,800
410	6.6	3,700	8.6	10,340		

monthly discharge of Tuolumne River, including Turlock and Modesto canals, at Lagrange, Cal., for 1903.

[Drainage area 1,501 square miles.]

nth.	Discharge in second-feet.			Total for month in acre-feet.	Run-off.	
	Maximum.	Minimum.	Mean.		Second-feet per square mile.	Depth in inches.
....	17,350	250	2,066	127,033	1.38	1.59
....	4,038	1,273	1,791	99,467	1.19	1.24
....	16,659	1,422	3,368	207,090	2.24	2.58
....	20,342	2,860	6,006	357,382	4.00	4.46
....	13,808	2,412	8,300	510,347	5.53	6.37
....	12,680	4,676	7,814	464,965	5.21	5.81
....	4,507	407	1,423	. 87,497	.95	1.101
....	491	134	263	16,171	.18	.21
er........	142	89	105	6,248	.07	.08
....	107	26	72	4,427	.05	.06 .
er........	3,451	46	1,038	61,765	.69	.77
r	607	293	432	26,563	.29	.33
e year......	20,342	26	2,723	1,968,955	1.82	24.60

TURLOCK CANAL AT LAGRANGE, CAL.

The Turlock canal, the property of the Turlock irrigation district, takes water from the left bank of the Tuolumne River at the Lagrange dam. This canal was designed to carry 1,500 second-feet and to irrigate a large area of fertile land in the vicinity of Turlock and Ceres, Stanislaus County, Cal. During 1898 water was first turned into the canal in small quantities and used for puddling the banks. A record of the gage height has been kept since July, 1899. Meter measurements are made when the gaging station on the Tuolumne River at Lagrange is visited, and Morgan flume, or flume No. 2, has been rated. The observer is J. L. Montgomery.

The observations at this station during 1903 have been made under the direction of J. B. Lippincott, supervising engineer.

Discharge measurements of Turlock canal at Lagrange, Cal., in 1903.

Date.	Hydrographer.	Gage height.	Discharge.
		Feet.	*Second-feet.*
February 9	S. G. Bennett	2.42	206
April 8	do	2.33	170
Do	do	1.58	95
Do	do	1.00	37
Do	do	3.00	286
April 9	do	3.11	276
June 16	do	4.10	383
September 23	do	1.25	33

Mean daily gage height, in feet, of Turlock canal at Lagrange, Cal., for 1903.

Day.	Jan.	Feb.	Mar.	Apr.	May.	June.	July.	Aug.	Sept.	Oct.
1	0.0	1.5	3.0	2.0	3.6	4.0	4.0	3.5	1.6	1.1
2	.0	1.5	2.3	3.0	3.6	4.0	4.0	3.4	1.5	1.1
3	.0	2.4	2.6	3.0	3.5	4.0	4.0	3.3	1.5	1.2
4	1.5	2.4	3.1	3.0	3.5	4.0	4.0	3.3	1.5	(a)
5	1.5	2.4	2.5	3.0	3.6	4.0	4.0	2.8	1.5	
6	1.5	2.4	3.0	3.0	3.6	4.0	4.0	2.7	1.4	
7	2.0	2.4	3.0	3.0	3.7	4.0	4.0	3.5	1.5	
8	2.0	2.4	1.9	3.0	3.8	4.0	4.0	2.7	1.5	
9	.0	.0	3.0	3.1	3.8	4.0	4.0	2.4	1.5	
10	.0	2.4	3.0	3.1	3.8	4.0	4.0	2.4	1.6	
11	.0	2.4	3.0	3.1	3.8	4.0	4.0	2.4	1.5	
12	1.9	2.4	3.0	3.1	3.8	4.0	4.0	2.4	1.4	
13	1.9	2.4	2.1	3.1	3.8	4.0	4.0	2.4	1.4	
14	2.1	2.4	2.5	3.1	3.8	4.0	4.0	2.4	1.3	
15	2.4	2.4	1.7	3.2	3.9	4.0	4.0	2.3	1.3	
16	2.4	2.4	3.0	3.3	3.9	4.0	4.0	1.8	1.3	
17	.0	.0	3.0	3.3	3.9	4.0	4.0	1.8	1.3	
18	2.0	.0	3.0	3.3	3.9	4.0	4.0	2.0	1.3	

a Canal dry from October 4 to December 31, inclusive.

Mean daily gage height, in feet, of Turlock canal at Lagrange, Cal., for 1903—Cont'd.

Day.	Jan.	Feb.	Mar.	Apr.	May.	June.	July.	Aug.	Sept.	Oct.
....	2.0	0.0	3.0	3.3	4.0	4.0	4.0	1.9	1.3	......
....	2.4	.0	3.0	3.3	4.0	4.0	4.0	1.8	1.3	......
....	1.5	.7	3.0	3.3	4.0	4.0	4.0	1.8	1.3	......
....	1.7	1.5	3.0	3.4	4.0	4.0	4.0	1.8	1.3	......
....	1.7	2.0	3.0	3.4	4.0	4.0	4.0	1.9	1.3	......
....	1.9	2.5	3.1	3.3	4.0	4.0	4.0	1.9	1.3	......
....	1.9	2.1	3.3	3.3	4.0	4.0	4.0	2.0	1.6	......
....	.0	3.0	3.3	3.3	4.0	4.0	4.0	1.9	1.6	......
....	.0	3.0	3.3	3.4	4.0	4.0	4.0	1.9	1.4	......
....	.0	3.0	2.1	3.5	4.0	4.0	4.0	1.8	1.4	......
....	.0	......	1.5	3.5	4.0	4.0	3.8	1.7	1.4	......
....	.0	......	1.6	3.5	4.0	4.0	3.8	1.7	1.4	......
....	.0	......	1.8	......	4.0	......	3.5	1.6	......	......

Rating table for Turlock canal at Lagrange, Cal., for 1903.

Gage height.	Discharge.	Gage height.	Discharge.	Gage height.	Discharge.	Gage height.	Discharge.
Feet.	*Second-feet.*	*Feet.*	*Second-feet.*	*Feet.*	*Second-feet.*	*Feet.*	*Second-feet.*
0.1	2	1.1	45	2.1	145	3.1	288
.2	4	1.2	52	2.2	158	3.2	304
.3	6	1.3	60	2.3	172	3.3	320
.4	8	1.4	69	2.4	186	3.4	336
.5	11	1.5	78	2.5	200	3.5	352
.6	15	1.6	88	2.6	214	3.6	368
.7	20	1.7	98	2.7	228	3.7	384
.8	26	1.8	109	2.8	243	3.8	400
.9	32	1.9	120	2.9	258	3.9	416
1.0	38	2.0	132	3.0	273	4.0	432

MODESTO CANAL AT LAGRANGE, CAL.

The Modesto canal is the property of the Modesto irrigation dis-
ct. The water is diverted from the right side of the Tuolumne
ver at the Lagrange dam. This canal was designed to carry 660
:ond-feet and to irrigate land in the vicinity of Modesto, Stanislaus
unty. The principal part of the construction work was done on
s canal prior to 1892, but on account of litigation the canal was not
npleted until April, 1903.

On April 26, 1903, a gage rod was set in and a rating made of
lian Hill flume, near Lagrange, Cal. From May 10 to June 3
d from June 10 to June 25, inclusive, boards were placed in the
mes to back the water up and keep the flumes saturated. During

this time gage heights were obtained by taking the depth of the water in the canal below Indian Hill flume. The observer is Annie P. McGinn.

The observations at this station during 1903 have been made under the direction of J. B. Lippincott, supervising engineer.

Discharge measurements of Modesto canal at Lagrange, Cal., in 1903.

Date.	Hydrographer.	Gage height.	Discharge.
		Feet.	*Second-feet.*
April 8	S. G. Bennett	2.14	127
April 9	....do	1.30	41
Do	....do	.70	14
April 7	....do	1.95	111
September 23	....do	1.00	3
November 27	....do	2.00	118

Mean daily gage height, in feet, of Modesto canal at Lagrange, Cal., for 1903.

Day.	Apr.	May.	June.	July.	Aug.	Sept.	Oct.	Nov.	Dec.
1		0.02	0.5	2.0	2.2	1.3	1.4	0.6	2.6
2		.05	.5	2.0	2.0	1.3	1.4	.6	2.5
3		2.70	.5	2.0	2.0	1.2	1.0	.6	2.5
4		.00	2.0	2.0	1.9	1.2	1.0	.6	2.5
5		.00	1.6	2.0	1.5	1.2	1.0	.6	2.5
6		.00	1.1	2.0	1.8	1.2	1.0	.6	2.5
7		.00	1.0	2.0	1.8	1.2	1.0	.6	2.5
8		2.0	1.1	2.0	.02	1.2	1.0	.1	2.5
9		2.0	1.3	1.5	1.8	1.2	1.0	.1	2.5
10		0.5	.7	1.0	1.8	1.2	1.0	.0	2.6
11		.5	.7	1.0	1.9	1.2	1.0	.0	2.5
12		.5	.7	1.4	1.9	1.2	1.0	.0	2.5
13		.5	.7	1.8	1.9	1.2	1.0	1.1	2.5
14		.5	.7	1.8	1.9	1.2	1.0	1.1	2.5
15		.5	.7	1.9	1.9	1.2	1.0	.0	2.6
16		.5	.7	2.2	1.8	1.2	1.3	.0	2.6
17		.5	.7	2.2	1.8	1.1	1.5	.1	2.5
18		.5	.7	2.2	1.7	1.1	.01	1.1	2.5
19		.5	.7	2.2	1.7	1.1	.01	1.2	2.5
20		.5	.7	2.0	1.7	1.1	1.6	1.0	2.5
21		.5	.7	1.0	1.6	1.1	1.6	.6	2.5
22		.5	.7	.1	1.6	1.1	1.6	.6	2.5
23		.5	.0	.3	1.7	1.1	1.6	.6	2.5
24		.5	.0	.3	1.6	1.1	1.4	1.6	2.5
25		.5	.0	1.9	1.5	1.4	1.4	1.6	2.5
26	0.30	.0	1.2	1.9	1.5	1.0	1.4	1.9	2.5
27	.20	.0	2.2	1.4	1.5	1.0	1.4	2.0	2.5
28	.05	.0	2.3	1.5	1.4	1.0	1.3	2.0	2.5
29	.05	.0	2.1	2.0	1.4	1.0	.68	2.0	2.5
30	.05	.0	2.0	.0	1.3	.9	.06	2.0	2.5
31		.5		2.0	1.3		.06		2.5

Rating table for Modesto canal at Lagrange, Cal., for 1903.

Gage height.	Discharge.	Gage height.	Discharge.	Gage height.	Discharge.	Gage height.	Discharge.
Feet.	*Second-feet.*	*Feet.*	*Second-feet.*	*Feet.*	*Second-feet.*	*Feet.*	*Second-feet.*
0.0	0	0.7	14	1.4	54	2.1	127
.1	1	.8	17	1.5	63	2.2	139
.2	2.5	.9	20	1.6	72	2.3	152
.3	4	1.0	24	1.7	82	2.4	165
.4	6	1.1	30	1.8	93	2.5	178
.5	8	1.2	38	1.9	104	2.6	191
.6	11	1.3	46	2.0	115		

MERCED RIVER ABOVE MERCED FALLS, CAL.

The measurement of this stream was undertaken in response to merous requests from mining and irrigation interests. The mid-nmer flow of the stream is less than the combined capacity of the igation and power canals taking water in the vicinity of Snelling. e gaging station at a point 1 mile above Merced Falls was estab-hed April 6, 1901. Meter measurements are made from a cable. e observer is Charles Siegfeldt.

The observations at this station during 1903 have been made under : direction of J. B. Lippincott, supervising engineer.

Discharge measurements of Merced River above Merced Falls, Cal., in 1903.

Date.	Hydrographer.	Gage height.	Discharge.
		Feet.	*Second-feet.*
)ruary 9	S. G. Bennett..................	9.80	771
ril 26	H. H. Henderson	11.48	2,945
y 24....................	.....do	10.55	2,106
ie 7....................	.....do	12.10	3,409
ie 21	.....do	10.62	2,174
y 5....................	.....do	9.50	1,204
y 19....................	.....do	8.80	606
;ust 4..................	.....do	8.40	503
;ust 23	.....do	8.00	339
)tember 15	.....do	7.90	293
)tember 27	.....do	7.70	269
tober 4..................	.....do	7.80	289
tober 18.................	.....do	7.80	285
vember 11	.....do	7.90	292

Mean daily gage height, in feet, of Merced River above Merced Falls, Cal., for 1903.

Day.	Jan.	Feb.	Mar.	Apr.	May.	June.	July.	Aug.	Sept.	Oct.	Nov.	D-
1..............	8.30	10.50	9.40	15.00	11.80	12.60	9.90	8.40	7.90	7.70	7.80	8
2..............	8.50	10.30	9.40	13.10	12.00	12.60	9.90	8.40	7.90	7.70	7.70	8
3..............	8.00	10.00	9.40	12.00	12.10	12.20	9.70	8.30	7.90	7.80	7.80	8
4..............	8.40	9.90	10.00	11.40	12.20	12.00	9.60	8.40	7.90	7.80	7.80	8
5..............	8.50	9.80	10.70	11.10	12.30	12.00	9.50	8.40	7.90	7.80	7.80	8
6..............	8.50	9.70	10.20	10.90	12.40	12.00	9.40	8.30	7.80	7.80	7.80	8
7..............	8.50	9.60	10.00	10.70	12.70	12.00	9.30	8.30	7.80	7.90	7.80	8
8..............	8.50	9.70	9.90	10.70	12.50	12.00	9.10	8.30	7.80	7.90	7.90	8
9..............	8.60	9.80	10.40	10.90	12.80	11.80	9.10	8.20	7.80	7.80	7.90	8
10..............	8.60	9.70	10.10	11.20	13.00	11.90	9.00	8.20	7.90	7.80	7.80	8
11..............	8.50	9.80	10.00	10.90	13.20	11.80	9.00	8.20	7.90	7.80	7.80	8
12..............	8.50	9.80	9.90	10.70	13.30	11.30	9.00	8.20	7.90	7.80	7.80	8
13..............	8.60	9.80	9.80	10.60	13.30	11.20	9.00	8.30	7.90	7.80	7.90	8
14..............	8.50	9.50	10.20	10.50	12.80	11.20	8.90	8.20	7.80	7.80	8.00	8
15..............	8.50	9.50	10.30	10.30	12.80	11.10	8.90	8.20	7.80	7.80	8.50	8
16..............	8.40	9.40	10.20	10.30	11.90	10.70	8.90	8.20	7.80	7.80	8.50	8
17..............	8.50	9.40	10.90	10.30	11.50	10.40	8.80	8.10	7.80	7.80	8.40	8
18..............	8.50	9.30	10.70	10.20	11.30	10.30	8.80	8.10	7.80	7.80	8.30	8
19...	8.40	9.30	10.70	10.10	11.30	10.50	8.80	8.00	7.80	7.80	8.10	8
20..............	8.40	9.30	10.60	10.10	11.10	10.80	8.80	8.00	7.80	7.80	8.30	8
21..............	8.40	9.30	10.40	10.20	10.70	10.70	8.70	8.00	7.80	7.80	8.80	8
22..............	8.40	9.40	10.10	10.50	10.80	10.70	8.60	8.00	7.80	7.80	8.80	8
23..............	8.50	9.40	10.00	10.80	10.60	10.60	8.50	8.00	7.80	7.80	8.60	8
24..............	8.50	9.40	10.20	11.20	10.60	10.50	6.50	8.10	7.80	7.80	8.50	8
25..............	9.20	9.40	11.10	11.70	10.60	10.50	8.50	8.00	7.80	7.80	8.40	8
26..............	10.70	9.40	10.70	11.50	10.60	10.60	8.50	8.10	7.80	7.80	8.50	8
27..............	12.00	9.40	10.50	11.40	11.80	10.50	8.50	8.00	7.70	7.80	8.40	8
28..............	13.80	9.40	10.60	11.20	11.90	10.40	8.40	8.00	7.70	7.80	8.30	8
29..............	11.10	....	11.30	11.30	12.10	10.30	8.50	8.00	7.70	7.80	8.30	8
30..............	10.50	....	11.40	11.70	12.50	10.00	8.40	7.96	7.70	7.80	8.30	8
31..............	10.70	....	13.50	....	12.70	....	8.40	7.90	....	7.80	....	8

Rating table for Merced River above Merced Falls, Cal., for 1903.

Gage height.	Discharge.	Gage height.	Discharge.	Gage height.	Discharge.	Gage height.	Discharge.
Feet.	Second-feet.	Feet.	Second-feet.	Feet.	Second-feet.	Feet.	Second-feet.
7.7	260	8.7	560	10.4	1,810	12.4	5,000
7.8	275	8.8	600	10.6	2,030	12.6	5,420
7.9	295	8.9	640	10.8	2,270	12.8	5,860
8.0	320	9.0	690	11.0	2,540	13.0	6,300
8.1	350	9.2	800	11.2	2,830	13.2	6,780
8.2	380	9.4	930	11.4	3,140	13.4	7,260
8.3	415	9.6	1,080	11.6	3,470	13.6	7,760
8.4	450	9.8	1,240	11.8	3,820	13.8	8,280
8.5	485	10.0	1,420	12.0	4,200	14.0	8,800
8.6	520	10.2	1,610	12.2	4,600	15.0	11,400

Estimated monthly discharge of Merced River above Merced Falls, Cal., for 1903.

[Drainage area, 1,090 square miles.]

Month.	Discharge in second-feet.			Total in acre-feet.	Run-off.	
	Maximum.	Minimum.	Mean.		Second-feet per square mile.	Depth in inches.
January	8,280	320	1,118	68,743	1.03	1.19
February	1,920	860	1,105	61,369	1.01	1.05
March	7,500	930	1,950	119,901	1.79	2.06
April	11,400	1,510	2,877	171,193	2.64	2.95
May	7,020	2,030	4,320	265,626	3.96	4.57
June	5,420	1,420	2,944	175,180	2.70	3.01
July	1,330	450	696	42,795	.64	.74
August	450	295	369	22,689	.34	.39
September	295	260	279	16,602	.26	.29
October	295	260	275	16,909	.25	.29
November	600	260	381	22,671	.35	.39
December	450	320	369	22,689	.34	.39
The year	11,400	260	1,390	1,006,367	1.28	17.32

KING RIVER AT KINGSBURG, CAL.

This station was established in 1879 by the engineering department of the Southern Pacific Company. No meter measurements have been made since 1898 except a low-water measurement in 1902, because it was found impossible to construct a satisfactory rating table, on account of the changes in gage heights caused by the raising and lowering of the head-gate of the Peoples canal, which takes water from King River, a few miles below the gaging station. The gage heights for 1903 have been furnished by William Hood, chief engineer of the Southern Pacific Company. Alf. Thompson was the observer.

Mean daily gage height, in feet, of King River at Kingsburg, Cal., for 1903.

Day.	Jan.	Feb.	Mar.	Apr.	May.	June.	July.	Aug.	Sept.	Oct.	Nov.	Dec.
1	5.0	5.9	5.2	6.7	7.8	·9.8	8.7	4.4	2.3	2.6	3.6	4.1
2	5.0	5.8	5.1	7.6	7.7	10.3	6.7	4.2	2.3	2.5	3.6	4.1
3	5.0	4.8	5.2	7.6	7.9	10.0	6.5	4.2	2.2	2.6	3.6	4.0
4	5.0	5.4	5.2	6.2	8.0	9.6	6.2	4.2	2.2	2.8	3.6	4.1
5	5.0	5.5	5.5	6.0	8.2	9.8	6.0	4.2	2.2	2.7	3.6	4.1
6	4.9	4.5	5.6	6.0	8.2	9.8	5.8	4.2	2.2	2.6	3.8	4.1
7	4.9	5.3	5.3	5.9	8.1	10.0	5.5	4.2	2.1	2.6	3.7	4.1
8	4.9	5.6	5.3	5.6	9.0	9.2	5.5	4.2	2.1	2.6	3.7	4.1
9	4.9	5.6	5.4	5.8	9.2	8.8	5.0	4.2	2.1	2.5	3.7	4.1
10	4.9	5.3	5.3	6.0	9.7	8.8	4.7	4.0	2.1	2.3	3.6	4.1
11	4.8	5.0	5.2	6.2	10.1	8.8	5.5	3.7	2.0	2.6	3.8	4.1
12	4.8	5.0	5.3	5.9	10.6	8.1	5.3	3.3	2.0	3.0	3.8	4.1
13	4.8	4.8	5.2	5.7	10.8	8.1	5.3	3.2	2.1	3.3	3.9	4.1
14	4.8	4.0	5.3	5.7	10.5	8.5	5.3	3.2	2.3	3.4	3.9	4.1
15	4.8	5.0	5.4	5.7	10.1	8.5	5.3	3.2	2.3	3.4	4.0	4.1
16	4.8	5.2	5.4	5.3	9.8	7.5	5.3	3.2	2.3	3.2	4.0	4.1
17	4.8	5.5	5.3	5.4	8.6	7.1	5.2	3.1	2.3	3.1	4.2	4.1
18	4.2	5.4	5.3	5.5	8.2	7.2	5.2	2.9	2.3	2.9	4.3	4.1
19	4.1	5.3	5.2	5.5	7.2	7.4	5.2	2.8	2.3	2.8	4.3	4.1
20	4.0	5.3	5.2	5.4	7.2	7.9	5.2	2.8	2.3	2.8	4.3	4.1
21	4.0	5.3	5.2	5.4	6.9	7.9	5.2	2.8	2.3	2.9	4.5	4.1
22	3.9	5.4	5.2	5.5	6.6	8.0	5.2	2.9	2.3	2.9	5.2	4.1
23	3.7	5.3	5.2	5.8	6.5	7.6	5.0	3.0	2.3	3.2	5.2	4.1
24	3.6	5.3	5.3	6.4	6.2	7.5	5.1	2.9	2.2	3.5	5.2	4.1
25	3.6	5.3	7.1	7.0	6.0	7.8	4.9	2.8	2.2	3.4	5.2	4.1
26	4.0	5.3	6.2	7.2	5.9	7.8	4.8	2.6	2.2	3.4	5.2	4.1
27	4.8	5.2	5.7	7.1	6.0	7.7	4.8	2.6	2.2	3.5	5.1	4.1
28	10.5	5.2	5.4	6.7	6.0	7.3	4.8	2.6	2.2	3.4	5.0	4.1
29	7.1	......	5.9	6.0	7.8	7.2	4.8	2.4	2.3	3.5	4.9	4.1
30	5.8	......	6.0	6.6	9.5	6.9	4.5	2.4	2.5	3.5	4.9	4.1
31	6.0	......	6.2	......	9.8	......	4.3	2.3	......	3.5	......	4.1

KING RIVER NEAR SANGER, CALIFORNIA.

This station was established September 3, 1895, by J. B. Lippincott. It is located 15 miles east of Sanger, Cal., near the mouth of the canyon, and is above all diversions. An automatic gage was installed April 18, 1903. There is also an inclined wooden gage near by from which readings were formerly taken and which is now used in checking the self-recording gage. Discharge measurements are made by means of a cable and car. The initial point for soundings is an eye-bolt embedded in concrete on right bank of river. The channel is nearly straight for 300 feet above and below the station, and has a width of 180 feet at ordinary stages. The bed of the stream is composed of gravel and small bowlders and changes but little. The right bank is high and not subject to overflow. The left bank is subject to overflow during extreme high water. The current is swift.

The observations at this station during 1903 have been made under the direction of J. B. Lippincott, supervising engineer.

Discharge measurements of King River near Sanger, Cal., in 1903.

Date.	Hydrographer.	Gage height.	Discharge.
		Feet.	*Second-feet.*
11	S. G. Bennett...................	5.35	1,025
....	.....do	7.30	3,155
....	.....do	6.90	2,370
....	.....do	9.35	6,680
r 18	.....do	3.80	176

:an daily gage height, in feet, of King River near Sanger, Cal., for 1903.

)ay.	Jan.	Feb.	Mar.	Apr.	May.	June.	July.	Aug.	Sept.	Oct.	Nov.	Dec.
....	4.2	6.0	5.1	9.0	8.6	10.4	8.4	5.2	4.1	3.9	3.8	4.0
....	4.2	5.6	5.0	8.1	8.9	10.9	8.3	5.1	4.1	4.0	3.8	4.0
....	4.2	5.5	5.0	7.5	9.2	10.7	7.9	5.1	4.1	4.0	3.8	4.0
....	4.2	5.5	5.4	7.5	9.5	10.1	7.7	5.0	4.1	4.1	3.7	3.9
....	4.2	5.4	5.8	7.1	9.4	10.5	7.4	5.0	4.2	4.0	3.7	3.9
....	4.2	5.3	5.6	7.1	9.7	10.8	7.2	4.9	4.1	4.0	3.7	3.9
....	4.1	5.1	5.3	6.9	10.1	10.7	6.9	4.9	4.1	4.0	3.7	3.9
....	4.1	5.4	5.3	7.2	10.2	10.0	6.6	4.8	4.0	4.0	3.8	3.9
....	4.2	5.3	5.6	7.5	10.6	9.8	6.4	4.8	4.0	3.9	3.8	3.8
....	4.2	5.2	5.5	8.2	10.7	9.8	6.3	4.8	4.0	3.9	3.8	3.8
....	4.1	5.4	5.5	7.4	11.1	9.9	6.3	4.8	4.0	3.9	3.8	3.8
....	4.1	5.4	5.5	7.1	11.6	9.6	6.4	4.7	4.0	3.9	3.8	3.8
....	4.1	5.2	5.5	7.1	(11.7)	9.7	6.4	4.7	4.0	3.8	3.8	3.8
....	4.0	5.1	5.5	7.1	(11.7)	9.9	6.4	4.7	3.9	3.8	3.8	3.8
....	4.1	5.0	5.7	6.9	(11.7)	9.6	6.3	4.7	3.9	3.8	3.9	3.8
....	4.1	5.5	5.6	6.7	11.6	8.7	6.2	4.7	3.9	3.8	4.0	3.8
....	4.0	5.0	5.8	7.1	10.9	8.6	6.2	4.6	3.9	3.8	3.9	3.8
....	4.0	5.0	5.6	6.6	10.2	8.9	6.3	4.5	3.9	3.8	3.9	3.9
....	4.0	5.0	5.5	6.6	9.9	9.1	6.4	4.5	3.9	3.8	3.9	3.9
....	4.0	5.0	5.5	6.6	10.0	9.2	6.3	4.4	3.9	3.8	3.9	3.9
....	4.1	5.1	5.4	6.8	9.6	9.4	6.2	4.4	3.8	3.8	4.4	3.9
....	4.0	5.1	5.5	7.3	9.2	9.2	6.1	4.4	3.8	3.8	4.3	3.9
....	4.1	5.1	5.5	7.9	9.3	9.1	6.1	4.4	3.8	3.7	4.2	3.9
....	4.1	5.0	6.3	8.5	(9.1)	9.2	6.0	4.4	3.8	3.7	4.2	3.9
....	4.3	5.1	8.2	8.4	(8.9)	9.3	5.4	4.3	3.8	3.7	4.2	3.9
....	4.5	5.0	6.9	8.6	(8.7)	9.4	5.4	4.3	3.7	3.7	4.1	3.9
....	7.3	5.0	6.6	8.5	(8.5)	9.2	5.3	4.2	3.8	3.7	4.1	3.9
....	11.0	5.0	6.5	8.0	(8.4)	8.9	5.3	4.2	3.8	3.8	4.0	3.9
....	6.7	......	6.9	8.0	8.4	8.7	5.2	4.2	3.8	3.7	4.0	3.9
....	6.0	......	7.0	8.4	10.6	8.5	5.2	4.1	3.9	3.7	4.0	3.8
....	5.9	......	7.8	......	10.7	......	5.2	4.1	......	3.8	......	3.9

Rating table for King River near Sanger, Cal, for 1903.

Gage height.	Discharge.	Gage height.	Discharge.	Gage height.	Discharge.	Gage height.	Discharge.
Feet.	*Second-feet.*	*Feet.*	*Second-feet.*	*Feet.*	*Second-feet.*	*Feet.*	*Second-feet.*
3.7	160	4.7	610	6.4	1,830	8.4	4,660
3.8	180	4.8	670	6.6	2,030	8.6	5,090
3.9	210	4.9	730	6.8	2,230	8.8	5,540
4.0	240	5.0	790	7.0	2,450	9.0	6,030
4.1	280	5.2	910	7.2	2,700	9.5	7,300
4.2	330	5.4	1,035	7.4	2,980	10.0	8,900
4.3	380	5.6	1,170	7.6	3,270	10.5	10,840
4.4	430	5.8	1,310	7.8	3,580	11.0	13,340
4.5	490	6.0	1,470	8.0	3,910	11.5	16,040
4.6	550	6.2	1,640	8.2	4,280		

Estimated monthly discharge of King River near Sanger, Cal., for 1903.

[Drainage area, 1,742 square miles.]

Month.	Discharge in second-feet.			Total in acre-feet.	Run-off.	
	Maximum.	Minimum.	Mean.		Second-feet per square mile.	Depth in inches.
January	13,240	240	930	57,183	0.53	0.61
February	1,470	790	930	51,650	.53	.55
March...............	4,280	790	1,470	86,513	.84	.97
April...............	6,030	2,030	3,287	195,590	1.89	2.11
May	17,290	4,680	9,546	586,961	5.48	6.32
June	12,250	4,880	7,876	468,655	4.52	5.04
July	4,680	910	1,948	119,778	1.12	1.29
August	910	280	560	34,433	.32	.37
September..........	330	160	224	13,329	.13	.14
October	280	160	195	11,990	.11	.13
November	430	160	227	13,507	.13	.14
December	430	180	203	12,482	.12	.14
The year......	17,290	160	2,283	1,652,071	1.31	17.81

KAWEAH RIVER BELOW THREERIVERS, CAL.

'his station was established April 29, 1903, by W. H. Stearns. It ocated at a point three-fourths of a mile below the confluence of North, Middle, and South forks. It is 17 miles from the Southern ific Railway station at Exeter, Tulare County, Cal., and one-fourth a mile west of the wagon road from Exeter to Threerivers. The ge consists of a vertical 2-inch pipe driven 8 feet into the river bed. is is used up to medium stages. For high-water readings a timber ge is securely nailed to a willow tree on the left bank of the stream. a gage is read twice each day by Miss Mary Lansdowne. Dis-rge measurements are made from a ¾-inch galvanized iron cable l car. The initial point for soundings is a sycamore tree on the left ik of the stream, to which the cable is fastened. The channel is iight for 400 feet above and below the station. The current is swift high stages, but sluggish at low water. There are rapids about 400 t above the cable. The right bank is low and subject to overflow high stages. The left bank is high enough to prevent overflow. ere are willow trees along the water's edge on both banks and a line willows, sycamores, and cottonwoods back from the water's edge on left bank. The bed of the stream is composed of sand, gravel, l bowlders. Some of the bowlders are 2 feet in diameter. The tion is probably permanent. The bench mark is a large rock 10 feet stream from the tree to which the cable is attached. It is marked s. M." in black paint. Its elevation is 13.95 feet above the zero of gage.

The observations at this station during 1903 have been made under direction of J. B. Lippincott, supervising engineer.

Discharge measurements of Kaweah River below Threerivers, Cal., in 1903.

Date.	Hydrographer.	Gage height.	Discharge.
		Feet.	*Second-feet.*
il 29	Stearns & Dean	6. 60	1, 297
y 4	W. F. Dean	7. 20	1, 951
y 11	.....do	7. 80	2, 083
y 28	.....do	6. 80	1, 306
ie 17	.....do	6. 50	1, 055
ie 25	.....do	6. 60	1, 186
y 20	A. J. Robertson..............	5. 40	215
gust 27	S. G. Bennett.................	4. 53	51
vember 22................	G. C. Morgan.................	4. 60	85

Mean daily gage height, in feet, of Kaweah River below Threerivers, Cal., for 1903.

Day.	Apr.	May.	June.	July.	Aug.	Sept.	Oct.	Nov.	Dec.
1		7.60	7.55	6.75	4.60	4.30	4.40	4.30	4.3
2		7.55	7.55	6.65	4.45	4.40	4.40	4.30	4.3
3		7.50	7.65	6.60	4.40	4.30	4.40	4.30	4.3
4		7.45	7.85	6.55	4.35	4.40	4.30	4.40	4.3
5		7.60	8.00	6.45	4.55	4.30	4.30	4.40	4.3
6		7.55	7.95	6.30	4.45	4.30	4.30	4.40	4.3
7		7.60	7.80	6.30	4.40	4.30	4.30	4.40	4.4
8		7.75	7.05	5.95	4.45	4.20	4.30	4.40	4.4
9		7.70	7.50	5.80	4.50	4.20	4.30	4.50	4.4
10		7.90	7.55	5.70	4.35	4.20	4.30	4.50	4.4
11		8.15	7.40	5.55	4.95	4.20	4.20	4.50	4.4
12		8.20	6.90	5.65	4.30	4.20	4.20	4.50	4.4
13		8.25	6.95	5.65	4.25	4.20	4.20	4.50	4.4
14		8.20	7.25	5.55	4.30	4.23	4.20	4.50	4.3
15		7.75	7.15	5.40	4.95	4.20	4.20	4.50	4.3
16		7.65	6.90	5.45	4.25	4.30	4.20	4.50	4.3
17		7.20	6.90	5.55	4.90	4.20	4.30	4.60	4.3
18		7.20	6.70	5.45	4.35	4.20	4.20	4.60	4.3
19		7.05	6.85	5.20	4.30	4.20	4.20	4.60	4.3
20		7.15	6.90	5.30	4.40	4.20	4.30	4.60	4.3
21		6.90	6.85	5.25	4.40	4.20	4.20	4.60	4.4
22		7.10	6.70	5.15	4.35	4.20	4.30	4.60	4.4
23		6.65	6.85	5.10	4.60	4.20	4.20	4.50	4.4
24		6.55	6.70	5.00	4.40	4.20	4.30	4.40	4.4
25		6.45	6.65	5.15	4.30	4.30	4.20	4.40	4.4
26		6.40	6.85	5.05	4.35	4.30	4.20	4.40	4.4
27		6.45	6.85	5.00	4.40	4.30	4.20	4.40	4.4
28		6.65	6.95	4.85	4.30	4.30	4.20	4.40	4.4
29	6.7	7.15	6.90	4.85	4.35	4.40	4.20	4.40	4.4
30	6.9	7.60	6.85	4.75	4.30	4.40	4.20	4.30	4.4
31		7.85	......	4.70	4.30	......	4.20	......	4.4

Rating table for Kaweah River below Threerivers, Cal., from April 29 to December 31, 1903.

Gage height.	Discharge.	Gage height.	Discharge.	Gage height.	Discharge.	Gage height.	Discharge.
Feet.	*Second-feet.*	*Feet.*	*Second-feet.*	*Feet.*	*Second-feet.*	*Feet.*	*Second-feet.*
4.2	40	5.2	225	6.2	795	7.2	1,750
4.3	45	5.3	265	6.3	875	7.3	1,870
4.4	50	5.4	305	6.4	960	7.4	1,990
4.5	60	5.5	350	6.5	1,050	7.5	2,110
4.6	70	5.6	400	6.6	1,140	7.6	2,230
4.7	85	5.7	460	6.7	1,235	7.7	2,360
4.8	105	5.8	520	6.8	1,330	7.8	2,490
4.9	130	5.9	580	6.9	1,430	7.9	2,620
5.0	160	6.0	650	7.0	1,530	8.0	2,750
5.1	190	6.1	720	7.1	1,640	8.1	2,890

mated monthly discharge of Kaweah River below Threerivers, Cal., for 1903.

[Drainage area, 520 square miles.]

Month.	Discharge in second-feet.			Total in acre-feet.	Run-off.	
	Maximum.	Minimum.	Mean.		Second-feet per square mile.	Depth in inches.
....	3,100	960	2,007	123,406	3.86	4.45
....	2,750	1,188	1,749	104,073	3.36	3.75
....	1,283	85	462	28,407	.89	1.03
....	70	42	50	3,074	.10	.11
)er..........	50	40	43	2,559	.08	.09
....	50	40	43	2,644	.08	.09
'er..........	70	45	56	3,332	.11	.12
er	50	45	48	2,951	.09	.10
he period....	3,100	40	557	270,446	1.07	9.74

TULE RIVER NEAR PORTERSVILLE, CAL.

gaging station is located about 8 miles east of Portersville at a
just below the wagon bridge near the McFarland ranch and
1 mile above the mouth of South Fork of Tule River. The
was established April 18, 1901. The gage rod is situated on
ht bank of the river 100 feet below the bridge. The zero of the
; 8 feet below a spike driven into a large cottonwood tree. The
er is Adah McFarland.

observations at this station during 1903 have been made under
ection of J. B. Lippincott, supervising engineer.

Discharge measurements of Tule River near Portersville, Cal., in 1903.

Date.	Hydrographer.	Gage height.	Discharge.
		Feet.	*Second-feet.*
y 13	S. G. Bennett...............	2.13	171
....	.....do	4.50	1,087
....	.....do	3.87	747
....	R. S. Hawley...............	2.35	185
)er 19	S. G. Bennett...............	1.16	28

Mean daily gage height, in feet, of Tule River near Portersville, Cal., for 1903.

Day.	Jan.	Feb.	Mar.	Apr.	May.	June.	July.	Aug.	Sept.	Oct.	Nov.	Dec.
1	1.40	3.0	2.1	4.7	3.0	2.5	1.60	1.10	0.95	0.95	1.0	1.1
2	1.35	2.6	2.1	4.0	3.1	2.5	1.50	1.08	.95	1.00	1.0	1.1
3	1.38	2.3	2.3	3.6	3.2	2.5	1.50	1.07	.95	1.00	1.0	1.1
4	1.40	2.4	2.8	3.4	3.2	2.5	1.40	1.06	.95	1.02	1.0	1.1
5	1.38	2.2	2.5	3.3	3.2	2.5	1.44	1.04	.95	1.01	1.0	1.1
6	1.40	2.1	2.3	3.0	3.2	2.5	1.42	1.03	.96	1.00	1.0	1.1
7	1.42	2.1	2.2	3.0	3.2	2.5	1.40	1.02	.98	1.00	1.1	1.1
8	1.40	2.5	2.2	3.0	3.2	2.4	1.40	1.01	.98	1.00	1.1	1.1
9	1.40	2.3	2.3	3.0	3.2	2.3	1.35	1.00	.90	1.00	1.1	1.1
10	1.40	2.3	2.3	3.7	3.2	2.5	1.35	1.00	.90	1.00	1.1	1.1
11	1.40	2.3	2.3	3.3	3.3	2.2	1.35	1.00	.95	1.00	1.1	1.1
12	1.33	2.2	2.3	3.2	3.4	2.2	1.34	1.00	.90	.98	1.1	1.1
13	1.38	2.1	2.3	3.1	3.2	2.2	1.32	1.00	.95	.98	1.1	1.1
14	1.38	2.1	2.5	3.0	3.2	2.2	1.30	1.00	.95	1.00	1.1	1.1
15	1.38	2.1	2.5	2.9	3.2	2.2	1.28	1.00	.95	1.00	1.1	1.1
16	1.38	2.0	2.4	2.9	3.0	2.1	1.25	.95	.95	1.00	1.1	1.1
17	1.38	2.0	2.4	2.8	2.9	2.1	1.24	.95	.95	1.00	1.1	1.1
18	1.38	2.0	2.2	2.8	2.9	2.1	1.22	.95	.90	1.00	1.1	1.1
19	1.38	2.1	2.3	2.7	2.8	2.0	1.20	.90	.95	.95	1.2	1.1
20	1.38	2.1	2.3	2.7	2.7	2.0	1.20	.90	.90	.98	1.2	1.1
21	1.37	2.1	2.3	2.8	2.7	1.9	1.20	.90	.90	.98	1.2	1.1
22	1.37	2.1	2.3	2.8	2.6	1.9	1.18	.90	.90	.95	1.2	1.1
23	1.37	2.1	2.4	3.0	2.6	1.9	1.16	.90	.95	.95	1.2	1.1
24	1.37	2.1	2.5	3.2	2.5	1.9	1.15	.90	.90	.95	1.2	1.1
25	1.45	2.1	2.7	3.3	2.4	1.8	1.15	.90	.90	.95	1.2	1.1
26	1.60	2.1	3.2	3.2	2.3	1.8	1.15	.90	.90	.95	1.1	1.1
27	5.00	2.1	2.8	3.1	2.2	1.7	1.14	.88	.92	.95	1.1	1.1
28	7.00	2.1	2.9	3.0	2.4	1.6	1.13	.88	.93	.95	1.1	1.1
29	3.50	......	3.2	2.9	2.5	1.6	1.12	.88	.95	.95	1.2	1.1
30	2.40	......	3.3	2.9	2.6	1.6	1.11	.88	.96	.95	1.2	1.1
31	3.00	......	3.3	......	2.6	......	1.11	.88	......	.95	......	1.1

Rating table for Tule River near Portersville, Cal., for 1903.

Gage height.	Discharge.	Gage height.	Discharge.	Gage height.	Discharge.	Gage height.	Discharge.
Feet.	*Second-feet.*	*Feet.*	*Second-feet.*	*Feet.*	*Second-feet.*	*Feet.*	*Second-feet.*
0.8	11	1.8	98	2.8	315	4.6	1,180
.9	14	1.9	115	2.9	345	4.8	1,300
1.0	18	2.0	132	3.0	380	5.0	1,420
1.1	23	2.1	150	3.2	450	5.5	1,770
1.2	30	2.2	170	3.4	530	6.0	2,170
1.3	39	2.3	190	3.6	625	6.5	2,620
1.4	48	2.4	210	3.8	725	7.0	3,120
1.5	58	2.5	235	4.0	830	7.5	3,670
1.6	70	2.6	260	4.2	940		
1.7	83	2.7	285	4.4	1,060		

Estimated monthly discharge of Tule River near Portersville, Cal., for 1903.

[Drainage area, 437 square miles.]

Month.	Discharge in second-feet.			Total in acre-feet.	Run-off.	
	Maximum.	Minimum.	Mean.		Second-feet per square mile.	Depth in inches.
anuary	3,790	45	254	15,618	0.58	0.67
'ebruary	380	132	173	9,608	.40	.42
{arch	675	150	258	15,864	.59	.68
₊pril	1,240	285	447	26,598	1.02	1.14
{ay	530	170	358	22,013	.82	.94
une ., ...	235	70	158	9,402	.36	.40
uly	70	24	38	2,337	.09	.10
₊ugust	23	13	17	1,045	.04	.05
ᴤeptember	16	13	14	833	.03	.03
)ctober	19	16	17	1,045	.04	.05
ᴎovember	30	18	24	1,428	.05	.06
December	30	23	28	1,722	.06	.07
The year	3,790	13	149	107,513	.34	4.61

KERN RIVER NEAR BAKERSFIELD, CAL.

This station, established in 1893 by Walter Jones, chief engineer of the Kern County Land Company, is located at what is known as "first point of measurement," 5 miles above Bakersfield and at the mouth of the canyon of the river. Miscellaneous meter measurements are taken, and an automatic gage records daily fluctuations of the river heights. A. K. Warren, the engineer in charge of this work for the Kern County Land Company, attends to the discharge measurements with accuracy and precision, and furnishes the Geological Survey with the final results.

Mean daily discharge, in second-feet, of Kern River near Bakersfield, Cal., in 1903.

[Drainage area, 2,345 square miles; observer, A. K. Warren.]

Day.	Jan.	Feb.	Mar.	Apr.	May.	June.	July.	Aug.	Sept.	Oct.	Nov.	Dec.
1	244	622	482	1,471	1,514	2,358	1,765	437	201	226	169	219
2	344	600	428	1,521	1,694	2,415	1,776	439	208	200	179	217
3	245	480	418	1,460	1,696	2,668	1,670	443	211	191	182	211
4	243	451	428	1,344	1,986	2,617	1,481	413	198	190	183	216
5	256	480	479	1,279	1,999	2,636	1,354	384	196	197	189	213
6	249	442	514	1,236	2,034	2,724	1,252	348	199	188	185	198
7	349	435	479	1,172	2,150	2,750	1,155	336	204	161	185	196
8	250	444	475	1,180	2,316	2,780	1,042	329	200	182	192	195
9	248	497	580	1,263	2,414	2,634	953	330	200	180	199	198
10	344	468	589	1,334	2,606	2,723	898	341	192	181	205	195
11	242	477	515	1,393	2,816	2,472	853	335	187	179	204	186
12	248	506	537	1,258	3,068	2,297	823	313	181	181	207	188
13	244	520	562	1,183	3,233	2,271	825	312	187	173	215	191
14	245	466	543	1,117	3,296	2,300	814	318	191	167	222	190
15	251	401	534	1,039	3,143	2,337	770	323	183	163	226	202
16	247	391	527	969	2,950	2,145	742	321	180	150	222	190
17	244	410	539	1,018	2,797	1,936	748	303	181	153	225	204
18	241	472	537	965	2,444	1,894	722	291	178	153	219	202
19	234	430	519	910	2,152	1,968	684	270	177	157	213	200
20	237	413	497	903	1,958	2,114	668	269	176	159	194	202
21	237	359	485	893	1,850	2,245	645	249	181	162	202	203
22	237	368	466	912	1,780	2,309	610	260	184	167	210	210
23	236	419	479	1,053	1,718	2,291	588	249	190	158	217	210
24	242	412	532	1,233	1,651	2,187	553	247	197	154	218	204
25	257	418	654	1,406	1,551	2,206	536	239	192	157	214	200
26	260	436	740	1,609	1,463	2,304	512	236	177	161	211	202
27	347	438	723	1,669	1,424	2,305	520	224	171	169	208	201
28	1,297	436	745	1,601	1,405	2,232	516	219	185	179	200	213
29	1,432	....	805	1,396	1,474	2,115	490	214	202	180	201	207
30	766	....	1,024	1,371	1,755	1,923	477	213	219	174	206	200
31	641	....	1,250	....	2,161	....	463	205	....	170	....	199
Total	10,856	12,701	17,935	37,465	66,599	70,191	26,900	9,389	5,727	5,385	6,106	6,227
Maximum	2,617	665	1,376	1,978	3,374	2,927	1,891	484	220	232	228	218
Minimum	229	337	413	872	1,380	1,756	445	198	165	150	166	176
Mean	350	454	579	1,249	2,148	2,340	868	303	191	174	203	201

Whole year: Total, 275,481; maximum, 3,374; minimum, 150; mean, 755.

Estimated monthly discharge of Kern River near Bakersfield, Cal., in 1903.

[Drainage area, 2,345 square miles.]

Month.	Discharge in second-feet.			Total in acre-feet.	Run-off.	
	Maximum.	Minimum.	Mean.		Second-feet per square mile.	Depth in inches.
January	2,617	229	350	21,521	0.15	0.17
February	665	337	454	25,214	.19	.20
March	1,376	413	579	35,601	.25	.29
April	1,978	872	1,249	74,321	.53	.59
May	3,374	1,380	2,148	132,075	.92	1.06
June	2,927	1,786	2,340	139,240	1.00	1.12
July	1,891	445	868	53,371	.37	.43
August	484	198	303	18,631	.13	.15
September	220	165	191	11,365	.08	.09
October	232	150	174	10,699	.07	.08
November	228	166	203	12,079	.09	.10
December	218	176	201	12,359	.09	.10
The year	3,374	150	755	546,476	.32	4.38

SAN JOAQUIN RIVER AT HERNDON, CAL.

The gage rod at this station was established by the engineering department of the Southern Pacific Railway Company in 1879. The old trestle bridge was torn down by the railroad company during 1899 and a new iron structure was erected in its place. A new gage rod, set to the datum of the old gage, was bolted to the western side of the central concrete pier. The bench mark is a nail in a post at the south end of the bridge on the west side, 0.2 foot above the ground, and marked "B. M." It is at an elevation of 24.12 feet above gage datum. The channel for some distance above and below the bridge is straight, and the water has a uniform velocity. The right bank is high, rocky, and steep. The bed of the stream is composed of small gravel and shifting sand. Because of the continual changes in the cross section, which were increased by a side channel breaking through the gravel pits on the left bank of the river just above the gaging station, meter measurements were discontinued at this station at the end of 1901.

The river stage record for 1903 has been furnished by William Hood, chief engineer of the Southern Pacific Company. G. G. Nelson was the observer.

Mean daily gage height, in feet, of San Joaquin River at Herndon, Cal., for 19

Day.	Jan.	Feb.	Mar.	Apr.	May.	June.	July.	Aug.	Sept.	Oct.	Nov.
1	3.0	4.0	3.3	7.0	7.0	9.0	5.0	2.0	2.0	2.0	2.0
2	3.0	4.5	3.3	7.0	7.2	9.0	5.0	2.0	2.0	2.0	2.0
3	3.0	4.2	3.3	6.0	7.5	8.8	5.4	2.0	2.0	2.0	2.0
4	3.0	3.8	3.3	5.4	7.7	8.5	5.4	2.0	2.0	2.0	2.0
5	3.0	3.8	3.5	5.2	8.0	8.8	5.4	2.0	2.0	2.0	2.0
6	3.0	3.7	3.5	5.0	7.6	8.2	4.5	2.0	2.0	2.0	2.0
7	3.0	3.6	3.4	5.0	8.2	8.5	4.5	2.0	2.0	2.0	2.0
8	3.0	3.6	3.4	5.0	8.3	8.0	4.0	2.0	2.0	2.0	2.0
9	2.9	3.7	3.4	5.6	8.8	7.8	4.0	2.0	2.0	2.0	2.0
10	2.9	3.6	3.4	5.8	9.0	7.7	4.0	2.0	2.0	2.0	2.2
11	2.9	3.7	3.4	6.2	9.5	8.2	4.0	2.0	2.0	2.0	2.2
12	2.9	3.5	3.3	5.4	10.0	7.6	4.0	2.0	2.0	2.0	2.2
13	2.8	3.5	3.3	5.0	10.0	7.2	4.0	2.0	2.0	2.0	2.0
14	2.8	3.5	3.3	5.0	9.7	7.0	4.0	2.0	2.0	2.0	2.0
15	2.9	3.5	3.3	4.8	9.2	7.0	4.0	2.0	2.0	2.0	2.2
16	2.8	3.6	3.3	4.7	8.8	6.8	4.0	2.0	2.0	2.0	2.2
17	2.8	3.5	3.5	4.5	8.2	6.4	4.0	2.0	2.0	2.0	2.2
18	2.8	3.5	3.6	4.2	8.0	6.0	3.5	2.0	2.0	2.0	2.2
19	2.8	3.5	3.6	4.2	7.0	6.2	3.5	2.0	2.0	2.0	2.2
20	2.8	3.5	3.6	4.2	6.8	6.2	3.5	2.0	2.0	2.0	2.2
21	2.8	3.5	3.5	4.2	6.5	6.2	3.5	2.0	2.0	2.0	2.2
22	2.8	3.4	3.5	4.2	6.5	6.7	3.5	2.0	2.0	2.0	2.0
23	2.8	3.4	3.5	5.5	6.0	6.5	3.5	2.0	2.0	2.0	2.0
24	2.8	3.3	3.7	6.2	5.8	6.5	2.8	2.0	2.0	2.0	2.0
25	2.8	3.3	4.1	6.5	5.7	6.6	2.2	2.0	2.0	2.0	2.0
26	2.8	3.3	4.5	6.8	5.5	6.5	2.0	2.0	2.0	2.0	2.0
27	3.2	3.3	4.5	6.5	5.5	6.5	2.0	2.0	2.0	2.0	2.2
28	10.0	3.3	4.2	6.0	5.8	6.8	2.0	2.0	2.0	2.0	2.2
29	6.0	......	4.6	6.0	6.3	6.3	2.0	2.0	2.0	2.0	2.2
30	4.5	......	5.0	6.0	8.0	6.0	2.0	2.0	2.0	2.0	2.2
31	4.0	......	5.8	......	9.3	......	2.0	2.0	......	2.0	.....

ANEOUS MEASUREMENTS IN SAN FRANCISCO BAY DRAINAGE
BASIN.

lowing miscellaneous measurements were made in the San
Bay drainage basin in 1903:

llaneous measurements in San Francisco Bay drainage basin in 1903.

Hydrographer.	Stream.	Location.	Discharge.
			Sec. ft.
I. E. Green.......	Klamath River	Klamathon........	2,000
S. G. Bennett......	Mad River	Vance.............	53
...do	Eel River	Stingleys station ...	189
I. G. Heisler......	Willow Creek (Susan River basin).	Merrillville........	20
S. G. Bennett......	Russian River	Healdsburg........	628
...do	Folsom canal (American River basin).	500 feet below headworks, Folsom.	1,466

PITT RIVER DRAINAGE.

I. G. Heisler......	Pitt River..........	Pittville	30
...do	.....do.............	Pecks Bridge, above mouth of Burney Creek.	2,617
...do	Fall River..........	Fall River mills....	1,510
...do	Hot Creek	Carbon............	657
...do	Burney River	Lower end of Burney Valley, below Burney Falls.	210
...do	Hatchet Creek	Near Montgomery..	10
...do	Montgomery Creek .	Montgomery.......	18

STANISLAUS RIVER DRAINAGE.

S. G. Bennett......	Stanislaus Water Co.'s canal.	Knights Ferry below penstock to power house.	61
...do	.....do.............	.....do............	41
...do	.....do.......	Knights Ferry above penstock to power house.	85

*Miscellaneous measurements in San Francisco Bay drainage basin in 1903—*Continued.

TUOLUMNE RIVER DRAINAGE.

Date.	Hydrographer.	Stream.	Location.	Dis-charge
				Sec.-ft.
Sept. 20	S. G. Bennett......	Clavey River.......	Above mouth of Twomile Creek, Carter-Lake Eleanor trail.	2.6
20	.....do	Twomile Creek.....	At mouth..........	.8
20	.....do	Hull Creek.........	Carter-Lake Eleanor trail crossing.	3.7
21	.....do	North Fork of Tuolumne River.	Above mouth of Basin Slope Creek.	2.9
19	.....do	Reed Creek	Rosasco ranch	2.6
19	.....do	Cherry River.......	Carter-Lake Eleanor trail crossing.	*.6
19	.....do	Elinor Creek	½ mile below Lake Eleanor.	1.0
18	.....do	Tuolumne River....	Above mouth, Hetch Hetchy Valley.	19.0
18	.....do	Rancheria Creek....	At mouth, Hetch Hetchy Valley.	1.3
17	.....do	Tiltill Creek........	.....do	.2
18	.....do	Tuolumne River....	Lower end of Hetch Hetchy Valley.	23
17	.....do	Middle Fork of Tuolumne River.	6 miles from Sequoia post-office.	.8
17	.....do	South Fork of Tuolumne River.	1 mile above Sequoia post-office.	7.7

MERCED RIVER DRAINAGE.

Date.	Hydrographer.	Stream.	Location.	Dis-charge
June 25	S. G. Bennett......	Teniah Creek.......	At Tassaack Avenue Bridge, Yosemite Valley.	159
Sept. 15	.....do	.....do............	.....do	3.0
June 25	.....do	Illilouette Creek....	Near mouth Yosemite Valley.	228
Sept. 15	.....do	.....do............	.....do	3.5
June 24	.....do	Merced River	Bridge near Yosemite Valley post-office.	1,135
Sept. 15	.....do	.....do............	.....do	27
June 24	.....do	Yosemite Creek	Wagon bridge below falls.	119
Sept. 15	.....do	.....do............	.....do	a.2
June 25	.....do	Bridal Veil Creek...	At Yosemite road bridge.	a 20

a Estimated.

zellaneous measurements in San Francisco Bay drainage basin in 1903—Continued.

MERCED RIVER DRAINAGE—Continued.

'ate.	Hydrographer.	Stream.	Location.	Discharge.
				Sec. ft.
t. 15	S. G. Bennett......	Bridal Veil Creek...	At Yosemite road bridge.	a 2
13	.....do	Alder Creek........	South Fork Merced.	1.0
12	.....do	South Fork Merced River.	1,000 feet below Wawona Bridge.	1.5
12	.....do	Washburn ditch....	South Fork Merced.	1.9
11	.....do	Big Creek..........	Summerdale........	3.2
11	.....do	Sugar Pine ditch....	South Fork Merced.	2.3

SAN JOAQUIN RIVER DRAINAGE.

t. 11	S. G. Bennett......	Soqual ditch........	At site of old mill of Madera Flume and Trading Co.	b 3.8
9	.....do	Chiquita San Joaquin.	At mouth..........	10
e 9	.....do	San Joaquin River..	Below mouth of Chiquita San Joaquin.	264

SALINAS RIVER DRAINAGE.

y 26	S. G. Bennett......	Vaquero Creek	At mouth..........	1.1
26	.....do	Spreckles's ditch ...	7 miles below headgate.	27
26	.....do	Arroyo Seco Irrigation Co's. ditch.	Road bridge, Joy's place.	36

KERN RIVER DRAINAGE.

r. 24	S. G. Bennett......	Little Kern..........	3 miles below mouth of Shotgun Creek.	5.3
22	.....do	.....do............	Above junction with Kern River.	25
22	.....do	North Fork Kern River.	Above junction with Little Kern.	278
· 21	.....do	Soda Creek.........	3 miles above mouth	2.4
21	.....do	Nameless Creek.....	....	2.4
20	.....do	Clark Creek	....	2.9
20	.....do	Jackson Creek......	....	2.8
20	.....do	Wade Creek........	....	2.6
19	.....do	Tobias Creek	At mouth..........	1.9
19	.....do	Salmon Creek	.....do	1.8
18	.....do	Bull Run Creek.....	Above mouth	.8

a Estimated.　　　b Total flow of North Fork of San Joaquin River at this point.

Miscellaneous measurements in San Francisco Bay drainage basin in 1903—Continued.

KERN RIVER DRAINAGE—Continued.

Date.	Hydrographer.	Stream.	Location.	Discharge.
				Sec. ft
Aug. 17	S. G. Bennett	Peterson ditch	Near head, above Kernville.	1.5
17	.....do	Thurston ditch	.....do	1.9
17	.....do	North Fork of Kern River.	3 miles above Kernville.	277
17	.....do	Big Blue ditch	At mill above Kernville.	25
17	.....do	Neal & Staverts's upper ditch.	Above Kernville	2.2
15	.....do	Brown's upper ditch.		15.3
15	.....do	Brown's ranch ditch.		14.5
15	.....do	Kernville town ditch.		.7
15	.....do	Cook's ditch	Kernville	.6
17	.....do	Hooper Mill ditch	Near Isabella	8.9
17	.....do	South Fork of Kern River.	Above mouth, near Isabella.	17
17	.....do	Kern River	Below mouth of South Fork, near Isabella.	251

KING RIVER DRAINAGE.

Date.	Hydrographer.	Stream.	Location.	Discharge.
Sept. 3	S. G. Bennett	King River	½ mile below mouth of North Fork.	332
3	.....do	North Fork of King River.	At mouth	32

KAWEAH RIVER DRAINAGE.

Date.	Hydrographer.	Stream.	Location.	Discharge.
Aug. 31	S. G. Bennett	Stony Creek	North Fork Kaweah trail crossing.	1.2
31	.....do	Dorst Creek	.....do	2.6
28	.....do	North Fork Kaweah.	3 miles above Three-rivers.	7.3
30	.....do	Marble Fork Kaweah	At bridge, Sequoia Park.	8.7
26	.....do	East Fork Kaweah	At Mineral King	5.3
26	.....do	.....do	Above headworks of Mount Whitney Power Co.'s canal.	23

TULE RIVER DRAINAGE.

Date.	Hydrographer.	Stream.	Location.	Discharge.
June 9	R. S. Hawley	South Fork Tule	Near mouth	35

SOUTHERN CALIFORNIA DRAINAGE.

ler the head of southern California drainage have been included
oncerning the streams of that part of the State south of the San
in basin. There are thus included the Mohave, which flows from
ountains north of San Bernardino into the Mohave Desert, a por-
f the great Interior Basin, as well as those flowing toward the
or southwest, whose waters, in times of flood at least, reach the
: Ocean.

oyo Seco rises on the eastern slope of the Santa Lucia Moun-
ihd flows east and empties into Salinas River at Soledad, Cal.

Lorenzo Creek drains the western slopes of the Gavilan Moun-
ind enters Salinas River near King City, Cal. There is a reser-
nd dam site 5 miles above its mouth. The flood waters are used
inter irrigation. Salinas River drains into the Pacific Ocean
;h Monterey Bay.

drainage basin of San Gabriel River lies on the southern slope
: Sierra Madre, being included in Los Angeles County, Cal.
irious tributaries join the river before it enters its lowest canyon,
e it appears finally on the plain in the vicinity of Azusa. The
;e waters of this valley appear lower down in the river and
enter the Pacific Ocean not far from the mouth of Los Angeles

All of the surplus waters of this stream are now used for
tion purposes, and it is only an occasional flood that passes the
; station.

ta Ana River has its source on the southern slope of the San
rdino Mountains and flows southerly, appearing from its canyon
's north of Redlands. Its waters are completely used in San
rdino Valley. At the lower part of the valley the water appears
in the vicinity of Rincon, where the river passes through a com-
vely narrow gorge; thence the general direction of the stream is
vesterly, emptying into the Pacific Ocean.

headwaters of Mohave River have their source on the northern
of the Sierra Madre. The river flows north, finally disappear-
the sands of the Mohave Desert.

following is a list of the stations in the southern California
ge basin:

Arroyo Seco near Piney, Cal.
an Lorenzo Creek near King City, Cal.
anta Ynez River near Santa Barbara, Cal.
Mono Creek at Mono dam site, near Santa Barbara, Cal.
Malibu Creek near Calabasas, Cal.
Triumpho Creek near Calabasas, Cal.
an Gabriel River and canals, Azusa, Cal.
an Luis Rey River near Pala, Cal.
anta Ana River below Warmsprings, Cal.
anta Maria River near Santa Maria, Cal.
Mohave River at Victorville, Cal.

The observations at this station during 1903 have been made the direction of J. B. Lippincott, supervising engineer.

Discharge measurements of Arroyo Seco near Piney, Cal., in 1903.

Date.	Hydrographer.	Gage height.	Discharge.
		Feet.	*Second-feet.*
March 21	S. G. Bennett	6.23	347
March 28	R. J. Love	8.6	1,707
April 1	do	10.8	3,663
May 26	S. G. Bennett	5.69	71
August 21	H. Hamlin	5.0	.6

Mean daily gage height, in feet, of Arroyo Seco near Piney, Cal., for 1903.

Day.	Jan.	Feb.	Mar.	Apr.	May.	June.	July.	Aug.	Sept.	Oct.	Nov.	Dec.
1	5.4	7.4	5.8	9.9	5.9	5.6	5.4	5.2	5.1	5.2	5.2	5.5
2	5.4	6.9	5.8	8.4	5.9	5.6	5.4	5.3	5.1	5.2	5.3	5.4
3	5.4	6.6	5.8	8.2	5.9	5.6	5.4	5.2	5.1	5.2	5.2	5.4
4	5.4	6.5	5.9	7.4	5.9	5.6	5.4	5.2	5.1	5.2	5.3	5.4
5	5.4	6.4	6.2	7.2	5.9	5.6	5.4	5.2	5.1	5.2	5.3	5.4
6	5.4	6.3	6.0	7.0	5.9	5.6	5.4	5.2	5.1	5.2	5.3	5.4
7	5.4	6.2	6.0	6.9	5.8	5.6	5.4	5.2	5.1	5.2	5.3	5.4
8	5.3	7.6	6.3	6.8	5.8	5.6	5.4	5.2	5.1	5.3	5.3	5.4
9	5.3	6.8	6.3	6.6	5.8	5.6	5.4	5.2	5.0	5.3	5.3	5.4
10	5.3	6.7	6.2	6.5	5.8	5.5	5.3	5.2	5.0	5.3	5.3	5.4
11	5.3	6.5	6.1	6.5	5.8	5.5	5.4	5.2	5.0	5.2	5.3	5.4
12	5.3	6.4	6.1	6.4	5.8	5.5	5.4	5.2	5.0	5.3	5.3	5.4
13	5.3	6.4	6.0	6.3	5.8	5.5	5.3	5.2	5.0	5.3	5.3	5.4
14	5.3	6.3	6.9	6.3	5.8	5.5	5.3	5.2	5.0	5.3	5.3	5.4
15	5.3	6.3	6.5	6.3	5.8	5.5	5.3	5.2	5.0	5.3	5.6	5.4
16	5.3	6.3	6.5	6.4	5.7	5.5	5.3	5.2	5.0	5.3	5.5	5.4
17	5.3	6.3	6.5	6.3	5.5	5.5	5.3	5.2	4.9	5.2	5.4	5.4
18	5.3	6.3	6.4	6.2	5.7	5.5	5.3	5.2	4.9	5.3	5.4	5.4
19	5.3	6.2	6.4	6.2	5.7	5.5	5.3	5.1	4.9	5.3	5.4	5.4
20	5.3	6.2	6.3	6.1	5.7	5.5	5.3	5.1	5.0	5.2	5.9	5.4
21	5.3	6.1	6.3	6.1	5.7	5.5	5.3	5.1	5.0	5.2	5.9	5.4
22	5.3	6.0	6.3	6.1	5.7	5.5	5.3	5.1	5.0	5.2	5.7	5.2
23	5.3	6.0	6.2	6.1	5.7	5.5	5.3	5.1	5.1	5.2	5.7	5.4
24	5.4	6.0	6.2	6.0	5.7	5.5	5.3	5.1	5.1	5.2	5.6	5.4
25	5.3	5.9	6.7	6.0	5.7	5.5	5.3	5.1	5.1	5.2	5.6	5.4

Mean daily gage height, in feet, of Arroyo Seco near Piney, Cal., for 1903—Continued.

Day.	Jan.	Feb.	Mar.	Apr.	May.	June.	July.	Aug.	Sept.	Oct.	Nov.	Dec.
26	5.4	5.9	6.3	6.0	5.7	5.5	5.3	5.1	5.1	5.2	5.5	5.4
27	8.4	5.9	6.2	6.0	5.6	5.4	5.3	5.1	5.1	5.2	5.5	5.45
28	8.6	5.9	7.8	6.0	5.6	5.4	5.3	5.1	5.1	5.2	5.5	5.45
29	7.2	......	7.8	6.0	5.6	5.4	5.3	5.1	5.1	5.8	5.5	5.45
30	7.5	......	9.2	5.9	5.6	5.4	5.3	5.1	5.1	5.8	5.5	5.45
31	7.0	......	9.8	......	5.6	......	5.2	5.1	......	5.8	......	5.45

Rating table for Arroyo Seco near Piney, Cal., for 1903.

Gage height.	Discharge.	Gage height.	Discharge.	Gage height.	Discharge.	Gage height.	Discharge.
Feet.	Second-feet.	Feet.	Second-feet.	Feet.	Second-feet.	Feet.	Second-feet.
5.0	1	6.0	150	7.0	600	8.2	1,430
5.1	6	6.1	190	7.1	660	8.4	1,610
5.2	13	6.2	230	7.2	720	8.6	1,790
5.3	22	6.3	270	7.3	780	8.8	1,970
5.4	32	6.4	310	7.4	840	9.0	2,170
5.5	44	6.5	350	7.5	900	9.2	2,380
5.6	56	6.6	400	7.6	970	9.4	2,600
5.7	72	6.7	450	7.7	1,040	9.6	2,830
5.8	92	6.8	500	7.8	1,110	9.8	3,080
5.9	120	6.9	550	8.0	1,270	10.0	3,350

Estimated monthly discharge of Arroyo Seco near Piney, Cal., for 1903.

[Drainage area, 217 square miles.]

Month.	Discharge in second-feet.			Total in acre-feet.	Run-off.	
	Maixmum.	Minimum.	Mean.		Second-feet per square mile.	Depth in inches.
January	1,790	22	205	12,605	0.94	1.08
February	970	120	313	17,383	1.44	1.50
March	3,080	92	466	28,653	2.15	2.48
April	3,210	120	483	28,740	2.23	2.49
May	720	56	84	5,165	.39	.45
June	120	32	46	2,737	.21	.23
July	56	13	25	1,537	.12	.14
August	32	6	10	615	.05	.06
September	6	0	4	238	.02	.02
October	22	13	17	1,045	.08	.09
November	120	13	40	2,380	.18	.20
December	44	32	33	2,029	.15	.17
The year	3,210	0	144	103,127	.66	8.91

SAN LORENZO CREEK NEAR KING CITY, CAL.

A gaging station was established December 16, 1900, at Hollenbeck ranch. To obtain a reliable observer it was necessary to move the gaging station to the Mathews dam site one-half mile above the Hollenbeck ranch on November, 1901. The stream is very flashy, and the rating curve for 1902 was completed by taking the cross section and slope with a level and calculating the discharge by the Kutter's formula. The observer is J. L. Mathews. This station was discontinued May 31, 1903.

The observations at this station during 1903 have been made under the direction of J. B. Lippincott, supervising engineer.

Discharge measurements of San Lorenzo Creek near King City, Cal., in 1903.

Date.	Hydrographer.	Gage height.	Discharge.
		Feet.	*Second-feet.*
June 1....................	S. G. Bennett..............	0.90	3.3
August 20..................	H. Hamlin	.65	.5

Mean daily gage height, in feet, of San Lorenzo Creek near King City, Cal., for 1903.

Day.	Jan.	Feb.	Mar.	Apr.	May.	Day.	Jan.	Feb.	Mar.	Apr.	May.
1	0.9	1.8	0.9	2.4	0.9	17.................	0.9	1.0	1.4	0.9	0.9
2	.9	1.6	.9	2.4	.9	18.................	.9	1.0	1.8	.9	.9
3	.9	1.5	1.0	2.0	.9	19.................	.9	1.0	1.6	.9	.8
4	.9	1.4	1.0	1.8	.9	20.................	.9	1.0	1.2	.9	.8
5	.9	1.2	1.0	1.8	.9	21.................	.9	1.0	1.0	.9	.5
6	.9	1.2	1.0	1.6	.9	22.................	.9	.9	1.0	.9	.5
7	.9	1.2	1.0	1.2	.9	23.................	.9	.9	1.0	.9	.5
8	.9	1.8	1.0	1.2	.9	24.................	.9	.9	1.0	.9	.5
9	.9	1.6	1.0	1.2	.9	25.................	1.0	.9	1.6	.9	.5
10	.9	1.6	1.0	1.1	.9	26.................	1.0	.9	1.8	.9	.5
11	.9	1.4	1.0	1.1	.9	27.................	1.2	.9	1.6	.9	.5
12	.9	1.2	1.0	.9	.9	28.................	1.9	.9	2.0	.9	.5
13	.9	1.0	1.0	.9	.9	29.................	1.8	....	2.0	.9	.5
14	.9	1.0	1.0	.9	.9	30.................	1.6	....	1.8	.9	.5
15	.9	1.0	1.8	.9	.9	31.................	2.0	....	2.4	....	.5
16	.9	1.0	1.6	.9	.9						

ible for San Lorenzo Creek near King City, Cal., from January 1 to May 31, 1903.

t.	Discharge.	Gage height.	Discharge.	Gage height.	Discharge.	Gage height.	Discharge.
	Second-feet.	*Feet.*	*Second-feet.*	*Feet.*	*Second-feet.*	*Feet.*	*Second-feet.*
}	3	1.4	88	2.0	274	2.6	585
)	5	1.5	110	2.1	317	2.7	650
)	15	1.6	135	2.2	364	2.8	720
	32	1.7	166	2.3	415	2.9	795
!	50	1.8	200	2.4	470	3.0	875
i	68	1.9	236	2.5	525		

iated monthly discharge of San Lorenzo Creek near King City, Cal., for 1903.

[Drainage area, 1,280 square miles.]

Month.	Discharge in second-feet.			Total in acre-feet.	Run-off.	
	Maximum.	Minimum.	Mean.		Second-feet per square mile.	Depth in inches.
....	274	5	34	2,091	0.14	0.16
y	200	5	52	2,888	.22	.23
....	470	5	89	5,472	.38	.44
....	470	5	69	4,106	.29	.32
....	5	3	4	246	.02	.02
he period ...	470	3	50	14,803	1.18	0.21

SANTA YNEZ RIVER NEAR SANTA BARBARA, CAL.

original station at which measurements were made during the
: part of 1903 is located about 1 mile above the mouth of Mono
It was established November 21, 1902, by W. B. Clapp,
1 by Howard Rankin. The gage is an inclined rod on the right
From January 25 to June 20, 1903, daily discharge measure-
were made by means of a cable and car. The channel is straight
) feet above and for 600 feet below the cable. The current is
ih at low stages. The right bank is low, but will overflow for
short distance. There is an overflow channel on the left bank.
d of the stream is composed of gravel and is permanent. The
mark is a nail in the root of the poplar tree on the left bank, to
the cable is attached. Its elevation is 7.18 feet above the zero
gage. On November 1, 1903, a new station was established at
braltar dam site by L. M. Hyde, assisted by H. W. Muzzall.
cated 5 miles below the original station and is below the mouth
10 Creek. It is 9 miles above the San Marcus ranch and halfway

between the old quicksilver mines. The gage is a 4 by 4-inch inch timber, spiked to a cottonwood tree on the right bank. The tree blazed above the gage rod for recording stages above the gage. Discharge measurements are made at high water by means of a cable feet above the gage. Measurements can usually be made by wad Weir measurements of the discharge were made daily from December 20 to 31, 1903. At the cable the initial point for soundings is a bl at the base of the cottonwood tree on the right bank, to which the c is attached. The channel is straight for 700 feet above and for feet below the station. The right bank is low, but is not liable overflow. The left bank rises abruptly about 20 feet beyond the s tree to which the cable is attached. It is not liable to overflow. bed of the stream is composed of sandy gravel, free from vegeta and bowlders. The cross section is regular, and is permanent. current is swift. The bench mark is a cross on a bench of a ledg the left bank, about 100 feet below the cable. Its elevation is 1 feet above the zero of the gage. The approximate elevation abov level, as estimated from topographic maps, is 1,200 feet.

The observations at this station during 1903 have been made u the direction of J. B. Lippincott, supervising engineer.

Discharge measurements of Santa Ynez River near Santa Barbara, Cal., in 190

Date.	Hydrographer.	Gage height.	Disc
		Feet.	Secon
January 12	H. Rankin	1.10	
January 25	do	1.10	
January 27	do	1.20	
January 28	do	3.55	
January 29	do	1.70	
January 30	do	1.50	
January 31	do	1.48	
February 1	do	1.50	
February 2	do	1.45	
February 3	do	1.42	
February 4	do	1.45	
February 5	do	1.45	
February 7	do	1.45	
February 8	do	1.58	
February 9	do	1.50	
February 10	do	1.49	
February 11	do	1.46	
February 12	do	1.45	
February 13	do	1.44	
February 14	do	1.44	
February 16	do	1.40	

neasurements of Santa Ynez River near Santa Barbara, Cal., in 1903—Cont'd.

Date.	Hydrographer.	Gage height.	Discharge.
		Feet.	*Second-feet.*
17	H. Rankin	1. 40	11. 3
18	....do	1. 40	11. 1
19	....do	1. 40	11
20	....do	1. 39	10. 8
21	....do	1. 39	11
23	....do	1. 38	10. 1
24	....do	1. 37	9
25	....do	1. 37	9. 1
26	....do	1. 36	8. 6
27	....do	1. 35	8. 7
28	....do	1. 35	8. 3
....	....do	1. 35	8. 2
....	....do	1. 35	8
....	....do	1. 35	7. 9
....	....do	1. 45	14. 8
....	....do	1. 65	32
....	....do	1. 50	20
....	....do	1. 50	20
....	....do	1. 43	13. 5
....	....do	1. 42	13. 2
....	....do	1. 42	13. 4
....	....do	1. 41	12. 7
....	....do	1. 42	13. 2
....	....do	1. 43	12. 8
....	....do	1. 43	13. 1
....	....do	1. 43	12. 8
....	....do	1. 42	11. 6
....	....do	1. 41	11. 3
....	....do	1. 40	11. 1
....	....do	1. 40	10. 9
....	....do	1. 40	10. 6
....	....do	1. 40	10. 5
....	....do	1. 40	10. 6
....	....do	1. 67	29
....	....do	1. 55	19. 6
....	....do	1. 50	17. 1
....	....do	1. 49	16. 9
....	....do	1. 50	18. 0
....	....do	2. 10	93
....	....do	2. 15	111
....	....do	3. 50	535
....	....do	2. 80	242

April 23	do		
April 24	do	1.75	
April 25	do	1.71	
April 27	do	1.68	
April 28	do	1.65	
April 29	do	1.60	
April 30	do	1.59	
May 2	do	1.56	
May 3	do	1.55	27
May 4	do	1.55	28
May 5	do	1.54	27
May 6	do	1.51	20
May 7	do	1.50	19.4
May 8	do	1.50	19.0
May 10	do	1.49	17.8
May 11	do	1.48	17.5
May 12	do	1.48	17.7
May 13	do	1.48	16.7
May 14	do	1.48	16.7
May 15	do	1.46	16.1
May 16	do	1.45	15.2
May 17	do	1.45	14.9
May 18	do	1.45	13.5

measurements of Santa Ynez River near Santa Barbara, Cal., in 1903—Cont'd.

Date.	Hydrographer.	Gage height.	Discharge.
		Feet.	*Second-feet.*
....	H. Rankin	1.45	15.2
....	.....do	1.45	14.4
....	.....do	1.45	13.5
....	.....do	1.45	13.6
....	.....do	1.45	13.4
....	.....do	1.43	12.2
....	.....do	1.42	11.9
....	.....do	1.40	11.4
....	.....do	1.40	11.2
....	.....do	1.40	10.7
....	.....do	1.40	10.4
....	.....do	1.38	8.7
....	.....do	1.38	8.5
....	.....do	1.37	7.8
....	.....do	1.35	7.1
....	.....do	1.35	7.0
....	.....do	1.35	7.0
....	.....do	1.33	6.0
....	.....do	1.32	5.7
....	.....do	1.32	4.9
....	.....do	1.31	4.9
....	.....do	1.30	4.9
....	.....do	1.30	4.9
....	.....do	1.30	4.7
....	.....do	1.30	4.4
....	.....do	1.30	4.4
....	.....do	1.30	4.4
....	W. B. Clapp.....................	1.28	3.1
....	L. M. Hyde.....................	1.20	1.2
r 20.................	H. W. Muzzall...............	(*a*)	.30
r 21.................	.....do	(*a*)	.35
r 22.................	.....do	(*a*)	.35
r 23.................	.....do	(*a*)	.40
r 24.................	.....do	(*a*)	.40
r 25.................	.....do	(*a*)	.40
r 26.................	.....do	(*a*)	.40
r 27.................	.....do	(*a*)	.40
r 28.................	.....do	(*a*)	.40
r 29.................	.....do	(*a*)	.42
r 30.................	.....do	(*a*)	.42
r 31.................	.....do	(*a*)	.42

a Weir.

Mean daily gage height, in feet, of Santa Ynez River near Santa Barbara, Cal., for 1903.

Day.	Jan.	Feb.	Mar.	Apr.	May.	June.	July.	Nov.	Dec.
1	1.10	1.50	1.35	3.50	1.58	1.38	.....	a 0.0	0.0
2	1.10	1.45	1.35	2.60	1.56	1.38	.....	0.0	0.0
3	1.10	1.42	1.35	2.20	1.55	1.37	.....	0.0	0.0
4	1.10	1.45	1.45	2.10	1.55	1.35	.....	0.0	0.0
5	1.10	1.45	1.65	1.90	1.54	1.35	.....	0.0	0.0
6	1.10	1.45	1.50	1.85	1.51	1.35	.....	0.0	0.0
7	1.10	1.45	1.50	1.80	1.50	1.34	.....	0.0	0.0
8	1.10	1.58	1.45	1.70	1.50	1.33	.....	0.0	0.0
9	1.10	1.50	1.43	1.65	1.50	1.32	.....	0.0	0.0
10	1.10	1.49	1.42	1.65	1.49	1.32	.....	0.0	0.0
11	1.10	1.46	1.42	1.62	1.48	1.31	.....	0.0	0.0
12	1.10	1.45	1.41	1.60	1.48	1.30	1.20	0.0	0.0
13	1.10	1.44	1.42	1.60	1.48	1.30	.....	0.0	0.0
14	1.10	1.44	1.43	1.56	1.48	1.30	.....	0.0	0.0
15	1.10	1.42	1.43	1.55	1.46	1.30	.....	0.0	0.0
16	1.10	1.40	1.43	2.50	1.45	1.30	.....	0.0	0.0
17	1.10	1.40	1.42	2.30	1.45	1.30	.....	0.0	0.0
18	1.10	1.40	1.41	2.10	1.45	1.30	.....	0.0	0.0
19	1.10	1.40	1.40	2.05	1.45	1.30	.....	0.0	0.0
20	1.10	1.39	1.40	2.00	1.45	1.28	.....	0.0	0.0
21	1.10	1.39	1.40	1.90	1.45	(b)	.....	0.0	0.0
22	1.10	1.39	1.40	1.85	1.45	.....	.....	0.0	0.0
23	1.10	1.38	1.40	1.80	1.45	.....	.....	0.0	0.0
24	1.10	1.37	1.45	1.75	1.44	.....	.....	0.0	0.0
25	1.10	1.37	1.67	1.71	1.43	.....	.....	0.0	0.0
26	1.10	1.36	1.55	1.70	1.42	.....	.....	0.0	0.0
27	1.20	1.35	1.50	1.68	1.40	.....	.....	0.0	0.0
28	3.55	1.35	1.49	1.65	1.40	.....	.....	0.0	0.0
29	1.70	.....	1.50	1.60	1.40	.....	.....	0.0	0.0
30	1.50	.....	2.10	1.59	1.40	.....	.....	0.0	0.0
31	1.48	.....	2.15	.....	1.39	.....	.....	.....	0.0

a On Nov. 1 new gaging station established at point 5 miles lower down on river at Gibraltar dam site.
b June 20 record was stopped for summer.

Estimated monthly discharge of Santa Ynez River near Santa Barbara, Cal., for 1903.[a]

[Drainage area, 71 square miles.]

Month.	Discharge in second-feet.			Total in acre-feet.	Run-off.	
	Maximum.	Minimum.	Mean.		Second-feet per square mile.	Depth in inches.
January	457	2	19	1,168	0.27	0.31
February	25	8	14	778	.20	.21
March	111	8	20	1,230	.28	.32
April	535	29	84	4,998	1.18	1.32
May	31	9	17	1,045	.24	.28
June 1 to 20	9	3	6	238	.08	.06
The period [b]	....	....	....	9,457	....	....

a Interpolated from discharge measurements.
b Mean discharge by weir measurement from Dec. 20 to 31 was 0.39 second-foot.

MONO CREEK AT MONO DAM SITE, CALIFORNIA.

his station was established November 22, 1902, by W. B. Clapp,
sted by Howard Rankin. It is located about one-half mile above
junction of Mono Creek and Santa Ynez River, 15 miles above the
a Marcus ranch and 17 miles by trail from Santa Barbara. The
er at this point traverses Mono flat. The gage is an inclined tim-
r hewn from a 4-inch sapling. It is spiked to a willow tree on the
ht bank. The tree has been blazed above the gage for use in record-
flood stages above the gage rod. Discharge measurements were
de daily from January 28 to June 20, 1903, by means of a cable
l by wading. The cable is located about 500 feet below the dam
and is just above the gage. The initial point for soundings is a
se at the base of the tree to which the cable is fastened on the left
k. The channel is slightly curved for about 500 feet above and
ow the station. Both banks are high and are not liable to overflow.
e bed of the stream is composed of sandy gravel, free from vegeta-
, and is not liable to shift. The bench mark is a standard United
tes Geological Survey bench-mark disk on a sandstone rock 100
t south of the large oak tree on the left bank. Its elevation is
92 feet above the zero of the gage and 1,410 feet above sea level.
a record of the discharge, kept by means of a weir, from November
o December 31, 1903, showed a constant discharge for this period
0.05 second-foot.
The observations at this station during 1903 have been made under
direction of J. B. Lippincott, supervising engineer.

Discharge measurements of Mono Creek at Mono dam site in 1902–3.

Date.	Gage height.	Discharge.	Date.	Gage height.	Discharge.
1902.	*Feet.*	*Second-feet.*	1903.	*Feet.*	*Second-feet.*
vember 11	....	8.7	February 10	1.28	8.4
1903.			February 11	1.27	8.8
nary 28.........	2.95	295	February 12	1.26	8.2
nary 29.........	1.55	20	February 13	1.25	7.6
nary 30.........	1.30	12.5	February 14	1.20	4.9
nary 31.........	1.25	9.0	February 16	1.20	5.1
ruary 1	1.35	14.8	February 17	1.20	5.0
ruary 2	1.30	10.1	February 18	1.20	5.1
ruary 3	1.20	6.9	February 19	1.20	5.2
ruary 4	1.28	8.8	February 20	1.19	4.8
ruary 5	1.28	8.2	February 21	1.19	4.7
ruary 6	1.21	8.1	February 23	1.17	4.6
bruary 8	1.35	12.0	February 24	1.15	4.6
bruary 9	1.29	8.8	February 25	1.15	4.8

Discharge measurements of Mono Creek at Mono dam site in 1902-3—Continued.

Date.	Gage height.	Discharge.	Date.	Gage height.	Discharge.
1903.	Feet.	Second-feet.	1903.	Feet.	Second-ft.
February 26	1.15	4.4	April 9	1.80	39
February 27	1.14	3.8	April 10	1.77	34
February 28	1.14	3.8	April 11	1.73	39
March 1	1.12	3.6	April 12	1.70	29
March 2	1.12	3.6	April 14	1.65	24
March 3	1.12	3.6	April 15	1.63	5
March 4	1.20	11.0	April 16	2.20	71
March 5	1.60	26	April 17	2.40	100
March 6	1.37	13.6	April 18	2.30	94
March 7	1.35	13.3	April 19	2.35	90
March 9	1.20	9.8	April 20	2.10	72
March 10	1.25	6.6	April 21	2.00	59
March 11	1.22	6.3	April 22	1.90	46
March 12	1.22	6.6	April 23	1.85	43
March 13	1.22	6.5	April 24	1.80	38
March 14	1.27	9.2	April 25	1.75	36
March 15	1.25	7.5	April 27	1.70	29
March 16	1.30	9.8	April 28	1.70	28
March 17	1.32	9.9	April 29	1.67	28
March 18	1.26	8.2	April 30	1.65	27
March 19	1.25	8.0	May 2	1.60	24
March 20	1.24	7.1	May 3	1.58	21
March 21	1.24	7.1	May 4	1.56	19.
March 22	1.23	7.2	May 5	1.55	18.
March 23	1.23	7.2	May 6	1.52	17.
March 25	2.50	153	May 7	1.50	17.
March 26	1.90	62	May 8	1.50	16.
March 27	1.70	36	May 10	1.50	15.
March 28	1.60	27	May 11	1.48	14.
March 29	1.61	30	May 12	1.48	14.
March 30	2.65	177	May 13	1.48	13.
March 31	2.80	227	May 14	1.47	13.
April 1	4.50	710	May 15	1.45	13.
April 2	3.00	270	May 16	1.45	11.
April 3	2.50	131	May 17	1.43	11.
April 4	2.40	92	May 18	1.43	11.
April 5	2.20	69	May 19	1.43	11.
April 6	2.00	52	May 20	1.42	10.
April 7	1.90	41	May 21	1.42	10.0
April 8	1.85	40	May 22	1.42	9.3

irge measurements of Mono Creek at Mono dam site in 1902-3—Continued.

ate.	Gage height.	Discharge.	Date.	Gage height.	Discharge.
)03.	*Feet.*	*Second-feet.*	1903.	*Feet.*	*Second-feet.*
....	1.42	9.1	June 8.............	1.35	5.1
....	1.41	8.8	June 9.............	1.40	8.2
....	1.41	8.4	June 10............	1.45	10.6
....	1.40	7.7	June 11............	1.40	8.6
....	1.40	7.5	June 12............	1.35	4.7
....	1.40	7.5	June 13............	1.33	4.4
....	1.40	7.4	June 15............	1.30	3.3
....	1.37	6.2	June 16............	1.30	3.2
....	1.35	5.7	June 17............	1.30	3.2
....	1.33	5.7	June 18............	1.29	3.0
....	1.30	4.4	June 20............	1.22	2.0
....	1.28	3.4	July 12............	.80	.4
....	1.28	3.2			

ily gage height, in feet, of Mono Creek at Mono dam site, California, for 1903.

Day.	Jan.	Feb.	Mar.	Apr.	May.	June.	July.
....	1.00	1.35	1.12	4.50	1.63	1.37	....
....	1.00	1.30	1.12	3.00	1.60	1.35	....
....	1.00	1.20	1.12	2.50	1.58	1.33	....
....	1.00	1.28	1.30	2.40	1.56	1.30	....
....	1.00	1.28	1.60	2.20	1.55	1.28	....
....	1.00	1.21	1.37	2.00	1.52	1.28	....
....	1.00	1.25	1.35	1.90	1.50	1.28	....
....	1.00	1.35	1.30	1.85	1.50	1.35	....
....	1.00	1.29	1.30	1.80	1.50	1.40	....
....	1.00	1.28	1.25	1.77	1.50	1.45	....
....	1.00	1.27	1.22	1.73	1.48	·1.40	....
....	1.00	1.26	1.22	1.70	1.48	1.35	c.80
....	1.00	1.25	1.22	1.68	1.48	1.33	....
....	1.00	1.20	1.27	1.65	1.47	1.32	....
....	1.00	1.20	1.25	1.63	1.45	1.30	....
....	1.00	1.20	1.30	2.20	1.45	1.30	....
....	1.00	1.20	1.32	2.40	1.43	1.30	....
....	1.00	1.20	1.26	2.30	1.43	1.29	....
....	1.00	1.20	1.25	2.35	1.43	1.26	....
....	1.00	1.19	1.24	2.10	1.42	a 1.22	....
....	1.00	1.19	1.24	2.00	1.42	....	....
....	1.00	1.18	1.23	1.90	1.42	....	....
....	1.00	1.17	1.23	1.85	1.42	....	....
....	1.00	1.15	1.30	1.80	1.42	....	....
....	1.00	1.15	2.50	1.75	1.41	....	....

a June 20, record was stopped for summer.

Mean daily gage height, in feet, of Mono Creek at Mono dam site, California, for 1903—
Continued.

Day.	Jan.	Feb.	Mar.	Apr.	May.	June.	July.
26	1.00	1.15	1.90	1.75	1.41		
27	1.00	1.14	1.70	1.70	1.40		
28	2.95	1.14	1.66	1.70	1.40		
29	1.55		1.61	1.67	1.40		
30	1.30		2.65	1.65	1.40		
31	1.25		2.80		1.38		

Record began again on November 1, 1903. Weir measurements were kept during November and December, 1903, and the discharge was constant at 0.05 second-feet.

Estimated monthly discharge of Mono Creek at Mono dam site, California, for 1903.[a]

[Discharge area, 119 square miles.]

Month.	Discharge in second-feet.			Total in acre-feet.	Run-off.	
	Maximum.	Minimum.	Mean.		Second-feet per square mile.	Depth in inches.
January	295	[b]0.17	11	676	0.09	0.10
February	15	4	7	389	.06	.06
March	227	4	30	1,845	.25	.29
April	710	25	81	4,820	.68	.76
May	26	6	13	799	.11	.13
June 1 to 20	11	2	5	198	.04	.03
November[c]						
December[c]						
The period				8,727		

a Interpolated from discharge measurements.
b Minimum discharge as determined by weir January 1 to 27.
c Weir measurements taken during November and December showed a constant discharge of 0.05 second-feet.

MALIBU AND TRIUMFO CREEKS NEAR CALABASAS, CAL.

These stations were established by S. G. Bennett at Chapman's ranch, November 29, 1901. They are 40 miles from Los Angeles by wagon road and 8 miles southwest of Calabasas. The Malibu is formed by the Triumfo and Las Virgines creeks, and is a short stream flowing through the Santa Monica Mountains and discharging into the Pacific Ocean about 15 miles above the city of Santa Monica, Cal. Sites for good storage reservoirs exist in both of these drainage basins, one on the Malibu a short distance below its gaging station, and one on the Triumfo above its gaging station. The water from these streams could be used for irrigation on the valuable land lying along the base of the mountains between the city of Los Angeles and Santa Monica, where it is greatly needed.

The channels of both streams are poor and are subject to change

ing high water. However, they are located at the only point
are an observer could be secured. The excessive cost of visiting
se stations has made it impossible to obtain as many meter meas-
ments as desired, but the observer is instructed to take float
ocities during floods at various gage heights, and these data, with
ss sections and grade of stream, are used in addition to meter
asurements for computing discharges for use in constructing rating
ves and tables. The gage rods for both stations are vertical 2 by
ich wooden rods graduated to feet and tenths.

he gaging station on the Malibu is located about one-fourth mile
ow the mouth of Las Virgines Creek. The channel above the
ion is straight for about 600 feet, and below the station is curved
about 300 feet, and the water is swift. Both banks are high.
right bank is rocky, and the bed of the stream is composed of
k and gravel. The initial point for soundings is on the right
k. The bench mark is a cross on a small projection on a rock
ff about 10 feet southwest of the gage rod. The assumed elevation
the bench mark is 530 feet. The zero of the gage rod is 524.57
t elevation.

he gaging station on Triumfo Creek is about one-half mile above
mouth of Las Virgines Creek. The channel is straight for about
feet above and 800 feet below the station, and the water is swift.
th banks are high and rocky. The bed of the stream is composed
gravel and sand, and is shifting. The initial point for soundings
m the right bank. The bench mark is a cross on a point of rock
feet above the bed of the creek. The assumed elevation is 550
t. The zero of the gage rod is 545.47 feet elevation. The observer
both stations is J. G. Chapman.

Discharge measurements of Malibu Creek near Calabasas, Cal., in 1903.

Date.	Hydrographer.	Gage height.	Discharge.
		Feet.	*Second-feet.*
ch 6	W. B. Clapp	1.30	23
ch 25	J. G. Chapman	2.45	a 292
ch 26	do	2.00	a 100
ch 27	do	1.90	a 72
ch 31	do	3.40	a 956
il 1	do	3.25	a 826
· 26	W. B. Clapp	1.12	2.1

a Velocity taken with floats and discharge calculated in office; approximate only.

Mean daily gage height, in feet, of Malibu Creek near Calabasas, Cal., for 1903.

Day.	Jan.	Feb.	Mar.	Apr.	May.	June.	July.	Nov.	Dec.
1	0.70	1.20	1.00	3.25	1.30	1.25	1.15	1.10	1.10
2	.70	1.20	1.00	2.55	1.30	1.25	1.15	1.10	1.10
3	.70	1.15	1.00	2.22	1.30	1.25	1.15	1.10	1.10
4	.70	1.15	1.40	2.10	1.30	1.25	1.15	1.10	1.10
5	.70	1.15	1.48	1.90	1.30	1.25	1.15	1.10	1.10
6	.70	1.15	1.40	1.90	1.30	1.25	1.15	1.10	1.10
7	.70	1.15	1.25	1.90	1.30	1.20	1.15	1.10	1.10
8	.70	1.15	1.20	1.60	1.30	1.20	1.15	1.10	1.10
9	.70	1.15	1.15	1.60	1.30	1.20	1.15	1.10	1.15
10	.70	1.15	1.15	1.50	1.30	1.20	1.15	1.10	1.15
11	.70	1.15	1.15	1.50	1.30	1.20	1.10	1.10	1.15
12	.75	1.15	1.10	1.50	1.30	1.20	1.10	1.10	1.20
13	.75	1.15	1.10	1.50	1.30	1.20	1.10	1.10	1.20
14	.75	1.15	1.10	1.45	1.30	1.20	1.10	1.10	1.20
15	.75	1.15	1.10	1.40	1.30	1.20	1.10	1.10	1.20
16	.75	1.15	1.10	3.20	1.30	1.20	1.10	1.10	1.20
17	.80	1.15	1.10	2.63	1.30	1.20	1.10	1.10	1.20
18	.80	1.15	1.10	2.15	1.30	1.20	1.10	1.10	1.20
19	.80	1.15	1.05	2.00	1.30	1.20	1.10	1.10	1.20
20	.80	1.15	1.05	1.80	1.30	1.20	1.10	1.10	1.20
21	.75	1.15	1.00	1.75	1.30	1.20	1.10	1.10	1.20
22	.75	1.15	1.00	1.70	1.30	1.20	1.10	1.10	1.20
23	.75	1.15	1.00	1.60	1.25	1.20	1.10	1.10	1.20
24	.75	1.15	1.18	1.50	1.25	1.20	1.10	1.10	1.20
25	.80	1.15	2.40	1.40	1.25	1.20	1.10	1.10	1.20
26	.80	1.15	1.95	1.40	1.25	1.20	1.10	1.10	1.20
27	.90	1.00	1.90	1.35	1.25	1.15	1.10	1.10	1.20
28	2.05	1.00	1.53	1.30	1.25	1.15	1.10	1.10	1.20
29	1.38	......	1.50	1.30	1.25	1.15	1.10	1.10	1.20
30	1.20	......	1.50	1.30	1.25	1.15	1.10	1.10	1.20
31	1.05	......	2.80	......	1.25	......	1.10	......	1.20

Rating table for Malibu Creek near Calabasas, Cal., from January 1 to March 30, 1903[a]

Gage height.	Discharge.	Gage height.	Discharge.	Gage height.	Discharge.	Gage height.	Discharge.
Feet.	*Second-feet.*	*Feet.*	*Second-feet.*	*Feet.*	*Second-feet.*	*Feet.*	*Second-feet.*
0.5	1.0	1.1	14	1.7	50	2.3	210
.6	1.5	1.2	18	1.8	62	2.4	265
.7	2.5	1.3	23	1.9	75	2.5	330
.8	4.5	1.4	28	2.0	98		
.9	7.0	1.5	34	2.1	128		
1.0	10.0	1.6	41	2.2	165		

[a] Two rating tables necessary on account of change in channel, March 31.

ting table for Malibu Creek near Calabasas, Cal., from March 31 to December 31, 1903.

Gage height.	Discharge.	Gage height.	Discharge.	Gage height.	Discharge.	Gage height.	Discharge.
Feet.	*Second-feet.*	*Feet.*	*Second-feet.*	*Feet.*	*Second-feet.*	*Feet.*	*Second-feet.*
1.0	1	1.7	44	2.4	265	3.1	720
1.1	2	1.8	57	2.5	330	3.2	785
1.2	5	1.9	72	2.6	395	3.3	850
1.3	10	2.0	95	2.7	460	3.4	915
1.4	16	2.1	128	2.8	525	3.5	980
1.5	24	2.2	165	2.9	590		
1.6	33	2.3	210	3.0	655		

Estimated monthly discharge of Malibu Creek near Calabasas, Cal., in 1903.

[Drainage area, 97 square miles.]

Month.	Discharge in second-feet.			Total in acre-feet.	Run-off.	
	Maximum.	Minimum.	Mean.		Second-feet per square mile.	Depth in inches.
.nuary	113	2.5	8.5	523	0.09	0.10
:bruary	18	10	15.7	872	.16	.17
arch	525	10	46.0	2,828	.47	.54
>ril	817	10	120.8	7,188	1.25	1.39
ay	10	8	9.4	578	.10	.12
me	8	4	5.5	327	.06	.07
ily	4	2	2.6	160	.03	.03
>vember	2	2	2.0	119	.02	.02
>cember	5	2	4.1	252	.04	.05
The period....	....	....	....	12,847	....	....

Discharge measurements of Triumfo Creek near Calabasas, Cal., in 1903.

Date.	Hydrographer.	Gage height.	Discharge.
		Feet.	*Second-feet.*
arch 6	W. B. Clapp...................	1.15	23
arch 25	J. G. Chapman	1.75	*a* 155
arch 26	.....do	1.15	*a* 40
arch 27	.....do	1.05	*a* 13
arch 31	.....do	2.75	*a* 1,075
>ril 1	.....do	2.35	*a* 625
>ril 16	.....do	2.30	*a* 551
ly 26	W. B. Clapp...................	.40	1.0

a Velocity taken with floats and discharge calculated in office; approximate only.

Mean daily gage height, in feet, of Triunfo Creek near Calabasas, Cal., for 1903.

Day.	Jan.	Feb.	Mar.	Apr.	May.	June.	July.	Nov.	Dec.
1	0.75	1.00	1.00	2.26	0.70	0.50	0.40	0.00	.15
2	.75	1.05	1.00	1.45	.70	.50	.40	.00	.15
3	.80	1.05	1.00	1.10	.70	.50	.40	.00	.15
4	.80	1.05	1.22	1.00	.70	.50	.40	.00	.15
5	.80	1.05	1.17	.90	.70	.50	.40	.00	.15
6	.80	1.05	1.10	.85	.70	.50	.40	.00	.15
7	.80	1.05	1.10	.80	.70	.50	.40	.00	.15
8	.80	1.05	1.10	.75	.70	.50	.40	.00	.15
9	.80	1.05	1.10	.75	.70	.50	.40	.00	.15
10	.80	1.05	1.05	.70	.70	.50	.40	.00	.15
11	.80	1.05	1.05	.65	.70	.50	.40	.00	.15
12	.75	1.05	1.05	.65	.70	.50	.40	.00	.15
13	.75	1.05	1.05	.65	.65	.50	.40	.00	.15
14	.75	1.05	1.05	.65	.65	.50	.40	.00	.15
15	.75	1.05	1.05	.65	.65	.50	.40	.00	.15
16	.75	1.05	1.05	2.15	.65	.50	.40	.00	.15
17	.80	1.05	1.05	1.90	.65	.50	.40	.00	.15
18	.80	1.05	1.05	1.38	.65	.50	.40	.00	.15
19	.80	1.05	1.05	1.30	.65	.50	.40	.00	.15
20	.80	1.05	1.05	1.20	.60	.45	.40	.00	.15
21	.75	1.05	1.00	1.10	.60	.45	.40	.40	.15
22	.75	1.05	1.00	1.05	.60	.45	.40	.40	.15
23	.75	1.05	1.00	1.00	.60	.45	.40	.40	.15
24	.80	1.05	1.10	.90	.60	.45	.40	.40	.15
25	.80	1.05	1.60	.80	.60	.45	.40	.40	.15
26	.80	1.05	1.13	.80	.60	.45	.40	.40	.15
27	.85	1.00	1.05	.75	.60	.40	.40	.40	.15
28	1.18	1.00	1.05	.70	.60	.40	.40	.35	.15
29	1.05	.....	1.05	.70	.60	.40	.40	.35	.15
30	1.00	.....	1.05	.70	.50	.40	.40	.35	.15
31	1.00	.....	2.13	.....	.50	.....	a.40	.....	.15

a Record closed July 31 for the summer, and began again November 1.

Rating table for Triumfo Creek near Calabasas, Cal., from January 1 to March 30, 1903.[a]

Gage height.	Discharge.	Gage height.	Discharge.	Gage height.	Discharge.	Gage height.	Discharge.
Feet.	Second-feet.	Feet.	Second-feet.	Feet.	Second-feet.	Feet.	Second-feet.
0.7	1	1.3	38	1.9	225	2.5	775
.8	2	1.4	55	2.0	280	2.6	895
.9	4	1.5	76	2.1	347	2.7	1,015
1.0	8	1.6	102	2.2	435		
1.1	16	1.7	136	2.3	545		
1.2	25	1.8	177	2.4	660		

a Two rating tables necessary on account of change in channel, March 31.

Rating table for Triumfo Creek near Calabasas, Cal., from March 31 to December 31, 1903.

Gage height.	Discharge.	Gage height.	Discharge.	Gage height.	Discharge.	Gage height.	Discharge.
Feet.	*Second-feet.*	*Feet.*	*Second-feet.*	*Feet.*	*Second-feet.*	*Feet.*	*Second-feet.*
0.3	0.5	0.9	19	1.5	101	2.1	373
.4	2.0	1.0	26	1.6	129	2.2	450
.5	4.0	1.1	35	1.7	162	2.3	550
.6	6.0	1.2	46	1.8	200	2.4	670
.7	9.0	1.3	60	1.9	250		
.8	14.0	1.4	78	2.0	305		

Estimated monthly discharge of Triumpho Creek near Calabasas, Cal., for 1903.

[Drainage area, 72 square miles.]

Month.	Discharge in second-feet.			Total in acre-feet.	Run-off.	
	Maximum.	Minimum.	Mean.		Second-feet per square mile.	Depth in inches.
January	23	1.5	3.3	203	0.05	0.06
February	12	8.0	11.6	644	.16	.17
March	373	8.0	27.5	1,691	.39	.45
April	510	8.0	60.0	3,570	.83	.93
May	9	4.0	7.5	461	.10	.12
June	4	2.0	3.5	208	.05	.06
July	2	2.0	2.0	123	.03	.03
November	2	0.0	.6	36	.01	.01
December	3	0.0	2.0	123	.03	.03
The period ...	510	0.0	....	7,059	....	....

SAN GABRIEL RIVER AND CANALS AT AZUSA, CAL.

Owing to the numerous diversions it has been difficult to obtain accurate discharge measurements at Azusa; but during 1898 the San Gabriel Electric Company completed its system, and measurements are now obtained with greater ease, and hence with greater accuracy. The headworks of this company are located about 6 miles above the mouth of the canyon; the water is carried along the left side by a series of tunnels and conduits, and a head of 400 feet is obtained where the electric power is generated. Weirs are placed on the conduit of the electric company, and the water is measured at this point. The capacity of the conduit is 90 second-feet.

The cable and gage are located about 1 mile from Azusa. During

the season of low water for a period of from six to eight months
canals above the station divert the entire flow, and there is no run
water at the station. The gage is a vertical 4 by 4 inch timber.
charge measurements are made by means of a cable. The channe
straight for 150 feet above and for several feet below the cable,
has a width of 280 feet at high water. At low stages there are
channels having different elevations, and accurate measurements
difficult to obtain. The bed of the stream is composed of cobblest
and bowlders, and the current is swift.

The total flow of the river is obtained by adding the daily disch
for the river to the figures for the corresponding dates for the can
The observer is H. F. Parkinson.

The observations at this station during 1903 have been made un
the direction of J. B. Lippincott, supervising engineer.

Discharge measurements of San Gabriel River and canals at Azusa, Cal., in 1903.

Date.	Hydrographer.	Gage height.	Discharge.	Remarks.
		Feet.	*Second-feet.*	
January 29	C. A. Miller...........	3. 20	467	River.
			64	Canal.
			531	Total.
February 5	W. B. Clapp	1. 70	56	River.
			70	Canal.
			126	Total.
March 23........	......do...............	1. 05	11	River.
			72	Canal.
			83	Total.
March 25........	......do...............	3. 50	570	River.
			72	Canal.
			642	Total.
April 1..........	......do...............	7. 95	8,432	River.
			72	Canal.
			8,504	Total.
May 23..........	......do...............	2. 20	97	River.
			70	Canal.
			167	Total.
June 10	......do...............	1. 70	38	River.
			70	Canal.
			108	Total.

Discharge measurements of San Gabriel River and canals at Azusa, Cal., in 1903—Con.

Date.	Hydrographer.	Gage height.	Discharge.	Remarks.
		Feet.	*Second-feet.*	
ʃune 13	W. B. Clapp	1. 50	20	River.
			8	Irrigation canal.
			73	Power canal.
			101	Total.
September 14 ...	......do.....................	....	27. 5	Flow over weir San Gabriel Power Co. canal equal total river.

Mean daily gage height, in feet, of San Gabriel River at Azusa, Cal., for 1903.

Day.	Jan.	Feb.	Mar.	Apr.	May.	June.
....	0. 00	2. 25	0. 00	7. 10	2. 90	1. 90
....	. 00	2. 20	. 00	4. 95	2. 85	1. 85
....	. 00	1. 80	. 00	4. 50	2. 85	1. 85
....	. 00	1. 90	1. 00	4. 00	2. 80	1. 80
....	. 00	1. 70	3. 00	3. 90	2. 80	1. 75
....	. 00	1. 65	2. 90	3. 60	2. 80	1. 70
7...	. 00	1. 60	2. 80	3. 30	2. 80	1. 70
8...	. 00	1. 65	2. 90	3. 20	2. 70	1. 70
9...	. 00	1. 45	2. 80	3. 10	2. 60	1. 70
0...	. 00	1. 40	2. 60	2. 90	2. 60	1. 70
1...	. 00	1. 45	2. 40	2. 80	2. 50	1. 60
2...	. 00	1. 45	2. 30	2. 90	2. 50	1. 50
3...	. 00	1. 55	2. 20	2. 70	2. 40	1. 50
4...	. 00	1. 45	2. 10	2. 65	2. 40	1. 48
5...	. 00	1. 45	2. 00	2. 70	2. 40	1. 45
6...	. 00	1. 30	1. 90	3. 00	2. 40	1. 40
7...	. 00	1. 20	1. 70	4. 25	2. 30	1. 38
8...	. 00	1. 10	1. 50	3. 50	2. 20	1. 35
9...	. 00	1. 00	1. 40	3. 50	2. 20	1. 30
0...	. 00	. 90	1. 30	3. 60	2. 20	1. 25
1...	. 00	. 60	1. 20	3. 50	2. 20	1. 22
2...	. 00	. 00	1. 10	3. 50	2. 20	1. 22
3...	. 00	. 00	1. 05	3. 50	2. 20	1. 09
4...	. 00	. 00	2. 30	3. 40	2. 25	1. 00
5...	. 00	. 00	3. 50	3. 30	2. 10	1. 00
6...	. 00	. 00	2. 35	3. 30	2. 10	. 00
7...	. 00	. 00	2. 40	3. 25	2. 10	. 00
8...	5. 65	. 00	2. 40	3. 00	2. 00	. 00
9...	3. 20	....	2. 40	2. 90	1. 90	. 00
10...	2. 30	....	2. 80	2. 90	1. 90	a . 00
51...	2. 30	....	4. 45	....	1. 90	....

a Dry remainder of year at rod (see canals).

Daily discharge, in second-feet, of San Gabriel canals at Azusa, Cal., for 1903.

Day.	Jan.	Feb.	Mar.	Apr.	May.	June.	July.	Aug.	Sept.	Oct.	Nov.	Dec.
1	30.5	68.0	56.0	72.0	70.0	70.0	58.5	34.8	23.5	31.5	21.5	21.5
2	30.0	68.0	57.0	22.0	70.0	70.0	57.0	34.6	23.0	31.0	22.5	24.0
3	29.0	68.0	57.0	20.0	70.0	70.0	56.0	34.5	23.0	30.0	22.5	24.0
4	28.5	70.0	70.0	20.0	70.0	70.0	53.0	32.5	22.0	29.0	22.5	24.0
5	28.0	70.0	72.0	52.0	70.0	68.0	52.0	31.5	21.0	26.5	22.5	24.5
6	27.0	70.0	72.0	54.0	70.0	70.0	50.0	30.7	24.0	26.5	23.0	24.0
7	27.5	70.0	72.0	54.0	70.0	70.0	50.0	30.7	27.0	26.0	23.0	24.5
8	26.0	70.0	72.0	62.0	70.0	68.0	49.0	30.7	23.0	26.0	24.5	24.5
9	25.0	70.0	72.0	62.0	70.0	68.0	47.0	30.2	23.0	26.0	24.0	24.5
10	25.0	70.0	72.0	62.0	70.0	70.0	46.0	30.0	22.5	27.8	24.0	24.5
11	24.5	70.0	72.0	62.0	70.0	84.0	45.5	30.0	22.5	24.8	24.0	24.7
12	24.5	70.0	72.0	62.0	70.0	84.0	43.0	30.0	24.7	22.8	24.0	24.5
13	24.0	70.0	72.0	62.0	70.0	81.0	43.0	30.0	26.0	22.5	25.0	25.0
14	23.5	70.0	72.0	62.0	70.0	84.0	43.0	30.0	27.5	22.5	25.0	25.0
15	23.0	70.0	72.0	62.0	70.0	84.0	43.0	29.5	25.5	22.5	25.5	25.0
16	22.5	74.0	72.0	62.0	70.0	83.0	42.0	28.0	23.5	22.0	25.5	25.0
17	22.0	74.0	72.0	62.0	70.0	82.0	42.0	26.7	22.5	21.6	25.5	24.5
18	22.0	74.0	72.0	62.0	70.0	81.0	42.0	25.8	23.0	20.5	24.5	24.5
19	22.0	74.0	72.0	16.0	70.0	80.0	42.0	26.3	23.0	21.5	24.5	24.5
20	22.0	74.0	72.0	18.0	70.0	79.5	41.0	26.3	23.0	21.5	24.5	24.6
21	22.0	74.0	72.0	40.0	70.0	79.0	40.0	25.8	23.0	21.0	25.0	24.8
22	22.0	72.0	72.0	52.0	70.0	79.0	38.0	25.3	22.5	21.0	24.5	24.8
23	22.0	70.0	72.0	62.0	70.0	81.0	38.0	25.8	21.7	21.0	22.4	24.3
24	22.5	68.0	72.0	62.0	50.0	80.0	37.5	25.4	21.7	21.0	24.0	24.5
25	25.0	65.0	72.0	62.0	70.0	73.0	37.0	26.9	24.0	21.0	24.0	24.5
26	23.0	63.0	72.0	00.0	60.0	66.0	36.5	27.0	24.5	21.0	23.3	24.5
27	32.0	61.0	72.0	68.0	60.0	64.0	36.0	27.0	29.0	21.0	23.3	24.0
28	64.0	61.0	72.0	68.0	68.0	63.0	36.0	26.0	34.5	21.0	23.3	24.0
29	64.0	......	72.0	68.0	70.0	63.0	35.5	25.0	31.5	21.0	23.3	24.0
30	68.0	......	72.0	68.0	70.0	62.0	35.3	25.0	31.5	21.0	21.5	24.0
31	68.0	......	72.0	......	70.0	......	34.8	24.5	......	21.0	......	24.0

Rating table for San Gabriel River at Azusa, Cal., from January 1 to March 30, 1903.[a]

Gage height.	Discharge.	Gage height.	Discharge.	Gage height.	Discharge.	Gage height.	Discharge.
Feet.	*Second-feet.*	*Feet.*	*Second-feet.*	*Feet.*	*Second-feet.*	*Feet.*	*Second-feet.*
0.6	1	1.6	45	2.6	211	4.2	1,110
.7	2	1.7	54	2.7	240	4.4	1,295
.8	4	1.8	65	2.8	273	4.6	1,500
.9	7	1.9	78	2.9	310	4.8	1,728
1.0	10	2.0	92	3.0	350	5.0	1,980
1.1	13	2.1	107	3.2	438	5.2	2,254
1.2	18	2.2	125	3.4	535	5.4	2,549
1.3	24	2.3	144	3.6	652	5.6	2,855
1.4	30	2.4	164	3.8	792		
1.5	37	2.5	186	4.0	945		

a Two rating tables necessary on account of change in channel March 31.

Rating table for San Gabriel River at Azusa, Cal., from March 31 to December 31, 1903.

Gage height.	Discharge.	Gage height.	Discharge.	Gage height.	Discharge.	Gage height.	Discharge.
Feet.	*Second-feet.*	*Feet.*	*Second-feet.*	*Feet.*	*Second-feet.*	*Feet.*	*Second-feet.*
0.8	0	1.8	44	3.6	652	5.6	2,855
.9	2	1.9	55	3.8	792	5.8	3,180
1.0	4	2.0	68	4.0	945	6.0	3,525
1.1	6	2.2	98	4.2	1,110	6.2	3,890
1.2	9	2.4	139	4.4	1,295	6.4	4,275
1.3	12	2.6	190	4.6	1,500	6.6	4,683
1.4	16	2.8	250	4.8	1,728	6.8	5,120
1.5	21	3.0	323	5.0	1,980	7.0	5,580
1.6	27	3.2	417	5.2	2,254	7.5	6,900
1.7	35	3.4	528	5.4	2,549	8.0	8,600

Estimated monthly discharge of San Gabriel River and canals at Azusa, Cal., for 1903.[a]

[Drainage area, 222 square miles.]

Month.	Discharge in second-feet.			Total in acre-feet.	Run-off.	
	Maximum.	Minimum.	Mean.		Second-feet per square mile.	Depth in inches.
anuary	2,999	22	148	9,100	0.67	0.77
ebruary	203	61	102	5,665	.46	.48
arch..............	1,417	57	257	15,802	1.16	1.34
pril..............	5,892	267	792	47,127	3.57	3.98
ay	355	125	217	13,343	.98	1.13
une	125	62	95	5,653	.43	.48
ıly	58	35	43	2,644	.19	.22
ıgust	35	24	29	1,783	.13	.15
ptember..........	34	21	25	1,488	.11	.12
tober	32	20	24	1,476	.11	.13
ıvember..........	26	22	24	1,428	.11	.12
cember	26	22	24	1,476	.11	.13
The year	5,892	20	148	106,985	.67	9.05

a Includes water in canals.

SAN LUIS REY RIVER NEAR PALA, CAL.

This station was established October 9, 1903, by W. B. Clapp. It
located at Sickler's mill, 4 miles above Pala, Cal. It is reached by
iving from Fallbrook or Escondido, stations on the Southern Cali-
rnia Railway, 18 and 25 miles distant, respectively. The gage is an

inclined 2 by 4 inch rod, graduated to feet and tenths, spiked to tree stumps and stakes at the left bank of the river. It is read once each day by M. M. Sickler. Discharge measurements are usually made by wading. During high water they are made from a car suspended from a ¼-inch galvanized iron wire cable stretched across the river at the gage. The initial point for soundings is the base of the oak tree to which the left end of the cable is fastened. The channel is straight for about 800 feet above and for 2,000 feet below the station. The rise and fall in the channel above and below the cable is 0.60 feet in 100 feet. The current is swift. The right bank rises abruptly about 15 feet beyond the oak tree to which the cable is fastened and is not liable to overflow. The left bank is low, but is not liable to overflow. It was once a portion of the river channel, but is now well above high-water marks. The bed of the stream is rocky in portions of flood channel, but the low-water channel is clear of rocks. There is a considerable growth of small timber in the channel, but this has been cleared the entire width of the cross section for a distance of 100 feet above and 50 feet below the station. This timber growth is not permanent, being washed out by floods every few years. The bench mark is a United States standard bronze-capped iron post set flush with the ground on the right bank of the river and the north side of the wagon road, and about 50 feet west from the line of the cable prolonged. Its elevation is 557 feet above mean sea level, and 26.98 feet above the zero of the gage.

Discharge measurements of San Luis Rey River near Pala, Cal., in 1903.

Date.	Hydrographer.	Gage height.	Discharge.
		Feet.	Second-feet.
August 16	W. B. Clapp	....	1.1
October 9	.....do	ª1.72	ª1.1

ª No rating table constructed, as the rod reads practically constant for balance of the year and discharge is 1.1 second-feet. No rain.

Mean daily gage height, in feet, of San Luis Rey River near Pala, Cal., for 1903.

Date.	Oct.	Nov.	Dec.	Date.	Oct.	Nov.	Dec.	Date.	Oct.	Nov.	Dec.
1............	......	1.73	1.72	12............	1.72	1.73	1.72	23............	1.72	1.73	1.73
2............	......	1.73	1.72	13............	1.71	1.73	1.72	24............	1.72	1.74	1.72
3............	......	1.72	1.71	14............	1.70	1.73	1.72	25............	1.72	1.74	1.73
4............	......	1.72	1.71	15............	1.69	1.74	1.73	26............	1.73	1.74	1.73
5............	......	1.72	1.71	16............	1.69	1.73	1.73	27............	1.72	1.73	1.73
6............	......	1.73	1.71	17............	1.70	1.73	1.73	28............	1.73	1.74	1.73
7............	......	1.73	1.71	18............	1.70	1.73	1.73	29............	1.73	1.73	1.73
8............	......	1.73	1.71	19............	1.70	1.74	1.73	30............	1.72	1.73	1.73
9............	1.72	1.73	1.72	20............	1.71	1.74	1.72	31............	1.72	......	1.73
10............	1.73	1.73	1.72	21............	1.71	1.74	1.73				
11............	1.72	1.73	1.72	22............	1.72	1.73	1.73				

SANTA ANA RIVER BELOW WARMSPRINGS, CAL.

This station was established in June, 1896. It is located 5 miles ortheast of Mentone, Cal., three-fourths of a mile below the head orks of the Mentone Power Company's canal, and opposite the warm ɔrings in the canyon:

The Edison Electric Company diverts the greater portion of the ·ater of Santa Ana River above the gaging station, but also returns ll of it above the station. They, however, allow only limited por- .ons of the water to pass out of their conduits during certain hours f the day, holding back the water for the purpose of obtaining addi- ional power when the greatest demand exists.

The Mentone Power Company's canal, formerly called the Santa ₁na canal, diverts water above the station, all of which is returned elow the point of measurement. During the low-water season the ntire flow of the river is diverted by the canals. The gage is an nclined 4 by 6 inch timber fastened to a large bowlder on the left ank. The channel was deepened by a flood March 31, 1903, and the age was accordingly lowered to reach low-water stages June 30, 1903. ᵀhe gage is read once each day by A. Laird. Discharge measurements re made by means of a cable and car, 100 feet below the gage. The iitial point for soundings is the bench-mark spike set in the north ide of a cottonwood tree on the left bank, 30 feet west of the tree to ·hich the cable is fastened. The channel is straight for 100 feet above nd below the station, and has a width of 22 feet at low and 125 feet t high stages. The current is swift at all stages. At flood stages the elocity is so high that measurements can be made only by means of ɔats. The right bank is low and is liable to overflow at flood stages ɔr about 100 feet. The left bank is low, but is not liable to overflow. ᵢoth banks are overgrown with alders. The bed of the stream is com- osed of firm sand and small bowlders; it is subject to considerable hange during flood stages. The bench mark, which is also used as ₁e initial point for soundings, is a spike in the north side of the ottonwood tree on the left bank, 30 feet west of the tree to which he cable is fastened. Its elevation is 7.29 feet above the zero of the ·age.

The observations of this station during 1903 have been made under he direction of J. B. Lippincott, supervising engineer.

Discharge measurements of Santa Ana River below Warmsprings, Cal., in 1903.

Date.	Hydrographer.	Gage height.	Discharge.	Remarks.
		Feet.	*Second-feet.*	
February 6	W. B. Clapp	1.30	37	River.
			3	Mentone Power Co. canal.
			40	Total river.
April 1	......do	7.00	4,906	Do.
April 2	......do	3.30	1,068	Do.
April 3.	......do	2.50	565	Do.
April 24	W. B. Clapp and J. C. Clausen.	1.10	132	Do.
May 16	W. B. Clapp	.85	95	Do.
June 30	......do	.20	56	Do.
August 31	......do	Weir.	41	Mentone Power Co. canal, total river.
November 24	......do	Weir.	28	Do.

Rating table for Santa Ana River below Warmsprings, Cal., from January 1, 1903, to March 30, 1903.[a]

Gage height.	Discharge.	Gage height.	Discharge.	Gage height.	Discharge.	Gage height,	Discharge.
Feet.	*Second-feet.*	*Feet.*	*Second-feet.*	*Feet.*	*Second-feet.*	*Feet.*	*Second-feet.*
0.9	7	1.5	63	2.1	208	2.7	440
1.0	12	1.6	80	2.2	240	2.8	485
1.1	18	1.7	102	2.3	277	2.9	540
1.2	26	1.8	125	2.4	315	3.0	600
1.3	36	1.9	150	2.5	355		
1.4	48	2.0	178	2.6	395		

a Two rating tables necessary on account of change in channel March 31.

ing table for Santa Ana River below Warmsprings, Cal., from March 31 to December 31,
1903.

Gage height.	Discharge.	Gage height.	Discharge.	Gage height.	Discharge.	Gage height.	Discharge.
Feet.	*Second-feet.*	*Feet.*	*Second-feet.*	*Feet.*	*Second-feet.*	*Feet.*	*Second-feet.*
0.0	48	1.0	112	3.0	856	5.0	2,550
.1	52	1.2	142	3.2	989	5.2	2,752
.2	56	1.4	179	3.4	1,136	5.4	2,961
.3	60	1.6	224	3.6	1,292	5.6	3,177
.4	65	1.8	278	3.8	1,455	5.8	3,399
.5	70	2.0	345	4.0	1,623	6.0	3,629
.6	76	2.2	425	4.2	1,799	6.2	3,868
.7	82	2.4	516	4.4	1,980	6.4	4,119
.8	90	2.6	618	4.6	2,165	6.6	4,378
.9	100	2.8	733	4.8	2,354	7.0	4,908

Estimated monthly discharge of Santa Ana River below Warmsprings, Cal., for 1903.[a]

[Drainage area, 182 square miles.]

Month.	Discharge in second-feet.			Total in acre-feet.	Run-off.	
	Maximum.	Minimum.	Mean.		Second-feet per square mile.	Depth in inches.
nuary	277	19	33	2,029	0.18	0.21
bruary	105	36	48	2,666	.26	.27
arch	2,451	38	148	9,100	.81	.93
pril	4,908	106	352	20,945	1.93	2.15
ay	113	79	92	5,657	.51	.59
ne	79	47	64	3,808	.35	.39
ly	57	44	51	3,136	.28	.32
ugust	59	50	55	3,382	.30	.35
ptember	64	45	52	3,094	.29	.32
tober	57	30	47	2,890	.26	.30
ovember	52	27	30	1,785	.16	.18
ecember	30	25	27	1,660	.15	.17
The year	4,908	19	83	60,152	.46	6.18

[a] Estimated monthly discharge includes Mentone Power Company's canal.

SANTA MARIA RIVER NEAR SANTA MARIA, CALIFORNIA.

This station was established October 22, 1903, by H. D. Clapp. It
located near the ranch house on Dutard's ranch, 21 miles above
anta Maria, Cal., a station on the Pacific Coast Railway. It is

reached by driving from Santa Maria. The gage is an inclined 2 by 6 inch timber graduated to feet and tenths and fastened to a rock ledge at the right bank. It is read once each day by Joseph A. Thompson. At low and medium water discharge measurements are made with meter by wading. During high water velocities are measured by means of floats. For this purpose two wires are stretched across the stream 254 feet apart. The measuring stations are marked on each wire. The initial points for soundings are blazes on the poplar trees on the left bank, to which the wires are attached. The channel is slightly curved for 300 feet above, and curved for 1,000 feet below the station. The water is swift at medium and flood stages. The rise in the channel above the upper wire is 0.40 foot in 100 feet, and below the upper wire it is 0.57 foot in 100 feet. The right bank is high and rocky, and not liable to overflow. The left bank is low, covered with scattering poplar trees, but not liable to overflow. The bed of the stream is composed of sand and gravel. A portion of the bed is covered with a light growth of low brush. The channel is not liable to much change. The bench mark is a spike driven near the ground, into the south side of the poplar tree to which the upper wire is fastened on the left bank of the stream. Its elevation is 9.65 feet above the zero of the gage.

Discharge measurements of Santa Maria River near Santa Maria, Cal., in 1903.

Date.	Hydrographer.	Gage height.	Discharge.
		Feet.	*Second-feet.*
August 26	W. B. Clapp	(a)	1.0
October 22	....do	1.97	.8

a No gage rod in at this date.

Mean daily gage height, in feet, of Santa Maria River near Santa Maria, Cal., for 1903.

Day.	Oct.	Nov.	Dec.	Day.	Oct.	Nov.	Dec.	Day.	Oct.	Nov.	Dec.
1	......	1.97	1.97	12	......	1.97	2.15	23	1.97	1.97	2.22
2	......	1.97	1.97	13	......	1.97	2.16	24	1.97	1.97	2.23
3	......	1.97	1.97	14	......	1.97	2.18	25	1.97	1.97	2.24
4	......	1.97	1.98	15	......	1.97	2.19	26	1.97	1.97	2.26
5	......	1.97	2.00	16	......	1.97	2.20	27	1.97	1.97	2.27
6	......	1.97	2.05	17	......	1.97	2.20	28	1.97	1.97	2.27
7	......	1.97	2.07	18	......	1.97	2.21	29	1.97	1.97	2.29
8	......	1.97	2.10	19	......	1.97	2.21	30	1.97	1.97	2.28
9	......	1.97	2.10	20	......	1.97	2.21	31	1.97	......	2.29
10	......	1.97	2.10	21	......	1.97	2.22				
11	......	1.97	2.13	22	1.97	1.97	2.22				

MOHAVE RIVER AT VICTORVILLE, CAL.

At Victorville, a station on the Atchison, Topeka and Santa Fe Railway, the river passes through a narrow gorge, locally known as The Narrows. This place has been under investigation as a possible dam site, and soundings for the depth of bed rock were made by the United States Geological Survey during the season of 1899. The greatest depth of bed rock was found to be 54 feet. The diamond drill showed the rock to be a fine granite. A more detailed account of this exploration will be found in the Twenty-first Annual Report, part 4. Above The Narrows the valley broadens into a large reservoir site, but as no surveys of it have been made the capacity is unknown. In order to determine the amount of water available for storage for this reservoir a gaging station was established February 27, 1899.

During 1902 no flood passed the gaging station. The channel is in sand, which is constantly shifting. The rod readings are of little value during low stages. Between January 1, 1902, and May 3, 1902, the discharge varied from 47 to 67 cubic feet per second, though the gage reading was 0.9 for the entire time. At the latter date the rod readings were discontinued.

The mean estimated discharge for each month was obtained by averaging the discharge as obtained by meter measurements made during the month.

The observations at this station during 1903 have been made under the direction of J. B. Lippincott, supervising engineer.

Discharge measurements of Mohave River at Victorville, Cal., in 1903.

Date.	Hydrographer.	Discharge.	Date.	Hydrographer.	Discharge.
		Sec.-feet.			*Sec.-feet.*
February 6	P. H. Leahy ..	59	July 11	P. H. Leahy ..	
February 14	.....do.......	57	July 18	......do.......	
February 21	.....do.......	67	July 31	......do.......	
February 28	.....do.......	74	August 8........	......do.......	
March 7........	.....do.......	75	August 15.......	......do.......	
March 14.......	.....do.......	69	August 24.......	......do.......	
March 21.......	.....do.......	74	September 14 ...	......do.......	
March 31.......	.....do.......	a13,413	September 21 ...	......do.......	
April 1	.....do.......	a3,760	September 28 ...	......do.......	
April 4	.....do.......	a1,135	October 8.......	......do.......	
April 15	.....do.......	a262	October 19......	......do.......	
April 18	.....do.......	a458	October 26......	......do.......	
April 25	.....do.......	a418	October 31......	......do.......	
May 9..........	.....do.......	54	November 8	......do.......	
May 16.........	.....do.......	50	November 14 ...	......do.......	57
May 23.........	.....do.......	47	November 21 ...	......do.......	54
May 30.........	.....do.......	45	November 28 ...	......do.......	50
June 6	.....do.......	45	December 5.....	......do.......	55
June 13	.....do.......	38	December 12....	......do.......	60
June 20	.....do.......	35	December 21....	......do.......	55
June 27	.....do.......	34	December 29....	......do.......	50
July 3.........	.....do.......	38			

a Float measurement.

Estimated monthly discharge of Mohave River at Victorville, Cal., for 1903. [a]

[Drainage area, 400 square miles.]

Month.	Discharge in second-feet.			Total in acre-feet.	Run-off.	
	Maximum.	Minimum.	Mean.		Second-feet per square mile.	Depth in inches.
uary	58	55	57	3,505	0.14	0.16
»ruary	74	57	63	3,499	.16	.17
rch...............	13,413	69	503	30,928	1.28	1.45
ril...............	3,760	262	765	45,521	1.91	2.13
y................	262	45	80	4,919	.20	.23
ιe	45	34	39	2,321	.10	.11
y................	40	32	37	2,275	.09	.10
gust	52	34	39	2,398	.10	.12
tember..........	53	36	41	2,440	.10	.11
ober	55	49	52	3,197	.13	.15
'ember	59	51	55	3,273	.14	.16
'ember	60	55	58	3,566	.15	.17
The year	13,413	32	149	107,842	.37	5.06

[a] Daily discharges obtained by interpolation between discharge measurements.

SEEPAGE MEASUREMENTS IN SOUTHERN CALIFORNIA.

3ince the United States Geological Survey has taken up the work
stream gaging in southern California the seasonal rainfall has been
:eedingly light, exceeding 11 inches of rain in any one season in
ly three instances during the last ten seasons previous to 1902-3,
: exceptions being the winters of 1894-95, with a precipitation of
10 inches; 1896-97, with 16.83 inches, and 1900-1901, with 16.38
:hes. These records are for Los Angeles, Cal., the mean rainfall
ing 16 inches for the past thirty years. These continued dry years,
th the great development of water from all streams in southern
lifornia, either by diversions on their upper reaches, by tunneling
d the building of submerged dams in the stream channels to save
e underflow, and the development from wells on the lower levels of
e stream washes, have lowered the water plane to such an extent that
en with a greatly increased seasonal rainfall, such as occurred during
e winter of 1902-3, and a correspondingly large stream discharge in
e mountain canyons, these flood waters extended only a short dis-
ice beyond the mouth of the canyons before being taken up in the
avel and sandy washes of the streams.

During the past winter, 1902-3, the rainfall at Los Angeles, Cal.,
s 19.32 inches. Until the latter portion of March this precipitation
ne in light rains, with a cool temperature, and was either taken

up by the dry soil or remained in the shape of snow on the higher elevations, producing no great amount of run-off. But on March 31 a warm and exceedingly heavy rain fell in southern California and the discharge of all streams at the mouth of the canyons where they leave the rough mountain region was greater than at any time since the winter of 1892–93. In many instances this flood discharge did not extend far into the valleys, much depending on the nature of the stream channels as they receded from the mountains. In some cases where these channels were flat and sandy the water soon disappeared entirely, but where the streams were confined with heavier grades and the channels composed of a cemented gravel the flood discharge extended a greater distance.

The rain of March 31 was followed by a lighter storm with a cool temperature on April 16. This rain produced no great flood discharge (much of it remaining as snow on the higher elevations), but was a great help in maintaining the flow during the spring and summer. As a result all streams where continuous records are kept show a discharge up to the present date, August 1, 1903, from 100 to 500 per cent greater than on corresponding dates in 1902. This resulted in a heavy spring irrigation throughout southern California, many irrigators taking advantage of this condition to thoroughly saturate the soil at a time when water was plenty and cheap, thereby saving much expense and contention which always accompanies a later irrigation when water is scarce, expensive, and in greater demand than can be supplied.

This heavy spring irrigation was practiced more extensively in the San Bernardino Valley, especially in the vicinity of Redlands, than in any other portion of southern California.

Immediately following the rain of April 16 a series of measurements were made by W. B. Clapp under the direction of the United States Geological Survey to determine the amount of water being absorbed in the sand and gravel washes of the larger tributary streams of the three principal river basins of southern California, viz, the Santa Ana, San Gabriel, and Los Angeles rivers. Measurements were taken at the mouths of the canyons where the streams leave the mountains, at all canal diversions, and at such intervals along the streams as time and the available force detailed for this work would allow, the location of the point where the stream entirely disappeared or left the valley being noted in all cases. This data has been compiled and shown by the following tables and by diagrams.[a]

The diagrams were not entirely satisfactory, as they show an even rate of loss from one point of measurement to another, in many cases long distances apart; thus, they only indicate the discharge at the

[a] Copies of these diagrams are on file at the United States Geological Survey Office at Washington, D. C.

mouth of canyon and at the point of no flow. If a larger force had been available and more time given to this work, numerous measurements could have been taken along the stream, which would have made it possible to have constructed a curve showing the loss in a much more comprehensive manner. These measurements, however, show the amount of water taken up in the sand and gravel washes of these river basins on the dates shown, and the tables which have been prepared, and which accompany this report, give this amount in cubic feet and acre-feet for a period of twenty-four hours.

Water discharged from tributary streams and sinking in the Los Angeles River basin above Burbank, Cal.

[Discharge for 24 hours.]

Stream.	Cubic feet.			Acre-feet.		
	Diversions.	Waste.	Total.	Diversions.	Waste.	Total.
April 18, 1903.						
Big Tujunga		13,564,800	13,564,800		311	311
Little Tujunga		2,332,800	2,332,800		54	54
Pacoima		8,467,200	8,467,200		194	194
Total		24,364,800	24,364,800		559	559
May 5, 1903.						
Big Tujunga		3,196,800	3,196,800		73	
Little Tujunga		345,600	345,600		8	
Pacoima		2,073,600	2,073,600		48	
Total		5,616,000	5,616,000		129	
June 4, 1903.						
Big Tujunga	259,200	803,520	1,062,720	6	18	24
Little Tujunga		60,480	60,480		1	1
Pacoima	259,200	432,000	691,200	6	10	16
Total	518,400	1,296,000	1,814,400	12	29	41

Water discharged from tributary streams and sinking in the San Gabriel River basin above El Monte, Cal.

[Discharge for twenty-four hours.]

Stream.	Cubic feet.			Acre-feet.			Passing El Monte.	
	Diversion.	Waste.	Total.	Diversion.	Waste.	Total.	Cubic feet.	Acre-feet.
April 26, 1903.								
San Gabriel...	3, 456, 000	24, 624, 000	28, 080, 000	79	565	644	19, 785, 600	454
San Dimas....	....	1, 641, 600	1, 641, 600	......	38	38	....	.....
Dalton	....	777, 600	777, 600	......	18	18	....	.....
Santa Anita ..	....	3, 456, 000	3, 456, 000	......	79	79	....	.....
Eaton Canyon.	....	2, 505, 600	2, 505, 600	......	57	57	....	.....
Total...	3, 456, 000	33, 004, 800	36, 460, 800	79	757	836	19, 785, 600	454
May 23, 1903.								
San Gabriel...	6, 048, 000	8, 380, 800	14, 428, 800	139	192	331	....	.....
San Dimas....	172, 800	43, 200	216, 000	4	4	5	....	.....
Dalton	112, 320	120, 960	233, 280	2	3	5	....	.....
Santa Anita ..	259, 200	864, 000	1, 123, 200	6	20	26	....	.....
Eaton Canyon.	259, 200	259, 200	518, 400	6	6	12	....	.....
Total...	6, 851, 520	9, 668, 160	16, 519, 680	154	225	379	....	.....

ter discharged from tributary streams and sinking in the Santa Ana River basin above Colton, Cal.

Discharge for twenty-four hours.

Stream.	Cubic feet.			Acre-feet.		
	Diversions.	Waste.	Total.	Diversions.	Waste.	Total.
April 24, 1903.						
nta Ana...............	1,382,400	10,022,400	11,404,800	32	230	262
ll Creek	432,000	2,937,600	3,369,600	10	67	77
unge Creek............	....	1,987,200	1,987,200	......	46	46
ty Creek	....	1,900,800	1,900,800	......	44	44
ist Twin Creek.......	....	864,000	864,000	......	20	20
est Twin Creek......	....	743,040	743,040	......	17	17
rtle Creek	604,800	4,838,400	5,443,200	14	111	125
Total	2,419,200	23,293,440	25,712,640	56	535	591
May 16, 1903.						
nta Ana...............	4,060,800	4,147,200	8,208,000	93	95	188
ill Creek	4,233,600	1,641,600	5,875,200	97	38	135
unge Creek..........	518,400	259,200	777,600	12	6	18
ty Creek	604,800	345,600	950,400	14	8	22
ist Twin Creek.......	172,800	259,200	432,000	4	6	10
est Twin Creek	172,800	172,800	345,600	4	4	8
rtle Creek	1,296,000	1,209,600	2,505,600	30	28	58
Total	11,059,200	8,035,200	19,094,400	254	185	439

ISCELLANEOUS DISCHARGE MEASUREMENTS IN SOUTHERN CALIFORNIA IN 1903.

The following miscellaneous measurements were made in southern alifornia in 1903:

Miscellaneous discharge measurements in southern California in 1903.

SANTA MARIA RIVER DRAINAGE.

Date.	Hydrographer.	Stream.	Location.	Discharge (sec.-ft.).
ug. 25	W. B. Clapp ...	Sisquoc River........	Sisquoc ranch........	1.1
ct. 23	......do	......do	.....do	0.0

VENTURA RIVER DRAINAGE.

ept. 19	W. B. Clapp ...	Ventura River	Ventura Light and Power Co. Upper diversion city supply.	7.0
		.....do	Below upper diversion.	3.7
		Total river.....	....	10.7
19	.....do	Power ditch	At old mill...........	3.3

Miscellaneous discharge measurements in southern California in 1903—Continued.

SANTA CLARA RIVER DRAINAGE.

Date.	Hydrographer.	Stream.	Location.	Discharge (sec.-ft.)
June 16	W. B. Clapp ...	Newhall ditch	Southern Pacific Rwy bridge, 3 miles below Saugus.	1.7
Sept. 16	.....do	.....do	.....do	1.8
Apr. 30	.....do	Santa Clara River	Road crossing Newhall ranch.	8.0
June 16	.....do	.....do	.....do	6.2
Sept. 16	.....do	.....do	.....do	5.4
Apr. 30	.....do	San Francisquito Creek	.....do	2.0
June 16	.....do	.....do	.....do	2.1
Sept. 16	.....do .:......	.....do	.....do	2.0
Apr. 30	.....do	Santa Clara River	Opposite head of Camulos ditch.	25.0
June 17	.....do	.....do	.....do	21.1
.		Camulos ditch	At head	7.6
		Total Santa Clara River.	....	28.7
Sept. 16	.....do	Santa Clara River	Camulos ditch	5.1
		Camulos ditch	At head	11.6
		Total Santa Clara River.	....	16.7
May 3	.....do	Santa Clara River	1 mile above Santa Paula.	127.0
June 22	.....do	.....do	.....do	22.0
		Richardson ditch.....	.....do	4.2
		Grees ditch	.....do	11.0
		Farmer's ditch	.....do	7.8
		Total Santa Clara River.	....	45.0
July 16	.....do	Santa Clara River	.....do	23.4
		Richardson ditch.....	.....do	2.3
		Grees ditch	.....do	3.8
		Farmer's ditch	.....do	8.4
		Total Santa Clara River.	....	39.9
Aug. 13	J. B. Lippincott and G. S. Power.	Santa Clara River	1 mile above Santa Paula.	20.8
		Richardson ditch.....	.....do	1.2
		Grees ditch	.....do	11.1
		Farmer's ditch	.....do	8.6
		Total Santa Clara River.	....	41.7

Miscellaneous discharge measurements in southern California in 1903—Continued.

SANTA CLARA RIVER DRAINAGE—Continued.

ite.	Hydrographer.	Stream.	Location.	Discharge (sec. ft.).
. 18	W. B. Clapp....	Santa Clara River	1 mile above Santa Paula.	18.2
		Richardson ditch.....	.....do	1.2
		Grees ditch...........	.....do	7.4
		Farmer's ditch.......	.....do	8.6
		Total Santa Clara River.	....	35.4
15	W. B. Clapp and G. S. Power.	Santa Clara Water and Irrigation Co. canal.	At head.............	30
16	.....do	....do	.....do	33
14	J. B. Lippincott and G. S. Power.	.....do	.....do	31.6
		Santa Clara River	Below head of canal...	3.3
		Total river.....	....	34.9
. 18	W. B. Clapp....	Santa Clara Water and Irrigation Co. canal.	Heading	26.7
		River	Below canal heading..	· 3.9
		Total river.....	....	30.6
1	.....do	Piru Creek...........	Dunton ranch	57
		.....do	Southern Pacific bridge	55
? 17	.....do	.....do	Upper diversion Piru Land and Water Co.	1.8
		.....do	1 mile above Esperanza.	3.9
		Total Piru Creek	....	5.7
. 17	.....do	Upper diversion......	1 mile above Esperanza.	1.2
		Lower diversion......	Southern Pacific bridge	1.0
		Piru Creek...........	.....do	.1
		Total Piru Creek	....	1.1
? 17	.....do	Lower diversion......	Southern Pacific bridge	3.7
		Piru Creek...........	.....do	2.7
		Total Piru Creek	....	6.4
2	.....do	Sespe Creek..........	Southern Pacific Rwy. bridge.	89

Miscellaneous discharge measurements in southern California in 1903—Continued.

SANTA CLARA RIVER DRAINAGE—Continued.

Date.	Hydrographer.	Stream.	Location.	Discharge (sec.-ft.)
Sept. 17	W. B. Clapp...	Sespe Land and Water Co. canal.	½ mile below head....	4.6
		Sespe Creek..........	Below canal heading..	.1
		Total Sespe Creek.	....	4.7
May 3	.....do	Santa Paula Creek....	5 miles above Santa Paula.	25
		.....do	At Santa Paula	19.0
		Santa Paula Water Co. ditch.	At Heading	5.6
June 22	.....do	.,...do	.....do.............	10.1
Sept. 18	.....do	.....do	.....do.............	2.7

SANTA YNEZ RIVER DRAINAGE.

Date.	Hydrographer.	Stream.	Location.	Discharge
Apr. 11	Howard Rankin	Amagusa Creek	At mouth	6.2
15	.....do ...,....	.....do	....'.do.............	4.5
21	.....do	.....do	.....do.............	7.6
23	.....do	.....do	.....do.............	6.9
28	.....do	.....do	.....do.............	6.2
10	.....do	Blue Canyon Creek...	.....do.............	11.1
22	.....do	.....do	.....do.............	11.7
29	.....do	.....do	.....do.............	6.3
10	.....do	Trail Creek	.....do.............	1.4
22	.....do	.....do	.....do.............	2.0
11	.....do	Ruiz Creek	.....do.............	5.4
15	.....do	.....do	.....do.............	4.3
21	.....do	.....do	.....do.............	7.0
23	.....do	.....do	.....do.............	5.7
28	.....do	.....do	.....do.............	4.5

SAN LUIS REY RIVER DRAINAGE.

Date.	Hydrographer.	Stream.	Location.	Discharge
Aug. 15	W. B. Clapp...	San Luis Rey River ..	Dam site, Warner's ranch reservoir.	2.8
14	.....do	.....do	Above head of Escondido canal.	.5
16	.....do	.....do	Pala, Cal.............	1.1
15	.....do	Sickler's canal	Sickler's ranch	2.1
Oct. 9	.....do	.....do	.....do.............	2.1

Miscellaneous discharge measurements in southern California in 1903—Continued.

LOS ANGELES RIVER DRAINAGE.

ate.	Hydrographer.	Stream.	Location.	Discharge (sec.-ft.).
9	J. C. Clausen ..	Pacoima Creek	Mouth of canyon	47
18	.....do	.....do	.....do.......	98
5	.....do	.....do	.....do.......	24
e 5	W. B. Clapp...	.....do	.....do.......	8.0
t. 26	.....do	.....do	.....do.......	0.0
9	J. C. Clausen ..	.....do	Southern Pacific Rwy.	32
19	.....do	.....do	.....do.......	57
5	.....do	.....do	.....do.......	16.0
e 5	W. B. Clapp...	.....do	...do.......	0.0
18	J. C. Clausen ..	.....do	3¼ miles below Southern Pacific Rwy.	0.0
5	.....do	.....do	1½ miles below Southern Pacific Rwy.	0.0
10	.....do	Little Tejunga Creek .	Mouth of canyon	10.0
17	.....do	.....do	.....do.......	27
5	.....do	.....do	.....do.......	4.0
e 4	W. B. Clapp...	.....do	.....do.......	1.0
t. 26	.....do	.....do	.....do.......	0.0
10	J. C. Clausen ..	Big Tejunga Creek ...	.....do.......	59
17	.....do	.....do	.....do.......	157
5	.....do	.....do	.....do.......	37
e 4	W. B. Clapp...	.....do	.....do.......	12.0
t. 26	.....do	.....do	.....do.......	.1
9	J. C. Clausen ..	.....do	Southern Pacific Rwy.	2.0
17	.....do	.....do	.....do.......	124
5	.....do	.....do	.....do.......	0.0
e 4	W. B. Clapp...	.....do	.....do.......	0.0
17	J. C. Clausen ..	.....do	1½ miles below Southern Pacific Rwy.	0.0
25	J. F. Danforth .	Los Angeles River....	Los Feliz Bridge......	109
8	J. C. Clausen ..	.....do	.....do.......	37
13	O. W. Peterson.	.....do	.....do.......	28
17	.....do	.....do	.....do.......	64
28	C. A. Miller ...	.....do	Aliso Street Bridge ...	66
11	J. C. Clausen ..	.....do	Ninth Street Bridge...	20
18	O. W. Peterson.	.....do	.....do.......	106
18	.....do	.....do	Boyle avenue.........	75
18	.....do	.....do	Opposite Bell station..	73
20	.....do	.....do	Above junction of Rio Hondo.	25
20	.....do	.....do	Below junction of Rio Hondo.	54

Miscellaneous discharge measurements in southern California in 1903—Continued.

LOS ANGELES RIVER DRAINAGE—Continued.

Date.	Hydrographer.	Stream.	Location.	Discharge (sec.-ft.)
Apr. 20	O. W. Peterson.	Los Angeles River....	One-half mile above Compton and Clearwater road.	82
20	.....do	.....do	Opposite Clearwater ..	53
21	.....do	.....do	One-half mile above Cerritos road.	39
21	.....do	.....do	1½ miles below Cerritos road.	77
21	.....do	.....do	Opposite Seabright....	78
13	W. B. Clapp...	Arroyo Seco	At mouth of canyon ..	24
17	.....do	.....do	.....do..............	100
May 21	.....do	.....do	.....do..............	10.0
Apr. 13	.....do	.....do	At Devils Gate	17.0
May 21	.....do	.....do	.....do..............	2.0
21	.....do	.....do	1 mile below Devils Gate.	0.0
Apr. 13	.....do	.....do	At Sheep Corral Spring, Pasadena.	4.0
17	.....do	.....do	.....do..............	80
13	.....do	.....do	Garvanza wagon bridge.	0.0
Jan. 28	C. A. Miller ...	.....do	At Avenue 26, Los Angeles.	274
Mar. 25	J. F. Danforth.	.....do	.....do..............	128
Apr. 17	O. W. Peterson.	.....do	.....do..............	77

SAN GABRIEL RIVER DRAINAGE.

Apr. 20	W. B. Clapp and J. C. Clausen.	Dalton Creek	At mouth of canyon ..	9.0
May 22	W. B. Clapp...	.....do	.....do..............	2.7
Sept. 14	.....do	.....do	.....do..............	0.0
May 22	.....do	.....do	1 mile below mouth of canyon.	0.0
Apr. 20	.....do	.....do	Southern California Rwy.	0.0
20	W. B. Clapp and J. C. Clausen.	San Dimas Creek......	Mouth of canyon	19.0
May 22	W. B. Clapp...	.....do	.....do..............	2.5
Sept. 14	.....do	.....do	.....do..............	0.0
Apr. 20	.....do	.....do	Base Line avenue.....	0.0
18	.....do	Santa Anita Creek....	Mouth of canyon	35

Miscellaneous discharge measurements in southern California in 1903—Continued.

SAN GABRIEL RIVER DRAINAGE—Continued.

Date.	Hydrographer.	Stream.	Location.	Discharge (sec.-ft.).
May 23	W. B. Clapp ...	Santa Anita Creek....	Mouth of canyon	10
		Baldwin diversion heading.	....	3
		Total creek	....	13
Sept. 9	.....do	Baldwin diversion heading.	....	1.4
Apr. 18	.....do	Santa Anita Creek....	Southern Pacific Railway, Monrovia branch.	0
May 23	.....do	.....do	White Oak avenue....	0
Apr. 18	.....do	Little Santa Anita Creek.	Above Santa Anita Creek.	5
18	.....do	Eaton Canyon Creek .	Mouth of canyon	29
May 21	.....do	.....do	.....do..............	3
		Do	Water company's diversion at heading, mouth of canyon.	3
		Total Creek....	....	6
Sept. 9	.....do	Eaton Canyon Creek .	Mouth of canyon	0
Apr. 18	.....do	.....do	At Southern California Railway	0
May 21	.....do	.....do	1 mile above Southern California Railway.	0
Apr. 14	.....do	San Gabriel River....	At upper road crossing, Duarte to Azusa.	264
May 26	.....do	.....do	Below head of Lexington wash.	29
		Do	Diversion below head of Lexington wash.	37
		Total river at same place.	....	66
Apr. 15	.....do	San Gabriel River....	At El Monte Bridge ..	50
15	.....do	.....do	Lexington wash at El Monte.	3
26	O. W. Peterson.	.....do	At El Monte Bridge ..	229
May 26	W. B. Clapp...	.....do	.....do..............	0
Apr. 25	O. W. Peterson.	.....do	At road crossing east of Durfee's ranch.	264
25	.....do	.....do	Whittier road crossing.	361
24	.....do	.....do	Southern California Railway at Rivera.	365
24	.....do	.....do	Southern Pacific Railway at Studebaker.	364

Miscellaneous discharge measurements in southern California in 1903—Continued.

SAN GABRIEL RIVER DRAINAGE—Continued.

Date.	Hydrographer.	Stream.	Location.	Discharge (sec.-ft.)
Apr. 23	O. W. Peterson.	San Gabriel River....	Opposite Alamitos	311
27	.....do	Rio Hondo	Old Mission Bridge ...	36
Oct. 2	W. B. Clapp...	.....do	.....do	29
Apr. 27	O. W. Peterson.	.....do	Whittier road crossing.	17
27	.....do	.....do	Road crossing west of Rivera.	12
27	.....do	.....do	Southern Pacific Railway west of Downey.	12
20	.....do	.....do	At S. P., L. A. & S. L. Railroad, above Workman.	29
25	.....do	San Jose Creek	Whittier and Puente road.	8
Oct. 3	W. B. Clapp...	Sheep Creek ditch....	Whittier flume crossing.	4.6
3	.....do	Rincon ditch	Rincon road crossing..	3.4
3	.....do	Durfee ditch	Road crossing above ranch house.	1.6
2	.....do........	Los Nietos or Santa ditch.	At heading...........	28
2	.....do	Ranchito or Standeferd ditch.	.....do..............	14
2	.....do	Cate ditch	In flume at road crossing.	5.5
Apr. 25	J. B. Lippincott	Arroyo ditch	Southern California Railway, Rivera.	21
Oct. 3	W. B. Clapp...	Whittier ditch	Pumping plant east of El Monte—developed water—discharge varies as needed.	5.1

SANTA ANA RIVER DRAINAGE.

Date.	Hydrographer.	Stream.	Location.	Discharge (sec.-ft.)
Apr. 1	W. B. Clapp...	San Antonio Creek ...	Ontario Power Co.'s power house, mouth of canyon.	854
21	W. B. Clapp and J. C. Clausen.	.....do	At power house, mouth of canyon.	42
		Do	San Antonio Water Co.'s diversion mouth of canyon.	15
		Total creek	....	57

Miscellaneous discharge measurements in southern California in 1903—Continued.

SANTA ANA RIVER DRAINAGE—Continued.

ate.	Hydrographer.	Stream.	Location.	Discharge (sec.-ft.).
y 25	W. B. Clapp...	San Antonio Creek ...	At power house, mouth of canyon.	14
		Do	San Antonio Water Co.'s diversion mouth of canyon.	20
		Total creek	....	34
y 3	.....do	San Antonio Creek ...	At power house, mouth of canyon. Total creek.	19. 5
)t. 13	.....do	.....do	.....do	11
r. 21	W. B. Clapp and J. C. Clausen.	.....do	1 mile below Southern Pacific Rwy.	.0
y 25	W. B. Clapp...	.....do	Southern Pacific Rwy.	.0
r. 21	W. B. Clapp and J. C. Clausen.	Cucamonga Creek....	Mouth of canyon	15
y 25	W. B. Clapp...	.....do	.....do	5. 5
)t. 13	.....do	.:...do	Above head works Hermosa Water Co.	3. 6
r. 21	W. B. Clapp and J. C. Clausen.	.....do	Southern California Rwy.	.0
y 25	W. B. Clapp...	.....do	Base line avenue......	.0
r. 1	.....do	Lytle Creek	At mouth of canyon ..	1,790
22	W. B. Clapp and J. C. Clausen.	.....do	.....do	56
		Rialto canal..........	Weir at head works...	7
		Total creek	....	63
y 19	W. B. Clapp...	Lytle Creek	Mouth of canyon	14. 6
		Rialto canal..........	Weir at head works...	14. 9
		Total creek	....	29. 5
y 2	.....do	Lytle Creek	Weir at head works Rialto canal.	16. 5
)t. 12	.....do	.....do	.....do	14. 7
r. 22	.....do	.....do	Highland avenue	.0
y 19	.....do	.....do	.....do	.0
r. 23	W. B. Clapp and J. C. Clausen.	West Twin Creek	Mouth of canyon	8. 6
y 18	W. B. Clapp...	.....do	.....do	2. 1
		Do	Ditch diversion.......	2. 3
		Total creek	....	4. 4

Miscellaneous discharge measurements in southern California in 1903—Continued.

SANTA ANA RIVER DRAINAGE—Continued.

Date.	Hydrographer.	Stream.	Location.	Discharge (sec.-ft.).
Sept. 11	W. B. Clapp	West Twin Creek	Mouth of canyon. Total creek.	0.3
Apr. 23	do	do	Southern California Rwy.	.0
May 18	do	do	1 mile above Southern California Rwy.	.0
Apr. 23	W. B. Clapp and J. C. Clausen.	East Twin Creek	Mouth of canyon	10
May 17	W. B. Clapp	do	do	3.2
		Do	Ditch diversion	1.6
		Total creek		4.8
Sept. 11	do	East Twin Creek	Ditch diversion. Total creek.	.4
Apr. 23	do	do	1 mile above Southern California Rwy.	.0
May 17	do	do	do	.0
Apr. 23	W. B. Clapp and J. C. Clausen.	City Creek	Mouth of canyon	22
May 17	W. B. Clapp	do	do	4.4
		Do	Canal diversion	6.8
		Total creek		11.2
Sept. 11	do	City Creek	Mouth of canyon. Total creek.	.2
Apr. 23	do	do	Road crossing south Harlem Springs.	.0
May 17	do	do	Base line avenue	.0
Apr. 23	W. B. Clapp and J. C. Clausen.	Plunge Creek	Mouth of canyon	23
May 17	W. B. Clapp	do	do	3
		Do	Diversion mouth of canyon.	6
		Total creek		9
Sept. 11	do	Plunge Creek	Mouth of canyon. Total creek.	.6
Apr. 23	do	do	Orange avenue	.0
May 17	do	do	do	.0
Apr. 1	do	Mill Creek	Head of Crafton zanja.	1,280

Miscellaneous discharge measurements in southern California in 1903—Continued.

SANTA ANA RIVER DRAINAGE—Continued.

ite.	Hydrographer.	Stream.	Location.	Discharge (sec.-ft).
24	W. B. Clapp and J. C. Clausen.	Mill Creek............	Head of Crafton zanja .	34
		Crafton zanja	....do	5
		Total Mill Creek.	....do	39
16	W. B. Clapp...	Mill Creek............	Head of Crafton zanja .	19
		Crafton zanja	....do	49
		Total Mill Creek.	....do	68
9	....do	Mill Creek............	....do	2
		Crafton zanja	....do	44
		Total Mill Creek.	....do	46
1	....do	Crafton zanja	At head; total Mill Creek.	29
1	....do	....do	....do	22
31	....do	Morton Canyon	Mouth of canyon	.2
31	....do	Redlands tunnel	At outlet.............	1.8
31	....do	Green Spot pipe line..	Weir at head.........	6.3
24	....do	Highland canal	....do	6
16	....do	....do	....do	18
31	....do	....do	....do	15.6
24	....do	Redlands canal	Sand-box weir........	9.6
16	....do	....do	....do	29
31	....do	....do	....do	28
5	K. Sanborn....	Castile ditch	1,000 feet below heading.	3
5	....do	Wilbur ditch	At Rogers' pipe trestle crossing.	1.7
5	....do	Newton ditch	West line section 28...	4.1
5	....do	Fuller ditch	At heading...........	13.4
5	....do	Roberts ditch	....do	2.6
23	W. B. Clapp...	Newberry ditch	Auburndale Bridge ...	3.3
23	....do	Durkee ditch	....do	3.4
23	....do	Gilliland ditch	....do	1.2
30	O. W. Peterson.	Santa Ana canal	1 mile below heading .	43
23	W. B. Clapp...	....do	....do	34
29	O. W. Peterson.	Anaheim and Fullerton canal.	Heading	2.8
25	W. B. Clapp...	....do	At Esperanza.........	32
25	....do	Yorba ditch	....do	.6

Miscellaneous discharge measurements in southern California in 1903—Continued.

SANTA ANA RIVER DRAINAGE—Continued.

Date.	Hydrographer.	Stream.	Location.	Discharge (sec. ft.)
May 15	W. B. Clapp and J. M. Mylne.	Santa Ana River	At Orange avenue	20
15	....do	....do	½ mile below Orange avenue.	.7
Apr. 22	W. B. Clapp and J. C. Clausen.	....do	Colton Bridge	13
May 18	W. B. Clapp	....do	....do	.7
Sept. 23	....do	....do	Auburndale Bridge	56
Mar. 31	....do	....do	Rincon wagon bridge	400
Apr. 29	O. W. Peterson	....do	....do	193
May 18	W. B. Clapp	....do	....do	101
June 12	....do	....do	....do	85
July 23	....do	....do	....do	50
Aug. 19	....do	....do	....do	60
Sept. 23	....do	....do	....do	76
Oct. 30	....do	....do	....do	93
Nov. 25	....do	....do	....do	102
Apr. 29	O. W. Peterson	....do	1 mile above heading of Santa Ana and Anaheim canal.	214
Sept. 23	W. B. Clapp	....do	Heading Santa Ana and Anaheim canals.	72
Apr. 29	O. W. Peterson	....do	2 miles below heading Santa Ana and Anaheim canals.	185
30	....do	....do	2 miles above Yorba	130
30	....do	....do	½ mile above Yorba	141
May 1	....do	....do	¾ mile above Southern California Rwy. crossing at Olive.	51
1	....do	....do	¾ mile below Southern California Rwy. crossing at Olive.	.0
Mar. 31	W. B. Clapp	Chino Creek	At Rincon wagon bridge.	146
Apr. 29	O. W. Peterson	....do	....do	23
May 18	W. B. Clapp	....do	....do	15
June 12	....do	....do	....do	6.5
July 23	....do	....do	....do	3.4
Aug. 19	....do	....do	....do	2.4
Sept. 23	....do	....do	....do	3.3
Oct. 30	....do	....do	....do	6
Nov. 25	....do	....do	....do	9.2

eral flow of return water to Santa Ana River, in second-feet, compared with developed water in San Bernardino Valley above Colton, Cal., 1903.

[Measurements by K. Sanborn, engineer Riverside Water Co.]

te.	Location.	Developed.	Natural.	Total.
		Sec.-ft.	*Sec.-ft.*	*Sec.-ft.*
9	Brown tract, artesian well......................	0.11	....	0.11
4	Barnhill pumping plant.........................	.80	....	.80
4	Bloomington pumping plant.....................	8.82	....	8.82
17	Beam ditch ....:	....	0.08	.08
4	City of Colton pumping plant..................	3.17	....	3.17
7	Cooley tract, artesian wells....................	2.70	....	2.70
22	Camp Carlton ditch...........................	2.20	....	2.20
9	Daley ditch	....	.00	.00
18	Flume pump No. 1, Riverside Water Co........	4.08	....	4.08
18	Flume pump No. 2, Riverside Water Co........	2.63	....	2.63
9	Garner tract, artesian well....................	.77	....	.77
21	Gage canal diversion...........................	....	.00	.00
21	Gage canal, Palm avenue.......................	30.82	....	30.82
9	Hurd tract, artesian wells	.21	....	.21
4	H. T. Hunter pumping plant...................	1.79	....	1.79
17	Haws & Talmadge ditch........................	....	.00	.00
4	Johnson & Hubbard pumping plant.............	.42	....	.42
4	Lawson Well Co. pumping plant	.59	....	.59
4	Lamb pumping plant	.24	....	.24
22	Logsden & Tarrell ditch.......................	....	.00	.00
1	Meeks & Daley ditch	....	12.87	12.87
9	McIntyre ditch................................	....	.00	.00
17	McKenzie ditch	....	.00	.00
18	Mill flume, Riverside Water Co	....	.00	.00
4	Orange Land and Water Co.....................	1.08	....	1.08
7	Rancheria pumping plant	1.09	....	1.09
1	Riverside Water Co., upper canal..............	23.89	32.01	55.90
4	Riverside Highland Water Co. pumping plant ..	8.98	....	8.98
17	Rabel dam ditch	....	.00	.00
19	River ditch pump, Riverside Water Co.........	5.79	....	5.79
19	C. W. Rogers pumping plant...................	4.44	....	4.44
17	Shay & Stout ditch...........................	.00	....	.00
7	Swamp ditch	....	.42	.42
9	Wozencraft tract, artesian well	.23	....	.23
22	Whitlock ditch................................	....	.00	.00
21	Ward & Warren ditch	1.44	.17	1.61
17	Whiting ditch.................................	....	.00	.00
19	West Riverside, 350 in pumping plant..........	3.41	....	3.41
	Total	109.70	45.55	155.25

Aug.	21	Evans ditch No. 1 at intake, Riverside County line.	....	.81
Aug.	22	Evans ditch No. 2 at intake, 1 mile below Riverside County line	....	.70
Aug.	20	Evans Island ditch at West Riverside bridge	....	4.71
Aug.	21	Lower Canal Riverside Water Co., head of canal.	....	2.05
	21	Bobidaux canal, head of canal	1.74	4.73
Aug.	25	Santa Ana River at Narrows	....	40.70
		Total	1.74	56.85

COLUMBIA RIVER DRAINAGE BASIN.

Next to the Colorado, Columbia River is the largest river in the region, its drainage basin including parts of Washington, Ore Idaho, and Montana, and a large area in Canada. The Columbia its numerous tributaries are of great importance, offering good for water-power development and an abundance of water for ir tion, while the main river is navigable for a considerable distance

A great part of the water of Columbia River and its tribut flows to waste, not being utilized. This is due to the fact that river has cut so deeply into the lava-covered plains that water ca be diverted except at points near the mountains, where the str are of small size and have not yet entered the deeply incised ca in the plateaus. The following rivers are tributary to the Colu

Umatilla River rises in the well-wooded country in northea Oregon and flows in a general westerly direction, entering Colu River below the mouth of Walla Walla River. The country nor Umatilla is high and rolling. A number of canals divert water the lower course of the stream to irrigate lands on either side.

Yakima River has its source in Keechelus Lake, on the eastern of the Cascade Mountains, in Kittitas County, Wash. Within a distance it receives the waters of Kachess Lake, and 2¼ miles Clealum it receives the outlet of the last of the three large head lakes. It enters Columbia River 23 miles below Kiona, Wash.

Naches River has its source on the eastern slope of the Ca Mountains, in Yakima County, Wash. It flows in a general s easterly direction, entering Yakima River a short distance above Yakima. Irrigation is practiced in the narrow valley along the course of the river, but its waters are of greater value for the i tion of lands west of North Yakima. The river has considerable

and the water can easily be diverted by means of comparatively short canals. For this reason it is of more value for irrigation purposes than Yakima River, which has less fall.

Teton River is the principal tributary of Naches River, discharging into the latter about 17 miles above its junction with Yakima River, near North Yakima. Its source is in the Cascade Mountains in the vicinity of Cowlitz Pass. A peculiar feature of the stream is the turbid, milk-white appearance of the water, it being similar in this respect to White River, on the western slope of the Cascade Range. The water of the South Fork of the Teton, 25 miles above the mouth, is, however, perfectly clear. The forks head in the glaciers of a peak of the Cascades known locally as Goat Rock.

Spokane River rises in the northern part of Idaho, being the outlet of Lake Coeur d'Alene. It passes into Washington, flows in a north-erly direction, and enters Columbia River near latitude 47° 52' north. It is about 120 miles long.

Missoula River has its source in Silverbow County, Mont., and flows northerly until it receives the waters of Little Blackfoot River, when it takes a more northwesterly course. The name Missoula is usually applied to that portion of the river between the junction of Blackfoot and Hellgate rivers and the mouth of Pend Oreille River. From that point to its junction with Columbia River it is called Clark Fork of Columbia.

The source of Bitterroot River is in the high mountains which form the boundary line between Montana and Idaho. It flows in a northerly direction, entering Missoula River a short distance below the city of Missoula. The tributaries on the east side drain comparatively low hills and contribute little to the supply of the river. The west side branches, on the contrary, are numerous, draining a precipitous and heavily wooded area. Their discharges are regulated by many small lakes fed by banks of snow, which continue far into the summer before disappearing altogether. From Hamilton to Missoula, a distance of 48 miles, the fall of the river is 350 feet, or 7.3 feet to the mile.

Snake River, which is the largest affluent of the Columbia, rises on the southern slope of the Continental Divide in the Yellowstone National Park, draining the country west and southwest of Yellow-stone Lake. From Shoshone, Lewes, and Hart lakes, near its head, the river flows in a southerly direction through a timbered and moun-tainous country, resulting in a long period of high water. After con-tinuing through this area for about 20 miles it broadens into Jackson Lake, a deep body of water about 3 miles wide and 8 miles long. Below the lake the river flows through Jackson Hole Valley—about 40 miles long and 8 miles wide—and then enters a long canyon near the Idaho-Wyoming line. All of the large tributaries come from the east, receiving their waters from the Wind River Range. The west side of the valley is bounded by the high Teton Mountains. from which most of the drainage flows westward through Teton River into North Fork of Snake River. It empties into Columbia River near Pasco Junction, in the State of Washington.

The headwater tributaries of Palouse River have their sources in western Idaho. After passing into Washington the streams unite to form Palouse River, which has a general southwesterly course, through a rolling country. Six miles below Hooper, Wash., the river bends suddenly to the south and enters its canyon, through which it flows until its junction with Snake River. A short distance above the mouth of the river are the Palouse Falls, approximately 180 feet high.

Weiser River drains Washington County, in the extreme western part of Idaho, and flows into Snake River at Weiser, Idaho.

The Boise drains a mountainous and well-wooded country in Elmore County, Idaho. The effects of the forests are shown in the high flow that is maintained throughout the summer season, in contrast to the discharge of Weiser River, farther to the west, which drains a more barren country. Below the gaging station, which is located in the canyon, a large number of canals divert water to irrigate lands in Boise Valley. The diversion of the water is now so great that frequent complaints of scarcity are heard.

Bruneau River rises in northern Nevada and flows in a general northerly course through southern Idaho, emptying into Snake River at a point south of Boise. Fall River is one of the small tributaries of Snake River at its headwaters in eastern Idaho.

The following rivers in the Columbia drainage basin are also worthy of mention: The Walla Walla, the drainage basin of which is one of the best irrigated and most productive localities in either Washington or Oregon; the Owyhee, which rises in northwestern Nevada and flows southwest through Idaho; the Malheur, rising in the mountains of east-central Oregon and emptying into Snake River west of Boise; the Grande Ronde, which drains the northern slope of the Blue Mountains and empties into Snake River in southeastern Washington.

The following list includes the stations in the Columbia River drainage basin:

Umatilla River near Umatilla, Oreg.
Umatilla River at Yoakum, Oreg.
Umatilla River at Pendleton, Oreg.
McKay Creek near Pendleton, Oreg.
Umatilla River at Gibbon, Oreg.
Walla Walla River at Milton, Oreg.
Walla Walla River (South Fork) near Milton, Oreg.
Yakima River at Kiona, Wash.
Yakima River at Union Gap, Wash.
Naches River near North Yakima, Wash.
Tieton River near North Yakima, Wash.
Clealum River near Roslyn, Wash.
Kaches River near Easton, Wash.
Yakima River near Martin, Wash.
Chelan River below Lake Chelan, Washington.
Methow River near Pateros, Wash.

Salmon River near Malott, Wash.
Johnson Creek near Riverside, Wash.
Sinlahekin Creek near Loomis, Wash.
Spokane River at Spokane, Wash.
Hangman Creek near Spokane, Wash.
Little Spokane River near Spokane, Wash.
Pend Oreille River at Priest River, Idaho.
Priest River at Priest River, Idaho.
Missoula River at Missoula, Mont.
Bitterroot River near Missoula, Mont.
Bitterroot River near Grantsdale, Mont.
Big Blackfoot River near Bonner, Mont.
Palouse River near Hooper, Wash.
Rock Creek near St. John, Wash.
Wallowa River near Elgin, Oreg.
Wallowa River near Wallowa, Oreg.
Wallowa River near Joseph, Oreg.
Grande Ronde River at Elgin, Oreg.
Grande Ronde River at Hilgard, Oreg.
Weiser River near Weiser, Idaho.
Malheur River near Ontario, Oreg.
Malheur River at Vale, Oreg.

Malheur River near Harper's ranch, above Vale, Oreg.
Bully Creek near Vale, Oreg.
Malheur Lake at The Narrows, Oreg.
Powder River near Baker City, Oreg.
Silvies River near Burns, Oreg.
Silvies River near Silvies, Oreg.
Silvies Creek near Riley, Oreg.
Boise River near Boise, Idaho.
Succor Creek near Homedale, Idaho.
Bruneau River near Grandview, Idaho.
Snake River near Minidoka, Idaho.
Owyhee River near Owyhee, Oreg.
Big Lost River near Mackay, Idaho.
Blackfoot River near Presto, Idaho.
Willow Creek near Prospect, Idaho.
Snake River (South Fork) near Lyon Idaho.
Snake River (South Fork) at Moran, Wyo.
Teton River near St. Anthony, Idaho.
Snake River (North Fork) near Ora, Idaho.
Fall River near Marysville, Idaho.

UMATILLA RIVER NEAR UMATILLA, OREG.

This station was established October 21, 1903, by John H. Lewis. It is located about 2 miles above Umatilla, Oreg., and about one-fourth mile below the diversion dam of the Oregon Land and Water Company. This dam diverts water into an irrigation ditch on the left bank. The inclined gage is on the left bank 45 feet below the cable and is in two sections. Both sections are made of 2 by 6 inch timber fastened by bolts which are cemented into the rock. The lower section reads from 1.2 to 3.5 feet. The upper section reads from 3.5 to 10.8 feet. The gage is read every other day by J. M. Griffith. Gage readings are taken every day during floods. Discharge measurements are made by means of a ⅜-inch wire cable, car, tagged wire, and stay wire. The cable has a span of 210 feet. The initial point for soundings is the zero mark on the tag wire, directly over the vertical portion of the left bank. The channel is straight for 500 feet above and for 1,000 feet below the cable. The current is swift. Both banks are high and rocky and will not overflow. The bed of the stream is composed of solid rock, free from vegetation, and is permanent. The bench mark is the head of a bolt cemented in the solid rock 1½ feet upstream from the gage rod. Its elevation is 10.3 feet above the zero of the gage. To obtain the total discharge of the river, the discharge of the irrigation ditch must be added to that of the river at the cable. The rapids just above Umatilla prevent backwater from Columbia River from affecting the gage heights at the station.

The observations at this station during 1903 have been made under the direction of J. T. Whistler, district engineer.

Discharge measurements of Umatilla River near Umatilla, Oreg., in 1903.

Date.	Hydrographer.	Gage height.	Discharge.
		Feet.	Second-feet.
October 21	John H. Lewis	2.70	15
November 29	do	3.95	1,300

Mean daily gage height, in feet, of Umatilla River near Umatilla, Oreg., for 1903.

Day.	Oct.	Nov.	Dec.	Day.	Oct.	Nov.	Dec.	Day.	Oct.	Nov.	Dec.
1			2.50	12			3.30	23			4.70
2			4.10	13			3.00	24		2.60	4.10
3			2.50	14			3.20	25			4.50
4			3.90	15			3.60	26		2.60	3.9
5			2.50	16			3.20	27			4.10
6			3.60	17			3.60	28		2.60	4.00	3.0
7			2.60	18			4.10	29			3.95
8			3.50	19			3.50	30		2.50	4.00	3.0
9			2.70	20			3.90	31			
10			3.40	21			3.30				
11			2.90	22			2.70	4.20				

UMATILLA RIVER AT YOAKUM, OREG.

This station was established May 5, 1903, by N. S. Dils. It is located one-half mile east of the Oregon Railroad and Navigation Company railroad station, at what is known as the Yoakum wagon bridge. The original gage is a vertical split rail spiked to the face of the south abutment on the upstream side. On September 5, 1903, a new gage, consisting of a 2 by 6 inch timber, 14 feet long, was spiked in a vertical position to the right abutment on the opposite side of the river from the original gage. The new gage was necessary to obtain gage readings at low stages. It is set at the same datum as the original gage and both gages read the same. During 1903 readings have been made once each day by Clyde Purner and Luther Dehaven. Discharge measurements are made from the single-span wagon bridge at which both gages are located. The initial point for soundings is the end of the lower chord of the upstream truss on the left bank. The channel is straight for 1,000 feet above and below the station. The current is swift and has a well-distributed velocity. Both banks are high and are composed of gravel. The right bank will not overflow. The left bank will overflow only at extreme flood stages. The bed of the stream is composed of gravel and is permanent. There is but one channel at all stages. The bench mark is a 60-penny nail and two 8-penny nails driven side by side into the second timber from the top

1e left abutment near the old gage. The elevation of ·the bench x is 13 feet above the zero of both gages.
1e observations at this station during 1903 have been made under direction of J. T. Whistler, district engineer.

Discharge measurements of Umatilla River at Yoakum, Oreg., in 1903.

Date.	Hydrographer.	Gage height.	Discharge.
		Feet.	*Second-feet.*
5......................	N. S. Dils....................	5.70	1,401
26......................	F. W. Huber	4.50	386
15......................	Whistler & Temple..........	3.80	218
3......................	J. H. Lewis.................	3.25	96
mber 5	.....do	2.95	58
er 9.....	.....do	3.90	248
er 23....................	.....do	3.52	146

Mean daily gage height, in feet, of Umatilla River at Yoakum, Oreg., for 1903.

Day.	May.	June.	July.	Aug.	Sept.	Oct.	Nov.	Dec.
....	....	4.20	3.30	2.90	....	3.20	3.40	5.70
....	....	4.20	3.30	2.90	....	3.20	3.40	5.90
....	....	4.20	3.20	2.90	....	3.10	3.40	5.60
....	....	4.10	3.20	2.90	....	3.80	3.40	5.40
....	5.70	4.10	3.30	2.90	2.95	3.40	3.40	5.20
....	5.70	4.00	3.20	2.90	3.00	3.50	3.40	4.90
....	5.60	3.90	3.20	2.90	3.00	3.90	3.40	4.60
....	5.50	3.80	3.20	2.90	3.10	3.90	3.60	4.50
....	5.30	3.70	3.20	2.80	3.10	3.90	3.70	4.40
....	5.30	3.70	3.20	2.80	3.10	3.90	3.80	4.30
....	5.10	3.70	3.20	2.80	3.20	3.80	4.00	4.30
....	5.10	3.70	3.10	2.80	3.30	3.80	4.20	4.30
... .	5.10	3.70	3.10	2.70	3.50	3.80	4.30	4.30
....	5.10	3.60	3.00	2.70	4.00	3.80	4.50	4.20
....	5.00	3.80	3.00	2.70	3.80	3.80	5.20	4.20
....	4.90	4.15	3.00	(a)	3.70	3.70	4.80	4.60
....	4.80	3.60	3.00	....	3.60	3.60	4.60	5.80
....	4.70	3.60	3.00	....	3.60	3.50	4.50	5.50
....	4.60	3.60	3.00	....	3.50	3.50	4.40	5.30
....	4.50	3.50	3.00	....	3.50	3.60	4.90	5.40
....	4.40	3.50	3.00	....	3.40	3.60	5.90	5.80
....	4.40	3.40	3.00	....	3.40	3.60	7.00	5.90
....	4.30	3.40	3.00	....	3.30	3.60	6.50	5.90
....	4.30	3.40	3.00	....	3.30	3.50	6.40	5.80
....	4.20	3.40	3.00	....	3.20	3.50	6.00	5.60
....	4.20	3.30	3.00	....	3.80	3.50	5.80	5.60
....	4.20	3.30	3.00	....	3.30	3.40	5.80	5.40
....	4.30	3.30	3.00	....	3.80	3.40	5.70	5.10
....	4.30	3.30	2.90	....	3.20	3.40	5.60	4.80
....	4.30	3.30	2.90	....	3.20	3.40	5.50	4.60
....	4.20	....	2.90	....	......	3.40	....	4.50

a Observer resigned August 16; new observer secured September 5, 1903.

Rating table for Umatilla River at Youkum, Oreg., from May 5 to December 31, 1903.

Gage height.	Discharge.	Gage height.	Discharge.	Gage height.	Discharge.	Gage height.	Discharge.
Feet.	*Second-feet.*	*Feet.*	*Second-feet.*	*Feet.*	*Second-feet.*	*Feet.*	*Second-feet.*
2.7	35	3.6	164	4.5	434	5.4	1,130
2.8	42	3.7	186	4.6	486	5.5	1,220
2.9	52	3.8	211	4.7	545	5.6	1,310
3.0	62	3.9	237	4.8	610	5.7	1,400
3.1	73	4.0	264	4.9	686	5.8	1,490
3.2	88	4.1	293	5.0	770	5.9	1,580
3.3	105	4.2	324	5.1	860		
3.4	123	4.3	357	5.2	950		
3.5	143	4.4	394	5.3	1,040		

Made from measurements between 2.95 and 5.70 feet gage height. Table extended above 5.70 and below 2.95 feet. Table well determined.

Estimated monthly discharge of Umatilla River at Yoakum, Oreg., for 1903.

Month.	Discharge in second-feet.			Total in acre-feet.
	Maximum.	Minimum.	Mean.	
May 5–31	1,400	324	676	36,202
June	324	105	189	11,246
July	105	52	73	4,489
August *a*	52	35	44	2,705
September *a*	264	42	107	6,367
October	237	73	160	9,838
November	2,570	123	775	46,116
December	1,580	324	903	55,708
The period	....	....	....	172,671

a Discharge estimated August 16 to September 4.

UMATILLA RIVER AT PENDLETON, OREG.

This station was established May 22, 1903, by F. W. Huber. It is located at the Main Street Bridge at Pendleton, Oreg. A short distance above the bridge at which the gage is located water is taken out of the river by the Farmers Mill ditch. This ditch carries from 30 to 50 second-feet, and at low stages the entire river is diverted into it. The water is returned to the river at a point about 4,000 feet below the Main Street Bridge and about 1,500 feet above the railroad bridge, at which discharge measurements are made. The original river gage was a vertical 1 by 5 inch board, 10 feet long, spiked to

the middle of the left, or south, side of the center pier of the Main
Street Bridge. On July 18, 1903, this gage was replaced by a 2 by 6
inch board, 10 feet long, fastened in the same position as the original
gage and on the same datum. During 1903 the gage has been read
irregularly by the Geological Survey office at Pendleton. Discharge
measurements are made from the Oregon Railroad and Navigation
Company's bridge, about 1 mile downstream from the gage. The
initial point for soundings is the face of the crib abutment on the left
bank. The railroad bridge consists of a single span of 145 feet, there
being 116 feet of trestle approach on the right bank and an approach of
34 feet on the left bank. The channel is straight for 250 feet above and
200 feet below the railroad bridge. The current is swift at this point,
but has a lower velocity than at the bridge at which the gage is located.
The right bank is low and will overflow under the trestle on this bank.
The left bank is high and is partly riprap. At low stages there will
be some backwater at the south bank. There is but one channel at all
stages. The bed of the stream is composed of gravel and may shift
slightly. The bench mark is the top of the south side of the steel
caisson at the east end of the middle pier of the Main Street Bridge.
Its elevation is 16.50 feet above the zero of the gage.

The flour mill ditch (Byer's) takes 100 feet or more of water about
a mile and a half above the gage and returns it to the river about 200
feet above the gage, so that it interferes with the measurements in
no way except for a time during very low water, when the water is
drawn down and then stored again. The gage was read only when the
flow was normal. This accounts for the few gage heights during the
summer.

The Farmers Mill ditch does not affect the reading of the gage when
above 4 feet. When the water is below that height it takes about 40
feet of water from the river above the gage and returns it above the
bridge, where measurements are made; so for low stages of the river
the amount of water in the ditch should be taken out from the measured
discharge in making curve and rating table and should then be added
to the discharge taken from the table. The amount of water in the
ditch is very uniform day and night, except for a short time in the
spring when repairs are being made.

Owing to the changes in the channel, one rating curve will not
answer for the whole year, but two must be made.

The observations at this station during 1903 have been made under
the direction of J. T. Whistler, district engineer.

Discharge measurements of Umatilla River at Pendleton, Oreg., in 1903.

Date.	Hydrographer.	Gage height.	Discharge.
		Feet.	*Second-feet.*
May 22	Huber and Davis	1.90	534
May 25	do	1.80	389
May 29	do	1.90	402
June 12	J. T. Whistler	1.30	204
June 13	F. Temple	1.35	227
July 10	J. H. Lewis	.90	94
July 20	do		63
July 21	J. T. Whistler		61
August 2	do	.30	39
September 6	do	.58	60
October 4	Lewis and Yates	1.10	92
October 26	do	1.28	116
November 24	do	3.70	1,606
December 2	H. A. Yates	3.20	1,462

Mean daily gage height, in feet, of Umatilla River at Pendleton, Oreg., for 1903.

Day.	May.	June.	July.	Aug.	Sept.	Oct.	Nov.	Dec.
1		1.80					1.20	3.1
2			0.80	0.30			1.28	3.2
3							1.25	2.91
4						1.08	1.24	2.60
5			.98				1.22	2.38
6		1.45	.95		0.58		1.28	2.30
7		1.35	.88				1.41	2.00
8		1.35	.85				1.45	1.90
9		1.30	.88				1.48	1.80
10		1.30	.90				1.66	1.71
11		1.27					1.71	1.70
12		1.30	a.70				2.19	1.65
13		1.32					2.17	1.65
14		1.22					2.51	1.65
15		1.15					2.90	1.65
16		1.10	.70				2.61	2.27
17		1.10					2.40	2.10
18		1.05					2.29	2.7
19								2.0
20		1.00			1.20			2.60
21		1.00					2.86	2.65
22	1.90	1.00				1.35	4.85	2.90
23	1.80	1.00				1.32	3.87	2.87
24	1.80	1.00				1.30	3.68	2.65
25	1.80	.90				1.28	3.28	2.50
26	1.80					1.28	3.00	2.25
27	1.90					1.28	3.03	2.15
28	1.90					1.27	2.95	
29	1.90	.85				1.25	2.80	
30	1.80			.60			2.76	1.90
31	1.80					1.23		1.80

a Between July 12 and October 21 gage was read only at time of normal flow.

M'KAY CREEK NEAR PENDLETON, OREG.

This station was established May 23, 1903, by E. I. Davis. It is located at the footbridge near the residence of C. W. Lyman, 2 miles west of Pendleton, Oreg. The gage is a vertical 1 by 6 inch board 11 feet long nailed to a post which is set in the bed of the stream and braced to a large poplar tree. It is about 200 feet north of C. W. Lyman's house and about 150 feet below the footbridge, on the left bank. It is read twice each day by C. W. Lyman. Discharge measurements are made from the footbridge, above the gage. It has a span of 65 feet. The initial point for soundings is the end of the log, of which the bridge consists, on the right bank. The channel is straight for 150 feet above and 100 feet below the bridge. The current has a good velocity at ordinary stages. Both banks are high, not liable to overflow, and are without trees. There is but one channel at all stages. The bed of the stream is composed of gravel, free from vegetation, and permanent. The bench mark is a nail in a blaze on a root of the poplar tree to which the gage is braced. Its elevation is 5.82 feet above the zero of the gage.

The observations at this station during 1903 have been made under the direction of J. T. Whistler, district engineer.

Discharge measurements of McKay Creek near Pendleton, Oreg., in 1903.

Date.	Hydrographer.	Gage height.	Discharge.
		Feet.	*Second-feet.*
May 23	F. W. Huber	1.10	46
May 25	....do	1.10	34
May 29	E. I. Davis	1.10	31
June 15	J. T. Whistler	.81	11
July 6	....do	.78	5
July 31	J. H. Lewis	.70	3.8
October 8	Lewis and Yates	1.10	25

Mean daily gage height, in feet, of McKay Creek near Pendleton, Oreg., for 1903.

Day.	May.	June.	July.	Aug.	Sept.	Oct.	Nov.	Dec.
....	....	1.00	0.80	0.70	0.70	0.90	1.00	1.0
....	....	1.00	.80	.70	.70	.90	1.00	1.0
....	....	1.00	.80	.70	.70	.90	1.00	1.0
....	....	1.00	.80	.70	.70	.90	1.00	1.0
....	....	1.00	.80	.70	.70	.90	1.00	1.0
....	....	.90	.75	.70	.70	.90	1.00	1.0
....	....	.90	.75	.70	.70	.90	1.10	1.0
....	....	.90	.75	.70	.70	1.00	1.10	1.0
....	....	.90	.75	.70	.70	1.00	1.20	1.0
....	....	.90	.75	.70	.70	1.10	1.20	1.0
....	....	.85	.75	.70	.70	1.10	1.20	1.0
....	....	.85	.75	.70	.70	1.10	1.25	1.0
....	....	.85	.75	.70	1.00	1.10	1.35	1.0
....	....	.85	.75	.70	1.35	1.10	1.45	1.0
....	....	.80	.75	.70	1.30	1.10	1.00	2.0
....	....	.80	.75	.70	1.20	1.10	1.60	2.0
....	....	.80	.75	.70	1.20	1.10	1.50	2.0
....	....	.80	.75	.70	1.20	1.10	1.50	1.0
....	....	.80	.75	.70	1.05	1.10	1.40	1.0
....	....	.80	.75	.70	1.05	1.10	1.40	1.0
21	....	.80	.75	.70	1.00	1.10	1.40	2.0
22	....	.80	.75	.70	1.00	1.10	2.45	2.0
23	....	.80	.75	.70	1.00	1.10	2.50	2.0
24	....	.80	.75	.70	1.00	1.00	2.50	2.0
25	1.10	.80	.75	.70	.95	1.00	2.40	2.0
26	1.10	.80	.75	.70	.95	1.00	2.40	2.0
27	1.10	.80	.70	.70	.95	1.00	2.30	2.0
28	1.10	.80	.70	.70	.95	1.00	2.00	2.0
29	1.10	.80	.70	.70	.90	1.00	2.00	2.0
30	1.10	.80	.70	.70	.90	1.00	1.70	1.9
31	1.10	....	.70	.70	....	1.00	....	1.8

UMATILLA RIVER AT GIBBON, OREG.

This station was established by C. C. Babb July 22, 1896. The original gage rod was located one-fourth mile below the railroad station. The gage, together with the cable from which discharge measurements were made, was carried away by a flood in May, 1902. The bench mark, consisting of a cross on the highest point of the rock to which the original gage was fastened, has also been destroyed. On September 10, 1902, the station was reequipped with a wire gage, cable, and car. The cable is located in its original position. To secure a better location the new gage was established a few hundred feet nearer the cable. It is located on a beam projecting over the water and spiked to the top of the cribwork on the left bank about 10 feet north of the railroad track. The wire gage was repaired July 29, 1903, but no change was made in the datum. The length of the wire from the marker to the bottom of the eye in the center of the web of the section of the rail used as a weight, is 16.30 feet. The initial

for soundings is the face of the tree on the right bank to which
ible is attached. The channel is straight for 150 feet above and
eet below the cable. Both the right and left banks are low. The
f the stream is composed of gravel and is permanent. Bench
No. 1 is the head of a 40-penny spike driven flush with the sur-
if the crib timber opposite the 1-foot mark of the gage scale.
evation is 8.66 feet above gage datum. Bench mark No. 2 is a
nny spike in a telegraph pole directly across the railroad track
the gage and 30 feet distant. It is about 5½ feet above the ground
ias an elevation of 10.35 feet above gage datum.
e gage is read once each day by Walter Swart.
e observations at this station during 1903 have been made under
irection of J. T. Whistler, district engineer.

Discharge measurements of Umatilla River at Gibbon, Oreg., in 1903.

Date.	Hydrographer.	Gage height.	Discharge.
		Feet.	*Second-feet.*
6......................	E. I. Davis......................	0.90	466
....	J. H. Lewis......................	.49	164
8......................	....do	.40	104
9......................	....do	.40	91
aber 29	....do	.50	95

Mean daily gage height, in feet, of Umatilla River at Gibbon, Oreg., for 1903.

Day.	Jan.	Feb.	Mar	Apr.	May	June.	July.	Aug.	Sept.	Oct.	Nov.	Dec.
....	1.90	1.30	0.70	2.50	1.65	0.90	0.45	0.40	0.60	0.60	0.50	1.80
....	2.40	1.10	.70	2.10	.60	.90	.45	40	55	.60	.50	1.80
....	3.60	.95	.70	1.90	.60	.80	.45	40	60	.60	.50	1.70
....	4.10	.90	.65	1.75	90	.80	.45	.40	65	.70	.55	1.60
....	3.50	.80	.60	1.60	1.50	.75	.50	.40	70	.90	.55	1.40
....	3.05	.70	.55	1.50	1.50	.75	.50	.40	80	.80	.60	1.40
....	2.80	.70	.50	1.45	1.50	.70	.50	.40	80	.80	.70	1.30
....	2.60	.70	.45	1.45	1.40	.65	.50	40	75	.80	.70	1.20
....	2.45	.70	.45	1.55	1.40	.65	47	40	70	.80	.70	1.00
....	2.35	.70	.45	1.45	50	.60	45	40	65	.70	1.00	1.00
....	2.25	.60	.45	1.40	1.40	.60	40	40	60	.70	1.00	1.00
....	2.15	.50	.45	1.35	40	.70	40	40	65	.70	1.20	1.00
....	2.00	.50	.60	1.30	40	.65	40	40	60	.70	1.30	1.00
....	1.90	.50	.60	1.25	30	.65	40	40	60	.70	1.60	1.00
....	1.80	.50	.60	1.25	20	.70	40	40	60	70	1.50	1.50
....	1.75	.45	.65	1.45	10	.65	40	40	60	70	1.30	1.90
....	1.70	.45	.65	1.70	1.01	.65	.40	.40	60	70	1.20	1.60
....	1.60	.45	.60	1.60	1.02	.60	.40	.40	60	70	1.10	1.60
....	1.60	.45	.60	1.80	1.00	.60	.40	.40	60	70	1.00	1.40
....	1.60	.15	.60	1.70	.95	.55	.40	.40	60	70	1.10	1.40
....	1.80	.45	.75	1.80	.95	.50	.40	.40	60	65	2.50	1.80

Mean daily gage height, in feet, of Umatilla River at Gibbon, Oreg., for 1903—Cont'd.

Day.	Jan.	Feb.	Mar.	Apr.	May.	June.	July.	Aug.	Sept.	Oct.	Nov.	Dec.
22	2.10	0.45	0.85	2.10	0.95	0.50	0.40	0.40	0.60	0.65	2.70	1.3
23	2.40	.45	1.05	2.00	1.00	.50	.40	.40	.60	.60	2.60	1.3
24	4.40	.55	1.45	2.00	1.00	.50	.40	.40	.60	.60	2.20	1.3
25	4.20	.60	1.85	2.04	.95	.50	.40	.70	.60	.55	2.10	1.3
26	3.00	.80	1.95	2.06	.90	.47	.40	.70	.60	.55	2.00	1.2
27	2.40	.65	2.20	2.08	.90	.47	.40	.60	.60	.55	2.00	1.3
28	2.00	.70	3.15	1.80	.90	.47	.40	.60	.60	.50	1.90	1.2
29	1.70	......	3.75	1.80	.95	.47	.40	.60	a.60	.50	1.80	1.3
30	1.55	......	3.10	1.70	.95	.47	.40	.60	.60	.50	1.80	1.3
31	1.40	......	2.90	......	.95	......	.40	.60	......	.50	......	1.3

a Interpolated.

WALLA WALLA RIVER AT MILTON, OREG.

This station was established February 14, 1903, by T. A. Noble. The gage is a vertical rod on the left bank one-half mile above the county bridge and just above the head-gate of an irrigation ditch. The gage is read once each day by William Wormington. During 1903 discharge measurements were made from a cable just above the gage. Measurements made at this point include the discharge of the irrigation ditch just below the gage. On October 29, 1903, a stay wire was installed about 35 feet above the county bridge to be used in making flood measurements. At the close of the season of 1903 the cable was abandoned, and discharge measurements will hereafter be made from the highway bridge. Measurements made at this point do not include the discharge of the irrigation ditch just below the gage. This ditch will have to be measured separately. There is a power ditch taken out on the left bank just below the bridge, which is included in the measurements made at the bridge. The initial point for soundings is the end of the lower chord on the upstream side of the bridge at the right bank. The bridge has a single span of 75 feet between abutments. The channel is straight for 80 feet above and for 150 feet below the station. The current is swift. Both banks are low, wooded, but not liable to overflow. There is but one channel at all stages. The bed of the stream is composed of gravel, free from vegetation, and liable to shift. The bench mark is the top of a sharp projecting rock 4 feet from the gage and 3 feet from the tree to which the gage is attached. Its elevation is 4.37 feet above the zero of the gage.

The observations at this station during 1903 have been made under the direction of J. T. Whistler, district engineer.

Discharge measurements of Walla Walla River at Milton, Oreg., in 1903.

Date.	Hydrographer.	Gage height.	Discharge.	Remarks.
		Feet.	*Sec.-feet.*	
February 13	T. A. Noble	1.11	247	At cable near gage rod.
Do.........	.....do	1.11	205	At county bridge, one-half mile below gage.
May 28.........	F. W. Huber	1.40	388	Do.
June 24	J. H. Lewis........	.98	171	Do.
September 8	.....do	1.05	152	At cable near gage rod.
Do.........	.....do	1.05	167	At county bridge, one-half mile below gage.
October 29......	Lewis and Yates ...	1.03	144	Gaging at cable, near gage.
Do.........	.....do	1.03	139	Gaging at county bridge, one-half mile below gage.
December 28....	H. A Yates........	1.15	224	Do.
Do.........	J. H. Lewis........	1.15	216	Do.
Do.........	.....do	1.15	237	Gaging at cable near gage rod.

Mean daily gage height, in feet, of Walla Walla River at Milton, Oreg., for 1903.

Day.	Feb.	Mar.	Apr.	May.	June.	July.	Aug.	Sept.	Oct.	Nov.	Dec.
1....................................		1.11	1.89	1.55	1.45	1.08	0.96	0.95	0.96	1.01	1.21
2....................................		1.11	1.74	1.55	1.44	1.13	.96	.94	.96	1.03	1.22
3....................................		1.11	1.61	1.55	1.37	1.10	.95	.94	.96	1.04	1.23
4....................................		1.11	1.54	1.60	1.32	1.07	.95	.94	1.02	1.05	1.20
5....................................		1.10	1.47	1.67	1.26	1.18	.95	.94	1.26	1.08	1.17
6....................................		1.09	1.41	1.70	1.24	1.13	.95	.98	1.62	1.10	1.12
7....................................		1.09	1.38	1.67	1.21	1.12	.94	1.07	1.37	1.25	1.08
8....................................		1.11	1.38	1.64	1.20	1.11	.94	1.05	1.27	1.19	.95
9....................................		1.10	1.42	1.58	1.19	1.11	.94	1.00	1.21	1.21	.96
10....................................		1.10	1.39	1.56	1.17	1.09	.94	1.08	1.18	1.24	.98
11....................................		1.10	1.35	1.54	1.14	1.07	.93	1.22	1.18	1.24	.99
12....................................		1.10	1.32	1.57	1.14	1.06	.93	1.23	1.11	1.27	.97
13....................................		1.07	1.32	1.65	1.18	1.05	.93	1.47	1.10	1.31	1.00
14....................................		1.06	1.32	1.70	1.14	1.04	.92	1.37	1.08	1.56	.96
15....................................	1.08	1.05	1.32	1.60	1.11	1.04	.92	1.26	1.07	1.57	.98
16....................................	1.07	1.06	1.35	1.56	1.11	1.04	.93	1.19	1.06	1.44	1.56
17....................................	1.07	1.06	1.41	1.49	1.10	1.02	.93	1.14	1.05	1.35	1.61
18....................................	1.07	1.07	1.42	1.46	1.07	1.01	.93	1.10	1.05	1.32	1.49
19....................................	1.07	1.06	1.40	1.43	1.04	1.02	.92	1.06	1.03	1.29	1.40
20....................................	1.07	1.06	1.43	1.39	1.04	1.02	.91	1.04	1.02	1.33	1.43
21....................................	1.07	1.24	1.53	1.35	1.02	.99	.91	1.02	1.01	1.62	1.45
22....................................	1.08	1.28	1.65	1.34	1.05	.97	.92	1.01	1.01	1.85	1.50
23....................................	1.10	1.29	1.67	1.33	1.03	.97	.93	1.00	.99	1.61	1.47
24....................................	1.10	1.31	1.70	1.32	1.02	.91	.93	.99	.98	1.47	1.37
25....................................	1.11	1.44	1.73	1.35	1.01	.92	.94	.99	.97	1.25	1.30
26....................................	1.11	1.48	1.83	1.38	1.10	.92	1.05	.99	.96	1.20	1.28

Mean daily gage height, in feet, of Walla Walla River at Milton, Oreg., for 1903—Cont'd

Day.	Feb.	Mar.	Apr.	May.	June.	July.	Aug.	Sept.	Oct.	Nov.	Dec.
27	1.11	1.52	1.75	1.35	1.09	0.91	1.17	0.99	0.95	1.26	1.9
28	1.11	1.85	1.64	1.49	1.14	.91	1.02	.95	.99	1.29	1.5
29		2.22	1.55	1.60	1.22	.90	.97	.96	.99	1.29	1.7
30		2.01	1.57	1.60	1.10	.95	.97	.96	1.02	1.19	1.8
31		2.00		1.62		.96	.96		1.08		1.6

Rating table for Walla Walla River at Milton, Oreg., from February 1 to December 31, 1903.

Gage height.	Discharge.	Gage height.	Discharge.	Gage height.	Discharge.	Gage height.	Discharge.
Feet.	*Second-feet.*	*Feet.*	*Second-feet.*	*Feet.*	*Second-feet.*	*Feet.*	*Second-feet.*
0.9	65	1.3	330	1.7	610	2.1	890
1.0	120	1.4	400	1.8	680	2.2	960
1.1	190	1.5	470	1.9	750		
1.2	260	1.6	540	2.0	820		

Table not well determined; curve extended below 1.05 and above 1.40 feet gage height.

Estimated monthly discharge of Walla Walla River at Milton, Oreg., for 1903.

Month.	Discharge in second-feet.			Total in acre-feet.
	Maximum.	Minimum.	Mean.	
February 15–28	197	169	181	5,026
March	981	155	310	19,061
April	743	344	482	28,681
May	610	344	471	28,961
June	435	127	228	13,567
July	246	70	142	8,731
August	239	70	93	5,718
September	449	85	164	9,759
October	554	90	177	10,883
November	715	127	324	19,279
December	547	90	266	16,356
Total for period				166,022

SOUTH FORK OF WALLA WALLA RIVER NEAR MILTON, OREG.

This station was originally established February 15, 1903, 6 miles above the mouth of the river and 12 miles from Milton, Oreg. The gage was read once each day from the date of establishment to October 31, 1903, by N. Redden. As there were no means for making flood measurements at this point, the station was moved to the highway bridge one-fourth mile above the junction of the North and South forks, 6 miles from Milton. The gage is a vertical 1¼ by 9 inch timber, secured to a stump on the right bank three-fourths mile above the highway bridge and directly back of the house of the observer, Harry Huber, who reads the gage once each day. Discharge measurements are made from the upstream side of the single-span highway bridge one-fourth mile above the mouth of South Fork. The initial point for soundings is the south side of a projecting beam which supports the north end of the lower chord of the bridge. The bridge has a span of 65 feet between abutments. The channel is straight for 100 feet above and for 150 feet below the bridge. The current is swift. The right bank is low, wooded, and liable to overflow. The left bank is low, but is not liable to overflow, and is without trees. The bed of the stream is composed of gravel, free from vegetation, and not liable to shift to any considerable extent. Bench mark No. 1 is a 20-penny nail driven into a cottonwood tree 1 foot in diameter 15 feet above the gage rod. Its elevation is 7 feet above the zero of the gage. Bench mark No. 2 is a 20-penny nail driven into the tree to which the gage is attached. Its elevation is 7 feet above the zero of the gage.

The observations at this station during 1903 have been made under the direction of J. T. Whistler, district engineer.

Mean daily gage height, in feet, of South Fork of Walla Walla River near Milton, Oreg., for 1903.

Day.	Feb.	Mar.	Apr.	May.	June.	July.	Aug.	Sept.	Oct.	Nov.	Dec.
1	......	1.57	2.33	2.11	2.18	1.64	1.56	1.55	1.68	·1.79	2.05
2	......	1.58	2.22	2.11	2.07	1.65	1.56	1.54	......	1.78	2.07
3	......	1.57	2.07	2.19	2.02	1.66	1.56	1.54	......	1.79	2.00
4	......	......	2.02	2.24	1.97	......	1.56	......	1.74	1.79	1.90
5	......	1.56	1.96	2.33	1.93	1.67	1.56	1.53	2.15	......	1.85
6	......	1.55	1.94	......	1.91	......	1.55	1.54	2.15	......	1.85
7	......	1.55	1.90	2.31	1.89	1.62	1.55	1.58	1.98	1.87	1.80
8	......	1.55	1.90	2.27	1.87	1.61	1.55	1.62	1.87	1.84	1.80
9	......	1.53	......	2.17	1.85	1.60	1.55	1.59	1.82	1.90	1.75
10	......	1.55	1.89	2.20	......	......	1.55	2.18	......	1.85	1.75
11	......	1.56	1.89	2.13	1.69	......	1.55	1.75	......	1.90	1.80
12	......	1.58	1.88	2.20	1.70	......	1.55	1.83	1.78	1.98	1.80
13	......	1.57	1.84	2.32	1.80	......	1.55	1.83	......	1.91	1.75
14	......	1.57	1.84	......	1.75	1.60	1.54	1.78	......	2.22	1.75
15	1.58	1.56	1.86	......	......	1.60	......	1.73	......	2.10	1.75
16	1.57	1.58	1.90	......	1.73	1.59	1.71	1.74	......	2.00	2.10
17	1.58	......	1.97	2.11	1.69	1.59	1.54	1.72	......	1.95	2.20
18	1.58	1.56	1.96	2.06	1.67	......	1.54	......	......	1.90	2.10
19	1.56	1.55	1.94	2.02	1.66	1.58	1.58	1.69	......	1.83	2.00
20	1.56	1.56	1.96	1.98	......	1.59	......	1.68	1.76	1.80	2.00
21	1.56	1.59	2.04	1.95	1.62	1.58	1.58	1.68	1.75	2.25	2.00
22	1.59	1.61	2.22	1.94	1.66	1.58	1.58	1.67	......	2.22	2.05

Mean daily gage height, in feet, of South Fork of Walla Walla River near Milton, Oreg., for 1903—Continued.

Day.	Feb.	Mar.	Apr.	May.	June.	July.	Aug.	Sept.	Oct.	Nov.	Dec.
23	1.59	1.74	2.21	1.94	1.64	1.58	1.54	1.67	......	2.12	2.05
24	1.60	1.93	2.22	1.96	1.63	1.58	1.54	......	1.74	2.02	2.00
25	1.60	2.00	2.34	2.00	1.63	1.58	1.59	1.70	......	1.95	1.92
26	......	2.04	2.41	2.10	1.61	1.58	1.83	1.69	......	1.95	1.87
27	......	1.90	2.26	2.07	......	1.57	1.67	1.68	......	......	1.81
28	1.59	......	2.18	2.11	1.68	1.57	1.61	1.68	1.75	1.97	......
29	......	2.86	2.17	2.04	1.61	1.56	1.57	......	1.76	1.92	1.80
30	......	2.82	2.14	2.07	1.62	1.56	1.56	1.68	1.76	1.96	1.80
31	......	2.50	......	2.21	......	1.56	1.55	......	(a)	......	......

a Readings February 15 to October 30, inclusive, on gage 12 miles above Milton, Oreg., by M. Redden. Readings November 1 to December 31, inclusive, on gage 6 miles above Milton, Oreg., by Harry Huber. Practically no water enters or leaves river between these points, and new gage was set to read same as old.

YAKIMA RIVER AT KIONA, WASH.

This station was established August 20, 1895. It is located at the highway bridge on the county road about 1,800 feet northwest of the Northern Pacific Railway station, at Kiona, Wash. It is about 23 miles above the mouth of the river. The original gage consisted of an inclined and a vertical section, spiked to the east end of the south pier of the bridge and anchored with rocks. The present gage is of the old wire type, and is located on the downstream side of the bridge between the fifth and seventh verticals from the right bank. The length of the wire from the end of the weight to the marker is 27.21 feet. The distance from the end of the scale board to the outside edge of the pulley is 2 feet. The gage is read once each day by W. A. Kelso. Discharge measurements are made from the upstream side of the bridge to which the gage is attached. There is a stay wire 70 feet above the bridge. The initial point for soundings is a point on the west side of the bridge 100 feet south of the center of the south pier of the main span. The channel is straight for 500 feet above and for 400 feet below the station. The current has a moderate velocity. The right bank is low, but is well protected by a levee, and is not subject to overflow. The left bank is somewhat higher and is not subject to overflow. The bed of the stream is composed of fine gravel, not subject to change. At low water the river flows beneath the middle, or main, span; at high water it passes under an additional shorter span at each end of the bridge. Bench mark No. 1 is the top of a spike in the east end of the cap of the first trestle bent on the right bank. Its elevation above gage datum is 20.94 feet. Bench mark No. 2 is a spike on the north side of the stay-wire post on the right bank. Its elevation above gage datum is 18.73 feet. Bench mark No. 3 is a nail in the upstream end of the first trestle bent on the left bank. Its elevation is 18.73 feet above gage datum. The top of the 2-inch pulley is 25.62 feet

above gage datum. The United States Geological Survey standard
iron bench-mark post, near the Northern Pacific Railway station, has
an elevation above sea level of 515 feet. The elevation of gage datum
above sea level is 453 feet. All bench marks are marked "B. M."
with black paint.

The observations at this station during 1903 have been made under
the direction of T. A. Noble, district engineer.

Discharge measurements of Yakima River at Kiona, Wash., in 1903.

Date.	Hydrographer.	Gage height.	Discharge.
		Feet.	*Second-feet.*
April 17	Sydney Arnold	6.80	4,661
June 4....................	....do	13.30	24,155
July 28	G. H. Ellis	5.65	2,741
September 23	....do	4.22	1,088

Mean daily gage height, in feet, of Yakima River at Kiona, Wash., for 1903.

Day.	Jan.	Feb.	Mar.	Apr.	May.	June.	July.	Aug.	Sept.	Oct.	Nov.	Dec.
1	4.95	5.55	4.65	8.18	9.25	11.95	9.65	4.16	3.80	4.30	4.30	7.23
2	4.67	5.45	4.70	8.92	9.52	12.92	6.95	5.00	3.67	4.55	4.32	5.19
3	5.50	5.52	5.72	8.70	9.19	12.75	8.40	4.97	3.80	4.30	4.34	3.40
4	8.25	5.85	5.68	8.50	9.42	13.35	8.00	4.85	3.85	4.25	3.36	2.65
5	11.30	5.64	5.66	8.60	10.02	13.00	7.82	4.78	3.31	4.30	5.42	9.65
6	11.72	5.41	5.65	8.05	10.70	12.05	7.65	4.65	3.25	4.40	5.57	8.47
7	11.22	5.84	5.65	7.86	11.11	11.57	7.62	4.56	3.17	5.45	5.65	5.00
8	10.32	5.92	5.64	7.83	11.00	12.30	7.50	4.48	3.10	6.75	5.83	7.35
9	9.56	6.15	5.65	7.80	10.80	12.55	7.45	4.44	3.12	7.15	5.98	7.50
10	8.98	6.05	5.65	7.50	10.10	13.13	7.40	4.38	3.20	7.04	6.08	6.80
11	8.47	5.96	5.62	7.34	10.05	13.00	7.25	4.36	3.47	6.75	6.20	6.80
12	7.98	5.95	6.40	7.23	10.05	13.02	7.28	4.35	3.72	6.75	6.26	6.65
13	7.68	5.80	6.13	7.16	10.40	13.82	7.23	4.30	4.08	6.85	6.29	6.70
14	7.34	5.80	5.95	7.08	10.95	13.83	7.15	4.24	4.38	6.84	6.30	6.65
15	7.10	5.80	5.90	7.02	11.50	13.30	7.08	4.20	4.52	6.83	6.02	6.33
16	6.92	5.80	5.88	6.82	11.55	13.28	6.82	4.24	4.48	6.74	5.85	6.25
17	6.77	5.60	5.72	6.85	11.45	13.35	6.70	4.18	4.42	6.52	5.68	6.35
18	6.65	5.58	5.76	6.94	10.82	13.12	6.60	4.13	4.35	6.40	5.50	6.45
19	6.58	5.55	5.71	7.08	10.42	12.60	6.57	4.05	4.28	6.10	5.45	6.70
20	6.53	5.53	5.72	7.36	9.98	11.82	6.58	3.92	4.42	5.93	5.40	6.30
21	6.61	5.50	5.73	7.40	9.65	11.50	6.62	3.90	4.15	5.83	5.40	6.15
22	6.72	5.58	5.72	7.51	9.30	11.30	6.60	3.87	4.16	5.62	5.40	6.18
23	6.83	5.50	5.85	7.65	9.45	10.90	6.60	3.88	4.22	5.52	5.00	6.10
24	6.84	5.50	6.07	8.00	8.80	10.38	6.52	3.88	4.32	5.44	5.62	6.07
25	7.45	5.54	6.40	8.45	8.83	10.06	6.40	3.78	4.43	5.32	5.65	5.80
26	7.85	5.55	6.90	8.93	9.35	9.88	6.17	3.73	4.50	5.10	5.64	5.80
27	7.40	5.65	(a)	9.45	10.00	9.62	5.93	3.80	4.55	5.10	5.62	5.80
28	7.20	5.72	7.70	9.58	10.30	9.65	5.02	3.74	4.50	4.97	5.90	5.70
29	7.06	......	8.95	9.65	10.20	9.70	5.47	3.63	4.48	5.02	6.45	5.70
30	6.90	......	9.40	9.50	10.20	9.50	5.33	3.53	4.45	5.16	7.00	5.60
31	6.72	......	9.40	......	10.55	......	5.22	3.50	......	5.24	......	5.55

a Observer absent.

Rating table for Yakima River at Kiona, Wash., for 1903.

Gage height.	Discharge.	Gage height.	Discharge.	Gage height.	Discharge.	Gage height.	Discharge.
Feet.	*Second-feet.*	*Feet.*	*Second-feet.*	*Feet.*	*Second-feet.*	*Feet.*	*Second-feet.*
3.1	320	4.4	1,250	6.6	4,350	10.0	12,400
3.2	370	4.6	1,450	6.8	4,740	10.5	13,720
3.3	430	4.8	1,670	7.0	5,160	11.0	15,100
3.4	490	5.0	1,900	7.2	5,600	11.5	16,580
3.5	555	5.2	2,150	7.4	6,040	12.0	18,080
3.6	620	5.4	2,420	7.6	6,480	12.5	19,580
3.7	690	5.6	2,700	7.8	6,940	13.0	21,080
3.8	760	5.8	3,000	8.0	7,400	13.5	22,580
3.9	835	6.0	3,320	8.5	8,590		
4.0	910	6.2	3,650	9.0	9,820		
4.2	1,070	6.4	3,990	9.5	11,100		

Table well defined to 9 feet gage height; uncertain above 9 feet. Rating table 1903 below 4.30 and above 11.20 same as that for 1902.

Estimated monthly discharge of Yakima River at Kiona, Wash., for 1903.

[Drainage area, 5,230 square miles.]

Month.	Discharge in second-feet.			Total in acre-feet.	Run-off.	
	Maximum.	Minimum.	Mean.		Second-feet per square mile.	Depth in inches.
January	17,180	3,000	6,938	426,601	1.33	1.53
February	4,350	2,420	2,985	165,778	.57	.59
March *a*	10,840	2,700	4,018	247,057	.77	.89
April	11,360	4,740	7,426	441,878	1.42	1.58
May	16,880	9,320	12,816	788,025	2.45	2.82
June	23,480	11,100	18,182	1,081,904	3.48	3.88
July	9,820	2,150	5,192	319,244	.99	1.14
August	2,150	555	1,119	68,805	.21	.24
September	1,450	320	924	54,982	.18	.20
October	5,600	1,070	3,006	184,832	.57	.66
November	5,160	2,280	2,979	177,263	.57	.64
December	11,360	2,700	5,004	307,684	.96	1.11
The year	23,480	320	5,882	4,264,053	1.12	15.28

a March 27 interpolated.

YAKIMA RIVER AT UNION GAP, WASHINGTON.

Yakima River enters Columbia River just above the town of Pasco. The first measurement of the river was made at this point on August 4, 1893. At that time there was an old vertical river rod attached to the central pier of the bridge. As the foot of this at low water was covered by rock and could not be read, an inclined gage was put in position at the west end of the county bridge. This consisted of two pieces of timber, having a total length of 24 feet. These were firmly secured to timbers, bedded, and loaded with rock. The gage rod was painted white and lettered in vertical feet and tenths of a foot. After this new gage was located it was ascertained that the readings on the old gage would be 1.13 feet higher than on the new. The zero of this new gage was 19.02 feet below the top of the rail of the Northern Pacific Railway immediately west of the west end of the bridge, which was about 40 feet from the gage. The high-water mark at that time showed that a flood had risen to 8.80 feet on the old gage. Readings at this point were begun on October 2, 1893, and continued during the following winter and spring until May 19, 1894. Owing to the destruction of the gage by floods the station was for a time abandoned.

During August, 1895, Arthur P. Davis visited the locality and found that the section was not favorable for making discharge measurements. He accordingly selected the present station, which is located at Union Gap, 6 miles below North Yakima, Wash., and 1,000 feet below the highway bridge. It is about 3 miles above the intake of the Sunnyside canal. The station is of value, as it is the only point near the large irrigated area above and below, which is unaffected by the taking out of water in irrigating canals. The gage rod is inclined and is attached to a willow stump and post set in the ground. It is read once each day by H. Kennedy, the section foreman. Discharge measurements are made by means of a cable, car, and tagged wire 150 feet above the gage and 1,000 feet below the highway bridge. The initial point for soundings is a cross chiseled on a rock 2.7 feet from the cable support on the right bank. The channel is straight for 1,000 feet above and below the station. The current has a moderate velocity. The right bank is high, not liable to overflow, and is covered with sagebrush. The left bank is a low gravel bar, liable to overflow. The bed of the stream is composed of gravel, free from vegetation, and is permanent. There is one channel at low water and two channels at ordinary and flood stages. Bench mark No. 1 is the top of a large bowlder between two other bowlders 43 feet north of the gage and 6.5 feet east of the fence. Its elevation is 17.52 feet above the zero of the gage. Bench mark No. 2 is the top of a large bowlder under the railroad fence 12 feet north of the gage. Its elevation is 21.29 feet above the zero of the gage. The elevation of the initial point for soundings is 17.45 feet

re the zero of the gage. The bench marks are marked "B. M."
ı black paint.

he observations at this station during 1903 have been made under
direction of T. A. Noble, district engineer.

Discharge measurements of Yakima River at Union Gap, Washington, in 1903.

Date.	Hydrographer.	Gage height.	Discharge.
		Feet.	*Second-feet.*
21......................	T. A. Noble..................	6. 40	5, 046
ımber 16...............	G. H. Bliss..................	4. 87	1, 663

ın daily gage height, in feet, of Yakima River at Union Gap, Washington, for 1903.

Day.	Jan.	Feb.	Mar.	Apr.	May.	June.	July.	Aug.	Sept.	Oct.	Nov.	Dec.
....	5. 40	6. 00	5. 40	7. 90	8. 20	11. 20	8. 90	5. 50	4. 10	4. 90	5. 80	8. 4
....	5. 50	5. 90	5. 50	7. 60	8. 20	11. 70	8. 00	5. 50	4. 10	4. 90	5. 70	8. 6
....	10. 50	5. 80	5. 50	7. 80	8. 30	11. 40	7. 90	5. 50	4. 00	4. 90	5. 70	8. 6
....	9. 90	5. 80	5. 40	7. 40	8. 60	10. 80	7. 90	5. 30	4. 00	5. 00	5. 70	8. 0
....	9. 10	5. 60	5. 30	7. 40	8. 60	10. 00	7. 90	5. 20	4. 00	5. 30	5. 70	7. 8
....	9. 10	5. 60	5. 30	7. 10	9. 60	9. 70	7. 90	5. 20	4. 00	5. 90	5. 80	7. 6
....	9. 00	5. 40	5. 30	6. 90	9. 60	9. 90	7. 40	5. 00	4. 10	6. 90	5. 80	7. 3
....	9. 00	5. 40	5. 40	6. 90	9. 40	10. 60	7. 00	4. 90	4. 10	6. 90	7. 10	6. 9
....	8. 60	5. 60	5. 40	6. 70	9. 40	11. 20	7. 00	4. 90	4. 10	6. 90	6. 70	6. 8
....	8. 20	5. 60	5. 30	6. 60	9. 00	11. 70	6. 90	4. 90	4. 30	6. 90	6. 50	6. 6
....	7. 80	5. 40	5. 30	6. 60	8. 70	11. 90	6. 90	4. 70	4. 30	6. 70	6. 40	6. 3
....	7. 60	5. 40	5. 50	6. 40	8. 60	11. 80	6. 90	4. 70	4. 90	6. 70	6. 40	6. 3
....	7. 00	5. 40	5. 70	6. 40	9. 00	11. 80	6. 90	4. 70	4. 90	6. 70	6. 40	6. 3
....	6. 80	5. 40	5. 90	6. 30	9. 70	11. 60	6. 70	4. 50	4. 90	6. 70	6. 40	6. 1
....	6. 60	5. 40	5. 60	6. 30	10. 00	11. 40	6. 70	4. 50	4. 70	6. 50	6. 00	6. 1
....	6. 30	5. 40	5. 40	6. 40	9. 80	10. 80	6. 50	4. 50	4. 70	6. 50	6. 00	6. 1
....	6. 20	5. 30	5. 30	6. 40	9. 30	11. 40	6. 50	4. 30	4. 70	6. 30	6. 00	6. 1
....	6. 20	5. 30	5. 30	6. 40	9. 10	10. 80	6. 50	4. 30	4. 70	6. 10	6. 00	6. 1
....	6. 00	5. 30	5. 30	6. 60	9. 00	10. 00	6. 30	4. 30	4. 70	6. 10	5. 80	6. 1
....	6. 00	5. 30	5. 30	6. 60	8. 60	9. 80	6. 30	4. 30	4. 70	6. 00	5. 80	6. 1
....	6. 00	5. 30	5. 30	6. 80	8. 60	9. 50	6. 30	4. 30	4. 70	6. 00	5. 80	6. 0
....	6. 30	5. 30	5. 30	6. 90	8. 00	9. 30	6. 30	4. 30	4. 80	6. 00	5. 40	6. 0
....	6. 30	5. 30	5. 30	7. 00	8. 00	9. 00	6. 30	4. 20	4. 80	5. 90	5. 90	6. 0
....	6. 60	5. 30	5. 80	7. 40	7. 80	9. 00	6. 10	4. 20	4. 90	5. 90	5. 90	5. 9
....	6. 60	5. 40	6. 00	7. 60	7. 80	8. 90	6. 10	4. 20	5. 00	5. 60	6. 00	5. 9
....	6. 30	5. 40	6. 30	8. 00	8. 00	8. 90	6. 00	4. 20	5. 00	5. 40	6. 00	5. 8
....	6. 30	5. 40	6. 90	8. 20	8. 60	8. 70	5. 90	4. 20	5. 00	5. 30	6. 10	5. 7
....	6. 10	5. 40	7. 00	8. 20	9. 00	8. 50	5. 70	4. 20	5. 00	5. 20	6. 10	5. 6
....	6. 30	......	8. 00	8. 10	9. 00	9. 00	5. 60	4. 20	5. 00	5. 60	7. 10	5. 6
....	6. 30	......	7. 90	8. 10	9. 20	9. 20	5. 60	4. 10	5. 00	5. 60	7. 80	5. 5
....	6. 30	......	7. 90	......	9. 70	......	5. 50	4. 10	......	5. 80	......	5. 5

April...............	10,000	4,210	6,417	381,838	1.94	2.16
May	17,700	8,500	12,706	781,261	3,85	4.44
June	26,250	11,200		1,142,955	5.82	6.49
July	12,800	2,650	5,499	338,120	1.67	1.93
August	2,650	920	1,501	92,293	.45	.52
September..........	1,940	820	1,446	86,043	.44	.49
October	5,700	1,810	3,671	225,721	1.11	1.28
November..........	8,500	2,500	3,920	233,256	1.19	1.33
December	11,600	2,650	5,026	309,037	1.52	1.75
The year	26,250	820	6,037	4,377,494	1,83	24.85

NACHES RIVER NEAR NORTH YAKIMA, WASH.

This station was established August 14, 1893, by F. H. Newell, at point a few hundred yards above the mouth of the Naches River, near the bridge of the Northern Pacific Railway. The vertical gage was nailed to the cribwork on the right-hand side of the river, above the railroad bridge, and could be read easily from the track. The 2-foot mark was 9.97 feet below the top of the rail on the bridge, the gage being about 60 feet easterly from the rail. The top of the iron pier, on the southeast end of the county bridge, was 5.87 feet above this 12-foot mark. Measurements were made from the county bridge. The locality was, however, not favorable for the purpose, as the water is very swift, and was broken by the piers of the bridge. Owing to the difficulty of securing accurate measurements the readings were discontinued on September 20, 1894, and not resumed until August 19, 1895. The flood of November, 1896, modified the channel very greatly, depositing a large mass of coarse gravel and small bowlders along the right side of the channel at the section, so that the rod was about 50 feet from the edge of the water at low stages. The current is swift, even at low water. On account of the instability of the channel the station was abandoned in February, 1897, although a number of discharge measurements were made during the season. The station was located below the heads of a number of ditches. May 19, 1897, a station was established on Yakima River, 5 miles above the mouth of the Naches, at the Northern Pacific Railway bridge near Selah, Wash., with the idea that the difference in discharge between this station and the one at Union Gap would give approximately the discharge of Naches River. Two ditches, those of the Moxee Valley, are taken out between the two points, but their amount is about counterbalanced by that received from Atanum Creek and the wastage at Old Town.

The North Yakima station was reestablished on February 1, 1898, and the station at Selah, on Yakima River, was discontinued. Since the reestablishment of the Naches station the river channel has been in a condition more favorable for meter observations than formerly. Discharge measurements are made from the lower side of the highway bridge. A new horizontal gage rod, with wire and weight, was attached to the main span of the Northern Pacific Railway bridge at the mouth of the Naches, a few hundred feet downstream from the highway bridge. The length of the gage wire from index to foot of weight was 30.41 feet. The pulley distance was 5.844 feet. The elevation of top of pulley was 24.57 feet. The bench mark was the top of the north end of east sill of clearance posts, about 150 feet north of Northern Pacific Railway bridge. Elevation, 23.766 feet above zero of gage. On December 27, 1898, Mr. Arnold connected the highway bridge with this bench mark. The top of the northeast concrete pier was found to be at an elevation of 22.09 feet, and the top of the bridge post at the 150-foot mark at an elevation of 26.76 feet above zero. The distance from the top of the post to the surface of the water has been carefully measured at each discharge measurement, so that the exact river height at the highway bridge is known.

On June 20, 1901, the gage rod and bench mark having been disturbed during alterations to the railroad bridge and approaches, a new 4 by 4 inch inclined gage rod was established on the left bank of the river, 30 feet downstream from the railroad bridge. This is the gage in present use. The lower section of the rod is inclined at an angle of 36° 30' with the horizontal. The upper section is inclined at an angle of 80° with the horizontal. On September 21, 1903, the station was equipped with a cable, car, tagged wire, and stay wire. The cable is located 180 feet above the Northern Pacific Railway bridge, and 170 feet below the highway bridge, at which discharge measurements were formerly made. The stay wire is 85 feet above the cable. The cable has a total span of 280 feet. The initial point for soundings is the south face of the cottonwood tree to which the cable is fastened, on the north bank. The channel is straight for 100 feet above and for 75 feet below the cable. The current is swift. Both banks are low and covered with gravel. At flood stages the rock-filled crib will prevent overflow on both banks. The bed of the stream is composed of gravel, free from vegetation, and somewhat shifting. Bench mark No. 1 is a cross in the top of the center of the down-stream end of the railroad bridge pier on the right bank. Its elevation above the zero of the gage is 19.94 feet, and above sea level is 1,090 feet. Bench mark No. 2 is a railroad spike driven into the south side of the telegraph pole, 23 feet north of the gage and 21 feet east of the railroad track. Its elevation is 17.81 feet above the zero of the gage.

The observations at this station during 1903 have been made under the direction of T. A. Noble, district engineer.

Discharge measurements of Naches River near North Yakima, Wash., in 1903.

Date.	Hydrographer.	Gage height.	Discharge.
		Feet.	*Second-feet.*
June 10	Sydney Arnold	10. 20	13, 064
September 21	G. H. Bliss	5. 25	461

Mean daily gage height, in feet, of Naches River near North Yakima, Wash., for 1903.

Day.	Jan.	Feb.	Mar.	Apr.	May.	June.	July.	Aug.	Sept.	Oct.	Nov.	Dec.
1	6.80	6.50	a6.00	a7.50	8.00	9.90	7.90	6.10	5.20	5.20	6.55	8.80
2	6.30	6.40	6.00	7.40	8.20	10.00	7.80	6.10	a5.10	5.10	6.50	8.60
3	7.20	6.30	6.00	7.30	8.35	a9.70	7.60	6.10	5.10	5.10	6.40	8.30
4	8.10	a6.30	6.00	7.20	8.50	9.20	7.60	6.00	5.10	5.60	6.40	8.00
5	9.00	6.30	6.00	7.15	a8.90	8.90	7.40	a5.90	5.00	6.10	6.50	7.70
6	8.50	6.30	6.00	7.10	9.10	9.00	7.60	5.90	5.00	a6.60	6.80	7.50
7	a8.00	6.30	6.00	7.00	9.00	9.40	a7.50	5.80	5.00	7.30	7.80	7.30
8	7.60	6.25	a6.00	a7.00	8.60	9.80	7.40	5.80	5.00	7.00	7.50	a7.10
9	7.80	6.20	6.00	7.00	8.40	10.00	7.40	5.80	a5.00	6.90	7.20	7.00
10	7.20	6.20	6.00	7.00	8.35	a10.20	7.40	5.80	5.00	6.70	7.10	6.90
11	7.10	a6.20	6.00	7.00	8.30	10.10	7.40	5.70	5.10	6.85	a7.00	6.80
12	7.00	6.20	6.00	6.95	a8.60	10.00	7.40	a5.70	5.10	7.00	7.00	6.70
13	6.90	6.20	6.00	6.90	9.00	10.00	7.40	5.60	5.15	a7.00	6.90	6.65
14	a6.80	6.20	6.00	6.80	9.40	9.80	a7.30	5.60	5.20	7.00	6.80	6.60
15	6.70	6.15	a6.00	a6.80	9.20	9.60	7.20	5.60	5.20	7.00	6.70	a6.60
16	6.60	6.10	6.00	6.80	8.70	10.10	7.00	5.55	a5.20	7.00	6.60	6.80
17	6.50	6.10	6.00	6.90	8.50	a10.00	7.00	5.50	5.20	6.90	6.60	6.90
18	6.50	a6.10	6.00	7.00	8.30	9.60	7.00	5.50	5.20	6.90	a6.60	6.80
19	6.50	6.10	6.00	7.00	a8.10	9.10	7.05	a5.50	5.20	6.90	6.60	6.80
20	6.50	6.00	6.00	7.00	7.90	9.00	a7.10	5.50	5.15	a6.60	6.60	6.75
21	a6.60	6.00	6.00	7.10	7.70	9.00	7.10	5.40	5.10	6.90	6.60	6.70
22	6.70	6.00	a6.15	a7.20	7.60	9.00	7.10	5.40	a5.30	6.80	6.60	a6.60
23	6.70	6.00	6.30	7.30	7.50	8.70	7.00	5.35	a5.30	6.80	6.60	·6.60
24	6.80	6.00	6.50	7.60	7.65	a8.60	6.90	5.30	5.40	6.70	6.60	6.50
25	6.80	a6.00	6.60	7.70	7.80	8.50	6.80	5.30	5.50	6.65	a6.60	6.40
26	6.80	6.00	6.90	7.70	a8.20	8.50	6.70	a5.30	5.50	6.60	6.70	6.30
27	6.80	6.00	7.00	7.70	8.40	8.40	6.60	5.30	5.40	a6.60	6.80	6.25
28	a6.70	6.00	7.40	7.90	8.30	8.35	a6.50	5.30	5.30	6.60	7.30	6.20
29	6.70	......	7.55	a7.90	8.30	8.80	6.40	5.30	5.30	6.70	7.40	6.10
30	6.60	......	7.70	7.80	8.70	8.30	6.30	5.20	a5.20	6.70	7.50	6.10
31	6.60	......	7.60	......	9.30	(a)	6.20	5.20	......	6.60	......	6.10

a Sundays interpolated.

Rating table for Naches River near North Yakima, Wash., for 1902 and 1903.

Gage height.	Discharge.	Gage height.	Discharge.	Gage height.	Discharge.	Gage height.	Discharge.
Feet.	*Second-feet.*	*Feet.*	*Second-feet.*	*Feet.*	*Second-feet.*	*Feet.*	*Second-feet.*
5.0	200	6.4	1,350	7.8	3,810	9.2	8,110
5.2	300	6.6	1,595	8.0	4,370	9.4	8,750
5.4	425	6.8	1,860	8.2	5,010	9.6	9,390
5.6	570	7.0	2,152	8.4	5,650	9.8	10,030
5.8	735	7.2	2,580	8.6	6,290	10.0	10,670
6.0	920	7.4	2,950	8.8	6,830	10.2	11,310
6.2	1,125	7.6	3,360	9.0	7,470		

Estimated monthly discharge of Naches River near North Yakima, Wash., in 1903.

[Drainage area, 1,000 square miles.]

Month.	Discharge in second-feet.			Total in acre-feet.	Run-off	
	Maximum.	Minimum.	Mean.		Second-feet per square mile.	Depth in inches.
January	7,470	1,235	2,352	144,619	2.35	2.71
February	1,470	920	1,084	60,202	1.08	1.12
March	3,580	920	1,355	83,316	1.36	1.57
April...............	4,050	1,860	2,623	156,079	2.62	2.92
May	8,750	3,150	5,704	350,725	5.70	6.57
June	11,310	5,330	8,459	503,345	8.46	9.44
July	4,050	1,125	2,491	153,166	2.49	2.87
August	1,020	300	585	35,970	.58	.67
September..........	495	200	296	17,613	.30	.33
October	2,760	250	1,673	102,869	1.67	1.93
November..........	3,810	1,350	1,972	117,342	1.97	2.20
December	6,290	1,020	2,244	137,978	2.24	2.58
The year	11,310	200	2,570	1,863,224	2.57	34.91

TIETON RIVER NEAR NORTH YAKIMA, WASH.

The gaging station on this stream was established April 14, 1902, at a point immediately below the mouth of Oak Creek, in sec. 3, T. 14 N., R. 16 E. of the Willamette meridian, and about 22 miles from North Yakima by road. The gage rod is inclined at an angle of 55° with the horizontal and is on the left bank of the stream. It consists of a 1 by 5 inch cedar plank supported and braced by stout logs. The equipment with which measurements are made consists of a ⅜-inch galvanized-iron cable supporting a wooden car, tag wire, and stay wire. The initial point for soundings is a cross chiseled in a ledge with a black ring painted around it. It is under the cable, feet east of the west shear legs. The observer is Omer Tetherow, a farmer. There are no side channels and the banks are not subject to overflow. The bed of the stream is rocky, with shifting gravel bars which make it difficult to find suitable cross sections for meter measurements. The point selected for the station is, however, a fairly good one, and the channel is straight both above and below the station for several hundred yards.

The observations at this station during 1903 have been made under the direction of T. A. Noble, district engineer.

Discharge measurements of Tieton River near North Yakima, Wash., in 1903.

Date.	Hydrographer.	Gage height.	Discharge.
		Feet.	Second-feet.
April 9	Sydney Arnold	8.00	718
April 27	....do	8.80	1,301
June 25	....do	9.70	2,409
September 18	G. H. Bliss	7.05	292

Mean daily gage height, in feet, of Tieton River near North Yakima, Wash., for 1903.

Day.	Jan.	Feb.	Mar.	Apr.	May.	June.	July.	Aug.	Sept.	Oct.	Nov.	Dec.
1	7.23	7.50	7.20	8.45	8.65	10.45	8.95	7.92	7.37	7.07	7.30	9.20
2	7.28	7.45	7.20	8.30	8.65	10.60	8.95	7.92	7.32	7.02	7.42	9.10
3	10.04	7.45	7.20	8.30	8.95	10.15	8.75	7.82	7.35	6.97	7.30	8.63
4	10.70	7.53	7.15	8.20	9.48	9.90	8.80	7.80	7.30	7.00	7.70	8.42
5	9.80	7.45	7.05	8.00	9.38	9.75	9.50	7.72	7.20	7.42	8.40	8.25
6	9.35	7.33	7.10	8.00	9.25	9.65	8.65	7.72	7.23	8.85	8.85	8.10
7	9.00	7.43	7.10	8.10	9.20	10.30	8.55	7.80	7.22	7.95	8.60	7.97
8	8.75	7.40	7.10	8.00	9.10	10.75	8.50	7.87	7.25	7.72	8.20	7.85
9	8.53	7.35	7.05	7.98	8.58	10.82	8.55	7.85	7.18	7.62	8.00	7.72
10	8.35	7.35	7.15	7.85	9.05	10.95	8.60	7.85	7.45	7.72	8.40	7.55
11	8.15	7.15	7.20	7.80	9.08	10.85	8.57	7.80	7.42	8.10	8.30	7.50
12	8.00	7.08	7.20	7.78	9.33	10.45	8.72	7.75	7.48	7.97	8.40	7.60
13		7.28	7.10	7.80	9.68	10.50	8.55	7.65	7.30	7.82	8.20	7.47
14		7.15	7.00	7.80	9.78	10.20	8.45	7.75	7.20	7.75	7.80	7.45
15	7.60	7.10	7.10	7.85	9.60	10.25	8.37	7.60	7.15	7.57	7.45	7.60
16	7.50	7.15	7.05	7.78	9.40	11.40	8.35	7.65	7.07	7.45	7.30	7.50
17	7.45	7.15	7.00	7.90	9.15	10.65	8.27	7.53	7.05	7.40	7.30	7.80
18	7.60	7.13	7.00	7.80	8.95	10.15	8.45	7.63	7.15	7.45	7.30	7.70
19	7.53	7.10	7.00	8.00	8.85	9.95	8.50	7.70	7.35	7.32	7.20	7.60
20	7.53	7.10	7.00	8.05	8.68	10.10	8.45	7.73	7.45	7.25	7.25	7.70
21	8.05	7.10	7.05	8.10	8.55	10.00	8.50	7.63	7.40	7.20	7.35	7.60
22	7.85	7.10	7.13	8.25	8.50	9.75	8.60	7.55	7.62	7.20	7.50	7.55
23	7.80	7.10	7.35	8.33	8.48	9.65	8.45	7.50	7.60	7.10	7.45	7.50
24	8.18	7.10	7.53	8.45	8.55	9.50	8.37	7.43	7.62	7.20	7.40	7.43
25	8.55	7.10	7.90	8.60	8.78	9.50	8.20	7.47	7.75	7.10	7.33	7.35
26	8.30	7.23	8.05	8.78	9.05	9.65	8.07	7.43	7.60	7.10	7.35	7.35
27	8.20	7.25	8.50	8.80	9.05	9.80	7.97	7.43	7.52	7.10	8.15	7.35
28	8.00	7.23	8.68	8.70	9.00	9.97	7.90	7.43	7.47	7.60	8.35	7.33
29	7.98		8.75	8.68	9.00	9.57	7.87	7.45	7.22	8.20	8.45	7.23
30	7.88		8.60	8.58	9.40	9.10	8.00	7.45	7.10	7.75	8.90	7.18
31	7.78		8.55		10.00		7.87	7.40		7.60		7.15

Rating table for Tieton River near North Yakima, Wash., for 1902 and 1903.

Gage height.	Discharge.	Gage height.	Discharge.	Gage height.	Discharge.	Gage height.	Discharge.
Feet.	*Second-feet.*	*Feet.*	*Second-feet.*	*Feet.*	*Second-feet.*	*Feet.*	*Second-feet.*
6.0	120	7.2	372	8.4	900	9.6	2,310
6.2	160	7.4	420	8.6	1,110	9.8	2,550
6.4	200	7.6	484	8.8	1,350	10.0	2,790
6.6	240	7.8	560	9.0	1,590	11.0	3,990
6.8	284	8.0	640	9.2	1,830		
7.0	328	8.2	750	9.4	2,070		

Estimated monthly discharge of Tieton River near North Yakima, Wash., for 1903.

[Drainage area, 269 square miles.]

Month.	Discharge in second-feet.			Total in acre-feet.	Run-off.	
	Maximum.	Minimum.	Mean.		Second-feet per square mile.	Depth in inches.
January [a]	3,630	383	966	58,905	3.31	3.83
February	467	350	365	21,382	1.33	1.38
March	1,290	328	500	30,744	1.73	1.99
April	1,350	560	789	46,949	2.73	3.04
May	2,790	1,000	1,682	103,422	5.82	6.71
June	4,470	1,710	2,960	176,132	10.24	11.42
July	2,190	580	1,011	62,164	3.50	4.04
August	600	420	507	31,174	1.75	2.02
September	541	339	410	24,397	1.42	1.58
October	1,170	110	470	28,899	1.63	1.88
November	1,470	372	646	38,440	2.24	2.50
December	1,830	361	608	37,384	2.10	2.42
The year	4,470	110	910	659,992	3.15	42.81

[a] January 13 and 14 interpolated.

CLEALUM RIVER NEAR ROSLYN, WASH.

This station was established October 10, 1903, by G. H. Bliss. It is located 1,000 feet below the outlet of Lake Clealum. It is 2¼ miles northwest of Roslyn and 6¼ miles northwest of Clealum, Wash. The gage is an inclined rod on the left bank, 20 feet upstream from the cable. It is read once each day by Charles M. Davis. Discharge measurements are made by means of a cable, car, tagged wire, and stay wire. The initial point for soundings is the south face of the black pine tree, 18 inches in diameter, to which the cable is fastened, on the left bank. The channel is straight for 300 feet above and for 900 feet below the station. The current is swift. Both banks are high, not liable to overflow, and are heavily timbered. The bed of the stream is composed of gravel, free from vegetation, and permanent. The bench mark is a large spike driven into the root of the tree to which the cable is fastened, on the left bank. The root is on the east or downstream side of the tree, and the tree is blazed. The elevation of the bench mark is 17.40 feet above the zero of the gage.

The observations at this station during 1903 have been made under the direction of T. A. Noble, district engineer.

The following discharge measurement was made by G. H. Bliss in 1903:

October 10: Gage height, 3.65 feet; discharge, 1,143 second-feet.

Mean daily gage height, in feet, of Clealum River near Roslyn, Wash., for 1903.

Day.	Oct.	Nov.	Dec.	Day.	Oct.	Nov.	Dec.	Day.	Oct.	Nov.	Dec.
1		3.20	5.20	12	4.05	3.05	3.10	23	2.90	2.33	2.55
2		3.10	5.90	13	4.31	2.90	2.90	24	2.83	2.30	2.50
3		3.00	6.70	14	4.32	2.90	2.90	25	2.75	2.25	2.40
4		2.90	5.20	15	4.15	2.65	2.80	26	2.64	2.30	2.35
5		3.10	4.70	16	3.90	2.65	2.75	27	2.56	2.70	2.28
6		3.15	4.25	17	3.65	2.45	2.70	28	2.58	3.45	2.25
7		3.25	3.95	18	3.47	2.45	2.65	29	2.96	3.85	2.18
8		3.20	3.70	19	3.30	2.45	2.63	30	3.35	4.45	2.15
9		3.20	3.45	20	3.18	2.40	2.65	31	3.30		2.10
10	3.65	3.17	3.30	21	3.06	2.40	2.60				
11	3.72	3.10	3.10	22	3.01	2.35	2.60				

KACHESS RIVER NEAR EASTON, WASH.

This station was established October 14, 1903, by G. H. Bliss. It is located 2 miles northwest of Easton, Wash., and one-half mile below the foot of Lake Kachess, at which a dam is being constructed by the Cascade Canal Company. The gage is an inclined rod on the left bank, directly under the cable. The gage is read once each day by W. W. Johnson. Discharge measurements are made by means of a cable, car, tagged wire, and stay wire. The cable is of one-half inch plow steel and has a span of 120 feet. The initial point for soundings is the south side of the aspen tree to which the cable is fastened, on the left bank. The channel is straight for 600 feet above and for 150 feet below the station. The current is swift at high stages only. Both banks are high, wooded, and not liable to overflow. The bed of the stream is composed of gravel and rocks, free from vegetation. The bench mark is the top of a large wire nail driven into the south side of the large aspen tree to which the cable is fastened, on the left or north bank. The tree is blazed and is marked "B. M." with black paint. The nail is near the base of the tree and has an elevation of 11.27 feet above the zero of the gage.

The observations at this station during 1903 have been made under the direction of T. A. Noble, district engineer.

The following discharge measurement was made by G. H. Bliss in 1903:

October 14: Gage height, 5.30 feet; discharge, 642 second-feet.

Mean daily gage height, in feet, of Kachess River near Easton, Wash., for 1903.

Day.	Nov.	Dec.	Day.	Nov.	Dec.	Day.	Nov.	Dec.		Nov.	De.
1		5.8	9		4.9	17		4.4	25	4.4	4.2
2		5.5	10		4.8	18		4.3	26	4.3	4.2
3		5.4	11		4.7	19		4.2	27	4.4	4.2
4		5.4	12		4.6	20	3.7	4.2	28	4.6	4.1
5		5.3	13		4.5	21	3.8	4.2	29	4.5	4.1
6		5.2	14		4.6	22	4.6	4.3	30	4.9	4.0
7		5.2	15		4.60	23	4.6	4.2	31		4.0
8		5.0	16		4.60	24	4.4	4.2			

YAKIMA RIVER NEAR MARTIN, WASH.

This station was established October 18, 1903, by G. H. Bliss. It is located 1,000 feet below the outlet of Lake Keechelus and 800 feet below the dam of the Cascade Lumber Company. It is 4 miles northwest of Martin, Wash. The gage is in two sections and is located on the right bank just above the cable. The lower inclined section reads from 5 to 7 feet. The upper vertical section reads from 7 to 13 feet. The gage is read once each day by Christian Hansen. Discharge measurements are made by means of a cable, car, tagged wire, and stay wire. The cable has a total span of 200 feet. The initial point for soundings is the north face of the tree on the south, or right, bank, to which the cable is fastened. The channel is straight for 500 feet above and for 350 feet below the station. Both banks are high, not liable to overflow, and are heavily timbered. The current is swift. There is but one channel at all stages. Bench mark No. 1 is the top of a spike in the root of a large cedar tree near the gage on the right bank. The root is on the south side of the tree and is blazed. The elevation of the bench mark is 15.63 feet above the zero of the gage. Bench mark No. 2 is the top of a spike in the root on the north side of a large cedar tree 50 feet south and 50 feet west of the gage. Its elevation is 21.67 feet above the zero of the gage. The bed of the stream is composed of gravel, free from vegetation, and permanent.

The observations at this station during 1903 have been made under the direction of T. A. Noble, district engineer.

The following discharge measurement was made by G. H. Bliss in 1903:

October 18: Gage height, 7.5 feet; discharge, 321 second-feet.

CHELAN RIVER BELOW LAKE CHELAN, WASHINGTON.

This station was established November 6, 1903, by G. H. Bliss.   It ɜ located at the highway bridge 3,000 feet below the outlet of the lake nd 4 miles northwest of Chelan Falls.   The gage is a vertical rod 16 eet long attached to the third pile bent of the northwestern approach o the bridge.   It is read once each day by G. L. Richardson.   Dis-barge measurements are made from the downstream side of the new ɩighway bridge, to which the gage is attached.   The initial point for ɔundings is the end vertical on the downstream side of the bridge at ɦe northwest approach.   The channel is straight for 50 feet above and ɔr 150 feet below the station.   The right bank can not overflow.   The eft bank is lower than the right, but is not liable to overflow.   Both ʙanks are without trees.   The bed of the stream is composed of rocks ·nd gravel, free from vegetation, and somewhat liable to shift.   The ʙench mark is a wire spike driven into the root of a large cottonwood ʀee, which is 40 feet downstream from the northwestern approach to ɦe bridge and 30 feet from the river.   The root is on the west side of ɦe tree.   The elevation of the bench mark is 11.86 feet above the ero of the gage.

The observations at this station during 1903 have been made under ɦe direction of T. A. Noble, district engineer.

The following discharge measurement was made by G. H. Bliss in 903:

November 6: Gage height, 6.60 feet; discharge, 1,764 second-feet.

Ƈean daily gage height, in feet, of Chelan River below Lake Chelan, Washington, for 1903.

Day.	Nov.	Dec.	Day.	Nov.	Dec.	Day.	Nov.	Dec.	Day.	Nov.	Dec.
....		6.18	9	6.60	6.20	17	6.40	6.00	25	6.15	5.97
....		6.35	10	6.55	6.10	18	6.35	5.90	26	6.16	6.00
....		6.30	11	6.60	6.00	19	6.31	5.90	27	6.18	5.95
....		6.28	12	6.50	5.80	20	6.30	5.90	28	6.20	5.90
....		6.28	13	6.48	5.90	21	6.30	5.88	29	6.20	5.90
....	6.60	6.30	14	6.45	5.90	22	6.30	5.90	30	6.18	5.85
....	6.59	6.28	15	6.45	5.95	23	6.28	5.92	31		5.85
....	6.61	6.25	16	6.43	6.00	24	6.22	5.95			

METHOW RIVER NEAR PATEROS, WASH.

This station was established May 3, 1903, by T. A. Noble.   It is ɔcated about 4,100 feet upstream from the county bridge.   The tem-ɷorary gage consisted of an inclined 1¼ by 8 inch board 12 feet ɷng.   It was fastened to sleepers buried in the left bank.   The per-ɴanent gage was installed June 17, 1903.   It is fastened between two ɔine trees on the left bank.   The lower section is inclined and reads ˈrom 0 to 10 feet.   The upper vertical section is a 1 by 3 inch timber ɪnd reads from 10 to 19 feet.   During 1903 the gage was read once

each day by Charles E. Nosler and K. K. Parker. Discharge meas
ments are made by means of a boat 10 feet downstream from the
of the old bridge, which has been demolished. This point is 4,100
downstream from the gage. It is expected that a new bridge wil
constructed near this point, from which measurements can be m
The initial point for soundings is a 30-penny spike driven in a
stake flush with the ground. It is on the left bank 10 feet downsti
from the old bridge. The distance from the initial point to the
graph pole on the opposite bank is 333 feet. The channel is str
for 3,000 feet above and for 400 feet below the gaging section.
point is about 400 feet above the junction of Columbia and Me
rivers. At high stages in the Columbia River backwater from
stream makes the current sluggish at the gaging section. The l
water does not extend far enough up the river to reach the gage.
right bank overflows only during periods of floods in the Colu
River. The left bank is high, not liable to overflow, and wil
trees. At low water the bed of the stream is composed of grave
small bowlders. That part of the channel which is covered ou
flood stages is mostly composed of sand. Bench mark No. 1 is a
Geological Survey standard iron post in front of the hotel at Pat
Wash. Its elevation is 26.05 feet above the zero of the gage and
feet above sea level. The initial point for soundings has an elev
of 20.08 feet above the zero of the gage. For a distance of 30 i
above the gaging section the Methow River is too swift and r
to be measured at high stages. It is only at a point just abov
mouth of the river, where the current is decreased by backwater :
the Columbia, that flood measurements can be made. Bench mark
2 is the top of a large granite bowlder marked "B. M." It is 30
northeast of the vertical portion of the gage on the left bank.
elevation is 15.21 feet above gage datum and 753.95 feet above r
sea level.

The observations at this station during 1903 have been made u
the direction of T. A. Noble, district engineer.

Discharge measurements of Methow River near Pateros, Wash., in 1903.

Date.	Hydrographer.	Gage height.	Disch:
		Feet.	*Second*
June 17	T. A. Noble	11.52	12
August 12	G. H. Bliss....................	5.29	1
November 8	.....do	5.20	1

Mean daily gage height, in feet, of Methow River near Pateros, Wash., for 1903.

Day.	May.	June.	July.	Aug.	Sept.	Oct.	Nov.	Dec.
1	......	(a)	8.50	5.69	4.90	4.90	5.10	4.90
2	......	(a)	8.30	5.60	4.80	4.90	5.10	4.95
3	5.90	(a)	8.10	5.49	4.70	4.80	5.10	5.00
4	5.92	(a)	7.80	5.40	4.70	4.90	5.10	4.90
5	5.93	(a)	7.70	5.40	4.65	4.90	5.10	4.75
6	5.94	(a)	7.40	5.40	4.70	4.90	5.20	4.80
7	5.94	(a)	7.30	5.40	4.70	4.80	5.20	4.90
8	5.90	(a)	7.20	5.40	4.70	4.90	5.20	4.90
9	6.00	(a)	7.10	(a)	4.60	4.90	5.20	4.90
10	5.88	(a)	7.00	(a)	4.60	4.80	5.10	4.90
11	5.88	(a)	6.90	(a)	4.70	4.90	5.13	4.90
12	5.88	(a)	6.87	(a)	4.80	5.00	5.07	4.90
13	5.93	(a)	6.85	(a)	4.80	5.00	5.00	4.90
14	6.00	(a)	6.77	(a)	4.80	5.00	5.05	4.85
15	6.03	(a)	6.52	(a)	4.80	5.10	5.00	4.85
16	6.01	(a)	6.55	5.20	4.80	5.10	4.95	4.80
17	5.99	11.60	6.54	5.10	4.70	5.10	4.90	4.80
18	5.96	11.05	6.50	5.10	4.80	5.10	4.90	4.80
19	5.94	11.08	6.30	4.90	4.80	5.20	4.90	4.75
20	5.93	10.07	6.05	4.80	4.80	5.20	4.90	4.75
21	5.92	10.80	6.02	4.90	4.80	5.20	4.90	4.75
22	5.88	10.50	5.99	4.90	5.00	5.20	4.90	4.75
23	5.88	10.20	5.96	4.90	5.10	5.20	4.85	4.75
24	5.88	9.70	5.93	4.90	5.10	5.20	4.80	4.70
25	5.92	9.70	5.90	5.00	5.10	5.20	4.80	4.65
26	5.93	9.60	5.87	5.00	5.10	5.10	4.80	4.65
27	5.96	9.50	5.84	5.00	5.10	5.10	4.77	4.65
28	6.01	9.40	5.81	4.90	5.00	5.10	4.77	4.65
29	6.05	9.00	5.78	4.90	5.00	5.10	4.77	4.60
30	6.05	8.80	5.75	4.80	5.00	5.10	4.80	4.60
31	(a)	....	5.72	4.80	....	5.10	....	4.60

a Missing.

SALMON CREEK NEAR MALOTT, WASH.

This station was established April 11, 1903, by T. A. Noble. It is
cated opposite R. D. Jones's house, which is on the county road half-
ay between Malott and Conconully, Okanogan County. It is reached
way of the Great Northern Railway to Wenatchee, thence by way
the Columbia River steamers to Brewster, and by the Conconully
ige from Brewster to Jones's ranch. The gage is vertical and con-
its of a 1 by 6 inch board graduated to feet and inches, and fastened
a small alder tree on the left bank of the river opposite the house
the observer, R. D. Jones, who reads the gage once each day. Dis-
arge measurements are made from the footbridge just above the
ge. The initial point for soundings is a large nail driven in a birch
b, 8 inches in diameter and 2 feet long, 4 feet east of the bridge and
feet north of the gage. It is on the left bank, 7 feet from the
ter's edge at ordinary stages. The channel is straight for 100 feet

above and for 200 feet below the station. There are rapids at the
bend in the river 100 feet above the station and at another bend 200
feet below the station. The current is swift. The right bank is low
and will overflow for about 100 feet, at which point it becomes steep.
The left bank is low and may overflow for 200 feet at extreme flood
stages. Both banks are without trees or brush with the exception of
a fringe of birch at the water's edge. The bed is rocky at the center
and sandy along the banks. It is without vegetation except near the
banks. The bench mark is the initial point for soundings. Its elevation is 4.58 feet above the zero of the gage.

The observations at this station during 1903 have been made under
the direction of T. A. Noble, district engineer.

Discharge measurements of Salmon Creek near Malott, Wash., in 1903.

Date.	Hydrographer.	Gage height.	Discharge.
		Feet.	*Second-feet.*
May 28	T. A. Noble	1.86	145
May 31	....do	2.31	226
June 4	C. Anderson	3.14	436
June 11	....do	2.23	302
June 20	E. A. Bailey	1.60	102
June 26	H. T. Jones	1.23	50
July 2	W. W. Schlecht	1.12	37
July 31	O. Laurgaard	.85	23
September 21	W. W. Schlecht	.94	22
Do	....do	.94	25
October 7	Calvin Casteel	1.02	26
Do	....do	1.02	29
Do	W. W. Schlecht	1.01	26

fean daily gage height, in feet, of Salmon Creek near Malott, Wash., for 1903.

Day.	April.	May.	June.	July.	Aug.	Sept.	Oct.	Nov.	Dec.
....		1.27	2.51	1.12	0.82	0.90	0.83	0.87	0.97
....		1.38	2.77	1.10	.92	.87	.84	.87	.93
....		1.48	3.06	1.14	.94	.84	.83	.90	.59
....		1.60	2.77	1.13	.92	.83	.85	.96	.62
....		1.67	2.50	1.23	.86	.83	.85	.93	.96
....		1.77	2.42	1.29	.81	.84	1.23	1.00	.85
....		1.67	2.44	1.50	.77	.84	1.01	.93	.85
....		1.62	2.47	1.46	.75	.83	.94	.94	.86
....		1.60	2.43	1.42	.68	.81	.92	1.00	.84
....		1.70	2.37	1.33	.71	.83	.94	.93	.85
....		1.59	2.17	1.23	1.00	.82	1.00	.85	.83
....	0.94	1.63	2.12	1.21	1.01	.94	.99	.69	.87
....	.93	1.83	2.34	1.11	.99	1.04	.96	.90	....
....	.90	2.04	2.02	1.12	.99	1.00	.94	.92	....
....	.92	1.96	1.93	1.01	.96	.99	.93	.92	....
....	.90	1.87	1.83	1.06	.96	.94	.92	.71	....
....	.96	1.83	1.76	1.00	.95	.92	.92	.67	....
....	.98	1.81	1.67	.98	.94	.92	.93	.73	....
....	.98	1.75	1.59	.97	.91	.87	.92	.83	....
....	1.04	1.72	1.60	.96	.90	.90	.92	.87	.91
....	1.10	1.74	1.52	.93	.90	.90	.90	.85	.80
....	1.12	1.63	1.45	.94	.90	1.04	.87	.83	.80
....	1.27	1.64	1.45	.92	.95	.95	.90	.87	.59
....	1.25	1.66	1.41	1.14	.98	.92	.90	.92	.77
....	1.31	1.75	1.26	1.06	1.04	.90	.92	.84	.80
....	1.42	1.87	1.25	.98	1.19	.90	.92	.96	.87
....	1.50	1.91	1.20	.96	1.04	.89	.87	1.00	.83
....	1.31	1.87	1.20	.95	.96	.85	.87	....	.85
....	1.32	1.87	1.66	.91	.94	.87	.89	.99	.87
....	1.30	1.92	1.66	.86	.93	.87	.88	.93	.85
....		2.25	....	.85	.92	....	.87		.83

g table for Salmon Creek near Malott, Wash., from April 12 to December 31, 1903.

ge ;ht.	Discharge.	Gage height.	Discharge.	Gage height.	Discharge.	Gage height.	Discharge.
ct.	*Second-feet.*	*Feet.*	*Second-feet.*	*Feet.*	*Second-feet.*	*Feet.*	*Second-feet.*
.6	16	1.3	57	2.0	166	2.7	318
.7	18	1.4	71	2.1	183	2.8	345
.8	20	1.5	86	2.2	202	2.9	373
.9	22	1.6	102	2.3	222	3.0	401
.0	27	1.7	118	2.4	244	3.1	429
.1	35	1.8	134	2.5	267		
.2	45	1.9	150	2.6	292		

le well defined. Curve extended below gage height 0.85 foot.

Estimated monthly discharge of Salmon River near Malott, Wash., for 1903.

Month.	Discharge in second-feet.			Total in acre-feet.
	Maximum.	Minimum.	Mean.	
April 12–30	86	22	41	1,56
May	212	51	124	7,63
June	429	45	170	10,11
July	86	21	38	2,37
August	45	18	24	1,67
September	31	20	23	1,39
October	51	21	24	1,47
November	27	17	22	1,30
December	27	16	21	1,29

JOHNSON CREEK NEAR RIVERSIDE, WASH.

This station was established by T. A. Noble, May 30, 1903. It is located at Sogle's ranch on the road from Riverside to Conconully, 1 mile from Riverside and 17 miles from Conconully. The equipment consists of a lip weir with an 8-foot opening and vertical sides. Below the level of the crest the weir consists of two 2-inch pine planks 12 inches wide, securely spiked together. Above the crest on each end are two planks 12 inches wide, which form the ends of the weir. The edges of the crest and ends are one-fourth inch wide and beveled on the downstream side to an angle of 60°. The pool above the weir is 10 feet long, 10 to 15 feet wide, and 1 foot deep. The velocity of approach is about 0.3 foot per second. The water has a fall of about 1 foot after passing the weir and then flows rapidly away.

The depth of water on the crest is determined by a hook gage and vernier reading to thousandths of a foot. The zero on the gage is level with the crest of the weir. Readings are made once each day by S. Sogle. The right bank of the stream is low for 10 feet back from the water's edge and then rises more abruptly. The left bank is steep and rocky. The bed of the creek is composed of small gravel. Bench mark No. 1 is the top of the fence post opposite the weir. Its elevation is 19.12 feet above gage datum. Bench mark No. 2 is a nail driven into a stake 3 feet northwest of the weir. Its elevation is 0.28 foot above gage datum.

The observations at this station during 1903 have been made under the direction of T. A. Noble, district engineer.

ean daily gage height, in feet, of Johnson Creek near Riverside, Wash., for 1903.

Day.	May.	June.	July.	Aug.	Sept.	Oct.	Nov.	Dec.
....	....	0.25	0.25	0.24	0.30	0.32	0.327	0.419
....	....	.25	.23	.21	.29	.34	.330	.395
....	....	.25	.23	.28	.29	.34	.440	.375
....	....	.26	.23	.26	.29	.35	.342	.450
....	....	.26	.29	.25	.34	.37	.340	.300
....	....	.27	.30	.27	.35	.39	.410	.371
....	....	.29	.32	.27	.35	.36	.374	.363
....	....	.29	.29	.27	.35	.36	.855	.878
....	....	.25	.29	.23	.36	.35	.371	.367
....	....	.25	.29	.21	.36	.36	.376	.367
....	....	.25	.28	.21	.36	.37	.370	.368
....	....	.22	.37	.22	.40	.36	.375	.371
....	....	.33	.35	.23	.38	.34	.378	.375
....	....	.32	.32	.22	.37	.32	.350	.374
....	....	.30	.33	.23	.37	.32	.330	.375
....	....	.20	.32	.24	.36	.33	.327	1.120
....	....	.25	.25	.26	.36	.33	.315	1.033
....	....	.24	.28	.22	.34	.32	.335	.553
....	....	.24	.24	.21	.32	.32	.348	.437
....	....	.29	.23	.20	.32	.33	.354	.437
....	....	.28	.25	.19	.37	.24	.380	.389
....	....	.29	.25	.18	.36	.30	.412	.436
....	....	.30	.25	.17	.36	.32	.400	.377
....	....	.30	.39	.24	.32	.32	.392	.370
....	....	.29	.30	.41	.28	.33	.381	.380
....	....	.28	.31	.38	.31	.32	.400	.412
....	....	.20	.31	.33	.31	.33	.415	.896
....	....	.19	.32	.32	.31	.33	.410	.385
....	0.27	.21	.32	.37	.30	.32	.400	.383
....	.26	.21	.29	.34	.30	.32	.421	.380
....	.25	....	.27	.34	....	.32	....	.380

table for Johnson Creek near Riverside, Wash., from May 29 to December 31, 1903.[a]

ge ght.	Discharge.	Gage height.	Discharge.	Gage height.	Discharge.	Gage height.	Discharge.
et.	Second-feet.	Feet.	Second-feet.	Feet.	Second-feet.	Feet.	Second-feet.
.1	0.9	.4	6.7	.7	15.3	1.0	26.0
.2	2.4	.5	9.3	.8	18.7	1.1	29.9
.3	4.4	.6	12.2	.9	22.3	1.2	34.0

computations for the rating table were based on the formula for contracted weirs given by
or Merriman.

e heights give directly depth of water over weir.

Estimated monthly discharge of Johnson Creek near Riverside, Wash., for 1903.

Month.	Discharge in second-feet.			Total in acre-feet.
	Maximum.	Minimum.	Mean.	
May 29–31	3.8	3.4	3.6	21
June	5.1	2.2	3.6	214
July	6.5	3.0	4.2	258
August	7.0	1.9	3.6	221
September	6.7	4.2	5.2	309
October	6.5	3.2	5.1	314
November	7.7	4.8	6.2	369
December	30.7	4.4	8.0	492

SINLAHEKIN CREEK NEAR LOOMIS, WASH.

This station was established July 1, 1903, by Charles E. Hewitt. It is located on the main road between Loomis and Conconully, Wash., 3 miles from Loomis and 19 miles from Conconully. The gage is a vertical staff driven into the ground and braced to the gatepost at the northeast corner of R. A. Garrett's yard. Mrs. Mary Garrett, the observer, reads the gage once each day. There is a highway bridge 500 feet north of Mr. Garrett's house, but discharge measurements are made from a plank footbridge near the gage. The initial point for soundings is a 1 by 2 inch iron bar 15 feet long driven flush with the ground at the northeast corner of the main part of Mr. Garrett's house, and 50 feet from the gage. The channel is straight for 20 feet above and for 50 feet below the station. Beyond these points are large bends in the stream. The right bank is low for about 15 feet back from the water's edge. Beyond this point it is not subject to overflow, as it rises more abruptly. The water's edge is covered with shrubbery. The left bank is low grass land, subject to overflow. The bed of the stream is a gravelly clay or loam, and is quite stable. The bench mark is the top of the iron bar used as the initial point for sounding. The zero of the gage has an elevation of 88.60 feet. The station is in charge of T. A. Noble. Its elevation is 11.40 feet above the zero of the gage.

The observations at this station during 1903 have been made under the direction of T. A. Noble, district engineer.

Discharge measurements of Sinlahekin Creek near Loomis, Wash., in 1903.

Date.	Hydrographer.	Gage height.	Discharge.
		Feet.	*Second-feet.*
ie 13......................	W. W. Schlecht	4. 10	71
ie 19......................	C. E. Hewitt	3. 50	44
y 17......................	.....do	2. 72	19
gust 8	W. W. Schlecht	2. 43	11

Mean daily gage height, in feet, of Sinlahekin Creek near Loomis, Wash., for 1903.

Day.	June.	July.	Aug.	Sept.	Oct.	Nov.	Dec.
....		2. 85	2. 50	2. 35	2. 50	2. 70	2. 75
....		2. 90	2. 50	2. 30	2. 50	2. 70	2. 75
....		2. 85	2. 50	2. 30	2. 50	2. 70	2. 75
....		2. 80	2. 50	2. 30	2. 55	2. 70	2. 70
....		2. 95	2. 50	2. 30	2. 62	2. 70	2. 65
....		3. 00	2. 45	2. 31	2. 80	2. 80	2. 60
....		3. 15	2. 45	2. 30	2. 75	2. 80	2. 60
....		3. 30	2. 40	2. 30	2. 70	2. 80	2. 55
....		3. 25	2. 40	2. 30	2. 60	2. 70	2. 50
....		3. 20	2. 40	2. 30	2. 60	2. 70	2. 55
....		3. 00	2. 40	2. 30	(a)	2. 80	2. 55
....		2. 90	2. 40	2. 40	2. 60	2. 90	2. 60
....	4. 10	2. 85	2. 35	2. 45	2. 65	2. 90	2. 60
....	4. 00	2. 80	2. 35	2. 45	2. 70	2. 80	2. 60
....	3. 90	2. 80	2. 35	2. 45	2. 70	2. 85	2. 60
....	2. 80	2. 75	2. 35	2. 40	2. 70	2. 70	2. 70
....	3. 60	2. 65	2. 35	2. 40	2. 70	2. 70	2. 65
....	3. 55	2. 60	2. 30	2. 40	2. 70	2. 70	2. 60
....	3. 50	2. 60	2. 30	2. 40	2. 70	2. 80	2. 60
....	3. 50	2. 55	2. 30	2. 40	2. 70	2. 85	2. 60
....	3. 55	2. 50	2. 30	2. 40	2. 70	3. 00	2. 55
....	3. 45	2. 50	2. 30	2. 50	2. 70	3. 00	2. 55
....	3. 35	2. 50	2. 40	2. 50	2. 70	3. 00	2. 55
....	3. 25	2. 70	2. 40	2. 50	2. 70	3. 00	2. 55
....	3. 15	2. 70	2. 50	2. 50	2. 70	2. 90	2. 55
....	3. 05	2. 60	2. 70	2. 50	2. 70	2. 95	2. 55
....	3. 00	2. 60	2. 50	2. 50	2. 70	2. 80	2. 50
....	3. 00	2. 55	2. 40	2. 50	2. 70	2. 80	2. 40
....	2. 90	2. 55	2. 40	2. 50	2. 70	2. 75	2. 40
....	2. 90	2. 55	2. 40	2. 50	2. 70	2. 75	2. 35
....		2. 50	2. 35	2. 50	2. 70	....	2. 20

a Missing.

SPOKANE RIVER AT SPOKANE, WASH.

Spokane River rises in Lake Coeur d'Alene, Idaho, and flows west-ly through eastern Washington. At Spokane its falls are used for)erating flour mills and manufacturing plants, and by the Washing-n Water Power Company for traction power and city lighting.

The gaging station was originally established by C. C. Babb, October 17, 1896, on the Oregon Railroad and Navigation Company's wooden bridge, about 1 mile above the falls, where discharge measurements and gage readings were taken until July 8, 1903. The distance from the end of the weight to the index of the first wire gage was 22 feet, and from the zero of the rod to the outside edge of the pulley 1.80 feet. The gage datum was found to be 1,880 feet above sea level by city datum and 1,865.9 feet by Government datum.

During 1901 new gages and bench marks were established. The bench mark is a railroad spike in an electric railway pole close to and on the south side of the railroad track, at the west end of bridge. Its elevation is 1,896.86 feet above city datum as determined by Mr. Fiskin, of the Washington Water Power Company, and 1,882.72 feet above Government datum, as determined by Mr. Bliss, of the United States Geological Survey, July 6, 1903, from a Government bench mark at the county court-house in Spokane.

A second wire gage was afterwards established on the north side of the west span of the bridge. The zero of this gage was at an elevation of 1,879.35 feet, coinciding with the position of the zero of the old gage. The distance from the end of the weight of the marker was also 22 feet, but the distance from the zero of the rod to the outside of the pulley was only 1.90 feet.

In July, 1903, the wooden bridge was torn out to be replaced by a steel structure, and the second gage board was destroyed. A third wire gage was established July 8, 1903, on the Olive avenue highway bridge, 950 feet below the railroad bridge. It is located on the south side of the bridge, between the fifth and seventh verticals from the west end, just outside a wooden conduit for pipes. It is 22.30 feet between the end of weight and marker, and tacks this distance apart have been driven into the top of the wooden conduit for checking the length of wire. One tack is in the middle of the conduit opposite the pulley which is set into the west end of the rod.

The bench mark is a railroad spike in the north face of the first telegraph pole west of the west approach of bridge. Its elevation is 1,881.052 feet above Government datum and 17.160 feet above rod datum.

This gage was established with the idea in mind that readings taken at this point would be a continuation of readings taken at the Oregon Railroad and Navigation Company's bridge, as the two gages were made to read the same when the new one was put in place and both sections are practically the same. For two weeks before the second gage (that on the railroad bridge) was destroyed, simultaneous readings were obtained, which showed no appreciable difference. The slope in the water surface between the two stations was 1.43 feet on July 8, 1904.

SPOKANE RIVER AT SPOKANE, WASH.

Mean daily gage height, in feet, of Spokane River at Spokane, Wash., for 1903.

Day.	Jan.	Feb.	Mar.	Apr.	May.	June.	July.	Aug.	Sept.	Oct.	Nov.	Dec.
1	3.25	5.50	3.00	6.00	8.50	8.30	6.00	3.10	1.85	1.70	2.20	3.40
2	3.35	5.30	3.00	6.40	8.50	8.55	5.85	3.05	1.85	1.70	2.15	3.70
3	3.50	5.25	2.95	6.65	8.40	8.90	5.70	3.00	1.85	1.70	2.15	4.00
4	4.20	4.95	2.95	6.75	8.35	9.25	5.50	2.90	1.85	1.70	2.15	4.30
5	5.25	4.85	2.95	6.75	8.25	9.55	5.40	2.85	1.80	1.70	2.15	4.50
6	5.90	4.65	2.95	6.75	8.25	9.85	5.40	2.80	1.80	a 1.75	2.15	4.65
7	6.25	4.60	2.95	6.75	8.50	9.80	5.15	2.75	1.80	a 1.75	2.15	4.60
8	6.30	4.45	2.90	6.70	8.60	9.75	5.05	2.70	1.80	a 1.80	2.15	4.50
9	6.30	4.40	2.85	6.65	8.75	9.65	4.90	2.65	1.80	a 1.80	2.60	4.50
10	6.25	4.40	2.85	6.60	8.90	9.55	4.80	2.60	1.80	a 1.80	2.75	4.50
11	6.25	4.20	2.85	6.55	8.90	9.55	4.75	2.55	1.80	1.80	2.85	4.50
12	6.05	4.10	2.85	6.35	8.80	9.50	4.65	2.50	1.80	1.85	2.90	4.50
13	5.85	3.90	2.85	6.30	8.70	9.45	4.55	2.45	1.80	1.90	2.95	4.30
14	5.75	3.80	2.95	6.05	8.60	9.25	4.45	2.40	1.80	1.95	2.95	4.20
15	5.65	3.75	2.95	6.00	8.80	9.05	4.35	2.35	1.80	2.00	3.00	4.20
16	5.30	3.70	2.95	5.85	9.00	8.90	4.25	2.30	1.80	2.00	3.00	4.10
17	5.15	3.60	3.00	5.85	9.30	8.85	4.25	2.30	1.80	2.05	3.00	4.05
18	5.15	3.50	3.00	5.85	9.45	8.65	4.05	2.25	1.80	2.05	2.95	4.00
19	4.90	3.45	3.00	5.95	9.45	8.60	3.95	2.20	1.80	2.05	2.90	4.00
20	4.85	3.45	3.05	6.00	9.40	8.30	3.85	2.20	1.80	2.05	2.90	4.00
21	4.75	3.45	3.05	6.10	9.25	8.00	3.75	2.00	1.80	2.05	2.90	3.95
22	4.65	3.35	3.05	6.25	9.00	7.80	3.65	2.00	1.80	2.00	2.90	3.95
23	4.65	3.25	3.05	6.50	8.85	7.60	3.60	1.95	1.80	2.00	3.00	4.00
24	4.50	3.20	3.10	6.75	8.60	7.40	3.55	1.80	1.80	1.90	3.10	4.00
25	4.65	3.15	3.20	7.05	8.45	7.20	3.50	1.80	1.80	1.90	3.20	4.00
26	5.15	3.10	3.30	7.25	8.40	7.00	3.40	1.85	1.80	1.90	3.25	3.90
27	5.65	3.10	3.50	7.50	8.30	6.80	3.30	1.85	1.75	1.95	3.30	3.90
28	5.65	3.05	3.75	7.95	8.15	6.60	3.20	1.85	1.75	2.30	3.30	3.85
29	5.65		4.25	8.20	8.20	6.40	3.20	1.85	1.75	2.25	3.30	3.80
30	5.55		4.85	8.50	8.25	6.20	3.15	1.85	1.70	2.25	3.35	3.75
31	5.55		5.65		8.25		3.10	1.85		2.20		3.65

a Taken from original record.

HANGMAN CREEK NEAR SPOKANE, WASH.

This is a miscellaneous station and is located on the highway bridge 6 miles southeast of Spokane, Wash. It is the fifth bridge from the mouth of the creek along the county road.

It was established July 10, 1903, by George H. Bliss. Measurements are taken from the highway bridge. No gage board was established and no gage-height readings have been taken. Bench mark No. 1 is on the root of a black pine tree, 4 feet in diameter, 100 feet northwest of the north end of the bridge. The assumed elevation of the nail is 25.766 feet above datum, which is the deepest point in the bed of the gaging section. Bench mark No. 2 is on the south side of a stump which is 35 feet west of the west side of the bridge. The elevation of the top of the nail is 16.719 feet above datum. The initial point for soundings is the most northerly bolt in the west guard rail (

id is surrounded by copper tacks. The channel above the station for
)0 feet is straight and the water sluggish. The channel below for
50 feet is straight and the water sluggish. The right bank is low,
overed with underbrush, and liable to overflow during high water.
he left bank is higher, covered with underbrush, and liable to over-
ow in extreme high water. The bed of the stream is compact, of mud
id sand, and not liable to shift.

The observations at this station during 1903 have been made under
le direction of T. A. Noble, district engineer.

Discharge measurements of Hangman Creek near Spokane, Wash., in 1903.

Date.	Hydrographer.	Gage height.	Discharge
		Feet.	*Second-feet.*
ily 10...................	G. H. Bliss...............	4.5	43
eptember 11 ....do	3.9	30

LITTLE SPOKANE RIVER NEAR SPOKANE, WASH.

This station was established August 3, 1903, by George H. Bliss. It
located about 2 miles above the mouth of the river at the second
ridge above the mouth. It is 9 miles northwest of Spokane, Wash.,
id 1½ miles northeast of what is known as the "9-mile bridge" over
Spokane River. The wire gage is located on the upstream side of the
bridge. The center of the pulley is 105 feet from the south end of
the bridge. The length of the wire from the end of the weight to the
marker is 13.25 feet. This distance has been laid off on the upper
surface of the bottom rail near the gage and is marked by copper
tacks inclosed in circles of black paint. These marks are used in
checking the length of the gage wire. The gage is read once each
day by Mary A. Keenan. Discharge measurements are made from
the upstream side of the bridge to which the gage is attached. The
initial point for soundings is the vertical end post on the upstream
side of the bridge at the south approach. The channel is straight for
100 feet above and 150 feet below the station. The current is swift.
Both banks are high, covered with underbrush, and liable to overflow
only at very high stages. The bed of the stream is composed of clean
gravel. The channel is broken by four bridge piers and has a width at
ordinary stages of about 125 feet. The bench mark is a wire nail
driven into the root of a black pine tree 2 feet in diameter. The root
is on the north side of the tree and extends toward the bridge. It is
60 feet distant from the south end of the bridge. The tree is blazed.
It is marked "B. M." with black paint. The bench mark has an
elevation of 21 feet above gage datum.

The observations at this station during 1903 have been made under the direction of T. A. Noble, district engineer.

Discharge measurements of Little Spokane River near Spokane, Wash., in 1903.

Date.	Hydrographer.	Gage height.	Discharge.
		Feet.	*Second-feet.*
November 16	G. H. Bliss..................	6.55	573
August 4	.....do	6.10	455

Mean daily gage height, in feet, of Little Spokane River near Spokane, Wash., for 1903.

Day.	Aug.	Sept.	Oct.	Nov.	Dec.	Day.	Aug.	Sept.	Oct.	Nov.	Dec.
1.....................		5.88	5.96	(a)	6.90	17.................	5.91	6.05	6.15	6.41	7.25
2.....................		5.94	6.06		6.90	18.................	5.97	6.09	6.30	6.39	7.40
3.....................	6.06	5.99	6.14		6.90	19.................	5.84	6.07	6.15	6.40	7.40
4.....................	6.10	5.96	6.10		6.60	20.................	5.88	6.05	6.55	6.55	7.20
5.....................	6.16	5.95	6.23		6.58	21.................	5.91	6.08	6.25	6.60	7.25
6.....................	6.05	5.97	6.30		6.50	22.................	6.01	6.14	6.12	6.74	7.30
7.....................	6.00	6.05	6.40		6.50	23......:.......	6.09	6.14	6.15	7.05	7.21
8.....................	6.00	6.15	6.30		6.50	24.................	6.21	6.10	6.10	6.79	7.20
9.....................	6.00	6.16	6.50		6.30	25.................	6.12	6.00	5.50	6.79	7.10
10.....................	6.05	6.10	6.40		6.40	26.................	6.18	6.04	6.40	6.53	7.00
11.....................	6.00	6.00	6.20		6.30	27.................	6.25	6.18	6.23	6.65	6.90
12.....................	5.92	6.25	6.30		6.30	28.................	6.25	6.10	6.60	6.69	6.85
13.....................	5.93	6.21	6.36		6.30	29.................	6.11	6.08	6.80	6.55	6.80
14.....................	5.80	6.20	6.30		6.40	30.................	6.00	6.04	6.40	6.67	6.70
15.....................	5.85	6.15	6.35	(a)	6.55	31.................	6.23		6.29		6.65
16..........	5.84	6.10	6.29	6.55	6.80						

a From November 1 to 15, inclusive, no records taken.

PEND OREILLE RIVER AT PRIEST RIVER, IDAHO.

This station was established June 26, 1903, by T. A. Noble, assisted by George H. Bliss. It is located about 1,000 feet west of Priest River railroad station and south of the railroad track, on the right bank. It is about 100 feet west of a sawmill. The stream at this point flows parallel to the railroad track, and both the platform to which the gage is attached and the ferry cable from which measurements are made are at right angles to the track and stream. The gage is of the wire and weight type, with horizontal scale board, fastened to the railing of the platform, which is built between two cottonwood and two black pine trees. The gage is adjusted to read the height of the water surface above sea level. Discharge measurements are made from a ferry cable about 400 feet downstream from the gage. The initial point for soundings is a stake on the left bank of the stream and the west side of the driveway. Its elevation is 2,062.11 feet above sea level. On the right bank 1,020 feet from the initial point is another stake, with

channel is straight for about 2,000 feet above and 4,000 feet below the station. The right bank is high, covered with underbrush, and not subject to overflow. The left bank is low, cleared, and liable to overflow. From the top of the left bank there is an upward slope of about 10 per cent. The water flows in one channel, and the bed of the stream is composed of sand, with occasional bowlders. The bench mark from which all elevations were obtained is the Geological Survey bench mark south of Priest River station, at the northeast corner of the hotel. Its elevation above sea level is 2,077 feet. A second bench mark, under the gage board platform, is at an elevation of 2,066.19 feet above sea level. A third, on a stump near the gage, is at an elevation of 2,073.02 feet above sea level. On July 16, 1903, the length of the gage wire from the end of the weight to the marker was measured and found to be 32.60 feet. On the bridge rail on the opposite side from the gage 2 copper tacks were driven 32.60 feet apart, to be used for future checking of the length of the wire by the observer.

The observations at this station during 1903 have been made under the direction of T. A. Noble, district engineer.

Discharge measurements of Pend Oreille River at Priest River, Idaho, in 1903.

Date.	Hydrographer.	Gage height.	Discharge.
		Feet.	Second-feet.
June 26	T. A. Noble	59.40	126,267
July 15	G. H. Bliss	53.85	61,422
August 6	do	48.48	38,862
September 3	do	45.15	17,805
October 28	do	44.81	16,183

Mean daily gage height, in feet, of Pend Oreille River at Priest River, Idaho, for 1903.

Day.	June.	July.	Aug.	Sept.	Oct.	Nov.	Dec.
1		58.35	49.71	45.40	44.41	44.75	44.71
2		58.11	49.55	45.30	44.28	44.71	44.70
3		57.80	49.30	45.15	44.38	44.70	44.86
4		57.46	49.01	45.15	44.41	44.69	44.96
5		57.11	48.75	45.16	44.40	44.67	45.00
6		56.82	48.50	44.95	44.40	44.79	44.99
7		56.49	48.34	44.99	44.50	44.85	45.00
8		56.17	48.20	44.95	44.55	44.81	45.01
9		55.91	48.02	44.85	44.59	44.95	45.01
10		55.62	47.85	44.81	44.49	44.84	45.02
11		55.25	47.65	44.80	44.52	45.06	44.96
12		55.00	47.56	44.90	44.60	44.90	45.01
13		54.76	47.46	44.75	44.61	44.99	44.98
14		54.65	47.39	44.70	44.62	44.21	44.98
15		53.85	47.09	44.66	44.69	44.89	44.98

daily gage height, in feet, of Pend Oreille River at Priest River, Idaho, for 1903—
Continued.

Day.	June.	July.	Aug.	Sept.	Oct.	Nov.	Dec.
....	....	53.69	46.97	44.61	44.74	44.89	45.00
....	....	53.41	46.81	44.59	44.78	44.94	44.99
....	....	53.11	46.77	44.59	44.79	44.96	44.93
....	....	52.80	46.61	44.52	44.79	44.70	44.95
....	....	52.46	46.45	44.54	44.82	44.69	44.90
....	....	52.30	46.34	44.52	44.85	44.72	44.90
....	....	51.93	46.24	44.50	44.89	44.68	44.92
....	....	51.68	46.15	44.49	44.88	44.74	44.84
....	....	51.41	46.10	44.43	44.89	44.68	44.82
....	....	51.22	46.00	44.40	44.90	44.70	44.81
....	....	50.86	45.90	44.41	44.84	44.68	44.76
....	....	50.75	45.80	44.40	44.82	44.65	44.70
....	59.14	50.85	45.72	44.40	44.81	44.65	44.70
....	58.90	50.20	45.59	44.39	44.71	44.67	44.74
....	58.64	50.10	45.61	44.40	44.70	44.69	44.60
....	....	49.85	45.48	....	44.78	....	44.58

PRIEST RIVER AT PRIEST RIVER, IDAHO.

his station was established June 28, 1903, by T. A. Noble, assisted
G. H. Bliss. It is located at the highway bridge, on the road
n the railroad station at Priest River to Priest Lake. The gage
vertical board nailed to a pile on the downstream side of the right
of the highway bridge. A ladder is attached to the pier to
litate reading the gage. It is read once each day by George
ing. Discharge measurements are made from the downstream
of the bridge, to which the gage is attached. The initial point
soundings is the bolt at the end of the guard rail at the right bank.
rcle has been painted around the bolt. The channel is straight for
feet above and for 300 feet below the bridge. Both banks are
1, wooded, and not liable to overflow. Extending from each pier
he single-span bridge to the bank is a breakwater composed of
s faced with planks. These breakwaters make the current slug-
ı between the piers and the banks. Under the main span of the
lge, a distance of 120 feet, the current is swift. The bed of the
am is composed of gravel. Bench mark No 1 is the bolt in
guard rail at the west end of the bridge, which is used as the initial
ıt for soundings. Its elevation is 29.04 feet above the zero of the
e, and 2,079.7 feet above sea level. Bench mark No. 2 is a spike
·en in a stump under the right approach to the bridge. Its eleva-
is 17.58 feet above the zero of the gage.
he observations at this station during 1903 have been made under
direction of T. A. Noble, district engineer.

Discharge measurements made on Priest River at Priest River, Idaho, in 1903.

Date.	Hydrographer.	Gage height.	Discharge.
		Feet.	*Second-feet.*
June 28	T. A. Noble	9.68	5,02?
July 14	G. H. Bliss	5.90	2,67?
August 7	do	4.48	1,53?
September 2	do	3.65	10?
October 27	do	3.75	97?

Mean daily gage height, in feet, of Priest River at Priest River, Idaho, for 1903.

Day.	June.	July.	Aug.	Sept.	Oct.	Nov.	Dec.
1		8.90	4.51	3.69	3.50	3.70	4.0
2		8.70	4.60	3.63	3.49	3.70	4.3
3		8.41	4.52	3.60	3.48	3.69	4.9
4		8.09	4.50	3.58	3.47	3.60	4.6
5		7.82	4.50	3.58	3.55	3.68	4.0
6		7.55	4.49	3.67	3.91	3.91	4.0
7		7.31	4.49	3.91	3.85	4.19	4.0
8		7.05	4.49	3.85	3.76	4.03	4.0
9		6.83	4.49	3.83	3.71	4.00	4.0
10		6.69	4.00	3.78	3.69	4.05	4.0
11		6.49	4.49	3.80	3.75	3.95	2.0
12		6.39	4.45	3.76	3.78	3.78	2.0
13		6.09	4.40	3.74	3.80	3.84	3.9
14		5.94	4.36	3.67	3.81	3.89	3.96
15		5.81	4.29	3.62	3.81	3.85	3.99
16		5.71	4.30	3.61	3.81	3.79	4.14
17		5.60	4.22	3.60	3.82	3.70	4.17
18		5.50	4.20	3.58	3.85	3.70	4.20
19		5.40	4.15	3.65	3.82	3.60	4.08
20		5.31	4.09	3.60	3.84	3.72	4.15
21		5.22	4.01	3.65	3.84	3.74	4.11
22		5.21	4.00	3.87	3.80	3.89	4.10
23		5.11	4.01	3.88	3.80	3.88	4.19
24		5.07	4.00	3.49	3.78	3.84	4.04
25		5.10	3.98	3.49	3.78	3.80	4.01
26		5.03	3.97	3.51	3.75	3.78	3.99
27		4.98	3.92	3.52	3.74	3.78	3.96
28	9.64	4.79	3.87	3.52	3.74	3.80	3.92
29	9.45	4.71	3.84	3.50	3.72	3.78	3.84
30	9.17	4.65	3.78	3.51	3.72	3.93	3.96
31		4.63	3.71		3.72		3.63

MISSOULA RIVER AT MISSOULA, MONT.

The original station was established July 10, 1898, by Cyrus C. Babb, and was located at Higgins Avenue Bridge in Missoula. As the river at this point flows in two channels, in which fluctuations occur, this location was abandoned and a new one found May 27, 1899, some distance downstream at the bridge of the Bitterroot Valley division of the Northern Pacific Railway. The river here is practically in one channel, except in times of flood, when some water passes through a slough 600 feet south of the bridge.

The measurements are made from the downstream side of the bridge, the initial point for soundings being over the northeast abutment opposite the center of the first angle block of the truss.

The riprapping around the crib piers of the bridge and remains of old cribs and piling in the channel under the bridge cause eddies which decrease the accuracy of measurements. At flood heights there is a visible difference in the elevation of the water surface above and below the station.

The gage is located on the right bank of the river some 400 feet above the station. It is of the wire type, and consists of a horizontal timber bolted to a cottonwood tree. Timbers above and below and guy wires brace it securely.

The correct length of wire from marker to bottom of weight is 21.95 feet. The elevation of the gage datum is 3,162.18 feet above sea level.

The gage is read twice each day by Thomas E. Westby. The bench mark consists of a United States Geological Survey iron post on Front, 200 feet west of McCormick street, 100 yards east of the gage, on the north side of the highway. Its elevation is 3,194.64 feet. The height of the river was read at the three rods during 1899, but all subsequent readings have been made from No. 3.

The observations at this station during 1903 have been made under the direction of C. C. Babb, district engineer.

Discharge measurements of Missoula River at Missoula, Mont., in 1903.

Date.	Hydrographer.	Gage height.	Discharge.
		Feet.	*Second-feet.*
April 24	F. M. Brown	5.22	3,776
June 30	C. D. Flaherty	6.40	6,580
July 13	do	5.55	3,661
September 12	do	3.96	1,670
October 6	J. H. Sloan	3.60	1,836

Mean daily gage height, in feet, of Missoula River at Missoula, Mont., for 1903.

Day.	Jan.	Feb.	Mar.	Apr.	May.	June.	July.	Aug.	Sept.	Oct.	Nov.	Dec.
1		(a)										
2		(a)										
3		(a)										
4		(a)							2.67			
5		(a)										
6		(a)										
7		(a)										
8		(a)							2.77			
9		(a)										
10		(a)							2.67			
11		(a)										
12		(a)										
13		(a)								3.77		
14	(a)	(a)								3.70		
15	(a)	(a)								3.70		
16	(a)	(a)								3.70		
17	(a)	(a)			7.09				(b)	3.70	(a)	
18	(a)	(a)			6.72					3.70	(a)	
19	(a)	(a)			6.55			3.90		3.70	(a)	
20	(a)	(a)	3.70	4.70	6.70	7.46				3.65	(a)	
21	(a)	(a)	3.75			7.25					(a)	
22	(a)	(a)					7.12	4.70				
23	(a)	(a)		5.15	5.90	7.08	4.70	3.70				
24	(a)	(a)	4.90	5.22	5.86	6.90	4.70	3.75				
25	(a)	(a)	4.90	5.30				4.70				
26	(a)	(a)	4.35	5.25	5.95	6.70	4.65	3.75		3.50		
27	(a)	(a)	4.50	5.43	6.13	6.63	4.56	3.87		3.50		
28	(a)	3.60	4.65	5.58	6.50	6.55	4.50	3.80		3.45		
29	(a)		4.75	5.65	6.85	6.58	4.50	3.73		3.45	(a)	
30	(a)		5.15	5.53	6.88	6.28	4.55	3.70		3.45	(a)	
31	(a)		5.58		7.25		4.50	3.70		3.45		

a Ice.   b Gage broken September 17 to October 11, 1903.   c To top of ice.

Rating table for Missoula River at Missoula, Mont., for 1903.

Gage height.	Discharge.	Gage height.	Discharge.	Gage height.	Discharge.	Gage height.	Discharge.
Feet.	*Second-feet.*	*Feet.*	*Second-feet.*	*Feet.*	*Second-feet.*	*Feet.*	*Second-feet.*
3.4	1,530	4.7	2,710	6.0	5,380	7.3	10,060
3.5	1,590	4.8	2,870	6.1	5,660	7.4	10,520
3.6	1,650	4.9	3,050	6.2	5,940	7.5	10,980
3.7	1,710	5.0	3,240	6.3	6,240	7.6	11,460
3.8	1,780	5.1	3,430	6.4	6,560	7.7	11,940
3.9	1,860	5.2	3,620	6.5	6,900	7.8	12,420
4.0	1,940	5.3	3,820	6.6	7,240	7.9	12,900
4.1	2,020	5.4	4,020	6.7	7,600	8.0	13,400
4.2	2,110	5.5	4,220	6.8	7,980	8.1	13,900
4.3	2,210	5.6	4,440	6.9	8,380	8.2	14,400
4.4	2,320	5.7	4,660	7.0	8,780	8.3	14,900
4.5	2,440	5.8	4,880	7.1	9,200		
4.6	2,570	5.9	5,120	7.2	9,620		

.ngent at 8 feet gage height, with differences of 500 per tenth.

Estimated monthly discharge of Missoula River at Missoula, Mont., for 1903.

[Drainage area, 5,960 square miles.]

Month.	Discharge in second-feet.			Total in acre-feet.	Run-off.	
	Maximum.	Minimum.	Mean.		Second-feet per square mile.	Depth in inches.
uary a			1,900	116,826	0.319	0.368
'ruary a			1,700	94,413	.285	.297
rch	4,440	1,530	2,018	124,082	.339	.391
il	4,550	2,110	2,949	175,476	.495	.556
y	9,840	3,430	5,817	357,673	.976	1.130
e	21,900	6,240	13,950	830,083	2.340	2.610
y	5,800	2,440	3,976	244,475	.667	.768
gust	2,440	1,710	1,995	122,667	.335	.386
tember a			1,800	107,107	.302	.337
ober a			1,700	104,529	.285	.329
;ember a			1,500	89,256	.252	.281
ember a			1,500	92,231	.252	.291
The year			3,400	2,458,820	.571	7.744

a Estimated.

BITTERROOT RIVER NEAR MISSOULA, MONT.

This station was established July 6, 1898, by C. C. Babb. The station is located at the Buckhouse wagon bridge, on the main road, 3 miles southwest of Missoula, Mont. As it is not far above its junction with the Missoula River, it will give the full discharge of the Bitterroot. The first gage established had a length of wire of 22.83 feet, and the center of the axle of the gage pulley was 3.801 feet below the bench mark. On April 8, 1901, the bridge was washed out, but was immediately replaced. The station was then reestablished with a gage wire of different length. Observations were discontinued November 1, 1901, but were begun again in 1903. The length of the gage wire on October 8, 1903, was found to be 30.40 feet, at which time the center of the gage pulley was 6.70 feet above the U. S. Geological Survey bench-mark post. The marker on the wire is taken as the end of the wire. Until October 10, 1903, the observer has been Donald Buckhouse. The present gage reader is Frank Mitchell. Discharge measurements are made from the downstream guard rail of the wagon bridge, to which the gage is attached. The initial point for soundings is a point marked 0 over the center of the northeast bridge pier. All distances are marked on the hand rail. The channel is nearly straight above and below the station. The right bank is low and liable to overflow. The left bank is high and rocky and juts out into the river at flood stages so that the channel is then congested at the bridge. The bed of the stream consists of gravel and is fairly constant. The depth varies from 3 to 6 feet.

The bench mark is a standard U. S. Geological Survey post set 25 yards northwest of the bridge. Its elevation above gage datum is 19.37 feet, and above sea level (Missoula datum) is 3,140 feet. The center of the gage pulley has an elevation of 26.07 feet above gage datum.

The observations at this station during 1903 have been made under the direction of C. C. Babb, district engineer.

harge measurements of Bitterroot River near Missoula, Mont., in 1903.

Date.	Hydrographer.	Gage height.	Discharge.
		Feet.	*Second-feet.*
....	F. M. Brown	....	4,131
....	J. S. Baker	3.40	3,377
....	C. D. Flaherty	7.05	9,982
....	....do	8.20	12,637
....	....do	6.86	9,121
....	....do	4.45	4,217
....	L. A. Cowan	2.30	1,546
13	....do	2.70	1,463
....	J. H. Sloan	....	1,683

ily gage height, in feet, of Bitterroot River near Missoula, Mont., for 1903.

Day.	June.	July.	Aug.	Sept.	Oct.	Nov.	Dec.
....	....	5.80	....	1.30	....	2.50	2.40
....	8.00	5.60	2.60	1.20	....	2.50	2.80
....	9.10	5.50	2.40	1.30	....	2.40	2.70
....	9.70	4.90	2.50	1.20	....	2.40	2.60
....	9.90	4.80	2.60	1.20	....	2.30	2.50
....	9.40	4.90	2.50	(a)	....	2.30	2.50
....	9.00	5.00	2.40	....	....	2.30	2.50
....	8.90	5.10	2.50	....	....	2.40	2.50
....	9.10	4.80	2.00	....	....	2.60	2.50
....	9.20	4.60	(b)	....	2.90	2.50	2.50
....	8.80	4.50	(b)	....	2.50	2.40	2.50
....	9.10	4.50	1.90	....	2.60	2.30	2.50
....	8.50	4.60	1.70	....	2.80	2.40	2.60
....	8.30	4.50	1.60	....	3.40	2.40	2.60
....	8.40	4.30	1.60	....	3.30	2.40	2.60
....	8.50	4.00	1.70	....	3.20	2.30	2.60
....	8.60	3.90	1.60	....	3.10	2.30	2.50
....	8.40	3.70	1.70	....	3.10	2.20	2.50
....	7.80	3.60	1.60	....	3.00	2.10	2.50
....	7.30	3.60	1.50	....	2.90	2.10	2.40
....	7.30	3.50	1.40	....	2.90	2.10	2.40
....	6.90	3.50	1.20	....	2.80	2.50	2.40
....	6.80	3.40	1.30	....	2.80	2.60	2.30
....	6.10	3.20	1.20	....	2.70	2.60	2.30
....	6.00	3.10	1.20	....	2.70	2.50	2.20
....	6.00	....	1.30	....	2.70	2.40	2.10
....	6.30	....	1.20	....	2.60	2.40	2.10
....	6.50	....	1.40	....	2.60	2.40	2.10
....	6.80	....	1.40	....	2.60	2.30	2.20
....	6.20	....	1.30	....	2.50	2.30	2.30
....	....	....	1.40	....	2.50	....	2.20

Observer absent from September 6 to October 10.   *b* Gage stolen.

BITTERROOT RIVER NEAR GRANTSDALE, MONT.

This station was established April 25, 1902, by H. B. Waters. It is ocated on the highway bridge 2 miles southwest of Grantsdale and 5 niles southwest of Hamilton, Mont. The gage is of the wire type nd fastened to the downstream truss of the bridge. The scale board s graduated to feet and tenths, and is read daily by T. J. Holt, who lives bout a quarter of a mile north of the gage.

Two large ditches, the new Hedge and the Republican, are taken out)f the river some distance above the station. They irrigate extensive 'arm lands and orchards in the vicinity of Hamilton.

The length of the wire from the end of the weight to the marker is $5.67 feet. Discharge measurements are made from the highway)ridge on the downstream side. The initial point for sounding is a jotch on the hand rail over the northeast bridge pier. The channel is straight both above and below the station.

The stream has a moderate velocity. The right bank has a gentle slope for about 100 feet, when it terminates in a high bank which is not liable to overflow. The left bank is high and is formed above the)ridge by a railroad fill. The bed of the stream is composed of gravel nd bowlders. Bench mark No. 1 is a wire nail driven in the northeast side of a large pine stump and marked "B. M. 24.40." The stump is across the road from the west end of the bridge. The elevation above the gage datum is 24.40 feet. Bench mark No. 2 is the northwest bolt in the northwest abutment plate of the bridge. It is marked "B. M. 19.36." Its elevation above the gage datum is 19.36 feet.

The observations at this station during 1903 have been made under the direction of C. C. Babb, district engineer.

Discharge measurements of Bitterroot River near Grantsdale, Mont., in 1903.

Date.	Hydrographer.	Gage height.	Discharge.
		Feet.	*Second-feet.*
April 28	F. M. Brown	3.59	3,283
May 29	C. D. Flaherty	4.05	4,763
June 16	...do	6.10	9,075
June 24	...do	4.50	3,914
July 9	...do	3.50	2,303
August 14	...do	1.25	322
September 9	...do	1.35	333
October 9	J. H. Sloan	2.36	647

Mean daily gage height, in feet, of Bitterroot River near Grantsdale, Mont., for 1903.

Day.	Jan.	Feb.	Mar.	Apr.	May.	June.	July.	Aug.	Sept.	Oct.	Nov.	Dec.
....	1.80	2.00	1.90	2.60	3.20	6.10	4.00	2.20	1.30	1.90	2.10	2.50
....	1.80	2.00	2.00	2.50	3.10	7.00	3.80	2.20	1.20	1.90	2.10	2.50
....	1.90	2.00	2.00	2.40	3.00	7.30	3.50	2.20	1.20	1.90	2.10	2.40
....	1.90	2.00	2.00	2.80	3.20	6.70	3.30	2.10	1.20	2.00	2.00	2.30
....	1.90	2.00	1.90	2.80	3.30	6.40	3.50	2.10	1.20	2.20	2.00	2.20
....	2.00	2.00	1.80	2.20	3.60	6.30	3.70	2.00	1.20	2.40	2.00	2.20
....	2.00	2.00	1.80	2.20	4.00	6.30	3.60	2.00	1.30	2.30	2.00	2.20
....	2.01	2.00	1.80	2.10	4.00	6.40	3.50	1.90	2.50	2.30	2.10	2.20
....	2.00	2.00	1.80	2.20	3.80	6.70	3.50	1.80	2.50	2.40	2.10	2.20
....	1.90	2.00	1.90	2.40	3.60	6.40	3.40	1.70	2.60	2.60	2.00	2.10
....	2.00	1.90	1.90	2.40	3.30	6.30	3.30	1.70	2.60	2.80	2.00	2.10
....	2.01	1.90	1.90	2.40	3.40	6.20	3.30	1.60	2.10	2.90	2.00	2.10
....	1.90	1.90	2.00	2.40	3.70	6.20	3.30	1.60	2.00	2.60	2.00	2.10
....	1.90	1.90	2.00	2.80	4.60	6.00	3.30	1.60	1.90	2.50	2.10	2.00
....	1.90	1.90	1.90	2.80	4.80	6.10	3.20	1.50	1.90	2.50	2.10	2.00
....	2.00	1.90	1.90	2.30	4.50	6.20	3.10	1.50	1.90	2.40	2.00	2.00
....	2.00	1.90	1.80	2.40	4.00	6.10	3.00	1.50	1.80	2.40	2.00	2.10
....	2.10	1.90	1.80	2.50	3.80	5.80	2.90	1.50	1.80	2.40	2.00	2.10
....	2.00	1.90	1.70	2.50	3.60	5.60	2.90	1.50	1.80	2.40	2.10	2.20
....	2.00	1.90	1.70	2.50	3.40	5.80	2.80	1.50	1.80	2.40	2.10	2.20
....	2.00	1.90	1.70	2.60	3.20	5.20	2.80	1.40	1.90	2.30	2.20	2.10
....	2.00	1.90	1.70	2.90	3.20	5.00	2.80	1.40	1.90	2.30	2.30	2.10
....	1.90	1.80	1.80	3.20	3.10	5.00	2.70	1.40	1.90	2.30	2.30	2.10
....	1.80	1.80	1.80	3.50	3.10	4.70	2.70	1.30	1.90	2.20	2.20	2.00
....	2.20	1.80	2.00	3.50	3.10	4.40	2.60	1.30	1.90	2.20	2.20	2.00
....	2.20	1.80	2.10	3.60	3.30	4.50	2.50	1.30	1.90	2.20	2.10	2.00
....	2.10	1.80	2.20	3.70	3.70	4.70	2.40	1.40	2.00	2.20	2.10	2.00
....	2.10	1.80	2.30	3.70	4.00	4.60	2.40	1.40	2.00	2.20	2.10	2.00
....	2.00	......	2.40	3.70	4.10	4.80	2.30	1.40	2.00	2.10	2.20	2.00
....	2.00	......	2.50	3.60	4.30	4.40	2.30	1.40	2.00	2.10	2.20	(a)
....	2.00	......	2.60	......	5.10	......	2.20	1.30	......	2.10	......	(a)

a Frozen.

Rating table for Bitterroot River near Grantsdale, Mont., for 1903.

Gage height.	Discharge.	Gage height.	Discharge.	Gage height.	Discharge.	Gage height.	Discharge.
Feet.	Second-feet.	Feet.	Second-feet.	Feet.	Second-feet.	Feet.	Second-feet.
1.0	125	2.2	700	3.4	2,450	5.2	6,380
1.1	135	2.3	790	3.5	2,650	5.4	6,855
1.2	150	2.4	900	3.6	2,850	5.6	7,325
1.3	175	2.5	1,020	3.7	3,050	5.8	7,790
1.4	205	2.6	1,150	3.8	3,255	6.0	8,340
1.5	240	2.7	1,290	3.9	3,470	6.2	8,970
1.6	280	2.8	1,430	4.0	3,700	6.4	9,670
1.7	340	2.9	1,585	4.2	4,140	6.6	10,370
1.8	395	3.0	1,745	4.4	4,575	6.8	11,075
1.9	460	3.1	1,910	4.6	5,005	7.0	11,795
2.0	530	3.2	2,085	4.8	5,440	7.3	12,875
2.1	615	3.3	2,265	5.0	5,900		

1903 rating table same as 1902.

Estimated monthly discharge of Bitterroot River near Grantsdale, Mont., for 1902.

[Drainage area, 1,550 square miles.]

Month.	Discharge in second-feet.			Total in acre-feet.	Run-off	
	Maximum.	Minimum.	Mean.		Second-feet per square mile.	Depth in inches.
January	700	395	518	31,850	0.334	1.26
February	530	395	471	26,156	.304	.27
March..............	1,150	340	519	31,912	.335	.3
April...............	3,050	615	1,409	83,841	.909	1.6
May	6,150	1,745	3,031	186,369	1.955	2.59
June	12,875	4,575	7,968	474,129	5.141	3.74
July	3,700	700	1,912	117,564	1.234	1.06
August	700	175	334	20,537	.215	.24
September.........	1,150	150	481	28,622	.310	.34
October	1,585	460	825	50,727	.532	.67
November	790	530	615	36,596	.397	.45
December	1,020	530	650	39,967	.419	.48
The year......	12,875	150	1,561	1,128,272	1.007	13.61

BIG BLACKFOOT RIVER NEAR BONNER, MONT.

This station was established for general information purposes in July, 1898, by C. C. Babb. It is situated a short distance above the junction of Big Blackfoot with Hellgate River, at the county highway bridge one-half mile west of Bonner and 6 miles east of Missoula. The power dam of the Big Blackfoot Milling Company is about 1,000 yards above the station, and interferes with the natural flow of the water. The opening and closing of the gates causes abrupt changes in the gage heights. The channel at the station is straight. Both banks are high and rocky. They are clothed with a vegetation of bushes and single trees. Neither bank is subject to inundation. The bed of the river is rocky and covered with cobbles and bowlders. It is not liable to change. The depth of water varies from 4 to 10 feet. The current is very swift and can seldom be gaged without guying the meter. The discharge measurements are made from the bridge, the distances being marked on the downstream hand rail. The initial point is a notch marked "0" at the left end of the downstream hand rail. The gage is attached to the upstream guard rail. It is of the wire type and reads to feet and tenths. The marker is a wire knot near the handle. The distance from the bottom of the weight to the marker is 22.75 feet. The observer, John McCormick, who lives at the bridge, reads the gage twice a day.

Bench mark No. 1 is a temporary bench mark of the topographic
vision of the Geological Survey. It is a crosscut in the northeast
rner of the top of the northeast abutment of the Northern Pacific Rail-
ay bridge near Bonner. Its elevation is 3,290.30 feet above sea level.
ench mark No. 2 is a standard United States Geological Survey iron
st located in front of McCormick's house, near the highway bridge.
s elevation is 3,246.038 feet above sea level. The elevation of the
le of the gage pulley, which serves as a temporary bench mark, is
251.68 feet above sea level.
The zero of the gage is 3,220.702 feet above sea level. All eleva-
ns refer to Missoula datum—25.34 feet above gage datum. The
vation of the axle of the gage pulley, which serves as a temporary
nch mark, is 30.98 feet above gage datum.
The observations at this station during 1903 have been made under
e direction of C. C. Babb, district engineer.

Discharge measurements of Big Blackfoot River near Bonner, Mont., in 1903.

Date.	Hydrographer.	Gage height.	Discharge.
		Feet.	*Second-feet.*
ril 26	F. M. Brown...................	2. 25	2, 352
ay 23.....................	J. S. Baker	2. 90	2, 979
ne 29.....................	C. D. Flaherty...............	3. 60	3, 830
ly 13.........."...........	......do	2. 55	2, 003
ptember 13	......do	1. 10	948
tober 8	J. H. Sloan...................	. 75	888

Mean daily gage height, in feet, of Big Blackfoot River near Bonner, Mont., for 1903.

Day.	May.	June.	July.	Aug.	Sept.	Oct.	Nov.	Dec.
1...................................		5. 35	3. 15	1. 70	0. 95	0. 95	1. 00	1. 10
2...................................		6. 30	3. 05	1. 65	. 95	. 90	. 90	1. 10
3...................................		6. 80	3. 10	1. 60	. 95	1. 00	. 90	1. 05
4...................................		7. 05	2. 90	1. 50	. 85	1. 25	. 90	1. 20
5...................................		6. 85	2. 90	1. 50	. 85	1. 00	. 90	a 1. 70
6...................................		6. 65	2. 85	1. 40	. 85	1. 10	1. 00	1. 10
7...................................		6. 50	2. 80	1. 40	. 90	1. 10	1. 00	. 90
8...................................		6. 35	2. 95	1. 40	. 90	1. 10	. 95	1. 10
9...................................		6. 25	2. 90	1. 35	. 95	1. 25	1. 00	1. 05
10...................................		5. 95	2. 85	1. 30	1. 20	1. 35	1. 10	1. 10
11...................................		5. 65	2. 70	1. 30	1. 20	1. 10	. 95	. 95
12...................................		5. 60	2. 60	1. 30	1. 10	1. 20	. 90	. 95
13...................................		5. 80	2. 60	1. 20	. 95	1. 20	. 90	. 60
14...................................		5. 85	2. 45	1. 15	1. 10	1. 20	. 90	1. 10
15...................................	4. 25	5. 50	2. 45	1. 10	1. 05	1. 20	1. 10	. 90
16...................................	4. 45	5. 30	2. 30	1. 15	. 95	1. 20	. 80	. 95
17...................................	4. 05	5. 20	2. 15	1. 20	. 90	1. 20	. 80	. 95
18...................................	3. 95	5. 20	2. 10	1. 20	1. 00	1. 20	. 80	. 90
19...................................	3. 65	4. 95	2. 05	1. 20	. 90	1. 20	. 80	. 90
20...................................	3. 40	4. 70	2. 00	1. 15	. 80	1. 20	1. 15	. 80

a Ice on bottom of river.

Mean daily gage height, in feet, of Big Blackfoot River near Bonner, Mont., in 1903—
Continued.

Date.	May.	June.	July.	Aug.	Sept.	Oct.	Nov.	De.
21	4.65	......	2.05	1.05	0.80	1.10	1.05	
22	4.55	......	1.90	1.00	.75	1.10	1.00	
23	4.25	......	1.85	1.00	.70	1.00	1.00	
24	2.80	......	1.85	1.00	.70	1.00	1.00	
25	2.80	......	1.75	1.00	.70	1.00	1.00	
26	2.85	......	1.80	1.20	.80	1.00	.95	
27	3.00	......	1.70	1.10	.85	1.00	.90	
28	3.60	3.60	1.70	1.15	.85	1.00	.85	
29	3.65	3.60	1.70	1.00	.70	1.00	.90	
30	4.00	3.20	1.65	.85	.70	1.00	1.00	
31	4.40	......	1.45	1.00	......	1.00	......	

Rating table for Big Blackfoot River near Bonner, Mont., for 1903.

Gage height.	Discharge.	Gage height.	Discharge.	Gage height.	Discharge.	Gage height.	Discharge.
Feet.	Second-feet.	Feet.	Second-feet.	Feet.	Second-feet.	Feet.	Second-feet.
0.5	700	1.7	1,470	2.9	2,870	4.1	4,620
.6	740	1.8	1,560	3.0	3,010	4.2	5,030
.7	780	1.9	1,660	3.1	3,150	4.3	5,250
.8	830	2.0	1,770	3.2	3,290	4.4	5,480
.9	890	2.1	1,880	3.3	3,440	4.5	5,710
1.0	950	2.2	1,990	3.4	3,590	4.6	5,940
1.1	1,010	2.3	2,100	3.5	3,740	4.7	6,180
1.2	1,070	2.4	2,210	3.6	3,900	4.8	6,420
1.3	1,140	2.5	2,330	3.7	4,070	4.9	6,660
1.4	1,220	2.6	2,460	3.8	4,250	5.0	6,900
1.5	1,300	2.7	2,590	3.9	4,430	5 1	7,140
1.6	1,380	2.8	2,730	4.0	4,620		

Tangent at 5 feet, with differences of 240 per tenth. Measurements made in 1903
between gage heights 0.75 and 3.60 feet. Rating curve and table extended above
and below these points by means of measurements made in 1899 and 1900.

Estimated monthly discharge of Big Blackfoot River near Bonner, Mont., for 1903.

[Drainage area, 2,465 square miles.]

Month.	Discharge in second-feet.			Total in acre-feet.	Run-off.	
	Maximum.	Minimum.	Mean.		Second-feet per square mile.	Depth in inches.
May 15–31	6,060	2,730	4,334	146,138	1.758	1.111
June[a]			7,575	450,744	3.073	3.428
July	3,220	1,425	2,191	134,719	.889	1.025
August	1,470	920	1,105	67,944	.448	.517
September	1,070	780	889	52,899	.361	.403
October	1,180	890	1,011	62,164	.410	.473
November	1,040	830	920	54,744	.373	.416
December	1,470	805	920	56,568	.373	.430

[a] June 21 to 27 estimated.

PALOUSE RIVER NEAR HOOPER, WASH.

For some distance above Hooper the Palouse River consists of a succession of deep pools, from 10 to 15 feet in depth, connected by short riffles. Its valley is about one-half a mile in width and is bordered with basaltic cliffs approximately 300 feet in height.

The measurements of Palouse River are of value in showing the amount of water that could be utilized for irrigation on the lands of Washtucna Valley and in the section north of Pasco. The gaging station was established April 1, 1897, by the land department of the Northern Pacific Railway. It is about 3 miles above the mouth of Cow Creek and 2 miles below the head of the ditch of the Palouse Irrigation Company, which carries 25 second-feet when full. No measurements were made during 1901, the rating table for 1900 being used for that year. The 1902 table gives larger discharges, and it is probable that the discharges as published for 1901 are too small.

On September 9, 1897, the Geological Survey took charge of the station and it was reestablished by C. C. Babb. An inclined gage was fastened to the right bank 1 mile below the Northern Pacific Railway gage and opposite the water tank. When this gage read 2.1 feet the Northern Pacific Railway gage read 6.1 feet. A wire gage was established about 20 feet above the cable by Sydney Arnold in 1903. The gage was repaired and checked from the original bench mark by G. H. Bliss, August 26, 1903. Its datum is the same as that of the original gage. The length of the wire from the end of the weight to the marker is 21.67 feet. It is read once each day by Frank Hill. Discharge measurements are made by means of a cable, car, tagged wire, and stay wire. The stay wire is located about 60 feet upstream from the cable. The initial point for soundings is the cable post on the left bank. The channel is straight for 200 feet above and for one-fourth of a mile below the cable. The current is swift. Both banks high, not liable to overflow, and covered with brush. The bed of the stream is rocky and free from vegetation. Bowlders on the bed of the stream make it difficult to obtain accurate results at this point at low water. There is but one channel at all stages. Bench mark No. 1 is the original bench mark established by C. C. Babb. It is the highest point on a ledge of rock on the left bank 200 feet below the cable directly opposite the point at which the inclined gage was located. It is painted black. Its elevation is 7.60 feet above gage datum. Bench mark No. 2 is the top of a large rock on the right bank 120 feet west of the cable post and 115 feet from the water's edge. Its elevation is 10.88 feet above gage datum. Bench mark No. 3 is a spike on the west side of the cable post on the right bank. It is near the surface of the ground, and has an elevation of 10.26 feet above gage datum.

The observations at this station during 1908 have been made under the direction of T. A. Noble, district engineer.

Discharge measurements of Palouse River near Hooper, Wash., in 1908.

Date.	Hydrographer.	Gage height.	Discharge.
		Feet.	*Second-feet.*
April 21	Sydney Arnold..............	5.28	1,546
August 26	G. H. Bliss	.95	25
November 18	.....do	1.87	133
December 18..................	.....do	2.75	318
Do......................	C. B. Cox..................	2.73	330
December 19..................	.....do	2.51	269
Do......................	.....do	2.51	274
December 20..................	.....do	2.51	256
Do......................	.....do	2.51	249
December 21..................	.....do	2.41	243

Mean daily gage height, in feet, of Palouse River near Hooper, Wash., for 1903.

Day.	Jan.	Feb.	Mar.	Apr.	May.	June.	July.	Aug.	Sept.	Oct.	Nov.	Dec.
1	4.20	5.30	4.20	8.10	4.80	3.40	0.85	0.30	0.95	1.05	1.20	2.00
2	4.20	4.70	4.00	7.90	4.45	3.30	.80	.30	.95	1.05	1.20	2.00
3	5.00	4.40	4.00	7.40	4.25	3.30	.80	.30	.95	1.05	1.20	2.15
4	11.40	4.10	3.90	6.40	4.15	3.20	.80	.30	1.00	1.05	1.20	2.55
5	11.10	4.00	3.70	5.50	4.00	3.20	.80	.30	1.10	1.05	1.20	2.70
6	11.60	3.90	3.70	5.10	3.95	3.00	.85	.25	1.10	1.10	1.20	2.40
7	10.40	4.00	3.60	4.90	4.00	2.90	.90	.25	1.05	1.10	1.20	2.20
8	8.10	4.00	3.60	4.80	4.00	2.80	.85	.25	1.05	1.10	1.20	2.10
9	6.80	4.10	3.60	4.60	4.00	2.70	.80	.20	1.05	1.10	1.20	2.10
0	6.20	4.30	3.50	5.00	3.95	2.65	.80	.20	1.05	1.10	1.20	2.10
1	5.80	4.10	3.40	5.00	3.75	2.60	.75	.20	1.05	1.10	1.25	2.05
2	4.90	4.10	3.60	4.90	3.70	2.45	.75	.20	1.05	1.40	1.30	1.85
3	4.70	3.50	4.70	4.80	3.60	2.45	.70	.15	1.05	1.30	1.40	1.85
4	4.60	3.40	4.80	4.50	3.55	2.40	.70	.15	1.05	1.25	1.80	1.85
5	4.40	3.20	4.75	4.30	3.40	2.40	.70	.15	1.05	1.45	1.90	1.85
6	4.10	3.10	4.50	4.30	3.40	2.35	.65	.15	1.05	1.40	1.80	2.05
7	3.70	3.20	4.50	4.30	3.40	2.30	.65	.15	1.05	1.35	2.00	2.10
8	3.40	3.25	4.45	4.30	3.40	2.25	.65	.15	1.20	1.35	1.85	2.75
9	2.80	3.25	4.40	4.50	3.55	2.10	.60	.15	1.20	1.35	1.70	2.55
10	2.80	3.25	4.30	4.75	4.70	2.10	.60	.15	1.15	1.25	1.75	2.50
11	2.90	3.40	4.30	5.25	5.00	1.95	.55	.15	1.15	1.25	1.70	2.40
12	5.50	3.70	4.20	5.15	4.50	1.20	.55	.10	1.10	1.30	1.70	2.75
13	5.40	4.00	4.30	5.20	4.25	1.90	.50	a .95	1.10	1.35	2.10	2.80
14	5.00	4.10	4.30	5.30	3.95	1.75	.50	.95	1.05	1.35	2.15	3.10
15	8.00	4.10	4.40	5.65	3.80	1.70	.45	.95	1.05	1.25	2.50	3.00
16	9.60	4.20	4.70	5.40	3.60	1.55	.45	.95	1.10	1.15	2.35	2.70
17	10.10	4.30	5.10	5.15	3.40	1.40	.45	.95	1.10	1.20	2.20	2.50
18	8.50	4.20	5.60	5.15	3.30	1.30	.40	.95	1.10	1.20	2.10	2.40
19	6.60	......	5.90	5.15	3.35	1.15	.35	.95	1.05	1.20	2.00	2.35
10	6.00	......	6.70	5.15	3.40	1.00	.35	.95	1.05	1.25	2.00	2.20
11	5.60	......	8.40	......	3.40	......	.35	.95	......	1.25	......	......

a An error of 0.85 in the gage readings was discovered at this point and adjusted.

ROCK CREEK NEAR ST. JOHN, WASH.

This station was established October 15, 1903, by G. H. Bliss. It s located at the highway bridge which crosses Rock Creek at the outlet of Rock Lake, three-fourths of a mile from the ranch of the observer, C. K. Remer. It is 9 miles northeast of St. John, Whitman County, Wash. The gage is a vertical rod fastened to the fifth pile bent of the southeast or left-bank approach. It is read once each day by C. K. Remer. At high stages discharge measurements are made from the downstream side of the highway bridge, to which he gage is attached. At low stages they are made by wading below. the bridge. The bridge is supported by pile bents and has a total span of 210 feet. The initial point for soundings is the end post of the downstream hand rail. The channel is straight for 200 feet above and for 75 feet below the station. The current is sluggish at the bridge at low stages. Both banks are low and rocky and liable to overflow at flood stages. The bed of the stream is covered with rocks and

ravel and is liable to shift at flood stages. Bench mark No. 1 is a
pike driven into the downstream side of the top of the sill of the ti.
Ile bent from the southeast or left-bank approach. Its elevation
1.52 feet above the zero of the gage. Bench mark No. 2 is the t:
a large rock on a point of rocks on the southeast side of the li
about 3,000 feet above the outlet. Its elevation is 16.70 feet ab
the zero of the gage.

The observations at this station during 1903 have been made und
le direction of T. A. Noble, district engineer.

The following discharge measurement was made by G. H. Bliss
: 903:

December 4: Gage height, 10.51 feet; discharge, 10 second-feet.

Mean daily gage height, in feet, of Rock Creek near St. John, Wash., for 1903.

Day.	Oct.	Nov.	Dec.	Day.	Oct.	Nov.	Dec.	Day.	Oct.	Nov.	
1		10.08	10.45	12		10.28	10.55	23		10.18	10.6	7
2		10.15	10.48	13		10.26	10.56	24		10.08	10.6	6
3		10.18	10.52	14		10.26	10.56	25		10.10	10.62	5
4		10.20	10.51	15	10.06	10.22	10.65	26		10.12	10.6	5
5		10.15	10.51	16	10.06	10.20	10.65	27		10.20	10.62	4
6		10.18	10.53	17	10.06	10.22	10.69	28		10.18	10.44	3
7		10.20	10.51	18	10.05	10.28	10.70	29		10.08	10.44	2
8		10.20	10.50	19	10.07	10.28	10.72	30		10.10	10.6	1
9		10.20	10.52	20	10.06	10.22	10.75	31		10.12		
10		10.22	10.55	21	10.07	10.25	10.75					
11		10.25	10.55	22	10.06	10.38	10.80					

WALLOWA RIVER NEAR ELGIN, OREG.

This station was established November 18, 1903, by John H. Lewis.
It is located at the county highway bridge just below the mouth of
Minam River and 12 miles from Elgin, Oreg. The gage is in 2 sections of 2 by 6 inch timber, located under the lower side of the bridge
on the left bank. Both sections are held in place by bolts, cemented
into the solid rock. The lower inclined section reads from 1.7 to 3 feet.
The upper vertical section reads from 3 feet to 10 feet. The gage
is read once each day by John McCulloch. Discharge measurements
are made from the downstream side of the single-span bridge, to which
the gage is attached. On account of the velocity of the water a stay
wire is used in making discharge measurements. The initial point for
soundings is the end of the bridge rail on the left bank. The channel
is straight for 100 feet above and for 1,000 feet below the station.
Both banks are high, rocky, not liable to overflow, and without timber or brush. The current is swift. The bed of the stream is composed of gravel, free from vegetation, and not liable to shift. There
is but one channel at all stages. Bench mark No. 1 is the highest
point of the rock to which the vertical section of the gage is fastened.
It is 6 inches from the gage rod and has an elevation of 9.60 feet

above the zero of the gage. Bench mark No. 2 is the center of the hole in the rock in which the bolt supporting the vertical rod is set. Its elevation is 7.07 feet above the zero of the gage.

The observations at this station during 1903 have been made under the direction of J. T. Whistler, district engineer.

The following discharge measurement was made by John H. Lewis in 1903:

November 18: Gage height, 2.60 feet; discharge, 540 second-feet.

Mean daily gage height, in feet, of Wallowa River near Elgin, Oreg., for 1903.

Day.	Nov.	Dec.	Day.	Nov.	Dec.	Day.	Nov.	Dec.	Day.	Nov.	Dec.
1		3.00	9		2.80	17	2.60	2.80	25	3.40	2.60
2		3.00	10		2.80	18	2.60	2.80	26	3.40	2.50
3		3.00	11		2.80	19	2.70	2.80	27	3.30	2.50
4		3.00	12		2.80	20	2.80	2.80	28	3.20	2.50
5		3.00	13		2.70	21	3.00	2.80	29	3.00	2.50
6		3.00	14		2.70	22	3.80	2.70	30	3.00	2.50
7		2.90	15		2.70	23	3.50	2.70	31		2.50
8		2.80	16		2.70	24	3.40	2.70			

WALLOWA RIVER NEAR WALLOWA, OREG.

This station was established November 14, 1903, by John H. Lewis. It is located at the county bridge, 1½ miles below Wallowa, Oreg., and one-fourth mile below the mouth of Bear Creek. A small irrigation ditch takes water from the river about 300 feet above the bridge on the right bank. The gage is a vertical 2 by 6 inch timber nailed to the downstream side of the timber crib pier on the right bank. Its location is such that it is protected by the pier from drift. It is read once each day by L. S. Johnson. Discharge measurements are made from the upstream side of the bridge to which the gage is attached. The initial point for soundings is the left end of the lower chord on the upstream side of the bridge. The channel is straight for 400 feet above and for 600 feet below the station. The current is swift. Both banks are low, wooded, and not liable to overflow. The bed of the stream is composed of gravel, free from vegetation, and is not liable to shift. There is but one channel at low water. At high water the channel is broken by the rock-filled timber crib pier, to which the gage is fastened. The bench mark is the head of a 30-penny wire nail driven nearly flush with the top of the crib, 2 inches from the edge, near the gage. Its elevation is 6.60 feet above the zero of the gage.

The observations at this station during 1903 have been made under the direction of J. T. Whistler, district engineer.

The following discharge measurement was made by John H. Lewis in 1903:

November 14: Gage height, 2.07 feet; discharge, 417 second-feet.

Mean daily gage height, in feet, of Wallowa River near Wallowa, Oreg., for 1901.

Day.	Nov.	Dec.	Day.	Nov.	Dec.	Day.	Nov.	Dec.	Day.	Nov.	Dec.
....	......	2.28	8..........	......	2.00	17..........	1.68	1.68	25..........	2.28	
....	......	2.25	10..........	......	2.00	18..........	1.65	1.65	26..........	2.18	
....	......	2.25	11..........	......	2.00	19..........	1.90	1.64	27..........	2.18	
....	......	2.18	12..........	......	1.98	20..........	1.62	2.00	28..........	2.18	
....	......	2.10	13..........	......	1.95	21..........	2.65	1.90	29..........	2.15	
....	......	2.10	14..........	2.00	1.92	22..........	2.50	1.90	30..........	2.10	
....	......	2.05	15..........	2.04	1.92	23..........	2.60	1.90	31..........		
....	......	2.00	16..........	1.98	1.95	24..........	2.35	1.89			

WALLOWA RIVER NEAR JOSEPH, OREG.

This station was established November 12, 1903, by John H. Lewis.
The gage is located on Wallowa Lake near its outlet. It is a vertical
2 by 6 inch board fastened to a log pier which extends into the lake.
It reads from zero to 5 feet. It is read once each day by G. F. Beal.
Discharge measurements are made from a footbridge about 500 feet
below the outlet of Wallowa Lake and 1½ miles above Joseph, Oreg.
The bridge has a single span of 50 feet. The initial point for soundings is the end of the upstream log supporting the footbridge on the
left bank. The channel is straight for 100 feet above and for 75 feet
below the station. The right bank is liable to overflow at high water
for about 30 feet, at which point it becomes steep. The left bank will
overflow for about 20 feet at high water. Both banks are timbered.
At the bridge the bed of the stream is composed of large bowlders.
free from vegetation, and is not liable to shift. There is but one channel at all stages. The direction of the wind is liable to affect both the
gage readings on the lake and the discharge. No bench marks have
been established.

The observations at this station during 1903 have been made under
the direction of J. T. Whistler, district engineer.

The following discharge measurement was made by John H. Lewis
in 1903:

November 12: Gage height, 148 feet; discharge, 79 second-feet.

Mean daily gage height, in feet, of Wallowa River near Joseph, Oreg., for 1903.

Day.	Nov.	Dec.	Day.	Nov.	Dec.	Day.	Nov.	Dec.	Day.	Nov.	Dec.
1..........	......	1.55	9..........	......	1.55	17..........	1.50	1.53	25..........	1.65	......
2..........	......	1.60	10..........	......	1.53	18..........	1.50	1.55	26..........	1.60	......
3..........	......	1.60	11..........	......	1.53	19..........	1.55	1.55	27..........	1.63	1.45
4..........	......	1.55	12..........	......	1.55	20..........	1.55	1.50	28..........	1.55	1.45
5..........	......	1.53	13..........	1.55	1.50	21..........	1.60	1.50	29..........	1.55	1.45
6..........	......	1.55	14..........	1.60	1.50	22..........	1.60	1.45	30..........	1.55	1.45
7..........	......	1.55	15..........	1.60	1.45	23..........	1.60	1.45	31..........	......	1.45
8..........	......	1.55	16..........	1.55	1.55	24..........	1.66	1.43			

GRANDE RONDE RIVER AT ELGIN, OREG.

This station was established November 20, 1903, by John H. Lewis. t is located at the county bridge on the road from Elgin to Wallowa,)reg., and is one-fourth mile east of the railroad station. It is at the ower end of the Grande Ronde Valley. The lower section of the ;age, reading from 0 to 2 feet, is a 2 by 4 inch rod driven into the nud on the upstream side of the vertical steel caisson of the left pier. *rom 2 to 9 feet the gage is painted on the side of the caisson. It is ·ead once each day by John Graham. Discharge measurements are nade from the downstream side of the bridge, to which the gage is .ttached. This bridge has a span between piers of 100 feet, with 130 eet of trestle approach from the left bank, and 30 feet of approach rom the right bank. The initial point for soundings is on the right ·ank directly over the center of the bent, 30 feet from the caisson. ?he channel is curved above a point 30 feet above the bridge, and is traight for 200 feet below. The right bank is high, rocky, free from ·egetation, and will not overflow. The left bank is low, free from ·egetation, and will overflow only under the trestle approach. The ·ed of the stream is uneven, covered with large bowlders, and is free rom vegetation. It is not liable to shift. The channel is broken by he piers and the trestle bents at high water. The bench mark is the op surface of the steel caisson directly over the gage. Its elevation s 15.14 feet above the zero of the gage.

Discharge measurements of Grande Ronde River at Elgin, Oreg., in 1903.

Date.	Hydrographer.	Gage height.	Discharge.
		Feet.	*Second-feet.*
November 20	J. H. Lewis...................	1.65	186
November 22	.....do	2.62	574

Mean daily gage height, in feet, of Grande Ronde River at Elgin, Oreg., for 1903.

Day.	Nov.	Dec.	Day.	Nov.	Dec.	Day.	Nov.	Dec.	Day.	Nov.	Dec.
....	.....	2.8	9...........	.....	2.2	17..........	.....	2.3	25.........	3.1	1.9
....	.....	3.0	10.........	.....	2.1	18..........	.....	2.2	26.........	3.2	1.8
....	.....	3.0	11.........	.....	2.2	19..........	.....	2.1	27.........	3.1	1.4
....	.....	2.8	12.........	.....	2.3	20..........	.....	2.2	28.........	2.9	1.4
....	.....	2.4	13.........	.....	2.3	21..........	2.1	2.2	29.........	2.8	1.4
....	.....	2.4	14.........	.....	2.3	22..........	2.6	2.3	30.........	2.8	1.1
....	.....	2.4	15.........	.....	2.3	23..........	2.9	2.2	31..........	.....	1.9
....	.....	2.4	16.........	.....	2.4	24..........	3.0	2.1			

GRANDE RONDE RIVER AT HILGARD, OREG.

This station was established November 6, 1903, by John H. Lewis. It is located at the county highway bridge one-half mile below the Oregon Railroad and Navigation Company station at Hilgard, Oreg. It is just below the mouth of Five Points Creek, which is the first important tributary above Grande Ronde Valley. There are two dams about 20 miles upstream, used to flood the river during the log driving season. The gage is a 1 by 4 inch board nailed to a vertical 4 by 6 inch timber, which is driven into the ground at the downstream end of the middle bridge pier, and which is bolted at the upper end to the log pier. J. D. Casey reads the gage once each day at ordinary stages and twice during floods. Discharge measurements are made from the downstream side of the two-span bridge to which the gage is attached. The bridge is supported by two timber crib abutments and by one middle crib pier. The initial point for soundings is at the point where the end post meets the lower chord of the bridge on the right bank. It is directly over the vertical outer edge of the abutment. The channel is straight for 100 feet above and for 200 feet below the station. At ordinary stages all the water passes under the right span, which has a length of 70 feet from the right abutment to the middle pier. At high water the water also passes under the shorter span, which has a length of 52 feet from the left abutment to the middle pier. The right bank is low, but is not liable to overflow. The left bank is low and will overflow only at a few points above the bridge. The bed of the stream is composed of sand and clay, free from vegetation and bowlders. It is permanent under the main span, but is liable to shift in the high-water channel under the shorter span. The bench mark is the head of a bolt through the lower chord of the bridge 7.5 feet from the timber to which the gage is attached. Its elevation is 13.8 feet above the zero of the gage.

The observations at this station during 1903 have been made under the direction of J. T. Whistler, district engineer.

Discharge measurements of Grande Ronde River at Hilgard, Oreg., in 1903.

Date.	Hydrographer.	Gage height.	Discharge.
		Feet.	*Second-feet.*
November 5	J. H. Lewis	2.07	44
November 6	do	2.00	46
November 23	do	3.02	406

Mean daily gage height, in feet, of Grande Ronde River at Hilgard, Oreg., for 1903.

Day.	Nov.	Dec.	Day.	Nov.	Dec.	Day.	Nov.	Dec.	Day.	Nov.	Dec.
....	....	3.10	9		2.10	2.50	17		2.50	25	3.10	2.50
....	....	3.10	10		2.50	18		2.50	26	3.30	2.50
....	....	2.90	11		2.10	2.50	19	2.20	2.50	27	3.20	2.50
....	....	2.70	12		2.10	2.50	20		2.50	28	3.00	2.50
....	....	2.60	13		2.00	2.50	21		2.50	29	2.90	2.50
....	2.00	2.60	14		2.50	22	2.90	2.50	30	3.00	2.50
....	2.10	2.60	15		2.50	23	2.90	2.50	31		2.50
....	2.00	2.60	16		2.50	24	2.90	2.50			

WEISER RIVER NEAR WEISER, IDAHO.

The drainage basin of this river is mountainous and rocky, in con-
ast to the well-wooded areas of the Boise and Payette basins, and
ɪe effect is shown in the high flood discharges and low summer flow.
number of small ditches utilize considerable water from this river,
ɪt the principal canal is the Galloway canal, which irrigates lands of
ɪe Weiser irrigation district north of Weiser. Above this canal is a
ɪnch country susceptible of cultivation if water could be brought to it.
The station was established December 6, 1894, by A. P. Davis. It
located on J. W. Lane's ranch in the canyon of the river about 10
iles above Weiser, Idaho. The gage rod, which was installed in
ɪ98, was covered during the process of grading for the roadbed of
ɪe Pacific and Idaho Northern Railroad, a line intended to run from
ᵀeiser to the mining country in the mountainous district to the north.
he present gage was installed October 31, 1899, at a point 100 feet
ɪove the old gage on the right bank. It is a 4 by 4 inch inclined
mber, 12 feet long, bolted to the rock bluff. One foot measured ver-
cally equals 1.15 feet measured along the gage. The gage is read
nce each day by Mrs. Annie T. Lane. Discharge measurements are
ɪade from a cable and car about 300 feet downstream from the gage.
ʰe initial point for soundings is 10 feet from the cable support on
he right bank. The channel is straight for 300 feet above and below
he station. The right bank is high and rocky and is not liable to
ᵥerflow. The left bank is low and will overflow for 100 feet. Both
anks are without trees, but brush grows on that part of the left bank
able to overflow. There is but one channel at all stages. The bed
f the stream is composed of gravel, free from vegetation, and not
able to change. Bench mark No. 1 is the highest point of a rock 40
ɪet southwest of the south anchorage of the cable. Its elevation is
ɪ.54 feet above the zero of the gage. Bench mark No. 2 is the
ighest point of a rock 60 feet southwest of the cable anchorage on
ɪe right bank. Its elevation is 25.16 feet above the zero of the gage.
he current is sluggish.

The observations at this station during 1906 have been made
the direction of D. W. Ross, district engineer.

Discharge measurements made on Weiser River near Weiser, Idaho, in 19

Date.	Hydrographer.	Gage height.	
		Feet.	
June 13	N. S. Dils	3.40	
August 3	F. Stockton	.30	
October 14	do	1.00	

Mean daily gage height, in feet, of Weiser River near Weiser, Idaho, for 1

Day.	Jan.	Feb.	Mar.	Apr.	May.	June.	July.	Aug.	Sept.	Oct.
1	2.35	3.60	1.00	7.00	4.00	4.00	1.60	5.50	3.60	0.70
2	2.30	3.00	1.00	6.16	4.00	4.40	1.80	.30	.80	.70
3	2.10	2.60	1.00	5.50	4.00	4.30	1.70	.50	.80	.90
4	1.90	2.60	1.00	5.00	4.30	4.00	1.60	.40	.80	1.00
5	1.80	2.50	1.00	4.60	4.40	4.20	1.50	.40	.40	1.10
6	1.80	2.40	1.00	4.40	4.40	4.20	1.40	.30	.40	1.20
7	1.80	2.20	1.00	4.10	4.50	4.00	1.50	.30	.40	1.20
8	1.80	2.00	1.00	3.90	4.50	4.00	1.70	.25	.40	1.20
9	1.70	1.90	1.00	3.50	4.80	3.90	1.60	.25	.40	1.10
10	1.60	1.80	1.00	3.90	4.10	3.50	1.60	.25	.60	1.00
11	1.60	1.60	2.00	4.20	4.00	3.60	1.40	.20	.50	1.10
12	1.60	1.40	3.60	3.90	4.00	3.40	1.40	.20	.70	1.10
13	1.60	1.30	5.60	3.60	4.10	3.40	1.30	.20	.70	1.10
14	1.50	1.20	8.00	3.50	4.30	3.30	1.30	.20	.80	1.10
15	1.50	1.20	9.60	3.40	4.50	3.20	1.20	.20	.90	1.00
16	1.40	1.20	9.60	3.40	4.30	3.00	1.20	.20	.90	1.00
17	1.30	1.20	9.10	3.60	4.10	2.90	1.20	.20	.90	1.00
18	1.30	1.10	7.50	3.80	3.70	2.80	1.20	.20	.80	.90
19	1.30	1.10	5.10	4.00	3.40	2.50	1.20	.20	.80	.90
20	1.40	1.10	5.00	3.90	3.10	2.40	1.20	.20	.80	1.00
21	1.50	1.00	5.10	3.90	3.00	2.30	1.10	.20	.70	1.00
22	1.70	1.00	5.20	4.10	3.00	2.20	1.10	.20	.70	1.00
23	2.80	1.00	5.20	4.40	3.10	2.20	1.00	.20	.70	1.00
24	4.80	1.00	5.50	4.60	3.20	2.10	.90	.20	.70	1.00
25	5.20	1.00	5.50	4.60	3.20	2.00	.80	.20	.60	.90
26	5.50	1.00	6.50	4.85	3.30	1.90	.80	.30	.60	.90
27	5.20	1.00	7.00	5.00	3.20	1.90	.80	.50	.60	.90
28	5.00	1.00	6.50	4.80	3.30	1.90	.80	.60	.60	.90
29	4.80	......	8.00	4.50	3.50	1.90	.80	.80	.70	.90
30	3.20	......	8.00	4.30	3.50	1.80	.70	.70	.70	.90
31	3.10	......	7.50	.....	3.80	......	.60	.70	......	.90

Rating table for Weiser River near Weiser, Idaho, for 1902 and 1903.

Gage height.	Discharge.	Gage height.	Discharge.	Gage height.	Discharge.	Gage height.	Discharge.
Feet.	Second-feet.	Feet.	Second-feet.	Feet.	Second-feet.	Feet.	Second-feet.
0.2	39	2.2	875	4.2	3,125	6.2	5,825
.4	71	2.4	1,050	4.4	3,395	6.4	6,095
.6	105	2.6	1,235	4.6	3,665	6.6	6,365
.8	150	2.8	1,425	4.8	3,935	6.8	6,635
1.0	215	3.0	1,625	5.0	4,205	7.0	6,900
1.2	290	3.2	1,840	5.2	4,475	7.5	7,575
1.4	375	3.4	2,075	5.4	4,745	8.0	8,250
1.6	470	3.6	2,325	5.6	5,015	8.5	8,925
1.8	585	3.8	2,585	5.8	5,285	9.0	9,600
2.0	720	4.0	2,855	6.0	5,555	9.5	10,275

Estimated monthly discharge of Weiser River near Weiser, Idaho, for 1903.

[Drainage area, 1,670 square miles.]

Month.	Discharge in second-feet.			Total in acre-feet.	Run-off.	
	Maximum.	Minimum.	Mean.		Second-feet per square mile.	Depth in inches.
nuary	4,880	330	1,369	84,176	0.82	0.95
·bruary .....:....	1,625	215	572	31,767	.34	.35
arch	10,410	215	4,246	261,076	2.54	2.93
)ril?	6,900	2,075	3,324	197,792	1.99	2.22
ıy	3,530	1,625	2,619	161,036	1.57	1.81
ne	3,530	585	1,838	109,369	1.10	1.23
ly	585	105	318	19,553	.19	.22
ıgust	150	39	60	3,689	.04	.05
ptember..........	180	71	115	6,843	.07	.08
tober	290	125	212	13,035	.13	.15
ıvember	4,407	150	828	49,269	.50	.56
cember	795	290	558	34,310	.33	.38
The year	10,410	39	1,338	971,915	.80	10.93

MALHEUR RIVER NEAR ONTARIO, OREG.

This station was established December 8, 1903, by John H. Lewis, sisted by William O'Brien. It is located at the new county bridge, out 2½ miles northwest of the town of Ontario, Oreg., and about 1½ iles from the junction of Malheur River with Snake River. The ·osnan ditch, 3 miles above the station, and the Nevada ditch, 12 miles above the station, divert some water at certain times. The gage consists of two 2 by 6 inch timbers, painted white and nailed to 4 by 6 inch timbers firmly secured to the left bank of the river, about 115 feet above the bridge. The lower sloping rod is graduated from 2 to 10 feet and the upper vertical section from 10 to 21 feet. The gage is read twice each day by William O'Brien. Discharge measurements are made from the county bridge, which has a single span of 120 feet. The initial point for soundings is at the end of the upstream railing over the pier on the left bank. Five-foot intervals are marked on the railing with 10-penny nails. The channel is straight for 150 feet above and for 200 feet below the station. The current is sluggish. Both banks are low, and liable to overflow at extreme high water. The right bank is covered with brush. The bed of the stream is composed of sand and gravel, free from vegetation, and may shift slightly during floods. There is one channel at low water, but several high-water sloughs. During extreme high water, generally caused by an ice jam, several high-water sloughs will have to be measured by wading. Floods in the Snake River are said to occur at different times from those in the Malheur River, and the water surface at the gage has not been affected by backwater during recent years.

The bench mark is the edge of the steel band on upstream bridge caisson at the left bank. Its elevation is 18.9 feet above the zero of the gage. A discharge measurement was made December 8, 1904, by John H. Lewis. A gage height of 6.46 feet and a discharge of 138 second-feet was registered.

The observations at this station during 1903 have been made under the direction of J. T. Whistler, district engineer.

Mean daily gage height, in feet, of Malheur River near Ontario, Oreg., for 1903.

Day.	Dec.	Day.	Dec.	Day.	Dec.	Day.	Dec.	Day.	Dec.
1		8	6.42	15	6.55	22	6.50	29	6.30
2		9	6.53	16	6.60	23	6.40	30	6.30
3		10	6.53	17	6.60	24	6.40	31	6.30
4		11	6.50	18	6.60	25	6.40		
5		12	6.50	19	6.60	26	6.30		
6		13	6.50	20	6.50	27	6.30		
7		14	6.50	21	6.50	28	6.30		

MALHEUR RIVER AT VALE, OREG.

This station was established as a temporary station May 20, 1903, by N. S. Dils. The permanent station was established by John H. Lewis, June 30, 1903. It is located at the steel highway bridge, one-eighth mile southeast of Vale, Oreg. The lower section of the gage is an inclined 2 by 6 inch timber under the downstream edge of the

₂, near the left pier. It reads from 0 to 14 feet. The upper sec-
reading from 14 to 20 feet, is painted on the downstream side of
ᵢwnstream caisson of the left abutment. The gage is read twice
lay by E. R. Murray. Discharge measurements are made from
ᵢpstream side of the bridge, at which the gage is located. The
point for soundings is the northwest side of the end post of the
at its base. The channel is straight for 250 feet above and for
ₑt below the station. The right bank is nearly vertical and is
ᵢped at the bridge. It is liable to overflow and is covered with
rush. The left bank is sloping and 'is the same height as the
ite bank. It iˢ riprapped at the bridge, and is liable to over-
The bed of the stream is sandy, free from vegetation, and is
to change. The bench mark iˢ the top surface of the down-
ₙ steel caisson on the left bank. The point is marked in black
and is a cross inclosed in a circle. Its elevation is 21.63 feet above
ᵢro of the gage. The highway bridge, which has a span of 150
ₑtween abutments, crosses the river at an angle of 20° with the
ᵢl to the direction of the current. This has been taken into
nt by laying off intervals of 5.1 feet on the hand rail of the
ₑ. These intervals correspond to 5-foot intervals of a section
ndicular to the direction of the current. A station has been
ained here at intervals since 1890.

ᵢ observations at this station during 1903 have been made under
rection of J. T. Whistler, district engineer.

Discharge measurements of Malheur River at Vale, Oreg., in 1903.

Date.	Hydrographer.	Gage height.	Discharge.
		Feet.	*Second-feet.*
₁.....................	N. S. Dils	5.00	327
....	H. D. Newell	4.88	349
2.....................	A. K. Sears...................	4.50	241
₃.....................	E. I. Davis	4.35	193
3.....................	.....do	3.95	87
₃.....................	J. H. Lewis...................	3.70	75
₍ 13	.....do	3.02	11
ᵢber 25	.....do	3.52	39
ber 16.................	.....do	4.28	197

Mean daily gage height, in feet, of Malheur River at Vale, Oreg., for 1903.

Day.	May.		Aug.	Sept.	Oct.	Nov.	Dec.
1...		4.50	3.70	3.30	3.50	3.50	4.00	4.0
2...		4.60	3.70	3.50	3.50	3.55	4.00	4.1
3...		4.70	3.70	3.30	3.50	3.60	4.00	4.6
4...		4.50	3.70	3.30	3.50	3.60	4.00	4.0
5...		4.50	3.70	3.50	3.50	3.70	4.00	4.8
6...		4.50	3.70	3.25	3.50	3.70	4.00	4.0
7...		4.65	3.65	3.30	3.50	3.70	4.00	4.1
8...		4.75	3.60	3.30	3.50	3.50	4.00	4.0
9...		4.65	3.60	3.50	3.50	3.30	4.00	4.0
10...		4.60	3.60	3.30	3.30	3.80	4.10	4.0
11...		4.60	3.70	3.15	3.50	3.80	4.15	4.0
12...		4.60	3.75	3.40	3.50	3.90	4.10	4.3
13...		4.50	3.80	3.00	3.50	3.90	4.15	4.3
14...		4.50	3.80	3.00	3.50	3.90	4.10	4.3
15...		4.60	3.80	3.00	3.00	3.90	4.10	4.5
16...		4.50	3.80	3.50	3.50	3.60	4.10	4.3
17...		4.75	3.80	3.00	3.50	3.00	4.30	4.3
18...		4.65	3.80	3.50	3.50	3.95	4.30	4.3
19...		4.60	3.80	3.00	3.30	4.00	4.30	4.3
20...	5.00	4.65	3.80	3.40	3.00	4.00	4.30	4.3
21...	5.00	4.50	3.80	3.30	3.50	4.00	4.30	4.3
22...	4.85	4.30	3.60	3.60	3.50	4.00	4.30	4.5
23...	4.60	4.15	3.70	3.00	3.00	4.00	4.00	4.3
24...	4.60	4.25	3.60	3.00	3.00	4.00	4.55	4.3
25...	4.70	4.10	3.50	3.00	3.00	4.00	5.25	4.4
26...	4.60	4.00	3.50	3.10	3.50	4.00	5.30	4.3
27...	4.55	3.90	3.50	3.10	3.50	4.00	5.15	4.5
28...	4.50	3.85	3.50	3.20	3.50	4.00	4.95	4.3
29...	4.50	3.75	3.50	3.35	3.50	4.00	4.75	4.1
30...	4.50	3.70	3.45	3.40	3.50	4.00	4.60	4.0
31...	4.50	....	3.35	3.50	....	4.00	....	4.0

Rating table for Malheur River at Vale, Oreg., from May 20 to December 31, 1903.

Gage height.	Discharge.	Gage height.	Discharge.	Gage height.	Discharge.	Gage height.	Discharge.
Feet.	Second-feet.	Feet.	Second-feet.	Feet.	Second-feet.	Feet.	Second-feet.
3.0	11	3.6	50	4.2	144	4.8	304
3.1	16	3.7	60	4.3	168	4.9	332
3.2	22	3.8	72	4.4	194	5.0	360
3.3	28	3.9	86	4.5	220	5.1	388
3.4	34	4.0	103	4.6	248	5.2	416
3.5	42	4.1	123	4.7	276	5.3	444

Table made from measurements between gage heights 3.02 and 5 feet. Table extended to 5.35 feet; not determined below gage height 3 feet.

Estimated monthly discharge of Malheur River at Vale, Oreg., for 1903.

Month.	Discharge in second-feet.			Total in acre-feet.
	Maximum.	Minimum.	Mean.	
1...............................	360	220	274	6,522
....	332	60	203	12,079
....	72	31	58	3,566
....	42	11	19	1,168
·r..............................	42	42	42	2,499
....	103	42	84	5,165
r	458	103	192	11,425
r	262	123	175	10,760
e period	....	....	....	53,184

.HEUR RIVER NEAR HARPER'S RANCH, ABOVE VALE, OREG.

nporary station was established December 15, 1903, by J. H.
it the old bridge site, 22 miles above Vale, Oreg., and 3 miles
ie Harper ranch. A permanent equipment was established
8, 1904. The gage established on the latter date consists of
al 2 by 6 inch timber attached to the frame bent of the old
)n the left bank. It is painted white and graduated with black
id tacks. It is read from the right bank at noon each day by
icott. Discharge measurements are made by means of a cable
about 150 feet above the gage. The cable is a $\frac{5}{8}$-inch galvan-
n rope, having a span of 233 feet, and carries a gaging car 2$\frac{1}{2}$
:t and 1 foot deep. The initial point for soundings is the zero
gaged wire, and is 20.6 feet from the lower end of the turn-
The channel is straight for 1,000 feet above and for 200 feet
he station. The right bank is low, covered with sagebrush,
ubject to overflow during high water. The left bank is high,
ind is not liable to overflow. The current is swift. The bed
itream is composed of gravel and sand, free from vegetation,
liable to shift. There is but one channel at low water. At
iter there is one main channel with two sloughs on the right
Bench mark No. 1 is the head of drift bolt projecting from the
of the frame of the bent to which the gage is attached at the
1. Its elevation is 17.64 feet above the zero of the gage.
mark No. 2 is a stake on the left bank near the gage and 75
m the river bank. Its elevation is 23.14 feet above the zero
rage. Bench mark No. 3 is a stake on the right bank directly
vith the cable and 185 feet from the deadman. Its elevation
leet above the zero of the gage. A single discharge measure-

ment was made at this station December 13, 1903, by J. H. Lewis. A gage height of 3.45 feet was registered, giving a discharge of 188 feet.

The observations at this station during 1903 have been made under the direction of J. T. Whistler, district engineer.

Mean daily gage height, in feet, of Malheur River at Harper's ranch, Oregon, for 1903.

Day.	Dec.	Day.	Dec.	Day.	Dec.	Day.	Dec.	Day.	Dec.
1...........	....	8...........	....	15...........	3.45	22...........	3.45	29........	3.9
2...........	....	9...........	....	16...........	3.45	23...........	3.40	30.........	3.9
3...........	....	10...........	....	17...........	3.50	24...........	3.30	31.........	3.9
4...........	....	11...........	....	18...........	3.45	25...........	3.90	-	
5...........	....	12...........	....	19...........	3.45	26...........	3.40		
6...........	....	13...........	....	20...........	3.40	27...........	3.30		
7...........	....	14...........	3.45	21...........	3.40	28...........	3.30		

BULLY CREEK NEAR VALE, OREG.

This station was established August 10, 1903, by John H. Lewis. It is located about one-eighth mile below the mouth of Cottonwood Creek and 13 miles above Vale, Oreg. It is about 2 miles below a proposed reservoir site on this creek. The gage is located on the left bank and consists of 2 by 6 inch timbers. The lower section is inclined and reads from 0 to 6 feet. The upper section is vertical and reads from 6 to 12 feet. The gage is about 200 feet from the house of F. O'Neill, who reads the gage twice each day. Discharge measurements are made by means of a cable and a car about 70 feet below the gage. The initial point for soundings is the zero on the tagged wire on the left bank, 25 feet from the lower end of the turnbuckle. The main channel at ordinary stages is straight for 200 feet above the cable. At flood stages the right bank overflows above the cable, causing a bend in the channel about 75 feet above. The channel is straight for 800 feet below the station. At the cable the right bank is low, will overflow at extreme flood stages, and is covered with sagebrush. The left bank is low, but is not liable to overflow. The bed of the stream is composed of gravel and sand, free from vegetation, and liable to shift. There is but one channel at all stages. Bench mark No. 1 is a projecting stone on the top of the wall in the front of the observer's house on the west side of the entrance. Its elevation is 16.92 feet above the zero of the gage. Bench mark No. 2 is a 2 by 4 inch timber projecting from the northeast corner of the granary, about 42 feet from the gage. Its elevation is 12.37 feet above the zero of the gage.

The observations at this station during 1903 have been made under the direction of J. T. Whistler, district engineer.

Discharge measurements of Bully Creek near Vale, Oreg., in 1903.

Date.	Hydrographer.	Gage height.	Discharge.
		Feet.	Second-feet.
August 11	J. H. Lewis	1.88	0.3
September 24	....do	2.03	2.0
December 11................	....do	2.60	12.0

Mean daily gage height, in feet, of Bully Creek near Vale, Oreg., for 1903.

Day.	Aug.	Sept.	Oct.	Nov.	Dec.	Day.	Aug.	Sept.	Oct.	Nov.	Dec.
1.....................	.....	2.00	2.03	2.10	2.55	17...................	1.86	2.10	2.10	2.15	2.62
2.....................	.....	2.00	2.03	2.14	2.50	18...................	1.86	2.10	2.10	2.20	2.60
3.....................	.....	2.00	2.10	2.18	2.37	19...................	1.86	2.10	2.10	2.36	2.60
4.....................	.....	2.00	2.10	2.18	2.34	20...................	1.86	2.10	2.10	2.40	2.60
5.....................	.....	2.00	2.10	2.18	2.38	21...................	1.86	2.10	2.10	2.40	2.60
6.....................	.....	2.00	2.10	2.18	2.40	22...................	1.86	2.10	2.10	2.40	2.60
7.....................	.....	2.00	2.10	2.18	2.40	23...................	1.86	2.10	2.10	3.25	2.60
8.....................	.....	2.00	2.10	2.18	2.40	24...................	1.86	2.06	2.10	3.40	2.60
9.....................	.....	2.00	2.10	2.18	2.51	25...................	2.37	2.03	2.10	3.00	2.60
10.....................		2.10	2.10	2.18	2.46	26...................	2.59	2.03	2.10	3.00	2.60
11.....................	1.88	2.10	2.10	2.45	2.55	27...................	2.15	2.03	2.10	2.90	2.60
12.....................	1.88	2.10	2.10	2.50	2.54	28...................	2.10	2.03	2.10	2.82	2.60
13.....................	1.87	2.10	2.10	2.40	2.55	29...................	2.10	2.03	2.10	2.75	2.60
14.....................	1.86	2.10	2.10	2.55	2.60	30...................	2.00	2.03	2.10	2.64	2.60
15.....................	1.86	2.10	2.10	2.45	2.60	31...................	2.00	......	2.10	......	2.60
16.....................	1.86	2.10	2.10	2.25	2.60						

MALHEUR LAKE AT THE NARROWS, OREG.

This station was established May 14, 1903, by N. S. Dils. It is located at the highway bridge across "The Narrows." The gage is a 1 by 4 inch board 10 feet long, fastened to a telegraph pole. It is read from the bridge once each day by C. A. Haines, the postmaster. The bench mark is a United States Geological Survey standard iron bench-mark post set at the east end of the bridge on the south side of the road near the fence and about 200 feet from the gage. Its elevation is 8.2 feet above the zero of the gage and 4,088 feet above sea level. This station is maintained for the purpose of obtaining data of the lake fluctuations. The lake has no outlet. No discharge measurements are made at this station.

The observations at this station during 1903 have been made under the direction of J. T. Whistler, district engineer.

Mean daily gage height, in feet, of Malheur Lake at The Narrows, Oreg., for 1903.

Day.	May.	June.	July.	Aug.	Sept.	Oct.	Nov.	Dec.
1..	....	5.60	5.20	4.60	3.30	1.90	1.80	1.6
2..	....	5.60	5.30	4.50	3.30	1.90	1.80	1.7
3..	....	5.60	5.30	4.50	3.20	1.80	1.80	..
4..	....	5.60	5.30	4.50	3.20	1.80	1.80	1.0
5..	....	5.60	5.20	4.40	3.10	1.80	1.90	1.9
6..	....	5.60	5.20	4.40	3.10	1.70	1.90	1.0
7..	....	5.50	5.20	4.40	3.00	1.70	1.80	1.0
8..	....	5.50	5.10	4.40	3.00	1.70	1.80	1.0
9..	....	5.50	5.10	4.30	2.90	1.70	1.90	1.0
10..	....	5.50	5.10	4.30	2.90	1.70	1.80	1.0
11..	....	5.50	5.10	4.30	2.80	1.60	1.80	1.0
12..	....	5.50	5.00	4.20	2.80	1.60	1.80	1.0
13..	....	5.50	5.00	4.20	2.70	1.60	1.80	1.0
14..	5.70	5.60	5.00	4.20	2.70	1.60	1.80	1.0
15..	5.70	5.60	5.00	4.20	2.60	1.50	1.80	1.0
16..	5.60	5.60	5.00	4.10	2.60	1.50	1.80	1.0
17..	5.60	5.50	5.00	4.10	2.50	1.50	1.80	1.8
18..	5.60	5.50	5.00	4.00	2.50	a1.40	1.80	1.8
19..	5.60	5.50	4.90	4.00	2.40	1.40	1.80	1.0
20..	5.00	5.50	4.90	3.90	2.40	1.40	1.80	2.0
21..	5.60	5.40	4.90	3.90	2.30	1.40	1.80	2.0
22..	5.00	5.40	4.80	3.90	2.30	1.40	1.50	2.0
23..	5.00	5.40	4.80	3.80	2.20	1.40	1.50	2.0
24..	5.60	5.40	4.80	3.70	2.20	1.40	1.50	2.0
25..	5.60	5.40	4.80	3.70	2.10	1.40	1.50	2.0
26..	5.60	5.40	4.70	3.60	2.00	1.40	1.50	2.0
27..	5.60	5.40	4.70	3.60	2.00	1.40	1.50	2.0
28..	5.60	5.30	4.70	3.50	2.00	1.40	1.50	2.0
29..	5.60	5.30	4.70	3.50	1.90	1.40	1.60	1.9
30..	5.60	5.30	4.60	3.40	1.90	1.40	1.60	1.9
31..	5.60	....	4.60	3.40	....	1.40	....	1.9

a Gage height 1.40 represents surface of dry bed. From about September 1, water at gage is disconnected with Malheur Lake and it has not yet been determined how closely subsequent records agree with fall of lake surface.

POWDER RIVER NEAR BAKER CITY, OREG.

This station was established December 20, 1903, by J. H. Lewis, assisted by R. M. Garrett. It is located 10 miles above Baker City, Oreg., and one-fourth mile below Salisbury station, on the Sumpter Valley Railroad. It is in SW. ¼ sec. 30, T. 10 S., R. 40 E. The gage consists of a vertical 2 by 6 inch timber painted white and graduated to feet and tenths with black paint and tacks. It is nailed to a tree on the left bank, about 400 feet below the house of the observer, and about 40 feet above a wagon bridge over the river. It is read twice each day by R. M. Garrett. Discharge measurements are made from the wagon bridge, which has a single span of 50 feet. The initial point for soundings is the end of the bridge on the left bank. Five-foot intervals are marked with 20-penny nails. The channel is straight for 75 feet above and for 100 feet below the station. The current is swift.

The right bank is low, timbered, and is said to have not been overflowed in recent years. The left bank is low, timbered, and not liable
to overflow. The bed of the stream is composed of gravel, free from
vegetation, and is not liable to shift. There is but one channel at all
stages. The bench mark consists of three 30-penny nails driven into
a cottonwood tree 3 feet in diameter on the left bank of the stream,
5 feet below the gage and 6 feet above the end of the bridge. Its
elevation is 8 feet above the zero of the gage. A single discharge
measurement was made at this station December 20, 1904, by J. H.
Lewis. A gage height of 2.24 feet and discharge of 35 second-feet
were registered.

The observations at this station during 1903 have been made under
the direction of J. T. Whistler, district engineer.

Mean daily gage height, in feet, of Powder River near Baker City, Oreg., for 1903.

Day.	Dec.	Day.	Dec.	Day.	Dec.	Day.	Dec.	Day.	Dec.
1		8		15		22	2.25	29	2.24
2		9		16		23	2.25	30	2.24
3		10		17		24	2.24	31	2.24
4		11		18		25	2.24		
5		12		19		26	2.25		
6		13		20	2.24	27	2.24		
7		14		21	2.24	28	2.24		

SILVIES RIVER NEAR BURNS, OREG.

This station was established as a temporary station by N. S. Dils,
May 10, 1903. The permanent station was established August 14,
1903, by W. T. Turner. It is located about 10 miles above Burns,
Oreg. On January 19, 1904, a new gage was installed together with
a cable for discharge measurements by John H. Lewis. The present
gage is on the left bank, 10 feet below the bridge, near Parker's house.
The lower inclined section reads from 2 to 12.6 feet. The upper vertical section reads from 12.7 to 17 feet. Both sections consist of
2 by 6 inch timbers, supported by a frame which is secured to the
bank. The gage is read once each day by Leonia Parker. Discharge
measurements are made from a $\frac{3}{8}$-inch galvanized iron cable with a
span of 390 feet. It is located at Lampshir's place, 1 mile above
the gage. Besides the timber supports at the ends of the cable it is
supported at a point 167 feet from the initial point for soundings by a
tree 10 inches in diameter. At high stages all water will pass beneath
the cable. At low water the current is sluggish at the cable and discharge measurements are made by wading. The initial point for
soundings is the zero of the tagged wire 22½ feet from the lower end
of the turn-buckle. The channel is straight for 75 feet above and
below the cable. The right bank is low and liable to overflow for a

:onsiderable distance. The left bank is high and rocky and will
overflow. Both banks have been cleared of brush. At the cable
bed of the stream is composed of sand, and is liable to shift. There
one channel at low water and two or more at high water. At the g
the bed of the stream is composed of clean sand and gravel. T
bench mark is the highest point of a large rock on the right bank
directly across the river from the gage, from which it is 135 feet
tant. Its elevation is 19.76 feet above the zero of the gage. T
present gage was established on the same datum as the one which
replaced.

The observations at this station during 1903 have been made under
the direction of J. T. Whistler, district engineer.

Discharge measurements of Silvies River near Burns, Oreg., in 1903.

Date.	Hydrographer.	Gage height.	Dischar.
		Feet.	*Second-feet.*
May 12.	N. S. Dils	6.30	
June 29.	M. D. Williams	2.85	34
July 17.	F. K. Lowry	2.43	4
July 26.	W. T. Turner	2.60	14
August 14	do	2.53	11
September 17	J. H. Lewis	2.62	4

Mean daily gage height, in feet, of Silvies River near Burns, Oreg., for 1903.

Day.	May.	June.	July.	Aug.	Sept.	Oct.	Nov.	De
1		4.10	2.80	2.50	2.30	2.20	2.50	3.
2		4.10	2.80	2.50	2.30	2.20	2.50	3.
3		4.10	2.80	2.50	2.30	2.20	2.50	3.
4		4.10	2.80	2.50	2.30	2.20	2.50	2.
5		4.10	2.70	2.50	2.30	2.20	2.50	2.
6		4.00	2.70	2.50	2.30	2.20	2.50	2.
7		4.00	2.70	2.50	2.30	2.20	2.50	2.
8		3.90	2.70	2.50	2.30	2.20	2.50	2.
9		3.80	2.70	2.50	2.20	2.20	2.50	2.
10	6.30	3.80	2.70	2.40	2.20	2.20	2.50	2.
11	6.30	3.60	2.70	2.40	2.20	2.20	2.50	2.
12	6.30	3.60	2.70	2.40	2.20	2.20	2.50	2.
13	6.20	3.50	2.70	2.40	(a)	2.20	2.50	2.
14	5.80	3.00	2.70	2.40		2.20	2.50	2.
15	5.40	3.00	2.60	2.40	....	2.20	2.60	2.
16	5.40	3.00	2.60	2.40	....	2.20	2.60	2.
17	5.40	3.00	2.60	2.40	2.62	2.20	2.60	2.
18	5.40	3.00	2.60	2.40	(a)	2.30	2.60	2.
19	5.40	3.00	2.60	2.40	....	2.30	2.70	2.
20	5.40	3.00	2.60	2.40	....	2.30	2.70	2.
21	5.40	3.00	2.60	2.40	....	2.30	2.70	2.
22	5.40	3.00	2.60	2.40	....	2.30	2.80	2.
23	5.30	3.00	2.60	2.30	....	2.30	2.90	2.

a Gage heights September 13–16 and 18–30, inclusive, are unreliable.

aily gage height, in feet, of Silvies River near Burns, Oreg., for 1903—Cont'd.

Day.	May.	June.	July.	Aug.	Sept.	Oct.	Nov..	Dec.
....	4.80	3.00	2.60	2.30	....	2.30	3.20	2.50
....	4.80	2.90	2.50	2.30	....	2.40	3.40	2.50
....	4.90	2.80	2.50	2.50	....	2.40	3.40	2.50
....	4.70	2.80	2.50	2.30	....	2.40	3.40	2.50
....	4.50	2.80	2.50	2.30	....	2.40	3.50	2.50
....	4.50	2.80	2.50	2.30	....	2.40	3.50	2.50
....	4.40	2.80	2.50	....	....	2.40	3.50	2.50
....	4.30	....	2.50	....	....	2.40	....	2.50

table[a] for Silvies River near Burns, Oreg., from May 10 to December 31, 1903.

Discharge.	Gage height.	Discharge.	Gage height.	Discharge.	Gage height.	Discharge.
Second-feet.	*Feet.*	*Second-feet.*	*Feet.*	*Second-feet.*	*Feet.*	*Second-feet.*
3	3.3	95	4.4	216	5.5	337
5	3.4	106	4.5	227	5.6	348
7	3.5	117	4.6	238	5.7	359
11	3.6	128	4.7	249	5.8	370
18	3.7	139	4.8	260	5.9	381
29	3.8	150	4.9	271	6.0	392
40	3.9	161	5.0	282	6.1	403
51	4.0	172	5.1	293	6.2	414
62	4.1	183	5.2	304	6.3	425
73	4.2	194	5.3	315	6.4	436
84	4.3	205	5.4	326	6.5	447

[a] This table is approximate: too few measurements for good curve.

Estimated monthly discharge of Silvies River near Burns, Oreg., for 1903.

Month.	Discharge in second-feet.			Total in acre-feet.
	Maximum.	Minimum.	Mean.	
o 31, inclusive	425	205	a 314	a 13,702
....	183	40	101	6,010
....	40	11	23	1,414
....	11	5	8	492
er 1 to 12, and 17 [b]	5	3	c 5	c 129
....	7	3	4	246
er	117	11	39	2,321
er 1 to 26, inclusive	95	11	d 20	d 1,031

[a] Mean in second-feet and total in acre-feet are for 22 days.
[b] Gage heights September 13 to 16 and 18 to 30, inclusive, are unreliable.
[c] Mean in second-feet and total in acre-feet are for 13 days.
[d] Mean in second-feet and total in acre-feet are for 26 days.

SILVIES RIVER NEAR SILVIES, OREG.

A temporary station was established May 8, 1903, by N. S. Dils, one-half mile above the present gage. Gage heights were read by John Craddock from May 8 to June 26, 1903. One discharge measurement was made at the temporary station. On June 16, 1903, a permanent gage was established by M. D. Williams at the proposed dam site one-half mile below the temporary gage. The station is 2 miles west of the stage road and three-fourths mile below the mouth of Trout Creek. There is a dam 1½ miles upstream. The gage is an inclined 2 by 6 inch plank on the left bank. It is read once each day by John Craddock. Discharge measurements are made by means of a ⅜-inch galvanized-iron cable having a total span of 494 feet. It is supported near the center. The cable is located 75 feet downstream from the gage. The initial point for soundings is the zero on the tagged wire, 29 feet from the cable support. The channel is straight for 150 feet above and for 200 feet below the station. The right bank is low and will overflow at high water for a considerable distance. The left bank is high, rocky, and is not liable to overflow. All water passes beneath the cable at all stages. The bed of the stream is composed of gravel, free from vegetation, and liable to shift. No bench mark has been established.

The observations at this station during 1903 have been made under the direction of J. T. Whistler, district engineer.

Discharge measurements of Silvies River near Silvies, Oreg., in 1903.

Date.	Hydrographer.	Gage height.	Discharge.
		Feet.	*Second-feet.*
May 9	N. S. Dils	4.50	200
June 16	M. D. Williams	3.78	78
June 17	F. K. Lowry	3.61	74
Do	do	3.58	74
June 19	do	3.45	53
June 26	do	3.00	28
September 11	J. H. Lewis	2.32	2

Daily gage height, in feet, of Silvies River near Silvies, Oreg., for 1903.

Day.	May.	June.	July.	Aug.	Sept.	Oct.	Nov.	Dec.
....	....	3.60	2.80	2.40	2.30	2.50	2.70	3.20
....	....	3.60	2.80	2.40	2.30	2.60	2.70	3.20
....	....	3.60	2.70	2.40	2.30	2.70	2.70	3.20
....	....	3.70	2.70	2.40	2.30	2.70	2.70	3.10
....	....	3.70	2.70	2.40	2.30	2.70	2.70	3.10
....	....	3.70	2.70	2.40	2.30	2.70	2.70	3.10
....	....	3.70	2.70	2.40	2.30	2.70	2.80	3.10
....	....	3.60	2.70	2.40	2.30	2.70	2.80	2.90
....	4.50	3.50	2.70	2.40	2.30	2.70	2.80	2.70
....	4.10	3.40	2.70	2.40	2.30	2.70	2.90	2.70
....	4.30	3.40	2.70	2.40	2.30	2.70	2.90	2.80
....	4.40	3.50	2.60	2.40	2.30	2.70	2.90	2.80
....	4.20	3.60	2.60	2.40	2.30	2.70	3.00	2.80
....	4.00	3.50	2.60	2.40	2.40	2.70	3.10	2.90
....	4.00	3.50	2.50	2.40	2.50	2.70	3.20	2.80
....	4.00	3.50	2.50	2.40	2.50	2.70	3.10	2.90
....	4.00	3.30	2.50	2.30	2.50	2.70	3.00	3.20
....	3.00	3.30	2.50	2.30	2.50	2.70	3.00	3.20
....	4.00	3.30	2.50	2.30	2.60	2.70	2.90	3.20
....	4.10	3.30	2.50	2.30	2.50	2.70	3.00	3.20
....	4.20	3.30	2.50	2.30	2.50	2.70	3.20	3.20
....	3.60	3.20	2.50	2.30	2.50	2.70	3.30	3.20
....	3.60	3.20	2.50	2.30	2.50	2.70	3.40	3.20
....	3.60	3.10	2.50	2.30	2.50	2.70	3.50	3.20
....	3.60	3.10	2.50	2.30	2.50	2.70	3.50	3.20
....	3.60	3.00	2.40	2.30	2.50	2.70	3.50	3.20
....	3.70	a3.00	2.60	2.30	2.50	2.70	3.40	3.20
....	3.80	3.00	2.60	2.30	2.50	2.70	3.40	3.20
....	3.60	3.00	2.50	2.30	2.50	2.70	3.30	3.20
....	3.60	2.90	2.40	2.30	2.50	2.70	3.20	3.20
....	3.40	....	2.40	2.30	....	2.70	....	3.20

gs. May 9 to June 26, 1903, inclusive, on temporary gage one-half mile above proposed
dings, June 27 to Dec. 31, 1903, inclusive, on permanent gage at proposed dam site.
nporary gage are reduced to datum of permanent gage.

eights are referred to gage established June 16.

le for Silvies River near Silvies, Oreg., from May 9 to December 31, 1903.

Discharge.	Gage height.	Discharge.	Gage height.	Discharge.	Gage height.	Discharge.
second-feet.	Feet.	Second-feet.	Feet.	Second-feet.	Feet.	Second-feet.
2	2.9	22	3.5	60	4.1	122
4	3.0	28	3.6	68	4.2	139
6	3.1	34	3.7	76	4.3	158
9	3.2	40	3.8	85	4.4	179
13	3.3	46	3.9	96	4.5	200
17	3.4	53	4.0	108		

y well defined.

Estimated monthly discharge of Silvies River near Silvies, Oreg., for 1903.

Month.	Discharge in second-feet.			Total acre-fe
	Maximum.	Minimum.	Mean.	
May 9–31............................	200	28	101	4,
June................................	76	22	52	3,
July	17	4	9	
August.............................	4	2	3	
September..........................	9	2	4	
October	13	6	13	
November	60	13	32	1,
December	40	13	33	2,
For period........................	....	....	....	14

SILVER CREEK NEAR RILEY, OREG.

This station is located on Silver Creek at Cecil's ranch, 12 m above Riley, Oreg., on the stage line from Burns to Shaniko. about 3 miles below the junction with Nichols Creek and about 6 n below the proposed reservoir site on Silver Creek. A perma: gaging station will be maintained at this point during 1904. Meas ments are made from three bridges over Silver Creek near the h of H. D. Cecil; also in one slough at high water, which is measure wading. The initial point for soundings is the west end of the at each of the three bridges. The channel is straight for about 80 above and 50 feet below the station. The current is sluggish. banks are low, covered with sagebrush, and are liable to overflo high water. The bed of the stream is composed of sand and clay, from vegetation, and is shifting. There are three channels at water and four or more at high water. A single discharge mea ment was made at this station September 21, 1903, by John H. L. A gage height of 2.15 feet was registered, giving a discharge o second-feet.

The observations at this station during 1903 have been made u the direction of J. T. Whistler, district engineer.

BOISE RIVER NEAR BOISE, IDAHO.

The station, established December 15, 1894, is located about 9 r above Boise, Idaho, at the mouth of the canyon. The original gag in two sections: The lower part is of 2 by 6 inch plank, inclined marked from 1 foot to 7.5 feet; the upper part is a 4 by 4 inch tin placed vertically, and marked from 7.5 to 12 feet, both portions pai white. The bench mark is a bridge spike driven into a cotton

ᵉᵉ 20 feet from gage and 20 feet from river. It is 3.4 feet above
ᵉ 8-foot mark on the gage. Measurements are made from a cable
ᵤₜ below the gage.

In the latter part of July, 1895, it was decided to locate a secondary
ᵃge on Boise River to obtain the slope of the water surface. This
ᵃs placed 425 feet below the old gage and carefully connected by
ᵉᵃⁿˢ of a wye level. Both were referred to the same datum. At
ᵃt time the lower end of the old gage was found to be warped and
ᵃs corrected. A gage was also placed on the lower Boise in order to
ᵗermine the water going by at the lowest stage during the irrigating
ᵃson. The meter can generally be used by wading, but in high water
ᵉasurements can be made from a wagon bridge.

April 18, 1897, the river cut into the right bank of the station, carry-
g out the cable and leaving the gage on a small island, so that the
ᵗcord after that date is unreliable. A temporary gage was therefore
ᵗablished May 12 at the Broadway Bridge, at Boise, and a record
ᵉpt by it until June 17, when a new gage, which is now used, was
ᵍain placed in the canyon 1 mile above the old location. The inclined
ᵈd is firmly attached to a cottonwood tree. The bench mark is a 20-
ᵉnny spike in the upstream face of the 6 by 8 inch cable support, about
feet above the ground. Its elevation is 15 feet above datum. Two
ᵖikes in same post are 14 feet above datum. Discharge measurements
ᵣe made from a cable and car 50 feet below the gage. An auxiliary
ᵃble for flood measurements is placed 117 feet above the main cable.
ʰe initial point for soundings is the face of the cable support on the
ight bank. At ordinary stages the channel is straight, both above and
ᵉlow the station. The banks are high and not liable to overflow.
ᵇout 300 feet below the cable is a gravel bar, reducing the width of
ʰe river at low water to about one-third of the channel and forcing
ʰe entire flow against the south bank. The channel is liable to change
ᵈuring extreme high floods. During 1900 the New York Canal Com-
ᵖany built a wing dam of timber and loose rock, headed about 150 feet
below the station and extending from the north bank diagonally down
ᵃnd across the stream a distance of about 50 feet, in order to protect
the north bank from erosion. The construction of this wing dam did
not seem to interfere with the flow of the river at the station. During
the year 1902 new cable supports were set and bench marks were carefully verified.

The observations at this station during 1903 have been made under
the direction of D. W. Ross, district engineer.

Discharge measurements of Boise River near Boise Idaho, in 1903.

Date.	Hydrographer.	Gage height.	Discharge
		Feet.	*Second-feet*
March 27	N. C. Dils	3.20	4,5
May 25	do	3.60	4,5
June 10	do	6.00	14,6
July 7	do	2.60	2,5
August 12	F. Stockton	.65	
August 34	do	.50	
October 12	do	1.00	1,1
October 26	do	.65	7
October 30	do	.65	

Mean daily gage height, in feet, of Boise River near Boise, Idaho, for 1903.

Day.	Jan.	Feb.	Mar.	Apr.	May.	June.	July.	Aug.	Sept.	Oct.	Nov.	Dec.
1	1.20	1.60	1.00	4.40	4.60	6.20	2.20	1.15	0.65	0.70	0.65	
2	.95	1.30	1.05	4.20	4.80	6.45	5.10	1.10	.55	.75	.62	
3	.80	1.20	1.10	5.00	4.50	6.20	2.90	1.00	.50	.75	.65	1.5
4	.85	1.20	1.10	3.90	4.80	6.00	2.65	.65	.50	.80		
5	.80	1.20	1.10	3.70	5.00	5.80	2.60	1.20	.50	.90		1.5
6	.80	1.25	1.00	3.70	5.20	5.70	2.62	.70	.50	1.10		
7	.80	1.10	.98	3.70	5.20	5.80	2.60	.75	.50	1.10		
8	.80	1.25	.90	3.30	5.20	5.90	2.50	.65	.50	1.00	.65	
9	.90	1.20	.90	3.40	5.10	6.00	2.35	.65	.50	1.00		
10	.80	1.10	1.00	4.60	5.10	6.00	2.20	.65	.50	.90	.65	
11	.97	1.00	1.10	4.30	5.10	5.90	2.20	.65	.50	.90		
12	.55	.80	1.10	3.90	5.10	5.70	2.10	.65	.65	1.00		7
13	.20	.70	1.30	3.60	5.40	5.50	2.20	.60	.80	.95		
14	.20	.90	1.40	3.70	5.70	5.30	2.15	.60	.80	.95		
15	.30	.60	1.50	3.75	5.70	5.20	2.10	.60	.80	.90		
16	.40	.80	1.50	3.90	5.50	5.30	2.20	.57	.75	.80		
17	.45	1.00	1.50	4.10	5.20	5.20	2.25	.65	.65	.80		
18	.50	.90	1.50	4.20	5.20	5.20	2.20	.55	.60	.70		
19	.60	.90	1.55	4.25	4.90	4.60	1.90	.55	.50	.70		
20	.90	.90	1.60	4.30	4.40	4.40	1.85	.50	.50	.70	1.00	
21	1.85	.90	1.60	4.60	3.90	4.20	1.90	.50	.50	.70		
22	1.60	1.00	1.80	5.50	3.70	4.10	2.00	.50	.55	.70		
23	1.30	1.00	1.90	5.40	3.60	4.10	1.90	.50	.52	.70	2.00	
24	2.20	1.00	2.20	5.50	3.60	3.80	1.80	.50	.50	.70		
25	3.20	1.00	2.80	5.60	3.70	3.70	1.80	.50	.50	.67	1.60	
26	2.60	.90	2.90	5.80	3.75	3.70	1.70	.55	.50	.60		
27	2.20	1.00	3.20	5.70	4.10	3.65	1.70	.60	.50	.60		
28	1.90	1.00	3.50	5.30	4.25	3.60	1.50	.70	.65	.65		
29	1.60		4.00	4.80	4.30	3.60	1.40	.75	.65	.67		
30	1.50		4.65	4.70	4.60	3.40	1.30	.65	.65	.67		
31	1.40		4.70		5.00		1.25	.60		.65		

Rating table for Boise River near Boise, Idaho, for 1903.

Gage height.	Discharge.	Gage height.	Discharge.	Gage height.	Discharge.	Gage height.	Discharge.
Feet.	*Second-feet.*	*Feet.*	*Second-feet.*	*Feet.*	*Second-feet.*	*Feet.*	*Second-feet.*
0.2	500	1.2	1,285	2.6	2,990	4.6	7,740
.3	570	1.3	1,375	2.8	3,340	4.8	8,410
.4	640	1.4	1,470	3.0	3,710	5.0	9,150
.5	715	1.5	1,570	3.2	4,100	5.2	10,010
.6	790	1.6	1,675	3.4	4,530	5.4	11,000
.7	865	1.7	1,785	3.6	4,980	5.6	12,110
.8	945	1.8	1,895	3.8	5,460	5.8	13,270
.9	1,025	2.0	2,130	4.0	5,970	6.0	14,430
1.0	1,110	2.2	2,400	4.2	6,520	6.2	15,590
1.1	1,195	2.4	2,680	4.4	7,110	6.4	16,750

Table well defined to 3.60 feet gage height. Above 3.60 feet gage height the rve is determined by one high-water measurement at 6 feet gage height.

Estimated monthly discharge of Boise River near Boise, Idaho, for 1903.

[Drainage area, 2,450 square miles.]

Month.	Discharge in second-feet.			Total in acre-feet.	Run-off.	
	Maximum.	Minimum.	Mean.		Second-feet per square mile.	Depth in inches.
nuary	4,100	500	1,316	80,918	0.54	0.62
bruary	1,675	790	1,136	63,090	.46	.48
arch	8,070	1,025	2,312	142,159	.94	1.08
pril	13,270	4,310	7,472	444,615	3.05	3.40
ay	12,690	4,980	8,396	516,250	3.43	3.95
ne	16,750	4,530	10,033	597,005	4.10	4.57
ly	4,310	1,330	2,379	146,279	.97	1.12
agust	1,285	715	854	52,510	.34	.39
ptember	945	715	.772	45,937	.32	.36
ctober	1,195	790	943	57,983	.39	.45
ovember[a]	2,130	790	1,127	17,883	.46	.14
ecember[b]	1,195	752	928	18,407	.38	.14

[a] November 1–3, 8, 10, 20, 23, 25.   [b] December 3, 5, 8, 12, 16, 19, 22, 25, 28, 31.

SUCCOR CREEK NEAR HOMEDALE, IDAHO.

This station was established March 19, 1903, by N. S. Dils. Records were discontinued September 30, 1903, but were begun again before water began flowing in the winter. It is located at a small truss bridge built to carry a flume and is about one-half mile above the mouth of the river. The gage is a vertical rod driven into the bed of the stream, fastened to the downstream side of the footbridge. The readings during the spring of 1903 were taken by Miss Mamie Mussel, and during the following winter by Mrs. Minnie Tracy. Discharge measurements are made from the footbridge. The initial point for soundings was taken at the west face of the sill on the right bank. The channel is straight for 100 feet above and for 300 feet below the bridge. The current is sluggish. Both banks are high and are not liable to overflow. The bed of the stream is sandy and free from vegetation. There is but one channel at all stages. The bench mark is an 8-penny nail driven into the lower chord of the bridge near the gage. Its elevation is 9.70 feet above the zero of the gage.

The observations at this station during 1903 have been made under the direction of D. W. Ross, district engineer.

Discharge measurements of Succor Creek near Homedale, Idaho, in 1903.

Date.	Hydrographer.	Gage height.	Discharge.
		Feet.	*Second-feet.*
March 19	N. S. Dils	2.50	43
March 30	do	3.40	306
June 3	do	2.30	3

Mean daily gage height, in feet, of Succor Creek near Homedale, Idaho, for 1903.

Day.	Mar.	Apr.	May.	June.	Dec.	Day.	Mar.	Apr.	May.	June.	Dec.
1		3.10	2.80	2.30	2.30	17		2.90	2.60	(a)	2.40
2		3.00	2.80	2.30	2.40	18		2.90	2.60	(a)	2.40
3		2.90	2.70	2.40	2.40	19	2.50	2.90	2.60	(a)	2.50
4		2.90	2.70	2.30	2.40	20	2.40	2.90	2.60	(a)	2.50
5		2.90	2.60	2.30	2.40	21	2.40	2.90	2.60	(a)	2.50
6		2.90	2.70	2.30	2.40	22	2.90	2.90	2.50	(a)	2.50
7		2.80	2.60	2.30	2.40	23	2.90	2.90	2.40	(a)	2.50
8		3.10	2.70	(a)	2.40	24	3.00	2.90	2.40	(a)	2.40
9		3.10	2.60	(a)	2.40	25	3.20	2.90	2.40	(a)	2.70
10		3.40	2.70	(a)	2.40	26	3.40	2.90	2.30	(a)	2.70
11		3.30	2.60	(a)	2.40	27	3.30	3.00	2.30	(a)	2.50
12		3.10	2.60	(a)	2.40	28	3.40	3.00	2.30	(a)	2.50
13		3.10	2.50	(a)	2.40	29	3.50	3.00	2.30	(a)	2.50
14		3.00	2.60	(a)	2.40	30	3.40	2.90	2.30	(a)	2.60
15		2.90	2.60	(a)	2.40	31	3.40		2.30	(a)	2.60
16		2.90	2.60	(a)	2.40						

a Dry.

ating table for Succor Creek near Homedale, Idaho, March 19 to December 31, 1903.

Gage height.	Discharge.	Gage height.	Discharge.	Gage height.	Discharge.	Gage height.	Discharge.
Feet.	*Second-feet.*	*Feet.*	*Second-feet.*	*Feet.*	*Second-feet.*	*Feet.*	*Second-feet.*
2.2	1	2.6	67	3.0	178	3.4	306
2.3	6	2.7	93	3.1	209	3.5	341
2.4	22	2.8	120	3.2	240		
2.5	43	2.9	148	3.3	272		

'able poorly defined; determined by only three points.

Estimated monthly discharge of Succor Creek near Homedale, Idaho, for 1903.

Month.	Discharge in second-feet.			Total in acre-feet.
	Maximum.	Minimum.	Mean.	
.rch 19–31	341	22	203	5,234
ril	306	120	171	10,175
ιy	120	6	57	3,505
ne 1–7 [a]			8	111
cember	67	6	32	1,968

[a] Creek dry after June 7.

BRUNEAU RIVER NEAR GRANDVIEW, IDAHO.

Systematic measurements on this river were maintained for four
ars by Andrew J. Wiley for the Owyhee Land and Irrigation Com-
.ny, immediately below the headworks of their canal system 10
iles east of Grandview, Idaho. The station was reestablished March
1903, by N. S. Dils. The station is located at the ford on the
d stage road, about 2 miles above the mouth of the river. The
.ge is in two sections, and is located on the left bank, 25 feet up-
ream from the old gage. The lower inclined section reads from zero
6.5 feet. The upper vertical section is driven into the bank and
ads from 6.5 to 9 feet. Both sections are of 2 by 4 inch timber.
ιe gage datum is the same as that of the old gage. Readings are
ken once each day by C. C. Gregg. Discharge measurements are
ade by means of a boat. The initial point for soundings is a piece
railroad iron set perpendicularly in the right bank. The channel
straight for 300 feet above and below the station, and the current is
ιggish. The right bank is low and liable to overflow at flood stages.
ιe left bank is high and is not liable to overflow. Both banks are
ithout trees or brush. The bed of the stream is sandy, free from
ιgetation, and shifting. There is but one channel at all stages. The
ιnch mark is the top of the northeast corner stone of the observer's

house. Its elevation is 13.14 feet above gage datum. This stati
was discontinued December, 1903.

The observations at this station during 1903 have been made u
the direction of D. W. Ross, district engineer.

Discharge measurements of Bruneau River near Grandview, Idaho, in 190.

Date.	Hydrographer.	Gage height.	
		Feet.	
March 6	N. S. Dils	2.00	
June 18	do	3.45	
July 31	F. Stockton	1.70	

Mean daily gage height, in feet, of Bruneau River near Grandview, Idaho, for 190.

Day.	Jan.	Feb.	Mar.	Apr.	May.	June.	July.	Aug.	Sept.	Oct.	Nov.	
1	1.75	2.00	2.00	4.85	3.15	3.55	2.05	1.70	1.60	1.70	1.85	
2	1.85	1.90	1.95	4.90	3.10	3.00	2.00	1.65	1.80	1.70	1.85	
3	1.90	1.85	1.95	2.90	3.10	4.05	2.05	1.65	1.80	1.70	1.85	
4	1.90	1.80	2.05	3.45	3.15	4.15	2.45	1.60	1.80	1.70	1.85	
5	1.90	1.80	2.10	3.20	3.15	4.10	2.05	1.60	1.50	1.75	1.85	
6	1.85	1.80	2.00	3.10	3.25	4.05	2.30	1.60	1.60	1.80	1.85	
7	1.85	1.75	2.50	2.90	3.35	4.00	2.30	1.60	1.80	1.75	1.85	
8	1.85	1.75	2.00	2.75	3.35	4.00	2.30	1.65	1.50	1.75	1.85	1.80
9	1.85	1.85	2.00	2.70	3.40	4.00	2.15	1.60	1.60	1.75	1.85	1.80
10	1.85	1.90	1.95	3.10	3.35	3.95	2.10	1.60	1.65	1.75	1.85	1.90
11	1.85	2.10	2.50	3.40	3.30	3.85	2.10	1.60	1.65	1.75	1.85	1.80
12	1.85	2.00	2.10	3.00	3.30	3.85	2.05	1.60	1.65	1.75	1.90	1.80
13	1.85	1.90	2.15	3.25	3.30	3.85	2.05	1.55	1.65	1.75	1.95	1.80
14	1.85	1.75	2.10	3.50	3.45	3.80	2.00	1.55	1.70	1.80	2.00	1.80
15	1.80	1.75	2.10	2.90	3.60	3.75	2.00	1.55	1.75	1.80	2.00	1.80
16	1.75	1.75	2.25	2.90	3.55	3.65	2.00	1.55	1.75	1.80	2.05	1.80
17	1.70	1.75	2.55	2.90	3.50	3.60	1.95	1.55	1.80	1.80	2.05	1.80
18	1.70	1.75	2.50	2.85	3.45	3.45	1.90	1.55	1.80	1.80	2.00	1.80
19	1.70	1.80	2.35	2.90	3.40	3.35	1.90	1.55	1.75	1.80	1.95	1.80
20	1.70	1.80	2.30	2.95	3.30	3.25	1.90	1.55	1.75	1.80	2.00	1.90
21	1.80	1.80	2.15	3.00	3.20	3.15	1.85	1.50	1.75	1.80	2.05	1.90
22	2.00	1.85	2.10	3.05	3.10	3.10	1.85	1.45	1.75	1.80	2.05	1.90
23	2.00	1.85	2.20	3.25	3.50	3.10	1.80	1.45	1.70	1.85	2.05	1.90
24	2.10	1.90	2.10	3.45	3.00	3.00	2.20	1.45	1.75	1.85	2.00	1.90
25	2.70	1.95	2.40	3.50	3.00	2.90	1.90	1.45	1.70	1.90	2.05	1.90
26	3.00	2.00	3.00	3.55	3.05	2.85	1.85	1.50	1.70	1.85	2.05	1.85
27	2.65	2.00	3.25	3.65	3.15	2.75	1.80	1.50	1.70	1.85	2.05	1.90
28	2.30	2.00	3.20	3.60	3.15	2.70	1.70	1.50	1.70	1.85	2.00	1.90
29	2.20		3.40	3.50	3.25	2.65	1.70	1.50	1.65	1.85	2.00	1.90
30	2.10		3.85	3.00	3.25	2.65	1.70	1.50	1.65	1.85	2.00	1.75
31	2.00		4.00		3.35		1.70	1.50		1.85		1.70

SNAKE RIVER NEAR MINIDOKA, IDAHO.

Ten miles above the gaging station at Montgomery Ferry occurs a natural dam site, from which, as a starting point, surveys for canal lines have been run, covering the large extent of rolling country susceptible of irrigation on both sides of the river. Measurements at Montgomery Ferry show the amount of water available for irrigation purposes there, and also the conditions that will exist for power purposes at Shoshone Falls, about 45 miles below, after the irrigable lands of Snake River Valley shall have been developed.

This station was originally established August 5, 1895, at Montgomery Ferry on the stage road from Minidoka to Albion, Idaho.

When the station was visited by N. S. Dils on October 14, 1899, a comparison was made of the gage rod with the bench mark, and it was found that the rod had moved to a considerable extent, owing to the action of the quicksand on the inclined portion of the rod. The heights as recorded by the observer, as well as the discharge measurements, were corrected. Gage readings were not taken during 1900.

The station was reestablished May 1, 1901, and the gage read morning and evening for the remainder of the year. Part of the inclined gage rod which had been moved by quicksand was corrected August 9, 1901, and all previous gage readings were carefully adjusted to correspond with the present position of the rod.

On October 16, 1903, the gage was carefully checked by means of a level by Fred Stockton. Different sections of the gage were found to be from 0.1 to 0.3 foot too high. The gage was corrected and the gage heights adjusted to conform with the old gage datum. The lower section of the gage is an inclined 4 by 4 inch timber and reads from zero to 6.8 feet. The upper section is vertical and reads from 6.8 to 13 feet. The gage is read once each day by George Montgomery, the ferryman. Discharge measurements are made from the ferryboat. A tagged wire has been stretched above the ferry cable. The initial point for soundings is the cable support on the right bank. The channel is straight for 300 feet above and below the station and has a width at the ferry of about 800 feet. Both banks are high, without vegetation, and are not liable to overflow. There is but one channel at all stages. The bench mark, established when the original gage was installed, is a spike in the east post of the tool house on the right bank. The spike is 1.2 feet above the ground, 52 feet west of the gage, and has an elevation of 17.50 feet above the zero of the gage. The letters "B. M." are marked in black paint on the post.

The observations at this station during 1903 have been made under the direction of D. W. Ross, district engineer.

Discharge measurements of Snake River near Minidoka, Idaho, in 1903.

Date.	Hydrographer.	Gage height.[a]	Discharge.
		Feet.	*Second-ft.*
June 20	N. S. Dils	8.40	2,90
July 9	do	5.20	11,07
July 30	Fred Stockton	2.90	5,36
August 21	do	1.20	2,04
October 16	do	2.80	6,000

[a] All gage heights refer to new rod.

Mean daily gage height, in feet, of Snake River near Minidoka, Idaho, for 1903.

Day.	Jan.	Feb.	Mar.	Apr.	May.	June.	July.	Aug.	Sept.	Oct.	Nov.	Dec.
1	6.00	4.00	3.70	3.00	4.55	5.20	7.15	2.60	1.20	2.50	2.90	1.25
2	5.60	4.00	3.70	3.60	4.35	5.40	7.00	2.50	1.20	2.50	2.90	1.3
3	4.35	4.25	3.70	3.00	4.15	5.90	6.90	2.40	1.20	2.60	2.90	1.3
4	3.85	4.05	3.70	3.00	4.05	6.40	6.70	2.40	1.20	2.60	2.90	1.3
5	2.95	3.75	3.35	3.60	3.95	6.75	6.40	2.30	1.20	2.70	2.90	1.15
6	2.75	3.55	3.05	3.60	4.80	7.00	6.20	2.10	1.20	2.80	3.00	1.10
7	2.70	3.85	3.28	3.60	3.95	7.25	5.80	1.90	1.20	2.90	3.00	1.05
8	2.70	3.25	2.65	3.60	4.05	7.40	5.50	1.80	1.20	2.90	3.00	1.15
9	2.70	3.10	2.60	3.60	4.30	7.50	5.20	1.70	1.30	3.00	3.00	2.10
10	2.70	3.55	2.60	3.60	4.65	7.75	5.05	1.50	1.40	3.00	3.00	2.40
11	2.65	3.70	2.70	3.60	4.85	7.95	4.85	1.60	1.50	3.00	3.00	2.90
12	2.60	3.55	2.80	3.60	4.85	8.00	4.65	1.50	1.70	3.00	3.00	2.95
13	2.50	3.30	2.90	3.60	4.85	8.10	4.60	1.40	1.75	3.00	3.00	3.00
14	2.50	3.10	3.00	3.60	4.80	8.10	1.50	1.40	1.90	3.00	3.10	3.10
15	2.60	3.10	2.60	3.60	4.80	8.10	4.40	1.40	2.00	3.00	3.10	3.10
16	2.60	3.40	2.60	3.60	5.10	8.10	4.25	1.30	2.15	3.00	3.20	3.10
17	3.00	3.40	2.60	3.60	5.50	8.10	4.15	1.30	2.33	2.90	3.20	3.00
18	4.20	3.30	2.60	3.60	5.90	8.15	4.05	1.20	2.40	2.90	3.10	3.00
19	4.55	3.30	2.70	3.60	6.05	8.20	3.95	1.20	2.40	2.90	2.95	2.90
20	3.85	3.20	2.50	3.10	5.90	8.20	3.75	1.20	2.40	2.90	2.80	2.90
21	3.60	3.25	2.50	3.10	5.50	8.10	3.60	1.20	2.50	2.90	2.70	3.00
22	3.60	3.45	2.50	3.10	5.20	8.00	3.55	1.20	2.50	2.90	2.85	3.00
23	3.60	3.65	2.50	3.20	......	7.90	3.45	1.20	2.50	2.90	3.00	3.00
24	2.95	3.60	2.80	3.35	......	7.80	3.40	1.20	2.50	2.90	3.10	3.10
25	3.25	3.60	2.60	3.55	......	7.75	3.30	1.20	2.50	2.80	3.20	3.15
26	3.90	3.60	2.60	3.75	......	7.05	3.10	1.20	2.50	2.80	3.25	3.20
27	3.70	3.60	3.00	3.85	......	7.50	3.10	1.20	2.50	2.80	3.35	3.00
28	3.60	3.70	3.00	4.05	......	7.50	3.00	1.20	2.50	2.80	3.50	3.90
29	4.20	......	3.00	4.35	......	7.60	2.90	1.20	2.50	2.80	3.20	4.70
30	4.10	......	3.60	4.50	......	7.55	2.90	1.20	2.50	2.80	3.20	3.50
31	4.00	......	3.60	......	4.95	......	2.50	1.20	......	2.80	......	5.20

Rating table for Snake River near Minidoka, Idaho, for 1903.

Gage height.	Discharge.	Gage height.	Discharge.	Gage height.	Discharge.	Gage height.	Discharge.
Feet.	Second-feet.	Feet.	Second-feet.	Feet.	Second-feet.	Feet.	Second-feet.
1.2	2,640	2.6	5,080	4.8	9,900	7.0	17,410
1.3	2,800	2.8	5,460	5.0	10,450	7.2	18,280
1.4	2,960	3.0	5,840	5.2	11,020	7.4	19,170
1.5	3,130	3.2	6,240	5.4	11,610	7.6	20,080
1.6	3,300	3.4	6,650	5.6	12,230	7.8	21,010
1.7	3,470	3.6	7,070	5.8	12,880	8.0	21,960
1.8	3,640	3.8	7,500	6.0	13,560	8.2	22,950
1.9	3,820	4.0	7,940	6.2	14,270	8.4	23,970
2.0	4,000	4.2	8,400	6.4	15,000		
2.2	4,360	4.4	8,880	6.6	15,770		
2.4	4,720	4.6	9,380	6.8	16,580		

Table well defined.

Estimated monthly discharge of Snake River near Minidoka, Idaho, for 1903.

Month.	Discharge in second-feet.			Total in acre-feet.
	Maximum.	Minimum.	Mean.	
anuary	13,560	4,900	6,919	425,433
'ebruary	8,520	6,040	6,918	384,206
[arch	7,280	4,900	5,773	354,968
.pril	9,130	5,840	7,023	417,898
[ay a	13,735	7,500	10,170	625,329
une	22,950	11,020	19,813	1,178,955
uly	18,060	5,460	9,674	594,831
.ugust	5,080	2,640	3,228	198,482
eptember	4,900	2,640	3,897	231,888
)ctober	5,840	4,900	5,553	341,441
[ovember	6,545	5,270	5,924	352,502
)ecember	11,920	5,460	6,489	398,993
The year	22,950	2,640	7,615	5,504,926

a May 23–30 interpolated.

OWYHEE RIVER NEAR OWYHEE, OREG.

This station was established August 27, 1903, by John H. Lewis. It is located at the county bridge 1½ miles from Owyhee, Oreg. A large irrigation ditch takes water from the river about 6 miles above. The gage is an inclined 2 by 6 inch timber reading from 1 to 5 feet. It is located at the upstream steel caisson of the left abutment of the ridge. The gage is painted on the vertical caisson from 5 to 18 feet. t is read once each day by D. T. Rigsby. Discharge measurements re made from the bridge at ordinary stages and by wading above the ridge at extreme low water. The initial point for soundings is the enter of the upstream caisson of the left abutment. The channel is traight for 200 feet above and for 400 feet below the station. The urrent is sluggish at low water. The right bank is high and rocky nd will not overflow. The left bank will overflow only at extreme igh water. The bed of the stream is composed of sand and gravel, nd is liable to shift during freshets. There are two channels at low water and one at high water. The bench mark is the top of the steel caisson, at its outer edge directly above the gage. Its elevation is 18.60 feet above the zero of the gage. The bench mark to which the old station, located at this point, was referred has been destroyed.

The observations at this station during 1903 have been made under the direction of J. T. Whistler, engineer.

Discharge measurements of Owyhee River near Owyhee, Oreg., in 1903.

Date.	Hydrographer.	Gage height.	Discharge.
		Feet.	*Second-feet.*
September 27	J. H. Lewis.	2.05	a 157
August 27	...do	1.87	b 115
December 17	...do	2.75	c 263

a 37 second-feet in Owyhee River; 120 second-feet in Owyhee ditch.
b 7 second-feet in Owyhee River; 108 second-feet in Owyhee ditch.
c No water in Owyhee ditch.

Mean daily gage height, in feet, of Owyhee River near Owyhee, Oreg., for 1903.

Day.	Aug.	Sept.	Oct.	Nov.	Dec.	Day.	Aug.	Sept.	Oct.	Nov.	Dec.
1		1.90	2.06	2.38	3.05	17		1.99	2.34	2.68	2.75
2		1.89	2.07	2.54	3.00	18		2.05	2.32	2.69	2.75
3		1.90	2.08	2.56	2.95	19		2.04	2.32	2.68	2.78
4		1.88	2.10	2.58	2.90	20		2.03	2.33	2.69	2.77
5		1.87	2.20	2.57	2.88	21		2.04	2.34	2.70	2.76
6		1.86	2.25	2.58	2.86	22		2.03	2.35	2.71	2.75
7		1.89	2.31	2.59	2.84	23		2.02	2.36	2.70	2.76
8		1.90	2.32	2.60	2.80	24		2.03	2.37	2.69	2.75
9		1.91	2.31	2.61	2.78	25		2.04	2.36	2.70	2.74
10		1.92	2.32	2.60	2.76	26		2.05	2.37	2.80	2.73
11		1.93	2.33	2.61	2.80	27		2.06	2.38	3.00	2.75
12		1.94	2.32	2.62	2.82	28	1.85	2.05	2.38	3.20	2.78
13		1.95	2.33	2.64	2.80	29	1.86	2.06	2.37	3.15	3.00
14		1.96	2.34	2.67	2.78	30	1.87	2.07	2.36	3.10	2.80
15		1.97	2.33	2.68	2.78	31	1.88		2.37		2.95
16		1.98	2.32	2.69	2.76						

BIG LOST RIVER NEAR MACKAY, IDAHO.

This station was established November 12, 1903, by Fred Stockton. It is located 3½ miles above Mackay, Idaho, above "The Narrows." The gage is a vertical 2 by 4 inch pine timber driven into the ground and spiked to a large willow. The observer is J. H. Haney.

The observations at this station during 1903 have been made under the direction of D. W. Ross, district engineer.

Mean daily gage height, in feet, of Big Lost River near Mackay, Idaho for 1903.

Day.	Dec.	Day.	Dec.	Day.	Dec.	Day.	Dec.	Day.	Dec.
....	0.70	8..........	0.70	15..........	0.70	22..........	0.70	29..........	0.70
....	.70	9..........	.70	16..........	.70	23..........	.70	30..........	.70
....	.70	10..........	.70	17..........	.70	24..........	.70	31..........	.70
....	.70	11..........	.70	18..........	.70	25..........	.70		
....	.70	12..........	.70	19..........	.70	26..........	.70		
....	.70	13..........	.70	20..........	.70	27..........	.70		
....	.70	14..........	.70	21..........	.70	28..........	.70		

BLACKFOOT RIVER NEAR PRESTO, IDAHO.

This station was established April 17, 1903, by N. S. Dils. It is located on the ranch of the observer, James Just, 2 miles west of Presto and about 15 miles from Blackfoot, Idaho. The gage consists of two vertical sections of 2 by 4 inch timber driven into the right bank. The lower section reads from 0 to 5 feet. The upper section reads from 5 to 9 feet. A cable will be installed from which to make discharge measurements. The initial point for soundings is at the gage rod. The channel is curved above and below the station. The current is sluggish. The right bank is high and will not overflow. The left bank will overflow at flood stages. Both banks are wooded. The bed of the stream is rocky, free from vegetation, and permanent. No bench marks have been established.

The observations at this station during 1903 have been made under the direction of D. W. Ross, district engineer.

Discharge measurements of Blackfoot River near Presto, Idaho, in 1903.

Date.	Hydrographer.	Gage height.	Discharge.
		Feet.	*Second-feet.*
April 17	N. S. Dils....................	3.40	293
June 22	......do	.80	150
August 8....................	Fred Stockton................	.30	151
September 3	......do	....	a 68

a Made at mouth of river.

Date.	Hydrographer.	Gage height.	
		Feet.	*Second-feet*
April 19	N. S. Dils	2.00	2.6
June 25	...do	6.10	21,8
August 3	F. Stockton	2.90	6,7
September 27	...do	1.85	3,7

Mean daily gage height, in feet, of South Fork of Snake River near Lyon, Idaho, for 190

Day.	Apr.	May.	June.	July.	Aug.	Sept.	Oct.	Nov.	Dec.
1	0.10	2.90	5.40	3.20	3.20	1.90	1.90	1.80	1.70
2	.10	3.30	5.30	3.15	3.15	1.85	1.95	1.80	1.70
3	.10	3.70	5.20	3.10	3.10	1.85	1.95	1.80	1.70
4	.20	4.00	5.10	2.85	2.85	1.80	1.95	1.80	1.70
5	.40	4.10	5.00	2.80	2.80	1.80	1.95	1.80	1.70
6	.50	4.20	4.90	2.75	2.75	1.90	1.90	1.80	1.70
7	.65	4.45	4.80	2.70	2.70	2.00	1.90	1.75	1.70
8	.80	4.70	4.70	2.65	2.65	2.05	1.90	1.75	1.70
9	.95	4.90	4.60	2.60	2.60	2.05	1.90	1.75	1.70
10	1.10	5.10	4.50	2.55	2.55	2.05	1.90	1.75	1.80
11	1.25	5.10	4.40	2.50	2.50	2.05	1.85	1.75	1.80
12	1.40	5.00	4.30	2.50	2.50	2.00	1.85	1.75	1.80
13	1.50	5.10	4.30	2.45	2.45	2.00	1.85	1.75	1.80

ily gage height, in feet, of South Fork of Snake River near Lyon, Idaho, for 1903—
Continued.

Day.	Apr.	May	June.	July.	Aug.	Sept.	Oct.	Nov.	Dec.
....	......	1.70	5.20	4.20	2.40	2.00	1.85	1.75	1.70
....	......	2.00	5.40	4.15	2.40	1.95	1.85	1.75	1.70
....	......	2.05	5.45	4.10	2.35	1.95	1.85	1.75	1.70
....	......	1.85	5.50	4.05	2.35	1.95	1.85	1.70	1.65
....	......	1.50	5.45	4.00	2.30	1.95	1.85	1.70	1.65
....	......	1.30	5.30	3.95	2.25	1.95	1.85	1.70	1.65
....	......	1.20	5.10	3.90	2.00	1.90	1.85	1.70	1.65
....	a0.05	1.20	6.90	3.85	2.00	1.90	1.80	1.70	1.65
....	.05	1.20	6.60	3.80	2.20	1.90	1.80	1.70	1.65
....	.10	1.20	6.30	3.75	2.15	1.90	1.80	1.70	1.65
....	.25	1.20	6.10	3.70	2.15	1.90	1.80	1.70	1.65
....	.30	1.20	6.00	3.65	2.10	1.90	1.80	1.70	1.65
....	.55	1.20	5.90	3.60	2.05	1.85	1.80	1.70	1.65
....	.75	1.20	5.80	3.55	2.00	1.85	1.80	1.70	......
....	.60	1.40	5.70	3.50	2.00	1.85	1.80	1.70	......
....	.20	1.80	5.60	3.40	1.95	1.85	1.80	1.70	......
....	......	2.10	5.50	3.30	1.95	1.90	1.80	1.70	......
....	......	2.45	......	3.25	1.90	......	1.80	......	......

ecord from April 21 to August 3 is somewhat doubtful, owing to the gage being washed out.

table for South Fork of Snake River near Lyon, Idaho, from June 25 to December 31, 1903.

e ht.	Discharge.	Gage height.	Discharge.	Gage height.	Discharge.	Gage height.	Discharge.
t.	*Second-feet.*	*Feet.*	*Second-feet.*	*Feet.*	*Second-feet.*	*Feet.*	*Second-feet.*
7	3,400	2.8	6,460	3.9	10,430	5.0	15,630
8	3,650	2.9	6,780	4.0	10,850	5.1	16,130
9	3,900	3.0	7,100	4.1	11,290	5.2	16,630
0	4,150	3.1	7,420	4.2	11,740	5.3	17,150
1	4,410	3.2	7,760	4.3	12,200	5.4	17,670
2	4,680	3.3	8,100	4.4	12,670	5.5	18,190
3	4,960	3.4	8,460	4.5	13,150	5.6	18,730
4	5,250	3.5	8,830	4.6	13,630	5.7	19,270
5	5,550	3.6	9,210	4.7	14,130	5.8	19,810
6	5,850	3.7	9,610	4.8	14,630	5.9	20,350
7	6,150	3.8	10,010	4.9	15,130	6.0	20,890

e is constructed from three points. Curve extended below 1.85 feet gage

Estimated monthly discharge of South Fork of Snake River near Lyon, Idaho, in 19..

Month.	Discharge in second-feet.			Total in acre-feet.
	Maximum.	Minimum.	Mean.	
June 25–30 a			19,540	232,54
July	17,670		11,906	732,07
August				324,92
September....				235,81
October				232,64
November				206,03
December b ...				205,82

a Record April 21 to August 3 somewhat doubtful owing to gage being washed out.
b December 27–31 interpolated.

FORK OF SNAKE RIVER   MORAN, WYO.

This station was established September   1903, by Fred Stockton.
It is located directly back of the post-office at Moran, Wyo., and
about three-fourths of a mile below the outlet of Jackson Lake. The
inclined gage is set on the left bank and nailed at the upper end to
            One foot of vertical height equals 1.29 feet measured
along the rod. The gage is read once each day by Marion Allen.
Discharge measurements are made by means of a small rowboat. The
initial point for soundings is the face of a cottonwood tree 6 inches in
diameter which grows on the left bank about 50 feet below the gage
and near the high-water mark. The channel is slightly curved for
about 300 feet above the station. Below the station the channel is
straight. At and above the station the current is smooth and has a
well distributed velocity. Below the station the current is broken by
small bowlders. Both banks are high and are not liable to overflow.
The right bank is wooded and the left bank is composed of gravel and
is without trees. The bed of the stream is composed of firm gravel.
There is but one channel at all stages. No bench mark has been
established.

The observations at this station during 1903 have been made under
the direction of D. W. Ross, district engineer.

The following discharge measurement was made by F. Stockton in
1903:

September 22: Gage height, 1.40 feet; discharge, 705 second-feet.

ily gage height, in feet, of South Fork of Snake River at Moran, Wyo., for 1903.

	Oct.	Nov.	Dec.	Day.	Oct.	Nov.	Dec.	Day.	Oct.	Nov.	Dec.
.....	1.40	1.40	1.40	12............	1.50	1.30	1.40	23............	1.40	1.30	1.20
.....	1.40	1.40	40	13......	1.50	1.30	1.30	24............	1.40	1.30	1.20
.....	1.50	1.40	40	14............	1.50	1.30	1.30	25............	1.40	1.40	1.20
.....	1.50	1.40	1.40	15....	1.50	1.30	1.30	26............	1.40	1.40	1.20
.....	1.50	1.40	1.40	16............	1.40	1.30	1.20	27............	1.40	1.40	1.20
.....	1.50	1.30	.40	17............	1.40	1.30	1.20	28............	1.40	1.40	1.20
.....	1.50	1.30	1.40	18............	1.40	1.30	1.20	29............	1.40	1.40	1.20
.....	1.50	1.30	1.40	19............	1.40	1.30	1.20	30............	1.40	1.40	1.20
.....	1.50	1.30	1.40	20............	1.40	1.30	1.20	31............	1.40	......	1.20
.....	1.50	1.30	1.40	21............	1.40	1.30	1.20				
.....	1.50	1.30	1.40	22............	1.40	1.30	1.20				

TETON RIVER NEAR ST. ANTHONY, IDAHO.

station was established April 23, 1903, by N. S. Dils. It is
l at the bridge on the stage road from St. Anthony to Teton,
The gage is a vertical 2 by 6 inch timber 8 feet long spiked
upstream side of the right crib abutment. It is read once each
Mrs. Kate Weaver. Discharge measurements are made from
o-span bridge to which the gage is attached. The bridge is
ted by crib abutments constructed of logs filled with lava rock
a similar middle pier. The initial point for soundings is a
)olt on the upstream side of the left abutment. The channel is
it for a short distance above and below the station. The right
s high and will not overflow. The left bank will overflow at
ie flood stages. The current is sluggish. Both banks and bed
nposed of hard gravel. The bed is permanent. Bench mark
is a cross on a large flat rock 40 feet northeast of the right end
bridge. It is marked "B. M." and has an elevation of 14.25
)ove the zero of the gage. Bench mark No. 2 is a similar rock
t northwest of the right end of the bridge. It is similarly
d and has an elevation of 14.41 feet above the zero of the gage.
observations at this station during 1903 have been made under
:ection of D. W. Ross, district engineer.

Discharge measurements of Teton River near St. Anthony, Idaho, in 1903.

Date.	Hydrographer.	Gage height.	Discharge.
		Feet.	*Second-feet.*
3.....................	.N. S. Dils....................	1.80	515
....do		3.80	2,085
6.....................	F. Stockton..................	1.90	724
ber 5 ....do		1.60	613

Estimated monthly discharge of South Fork of Snake River near Lyon, Idaho, in 19

Month.	Discharge in second-feet.			Total i
	Maximum.	Minimum.	Mean.	acre-fe
June 25–30 a	20,890	18,190	19,540	212,
July	17,670	7,930	11,906	732,
August	7,760	3,900	5,285	324,
September	4,280	3,650	3,963	235,
October	4,025	3,650	3,787	232,
November	3,650	3,400	3,496	208
December b	3,400	3,280	3,342	205

a Record April 21 to August 3 somewhat doubtful owing to gage being washed out.
b December 27–31 interpolated.

SOUTH FORK OF SNAKE RIVER AT MORAN, WYO.

This station was established September 21, 1903, by Fred Stockt
It is located directly back of the post-office at Moran, Wyo.,
about three-fourths of a mile below the outlet of Jackson Lake.
inclined gage is set on the left bank and nailed at the upper en
some willows. One foot of vertical height equals 1.29 feet meas
along the rod. The gage is read once each day by Marion Al
Discharge measurements are made by means of a small rowboat.
initial point for soundings is the face of a cottonwood tree 6 inche
diameter which grows on the left bank about 50 feet below the g
and near the high-water mark. The channel is slightly curved
about 300 feet above the station. Below the station the channe
straight. At and above the station the current is smooth and h
well distributed velocity. Below the station the current is broker
small bowlders. Both banks are high and are not liable to overfl
The right bank is wooded and the left bank is composed of gravel
is without trees. The bed of the stream is composed of firm gra
There is but one channel at all stages. No bench mark has b
established.

The observations at this station during 1903 have been made un
the direction of D. W. Ross, district engineer.

The following discharge measurement was made by F. Stockton
1903:

September 22: Gage height, 1.40 feet; discharge, 705 second-feet.

gage height, in feet, of South Fork of Snake River at Moran, Wyo., for 1903.

Oct.	Nov.	Dec.	Day.	Oct.	Nov.	Dec.	Day.	Oct.	Nov.	Dec.
1.40	1.40	1.40	12............	1.50	1.30	1.40	23............	1.40	1.30	1.20
1.40	1.40	1.40	13......	1.50	1.30	1.30	24............	1.40	1.30	1.20
1.50	1.40	1.40	14............	1.50	1.30	1.30	25............	1.40	1.40	1.20
1.50	1.40	1.40	15............	1.50	1.30	1.30	26............	1.40	1.40	1.20
1.50	1.40	1.40	16............	1.40	1.30	1.20	27............	1.40	1.40	1.20
1.50	1.30	1.40	17............	1.40	1.30	1.20	28............	1.40	1.40	1.20
1.50	1.30	1.40	18............	1.40	1.30	1.20	29............	1.40	1.40	1.20
1.50	1.30	1.40	19............	1.40	1.30	1.20	30............	1.40	1.40	1.20
1.50	1.30	1.40	20............	1.40	1.50	1.20	31............	1.40		1.20
1.50	1.30	1.40	21............	1.40	1.30	1.20				
1.50	1.30	1.40	22............	1.40	1.30	1.20				

TETON RIVER NEAR ST. ANTHONY, IDAHO.

tation was established April 23, 1903, by N. S. Dils. It is
it the bridge on the stage road from St. Anthony to Teton,
The gage is a vertical 2 by 6 inch timber 8 feet long spiked
ostream side of the right crib abutment. It is read once each
Mrs. Kate Weaver. Discharge measurements are made from
span bridge to which the gage is attached. The bridge is
d by crib abutments constructed of logs filled with lava rock
, similar middle pier. The initial point for soundings is a
lt on the upstream side of the left abutment. The channel is
for a short distance above and below the station. The right
high and will not overflow. The left bank will overflow at
flood stages. The current is sluggish. Both banks and bed
oosed of hard gravel. The bed is permanent. Bench mark
a cross on a large flat rock 40 feet northeast of the right end
ridge. It is marked "B. M." and has an elevation of 14.25
ve the zero of the gage. Bench mark No. 2 is a similar rock
northwest of the right end of the bridge. It is similarly
and has an elevation of 14.41 feet above the zero of the gage.
oservations at this station during 1903 have been made under
ction of D. W. Ross, district engineer.

scharge measurements of Teton River near St. Anthony, Idaho, in 1903.

Date.	Hydrographer.	Gage height.	Discharge.
		Feet.	*Second-feet.*
....	S. S. Dils....................	1.80	515
....do		3.80	2,085
....	F. Stockton....................	1.90	724
r 5 ....do		1.60	613

Mean daily gage height, in feet, of Teton River near St. Anthony, Idaho, for 1903.

Day.	Apr.	May.	June.	July.	Aug.	Sept.	Oct.	Nov.	Dec.
1		1.80	4.00	3.80	2.10	1.80	1.70	1.70	
2		1.70	4.20	3.50	2.00	1.70	1.70	1.70	
3		1.60	4.40	3.20	2.00	1.70	1.80	1.80	
4		1.60	4.50	3.00	2.00	1.70	1.80	1.60	
5		1.80	4.60	2.80	2.00	1.70	1.80	1.80	
6		2.00	4.60	2.50	1.90	1.70	1.80	1.60	
7		2.20	4.60	2.70	1.90	1.70	1.80	1.60	
8		2.40	4.80	2.70	1.90	1.70	1.80	1.80	
9		2.40	4.60	2.60	1.90	1.70	1.80	1.90	
10		2.50	4.80	2.60	1.90	1.70	1.80	1.80	
11		2.30	4.80	2.60	1.90	1.70	1.80	1.80	
12		2.20	4.50	2.50	1.90	1.80	1.70	1.70	
13		2.70	4.40	2.60	1.90	1.80	1.70	1.70	
14		3.40	4.60	2.50	1.90	1.80	1.70	1.60	
15		3.80	4.40	2.40	1.90	1.70	1.70	1.60	
16		3.60	4.40	2.40	1.90	1.70	1.70	1.70	
17		2.90	4.50	2.40	2.40	1.70	1.70	1.70	
18		2.70	4.30	2.30	2.30	1.70	1.70	1.80	
19		2.60	4.10	2.30	1.90	1.70	1.70	1.70	
20		2.40	4.10	2.20	1.80	1.70	1.70	1.60	
21		2.30	3.90	2.30	1.80	1.70	1.70	1.60	
22		2.30	3.70	2.30	1.90	1.70	1.70	1.70	
23	1.80	2.20	3.70	2.30	1.90	1.70	1.70	1.80	
24	1.80	2.20	3.50	2.30	1.80	1.70	1.70	1.90	
25	1.80	2.30	3.40	2.30	1.80	1.70	1.70	1.80	
26	1.90	2.40	3.70	2.20	1.80	1.70	1.70	1.70	
27	2.00	2.50	3.60	2.20	1.80	1.70	1.70	1.60	
28	2.00	2.80	3.70	2.20	1.80	1.70	1.70	1.60	
29	1.90	3.00	3.80	2.10	1.80	1.70	1.70	1.60	
30	1.90	3.20	3.80	2.10	1.80	1.70	1.70	1.50	
31		3.70		2.10	1.80		1.70		

Rating table for Teton River near St. Anthony, Idaho, from April 23 to December 31, 1903.

Gage height.	Discharge.	Gage height.	Discharge.	Gage height.	Discharge.	Gage height.	Discharge.
Feet.	Second-feet.	Feet.	Second-feet.	Feet.	Second-feet.	Feet.	Second-feet.
1.4	550	2.3	940	3.2	1,570	4.1	2,355
1.5	581	2.4	1,000	3.3	1,650	4.2	2,445
1.6	613	2.5	1,065	3.4	1,735	4.3	2,535
1.7	648	2.6	1,130	3.5	1,820	4.4	2,625
1.8	685	2.7	1,200	3.6	1,905	4.5	2,715
1.9	725	2.8	1,270	3.7	1,995	4.6	2,805
2.0	770	2.9	1,340	3.8	2,085	4.7	2,895
2.1	823	3.0	1,415	3.9	2,175	4.8	2,985
2.2	880	3.1	1,490	4.0	2,265	4.9	3,075

Table determined by three points—extended above 3.80 feet gage height and below 1.60 feet gage height.

d monthly discharge of Teton River near St. Anthony, Idaho, for 1903.

Month.	Discharge in second-feet.			Total in acre-feet.
	Maximum.	Minimum.	Mean.	
....	770	685	721	11,440
....	2,085	613	1,090	67,021
....	3,075	1,735	2,454	146,023
....	2,085	823	1,094	67,267
....	1,000	685	737	45,316
....	685	648	653	38,856
....	685	648	659	40,520
....	725	581	636	37,845
....	1,270	550	668	41,074

NORTH FORK OF SNAKE RIVER NEAR ORA, IDAHO.

tion was established August 20, 1902, by N. S. Dils. It is
:he North Fork Bridge, 2 miles south of Ora and 10 miles
Anthony, Idaho. A temporary vertical gage was set on the
of the first pier from the north shore, at which point the
gage will be located. The observer is Mrs. Martha J.
e initial point for soundings is a bolt through the toe of the
on the north end of the lower side of the bridge. The
10 feet long. It consists of four spans resting on three
crib piers. Measurements are made from the lower side of
. The channel is straight both above and below this station.
are high and do not overflow. The bed of the stream is
l, quite smooth, and not liable to change. The current is
e bench mark is a cross on a large flat rock 25 feet north-
he north or right end of the bridge. It is marked "B. M."
n is 11.10 feet above the zero of the gage.
rvations at this station during 1903 have been made under
n of D. W. Ross, district engineer.

: measurements of North Fork of Snake River near Ora, Idaho, in 1903.

Date.	Hydrographer.	Gage height.	Discharge.
		Feet.	Second-feet.
....	N. S. Dils..............	2.70	2,133
....	....do	2.30	1,367
....	....do	2.35	1,540
....	F. Stockton.............	2.15	1,277
....	....do	2.00	1,192

Mean daily gage height, in feet, of North Fork of Snake River near Ora, Idaho, for 1903.

Day.	Jan.	Feb.	Mar.	Apr.	May.	June.	July.	Aug.	Sept.	Oct.	Nov.	Dec.
1	2.00	2.00	2.00	2.30	2.50	2.70	2.30	2.10	2.00	2.00	2.00	1.9
2	2.00	2.00	2.00	2.10	2.40	2.70	2.30	2.10	2.00	2.00	2.00	1.9
3	2.00	2.00	2.00	2.00	2.70	2.80	2.30	2.10	2.00	2.10	2.00	1.9
4	2.00	2.00	2.00	2.00	2.90	2.80	2.30	2.10	2.00	2.10	2.00	1.9
5	2.00	1.90	2.00	2.00	3.00	2.80	2.30	2.10	2.00	2.10	2.00	1.9
6	2.00	1.90	2.00	2.00	3.20	2.70	2.30	2.10	2.00	2.10	2.00	1.9
7	2.00	1.90	2.00	2.00	3.40	2.70	2.20	2.10	2.00	2.00	2.00	1.9
8	1.90	1.90	2.00	2.00	3.50	2.70	2.20	2.10	2.00	2.10	2.00	1.9
9	1.90	1.90	2.00	2.00	3.40	2.60	2.20	2.10	2.00	2.10	2.00	1.9
10	1.90	1.90	1.90	2.10	3.10	2.60	2.20	2.10	2.00	2.10	2.00	1.8
11	1.90	1.90	1.90	2.00	3.00	2.60	2.20	2.10	2.00	2.10	2.00	1.8
12	2.00	1.90	1.90	2.00	3.00	2.50	2.20	2.00	2.00	2.00	2.00	1.8
13	2.00	1.90	1.90	2.00	3.00	2.50	2.20	2.00	2.00	2.00	2.10	1.8
14	2.00	1.90	1.90	2.00	3.10	2.50	2.20	2.00	2.10	2.00	2.30	1.8
15	2.00	1.90	1.90	2.00	3.10	2.50	2.20	2.00	2.10	2.00	2.30	1.8
16	2.00	1.90	1.90	2.00	3.10	2.50	2.10	2.00	2.10	2.00	2.10	2.0
17	2.00	1.90	1.90	2.10	3.00	2.40	2.10	2.00	2.10	2.00	2.00	2.0
18	2.00	2.00	1.80	2.10	2.90	2.40	2.10	2.00	2.10	2.00	2.00	1.9
19	2.00	2.00	1.80	2.10	2.70	2.40	2.10	2.00	2.10	2.00	2.00	1.9
20	2.00	2.00	1.90	2.10	2.70	2.30	2.10	2.00	2.00	2.00	2.00	1.9
21	2.00	2.00	1.90	2.20	2.60	2.30	2.10	2.00	2.00	2.00	2.00	1.9
22	2.00	2.00	1.90	2.30	2.60	2.30	2.10	2.00	2.00	2.00	2.00	1.9
23	2.00	2.00	1.90	2.40	2.60	2.30	2.10	2.00	2.00	2.00	2.00	1.9
24	2.00	2.00	1.90	2.60	2.50	2.30	2.10	2.00	2.00	2.00	2.00	1.9
25	2.10	2.00	1.90	2.70	2.50	2.30	2.10	2.00	2.00	2.00	2.00	1.9
26	2.10	2.00	1.90	2.50	2.50	2.30	2.10	2.00	2.00	2.00	2.00	1.9
27	2.00	2.00	1.90	3.00	2.60	2.30	2.10	2.00	2.00	2.00	2.00	1.9
28	2.00	2.00	1.90	2.90	2.70	2.30	2.10	2.00	2.00	2.00	2.00	1.9
29	2.00	.....	2.00	2.70	2.70	2.30	2.10	2.00	2.00	2.00	2.00	2.0
30	2.00	.....	2.00	2.70	2.70	2.30	2.10	2.00	2.00	2.00	2.00	2.0
31	2.00	.....	2.10	.....	2.70	.....	2.10	2.00	.....	2.00	.....	2.0

Rating table for North Fork of Snake River near Ora, Idaho, from August 20, 1902, to December 31, 1903.

Gage height.	Discharge.	Gage height.	Discharge.	Gage height.	Discharge.	Gage height.	Discharge.
Feet.	Second-feet.	Feet.	Second-feet.	Feet.	Second-feet.	Feet.	Second-feet.
1.8	1,135	2.3	1,440	2.8	2,360	3.3	3,510
1.9	1,170	2.4	1,570	2.9	2,590	3.4	3,740
2.0	1,210	2.5	1,730	3.0	2,820	3.5	3,970
2.1	1,265	2.6	1,915	3.1	3,050		
2.2	1,340	2.7	2,130	3.2	3,280		

Table not well defined. Curve extended above 2.70 gage height and below 2.00 feet gage height.

onthly discharge of North Fork of Snake River near Oro, Idaho, for 1902–3.

Month.	Discharge in second-feet.			Total in acre-feet.
	Maximum.	Minimum.	Mean.	
1902.				
31	1,210	1,210	1,210	28,800
....	1,210	1,170	1,209	71,941
....	1,265	1,210	1,215	74,707
....	1,265	1,170	1,210	72,000
....	1,210	1,170	1,196	73,539
1903.				
....	1,265	1,170	1,208	74,277
....	1,210	1,170	1,191	66,145
....	1,265	1,135	1,185	72,863
....	2,820	1,210	1,518	90,327
....	3,970	1,570	2,493	153,289
....	2,360	1,440	1,756	104,489
....	1,440	1,265	1,321	81,225
....	1,265	1,210	1,230	75,630
....	1,265	1,210	1,221	72,655
....	1,265	1,210	1,226	75,384
....	1,340	1,210	1,222	72,714
....	1,265	1,210	1,215	74,707
year	3,970	1,135	1,399	1,013,705

FALL RIVER NEAR MARYSVILLE, IDAHO.

ation was established August 21, 1902, by N. S. Dils. It is
. P. Wilson's sawmill, 12 miles southeast of Marysville, Idaho.
is a plain staff graduated to feet and tenths and is firmly
the lower side of the first bent from the south shore. It is
teorge Harringfield, a local civil engineer. The initial point
ings is the face of the south abutment and measurements are
n the bridge. The bridge rests on crib abutments and framed
'he river is straight both above and below the station. The
e river is hard gravel with occasional lava bowlders. Both
high and rocky, are not liable to overflow, and are free from
d timber. The bench mark is a 40-penny spike driven into
9 inch cap on the bridge support to which the gage is attached.
site the 8-foot mark on the gage. A cable will be installed
e-half mile above the bridge. Mr. Harringfield states that
in this vicinity has a fall of from 37½ to 39 feet per mile.
ingfield canal is being built to divert the water from above
n.

The observations at this station during 1903 have been made under the direction of D. W. Ross, district engineer.

Discharge measurements of Fall River near Marysville, Idaho, in 1903.

Date.	Hydrographer.	Gage height.	Discharge.
		Feet.	*Second-feet.*
July 1.....................	N. S. Dils	4.15	2.48
September 11	F. Stockton.................	2.10	65

Mean daily gage height, in feet, of Fall River near Marysville, Idaho, for 1903.

Day.	Jan.	Feb.	Mar.	Apr.	May.	June.	July.	Aug.	Sept.	Oct.	Nov.	Dec.
1.............	1.80	1.80	1.65	2.00	2.15	4.15	4.15	2.40	2.30	2.40	2.10	2.00
2.............	1.80	1.80	1.65	1.87	2.20	4.20	4.00	2.40	2.20	2.40	2.10	2.00
3.............	1.80	1.80	1.65	1.85	2.30	4.20	3.55	2.40	2.30	2.40	2.00	2.00
4.............	1.80	1.80	1.65	1.85	2.60	4.32	3.30	2.35	2.20	2.40	2.00	2.00
5.............	1.80	1.80	1.65	1.85	2.00	4.35	3.30	2.30	2.20	2.35	2.00	2.00
6.............	1.80	1.80	1.65	1.80	3.07	4.25	3.30	2.30	2.20	2.30	2.00	2.00
7.............	1.80	1.80	1.65	1.80	3.30	4.30	3.27	2.30	2.20	2.25	2.00	2.00
8.............	1.80	1.80	1.65	1.80	3.85	4.45	3.12	2.30	2.20	2.25	2.05	2.00
9.............	1.80	1.80	1.65	1.87	3.20	4.70	3.12	2.30	2.20	2.25	2.05	2.00
10.............	1.80	1.80	1.65	1.82	3.07	4.70	3.10	2.30	2.20	2.25	2.05	2.00
11.............	1.80	1.80	1.65	1.80	3.07	4.70	3.10	2.25	2.20	2.25	2.10	2.00
12.............	1.80	1.80	1.65	1.80	3.25	4.70	3.05	2.25	2.20	2.25	2.10	2.00
13.............	1.80	1.80	1.65	1.80	3.75	4.70	2.95	2.25	2.20	2.25	2.10	2.00
14.............	1.80	1.80	1.65	1.80	4.10	4.70	3.00	2.25	2.20	2.20	2.10	2.00
15.............	1.80	1.65	1.65	1.80	4.40	4.70	2.90	2.25	2.20	2.15	2.10	2.00
16.............	1.80	1.65	1.65	1.80	3.95	4.70	2.85	2.25	2.20	2.15	2.05	2.00
17.............	1.80	1.65	1.65	1.80	3.50	4.70	2.80	2.25	2.20	2.15	2.00	2.00
18.............	1.80	1.65	1.65	1.80	3.00	4.70	2.80	2.25	2.20	2.10	2.00	2.00
19.............	1.80	1.65	1.65	1.90	2.90	4.70	2.65	2.25	2.20	2.10	2.00	2.00
20.............	1.80	1.65	1.65	1.92	2.75	4.60	2.55	2.25	2.20	2.10	2.00	2.00
21.............	1.80	1.65	1.65	1.97	2.80	4.40	2.55	2.25	2.20	2.10	2.00	2.00
22.............	1.80	1.65	1.65	2.15	2.80	4.30	2.50	2.25	2.20	2.10	2.05	2.00
23.............	1.80	1.65	1.65	2.25	2.80	4.25	2.55	2.25	2.20	2.10	2.05	2.00
24.............	1.80	1.65	1.65	2.32	3.00	3.95	2.50	2.25	2.20	2.10	2.05	2.00
25.............	1.80	1.65	1.65	2.52	3.20	3.90	2.50	2.25	2.20	2.05	2.05	2.00
26.............	1.80	1.65	1.67	2.75	3.10	3.90	2.50	2.25	2.20	2.05	2.05	2.00
27.............	1.80	1.65	1.72	2.52	3.30	4.05	2.50	2.20	2.20	2.00	2.05	2.00
28.............	1.80	1.65	1.75	2.35	3.65	4.10	2.50	2.20	2.25	2.00	2.05	2.00
29.............	1.80	......	1.82	2.20	3.70	4.20	2.40	2.20	2.30	2.00	2.05	2.00
30.............	1.80	......	1.90	2.15	3.80	4.30	2.40	2.20	2.40	2.00	2.03	2.00
31.............	1.80	......	2.00	......	4.00	......	2.40	2.20	......	2.00	......	2.00

SEEPAGE MEASUREMENTS IN BOISE VALLEY.

The following measurements were made between the gaging station near Boise and the canyon 2½ miles above Caldwell, by Fred Stockton. The conditions under which they were made were very favorable. The irrigation season was practically over. Very little of the diverted water found its way back to the river in small streams that would have

a difficult to measure with a meter, but some of the waste water s running in good-sized streams. Some of the canals had all the er turned out of them for repairs, and others were carrying only art of their normal flow. The river at the gaging station was ed at the beginning and at the end of the measurements, and the e reading remained constant during this period.

The net seepage flow is the difference between two sums. The first these is the quantity flowing in the river at the gaging station in upper canyon plus the natural drainages and the waste from canals ering between the two points at which the river was gaged. The ond sum is the quantity flowing in the lower canyon plus all that is erted between the two points at which the river was gaged.

A similar determination of the seepage was made in the years 1898 1902, inclusive.

harge of Boise River at upper point of measurement, plus tributary drainage and waste, in 1903.

ate.	Stream.	Locality of heading or place of measurement.	Discharge.
			Sec.-feet.
. 24	Boise River	Gaging station (upper point of measurement).	771.6
24	Tailrace from Goodwin's mill	At river	29.8
24	Small stream	Near Boise	.2
24	Waste through natural drainage way.	....do	11.0
26	Small waste	Below Farmers' Union canal	1.0
26	Small stream	....do	.2
26	....do	By Baptist church	1.6
26	....do		.3
27	Dry Creek	At valley road	5.1
27	Small waste	J. Wood's ranch	.5
27	....do	Levi Smith's ranch	.5
27	Small stream	Railroad track	1.0
27	Natural drainage	Middleton	114.4
27	Willow Creek	Below Hartley lateral	26.3
28	Tenmile Creek	Bridge	27.6
28	Three small wastes	Near Boise	3.0
29	Rossi waste	In South Boise	134.2
29	Tailrace from car-line power house.	Near Boise	5.9
	Total		362.6

Second-
feet.

Total of Boise River at gaging station, plus all tributaries between upper and
lower points of measurement... 1,134.2

Total of Boise River 2½ miles above Caldwell, plus all diversions between
upper and lower points of measurement................................ 1,389.5

Seepage between Boise and Caldwell, Idaho, 37 miles...................... 255.3

MISCELLANEOUS MEASUREMENTS IN COLUMBIA RIVER DRAINAGE BASIN.

The following discharge measurements were made in Columbia
River drainage basin in 1903:

Date.	Hydrographer.	Stream.	Locality.	Dis-charge.
				Sec.ft.
Apr. 27	....	Rattlesnake Creek..	Missoula, Mont.........	274

TRIBUTARIES OF THE SNAKE RIVER.

Date.	Hydrographer.	Stream.	Locality.	Discharge.
				Sec.-ft.
t. 3	Fred Stockton ...	Blackfoot River	At mouth, Idaho	68
9	.....do	North Fork of Snake River.	Outlet of Henrys Lake, Idaho.	27
8	.....do ....do		10 miles below Henrys Lake, Idaho.	206
9	....do	Buffalo Creek	5 miles below Flat Rock, Idaho.	183
10	.....do	Warm River	At mouth, near Marysville, Idaho.	166
11	.....do	Conant Creek	Just below mouth of Squirrel Creek, Idaho.	20
17	.....do	Bechler Creek	Southwestern part of Yellowstone Park.	312
19	.....do	Lewis River	Yellowstone Park, at mouth.	119
22	.....do	South Fork of Snake River.	Moran, Wyo., just below outlet Jackson Lake.	705
22	.....do	Pacific Creek	At mouth, Wyoming....	84
22	.....do	Buffalo Creek......	At mouth, near Pacific Creek, Wyo.	258
23	.....do	Lake Creek........	Near outlet of Jennie Lake, Wyo.	41
24	.....do	Gros Ventre River .	5 miles above mouth, in Wyoming.	117
25	.....do	Hoback River	At mouth, Wyoming....	195
26	.....do	John Days River...	At ford 1 mile above mouth, Wyoming.	458
26	.....do	Salt River	1 mile above mouth, in Wyoming.	833

PUGET SOUND DRAINAGE BASIN.

For convenience in arrangement the smaller rivers which have their
headwaters on the western slope of the Cascade Range and which flow
to Puget Sound north of Seattle have been grouped as the Puget
Sound drainage. Of these, White River has its source near Mount
Rainier and flows into Puget Sound near Seattle, Wash. Cedar River
a tributary of White River. Snoqualmie and Skykomish rivers
unite to form the Snohomish, which flows into the Sound about 10
miles beyond the junction near Everett, Wash. The Stilaguamish lies
north of the Skykomish and has a parallel course. The following list
includes the stations in the Puget Sound drainage basin:

White River at Buckley, Wash.
Cedar River near Ravensdale, Wash.

Cedar River at Cedar Lake, near Northbend, Wash.
Snoqualmie River near Snoqualmie Falls, Wash.
Skykomish River (South Fork) near Index, Wash.
Stilaguamish River near Robe, Wash.

WHITE RIVER AT BUCKLEY, WASH.

In order to determine the amount of power that could be develop[ed] on White River, a station was established by Sydney Arnold, A[pril] 22, 1899, at the new highway bridge, 500 feet above the North[ern] Pacific Railway bridge and one-half mile north of the town of Buckl[ey] Wash. The location is in sec. 34, T. 20 N., R. 6 E. of the Willame[tte] meridian. The rod of the wire gage is fastened to the guard rail [of] the highway bridge. The length of the wire gage is 20 feet. Th[e] elevation of the top of the guard rail at the pulley is 19.56 feet abo[ve] gage datum. The bridge has a clear span of 180 feet, crossing t[he] river a trifle obliquely. The channel is straight for some distan[ce] above and below the bridge, but the section is not a very good on[e] owing to the sudden fall a short distance below. About 300 feet abo[ve] the bridge is a good section, with fine gravel bottom in calmer wat[er] but in order to utilize it it would be necessary to install a cable. A[n] examination of the river channel for some distance above and bel[ow] this point was made, but the present location is about the only o[ne] available for accurate results.

No measurements of discharge were made at this station during 19[02] or 1903.

This station was discontinued August 31, 1903.

The observations at this station during 1903 have been made und[er] the direction of T. A. Noble, district engineer.

Mean daily gage height, in feet, of White River at Buckley, Wash., for 1903.

Day.	Jan.	Feb.	Mar.	Apr.	May.	June.	July.	A[ug.]
1	1.80	2.25	2.15	2.20	4.95	6.20	5.75	
2	3.50	2.25	2.15	2.20	4.95	6.05	5.75	
3	8.75	2.25	2.15	2.20	5.10	5.75	5.55	
4	7.00	2.20	2.15	2.15	5.25	5.45	5.60	
5	4.90	2.20	2.15	(a)	5.45	5.35	5.70	
6	3.75	2.20	2.15	(a)	5.70	5.60	5.50	
7	2.70	2.20	2.15	(a)	5.95	6.15	5.55	
8	2.50	2.20	2.15	4.00	4.80	6.65	5.50	
9	2.60	2.25	2.15	8.95	4.75	6.95	5.50	
10	2.45	2.20	2.15	3.95	4.70	7.05	5.30	
11	2.45	2.20	2.15	3.90	4.65	7.05	5.55	
12	2.40	2.30	2.15	3.90	4.75	7.45	5.50	
13	2.40	2.35	2.15	4.00	5.20	6.85	5.40	
14	2.35	2.40	2.15	4.00	5.60	6.40	5.25	
15	2.35	2.20	2.15	4.00	5.70	6.80	5.20	
16	2.35	2.20	2.15	4.05	5.50	7.35	5.10	
17	2.30	2.15	2.15	4.05	5.30	6.85	5.10	
18	2.30	2.15	2.15	4.10	5.20	6.35	5.10	

a Missing.

ily gage height, in feet, of White River at Buckley, Wash., for 1903—Continued.

Day.	Jan.	Feb.	Mar.	Apr.	May.	June.	July.	Aug.
....	2.30	2.15	2.15	4.10	5.15	6.25	5.20	4.65
....	2.35	α2.15	2.15	4.10	5.05	6.20	5.30	4.65
....	2.50	2.15	2.15	4.20	4.90	6.25	5.40	4.60
....	2.50	2.10	2.15	4.40	4.75	6.15	5.30	4.60
........................:	2.40	2.10	2.15	4.60	4.50	6.10	5.30	4.55
....	3.10	2.10	2.15	4.65	4.60	5.90	5.05	4.55
....	2.60	2.10	2.15	4.85	4.70	6.05	4.90	4.55
....	2.40	2.15	2.15	5.35	4.85	6.00	4.80	4.65
....	2.35	2.15	2.15	5.30	4.85	6.55	4.75	4.65
....	2.30	2.15	2.20	5.15	4.85	6.70	4.75	4.65
....	2.30	....	2.20	5.05	5.00	6.20	4.80	4.50
....	2.30	....	2.20	5.00	5.35	5.65	4.75	4.50
....	2.25	....	2.20	....	6.30	....	4.75	4.45

α Interpolated.

CEDAR RIVER NEAR RAVENSDALE, WASH.

station is located at the intake of the Seattle waterworks and
iiles below Cedar Lake, 4 miles from the Northern Pacific Rail-
, Ravensdale, and 6 miles from the Columbia and Puget Sound
y at Maple Valley. The station was established September 27,
by T. A. Noble. The gage is a plain staff graduated to feet and
dths, to which is attached a hook gage and vernier reading to
ndths. When this gage reads zero the hook is level with the
f the dam. It is fastened securely to the head gates above the
nd is read daily by George Landsburg. The bench mark is
st of the dam. The elevation, from city levels, of the south end
831 feet and of the north end 535.840 feet. The gagings at this
are made at two points. The first is 142 feet below the dam,
the cross section is small, the current rapid and suitable for
· the stream at stages below 1 foot on the gage. The measure-
at this point are made from a cable. The initial point for
ngs is on the right bank. The channel is straight. The right
s steep, the left bank has a sloping gravelly beach, and the bed of
eam is rocky. At all stages of the river above 1 foot on the gage
asurements are made from a cable located 600 feet above the
vhere the cross section is large and suitable for gaging the higher
of the river. The initial point for soundings is a boat spike
into the top of a hemlock stump about 12 inches in diameter
ie edge of the water. The right bank is a sloping sandy beach;
t bank is steep and of hardpan formation. The bed is perma-
rocky near the right bank, and of sand and gravel near the left

ie discharge measurements made below the dam should be added
ount of water flowing into the gravity system which supplies the
Seattle. This varies from 34 to 37 second-feet. The discharge

3..............................	5.10	1.29	.90	1.29	1.53	1.62	1.33	.68	......	1.15	.858	2.13
4..............................	5.10	1.24	.89	1.28	1.58	1.68	1.38	.68	......	1.20	.943	1.78
5..............................	5.56	1.19	.88	1.20	1.41	1.55	1.48	.62	......	1.40	.980	1.60
6..............................	4.12	1.15	.86	1.18	1.51	1.52	1.45	.60	0.45	2.05	1.543	1.78
7..............................	2.97	1.12	.91	1.20	1.64	1.66	1.46	.59	.47	2.20	1.575	1.77
8..............................	2.87	1.05	.89	1.31	1.75	1.90	1.40	.54	.52	2.00	1.46	1.65
9..............................	2.56	1.19	.87	1.27	1.79	1.98	1.32	.45	.62	1.80	1.64	1.62
10..............................	2.31	1.35	1.02	1.26	1.83	2.04	1.27	.42	.88	1.69	1.52	1.59
11..............................	2.07	1.30	1.16	1.24	1.76	2.05	1.18	.41	.87	1.61	1.57	1.57
12..............................	1.91	1.24	1.11	1.17	1.68	2.01	1.15	.41	1.30	1.55	1.51	1.60
13..............................	1.76	1.20	1.07	1.12	1.73	1.91	1.11	.39	1.17	1.50	1.46	1.62
14..............................	1.72	1.14	1.04	1.11	1.96	1.78	1.10	.38	1.13	1.42	1.41	1.64
15..............................	1.55	1.10	1.02	1.09	2.10	1.68	1.04	.38	1.08	1.36	1.39	1.65
16..............................	1.43	1.07	.98	1.04	2.12	1.72	1.01	.38	1.04	1.27	1.35	1.73
17..............................	1.37	1.03	.95	1.02	2.08	1.79	.96	.39	.99	1.19	1.32	1.75
18..............................	1.33	1.00	.93	1.00	1.92	1.65	.92	.38	.96	1.14	1.26	1.82
19..............................	1.38	.96	.90	1.02	1.90	1.53	.90	.38	.92	1.08	1.22	1.86
20..............................	1.43	.97	.87	1.01	1.81	1.48	.88	.37	.90	1.05	1.22	1.85
21..............................	1.94	.96	.87	.98	1.72	1.44	.85	.37	1.10	1.09	1.30	1.91
22..............................	2.10	.95	.86	1.09	1.70	1.40	.83	.37	1.06	.98	1.32	1.95
23..............................	2.27	.96	.85	1.17	1.65	1.38	.81	.38	1.02	.93	1.29	1.99
24..............................	2.39	.96	.87	1.21	1.50	1.33	.80	.39	1.03	.88	1.27	1.62
25..............................	2.87	.96	.89	1.25	1.48	1.31	.78	.50	1.22	.87	1.31	1.75
26..............................	2.55	.96	.89	1.40	1.50	1.27	.76	.53	1.20	.84	1.33	1.70
27..............................	2.27	.94	.92	1.54	1.56	1.29	.73	.51	1.16	.84	2.11	1.73
28..............................	1.92	.93	1.03	1.59	1.55	1.52	.72	.50	1.14	.62	1.85	1.60
29..............................	1.73	......	1.24	1.47	1.53	1.43	.70	.49	1.26	.60	1.77	1.65
30..............................	1.66	......	1.31	1.42	1.92	1.33	.67	......	1.22	.54	1.86	1.60
31..............................	1.55	......	1.40	......	1.92	......	.67	......	......	.53	......	1.56

ng table for Cedar River near Ravensdale, Wash., from September 27, 1902, to December 31, 1903.

Gage height.	Discharge.	Gage height.	Discharge.	Gage height.	Discharge.	Gage height.	Discharge.
Feet.	*Second-feet.*	*Feet.*	*Second-feet.*	*Feet.*	*Second-feet.*	*Feet.*	*Second-feet.*
0.4	177	1.4	795	2.4	1,640	3.8	3,300
.5	235	1.5	861	2.5	1,750	4.0	3,540
.6	294	1.6	927	2.6	1,860	4.2	3,780
.7	354	1.7	995	2.7	1,980	4.4	4,020
.8	414	1.8	1,065	2.8	2,100	4.6	4,260
.9	476	1.9	1,145	2.9	2,220	4.8	4,500
1.0	538	2.0	1,235	3.0	2,340	5.0	4,740
1.1	600	2.1	1,335	3.2	2,580	5.2	4,980
1.2	664	2.2	1,435	3.4	2,820	5.4	5,220
1.3	729	2.3	1,535	3.6	3,060	5.6	5,460

Table fairly well defined to 2.7 feet gage height; curve extended above 2.7 feet ge height.

timated monthly discharge of Cedar River near Ravensdale, Wash., for 1902 and 1903.

Month.	Discharge in second-feet.			Total in acre-feet.
	Maximum.	Minimum.	Mean.	
1902.				
September 27-30	538	354	445	3,531
October	354	148	230	14,142
November	1,145	324	811	48,258
December	2,700	632	1,113	68,436
1903.				
January	5,400	762	1,727	106,189
February	828	507	607	33,711
March	795	445	521	32,035
April	894	538	676	40,225
May	1,335	729	999	61,426
June	1,285	696	969	57,659
July	861	324	566	34,802
August	324	148	221	13,589
September	729	206	476	28,324
October	1,435	264	681	41,873
November	1,335	414	786	46,770
December	1,335	894	1,040	63,947
The year	5,400	148	772	560,550

CEDAR RIVER AT CEDAR LAKE, NEAR NORTHBEND, WASH.

This station was established October 17, 1902, by T. A. Noble. It is located 9 miles southeast of Northbend, Wash. The gage is a plain staff graduated to feet and tenths. A hook and vernier are used in reading to hundredths. It is securely nailed to a large cedar stump. The bench mark is located on the stump which holds the gage. Its elevation is 1,542.07 feet above the datum of the city of Seattle, Wash. The initial point for soundings is a nail driven into the root of a hemlock stump on the right bank. The gagings were made from a cable. The cable has been removed on account of the commencement of work on a new dam by the city of Seattle, Wash., for water power and water supply. The channel is straight both above and below the station. The right bank is steep, high, and never overflows; the left bank is flat and overflows at extreme high water. The bed of the stream is rock near the left bank, and of gravel and sand at the right bank. Underlying the bed is a formation of very compact hardpan, which shows no sign of washing away. This formation also underlies the river up to and including the lower portion of Cedar Lake, which is about 8,000 feet upstream from the gaging station. Cedar Lake contains about 1,200 acres at low water.

The observations at this station during 1903 have been made under the direction of T. A. Noble, district engineer.

Discharge measurements of Cedar River at Cedar Lake, Washington, in 1902 and 1903.

Date.	Hydrographer.	Gage height.	Discharge
1902.		*Feet.*	*Sec-ft.*
October 17	T. A. Noble	2.21	36
October 18	do	2.17	35
Do	do	1.70	15
October 19	do	1.50	9
Do	do	1.13	5
1903.			
May 1	H. W. Quinan	3.28	56
May 11	do	4.29	57
June 8	do	4.54	90
July 3	do	2.98	40

an daily gage height, in feet, of Cedar River at Cedar Lake, Washington, for 1903.

Day.	Jan.	Feb.	Mar.	Apr.	May.	June.	July.	Aug.
....	4.25	2.95	2.08	3.26	3.28	5.30	3.27	(a)
....	4.90	2.80	2.07	3.11	3.18	4.97	3.33	1.75
....	12.80	2.65	2.05	3.05	3.16	4.52	3.19	1.72
....	12.30	2.54	2.02	2.92	3.34	4.09	3.10	1.70
....	9.60	2.44	1.99	2.83	3.62	3.81	3.38	1.67
....	7.90	2.35	1.98	2.75	4.05	3.97	3.34	1.62
....	6.85	2.30	2.09	2.87	4.19	4.56	3.31	1.61
....	5.65	2.32	2.08	3.02	4.34	5.26	3.20	1.59
....	5.10	2.50	2.09	2.96	4.57	5.70	3.06	....
....	4.50	2.90	2.20	2.83	4.56	5.88	2.96	....
....	4.10	2.85	2.45	2.75	4.31	5.75	2.86	....
....	3.75	2.80	2.45	2.59	4.23	5.65	2.75	....
....	3.50	2.65	2.52	2.47	4.60	5.22	2.71	....
....	3.25	2.50	2.33	2.41	5.20	4.61	2.70	....
....	3.05	2.40	2.27	2.35	5.50	4.39	2.55	....
....	2.88	2.31	2.20	2.29	5.68	4.73	2.44	....
....	2.80	2.25	2.15	2.27	5.34	4.60	2.39	....
....	2.78	2.19	2.09	2.26	4.90	4.19	2.33	....
....	2.80	2.14	2.05	2.27	4.76	3.86	2.27	....
....	3.05	2.11	2.00	2.25	4.49	3.92	2.22	....
....	4.70	2.24	1.96	2.27	4.17	3.74	2.16	....
....	4.97	2.24	1.94	2.49	3.87	3.57	2.12	....
....	5.10	2.07	1.94	2.74	3.62	3.51	2.09	....
....	5.50	2.08	1.96	2.85	3.48	3.36	2.06	....
....	6.65	2.09	2.01	3.01	3.33	3.29	2.02	....
....	5.90	2.10	2.06	3.53	3.72	3.16	(a)	....
....	5.10	2.10	2.10	3.73	3.82	3.02	(a)	....
....	4.30	2.09	2.32	3.63	3.76	3.70	(a)	....
....	3.90	....	2.85	3.44	3.82	3.48	(a)	....
....	3.47	....	3.09	3.34	4.32	3.22	(a)	....
....	3.20	....	3.27	....	5.18	....	(a)	....

a Missing.

g table for Cedar River at Cedar Lake, Washington, from October 17, 1902, to August 8, 1903.

Gage height.	Discharge.	Gage height.	Discharge.	Gage height.	Discharge.	Gage height.	Discharge.
Feet.	Second-feet.	Feet.	Second-feet.	Feet.	Second-feet.	Feet.	Second-feet.
1.0	23	2.0	186	4.0	812	7.5	2,190
1.1	32	2.2	234	4.2	888	8.0	2,390
1.2	42	2.4	288	4.4	964	8.5	2,590
1.3	54	2.6	344	4.6	1,040	9.0	2,790
1.4	68	2.8	402	4.8	1,116	9.5	2,990
1.5	84	3.0	466	5.0	1,192	10.0	3,190
1.6	102	3.2	530	5.5	1,390	10.5	3,390
1.7	122	3.4	596	6.0	1,590	11.0	3,590
1.8	142	3.6	664	6.5	1,790	12.0	3,990
1.9	164	3.8	736	7.0	1,990	13.0	4,390

his table will not apply after the dam is built at outlet of Cedar Lake. Curve nded above 5.15 feet gage height.

Estimated monthly discharge of Cedar River at Cedar Lake, Washington, in 1902 and [...]

Month.	Discharge in second-feet.			Total in acre-feet
	Maximum.	Minimum.	Mean.	
1902.				
October 17–31 [a]	288	84	173	5,1[...]
November	1,173	198	668	39,7[...]
December	2,650	274	797	49,0[...]
1903.				
January	4,310	402	1,263	77,6[...]
February	450	198	289	16,0[...]
March..............................	546	175	243	14,9[...]
April..............................	718	247	414	24,6[...]
May	1,470	514	894	54,9[...]
June	1,550	466	931	55,3[...]
July [b]	596	142	340	20,9[...]
August 1–8 [c]	132	102	116	1,8[...]

[a] October 20 and 21, 1902, estimated.   [c] Station discontinued August 9, 1903.
[b] July 26–August 1, 1903, estimated.

SNOQUALMIE RIVER NEAR SNOQUALMIE FALLS, WASH.

This station was originally established by T. A. Noble on September 14, 1902. The gage was then located below the falls, but was destroyed by the flood of December 1, 1902. On November 2 another gage was placed about 3 miles above Snoqualmie Falls post-office. The gage consisted of a plain staff graduated to feet and tenths and reading to hundredths of a foot by means of a hook and vernier. The elevation of the zero on the gage was assumed to be 100 feet. The gage was fastened to an alder tree. January 3, 1903, this gage was washed out and on January 7 was replaced by a gage in two parts, 7 and 8 feet long, respectively. The elevation of the zero of the gage is 100.06 feet. The bench mark is on a large maple stump on the right bank. It consists of a spike driven into the stump about 4 feet from the ground. Its elevation is 127.89, or 27.83 feet above the zero of the gage. This is also the initial point for the soundings. The right bank is high and never overflows; the left bank overflows at extreme high water. The bed of the stream is of gravel and sand and is not liable to shift. The station is located below the junction of the north, south, and middle forks of Snoqualmie River. At Snoqualmie Falls, about 4 miles below this station, the river flows over a precipice 268 feet high. Above the falls the Snoqualmie Falls Power Company has built a dam and water-power plant. The slack water from this dam reaches back from the falls about 3 miles and probably affects the

of the river slightly at the gaging station. This is the only possible location for a gaging station which will include all three forks of
the river.

On August 14, 1902, the dam below the station was raised 4 feet,
backing up, the water on the gage. This affected the gage height
from August 14 to October 5, 1903, at which time the dam was
washed out.

The observations at this station during 1903 have been made under
the direction of T. A. Noble, district engineer.

Discharge measurement of Snoqualmie River near Snoqualmie Falls, Wash., in 1903.

Date.	Hydrographer.	Gage height.	Discharge.
		Feet.	*Second-feet.*
anuary 29	T. A. Noble...................	3.86	2,039
April 30	.....do	4.10	2,467
May 10........................	H. W. Quinan	5.70	4,049
May 30........................	.....do	7.56	6,423
June 9	.....do	8.70	8,275
June 6........................	.....do	6.49	5,210
June 5........................	.....do	5.32	3,685
August 26.....................	.....do	3.97	693
August 28.....................	.....do	3.94	637
December 29.................	G. H. Bliss	3.13	1,690
Do	.....do	3.13	1,600

Mean daily gage height, in feet, of Snoqualmie River near Snoqualmie Falls, Wash., for 1903.

Day.	Jan.	Feb.	Mar.	Apr.	May.	June.	July.	Aug.	Sept.	Oct.	Nov.	Dec.
1.......................	4.94	3.28	2.18	4.20	3.95	7.74	4.76	2.23	3.75	4.64	2.82	8.72
2.......................	5.42	3.05	2.10	3.69	3.83	6.93	5.35	2.12	3.68	4.41	2.78	9.43
3.......................	19.62	2.77	2.08	3.50	4.28	5.77	4.90	1.98	3.60	4.65	3.43	6.48
4.......................	(a)	2.63	1.89	3.42	5.10	5.18	4.63	1.91	3.55	4.68	3.98	5.28
5.......................	(a)	2.51	1.82	3.20	5.48	5.32	4.77	1.89	3.56	7.64	4.46	5.32
6.......................	(a)	2.44	1.81	3.00	6.14	6.50	4.92	1.86	3.88	69.12	8.65	4.15
7.......................	6.16	2.35	2.13	3.41	5.40	7.60	4.58	1.89	9.40	7.34	6.52	3.65
8.......................	5.93	2.38	2.00	3.86	5.41	8.94	4.69	1.83	7.39	5.23	5.00	3.32
9.......................	6.09	2.84	1.89	3.28	6.64	8.95	4.57	1.81	5.33	4.25	4.86	3.24
10.......................	5.60	4.35	2.65	2.99	5.91	8.86	4.49	1.81	9.50	4.53	4.49	3.15
11.......................	4.92	3.41	4.79	2.89	5.10	8.83	4.42	1.79	8.23	5.36	4.22	3.28
12.......................	4.37	2.96	3.72	2.60	5.82	8.38	4.15	1.73	7.08	6.22	3.91	3.42
13.......................	4.02	2.71	2.85	2.54	6.83	7.02	3.94	1.72	6.83	5.63	3.60	4.63
14.......................	3.89	2.50	2.70	2.51	7.25	7.21	4.18	1.71	5.10	4.21	3.39	4.73
15.......................	3.86	2.37	2.49	2.47	6.91	6.91	3.85	3.30	4.72	3.67	3.24	5.14
16.......................	3.72	2.25	2.32	2.45	6.37	6.50	3.83	3.81	4.43	3.22	3.10	6.10

a Gage washed out.　　　　　　c Dam at falls 2 miles below station raised about 4 feet.
b 4 feet false work washed out.

Mean daily gage height, in feet, of Snoqualmie River near Snoqualmie Falls, Wash., for 1903—Continued.

Day	Jan.	Feb.	Mar.	Apr.	May.	June.	July.	Aug.	Sept.	Oct.	Nov.	De.
17	3.43	2.15	2.19	2.46	5.60	6.42	3.83	3.80	4.21	2.01	3.84	4.5
18	3.72	2.11	2.12	6.09	5.13	5.71	3.74	3.77	6.14	2.83	3.67	4.6
19	4.66	2.19	2.04	2.64	5.64	5.94	3.69	3.81	4.12	2.68	3.62	4.6
20	4.16	2.08	2.02	2.66	5.13	6.10	3.68	3.87	4.23	2.44	2.77	4.2
21	4.12	2.13	1.98	3.22	4.99	5.89	3.67	3.82	3.81	2.31	3.48	4.4
22	6.99	2.34	1.97	4.18	4.73	5.89	3.32	3.75	6.78	2.39	4.88	4.3
23	4.38	2.47	2.14	4.23	4.71	5.73	3.26	3.86	6.38	2.14	4.79	4.2
24	7.72	2.50	2.23	3.98	4.75	5.45	3.11	3.91	5.41	2.03	4.67	4.4
25	7.77	2.45	2.82	4.44	5.34	5.71	2.90	3.96	8.22	1.97	4.73	4.6
26	5.79	2.43	2.81	5.67	7.38	5.09	2.58	3.99	7.41	1.88	4.90	3.6
27	5.65	2.35	2.65	(a)	5.80	6.36	2.54	3.95	4.93	1.79	3.69	3.2
28	4.20	2.25	4.10	4.64	3.13	7.21	2.46	3.96	4.86	2.41	3.8	3.2
29	3.89	.....	5.65	4.16	5.48	5.50	2.39	3.92	6.10	4.55	6.84	3.7
30	3.74	.....	4.43	4.00	7.55	4.81	2.32	3.90	5.45	3.76	10.70	3.2
31	3.50	.....	4.82	.....	7.68	.....	2.33	3.88	.....	2.85	.....	3.5

a Missing.

Rating table for Snoqualmie River near Snoqualmie Falls, Wash., from November 3, 1901, to August 14, 1903, and from October 7, 1903, to December 31, 1903.

Gage height.	Discharge.	Gage height.	Discharge.	Gage height.	Discharge.	Gage height.	Discharge.
Feet.	Second-feet.	Feet.	Second-feet.	Feet.	Second-feet.	Feet.	Second-feet.
1.7	870	3.6	1,990	7.0	5,770	12.5	13,470
1.8	910	3.8	2,150	7.5	6,470	13.0	14,170
1.9	950	4.0	2,325	8.0	7,170	13.5	14,870
2.0	1,000	4.2	2,505	8.5	7,870	14.0	15,570
2.2	1,100	4.4	2,690	9.0	8,570	14.5	16,270
2.4	1,200	4.6	2,890	9.5	9,270	15.0	16,970
2.6	1,300	4.8	3,090	10.0	9,970	16.0	18,370
2.8	1,420	5.0	3,290	10.5	10,670	17.0	19,770
3.0	1,540	5.5	3,850	11.0	11,370	18.0	21,170
3.2	1,680	6.0	4,450	11.5	12,070	19.0	22,570
3.4	1,830	6.5	5,100	12.0	12,770	20.0	23,970

Table well defined to 8.70 feet gage height. Curve extended above 8.70 feet gage height. From August 14 to October 7, 1903, subtract 670 from discharge given in table.

mated monthly discharge of Snoqualmie River near Snoqualmie Falls, Wash., in 1902 and 1903.

Month.	Discharge in second-feet.			Total in acre-feet.
	Maximum.	Minimum.	Mean.	
1902.				
*mber 3-30	4,775	1,540	2,673	148,451
*mber	16,970	1,250	3,711	228,180
1903.				
*ary 1-3 and 7-31 a			4,218	234,256
*ruary	2,640	1,050	1,320	73,309
*rch.............................	4,030	910	1,500	92,231
*ril b............................	4,030	1,225	1,951	116,093
*ay	6,750	1,645	4,065	249,947
*ne	8,500	3,090	5,298	315,253
*ly...............................	3,675	1,150	2,233	137,302
*gust c	1,655	870	1,255	77,167
*ptember c	8,600	1,280	3,616	215,167
*ctober c	8,040	910	2,497	153,534
*ovember	10,950	1,250	3,250	193,388
*cember	9,200	1,645	3,237	199,035

a January 4-6 gage board washed out.
b April 27 es*imated.
c August 15 to October 6 approximate, owing to influence on gage of a temporary dam 2 miles elow the station.

SOUTH FORK OF SKYKOMISH RIVER NEAR INDEX, WASH.

This station was established October 6, 1902, by T. A. Noble. It s located about 2 miles above Index and about 300 feet from the rail-oad track. The gage is a plain staff graduated to feet and tenths. A hook and vernier are used for reading to hundredths of a foot. Readings are made daily by Louis G. Heybrock. The gage is fas-ened by means of plugs driven in drill holes in the solid rock. The ench mark is a cross cut in the rock about 6 feet above low water und 40 feet downstream from the gage. Its elevation, as obtained rom the Great Northern Railway, is 679.153 feet. The elevation of he zero of the gage is 669.926 feet. The initial point for soundings is on the left bank at a plug driven in the solid rock 10 feet from the edge of stream at low water. The gagings are made from a cable. The channel is straight for 500 feet above and 300 feet below the sta-tion. Both banks are of solid rock and are not liable to overflow. The bed of the stream is of sand and gravel, not liable to change except near the left bank, where the sand shifts at high water. This does not cause any important change in the cross section. This sta-tion is 300 feet upstream from Sunset Falls, where the river plunges

June 20 ....do **8. 55**   6, 134

June 21 ....do **8. 19**   5, 627

June 22 ....do 7. 80   5. 237

Mean daily gage height, in feet, of South Fork of Skykomish River near Index, Wash., for 1903.

Day.	Jan.	Feb.	Mar.	Apr.	May.	June.	July.	Aug.	Sept.	Oct.	Nov.	Dec.
1......................	4.95	3.11	2.41	4.48	4.45	9.10	7.18	3.15	1.74	2.98	3.34	11.60
2.........	5.52	2.44	2.30	4.50	4.99	8.20	6.28	3.18	1.55	3.07	3.74	9.60
3......................	a17.01	2.56	2.20	4.21	4.99	7.25	6.05	2.95	1.55	3.02	3.82	6.95
4......	a12.00	2.45	2.46	4.60	5.92	6.85	5.84	2.74	1.52	5.61	3.48	5.54
5......................	9.25	2.41	2.00	3.82	6.45	6.23	6.32	2.72	1.43	9.07	5.65	4.64
6......................	8.12	2.35	2.00	3.40	7.10	8.10	5.68	2.75	1.65	9.54	7.90	4.60
7......................	6.70	2.31	2.20	3.85	6.88	9.10	5.42	2.80	5.83	7.14	5.68	4.51
8......................	5.81	2.19	2.00	4.32	6.03	10.50	5.44	2.72	4.00	5.62	4.71	4.12
9......................	5.61	3.00	2.00	3.58	6.35	10.75	5.70	2.70	3.18	5.65	4.25	3.75
10......................	5.22	4.20	3.60	3.23	6.21	10.28	5.54	2.68	5.85	5.70	4.42	3.66
11......................	4.79	3.22	4.45	3.52	5.82	10.50	5.59	2.68	4.51	6.10	4.45	2.55
12......................	4.39	2.72	3.30	2.69	7.59	9.22	5.62	2.68	4.38	6.48	4.48	2.55
13......................	4.12	2.65	2.95	2.59	8.21	9.00	5.35	2.62	4.35	5.26	3.85	4.15
14......................	3.82	2.35	2.82	2.60	7.45	8.25	5.15	2.52	4.88	5.10	3.54	3.70
15......................	3.56	2.41	2.60	2.66	7.51	9.35	4.95	2.51	3.88	4.34	3.39	3.72
16......................	3.47	2.20	2.20	2.70	7.22	10.35	4.84	2.22	2.74	3.95	3.15	6.40
17......................	3.34	2.20	2.10	3.00	6.35	9.10	5.10	2.15	2.88	3.64	2.92	5.20

a Approximate, above gage board.

an daily gage height, in feet, of South Fork of Skykomish River near Index, Wash., for 1903—Continued.

Day.	Jan.	Feb.	Mar.	Apr.	May.	June.	July.	Aug.	Sept.	Oct.	Nov.	Dec.
....	3.85	2.10	2.00	3.10	5.83	8.75	5.10	2.13	2.32	3.33	2.72	4.46
....	4.12	2.10	1.95	3.15	5.45	7.08	5.19	2.16	2.34	3.11	2.64	4.91
....	5.61	2.05	2.00	3.14	5.10	8.60	4.93	2.09	2.55	3.00	2.52	5.11
....	7.35	2.10	1.90	3.35	4.60	8.22	5.22	1.98	2.74	2.85	3.67	4.64
....	6.55	2.15	2.10	4.44	4.48	7.15	5.81	1.87	5.65	2.82	4.41	4.60
....	5.91	2.45	2.18	4.65	4.52	7.17	4.86	1.98	4.10	2.61	3.86	4.20
....	6.15	2.53	2.16	4.30	4.99	7.10	4.92	2.00	3.14	2.42	3.45	3.64
....	6.75	2.55	2.96	4.95	5.98	7.70	4.12	1.96	5.10	2.35	3.14	3.75
....	6.25	2.52	3.30	5.82	6.70	8.82	3.91	1.93	3.78	2.22	4.47	3.46
....	4.55	2.51	2.95	5.20	6.74	9.18	3.65	1.94	3.82	2.32	10.80	3.36
....	4.10	2.43	4.98	4.87	6.52	8.16	3.65	1.85	3.22	4.85	7.22	3.21
....	3.65	......	5.46	4.68	6.30	7.10	3.41	1.84	3.68	5.22	6.92	3.00
....	3.32	......	4.74	4.63	8.10	7.21	3.33	1.85	3.38	4.33	9.80	3.39
....	3.25	......	5.10	......	10.00	......	3.23	1.85	......	3.45	......	3.21

ating table for South Fork of Skykomish River near Index, Wash., from October 7, 1902, to December 31, 1903.

Gage height.	Discharge.	Gage height.	Discharge.	Gage height.	Discharge.	Gage height.	Discharge.
Feet.	Second-feet.	Feet.	Second-feet.	Feet.	Second-feet.	Feet.	Second-feet.
0.9	462	2.0	843	4.2	2,090	8.5	6,050
1.0	493	2.2	931	4.4	2,230	9.0	6,620
1.1	524	2.4	1,025	4.6	2,370	9.5	7,220
1.2	555	2.6	1,125	4.8	2,510	10.0	7,820
1.3	586	2.8	1,225	5.0	2,670	11.0	9,020
1.4	618	3.0	1,335	5.5	3,070	12.0	10,220
1.5	652	3.2	1,450	6.0	3,520	13.0	11,420
1.6	687	3.4	1,570	6.5	3,980	14.0	12,620
1.7	724	3.6	1,690	7.0	4,480	15.0	13,820
1.8	762	3.8	1,810	7.5	4,980	16.0	15,020
1.9	802	4.0	1,950	8.0	5,500	17.0	16,220

Table extended above 8.60 feet gage height. The channel at this station is partly n solid rock, and is permanent. This table will be correct for all gage heights as age is located at present (June 30, 1903).

stimated monthly discharge of South Fork of Skykomish River near Index, Wash., from October 7, 1902, to December 31, 1903.

Month.	Discharge in second-feet.			Total in acre-feet.
	Maximum.	Minimum.	Mean.	
1902.				
October 7–31	1,880	462	671	33,273
November	3,610	1,335	2,172	129,243
December	9,500	887	2,521	155,010
1903.				
January	16,220	1,480	3,561	218,957
February	2,090	865	1,090	60,536
March	3,030	802	1,300	79,934
April	3,340	1,125	1,892	112,582
May	7,820	2,265	3,856	237,006
June	8,720	3,745	6,087	362,202
July	4,680	1,480	2,783	171,120
August	1,450	782	1,017	62,533
September	3,385	635	1,654	98,420
October	7,280	931	2,435	149,722
November	8,780	1,075	2,585	153,818
December	9,020	1,100	2,560	157,962
The year	16,220	635	2,569	1,864,882

STILAGUAMISH RIVER NEAR ROBE, WASH.

This station was established December 3, 1902, by T. A. Noble. It was located at the bridge of the Everett and Monte Cristo Railway, 1½ miles east of Robe, Wash. The gage is a plain staff graduated to feet and tenths, and read by means of a hook gage and vernier to hundredths. It is fastened to the east bridge pier, and is read daily by William A. Dobson. The bench mark is the top of the rail above the east pier of the bridge. Its elevation is 901.06 feet above sea level. The elevation of the zero of the gage is 873.184 feet. The elevation of the bridge to which the gage is attached was furnished by the Everett and Monte Cristo Railway Company. No discharge measurements were made in 1903. This station was discontinued July 31, 1903.

The observations at this station during 1903 have been made under the direction of T. A. Noble, district engineer.

daily gage height, in feet, of Stilaguamish River near Robe, Wash., for 1903.

Day.	Jan.	Feb.	Mar.	Apr.	May.	June.	July.
....	2.89	1.59	1.44	2.72	2.46	3.54	4.59
....	4.58	1.57	1.39	2.23	2.19	3.06	3.32
....	10.25	1.51	1.35	2.09	2.49	2.66	3.17
....	6.36	1.46	1.32	1.99	2.97	2.65	2.75
....	4.95	1.34	1.29	1.83	3.08	2.68	2.51
....	4.87	1.33	1.25	1.98	3.47	3.28	2.48
....	3.22	1.44	1.49	2.59	2.94	3.87	2.23
....	3.01	2.09	1.32	2.33	2.94	4.25	2.15
..............................:..	3.95	2.96	1.26	1.97	3.03	4.26	2.21
....	2.99	2.51	1.92	1.78	2.76	4.21	2.23
....	2.51	1.81	2.56	1.46	2.58	3.97	2.24
....	2.13	1.73	1.94	1.47	3.07	3.97	2.15
....	2.09	1.55	1.91	1.45	3.74	3.44	2.12
....	2.07	1.44	1.76	1.48	3.93	2.79	2.16
....	2.07	1.41	1.53	1.52	4.05	2.85	1.85
....	1.95	1.25	1.44	1.51	4.00	3.61	2.06
....	1.84	1.30	1.36	1.66	3.11	3.09	2.02
....	2.11	1.29	1.37	1.74	2.76	2.77	2.06
....	3.28	1.28	1.34	1.86	2.61	2.74	2.04
....	3.76	1.28	1.34	1.75	2.34	2.67	2.03
....	4.94	1.43	1.33	1.93	2.17	2.79	2.03
....	4.47	1.56	1.36	3.28	2.07	2.78	2.02
....	3.79	1.66	1.53	2.92	2.09	2.77	1.95
....	3.51	1.67	1.79	2.53	2.31	2.61	2.01
....	4.06	1.63	1.78	2.79	2.76	3.21	1.69
....	2.88	1.61	1.90	2.34	2.45	3.60	1.54
....	2.49	1.56	1.95	2.81	2.83	6.64	1.49
....	2.35	1.49	2.81	2.45	2.54	4.24	1.54
....	1.86	....	3.72	2.28	2.88	3.02	1.53
....	1.88	....	2.58	2.66	3.77	2.49	1.44
....	1.78	....	4.13	....	4.38	....	1.41

...ANEOUS MEASUREMENTS IN THE PUGET SOUND DRAINAGE BASIN.

r River at Renton, Wash., was measured by H. W. Quinan on
., 1903. It had a gage height of 44.87 feet and a discharge of
second-feet.

HUDSON BAY DRAINAGE BASIN.

River drains a large basin in the United States, covering por-
Minnesota and North and South Dakota, characterized by a
topography, broken up in places by moraines and other glacial
s. The major part is prairie, and its eastern half has an
nce of lakes and some woods. The main river flows nearly due
cutting a deep channel in its broad, level valley, and is subject
len rises caused by heavy spring rains, entailing frequently
rable loss of life and property. The valley of Red River com-
about 9,000,000 acres of excellent agricultural lands, which to a
xtent still await settlement. A number of water powers have

een developed during recent years on the tributaries entering from both sides.

Red River drains into Hudson Bay through Lake Winnipeg and elson River. Red Lake River, one of its principal tributaries in linnesota, drains Red Lake in the northern part of the State. The heyenne River, its principal tributary from North Dakota, joins the ed River about 10 miles below Fargo, N. Dak.

St. Mary River heads in northern Montana, near the Canadian oundary line, on the eastern slope of the main range of the Rocky lountains, in a region of perpetual snow and in the midst of numer us glaciers. It starts from the great Blackfoot Glacier (probably the rgest in the Rocky Mountains within the United States) and receives flluents from at least a dozen lesser ones. These small streams unite ithin a short distance from their sources and flow into a lake hemmed by high mountains, known as Upper St. Mary Lake. Below this, parated by a narrow strip of land, is Lower St. Mary Lake. The ggregate length of these two lakes is about 22 miles. The river flows ut of the lower lake, the elevation of which is 4,460 feet above sea level, and within 2 miles is joined by a stream nearly, if not quite, large as itself, known as Swift Current Creek, which receives the ater of the Grinnell Glacier and four lesser ones. From the confluence of these streams to the boundary of the British possessions, a distance of 12 miles, the river flows in a northerly direction. Entering Alberta, it empties into Belly River, its waters at length finding their way through Saskatchewan River into Hudson Bay. A canal has been constructed in Canada, by the Canadian Northwest Irrigation Company, which diverts water from the right bank of St. Mary River about 5 miles below the international boundary line.

The following list includes the stations of the Hudson Bay drainage basin:

St. Mary River at international line near Cardston, Alberta.
Swiftcurrent Creek near Wetzel, Mont.
Kennedy Creek near Wetzel, Mont.
St. Mary River at dam site near St. Mary, Mont.
Mouse River at Minot, N. Dak.
Pembina River at Neche, N. Dak.
Red Lake River at Crookston, Minn.
Red River at Grand Forks, N. Dak.
Sheyenne River at Haggart, N. Dak.
Red River at Fargo, N. Dak.

ST. MARY RIVER AT INTERNATIONAL LINE NEAR CARDSTON, ALBERTA. B. C.

This station was established September 4, 1902, by C. T. Prall. It is located at the ranch of L. C. Shaw, about 1,000 feet from the house It is one-fourth of a mile north of the boundary line between the United States and Canada, and 17 miles south of Cardston, Alberta

The chain gage is located at the crossing near L. C. Shaw's house, about 1.200 feet above the cable. It is fastened to a tree and post on the bank. The length of the chain from the end of the weight to the marker is 10.85 feet. The marker is the outside end of the ring at the end of the chain. The gage is read once each day by Vernon Shaw. Discharge measurements are made by means of a cable, car, and tagged wire. The initial point for soundings is the zero of the tagged wire on the left bank. The channel is straight for 300 feet above and 150 feet below the station. The right bank is high and not liable to overflow. The left bank is sloping and is liable to overflow at very high water. The bed of the stream is of sand and gravel and is not liable to scour. There is but one channel and the current is swift near the right bank.

Bench mark No. 1 is a 60-penny spike driven in a cottonwood tree located directly back of the rod. Its elevation is 12.92 feet above the zero of the gage. Bench mark No. 2 is a 60-penny spike driven in a post 1 foot in diameter and 2 feet high. The post is set in the ground 92 feet northwest of the gage rod. Its elevation is 17.56 feet above the zero of the gage.

The observations at this station during 1903 have been made under the direction of C. C. Babb, district engineer.

Discharge measurements of St. Mary River at international line near Cardston, Alberta, B. C., in 1903.

Date.	Hydrographer.	Gage height.	Discharge.
		Feet.	*Second-feet.*
May 11.....................	F. M. Brown	4.76	1,130
Do.....................	.....do	4.76	1,125
June 5	C. T. Prall	7.05	6,942
June 23	.....do	6.55	·5,212
August 3	.....do	5.20	1,730
October 5...................	.....do	4.80	1,325

Mean daily gage height, in feet, of St. Mary River at international line near Cardston, Alberta, B. C., for 1903.

Day.	Jan.	Feb.	Mar.	Apr.	May.	June.	July.	Aug.	Sept.	Oct.	Nov.	Dec.
1	4.65	4.40	4.70	5.70	4.40	6.45	6.70	5.80	4.55	5.15	4.00	4.0
2	4.60	(a)	4.60	5.55	4.40	6.80	6.50	5.30	4.50	5.00	3.95	4.0
3	4.50	4.25	4.68	5.25	4.45	7.00	(a)	5.20	4.45	4.90	3.90	4.0
4	4.60	4.35	4.70	5.10	4.60	7.00	6.30	5.15	4.40	4.85	3.90	4.0
5	4.65	4.35	4.60	5.00	4.60	7.05	(a)	5.00	4.40	4.80	3.90	4.0
6	4.60	4.50	4.50	5.00	4.70	7.05	(a)	5.00	4.45	4.75	3.95	4.0
7	4.30	4.40	4.55	4.90	4.80	7.05	6.35	5.00	(a)	4.60	3.95	4.0
8	4.50	4.40	4.58	4.80	4.85	7.15	6.20	4.95	4.45	(a)	4.15	4.0
9	(a)	4.35	4.60	4.85	4.80	7.00	6.10	4.95	4.30	(a)	4.05	4.0
10	4.35	4.35	4.50	4.50	(a)	7.00	6.05	5.05	4.55	4.40	4.15	4.0
11	4.20	4.45	4.48	4.25	(a)	6.00	(a)	4.50	4.50	4.10	4.10	4.0
12	4.50	4.40	4.35	3.80	4.80	6.80	5.95	5.00	4.50	4.45	4.00	4.0
13	4.30	3.70	(a)	3.60	4.95	6.90	5.90	4.90	4.45	4.40	3.95	4.0
14	4.45	4.40	4.80	3.63	4.90	6.90	5.90	4.95	4.35	4.40	3.95	4.0
15	4.50	4.30	4.80	3.90	5.00	(a)	5.85	4.90	4.30	4.45	3.75	4.0
16	4.50	4.65	5.00	4.50	5.10	6.80	5.65	4.85	4.25	4.45	3.60	4.0
17	4.25	4.55	4.90	4.55	5.30	7.00	5.70	4.85	4.20	(a)	3.60	4.0
18	4.05	(a)	4.70	4.60	(a)	6.95	5.65	4.80	4.20	4.30	3.60	4.0
19	4.10	4.65	(a)	4.50	(a)	6.90	5.60	4.80	4.20	4.30	3.60	4.0
20	4.00	4.50	(a)	4.40	5.40	6.90	(a)	4.75	4.25	4.30	3.65	4.0
21	(a)	4.55	(a)	4.30	5.40	6.80	5.50	4.75	4.25	4.00	(a)	4.0
22	4.20	4.65	4.00	4.35	5.30	6.70	5.55	4.70	4.65	4.30	4.00	4.0
23	4.00	4.60	4.05	4.35	5.40	6.55	5.50	(a)	4.75	4.25	3.95	4.0
24	4.25	4.70	4.60	4.30	5.45	6.55	5.65	(a)	4.85	4.25	3.90	4.0
25	4.30	4.65	4.60	4.45	5.45	6.40	5.65	4.80	5.10	4.15	3.95	4.0
26	4.20	4.00	4.70	4.50	5.70	6.40	5.60	5.00	5.55	4.10	3.90	4.0
27	4.30	4.45	4.80	4.40	5.95	6.40	5.50	4.95	5.45	4.10	4.00	3.5
28	4.55	4.70	4.83	4.40	5.75	6.65	5.45	4.90	5.30	4.05	3.85	3.4
29	4.60	.....	5.70	4.40	5.60	6.85	5.40	4.80	5.25	4.05	3.90	3.4
30	4.60	.....	5.80	4.35	5.70	6.70	5.30	4.70	5.15	4.00	4.00	3.4
31	4.50	.....	5.75	.....	5.75	.....	5.35	4.65	.....	4.00	.....	...

a No reading.

Rating table for St. Mary River at international line near Cardston, Alberta, B. C., for 1903.

Gage height.	Discharge.	Gage height.	Discharge.	Gage height.	Discharge.	Gage height.	Discharge.
Feet.	Second-feet.	Feet.	Second-feet.	Feet.	Second-feet.	Feet.	Second-feet.
3.0	150	4.0	575	5.0	1,510	6.0	3,800
3.1	175	4.1	645	5.1	1,640	6.1	4,060
3.2	205	4.2	720	5.2	1,800	6.2	4,320
3.3	239	4.3	800	5.3	2,000	6.3	4,580
3.4	275	4.4	880	5.4	2,240	6.4	4,840
3.5	315	4.5	965	5.5	2,500	6.5	5,120
3.6	360	4.6	1,060	5.6	2,760	6.6	5,420
3.7	410	4.7	1,165	5.7	3,020	6.7	5,740
3.8	460	4.8	1,275	5.8	3,280	6.8	6,060
3.9	515	4.9	1,390	5.9	3,540	7.0	6,700

Tangent above 6.6 feet gage height with a difference of 320 per tenth.

mated monthly discharge of St. Mary River at international line near Cardston, Alberta, B. C., for 1903.

[Drainage area, 482 square miles.]

Month.	Discharge in second-feet.			Total in acre-feet.	Run-off.	
	Maximum.	Minimum.	Mean.		Second-feet per square mile.	Depth in inches.
1902.						
:ember *a*	760	175	607	36,118	1.343	1.503
)ber *b*	610	295	475	29,206	1.051	1.211
ember	410	150	336	19,993	.743	.823
:mber	1,800	240	1,105	67,943	2.444	2.814
1903.						
iary *c*	1,112	575	862	53,002	1.907	*2.198
ruary *d*	1,165	410	921	51,150	2.038	2.122
:h *e*	3,280	840	1,327	81,594	2.936	3.385
l...............	3,020	360	1,106	65,812	2.447	2.728
f	3,670	880	1,875	115,290	4.148	4.781
·g	7,180	4,840	6,116	363,927	13.531	15.100
h	5,740	2,000	3,424	210,535	7.575	8.734
ıst *i*	2,000	1,112	1,426	87,682	3.154	3.634
ember *j*	2,630	720	1,145	68,133	2.533	2.823
ber *k*	1,720	575	928	57,061	2.053	2.363
:mber	880	360	536	31,894	1.185	1.326
:mber *l*	682	275	436	26,809	.965	1.116
The year	7,180	275	1,675	1,212,889	3.706	50.310

a September 1 and 2 estimated.
b October 5 to 11 estimated.
c January 9 and 21 estimated.
d February 2 and 18 estimated.
e March 13, 19, 20, and 21 estimated.
f May 10, 11, 18, and 19 estimated.

g June 11 and 15 estimated.
h July 3, 5, 6, and 20 estimated.
i August 11, 23, and 24 estimated.
j September 7 estimated.
k October 8, 9, and 17 estimated.
l December 21 and 31 estimated.

SWIFTCURRENT CREEK NEAR WETZEL, MONT.

his station was originally established April 8, 1902, by J. S. Baker. was located one-half mile northwest of Henkel's ranch and 36 :s northwest of Browning, Mont. The nearest post-office is Wet- Henkel's ranch is reached by regular stage from Wetzel, Mont. channel is straight for 500 feet above and 200 feet below the sta- . The right bank is low and liable to overflow; the left bank is ı and rocky. The bed of the stream is rocky. The station as :inally established by J. S. Baker was washed away by the high :r in June, 1902, and was reestablished July 30, 1902, by W. W. lecht. The original gage was placed on the right bank of the am. The new gage was located 1,800 feet above the first gage **and** on the same bank of the stream. The gage was again moved on September 27, 1902, by C. T. Prall, and located about 900 feet above the former station, as the second location was directly above a dam. The length from the pointer to the bottom of the weight was 14.60 feet. The distance from the zero of the rod to the outside of the pulley was 2.15 feet.

This gage was washed out about June 10, 1903, and replaced June 17 in practically its original position. It is firmly nailed to a horizontal support. On September 30, 1903, the length of the wire from the end of the weight to the marker was found to be 14.6 feet, and the distance from the zero of the scale to the outside edge of the pulley was found to be 2.15 feet, the same as when the previous gage was established. Discharge measurements are made by means of a cable, car, and tagged wire. At low water measurements are made by wading. The initial point for soundings is on the left bank.

Bench mark No. 1 is a point chipped on a large bowlder 32.4 feet south of the gage. Its elevation is 12.52 feet above the zero of the gage. Bench mark No. 2 is the head of a 20-penny wire nail driven in a cottonwood tree 64.4 feet east of the gage. Its elevation is 12.86 feet above the zero of the gage.

The gage is read once each day by Henry Henkel.

The observations at this station during 1903 have been made under the direction of C. C. Babb, district engineer.

Discharge measurements of Swiftcurrent Creek near Wetzel, Mont., in 1903.

Date.	Hydrographer.	Gage height.	Discharge.
		Feet.	*Second-feet.*
May 9	F. M. Brown	2.87	430
June 3	C. T. Prall	5.40	2,473
June 20	do	4.75	1,546
June 26	do	4.35	1,290
July 3	do	3.90	1,030
July 9	do	3.85	964
July 18	do	3.50	745
July 25	do	3.50	718
August 1	do	3.20	500
August 8	do	3.10	411
August 15	do	3.10	407
August 22	do	3.00	356
August 27	do	3.10	413
September 26	do	4.23	1,296
October 3	do	2.90	348
October 10	do	2.70	262
October 17	do	2.90	368

Mean daily gage height, in feet, of Swiftcurrent Creek near Wetzel, Mont., for 1903.

Day.	Jan.	Feb.	Mar.	Apr.	May.	June.	July.	Aug.	Sept.	Oct.	Nov.	Dec.
...	1.50	1.30	1.20	1.30	2.40	4.80	4.00	3.20	3.00	3.10	2.50	2.30
...	1.50	1.30	1.20	1.20	2.50	5.20	3.80	3.10	2.90	3.10	2.50	2.60
...	1.50	1.30	1.20	1.20	2.30	5.50	3.90	3.20	2.80	3.00	2.40	2.50
...	1.40	1.20	1.20	1.20	2.40	5.30	3.80	3.10	2.80	2.90	2.40	2.40
...	1.40	1.20	1.40	1.20	2.50	5.30	3.90	3.00	2.80	2.90	2.50	2.30
...	1.30	1.20	1.60	1.20	2.50	5.40	3.90	3.00	2.70	2.80	2.50	2.40
...	1.30	1.20	1.50	1.20	2.60	5.60	4.10	3.00	2.70	2.70	2.40	2.60
...	1.30	1.20	1.30	1.20	2.70	(a)	3.90	3.10	2.60	2.70	2.40	2.50
...	1.30	1.20	1.30	1.20	2.80	(a)	3.70	3.10	2.60	2.60	2.40	2.40
...	1.30	1.20	1.20	1.20	2.90	(a)	3.60	3.20	2.80	2.70	3.40	2.40
...	1.30	1.20	1.20	1.20	2.90	(a)	3.50	3.10	2.80	2.70	3.60	b 2.50
...	1.30	1.20	1.20	1.10	3.10	(a)	3.60	3.10	2.90	2.90	3.60	2.50
...	1.30	1.20	1.20	1.10	3.20	(a)	3.60	3.10	2.80	2.90	3.20	2.50
...	1.20	1.20	1.20	1.10	3.50	(a)	3.60	3.10	2.70	2.90	3.20	2.50
...	1.20	1.20	1.20	1.10	3.50	(a)	3.50	3.00	2.60	3.30	3.00	2.50
...	1.20	1.20	1.20	1.20	3.70	(a)	3.50	3.00	2.60	2.90	2.90	2.50
...	1.20	1.20	1.20	1.20	3.30	4.50	3.40	2.90	2.60	2.90	2.80	2.50
...	1.20	1.20	1.20	1.20	3.00	4.90	3.40	3.00	2.60	2.90	2.80	2.50
...	1.20	1.20	1.20	1.20	2.60	5.00	3.80	3.00	2.50	2.70	2.60	2.50
...	1.20	1.20	1.20	1.30	2.10	4.80	3.80	3.00	2.80	2.60	2.50	2.50
...	1.20	1.20	1.20	1.40	2.50	4.50	3.40	3.00	2.90	2.60	2.40	2.50
...	1.20	1.20	1.20	1.70	2.30	4.40	3.40	2.90	3.00	2.60	2.30	c 2.70
...	1.20	1.20	1.20	1.80	2.40	4.40	3.50	2.90	3.40	2.60	2.40	2.70
...	1.20	1.20	1.30	2.00	2.50	4.20	3.50	2.90	3.80	2.60	2.40	2.70
...	1.20	1.20	1.30	2.40	2.60	4.20	3.50	3.00	4.40	2.50	2.40	2.70
...	1.20	1.20	1.30	2.50	2.80	4.30	3.40	3.00	4.40	2.50	2.40	2.70
...	1.20	1.20	1.40	2.50	2.90	4.30	3.30	3.10	3.90	2.50	2.30	......
...	1.20	1.20	1.40	2.60	2.90	4.90	3.30	3.00	3.30	2.50	2.20	......
...	1.20	......	1.30	2.50	3.00	5.10	3.20	3.00	3.30	2.50	2.20	......
...	1.20	......	1.30	2.40	3.00	4.40	3.10	3.00	3.20	2.60	2.20	......
...	1.20	......	1.30	......	3.90	......	3.10	3.00	......	2.60	......	......

ᵃ Gage washed out. ᵇ To top of ice December 11–22. ᶜ To top of ice December 23–26; ice 6 inches thick.

Rating table for Swiftcurrent Creek near Wetzel, Mont., from April 25 to December 31, 1903.ᵃ

Gage height.	Discharge.	Gage height.	Discharge.	Gage height.	Discharge.	Gage height.	Discharge.
Feet.	Second-feet.	Feet.	Second-feet.	Feet.	Second-feet.	Feet.	Second-feet.
2.0	99	2.9	321	3.8	910	4.7	1,550
2.1	113	3.0	369	3.9	980	4.8	1,650
2.2	129	3.1	425	4.0	1,050	4.9	1,770
2.3	147	3.2	490	4.1	1,120	5.0	1,900
2.4	167	3.3	560	4.2	1,190	5.1	2,040
2.5	189	3.4	630	4.3	1,260	5.2	2,180
2.6	215	3.5	700	4.4	1,330	5.3	2,320
2.7	245	3.6	770	4.5	1,400	5.4	2,460
2.8	280	3.7	840	4.6	1,470		

ᵃ Use 1902 rating table from January 1 to April 24, 1903.

Measurements made in 1903 extend from gage height 2.70 feet to 5.40 feet. Curve and table are extended below 2.70 feet and above 5.40 feet gage height.

Estimated monthly discharge of Swiftcurrent Creek near Wetzel, Mont., for 1903.

[Drainage area, 101 square miles.]

Month.	Discharge in second-feet.			Total in acre-feet.	Run-off.	
	Maximum.	Minimum.	Mean.		Second-feet per square mile.	Depth in inches
January	67	45	50	3,074	0.50	0.55
February	52	45	46	2,555	.46	.53
March	78	45	50	3,074	.50	.57
April	215	40	81	4,820	.80	.89
May	990	113	335	20,596	3.32	3.83
June *a*	2,740	1,190	1,902	113,177	18.83	21.03
July	1,120	425	734	45,132	7.27	8.38
August	490	321	391	24,042	3.87	4.46
September	1,330	189	422	25,111	4.18	4.68
October	560	189	275	16,909	2.72	3.14
November	770	129	264	15,709	2.61	2.91
December *b*	215	147	170	10,453	1.68	1.94
The year	2,740	40	393	284,654	3.90	52.96

a June 6 to 16 estimated.   *b* December 11 to 31 estimated.

KENNEDY CREEK NEAR WETZEL, MONT.

This station was established October 17, 1903, by C. T. Prall. It is located 50 feet above the road from Altyn, Mont., to Cardston, Alberta, B. C. It is 35 miles northwest of Browning, Mont., and 20 miles from Wetzel, Mont., and about 5 miles north of the St. Mary dam site. The station is at the mouth of the canyon and about 1 mile above the mouth of Kennedy Creek. The horizontal wire gage is bolted to a tree and post on the bank of the river. The length of the wire from the end of the weight to the marker is 8.45 feet.

Discharge measurements are made by means of a cable, car, tagged wire, and stay wire. The initial point for soundings is the zero of the tagged wire on the left bank. The channel is straight for 200 feet above and 100 feet below the station. The current is very rapid at high stages and moderate at low stages. The right bank is rocky and may overflow at extreme high water. The left bank is high and rocky and not liable to overflow. The bed of the stream is composed of bowlders and gravel, and is probably subject to some change at high floods. Bench mark No. 1 is a chipped point on a large bowlder at the strut near the north end of the cable. Its elevation is 12.88 feet above gage datum. Bench mark No. 2 is a 60-penny spike in a cottonwood tree 90 feet north of the gage. Its elevation is 10.49 feet above gage datum.

The observations at this station during 1903 have been made under the direction of C. C Babb, district engineer.

Discharge measurements of Kennedy Creek near Wetzel, Mont., in 1903.

Date.	Hydrographer.	Gage height.	Discharge.
		Feet.	*Second-feet.*
August 22	C. T. Prall...................	6. 00	114
October 3......................	.do	5. 95	102
October 10...................	....do	5. 85	84
July 25........................	....do	6. 30	208
August 1.......................	....do	6. 20	164
August 8......................	...do	6. 10	136
August 15....................	....do	6. 10	138
October 17...................	....do	5. 85	91

ST. MARY RIVER AT DAM SITE NEAR ST. MARY, MONT.

This station was established April 9, 1902, by J. S. Baker. It is located on Henry Henkel's ranch, about 1 mile east of his house, and is 35 miles northwest of Browning, Mont. It is also about 4,500 feet below Lower St. Mary Lake. The gage, which is located about 1,000 feet above the cable from which the measurements are made, is of the wire type. It is supported upon a horizontal arm which extends over the river and is fastened to a cottonwood stump. The scale board is graduated to feet and tenths. The length of the wire from the bottom of the weight to the marker is 11.40 feet. A new wire gage was installed by C. T. Prall June 17, 1903. It reads the same and has the same length of wire as the 1902 gage. The observer is Henry Henkel, who lives about 1 mile west of the gage and who reads the gage once each day. Discharge measurements are made by means of a cable, car, and tagged wire. The cable is fastened to a cottonwood tree on each bank. The initial point for soundings is on the left bank at the middle of the cottonwood tree to which the cable is fastened. The channel is straight for 500 feet above and for 200 feet below the station. Both banks are high and rocky and have gentle slopes. The current has a moderate velocity. The bed of the stream is composed of gravel and bowlders. Bench mark No. 1 is a spike in the foot of a cottonwood tree 30 feet north of the gage. It is marked B., M. 9.24. Its elevation above gage datum is 9.24 feet. Bench mark No. 2 was a wire spike in the rear post of the gage. It was marked B. M. 6.97 feet. Its elevation above gage datum was 6.97 feet. It has been destroyed. Bench mark No. 3 is a 60-penny spike in the base of a cottonwood tree 125 feet above the gage. Its elevation is 9.61 feet above gage datum.

The observations at this station during 1903 have been made under the direction of C. C. Babb, district engineer.

Discharge measurements of St. Mary River at dam site near St. Mary, Mont., in 1903.

Date.	Hydrographer.	Gage height.	Discharge.
		Feet.	*Second-feet.*
May 10	F. M. Brown	2.62	484
June 3	C. T. Prall...................	4.70	2,431
June 18	.....do	6.00	3,683
June 25	.....do	5.25	2,756
June 30	.....do	5.85	3,418
July 6......................	.....do	4.75	2,212
July 8......................	F. P. Sherburne...........	4.59	2,082
July 14...........	C. T. Prall................	4.25	1,831
July 18.....................	.....do	4.05	1,639
July 25.....................	.....do	4.02	1,571
July 31.....................	.....do	3.64	1,291
August 8....................	.....do	3.14	90
August 14...................	.....do	3.05	84
August 22...................	.....do	2.97	80
August 27...................	.....do	2.92	75
September 2	.....do	2.70	61
September 26	.....do	3.20	97
October 3...................	.....do	3.22	96
October 10..................	.....do	2.65	63
October 17..................	.....do	2.70	66

Mean daily gage height, in feet, of St. Mary River at dam site near St. Mary, Mont., for 1903.

Day.	Jan.	Feb.	Mar.	Apr.	May.	June.	July.	Aug.	Sept.	Oct.	Nov.	Dec.
1......................	1.60	1.30	1.30	1.30	1.80	4.60	5.70	3.50	2.90	3.40	2.40	2.60
2......................	1.60	1.30	1.30	1.40	1.80	4.60	5.50	3.50	2.80	3.30	2.40	2.10
3......................	1.60	1.30	1.20	1.40	1.90	4.70	5.30	3.50	2.80	3.20	2.40	2.10
4......................	1.50	1.30	1.20	1.40	1.90	4.90	5.00	3.40	2.70	3.10	2.30	2.10
5......................	1.50	1.30	1.20	1.40	2.00	5.40	4.70	3.30	2.70	3.10	2.30	2.10
6......................	1.50	1.30	1.20	1.40	2.20	5.90	4.60	3.20	2.70	3.00	2.30	2.10
7......................	1.40	1.30	1.20	1.40	2.40	5.80	4.60	3.20	2.70	2.90	2.20	2.10
8......................	1.40	1.30	1.20	1.40	2.50	5.90	4.50	3.10	2.60	2.80	2.20	2.10
9......................	1.40	1.30	1.20	1.40	2.60	5.90	4.50	3.10	2.60	2.70	2.20	2.10
10.....................	1.40	1.30	1.20	1.40	2.60	5.90	4.40	3.10	2.50	2.60	2.20	2.00
11.....................	1.40	1.30	1.20	1.40	2.50	5.80	4.30	3.20	2.50	2.60	2.30	a2.10
12.....................	1.40	1.30	1.20	1.40	2.50	5.90	4.30	3.20	2.50	2.70	2.30	a2.10
13.....................	1.40	1.30	1.20	1.40	2.60	5.90	4.30	3.10	2.50	2.70	2.20	a2.10

a To top of ice.

Mean daily gage height, in feet, of St. Mary River at dam site near St. Mary, Mont., for 1903—Continued.

Day.	Jan.	Feb.	Mar.	Apr.	May.	June.	July.	Aug.	Sept.	Oct.	Nov.	Dec.
14	1.40	1.30	1.20	1.40	2.60	5.90	4.20	3.00	2.50	2.70	2.20	a2.10
15	1.40	1.30	1.20	1.40	2.70	5.90	4.20	3.00	2.40	2.70	2.20	a2.10
16	1.30	1.30	1.20	1.40	2.80	5.80	4.20	3.00	2.40	2.70	2.20	a2.10
17	1.30	1.30	1.20	1.40	2.80	5.90	4.10	2.90	2.40	2.70	2.20	a2.10
18	1.30	1.30	1.20	1.40	2.80	5.90	4.00	3.00	2.40	2.70	2.20	a2.10
19	1.30	1.30	1.20	1.40	2.70	6.00	4.00	3.00	2.40	2.60	2.20	a2.10
20	1.30	1.30	1.20	1.50	2.80	5.90	4.00	3.00	2.40	2.60	2.20	a2.10
21	1.30	1.30	1.20	1.50	2.70	5.80	4.00	3.00	2.50	2.50	2.20	2.00
22	1.30	1.30	1.20	1.60	2.90	5.60	3.90	2.90	2.50	2.60	2.20	2.00
23	1.30	1.30	1.20	1.60	3.00	5.60	3.90	2.90	2.60	2.60	2.10	2.00
24	1.30	1.30	1.30	1.70	3.20	5.40	3.90	2.90	2.60	2.60	2.10	2.00
25	1.30	1.30	1.30	1.80	3.40	5.30	3.90	2.90	2.70	2.50	2.10	2.00
26	1.30	1.30	1.30	1.80	3.50	5.20	3.90	2.90	3.20	2.50	2.10	2.00
27	1.30	1.30	1.30	1.80	3.50	5.20	3.90	2.90	3.40	2.50	2.10	2.00
28	1.30	1.30	1.30	1.80	3.80	5.40	3.80	2.90	3.40	2.50	2.10	2.00
29	1.30	......	1.30	1.70	3.90	5.60	3.70	2.90	3.50	2.50	2.10	2.00
30	1.30	......	1.30	1.70	4.10	5.80	3.70	2.90	3.50	2.50	2.00	......
31	1.30	......	1.30	......	4.60	......	3.50	2.80	......	2.40	......	......

a To top of ice.

Rating table for St. Mary River at dam site near St. Mary, Mont., for 1903.

Gage height.	Discharge.	Gage height.	Discharge.	Gage height.	Discharge.	Gage height.	Discharge.
Feet.	*Second-feet.*	*Feet.*	*Second-feet.*	*Feet.*	*Second-feet.*	*Feet.*	*Second-feet.*
1.2	37	2.5	525	3.8	1,395	5.1	2,595
1.3	46	2.6	575	3.9	1,480	5.2	2,700
1.4	67	2.7	630	4.0	1,570	5.3	2,810
1.5	97	2.8	690	4.1	1,660	5.4	2,920
1.6	140	2.9	750	4.2	1,750	5.5	3,040
1.7	182	3.0	810	4.3	1,840	5.6	3,160
1.8	225	3.1	880	4.4	1,930	5.7	3,280
1.9	267	3.2	950	4.5	2,020	5.8	3,400
2.0	310	3.3	1,020	4.6	2,110	5.9	3,520
2.1	352	3.4	1,090	4.7	2,200	6.0	3,640
2.2	395	3.5	1,160	4.8	2,295		
2.3	437	3.6	1,235	4.9	2,390		
2.4	480	3.7	1,315	5.0	2,490		

No discharge measurements have been made below 1.50 feet on the gage. Upper part of table is well determined.

Estimated monthly discharge of St. Mary River at dam site near St. Mary, Mont., for 1903.

[Drainage area, 177 square miles.]

Month.	Discharge in second-feet.			Total in acre-feet.	Run-off.	
	Maximum.	Minimum.	Mean.		Second-feet per square mile.	Depth in inches.
January	140	46	66	4,058	0.373	0.430
February............	46	46	46	2,555	.259	.270
March...............	46	37	40	2,460	.226	.351
April................	225	46	106	6,307	.598	.689
May	2,110	225	750	46,116	4.237	4.883
June	3,640	2,110	3,154	187,676	17.819	19.886
July	3,280	1,160	1,832	112,645	10.350	11.934
August	1,160	690	864	53,125	4.881	5.621
September..........	1,160	480	653	38,854	3.689	4.120
October	1,090	480	638	39,229	3.604	4.154
November	480	310	398	23,683	2.248	2.509
December	352	310	321	19,737	1.814	2.065
The year	3,640	37	739	536,445	4.175	56.822

MOUSE RIVER AT MINOT, N. DAK.

This station was established May 5, 1903, by F. E. Weymouth
It is located at the footbridge, 150 feet northwest of the Great North
ern Railroad roundhouse at Minot, N. Dak. The gage is a vertica
1 by 6 inch board, 20 feet long, nailed to a pile of the center pier o
the bridge on the downstream side. It is read once each day b;
H. E. Wheeler. Discharge measurements are made from the down
stream side of the bridge, at which the gage is located. The bridg
makes an angle of 15° with the normal to the direction of the current
which has to be taken into account in computing discharge measure
ments. The initial point for soundings is the zero mark on the down
stream guard rail at the electric-light pole on the right bank. The
channel is straight for 100 feet above and below the station. The
current has a moderate velocity. Both banks are high, covered with
trees and shrubs, and will not overflow. The bed of the stream is
composed of sand and is fairly constant. There are some snags and
brush in the bottom of the channel. Bench mark No. 1 is the top of
the hydrant at the corner of the street, about 150 feet north of the
bridge. Its elevation is 21.83 feet above the zero of the gage. Bench
mark No. 2 is the top of the hydrant one block west of bench mark
No. 1. This elevation is 21.85 feet above the zero of the gage.
Bench mark No. 3 is the top of the rail of the side track at the southwest corner of the roundhouse. Its elevation is 24.18 feet above the
zero of the gage. The top of the rail of the main track of the Great

Northern Railroad station has an elevation of 24.49 feet above the zero of the gage. As determined by connection with railroad levels, the zero of the gage has an elevation of 1,540 feet above sea level.

The observations at this station during 1903 have been made under the direction of E. F. Chandler, district hydrographer.

Discharge measurements of Mouse River at Minot, N. Dak., in 1903.

Date.	Hydrographer.	Gage height.	Discharge.
		Feet.	*Second-feet.*
April 17	C. C. Babb	5.45	336
May 5	F. E. Weymouth	4.85	255
May 30	E. F. Chandler	5.20	359
June 25	....do	4.62	193
September 25	....do	9.70	1,117
October 5	....do	5.48	346

Mean daily gage height, in feet, of Mouse River at Minot, N. Dak., for 1903.

Day.	May.	June.	July.	Aug.	Sept.	Oct.	Nov.	Dec.
1		5.10	4.50	4.90	4.40	6.40	4.50	(a)
2		5.30	4.50	4.80	4.60	6.00	4.50	....
3		5.00	4.50	4.80	4.90	5.80	4.50	....
4		5.40	4.50	4.70	5.70	5.70	4.50	....
5	4.80	5.70	4.50	4.70	5.90	5.60	4.50	....
6	4.80	5.90	4.50	4.60	5.90	5.60	4.50	....
7	4.80	5.90	4.50	4.50	5.50	5.50	4.50	(a)
8	4.70	6.30	4.40	4.50	5.40	5.50	4.40	....
9	4.70	6.40	4.40	4.50	5.70	5.40	4.40	....
10	4.70	6.50	4.40	4.50	6.00	5.40	(a)	....
11	4.60	6.80	4.40	4.40	6.20	5.40	....	....
12	4.60	6.90	5.00	4.40	6.40	5.30	(a)	(a)
13	4.50	5.60	5.00	4.40	7.00	5.30	....	....
14	4.50	5.60	5.00	4.40	7.50	5.20	....	....
15	4.50	5.60	5.00	4.40	7.60	5.20	....	....
16	4.60	5.40	5.00	4.40	7.80	5.00	....	....
17	4.60	5.00	5.00	4.40	8.20	5.00	(a)	(a)
18	4.60	5.00	5.00	4.40	8.90	5.00	....	....
19	4.70	4.90	5.00	4.30	9.00	5.00	....	....
20	4.80	4.90	5.20	4.30	9.40	5.00	....	....
21	4.80	4.90	5.40	4.30	9.60	4.90	....	....
22	5.30	4.90	5.60	4.30	9.60	4.80	(a)	(a)
23	5.30	4.80	5.80	4.30	9.70	4.80	....	....
24	5.30	4.70	5.80	4.30	9.80	4.70	....	....
25	5.50	4.60	5.80	4.30	9.80	4.70	....	....
26	5.50	4.60	5.80	4.40	9.70	4.70	....	....
27	5.30	4.60	5.70	4.40	9.30	4.70	(a)	(a)
28	5.20	4.50	5.60	4.40	8.30	4.60	....	....
29	5.20	4.60	5.50	4.40	7.40	4.60	....	....
30	5.20	4.60	5.20	4.40	6.90	4.60	....	....
31	5.20	....	5.00	4.40	....	4.60	....	....

a Frozen November 10 to December 31, 1903.

Rating table for Mouse River at Minot, N. Dak., from April 17 to December 31, 1

Gage height.	Discharge.	Gage height.	Discharge.	Gage height.	Discharge.	Gage height.	Disch
Feet.	*Second-feet.*	*Feet.*	*Second-feet.*	*Feet.*	*Second-feet.*	*Feet.*	*Secon*
4.2	153	5.5	347	6.8	581	8.2	
4.3	165	5.6	365	6.9	599	8.4	
4.4	178	5.7	383	7.0	617	8.6	
4.5	191	5.8	401	7.1	635	8.8	
4.6	205	5.9	419	7.2	653	9.0	
4.7	219	6.0	437	7.3	671	9.2	1.
4.8	233	6.1	455	7.4	689	9.4	1,
4.9	248	6.2	473	7.5	707	9.6	1.
5.0	263	6.3	491	7.6	725	9.8	1.
5.1	279	6.4	509	7.7	743	10.0	1,
5.2	296	6.5	527	7.8	761		
5.3	313	6.6	545	7.9	779		
5.4	330	6.7	563	8.0	797		

Estimated monthly discharge of Mouse River at Minot, N. Dak., for 1903.

Month.	Discharge in second-feet.			T ac
	Maximum.	Minimum.	Mean.	
May 5–31	347	191	251	
June	599	191	324	
July	401	178	271	
August	248	165	186	
September	1,134	178	695	
October	509	205	293	
November 1–9	191	178	188	

PEMBINA RIVER AT NECHE, N. DAK.

This station was established April 29, 1903, by F. E. Weyr
assisted by E. F. Chandler. At this time a temporary gage w
in. The permanent gage was installed by E. F. Chandler M
1903. It is located at the Great Northern Railroad bridge, two-
of a mile north of the railroad station at Neche, N. Dak. The
near the northeast corner of sec. 36, T. 164 N., R. 54 W. It c
of two sections of 1 by 6 inch plank. The lower section, reading
0 to 5 feet, is driven 2 feet into the bed of the stream, and t
upper end spiked to the bridge abutment timbers on the left
The upper section, reading from 5 to 24 feet, is spiked to the

pier on the right bank. It is read once each day by P. J. Horgan. Discharge measurements are made from the single-span highway bridge, 400 feet below the gage. The initial point for soundings is a point on the downstream hand rail 3 feet from its right end. The bridge crosses the river obliquely. To take this fact into account, the hand rail has been divided into intervals of 10.4 feet. This is equivalent to 10-foot intervals of a cross section normal to the direction of the current. The channel is straight for 100 feet above and below the station. Both banks extend about 20 feet above the zero of the gage. They are not liable to overflow except at exceptionally high stages. The right bank is densely wooded and the left bank is covered with brush. The bed of the stream is composed of sand and of mud, in which there are some sunken snags. The bed may change slightly. About one-third of a mile below the gage there is a loose-rock dam 4 feet high. This raises the water 1 or 2 feet at the bridge, but, as the dam is not tight, the water may fall at low stages. Bench mark No. 1 is the top of the horizontal timber on the upstream side of the right pier, near the gage. It is marked with brass-headed nails, and has an elevation of 23.70 feet above the zero of the gage. Bench mark No. 2 is the west rail of the track at the rail joint at the crossing about 550 feet south of the gage. Its elevation is 26.54 feet above the zero of the gage. Bench mark No. 3 is a spike driven in the north face of the telephone pole inside the fence in the field west of the track and about 550 feet south of the gage. It is about 1½ feet above the ground and has an elevation of 24.28 feet above the zero of the gage. The elevation of the top of the 12 by 12 inch timber, to which the low-water gage is fastened, is 4.93 feet. The elevation of the zero of the gage above sea level, as determined by hand level from the railroad station at Neche, N. Dak., is 815 feet.

The observations at this station during 1903 have been made under the direction of E. F. Chandler, district hydrographer.

Discharge measurements of Pembina River at Neche, N. Dak., in 1903.

Date	Hydrographer.	Gage height.	Discharge.
		Feet.	*Second-feet.*
April 29	F. C. Weymouth	4.06	240
May 18	E. F. Chandler	3.59	166
July 8	....do	2.78	58
September 28	....do	2.76	a 27
November 9	....do	4.31	b 46

a Taken while dam below was being changed and rebuilt. Gage height of no value.
b Taken after dam was rebuilt into its permanent form; should be the same next season.

Mean daily gage height, in feet, of Pembina River at Neche, N. Dak., for 1903.

Day.	May.	June.	July.	Aug.	Sept.	Oct.	Nov.	Dec.
1	4.00	3.80	3.20	2.60			4.30	(c)
2	4.00	3.80	3.10	2.60			4.30	
3	(b)	3.80	3.10	2.60			4.30	
4		3.70	3.10	2.60			4.30	
5		3.70	3.10	2.60			4.30	
6		3.70	3.10	2.60			4.20	
7		3.70	3.10	2.60			4.20	(c)
8		3.70	2.90	2.60			4.20	
9		3.60	2.90	(c)			4.30	
10		3.50	2.80				4.30	
11		3.50	2.80			4.30	4.20	
12		3.50	2.80			4.30	4.20	(c)
13		3.50	2.80			4.30	4.10	
14		3.50	2.80			4.30	4.10	
15		3.50	2.80			4.30	(c)	
16		3.40	2.70			4.30		
17		3.40	2.70			4.30		(c)
18	3.60	3.40	2.70			4.30		
19	3.60	3.40	2.70			4.30		
20	3.70	3.40	2.70			4.30		
21	3.80	3.30	2.70			4.30		
22	3.80	3.30	2.70			4.30	(c)	(c)
23	3.80	3.30	2.70			4.30		
24	3.90	3.20	2.70			4.30		
25	3.90	3.20	2.70			4.30		
26	4.00	3.20	2.60			4.20		
27	3.90	3.30	2.60			4.25	(c)	(c)
28	3.90	3.30	2.60			4.20		
29	3.90	3.30	2.60			4.20		
30	3.90	3.30	2.60			4.20		
31	3.80		2.60			4.20		

a Frozen November 15 to December 31.

b No readings made May 3 to 17.

c No readings from August 9 to October 1 on account of rebuilding dam.

Rating table for Pembina River at Neche, N. Dak., from April 29 to August 8, 1903.

Gage height.	Discharge.	Gage height.	Discharge.	Gage height.	Discharge.	Gage height.	Discharge.
Feet.	Second-feet.	Feet.	Second-feet.	Feet.	Second-feet.	Feet.	Second-feet.
2.6	35	3.0	84	3.4	138	3.8	198
2.7	47	3.1	97	3.5	152	3.9	214
2.8	59	3.2	110	3.6	167	4.0	230
2.9	71	3.3	124	3.7	182		

Rating table for Pembina River at Neche, N. Dak., from October 11 to November 14, 1903.

Gage height.	Discharge.	Gage height.	Discharge.	Gage height.	Discharge.
Feet.	Second-feet.	Feet.	Second-feet.	Feet.	Second-feet.
4.1	21	4.2	32	4.3	45

Estimated monthly discharge of Pembina River at Neche, N. Dak., for 1903.

Month.	Discharge in second-feet.			Total in acre-feet.
	Maximum.	Minimum.	Mean.	
May a			202	12,420
June	198	110	149	8,866
July	110	35	60	3,689
August 1 to 8			35	555
September b				
October 11 to 31..			42	1,749
November 1 to 14			42	1,156

a May 3 to 17 estimated.   b No readings from August 9 to October 10.

RED LAKE RIVER AT CROOKSTON, MINN.

This station was established May 19, 1901, by C. M. Hall. It is located at the bridge which connects Robert and St. Paul streets, and which is known as the Sampson addition bridge. It is about one-sixth mile west of the Great Northern Railroad station. The low-water vertical gage is a 1 by 6 inch board fastened to the piling of the left abutment under the bridge. It reads from 0 to 9.6 feet. The high-water vertical gage is attached to the piling of the pier at the right end of the bridge. It reads from 7.5 to 15 feet. There is also a wire gage near the middle of the single-span bridge. Its horizontal scale reads from 3 to 19 feet. The gage is read once each day by J. E. Carroll, the city engineer. Discharge measurements are made from the lower side of the single-span bridge at which the gages are located. The initial point for soundings is the post of the hand rail at the left end of the bridge at a point where the diagonal member of the bridge truss meets the floor timbers. The channel is straight for 250 feet above and 200 feet below the station. The current is swift. The right bank is low and is covered with trees and brush. It is liable to overflow at very high stages, but there would be little current in the flooded section, owing to the trees and brush. The left bank is high, wooded, and not liable to overflow. The bed of the stream is sandy, free from vegetation, and shifting. The dam and power house for the city waterworks are located about 1,000 feet above the gage. The opening and closing of the sluices cause some variation in the flow, but the gage readings represent a close average of the daily river height. Bench mark No. 1 is the top of the hydrant at the corner of St. Paul and Robert streets, 30 feet south of the bridge. Its elevation is 24.23 feet above gage datum. Bench mark No. 2 is the top of a hydrant 00 feet north of the bridge at the corner of St. Paul and Woodlawn

streets.   Its elevation is 27.46 feet above the zero of the gage. Th
city datum of Crookston has an elevation of 4.10 feet above ga
datum.   The top of the rail of the main track at the Great Northei
Railroad station at Crookston has an elevation of 37.63 feet above ga
datum.   The gage datum has an elevation of 825 feet above sea leve

The observations at this station during 1903 have been made und
the direction of E. F. Chandler, district hydrographer.

Discharge measurements of Red Lake River at Crookston, Minn., in 1903.

Date.	Hydrographer.	Gage height.	Discharg
		Feet.	*Second-fe*
February 16	W. R. Hoag	....	6
April 28	C. C. Babb	8.80	4,0
May 25	E. F. Chandler	8.30	3,0
June 8	.....do	6.77	2,4
June 22	.....do	5.84	1,0
July 18	.....do	4.49	
Do	.....do	5.30	1,
October 12	.....do	7.02	2,
November 16	.....do	4.12	

Mean daily gage height, in feet, of Red Lake River at Crookston, Minn., for 1903.

Day.	May.	June.	July.	Aug.	Sept.	Oct.	Nov.	De
1	....	7.60	4.90	4.55	5.10	4.30	6.10	
2	....	7.30	4.70	4.35	5.00	4.55	5.80	
3		7.20	4.60	4.40	4.50	4.80	5.80	
4	....	7.20	4.70	4.45	5.30	4.50	6.35	
5		7.15	4.90	4.30	5.50	5.30	5.80	
6		7.05	5.40	4.15	5.00	5.50	5.70	
7		7.05	5.05	4.20	5.30	6.25	5.60	
8		6.85	4.95	4.25	5.00	6.65	6.15	
9		6.45	4.80	4.25	5.60	6.80	5.70	
10		6.45	4.50	4.25	5.50	6.90	5.65	
11		6.40	4.50	4.10	5.90	6.90	5.65	
12		6.30	5.10	4.00	....	7.00	5.55	
13		6.10	5.05	4.05	6.10	7.00	5.20	
14		6.20	5.30	4.00	6.90	7.00	4.80	
15		6.00	5.15	3.95	5.85	6.95	a 4.50	
16		5.75	4.25	3.85	5.80	6.95	4.10	
17		5.75	4.40	3.95	5.60	6.90	4.40	
18		5.70	5.10	3.95	5.50	6.55	4.05	
19		5.45	4.90	3.70	5.90	6.85	5.35	
20		5.55	4.85	3.95	6.30	6.10	5.10	
21		5.40	4.90	3.65	6.00	6.40	5.10	
22		5.35	4.85	3.95	5.70	6.80	4.90	
23		5.40	4.75	3.50	5.35	6.45	4.85	

a Margin of river frozen, channel continuously narrowing till December 31.

Mean daily gage height, in feet, of Red Lake River at Crookston, Minn., for 1903—
Continued.

Day.	May.	June.	July.	Aug.	Sept.	Oct.	Nov.	Dec.
14.	....	5.50	4.40	3.95	5.10	6.30	4.35	6.00
15.	8.45	5.30	4.95	4.05	5.15	6.95	3.75	6.10
16.	8.25	5.05	4.70	3.65	4.90	6.10	4.65	6.10
17.	8.40	5.50	4.50	4.65	5.10	6.45	5.95	6.10
18.	8.25	5.10	4.45	4.65	5.10	5.90	5.95	6.10
19.	7.95	4.90	4.85	5.00	4.60	6.45	5.90	6.00
20.	7.80	5.05	4.55	4.30	4.72	5.90	6.10	6.30
21.	7.60	....	4.60	4.90	....	6.00	....	a 6.10

a Frozen over.

Rating table for Red Lake River at Crookston, Minn., from May 19, 1901, to December 31, 1903.

Gage height.	Discharge.	Gage height.	Discharge.	Gage height.	Discharge.	Gage height.	Discharge.
Feet.	*Second-feet.*	*Feet.*	*Second-feet.*	*Feet.*	*Second-feet.*	*Feet.*	*Second-feet.*
3.6	430	4.9	1,050	6.2	1,880	7.8	3,105
3.7	463	5.0	1,110	6.3	1,950	8.0	3,285
3.8	500	5.1	1,170	6.4	2,020	8.2	3,465
3.9	540	5.2	1,230	6.5	2,090	8.4	3,645
4.0	585	5.3	1,290	6.6	2,160	8.6	3,825
4.1	635	5.4	1,350	6.7	2,230	8.8	4,005
4.2	685	5.5	1,410	6.8	2,300	9.0	4,185
4.3	735	5.6	1,470	6.9	2,380	9.2	4,370
4.4	785	5.7	1,530	7.0	2,460	9.4	4,570
4.5	835	5.8	1,600	7.2	2,620	9.6	4,770
4.6	885	5.9	1,670	7.4	2,780	9.8	4,970
4.7	935	6.0	1,740	7.6	2,940	10.0	5,170
4.8	990	6.1	1,810				

Table not well established below 4.3 feet gage height.

Estimated monthly discharge of Red Lake River at Crookston, Minn., for 1902-

| Month. | Discharge in second-feet. | | | Tot acre |
	Maximum.	Minimum.	Mean.	
1902.				
March 13–31 a	5,020	2,780	3,516	13
April...	3,330	1,200	2,188	13
May	5,170	3,420	4,120	23
June	4,970	2,230	3,901	23
July b	2,460	885	1,916	11
August	1,530	735	1,165	7
September..............................	1,170	910	1,028	6
October	1,350	1,050	1,181	7
November c	1,470	1,080	1,346	8
1903.				
May 25–31	....	....	3,384	4
June	2,940	1,050	1,815	10
July	1,350	710	995	6
August	1,110	401	671	4
September d	2,380	835	1,388	8
October	2,400	735	1,935	11
November e	1,985	430	1,273	7
December e	1,870	1,290	1,489	9

a March 16 estimated.
b July 9, 10, and 11 estimated.
c November 30 estimated.

d September 12 estimated.
e Estimates for November and December
corrected to allow for the effect of ice.

RED RIVER AT GRAND FORKS, N. DAK.

This station was established May 26, 1901, by C. M. Hall.
located at the Northern Pacific Railroad bridge at Grand Fork:
Dak. The original gage is a vertical 1 by 8 inch board attached to
north end of the breakwater of the middle bridge pier. The zer
this gage was placed 5 feet below the zero of the U. S. Army engin
gage, which is attached to the same breakwater. A standard
gage has been established with the same datum as the vertical
and is attached to the downstream side of the bridge. The lengt
the chain from the end of the weight to the marker is 50.44 feet.
gage is read once each day by Philip Hayes. Discharge meas
ments are made from the Great Northern Railroad bridge, about
fifth mile above the gage. A highway bridge crosses the river bet
the two railroad bridges. The initial point for soundings is ma
in red paint on the downstream guard rail at the left end of the
span. The channel is straight for 500 feet above and for 150

ѵ the Great Northern Railroad bridge. The current has a mod-
velocity. The right bank is liable to overflow at high stages
s wooded. The left bank will overflow only at very high stages
or a short distance. The water at all stages will pass beneath the
n bridge and its trestle approaches. The bed of the stream is
osed of sand and mud and is subject to some change. The water
ially heavily laden with sediment from Red Lake River, which
s one-half mile above the station. Bench mark No. 1 is the north
er of the iron plates of the turntable near the center of the middle
of the Northern Pacific Railroad bridge. The bench mark is on
orth side, about 1 inch above the surface of the stone pier. Its
tion is 43.95 feet above gage datum. Bench mark No. 2 is a
in a telegraph pole in the lumber yard southwest of the left end
e bridge, from which it is 200 feet distant. Its elevation is 48.50
above gage datum. The top of the pulley of the chain gage is
l feet above gage datum. Gage datum is 45.58 feet above the
datum of Grand Forks and is 777.9 feet above sea level.
e observations at this station during 1903 have been made under
irection of E. F. Chandler, district hydrographer.

Discharge measurements of Red River at Grand Forks, N. Dak., in 1903.

Date.	Hydrographer.	Gage height.	Discharge.
		Feet.	*Second-feet.*
15	C. C. Babb	23.20	14,009
27	do	14.00	6,223
14	E. F. Chandler	11.91	4,783
9	do	10.68	4,005
18	do	8.74	2,878
7	do	6.58	1,695
st 1	do	5.52	1,134
st 4	do	5.87	1,189
er 29	do	8.76	2,648

26									5.20	6 80		....	
27				13.75					5.20	6 80		(a)	(a)
28				13.90	٩.	7.20	5.45	5.25	6.60			....
29				13.75	15 25	7.40	5.95	5 75	6 50			....
30			(a)	13.55	14.60	7.20	5.45	5 80	6.85		8.00	......
31	7.85		10.50		14.10		5.60	5.95		8.70		8 05

a Frozen January 1 to March 31 and from November 16 to December 31.

g table for Red River at Grand Forks, N. Dak., from September 1, 1902, to December 31, 1903.

e it.	Discharge.	Gage height.	Discharge.	Gage height.	Discharge.	Gage height.	Discharge.
.	*Second-feet.*	*Feet.*	*Second-feet.*	*Feet.*	*Second-feet.*	*Feet.*	*Second-feet.*
)	870	6.9	1,775	10.6	3,935	19.5	10,650
1	910	7.0	1,830	10.8	4,065	20.0	11,090
2	955	7.2	1,940	11.0	4,195	20.5	11,540
3	1,000	7.4	2,050	11.5	4,520	21.0	11,990
4	1,045	7.6	2,160	12.0	4,845	21.5	12,440
5	1,090	7.8	2,270	12.5	5,170	22.0	12,900
6	1,135	8.0	2,380	13.0	5,520	22.5	13,375
7	1,180	8.2	2,490	13.5	5,870	23.0	13,850
8	1,225	8.4	2,600	14.0	6,220	23.5	14,325
9	1,270	8.6	2,715	14.5	6,575	24.0	14,800
)	1,320	8.8	2,835	15.0	6,950	24.5	15,275
1	1,370	9.0	2,955	15.5	7,325	25.0	15,760
2	1,420	9.2	3,075	16.0	7,720	25.5	16,260
3	1,470	9.4	3,195	16.5	8,120	26.0	16,760
4	1,520	9.6	3,315	17.0	8,525	26.5	17,260
5	1,570	9.8	3,435	17.5	8,950	27.0	17,770
6	1,620	10.0	3,555	18.0	9,375	27.5	18,295
7	1,670	10.2	3,675	18.5	9,800	28.0	18,820
8	1,720	10.4	3,805	19.0	10,225		

Estimated monthly discharge of Red River at Grand Forks, N. Dak., for 19

Month.	Discharge in second-feet.			T ac
	Maximum.	Minimum.	Mean.	
1902.				
September	1,995	1,470	1,703	
October	2,655	1,320	1,637	
November	3,075	1,495	2,544	
1903.				
January a			1,600	
February a			1,420	
March a			2,100	
April	18,767	4,455	10,626	
May	7,137	4,260	5,388	
June	5,870	1,940	3,342	
July	2,105	1,067	1,443	
August	1,295	870	1,050	
September	2,655	1,180	1,891	
October	3,935	1,395	2,882	
November a			2,204	
December a			1,960	
The year			2,997	2,

a January 1 to March 31 and November 16 to December 31, inclusive, river frozen ove observations not made, but discharge estimated from occasional observations.

SHEYENNE RIVER NEAR HAGGART, N. DAK.

This station was established March 22, 1902, by Charles M It is located near the way station of Haggart, on the Northern Railway, 5 miles west of Fargo, N. Dak., at a private wagon about one fourth of a mile north of the railroad. The gage is a cal 1 by 6-inch board fastened to the piling pier at the middle bridge, and reads up to 17 feet. It is read once each day by J Haggart.

Discharge measurements are made from the same bridge. The point for soundings is the end of the hand rail on the lower s the bridge, right bank. The channel is straight for 30 feet abo 200 feet below the station and the current moderate. Both bar steep and not liable to overflow, except in unusual floods. Th a single channel at all stages and its bed is of clay and only sl shifting.

The observations at this station during 1903 have been made the direction of E. F. Chandler, district hydrographer.

Discharge measurements of Sheyenne River near Haggart, N. Dak., in 1903.

Date.	Hydrographer.	Gage height.	Discharge.
		Feet.	*Second-feet.*
April 13	C. C. Babb	14.07	1,477
May 9	F. E. Weymouth	5.32	310
June 3	D. E. Willard	5.50	266
July 11	do	3.70	77
July 30	E. F. Chandler	3.76	94
October 24	do	4.00	77

Mean daily gage height, in feet, of Sheyenne River near Haggart, N. Dak., for 1903.

Day.	April.	May.	June.	July.	Aug.	Sept.	Oct.	Nov.	Dec.
1		6.20	5.60	3.90	3.30	2.60	4.00	3.90	(a)
2		6.00	5.60	3.50	3.60	3.60	3.90	3.90	
3		5.80	5.50	3.90	3.30	3.50	4.00	3.80	
4		5.70	5.30	3.80	3.20	3.50	4.10	3.60	
5		5.60	5.20	3.80	3.00	3.40	4.20	3.90	
6		4.90	5.20	3.60	3.20	3.50	4.00	4.00	
7		5.80	5.00	3.70	3.30	3.50	4.20	4.00	
8	11.50	5.80	4.80	3.90	3.50	3.70	4.20	4.00	
9	12.60	5.50	4.50	3.80	3.40	3.80	4.10	4.00	
10	14.20	5.20	4.80	3.90	3.00	3.80	4.10	3.70	
11	14.70	4.90	4.70	3.70	3.30	3.90	3.90	4.20	
12	14.50	4.90	4.70	3.70	3.30	3.90	4.00	4.10	(a)
13	14.00	5.00	4.60	3.80	3.10	4.00	3.70	a 4.40	
14	13.80	5.00	4.50	3.40	3.50	4.10	3.50	a 4.00	
15	13.00	4.80	4.40	3.40	3.40	3.90	3.80	(a)	
16	12.60	4.70	4.10	3.60	3.10	3.50	3.90	(a)	
17	11.80	4.50	4.20	3.80	2.70	3.50	4.00	(a)	
18	11.20	4.50	4.30	3.70	2.60	3.80	4.10		
19	10.80	4.50	4.20	3.70	3.10	3.90	4.00		
20	10.40	4.60	4.10	3.80	3.60	3.80	3.80		
21	10.00	4.60	1.10	3.30	3.50	3.80	4.00		
22	9.60	5.20	4.10	3.60	3.60	3.90	4.10	(a)	(a)
23	8.90	6.80	4.00	3.80	3.40	4.10	4.00		
24	8.60	8.20	4.10	3.80	3.30	4.10	4.00		
25	8.10	9.40	4.10	3.50	3.20	4.00	4.00		
26	7.60	11.50	4.00	3.80	3.00	3.90	3.90		
27	7.30	10.20	4.00	3.90	3.10	3.90	3.80	(a)	
28	7.00	8.90	4.00	3.80	3.30	3.90	3.40		
29	6.70	7.70	3.90	3.70	3.40	3.80	3.90		
30	6.40	6.90	3.80	3.50	3.50	3.40	3.90		
31		6.30		4.40	3.50		4.00		

(a) Frozen November 13 to December 31.

March 29–31	1,204	840	994	8
April	2,030	616	1,492	8
May	644	299	535	3
June	448	124	262	1
July	158	67	106	6
August	113	53	82	
September	113	67	84	
October	146	53	92	
November 1–8	103	93	97	1
1903.				
April 8–30	1,568	406	1,003	45
May	1,120	158	369	22
June	299	84	163	9
July	146	47	80	4
August	67	19	46	2
September	113	53	82	4,
October	124	53	99	6,
November 1–14	146	67	100	2,

RED RIVER AT FARGO, N. DAK.

This station was established by C. M. Hall May 27, 1901, and is
:ated at the bridge connecting Front street, Fargo, N. Dak., and
ain street, Moorhead, Minn. The gage consists of a 1 by 8 inch
ard, painted white, graduated to feet and tenths in black, and
:ached to the east side of the breakwater for the center pier of the
idge. The zero of the gage is 44.45 feet below the top of the plank
lk of the bridge over the gage and is 860.9 feet above sea level, the
vation having been determined by leveling from top of rail of
rthern Pacific station, Moorhead, Minn. The danger line is at 26.5
t. Above the station the river curves to the west. The west bank
high and steep; the east bank low and subject to overflow at times
high water. Measurements are made from the bridge. The river
d consists of soft mud. The observer is H. W. Grasse, United
ates Weather Bureau, Moorhead, Minn.
The observations at this station during 1903 have been made under
e direction of E. F. Chandler, district hydrographer.

Discharge measurements of Red River at Fargo, N. Dak., for 1903.

Date.	Hydrographer.	Gage height.	Discharge.
		Feet.	*Second-feet.*
il 14	C. C. Babb	9.97	1,014
il 25	do	8.88	709
9	F. E. Weymouth	8.35	636
ə 3	D. E. Willard	8.30	574
11	do	7.80	330
30	E. F. Chandler	7.20	276
Do	do	7.19	254
·ber 24	do	8.09	471
Do	do	8.08	461

Mean daily gage height, in feet, of Red River at Fargo, N. Dak., for 1903.

Day.	Mar.	Apr.	May.	June.	July.	Aug.	Sept.	Oct.	Nov.	Dec.
		12.70	8.60	8.40	7.80	7.20	7.40	7.50	8.10	(a)
		13.30	8.60	8.40	7.90	7.20	7.40	7.50	8.10	(a)
		13.50	8.50	8.30	.90	7.20	7.40	7.60	8.10	(a)
		13.70	8.50	8.30	7.90	7.20	7.40	7.60	8.10	(a)
		13.80	8.50	8.30	7.90	7.30	7.40	7.70	8.10	(a)
		13.90	8.50	8.30	8.00	7.30	7.40	7.80	8.10	(a)
		13.00	8.50	8.20	8.00	7.30	7.30	8.00	8.10	(a)
		12.00	8.40	8.20	8.00	7.30	7.30	8.10	8.10	(a)
		12.20	8.40	8.10	7.90	7.30	7.20	8.20	8.10	(a)

a Frozen.

Gage height.	Discharge.	Gage height.	Discharge.	Gage height.	Discharge.	Gage height.	Discharge.
Feet.	Second-feet.	Feet.	Second-feet.	Feet.	Second-feet.	Feet.	Second-feet.
7.0	225	8.3	513	9.6	880	11.6	1,560
7.1	244	8.4	538	9.7	910	11.8	1,630
7.2	263	8.5	564	9.8	940	12.0	1,700
7.3	283	8.6	590	9.9	970	12.2	1,770
7.4	304	8.7	616	10.0	1,000	12.4	1,850
7.5	325	8.8	643	10.2	1,070	12.6	1,930
7.6	347	8.9	671	10.4	1,140	12.8	2,010
7.7	369	9.0	700	10.6	1,210	13.0	2,090
7.8	392	9.1	730	10.8	1,280	13.2	2,170
7.9	415	9.2	760	11.0	1,350	13.4	2,250
8.0	439	9.3	790	11.2	1,420	13.8	2,410
8.1	463	9.4	820	11.4	1,490	14.0	2,490
8.2	488	9.5	850				

Curve not well established above gage height 11 feet.

monthly discharge of Red River at Fargo, N. Dak., for 1902 and 1903.

Month.	Discharge in second-feet.			Total in acre-feet.
	Maximum.	Minimum.	Mean.	
1902.				
28	616	347	497	3,943
....	1,105	513	737	45,316
....	700	392	469	27,907
....	1,175	439	733	45,070
....	1,035	730	904	53,792
....	850	463	676	41,566
....	564	369	463	28,469
....	369	283	328	19,517
....	392	207	275	16,909
....	392	172	298	17,732
2..........................	189	189	189	750
1903.				
....	1,560	225	965	9,570
....	2,450	564	1,260	74,975
....	643	513	566	34,802
....	538	392	460	27,372
....	439	283	363	22,320
....	325	225	268	16,479
....	369	263	317	18,863
....	564	325	471	28,961
-26, inclusive)............	590	439	474	24,444

June 18	.....do ....do ....do	
July 7	.....do ....do ....do	1,301
Aug. 4	.....do ....do ....do	913
June 18	.....do .... Red River of the North.	Above the forks, ... Grand Forks, N. Dak.	1,164

INDEX.

505

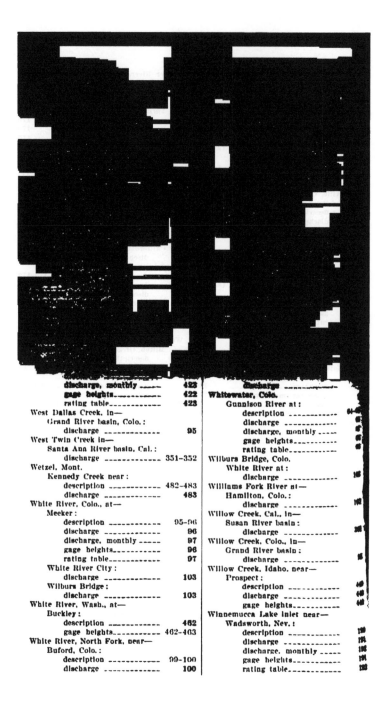

O

Supply and Irrigation Paper No. 101

Series O, Underground Waters, 23

DEPARTMENT OF THE INTERIOR

UNITED STATES GEOLOGICAL SURVEY

CHARLES D. WALCOTT, Director

UNDERGROUND WATERS

OF

)UTHERN LOUISIANA

BY

GILBERT DENNISON HARRIS

WITH DISCUSSIONS OF THEIR USES FOR WATER SUPPLIES
AND FOR RICE IRRIGATION

BY

M. L. FULLER

WASHINGTON

GOVERNMENT PRINTING OFFICE

1904

CONTENTS.

4 CONTENTS.

ILLUSTRATIONS.

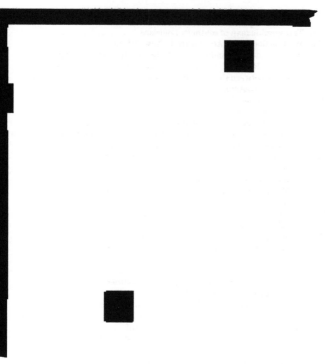

LETTER OF TRANSMITTAL.

DEPARTMENT OF THE INTERIOR,
UNITED STATES GEOLOGICAL SURVEY,
HYDROGRAPHIC BRANCH,
Washington, D. C., December 4, 1903.

SIR: I have the honor to transmit herewith a manuscript by Prof.
D. Harris on the "Underground Waters of Southern Louisiana,"
which has been added short discussions of certain economic features,
luding the uses of underground waters for water supplies and for
e irrigation, by Mr. M. L. Fuller. The part by Professor Harris
n elaboration of a portion of an earlier paper published in the
orts of the geological survey of Louisiana, and by means of its
scriptions and illustrations brings out clearly the nature of the
urrence and the importance of the underground water resources of
region considered. It is believed that the discussion of water sup-
es, by reiterating the importance of pure sources, will hasten their
roduction. The irrigation of rice, though yet in its earlier stages,
s already increased tenfold the value of land over large areas, and
publication of information which will in any way call attention to
importance of underground waters in its development will be of
nsiderable value. I would recommend that the report be published
the series of Water-Supply and Irrigation Papers.

Very respectfully,

F. H. NEWELL, *Chief Engineer.*

Hon. CHARLES D. WALCOTT,
Director, United States Geological Survey.

9

DERGROUND WATERS OF SOUTHERN LOUISIANA.

By Gilbert D. Harris.

PREFATORY REMARKS.

the writer's studies of the geology of southern Louisiana during
ast three years, opportunities have presented themselves for col-
ng data relating to the underground waters of this section of the
:. A brief summary of the data so collected was given in Part
f the report of the State geological survey for the year 1902, in
ial Report No. 6, "The Subterranean Waters of Louisiana."
e the publication of this work one winter season has been spent
outhern Louisiana in general geological work, and one month
e 20 to July 20, 1903) has been devoted to field work bearing
usively on water supplies. This report, therefore, may be con-
red as an enlarged and revised edition of the special report named,
d, in large measure, on facts gathered by the writer while he
employed by the State of Louisiana. After this explanation it
is scarcely necessary to use quotation marks or to give precise
rences in every case where facts have been taken from the earlier
rt.

many instances the height above tide (mean sea level) of stations
g the Southern Pacific Railroad will be found to vary as much as
et in the two reports. This is due to the fact that early eleva-
s furnished by this road to the United States Geological Survey,
published in Bulletin 160, were different from those now posted on
stations along the line throughout southern Louisiana (see Pl. I).

ORIGIN OF ARTESIAN AND DEEP-WELL WATERS IN SOUTH-ERN LOUISIANA.

PRECIPITATION.

Last year's Weather Bureau report gives the following figures regarding precipitation at several stations in southern Louisiana:[a]

Precipitation at stations in southern Louisiana.

Station.	1902.	Aver'g.
Alexandria	45. 34	53. 95
Amite	41. 44	60. 41
Cheneyville	40. 74	53. 13
Clinton	52. 29	55. 01
Hammond	47. 01	58. 13
Lafayette	36. 36	53. 44
Lake Charles	41. 19	54. 94
Opelousas	39. 77	54. 64
Sugartown	48. 12	54. 55

From this it appears that the average annual precipitation in this part of the State is about 55 inches. This means that each acre of land receives more than double enough rain water to irrigate it properly if planted in rice. But much of this water is lost, so far as agricultural purposes are concerned, by flowing away in surface streams to the Gulf. Much, too, that descends into the soil and lower strata of the earth, doubtless leaches out into the Gulf underground. Unfortunately for our present study, the main local streams of southern Louisiana have never been gaged, and consequently the amount of water that reaches the sea, even by surface streams, is not known. The extent, therefore, to which the total amount of rainfall may be utilized as deep-well water can not at present be even approximately estimated. That much rain water is absorbed and transported to distant places through underground porous layers is evident from the existence of many satisfactory deep and artesian wells throughout the southernmost parishes of the State. Yet it is often held that the supply of deep waters may be derived from large bodies of neighboring water—for example, from lakes and rivers and small streams that have a greater altitude than the surface of the water in the deep wells. This may, indeed, be the case in a region in which there are limestone formations, or in a region where the gradient of the streams is considerable and erosion is scouring and cleaning the sides and bottom of the channels and where practically no silt is being deposited, but in

[a] U. S. Dept. Agr., Ann. Summary, 1902, Louisiana Section, Weather Bureau Office, New Orleans, La.

uisiana none of these conditions exist, so far as the larger streams
1 other large bodies of water are concerned.   However, we will
nsider with all necessary detail two of the common theories advanced
 account for the presence of water in such apparently immense
antities beneath the surface in southern Louisiana.

GULF WATER AS A SOURCE OF SUPPLY OF DEEP WELLS.

It is frequently asserted that the continuance of southerly winds or
igh tides causes an appreciable rise in the level of the water in wells
ot far from the coast; that when wells are vigorously pumped the
ater level descends below tide; that therefore there is an intimate
onnection between the waters of the Gulf and those encountered so
oundantly in deep wells.

That there is more or less connection between the fresh water under
e ground and the salt water of the Gulf there can be no doubt.   A
ariation in the height of the water in a few wells coincident with that
 a neighboring body of water in which there is a perceptible tide
as long ago recorded by members of the Louisiana State geological
rvey and others.   That there is no underground current from the
ulf landward is evident from the facts (1) that when pumping ceases
r a few hours the water level in the wells quickly rises above tide,
d (2) that any water derived from the Gulf would possess a salti-
ess that has not thus far been recorded in any deep irrigation well.
ny impediment tending to retard the escape of the underground
aters Gulfward, as the weight of water collected from long-continued
avy showers or the backing up of the Gulf's waters from the south,
ll necessarily raise the level of the water in the deep wells or cause
e artesian wells to flow more strongly.

RIVER WATERS AS A SOURCE OF SUPPLY.

In 1860 Raymond Thomassy[a] published his Géologie pratique de
Louisiane.   He seems to have been greatly captivated with the
ea that a large amount of the water flowing into the Mississippi from
 various tributaries never reaches the Gulf by surface streams, but
absorbed by the pervious layers that form the banks and bottom of
e. river, and is carried thence through underground passages and
rous layers to the Gulf coast, or beneath its waters.

Thomassy was of course not aware of the great possibilities of irri-
tion in southern Louisiana, but had he lived to see hundreds of 10
 12 inch wells yielding almost rivers of deep, cool water, he would
ubtless have felt that his absorption theory was at last fully proved,
e whence could all this underground water come?

Géologie pratique de la Louisiane, par R. Thomassy (accompagné de 6 planches), chez l'auteur,
ı Nouvelle-Orléans et à Paris, 1860.

No definite statements can be made regarding the amount of water furnished by the rivers of Louisiana to the general underground supply until the topography and stratigraphy have been determined in detail. Yet it may be shown here that the oft-repeated popular statement that waters of the Mississippi River supply the wells in southern Louisiana is but partly, if at all, correct. Certainly no "veritable river" is leaving the Mississippi in its lower reaches to force its way laterally for long distances underground. The process of transferring discharge measurements from one point on the river to another, as employed by Humphreys and Abbot[a] in their delta survey, has shown that the difference in discharge at two stations at equal stages in the river is due to increment of water from tributaries and loss in distributary bayous and crevasses between the two places. Daily discharge measurements made at Vicksburg were compared with discharge measurements made at stations up and down the river, and these agreed in a remarkable way.

In other words, there is no difference in the amounts of discharge at Vicksburg and Carrollton, for example, that can not be explained by taking into account the difference between water received and that given up by surface channels. The absorption, therefore, of the Mississippi's waters by underground porous layers is a subject that is of no importance in the present report.

The impropriety of assuming that variations in "head" noticed in deep wells located at any considerable distance from the Mississippi are due to difference in the stage or height of the river, is evident from facts presented farther on in this report. It is fortunate that the measurements of well stages here recorded were made mostly in the spring of 1901, especially in April and May. The wells showed a slight temporary rise about April 22, due to local showers, but thereafter the usual marked decline for the summer went steadily on. Not so the river; it gradually rose till it reached the highest point of the season on the dates which follow, at the localities designated:[b] May 16, Vicksburg, Miss.; May 15–16, St. Joseph, La.; May 16–17, Natchez; May 17, Red River Landing; May 17, Bayou Sara; May 17, Baton Rouge; May 16, Plaquemine; May 19, Donaldsonville; May 15, College Point; May 17 and 20, Carrollton. After these dates, at the stations named, the river began to decline.

The cross sections presented in figs. 8 and 9 (pp. 29, 30) show clearly the behavior of deep waters in the vicinity of large stream channels. There is therefore reason to suppose that the Mississippi and other large streams serve as drains on the underground-water supply rather than as feeders.

[a] Report upon the physics and hydraulics of the Mississippi River; upon the protection of the alluvial region from overflow, etc.: Professional Paper No. 13, Corps of Engineers, U. S. Army, 1861. See reprint of 1876, pp. 260, 358–363.

[b] Stages of the Mississippi, etc.: Miss. Riv. Comm., 1901, St. Louis, Mo., Mississippi River Commission Print, 1902.

TOPOGRAPHY OF SOUTHERN LOUISIANA.

Since the cause of flow of underground waters must be due mainly
the action of gravity, it follows that the surface features of the land
e a marked influence on the rate of underground as well as of over-
und flow.   Southern Louisiana has only just begun to cooperate
h the General Government in the construction of detailed topo-
phic maps, so it is not possible to show the surface features as well
could be desired; yet private individuals, corporations (such as
lroad and canal companies), United States engineers, and members
the State geological survey have done a large amount of spirit
eling throughout the area, and from such data it has been found

Fig. 1.—Map showing topographic subdivisions of southern Louisiana.

ssible to compile a small-scale contour map (Pl. I) and a still smaller
lex map (fig. 1) to the topography of this part of the State.

TOPOGRAPHIC SUBDIVISIONS.

SWAMP-LAKE AREA.

To this subdivision may be assigned in general that portion of the
te having an elevation above tide of less than 20 feet (see fig. 1).
size is surprisingly great when compared with that of the more
vated areas.   Pl. I represents an area in Louisiana, exclusive of
ge lakes, bays, etc., covering 28,900 square miles, of which 15,800
· below the 20-foot contour.   The Five Islands in Iberia and St.
ry parishes are the only areas furnishing what might be called
able relief in this subdivision of southern Louisiana.   One "island"
·s to a height of 150 feet above the surrounding marsh land; others
but two-thirds or half as high.   Since, however, the diameter of
largest is only approximately 2 miles, their total area is extremely
ignificant when compared with the vast extent of low land shown

on the map.   Southern Cameron and Vermilion parishes contain extensive swamp tracts that lie several miles back from the Gulf border, but close to the Gulf there are several remarkably persistent dry, sandy ridges that rise from 5 to 10 feet above mean tide (see Pl. 1).   In the swampy areas there are several broad and very shallow lakes or bays, as may be seen by consulting the same plate.   They rarely show a depth of more than 15 or 20 feet, usually much less.   The bayous and rivers, however, have cut very deep channels through these lowlands.   Depths of 30 to 40 feet are by no means unusual, while the Mississippi has long stretches of channel that range in depth from 72 to 90 feet, and occasional pools 200 feet deep.   The manner in which the ground slopes above and below Gulf level, the basin-like character of the lakes, and the deepness of the river channels are typically shown in fig. 2.

The topography of the region lying between Lake Pontchartrain and the Atchafalaya River—the so-called delta region of the Mississippi—deserves a few additional remarks.

Large areas in this tract are scarcely above sea level. The figures shown on Pl. 1, along the Southern Pacific Railroad from New Orleans to Morgan City, indicate feet above tide.   Here, as in all mature river flood plains, there is a tendency to deposit sediment along the immediate banks of the streams so as to form low, natural levees.   This feature is indicated to some extent by the figures just referred to, but in the lower delta region it is clearly seen along the sea-level line.   Nearly all the streams are leveed, as it were, out into the Gulf, especially the Mississippi.

The large quantities of water that have passed over this delta region in comparatively recent geologic times have kept its surface from rising above sea level at the same rate as did adjoining portions of the State lying north, east, and west.   The result is that this region has been eroded by waters coming from nearly all the middle Western States, whereas the adjoining tracts have been worn down only by the results of the precipitation upon their own area.   Now, river action is gradually building up this delta region, whereas to the east and west the land surface is being gradually degraded.

REGION OF PRAIRIES AND LOW, ROLLING HILLS.

There is naturally no sharp line of demarcation between this topographic division and the one just described.   The swamp and lake regions gradually become drier to the north, and the former Gulf, lake, or swamp bottoms assume the rôle of "crawfish"

Fig. 2.—North-south section from the Kansas City Southern Railroad, in Texas, through Sabine Lake and southwestern Cameron Parish, La. Length, 55 miles; extreme elevations, 30 to 40 feet.

iries. This is specially true of the low plains west of the Atchafa-
a. In general, this region may be approximately defined as extend-
from the 20-foot to the 100-foot contour line. For a stretch of about
miles in width west of the Mississippi the general appearance of this
ion is somewhat changed by the erosion and the alluvial deposits
this great stream and its tributaries or distributaries. East of
Mississippi, however, the prairies again appear here and there,
ugh the forests often descend to the very edge of the swamp lands.
'the 100-foot contour is approached, the land becomes dissected by
merous small streams, and when cleared of its forest growth pre-
ts a decidedly rolling surface. East of the Mississippi the plains
ng near the level of the 20-foot contour are still in places thickly
ded with graceful, palm-like "long-leaf" pines. Their years are
mbered, though, as the many huge sawmill plants in their midst
l attest.

The soil of the region or zone that lies nearly at the level of the 20-
t contour is decidedly clayey and "tight-bottomed," a feature of
at economic importance to the rice planter. Farther up, toward
100-foot contour, the soil is more sandy and is, therefore, more
rvious to surface waters. This, too, as we shall see later on, is an
xremely fortunate circumstance so far as the supply of underground
ter farther south is concerned.

HILL LANDS.

As the low lake and swamp lands pass gradually into the prairies, so
upper undulating prairie and timber lands pass gradually into the
re abrupt dissected area. The chief difference to be noted is that
this last subdivision the streams are so numerous and their valleys
deep that there is little left of the old sea-bottom plain out of which
is rugged topography was carved. As the surface of the land in
is area rises from 100 feet to over 400 in a distance usually less than
om the sea margin to the 20-foot contour, it is no wonder that the
lects of erosion are well marked. Here the soil is still more gravelly
sandy than in the belt below the 100-foot contour. This fact, too,
s much to do with the rapid erosion that is apparent on every hand.
small exception to the general appearance of these "long-leaf pine
ll lands" is to be seen in the calcareous prairies (Anacacho) near
eesville, Vernon Parish.

STRATIGRAPHY OF SOUTHERN LOUISIANA.

GENERAL CONSIDERATIONS.

So far as underground waters are concerned the stratigraphy of
uthern Louisiana is very simple, for nearly all of the wells dis-
ssed in this report are in very young or Quaternary deposits.

Here and there, to be sure, peaks and uplifts of the older be
approach the surface, or even protrude above the general level of t
land, but such uplifts are generally of ex-
tremely local nature.   The Five Islands,
for example, stretch along the coast for a
distance of over 35 miles, but the greatest
diameter of the largest one is only 2 miles.

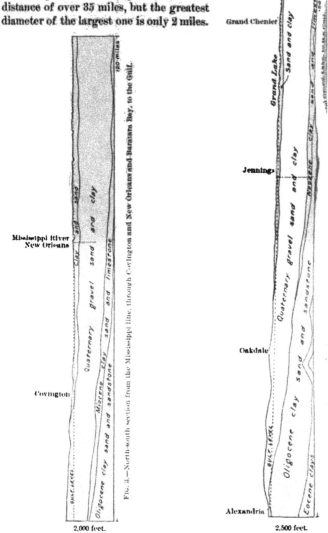

FIG. 3.—North-south section from the Mississippi line, through Covington and New Orleans and Barataria Bay, to the Gulf.

Again, these Five Islands are separated by a stretch of 25 miles fr
the truncated cone at Anse la Butte, or by 75 or 80 miles from si

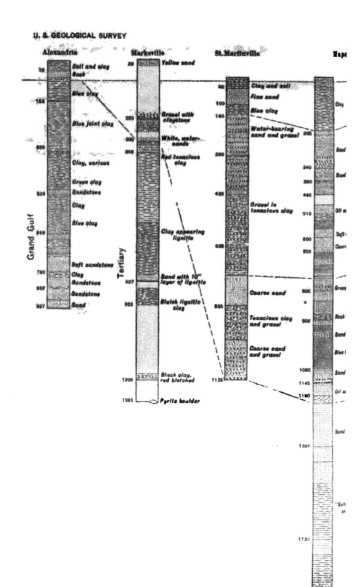

WELL SECTIONS FROM

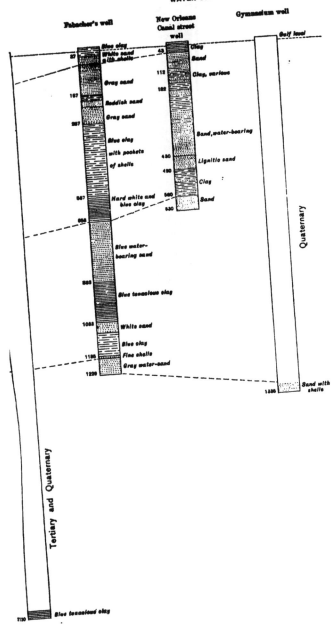

Fabacher's well

New Orleans
Canal street
well

Gymnasium well

Gulf level

Blue clay
87 White sand
with shells

Clay
49 Sand

112 Clay, various

Gray sand

167 182
Reddish sand
Gray sand
287

Sand, water-bearing

Blue clay
with pockets
of shells

430 Lignitic sand
490 Clay

547 Hard white and
blue clay

560 Sand
530

696

Blue water-
bearing sand

Blue tenacious clay

885

1083 White sand

Blue clay
Fine shells

1196 Gray water-sand
1229

Sand with
shells
1335

Quaternary

Tertiary and Quaternary

Blue tenacious clay
1730

XANDRIA TO NEW ORLEANS.

structures at Sulphur and Vinton. The Cretaceous limestone of
ou Chicot is 60 miles north of the northernmost island. There
doubtless other and undiscovered irregularities in the underlying
ks in southern Louisiana, but they are so evenly blanketed over
Quaternary clays and sands that there is no evidence of their
stence.

The two cross sections herewith given show the general stratigraphy
the water-bearing sands in southern Louisiana (see figs. 3 and 4).
her sections can be constructed by placing in juxtaposition well sec-
ons that have been taken along some one general trend. On Pl. II is
own the stratigraphic relation of the beds encountered in well sec-

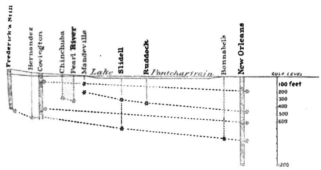

FIG. 5.—Correlation of water-bearing sands north and south of Lake Pontchartrain.

ns at Alexandria, Marksville, St. Martinville, Napoleonville, and
w Orleans. Fig. 5 is a similar section, extending from a point 9
es northwest of Covington to New Orleans.

TERTIARY.

OLIGOCENE.

n considering the Quaternary sands of this region, it seems proper
ake some notice of the beds upon which they lie. Again, if the
ntry around Alexandria, for example, be included in the region
e called southern Louisiana, this discussion should embrace a con-
ration of the outcrops of Tertiary (Oligocene) rocks in that neigh-
hood, which are of considerable importance in connection with the
ply of underground potable waters of the State. The well of the
xandria Ice and Storage Company, recently put down, will give a
idea of the character of the Oligocene (Grand Gulf) material in
part of the State. Its section is as follows:

Section of well of Alexandria Ice and Storage Company, Alexandria, La.

	Thickness in feet.	Depth in feet.
Surface ground clay	21	21
Sand	2	23
Clay	15	38
Rock	27	65
Blue clay	88	153
Hard rock	2	155
Blue clay	20	175
Rock	8	183
Blue joint clay	145	328
Limestone	3	331
Clay	43	374
Hardpan	90	464
Hard limestone	2.5	466.5
Green clay	12	478.5
Hard rock	1.5	480
Blue clay	10	490
Sandstone	14	504
Clay	30	534
Sand	3	537
Rock	2	539
Clay	10	549
Sand	1	550
Clay	8	558
Sand	2	560
Blue clay	89	649
Sand	16	665
Clay	28	693
Sand	10	703
Blue clay	24	727
Soft sandstone	53	780
Clay	24	804
Sand	5	809
Soft sandstone	42	851
Clay	2	853
Sandstone	44	897
Sand	30	927

This well is provided with a 70-foot strainer, and before it was cleaned had a flow of 125 gallons a minute, according to the report of a local paper.

The somewhat misleading information received at Alexandria two years ago regarding the material passed through in sinking the water-works well would have inclined one to place the water-bearing sands here at a horizon that is manifestly far too low. The great thickness of the Grand Gulf beds here is surprising, but the description of the material penetrated certainly places it in this division of the Tertiary.

The several fine flowing wells at Boyce are evidently mainly if not wholly in this horizon, though perhaps the one 810 feet deep, which yields gas with the water, may have a somewhat lower origin than the shallower ones.

Similar water-bearing Grand Gulf sands on the Ouachita River near Catahoula Shoals have already been described.[a]

It is doubtful whether these beds have ever been encountered in drilling for water or oil farther south in Louisiana, except perhaps around some of the local upheavals or buried cones already referred to. The most probable exception is the Spring Hill oil well, not far east of Kinder, where, at a reported depth of 1,500 feet (probably about 1,200), the writer observed that the drill was passing through sharp quartz sand, mixed with flakes of green clay. It was reported that a soft sandstone, about 14 feet thick, was penetrated by the drill just before the writer's visit.

MIOCENE.

There is little if anything in the stratigraphy of these rocks that concerns us here. Their position must be such (see figs. 3 and 4) that their water supply would be very uncertain, both as to quantity and quality. They probably contain salt water, and this has been found in them by many of the oil-well drillers. From samples of well borings already studied, it seems probable that where there are no local disturbances these beds in southern Louisiana, say along the thirtieth parallel, scarcely ever rise above a plane that lies 2,000 feet below sea level.

QUATERNARY.

SUBDIVISIONS.

The longer the geology of southern Louisiana is studied the more futile appears the attempt to make satisfactory subdivisions in the Quaternary deposits—subdivisions that have any definite time or structural limits. Differences in conditions of deposition during the same period of time have produced results that vary greatly in different localities. The same differences in conditions of deposition that we see to-day at different places in southern Louisiana, producing the sea-marsh clays with vegetable and brackish-water organic inclusions,

---

[a] Report Geol. Survey Louisiana for 1902, p. 214.

It seems, therefore, that if there is anything to be gained by applying a name to clays that were evidently deposited in brackish-water bays, estuaries, and lakes along the Gulf border, some such term as "Pontchartrain clays" may be used, with the understanding, however, that the name shall denote a particular kind of deposit or phase of deposition having no special time value. So, too, the deposits, mainly alluvial, containing a large amount of vegetable matter, especially stumps and trunks of trees, may, if necessary, be classed as Port Hudson clays; and marine sands may be referred to as Biloxi sands; but in all cases the terms must be understood as denoting mere phases of deposition, not stratigraphic units.

But the names that may be applied to the different portions of the Quaternary deposits of this State are of little importance so far as the present work is concerned. The important facts are these: Pervious material, such as sand (coarse and fine) and gravel, alternate with impervious clay beds of various thicknesses throughout the Quaternary deposits of southern Louisiana; these beds vary greatly as regards inclosed organic remains and products of decomposition and in different localities are inclined at different angles, the "dip" being, roughly speaking, in the same general direction as the slope of the surface of the land, though somewhat greater in amount; the character of the water is greatly modified by the medium through which it passes; the position or state of the water, i. e., whether "deep well" or artesian," is dependent largely on local topography.

A. REMNANT OF GRAND CHENIER RIDGE AT THE FERRY LANDING ON
MERMENTAU RIVER.

LOCATION OF SPRINGS AMONG THE LIVE OAKS ON THE BORDER BETWEEN
THE SEA MARSH AND THE SOUTH SIDE OF GRAND CHENIER ISLAND
ABOUT 2 MILES EAST OF THE VILLAGE

the generalized sections here given, running north and south
this portion of the State (see figs. 3 and 4), no attempt has been
to show the many and various clay, sand, and gravel beds that
the Quaternary series of this region. The fact has been indi-
however, that generally, where the land is flat and erosion has
light, the latest (uppermost) layers consist of fine sand and clays.

GENESIS OF DEPOSITS.

statistics upon which the above-mentioned general statements
sed are mainly of two kinds—first, well sections and the fossils
)ck material accompanying them, and second, facts noted in a
/hat detailed study of present areas of deposition along the
:rn border of the State. Well statistics will form an important
n of this report. Their interpretation, however, depends on an
ite knowledge of present conditions of sedimentation. The
'ing remarks and illustrations will therefore serve to throw light
: general statements already made and give a meaning to the
:d well records which follow.
shore line of southern Louisiana is generally sandy, and there
ften sand and shell ridges extending for miles parallel to the
either in close proximity to the Gulf or some distance inland.
more distant from the present southern border of the land often
axial directions not in accord with those nearer the Gulf, as may
n by observing the direction of the ridges toward the eastern
n of Pecan "Island." These peculiar forms are not due to any
lerable extent to erosion. Ridges of the same character are now
formed along the Gulf border, just above and just below mean
Off Cameron and Mud lakes, for example, one can see how,
g storms, the waves have beaten up the sand and shells in ridges
in some places to a height of 10 feet above mean tide. Out in
ulf some distance from the land the same force is at work making
ibine Shoals (see Pl. I.) The curve of Point au Fer, off Atchafa-
3ay, gives a strong hint as to the formation of ridges with a
somewhat at variance to the general direction of the shore line.
)ernière, Timbalier, Ship, and Cat islands, and the Chandeleurs
robably become inland ridges like Pecan and Grand Chenier, in
west Louisiana, or like the less elevated and less conspicuous
·idge encountered in sinking the foundation for pumping station
for the drainage of New Orleans.
: dimensions and general character of these ridges are well shown
;. 6A. Pl. III, .1, taken from the Louisiana report of 1902
it., Pl. IX), shows Mermentau River flowing in a westerly direc-
)r some distance before it finally breaks through the ridge on its
o the Gulf. To the right the sea marsh stretches away to the

east and south, with its waters practically at Gulf level. At the border of the Gulf, another though less important ridge is being formed at the present day. This statement, however, is not meant to imply that the large ridge shown in Pl. III, *A*, is of any other than Pleistocene origin. In fact, the white objects shown in the foreground of the view are remains of large molluscan species which are now living in the Gulf and along the Atlantic coast. No doubt should be taken of the fact, however, that these shells are still in place. The few *Rangia* mixed with them show considerable erosion, and have evidently been washed into this place by Gulfward flowing streams.

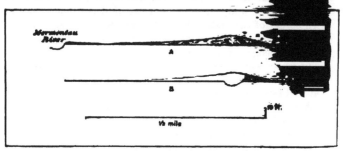

The sands and shells forming these ridges absorb enough rain water to furnish a continuous supply to many springs that flow out at sea level on either flank. Pl. III, *B*, shows the location of such springs along the line of and between the great live oaks that have given the name of Chenier to this island.

The abrupt transition from the firmer ridge material to the softer marsh ground to the south is well shown by the fact that the aged oaks nearly always incline toward and finally fall into the marsh (to the left in the plate). In drilling for water similar abrupt changes are often met with in wells but a short distance apart.

On the north or opposite side of the ridge, scarcely three-fourths of a mile away, the character of the vegetation and deposition is very different (see Pl. IV, *A*). The marshy land is less even or is slightly undulating, showing accordingly all stages of transition from moist to wet lands through occasionally inundated swamps to areas nearly always beneath the water. These areas are receiving sediment from the flood stages of the bayous and hence are gradually filling in and presumably rising, irrespective of any uplifting movement that may be affecting the coast as a whole.

Such areas explain the way that the deposits encountered in the various wells to the north were formed. The occurrence of decayed

A. NORTH SIDE OF GRAND CHENIER ISLAND.

The Mermentau River in the background to the right, near the trees.

JTH SHORE OF LAKE PORTCHARTRAIN. 1½ MILES WEST OF WEST END.

Black clay s   to acc ral r aps of white weathered *Rangia* shells.

ves, wood, fresh-water and land shells, together with fragments of
arine shells in many borings, thus receives a natural explanation.

Equally interesting and important in the formation of this portion
the State are the shallow lakes, reference to which has been made
discussing fig. 2, such as Sabine, Calcasieu, Grand, White, Maure-
s, Pontchartrain, and Borgne lakes, as well as the bodies of water
lled bays simply because they are not so completely surrounded by
nd. Of these bays Vermilion, Côte Blanche, Atchafalaya, Caillou,
errebonne, Timbalier, as well as the still more open Chandeleur and
ississippi sounds, are good examples. In these there is a complete
ries of beds showing transition from purely marine to brackish or
en fresh water. The fulgurs, naticas, arcas, oysters, tellinas, and
actras in the open sounds give place in the more inclosed bays to
sters, mactras, and rangias, while in the still more inland lakes the
ngias lose their fellowship with the salt-water forms and live in com-
rt and harmony with the purely fresh-water unios. This condition
ay be seen in Lake Charles, a small swelling in Calcasieu River about
miles from the coast.

Marks of wave action and heaps of brackish-water *Rangia* shells may
seen along the low shore of Lake Pontchartrain, shown in Pl. IV,
'. The characteristics of this vast expanse of shallow brackish water
serve more than passing notice by one who would understand the
neral geological history of southern Louisiana. It would scarcely
an exaggeration to assert that during some period of Pleistocene
me practically all of the land area of this part of the State passed
rough a Pontchartrain stage. By this it is not meant that the
hole of this area was one great brackish, inland lake at the same
me; far from it. There are now in this region open sounds, more
closed bays, still more landlocked lakes, growing smaller, usually,
e farther inland the body of water lies. North of Lake Charles
ere is an extensive swamp area that has the appearance of being an
d lake bed from which the waters are nearly drained off. Little
ake Charles is a remnant of a corner of this former extensive body
' water. Still farther up are The Bays, low, flat, level, hard, wet-
ottomed areas, embracing several thousand acres of land. The water
d oil wells that have been drilled during recent years in southern
ouisiana seldom fail to encounter masses of *Rangia* shells at some
pth. Water wells, at Jennings for example, sometimes pass through
bed of such shells 10 feet thick, lying at depths ranging from 50 to
0 feet below the surface. On the shores of Lake Charles, Lake
rthur, Grand Lake, Berwick Bay, Lake Pontchartrain, and else-
here in countless localities the same recent *Rangia* can be seen
aped up in ridges. At Jennings again similar Pontchartrain clays,
th the same *Rangia*, are generally encountered just above the oil in
ells, at a depth of about 1,800 feet.

EFFECT OF THE MISSISSIPPI ON STRATIGRAPHY OF SOUTHERN LOUISIANA.

The Quaternary material of Louisiana was evidently brought to its present place by Mississippi River and other smaller streams emptying into the Gulf, as it then was, throughout a stretch of perhaps 250 miles. In Tertiary as well as Quaternary times the Mississippi has had a marked influence on the character of deposition and the character of life to be found in this section. In several stages of the Eocene the deposits along the Mississippi axis are decidedly lignitic; farther to the east and west they are more marine. Certain stages in the Oligocene show similar conditions and differences. It must not surprise us, then, to find that the greater part of the Quaternary deposits of southern Louisiana are composed of beds that bespeak clearly the proximity of brackish or fresh water or land conditions. We attribute the presence of tenacious clays in the wells of southern Louisiana, to a depth of 1,800 feet in places, indirectly to the rapid filling in of the Gulf border by the Mississippi sedimentation. In some places there has been a continual loading and consequent depression of the Gulf's border; this has given rise to uplifts in regions not far distant. The shifting of the mouth of the river and the consequent change of loading point has caused a shifting of regions of depression and upheaval. If the region of uplift is some distance from the coast, then shallow sounds, bays, or lakes result, according to the extent of the uplifting. These, when finally filled with Pontchartrain clays derived from the sediment of inflowing rivers, pass through the sea-marsh stage into "crawfish" prairies, when the region in which they occur has been elevated a few feet.

Wave action, to be sure, performs a significant part in the formation of certain ridges that will eventually act as temporary borders to these landlocked bodies of water; but, after all, it is mainly the action of the Mississippi that causes the many changes of level that are so well recorded in the Quaternary deposits throughout south Louisiana.

In the immediate vicinity of the present course of the larger rivers, especially the Mississippi, Biloxi conditions can scarcely be expected to prevail. Here and there will be ridges of sand containing a purely marine fauna, but they will be notably local. Perhaps no better example of such a marine oasis in beds generally of a somewhat brackish water origin can be referred to than the sands containing beautifully preserved seashells at pumping station No. 7 of the New Orleans Drainage Works. This is evidently one of those ridges caused by wave action and slight upheaval that have served to cut off portions of the Gulf in the manner described above. The mouth of the Mississippi was at that time doubtless as far up as Bayou Sara, and its waters would not materially modify the life at a point then so far out at sea. So, too, at Napoleonville, the fauna at a depth of 2,100 feet seems purely marine. When this fauna lived doubtless the mouth of the Mississippi was as far north as the point named above, hence no great modification was brought about by the fresh waters of that great stream. Later, however, the fauna became brackish, with a preponderance of *Rangia* at 1,200 feet, and the drillers brought out a large tooth, equine in appearance, from a depth of 800 feet. Large quantities of *Rangia* have been found in a well near Morgan City at a depth of about 500 feet, and specimens of the same species were obtained at 400 feet at the Istrouma Hotel well at Baton Rouge.

SUBDIVISIONS OF SOUTHERN LOUISIANA, BASED ON UNDERGROUND WATER CONDITIONS.

MODIFICATION OF KIND AND CONDITIONS OF WATER BROUGHT ABOUT BY LOCAL TOPOGRAPHY AND STRATIGRAPHY.

For a somewhat detailed exposition of the topographic features of the southern part of Louisiana the reader is referred to the map herewith published as Pl. I, but a clearer and more general idea of the subject can be obtained more quickly by referring to fig. 1. Topography alone may have little bearing on the subject of underground water supplies, but when considered in connection with stratigraphy its significance may be great. Where the different formations or beds slope coastwise at an angle slightly greater than that of the surface of the land, where there are more or less extended beds of pervious material alternating with impervious, and where there is an abundant rainfall back in the country the conditions are favorable for an underground supply of water. The pressure head of this supply will depend largely upon the topography and stratigraphy, while the kind of water will depend upon the kinds of rock the water has to penetrate and the length of time consumed in its penetration. Kind of water is, therefore, indirectly more or less influenced by topography and stratigraphy. As a result of all these influences underground water occurs in southern Louisiana approximately as indicated by the accompanying fig. 7.

The influence of topography on pressure head will be evident to anyone who will study the outline cross section given in fig. 8 in connection with the topographic map, Pl. I. The section extends from Pearl River to Oberlin, passing through Covington, Hammond, Baton Rouge, Opelousas, and the country lying westward, to Oberlin. The situation at Opelousas is interesting. Here the surface of the ground is 67 feet above the Gulf level, but water rises only 22 feet above that level. A glance at the topographic map will show the cause of this low pressure head. Northward, in the direction of the uprising strata or bedding planes, there is no higher ground than at Opelousas. The water there present must work sidewise along the pervious strata

from the somewhat distant hill land lying west and northwest. East
of the Mississippi the conditions are different, for only a few miles

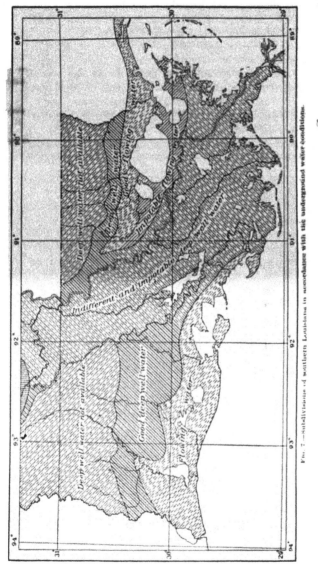

FIG. 7.—Subdivisions of southern Louisiana in accordance with the underground water conditions.

east of Baton Rouge the pressure head is considerably above the surface of the ground, the hill land to the north being close by. The

ʳked decrease in pressure head shown at Baton Rouge
ᵉ to the nearness of the Mississippi River Valley. Dᴄ
is same valley has something to do with the
w stand of the water at Opelousas.

Lesser depressions than the Mississippi Val-
y have their influence on the head of subter-
nean waters, as may be seen by the section
ɔng the line of the Southern Pacific Railroad
om Lafayette west, shown in fig. 9. Mer-
entau River, with its tributaries, has degraded
is central portion of southwest Louisiana, and
e pressure head of the deep-well waters re-
onds to this topographic feature. Calcasieu
iver seems to have the opposite effect on the
·essure head about Lake Charles. Here, how-
·er, we are dealing with a region that is im-
ediately south of some of the highest land in
e State, and it is doubtless this condition that
·unteracts any reverse influence the Calcasieu
alley might possess.

East of the Mississippi, in the neighborhood
New Orleans, the low pressure shown by the
.rious water-bearing layers penetrated in wells
ss than 1,300 feet deep is probably due to the
ide, low stretch just to the north—i. e., the
ıke Pontchartrain depression. On Ship Is-
nd good water flows freely, and with much
ore force than is exhibited by the New Or-
ıns wells. The narrowness of Mississippi
ɔund, as compared to Lake Pontchartrain,
fers a ready explanation of this fact.

REMARKS ON SPECIAL AREAS.

On the small map (fig. 7) there is indicated
small artesian area about Alexandria. The
:tent of this area must of necessity be very
nited, for the Grand Gulf formation usually
ps rapidly Gulfward, so that the water-
·aring strata would soon be below practicable
·pths so far as ordinary water supply and irri-
ɪtion are concerned (see stratigraphy indi-
.ted by fig. 4).

To what extent water would flow close to the
ulf border from Sabine River to Atchafalaya
ay can scarcely be conjectured, though from the fact
·ighborhood of Lake Charles, Gueydan, and places

FIG. 8.—Relations of land surface to pressure head of artesian and deep-well wat ᵉʳ approximately along parallel 30° 27'.

Light-house well.—Mouth of well perhaps 10 feet above tide; flow, vigorous; estimated at 50 gallons per minute from a 2-inch pipe; depth, 750 feet.

Section of well at light-house on Ship Island.

[Section by Dr. Murdock.]

Soil.	Thickness in feet.	Depth in feet.
Sand	250	250
Yellow clay	100	350
Blackish mud..	50	400
Fine sand, with shells..................................	60	410
Blue clay..	250	700
Water-bearing sand	50	750

MISSISSIPPI CITY.

C. Clemenshaw's well.—Depth, 925 feet; mouth of the well about 12 feet above tide. Regarding the well Mr. Clemenshaw remarks:

Passed through no hard rock, no quicksand, but clay and blue sand, the latter often highly micaceous. A 60-gallon per minute flow was found at a depth of 600 feet, a 200-gallon flow at 925 feet.

E. P. Ellis's well.—Depth, 850 feet; 3-inch pipe; flow, 80 gallons per minute; 55 feet above tide.

Court-house well.—Pipe, 2½ inches; reduced to 1¼ inches; flow, 30 gallons per minute; 28 feet above ground, or about 50 feet above tide.

GENERAL SECTION FROM PASS CHRISTIAN TO BILOXI.

According to Mr. A. Dixon, who has accompanied a well-drilling outfit for several years in this part of the State, the majority of the wells show approximately the following section:

General section of wells between Pass Christian and Biloxi.

Soil.	Thickness in feet.	Depth in feet.
Sand	80	80
Clay		125
Sand, and clay..		425
Light-gray fine sand..................................		500
Clay		600
Water-bearing sand		685

ich a fair quantity of flowing water may be obtained, but as a rule
best wells are sunk to a much greater depth here than in southern
ᴜuisiana.

HARRISON COUNTY.

BILOXI.

Section of well one-half mile east of railroad station.

[Section by Brown.]

Soil.	Thickness in feet.	Depth in feet.
il and clay	4	4
nd, bearing good pumping water..........................	61	65
hitish clay	35	100
eenish clay...	390	490
ᴅd, extremely fine at first, becoming coarser below, coarse gravel...	428	918

Pipe, 6 and 4 inches; flow, at surface of the ground, 1,000 gallons
ʳ minute; 500 gallons at elevation of 35 feet, 250 gallons at elevation
55 feet. This indicates that the pressure head is not far from 75
ᵗ above tide.

City waterworks wells.—No notes were obtained regarding the
ᴘths of these wells. It was observed, however, that the large 6-inch
ᵉs carried the water up rapidly and filled the elevated tanks to a
ght of 40 feet above the general surface of the ground.

Ice-factory wells.—At these wells the difference in temperature of
ᵉ shallow and deep well was specially noted, viz: Water (flowing)
ᴍ 500-foot stratum, 79.5° F.; from 900-foot stratum, 82.5° F.

SHIP ISLAND.

Quarantine station well.—Depth, 730 feet; mouth of well about 10
ᵗ above tide.

Section of well at Quarantine station, Ship Island.

[Section by Dr. P. C. Kallock.]

Soil.	Thickness.		Depth.	
	Feet.	In.	Feet.	In.
hite sand..	45	0	45	0
ft clay and mud.......................................	155	0	200	0
ᴀnd blue clay...	100	0	300	0
hite sand...	5	0	305	0
ᴜe clay...	60	0	565	0
ᴀndstone		5	565	5
ᴜe clay...	156	0	721	5
ᴀter-bearing sand.....................................	9	0	730	5

Light-house well.—Mouth of well perhaps 10 feet above tide; flow, vigorous; estimated at 50 gallons per minute from a 2-inch pipe; depth, 750 feet.

Section of well at light-house on Ship Island.

[Section by Dr. Murdock.]

Soil.	Thickness in feet.	Depth in feet.
Sand	250	250
Yellow clay	100	350
Blackish mud	50	400
Fine sand, with shells	50	450
Blue clay	250	700
Water-bearing sand	50	

MISSISSIPPI CITY.

C. Clemenshaw's well.—Depth, 925 feet; mouth of the well about 18 feet above tide. Regarding the well Mr. Clemenshaw remarks:

Passed through no hard rock, no quicksand, but clay and blue sand, the latter often highly micaceous. A 80-gallon per minute flow was found at a depth of 600 feet, a 200-gallon flow at 925 feet.

E. P. Ellis's well.—Depth, 850 feet; 3-inch pipe; flow, 80 gallons per minute; 55 feet above tide.

Court-house well.—Pipe, 2¼ inches; reduced to 1¼ inches; flow, 30 gallons per minute; 28 feet above ground, or about 50 feet above tide.

GENERAL SECTION FROM PASS CHRISTIAN TO BILOXI.

According to Mr. A. Dixon, who has accompanied a well-drilling outfit for several years in this part of the State, the majority of the wells show approximately the following section:

General section of wells between Pass Christian and Biloxi.

Soil.	Thickness in feet.	Depth in feet.
Sand	80	80
Clay		125
Sand, and clay		425
Light-gray fine sand		500
Clay		600
Water-bearing sand		685

.I. WELL IN BARN LOT OF THE HERNANDEZ PLACE. 2½ MILES NORTH OF
COVINGTON LA.

B. WELL IN MR AN ERL : : BARNYARD THREE FOURTHS . A M LE NORTH.
AEST F HAMMOND STAT ON LA.

Section of Dummet well, St. Tammany Parish—Continued.

	Thickness in feet.	Depth feet.
Red sand and gravel.......................................	20	10
....	32	42
Red sand	38	80
	25	105
	4	109
White clay	18	127
Blue clay ..	183	310
Fine bluish and greenish water-bearing sand..................	7	317
Blue clay ..	71	388
Gray sand .	0	388
Fine blue and greenish sand ...	8	396

John Dutch's well.—In north-central part of Covington; depth, 600 feet; flow, 20 gallons per minute; temperature, 74° F., April 5, 1901; elevation of flow, 35.6 feet above tide.

Mrs. Flower's wells.—These records were furnished by Mr. Wallbillick, and show that here, as elsewhere, there are sandy strata bearing water at far less depths than the beds furnishing the water that **will flow above** the surface of the ground. Such wells are termed shallow or pumping wells.

Sections of Mrs. Flower's wells, St. Tammany Parish.

	Thickness.		Depth.	
	Ft.	*in.*	*ft.*	*in.*
Well No. 1:				
White clay	30	6	30	6
Blue clay	18	6	49	0
White sand...	2	0	51	0
Blue clay	17	8	68	8
Shells mixed with blue clay...............................	1	6	70	2
Fine white sand..	27	6	97	8
Coarse white sand (pumping stratum)......................	5	0	102	8
Well No. 2:				
White clay...	40	0	40	0
Blue clay	2	6	42	6
White clay	21	0	63	6
Shells mixed with black clay..............................	0	0	63	6
Dark clay..	9	6	73	0
White sand	21	0	94	0

These wells are but 300 feet apart.

.1. WELL IN BARN LOT OF THE HERNANDEZ PLACE, 2¼ MILES NORTH OF COVINGTON, LA.

B. WELL IN MR. ANDERSON'S BARNYARD, THREE-FOURTHS OF A MILE NORTH-WEST OF HAMMOND STATION, LA.

H. Haller's well.—Southwestern part of Covington; depth, 520
et; pipe, 2 inches; flow, 30 gallons per minute; temperature, 72 F.,
ne, 1903.

Hernandez place, well by house.   About 2 miles north of Covington;
pth, 610 feet; pipe, 2.5 inches; flow from 1-inch pipe, January,
1, 38½ gallons per minute; April, 1901 (from whole pipe?), 60
llons per minute; temperature, 1901, 73 F.; elevation of ground,
.1 feet above tide; top of basin, 47.3 feet; of pipe, 48.5 feet.

Hernandez well, by barn.—About 2½ miles north of Covington;
pth approximately as in last well; pipe, 2½ inches; flow, January,
1, 35½ gallons per minute; March, 1902, 54½ gallons per minute;
ne, 27, 1903, 40 gallons per minute; temperature, 72.25 ; elevation
ground, 47.4 feet; of pipe, 52; pressure head considerably above 60
ove tide.

This is the well shown in Pl. V, .1, and is usually considered one of
· best in this part of the State, but it has not the capacity of the well
 the house, which is so piped that satisfactory measurements of its
w are hard to obtain.   This beautiful summer residence is now the
operty of Louis P. Rice, of Covington and New Orleans.

Ice factory wells.- -Three wells of the "shallow" type before men-
ned, two 2 inch and one 2½ inch, furnish, when pumped, sufficient
ater for the ice factory.   The water rises to within about 8 feet of
e surface.

Lyon well.—At Claiborne, 1 mile east of Covington; depth, 630 feet;
pe, 2 inches; flow, 30 gallons per minute; temperature, 73 . April,
01; flow, 26 gallons per minute; temperature, 74 . June 26, 1903;
evation, 26.6 feet above tide.

Maison Blanche.—Depth, 480 feet; pipe, 2 inches, reduced to 1 inch;
w per minute, April, 1901, 20½ gallons; March, 1902, 23½ gallons;
ane 26, 1903, 16½ gallons; temperature, 72.25 ; elevation of ground,
 feet 6 inches of top of basin, 33.6 feet of flow, 35.5 feet above tide.

Other wells.—There are many other flowing wells about Covington,
at the data presented above will give a fair idea of their general
aracter.   It will be seen that as the depth increases the temperature
so increases, as might well be expected.   For a 600-foot well a tem-
rature of 74 is about normal here.   Compare these in this respect
ith the Hammond and Ponchatoula wells.

There is a flow about Covington at present, within a radius of 3
iles, of about 300 gallons per minute; and as is generally the case the
ater is mainly wasted, i. e., allowed to flow to no purpose.

ABITA SPRINGS.

Abita Hotel and Cottage Company well.- Half mile east of Abita
prings Station (elevation of station 38.3 feet above tide); depth, given
y some as 545, by others, 525 feet; pipe, 2 inches; flow through stop-

cock, 54 gallons per minute; temperature, 73°; no screen. This is a new well, put down this season (1903). When allowed to flow freely it reduces the pressure of neighboring wells materially, especially those to the west and south.

Aubert Hotel well.—About one-third mile southeast of station; depth, 585 feet; pipe, 1¼ inches; flow from a faucet, 2½ feet above the ground; 38.3 feet above tide January, 1901; 12½ gallons per minute through a network of pipes 60 feet long; June 26, 1903, 22 gallons per minute, direct from well at a height of about 38 feet above tide.

See analysis given on p. 78. Pressure head at least 50 feet above tide.

Frank Brinker's well.—One-fourth mile northwest of the station; pipe, 2 inches; depth, 574 feet; flow through stopcock about 2 feet above the surface of the ground, 27 gallons per minute; no screen; temperature, 73° F.

Labat Hotel well.—One-fourth mile north of the station; depth, 536 feet; pipe, 1¼ inches; original flow, seven or eight years ago, said to be 45 gallons per minute; flow, January, 1901, from faucet, 45.2 feet above tide, 37 gallons; flow from pipe with stopcock but without faucet, June 26, 1903, 56 gallons per minute; temperature, 74°. When the size of the pipe is taken into consideration this is the most freely flowing well in St. Tammany Parish.

Chas. W. Schmidt's well.—A few yards south of the station; depth supposed to be 800 feet; pipe, 1¼ inches; flow through a one-half inch faucet, in 1901 and 1903, 4 gallons per minute; temperature, 72°; elevation of ground, 35.6 feet; of faucet, 36.6 feet above tide. This was perhaps the earliest artesian well in this vicinity. It was not decidedly successful, doubtless on account of the novelty of the undertaking. The temperature indicates that its flow of water comes from a depth much short of 800 feet.

Simon's Hotel well.—Just east of the station; hotel building burned; pipe, 1¼ inches; flow through two elbows and a horizontal pipe 2 feet in length, January, 1901, 12 gallons per minute; April, 1901, 11 gallons; June, 1903, 10 gallons; temperature, 1901, 72°; 1903, 73°; elevation of ground, 38.3 feet of top of basin, 41.7 feet of top of pipe, 43.6 feet above tide.

Limit of supply.—The present flow of water from artesian wells about Abita Springs is not far from 200 gallons per minute. The sensitiveness, especially on the part of the smaller wells, to the flow from the new, large well would seem to indicate that the supply, though ample for all legitimate uses, should not be unduly drawn upon, else pumping in some instances will have to be resorted to.

PEARL RIVER JUNCTION.

When compared with most of the wells in this part of the State the well at P   ' River Junction appears remarkable for the great amount of water it furnishes at a shallow depth. The water is not regarded as suitable for boiler and drinking purposes, though for common household uses it serves excellently. Depth, 350 feet; pipe, 2½ inches; flow through a stopcock at the rate of 72 gallons per minute; flow from open 2½-inch pipe said to be 90 gallons per minute; pressure head, 54 feet above tide. The elevation of station is 31 feet above tide.

MANDEVILLE JUNCTION.

At Mandeville Junction there is an excellent well that furnishes the railroad tank with water, flowing up freely 27 feet above the ground. Since no levels have ever been run over this road it is not possible to state the exact height of the well above tide.

MANDEVILLE.

The elevation of station at Mandeville is 6.8 feet above tide.

Dessone well.—Northeastern part of the village, in flower garden; depth, 217 feet; pipe, 2 inches; flow per minute, March, 1901, 28.1 gallons; March, 1902, 26 gallons; June 27, 1903, 28 gallons; temperature in 1902, 69.5°; in June, 1903, 69.8° F. Flows from pipe 9 feet above tide; pressure head, 14½ feet above tide.

Mrs. John Hawkins's well.—Western part of the village; pipe, 2 inches, reduced to 1¼ inches; flow per minute, 1902, 40 gallons; in 1903, 13 gallons; temperature in 1902, 68.5° ; in 1903, 70° F. Flow from a pipe 7.35 feet above tide.

C. H. Jackson's well.—Depth, 135 feet; pipe, 1½ inches, reduced to 1 inch; flow, 0.97 gallon per minute; height of flow, 13.8 feet above tide.

Dr. Paine's well.—Flow, open 2-inch pipe, 10.6 gallons per minute; reduced to 1 inch, 10½ gallons per minute; through inch pipe with stopcock attached, 9.1 gallons per minute. Elevation of ground, 3.85 feet; of flow, 6.80 feet above tide.

Ribava well.—Depth, 247 feet; flow, from open 1½ inch pipe, 12 gallons per minute, 1901; through stopcock, 9½ gallons per minute in 1902; through stopcock, 1903, 7 gallons per minute; temperature, 71° , February, 1902; 72°, June, 1903; elevation of ground, 3.42 feet; of flow, 4.9 feet above tide.

Rush well.—North of station, perhaps one-third mile; depth, 252 feet; pipe, 2 inches, reduced to 1 inch; flow, 7 gallons per minute; 4 feet above the general level of the ground.

Depths.—As in several other regions already described, there are here to be found beds yielding water at a depth considerably less than that attained by most of the artesian wells. The water in the shallower wells usually stands, in the vicinity of Covington, as well as about Hammond, from 2 to 10 feet below the surface. Here such shallow wells, about 90 feet deep, actually flow, though not vigorously.

It will be noticed that the wells about Mandeville are very much shallower than at Covington, 9 miles to the north. They are also 3° cooler, and have a less ferruginous taste and appearance. The wells about Slidell have not been examined, but Mr. Blakemore, of New Orleans, says that there the Mandeville water (300 feet), and a decidedly bad "yellow" water (perhaps 700 feet down), are met with. The latter is described as the same water that is found at the same depth in the city of New Orleans.

CHINCHUBA.

Depth, 325 feet; pipe, 2 inches; flow reduced to one-third inch pipe, hence with pressure head of but 7.3 feet; temperature, 73° F.; elevation of ground, 19 feet above tide.

Other wells at a brickyard to the north, and at a locality 4 miles to the northwest, are reported to have satisfactory artesian wells, but they were not visited.

TANGIPAHOA PARISH.

SINGLETRY'S STILL.

This well is about 9 miles northwest of Covington, or in the SW. ¼ NW. ¼ sec. 31, 5 S., 10 E. It is so distant from any other flowing well that the following statistics and section, though imperfect, will be of considerable interest to landowners and well men in this section of the State.

Section of well at Singletry's still.

[Section given by E. P. Singletry.]

	Thickness in feet.	Depth in feet.
Sand and clay	100	100
Quicksand	120	220
Red clay	170	390
Pipe clay	160	550
Blue sand	10	560

Depth, 560 feet; pipe, about 2 inches; flow, 18 gallons per minute, with several small leaks; height of pipe where measured, 78 feet above tide. (See analysis, p. 78.) The elevation was determined in 1901 by J. Pacheco and G. D. Harris, who ran a spirit-level line out from Covington.

HAMMOND.

The elevation of the railroad station at Hammond is 43.3 feet above tide.

Captain Anderson's well.—For general appearance of the well see

V, *B;* depth, 272 feet; size, 2 inches; flow, 20 gallons per minute; nperature of water, 70.5°; strainer or point, 10 feet.

Section: Sand to 40 feet, a thick bed of blue clay, then sand and avel to the bottom.

Well sunk and cased with galvanized pipe for 55 cents per foot; nce total cost of well, approximately, $150.

Baltzell & Thomas livery stable well.—Depth, 330 feet; size, 2 ches; temperature, 71.5°; flow, 24 gallons per minute, June 23, 1903; reen, 7 feet long.

B. F. Bauerle's well.—One and one-half miles south-southwest of ammond; depth, 212 feet; size, 2 inches; temperature, 69°; flow, ine 23, 1903, 8¾ gallons per minute.

Durkee well.—Depth, 297 feet; size, 2 inches reduced to 1¼ inches; w in March, 1901, 24 gallons per minute, besides two small distrib- ing pipes that could not be closed; flow, same conditions, June 23, 03, 24 gallons per minute.

Eastman well.—One and one-half miles south of Hammond; depth, 9 feet; pipe, 2 inches; flow, 30 gallons per minute, in 1901; pressure, pounds per square inch; temperature, 72° F.

Forbes well.—One mile east of Hammond, NE. ¼ NW. ¼, section 30; pth, 250 feet; flow, June 23, 1903, 7¼ gallons per minute; size, 1¼ ches; pressure head, 17 feet above surface of ground; age, 8 years. Three water-bearing beds were encountered in sinking this well:) Depth, 52 feet, water coming to within 2 feet of surface; (2) 150 et, coming to within 8 feet of surface; (3) 250 feet, with head of 17 et.

Hermann well.—Two miles south-southwest of Hammond. Impos- ble to obtain accurate data, except pressure, 8½ pounds per square ch.

Hammond Ice Company's well.—Depth, 340 feet; pipe, 2 inches; w in 1901, about 50 feet above tide, 15 gallons per minute; same nditions, June, 1903, 11 gallons per minute; temperature, both ars, 72° F.

Hammond Mineral Water Company (Limited).—Well, 460 feet deep; ipe, 3 inches; flow, about 46 feet above tide, 65 gallons per minute.

C. H. Hommel's well.—One-half mile southeast of Hammond; depth, 8 feet; flow, impossible to measure now; said to have been, when ell was first put down, 45 gallons per minute; temperature, 70.5° F.

Alfred Jackson's well.—Depth, 265 feet; pipe, 1¼ inches; flow, 3 et above surface of the ground, June, 1903, 6¾ gallons per minute; mperature, 71° F.

June Brothers' sawmill well.—Depth, 377 feet; pipe, 2 inches; w, at a point 5 feet above the ground, open pipe with one elbow, ine 22, 1903, 24 gallons per minute; temperature, 71° F.

Kate well.—In western part of the village, on Morris avenue; size

of pipe, 2 inches; free flow, perhaps 3 feet above the general surface
of the ground, 30 gallons per minute, June 23, 1903; temperature,
70.6° F. A new well, just finished.

Fred Karlton's well.—One-half mile southeast of Hammond; depth,
302 feet; pipe, 2 inches below, reduced to 1½ above surface of ground;
screen, 10 feet; flow, June, 1903, 24 gallons per minute; cost, $150;
temperature, 70.5° F.

Kemp well.—Three-fourths mile southeast of Hammond; pipe, 1½
inches; flow, June, 1903, 5 gallons per minute; temperature, 70° F.

Merritt Miller's well.—Depth, 265 feet; pipe, 2 inches, reduced
to 1½; flow in 1901, 28½ gallons per minute; elevation of flow, 44
feet above tide; pressure head, 56.6 feet above tide; temperature,
71° F.

Morrison well.—Pipe, 2 inches; flow, 46 feet above tide, 1901, 30
gallons per minute; June, 1903, same flow; pressure head, 51.7 feet
above tide; temperature, 72° F.

Oaks Hotel well.—Depth, 300 feet; pipe, 2 inches; flow, 25 gallons
per minute; age, ten years; temperature, 71° F. See analysis,
p. 78.

Oil well.—The following section was obtained from samples in
1901:

Section of oil well at Hammond, Tangipahoa Parish.

	Depth in feet
Clay	45-55
Sand and gravel	85-100
Yellow loam	173
Water-bearing sand	294
Coarse sand	368
Coarse sand and gravel	475
The same, more sandy	500-512
5-foot bed of hard blue clay, about	570
"Pepper and salt sand"	570+

The new well, June, 1903, was over 760 feet deep. It was generally understood that its section tallied with the old one fairly closely
so far as the latter went down. The "5-foot bed of clay" of the old
well showed only 3 feet in the new. From approximately 570 feet in
the new, gravel was abundant to 760. Below, a hard bed of clay had
been encountered, light colored above, but growing much darker
below.

Pushee well.—One mile south of Hammond; west of the railroad;
depth, 380 feet; 340 feet of 1½-inch pipe, 40 feet of 1¼-inch pipe; no
screen; flow recorded by Mr. Pacheco March, 1901, 14½ gallons per
minute; April, 1901, 15½ gallons; by G. D. Harris, June 23, 1903, 11
gallons per minute; temperature, 70.6° F.

Robinson well.—Northwest quarter of the town (see analysis, p. 78); depth, 356 feet; size of pipe, 2 inches; flow not ascertained because it is piped to various places quite inaccessible.

Rogers's (Ben) well.—West of Hammond, ⅛ mile; depth, 284 feet; 2-inch pipe reduced to 1 inch; flow, 17 gallons per minute; temperature, 70.5° F.; cost, $142.

Section of Rogers's well, Hammond, Tangipahoa Parish.

Clay.
Quicksand to 75 feet.
Clay.
Sand, last 50 feet.
Lower end of screen (10 feet) stuck in clay bed.

Erastus Rogers's well.—Depth, 225 feet; pipe, 1¼ inches; flow, 5 feet from ground, 2½ gallons per minute; temperature, 70°; strainer (screen), 8 feet.

J. T. Smith's well.—One mile east of Hammond; depth, 235 feet; temperature, 69° F.; pipe, 1¼ inches; flow, 6.5 feet above ground, 7¼ gallons per minute; age, one year; cost, $108.

W. B. Smith's well.—One-half mile southeast of Hammond; depth said by some to be 260, by others 305, feet; pipe, 2 inches reduced to 1¼; temperature, 70.5°; flow, 11¼ gallons per minute; age, eight years.

Tigner well.—Two miles southeast of Hammond; flow, 20 gallons per minute; pipe, 2 inches; temperature, 70° F.

W. J. Wilmot's well.—Depth, about 370 feet; pipe, 2 inches reduced to 1 inch; flow, said to be 40 gallons per minute; pressure, 2 feet above the ground, 7.7 pounds per square inch; flows readily 14 feet above ground, with small leaks in pipe; would doubtless flow 20 feet above ground.

H. Walsh's well.—One and one-half miles south-southeast of Hammond, in section 31; depth, 298 feet; pipe, 1½ inches reduced to 1 inch; flows through 30 feet horizontal pipe, with stopcock, 5½ gallons per minute; temperature, 70.75° F.; age, five years.

Way well.—One and one-half miles south-southwest of Hammond; depth, 140 feet; flow, 3 gallons per minute; temperature, 69° F.

Summary of wells about Hammond.—Water may be had by pumping, from wells ranging in depth from 30 to 100 feet; a sand, or quicksand, furnishes a slight flow generally at 140 to 150 feet, flow or not depending on topography; temperature, 69° F.; after passing more clay, to depths ranging from 230 to 380 feet, coarser sand or gravel is encountered, furnishing an artesian flow above the ground of from 10 to 20 feet, according to topography; temperature, 69° to 72° F.

The well of the Mineral Water Company, with 3-inch pipe, and a depth of about 460 feet, with a flow of 65 gallons per minute, as well as the log of the oil well, shows conclusively that better and larger wells may be expected in this vicinity. The Morrison and Durieu wells, some distance apart, in the central portion of the town, have shown no change whatever in flow for the past two years. Since they are of the normal size and depth, it is evident that the available supply is as yet far greater than the demand.

In a radius of two miles of Hammond there are already about 50 flowing wells, yielding about 1,000 gallons of water per minute, or half a billion gallons annually, nearly all of which is wasted.

Decrease in the flow of certain wells in this neighborhood is due solely to increased obstruction in the lower end of the pipe.

The cost of these wells is not far from 50 cents a foot, labor, casing, etc., being furnished by the driller. The usual size pipe is 2 inches in diameter; in case smaller pipe is used the cost of the well is somewhat less. See notes on J. T. Smith's well, above.

Age of the wells examined, from two months to ten years. When properly screened, or put down into coarse gravel, these wells seem to flow as freely now as when first put down.

Local well drillers: Bacon and Gamble, Edwin Way, John Blumquist.

PONCHATOULA.

The elevation of the railroad station at Ponchatoula is 29 feet above tide.

Alber well.—Two hundred feet from the town well; depth, 413 feet; pipe, 2 inches; flow, 25 gallons per minute; head about 30 feet above the surface of the ground. Bacon and Gamble, drillers.

G. H. Biegel's well.—At Pelican Hotel; depth, 232 feet; flow, 4½ gallons per minute, 1901; 2¼ gallons per minute, 1903; temperature, 71° in 1901; 69.5° in 1903; pipe, 1½ inches; height of flow about 31 feet above tide.

Mrs. Bishop's well.—Old, deserted place, 3 miles north of Ponchatoula, 2 miles south of Hammond; depth, 170 feet; pipe, 1½ inches; temperature, 69.5° F.; flow, 10 gallons per minute; age, about nine years.

The section of this well, according to John Blumquist, who drilled it, is as follows:

Section of Bishop well, Ponchatoula.

	Thickness in feet.	Depth in feet.
Clay	50	50
Sand, with some water	20	70
Blue clay	94	164
Coarse sand	6	170

C. A. McKinney's well.—About ¼ mile southwest of Ponchatoula; pth, 199¾ feet; flow said to be variable, caving in evidently taking ᴀce below; on June 24, 1903, 12 gallons per minute; pipe, 1½ inches; ᴇ, four years.

Moon well.—Same general vicinity as preceding; depth, 200 feet; pe, 1¼ inches; flow, 12 gallons per minute; age, seven years. Near ʳ this is the Fisher well with a flow of 10 gallons per minute.

Railroad well.—At Chester, 100 feet north of fiftieth milepost from ᴊw Orleans, west of track and 5 feet below the level of rails; flow, gallons per minute; pipe, 1¼ inches; temperature, 70 F.

Town well.—In public square; flow, 1901, 2½ gallons; in 1903, 2¾ llons per minute; temperature, 71°, 1901; 70°, 1903. See table of ᴀlyses for further information regarding this and the Biegel well.

Sawmill well.—Depth, 332 feet; flow, 5 gallons per minute.

Section of sawmill well, Ponchatoula.

[Section given by Bacon and Gamble.]

	Thickness in feet.	Depth in feet.
llow and gray blue clay....................................	75	75
ᴀy sand and gravel	15	90
ᴇ clay, about.................................	35	125
ᴇ blue sand..	105	230
ᴀrse white sand..	30	260
ᴇ blue sand, with thin beds of clay...	40	300
ᴀd a little coarser, weak flow of water.....................	32	332

ORLEANS PARISH.

The fact that there are two well-defined water-bearing strata[a] under ᴊw Orleans has already been mentioned. A number of additional ᴇts can now be presented.

The old Canal street well of 1854, so often referred to in geological ᴇrature, both on account of its great depth, as borings then went, ᴅ, more especially, on account of the careful record kept by Mr. ᴀnchard of the beds passed through, including many fossils, still ᴍains the type section for this general region of the country down ᴀ depth of 630 feet. No recent boring has been recorded with ᴇ interest and painstaking care that was displayed in this well. ᴛhis is most seriously to be regretted, as the number of wells sunk ᴀs been very large, and their records, if carefully kept, would furnish ᴀterial for an interesting chapter in the geological history of the

[a] Rept. Geol. Survey Louisiana for 1902, p. 221.

Chloride magnesium	75.7	44.9
Chloride ammonia	1.3	
Chloride potash	Trace.	Trace.
Carbonate calcium	86.8	40.8
Oxides of Fe and Al	4.7	2.8
Phosphate	Trace.	Trace.

a Ounces.

Fabacher's well.—At Fabacher's "Casino," corner Nashville avenue and St. Charles street; depth, 1,229 feet; pipe, 4 inches; flow 1 foot above ground, 55 gallons per minute; flow, reduced to 2 inches and raised 10 feet above the ground, 6 gallons per minute; flow stops at 12 feet above ground; temperature, 81.5° F.

Section of Fabacher's well, New Orleans.

[Furnished by Mr. Blakemore.]

Character of material.	Thickness in feet.	Depth in feet.
Bue clay	37	37
White sand with shells	20	57
Yellowish-white clay	5	62
Gray sand	105	167
Bue clay	20	187
Reddish sand	20	207
Gray sand	80	287
Bue clay, with pockets of shells	280	567
Gray sand	2	569
Blue clay	40	609
Hard white clay	19	628
Hard blue clay	30	658
Blue water sand (fresh water)	225	883
Blue tenacious clay	150	1,033
White sand (resembling white sugar)	40	1,073
Blue clay	85	1,158
Fine shells	6	1,164
Gray water sand	65	1,229

A forthcoming report will deal with the fossil remains saved from this well by Mr. Fabacher, and similar ones preserved by Mr. John Kracke, from the gymnasium well. They appear to be of Pleistocene or Quaternary age.

THE COMMON "YELLOW-WATER" WELLS.

These include the 600 to 900 foot wells bored at frequent intervals over the city. One of the earliest wells of this class sunk in New Orleans was in the neutral grounds on Canal street, between Caron-dalet and Baronne streets, in the year 1854. A colored section of this well, as originally kept by A. G. Blanchard, C. E., of New Orleans, appears in the report of the board of health of Louisiana for 1890–91.[a] From this it will be observed that the strata penetrated to a depth of 330 feet consist of light yellowish and bluish sands and clays with some light greenish layers and occasional shell sands.[b]

One of the most recent wells of this class is that at the Marine Hospital, Audubon Park. This is 765 feet deep. The first 600 feet are reported as sand, silt, and clay beds; a bed of yellow sand, perhaps

[a] Biennial Rept. Board of Health to the general assembly of the State of Louisiana for 1890–91, plate opposite p. 148. Baton Rouge, 1892.
[b] Rept. Geol. Survey Louisiana for 1903, p. 221.

40 feet thick, was encountered some distance below, and continued to 705 feet. From there on for 60 feet the material consists of white sand. The water rises to within about 3 feet of the surface at present. This 6-inch well is capable of furnishing 300 gallons per minute. The water is classed as excellent for washing purposes, requiring but half as much soap as the river water. It is also excellent for boiler purposes, but is impotable.

The flow from this shallower class of wells has always been weak, and the large number of such wells has still further weakened the flow. There is a tendency now, when more water is required, to seek the lower level. This water is excellent for bathing purposes, containing, as the above analysis shows, a large amount of common salt.

The great range in depth here given really includes two or more water-bearing horizons, though at various localities but one may be represented.

THE 400-FOOT SANDS.

In the old well on the neutral grounds, just referred to, a sand bed was passed through from 335 to 480 feet below the surface that furnished artesian water at the rate of 350 gallons an hour.

SHALLOW WELLS.

Very close by Mr. Fabacher's deep well, above described, is a well but 180 feet in depth, fitted with a 3-inch casing, that flows 12 gallons per minute 1 foot above the surface of the ground. It is brackish. Temperature, 70° F.

Small driven wells in the city limits, at varying shallow depths, reach sandy, coarse material that bears water, evidently closely connected with the river.

BONNABEL WELL.

One of the most interesting wells that has ever been put down in the vicinity of New Orleans is that on the shore of Lake Pontchartrain, about 1 mile west of West End. An attempt was here made to start a summer resort under the name of Lake City, and this well was sunk for a supply of fresh water. According to Mr. Bonnabel, the well is 1,200 feet deep, but a letter from the driller indicates that it is not over 900 feet deep. It now flows from a 2¾-inch pipe, standing about 8 feet above tide, 12 gallons per minute, with a temperature of 79° (measured July 5, 1903).

Mr. Bonnabel makes the following remarks regarding the well section:

Five-inch casing to 600 feet, hitting rock; 3-inch casing to 700 feet; then 1½-inch casing to 1,200 feet. Compact, ferruginous conglomerate, 60 feet thick, was passed through about 700 feet down; then a black, hard clay was encountered, giving way to bluish sand; water in pale blue clay.

he analysis of the water by Mr. Joseph Albrecht, as given in a
dbook regarding "Lake City," is as follows:

Analysis of water from Bonnabel well.

	Grains per gallon.
ium chloride	27.74
ium carbonate	34.39
sium carbonate	4.49
ca carbonate	1.69
anic matter free of nitrogen	0.46
bonic acid combined as bicarbonates	13.33
Total	82.10

The features of the section outlined by Mr. Bonnabel are in some
ys remarkable, and if it were certain that there is no error in the
tter there might be grounds for supposing that there had been
ne orogenic movement in this region that brought up rocks belong-
; to a horizon beneath the Pleistocene to an elevation of but 600
t below tide level.   It is probable that the water comes from the
ne stratum that is found at a depth of 500 to 600 feet about Coving-
ι and Abita Springs, and that it is the same as the 600 to 900 foot
ιd beds penetrated and so largely drawn from throughout the city of
w Orleans.   The fact that the water may be potable at Covington,
rely so at Lake City, and quite impotable in New Orleans, is readily
plained by the very slight slope of the water-bearing stratum, and
nce the very slow movement of the underground waters.   A slope
perhaps 150 feet in 35 or 36 miles can scarcely give an appreciable
ly motion through sand that is generally very fine.   When we
ιsider also the rapid formation of this coastal region of Louisiana,
ι the great amount of organic matter that was brought Gulfward
ιn as well as now and deposited along in the sand and clay beds of
iocene times, it is no wonder that the slowly moving waters should
ιome strongly impregnated with various salts and so-called impuri-
ι as they pass Gulfward (see fig. 4).
Since such is the condition of affairs in and about New Orleans,
ιre seems to be no valid reason for supposing that the city will ever
supplied with potable artesian water derived from local wells.

ST. JOHN THE BAPTIST PARISH.

RUDDOCK.

Mr. John Blumquist, of Hammond, says that the well at this place
posite the railroad station is 338 feet deep.   It flows strongly, but
ι water stains everything red, even glass.

EAST BATON ROUGE PARISH.

BATON ROUGE AND VICINITY.

Waterworks, two wells.—Old well put down in 1892; depth, 758 feet; water rises to within 6 feet of surface, i. e., approximately 30 feet above tide; capacity given as 500,000 gallons daily.

Analysis of water of waterworks well at Baton Rouge.

[B. B. Ross, analyst.]

	Grains per gallon.
Total solid matter	14.2176
Mineral matter	12.1357
Organic and volatile matter	2.1573
Silica	1.3412
Potash	.2251
Soda	5.1929
Lime	.5009
Magnesia	.2553
Oxides of Fe and Al	.5074
Phosphoric acid	.0019
Sulphuric acid	1.8079
Chlorine	.4595
Oxygen oxidizing organic matter	.0423
Nitrogen, albuminoid ammonia	.0056
Nitrogen as free ammonia	.00519
Nitrogen as nitrates	.00199
Sulphuric acid and chlorine combined as—	
Potassium sulphate	.4171
Sodium sulphate	3.0022
Sodium chloride	.7404

This well has an 8-inch pipe for 386 feet; 6-inch pipe for 304 feet; 4¼-inch pipe for 68 feet. New well starts with 10-inch pipe, and is 6 inches the rest of the way down; depth, 800 feet; flow at surface about 35 feet above tide.

The two wells are said to have a capacity of 1,000,000 gallons a day. Pumped with compressed air.

Istrouma Hotel well.—Depth, according to the Blakemore Well Company, of New Orleans, 770 feet; water stands 18 feet below the surface of the ground. It is of the same quality as the water obtained at the waterworks, and pumps with a suction pump at the surface about 80 gallons per minute.

Well 4 miles east of Baton Rouge.—Pipe 4-inch, flow from 2-inch hole 4¼ feet above ground, 5 gallons per minute; from 2-inch hole 1¼ feet above ground, 30 gallons per minute; temperature 71° F. Pressure head about 50 feet above tide.

BAKER.

Well at old mill, one-fourth mile south of station.—Depth, 850 feet; 2-inch pipe; has flowed freely 16 feet above present faucet. It now furnishes large quantities of water.

Elevation of pressure head, about 100 feet above tide. (Elevation Baker station given by Gannett as 82 feet above tide.) Driven wells, 150 feet deep, furnish fair water. Bored wells, 25 to ▸ feet deep, yield very impure water.

ZACHARY.

Wells here, some as deep as 200 feet, have to be pumped. Most of the water used is from shallow bored wells.

WEST FELICIANA PARISH.

BAYOU SARA.

Well just southeast of railroad station, 240 feet deep; passed through gravel at 100 feet. It is pumped. Darton gives the following data from one well at this place: Depth, 736 feet; pipe 4-inch; yield, 347 gallons; height of water [above mouth of well?] +2 feet; temperature 8°. For another he gives simply depth 450 feet and "height" +1 foot.

ARTESIAN AND DEEP WELLS IN LOUISIANA WEST OF THE MISSISSIPPI.

LA FOURCHE PARISH.

THIBODAUX.

Ice factory well.—Depth, 225 feet; passes through moderately fine bluish sand all the way down; water impotable on account of various salts; stands 13 feet below the surface of the ground; used for condensing.

ASSUMPTION PARISH.

NAPOLEONVILLE.

City waterworks.—Two wells, an 8-inch, 190 feet deep; a 6-inch, 210 feet deep. Both said to furnish 25.000 gallons per hour; the smaller, and deeper, with 9-foot strainer, furnishes more water than the larger, with 20-foot strainer.

Several such wells around the town furnish a similar water, i. e., very ferruginous, staining bath tubs and connections an orange yellow.

ST. JAMES PARISH.

St. James well.—Mr. Weasel, contractor for well drilling on the Texas and Pacific Railroad, says that at St. James he found good water at a depth of 285 feet, passing through a bed of shells (probably *Rangia* shells).

Cement well. — Mr. C. Oley, of the Blakemore Well Drilling Company, states that here he put a well down to the depth of 190 feet and procured good water. It rises and falls with the Mississippi.

ST. MARY PARISH.

MORGAN CITY.

Well penetrated a very coarse gravel bed at a depth of 500 feet.

GLENCOE.

Clendenin[a] gives a section of an artesian well at this place furnished by Doctor Simmons. It shows coarse sand and water at a depth of 615 feet.

Section of well at Glencoe, St. Mary Parish.

	Thickness in feet.	Depth in feet.
Soil	1	1
Yellow clay	11	12
Quicksand	12	24
Blue clay	160	184
Shale	} Undetermined. {	
Tough, gray clay		
Coarse sand and gravel and water at		615

IBERIA PARISH.

JEANERETTE AND VICINITY.

Moresi's barnyard well.[b] — Depth, 140 feet; pipe, 1½-inch; flow, February 16, 1901, 7½ gallons per minute; temperature, 70°. See table of analyses given on page 78.

Elevation of station, 18 feet above tide; well, 13.2 feet below station; hence flow is about 5 feet above tide.

Moresi's foundry well. — Depth, 700 feet. See table of analyses given on page 78. Section given as follows:

Section of Moresi's foundry well at Jeanerette, Iberia Parish.

	Thickness in feet.	Depth in feet.
Clay	40	40
Sand and gravel	160	200
Blue and gray clay, shells, and red water	460	660
Gravel	40	700

a Part III, Geol. and Agric. State Exp. Sta., 1896, p. 243.
b Survey Louisiana for 1902, p. 232.

Elevation, 5.5 feet below railroad station; water stands within 5 or
eet of the surface; hence water is about 8 feet above tide.

Ice factory well.—Pipe, 8-inch. Clendenin gives this well section as
lows:

Section of well at ice factory, Jeanerette.

	Thickness in feet.	Depth in feet.
1 clay	15	15
ttled clay and sand	80	95
anic bed	10	105
d and gravel	70	175
llow clay	175	350

Flow from base of cap, 7.69 feet below railroad station or about 10.5
t above tide.

Old Moresi plantation.—One mile southeast of Jeanerette; depth,
0 feet; flows freely about 8 feet above tide; stains pipes and con-
ctions bright reddish yellow.

S. B. Roane's well.—Three miles south of Jeanerette; depth, 420
et; pipe, 10-inch; water flows over the top of pipes perhaps 10
et above tide when wells are not pumped for a time; water seems
od for general family use; potable; wells pumped for rice irrigation.
his is known as the Kilgore plantation. The section is as follows:

Section of Roane's well at Kilgore plantation, near Jeanerette, Iberia Parish.

	Thickness in feet.	Depth in feet.
ay	80	80
avel	6	86
ty, full of shells	150	236
avel and sand	184	420

NEW IBERIA.

Ice works wells.—Depth, about 230 feet; quality and quantity not as
ired for general use; pipes soon cake and clog up.

John Emms's well.—Depth, about 260 feet; extremely ferruginous;
t potable; rises 5 feet above the bayou at mid-stage.

Oil well.—North of New Iberia; depth, said to be about 500 feet,
th pockets of oil, and one "rock" 2 feet thick; good water also
ported at a depth of about 400 feet.

The quality of the water at the Roane well, mentioned above, is such
to seem to bear out ex-Mayor Moresi's statement that good water is
be found only at the usual depths some distance back from the

bayou. It is probable that the problem of furnishing New Iberia with good water will be solved by pumping it from a station a few miles to the west. Mr. Caldwell, the machinist, vouches for the statement that good water was found in the "Oil" well.

ST. MARTIN PARISH.

ST. MARTINVILLE AND VICINITY.

Oil well.—About 1½ miles northwest of St. Martinville.

Section of oil well near St. Martinville, St. Martin Parish.

(According to Mr. William Kennedy.)

	Thickness in feet	Depth in feet
Clay and soil	40	40
Fine sand	60	100
Blue clay	40	140
Water-bearing sand and gravel	150	290
Tenacious clay and gravel	25	315
Water-bearing sand and gravel	120	435
Tenacious clay, with gravel	200	635
Coarse sand	200	835
Tenacious clay and gravel	150	985
Coarse sand and gravel	150	1,135

It will be observed that two beds of water-bearing sand and gravel are mentioned. Doubtless other sand and gravel beds, like the lowest penetrated, would furnish an ample supply of water, though very likely to be salty.

The Southern Pacific Railroad station is marked 25 feet above tide: a spirit-level line to the well shows that the floor of the derrick is 16.3 feet above tide. For diagram of this well see Pl. II.

In an irrigation well close by the water surface stood at a height of 11.6 feet above tide January 13, 1903.

Labbé's well.—Four miles south of St. Martinville; spirit-level line from St. Martinville showed surface of water to be 11.13 feet above tide; surface of the land 17.3 feet above tide January 14, 1903.

BREAUX BRIDGE.

Gilbeaux place.—Three-fourths mile west of station, on Gilbeaux plantation; elevation of railroad station 27.5 feet above tide; well pipe 12 feet above tide; water said to have flowed over the top of this pipe when well was first put down.

LAFAYETTE PARISH.

LAFAYETTE AND VICINITY.

Waterworks wells.—We have here an instance of lack of care in ving the orifice of the wells accessible, so that the wells may at any ւe be cleaned, or rather flushed, when clogged with sand. Three lls have been put down here in succession, because, after a few ưrs, they became clogged np. The depth of the new and conse- ⺟ntly best well was given as 226 feet. Its casing is 6 inches; screen, 5 feet long. This well supplies Lafayette, besides 220,000 gallons the Southern Pacific Railroad daily; height of surface of ground, ҈koning Lafayette station as 40 feet, about 34.6 feet; water said to between 20 to 25 feet below, hence about 10 or 15 feet above tide; ⺟en cleaned occasionally it is as good as when first put down; screen very coarse gravel; C. H. Melchert, engineer in charge.

Lafayette Compress and Storage Company's well.—Depth, 125 feet; ⺟ter surface about 25 feet below surface of the ground, i. e., about feet above tide.

ST. LANDRY PARISH.

OPELOUSAS.

Waterworks well.—Depth, 184 feet; pipe, 10 inches; screen, 64 feet; ⺟ been pumped to the extent of 300 gallons per minute, guaranteed 0; elevation of water in well, 22.28 feet above tide, i. e., considering ⺟ station as 67.5 feet, as given by the Southern Pacific Railroad.

Section of waterworks well at Opelousas, St. Landry Parish.

	Thickness in feet.	Depth in feet.
⺟y	83	83
⺟e sand	37	120
⺟vel to bottom of well	64	184

Oil Mill well.—Depth, 208 feet; pipe, 8 inches, with 40 feet of screen.

WASHINGTON.

Washington well.—The following section was given for the well at ⺟shington:

Section of well at Washington, St. Landry Parish.

	Thickness in feet.	Depth in feet.
⺟icksand	18	18
⺟d	52	70
⺟vel	124	194

Water said to rise to within 11 feet of surface of ground, or about feet above tide.

WEST BATON ROUGE PARISH.

BATON ROUGE JUNCTION.

Mr. Weasel says he found good water here at 160 feet.

LOBDELL.

The same authority just quoted says that good water is found here at a depth of 150 feet. The surface of the water is 21 feet below the general level of the ground.

POINTE COUPEE PARISH.

NEW ROADS.

Mr. Weasel reports poor water at 120 feet.

BATCHELOR.

The same condition exists here as at New Roads.

AVOYELLES PARISH.

BUNKIE.

Railroad wells.—One 90, the other 142 feet deep; the water from both impotable. Water stands in both about 13 feet below station.

W. D. Haas's well.—One 4-inch well, 180 feet deep, furnishes enough water to run four large boilers in Haas's cotton compress works; water stands about 10 feet below surface of ground or about 11.5 feet below the station.

Gannett gives Bunkie an elevation of 66 feet. Hence water stands in these wells about 52 or 54 feet above tide.

MARKSVILLE.

Court-house well.—This well is reported to have a depth of 800 feet, encountering salt water. At a depth of 230 feet a 5-foot stratum of lignite was penetrated. Mouth of well 0.3 foot above railroad station, hence approximately 82 feet above tide.

VERMILION PARISH.

ABBEVILLE AND VICINITY.

Court-house well.—Well about 16.6 feet above tide with section as follows:

Section of well at court-house, Abbeville, Vermilion Parish.

	Thickness in feet.	Depth in feet.
Clay	15	15
Fine sand	65	80
Clay	2	82
Hard layers of clay alternating with sand	57	139
Coarse white sand with white pebbles	21	160
Reddish clay and "rock"	60	220

The upper bed here alone furnishes water. Exact height of water
uld not be told; certainly it lacks several feet of overflowing.

Well 9 miles west of Abbeville.—On Mr. John Waltham's place W.
ᴴE. ¼ sec. 32, 12 S., R. 3 E., are several wells. The land is here 10
ᵗt above tide and the general well section, according to Mr. Moresi,
about as follows:

Section of well on Waltham's place, 9 miles west of Abbeville, Vermilion Parish.

	Thickness in feet.	Depth in feet.
ᴸy	30	30
ᵃy sand	10	40
ᴸy	5	45
ᴺite sharp sand and gravel....................................	30+	75+

Even at this low level the water does not overflow.

SHELL BEACH.

Wells that have a feeble flow above the surface of the ground were
ard of at this place, but were not visited.

GUEYDAN.

Wilkinson's well, 3 miles southwest of Gueydan.—Depth, 190 feet;
pe, 8-inch; flow, 8+ gallons per minute; temperature, 73°. Eleva-
ᵒn of flow, 6.9 feet above tide, determined by spirit-level line from
ueydan; bench mark on station; according to Southern Pacific Rail-
ad, 9.07 feet above tide.

Donnelly place, 6 or 7 miles east of Gueydan.—Two 8-inch and two
inch wells. Water said to rise 8 inches above the surface.

ACADIA PARISH.

RAYNE AND VICINITY.

Chapuis's well.—Depth, 210 feet, with 10-foot strainer; water stands
feet below surface. Elevation of station, according to Southern
ᴸcific Railroad, 37.5 feet above tide, well about 2 feet below; hence,
ᵗter in well about 19.5 feet above tide.

Hippolite Richard's well.—This is 3 miles east-northeast of Rayne.
ᵉpth, 200 feet; water stands within 17.5 feet of surface. Elevation
surface of water in well about 20 feet above tide, based on spirit-
ᵛel line run from Rayne to mouth of well.

CROWLEY AND VICINITY.

Railroad well.—Depth, 173 feet. Water usually rises to within 5
ᵒ 6 feet of surface. Elevation of water, about 19 feet above tide.

Ice factory well.—Depth, 600 feet; water unsatisfactory; pipe withdrawn to the usual depth, 170–180 feet.

Long Point, 15 miles northeast of Crowley.—One 8-inch and three 6-inch wells. Water at 180 feet; rises to within 26 feet of the surface.

Three miles east of Crowley.—Two wells pass through logs at depth of 168 and 202 feet. In the first, beneath the 168-foot log, 7 feet of water-bearing sand was encountered, water rising to within 7 feet of surface.

Sol Wright's well.—About 3 miles southwest of Crowley, or in center of sec. 19, T. 10 S., R. 1 E.; depth, 293 feet; surface of ground, 19.37 feet above tide; of water, 9.37 feet above tide January 29, 1903. Strainer, 70 feet long.

L. J. Bowen's well.—Middle of NE. ¼ sec. 19; depth, 196.6 feet; top of pipe, 21.39 feet above tide; of water, 9.49 feet above tide.

MIDLAND.

Water stands in this well, February 5, 1903, 10.5 feet below station; hence 7.5 feet above tide.

ORIZA AND VICINITY.

John Wendling's well, 1 mile southwest of Oriza.—Pipe, 6-inch; flow, 1.2 feet above surface; 20 gallons per minute. Elevation of Oriza (Southern Pacific Railroad), 24 feet above tide. By spirit-level line, top well is 11.4 feet above tide.

D. J. Scanlin's well, 2 miles southwest of Oriza.—Elevation of surface of water, 12.2 feet above tide; line from Oriza.

F. Scanlin's well, 2 miles south-southwest of Oriza.—Elevation of surface of water, 12 feet above tide; leveled from Oriza.

CALCASIEU PARISH.

It is in the eastern half of this parish that perhaps two-thirds of all the large irrigation wells of southwestern Louisiana are located. Not that this particular area is better adapted to the growing of rice than many other sections of southern Louisiana, but by a glance at any map of this part of the State it will be seen that east Calcasieu has comparatively few large rivers, creeks, or bayous from which water may be had for irrigation purposes. The result is that here are found the most advanced methods of sinking wells and lifting the water from them.

It is entirely out of the question to refer to even a tenth part of the wells now in operation in this section; of late years their number has gone up into the hundreds, and will soon reach a thousand or more. A few statistics regarding some of these wells will show the general characters of all of them. Welsh may be taken as the central point of interest in deep-well activity.

C. L. Bower's well.—About one-half mile northeast of Welsh, cen-
of sec. 30, called in the last report of the Geological Survey of
iisiana (1902) "E. L. Brown's well;" depth, 130 feet; pipe, 8 inches;
iiner, 38 feet; surface of water above tide February 26, 1901, 16.68
:; March 21, 1903, 13.92; July 13, 1.18 feet. The section shows
r to 65 feet and sand, growing coarser below, to 130 feet.

Ir. Bower has recently put down another well 92 feet north of
i well; it has a 10-inch casing, is 175 feet deep, and has a 56-foot
iiner. From top of pipe to water surface, March 21, 1903, 6.73
t of the water stood 13.26 above tide; July 13, 1903, 0.5 foot
ve tide.

Cooper's well, ¼ mile east of Welsh.—The section shows clay to 90
t, coarse sand, clay, sand, and finally blue sand at a depth of
–145 feet.

Field's well, ¼ mile east of station.—The section shows clay to 90
t; sand, clay, coarse below, to 164 feet.

Welsh planing mill well.—Pipe, 3 inches; top of pipe, 20.33 feet;
face of water, March 19, 13.86 feet; March 21, 13.93 feet above
e.

Section of well at Welsh planing mill, Welsh, Calcasieu Parish.

	Thickness in feet.	Depth in feet.
y	12	12
d	4	16
y	182	198
rse, light sand	40	238

S. R. May's well, ¼ mile north of station.—Top of flume 20.3 feet
ove tide, of water; March 19, 1903, 14.3 feet; March 21, 15.16 feet;
ly 12, 1903, 0.8 foot above tide; July 13, after pump had been
rking 1 hour, but had stopped 5 minutes before the measurement
s taken, 1.7 foot above tide; same conditions except pump had been
pped for about 20 minutes, 1.1 foot above tide; lowest level in 1902
d to be −8 feet; depth, 190 feet; pipe, 8 inches; temperature
water, 71.5° F.; supplies 1,200 gallons per minute when pumped
a 20-horsepower Erie engine.

Abbot's well, 2 miles southeast of Welsh.—Elevation of water surface,
bruary 26, 1901, 16.42 feet above tide; that is, 7.08 feet below the
lroad station.

Herald's well, perhaps 1½ miles east-southeast of the station.—Eleva-
n of water, February 26, 1901, 16.6 feet above tide, or 6.9 feet
ow the railroad station.

Well 9 miles north-northwest of Welsh.—The section shows clay to .92 feet and sand to 235 feet.

LAKE ARTHUR.

Wells at mills reported as flowing 5 feet above tide.

R. E. Camp's well, 1½ miles northwest of Lake Arthur.—Southeast ¼ sec. 8, 11 S, 3 W.; depth, 215.7 feet, water-bearing sand 40 feet thick; elevation of top of pipe as determined by a spirit-level line from the lake, 17.5 feet above tide; elevation of water surface, 8 feet above tide.

JENNINGS AND VICINITY.

Anderson's wells, about 1 mile west-southwest of Jennings.—Three 10-inch wells, connected to a 14-inch main and pumped with a 50 horse-power engine. Depth, approximately, 300 feet; wells about 20 feet apart, furnishing, with engine running at perhaps half-rated power, ,800 gallons per minute.

These wells are furnishing now (1903) about half as much water as they did last year owing to clogging of the strainer with fine sand. The fireman at the plant says the 150-foot well, about 50 feet north of the three, is capable of furnishing nearly as much as these three are furnishing now. Though so various in depth, when the deeper wells are pumped, the amount obtainable from the shallower one is materially diminished.

Carey's wells.—In this same vicinity are the three Carey wells, a general log of which is herewith given:

General section of three wells near Jennings, La.

	Thickness in feet.	Depth in feet.
Clay, with shells at about 50 feet, and with vegetable matter and a log below	.15	115
Quicksand above, gravelly below	45	160
Bluish, sandy gravel	20	180
Sandy clay	50	230
Gravel	30	260

City waterworks well.—When measured, March 19, 1903, the water in this new well stood 18 feet below the mouth of the pipe or, perhaps, 12 feet above tide. The capacity of the tank is 65,000 gallons. The engineer informed us that the well seemed to lower none while the tank was being filled, the operation lasting about three hours.

Well 3 miles east-southeast of Jennings.—This well was being sunk on February 24, 1900, by the Brechner outfit. The beds penetrated showed reddish, yellow, and gray mottled clay for 30 feet, becoming

A. ARTESIAN WELL OF BRADLEY AND RAMSAY LUMBER COMPANY, 1 MILE
NORTH OF LAKE CHARLES LOUISIANA.

Flow in March, 190? 2?0 gallons a minute

B. SCREEN WOUND AT THE MOREJI BROTHERS SHOP JEANERETTE, LA.

s tenacious, with fossil fragments, *Rangia, Helix, Balanus,* etc.,
til a depth of 90 feet was reached, when blue sand, with thickness
determined, was struck.

KINDER AND VICINITY.

McRill's well, 1 mile north of Kinder.—Depth, 150 feet; elevation
f water surface as determined by leveling, from Kinder Station,
larch 8, 1902, 27.1 feet above tide, assuming that the station is 49.3
et above tide.

Tillotson's well.—Depth, 138 feet; depth of water from top of pipe,
l feet, 10 inches; temperature, 68° F.; elevation of water surface
larch 7, 1902, 25.4 feet above tide.

CHINA.

McBirney's wells.—A number of wells in this vicinity, ranging in
epth from 140 to 175 feet and in size from 6 inches to 8 inches, in
hich water rises to within 14 to 23 feet of surface, depending on local
pography.

OBERLIN.

Mr. Dennis Moore says that the railroad tank well is 190 feet in
pth, and that water rises to within 10 feet of the surface, or about
feet above tide.

In general the water level would probably be somewhat lower than
s. No hopes can be entertained of obtaining a flowing well at this
mparatively shallow depth.

LAKE CHARLES.

Well 1 mile north of lake.—The Bradley and Ramsay Lumber
mpany's well, about 500 feet deep, has the greatest flow of any
ll measured in the State, 210 gallons per minute; pipe, 6 inches.
e analysis given below. (See Pl. VI, 1, for view at well.) Ele-
tion, 10.5 feet above tide. Based on tide gage reading at Lake
arles, by G. D. Harris.

Reiser's machine-shop well.—The following is a section of the well:

Section at well at Reiser's machine shop, near Lake Charles.

	Thickness in feet.	Depth in feet.
td	96	96
l sand with pebbles..	6	102
ty sand and clay alternating..................................	98	200

Water with iron taste. See analysis given below. Elevation of well about 13 feet; known to flow to 17 feet and said to have flowed to 27 feet above tide.

Judge Miller's well.—Pressure of 5.25 pounds per square inch; flows 12 gallons per minute. Elevation of present flow, 12.72 feet above tide; would flow at 24.79 feet above tide.

WEST LAKE.

Perkins and Miller Lumber Company's well.—Pipe, 4 inches; elevation of flow, 10 feet above tide, and would doubtless flow to 16 feet or more above tide.

Well 3 miles northwest of lake.—Pipe, 8 inches. Following is a partial section of this well:

Partial section of well at West Lake, Calcasieu Parish.

	Feet.
Hard clay between	250–300
Shells	300
Gravel	300

This is a very strong flowing well.

RAPIDES PARISH.

BLOWING WELLS.

It would doubtless be an unpardonable omission, if in enumerating the various classes of wells in southern Louisiana, with their depths, kinds of water, and other characteristics, no mention were made of the "blowing" wells of Rapides Parish, that have attracted much attention, at least locally.

Judge Blackman, of Alexandria, has frequently called attention to a certain well of this character, and has recently sent, through Mr. Kennedy, of the Southern Pacific geological survey, a clipping from the Alexandria Town Talk, of September 19, 1903, relating to this subject.

Though Judge Blackman knows of two other wells having similar characteristics, the one best known is located on the farm of Mr. Frank Melder, Melder post-office, between Spring Creek and Calcasieu River, 2 miles east of the river, and 3 miles east of Strothers Crossing, on the Calcasieu.

It was in 1892 that Mr. Frank Melder started to bore a 12-inch well, but had to give it up after reaching a depth of 80 feet. The air would come rushing from the well, sometimes for a period of three or four days, and again at shorter periods. When the air was not rushing from the well, it would turn the other way and be sucked into the well with great force * * * The force of the air coming from the well would keep a man's hat suspended over it.

In boring the well a stratum of about 1 foot of pipe clay was penetrated, and for the remainder of the distance, over 75 feet, a bed of yellow sand was penetrated.

While boring it was discovered that every foot deeper the well was sunk, the harder the air would blow from it. When the well was first completed, it would blow a day and then air would be sucked in for a day. No water ever appeared in the well at any period.

The subject of "blowing wells" has been discussed in Water-Supply and Irrigation Paper No. 29, by Mr. Barbour.[a] He attributes such phenomena, doubtless correctly, to changes of atmospheric pressure at the surface of the earth. Those interested in this subject will find, without doubt, that when the wells are "blowing," the barometer reading as recorded by the nearest weather station is low; when the wells are "sucking in," the barometer is rising.

It seems from the above statement regarding the section of the Melder well that its great capabilities as a "blowing" well are due to the absence of water between the grains of sand.

When such interstices are mainly filled with water, as is usually the case, the phenomenon of "blowing" is much less noticeable.

VARIATION IN FLOW AND PRESSURE HEAD SHOWN BY WELLS IN SOUTH LOUISIANA.

WELLS EAST OF THE MISSISSIPPI.

As a result of investigations already carried on, it is safe to say that the total amount of water obtained from deep and artesian wells in this part of the State north of Lake Pontchartrain does not exceed 3,000 gallons per minute. South of the lake, in the city of New Orleans, there are a number of 6-inch wells, but they are pumped so irregularly, both as to time and amount, and are so "connected up," that no safe estimate can be given as to their total yield. The water-bearing sands, ranging from 600 to 900 feet below the surface throughout the city, have been penetrated in so many places that the water rarely overflows from these wells. All admit that the head has been gradually lowered somewhat in proportion to the number of new wells put down. (For a record of the present stand of the waters in these wells, see pp. 44–47.)

There seems to have been a slight decline in the waters of the Mandeville region, if we may trust occasional-measurements, yet by referring to the data presented under Mandeville (p. 37), it will be seen that some of the important wells are flowing now almost as much as two years ago. Some have become practically clogged up and of little or no value. The presumption is that, were new wells put down or were those now in existence occasionally flushed, the supply would be as great as ever from each well. Very few new wells have been put down in this vicinity during-recent years.

---

a Barbour, E. H., Wells and windmills in Nebraska: Water-Sup. and Irr. Paper No. 29, U. S. Geol. Survey, 1899, pp. 73–82.

About Covington the new wells seem to show the same head as those put down two or more years showed at that time. Here, too, there is a suspicion that the marked falling off of head in several of the wells is to be accounted for by the clogging of the pipes.

At Abita Springs it has been noticed that the flowing of the last new large well put down decreases to a marked extent the head in the wells close by, especially to the south and west. Some of the better wells, however, have shown an increase rather than a decrease, so that with care in properly spacing the wells and judgment in using the water no one need expect to be obliged to resort to pumping for a long time to come.

At Hammond the better wells have shown no decrease of flow or pressure head for the last two years, even though their number has greatly increased during this interval.

When the extent of catchment area is taken into account, reaching, as it must, northward as far as Crystal Springs, Miss., and when the total amount of waters obtained from deep sources in this section of the State is considered, it is no wonder that there seems to be no general variation in flow or pressure head thus far recorded. Two moderate-sized rice plantations in southwest Louisiana would call for more water during the summer months than flows from all these wells combined. Until irrigation is practiced far more generally in this section of the country there will probably be no marked decline in the flow of the carefully constructed artesian and deep wells.

WELLS WEST OF THE MISSISSIPPI.

The statement is often made that the wells along the Mississippi and in the alluvial or delta region to the west vary as to head according to the different stages of the river. In the lowest regions, close to the river channel, this probably means that when the river is very high, held far above the wells by the great levee system, some of the river water gradually seeps through the intervening soils and enters the wells. Many instances are on record of the pressure of the river water becoming so great as to cause a spring to burst forth from the ground several hundred yards from the river's border. When such waters are welled up to a height corresponding to that of the surface of the river, they cease to flow.

However, if it is assumed that the motion of most underground waters is but a few feet a day, or only a mile or two a year, it is evident that the underground transmission of water from the Mississippi eastward, westward, or Gulfward is not sufficiently rapid to be detected and correlated with stages of the river except for a distance of a few hundreds yards from the channel.

It is obvious, however, that there may be a transmission of pressure, affecting the flow of wells more promptly and at a greater distance

n would the actual translation of the water itself. Data touching
m this interesting question are in the delta region unfortunately
cing, and this for two reasons: (1) Since the water there obtained
m wells is usually of poor quality, their number is not great, and
when they are put down they are nearly always on the bank of
ie navigable bayou where the villages and sugarhouses are to be
.nd. The fluctuations of such wells may be due, as explained
ve, mainly to the lateral transmission of river or bayou water, and
, to the simple transmission of pressure.

Vells farther west, some distance from the Mississippi and its
tributaries, show, as will be seen below, no appreciable effect of
nsmission of either water or pressure from the Mississippi.

Vo observations continuing throughout the whole year have been
de, so far as the writer is aware, of the height of water in the vari-
s deep wells in the southwest part of the State. As explained in the
:fatory notice to this paper, the facts upon which this report is based
re collected by the writer during the winter months, while engaged
general work of the State geological survey. However, several
ort series of observations have been made, covering intervals in three
ccessive years. In 1901 Mr. Pacheco, of the State survey, was kept
the field nearly two months for the sole purpose of making such
servations. The results of his observations, as published by the
ate survey, are as follows:

ariation of height of water in Hammill's well, 2½ miles south of station, Jennings, La.

1901.	Hour.	Feet.	Inches.	1901.	Hour.	Feet.	Inches.
Feb. 21	....	13	4.0	Apr. 29	a. m.	13	7.2
Apr. 20	....	13	9.5		p. m.	13	7.0
21	a. m.	13	9.0	30	a. m.	13	7.16
	p. m.	13	8.5		p. m.	13	7.12
22	9 a. m.	13	7.25	May 1	2 p. m.	13	7.0
	11 a. m.	13	7.0		4 p. m.	13	6.9
	12 m.	13	6.9		5 p. m.	13	6.8
	2 p. m.	13	6.87	5	....	13	7.75
	3 p. m.	13	6.75	6 { a. m. / p. m. }		13	7.75
	5 p. m.	13	6.75				
24	....	13	8.75	14	....	13	10.25
25	....	13	8.0	15	....	13	11.0
26	a. m.	13	8.33	16	....	13	11.75
	p. m.	13	8.25	17	....	14	0.125
27	10 a. m.	13	8.5	18	....	14	2.0
	11 a. m.	13	8.4	20	....	14	2.0
28	....	13	7.0	Water dropped below pump.			

2 p. m.

· · · · · · ● ۶ ۶

11 a. m.

Variation of height of water in Bower's well, Welsh, La.

1901.	Hour.	Feet.	Inches.	1901.	Hour.	Feet.	Inches
Feb. 26	....	4	6.0	May 12	....	7	2.5
Mar. 21	....	4	3.0	13	....	7	2.75
Apr. 20	.....۰........	7	1.25	14	....	7	2.75
23	....	7	1.5	15	....	4	3.5
24	8 a. m.	4	1.4	16	....	7	3.75
	10 a. m.	4	1.5	17	....	7	3.75
	11 a. m.	4	1.6	18	....	7	4.0
	12 m.	4	1.75	19	....	7	4.25
May 5	....	7	1.75	20	....	7	4.5
5	....	7	2.0	21	....	7	5.0
6	....	7	2.0	22	....	7	5.0
7	....	7	1.75	25	....	7	7.0
8	....	7	2.12	26	....	4	9.0
9	....	7	2.12	28	....	5	5.0
10	....	4	2.12	30	....	5	9.0
11	....	4	2.25				

Variation of heighth of water in Hawkeye rice mill well, Fenton, La.

1901.	Hour.	Feet.	Inches.	1902.	Hour.	Feet.	Inches.
Mar. 31	....	14	10	Mar. 7	....	18	3
May 5	.....:......	15	....	8	....	18	2

It will be observed that in these measurements the numbers under
et and inches indicate distances downward from some datum plane,
:nerally the top of the casing or the floor of the discharge trough.
s the season advances, the surface of the water in the wells gradually
·wers. The rate of lowering is not constant, but the total result of
le various fluctuations is to materially lower the water surface as
immer approaches. The noticeable acceleration in the rate of lower-
ig after May 15 is due to the beginning of pumping for rice irriga-
on. Perhaps there is nothing new or unexpected in these results
ius far. The variations shown throughout different hours of the day
re much more difficult of explanation. Very possibly, though, care-
ully kept barometric readings would give a clew to their meaning.

By far the most interesting and unexpected variations are those of
bout April 22, 1901, and February 25 to 27, 1902. Instead of the
radual downward course, there is indicated for these dates a notice-
ble rise. The Weather Bureau reports show that heavy showers
rere abundant on the 16th, 17th, and 18th of April, 1901, in this part
f the State, and from the 19th to the 26th of February, 1902.

Again, these same tendencies toward a lowering in summer and a
[uick response to local showers has been observed this year (1903), as
s shown by the following table:

Date.	May well.	Rice mill.	Planing mill.	Bower's wells.	
				North.	South.
	Feet.	Feet.	Feet.	Feet.	Feet.
Mar. 19....	6.00	....	6.4	....	....
21....	5.14	....	6.33	6.73	6.26
25....	5.85	6.70	6.50	6.87	6.63
July 12....	19.5	18.8	....	19.5	....
13....	19.2	....	....	....	....

Over 2 inches of rain fell on the 19th and 20th of March in this
vicinity, and from the changes in level noted in the foregoing tables
for previous years it is only to be expected that these wells would
show a similar change for a similar cause. Observe especially in the
May well how the water level rose on the 21st, but went back again on

the 25th. Notice, too, the effect of the summer with its pumping season, under July 12.

The marked effect of copious showers on the water level in the deep wells of southwestern Louisiana has not escaped the general observation of planters.[a]

The extent to which very local heavy showers affect the territory just without their limits is an interesting topic that thus far has not been investigated, nor have time and circumstances permitted the observation of effects produced by local or extensive rainfall in different directions from any given well or group of wells, though the importance of such observations, when a full explanation of the occurrence and conditions of the underground waters of this part of the State is attempted, can not be too much emphasized.

From what has already been said, it is evident that in some respects the waters of this section behave like the common "ground water" of this or any other well-watered land; but that, ordinarily, there is no very direct connection between the water of these deep wells and the ordinary soil supply is evident from the fact that at a number of places the deep waters flow several feet above the surface of the soil for miles around; and, again, the water in the casing of the deep wells never, so far as observed, stands at the same level as the water in the pit outside. Again, the supply of deep water is not obtained until one or, more generally, several, thick, impervious layers of clay have been penetrated.

Since the thickness and character of the sand and clay beds encountered in sinking wells but a short distance from one another may vary greatly, and since the position of a clay bed in one well may be taken by a sand bed in another it is very evident that, in southern Louisiana, the artesian and deep-well conditions are somewhat different from those encountered in regions where there is one great extensive underlying formation, sharply defined from overlying and underlying beds, and alone transmitting the deep underground flow. Yet some typical or ideal artesian features are represented in this part of the State. The first hundred or two hundred feet passed through in sinking deep wells contains comparatively few very porous layers; below, the sand usually becomes coarser, and sometimes thick beds of gravel are found. Gravel deposits are by no means uncommon to a depth of 1,000 feet, as will be seen by inspecting the logs of the wells put down in search for oil or deep artesian water and published herewith as Pl. II. Very coarse gravel is reported in the bottom of many of the best water wells throughout the Gulf border. As will be seen by referring to the record of a well just completed in Biloxi, Miss., the casing, over 900 feet down, is in extremely coarse gravel (see p. 31).

---

[a] For remarks on this point, see Rept. Geol. Survey Louisiana for 1902, p. 246.

Water naturally flows much more readily through coarse than
rough fine material. The best flowing or deep wells of southwest
ɔuisiana obtain their waters from very coarse sand or gravel beds.
ɪch beds are generally below 150 or 200 feet from the surface.
round-water features or characteristics decrease in this region down-
ard, according as those more typically artesian increase.

There is one more somewhat interesting fact connected with varia-
on in pressure head as noticed in the May well at Welsh, though
robably it is common to all others in this part of the State. On the
ᴵth of July no pumping was done, and from all appearances none
ɪd been done for several days. At 5 o'clock in the evening the water
ood 19.5 feet below the top of the mouth of the casing. Next morn-
ɪg the pump had run but an hour when, at the writer's request, it
as stopped in order that the stage of the water might be measured.
he surface of the water, after dropping suddenly, balanced up and
ɔwn for a moment and then appeared to have come to rest. Five
ɪinutes after the pump had ceased working, the water stood 18.6 feet
ɛlow the mouth of the casing. After the pump had been stopped for
venty minutes the water stood at 19.2 below the same datum plane.
: thus appears that the pumping, which was equivalent to a flow of
200 gallons per minute, or 72,000 gallons per hour, had not in one
ɔur's time materially lowered the water level--in fact, had actually
ised it temporarily.

That long-continued pumping does lower the level of the water in
ells is understood by all who are connected with deep-water supplies.
ɔr example, in July, 1903, Mr. Roanes's place was visited, and,
though under ordinary circumstances his wells are flowing, at that
ne, owing to several hours of intermittent pumping, continuing for
period of several days, the water stood just below the tops of the
pes.

The Fabacher well in New Orleans (see p. 44), which ordinarily flows
ntinuously from a 4-inch pipe but 2 feet or less above the general
ᵛel of the ground, will, if suddenly turned into a smaller pipe, rise up
d overflow for a few minutes to a height of 11 or 11.5 feet above
e ground. Then the water gradually descends to a permanent head
about 10 feet above the ground. The cause of the temporary,
ɪusually high head in the above-mentioned cases is doubtless attrib-
able to the momentum of the water in the porous sand or gravel bed
low. What seems worthy of special note is the length of time
quired for the water to descend to its normal head, especially in the
se of wells that have just been pumped.

WELL DRILLING AND PUMPING.

METHODS OF DRILLING.

In southern Louisiana practically but one method is used in sinking wells, either for water or oil. This consists primarily of loosening the earth with a hollow revolving bit and bringing it to the surface by the upward current of water obtained by forcing water down through the hollow bit. There are, to be sure, many different devices for producing the necessary rotating motion, many differently shaped bits, and many different sized and shaped derricks used; but the fundamental principle of drilling is the same with all.

Preferences as to kind and size of well desired differ considerably in different localities. East of the Mississippi and north of Lake Pontchartrain most of the wells are furnished with a 2-inch casing, and the water is expected to flow at the surface of the ground or even some feet above. The wells are used for dairy or ordinary household purposes. West of the delta region the wells are usually 6, 8, 10, or 12 inches in diameter, the water is not expected to rise to the surface, and irrigation is the main object for which the wells are put down. As a result of the number, kind, and size of wells required in different sections of the State, methods of drilling varying somewhat in detail are resorted to by local drillers.

JETTING.

Fig. 10 shows what is usually called the jetting process. The traction engine furnishes steam to run the small force pump (A), which obtains water from a local source and pumps it through a strong hose (B) to the drill pipe (C). The rotating of the pipe with bit attached is here accomplished by the simple method of temporarily attaching a Stilson wrench (H) and moving it to and fro. The pipe carrying the downward current of water with the bit is held up by a block (D) and ropes (E) and is moved up and down every few seconds by power from the engine transmitted by a rope and the force of gravity. The rope going to the engine in this case simply passed over a drum or large spool about 6 inches in diameter, on the outer end of the flywheel shaft. Two or three turns only were made around this drum, and when no work was required of the rope the engine continued turning, but the coils were allowed to slip loosely on the drum. By tightening the coils the drill pipe was immediately raised. The drill pipe in this instance is about 1¼ inches in diameter, while the casing is about 2½ inches. The casing is sunk nearly as far as the drill has penetrated, and the return water, laden with drillings, comes up between the pipes. Its exit is shown at F.

:arning back and forth on the long-handled wrench at **K** the
is loosened from the outside sand and clay and ordinarily readily
ls by its own weight about as rapidly as the jet clears the way,
some instances is forced down by driving.

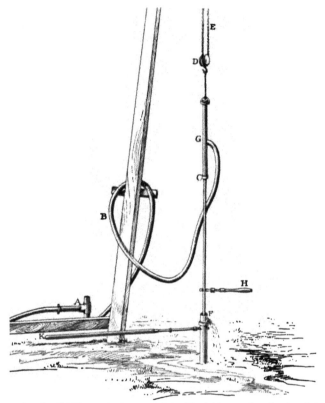

Portion of well-drilling outfit of Bacon and Gamble, sinking a well at Ponchatoula by
the jetting process.

he drill descends and the swivel coupling (G) approaches the top
casing, the coupling is unscrewed and another length of drill
2 to 20 feet long, is put in, and the drilling is continued.

ROTARY PROCESS.

:re many wells of large diameter are to be put down, as in the
estern part of the State, much of the manual labor required by
ove-described process is done away with by the use of a mechan-

After the desired depth has been attained, the 3-inch pipe is removed, section by section, and the 6-inch, 10-inch, or 12-inch casing is hoisted up and sunk into the hole made by the 3-inch pipe and its arrow-head bit. The hole is often nearly 14 inches in diameter.

The first one, two, or three sections of this large pipe or "casing" are perforated and form the strainer, near the bottom of the completed well. If the strainer is to be three lengths long, say 60 feet, care is taken to insert in the casing three lengths of 3-inch pipe and to fill the space between this inner and the outside pipe with shavings so that it can not fill with earthy matter while descending. Length after length of casing is screwed on and lowered until the desired amount is sunk into the ground. In case it does not descend readily of its own accord, resort is had to rotating the casing by machinery precisely as the 3-inch pipe was rotated in the beginning. The lower margin of the casing is cut with points like saw teeth, so that it answers fairly well as a drill or auger.[b] The upper end of the 3-inch pipe within carries a conical sleeve, so that it can be caught readily by the thread end of other lengths that are lowered afterwards and coupled up with the three lengths already spoken of as being in the strainer part of the casing. The shavings can now be jettied out, the interior pipe withdrawn, and the well "pumped" to withdraw all the muddy impurities forced down while drilling, as well as fine sand that might eventually fill up the strainer.

One of the most satisfactory methods of drilling is by portable outfits, in which the derrick, traction and dummy engines, pumps, etc.,

[a] Harris, G. D., and Pacheco, J., The subterranean waters of Louisiana: Rept. Geol. Survey Louisiana for 1902, pt. 6, Special Report No. 6, pp. 236–238.

[b] Bond, Frank, Irrigation of rice in the United States: U. S. Dept. Agr. Exp. Sta. Bull. No. 113, p.

A. MAY PUMPING PLANT, WELSH, LA.

Shows general appearance of small stations throughout the rice district of southwestern Louisiana.

B. PUMPING FROM A 12-INCH WELL ON THE FARM OF A. E. LEE, 4 MILES
NORTHWEST OF CROWLEY, LA.

: loaded on special carriages. The lightness of this rig and the
1sequent facility with which it can be moved from place to place
1d to make it popular in regions where depths no greater than 300
't are to be drilled. For various styles of light derricks see the State
:vey report already referred to (Pls. XLII and XLIII).

If it is expected that drilling will be carried to a depth of 500 or
1000 feet, larger, stronger outfits are called for. The great advantage
the taller form of derrick is that in hoisting the drill pipe or casing,
1enever necessary, it can be uncoupled two lengths at a time instead
length by length, so that nearly half the labor required to remove
replace the pipe is thus avoided.

Oil wells that reach depths of 1,000, 2,000, or even 3,000 feet are put
wn by similar but heavier outfits. The derrick is sufficiently high
allow the pipes and casing to be removed three lengths at a time.

SCREENS.

Nearly every driller has his own ideas as to the proper manner of
·ating or placing the lower end of the casing so that a well may have
free inflow of water and at the same time may not be liable to clog
. Many assert that all ordinary screens are liable to give out and
in the wells they are in. No screen at all is most satisfactory if the
ver end of the pipe is set in very coarse gravel with no mixture of
.y or fine sand. Some advocate the pumping out of several tons of
er material from around the bottom of the pipe and the forcing
wn, in its stead, of several wagonloads of gravel, so as to make a
bble screen.

As a rule, however, some kind of metallic screen is used. Mr.
1nd, in the bulletin referred to above, thus describes a common type
use in southwestern Louisiana:

In the screens now generally used perforations in the well casing are three-fourths
seven-eighths of an inch in diameter, and the distance between centers averages
1ut 1½ inch, the perforated portion being carefully wound with galvanized-iron
'e. On 10-inch pipe No. 14 wire is wound nine wires to the inch; on 18-inch
e No. 16 wire is wound eleven wires to the inch; on 6-inch pipe No. 17 wire is
und fourteen wires to the inch. A common machine-shop lathe is used for wind-
· the wire upon the casing, and the wire is not only wound on tightly, but is
dered in place to prevent its sliding, so as to close openings between strands.
·en rows of solder are placed upon a 10-inch pipe, the number increasing with
ger pipe and decreasing with smaller pipe.

Fig. 11 is taken from Bond's work, and represents the casing, holes,
re, rows of solder as he has just described them.

Pl. VI, B. shows a different method of constructing a screen. The
res are wound much farther apart than in the type above described.
·er the wires is placed fine brass gauze. The pipe is then wound

again over the gauze in the opposite oblique direction. The outside coarse wire is mainly to protect the brass gauze, while the inner coarse

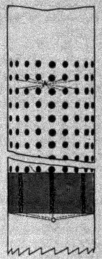

wire is to hold the same from fitting down tight upon the exterior of the pipe, thus shutting out all ingress of water except immediately over the bored holes.

Machinists very quickly find it to their advantage to have three to five strands winding at once, side by side, not simply one at a time, as represented in Bond's figure.

The lower end of these pipes is generally closed by a ball valve that is so constructed as to allow the jet of water to pass down and out, but immediately closes against any pressure from below. This is to prevent the entrance of fine sand or other foreign substance.

PUMPING.

As would naturally be supposed, water is pumped from deep wells of southwestern Louisiana by steam power. Formerly the fuel used for generating steam was wood from the nearby lowlands or banks of the bayous or coal brought

Fig. 11.—A common method of constructing a screen.

from Alabama or Kansas City by rail. Since the discovery of oil in such quantities at Beaumont, Tex., nearly all the pumping plants have

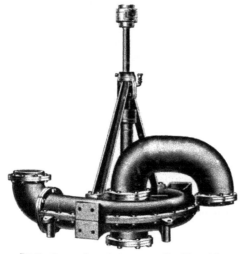

Fig. 12.—Common form of rotary pump. Van Wie model. -

erected tanks at an elevation of from 8 to 10 feet above the boiler furnace, and so are able to store and use oil in a very easy and economcal manner. However, as the price of oil gradually rises above 80 cents per barrel there is a tendency to return to the old methods and materials for making steam. Pl. VII, A. shows a typical small pumping plant of to-day, with its fuel tank and cheap board structure with engine inside.

Pl. VII, B, shows the rear of a similar plant. A centrifugal pump (see fig. 12) is on the lower end of the same shaft that carries the band wheel. It is placed in a wooden-curbed well sufficiently low to be beneath the surface of the water at the driest season of the year. When it is not so placed resort must be had to priming every time the pump is started. Around Kinder and China, where the usual head of the water is 25 feet below the surface of the ground, the pumps are depressed to a depth of 25 or 30 feet.

WATER SUPPLIES FROM WELLS IN SOUTHERN LOUISIANA.

By M. L. FULLER.

INCREASED USE OF UNDERGROUND WATER.

The past decade has witnessed a great impetus to well drilling in southern Louisiana, and as the advantages of underground water supplies become better understood, more and more attention will be given to such sources. The use of underground waters for the irrigation of rice has led to the sinking of an unusually large number of wells, especially in the region along the coast, where values in some localities have increased five to ten fold within the last ten years through the reclamation of the land by irrigation. The use of water for this purpose will be considered in detail in the section on "Rice irrigation in southern Louisiana," the present discussion being limited to town, domestic, farm, railroad, and manufacturing supplies.

TOWN AND DOMESTIC SUPPLIES.

Increasing attention is always given to the quality of water supplies as a country becomes older. In the early stages of development the settlements are of small size, and are more or less remote from one another. Even within the villages themselves the houses are generally scattering. Under such conditions a sufficient water supply can usually be had near at hand, either from surface streams, or from springs, or shallow wells, though in some instances a deep supply must be sought from the start. As the country develops and the villages and towns become more crowded the original sources of supply are frequently either exhausted or become too contaminated for use.

Contamination of the shallow wells, where arising from local sources, can frequently be prevented by proper systems of drainage or sewage disposal, but in small communities such systems are often more expensive than a new and deeper system of water supply. Moreover, shallow wells of the open type are not only liable to pollution by the entrance of surface water or of ground water of the surface zone, both of which are often charged with matter derived from stables, privies, cesspools, etc., but receive more or less refuse blown in by the wind from the adjoining yards or streets, while small

74

imals not uncommonly fall into such wells and the water is contamated by their decaying bodies. The odor and taste of the water in me instances, and the odorous masses of muck removed in cleaning others, attest the occurrence of large amounts of decaying organic atter. In fact, an open well can seldom be so guarded as to entirely revent pollution, and although often not especially deleterious to ealth, the water is rarely equal to that of a driven well of the same epth, from which it is possible to shut out all waters from or near e surface.

In the case of streams, the contamination may not, and in fact usuly does not, rest with the community using the water, but with her villages or cities farther up the stream, perhaps in another ate. Little can be done in such cases toward removing the sources pollution, and to secure even a moderately pure supply resort must had to filtration or other processes of purification, or to deep wells. In many parts of the country little water can be obtained from deep lls, but in southern Louisiana the conditions are exceptionally orable for obtaining satisfactory supplies in this manner. Fig. 7, ge 28, shows graphically the subdivisions of this portion of the ate as regards the occurrence of underground water. It will be ted that there are three definite east-west belts, each of which is ected by the northwest-southeast belt along Mississippi River. e latter belt, which in area is the largest of them all, is made up of ds consisting mainly of materials deposited by the river in recent ological times, though older deposits sometimes show at the surface. general it consists of alternations of sands and mucks, all of which rry more or less organic matter. In this area water can be obtained almost any depth, but it carries a large amount of iron and organic atter, and although used for drinking purposes and for watering ock, has a decidedly deleterious action on health and is a great hinrance to the proper development of the region, especially as the vailable surface supplies except along Mississippi, Red, and Atchalaya rivers, Bayou Lafourche, Bayou Teche, etc., are mainly from uggish streams and bayous (Pl. VIII), which are generally equally d. Certain of the waters of this belt are, however, sometimes aced on the market as mineral waters, and are used for bathing at veral resorts.

The most southerly of the three east-west belts affords the best underound supplies. In this area the water can be obtained at a moderate pth, flows without pumping, and is of good quality. Most of the wns depending for their public supplies on wells (see table on p. 77) e located in this belt. In the middle east-west belt pure supplies are so obtained at no great depths, but in general the wells do not flow. e towns listed in the table mentioned and not situated in the southn belt occur in the middle belt. In the northern belt the land is

barrier to the passage of surface waters, will be entirely satisfactory as far as freedom from contamination is concerned. In flat areas a source of supply 20 to 30 feet below the nearest source of pollution would probably be safe to use if all access to surface waters were cut off by proper casing.

Among the disadvantages of underground supplies are (1) their uncertain distribution and depth, (2) their uncertain quality, (3) the cost of deep wells, (4) the cost of pumping nonflowing wells, and (5) the insufficiency of supply in certain crowded communities and in some irrigable areas. The first two objections are of great importance. The conditions of the occurrence of waters are, however, well understood by those who have investigated them, and valuable information can usually be obtained from the numerous State or national bureaus engaged in the study of the subject. The cost of drilling and pumping deep wells is nearly always greater than the cost of obtaining surface supplies where the latter are at hand, but this is offset in many parts of southern Louisiana by the greater purity and the consequent greater number of uses to which well water can be put, and by the greater number of points at which it can be obtained. Good wells can frequently be obtained at localities far removed from surface sources and in such instances afford the only means of development of the country. The fifth objection is one that is less readily met, but

A CHARACTERISTIC BAYOU OF THE MORE SLUGGISH TYPE IN THE GULF COASTAL REGION.

Analyses of artesian waters from southern Louisiana.

[Parts per million.]

Name of well.	Locality.	Solid matter.	Ash.	Organic matter.	Albu-minoids.	Free ammonia.	Nitrites.	Nitrates.	CaO.	K₂O.	?	Remarks.
A. A. Bayer	Mandeville	208.40	178.4	30.0	0.14	0.04	Trace.	0.8	4.0	4.14	1.19	Colorless, with little suspended matter.
Moresi	Jeanerette	480.0	400.0	80.0	.96	.10	.02	.50	30.40	8.62	.64	Very cloudy.
Moresi (barnyard)	...do	417.8	366.8	51.0	.02	1.08	Trace.	1.0	91.0	7.93	4.84	Cloudy.
E. Dessome, at flower garden.	Mandeville	179.4	150.4	29.0	.06	.09	Trace.	.2	2.30	6.40	.59	Perfectly clear.
Mrs. Aubert	Abita Springs	184.0	154.0	30.0	None.	.05	.06	.24	.7	4.80	.74	Do.
Hernandez, by house.	Covington	161.6	133.0	28.6	None.	.08	Trace.	.4	15.4	8.27	.69	Colorless, with little suspended matter.
Singletry's still	...do	139.0	117.0	22.0	None.	.01	None.	.72		7.94	.48	Colorless, with suspended matter.
Lockmore & Co	{West Lake / Lake Charles}	288.0	214.0	54.0	.10	.01	None.	.20	44.0	3.16	.42	Colorless, with suspended matter (fishy smell).
Bradley & Ramsay Co.	Lake Charles	245.0	219.0	26.0	None.	.09	None.	.20	36.0	4.33	.21	Whitish cloudiness.
Reiser machine shop	...do	280.0	235.0	25.0	None.	.08	Trace.	.32	46.5	3.94	.26	Do.
Menafee Lumber Co	...do	271.4	235.0	38.4	None.	.18	None.	.32	33.3	2.52	.42	Slightly cloudy.
Judge E. D. Miller	...do	277.0	229.0	48.0	None.	.11	None.	.32	48.2	2.56	.42	Do.
Oaks Hotel	Hammond	185.0	144.6	40.4	.16	.14	.001	.80	5.0	0.98	(?)	Colorless, with suspended matter.
Doctor Robinson	...do	187.0	152.0	35.0	.10	.014	Trace.	.29	6.0	216	3.40	

rface supplies are often subject to the same drawbacks and fail
here wells succeed.

In the following table is given a list of the towns and cities in
uthern Louisiana depending in whole or in part on deep wells for
eir supplies. Doubtless all of the inhabitants of a given community
o not draw upon the public supply; but, on the other hand, the
aters are frequently piped beyond the corporate limits. The total
umber of persons using water from the deep wells in the localities
entioned is, therefore, probably not far from the number indicated
· the figures of population, aggregating about 45,000. To these
ust be added the large but unknown number living in the smaller
wns, or scattered throughout the country, who draw their supplies
om deep private wells. Many of the more important hotels possess
ch wells. When it is borne in mind that the amount of sickness
d number of deaths is much lower among those using deep waters,
d that the productiveness of the State is thereby increased by
ndreds of thousands of dollars, the importance of pure water
pplies will be appreciated.

Cities and towns depending on deep wells for public water supplies.

[Compiled from Insurance Maps of Sanborn Map Company and other sources.]

Town.	Parish.	Date of information.	Population. 1900.	Number of wells.	Depth of wells.	Method of storage.
xandria	Rapides	1900	5,640	2	500,760	Standpipe.
ton Rouge	East Baton Rouge	1903	11,269	2	758,800	Do.
wley	Acadia	1902	4,214	2	170,270	Do.
anklin	St. Mary	1899	2,692	2		Do.
nerette	Iberia	1898	1,905	1		Tanks 60 feet deep.
nnings	Calcasieu	1903	1,539	1	240	steel tank on trestle.
ayette	Lafayette		3,314	3	150,150,225	Brick reservoir.
ke Charles	Calcasieu	1903	6,680	2		Standpipe.
Slousas	St. Landry	1899	2,951	2	175,184	standpipe and reservoir.
aquemine	Iberville	1900	3,590	2		Elevated tank.
nea	Acadia	1903	1,007	1	250	steel tank on trestle.

a System proposed.

The interest exhibited in the problems of pure water supplies is made
anifest by a constantly increasing number of analyses, especially
ose of a sanitary character. A number of analyses, in part sanitary
d in part purely chemical, made by the State experiment station[a]
e given in the following table:

a Rept. Geol. Survey Louisiana for 1902, pt. 6, special report No. 6, pp. 251-252.

Analyses of artesian waters from southern Louisiana.

[Parts per million.]

Name of well.	Locality.	Solid matter.	Ash.	Organic matter.	Albu-minoids.	Free-ammonia.	Nitrites.	Nitrates.	CaO.	K₂O.	P₂O₅.	Remarks.
A. A. Bayer	Mandeville	208.40	178.4	30.0	0.14	0.04	Trace.	0.8	4.0	4.11	1.19	
Moresi	Jeanerette	480.0	400.0	80.0	.06	.10	.02	.56	33.40	6.62	.64	
Moresi (barnyard)	do	417.8	366.8	51.0	.02	1.66	Trace.	1.0	91.0	7.65	4.84	
E. Dessonic, at flower garden.	Mandeville	179.4	150.4	29.0	.06	.09	Trace.	.2	2.30	6.46	.59	
Mrs. Aubert	Abita Springs	184.0	154.0	30.0	None.	.05	.06	.24	.7	4.80	.74	
Hernandez, by house.	Covington	161.6	133.0	28.6	None.	.08	Trace.	.4	15.4	8.27	.69	
Singletry's still	do	139.0	117.0	22.0	None.	.01	None.	72	......	7.34	.48	
Lockmore & Co	{ West Lake Lake Charles	288.0	214.0	54.0	.10	.01	None.	20	44.0	3.16	.42	
Bradley & Ramsay Co.	Lake Charles	245.0	219.0	26.0	None.	.09	None.	.20	36.0	4.33	.21	itish
Reiser machine shop	do	280.0	235.0	25.0	None.	.08	Trace.	.52	46.5	3.01	.25	Do.
Menafee Lumber Co.	do	271.4	235.0	36.4	None.	.18	None.	.32	33.3	2.52	.42	Slightly
Judge E. D. Miller	do	277.0	229.0	48.0	None.	.11	None.	.32	48.2	2.54	.42	
Oaks Hotel	Hammond	185.0	144.6	40.4	.16	.14	.001	.80	5.0	0.96	(*)	
Doctor Robinson	do	187.0	152.0	35.0	.19	.014	Trace.	.25	6.0	.31	3.40	

Owl Bayou Cypress Co. Strater station	619.0	583.4	85.6	.50	3.08	Trace.	.05	2.0	1.28	1.02
Ponchatoula town well. Ponchatoula	265.0	179.0	26.0	None.	.07	Trace.	.44	11.50	1.48	6.40
...do...	237.0	198.0	39.0	.205	Trace.	.60	.80			11.64
Biegel's overflow (232 feet deep).								2.0	10.24	
Biegel's pump well (100 feet deep).	512.6	450.0	62.6	.20	1.98	.40	1.0			

a Not enough to determine.

In the above analyses the total amount of solid matter held by the water is indicated in the third column. This solid matter is the residue which is left after the water has been evaporated to dryness, and includes both the suspended matter noted in the last column and matter held in solution. The waters of the table show less foreign matter than is generally found in ground waters, except in the New England States. Iowa ground waters, for example, usually contain from 1,000 to 12,500 parts of residue per million; Illinois waters run from 300 to 1,200, while those of Kansas vary from 1,000 to 7,000 parts. The "ash" column indicates the amount of residue left after the solid matter of the previous column has been heated to a red heat. In the table the difference between the amounts of the ash and the solid matter is given in the third column as organic matter. In reality this is not all organic matter, but includes the volatile parts of carbonates, nitrates, etc., the amounts being, therefore, considerably too large. The figures in the sixth or albuminoid column indicate the amount of ammonia in actual organic combination, while the figures in the free ammonia, nitrite, and nitrate columns represent the amounts in each of the progressive stages through which organic matter passes during its oxidation to mineral matter. No single determination is an absolute indication of the quality of the water, but in general an association of high ammonia and nitrites, especially when associated with high chlorine, indicates pollution by sewage.

RAILROAD SUPPLIES.

ѕ of the most common and important uses of underground water
· the locomotive supplies of railroads. On every line, usually
few miles apart, are located the familiar water tanks, each of
hundreds or thousands of barrels capacity. For these, pure sup-
must be had. The waters of the bayous and streams are in many
ices unsatisfactory for locomotive use, and wells are commonly
ed to. Relatively little difficulty is encountered in southern
iana in obtaining water in this way. Except in the Mississippi
nds, water of a satisfactory quality may usually be obtained in
ᵻ amounts at moderate depths. In general, the waters give only
amounts of boiler scale.

MANUFACTURING SUPPLIES.

der this head are included both the boiler supplies of manufac-
ᵷ establishments and the supplies used directly in manufacturing
sses. The statements in regard to waters required for railroad
ιotives apply equally to boiler waters in other lines. Taken as a
ᵻ, such supplies are of great importance, the very existence of
ιdustries of certain localities being largely dependent upon them.
lumber business, with its numerous saw and planing mills,
nds supplies of pure water at a great number of points. This
· is generally obtained from deep wells. Well waters are also of
importance in many other industries, and as these increase in
er, variety, size, and output the economic value of water will
ırtionally increase.

the processes in which water plays a direct and leading part the
facture of ice is the most important. Many of the cities and
ι, including Baton Rouge. Crowley, Covington, Hammond,
ιrette, and New Iberia, employ well water for this purpose.

ιʀʀ 101—04——6

Rice land at a distance from railroads and not under canals may still be had for about $15 an acre, and will doubtless yield fair profits if carefully developed. It is with a view of calling attention to the possibilities of rice irrigation that the present description has been prepared. Many of the main facts here presented are taken from the descriptions of Mr. Frank Bond, who investigated the subject for the Office of Experiment Stations, United States Department of Agriculture, and published a report of his investigations as Bulletin 113 of that Bureau. The statistics of the use of wells for rice irrigation, which are brought up to the end of 1902, are presented through the kindness of the Bureau of the Census, Department of Commerce and Labor.

The first people to plant rice in southern Louisiana were the Acadians, who, after their expulsion from Nova Scotia by the English in 1755, settled in considerable numbers in Louisiana and planted small areas to rice. The cultivation, primitive in its methods, was confined to the lowlands along the bayous, the prairies affording pasturage for their herds of cattle. The lowland areas seldom admitted of satisfactory drainage and were too small for profitable cultivation. The crops frequently failed in years of deficient rainfall. Attempts were made to create additional water supplies by building levees across low sags

A. PUMPING STATION ON ONE OF THE LARGER STREAMS OF THE GULF
COASTAL REGION.

B. DISCHARGE OF A HEAVY PUMP SYSTEM.

coulees at poin..s higher than the cultivated areas, but in most cases
her the rainfall proved deficient or the capacity of the re ervoirs too
ited.

Little advance was made over the Acadian methods until a very
:ent date. Experiments in unusually wet years had served to show
it the soils of the prairies were adapted to the growth of rice if suf-
ient water was at hand. This led to the trial of pumps as a means
raising water from the bayous to the rice fields. So successful was
: test that pumps were at once installed at many points, and in a few
urs tens of thousands of acres of previously nearly worthless land
ng from 10 to 70 feet above the bayous were put under cultivation.
The first important pump was installed in 1894. It was a vacuum
mp of the pattern used in the mining camps of the Northwest, and
s established on the Bayou Plaquemine, in Acadia Parish, near
owley. Although its failure at a critical time involved the partial
s of the crop, it showed the possibilities of pumping methods. In
: following year a centrifugal pump was introduced, but was too
all to meet the demands made upon it, and was succeeded in 1896
a pump having a capacity of 5,000 gallons per minute, which by its
:cess opened a new era in rice cultivation. Still larger pumps have
ice been introduced, both of the centrifugal and rotary types. These
ve discharge pipes ranging from 12 to 60 inches in diameter, and
se 20 to 100 cubic feet of water per second through a distance of
veral feet. In the larger plants batteries of pumps operated by
mpound Corliss engines of 400 to 800 horsepower are in common
e. Pl. IX, A, shows the surroundings of a typical pumping plant
awing its supply from a stream or bayou, while B of the same plate
res a good idea of the volume of discharge from a powerful battery
pumps.

SOURCES OF WATER.

Bayous.—In the early stages of rice irrigation practically all the
iter was drawn from the bayous. In the portion of Louisiana
voted to the cultivation of rice these are the channels of the slug-
sh streams draining the prairies or marshes and not the abandoned
distributary channels of a river, such as those near the Mississippi.
physical aspect, however, they are very similar. The current,
ough fairly strong at certain seasons, is very weak at others; the
w moving waters resting in channels sunk below the prairies are
ore or less clogged in many instances by snags of waterlogged
mps, logs, trees, etc., and bordered by dense vegetation, including
: constantly encroaching cypress. Notwithstanding the sluggish-
ss of the currents throughout the greater part of the year, however,
: bayous maintain deep channels, the bottoms of which, in the region
ar the coast, are often many feet below the level of the sea.

Wells.—The bayous for a number of years furnished an adequate supply of water for the areas under cultivation, but the increase of acreage, combined with a deficiency of rainfall, as in 1901, brought to the attention of everyone the inadequacy of the supply under such conditions. In that year considerable areas planted to rice had to be abandoned, as the water supply failed, and in many localities the bayous were so lowered that salt water entered and by rendering the supply brackish still further reduced the production. Emphasis was added to the fact already predicted that in dry years considerable areas remote from the bayous would either have to be abandoned or a new supply obtained. Deep wells, however, had already yielded abundant supplies at several points and were now regarded as the key to the situation. A number of the early wells are indicated on a map issued by the Office of Experiment Stations of the Department of Agriculture, and in part reproduced as Pl. XI. To these have been added a number of new wells.ᵃ The wells shown, however, should be regarded as indicating the locations rather than the exact number, as at the close of 1903 several hundred wells existed where only a few are shown on the map.

IRRIGATION SYSTEMS IN OPERATION.

The following tables give an idea of the extent and importance of the use of well and combined well and bayou systems for the irrigation of rice in Louisiana:

Owners, acreage, and cost of rice irrigation from wells in 1902.

[As reported to the Bureau of the Census.]

ACADIA PARISH.

Owner of well system.	Post-office.	Acres irrigated in 1902.	Farms irrigated.	Length of main canal in miles.	Total cost.
Cromwell, Wm.	Abbott	160	1	⁰ ¾	$25.00
Scanlan, Denis J.	....do	225	1	¼ ¾	25.00
Baronsse, E.	Branch	200	1		24.30
Burns, J. W.	....do	200	1		30.00
Edgar, H. F.	....do	160	1	¾	35.00
Prosper Bros. & Edgar	....do	40	1		15.00
Allen, Roble	Crowley	200	1	¾	42.00
Black, R. J.	....do	550	2		56.00
Carper, Benjamin F.	....do	450	1	¼	67.60
Cromwell, E.	....do	450	2		32.00
Cullumber, Chas.	....do	300	1		25.00
Gluters, Oscar	....do	170	2		19.48
Hoag, Philip H.	Jennings, **Calcasieu** Parish.	150	1	1	25.20
Jamison, Thomas	Crowley	130	1		10.40
Kraus, George	....do	120	1		18.40
Lee, Alonzo E	....do	200	1		27.40
Lineberger, Jacob	....do	120	1		20.45
Linscombe, John (manager)	....do	80	1		30.46
Minga, J. A.	....do	430	2	1	42.62
Omealy, Geo. H.	....do	520	2	¼	25.22

ᵃ Information furnished by Mr. A. C. Veatch.

PUMPING PLANT, BAYOU DES CANNES.

…s, acreage, and cost of rice irrigation from wells in 1902—Continued.

ACADIA PARISH—Continued.

…of well system.	Post-office.	Acres irrigated in 1902.	Farms irrigated.	Length of main canal in miles.	Total cost.
....	Crowley	550	3	2¼	$52.00
....	.....do	140	1	¼	26.50
....	.....do	200	1	2	45.00
…nd D.	.....do	200	1	....	33.00
W.	.....do	90	1	¼	21.88
d	.....do	500	1	....	40.00
…m.	.....do	300	2	....	30.00
F. (manager Crowley …).	.....do	480	1	2	58.50
…nas and Walter	.....do	300	1	....	44.00
and Isaac	.....do	218	2	....	35.00
....	Gervais	220	2	¼	7.75
....	Iota	150	1	....	24.00
…ey J	.....do	160	1	....	25.00
…rell	.....do	130	1	....	6.00
…S. and J. J	.....do	275	2	1	53.00
....	Mermenton	80	1	....	24.45
....	Rayne	100	1	1	36.75
....	.....do	200	1	¼	24.00
....	.....do	150	1	....	1.15
....	.....do	515	5	1	47.00
....	.....do	300	4	1	26.00
…nie	.....do	450	4	2	25.20
…rad	.....do	275	4	....	25.00
....	.....do	100	1	¼	35.00
…ois	.....do	152	1	....	24.00
....	.....do	300	3	1	21.00
....	.....do	175	1	1½	20.50
....	.....do	60	1	....	8.22
…d.	.....do	100	1	....	30.68
…J	.....do	200	1	1	31.80
K	.....do	100	1	....	48.00
…K. Andrews, tenant) ..	Moweaqua, Ill	300	1	1	50.00
H	Santo	30	1	1	68.25
....	.....do	200	1	¼	32.00
…on	.....do	150	1	2	37.00
....	Star	150	2	1	27.00
....	.....do	140	1	¼	25.50
....	.....do	125	1	....	31.00
y	Duson, Lafayette Parish.	100	1	¼	24.00
....	Crowley	400	1	1½	55.00
…obert	.....do	200	1	2¼	82.00
....	Eunice	250	3	¼	31.00
....	Mermenton	800	4	3	18.00
…ppins	Rayne	1,000	1	4	15.00
....	.....do	150	1	....	26.00
....	.....do	500	1	2	19.00
....	.....do	700	1	3	70.00
....	.....do	480		1½	35.00
....	Star	200	1	1	5.00
....	.....do	200	1	1	5.00

Name	Location			
Maund, George E	do		2	......	31.00
Eastman, W. W.	do		1	......	25.00
Garlick, Geo. W	do	---	1	1	26.50
Harris, W. E.	do	300	1	......	37.00
Anderson, Albert	do	830	2	1	65.00
Jones, Augustus	do	175	1	......	25.50
Jones, Perry B.	do	320	2		40.50
Kenny, J. A. (manager)	do	800	1		32.00
Marsh, Martin V.	do	50	1	2	25.00
Maund, James	do	150	1	......	26.00
Meyers, John R	do	120	1	1½	22.30
Pearl, John	do	575	3	......	62.00
Remage, Dr. G. W	do	140	1	1	29.00
Roberts, John H	do	120	1	......	20.00
Twitchell, V. M	do	120	1	......	29.00
White, H	do	300	1	......	30.00
Oden, R. E.	Kinder	600	3	1	69.00
Cary, Howard L	Jennings	600	5	4	208.00
Garlick, G. W	do	240	1	......	27.90
Braden, John E.	Lake Arthur	300	1		55.00
Camp, R. E	do	200	2	......	33.10
Traham, Euzebe	do	150	3		29.30
Winn, T. H	do	850	1	3	131.00
Baker, M. S	do	120	1	......	32.20
Wilcason, Dr.	Jennings	200	1		16.00
Demeist, John B.	do	122	1		26.00
Hamond	Lake Charles	240	1	1½	48.00
Harlan, A. D.	do	200	1	......	34.00
Sherman, Mark	do	90	1	......	2.00
Raymond, Charles	Raymond	100	1	......	19.00

mers, acreage, and cost of rice irrigation from wells in 1902—Continued.

CALCASIEU PARISH—Continued.

rner of well system.	Post-office.	Acres irrigated in 1902.	Farms irrigated.	Length of main canal in miles.	Total cost.
....	Roanoke	300	1	....	$29.00
....	....do	300	1	....	75.00
. A. (Auburn, Ala)	....do	100	1	⅛	20.00
homas	....do	300	1	....	25.00
n	....do	150	1	....	27.00
O. R.....................	....do	145	1	....	30.00
H	....do	100	1	....	17.45
L. N.....................	....do	155	1	....	31.00
J. B	....do	80	1	....	33.00
'. B.....................	....do	110	1	....	18.00
C	....do	200	2	....	25.00
E. T.....................	....do	130	1	....	12.00
ianders Plantation (W. H. ianager).	....do	200	1	....	35.00
avid and John.............	....do	125	1	1	22.25
on, C. H	....do	850	1	1	135.00
8........................	....do	500	1	....	57.00
. M......................	....do	740	1	1½	95.00
A........................	Welsh	450	1	....	65.00
ner......................	....do	250	2	1½	37.00
N	....do	90	2	⅛	19.90
Spaulding	....do	200	1	....	25.00
ieo......................	....do	160	2	....	16.70
., and Patterson, A. D	....do	110	1	....	18.00
s........................	....do	300	1	....	47.25
I. B	....do	100	1	....	28.00
....	....do	30	1	....	10.00
H. A	....do	80	1	....	21.00
....	....do	350	1	1	47.18
ons, M	....do	100	1	⅛	25.55
est......................	....do	200	1	2	42.00
irles....................	....do	140	1	....	31.00
....	....do	300	1	....	26.00
'rentice.................	....do	320	1	⅛	29.40
, Wm. and A. R...........	....do	900	1	2½	18.50
I........................	....do	450	2	1	36.02
i........................	....do	180	2	....	25.00
.ward....................	....do	500	1	1	47.00
....	....do	65	1	....	24.00
, S. A	....do	200	3	⅛	24.28
'red.....................	....do	200	1	....	22.00
....	Lake Arthur..........	3,300	10	3	80.00
Managan..................	West Lake............	150	1	3	42.00

VERMILION PARISH.

Adam	Abbeville	150	1	....	$22.93
....	....do	300	2	....	33.00
'm.......................	....do	207	3	....	16.25
Ion. R. P.................	....do	240	1	....	41.00
we, Edmond	....do	225	1	1	21.35
Planting and Milling Co. d).	....do	1,200	1	4	100.00

Laurents, Mrs. G	Laurents	100	1	1	25.50
Laurents, Jules G	do		3	3½	45.00
Laurents, P	do	100	1		28.50
Huber, J. F	Perry	700	5		60.00

Owners, acreage, and cost of rice irrigation from wells in 1902.

[As reported to the Bureau of Census.]

MISCELLANEOUS PARISHES.

Parish.	Owner of system.	Post-office.	Acreage.	Farms.	Length of main canal.	Total cost.
					Miles.	
Cameron	Hamblin, Albert F	Laurents	60	1		$12.00
Do	Bridgeford, Walter F	Lakeside	200	1		24.00
Do	Pomeroy and Sons	do	100	1		28.00
Do	Monroe Rice Plantation	do	800	2	5	101.53
Do	Lakeside Irrigation Co	Jennings	3,000	12	12	430.00
Iberia	Loard, Mrs. M	New Iberia	150	1	½	29.10
Do	Poirson and Roane	Jeanerette	400	1	1½	54.40
Do	Poirson and Hebert	do	200	1	½	41.05
Lafayette	Avant, Berr	Duson	50	1	½	21.90
Orleans	Funk, John	New Orleans	3	1		3.00
Do	Sarradet, J. M	do	4	1		1.00
Do	Schenck and Son, Michael	do	6	1		2.70
Do	Seyer, Chas	do	1	1		.25
Do	Witzel, Mrs. Chas	do	3	1		2.00

Owners, acreage, and cost of rice irrigation from wells in 1902—Continued.

MISCELLANEOUS PARISHES—Continued.

Parish.	Owner of system.	Post-office.	Acreage.	Farms.	Length of main canal.	Total cost.
					Miles.	
St. Landry.....	Fruge, Azalier	Mamon...........	405	6	4	52.50
Do.........	Lafleur, Dorville.............	Chataignier	60	3	....	20.00
Do.........	Bordelon and E. B. Dubuson	Opelousas				
Do.........	Gus Fusilier and Gaumay ..	Eunice	200	4	....	40.00
Do.........	Helms, Lafayette A	......do	100	4	....	20.00
Do.........	Miller, Durel................	.....do	160	4	....	26.00
Do.........	Tate, Theodore...............	.....do	150	4	....	25.00
Do.........	Woolf, Leon.................	Washington	130	4	....	21.00
St. Martin	Smedes, C. E	Cades.............	160	4	....	27.55
Do.........	Long and Son...............	St. Martinsville...	109	1	4	16.00
Do.........	Martin, Dr. J. S..............	.....do	80	1	....	8.50
Tangipahoa ...	Hammel, C. H..............	Hammond	12	1	....	1.85

Number of rice irrigation systems.

[As reported to the Bureau of the Census.]

Parish.	Number of irrigation systems—			
	Supplied with water from—			Total.
	Streams.	Wells.	Streams and wells.	
Acadia	37	59	11	107
Calcasieu	37	93	2	132
Iberia..................	16	3	....	19
Plaquemines.............	413	....	....	413
Vermilion	13	27	6	46
Cameron		1	4	
Lafayette...............		1	....	
Orleans.................		5	....	
St. Landry...............	95	6	1	118
St. Martin		3	....	
Tangipahoa..............		4	....	
All other parishes.......		1	....	
Total.............	611	200	24	835

Acadia	543	86	22	651
Calcasieu	419	116	11	546
Iberia	54	3	....	57
Plaquemines	432	....		432
Vermilion	450	43	17	510
Cameron		1	16	
Lafayette		1	....	
Orleans		5	....	
St. Landry	186	8	6	237
St. Martin		3	....	
Tangipahoa		1	....	
All other parishes		11	....	
Total	2,084	278	72	2,433

Acreage under irrigation for rice.

Parish.	Streams.	Wells.	Streams and wells.	Total.
Acadia	87,666	13,460	4,880	106,006
Calcasieu	111,636	23,117	3,450	138,203
Iberia	9,376	750	....	10,126
Plaquemines	14,015	....	....	14,015
Vermilion	53,875	9,248	3,215	66,338
Cameron		60	4,100	
Lafayette		50	....	
Orleans		17	....	
St. Landry	46,191	800	405	52,769
St. Martin		340	....	
Tangipahoa		12	....	
All other parishes		794	....	
Total	322,759	48,648	16,050	387,457

Lengths of rice canals and ditches in 1902.

[As reported to the Bureau of the Census.]

Parish.	Total length in miles of main canals from well and well and stream systems.	Total length in miles of ditches of all systems.
Acadia	53	239
Calcasieu	47	291
Iberia	3	12
Plaquemines	....	8
Vermilion	25	108
Cameron	17	
St. Landry	4	50
All other parishes	....	
Total	149	708

PUMPING.

The different types of pumps in common use have already been mentioned. The centrifugal, which is the prevailing type, is lighter, simpler, more readily established, and cheaper than the rotary pumps, although the latter are more efficient when carefully installed. The total lift of such pumps in raising waters from the bayous to the canals varies from 7 to 35 feet, 20 feet being an average lift. Higher levels require supplementary lifts. Pl. X shows a typical pumping plant on the Bayou des Cannes, Louisiana.

In the case of some of the flowing wells the water can be turned directly into the canals for distribution, but where applied at a higher level than the wellhead, pumps are used for lifting the supply. Where the water rises within a few feet of the surface an excavation is made to such depth that the pump is submerged by the water. In the case of wells of small bore the pumping is generally conducted on batteries of wells located 12 to 20 feet apart, though single wells are sometimes pumped. Great numbers of such batteries have been installed in Calcasieu Parish, where their success has been very marked. Much trouble is caused by sand entering the wells, but this can be largely remedied by screening devices, such as are described on pages 71-72.

The fuel used in pumping is of three kinds: Coal, wood, and oil. In 1901 bituminous coal cost as high as $4.75 per ton; wood, $1.50 to $3 per cord, and oil from 48 to 62½ cents per barrel. The cost of oil was at that time said to be about $1 per acre for the season, while that of coal and wood was from $2 to $3 per acre. Oil now commands a much higher price and there is much less money saved through its use. Coal will doubtless continue to be extensively used in the plants near the railroads, but in localities remote from transportation facilities wood will probably afford the most available supply. Pl. X shows the process of unloading wood from a flatboat by means of a moving belt.

APPLICATION OF THE WATER.

CANALS.

The water received from the pumps or directly from the flowing wells is conducted to the rice fields by canals. These consist of two parallel levees constructed of wet, impervious clays, or clayey loams, free from roots and twigs, between which the water is conducted. Fig. 13 is a cross section of the type of canal which has

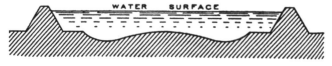

FIG. 13.—Cross section of rice canal.

been found to yield the best results. Care should be taken to remove stumps and to keep out all growth of weeds or other sources of obstruction to the flow of the water. Pl. XI shows the distributary system in the leading rice district of Louisiana.

FIELD LEVEES.

The best form of field levees are low swells, from 15 to 20 feet in width, having the shape shown in fig. 14. They are used to regu-

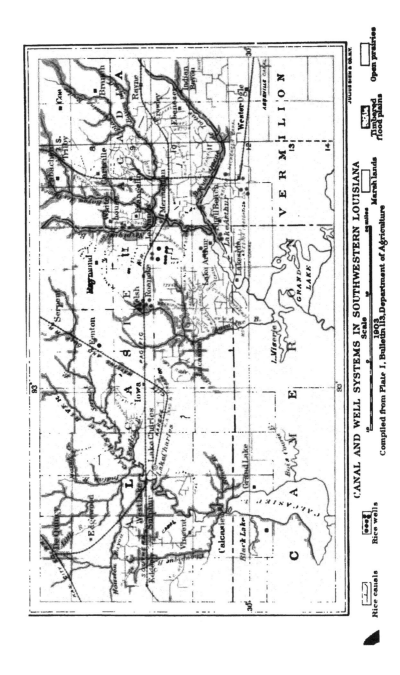

CANAL AND WELL SYSTEMS IN SOUTHWESTERN LOUISIANA

Scale

1903
Compiled from Plate I, Bulletin 113, Department of Agriculture

Rice canals   Rice wells   Marsh lands   Timber and flood plains   Open prairies

e application of water in irrigation. The advantages of levees
s type over the old high and narrow variety are: (1) They
sily crossed, and without damage, by farm machinery; (2) no
s withdrawn from cultivation by them; (3) the growth of the
less red rice and of undesirable grasses and plants is largely pre-
l because of the cultivation of the entire area; (4) they are
d to the varying slopes of the different types of rice fields.
kind of levee is more difficult to construct, and before its intro-
n the fields developed under the old system must be releveled.
levees are also required on sloping ground. In the end, how-
ts use will probably prove the most economical of the various

METHODS OF FARMING.

type of soil best adapted for the growing of rice is a medium
the materials of which are clayey enough to form resistant levees
support heavy harvesting machinery. Organic matter tends to
the material more porous, and is undesirable where it is to be
or levees.
land is plowed with gang plows in the fall or spring, sometimes
then disked and harrowed thoroughly. Planting is done with
oadcast machine attached to an ordinary farm wagon, or the

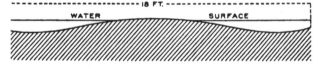

FIG. 14.—Cross section of correct form of field levee.

drilled in rows from 7 to 8 inches apart, the latter method
ng a better crop. During the planting season, which extends
April 1 to June 15, or later, no water is put upon the land,
dence being placed upon rainfall to sprout the seed and promote
owth of the plant for a period varying between one and two
s, depending upon the season and water supply. Flooding
y begins when the rice reaches a height varying between 6 and
hes, and from this time on until the grain is in the milk and well
d—a period of about seventy days—the fields are kept flooded.
ler cases much less water is used. The accompanying diagram
5) shows the depths of water and dates of flooding of such a field
wley.
ut ten days before harvest the levees are cut and the fields are
d. The grain rapidly hardens and matures, and by the time it
ly to cut the field is sufficiently dry to permit the use of the
and binder. This machine is identical with that used in the
fields elsewhere in the United States. The sheaves of rice are

shocked in the field immediately after the binder, ten sheaves to a
shock being the rule, in order that there may be a free circulation of
air to dry the straw.   When harvesting begins the stalks and leaves of
the rice are still green, in the main, but the head is golden yellow on
the terminal two-thirds.   The green straw properly cured is a valuable
substitute for hay, and is baled and fed to live stock, including the
work horses and mules, which become accustomed to it, often prefer-
ring it to prairie hay.   Harvesting begins in September and continues
through October and part of November, often until the 1st of Decem-

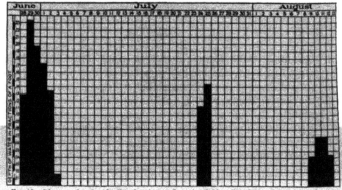

Fig. 15.—Diagram showing depths of water used on rice field at Crowley and dates of irrigation.

ber, and thrashing the rice from the shock begins after it has been
allowed to cure and dry for a period of two weeks at least.   The
machines used are the modern styles of wheat thrashers using steam
power, revolving knives for cutting the binding twine, and a blower
to remove and stack the straw.   The rough rice as it comes from the
thrasher is put in large gunny sacks, weighing, when filled, an average
of 185 pounds each.   The sacked rice is hauled either to the warehouses
or directly to the mills.

INDEX.

O